NEUROBIOLOGY OF ADDICTION

NEUROBIOLOGY OF ADDICTION

GEORGE F. KOOB

Molecular and Integrative Neurosciences Department,
The Scripps Research Institute,
La Jolla, California, USA

MICHEL LE MOAL

Physiopathologie du Comportement
Institut National de la Santé et de la Recherche Médicale,
Institut François Magendie,
Université Victor Ségalen Bordeaux 2
Bordeaux, France

AMSTERDAM • BOSTON • HEIDELBERG • LONDON • NEW YORK • OXFORD
PARIS • SAN DIEGO • SAN FRANCISCO • SINGAPORE • SYDNEY • TOKYO
Academic Press is an imprint of Elsevier

Academic Press is an imprint of Elsevier
84 Theobald's Road, London WC1X 8RR, UK
30 Corporate Drive, Suite 400, Burlington, MA 01803, USA
525 B Street, Suite 1900, San Diego, California 92101-4495, USA

This book is printed on acid-free paper ∞

Library of Congress Catalog Number: 2005933582

British Library Cataloguing in Publication Data
A catalogue record for this book is available from the British Library

ISBN–13: 978-0-12-419239-3
ISBN–10: 0-12-419239-4

For information on all Academic Press publications
visit our web site at http://books.elsevier.com

Printed and bound in China

06 07 08 09 10 10 9 8 7 6 5 4 3 2 1

"Drugs don't affect me"
Dr. Theodora J Koob

Contents

7. Cannabinoids

8. Imaging

9. Neurobiological Theories of Addiction

10. Drug Addiction: Transition from Neuroadaptation to Pathophysiology

Preface

This book began with an idealistic goal to summarize, integrate, and synthesize the world's literature on the Neurobiology of Addiction under one conceptual framework, in one volume. As we embarked on this journey, it became increasingly evident that this was a Herculean task that required an enormous commitment to find original sources, review diverse topics, search under conceptual stones, and select relevant facts, papers and frameworks, and ultimately limit our appetite for citing every paper.

The journey took over two years and led to the uncovering of heretofore unknown intellectual gems (to us anyway) and some contradictions—but also a surprising number of neurobiological consistencies and commonalities across the spectrum of addictive drugs. What evolved was, what we think, is a reasonably objective view of the field, with a concerted attempt to accomplish what we set out to do.

As a result, there is a chapter on *What is Addiction?* and another on *Animal Models of Addiction* to guide the reader through the conceptual and technical framework of the book. What follows are five chapters on the major classes of drugs of addiction: *Psychostimulants, Opioids, Alcohol, Nicotine, and Cannabinoids.* Each chapter stands on its own and integrates human use and addiction patterns and behavioral mechanisms with the neurobiology explored at three levels: neurocircuitry (neuropharmacology), cellular (electrophysiology), and molecular (molecular measures and molecular genetic approaches). Appendices 1–5 provide human case histories and anecdotes describing addiction profiles and personal experiences relevant to each drug chapter.

These levels of analysis are arbitrary and overlap but provide a succinct framework for integration across disciplines. Human neuroimaging has been incorporated as a separate chapter to allow a view of common elements in the imaging field and to reduce redundancy of technique, approach, and methodology.

Towards the end of the book, there is a chapter on *Neurobiological Theories of Addiction* which explores the different conceptual views of prominent investigators in the field, from both a neurobiological and evolving historical perspective. An attempt is made to integrate the different theories into a heuristic model to account for most of the stages of the addiction cycle. The last chapter, *Drug Addiction: Transition from Neuroadaptation to Pathophysiology*, is what we unabashedly admit is our world view of the neurobiology of addiction; as such, this final chapter may be considered parochial. Nevertheless, we believe, that after two years of intense research, we can bring a unique view to the field, as if we were in a zeppelin floating above the vast sea of data.

Several facts are worth noting. First, an attempt was made to trace every single statement to its original source, and not to use secondary references. If, perchance, we have failed, we welcome corrections. Second, we made a valiant attempt to fully cover the field. If we have left out an important component of a given piece of the field, we also welcome input. Third, we restricted our journey to the major drugs of addiction and left out numerous other drugs of abuse and drugs of dependence (with a little "d"; see Chapter 1). We did not cover psychedelics, inhalants, steroids, caffeine, benzodiazepines, gambling, etc. Such a broadened perspective will be saved for another day.

George F Koob

Michel Le Moal

Acknowledgments

This book would not have been possible without the following contributions. First and foremost, without the tireless efforts of Michael A. Arends, we would still be writing. Mike found and tracked down references (sometimes literally hundreds of years old) from the world over, coordinated all tables, figures, figure legends, and references, and leant his incredibly thorough editing skill.

All figures for the book were redrawn from their original sources by Janet Hightower of The Scripps Research Institute Biomedical Graphics department.

We cannot thank them enough.

We appreciate Isabelle Batby and Mellany Santos for their editorial contributions and many excursions to the library. We thank Marisela Perez-Meza (The Scripps Research Institute Kresge Library) and Hélène Renaud (Institut François Magendie Library) for their diligent interlibrary loan support. We acknowledge the following fellows and students for their varied editing contributions: Sheila Drnec, Thomas Greenwell, Simon Katner, Maegan Mattock, Beth Maxwell Boyle, Cindy Reiter-Funk, Bryant Silbaugh, and Brendan Walker. We also thank K. Noelle Gracy for soliciting this work, and Johannes Menzel, Pauline Sones, and Maureen Twaig of Elsevier for their support and enduring patience.

Finally, we owe a debt of gratitude to many colleagues for providing encouragement, suggestions, references, and their own personal interpretations of studies:

Nora Abrous
Serge Ahmed
James Anthony
Gary Aston-Jones
Tamas Bartfai
Floyd Bloom
Al Collins
Véronique Deroche-Gamonet
Barry Everitt
Eliot Gardner
Jacques Glowinski
Howard Gutstein
R. Adron Harris
Markus Heilig
Jack Henningfield
Ralph Hingson
Reese Jones
Pierre Karli
Conan Kornetsky
Robin Kroft
Charles Ksir
Jean-Paul Laulin
Rong Lee
Ting-Kai Li
Athina Markou
Remi Martin-Fardon
Barbara Mason
Charles O'Brien
Loren Parsons
Pier Vincenzo Piazza
John Pierce
Linda Porrino
Heather Richardson
Trevor Robbins
Marisa Roberto
Bernard Roques
Saul Shiffman
George Siggins
Hervé Simon
Guy Simonnet
Luis Stinus
Jean-Pol Tassin
Anne-Marie Thierry
Tamara Wall
Friedbert Weiss
Sam Zakhari
Eric Zorrilla

CHAPTER

1

What is Addiction?

OUTLINE

DEFINITIONS OF ADDICTION

Drug Use, Drug Abuse, and Drug Addiction

Drug addiction, also known as Substance Dependence (American Psychiatric Association, 1994), is a chronically relapsing disorder that is characterized by (1) compulsion to seek and take the drug, (2) loss of control in limiting intake, and (3) emergence of a negative emotional state (e.g., dysphoria, anxiety, irritability) when access to the drug is prevented (defined here as dependence) (Koob and Le Moal, 1997). The occasional but limited use of an abusable drug clinically is distinct from escalated drug use, loss of control over limiting drug intake, and the emergence of chronic compulsive drug-seeking that characterizes addiction. Modern views have focused on three types of drug use: (1) occasional, controlled or social use, (2) drug abuse or harmful use, and (3) drug addiction. An important goal of current neurobiological research on addiction is to understand the neuropharmacological and neuroadaptive mechanisms within specific neurocircuits that mediate the transition between occasional, controlled drug use and the loss of behavioral control over drug-seeking and drug-taking that defines chronic addiction (Koob and Le Moal, 1997).

The critical nature of the distinction between drug use, abuse and dependence has been illuminated by data showing that approximately 15.6 per cent (29 million) of the U.S. adult population will go on to engage in nonmedical or illicit drug use at some time in their lives, with approximately 3.1 per cent (5.8 million) of the U.S. adult population going on to drug abuse and 2.9 per cent (5.4 million) going on to Substance Dependence on illicit drugs (Grant and Dawson, 1998; Grant *et al.*, 2005). For alcohol, 51 per cent (120 million) of people over the age of 12 were current users, 23 per cent (54 million) engaged in binge drinking, and 7 per cent (16 million) were defined as heavy drinkers. Of these current users, 7.7 per cent (18 million) met the criteria for Substance Abuse or Dependence on Alcohol (see *Alcohol* chapter). For tobacco, 30 per cent (71.5 million) of people aged 12 and older reported past-month use of a tobacco product. Also, 19 per cent (45 million) of persons in the U.S. smoked every day in the past month. From the 1992 National Comorbidity Survey, 75.6 per cent of 15–54-year-olds ever used tobacco, with 24.1 per cent

Neurobiology of Addiction, by George F. Koob and Michel Le Moal.
ISBN – 13: 978-0-12-419239-3 ISBN – 10: 0-12-419239-4

TABLE 1.1 Estimated Prevalence Among 15–54-Year-Olds of Nonmedical Use and Dependence Among Users (1990–1992) from The National Comorbidity Survey

	Ever used (%)	Prevalence of dependence (%)	Dependence among users (%)
Tobacco	75.6	24.1	31.9
Alcohol	91.5	14.1	15.4
Illicit Drugs	51.0	7.5	14.7
Cannabis	46.3	4.2	9.1
Cocaine	16.2	2.7	16.7
Stimulants	15.3	1.7	11.2
Anxiolytics	12.7	1.2	9.2
Analgesics	9.7	0.7	7.5
Psychedelics	10.6	0.5	4.9
Heroin	1.5	0.4	23.1
Inhalants	6.8	0.3	3.7

[Reproduced with permission from Anthony *et al.*, 1994.]

meeting the criteria for Dependence (Anthony *et al.*, 1994) (see *Nicotine* chapter).

The number of individuals meeting the criteria for Substance Dependence on a given drug as a function of ever having used the drug varies between drugs. According to data from the 1990–1992 National Comorbidity Survey, the percentage addicted to a given drug, of those people who ever used the drug, decreased in the following order: *tobacco* > *heroin* > *cocaine* > *alcohol* > *marijuana* (Anthony *et al.*, 1994) (**Table 1.1**). More recent data derived from the National Household Survey on Drug Abuse (Substance Abuse and Mental Health Services Administration, 2003) showed that the percentage addicted to a given drug, of those who ever used, decreases in the following order: *heroin* > *cocaine* > *marijuana* > *alcohol* (**Fig. 1.1**). These more recent data suggest unsettling evidence of an overall trend for a significant increase in Substance Dependence with marijuana (see *Cannabinoids* chapter).

The cost to society of drug abuse and drug addiction is prodigious in terms of both direct costs and indirect costs associated with secondary medical events, social problems, and loss of productivity. In the United States alone, it is estimated that illicit drug abuse and addiction cost society $161 billion (Office of National Drug Control Policy, 2001; see also Uhl and Grow, 2004). It is estimated that alcoholism costs society $180 billion per year (Yi *et al.*, 2000), and tobacco addiction $155 billion (Centers for Disease Control and Prevention, 2004). In France, the total cost of drug use is USD 41 billion (including $22 billion for alcohol, $16 billion for tobacco, and nearly $3 billion for illicit drugs) (Kopp and Fenoglio, 2000).

Addiction and *Substance Dependence* will be used interchangeably throughout this text and will refer to a final stage of a usage process that moves from drug use to abuse to addiction. Drug addiction is a disease and, more precisely, a *chronic* disease (Meyer, 1996). As such, it can be defined by its diagnosis, etiology, and pathophysiology as a chronic relapsing disorder (**Fig. 1.2**). The associated medical, social, and occupational difficulties that usually develop during the course of addiction do not disappear after detoxification. Addictive drugs are hypothesized to produce changes in brain pathways that endure long after the person stops taking them. These protracted brain changes and the associated personal and social difficulties put the former patient at risk of relapse (O'Brien and McLellan, 1996), a risk higher than 60 per cent within the year that follows discharge (Finney and Moos, 1992; Hubbard *et al.*, 1997;

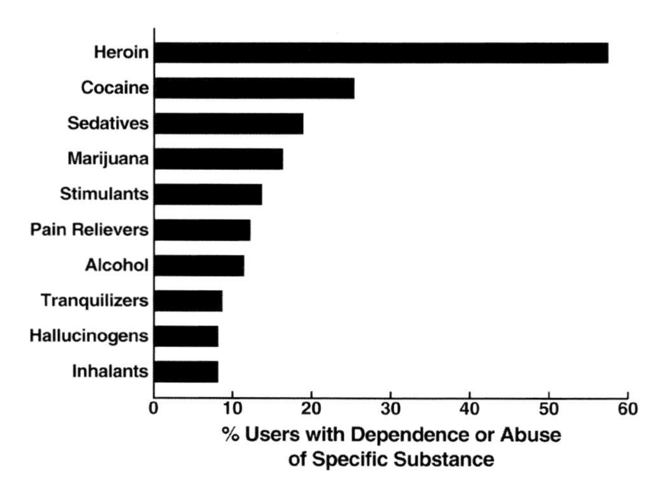

FIGURE 1.1 Dependence or abuse of specific substances among past-year users of substance (Substance Abuse and Mental Health Services Administration, 2003). Heroin: 57.4% (0.2 million), Cocaine: 25.6% (1.5 million), Marijuana: 16.6% (4.2 million).

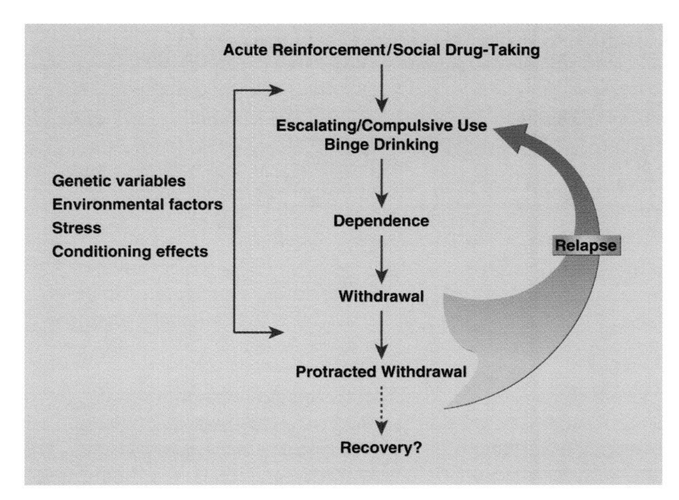

FIGURE 1.2 Stages of addiction to drugs of abuse. Drug-taking invariably begins with social drug-taking and acute reinforcement and often, but not exclusively, then moves in a pattern of use from escalating compulsive use to dependence, withdrawal, and protracted abstinence. During withdrawal and protracted abstinence, relapse to compulsive use is likely to occur with a repeat of the cycle. Genetic factors, environmental factors, stress, and conditioning all contribute to the vulnerability to enter the cycle of abuse/dependence and relapse within the cycle.

McLellan and McKay, 1998; McLellan *et al.*, 2000). While much of the initial study of the neurobiology of drug addiction focused on the acute impact of drugs of abuse (analogous to comparing no drug use to drug use), the focus is now shifting to chronic administration and the acute and long-term neuroadaptive changes in the brain that result in relapse. Cogent arguments have been made which support the hypothesis that addictions are similar in their chronic relapsing properties and treatment efficacy to other chronic relapsing disorders such as diabetes, asthma, and hypertension (McLellan *et al.*, 2000). The purpose of current neuroscientific drug abuse research is to understand the cellular and molecular mechanisms that mediate the transition from occasional, controlled drug use to the loss of behavioral control over drug-seeking and drug-taking that defines chronic addiction (Koob and Le Moal, 1997).

Diagnostic Criteria of Addiction

The diagnostic criteria for addiction as described by the *Diagnostic and Statistical Manual of Mental Disorders*, 4th edition (DSM-IV) (American Psychiatric Association, 1994), also have evolved over the past 30 years with a shift from the emphasis and necessary criteria of tolerance and withdrawal to other criteria directed more at compulsive use. In the DSM-IV, tolerance and withdrawal form two of seven potential criteria. The criteria for Substance Dependence outlined in the DSM-IV closely resemble those outlined by the *International Statistical Classification of Diseases and Related Health Problems* (ICD-10) (World Health Organization, 1992) (**Tables 1.2** and **1.3**). The number of criteria met by drug addicts vary with the severity of the addiction, the stage of the addiction process, and the drug in question (Chung and Martin, 2001). For example, in adolescents, the most frequently observed criteria are *much time getting or recovering from use* (DSM-IV criteria #5 and #7), *continued use despite problems in social and occupational functioning* (DSM-IV criterion #6), and *tolerance or withdrawal* (DSM-IV criteria #1 and #2) (Crowley *et al.*, 1998) (see *Cannabinoids* chapter).

TABLE 1.2 DSM-IV and ICD-10 Diagnostic Criteria for *Alcohol and Drug Abuse/Harmful Use*

DSM-IV Alcohol and drug abuse	ICD-10 Harmful use of alcohol and drugs
A. A maladaptive pattern of substance use leading to clinically significant impairment or distress, as manifested by one (or more) of the following occurring within a 12-month period: 1. recurrent substance use resulting in a failure to fulfil major role obligations at work, school, or home. 2. recurrent substance use in situations in which use is physically hazardous. 3. recurrent substance-related legal problems. 4. continued substance use despite having persistent or recurrent social or interpersonal problems caused or exacerbated by the effects of the drug. B. The symptoms have never met the criteria for substance dependence for the same class of substance.	A. A pattern of substance use that is causing damage to health. The damage may be physical or mental. The diagnosis requires that actual damage should have been caused to the mental or physical health of the user. B. No concurrent diagnosis of the substance dependence syndrome for same class of substance.

TABLE 1.3 DSM-IV and ICD-10 Diagnostic Criteria for *Alcohol and Drug Dependence*

	DSM-IV	ICD-10
Clustering criterion	A. A maladaptive pattern of substance use, leading to clinically significant impairment or distress as manifested by three or more of the following occurring at any time in the same 12-month period:	A. Three or more of the following have been experienced or exhibited at some time during the previous year:
Tolerance	1. Need for markedly increased amounts of a substance to achieve intoxication or desired effect; or markedly diminished effect with continued use of the same amount of the substance.	1. Evidence of tolerance, such that increased doses are required in order to achieve effects originally produced by lower doses.
Withdrawal	2. The characteristic withdrawal syndrome for a substance or use of a substance (or a closely related substance) to relieve or avoid withdrawal symptoms.	2. A physiological withdrawal state when substance use has ceased or been reduced as evidenced by the characteristic substance withdrawal syndrome, or use of substance (or a closely related substance) to relieve or avoid withdrawal symptoms.
Impaired control	3. Persistent desire or one or more unsuccessful efforts to cut down or control substance use.	3. Difficulties in controlling substance use in terms of onset, termination, or levels of use.
	4. Substance used in larger amounts or over a longer period than the person intended.	
Neglect of activities	5. Important social, occupational, or recreational activities given up or reduced because of substance use.	4. Progressive neglect of alternative pleasures or interests in favor of substance use; or
Time spent	6. A great deal of time spent in activities necessary to obtain, to use, or to recover from the effects of substance used.	A great deal of time spent in activities necessary to obtain, to use, or to recover from the effects of substance use.
Inability to fulfil roles	None	None
Hazardous use	None	None
Continued use despite problems	7. Continued substance use despite knowledge of having a persistent or recurrent physical or psychological problem that is likely to be caused or exacerbated by use.	5. Continued substance use despite clear evidence of overtly harmful physical or psychological consequences.
Compulsive use	None	6. A strong desire or sense of compulsion to use substance.
Duration criterion	B. No duration criterion separately specified. However, several dependence criteria must occur repeatedly as specified by duration qualifiers associated with criteria (e.g., 'often', 'persistent', 'continued').	B. No duration criterion separately specified.
Criterion for subtyping dependence	*With physiological dependence:* Evidence of tolerance or withdrawal (i.e., any of items A-1 or A-2 above are present). *Without physiological dependence:* No evidence of tolerance or withdrawal (i.e., none of items A-1 or A-2 above are present).	None

Dependence View of Addiction

Historically, definitions of addiction began with definitions of dependence. Himmelsbach defined physical dependence as:

> '... an arbitrary term used to denote the presence of an acquired abnormal state wherein the regular administration of adequate amounts of a drug has, through previous prolonged use, become requisite to physiologic equilibrium. Since it is not yet possible to diagnose physical dependence objectively without withholding drugs, the *sine qua non* of physical dependence remains the demonstration of a characteristic abstinence syndrome' (Himmelsbach, 1943).

Eventually this definition evolved into the definition for physical dependence or 'intense physical disturbances when administration of a drug is suspended' (Eddy *et al.*, 1965). However, this terminology clearly did not capture many of the aspects of the addictive process where no *physical* signs were observed, necessitating a second definition of *psychic dependence* to capture the more *behavioral* aspects of the symptoms of addiction: 'A condition in which a drug produces "a feeling of satisfaction and a psychic drive that require periodic or continuous administration of the drug to produce pleasure or to avoid discomfort"...' (Eddy *et al.*, 1965). Modern definitions of addiction resemble a combination of physical and psychic dependence with more of an emphasis on the psychic or motivational aspects of withdrawal, rather than on the physical symptoms of withdrawal:

> '*Addiction* from the Latin verb "addicere", to give or bind a person to one thing or another. Generally used in the drug field to refer to chronic, compulsive, or uncontrollable drug use, to the extent that a person (referred to as an "addict") cannot or will not stop the use of some drugs. It usually implies a strong (Psychological) Dependence and (Physical) Dependence resulting in a Withdrawal Syndrome when use of the drug is stopped. Many definitions place primary stress on psychological factors, such as loss of self-control and over powering desires; i.e., addiction is any state in which one craves the use of a drug and uses it frequently. Others use the term as a synonym for physiological dependence; still others see it as a combination (of the two)' (Nelson *et al.*, 1982).

Unfortunately, the word *dependence* has multiple meanings. Any drug can produce dependence if dependence is defined as the manifestation of a withdrawal syndrome upon cessation of drug use, but meeting the DSM-IV criteria for *Substance Dependence* is much more than a manifestation of a withdrawal syndrome, but rather is equivalent to addiction. For the purposes of this book, *dependence* with a lower-case 'little d' will refer to the manifestation of a withdrawal syndrome, whereas *Dependence* with a capital 'big D' will refer to Substance Dependence as defined by the DSM-IV or addiction. The words *Substance Dependence* (as defined by the DSM-IV), *addiction* and *alcoholism* will be held equivalent for this book.

Psychiatric View of Addiction

From a psychiatric perspective, drug addiction has aspects of both impulse control disorders and compulsive disorders. Impulse control disorders are characterized by an increasing sense of tension or arousal before committing an impulsive act—pleasure, gratification or relief at the time of committing the act—and there may or may not be regret, self-reproach or guilt following the act (American Psychiatric Association, 1994). In contrast, compulsive disorders are characterized by anxiety and stress before committing a compulsive repetitive behavior and relief from the stress by performing the compulsive behavior. As an individual moves from an impulsive disorder to a compulsive disorder, there is a shift from positive reinforcement driving the motivated behavior to negative reinforcement driving the motivated behavior (Koob, 2004) (**Fig. 1.3**). Drug addiction has been conceptualized as a disorder that progresses from impulsivity to compulsivity in a collapsed cycle of addiction comprised of three stages: preoccupation/anticipation, binge/intoxication, and withdrawal/negative affect. Different theoretical perspectives ranging from experimental psychology, social psychology, and neurobiology can be superimposed on these three stages which are conceptualized as feeding into each other, becoming more intense, and ultimately leading to the pathological state known as addiction (Koob and Le Moal, 1997) (**Fig. 1.4**).

Psychodynamic View of Addiction

A psychodynamic view of addiction that integrates well with the neurobiology of addiction is that of Khantzian and colleagues (Khantzian, 1985, 1990, 1997) with a focus on the factors leading to vulnerability for addiction. This perspective is deeply rooted in clinical practice and in psychodynamic concepts developed in a contemporary perspective in relation to substance use disorders. The focus of this approach is on developmental difficulties, emotional disturbances, structural (ego) factors, personality organization, and the building of the self. It is important to note that this contemporary perspective contrasts with a classic but not abundant psychoanalytic literature on the subject which emphasizes the pleasurable aspects of drugs and the regressive aspects of drug use.

Two critical elements (disordered emotions and disordered self-care) and two contributory elements (disordered self-esteem and disordered relationships) have been identified, which have evolved into a modern self-medication hypothesis, where individuals with substance use disorders are hypothesized to take drugs as a 'means to cope with painful and threatening emotions.' In this conceptualization, addicted individuals experience states of subjective distress and suffering that may or may not be associated with conditions meeting DSM-IV criteria for a psychiatric diagnosis (American Psychiatric Association, 1994).

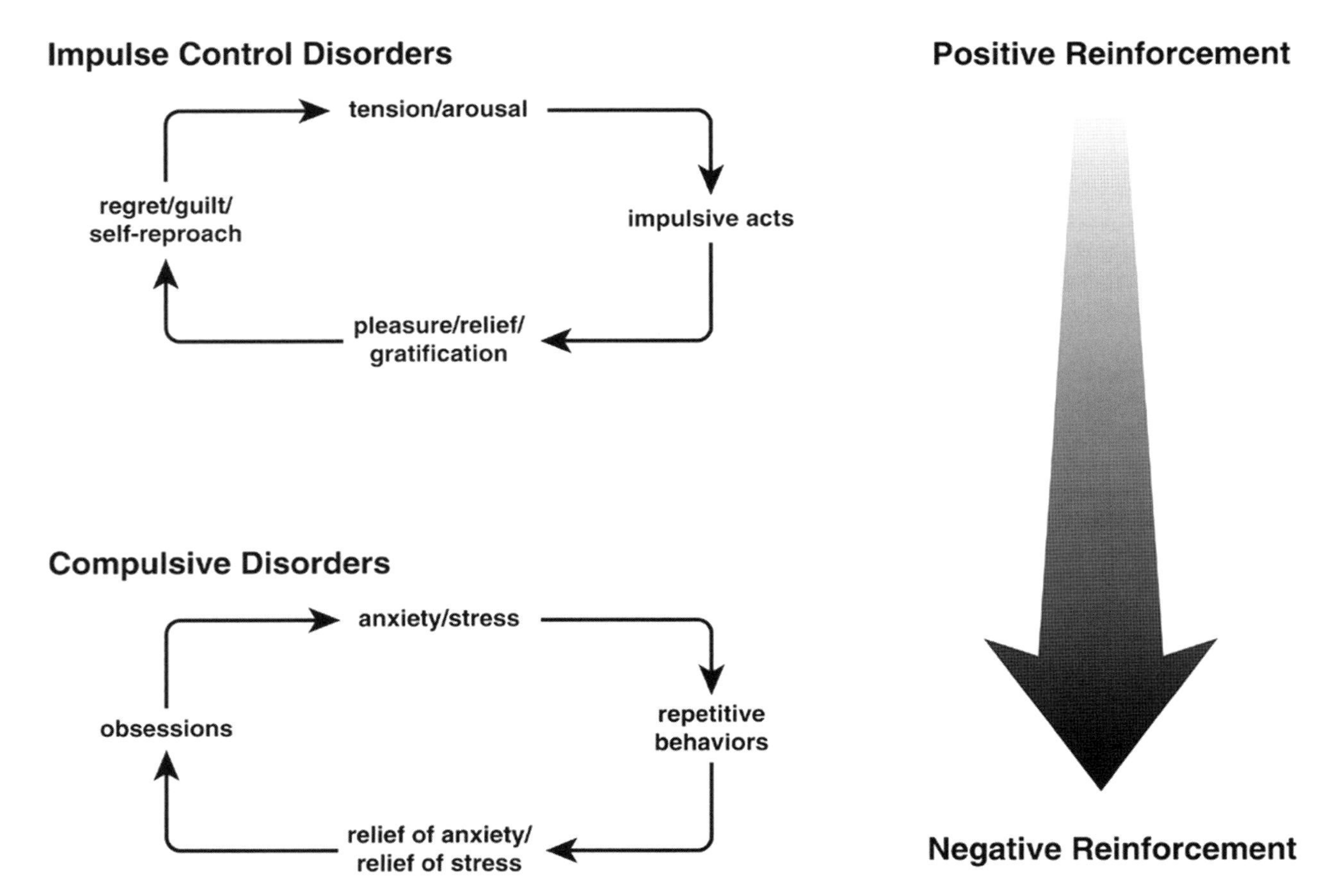

FIGURE 1.3 Diagram showing stages of impulse control disorder and compulsive disorder cycles related to the sources of reinforcement. In impulse control disorders increasing tension and arousal occur before the impulsive act, with pleasure, gratification, or relief during the act. Following the act there may or may not be regret or guilt. In compulsive disorders, there are recurrent and persistent thoughts (obsessions) that cause marked anxiety and stress followed by repetitive behaviors (compulsions) that are aimed at preventing or reducing distress (American Psychiatric Association, 1994). Positive reinforcement (pleasure/gratification) is more closely associated with impulse control disorders. Negative reinforcement (relief of anxiety or relief of stress) is more closely associated with compulsive disorders. [Reproduced with permission from Koob, 2004.]

Addicts have feelings that are overwhelming and unbearable and may consist of an affective life that is absent and nameless. From this perspective, drug addiction is viewed as an attempt to medicate such a dysregulated affective state. The suffering of the patient is deep-rooted in disordered emotions characterized at their extremes either by unbearable painful affect or by a painful sense of emptiness. Others cannot express personal feelings or cannot access emotions and are hypothesized to suffer from alexithymia, defined as 'a marked difficulty to use appropriate language to express and describe feelings and to differentiate them from bodily sensation' (Sifneos, 2000).

Such self-medication may be drug-specific in that patients may have a preferential use of drugs that fits with the nature of the painful affective states that they are self-medicating. Opiates might be effective in reducing psychopathological states of violent anger and rageful feelings. Others suffering from anhedonia, anergia, or lack of feelings, will prefer the activating properties of psychostimulants. Some flooded in their feelings, or cut off from feelings, will welcome repeated moderate doses of alcohol or depressants as medicine to express feelings that they are not able to communicate. Thus, in some cases, the subjects operate to relieve painful feelings, in others, the operative motive is to control or express feelings (Khantzian, 1995, 1997; Khantzian and Wilson, 1993). The common element to this hypothesis is that each class of drugs serves as an antidote to dysphoric states and acts as a 'replacement for a defect in the psychological structure' of such individuals (Kohut, 1971). The paradox is that the choice of drugs to self-medicate such emotional pain will later by itself perpetuate it, thereby continuing a life revolving around drugs.

Disordered self-care is hypothesized to combine with a disordered emotional life to become a principal

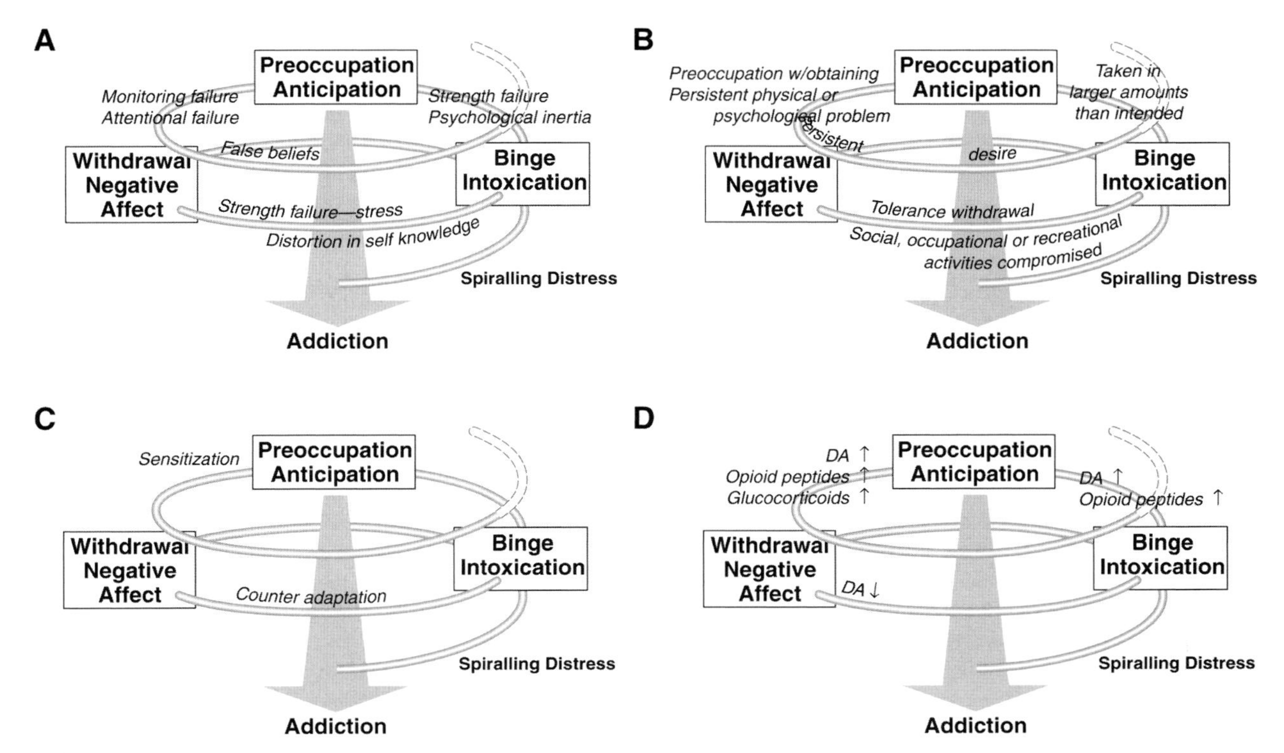

FIGURE 1.4 Diagram describing the spiraling distress—addiction cycle from four conceptual perspectives: social psychological, psychiatric, dysadaptational, and neurobiological. Note that the addiction cycle is conceptualized as a spiral that increases in amplitude with repeated experience, ultimately resulting in the pathological state known as addiction. **(A)** The three major components of the addiction cycle—preoccupation/anticipation, binge/intoxication, and withdrawal/negative affect—and some of the sources of potential self-regulation failure in the form of underregulation and misregulation. **(B)** The same three major components of the addiction cycle with the different criteria for substance dependence incorporated from the DSM-IV. **(C)** The places of emphasis for the theoretical constructs of sensitization and counteradaptation. **(D)** The hypothetical role of different neurochemical and endocrine systems in the addiction cycle. Small arrows refer to increased functional activity. DA, dopamine; CRF, corticotropin-releasing factor. [Reproduced with permission from Koob and Le Moal, 1997.]

determinant of substance use disorders. Self-care deficits reflect an inability to ensure one's self-preservation and are characterized by an inability to anticipate or avoid harmful and dangerous situations, and an inability to use appropriate judgment and feeling as guides in the face of danger. Thus, self-care deficits reflect an inability to appropriately experience emotions and appreciate the consequences of dangerous behaviors, and the core element of this psychodynamic perspective is a dysregulated emotional system or systems in individuals vulnerable to addiction.

This psychodynamic approach integrates well with a growing amount of evidence for a critical role of dysregulated brain reward and stress systems, from studies on the neurobiology of addiction using animal models that have developed from a physiological framework (see chapters that follow). However, from a neurobiological perspective, there is the additional insult to the personality produced by the direct effects of the drugs themselves to perpetuate, and actually *create*, such character flaws (Koob, 2003).

Social Psychological / Self-regulation View of Addiction

At the social psychology level, self-regulation failure has been argued as the root of the major social pathology in present times (Baumeister *et al.*, 1994). From this perspective there are important self-regulation elements that may be involved in the different stages of addiction to drugs, as well as in other pathological behaviors such as compulsive gambling and binge eating (Baumeister *et al.*, 1994). Such self-regulation failures ultimately may lead to addiction in the case of drug use or an addiction-like pattern with nondrug behaviors. Underregulation as reflected in strength deficits, failure to establish standards or conflicts in standards, and attentional failures as well as misregulation (misdirected attempts to self-regulate) can contribute to the development of addiction-like patterns of behavior (**Fig. 1.4**). The transition to addiction can be facilitated by lapse-activated causal patterns. That is, patterns of behavior that contribute to the transition from an initial lapse in self-regulation to a large-scale breakdown in self-regulation can lead to spiraling distress (Baumeister *et al.*, 1994). In some cases, the first self-regulation failure can lead to emotional distress which sets up a cycle of repeated failures to self-regulate and where each violation brings additional negative affect, resulting in spiraling distress (Baumeister *et al.*, 1994). For example, a failure of strength may lead to initial drug use or relapse, and other self-regulation failures can be recruited to produce an entrance to, or prevent an exit from, the addiction cycle.

At a neurobehavioral level, such dysregulation again may be reflected in deficits of information-processing, attention, planning, reasoning, self-monitoring, inhibition, and self-regulation, many of which involve functioning of the frontal lobe (Giancola *et al.*, 1996a,b) (see chapters that follow). Executive function deficits, self-regulation problems, and frontal lobe dysfunctions or pathologies constitute a risk factor for biobehavioral disorders including drug abuse (Dawes *et al.*, 1997). Deficits in frontal cortex regulation in children or young adolescents predict later drug and alcohol consumption, especially for children raised in families with drug and biobehavioral disorders histories (Dawes *et al.*, 1997; Aytaclar *et al.*, 1999).

Vulnerability to Addiction

Drug abuse is a far more complex phenomenon than previously thought, and it is now recognized that drug abusers represent a highly heterogeneous group, and the patterns leading to dependence are diverse. Individual differences in temperament, social development, comorbidity, protective factors, and genetics are areas of intense research, and a detailed discussion of these contributions to addiction are beyond the scope of this book. However, each of these factors presumably interacts with the neurobiological processes discussed in this book. A reasonable assertion is that the initiation of drug abuse is more associated with social and environmental factors, whereas the movement to abuse and addiction are more associated with neurobiological factors (Glantz and Pickens, 1992).

Temperament and personality traits and some temperament clusters have been identified as factors of vulnerability to drug abuse (Glantz *et al.*, 1999) and include disinhibition (behavioral activation) (Windle and Windle, 1993), negative affect (Tarter *et al.*, 1995), novelty- and sensation-seeking (Wills *et al.*, 1994), and 'difficult temperament' (conduct disorder) (Glantz *et al.*, 1999).

From the perspective of comorbid psychiatric disorders, some of the strongest associations are found with mood disorders, anxiety disorders, antisocial personality disorders, and conduct disorders (Glantz and Hartel, 1999). Data from the International Consortium in Psychiatric Epidemiology (representing six different sites in the United States, Germany, Mexico, The Netherlands, Ontario, and Canada) and the National Comorbidity Study (United States; approximately 30 000 subjects) have revealed that approximately 35 per cent of the sample with drug dependence met lifetime criteria for a mood disorder. About 45 per cent met criteria for an anxiety disorder, and 50 per cent met criteria for either conduct or antisocial personality disorder (Merikangas *et al.*, 1998). More recent data on 12-month prevalence of comorbidity from the National Institute on Alcohol Abuse and Alcoholism's National Epidemiologic Survey on Alcohol and Related Conditions represents over 43 000 respondents and shows similar results (21–29 per cent for comorbidity of mood disorders; 22–25 per cent comorbidity for anxiety disorders; 32–70 per cent comorbidity for personality disorders) (Grant *et al.*, 2004a,b,c) (**Table 1.4**). The association of Attention Deficit Hyperactivity Disorder (ADHD) with drug abuse can be explained largely by the higher comorbidity with conduct disorder in these children (Biederman *et al.*, 1997). Independent of this association, there is little firm data to support a risk due to treatment of ADHD with stimulants (Biederman *et al.*, 1999), and no preference for stimulants over other drugs has been noted (Biederman *et al.*, 1997).

Developmental factors are important components of vulnerability, with strong evidence developing that adolescent exposure to alcohol, tobacco, or drugs of abuse leads to significant vulnerability for alcohol

TABLE 1.4 12-Month Prevalence of Comorbid Disorders Among Respondents with Nicotine Dependence, Alcohol Dependence, or Any Substance Use Disorder

	Mood	Anxiety	Personality
Alcohol	27.6%	23.5%	39.5%
Nicotine	21.1%	22.0%	31.7%
Substance Dependence (including alcohol but not nicotine)	29.2%	24.5%	69.5%

[Data from Grant *et al.*, 2004a,b,c.]

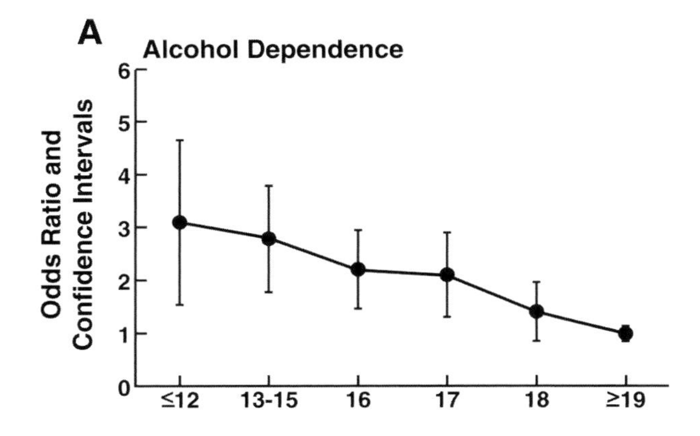

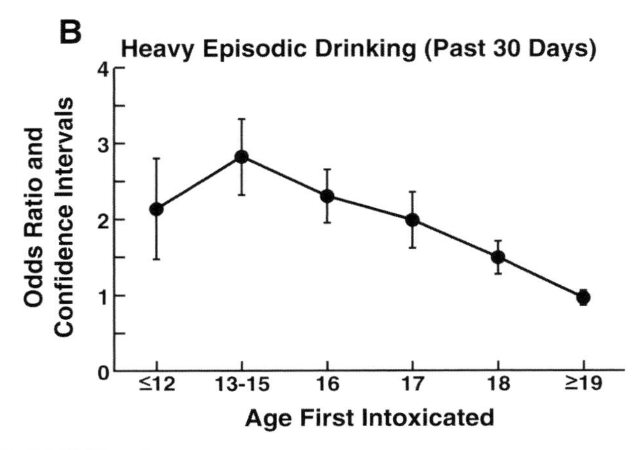

FIGURE 1.5 1999 College Alcohol Survey. **(A)** Alcohol Dependence according to age first intoxicated. **(B)** Past 30 days heavy episodic drinking according to age first intoxicated. After controlling for personal and demographic characteristics and respondent age, the odds of meeting alcohol dependence criteria were 3.1 times greater for those first drunk at or prior to age 12 compared with drinkers who were first drunk at age 19 or older. The relationship between early onset of being drunk and heavy episodic drinking in college persisted even after further controlling for alcohol dependence. Respondents first drunk at or prior to age 12 had 2.1 times the odds of reporting recent heavy episodic drinking than college drinkers first drunk at age 19 or older. [Reproduced with permission from Hingson *et al.*, 2003.]

dependence and alcohol problems in adulthood. Persons first intoxicated at 16 or younger were more likely to drive after drinking, to ride with intoxicated drivers, to be injured seriously when drinking, to be more likely to become heavy drinkers, and to be 2–3 times more likely to develop substance dependence on alcohol (Hingson *et al.*, 2003) (**Fig. 1.5**). Similarly, persons who smoked their first cigarette during 14–16 years of age were 1.6 times more likely to become dependent than those who initiated at a later age (Breslau *et al.*, 1993; Everett *et al.*, 1999). Others have argued that regular smoking during adolescence raises the risk for adult smoking by a factor of 16 compared to nonsmoking during adolescence (Chassin *et al.*, 1990). Most smoking initiation occurs in the United States during the transition from junior high school to high school (14–15 years of age) (Winkleby *et al.*, 1993). The age at which smoking begins influences the total years of smoking (Escobedo *et al.*, 1993), the number of cigarettes smoked in adulthood (Taioli and Wynder, 1991), and the likelihood of quitting (Ershler *et al.*, 1989; Chassin *et al.*, 1990) (**Fig. 1.6**). When prevalence of lifetime illicit or nonmedical drug abuse and Substance Dependence was estimated for each year of onset of drug use from ages 13 and younger to 21 and older, early onset of drug use was a significant predictor of the subsequent development of drug abuse over a lifetime (Grant and Dawson, 1998) (**Fig. 1.7**). Drugs included sedatives, tranquilizers, opioids other than heroin, amphetamines, cocaine and crack cocaine, cannabis, heroin, methadone, hallucinogens, and inhalants. Overall, the lifetime prevalence of

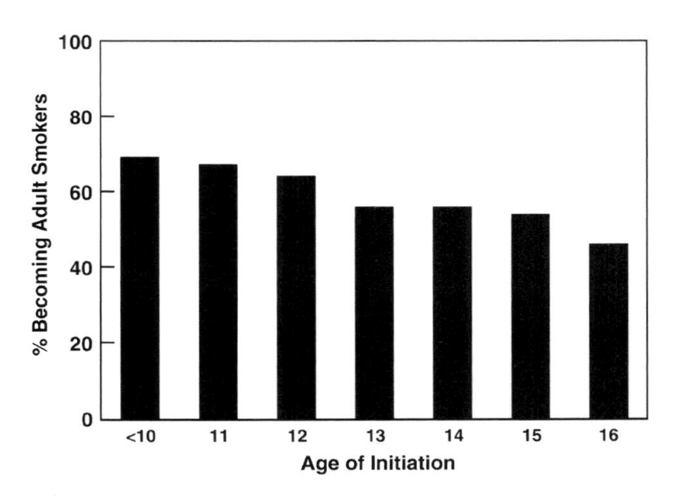

FIGURE 1.6 Percentages of adolescent regular smokers who became adult regular smokers as a function of grade of smoking initiation. Subjects consisted of all consenting 6th to 12th graders in a Midwestern county school system who were present in school on the day of testing. All 6th to 12th grade classrooms (excluding special education) were surveyed annually between 1980 and 1983. There was a potential pool of 5799 individuals who had been assessed at least once during their adolescence between 1980 and 1983. At the time of follow-up, 25 of these subjects were found to be deceased, and 175 refused participation. 4156 provided data (72%). The subjects were predominantly Caucasian (96%), were equally divided by sex (49% male; 51% female), and were on an average 21.8 years old. 71% had never been married, and 26% were currently married. 58% had completed at least some college by the time of follow-up. 32% were still students. 43% had a high school education. For nonstudents, occupational status ranged from 29% in factory, crafts, and labor occupations, to 39% in professional, technical, and managerial occupations. At follow-up, the overall rate of smoking at least weekly was 26.7%. [Reproduced with permission from Chassin *et al.*, 1990.]

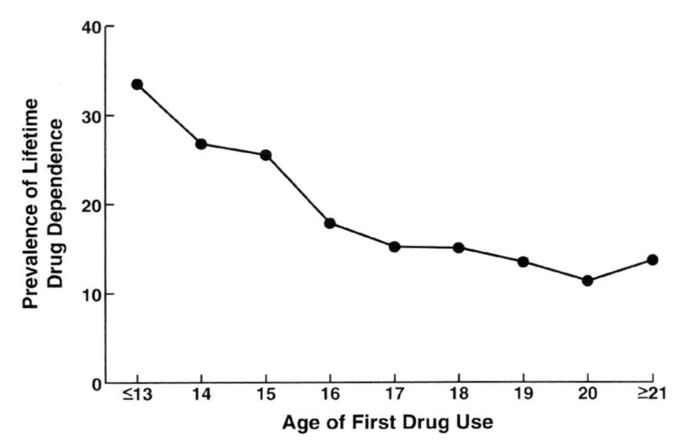

FIGURE 1.7 Prevalence of lifetime drug dependence by age at first drug use. The prevalence of lifetime dependence decreased steeply with increasing age of onset of drug use. Overall, the prevalence of lifetime dependence among those who started using drugs under the age of 14 years was about 34%, dropping sharply to 15.1% for those initiating use at age 17, to about 14% among those initiating use at age 21 or older. [Reproduced with permission from Grant and Dawson, 1998.]

dependence among those who started using drugs under the age of 14 years was 34 per cent; this percentage dropped to 14 per cent for those who started using at age 21 or older (Grant and Dawson, 1998).

In adolescents, it has been proposed that there are stages and pathways of drug involvement (Kandel and Jessor, 2002). There is considerable support for the hypothesis that initiation begins with legal drugs, alcohol and tobacco, and involvement with illicit drugs occurs later in the developmental sequence, marijuana often being the bridge between licit and illicit drugs. However, although this sequence is common, this does not represent an inevitable progression. Only a very small percentage of youths progress from one stage to the next and on to late stage illicit drug use or Dependence.

Genetic contributions to addiction have long been postulated and can result from complex genetic differences that range from alleles that control drug metabolism to hypothesized genetic control over drug sensitivity and environmental influences. Complex genetic influences are those that are genetic but are not due to single-gene effects that produce Mendelian inheritance patterns, as stressed by Uhl and Grow (2004). The classical approaches to complex trait genetics have been the examination of co-occurrence or comorbidity for the trait in monozygotic versus dizygotic twins, reared together or apart, and in analogous family studies with other sorts of biological relatives. Twin and adoption studies can provide researchers with estimates of the extent of genetic effects, termed *heritability* (the proportion of observed variation in a particular trait that can be attributed to inherited genetic factors in contrast to environmental factors). Using such estimates, genetic studies have demonstrated that genetic factors can account for approximately 40 per cent of the total variability of the phenotype (**Table 1.5**). Twin studies suggest significant overlap between genetic predisposition for Dependence on most classes of addictive substances (Karkowski *et al.*, 2000). Clearly, in no case is heritability 100 per cent, which argues strongly for gene–environment interactions, including the stages of the addiction cycle, developmental factors, and social factors.

TABLE 1.5 Heritability Estimates for Drug Dependence

	Males	Females
Cocaine	44%	65%
Heroin (opiates)	43%	—
Marijuana	33%	79%
Tobacco	53%	62%
Alcohol	49% (40–60%)	64%
Addiction overall	40%	

Male cocaine, heroin, marijuana: Tsuang *et al.*, 1996. *Male nicotine:* Carmelli *et al.*, 1990. *Female cocaine*: Kendler and Prescott, 1998b. *Female marijuana*: Kendler and Prescott, 1998a. *Female nicotine*: Kendler *et al.*, 1999. *Male alcohol*: Liu *et al.*, 2004; Prescott and Kendler, 1999; McGue *et al.*, 1992. *Female alcohol*: McGue *et al.*, 1992. *Addiction overall*: Uhl and Grow, 2004.

It also should be emphasized that genetic and environmental factors can convey not only vulnerability, but also protection against drug abuse. Certain Asian populations missing one or more alleles for acetaldehyde dehydrogenase show significantly less vulnerability to alcoholism (Goedde *et al.*, 1983a,b; Mizoi *et al.*, 1983; Higuchi *et al.*, 1995). There is also similar evidence developing for individuals with a genetic defect in metabolizing nicotine (Tyndale and Sellers, 2002; Sellers *et al.*, 2003). Clearly, there are also protective factors within the social environment that can promote competent adaptation and as a result prevent drug abuse (Dishion and McMahon, 1998).

NEUROADAPTATIONAL VIEWS OF ADDICTION

Behavioral Sensitization

Repeated exposure to many drugs of abuse results in a progressive and enduring enhancement in the motor stimulant effect elicited by a subsequent challenge. The phenomenon of *behavioral sensitization* has been thought to underlie some aspects of drug addiction

(Vanderschuren and Kalivas, 2000). Behavioral or psychomotor sensitization, as defined by increased locomotor activation produced by repeated administration of a drug, is more likely to occur with intermittent exposure to drugs, whereas tolerance is more likely to occur with continuous exposure. This phenomenon was observed and characterized in the 1970s and 1980s for various drugs (Babbini *et al.*, 1975; Eichler and Antelman, 1979; Bartoletti *et al.*, 1983a,b; Kolta *et al.*, 1985). Another intriguing aspect is that it has been suggested that the sensitization grows with the passage of time (Antelman *et al.*, 1983, 1986, 2000). Moreover, stress and stimulant sensitization effects show cross-sensitization (Antelman *et al.*, 1980). Psychomotor sensitization is linked invariably to a sensitization of the activity of the mesolimbic dopamine system (Robinson and Berridge, 1993).

A conceptualization of the role of psychomotor sensitization in drug addiction has been proposed where a shift in an incentive-salience state described as *wanting*, as opposed to *liking*, was hypothesized to be progressively increased by repeated exposure to drugs of abuse (Robinson and Berridge, 1993) (**Fig. 1.8**). The transition to pathologically strong *wanting* or craving was proposed to define compulsive use.

The theory posits that there is no causal relationship between the subjective pleasurable effects of the drugs (drug *liking*) and the motivation to take drugs (drug *wanting*). The brain systems that are sensitized do not mediate the pleasurable or euphoric effects of drugs, but instead they mediate a subcomponent of reward termed *incentive salience* (i.e., motivation to take drug or drug *wanting*). It is the psychological process of incentive-salience specifically that is responsible for instrumental drug-seeking and drug-taking behavior (*wanting*) (Robinson and Berridge, 2003). When sensitized, this incentive-salience process produces compulsive patterns of drug use. By means of associative learning, the enhanced incentive value becomes oriented specifically toward drug-related stimuli, leading to escalating compulsion for seeking and taking drugs. The underlying sensitization of neural structures persists, making addicts vulnerable in the long-term to relapse.

The theory posits:

> '... it is specifically sensitization of incentive salience attribution to representation of drug cues and drug-taking that cause the pursuit of drugs and persisting vulnerability to relapse and addiction ... Individuals are guided to incentive stimuli by the influence of Pavlovian stimulus-stimulus (S-S) associations on motivational systems, which is psychologically separable from the symbolic cognitive systems that mediate conscious desire, declarative expectancies of reward, and act-outcome representations' (Robinson and Berridge, 2003).

Counteradaptation–Opponent-Process

Counteradaptation hypotheses have long been proposed to explain tolerance and withdrawal and the motivational changes associated with the development of addiction. Here, the initial acute effect of the drug is opposed or counteracted by homeostatic changes in systems that mediate primary drug effects (Solomon and Corbit, 1974; Siegel, 1975; Poulos and Cappell, 1991). The origins of such counteradaptive hypotheses can be traced to some of the earlier work on physical dependence (Himmelsbach, 1943), and the counter-adaptive changes associated with acute and chronic opioid administration on physiological measures.

Martin (1968) proposed a homeostatic and redundancy theory of tolerance and dependence to opioids that had a striking resemblance to what was to follow as a more general *opponent process* theory by other researchers (see below). Based on studies of acute tolerance and physical dependence produced in dogs by infusing 8 mg/kg of morphine per hour for 7–8 h, and subsequently precipitating abstinence with 20 mg/kg of the mixed agonist/antagonist nalorphine, a regular sequence of changes in physiological parameters such as temperature took place over time. Martin argued that the following sequence of events transpired in the development of *acute tolerance* (**Fig. 1.9**, left side). Morphine lowered the homeostat—that is, lowered the thermoregulatory set point (A). The difference between the homeostatic level and the level of the internal environment or existing state (B) gave rise to an error force (C) which in turn drove a physiological system (D) to rectify the error. In the case of a temperature change, this was effected by panting in the dog. As the error force was diminished, the level of function of the physiological system rectifying the error also diminished, and acute tolerance developed.

A similar scheme explained *acute physical dependence* except that initially, nalorphine rapidly reversed the effects of morphine on body temperature (A), restoring the homeostat to a control level. However, a new error force of the opposite valence was established by the nalorphine (C) which recruited heat-generating mechanisms (D) (**Fig. 1.9**). These error forces remained until a new equilibrium state was achieved (D).

Similar shifts in the homeostatic level of control were hypothesized for signs of withdrawal and precipitated abstinence in *chronically dependent subjects*. The mechanism involved was the same, except that with chronic physical dependence, chronic

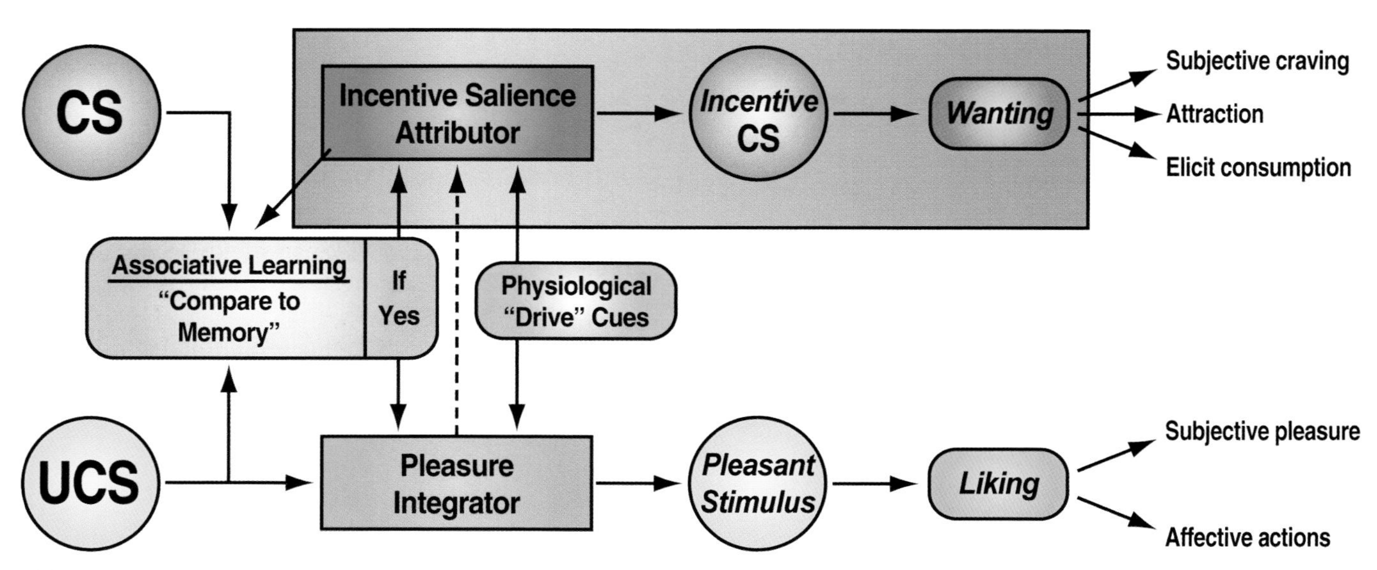

FIGURE 1.8 A schematic illustration of a model which proposes a process of incentive salience and accounts for the consequences of drug-induced sensitization. In this modified model of incentive motivation, the psychological process (and neural substrate) for pleasure (*liking*) is separate from the psychological process (and neural substrate) responsible for incentive salience (*wanting*). Robinson and Berridge (1993) proposed that the activation of esotelencephalic dopamine systems plays a *direct* role only in the process of *wanting* via the attribution of incentive salience to the perception and representation of conditioned stimuli (CS). This portion of the model (i.e., the psychological process) is sensitized by repeated drug administration. It is the hyperactivation of this specific psychological process (incentive salience), due to sensitization of its neural substrate by drugs, that results in the excessive attribution of incentive salience to drug-related stimuli. Whereas normal levels of incentive salience attribution results in normal *wanting*, hyperactivation of this system is hypothesized to result in excessive incentive salience attribution, which is experienced as craving. Craving is pathologically intense *wanting*. The major difference between this model of incentive motivation and the traditional model is that psychological processes and neural substrates responsible for pleasure (*liking*) are separate from those for incentive salience (*wanting*). Thus, natural incentives (unconditioned stimuli [UCS]) produce pleasure directly, but produce incentive salience and elicit goal-directed approach behavior only indirectly (as indicated by the dashed arrow from 'pleasure integrator' to the 'incentive salience attributor'). The direction of incentive salience attribution to stimuli that preceded or accompanied incentive salience activation is determined by associative learning. Thus, activation of the incentive salience attributor by an unconditioned stimulus results in incentive salience being assigned to the perception of conditioned stimuli that were originally neutral (such as the sight of a syringe) and to their mental representations. This is what makes conditioned stimuli attractive and 'wanted' and able to elicit approach. Conditioned stimuli (and unconditioned stimuli) are always compared against past associative memories. Without the direction provided by associative learning, incentive salience could not be focused on any single target. Although diffuse attribution of incentive salience would be both psychologically and behaviorally activating, without associative direction it would not be sufficient to guide behavior toward a specific goal. Familiar conditioned stimuli that have been paired with incentive salience attribution in the past are the target of incentive salience when encountered again, especially when an animal is in particular physiological states (indicated by the arrow from *Physiological 'Drive' Cues*). Incentive salience assigned to conditioned stimuli must be further 'reboosted' each time they are paired again with salience activation (indicated by the dashed arrow from the incentive salience attributor to associative learning). Disruption of this reboosting, by neuroleptics for example, can produce 'extinction mimicry' or decay of incentive value. Ordinarily, incentive salience is assigned only to stimuli that have been paired with pleasure. But brain manipulations (such as drugs or electrical brain stimulation) may circumvent pleasure, by activating the neural substrate of incentive salience directly. This will result in the attribution of incentive salience to associated stimuli and actions and result in their becoming 'wanted,' even in the absence of pleasure. This can be considered a kind of 'sham reward.' Sensitization of the neural substrate for incentive salience will lead to pathological *wanting* (craving) for stimuli associated with its excessive activation (e.g., those involved in drug taking), even if this produces little or no pleasure. As mentioned, the direction of incentive salience by associative learning is the primary determinant of exactly which stimuli become craved. Thus, in the addict, drug-paired stimuli which have been experienced repeatedly in association with the excessive stimulation of dopamine systems become the nearly exclusive targets for the attribution of incentive salience. Other contributions of associative learning are also possible in this model. For example, the pleasure elicited by an unconditioned stimulus can change with repeated experience, as when one develops an appreciative palate for Scotch whiskey (this is indicated in the model by the arrow from learning to the 'pleasure integrator'). Also, a conditioned stimulus that has been repeatedly paired with pleasure can itself come to elicit subjective pleasure, as in the example of a conditioned 'high' reported by 'needle freaks' (arrows from the conditioned stimulus to the 'pleasure integrator' via associative learning). But these effects are separate from the attribution of incentive salience, and they have only a relatively weak influence on motivated behavior compared to the craving produced by the attribution of excessive incentive salience. Finally, none of the psychological processes described in this model, except for subjective *wanting* (craving) and subjective pleasure, are apparent to conscious awareness. The interaction among incentive salience, pleasure and associative learning is not available to introspection. Only the final products of the interaction are interpreted by cognitive mechanisms as subjective *wanting* and *liking*. For an addict whose neural substrates of incentive salience have been sensitized, the subjective product is dominated by the intense experience of drug craving. [Reproduced with permission from Robinson and Berridge, 1993.]

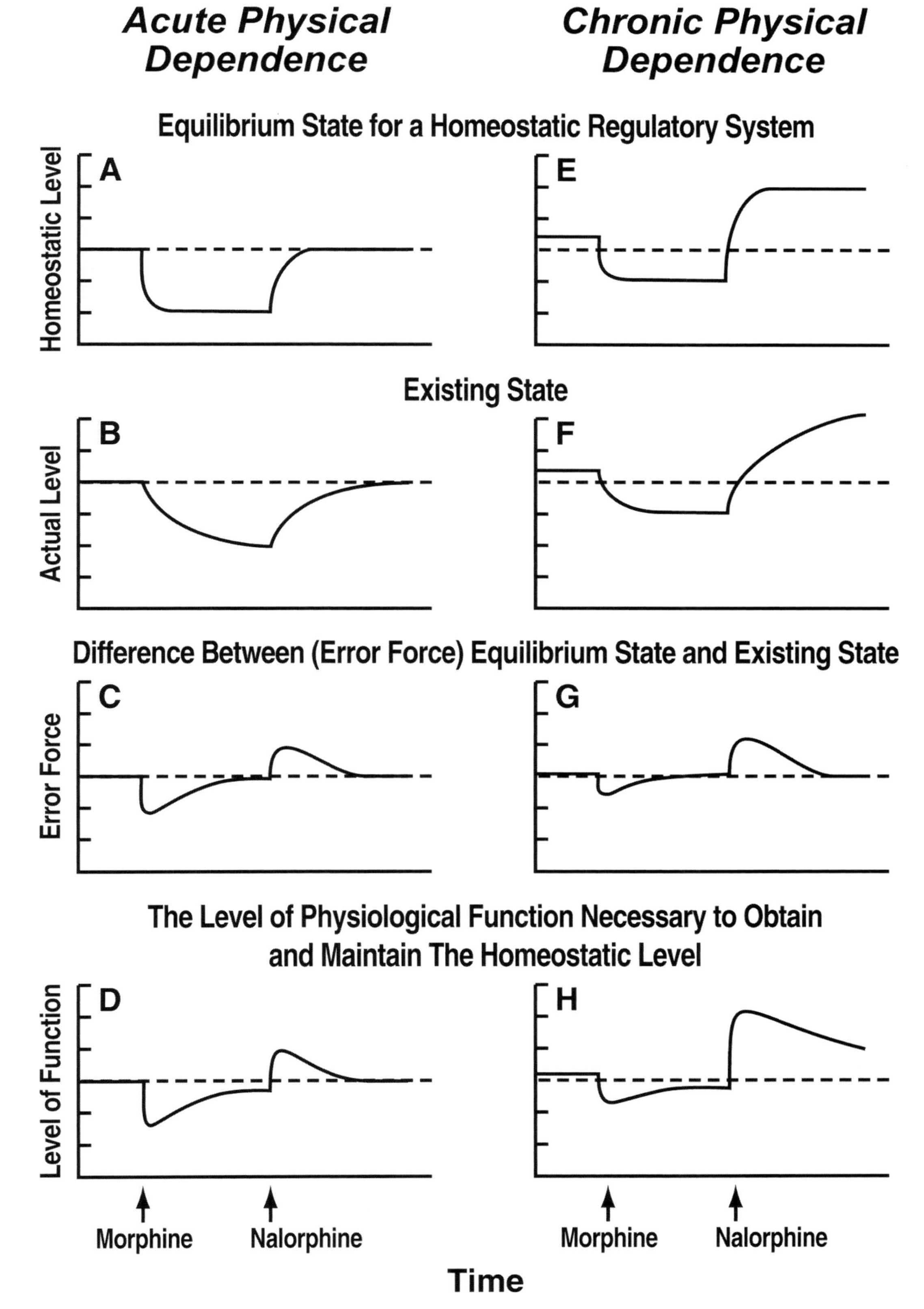

FIGURE 1.9 A general and theoretical formulation of the homeostatic theory of acute and chronic tolerance and physical dependence (see text for details). [Reproduced with permission from Martin, 1968.]

morphine elevated the homeostatic level above the pre-addiction level. In other words, at the time when morphine was administered, the animal was already in early abstinence and the homeostatic level was already slightly above the control level (E) (**Fig. 1.9**). When morphine was administered, the level to which the homeostat was depressed was smaller than it was in the nondependent state. Martin hypothesized that it proceeded from an elevated baseline and the absolute magnitude of the depression was smaller. As a consequence, the error force was smaller (G) (**Fig. 1.9**), and the level of function of the restorative system was lower (H) (**Fig. 1.9**). When nalorphine was administered, a very large error force was generated (G) (**Fig. 1.9**) revealing the true level of the hypersensitized homeostat.

Martin (1968) went on to speculate that the changes observed in homeostatic set point could be explained by redundancy theory where two separate neurochemical systems mediate a given function (Martin and Eades, 1960). When applied to tolerance and dependence, it is assumed that morphine interrupts one of the redundant systems (pathway B) but does not disrupt the other (pathway A). Eventually, pathway A will develop hypertrophy and take over the previous function of pathway B. The tolerance that develops is a consequence of the hypertrophy of the redundant pathway A, not a decrement in the effect on pathway B. When the drug is withdrawn, pathway B returns to its normal level of excitability, but the total system functions at a much higher level because of the contribution of the hypertrophied pathway A. One means of integrating the redundancy theory with the original contra-adaptive theory of Himmelsbach (1942) was to argue that there exists a negative feedback mechanism on pathways A and B that is diminished when pathway A is hypertrophied.

The views of Martin (1968) significantly predate *opponent process* theory (Solomon and Corbit, 1974) and within-system (hypertrophy of pathway A) and between-system (decreased negative feedback of pathway A) neuroadaptations (Koob and Bloom, 1988), but certainly contained elements of both. In addition, as we will see later in the book, Martin's concepts of acute tolerance and acute dependence apply not only to temperature regulation but also to analgesia and the hedonic effects of drugs in humans and animals.

Opponent-process theory was developed during the 1970s by Solomon and colleagues (Solomon and Corbit, 1973, 1974; Hoffman and Solomon, 1974; D'amato, 1974). Since then, it has been applied by many authors to various situations such as drugs (opiates, nicotine, alcohol) to adjunctive drinking, fear conditioning, tonic immobility, ulcer formation, eating disorders, jogging, peer separation, glucose preference, and parachuting (Solomon and Corbit, 1973, 1974; Hoffman and Solomon, 1974; Solomon, 1980).

The theory assumes that the brain contains many affect control mechanisms, working as though they were affect immunization systems that counter or oppose all departures from affective neutrality or equilibrium, whether they be aversive or pleasant (Solomon and Corbit, 1974). The theory is a negative feed-forward control construct designed to keep affect in check even though stimulation is strong. The device is composed of three subparts organized in a temporal manner. Two opposing processes control a summator, which determines the controlling affect at a given moment. First, an unconditional arousing stimulus triggers a primary affective process, termed the *a-process*. It is an unconditional reaction that translates the intensity, quality, and duration of the stimulus (for example, a first opiate intake). Second, as a consequence of the *a-process*, and inherently linked to it on a biological basis, the *b-process* is evoked after a short delay, an opponent process. Empirically, the *b-process* feeds a negative signal into the summator, subtracting from the impact on the summator the already existing *a-process*. The two responses are consequently and temporarily linked (*a* triggers *b*) but were hypothesized to depend on different neurobiological mechanisms. The *b-process* has a longer latency, but some data show that it may appear soon after the beginning of the stimulus in the course of the stimulus action (Larcher *et al.*, 1998). The *b-process* also has more inertia, a slower recruitment, and a more sluggish decay. At a given moment, the pattern of affect will be the algebraic sum of these opposite influences and the dynamics reveal, with the passage of time, the net product of the opponent process (Solomon, 1980) (**Fig. 1.10**).

In this opponent-process theory from a drug addiction perspective, tolerance and dependence are inextricably linked (Solomon and Corbit, 1974). Solomon argued that the first few self-administrations of an opiate drug produce a pattern of motivational changes where the onset of the drug effect produces a euphoria that is the *a-process*, and this is followed by a decline in intensity. Then, after the effects of the drug wear off, the *b-process* emerges as an aversive craving state. The *b-process* gets larger and larger over time, in effect contributing to or producing more complete tolerance to the initial euphoric effects of the drug (**Fig. 1.10**).

What is important to understand is that the dynamics, with the repetition of the stimulus, is the result of a progressive increase in the *b-process*. In other words, the *b-process* sensitizes through drug use, appears more and more rapidly after the unconditional stimulus onset, lasts longer and longer (the conditional effect), and masks the unconditional effect (*a-process*), resulting in

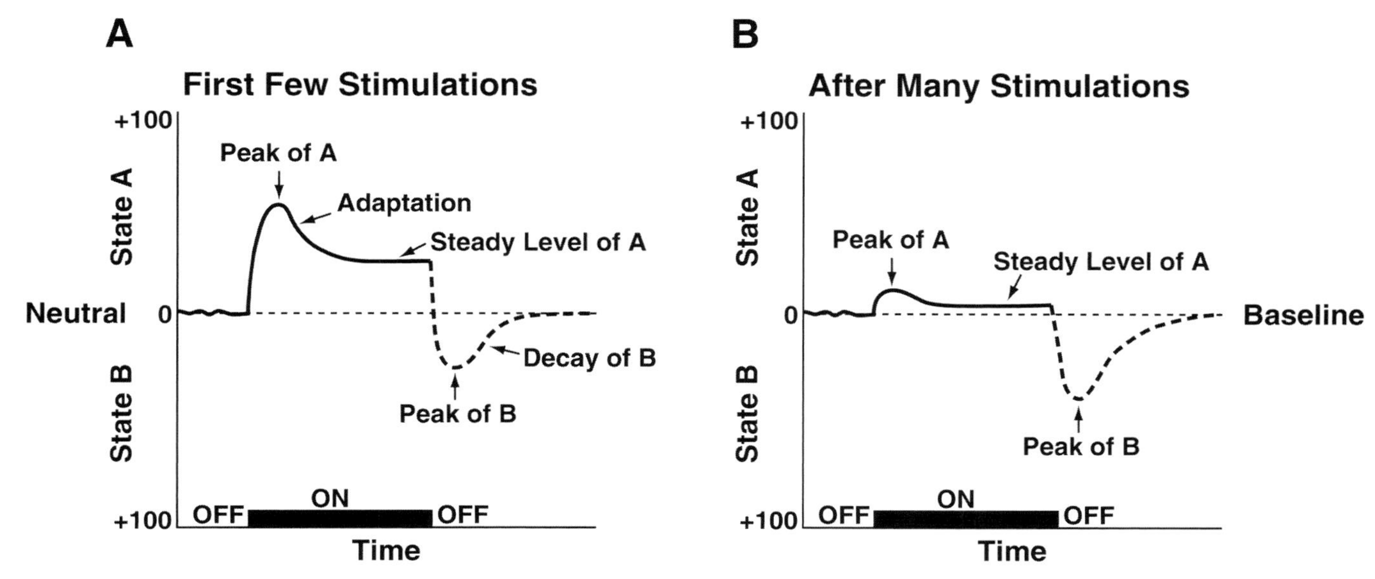

FIGURE 1.10 **(A)** The standard pattern of affective dynamics produced by a relatively novel unconditioned stimulus. **(B)** The standard pattern of affective dynamics produced by a familiar, frequently repeated unconditioned stimulus. [Reproduced with permission from Solomon, 1980.]

an apparent tolerance (Laulin *et al.*, 1999). Experimental data show that if the development of the *b-process* is blocked, no tolerance appears. The unconditioned effect of the drug does not change with repeated drug administration. The development of the *b-process* equals the development of a negative affective state and withdrawal symptoms, in opposition to the hedonic quality of the unconditioned stimulus. Importantly, the nature of the acquired motivation is specified by the nature of the *b-process*, that is, an aversive affect in the case of drug abuse. The subject will work to reduce, terminate, or prevent the negative affect.

Motivational View of Addiction

Rather than focusing on the *physical* signs of dependence, our conceptual framework has focused on *motivational* aspects of addiction. Emergence of a negative emotional state (e.g., dysphoria, anxiety, irritability) when access to the drug is prevented (defined here as dependence) (Koob and Le Moal, 2001), has been associated with this transition from drug use to addiction. Indeed, some have argued that the development of such a negative affective state can define dependence as it relates to addiction:

> 'The notion of dependence on a drug, object, role, activity or any other stimulus-source requires the crucial feature of negative affect experienced in its absence. The degree of dependence can be equated with the amount of this negative affect, which may range from mild discomfort to extreme distress, or it may be equated with the amount of difficulty or effort required to do without the drug, object, etc.' (Russell, 1976).

A key common element that has been identified in animal models is the dysregulation of brain reward function associated with removal from chronic administration of drugs of abuse, and this observation lends credence to the motivational view (see subsequent chapters).

Rapid acute tolerance and opponent process-like effects to the hedonic effects of cocaine have been reported in human studies of smoked coca paste (Van Dyke and Byck, 1982) (**Fig. 1.11**). After a single smoking session, the onset and intensity of the 'high' are very rapid via the smoked route of administration, and a rapid tolerance is manifest in that the 'high' decreases rapidly despite significant blood levels of cocaine. Even more intriguing is that human subjects also actually report a subsequent 'dysphoria', again despite significant blood levels of cocaine. Intravenous cocaine produced similar patterns of a rapid 'rush' followed by an increased 'low' in human laboratory studies (Breiter *et al.*, 1997) (**Fig. 1.12**).

The hypothesis that compulsive use of cocaine is accompanied by a chronic perturbation in brain reward homeostasis has been tested in an animal model of escalation in drug intake with prolonged access. Animals implanted with intravenous catheters and allowed differential access to intravenous self-administration

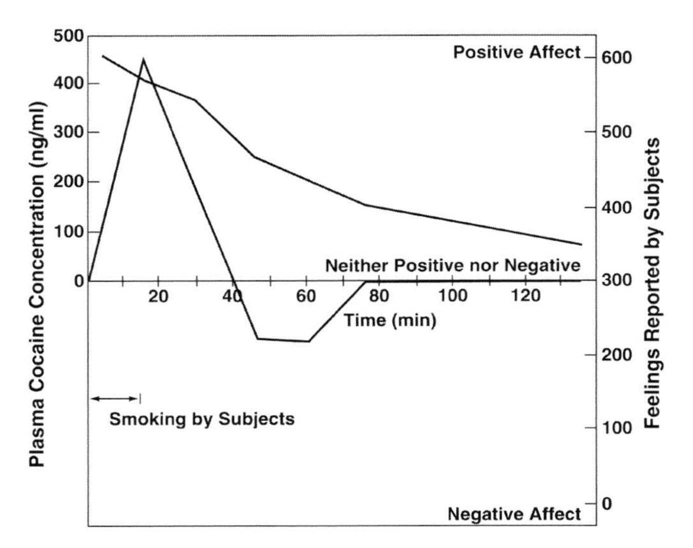

FIGURE 1.11 Dysphoric feelings followed the initial euphoria in experimental subjects who smoked cocaine paste, even though the concentration of cocaine in the plasma of the blood remained relatively high. The dysphoria is characterized by anxiety, depression, fatigue, and a desire for more cocaine. The peak feelings for the subjects were probably reached shortly before the peak plasma concentration, but the first psychological measurements were made later than the plasma assay. Hence, the temporal sequence of the peaks shown cannot be regarded as definitive. [Reproduced with permission from Van Dyke and Byck, 1982.]

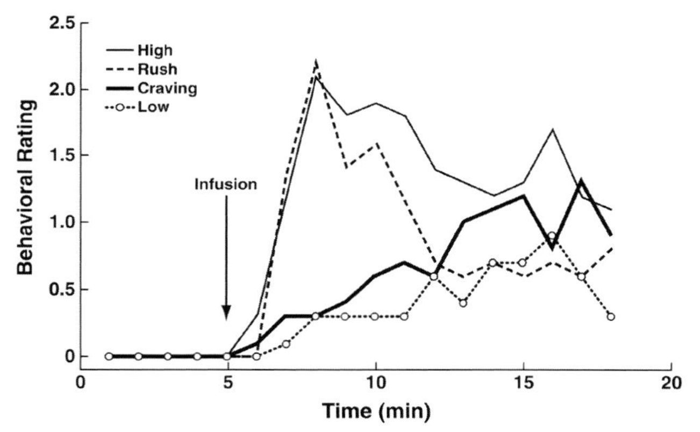

FIGURE 1.12 Average behavioral ratings after an infusion of cocaine (0.6 mg/kg over 30 s; $n = 9$). The rush, high, low, and craving ratings were averaged within each category for the subjects who had interpretable cocaine functional magnetic resonance imaging data after motion correction and behavioral ratings time-locked to the scanner. Both peak rush and peak high occurred 3 min post-infusion. Peak low (primary reports of dysphoria and paranoia) occurred 11 min post-infusion. Peak craving occurred 12 min post-infusion. No subject reported effects from the saline infusion on any of the four measures. Ratings obtained for rush, high, low, and craving measures were higher in subjects blinded to the 0.6 mg/kg cocaine dose compared to subjects unblinded to a 0.2 mg/kg cocaine dose. [Reproduced with permission from Breiter *et al.*, 1997.]

of cocaine show increases in cocaine self-administration from day to day in the long-access group (6 h; LgA) but not in the short-access group (1 h; ShA) (Ahmed and Koob, 1998; Deroche-Gamonet *et al.*, 2004; Mantsch *et al.*, 2004). The differential exposure to cocaine self-administration had dramatic effects on intracranial self-stimulation (ICSS) reward thresholds. ICSS thresholds progressively elevated for LgA rats, but not for ShA or control rats across successive self-administration sessions (Ahmed *et al.*, 2002) (see *Psychostimulants* chapter). Elevation in baseline ICSS thresholds temporally preceded and was highly correlated with escalation in cocaine intake. Post-session elevations in ICSS reward thresholds failed to return to baseline levels before the onset of each subsequent self-administration session, thereby deviating more and more from control levels. The progressive elevation in reward thresholds was associated with the dramatic escalation in cocaine consumption that was observed previously. After escalation had occurred, an acute cocaine challenge facilitated brain reward responsiveness to the same degree as before but resulted in higher absolute brain reward thresholds in LgA when compared to ShA rats.

With intravenous cocaine self-administration in animal models, such elevations in reward threshold begin rapidly and can be observed within a single session of self-administration (Kenny *et al.*, 2003) (**Fig. 1.13**), bearing a striking resemblance to human subjective reports. These results demonstrate that the elevation in brain reward thresholds following prolonged access to cocaine failed to return to baseline levels between repeated, prolonged exposure to cocaine self-administration (i.e., residual hysteresis), thus creating a greater and greater elevation in 'baseline' ICSS thresholds. These data provide compelling evidence for brain reward dysfunction in escalated cocaine self-administration that provide strong support for a hedonic allostasis model of drug addiction.

Allostasis and Neuroadaptation

More recently, opponent process theory has been expanded into the domains of the neurocircuitry and neurobiology of drug addiction from a physiological perspective. An allostatic model of the brain motivational systems has been proposed to explain the persistent changes in motivation that are associated with vulnerability to relapse in addiction, and this model may generalize to other psychopathology associated with dysregulated motivational systems. Allostasis from the addiction perspective has been defined as the process of maintaining apparent reward

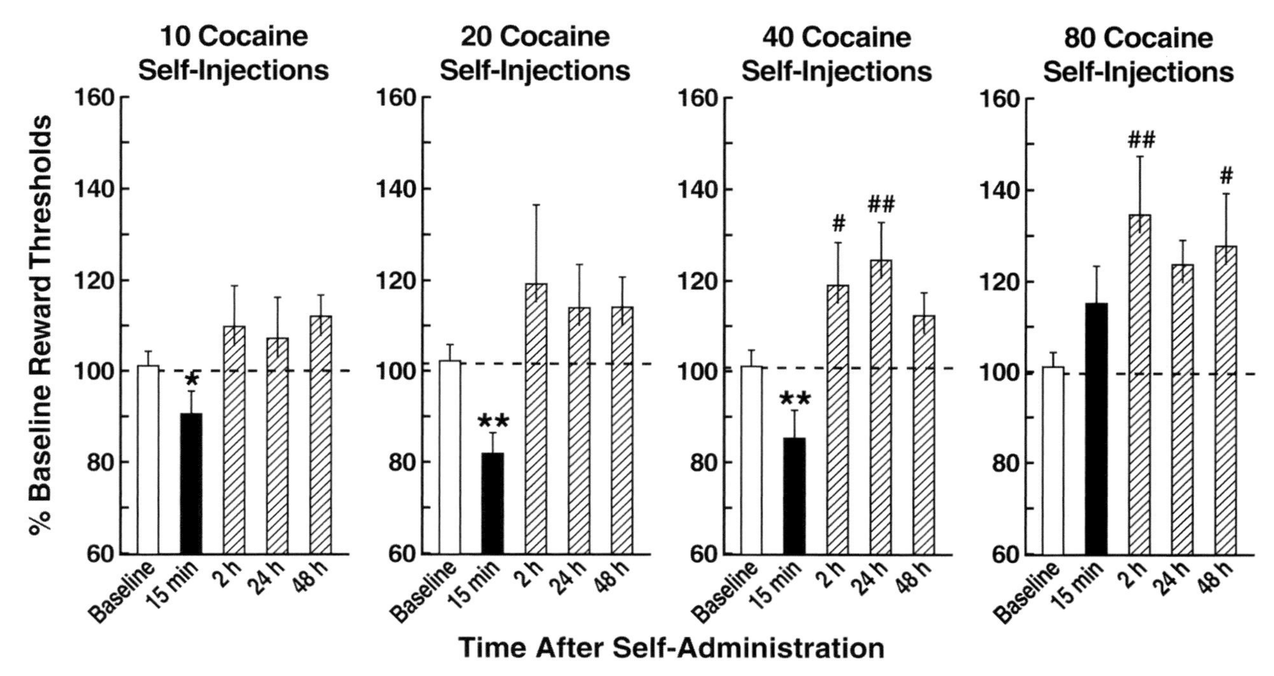

FIGURE 1.13 Rats ($n = 11$) were allowed to self-administer 10, 20, 40, and 80 injections of cocaine (0.25 mg per injection), and ICSS thresholds were measured 15 min and 2, 24, and 48 h after the end of each intravenous cocaine self-administration session. The horizontal dotted line in each plot represents 100% of baseline levels. All data are presented as mean + SEM percentage of baseline ICSS thresholds. *$p < 0.05$, **$p < 0.01$ compared to baseline; paired t-test. #$p < 0.05$, ##$p < 0.01$ compared to baseline; Fisher's LSD test after a statistically significant effect in the repeated-measures analysis of variance. [Reproduced with permission from Kenny *et al.*, 2003.]

function stability through changes in brain reward mechanisms (Koob and Le Moal, 2001). The allostatic state represents a chronic deviation of reward set point that often is *not* overtly observed while the individual is actively taking the drug. Thus, the allostatic view is that not only does the *b-process* get larger with repeated drug taking, but the reward set point from which the *a-process* and *b-process* are anchored gradually shifts downward creating an allostatic state (Koob and Le Moal, 2001) (**Fig. 1.14**).

The allostatic state is fueled not only by dysregulation of neurochemical elements of reward circuits per se, but also by the activation of brain and hormonal stress responses (see *Neurobiological Theories of Addiction* chapter). From the perspective of a given drug, it is unknown whether the hypothesized reward dysfunction is specific to that drug, common to all addictions, or a combination of both perspectives. However, from the data generated to date, and the established anatomical connections, the manifestation of this allostatic state as compulsive drug-taking and loss of control over drug-taking is hypothesized to be critically based on dysregulation of specific neurotransmitter function in the central division of the extended amygdala (a basal forebrain macrostructure comprised of the central nucleus of the amygdala, bed nucleus of the stria terminalis, and a transition area in the region of the shell of the nucleus accumbens) (Koob *et al.*, 1998) (see *Neurobiological Theories of Addiction* chapter). Decreases in the function of γ-aminobutyric acid, dopamine, serotonin, and opioid peptides, as well as dysregulation of brain stress systems such as corticotropin-releasing factor and neuropeptide Y are hypothesized to contribute to a shift in reward set point. Thus, a chronic elevation in reward thresholds as elaborated in Koob and Le Moal (2001) is viewed as a key element in the development of addiction and as setting up other sources of self-regulation failure and persistent vulnerability to relapse (protracted abstinence).

It is hypothesized further that the pathology of this neurocircuitry is the basis for the emotional dysfunction long associated with drug addiction and alcoholism in humans. Some of this neurocircuitry pathology persists into protracted abstinence, thereby providing a strong motivational basis for relapse. The view that drug addiction and alcoholism are the pathology that results from an allostatic mechanism that usurps the circuits established for natural rewards provides a realistic approach to identifying the

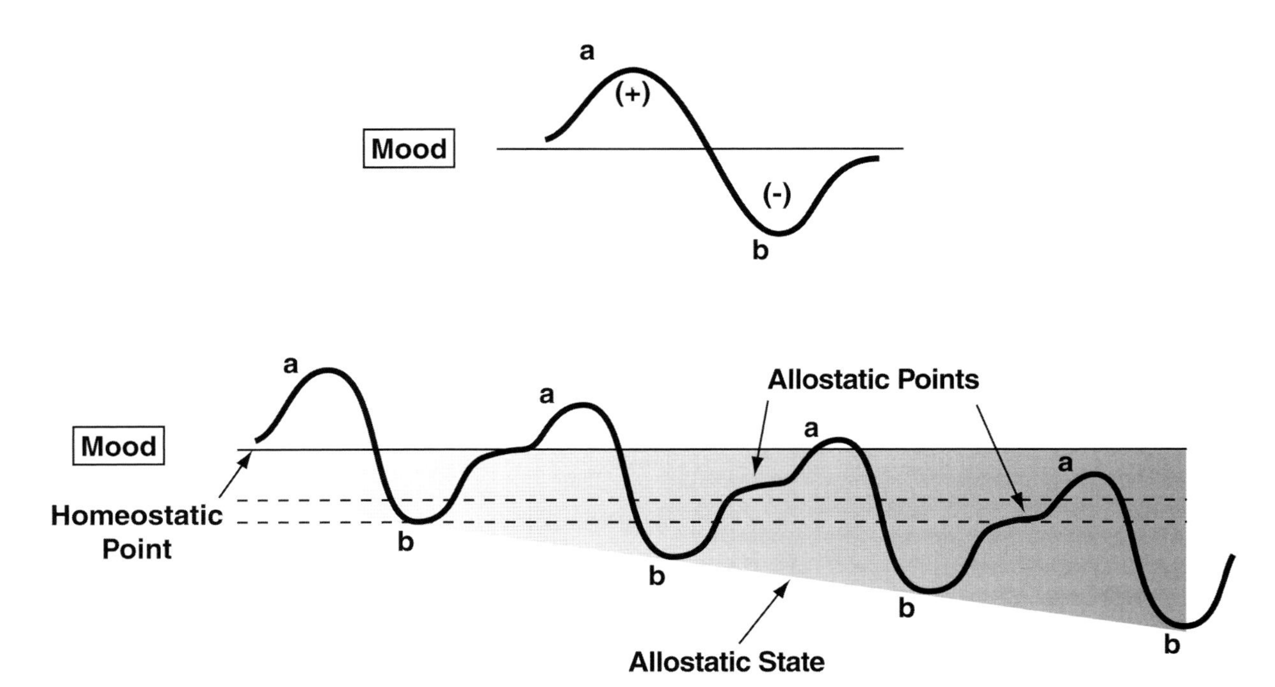

FIGURE 1.14 Diagram illustrating an extension of Solomon and Corbit's (1974) opponent-process model of motivation to outline the conceptual framework of the allostatic hypothesis. Both panels represent the affective response to the presentation of a drug. (Top) This diagram represents the initial experience of a drug with no prior drug history. The *a-process* represents a positive hedonic or positive mood state, and the *b-process* represents the negative hedonic or negative mood state. The affective stimulus (state) has been argued to be a sum of both an *a-process* and a *b-process*. An individual experiencing a positive hedonic mood state from a drug of abuse with sufficient time between re-administering the drug is hypothesized to retain the *a-process*. In other words, an appropriate counteradaptive opponent process (*b-process*) that balances the activational process (*a-process*) does not lead to an allostatic state. (Bottom) The changes in the affective stimulus (state) in an individual with repeated frequent drug use that may represent a transition to an allostatic state in the brain reward systems and, by extrapolation, a transition to addiction. Note that the apparent *b-process* never returns to the original homeostatic level before drug-taking is reinitiated, thus creating a greater and greater allostatic state in the brain reward system. In other words, the counteradaptive opponent process (*b-process*) does not balance the activational process (*a-process*) but in fact shows a residual hysteresis. While these changes are exaggerated and condensed over time in the present conceptualization, the hypothesis here is that even during post-detoxification, a period of protracted abstinence, the reward system is still bearing allostatic changes. In the nondependent state, reward experiences are normal, and the brain stress systems are not greatly engaged. During the transition to the state known as addiction, the brain reward system is in a major underactivated state while the brain stress system is highly activated. The following definitions apply: *allostasis*, the process of achieving stability through change; *allostatic state*, a state of chronic deviation of the regulatory system from its normal (homeostatic) operating level; *allostatic load*, the cost to the brain and body of the deviation, accumulating over time, and reflecting in many cases, pathological states and accumulation of damage. [Reproduced with permission from Koob and Le Moal, 2001.]

neurobiological factors that produce vulnerability to addiction and relapse.

The neurobiological view of drug addiction presented in this book represents a neuroadaptational perspective that is shared by most current neurobiological theories. Controversies exist, however, over the importance of the phenomenon of psychomotor sensitization associated with the mesolimbic dopamine system (see *Neurobiological Theories of Addiction* chapter). According to the psychomotor sensitization conceptual framework, the *wanting* and *liking* of drugs are separate phenomena with separate neurobiological substrates, and a shift in an incentive-salience state described as *wanting* was hypothesized to be progressively increased by repeated exposure to drugs of abuse (Robinson and Berridge, 1993). However, the allostatic-neuroadaptational position is that locomotor sensitization may play a role in initial sensitivity to a drug, but that it disappears or becomes irrelevant with the development of motivational dependence. Intertwined with the psychomotor sensitization hypothesis is a prominent or even critical role for dopamine in the motivational effects of drugs of abuse. The allostatic- neuroadaptational position is that dopamine has a role in addiction, particularly for psychomotor stimulants, but is not critical nor sufficient for the development of addiction to many drugs of abuse such as opiates, alcohol, phencyclidine, and others. An extension of the psychomotor sensitization and dopamine theories of addiction is the dismissal of drug withdrawal as a motivating factor in drug addiction. The allostatic-neuroadaptational position is that drug withdrawal is largely misunderstood by the neurobiology of drug addiction research community. The focus should not be on *physical* withdrawal, which for the allostatic-neuroadaptational position is largely a marker for dependence, but rather on *motivational* withdrawal which allostatic-neuroadaptational hypotheses hold as one of the key elements of drug addiction (see *Drug Addiction: Transition from Neuroadaptation to Pathophysiology* chapter).

SUMMARY

This chapter defines addiction as a chronic relapsing disorder characterized by compulsive drug seeking, a loss of control in limiting intake, and emergence of a negative emotional state when access to the drug is prevented. The definition of addiction is derived from the evolution of the concept of dependence and the nosology of addiction diagnosis, and a distinction is made between drug use, drug abuse, and drug addiction. Addiction affects overall a large percentage of society, including illicit drugs, licit drugs, alcohol, and tobacco, and has with it enormous monetary costs. Addiction is further conceptualized as a condition that evolves, moving from impulsivity to compulsivity, and ultimately being comprised of three major stages: binge/intoxication, withdrawal/negative affect and preoccupation/anticipation. Motivational, psychodynamic, social psychological, and vulnerability factors all contribute to the etiology of addiction, but the focus of the conceptualization for this book is placed on the neuroadaptational changes that occur during the addiction cycle. A theoretical framework is described that derives from early homeostatic theories and subsequent opponent process theories that provides a heuristic framework for understanding the neurobiology of addiction. This framework is followed in each subsequent major drug class, each covered by a separate chapter (*Psychostimulants, Opioids, Alcohol, Nicotine, and Cannabis*) and is tied together in the *Imaging, Neurobiological Theories of Addiction* and *Drug Addiction: Transition from Neuroadaptation to Pathophysiology* chapters.

REFERENCES

Ahmed, S. H., and Koob, G. F. (1998). Transition from moderate to excessive drug intake: Change in hedonic set point. *Science* **282**, 298–300.

Ahmed, S. H., Kenny, P. J., Koob, G. F., and Markou, A. (2002). Neurobiological evidence for hedonic allostasis associated with escalating cocaine use. *Nature Neuroscience* **5**, 625–626.

American Psychiatric Association (1994). *Diagnostic and Statistical Manual of Mental Disorders*, 4th ed. American Psychiatric Press, Washington DC.

Antelman, S. M., Eichler, A. J., Black, C. A., and Kocan, D. (1980). Interchangeability of stress and amphetamine in sensitization. *Science* **207**, 329–331.

Antelman, S. M., DeGiovanni, L. A., Kocan, D., Perel, J. M., and Chiodo, L. A. (1983). Amitriptyline sensitization of a serotonin-mediated behavior depends on the passage of time and not repeated treatment. *Life Sciences* **33**, 1727–1730.

Antelman, S. M., Kocan, D., Edwards, D. J., Knopf, S., Perel, J. M., and Stiller, R. (1986). Behavioral effects of a single neuroleptic treatment grow with the passage of time. *Brain Research* **385**, 58–67.

Antelman, S. M., Levine, J., and Gershon, S. (2000). Time-dependent sensitization: the odyssey of a scientific heresy from the laboratory to the door of the clinic. *Molecular Psychiatry* **5**, 350–356.

Anthony, J. C., Warner, L. A., and Kessler, R. C. (1994). Comparative epidemiology of dependence on tobacco, alcohol, controlled substances, and inhalants: Basic findings from the National Comorbidity Survey. *Experimental and Clinical Psychopharmacology* **2**, 244–268.

Aytaclar, S., Tarter, R. E., Kirisci, L., and Lu, S. (1999). Association between hyperactivity and executive cognitive functioning in childhood and substance use in early adolescence. *Journal of the American Academy of Child and Adolescent Psychiatry* **38**, 172–178.

Babbini, M., Gaiardi, M., and Bartoletti, M. (1975). Persistence of chronic morphine effects upon activity in rats 8 months after ceasing the treatment. *Neuropharmacology* **14**, 611–614.

Bartoletti, M., Gaiardi, M., Gubellini, C., and Babbini, M. (1983a). Further evidence for a motility substitution test as a tool to detect the narcotic character of new drugs in rats. *Neuropharmacology* **22**, 177–181.

Bartoletti, M., Gaiardi, M., Gubellini, G., Bacchi, A., and Babbini, M. (1983b). Long-term sensitization to the excitatory effects of morphine: a motility study in post-dependent rats. *Neuropharmacology* **22**, 1193–1196.

Baumeister, R. F., Heatherton, T. F., and Tice, D. M. (Eds.), (1994). *Losing Control: How and Why People Fail at Self-Regulation*, Academic Press, San Diego.

Biederman, J., Wilens, T., Mick, E., Faraone, S. V., Weber, W., Curtis, S., Thornell, A., Pfister, K., Jetton, J. G., and Soriano, J. (1997). Is ADHD a risk factor for psychoactive substance use disorders? Findings from a four-year prospective follow-up study. *Journal of the American Academy of Child and Adolescent Psychiatry* **36**, 21–29.

Biederman, J., Wilens, T., Mick, E., Spencer, T., and Faraone, S. V. (1999). Pharmacotherapy of attention-deficit/hyperactivity disorder reduces risk for substance use disorder. *Pediatrics* **104**, e20.

Breiter, H. C., Gollub, R. L., Weisskoff, R. M., Kennedy, D. N., Makris, N., Berke, J. D., Goodman, J. M., Kantor, H. L., Gastfriend, D. R., Riorden, J. P., Mathew, R. T., Rosen, B. R., and Hyman, S. E. (1997). Acute effects of cocaine on human brain activity and emotion. *Neuron* **19**, 591–611.

Breslau, N., Fenn, N., and Peterson, E. L. (1993). Early smoking initiation and nicotine dependence in a cohort of young adults. *Drug and Alcohol Dependence* **33**, 129–137.

Carmelli, D., Swan, G. E., Robinette, D., and Fabsitz, R. R. (1990). Heritability of substance use in the NAS-NRC Twin Registry. *Acta Geneticae Medicae et Gemellologiae* **39**, 91–98.

Centers for Disease Control and Prevention (2004). *Targeting Tobacco Use: The Nation's Leading Cause of Death*, Centers for Disease Control and Prevention, Atlanta.

Chassin, L., Presson, C. C., Sherman, S. J., and Edwards, D. A. (1990). The natural history of cigarette smoking: predicting young-adult smoking outcomes from adolescent smoking patterns. *Health Psychology* **9**, 701–716.

Chung, T., and Martin, C. S. (2001). Classification and course of alcohol problems among adolescents in addictions treatment programs. *Alcoholism: Clinical and Experimental Research* **25**, 1734–1742.

Crowley, T. J., Macdonald, M. J., Whitmore, E. A., and Mikulich, S. K. (1998). Cannabis dependence, withdrawal, and reinforcing effects among adolescents with conduct symptoms and substance use disorders. *Drug and Alcohol Dependence* **50**, 27–37.

D'Amato, M. R. (1974). Derived motives, *Annual Review of Psychology* **25**, 83–106.

Dawes, M. A., Tarter, R. E., and Kirisci, L. (1997). Behavioral self-regulation: correlates and 2 year follow-ups for boys at risk for substance abuse. *Drug and Alcohol Dependence* **45**, 165–176.

Deroche-Gamonet, V., Belin, D., and Piazza, P. V. (2004). Evidence for addiction-like behavior in the rat. *Science* **305**, 1014–1017.

Dishion, T. J., and McMahon, R. J. (1998). Parental monitoring and the prevention of child and adolescent problem behavior: a conceptual and empirical formulation. *Clinical Child and Family Psychology Review* **1**, 61–75.

Eddy, N. B., Halbach, H., Isbell, H., and Seevers, M. H. (1965). Drug dependence: its significance and characteristics. *Bulletin of the World Health Organization* **32**, 721–733.

Eichler, A. J., and Antelman, S. M. (1979). Sensitization to amphetamine and stress may involve nucleus accumbens and medial frontal cortex. *Brain Research* **176**, 412–416.

Ershler, J., Leventhal, H., Fleming, R., and Glynn, K. (1989). The quitting experience for smokers in sixth through twelfth grades. *Addictive Behaviors* **14**, 365–378.

Escobedo, L. G., Marcus, S. E., Holtzman, D., and Giovino, G. A. (1993). Sports participation, age at smoking initiation, and the risk of smoking among US high school students. *Journal of the American Medical Association* **269**, 1391–1395.

Everett, S. A., Warren, C. W., Sharp, D., Kann, L., Husten, C. G., and Crossett, L. S. (1999). Initiation of cigarette smoking and subsequent smoking behavior among U.S. high school students. *Preventive Medicine* **29**, 327–333.

Finney, J. W., and Moos, R. H. (1992). The long-term course of treated alcoholism: II. Predictors and correlates of 10-year functioning and mortality. *Journal of Studies on Alcohol* **53**, 142–153.

Giancola, P. R., Moss, H. B., Martin, C. S., Kirisci, L., and Tarter, R. E. (1996a). Executive cognitive functioning predicts reactive aggression in boys at high risk for substance abuse: a prospective study. *Alcoholism: Clinical and Experimental Research* **20**, 740–744.

Giancola, P. R., Zeichner, A., Yarnell, J. E., and Dickson, K. E. (1996b). Relation between executive cognitive functioning and the adverse consequences of alcohol use in social drinkers. *Alcoholism: Clinical and Experimental Research* **20**, 1094–1098.

Glantz, M. D., and Hartel, C. R. (Eds.), (1999). *Drug Abuse: Origins and Interventions*. American Psychological Association, Washington DC.

Glantz, M. D., and Pickens, R. W. (Eds.), (1992). *Vulnerability to Drug Abuse*. American Psychological Association, Washington DC.

Glantz, M. D., Weinberg, N. Z., Miner, L. L., and Colliver, J. D. (1999). The etiology of drug abuse: mapping the paths. *In Drug Abuse: Origins and Interventions* (M. D. Glantz, and C. R. Hartel, Eds.), American Psychological Association, Washington DC, pp. 3–45.

Goedde, H. W., Agarwal, D. P., and Harada, S. (1983a). The role of alcohol dehydrogenase and aldehyde dehydrogenase isozymes in alcohol metabolism, alcohol sensitivity and alcoholism. In *Cellular Localization, Metabolism, and Physiology* (series title: *Isozymes: Current Topics in Biological and Medical Research*, vol. 8 (M. C. Rattazzi, J. G. Scandalios, and G. S. Whitt Eds.), pp. 175–193. Alan R. Liss, New York.

Goedde, H. W., Agarwal, D. P., Harada, S., Meier-Tackmann, D., Ruofu, D., Bienzle, U., Kroeger, A., and Hussein, L. (1983b). Population genetic studies on aldehyde dehydrogenase isozyme deficiency and alcohol sensitivity. *American Journal of Human Genetics* **35**, 769–772.

Grant, B. F., and Dawson, D. A. (1998). Age of onset of drug use and its association with DSM-IV drug abuse and dependence: results from the National Longitudinal Alcohol Epidemiologic Survey. *Journal of Substance Abuse* **10**, 163–173.

Grant, B. F., Hasin, D. S., Chou, S. P., Stinson, F. S., and Dawson, D. A. (2004a). Nicotine dependence and psychiatric disorders in the United States: results from the national epidemiologic survey on alcohol and related conditions. *Archives of General Psychiatry* **61**, 1107–1115.

Grant, B. F., Stinson, F. S., Dawson, D. A., Chou, S. P., Dufour, M. C., Compton, W., Pickering, R. P., and Kaplan, K. (2004b). Prevalence and co-occurrence of substance use disorders and independent mood and anxiety disorders: results from the National Epidemiologic Survey on Alcohol and Related Conditions. *Archives of General Psychiatry* **61**, 807–816.

Grant, B. F., Stinson, F. S., Dawson, D. A., Chou, S. P., Ruan, W. J., and Pickering, R. P. (2004c). Co-occurrence of 12-month alcohol and drug use disorders and personality disorders in the United States: results from the National Epidemiologic Survey on Alcohol and Related Conditions. *Archives of General Psychiatry* **61**, 361–368.

Grant, B., Dawson, D., Stinson, F., Chou, P., Dufour, M., and Pickering, R. (2005). The 12-month prevalence and trends in DSM-IV alcohol abuse and dependence: United States, 1991–1992 and 2001–2002. *Drug and Alcohol Dependence*, in press.

Higuchi, S., Matsushita, S., Murayama, M., Takagi, S., and Hayashida, M. (1995). Alcohol and aldehyde dehydrogenase polymorphisms and the risk for alcoholism. *American Journal of Psychiatry* **152**, 1219–1221.

Himmelsbach, C. K. (1942). Clinical studies of drug addiction: Physical dependence, withdrawal and recovery. *Archives of Internal Medicine* **69**, 766–772.

Himmelsbach, C. K. (1943). Can the euphoric, analgetic, and physical dependence effects of drugs be separated? IV With reference to physical dependence. *Federation Proceedings* **2**, 201–203.

Hingson, R., Heeren, T., Zakocs, R., Winter, M., and Wechsler, H. (2003). Age of first intoxication, heavy drinking, driving after drinking and risk of unintentional injury among U.S. college students. *Journal of Studies on Alcohol* **64**, 23–31.

Hoffman, H. S., and Solomon, R. L. (1974). An opponent-process theory of motivation: III. Some affective dynamics in imprinting. *Learning and Motivation* **5**, 149–164.

Hubbard, R. L., Craddock, G., Flynn, P. M., Anderson, J., and Etheridge, R. M. (1997). Overview of 1-year follow-up outcomes in the Drug Abuse Treatment Outcome Study (DATOS). *Psychology of Addictive Behaviors* **11**, 261–278.

Kandel, D. B., and Jessor, R. (2002). The gateway hypothesis revisited. *In Stages and Pathways of Drug Involvement: Examining the Gateway Hypothesis* D. B. Kandel (Ed.), pp. 365–372. Cambridge University Press, New York.

Karkowski, L. M., Prescott, C. A., and Kendler, K. S. (2000). Multivariate assessment of factors influencing illicit substance use in twins from female–female pairs. *American Journal of Medical Genetics* **96**, 665–670.

Kendler, K. S., and Prescott, C. A. (1998a). Cannabis use, abuse, and dependence in a population-based sample of female twins. *American Journal of Psychiatry* **155**, 1016–1022.

Kendler, K. S., and Prescott, C. A. (1998b). Cocaine use, abuse and dependence in a population-based sample of female twins. *British Journal of Psychiatry* **173**, 345–350.

Kendler, K. S., Neale, M. C., Sullivan, P., Corey, L. A., Gardner, C. O., and Prescott, C. A. (1999). A population-based twin study in women of smoking initiation and nicotine dependence. *Psychological Medicine* **29**, 299–308.

Kenny, P. J., Polis, I., Koob, G. F., and Markou, A. (2003). Low dose cocaine self-administration transiently increases but high dose cocaine persistently decreases brain reward function in rats. *European Journal of Neuroscience* **17**, 191–195.

Khantzian, E. J. (1985). The self-medication hypothesis of affective disorders: focus on heroin and cocaine dependence. *American Journal of Psychiatry* **142**, 1259–1264.

Khantzian, E. J. (1990). Self-regulation and self-medication factors in alcoholism and the addictions: similarities and differences. *In Combined Alcohol and Other Drug Dependence* (series title: *Recent Developments in Alcoholism*, vol. 8), (M. Galanter Ed.), pp. 255–271. Plenum Press, New York.

Khantzian, E. J. (1995). The 1994 distinguished lecturer in substance abuse. *Journal of Substance Abuse Treatment* **12**, 157–165.

Khantzian, E. J. (1997). The self-medication hypothesis of substance use disorders: a reconsideration and recent applications. *Harvard Review of Psychiatry* **4**, 231–244.

Khantzian, E. J., and Wilson A. (1993). Substance abuse, repetition, and the nature of addictive suffering. *In Hierarchical Concepts in Psychoanalysis: Theory, Research, and Clinical Practice* (A. Wilson, and J. E. Gedo, Eds.), pp. 263–283. Guilford Press, New York.

Kohut, H. (1971). *The Analysis of the Self* (series title: *The Psychoanalytic Study of the Child*, vol. 4), International Universities Press, New York.

Kolta, M. G., Shreve, P., De Souza, V., and Uretsky, N. J. (1985). Time course of the development of the enhanced behavioral and biochemical responses to amphetamine after pretreatment with amphetamine. *Neuropharmacology* **24**, 823–829.

Koob, G. F. (2003). The neurobiology of self-regulation failure in addiction: an allostatic view [commentary on Khantzian, 'Understanding addictive vulnerability: An evolving psychodynamic perspective'], *Neuro-Psychoanalysis* **5**, 35–39.

Koob, G. F. (2004). Allostatic view of motivation: implications for psychopathology. *In Motivational Factors in the Etiology of Drug Abuse* (series title: *Nebraska Symposium on Motivation*, vol. 50), (R. Bevins, and M.T. Bardo Eds.), pp. 1–18. University of Nebraska Press, Lincoln NE.

Koob, G. F., and Bloom, F. E. (1988). Cellular and molecular mechanisms of drug dependence. *Science* **242**, 715–723.

Koob, G. F., and Le Moal, M. (1997). Drug abuse: Hedonic homeostatic dysregulation. *Science* **278**, 52–58.

Koob, G. F., and Le Moal, M. (2001). Drug addiction, dysregulation of reward, and allostasis. *Neuropsychopharmacology* **24**, 97–129.

Koob, G. F., Sanna, P. P., and Bloom, F. E. (1998). Neuroscience of addiction. *Neuron* **21**, 467–476.

Kopp, P., and Fenoglio, P. (2000). *Le cout Social des Drogues Licites (Alcool et Tabac) et Illicites en France*, etude 22, Observatoire Francais des Drogues et des Toxicomanies, Paris.

Larcher, A., Laulin, J. P., Celerier, E., Le Moal, M., and Simonnet, G. (1998). Acute tolerance associated with a single opiate administration: Involvement of N-methyl-D-aspartate-dependent pain facilitatory systems. *Neuroscience* **84**, 583–589.

Laulin, J. P., Celerier, E., Larcher, A., Le Moal, M., and Simonnet, G. (1999). Opiate tolerance to daily heroin administration: An apparent phenomenon associated with enhanced pain sensitivity. *Neuroscience* **89**, 631–636.

Liu, I. C., Blacker, D. L., Xu, R., Fitzmaurice, G., Lyons, M. J., and Tsuang, M. T. (2004). Genetic and environmental contributions to the development of alcohol dependence in male twins. *Archives of General Psychiatry* **61**, 897–903.

Mantsch, J. R., Yuferov, V., Mathieu-Kia, A. M., Ho, A., and Kreek, M. J. (2004). Effects of extended access to high versus low cocaine doses on self-administration, cocaine-induced reinstatement and brain mRNA levels in rats. *Psychopharmacology* **175**, 26–36.

Martin, W. R. (1968). A homeostatic and redundancy theory of tolerance to and dependence on narcotic analgesics. In *The Addictive States* (series title: *Its Research Publications*, vol. 46), A. Wikler pp. 206–225. Williams and Wilkins, Baltimore.

Martin, W. R., and Eades, C. G. (1960). A comparative study of the effect of drugs on activating and vasomotor responses evoked by midbrain stimulation: atropine, pentobarbital, chlorpromazine and chlorpromazine sulfoxide. *Psychopharmacologia* **1**, 303–335.

McGue, M., Pickens, R. W., and Svikis, D. S. (1992). Sex and age effects on the inheritance of alcohol problems: a twin study. *Journal of Abnormal Psychology* **101**, 3–17.

McLellan, A. T., and McKay, J. (1998). The treatment of addiction: what can research offer practice? *In Bridging the Gap Between Practice and Research: Forging Partnerships with Community-Based Drug and Alcohol Treatment*, S. Lamb, M. R. Greenlick, D. McCarty (Eds.), pp. 147–185. National Academy Press, Washington DC.

McLellan, A. T., Lewis, D. C., O'Brien, C. P., and Kleber, H. D. (2000). Drug dependence, a chronic medical illness: implications for treatment, insurance, and outcomes evaluation. *Journal of the American Medical Association* **284**, 1689–1695.

Merikangas, K. R., Mehta, R. L., Molnar, B. E., Walters, E. E., Swendsen, J. D., Aguilar-Gaziola, S., Bijl, R., Borges, G., Caraveo-Anduaga, J. J., DeWit, D. J., Kolody, B., Vega, W. A., Wittchen, H. U., and Kessler, R. C. (1998). Comorbidity of substance use disorders with mood and anxiety disorders: results of the International Consortium in Psychiatric Epidemiology. *Addictive Behaviors* **23**, 893–907.

Meyer, R. E. (1996). The disease called addiction: emerging evidence in a 200-year debate. *Lancet* **347**, 162–166.

Mizoi, Y., Tatsuno, Y., Adachi, J., Kogame, M., Fukunaga, T., Fujiwara, S., Hishida, S., and Ijiri, I. (1983). Alcohol sensitivity related to polymorphism of alcohol-metabolizing enzymes in Japanese. *Pharmacology Biochemistry and Behavior* 18(Suppl. 1), 127–133.

Nelson, J. E., Pearson, H. W., Sayers, M., and Glynn, T. J. (Eds.), (1982). *Guide to Drug Abuse Research Terminology*, National Institute on Drug Abuse, Rockville MD.

O'Brien, C. P., and McLellan, A. T. (1996). Myths about the treatment of addiction. *Lancet* **347**, 237–240.

Office of National Drug Control Policy, *The Economic Costs of Drug Abuse in the United States: 1992–1998*, Office of National Drug Control Policy, Washington DC, 2001.

Poulos, C. X., and Cappell, H. (1991). Homeostatic theory of drug tolerance: A general model of physiological adaptation. *Psychological Reviews* **98**, 390–408.

Prescott, C. A., and Kendler, K. S. (1999). Genetic and environmental contributions to alcohol abuse and dependence in a population-based sample of male twins. *American Journal of Psychiatry* **156**, 34–40.

Robinson, T. E., and Berridge, K. C. (1993). The neural basis of drug craving: An incentive-sensitization theory of addiction. *Brain Research Reviews* **18**, 247–291.

Robinson, T. E., and Berridge, K. C. (2003). Addiction. *Annual Review of Psychology* **54**, 25–53.

Russell, M. A. H. (1976). What is dependence? *In Drugs and Drug Dependence* (G. Edwards, Ed.), pp. 182–187. Lexington Books, Lexington, MA.

Sellers, E. M., Tyndale, R. F., and Fernandes, L. C. (2003). Decreasing smoking behaviour and risk through CYP2A6 inhibition. *Drug Discovery Today* **8**, 487–493.

Siegel, S. (1975). Evidence from rats that morphine tolerance is a learned response. *Journal of Comparative and Physiological Psychology* **89**, 498–506.

Sifneos, P. E. (2000). Alexithymia, clinical issues, politics and crime. *Psychotherapy and Psychosomatics* **69**, 113–116.

Solomon, R. L. (1980). The opponent-process theory of acquired motivation: the costs of pleasure and the benefits of pain. *American Psychologist* **35**, 691–712.

Solomon, R. L., and Corbit, J. D. (1973). An opponent-process theory of motivation. II. Cigarette addiction. *Journal of Abnormal Psychology* **81**, 158–171.

Solomon, R. L., and Corbit, J. D. (1974). An opponent-process theory of motivation: 1. Temporal dynamics of affect. *Psychological Reviews* **81**, 119–145.

Substance Abuse and Mental Health Services Administration (2003). *Results from the 2002 National Survey on Drug Use and Health: National Findings* (Office of Applied Studies, NHSDA Series H-22, DHHS Publication No. SMA 03-3836), Rockville MD.

Taioli, E., and Wynder, E.L. (1991). Effect of the age at which smoking begins on frequency of smoking in adulthood. *New England Journal of Medicine* **325**, 968–969.

Tarter, R. E., Blackson, T., Brigham, J., Moss, H., and Caprara, G. V. (1995). The association between childhood irritability and liability to substance use in early adolescence: a 2-year follow-up study of boys at risk for substance abuse. *Drug and Alcohol Dependence* **39**, 253–261.

Tsuang, M. T., Lyons, M. J., Eisen, S. A., Goldberg, J., True, W., Lin, N., Meyer, J. M., Toomey, R., Faraone, S. V., and Eaves, L. (1996). Genetic influences on DSM-III-R drug abuse and dependence: a study of 3,372 twin pairs. *American Journal of Medical Genetics* **67**, 473–477.

Tyndale, R. F., and Sellers, E. M. (2002). Genetic variation in CYP2A6-mediated nicotine metabolism alters smoking behavior. *Therapeutic Drug Monitoring* **24**, 163–171.

Uhl, G. R., and Grow, R. W. (2004). The burden of complex genetics in brain disorders. *Archives of General Psychiatry* **61**, 223–229.

Van Dyke, C., and Byck, R. (1982). Cocaine, *Scientific American* **246**, 128–141.

Vanderschuren, L. J., and Kalivas, P. W. (2000). Alterations in dopaminergic and glutamatergic transmission in the induction and expression of behavioral sensitization: a critical review of preclinical studies. *Psychopharmacology* **151**, 99–120.

Wills, T. A., Vaccaro, D., and McNamara, G. (1994). Novelty seeking, risk taking, and related constructs as predictors of adolescent substance use: an application of Cloninger's theory. *Journal of Substance Abuse* **6**, 1–20.

Windle, M., and Windle, R. C. (1993). The continuity of behavioral expression among disinhibited and inhibited childhood subtypes. *Clinical Psychology Review* **13**, 741–761.

Winkleby, M. A., Fortmanm, S. P., and Rockhill, B. (1993). Cigarette smoking trends in adolescents and young adults: the Stanford Five-City Project. *Preventive Medicine* **22**, 325–334.

World Health Organization (1992). *International Statistical Classification of Diseases and Related Health Problems*, 10th revision, World Health Organization, Geneva.

Yi, H., Williams, G. D., and Dufour, M. C. (2000). *Trends in Alcohol-Related Fatal Traffic Crashes*, United National Institute on Alcohol Abuse and Alcoholism, *10th Special Report to the U.S. Congress on Alcohol and Health: Highlights from Current Research*, National Institute on Alcohol Abuse and Alcoholism, Bethesda MD.

C H A P T E R

2

Animal Models of Drug Addiction

OUTLINE

DEFINITIONS AND VALIDATION OF ANIMAL MODELS

Definitions of Drug Addiction Relevant to Animal Models

The definition of drug addiction used in the present book focuses on several different meanings of drug addiction (see *What is Addiction?* chapter). Addiction has been viewed from psychiatric, psychodynamic, and social psychological perspectives, and several common elements have been identified. Drug addiction, also known as Substance Dependence (American Psychiatric Association, 1994), was defined as a chronically relapsing disorder that is characterized by: (1) compulsion to seek and take the drug, (2) loss of

Neurobiology of Addiction, by George F. Koob and Michel Le Moal.
ISBN – 13: 978-0-12-419239-3 ISBN – 10: 0-12-419239-4

control in limiting intake, and (3) emergence of a negative emotional state (e.g., dysphoria, anxiety, irritability) when access to the drug is prevented.

Much of the recent progress in understanding the mechanisms of addiction has been derived from the study of animal models of addiction on specific drugs such as opiates, stimulants, and alcohol. While no animal model of addiction fully emulates the human condition, animal models do permit investigation of specific elements of the process of drug addiction. Such elements can be defined by models of different systems, models of psychological constructs such as positive and negative reinforcement, models of different stages of the addiction cycle, and models of actual symptoms of addiction as outlined by psychiatric nosology (Koob and Le Moal, 1997; Koob *et al.*, 1998) (see *What is Addiction?* chapter). Drug addiction has been conceptualized as a disorder that progresses from impulsivity to compulsivity in a collapsed cycle comprised of three stages: (1) preoccupation/anticipation, (2) binge/intoxication, and (3) withdrawal/negative affect. While much focus in animal studies has been on the synaptic sites and transductive mechanisms in the nervous system on which drugs of abuse act initially to produce their positive reinforcing effects, new animal models of the negative reinforcing effects of dependence have been developed for exploring how the nervous system adapts to drug use.

The construct of *reinforcement*, or *motivation*, is a crucial part of this definition. A reinforcer can be defined operationally as 'any event that increases the probability of a response'. This definition also can be used to signify a definition for reward, and the two words are often used interchangeably. However, *reward* often connotes some additional emotional value such as pleasure. Multiple powerful sources of reinforcement have long been identified during the course of drug dependence that provide the motivation for compulsive use and loss of control over intake (Wikler, 1973).

Motivation as a concept also has many definitions. Donald Hebb argued that motivation is 'stimulation that arouses activity of a particular kind' (Hebb, 1949), and C.P. Richter argued that 'spontaneous activity arises from certain underlying physiological origins and such "internal" drives are reflected in the amount of general activity' (Richter, 1927). Dalbir Bindra defined motivation as a 'rough label for the relatively persisting states that make an animal initiate and maintain actions leading to particular outcomes or goals' (Bindra, 1976). A more behavioristic view is that motivation is 'the property of energizing of behavior that is proportional to the amount and quality of the reinforcer' (Kling and Riggs, 1971). Finally, a more neurobehavioral view is that motivation is a 'set of neural processes that promote actions in relation to a particular class of environmental objects' (Bindra, 1976).

The primary pharmacological effect of a drug can produce a direct effect through positive reinforcement or negative reinforcement (e.g., self-medication or relief from aversive abstinence signs). The secondary pharmacological effects of the drug also can have motivating properties. Conditioned positive reinforcement involves the pairing of previously neutral stimuli with acute positive reinforcing effects of drugs, and conditioned negative reinforcement involves the pairing of previously neutral stimuli with the aversive stimulus effects of withdrawal or abstinence (**Table 2.1**).

An approach to the development of animal models that has gained wide acceptance is that animal models are most likely to have construct or predictive validity when the model mimics only the specific signs or symptoms associated with the psychopathological condition (Geyer and Markou, 1995). Animal models for a complete syndrome of a psychiatric disorder are unlikely to be possible either conceptually or practically. Certain areas of the human condition obviously are difficult to model in animal studies (e.g., comorbidity, polydrug addictions, child abuse, etc.). From a practical standpoint, psychiatric disorders are based on a nosology that is complex and constantly evolving, and most certainly involves multiple subtypes, diverse etiology, and constellations of many different disorders. In addition, models that attempt to reproduce entire syndromes require multiple endpoints, thereby making it very difficult in practice to study underlying mechanisms.

Under such a framework of mimicking only the specific signs or symptoms associated with the psychopathological condition, specific 'observables' (dependent variables) (Geyer and Markou, 1995) that have been identified in addiction provide a focus for study in animals. The reliance of animal models on a given observable also eliminates a fundamental problem associated with animal models of psychopathology, that of the frustration of attempting to provide complete validation of the whole syndrome. More definitive information related to a specific domain of addiction can be generated, and thus one can increase the confidence of cross-species validity. This framework also leads to a

TABLE 2.1 Relationship of Addiction Components and Behavioral Constructs

Addictive component	Behavioral construct
Pleasure	Positive reinforcement
Self-medication	Negative reinforcement
Habit	Conditioned positive reinforcement
Habit	Conditioned negative reinforcement

more pragmatic approach to the study of the neurobiological mechanisms of the behavior in question.

In the present chapter, these observables are organized by the binge/intoxication, withdrawal/negative affect, and preoccupation/anticipation (craving) stages of addiction. However, later in the chapter these observables are linked to the actual *Diagnostic and Statistical Manual of Mental Disorders*, 4th edition (DSM-IV) (American Psychiatric Association, 1994) criteria for addiction. The particular behavior being used for an animal model may or may not even be symptomatic of the disorder, but must be defined objectively and observed reliably. Indeed, the behavior being used may be found both in pathological and nonpathological states but still have predictive validity. A good example of such a situation would be the widespread use of drug reinforcement or reward as an animal model of addiction. Drug reinforcement does not necessarily lead to addiction (e.g., social drinking of alcohol), but the self-administration of alcohol has major predictive validity for the binge/intoxication stage of addiction, and it is difficult to imagine alcohol addiction without alcohol reinforcement.

Simultaneously with the movement of animal models to an observable-based framework has been the movement of the human addiction field in the domain of genetics to the concept of endophenotype. Mainly evolved from genetic studies where multiple genes have been hypothesized to mediate psychopathology, an endophenotype has been defined as 'some measurable characteristic of a person that can be detected with a laboratory procedure and that itself is a product of the gene or genes predisposing to alcoholism risk' (Iacono *et al.*, 2000). It has been argued that a valid endophenotype would lessen etiologic heterogeneity inherent in clinical phenotypes and would reduce both false positives and false negatives. Such an approach in genetics would make it possible to determine the mode of genetic transmission and eliminate the requirement that only certain kinds of families, which may not be representative of families with an alcoholic proband, serve as the study population (Iacono, 1998). Endophenotypes used in genetic studies closely resemble the 'observable behavior' argued by Geyer and Markou (1995) that provides a more easily validated construct for the study of the neurobiology of addiction.

Validation of Animal Models of Drug Addiction

Animal models are critical for understanding the neuropharmacological mechanisms involved in the development of addiction. While there are no complete animal models of addiction, animal models do exist for many elements of the syndrome. An animal model can be viewed as an experimental preparation developed for the purpose of studying a given phenomenon found in humans. The most relevant conceptualization of validity for animal models of addiction is the concept of *construct validity* (Ebel, 1961). Construct validity refers to the interpretability, 'meaningfulness,' or explanatory power of each animal model and incorporates most other measures of validity where multiple measures or dimensions are associated with conditions known to affect the construct (Sayette *et al.*, 2000). An alternative conceptualization of construct validity is the requirement that models meet the construct of functional equivalence, defined as 'assessing how controlling variables influence outcome in the model and the target disorders' (Katz and Higgins, 2003). The most efficient process for evaluating functional equivalence has been argued to be through common experimental manipulations which should have similar effects in the animal model and the target disorder (Katz and Higgins, 2003). This process is very similar to the broad use of the construct of *predictive validity* (see below). *Face validity* often is the starting point in animal models where animal syndromes are produced which resemble those found in humans in order to study selected parts of the human syndrome but is limited by necessity (McKinney, 1988). *Reliability* refers to the stability and consistency with which the variable of interest can be measured and is achieved when, following objective repeated measurement of the variable, small within- and between-subject variability is noted, and the phenomenon is readily reproduced under similar circumstances (for review, see Geyer and Markou, 2002). The construct of *predictive validity* refers to the model's ability to lead to accurate predictions about the human phenomenon based on the response of the model system. Predictive validity is used most often in the narrow sense in animal models of psychiatric disorders to refer to the ability of the model to identify pharmacological agents with potential therapeutic value in humans (Wilner, 1984; McKinney, 1988). However, when predictive validity is more broadly extended to understanding the physiological mechanism of action of psychiatric disorders, it incorporates other types of validity (i.e., etiological, convergent or concurrent, discriminant) considered important for animal models, and approaches the concept of construct validity (Markou *et al.*, 1993). The present chapter will describe animal models that have been shown to be reliable and in many cases to have construct validity for various stages of the addictive process.

ANIMAL MODELS FOR THE BINGE/INTOXICATION STAGE OF THE ADDICTION CYCLE

Animals and humans will readily self-administer drugs in the nondependent state. Drugs of abuse have powerful reinforcing properties in that animals will perform many different tasks and procedures to obtain drugs, even when not dependent. The positive reinforcing or rewarding effects of drugs are generally considered to be an important part of the beginning of the addiction cycle, and thus are included here. The drugs that have positive reinforcing effects as measured by direct self-administration, lowering of brain stimulation reward thresholds, and conditioned place preference in rodents and primates correspond very well with the drugs that have high abuse potential in humans (Kornetsky and Esposito, 1979; Collins *et al.*, 1984; Carr *et al.*, 1989) (**Table 2.2**).

Intravenous Drug Self-Administration

Drugs that are self-administered by animals correspond well with those that have high abuse potential in humans, and intravenous drug self-administration is considered an animal model that is predictive of abuse potential (Collins *et al.*, 1984). Intravenous drug self-administration also has proven to be a powerful tool for exploring the neurobiology of drug positive reinforcement as will be seen in subsequent chapters (Koob and Goeders, 1989). Self-administration of cocaine and heroin intravenously in rodents produces a characteristic pattern of behavior that lends itself to pharmacological and neuropharmacological study. Rats on a simple schedule of continuous reinforcement, such as a fixed ratio-1 schedule where one press of a lever or one nosepoke delivers one drug delivery, will develop a highly stable pattern of drug self-administration in a limited access situation (Caine *et al.*, 1993) (**Fig. 2.1**). However, as the unit dose is decreased, animals increase their self-administration rate, apparently compensating for decreases in the unit dose. Conversely, as the unit dose is increased, animals reduce their self-administration rate. Thus, manipulations which increase the self-administration rate on this fixed-ratio schedule resemble decreases in the unit dose, and may be interpreted as decreases in the reinforcing potency of the drug under study.

As would be predicted by the unit dose–response model, low to moderate doses of dopamine receptor antagonists increase cocaine self-administration maintained on this schedule in a manner similar to decreasing the unit dose of cocaine, suggesting that partial blockade of dopamine receptors by competitive antagonists reduces the reinforcing potency of cocaine. Conversely, dopamine receptor agonists decrease cocaine self-administration in a manner similar to increasing the unit dose of cocaine, suggesting that the effects of dopamine agonists together with cocaine self-administration can be additive, perhaps due to their mutual activation of the same neural substrates (Caine and Koob, 1993).

The use of different schedules of reinforcement in intravenous self-administration can provide important control procedures for nonspecific motor and motivational actions such as increases in exploratory activity and locomotion. Increasing the fixed-ratio value requirement for obtaining a reinforcer provides evidence that the animal is not simply involved in superstitious behavior; up to a certain point the animal will increase the number of lever presses to maintain the previous rate of drug infusion. In addition, a second, inactive lever can be introduced into the testing chamber, so that selective increases in responding on the active lever can be measured. Another approach is to increase the fixed-ratio value requirement for obtaining a reinforcer.

Other schedules called second-order schedules are used to test the motivational effects of drugs of abuse and have proven highly useful for the study of the

TABLE 2.2 Drugs Which are Self-Administered by Rats or Monkeys

Class	Drug name
Psychomotor stimulants	Cocaine
	D-amphetamine
	Methamphetamine
	Methylenedioxymethamphetamine
	Phenmetrazine
	Methylphenidate
	Diethylpropion
Opiates	Morphine
	Meperidine
	Codeine
	Pentazocine
	Heroin
Barbiturates	Amobarbital
	Secobarbital
	Pentobarbital
	Hexobarbital
Benzodiazepines	Chlordiazepoxide
	Diazepam
Other	Ethanol
	Nicotine
	Phencyclidine

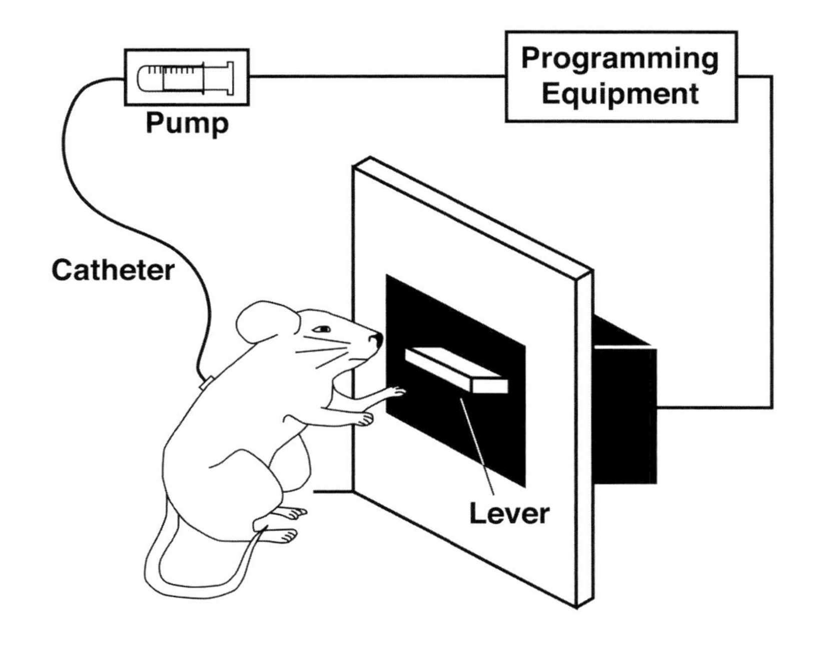

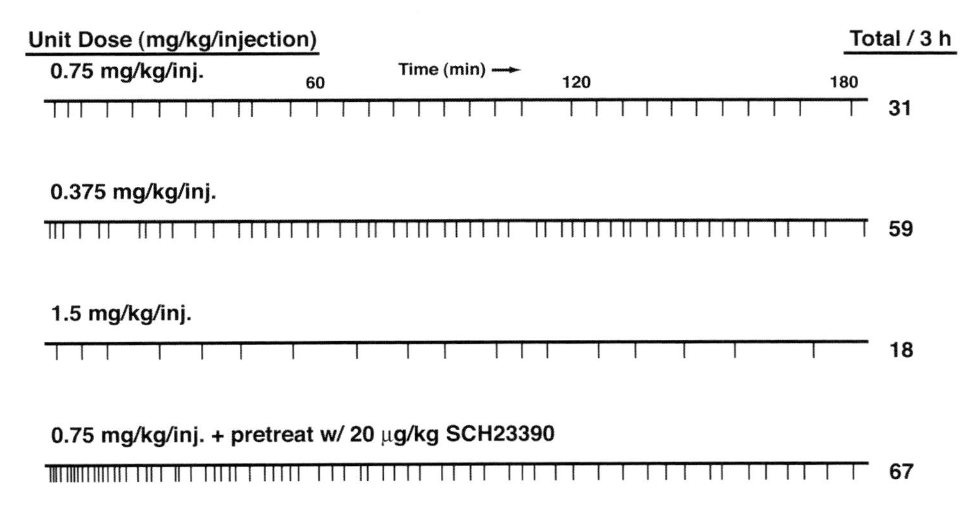

FIGURE 2.1 (Top) Drawing describing the procedure for intravenous self-administration in the rat. (Bottom) Event record and dose–response relationship relating dose of cocaine to the number of infusions. Rats implanted with intravenous catheters and trained to self-administer cocaine with limited access (3 h/day) will show stable and regular drug intake over each daily session. No obvious tolerance or dependence develops. Rats are generally maintained on a low-requirement, fixed-ratio (FR) schedule for intravenous infusion of the drug, such as an FR-1 or FR-5. In an FR-1 situation, one lever press is required to deliver an intravenous infusion of cocaine. In an FR-5 situation, five lever presses are required to deliver an infusion of cocaine. A special aspect of using an FR schedule is that the rats appear to regulate the amount of drug self-administered. Lowering the dose from the training level of 0.75 mg/kg/injection increases the number of self-administered infusions and vice versa. [Reproduced with permission from Caine *et al.*, 1993.]

neuropharmacological bases of the positive reinforcing effects of drugs of abuse and the conditioned rein-for-cing effects of drugs of abuse (Katz and Goldberg, 1991; Schindler *et al.*, 2002). In a second-order schedule of reinforcement, each *n*th response of a fixed-ratio schedule produces a brief stimulus, usually visual, and the first fixed-ratio completed after the completion of a fixed-interval produces this same stimulus accompanied by a drug injection (Katz and Goldberg, 1987). This stimulus comes to acquire secondary reinforcing properties and is used as an animal model of relapse. In simple fixed-ratio schedules, self-administration rate is inversely related to dose. In contrast, response rates in second-order schedules have been shown to increase with increasing drug doses (Goldberg *et al.*, 1975; Kelleher, 1975; Goldberg and Gardner, 1981; Katz and Goldberg, 1991) (**Fig. 2.2**) and are discussed in detail below. Further increases in dose lead to a decrease or leveling-off of response rates and an inverted-U or sigmoidal dose–response function.

In a multiple schedule, self-administration is one component, and another component may involve a nondrug natural reinforcer. Behavior maintained by food or cocaine alternately in the same test session and with identical reinforcement requirements has been reported for various species (Balster and Schuster, 1973; Kleven and Woolverton, 1990; Caine and Koob, 1994). As depicted in **Fig. 2.3,** these schedules may be used to evaluate the selectivity of manipulations which apparently selectively reduce the reinforcing efficacy of cocaine (Caine and Koob, 1994).

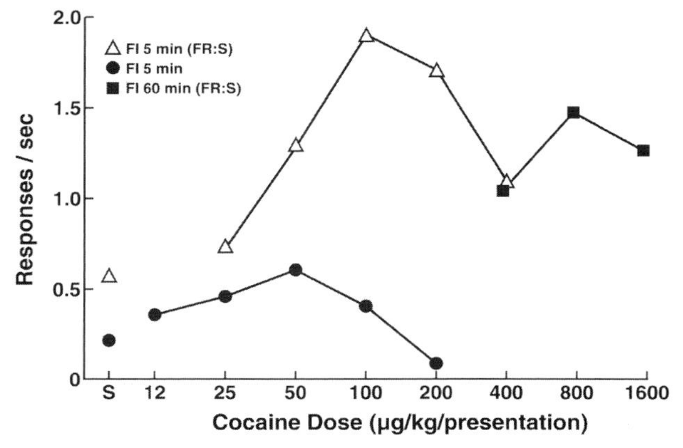

FIGURE 2.2 Effects of dose of cocaine per presentation on average rates of responding maintained with squirrel monkeys under different schedules of cocaine injection. Circles: Fixed-interval 5 min schedule with a 1 min timeout following each injection and 15 injections per experimental session (squirrel monkeys S-467 and S-474 from Goldberg and Kelleher, 1976). Triangles: Second-order schedule, fixed-interval 5 min (FR:S), with a 1 min timeout following each injection and 15 injections per experimental session. Squares: Second-order schedule, fixed-interval 60 min (FR:S), with 15 injections of cocaine spaced 2 s apart following the reinforced response concluding the experimental session. Abscissa: Dose of cocaine per presentation, log scale. [Reproduced with permission from Katz and Goldberg, 1991.]

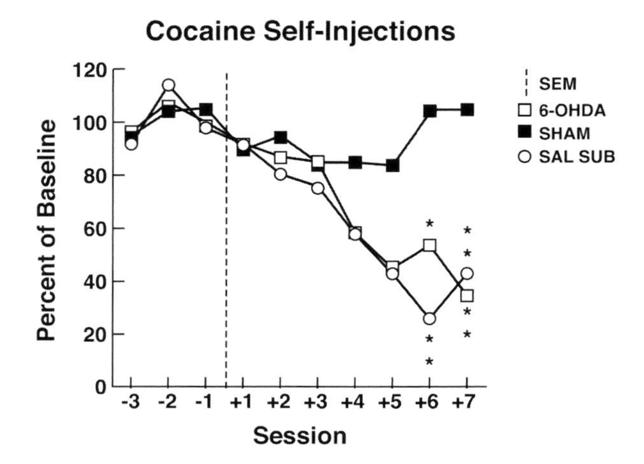

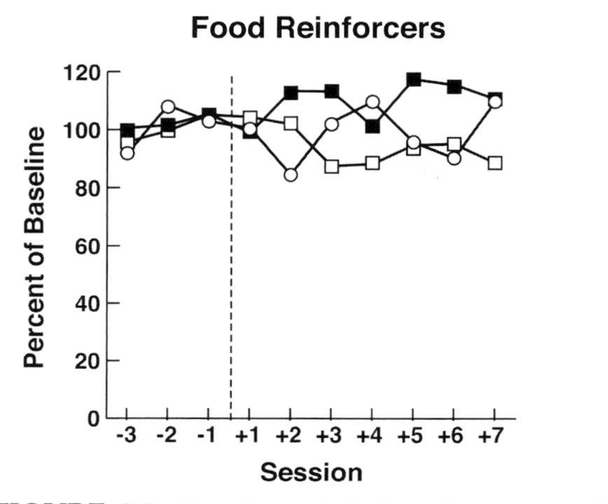

FIGURE 2.3 The effects of 6-hydroxydopamine (6-OHDA) or vehicle infusion into the nucleus accumbens and olfactory tubercle, or substitution of saline (SAL SUB) for cocaine, on the number of self-injections (top panel) or food pellets (bottom panel) earned in daily multiple-schedule sessions in rats. Values are the mean of the individual subjects' percentage of baseline (6-OHDA and sham, $n = 6$ per group; SAL SUB, $n = 4$). Inset shows the average SEM for the data points. [Reproduced with permission from Caine and Koob, 1994.]

To directly evaluate the reinforcing efficacy of a self-administered drug, one can use a progressive-ratio schedule of reinforcement. Here, the response requirements for each successive reinforcer delivered, in the case of drug self-administration, is increased, and a 'break point' is determined where an animal no longer responds or suspends responding at a certain response requirement. Alternatively, one can simply measure the highest ratio obtained in a given session (Grasing *et al.*, 2003). This schedule is effective in determining relative reinforcement strength for different reinforcers, including drugs. The unit dose–response model demonstrates that increasing the unit dose of self-administered drugs increases the break point on a progressive-ratio schedule (Griffiths *et al.*, 1978; Roberts *et al.*, 1989), and dopamine receptor antagonists have been shown to decrease the break point for cocaine self-administration (Roberts *et al.*, 1989; Depoortere *et al.*, 1993; Ward *et al.*, 1996) (**Figs. 2.4** and **2.5**).

Oral Drug Self-Administration

Oral self-administration almost exclusively involves alcohol because of the obvious face validity of oral alcohol self-administration. Two procedures are largely used to explore the neurobiological basis of ethanol reinforcement: two-bottle choice and operant self-administration. Historically, home cage drinking and preference have been used for characterizing genetic differences in drug preference, most often alcohol preference (Li, 2000), and for exploring the effects of pharmacological treatments on drug intake and preference. Here, a choice is offered between a drug solution and alternative solutions, one of which often is water, and the proportion of drug intake relative to total intake is calculated as a preference ratio. For two-bottle choice testing of alcohol in mice or rats, animals are singly housed, and a bottle containing alcohol and one containing water are placed on each cage.

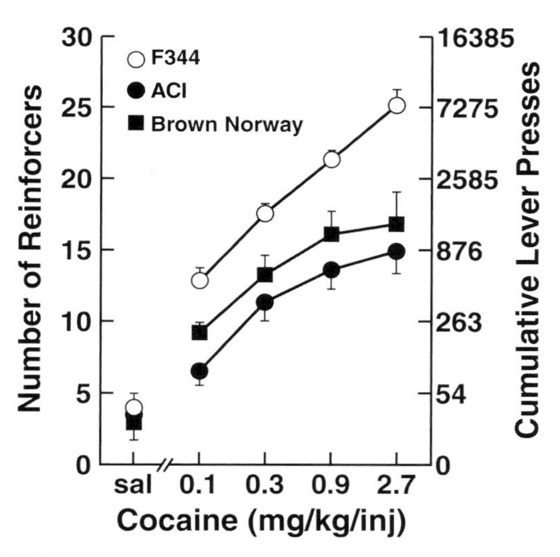

FIGURE 2.4 Baseline dose–effect curves for cocaine self-administration on a progressive ratio schedule in rats. Left ordinate: Average number of cocaine injections self-administered during the session. Right ordinate: Cumulative number of lever presses emitted during the session. Fischer 344 rats (○) maintained a higher average breaking point at the medium (0.3 mg/kg) and high dose (0.9 mg/kg) of cocaine than August × Copenhagen Irish (●) or Brown Norway rats (■). There was no difference between strains at the low dose of cocaine (0.1 mg/kg). Data are represented as mean ± SEM. [Reproduced with permission from Ward *et al.*, 1996.]

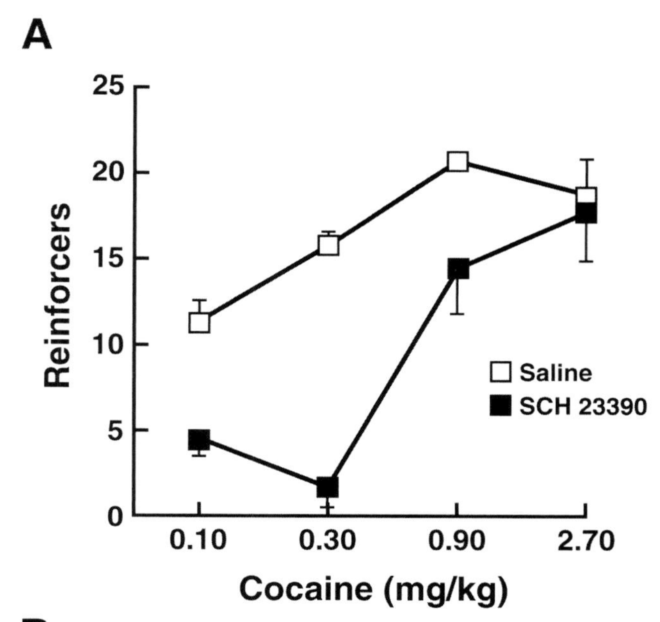

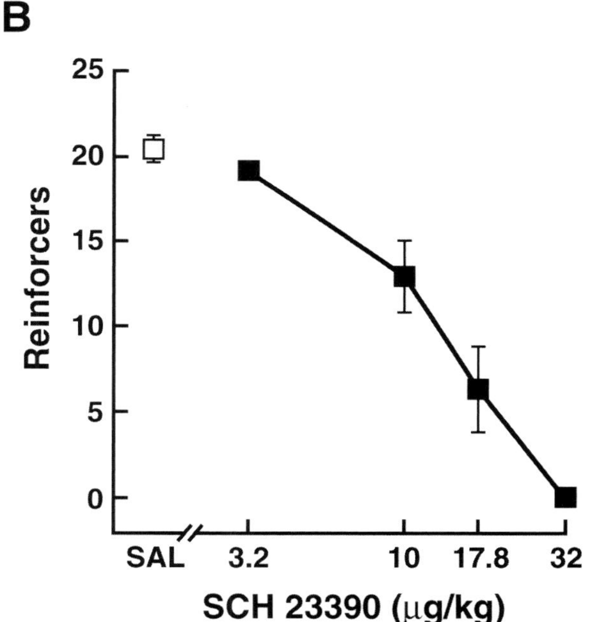

FIGURE 2.5 Effect of SCH 23390 pretreatment on the number of reinforcers obtained in intravenous self-administration of cocaine in rats. **(A)** Each point represents the average number of reinforcers obtained in a session with pretreatment of saline or SCH 23390 (10 μg/kg, s.c.) ($n = 7$). **(B)** Each point represents the average number of reinforcers obtained in a session with s.c. pretreatment with saline or SCH 23390. Saline or SCH 23390 were tested against the training dose of cocaine (0.90 mg/kg) ($n = 10$). [Reproduced with permission from Depoortere *et al.*, 1993.]

Most commonly, animals are allowed free choice of these drinking solutions for successive 24-h periods with simultaneous free access to food. However, limited access to the drug can induce high drug intakes in short periods of time (Files *et al.*, 1994). While alcohol is most often studied with these procedures, similar studies have been done with other drugs (Meisch *et al.*, 1992; Stewart *et al.*, 1994, 1996; Jentsch *et al.*, 1998).

For operant self-administration of alcohol, rats can be trained to lever press for alcohol using a variety of techniques. Many of these approaches are designed to overcome the aversive effects of initial exposure to alcohol, either by slowly increasing the concentration of alcohol or by adding a sweet solution. Using a sweetened solution fading procedure (Samson, 1986; Weiss *et al.*, 1990), alcohol concentrations are increased to a final concentration of 10%v/v ethanol over 20 days, with each concentration being mixed first with saccharin or sucrose and then presented alone. Using this approach, animals can be trained to lever press for concentrations of alcohol up to 40%v/v (Samson, 1986) (**Table 2.3**). They will perform on fixed-ratio schedules and progressive-ratio schedules and obtain significant blood alcohol levels in a 30-min session. Operant self-administration of oral alcohol also has been validated as a measure of the reinforcing effects of alcohol in primates (Stewart *et al.*, 1996).

TABLE 2.3 Mean Responding and Fluid Intakes for Various Operant Conditions

Fluid presented in dipper	Number of sessions	Reinforcement schedule	Number of responses	Ethanol (g/kg)
Single lever				
20% sucrose	5	FR4	219 ± 16	—
10% sucrose	2	FR4	308 ± 15	—
10% sucrose + 2% ethanol	3	FR4	354 ± 17	0.30 ± 0.01
10% sucrose + 5% ethanol	3	FR4	397 ± 21	0.84 ± 0.04
5% sucrose + 5% ethanol	6	FR4	315 ± 30	0.66 ± 0.05
5% ethanol	3	FR4	140 ± 15	0.28 ± 0.03
5% sucrose + 10% ethanol	5	FR4	238 ± 29	0.51 ± 0.07
10% ethanol	10	FR4	124 ± 16	0.51 ± 0.07
15% ethanol	5	FR4	121 ± 9	0.74 ± 0.06
20% ethanol	5	FR4	87 ± 8	0.71 ± 0.06
30% ethanol	5	FR4	77 ± 7	0.90 ± 0.05
40% ethanol	5	FR4	60 ± 4	0.94 ± 0.05
10% ethanol	10	FR4	148 ± 15	0.59 ± 0.05
Two-lever concurrent				
10% ethanol	24	FR4	151 ± 19	0.56 ± 0.05
Water		FR4	37 ± 11	—
10% ethanol	11	FR4	156 ± 22	0.57 ± 0.06
1% sucrose		FR4	147 ± 51	—
10% ethanol	15	FR4	109 ± 24	0.40 ± 0.10
5% sucrose		FR4	356 ± 53	—
10% ethanol	5	FR4	154 ± 26	0.55 ± 0.09
5% sucrose		FR64	113 ± 23	—
10% ethanol	4	FR4	143 ± 24	0.51 ± 0.09
Water		FR4	39 ± 13	—
Single lever				
10% ethanol	5	FR4	194 ± 25	0.69 ± 0.08

Reproduced with permission from Samson, 1986.

The advantages of the operant approach are that the effort to obtain the substance can be separated from the consummatory response (e.g., drinking) and intake can be charted easily over time. In addition, different schedules of reinforcement can be used to change baseline parameters.

Acquisition of Drug Self-Administration

Acquisition studies involve subjects that are naive to drug learning a simple operant response for intravenous delivery of the drug in a limited-access situation. Individual differences in the response to psychostimulants and other drugs of abuse in general have been widely demonstrated in humans (De Wit *et al.*, 1986; O'Brien *et al.*, 1986) and laboratory animals (Piazza *et al.*, 1989, 1996; Crabbe *et al.*, 1994). An important issue is to better understand the vulnerable phenotype that predisposes to drug intake. The concept of subjects at risk supposes the existence of a vulnerable phenotype and of course a specific brain abnormality, either inherent or acquired. Although the importance of individual differences in humans is well accepted in clinical practice, it has generally been neglected in animal studies. The most sensitive model to test vulnerability to drugs of abuse experimentally is to provide naive animals with very low unitary doses of drugs in an acquisition paradigm such that only the more sensitive individuals develop self-administration. The differences are hypothesized to reflect differential reactivity of neurotransmitters (Piazza *et al.*, 1989) (**Fig. 2.6**). Such types of differential responses have been shown for cocaine, amphetamine, and heroin (Piazza and Le Moal, 1996; Piazza *et al.*, 2000). The difference between animals can be studied further by dividing the population in two or by the median (i.e., 50 per cent–50 per cent), or by comparing the lowest and highest interquartiles to maximize the phenotypic differences (Ambrosio *et al.*, 1995; Elmer *et al.*, 1995). Such a model

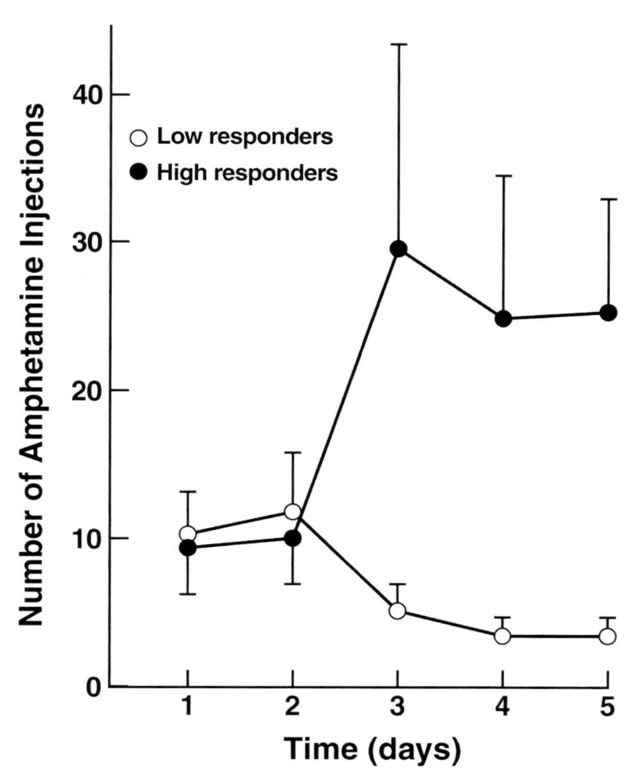

FIGURE 2.6 Acquisition of amphetamine self-administration of rats in high-responder and low-responder groups after repeated intraperitoneal administration of saline. After saline treatment, the groups (n = 10/group) differed in their acquisition of self-administration both in terms of total amphetamine administered over the five days and in terms of the number of injections over the different days. [Reproduced with permission from Piazza *et al.*, 1989.]

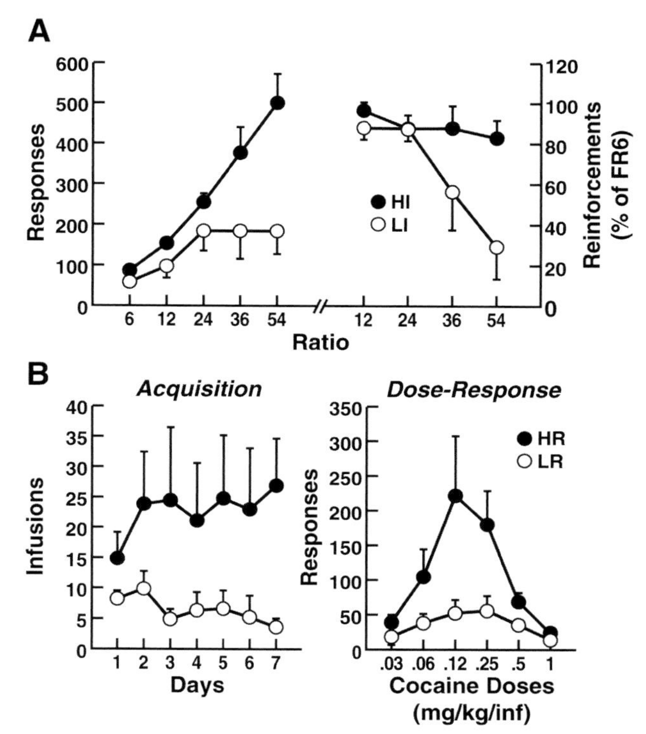

FIGURE 2.7 **(A)** Ratio-response function for intravenous cocaine self-administration in rats with high and low cocaine intake. Results are expressed as number of responses (A) and reinforcements (number of infusions) earned (B) over ratios. HI animals reacted at the increase in ratio (the response requirement necessary to obtain one drug infusion) with a proportional increase in responding, whereas after a fixed-ratio 24 schedule (24 responses for one infusion), responding did not increase any more in LI animals. As a consequence, the intake of cocaine remained stable for HI but rapidly decreased in LI at the increase in ratio. **(B)** Intravenous cocaine self-administration acquisition (A) and dose–response function (B) in high and low responders to novelty. HRs compared with LRs were the only group to acquire self-administration for a low dose of cocaine (100 μg/infusion), and their dose–response function was shifted upward. [Reproduced with permission from Piazza *et al.*, 2000.]

leads to at least two avenues of research: (1) to investigate the biological and brain parameters that differentiate behavioral phenotypes, and (2) to characterize as a syndrome the vulnerable versus resistant phenotype in such a way that the knowledge or the measure of a behavioral characteristic is predictive of the type of response to the drug (Piazza *et al.*, 1989, 2000). The validity of the vulnerability approach with low doses is demonstrated after dose–response, dose-intake, and ratio-intake studies for drug (e.g., cocaine) self-administration. Drug potency does not significantly vary among individuals, but instead large variations in cocaine-reinforcing efficacy exist, producing dramatic differences in drug intake across doses (Piazza *et al.*, 2000) (**Fig. 2.7**). Vulnerability to drugs of abuse is independent of dose, confirming—no matter the dose—that drug intake is very high in so-called 'vulnerable' rats and very low in 'resistant' rats. This model has face validity. Vulnerable subjects would have a higher chance of developing addiction independent of the quantity of drug available, as appears to occur in the real world (Piazza *et al.*, 2000). Differential sensitivity to the drug and to self-administration correlate with reduced dopamine activity in the prefrontal cortex, increases in dopamine in the nucleus accumbens (Piazza *et al.*, 1991), an exaggerated reaction to stressful situations, and an overactivation of the stress hormone axis (Piazza and Le Moal, 1996; Piazza *et al.*, 1996).

Brain Stimulation Reward

Animals will reliably self-administer electrical self-stimulation of certain brain areas, and humans have described stimulation in some of these areas as pleasurable (Olds and Milner, 1954). Animals will perform a variety of tasks to self-administer short electrical trains of stimulation (approximately 250 ms) to many different brain areas, but the highest rates and preference for stimulation follow the course of the medial forebrain bundle coursing bidirectionally from the midbrain to basal forebrain (for review, see Gallistel, 1983). The study of the neuroanatomical and neurochemical substrates of intracranial self-stimulation (ICSS) has led to the hypothesis that ICSS directly activates neuronal circuits that are activated by conventional reinforcers (e.g., food, water, sex) and that ICSS may reflect the direct electrical stimulation of the brain systems involved in motivated behavior. Drugs of abuse decrease thresholds for ICSS, and there is a good correspondence between the ability of drugs to decrease ICSS thresholds and their abuse potential (Kornetsky and Esposito, 1979; Kornetsky and Bain, 1990) (**Table 2.4**).

Early ICSS procedures involved simple rate-of-responding measures (for review, see Stellar and Stellar, 1985), but important methodological advances provided a valid measure of reward threshold unconfounded by influences on motor/performance capability. Two ICSS procedures that have been used extensively to measure the changes in reward threshold produced by drugs are the rate-frequency curve-shift procedure (Campbell *et al.*, 1985) and the discrete-trial, current-intensity procedure (Kornetsky and Esposito, 1979; Markou and Koob, 1992). The rate-frequency procedure involves the generation of a stimulation-input response-output function and provides a frequency threshold measure (Campbell *et al.*, 1985; Miliaressis *et al.*, 1986). In this procedure, the frequency of the stimulation (input) is varied, and the subject's response rate (output) is measured as a function of frequency. Rate-frequency curves are collected by allowing the rats to press a lever for an ascending series of pulse frequency stimuli, delivered through an electrode in the medial forebrain bundle or other rewarding brain site. A runway apparatus also can be used with running speed as the dependent measure (Edmonds and Gallistel, 1974). Frequencies can be presented in a descending or random order or in alternating descending and ascending series, and are changed in 0.05 or 0.1 log-unit steps. The rate-frequency function is a sigmoid curve which offers two measures. The locus of rise is a frequency threshold measure and is presumed to be a measure of ICSS reward threshold (Campbell *et al.*, 1985). Locus of rise refers to 'location' (that is, frequency) at which the function rises from zero to an arbitrary criterion level of performance. The most frequently used criterion is 50 per cent of maximal rate. The behavioral maximum measure is the asymptotic maximal response rate, and changes in the maximum measure are thought to reflect motor or performance effects (Edmonds and Gallistel, 1974). The procedure has been validated and investigated extensively, and studies indicate that

TABLE 2.4 Effects of Various Drugs on Intracranial Self-Stimulation Reward Threshold

Lowers	No change	Raises
Morphine	Tetrahydrocannabinol	Haloperidol
6-acetylmorphine	Lysergic acid dithalymide	Pimozide
Buprenorphine	Naloxone	Chlorpromazine
Nalbuphine[a]	Naltrexone	Imipramine
Methamphetamine	Cyclazocine	Atropine
Amfonelic acid	Ethylketocyclazocine	Scopolamine
Tripelennamine[b]	Nisoxetine	
Methylenedioxymethamphetamine	Apomorphine	
Heroin	U50, 488	
Cocaine	Pentobarbital	
Pentazocine[a]	Procaine	
D-amphetamine		
Phencyclidine		
Bromocriptine		
Nicotine		
Ethanol[c]		

[a] Especially in combination with tripelennamine.
[b] Especially in combination with pentazocine or nalbuphine.
[c] Only under conditions of self-administration (Moolten and Kornetsky, 1990).
Note: The drugs listed are only those that have been tested in the laboratory of the original source of this work using the rate-independent threshold procedure. Some drugs that caused no change, especially pentobarbital, might lower the threshold under drug self-administration conditions, as observed with ethanol. [Reproduced with permission from Kornetsky and Bain, 1990.]

changes in the reward efficacy of the stimulation (i.e., intensity manipulations) shift the rate-frequency functions laterally which translates into large changes in the locus of rise value but produce no alterations in the asymptote or in the shape of the function (Edmonds and Gallistel, 1974). In contrast, performance manipulations (e.g., weight on the lever, curare, etc.), including changes in motivation (that is, priming), alter the maximum measure value and the shape of the function (Edmonds and Gallistel, 1974; Miliaressis *et al.*, 1986; Fouriezos *et al.*, 1990). Drugs of abuse such as cocaine or amphetamine shift the dose–response curve to the left (Bauco and Wise, 1997) (**Fig. 2.8**).

A second procedure that controls for rate of responding and nonspecific performance deficits is a discrete-trial procedure. The discrete trial procedure is a modification of the classical psychophysical method of limits and provides a current intensity threshold measure (Kornetsky and Esposito, 1979; Markou and Koob, 1992). Here, at the start of each trial, rats receive a noncontingent, experimenter-administered electrical stimulus. The subjects then have 7.5 s to turn the wheel manipulandum one-quarter of a rotation to obtain a contingent stimulus identical to the previously delivered noncontingent stimulus (positive response) (Markou and Koob, 1992) (**Fig. 2.9**). If responding does not occur within the 7.5 s after the delivery of the noncontingent stimulus (negative response), the trial is terminated and an intertrial interval follows. Stimulus intensities vary according to the psychophysical method of limits.

The procedure provides two measures for each test session. The threshold value is defined as the midpoint in microamperes between the current intensity level at which the animal makes two or more positive responses out of the three stimulus presentations, and the level where the animal makes less than two positive responses at two consecutive intensities. Response latency is defined as the time in seconds that elapses between the delivery of the noncontingent electrical stimulus (end of the stimulus) and the animal's response on the wheel. Lowering of thresholds can be interpreted as an increase in the reward value of the stimulation, while increases in threshold reflect decreases in reward value. Increases in response latency can be interpreted as motor/performance deficit (**Fig. 2.10**).

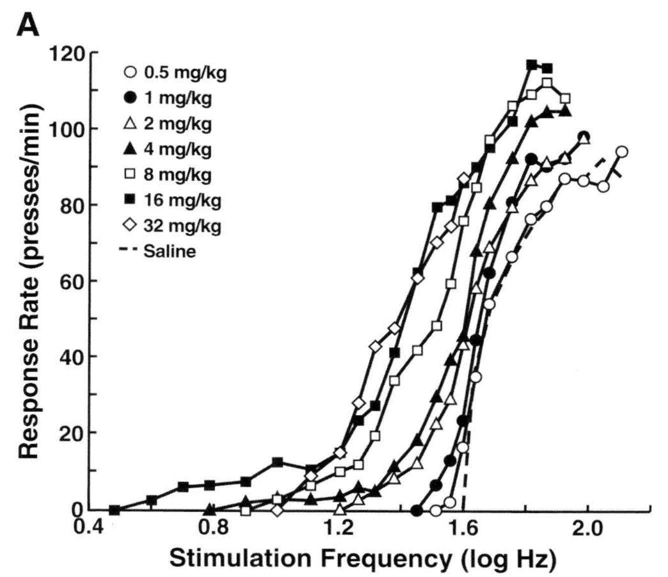

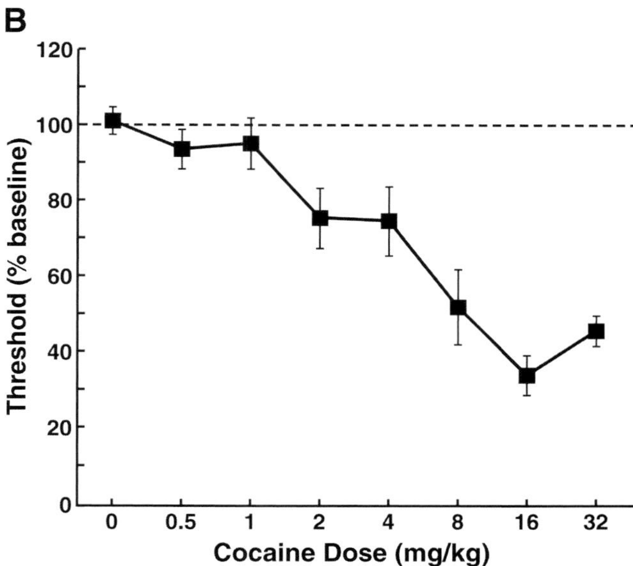

FIGURE 2.8 **(A)** Mean lever-pressing rate in rats during the first 15 min after cocaine injection as a function of stimulation frequency and cocaine dosage. Frequency data were transformed to log difference-from-baseline values for each animal before averaging so that the slope of the mean rate-frequency functions would not be contaminated by between-animal differences in threshold. **(B)** Mean (± SEM) self-stimulation frequency threshold (expressed as percentage of baseline) as a function of cocaine dosage. Values are the means from the first threshold determination in the first hour after injection. Each reference (baseline) value is the mean from the two threshold determinations taken just before the respective drug test. [Reproduced with permission from Bauco and Wise, 1997.]

Conditioned Place Preference

Conditioned place preference, or place conditioning, is a nonoperant procedure for assessing the reinforcing efficacy of drugs using a classical or Pavlovian conditioning procedure. In a simple version of the place preference paradigm, animals experience two distinct neutral environments that are paired spatially and temporally with distinct drug or nondrug states. The animal then is given an opportunity to choose to

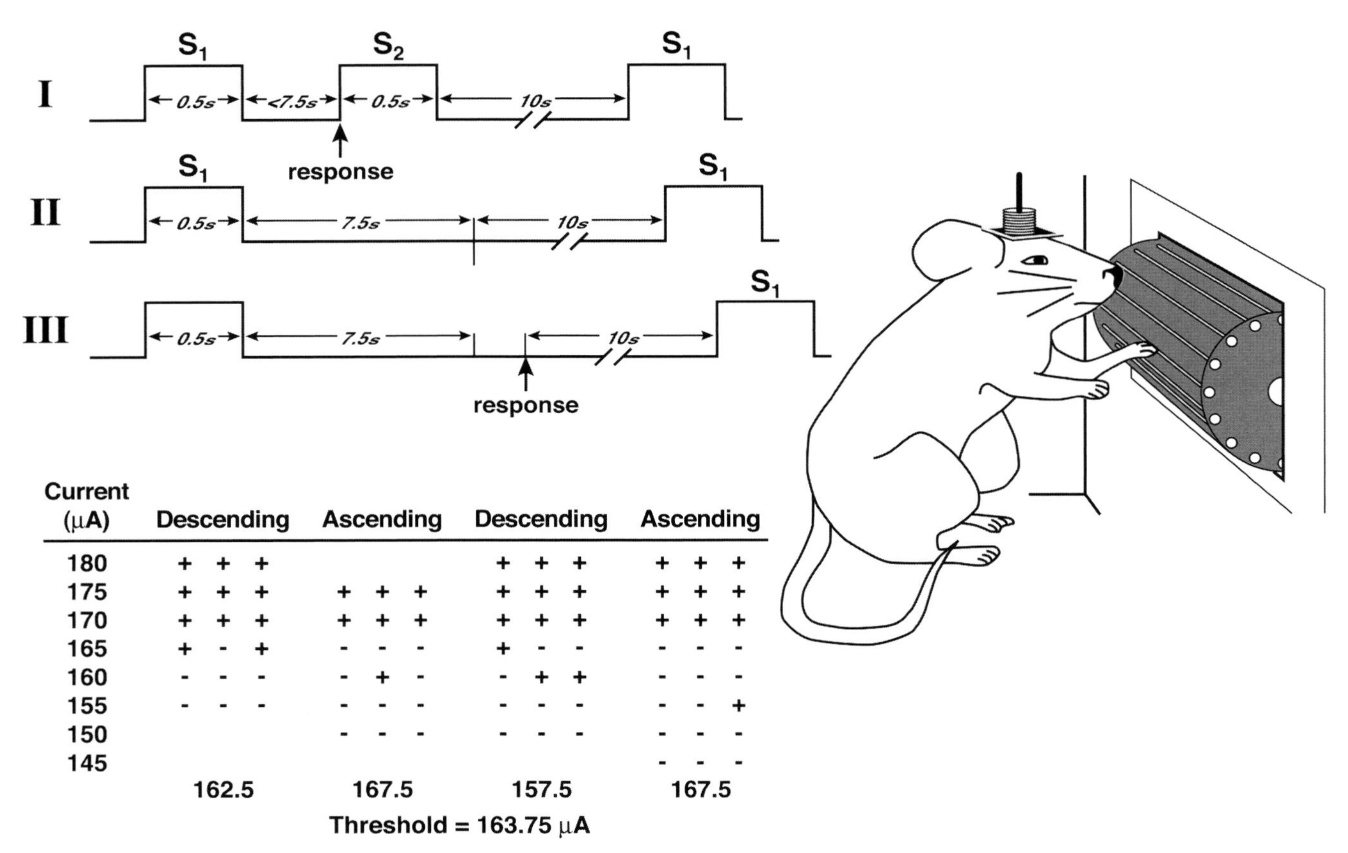

Current (μA)	Descending	Ascending	Descending	Ascending
180	+ + +		+ + +	+ + +
175	+ + +	+ + +	+ + +	+ + +
170	+ + +	+ + +	+ + +	+ + +
165	+ - +	- - -	+ - -	- - -
160	- - -	- + -	- + +	- - -
155	- - -	- - -	- - -	- - +
150		- - -	- - -	- - -
145				- - -
	162.5	167.5	157.5	167.5

Threshold = 163.75 μA

FIGURE 2.9 Intracranial self-stimulation threshold procedure. Panels I, II, and III illustrate the timing of events during three hypothetical discrete trials. Panel I shows a trial during which the rat responded within the 7.5 s following the delivery of the noncontingent stimulus (positive response). Panel II shows a trial during which the animal did not respond (negative response). Panel III shows a trial during which the animal responded during the intertrial interval (negative response). For demonstration purposes, the intertrial interval was set at 10 s. In reality, the interresponse interval had an average duration of 10 s and ranged from 7.5–12.5 s. The table depicts a hypothetical session and demonstrates how thresholds were defined for the four individual series. The threshold of the session is the mean of the four series' thresholds. [Reproduced with permission from Markou and Koob, 1992.]

enter and explore either environment, and the time spent in the drug-paired environment is considered an index of the reinforcing value of the drug. Animals exhibit a conditioned preference for an environment associated with drugs that function as positive reinforcers (i.e., spend more time in the drug- compared to placebo-paired environment) and avoid those environments that induce aversive states (i.e., conditioned place aversion). This procedure permits assessment of the conditioning of drug reinforcement and can provide indirect information regarding the positive and negative reinforcing effects of drugs (Carboni and Vacca, 2003) (**Fig. 2.11**).

The apparatus used in conditioning experiments consists of two or three environments that are differentiated from each other on the basis of color, texture, and/or lighting (Swerdlow *et al.*, 1989) (**Fig. 2.12**). The distinctiveness of the environments is essential for the development of conditioning. There are a number of critical independent and dependent variables that can affect place conditioning. Dependent variables include the duration of the post-training testing, the method for calculating preference (e.g., difference score, percentage of pretraining, etc.), and the actual measures used (e.g., number of entries, mean duration of time per entry). A critical independent variable is the use of a 'biased' or 'unbiased' training schedule. In an unbiased design, the environments are adjusted so that animals differentiate one environment from the other but do not exhibit an innate preference for either environment. Pairing of a drug with a particular environment is counterbalanced, and changes in the time spent in the drug-paired environment can be attributed directly to the conditioned reinforcing effects of the drug. Because it is an unbiased design, a preconditioning phase to assess pretest preferences is not

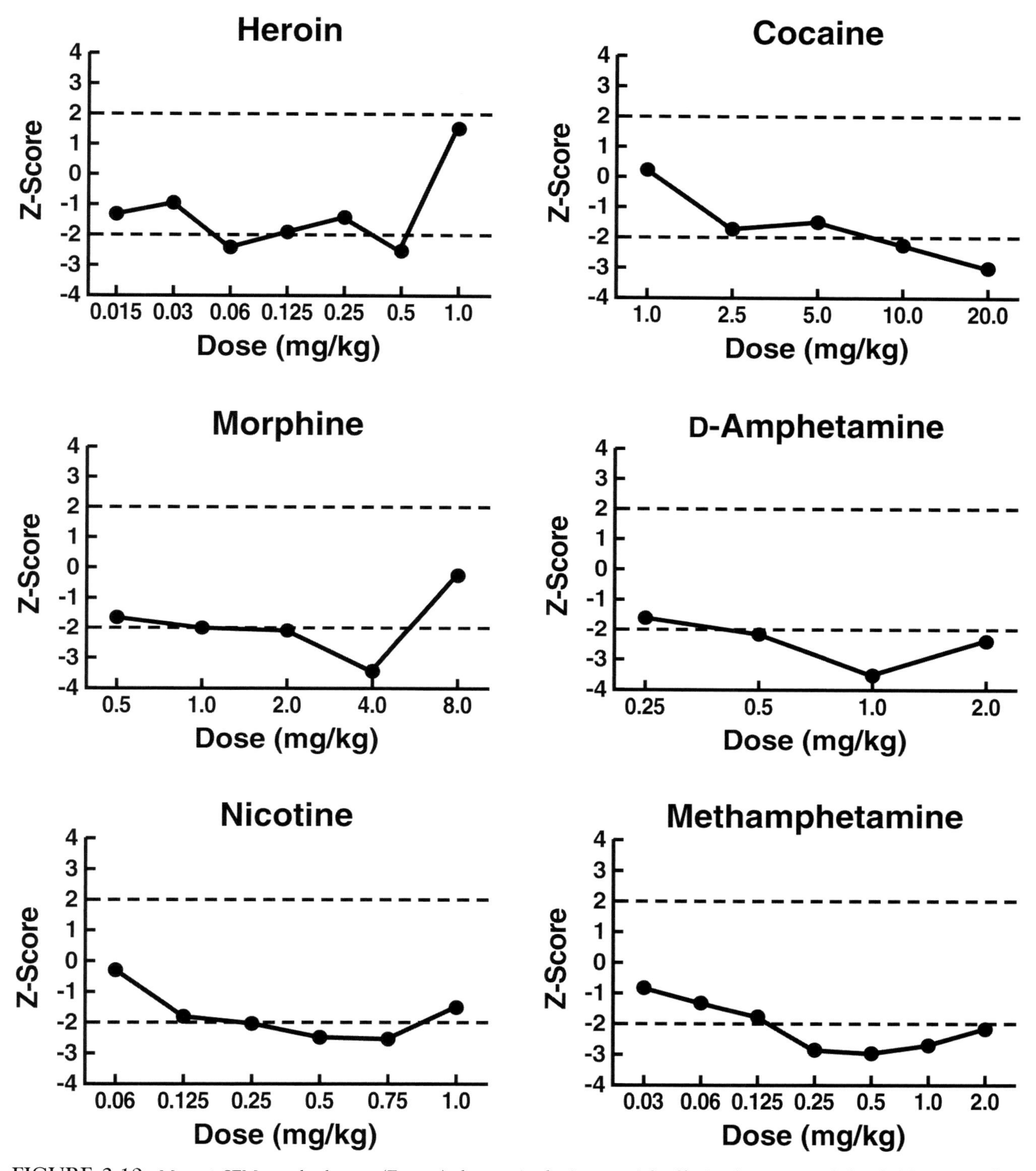

FIGURE 2.10 Mean ± SEM standard score (Z-score) changes in the intracranial self-stimulation reward thresholds in rats after administration of various doses of heroin, morphine, nicotine, cocaine, D-amphetamine, and methamphetamine. A Z-score is based on the pre- and post-drug changes in threshold, and a Z-score of ± 2.0 indicates the 95 per cent confidence limit based on the mean and standard deviation for all saline days. [Reproduced with permission from Kornetsky, 1985 (D-amphetamine), Hubner and Kornetsky, 1992 (heroin, morphine), Huston-Lyons and Kornetsky, 1992 (nicotine), Izenwasser and Kornetsky, 1992 (cocaine), Sarkar and Kornetsky, 1995 (methamphetamine).]

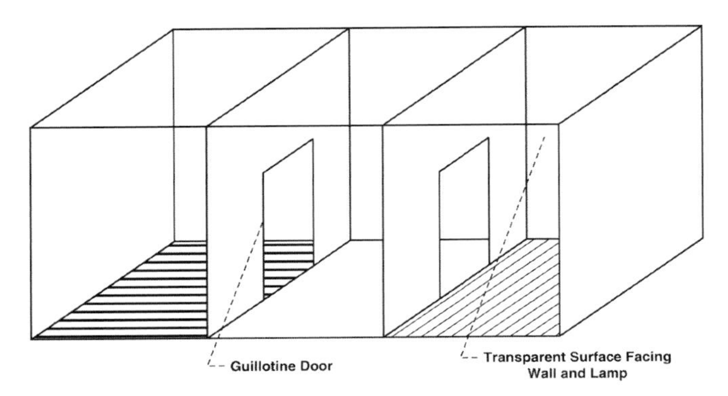

FIGURE 2.11 Three-compartment conditioned place preference apparatus. A three-compartment apparatus may consist of two square base compartments (38 × 30 × 30 cm), one with white and the other with gray walls. Between two compartments, there is a smaller compartment (38 × 15 × 30 cm) with clear gray walls and floor. Covers and doors are similar to the two-compartment apparatus. The apparatus can be made of 4 mm thick Plexiglas. Covers can be made of transparent Plexiglas which is sheltered with a semitransparent film to allow the experimenter to see the rat but not vice versa. Six apparatuses can be placed 50 cm from each other and 50 cm from a wall where one fluorescent lamp (40 W) per apparatus is placed. To avoid dark corners in the apparatus, each compartment has a transparent wall facing the lamp to provide uniform illumination. It is advisable to keep the apparatus in a soundproof room with 60 dB white noise. When a video camera is used to record the sessions, it is placed above two apparati. It is suggested to have the experimental room attached to a small ante room provided with washing facilities. In some tests, tactile cues can be included to allow distinction among compartments. One floor can be made with 2 mm diameter rods placed in a frame at 1.5 cm distance. The second floor can be made with 4 mm high squares (1 × 1 cm) placed on a base at 1 cm distance from each other. Alternatively, a wire mesh floor can be used. Ease of cleaning has to be considered in choosing the type of floor. When a third compartment is used, the floor is left smooth. It is possible to include in the conditioned place preference procedure olfactory cues by using different bedding or other odors. [Reproduced with permission from Carboni and Vacca, 2003.]

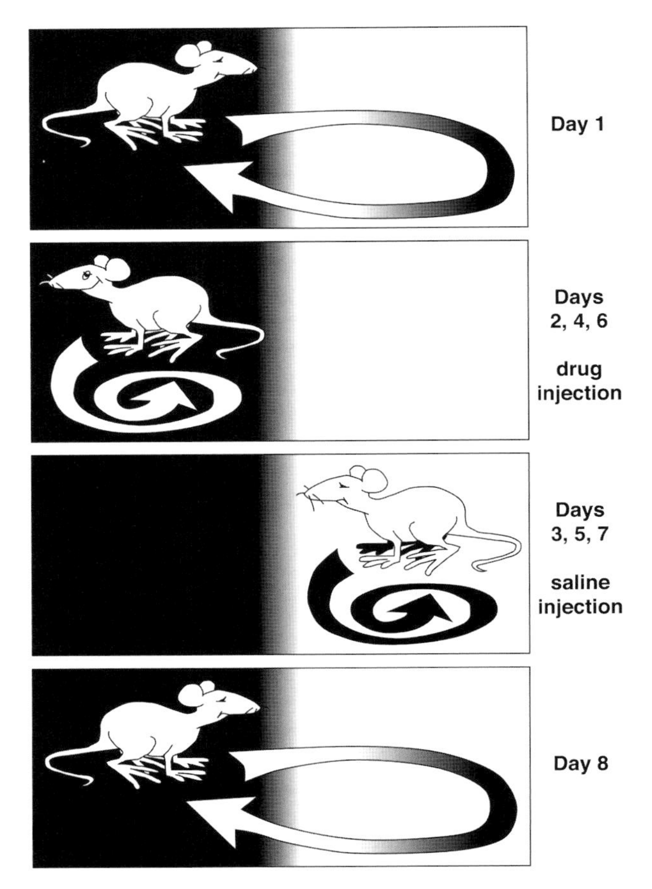

FIGURE 2.12 Diagram illustrating the place conditioning procedure in the rat. Animals experience two distinct neutral environments (here, black and white shaded) paired spatially and temporally with distinct unconditioned stimuli (here, drug on Days 2, 4 and 6, and saline on Days 3, 5, and 7). On Day 8, animals are given an opportunity to enter either environment, and the time spent in each environment is used as an index of the reinforcing value of each unconditioned stimulus. These time values often are compared to a baseline preference for each environment (here, measured at Day 1). [Reproduced with permission from Swerdlow *et al.*, 1989.]

required, which can save time in experimentation. Other independent variables are numerous and include housing of the animals, age of the animals, familiarity with the training environment, and the physical structure of the training and testing environment. Detailed discussion of these issues is beyond the scope of this chapter (for more information, see Mucha *et al.*, 1982; Phillips and Fibiger, 1987; Stewart and Eikelboom, 1987; van der Kooy, 1987; Carr *et al.*, 1989; Swerdlow *et al.*, 1989).

A procedure with multiple choices of environments (more than two) simply adds additional environmental choices such as three distinct environments (Stinus *et al.*, 1990) or multiple spatial locations such as on an elevated radial maze (McDonald and White, 1993). In either case, an additional choice allows for additional controls for nonspecific effects and permits easier balancing between two locations being used for subsequent pairings. For details of two types of apparatuses and procedures, see Stinus *et al.* (1990) and McDonald and White (1993).

Drug Discrimination

Drug discrimination procedures developed in animals have provided a powerful tool for identifying the relative similarity of the discriminative stimulus effects of drugs and, by comparison with known drugs of abuse, the generation of hypotheses regarding the abuse potential of these drugs (Holtzman, 1990). Drug discrimination typically involves training an animal to produce a particular response in a given drug state for a food reinforcer, and to produce a different response in the placebo or nondrug state. The interoceptive cue state (produced by the drug) controls the behavior as a discriminative stimulus or cue, which informs the animal to make the appropriate response to gain reinforcement. The choice of response which follows administration of an unknown test compound can provide valuable information about the similarity of that drug's interoceptive cue properties to those of the training drug. Some of the original drug discrimination procedures utilized a T-maze escape procedure (Overton, 1974). However, high drug doses are required, and the T-maze is not easily automated.

More commonly, an appetitively motivated operant procedure is used where the rat has access to two levers (Colpaert *et al.*, 1975). One lever (e.g., left lever) is reinforced on a fixed-ratio 10 schedule for food following injection of the training drug. The other lever is reinforced on a fixed-ratio 10 schedule for food in sessions that follow the injection of drug vehicle. No more than nine responses can be emitted on the inappropriate lever or the contingency is terminated without reinforcement. Other schedules of reinforcement can be used, although simpler schedules tend to produce a lower level of performance accuracy than the fixed-ratio schedules (Shannon and Holtzman, 1976; Stolerman *et al.*, 1989; Holtzman, 1990). Training continues until the animal reaches a certain predetermined performance level, usually 10 consecutive sessions during which 85–95 per cent of responses prior to delivery of the first reinforcer are emitted on the choice lever appropriate for the drug state of the animal (this may require 30–80 training sessions) (Holtzman, 1990).

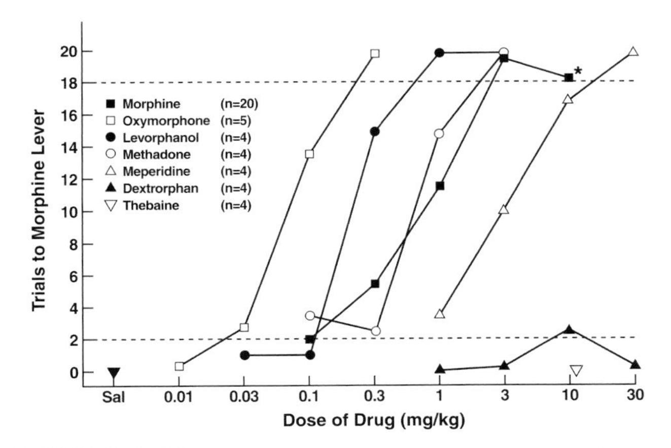

FIGURE 2.13 Dose–response curves of narcotic analgesics from diverse chemical families and the structural analogs dextrorphan and thebaine, which lack narcotic activity, in rats trained to discriminate between saline and 3.0 mg/kg of morphine. The number of trials completed on the morphine-appropriate lever after the administration of saline is indicated by ▼ at Sal (Saline). The ordinate is the mean number of trials out of a maximum of 20 which were completed on the morphine-appropriate lever. The abscissa is the dose of drug. Four of the 20 rats completed only 10, 13, 14, and 18 trials, respectively, out of 20 trials after administration of 10 mg/kg morphine. [Reproduced with permission from Shannon and Holtzman, 1976.]

Tests of generalization to a different drug can be interspersed among the training sessions once performance has stabilized (**Fig. 2.13**). Alternatively, an entire drug generalization function can be generated in a few hours on a single day using a cumulative dosing method where animals are tested in a series of short sessions with a time-out between each session (Bertalmio *et al.*, 1982). In tests of stimulus generalization, data often are collected only until the delivery of the first reinforcer to eliminate the influence of reinforcement on subsequent choice responding (which would effectively place the animal in a new training situation), or the sessions can be conducted as extinction sessions where no reinforcers are delivered.

An alternative drug discrimination training procedure used extensively in both rats and squirrel monkeys (Schuster and Woods, 1968; Shannon and Holtzman, 1976) involves a discrete-trial procedure using avoidance or escape from shock. Here, animals are trained to lever-press on one of two levers to avoid or escape electric shocks which are delivered intermittently through the grid floor of the test cage. A trial is signaled by the illumination of a house light. A third lever (called the observing lever) must be pressed before the choice is made to prevent the rat from perseverating on the appropriate choice lever. A major advantage of this aversively maintained responding is that no food restriction is necessary and there is no confound from the anorexic effects of the drug in question (Holtzman, 1990).

Summary of Animal Models for the Binge/Intoxication Stage

The procedures outlined above have proven reliable and to have predictive validity in their ability to understand the neurobiological basis of the acute reinforcing effects of drugs of abuse. One could reasonably argue that drug addiction mainly involves counteradaptive mechanisms that go far beyond the acute reinforcing actions of drugs. However, understanding the neurobiological mechanisms for positive reinforcing actions of drugs of abuse also provides a framework for understanding the motivational effects of counteradaptive mechanisms (see *What is Addiction?* chapter). Many of the operant measures used as models for the reinforcing effects of drugs of abuse lend themselves to within-subjects designs, limiting the number of subjects required. Indeed, once an animal is trained, full dose–effect functions can be generated for different drugs, and the animal can be tested for weeks and months. Pharmacological manipulations can be conducted with standard reference compounds to validate any effects. In addition, a rich literature on the experimental analysis of behavior is available for exploring the hypothetical constructs of drug action as well as for modifying drug reinforcement by modifying the history and contingencies of reinforcement.

The advantage of the ICSS paradigm as a model of drug effects on motivation and reward is that the behavioral threshold measure provided by ICSS is easily quantifiable. ICSS threshold estimates are very stable over periods of several months (for review, see Stellar and Stellar, 1985). Another considerable advantage of the ICSS technique is the high reliability with which it predicts the abuse liability of drugs. For example, there has never been a false positive with the discrete trials threshold technique (Kornetsky and Esposito, 1979).

The advantages of place conditioning as a model for evaluating drugs of abuse include its high sensitivity to low doses of drugs, its potential utility in studying both positive and negative reinforcing events, the fact that testing for drug reward is done under drug-free conditions, and its allowance for precise control over the interaction of environmental cues with drug administration (Mucha *et al.*, 1982).

ANIMAL MODELS FOR THE DRUG WITHDRAWAL/NEGATIVE AFFECT STAGE OF THE ADDICTION CYCLE

Drug withdrawal from chronic drug administration is usually characterized by responses opposite to the acute initial actions of the drug. For example, if a drug acutely decreases sympathetic activity such as blood pressure, during withdrawal blood pressure may increase. Many of these overt physical signs associated with withdrawal from drugs (e.g., alcohol and opiates) can be easily quantified. Standard rating scales exist for opiate, nicotine, and alcohol withdrawal (Gellert and Holtzman, 1978; Malin *et al.*, 1992; Macey *et al.*, 1996) (**Table 2.5**). However, while the physical signs of withdrawal are measured easily and may provide a marker for the study of neurobiological mechanisms of dependence, motivational measures of withdrawal have more validity for understanding the counteradaptive mechanisms that drive addiction (see *What is Addiction?* chapter). Such motivational measures have proven to be extremely sensitive to drug withdrawal and provide powerful tools for exploring the

TABLE 2.5 Measures of Physical Withdrawal in Rats

Opiates

Counted signs	*Counted signs*
Escape attempts	Eye blinks
Wet-dog shakes	Grams of weight loss
Facial fasciculations	*Checked signs*
Cheek tremors	Diarrhea
Chews	Vocalization
Teeth-chattering	Irritability on handling
Abdominal constrictions	Abnormal posture
Genital licks	Ptosis

Methods: After a 10 min initial period of habituation to a separate testing room, rats are injected with naloxone (0.03–3.0 mg/kg, s.c.) and placed for 10 min into a clear square-shaped container measuring 1 cubic foot. With the exception of weight measurements, observations of somatic signs of opiate withdrawal last 10 min, starting 1 min after naloxone injection. To obtain a withdrawal score, counted and checked signs are given a weighting factor based on the observation period (5 min each) in which the sign is observed (e.g., ptosis observed in the 2nd observation period is assigned a weighting factor of 2). Total withdrawal scores are obtained by combining the weighted scores (Gellert and Holtzman, 1978).

Alcohol

Observed signs
- Ventromedial distal limb flexion response
- Tail stiffness
- Abnormal body posture

Methods: A subjective 0–2 point scale of severity is utilized for each of three withdrawal signs, with 0 representing undetectable, 1 representing moderately severe, and 2 representing severe withdrawal signs. The ventromedial distal limb flexion response is measured by grasping the rat by the scruff of the neck and checking for retraction of the limbs toward the body. Tail stiffness is characterized by the presence of a rigid, awkwardly bent tail. Abnormal body posture in the rat is indicated by the presence of a broad-based stance or abnormal gait. Ratings for each individual withdrawal sign are analyzed for percentage of subjects showing the trait, either 1 or 2 on the rating scale. In addition, overall withdrawal severity scores are determined by combining all individual withdrawal sign ratings, yielding a range of 0–6. Using continuous ethanol vapor exposure which yields blood alcohol levels of 150–200% for two weeks, withdrawal scores range from 3–5 in individual animals (Macey *et al.*, 1996).

Nicotine

Observed signs

Blinks	Yawns	Teeth-chattering
Body shakes	Scratches	Ptosis
Chews	Head shakes	Hops
Cheek tremors	Genital licks	Writhes
Escape attempts	Gasps	Foot licks

Methods: Each rat is placed in a plastic opaque cylindrical container (30 × 29 cm) and observed for 10 min for somatic signs of nicotine withdrawal 30 min after administration of saline or the nicotinic antagonist mecamylamine (1.5 mg/kg, s.c.). An observer records the frequency and time of occurrence of withdrawal signs. Multiple successive counts of any sign require a distinct pause between episodes. The total number of somatic signs per 10 min observation period is defined as the sum of the number of occurrences of all signs. Both frequency of individual signs and total number of signs are analyzed statistically (Malin *et al.*, 1992).

neurobiological bases for the motivational aspects of drug dependence.

Animal models for the motivational effects of drug withdrawal have included operant schedules, place aversion, ICSS, the elevated plus maze, and drug discrimination. While each of these models can address a different theoretical construct associated with a given motivational aspect of withdrawal, some reflect more general malaise, while others reflect more specific components of the withdrawal syndrome.

Intracranial Self-stimulation

ICSS thresholds have been used to assess changes in systems mediating reward and reinforcement processes during the course of drug dependence. Acute administration of psychostimulant drugs lowers ICSS thresholds (i.e., increases reward) (for reviews, see Kornetsky and Esposito, 1979; Stellar and Rice, 1989), and withdrawal from chronic administration of virtually all major drugs of abuse elevates ICSS thresholds (i.e., decreases reward) (Leith and Barrett, 1976; Kokkinidis and McCarter, 1990; Markou and Koob, 1991, 1992; Schulteis *et al.*, 1994, 1995; Epping-Jordan *et al.*, 1998; Gardner and Vorel, 1998; Paterson *et al.*, 2000) (**Fig. 2.14**).

Conditioned Place Aversion

Place aversion has been used to measure the aversive stimulus effects of withdrawal (Hand *et al.*, 1988; Stinus *et al.*, 1990). In contrast to conditioned place preference discussed above, rats exposed to a particular environment while undergoing precipitated withdrawal to opiates, spend less time in the withdrawal-paired environment when subsequently presented with a choice between that environment and an unpaired environment. These aversive stimulus effects can be measured from 24 h to 16 weeks later (Hand *et al.*, 1988; Stinus *et al.*, 1990, 2000) (**Fig. 2.15**). The place aversion does not require maintenance of opiate dependence for its manifestation, in that such an association continues to be manifested weeks after animals are 'detoxified' (e.g., after the morphine pellets are removed) (see Baldwin and Koob, 1993; Stinus *et al.*, 2000). Also, a place aversion in opiate-dependent rats can be observed with doses of naloxone below which somatic signs of withdrawal are observed (Schulteis *et al.*, 1994). Although naloxone itself will produce a place aversion in nondependent rats, the threshold dose required to produce a place aversion decreases significantly in dependent rats (Hand *et al.*, 1988).

Operant Schedules in Dependent Animals

Several operant schedules have been used to characterize the response-disruptive effects of drug withdrawal (Denoble and Begleiter, 1976; Gellert and Sparber, 1977; Koob *et al.*, 1989), providing a readily quantifiable measure of withdrawal (e.g., response rate). These include high rate schedules such as fixed-ratios (Gellert and Sparber, 1977), and schedules that produce a low but stable rate of responding (Denoble and Begleiter, 1976). However, response disruption can be caused by any number of variables, from motor problems to malaise and decreases in appetite, and thus other measures must be used to rule out nonspecific effects.

Drug Discrimination

Drug discrimination can be used to characterize both specific and nonspecific aspects of withdrawal. Animals have been trained to discriminate the anxiogenic substance pentylenetetrazol from saline (Gauvin and Holloway, 1991), and opiate-dependent animals have been trained to discriminate an opiate antagonist from saline (Emmett-Oglesby *et al.*, 1990). Generalization to the pentylenetetrazol cue during withdrawal has suggested an anxiogenic-like component to the withdrawal syndrome. In contrast, generalization to an opiate antagonist provides a more general nonspecific measure of opiate withdrawal intensity and time-course (Gellert and Holtzman, 1979; France and Woods, 1989). Hypothetically, similar experiments could be done with spontaneous withdrawal at discrete time points. However, as of this writing, the authors are unaware of any published results demonstrating stimulus generalization effects of spontaneous withdrawal using operant drug discrimination.

Summary of Animal Models for the Withdrawal/Negative Affect Stage

Motivational measures of drug withdrawal have much the same value for the study of the neurobiological mechanisms of addiction as procedures used to study the positive reinforcing effects of drugs. ICSS threshold procedures have high predictive validity for changes in reward valence. Disruption of operant responding during drug abstinence is very sensitive. Place aversion is hypothesized to reflect an aversive unconditioned stimulus. Drug discrimination allows a powerful and sensitive comparison to other drug states. The use of multiple dependent variables for the study of the motivational effects of withdrawal may

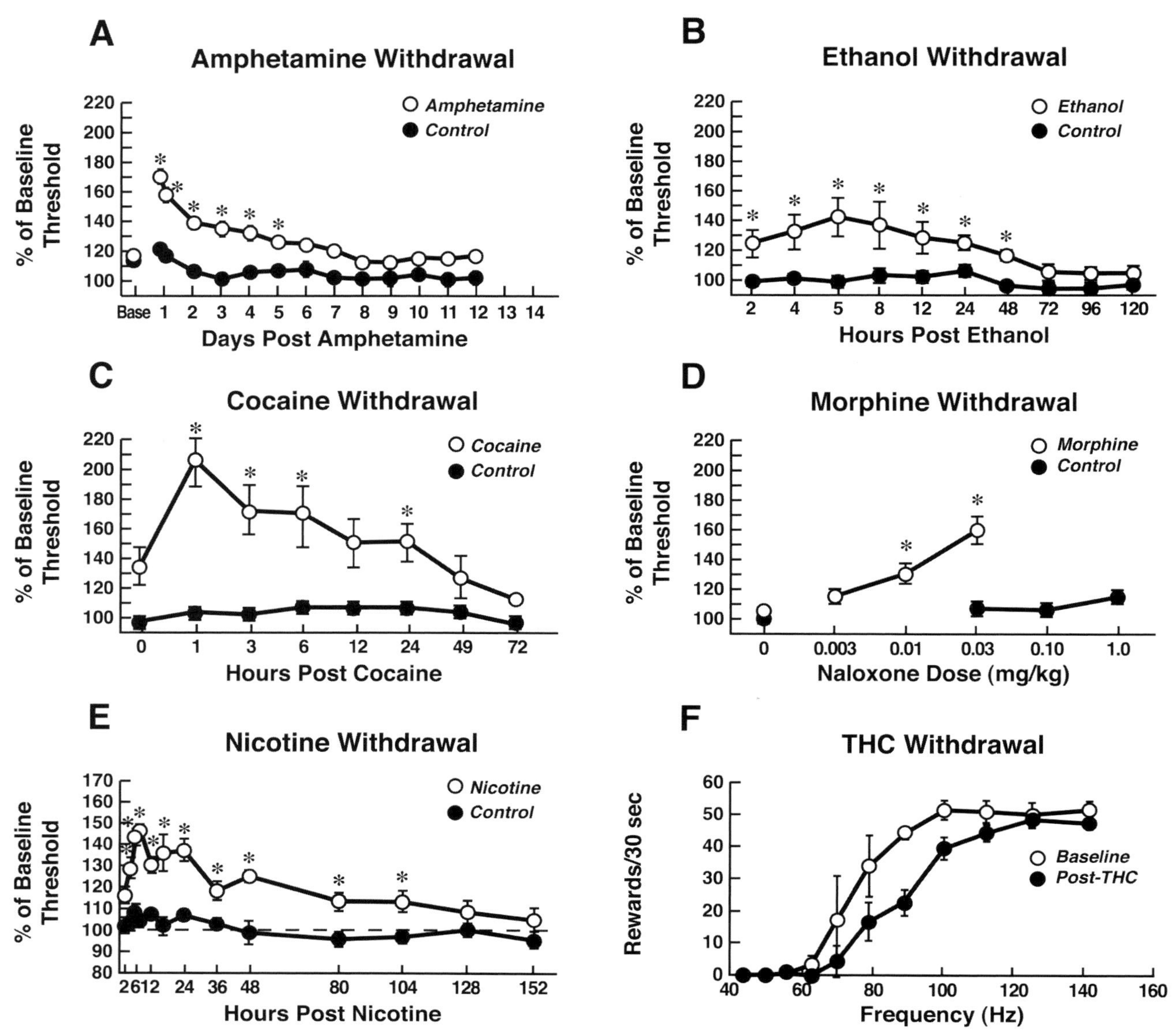

FIGURE 2.14 Intracranial self-stimulation in rats during drug withdrawal. **(A)** Mean intracranial self-stimulation reward thresholds (± SEM) during amphetamine withdrawal (10 mg/kg/day for six days). Data are expressed as a percentage of the mean of the last five baseline values prior to drug treatment. Asterisks (*) indicate statistically significant differences from the saline control group ($p < 0.05$). [Reproduced with permission from Paterson *et al.*, 2000.] **(B)** Mean ICSS thresholds (± SEM) during ethanol withdrawal (blood alcohol levels achieved: 197.29 mg%). Elevations in thresholds were time-dependent. Asterisks (*) indicate statistically significant differences from the control group ($p < 0.05$). [Reproduced with permission from Schulteis *et al.*, 1995.] **(C)** Mean ICSS thresholds (± SEM) during cocaine withdrawal 24 h following cessation of cocaine self-administration. Asterisks (*) indicate statistically significant differences from the control group ($p < 0.05$). [Reproduced with permission from Markou and Koob, 1991.] **(D)** Mean ICSS thresholds (± SEM) during naloxone-precipitated morphine withdrawal. The minimum dose of naloxone that elevated ICSS thresholds in the morphine group was 0.01 mg/kg. Asterisks (*) indicate statistically significant differences from the control group ($p < 0.05$). [Reproduced with permission from Schulteis *et al.*, 1994.] **(E)** Mean ICSS thresholds (± SEM) during spontaneous nicotine withdrawal following surgical removal of osmotic minipumps delivering nicotine hydrogen tartrate (9 mg/kg/day) or saline. Asterisks (*) indicate statistically significant differences from the control group ($p < 0.05$). [Data adapted with permission from Epping-Jordan *et al.*, 1998.] **(F)** Mean rewards/30 s (± SEM) measured 24 h following an acute 1.0 mg/kg dose of Δ^9-tetrahydrocannabinol (THC). Withdrawal significantly shifted the reward function to the right (indicating diminished reward). [Reproduced with permission from Gardner and Vorel, 1998.] Note that because different equipment systems and threshold procedures were used in the collection of the above data, direct comparisons among the magnitude of effects induced by these drugs cannot be made.

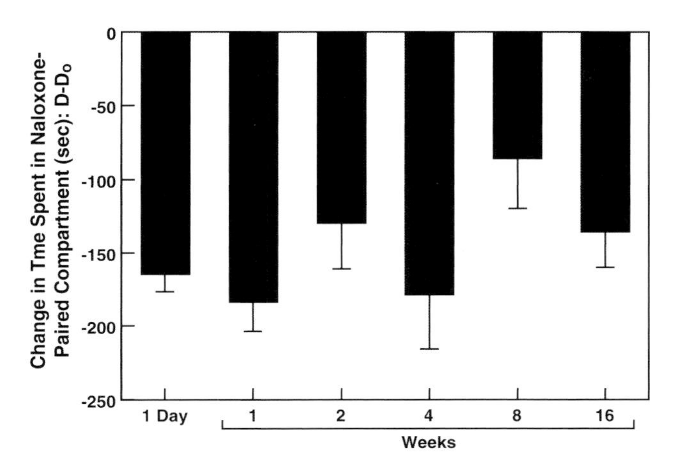

FIGURE 2.15 Place conditioning with naloxone (15 μg/kg, s.c.) in morphine-dependent rats. D_0 indicates time spent in the naloxone-paired compartment before conditioning. Rats then were evaluated 1 day, and 1, 2, 4, 8, and 16 weeks later. Values represent the mean ± SEM time spent in the naloxone compartment after conditioning minus the time spent in the naloxone compartment before conditioning. [Data from Stinus *et al.*, 2000.]

provide a powerful means of assessing overlapping neurobiological substrates and provide a heuristic framework for the counteradaptive mechanisms hypothesized to drive addiction (see *What is Addiction?* chapter).

ANIMAL MODELS FOR THE PREOCCUPATION/ANTICIPATION (CRAVING) STAGE OF THE ADDICTION CYCLE

The chronic, relapsing nature of addiction is one of its defining characteristics. Animal models of relapse fall into three categories of a broad-based conditioning construct: drug-induced reinstatement, cue-induced reinstatement, and stress-induced reinstatement. The general conceptual framework for the conditioning construct is that cues, either internal or external, become associated with the reinforcing actions of a drug by means of classical conditioning and can elicit subjective states that trigger drug use. Environmental cues repeatedly paired with primary reinforcers can acquire reinforcing properties via classical conditioning processes (McFarland and Ettenberg, 1997; Arroyo *et al.*, 1998; See *et al.*, 1999; Weiss *et al.*, 2000; Deroche-Gamonet *et al.*, 2004), and these conditioned re-inforcing effects have been hypothesized to contribute to drug craving and relapse. Human studies have shown that the presentation of stimuli previously associated with drug delivery or drug withdrawal increase the likelihood of relapse as well as self-reports of craving and motivation to engage in drug-taking (O'Brien *et al.*, 1977, 1992; Childress *et al.*, 1988).

A drug-predictive discriminative stimulus can signal the availability of a reinforcer, and thereby provides motivation to engage in behavior that brings the organism into contact with the reinforcer. A condition often associated with drug craving in humans is cognitive awareness of drug availability (e.g., Meyer and Mirin, 1979). Discriminative stimuli, therefore, may have a prominent role in craving and the resumption of drug-seeking behavior in abstinent individuals. Moreover, the response-contingent conditioned stimulus, acting as a conditioned reinforcer, may contribute to the maintenance of subsequent drug-seeking behavior once it is initiated. In fact, these contingencies can be conceptualized to resemble those associated with the relapse process in humans—certain drug-related cues may provide the initial central motivational state to engage in drug-seeking behavior while others may maintain this behavior until the primary reinforcer is obtained.

Resistance to Extinction Associated with Drug Self-administration

Extinction procedures can provide measures of the motivational properties of drugs by assessing the persistence of drug-seeking behavior in the absence of response-contingent drug availability. In an extinction paradigm, subjects are trained to self-administer a drug until stable self-administration patterns are achieved, and then the drug is removed (Schuster and Woods, 1968) (**Fig. 2.16**). Extinction testing sessions are identical to training sessions except that no drug is delivered after completion of the response requirement. Measures provided by an extinction paradigm reflect the degree of resistance to extinction and include the duration of extinction responding, the total number of responses emitted during the entire extinction session, and the probability of reinitiating responding under extinction conditions at a later time after successful extinction of the self-administration behavior (i.e., propensity to relapse).

Drug-Induced Reinstatement

Additional measures of the motivational properties of drugs can be derived from an extinction paradigm if noncontingent drug injections are administered after extinction. After priming with the drug, the latency to reinitiate responding, or the amount of responding on the previously extinguished lever, are used to reflect

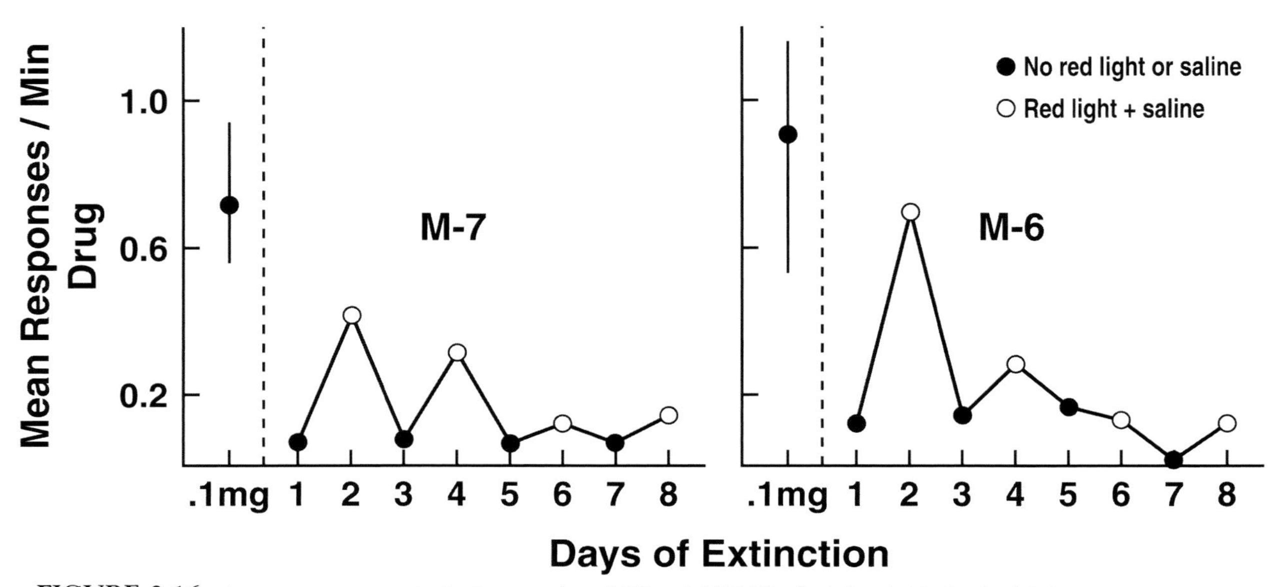

FIGURE 2.16 Average response rate for two monkeys M#6 and M#7. The far left point is the final 5 day average response rate for morphine reinforcement. The monkeys were placed on a 2.5 min variable internal schedule of reinforcement for both food and morphine. Each 24 h day was broken into four cycles of 6 h each. Brackets indicate the range of daily response rate over the 5 days. Points to the right of the dashed line show extinction data following 15 drug-free days of rest. The closed circles indicate the absence of response consequences for drug responding, and the open circles indicate the consequences of infusion of saline and the presentation of a red light. [Reproduced with permission from Schuster and Woods, 1968.]

the motivation for drug-seeking behavior. This type of paradigm has been used primarily to investigate drug-induced initiation of responding for drugs, often referred to as priming or drug-induced reinstatement (Stewart and de Wit, 1987). Both stimulant and opiate self-administration have been consistently reinstated following extinction in animals with systemic or intracerebral noncontingent drug infusions (Gerber and Stretch, 1975; Stewart and Wise, 1992). Animal models for drug-primed reinstatement show that exposure to cocaine can elicit strong recovery of extinguished drug-seeking behavior in the absence of further drug availability (Gerber and Stretch, 1975; de Wit and Stewart, 1981) (**Fig. 2.17**). Systemic injections of compounds of drug classes other than the training drug are generally less effective in reinstating drug self-administration behavior in rats (Stewart and de Wit, 1987) and monkeys (Gerber and Stretch, 1975). The effectiveness of compounds in reinstating drug self-administration decreases as their discriminative stimulus similarity to the training drug decreases.

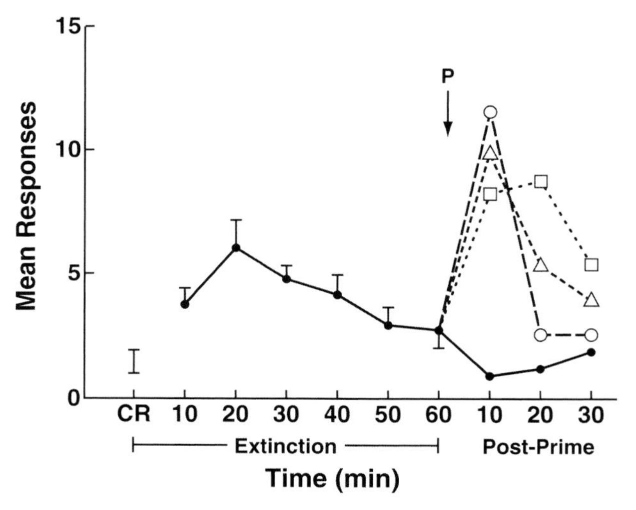

FIGURE 2.17 Mean number of responses in rats per 10 min during cocaine self-administration (continuous reinforcement, CR), during extinction, and after cocaine priming injections of 2.0, 1.0, and 0.5 mg/kg or after a 'dummy trial' (0.0 mg/kg). 'P' indicates the point at which the priming injection was given. The mean values during extinction are based on eight determinations for each of five rats; the means after the priming infusions are based on two determinations per rat for each of five rats. Standard errors of the mean are indicated for the mean values during extinction. Closed circles, 0.0 mg/kg; open circles, 0.5 mg/kg; triangles, 1.0 mg/kg; squares, 2.0 mg/kg. [Reproduced with permission from de Wit and Stewart, 1981.]

Cue-induced Reinstatement

A number of animal models are available for characterizing the reinforcing value imparted on formerly neutral environmental stimuli that have been associated repeatedly with drug self-administration. A conditioned reinforcer can be defined as any motivationally neutral stimulus which acquires motivational properties through association with a primary reinforcer (Hilgard and Marquis, 1961). In a conditioned reinforcement paradigm, subjects usually are trained in an operant box containing two levers where responses on one lever result in presentation of a brief stimulus followed by a drug injection (active lever), while responses on the other lever have no consequences throughout the experiment (inactive lever) (Schuster and Woods, 1968; Davis and Smith, 1987). Previously neutral stimuli also can acquire conditioned reinforcement properties when the drug administration is not contingent on the animal's behavior, as long as the stimulus precedes the drug injection (Davis and Smith, 1987). Subsequently, the ability of the previously neutral, drug-paired stimuli to maintain responding in the absence of drug injections provides a measure of the motivational value of these stimuli.

Early evidence for the ability of drug-paired stimuli to function as conditioned reinforcing stimuli was provided by a study in which an anise-flavored solution of etonitazene (an opiate agonist), was provided as the sole drinking solution to nonopiate-dependent rats (Wikler *et al.*, 1971). Several months later, in a two-bottle choice situation, the animals consumed twice as much anise-flavored water as control rats. Furthermore, opiate-dependent rats that were given access to the anise-flavored etonitazene solution during morphine withdrawal several months later consumed twice as much anise-flavored water as the rats with similar anise-flavored etonitazene experience that were never opiate-dependent. These results suggest that previously neutral stimuli can acquire reinforcing properties when paired with reinforcing drugs, and that the development of dependence and the relief of withdrawal also can contribute to the motivational efficacy of such secondary reinforcers.

Drug-associated stimuli (S^+/S^-) that signal response-contingent availability of intravenous cocaine versus saline (Weiss *et al.*, 2000) reliably elicit drug-seeking behavior in experimental animals, and responding for these stimuli is highly resistant to extinction (See *et al.*, 1999; Weiss *et al.*, 2000). Subsequent re-exposure after extinction to a cocaine S^+, but not a nonreward S^-, produces strong recovery of responding at the previously active lever in the absence of any further drug availability. Cues associated with availability of oral alcohol self-administration also can reinstate responding in the absence of the primary reinforcer (Katner *et al.*, 1999; Ciccocioppo *et al.*, 2001). Consistent with the well-established conditioned cue reactivity in human alcoholics (Drummond, 2000), the motivating effects of alcohol-related stimuli are highly resistant to extinction in that they retain their efficacy in eliciting alcohol-seeking behavior over more than one month of repeated testing (Katner *et al.*, 1999; Ciccocioppo *et al.*, 2001). Animal models for cue reinstatement show that discriminative stimuli predictive of cocaine reliably elicit strong recovery of extinguished drug-seeking behavior in the absence of further drug availability (Weiss *et al.*, 2001) (**Fig. 2.18**).

Stress-induced Reinstatement

Many human studies show that relapse to drug-taking is more likely to occur under situations of stress (Kosten *et al.*, 1986; Brown *et al.*, 1995). Animal models for stress reinstatement show that stressors elicit strong recovery of extinguished drug-seeking behavior in the absence of further drug availability (Erb *et al.*, 1996; Ahmed and Koob, 1997). Administration of acute intermittent footshock induced reinstatement of cocaine-seeking behavior after prolonged extinction, and this was as effective as a priming injection of cocaine (Erb *et al.*, 1996; Ahmed and Koob, 1997; Deroche *et al.*, 1997) (**Figs. 2.19** and **2.20**). Such effects are observed even after a 4–6 week drug-free period (Erb *et al.*, 1996) and appear to be selective for drug, in that food-seeking behavior was not reinstated (Ahmed and Koob, 1997).

The specificity of conditioning in animal models may have predictive validity with the specificity of conditioned responses in human drug users. An experimental study in human drug users indicated that cocaine-related stimuli were effective in eliciting conditioned physiological responses and self-reported cocaine craving in cocaine users but not in opiate or nondrug users. Furthermore, self-reported cocaine withdrawal-related or neutral stimuli were ineffective in eliciting any conditioned responses (Ehrman *et al.*, 1992). This effect has been confirmed in animal studies. Thus, the noncontingent presentation of drug-paired stimuli, in the absence of drug infusions, also can reinstate drug-seeking behavior after a period of extinction (de Wit and Stewart, 1981). In addition to general physiological responses, conditioned stimuli associated with drug administration also induce the 'psychological' phenomenon of drug craving in humans, even after a period of abstinence (McLellan *et al.*, 1986; Childress *et al.*, 1988).

Second-order Schedules of Reinforcement

In second-order schedules of reinforcement, animals can be trained to work for a previously neutral stimulus that ultimately predicts drug availability (Katz and Goldberg, 1991). In such a paradigm, completion of the

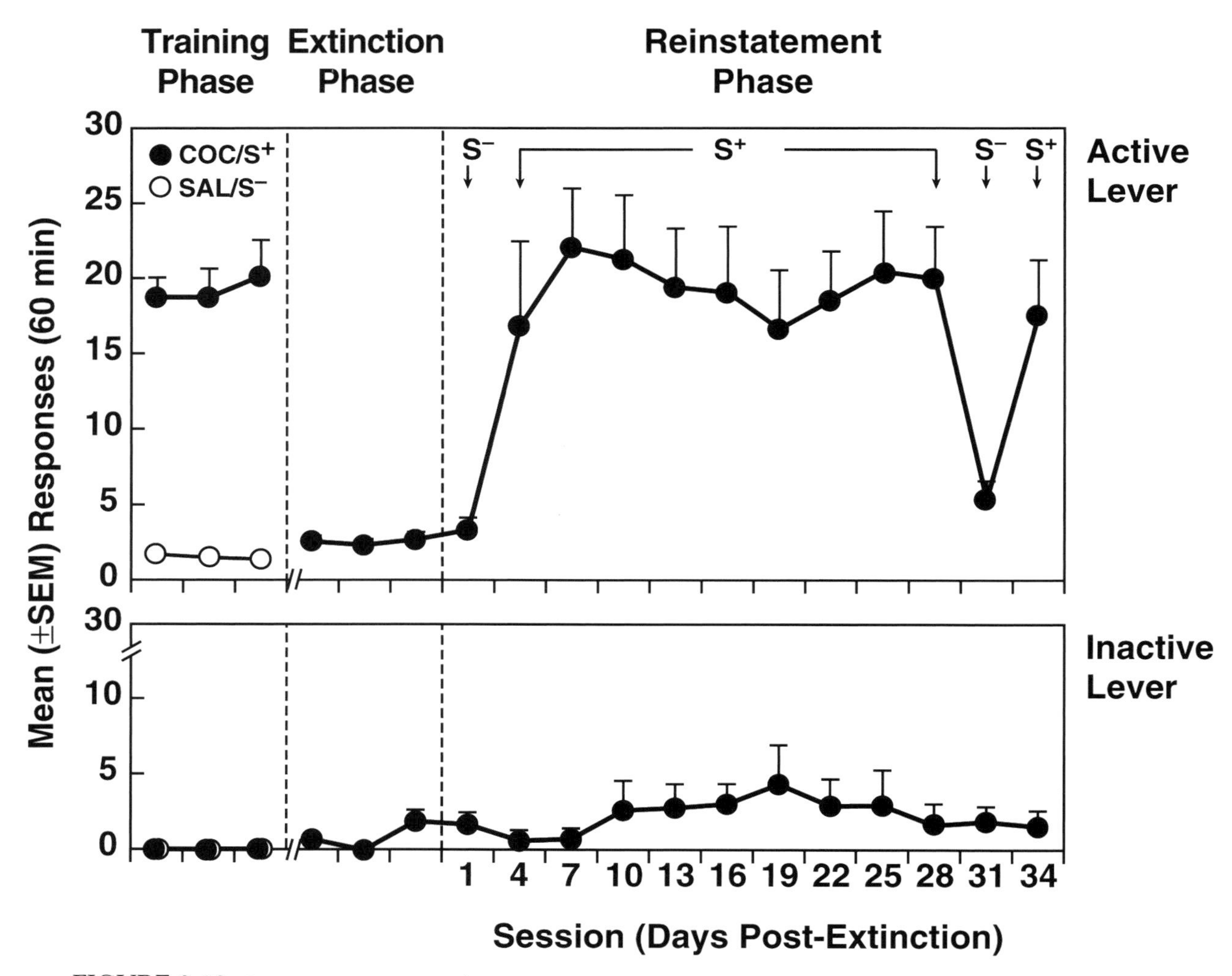

FIGURE 2.18 Lever-press responses of rats during cocaine self-administration training, extinction, and reinstatement sessions at an active (top) and inactive (bottom) lever. Training Phase: Cocaine-reinforced (○) and saline/nonreinforced (●) responses during the final three days of the self-administration phase in rats trained to associate discriminative stimuli with the availability of intravenous cocaine (S^+) or saline (S^-). Rats were designated for tests of the resistance to extinction of cocaine-seeking behavior induced by the cocaine S^+ during the Reinstatement Phase. Extinction Phase: Extinction responses at criterion (< 4 responses/session over three consecutive days). The number of days required to reach the criterion was 15.3 ± 3.9. Reinstatement Phase: Responses in the presence of the S^+ and S^-. Exposure to the S^+ elicited significant recovery of responding in the absence of further drug availability, while responding in the presence of the S^- remained at extinction levels. [Reproduced with permission from Weiss *et al.*, 2001.]

first component or unit of the schedule usually results in the presentation of a brief stimulus (often a light), and completion of the overall schedule produces the stimulus and the primary reinforcer. For example, each *n*th response produces a brief visual stimulus, and the first fixed-ratio completed after a fixed-interval produces the visual stimulus and the primary reinforcer (a drug injection). A second-order schedule can be repeated several times during a session, which would result in multiple drug administrations. Responses occurring before any drug administration can be used as measures of the conditioned reinforcing properties of drugs.

Studies in nonhuman primates (e.g., macaque monkeys, baboons) and dogs indicated that subjects will readily and reliably perform long behavioral sequences in second-order schedules, which are several minutes to hours in duration, in order to receive a variety of drugs, including psychomotor stimulants, opiates, nicotine and barbiturates, even when the drug is administered only at the end of the schedule (Goldberg, 1973; Goldberg *et al.*, 1975; Kelleher and Goldberg, 1977; Goldberg and Gardner, 1981). Rats also have been successfully trained to work on second-order schedules for drugs (Arroyo *et al.*, 1998) (**Fig. 2.21**). Performance in second-order schedules is

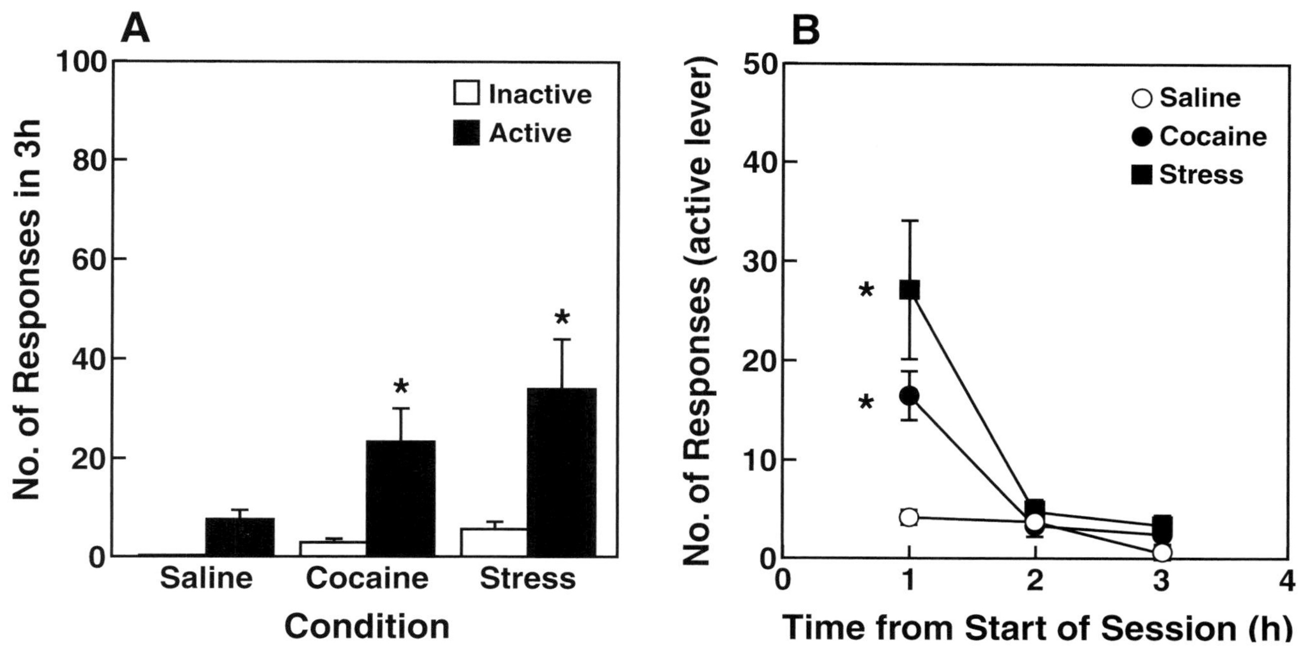

FIGURE 2.19 **(A)** Mean (± SEM) number of responses by rats on the previously inactive and active levers in a 3 h test for reinstatement after a noncontingent intravenous injection of saline, a noncontingent intravenous priming injection of cocaine (2.0 mg/kg), and intermittent footshock stress (10 min, 0.5 mA, 0.5 s on, mean off period of 40 s). **(B)** Mean (± SEM) number of responses on the previously active lever during each hour following the saline, cocaine, and footshock primes. $*p < 0.05$, significantly different from the saline condition. [Reproduced with permission from Erb *et al.*, 1996.]

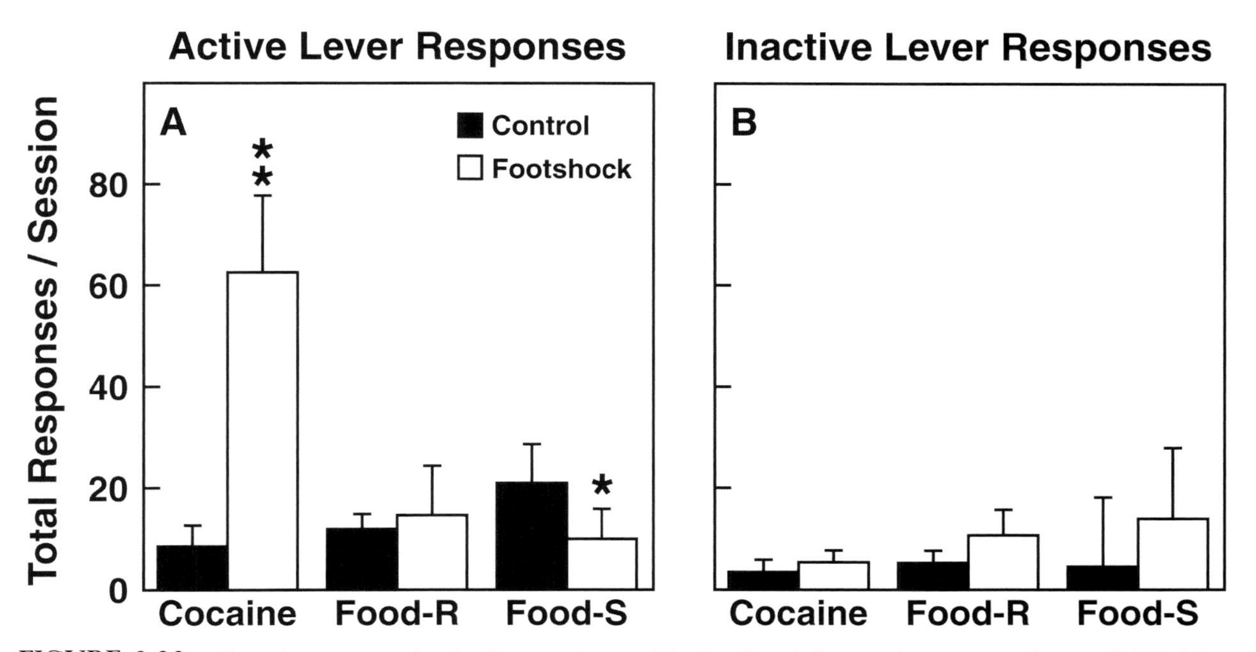

FIGURE 2.20 Effect of intermittent footshock on cocaine- and food-seeking behavior after extinction by rats. **(A)** Each bar represents the mean ± SEM total presses on the active lever in a 2 h extinction session following a 15 min shock-free period (control testing session) or 15 min of intermittent footshock (0.86 mA; 0.5 s on, with a mean off period of 40 s) in either cocaine-trained (n = 6) or food-trained rats (food-restricted [Food-R] $n = 8$ or food-sated [Food-S] $n = 8$ at the time of testing). During these testing sessions, presses were without consequences. Asterisks (*) indicate statistically significant differences from the control testing session ($**p < 0.01$; $*p < 0.05$). **(B)** Same as in A, but for presses on the inactive lever. [Reproduced with permission from Ahmed and Koob, 1997.]

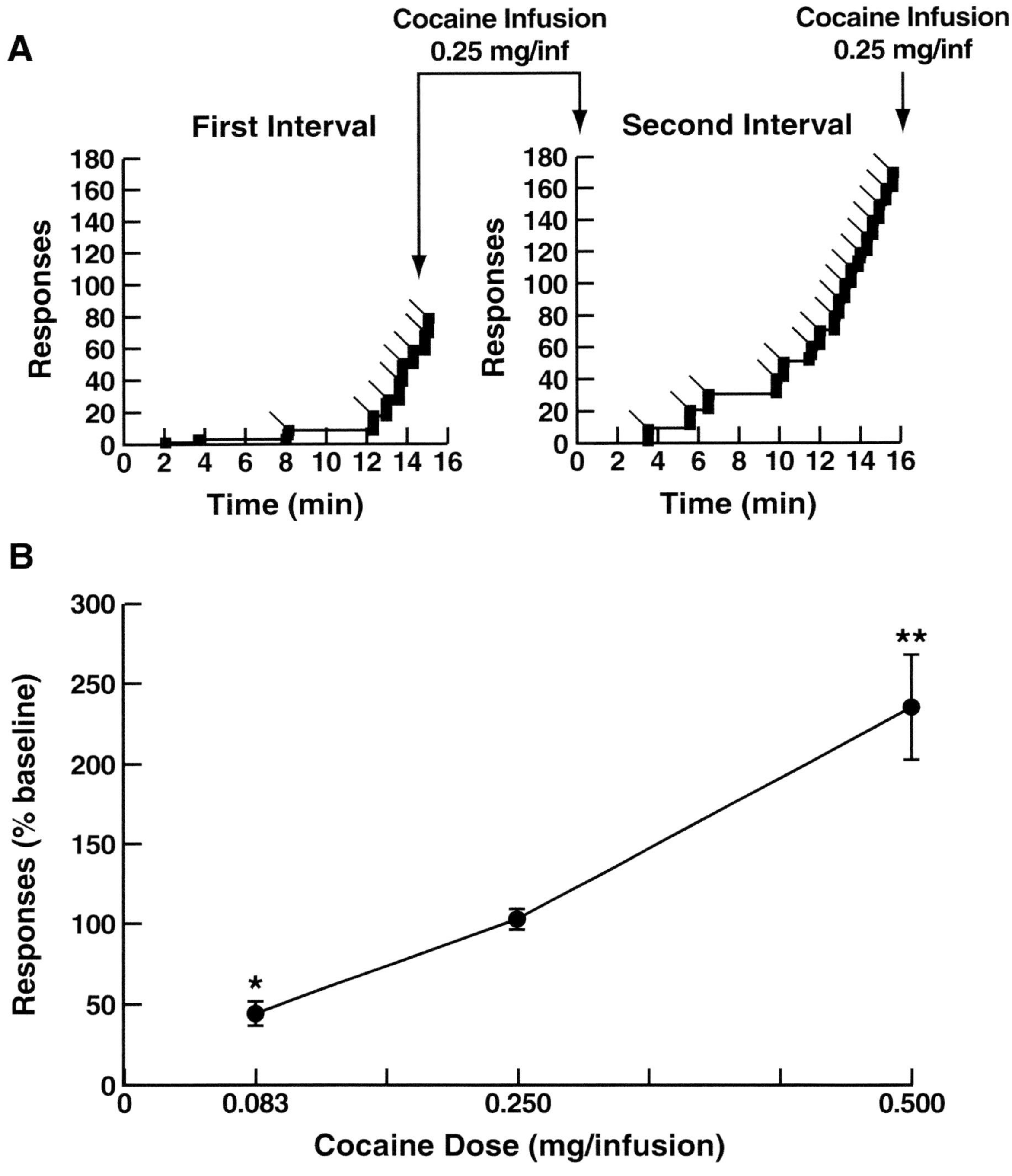

FIGURE 2.21 **(A)** Example of pattern of responding during the first two intervals of a typical intravenous cocaine self-administration session in rats under a fixed-interval 15 min (FR10:S) schedule of reinforcement. The cumulative number of responses on the active lever is plotted against the duration of the interval. Each small diagonal line represents the presentation of a 1 s stimulus light (conditioned stimulus). The completion of a 10-response unit after 15 min had elapsed resulted in the end of the first interval, the delivery of an intravenous cocaine infusion (0.25 mg/infusion over 4 s), 20 s lever retraction, 20 s conditioned stimulus presentation, and the beginning of the second interval. Each daily session consisted of five intervals occurring consecutively in the same manner. **(B)** Effect of changes in unit dose of cocaine on responding under a fixed-interval 15 min (FR10:S) schedule of intravenous cocaine reinforcement. The number of responses for the first interval of the session at the training dose of cocaine (0.25 mg/infusion) was compared to responding on the day immediately after the three consecutive sessions when the self-administered dose of cocaine was changed to double (0.50 mg/infusion) or a third of (0.083 mg/infusion) the training dose. Response levels are expressed as percentage of the average baseline levels of responding reached during the three days preceding the change of dose. Asterisks indicate significantly different responding with the 0.083 mg/infusion (* $p < 0.05$) and 0.50 mg/infusion (** $p < 0.01$) doses compared to the training dose. [Reproduced with permission from Arroyo *et al.*, 1998.]

maintained by injections (intravenous, intramuscular, or oral) of a variety of drugs that are abused by humans, with the animals exhibiting similar behavioral patterns in second-order schedules that terminate in drug injections (Goldberg *et al.*, 1976; Goldberg and Gardner, 1981).

Alteration of the stimuli that maintain a second-order schedule can alter acquisition, maintenance, resistance to extinction, and recovery from extinction in second-order schedules (Goldberg and Gardner, 1981). For example, replacement of drug-paired stimuli with nondrug-paired stimuli decreases response rates (Spear and Katz, 1991). These findings suggest that drug-paired stimuli function as conditioned reinforcers, and as such they are essential in maintaining performance in these long schedules. This maintenance of performance in second-order schedules with drug-paired stimuli appears to be analogous to the maintenance and reinstatement of drug-seeking behavior in humans with the presentation of drug-paired stimuli (McLellan *et al.*, 1986; Childress *et al.*, 1988).

Second-order schedules maintain high rates of responding (e.g., thousands of responses per session in monkeys) and extended sequences of behavior before any drug administration. Thus, potentially disruptive, nonspecific, acute drug and treatment effects on response rates are minimized. High response rates are maintained even for doses that decrease rates during a session on a regular fixed-ratio schedule, indicating that performance on the second-order schedule is unaffected by those acute effects of the drug which disrupt operant responding (Katz and Goldberg, 1987).

Protracted Abstinence

Relapse to drugs of abuse often occurs even after physical and motivational withdrawal signs have ceased, suggesting perhaps that the neurochemical changes which occur during the development of dependence can persist beyond the overt signs of acute withdrawal. Indeed, animal work has shown that prior dependence lowers the 'dependence threshold' such that previously ethanol-dependent animals made dependent again display more severe withdrawal symptoms than groups receiving alcohol for the first time (Branchey *et al.*, 1971; Baker and Cannon, 1979; Becker and Hale, 1989; Becker, 1994). This supports the notion that alcohol experience, and the development of dependence in particular, can lead to relatively long-lasting motivational alterations in responsiveness to alcohol.

In human alcoholics, numerous symptoms that can be characterized as components of negative affect can persist long after acute physical withdrawal from ethanol. Fatigue and tension have been reported to persist up to 5 weeks postwithdrawal (Alling *et al.*, 1982). Anxiety has been shown to persist up to 9 months (Roelofs, 1985), and anxiety and depression have been shown to persist in up to 20–25 per cent of alcoholics for up to two years postwithdrawal. These symptoms, postacute withdrawal, tend to be affective in nature, subacute and often precede relapse (Hershon, 1977; Annis *et al.*, 1998). A recent factor analysis of Marlatt's relapse taxonomy found that negative emotion, which includes elements of anger, frustration, sadness, anxiety, and guilt, was a key factor in relapse (Zywiak *et al.*, 1996), and the leading precipitant of relapse in a large-scale replication of Marlatt's taxonomy was negative affect (Lowman *et al.*, 1996). In secondary analyses of patients not meeting criteria for any DSM-IV mood disorder in a 12-week clinical trial with alcohol dependence, the association with relapse and a subclinical negative affective state was particularly strong (Mason *et al.*, 1994). This state has been termed *protracted abstinence* and has been defined in humans as showing a Hamilton Depression rating ≥8 with the following three items consistently reported by subjects: depressed mood, anxiety, and guilt (Mason *et al.*, 1994).

Prolonged increases in ethanol self-administration can be observed in rats after acute withdrawal and detoxification (Roberts *et al.*, 2000). This increase in self-administration of ethanol is accompanied by increases in blood alcohol levels and persists for up to 8 weeks postdetoxification (Roberts *et al.*, 2000) (**Fig. 2.22**). The increase in self-administration also is accompanied by increased responsivity to stressors and increased responsivity to antagonists of the brain corticotropin-releasing factor stress systems (Valdez *et al.*, 2003) (see *Alcohol* chapter). The persistent increase in ethanol self-administration has been hypothesized to involve an allostatic-like adjustment, such that the set point for ethanol reward is elevated (Roberts *et al.*, 2000; Koob and Le Moal, 2001). These persistent alterations in ethanol self-administration and residual sensitivity to stressors has been arbitrarily defined as a state of 'protracted abstinence.' Protracted abstinence so defined in the rat spans a period, after acute physical withdrawal has disappeared, where elevations in ethanol intake over baseline and increased stress responsivity persist 2–8 weeks after withdrawal from chronic ethanol.

Animal Models for Conditioned Withdrawal

Motivational aspects of withdrawal also can be conditioned, and conditioned withdrawal has been repeatedly observed in opiate-dependent animals

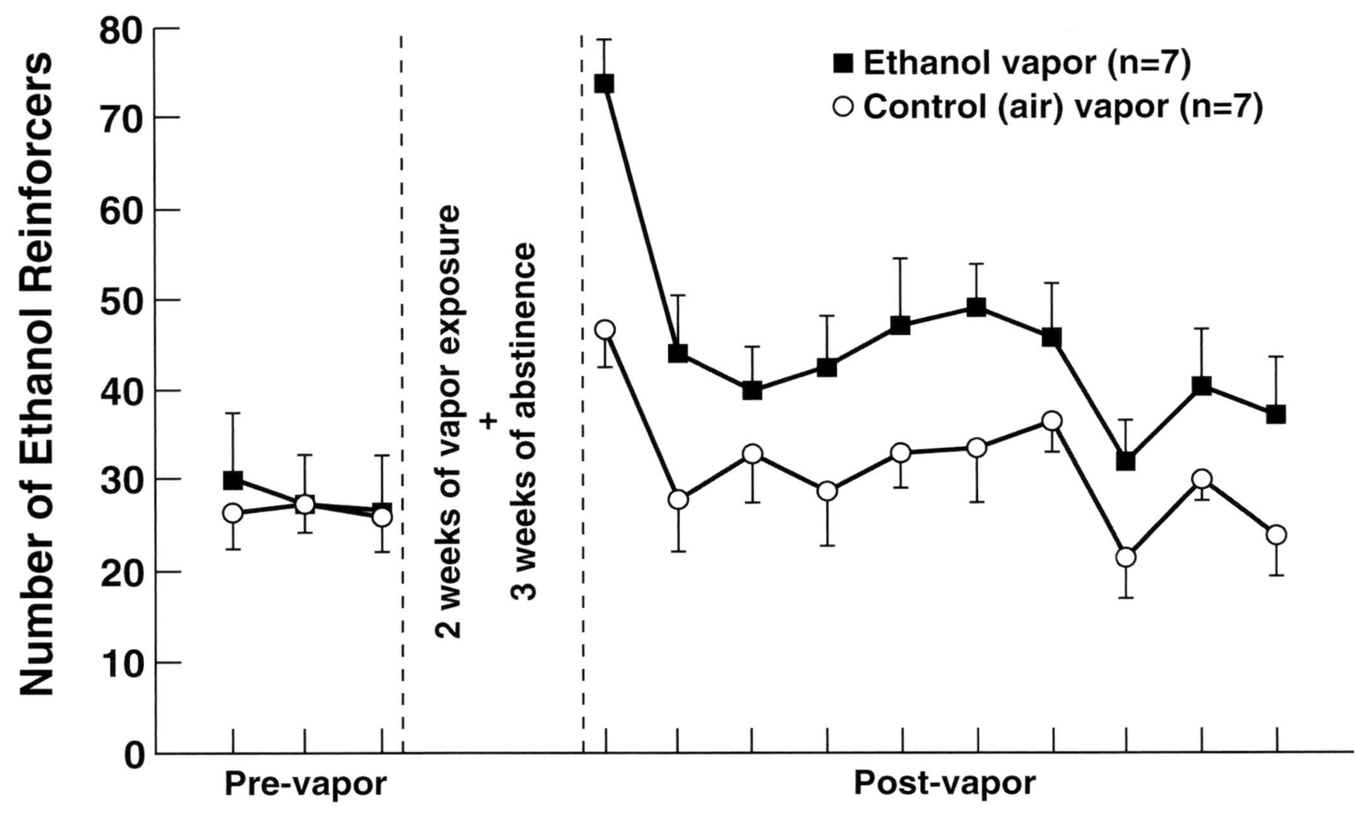

FIGURE 2.22 A prior history of chronic ethanol exposure enhances the alcohol deprivation effect in rats. Operant responding for oral ethanol across 10 days of 30 min test sessions in rats exposed to two weeks of ethanol vapor ($n = 6$) or air ($n = 6$). The three prevapor test sessions were used for group selection. Daily operant test sessions were resumed two weeks following removal from the vapor chambers. Number of deliveries are represented as means ± SEM. Asterisks (*) indicate a significant difference between the ethanol and control groups ($p < 0.05$). [Reproduced with permission from Roberts *et al.*, 2000.]

and humans. Human subjects report signs and symptoms resembling opiate abstinence when returning to environments similar to those associated with drug experiences, even following detoxification (O'Brien *et al.*, 1977). In an experimental laboratory situation, opiate addicts maintained on methadone (see *Opioids* chapter) were subjected to naloxone injections repeatedly paired with a compound stimulus of a tone and a peppermint smell (O'Brien *et al.*, 1977). Following these pairings, presentation of only the tone and odor elicited both subjective reports of sickness and discomfort and objective physical signs of withdrawal.

In early animal studies, rats made dependent by gradually increasing daily doses of morphine were exposed to a distinct environment each evening while experiencing the gradual and progressive onset of morphine abstinence. After six weeks of such pairings, rats exposed to this same distinct environment showed signs of withdrawal up to 155 days after the last morphine injection (Wikler and Pescor, 1967; Wikler, 1973). Motivational signs of withdrawal also have been conditioned in animals. Morphine-dependent rhesus monkeys, trained to lever press for food on a fixed-ratio 10 schedule, showed an immediate suppression of food-maintained responding in addition to clear physical signs of withdrawal after injection of the mixed opiate agonist/antagonist nalorphine (Goldberg and Schuster, 1967, 1970; Goldberg, 1975). Presentation of the conditioned stimuli resulting from repeated pairings of a light or tone with the nalorphine injection also produced a complete suppression of this food responding (Goldberg and Schuster, 1967, 1970). Similar results have been observed with rats. Morphine-dependent rats trained to lever press for food on a fixed-ratio 15 schedule showed effects similar to those detected in monkeys (Baldwin and Koob, 1993). Four pairings of a compound stimulus of a tone and smell with an injection of naloxone resulted in a reduction in operant responding in response to the tone and smell alone, and this conditioned response persisted for one month, even after pellet removal (Goldberg and Schuster, 1967; Sparber and Meyer, 1978; Baldwin and

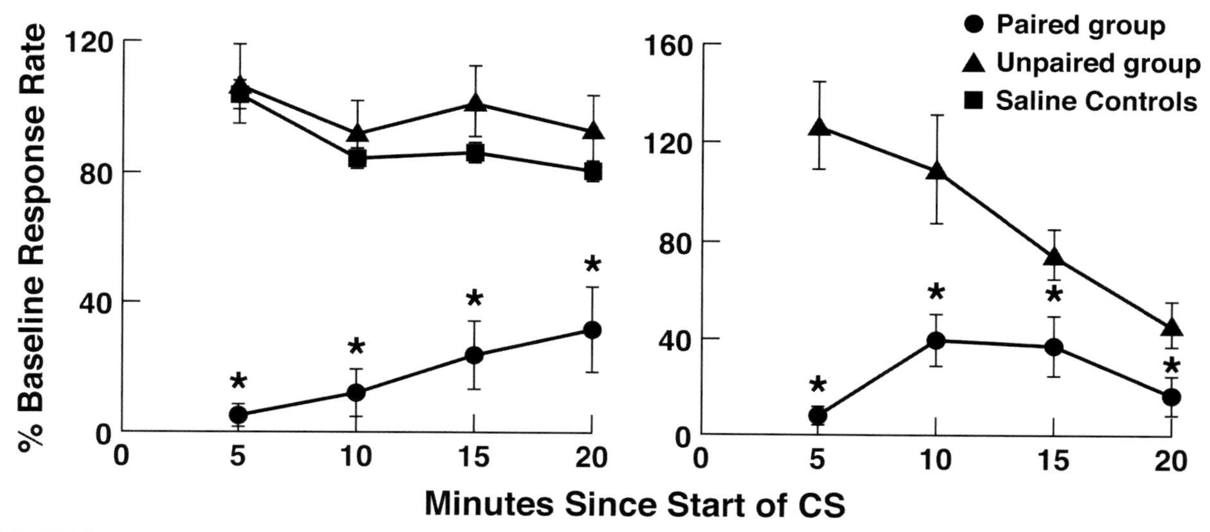

FIGURE 2.23 Left panel: Test for conditioned withdrawal in morphine-dependent (Left Panel) and postdependent (Right Panel) rats. Values are mean ± SEM percentages of baseline rates of lever-pressing per minute since the start of the conditioned stimulus (CS). Mean ± SEM pretreatment baseline response rates on the test day for morphine-dependent rats were as follows: paired group = 60.0 ± 8.4; unpaired group = 74.7 ± 8.1; saline controls = 95.1 ± 8.7 lever-presses/min. Asterisks (*) indicate significantly different from the unpaired group and saline controls ($p < 0.05$). Mean ± SEM pretreatment baseline response rates on the test day for postdependent rats were as follows: paired group = 66.0 ± 10.8; unpaired group = 81.9 ± 9.3 lever-presses/min. Asterisks indicate significantly different from the unpaired group ($p < 0.05$). [Reproduced with permission from Baldwin and Koob, 1993.]

Koob, 1993) (**Fig. 2.23**). Perhaps even more interesting is that the animals showed no obvious conditioned physical signs of withdrawal. These results suggest that a conditioned stimulus can acquire aversive stimulus effects that persist even in the absence of opioid receptor occupancy, and that motivational signs of conditioned withdrawal can occur in the absence of withdrawal-like somatic symptoms.

Paradigms demonstrating the motivational significance of conditioned withdrawal were provided in a series of elegant operant studies by Goldberg and colleagues (Goldberg *et al.*, 1969, 1971). Monkeys allowed to intravenously self-administer morphine 24 h per day were injected once per day with nalorphine, preceded by 10 min with a light cue. After repeated pairings of nalorphine and the light, presentation of the light alone (with injection of saline) resulted in a conditioned increase in responding for morphine, presumably to avoid the onset of withdrawal (Goldberg *et al.*, 1969). An even more compelling demonstration of the negative reinforcing properties of conditioned withdrawal was provided by a study in which opiate-dependent monkeys were given daily 2 h sessions in which a green light signaled an intravenous infusion of nalorphine or naloxone (Goldberg *et al.*, 1971). Lever-pressing by the monkey terminated the green light and prevented the injections of opiate antagonists for 60 s. Initially, most responses occurred after the onset of injection, but with repeated pairings of the light cue and the antagonist, most of the responding occurred during the period when the light cue was illuminated, but before the antagonist infusion. These results are a powerful demonstration of the negative reinforcing properties of antagonist-precipitated drug withdrawal.

Attempts to modify such conditioned effects hypothetically could contribute to knowledge of the factors that contribute to relapse or craving. One also could envisage the use of cue- and drug-induced reinstatement of an extinguished place conditioning response as a measure of relapse (Mueller and Stewart, 2000). These results support the hypothesis that learned responses to drug-related environmental stimuli can be important factors in the reinstatement of drug-seeking in animals and provide a powerful model for elucidating the neuropharmacological basis for such effects that are related to the human concepts of relapse and craving (Weiss *et al.*, 2000).

Summary of Animal Models for the Preoccupation/Anticipation Stage

Each of the models outlined above has face validity to the human condition and ideally heuristic value for understanding the neurobiological bases for different aspects of the craving stage of the addiction cycle. The DSM-IV criteria that apply to the craving stage and loss of control over drug intake include 'any unsuccessful effort or persistent desire to cut down or control substance use.' The extinction paradigm has predictive validity, and with the reinstatement procedure, it can be a reliable indicator of the ability of conditioned stimuli to reinitiate drug-seeking behavior. The conditioned-reinforcement paradigm has the advantage of assessing the motivational value of a drug infusion in the absence of acute effects of the self-administered drug that could influence performance or other processes which interfere with motivational functions. For example, nonspecific effects of manipulations administered before the stimulus–drug pairings do not directly affect the assessment of the motivational value of the stimuli because the critical test can be conducted several days after the stimulus–drug pairings. Also, the paradigm contains a built-in control for nonspecific motor effects of a manipulation by its assessment of the number of responses on an inactive lever.

The animal models for the conditioned negative reinforcing effects of drugs are reliable measures and have good face validity. Work in this area, however, has largely been restricted to the opiate field where competitive antagonists reliably precipitate a withdrawal syndrome. There is general consensus that the animal reinstatement models have face validity (Epstein and Preston, 2003; Katz and Higgins, 2003). However, predictive validity remains to be established. To date, there is some predictive validity for the stimuli that elicit reinstatement in the animal models, but little evidence of predictive validity from studies of the pharmacological treatments for drug relapse (Katz and Higgins, 2003). Very few clinical trials have tested medications that are effective in the reinstatement model, and very few anti-relapse medications have been tested in the animal models of reinstatement (see *Alcohol* chapter). From the perspective of functional equivalence or construct validity there is some evidence of functional commonalities. For example, drug re-exposure or priming, stressors, and cues paired with drugs all produce reinstatement in animal models and promote relapse in humans.

ANIMAL MODELS FOR THE TRANSITION TO ADDICTION

Another conceptual framework upon which animal models can be directly related to the compulsive behavior and loss of control over intake that is the hallmark of addiction, is to specifically relate a given animal model to a specific symptom of the DSM-IV criteria for addiction (**Table 2.6**). Several recent studies have emphasized animal models that contribute to specific elements of the DSM-IV criteria with strong face validity, and at the same time may represent specific endophenotypes of the compulsive nature of the addiction process.

Escalation in Drug Self-Administration with Prolonged Access

A progressive increase in the frequency and intensity of drug use is one of the major behavioral phenomena characterizing the development of addiction and has face validity with the DSM-IV criteria: 'The substance is often taken in larger amounts and over a longer period than was intended' (American Psychological Association, 1994). A framework with which to model the transition from drug use to drug addiction can be found in recent animal models of prolonged access to intravenous cocaine self-administration. Historically, animal models of cocaine self-administration involved the establishment of stable behavior from day to day to allow the reliable interpretation of data provided by within-subject designs aimed at exploring the neuropharmacological and neurobiological bases of the reinforcing effects of acute cocaine. Typically, after acquisition of self-administration, rats that are allowed access to cocaine for 3 h or less per day establish highly stable levels of intake and patterns of responding between daily sessions. To explore the possibility that differential access to intravenous cocaine self-administration in rats may produce different patterns of drug intake, rats were allowed access to intravenous self-administration of cocaine for 1 or 6 h per day. One-hour access (short access or ShA) to intravenous cocaine per session produced low and stable intake as observed previously. In contrast, with 6-h access (long access or LgA) to cocaine, drug intake gradually escalated over days (Ahmed and Koob, 1998) (**Fig. 2.24**). In the escalation group, there was increased intake during the first hour of the session as well as sustained intake over the entire session and an upward shift in the dose–effect function, suggesting an increase in hedonic set point. When animals were allowed access to different doses of cocaine, both the LgA and ShA animals titrated their cocaine intake, but the LgA rats consistently self-administered almost twice as much cocaine at any dose tested, further suggesting an upward shift in the set point for cocaine reward in the escalated animals (Ahmed and Koob, 1999). Escalation also is associated with an increase in break point for cocaine in a progressive-ratio schedule,

TABLE 2.6 **Animal Models for DSM-IV Criteria**

DSM-IV	Animal models
A. A maladaptive pattern of substance use, leading to clinically significant impairment or distress as manifested by three or more of the following occurring at any time in the same 12-month period:	
1. Need for markedly increased amounts of a substance to achieve intoxication or desired effect; or markedly diminished effect with continued use of the same amount of the substance.	Tolerance to stimulus effects—drug discrimination Psychostimulants Opioids Alcohol
2. The characteristic withdrawal syndrome for a substance or use of a substance (or a closely related substance) to relieve or avoid withdrawal symptoms.	Increased reward thresholds during drug withdrawal Psychostimulants Opioids Alcohol Nicotine Δ^9-tetrahydrocannabinol Somatic signs of withdrawal Opioids Alcohol Nicotine Δ^9-tetrahydrocannabinol Self-administration in dependent animals Opioids Alcohol Nicotine
3. Persistent desire or one or more unsuccessful efforts to cut down or control substance use.	Conditioned positive reinforcing effects Psychostimulants Opioids Alcohol Conditioned place preference Psychostimulants Opioids Alcohol Nicotine
4. Substance use in larger amounts or over a longer period than the person intended.	Escalation in intake Psychostimulants Opioids Alcohol Deprivation Effect Self-administration in dependent animals Opioids Alcohol Nicotine
5. Important social, occupational, or recreational activities given up or reduced because of substance use.	Choice paradigms Behavioral economics—loss of plasticity Self-administration in presence of aversive stimuli
6. A great deal of time spent in activities necessary to obtain, to use, or to recover from the effects of substance used.	Runway model Second-order schedules of reinforcement
7. Continued substance use despite knowledge of having a persistent or recurrent physical or psychological problem that is likely to be caused or exacerbated by use.	Progressive-ratio schedules Cocaine binge toxicity

suggesting an enhanced motivation to seek cocaine or an enhanced efficacy of cocaine reward (Paterson and Markou, 2003) (**Fig. 2.25**). Similar changes in the reinforcing and incentive effects of cocaine have been observed in an experimental model of intravenous drug self-administration in rats studied with cocaine intake over 14–29 sessions. A progressive escalation in cocaine intake was observed at each dose tested in animals that are allowed 14 sessions of extended access to cocaine, and a shift upward of the dose–response function was observed (Deroche *et al.*, 1999) (**Fig. 2.26**).

To further assess the changes in the reinforcing and incentive properties of cocaine associated with such an escalated drug intake, rats previously allowed to self-administer cocaine for either a short (6 session) or

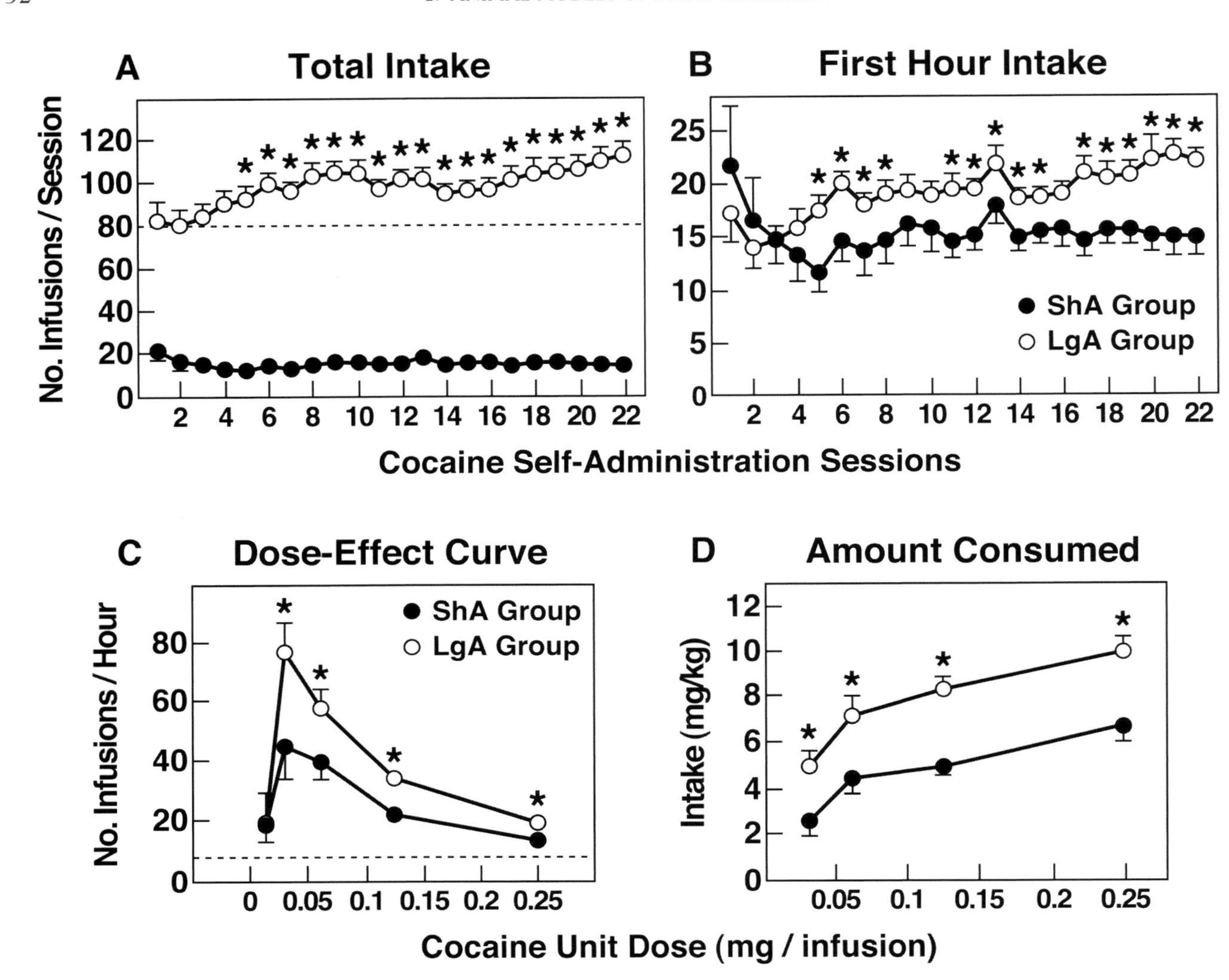

FIGURE 2.24 Effect of drug availability on cocaine intake (mean ± SEM). **(A)** In long-access (LgA) rats ($n = 12$) but not in short-access (ShA) rats ($n = 12$), mean total cocaine intake started to increase significantly from session 5 ($p < 0.05$; sessions 5–22 compared to session 1) and continued to increase thereafter ($p < 0.05$; session 5 compared to sessions 8–10, 12, 13, 17–22). **(B)** During the first hour, LgA rats self-administered more infusions than ShA rats during sessions 5–8, 11, 12, 14, 15, and 17–22 ($p < 0.05$). **(C)** Mean infusions (± SEM) per cocaine dose tested. LgA rats took significantly more infusions than ShA rats at doses of 31.25, 62.5, 125, and 250 µg/infusion ($p < 0.05$). **(D)** After escalation, LgA rats took more cocaine than ShA rats regardless of the dose ($p < 0.05$). *$p < 0.05$ (Student's t test after appropriate one-way and two-way analysis of variance). [Reproduced with permission from Ahmed and Koob 1998.]

long (29 session) period were compared in three different tests of the incentive and reinforcing effects of cocaine (Deroche *et al.*, 1999) (**Fig. 2.27**). In the first test, cocaine-induced reinstatement of the extinguished self-administration behavior was used to measure the motivational properties of priming doses of cocaine. In the second test, the cocaine-induced runway model allowed evaluation of the reinforcing properties of cocaine using an operant response that was independent of the response used during self-administration. In this test, the motivation to obtain the drug was assessed daily in a drug-free state by measuring how fast the animal ran along an alley to receive cocaine infusions. In the third test, cocaine-induced place conditioning was used as an index of the appetitive properties of a drug that was not based on the measure of an operant response, but rather on the evaluation of the degree of preference for an environment previously paired with the effects of the drug in a drug-free state.

Escalating drug intake during cocaine self-administration is paralleled by an increase in the responsiveness to the reinforcing and incentive effects of the drug. Animals with prolonged training for cocaine self-administration showed a greater sensitivity to both

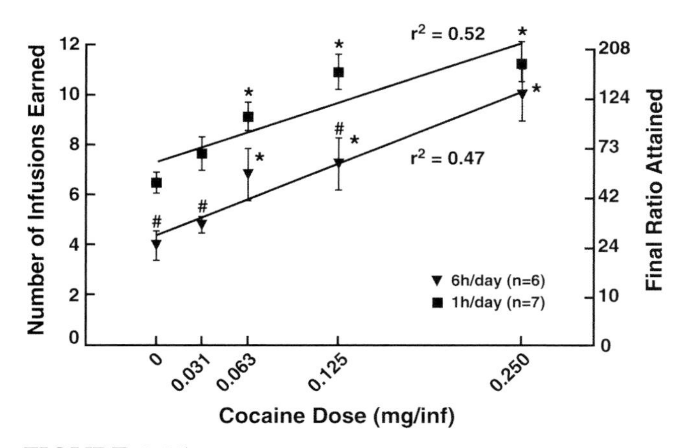

FIGURE 2.25 The effect of escalation in cocaine intake on the dose–response functions obtained under a progressive-ratio schedule of reinforcement in rats. The graph shows the dose–response functions obtained in the short-access ($n = 7$) and long-access ($n = 6$) groups. *Significant differences ($p < 0.05$) from vehicle-maintained responding (within-group comparisons). #$p < 0.05$ between groups. [Reproduced with permission from Paterson and Markou, 2003.]

cocaine-induced runway behavior and cocaine-induced reinstatement of self-administration. In contrast, the threshold for cocaine-induced place conditioning was not affected. This suggested to the authors that over repeated drug taking it was not the motivational impact of the drug that was increased, but only the likelihood that such a state was translated into drug-directed behavior (Deroche *et al.*, 1999).

Operant Drug Self-administration in Dependent Animals

While physical symptoms of withdrawal are not always observed in alcoholics, motivational signs of withdrawal such as anxiety, dysphoria, and malaise are considered an important factor in the continued use of alcohol or relapse. Thus, alcoholics drink alcohol not simply for its euphorigenic effects, but also to self-medicate existing negative affective states or to avoid or reverse the negative affective states associated with withdrawal (Cappell and LeBlanc, 1981; Edwards, 1990; Lowman *et al.*, 1996; Zywiak *et al.*, 1996). Indeed, such motivational withdrawal symptoms (especially depression and anxiety) were found to provoke drinking in alcoholics who experienced these symptoms (Hershon, 1977). Recent conceptualizations of the contribution of different sources of reinforcement to the addictive process have been elaborated where negative reinforcement and positive reinforcement combine to produce powerful motivation to drink in

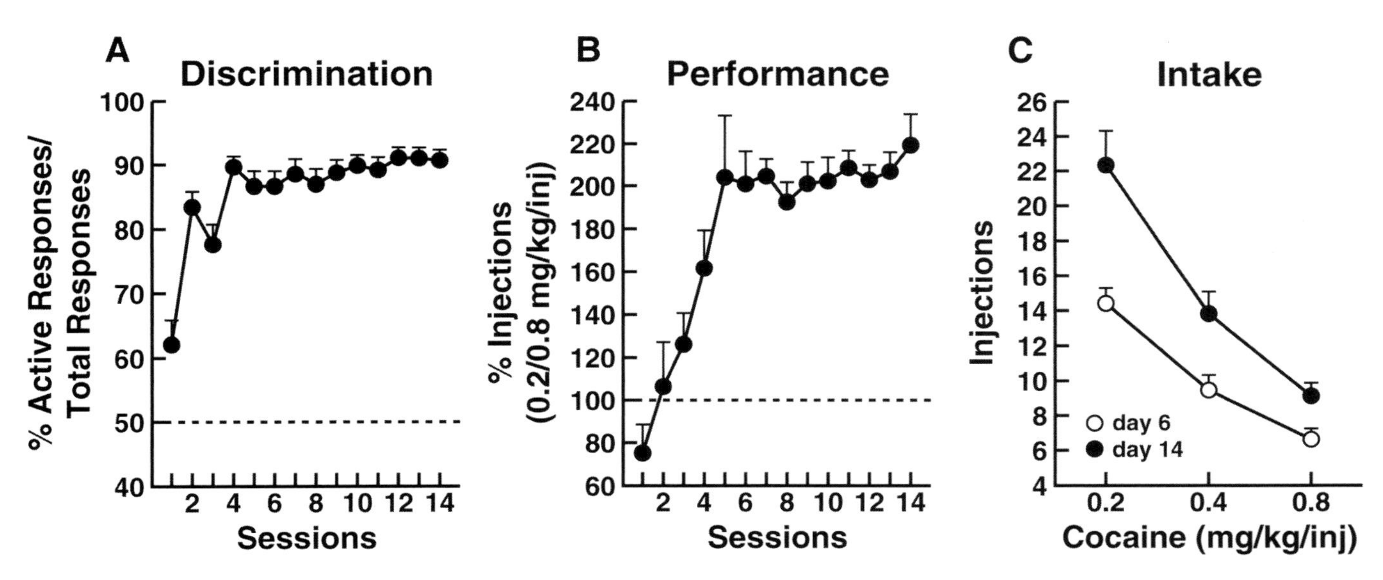

FIGURE 2.26 Temporal evolution of intravenous cocaine self- administration in rats allowed access for 14 days to 0.2, 0.4, and 0.8 mg/kg/inj. Two holes, located on opposite sides of the box, were used as devices to record responding. Nosepokes into the active hole activated the infusion pump (20 µl over 1 s). Nosepokes into the inactive hole produced no consequence. A fixed-ratio 1 timeout 20 s schedule was used. Doses of cocaine were tested in descending order. The drug components (30 min each) were separated by 20 min drug-free periods which were signalled by illumination of a cue light above the box. **(A)** The discrimination between the active (the one delivering cocaine) and inactive (the one with no scheduled consequence) lever progressively increased from Day 1 to Day 4, reaching a stable plateau thereafter. **(B)** The slope of the dose–response curve (performance) expressed as the percentage of self-injections at the lowest dose (0.2 mg/kg/inj) over self-injections at the highest dose (0.8 mg/kg/inj) quickly increased over the initial days of testing and reached a stable plateau on the fifth session. **(C)** Though self-administration behavior was stable between Day 6 and Day 14, the number of self-injections increased for all doses of cocaine tested. [Reproduced with permission from Deroche *et al.*, 1999.]

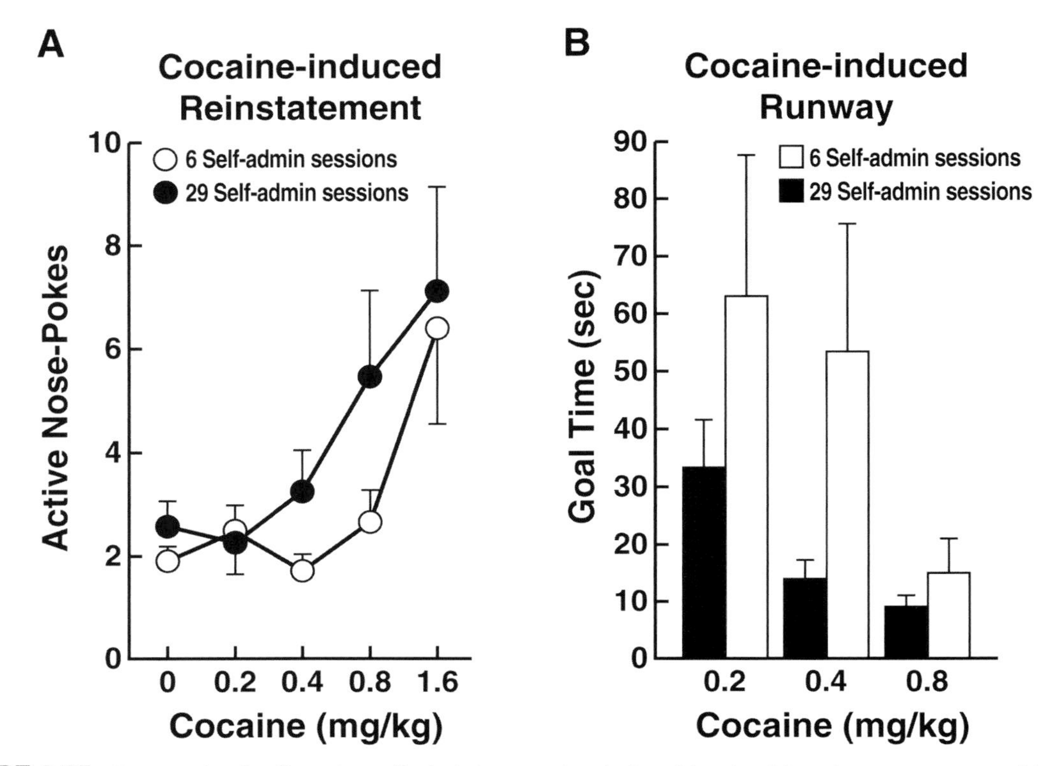

FIGURE 2.27 Rats previously allowed to self-administer cocaine during either 6 or 29 sessions were compared in two tests of the incentive and reinforcing effects of the drug: (A) cocaine-induced reinstatement, and (B) cocaine-induced runway behavior. A daily within-session paradigm was applied using three doses of cocaine (0.8, 0.4, and 0.2 mg/kg/inj). **(A)** Cocaine-induced reinstatement of nonreinforced responding for the drug in rats trained for either 29 or 6 cocaine self-administration sessions. Data are cumulated over the 10 min following each cocaine priming injection. Cocaine priming dose-dependently increased responding in the nosepoke device previously associated with cocaine self-infusion (i.e., active nose pokes) (see Fig. 2.26). The 29-session group showed increased sensitivity to cocaine-induced reinstatement compared to the 6-session group (significant only for the 0.4 and 0.8 mg/kg doses of cocaine). **(B)** Cocaine-induced runway behavior in rats trained for either 29 or 6 cocaine self-administration sessions and tested 15 days after the last self-administration session. The apparatus was a straight alley (200 cm long and 11 cm wide) connecting a start box and a goal box. Each daily trial consisted of reinforcing the run with three intravenous cocaine injections spaced 20 s apart. Three unit doses of cocaine were successively tested (0.8, 0.4, and 0.2 mg/kg/inj in a volume of 40 μl for 8, 7, and 5 days, respectively). Means of the last two days of testing for each dose are presented. The two groups did not differ for the basal goal time measured during the first trial (i.e., before any drug had been administered in the apparatus). For both experimental groups, the goal time increased as the dose of cocaine administered in the goal box decreased. The 29-session group exhibited a shorter dose-dependent goal time compared to the 6-session group. [Reproduced with permission from Deroche *et al.*, 1999.]

dependent or post-dependent individuals (see *What is Addiction* Chapter). Therefore, the development of animal models of ethanol intake in dependent animals, and elucidation of the neurobiological mechanisms mediating ethanol intake in dependent versus nondependent animals, is important for understanding alcoholism or substance dependence on alcohol. Excessive drug intake associated with withdrawal has face validity with the DSM-IV criteria of 'substance is taken to relieve or avoid withdrawal symptoms.'

Historically, animal models for the negative reinforcement associated with ethanol dependence have proven difficult, especially with rodents. There is evidence that manipulations of physical dependence can enhance preference for ethanol (Veale and Myers, 1969; Deutsch and Koopmans, 1973; Hunter *et al.*, 1974; Samson and Falk, 1974; Deutsch and Walton, 1977; Wolffgramm and Heyne, 1991; Roberts *et al.*, 1996; Schulteis *et al.*, 1996), yet other reports have not supported an enhanced preference for ethanol in dependent animals (Myers *et al.*, 1972; Begleiter, 1975; Winger, 1988). Recently, reliable and useful models of ethanol consumption in dependent rats and mice have been developed. A critical issue appears to be the learning of the association between drinking and the alleviation of withdrawal symptoms. Alcohol first was established was a reinforcer in nondependent, limited-access situations, and then the animals were made dependent. Animals are maintained on liquid diet or continuous alcohol vapor exposure at blood alcohol levels of 100–200 mg% (LeBourhis, 1975). Upon termination of the alcohol exposure, which lasted at least 2–3 weeks, animals have shown mild-to-moderate physical withdrawal symptoms when the ethanol was removed, but significant motivational signs were observed as changes in brain stimulation reward during acute withdrawal from ethanol (Schulteis *et al.*, 1995). Therefore, any physical withdrawal symptoms that the rats did experience predictably would be quite mild and would not be expected to interfere with the ability of the rats to respond. Finally, the rats were tested repeatedly following the induction of dependence to allow for the examination of learning and/or practice effects. Animals showed reliable increases in self-administration of ethanol during withdrawal where the amount of intake approximately doubled and the animals maintained blood alcohol levels from 100–150 mg% for up to 12 h of withdrawal (Roberts *et al.*, 1996; Aufrere *et al.*, 1997; Naassila *et al.*, 2000; Lallemand *et al.*, 2001) (**Fig. 2.28**). Animals exposed intermittently to the same amount of ethanol as continuously exposed animals showed an even more dramatic increase in self-administration (O'Dell *et al.*, 2004). Animals allowed access to alcohol with a liquid diet also showed increased alcohol self-administration during withdrawal (Macey *et al.*, 1996; Schulteis *et al.*, 1996; Weiss *et al.*, 1996; Valdez *et al.*, 2004).

Opioid self-administration also has been conducted in dependent animals, and the procedures are very similar to those discussed earlier regarding drug self-administration in nondependent rats. Indeed, significant evidence suggests that the reinforcing efficacy of a drug can increase with dependence. Monkeys made dependent on morphine showed enormous increases in their progressive-ratio performance compared to their performance in the nondependent state (Yanagita, 1973). In studies in rats, several animal models of heroin self-administration in dependent rats have been established. Rats trained to self-administer heroin and then made dependent on morphine using morphine pellets (2 × 75 mg morphine base, s.c.) showed dramatically increased sensitivity to naloxone (Carrera *et al.*, 1999) and escalation in heroin intake (Walker *et al.*, 2003). Also, exposure to heroin either continuously via a minipump for 6 days or intermittently (12 h on/12 h off) at doses sufficient to elicit a robust precipitated withdrawal, produced an increase in heroin intake (Walker *et al.*, 2003).

Independent of passive administration of opiates, rats allowed intravenous access to heroin for 11 h also escalate their intake of heroin and show a resistance to extinction, suggesting an increased motivation for heroin intake (Ahmed *et al.*, 2000). Perhaps most compelling, the studies with passive administration

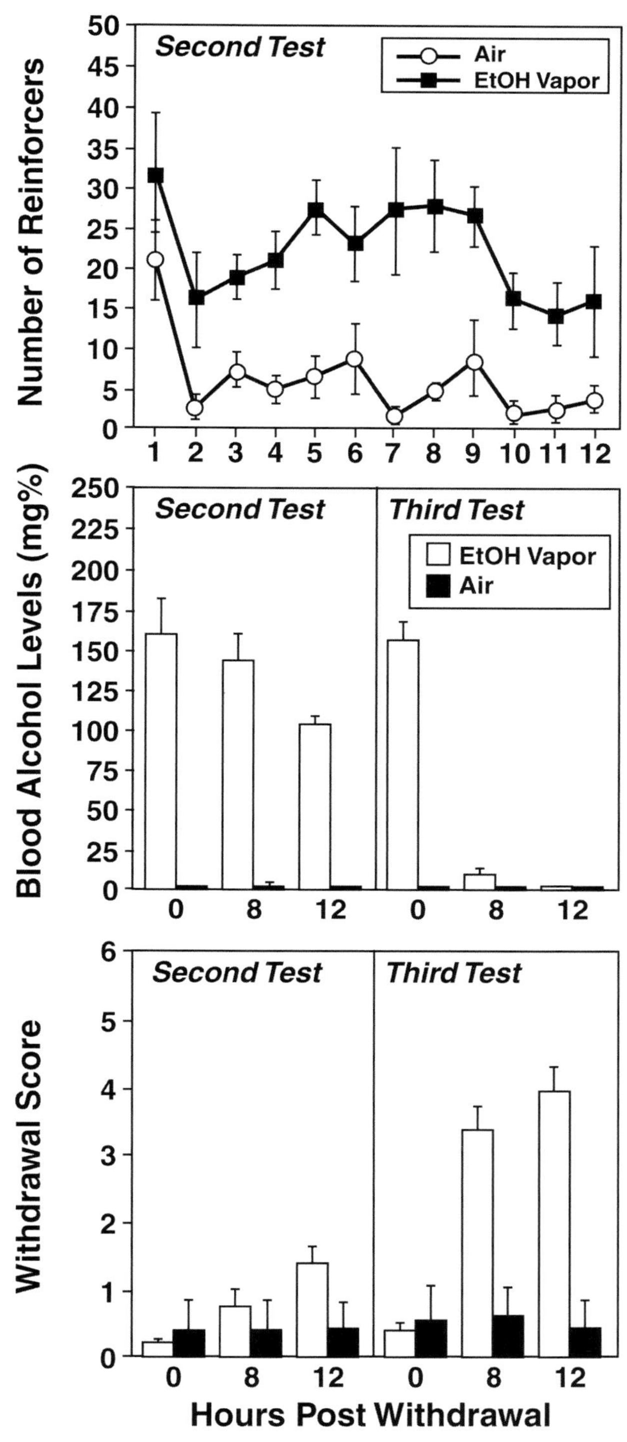

FIGURE 2.28 Operant responding for ethanol (EtOH) across a 12 h test period by air-exposed and ethanol vapor-exposed rats (top). In addition, blood alcohol levels (middle) and ethanol withdrawal severity (bottom) obtained during test 2 (while rats were allowed access to ethanol in the operant boxes) and test 3 (while in home cages) are shown. Data are expressed as means ± SEM. [Reproduced with permission from Roberts *et al.*, 1996.]

and active prolonged intravenous self-administration led to the exploration of an unlimited-access model of intravenous heroin self-administration. Access to heroin for 23 h via intravenous self-administration with continuous concomitant access to food and water via a nosepoke response led to a more rapid escalation of heroin intake than previous models and even more rapid development of dependence. Animals showed signs of dependence as early as 7 days into continuous access as measured by changes in the pattern and distribution of heroin and food intake. In addition, animals with 23-h access to heroin self-administration that have escalated intake show increases in heroin self-administration to very low doses of naloxone and decreases in heroin self-administration to administration of buprenorphine, validating pharmacologically this procedure as a measure of dependence (**Fig. 2.29**).

Thus, the reinforcing value of drugs may change with dependence. Much evidence has been generated to show that drug dependence can produce an aversive or negative motivational state that is manifested in changes in a number of behavioral measures such as response disruption, changes in reward thresholds, and conditioned place aversions, and that contributes to the increased value of reinforcing drugs (Koob *et al.*, 1993).

Alcohol Deprivation Effect

A robust and reliable feature of animal models of alcohol drinking is an increase in consumption observed after a period of deprivation. Termed the 'alcohol deprivation effect', the increase in consumption has been observed in mice (Salimov and Salimova, 1993), rats (Le Magnen, 1960; Sinclair and Senter, 1967, 1968; Spanagel *et al.*, 1996), monkeys (Kornet *et al.*, 1991), and human social drinkers (Burish *et al.*, 1981). Deprivation typically involves 3 days to several weeks and can result in temporary increases in alcohol intake usually for only 1–2 days. Maximum increases usually range between 50–100 per cent over baseline and can be seen in nonphysically dependent animals and animals with a history of physical dependence (Roberts *et al.*, 2000). The alcohol deprivation effect in nondependent animals is more robust in animals with extended access to ethanol (>1 month) in daily limited-access situations (Heyser *et al.*, 1997). The alcohol deprivation effect is exaggerated in alcohol-preferring rats where deprivations can increase intakes by 200–300 per cent and over more prolonged periods (Rodd-Henricks *et al.*, 2001).

There is some face validity to the alcohol deprivation effect in nondependent animals as a model of relapse, in that it is well-established that even after withdrawal and having been alcohol free for some time, relapses to problem drinking occur (Hunt *et al.*, 1971; Marlatt and George, 1984; Dole, 1986). Predictive validity is under test currently in the field of alcohol research, but preliminary evidence shows that the alcohol deprivation effect is particularly sensitive to naltrexone and acamprosate, two medications currently available for the treatment of relapse prevention in alcoholism (Spanagel *et al.*, 1996; Heyser *et al.*, 1997, 1998) (see *Alcohol* chapter).

Genetic Animal Models of Alcoholism

Many genetically selected lines of rats have been developed for differences in alcohol consumption and include the University of Chile A and B rats (Mardones and Segovia-Requelme, 1983), Alko alcohol (AA) and Alko nonalcohol (ANA) rats (Erikson, 1968), University of Indiana alcohol-preferring (P) and alcohol nonpreferring (NP) rats (Lumeng *et al.*, 1977), University of Indiana high-alcohol-drinking (HAD) and low-alcohol-drinking (LAD) rats (Li *et al.*, 1993), Sardinian alcohol preferring (sP) and Sardinian-nonpreferring (sNP) rats (Fadda *et al.*, 1989) (**Table 2.7**).

The P rats have been the most extensively studied and provide a model of some of the most salient features of excessive consumption of ethanol (Li and McBride, 1995; Murphy *et al.*, 1986, 2002; Waller *et al.*, 1984; Rodd-Henricks *et al.*, 2001) (**Table 2.8**). P rats voluntarily consume 6.5 g/kg ethanol per day in free choice drinking and attain blood alcohol levels in the 50–200 mg% range. When deprived of alcohol, these animals can attain intakes of up to 4 g/kg in the first hour of re-exposure and blood alcohol levels of over 150 mg% even in limited-access situations (Rodd-Henricks *et al.*, 2001; Rodd *et al.*, 2004). In an operant situation, repeated alcohol deprivations doubled break point measures in a progressive-ratio test (Rodd *et al.*, 2003). P rats will work to obtain alcohol in operant situations and will readily drink much higher concentrations than the 10 per cent ethanol for which they were selected, including concentrations as high as 35–40 per cent (Murphy *et al*, 2002). P rats drink for the pharmacological effects of alcohol as demonstrated not only by numerous control experiments, but also by the observation that they will readily self-administer alcohol in intoxicating amounts intragastrically (Waller *et al.*, 1984). P rats also will self-administer alcohol directly into the ventral tegmental area of the brain (Gatto *et al.*, 1994). Finally, P rats drink sufficient amounts of alcohol to attain metabolic and pharmacodynamic tolerance and physical dependence.

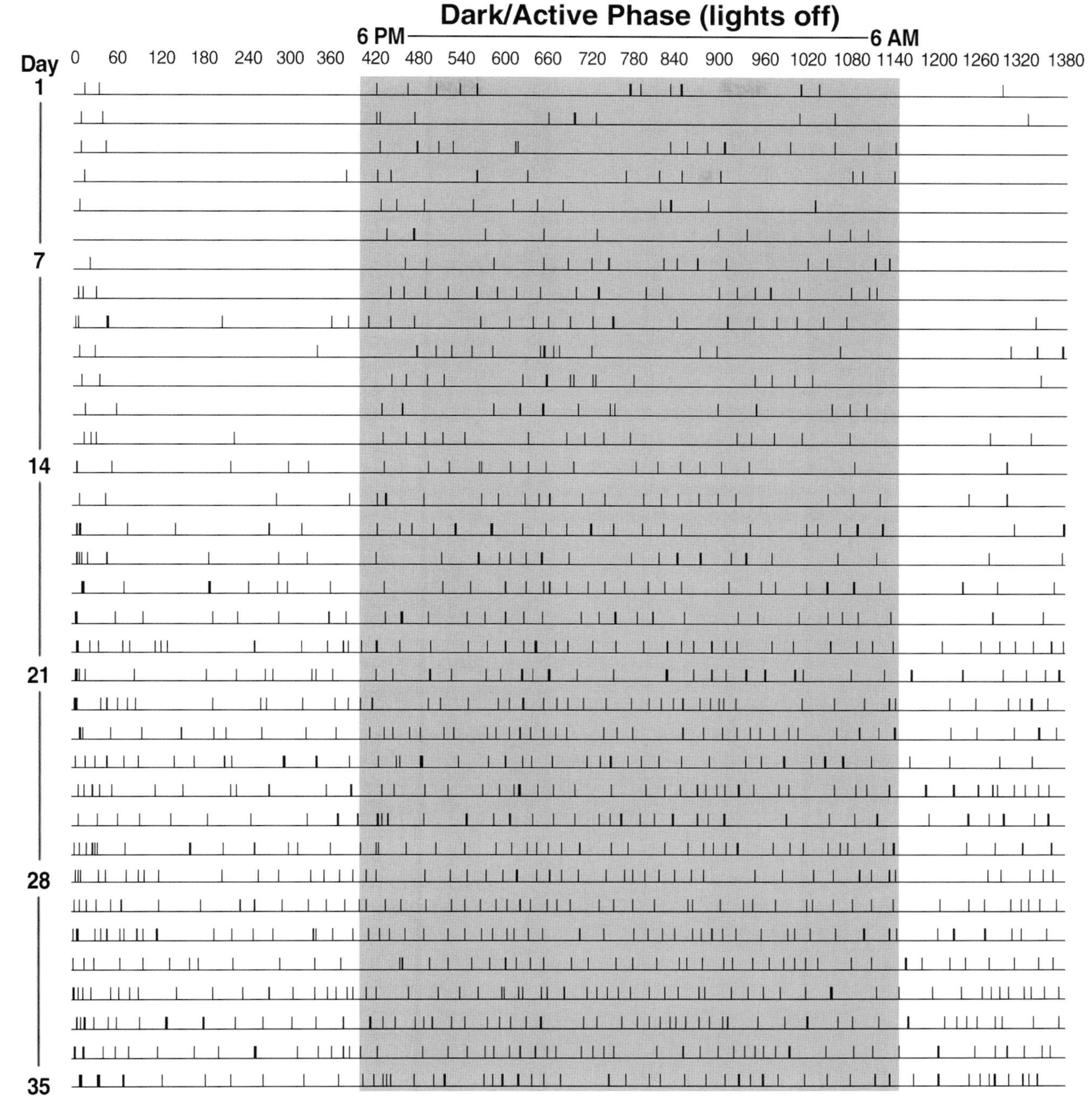

FIGURE 2.29 Event recording of heroin self-administration for one representative rat. Early sessions show drug intake primarily in the dark phase (6 PM — 6 AM) of the light/dark cycle. However, with more sessions, the rat began to increase its drug intake beyond the dark phase and into the light phase (6 AM — 6 PM) to the point where it was self-administering heroin throughout the 23-h session. [S. A. Chen, L. O'Dell, M. Hoefer, T. N. Greenwell, E. P. Zorrilla, and G. F. Koob, unpublished results.]

TABLE 2.7 Average Intake of Various Alcohol-Preferring and -Nonpreferring Rat Strains

Strain	Abbreviation	Ethanol intake	Reference
ALKO alcohol	AA	4.8–9.7 g/kg/day	Eriksson, 1968
ALKO nonalcohol	ANA	1.8–2.9 g/kg/day	Eriksson, 1968
University of Chile B (preferring)	UChB	3.6 g/kg/day	Mardones and Segovia-Riquelme, 1983
University of Chile A (nonpreferring)	UchA	0.5 g/kg/day	Mardones and Segovia-Riquelme, 1983
Sardinian alcohol-preferring	sP	7 g/kg/day	Fadda *et al.*, 1989; Colombo, 1997
Sardinian alcohol-nonpreferring	Snp	< 1 g/kg/day	Fadda *et al.*, 1989; Colombo, 1997
Indiana alcohol-preferring	P	6.5 g/kg/day	Lumeng *et al.*, 1977
Indiana alcohol-nonpreferring	NP	0.5 g/kg/day	Lumeng *et al.*, 1977
Indiana high alcohol drinking	HAD	9.5 g/kg/day	Li *et al.*, 1993
Indiana low alcohol drinking	LAD	0.5 g/kg/day	Li *et al.*, 1993

Drug-taking in the Presence of Aversive Consequences after Extended Access

Drug-taking or drug-seeking behavior that is impervious to environmental adversity, such as signals of punishment, has been hypothesized to capture elements of the compulsive nature of drug addiction. From a DSM-IV perspective, this observable may fit well with 'continued substance use despite knowledge of having a persistent physical of psychological problem'. Presentation of an aversive stimulus suppressed cocaine-seeking behavior in rats with limited-access and sucrose-seeking in rats, but extended access to cocaine and sucrose produced differential effects (Vanderschuren and Everitt, 2004) (**Fig. 2.30**). Rats with extended access to cocaine did not suppress drug-seeking in the presence of an aversive conditioned stimulus, but in rats with extended access to sucrose the conditioned aversive stimulus continued to suppress responding for sucrose. The authors provided controls to show that this effect was not due to impaired fear conditioning or an increased incentive value of cocaine and hypothesized that this behavioral change reflected 'the establishment of compulsive behavior' (Vanderschuren and Everitt, 2004).

TABLE 2.8 Summary of Blood Alcohol Levels Attained by University of Indiana Alcohol-Preferring (P) Rats Under Different Conditions of Ethanol Intake

Condition	Blood alcohol level (mg%)	Ethanol intake (g/kg)	Reference
24 h free access 10% ethanol	90 ± 10[a]	6.0 ± 0.2 (24 h)	Murphy *et al.*, 1986
4 h limited access 10% ethanol	120 ± 15[b]	2.1 ± 0.2 (4 h)	Murphy *et al.*, 1986
1 h limited access 10% ethanol	76 ± 13[b]	1.3 ± 0.1 (1 h)	Murphy *et al.*, 1986
24 h free access 20% ethanol, intragastric	200 (range: 115–300)[c]	5.5 ± 0.2 (24 h)	Waller *et al.*, 1984
24 h free access 40% ethanol, intragastric	230 (range: 90–415)[c]	9.4 ± 1.7 (24 h)	Waller *et al.*, 1984
24 h relapse 10, 20, or 30% ethanol	180 (range: 160–205)[d]	5.3 ± 0.6 (2 h)	Rodd-Henricks *et al.*, 2001

[a] Samples taken from retro-orbital sinus at set times throughout a 24 h period. Peak value was attained 3 h into the dark cycle.
[b] Samples taken from retro-orbital sinus 1 h into the session.
[c] Samples taken from retro-orbital sinus 30–40 min after completing an intragastric self-administration episode.
[d] Trunk blood sampled 2 h after ethanol solutions were restored following 2 weeks of alcohol deprivation.

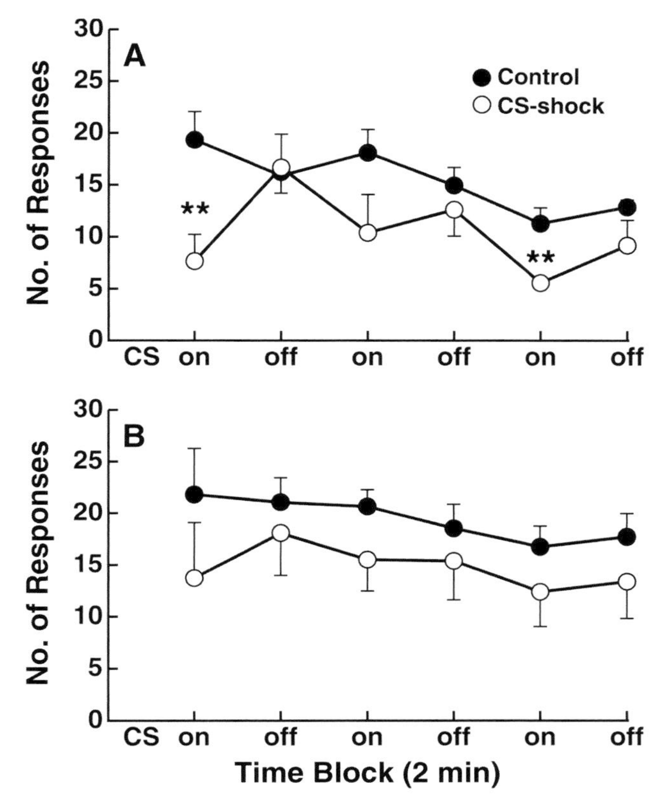

FIGURE 2.30 Presentation of an aversive conditioned stimulus (CS) suppresses cocaine-seeking behavior in rats after limited (A) but not prolonged (B) cocaine self-administration. **(A)** Mean (± SEM) cocaine-seeking responses per 2 min interval in the CS-shock and control groups after limited cocaine exposure, with the aversive CS on or off during alternating 2 min periods. $^{**}p < 0.01$ (Student-Newman-Keuls). **(B)** Mean (± SEM) cocaine-seeking responses per 2 min interval in the CS-shock and control groups after extended cocaine exposure, with the aversive CS on or off during alternating 2 min periods. [Reproduced with permission from Vanderschuren and Everitt, 2004.]

Drug-taking on a Progressive-Ratio Schedule with Extended Access

In a study of the behavioral effects of drug-taking in animals with access to cocaine for 3 months, a number of behavioral tests were administered that were hypothesized to capture DSM-IV criteria of addiction (Deroche-Gamonet *et al.*, 2004). 'Unsuccessful effort or a persistent desire to cut down or control substance use' was linked to the persistence of cocaine seeking during a period of signaled nonavailability. 'A great deal of time spent in activities necessary to obtain the substance' was linked to performance on a progressive-ratio schedule, and 'continued substance use despite knowledge of having a persistent physical of psychological problem' was linked to the persistence in responding for drug by animals when drug delivery was associated with punishment. Rats were trained to self-administer cocaine intravenously and then separated by groups based on a test for reinstatement to small doses of cocaine administered after 5 days of extinction. The animals with the high tendency to show reinstatement showed progressively increased responding during signaled nondrug periods, higher break points on the progressive-ratio test, and higher responding after punishment (Deroche-Gamonet *et al.*, 2004) (**Fig. 2.31**). Further study of rats subjected to all three tests above revealed that the animals that met all three positive criteria represented 17 per cent of the entire population, a percentage noted by the authors to be similar to the number of human cocaine users meeting the DSM-IIIR criteria for addiction (Anthony *et al.*, 1994). These models highlight the importance of differential vulnerability to addiction.

Summary of Animal Models for the Transition to Addiction

The studies outlined in this section illustrate how animal models for addiction have progressed from simple drug reinforcement models to sophisticated models with solid face validity. Escalation in drug intake with extended access has been observed now in numerous laboratories (Ahmed and Koob, 1998, 1999; Deroche *et al.*, 1999; Ahmed *et al.*, 2000, 2002, 2003; Mantsch *et al.*, 2001, 2003, 2004; Morgan *et al.*, 2002, 2005; Deroche-Gamonet *et al.*, 2003; Paterson and Markou, 2003; Walker *et al.*, 2003; Ben-Shahar *et al.*, 2004; Roth and Carroll, 2004; Liu *et al.*, 2005), and this escalation in intake has been linked to between-system tolerance and reward allostatic mechanisms (see *Neuroadaptational Theories of Addiction* chapter). Increased drug-taking during dependence has been well established with alcohol and opiates and can produce sufficient intake to maintain dependence. More recently, animal models for the criteria of Substance Dependence (addiction) that reflect the channeling of behavior toward drug seeking at the expense of other environmental contingencies (Deroche-Gamonet *et al.*, 2004; Vanderschuren and Everitt, 2004) have been developed and linked to the compulsive loss of control over intake that forms the hallmark of addiction. Such models may prove particularly sensitive to the transition to dependence in otherwise vulnerable individuals.

It is particularly important for the study of the neurobiology of addiction that animal models have moved from simple measures of drug reinforcement to measures with strong face validity for the 'observable symptoms' of addiction. What remains to be

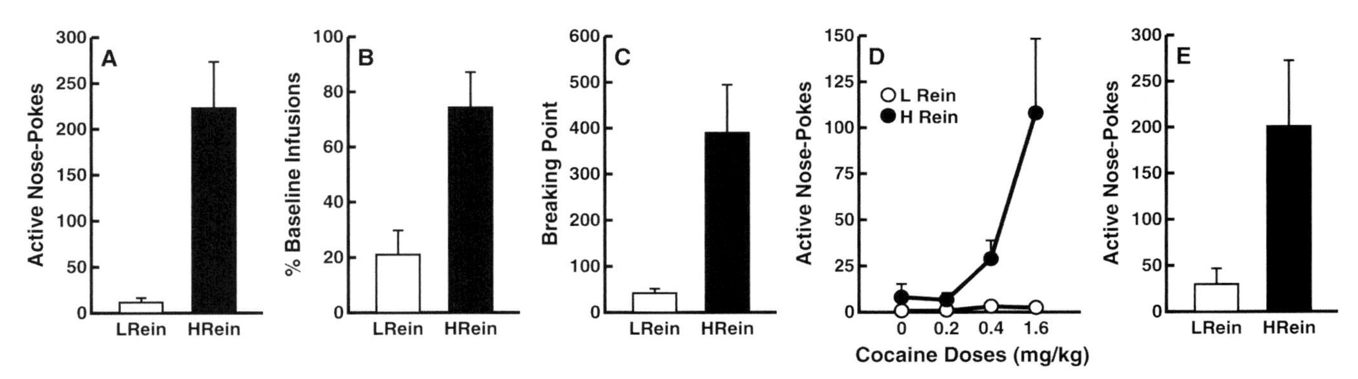

FIGURE 2.31 Development of addiction-like behaviors over subsequent cocaine self-administration sessions in rats showing high (HRein) or low (LRein) cocaine-induced reinstatement induced by cocaine infusion at 1.6 mg/kg after a 30-day withdrawal period. **(A)** Persistence in drug-seeking behavior, as measured by number of nose-pokes in the cocaine-associated device during the no-drug period of the 54th self-administration session. **(B)** Resistance to punishment, as measured by change in the number of cocaine self-infusions (expressed as percentage of baseline self-administration) when cocaine delivery was associated with an electric shock during the 72nd self-administration session. **(C)** Motivation for the drug, as measured by the break point during the progressive-ratio schedule conducted during the 60th self-administration session. **(D)** Drug-induced reinstatement, as measured by number of nose-pokes in the drug-associated device as a function of the priming dose of cocaine. **(E)** Reinstatement induced by a CS, as measured by the number of nose-pokes in the drug-associated device when responding was associated with the contingent presentation of the CS. Tests for cocaine- and CS-induced reinstatement were performed after 30 and 32 days of withdrawal, respectively, using a Latin square design. [Reproduced with permission from Deroche-Gamonet *et al.*, 2004.]

accomplished is to show that both construct validity (functional equivalence) and predictive validity exist for these models. A further challenge for future studies will be the development of 'endophenotypes' that cross species from animal to human that will allow further construct validity (functional equivalence) for studies of genetic and environmental vulnerability and the neurobiological mechanisms therein.

SUMMARY

Most of the animal models discussed above have predictive validity for some component of the addiction cycle (compulsive use, withdrawal or craving) and are reliable. For the positive reinforcing effects of drugs, drug self-administration, ICSS, and conditioned place preference have been shown to have predictive validity. Drug discrimination has predictive validity indirectly through generalization to the training drug. Animal models of withdrawal are focused on motivational constructs as opposed to the physical or somatic signs of withdrawal. Animal models of conditioned drug effects are successful in predicting the potential for conditioned drug effects in humans. Predictive validity is more problematic for such concepts as craving, largely due to the inadequate formulation of the concept of craving in humans (Markou *et al.*, 1993; Sayette *et al.*, 2000; Tiffany *et al.*, 2000). Virtually all of the measures described above have demonstrated reliability. Consistency and stability of the measures, small within-subject and between-subject variability, and reproducibility of the phenomenon are characteristic of most of the measures employed in animal models of dependence.

Clearly, much remains to be explored about the face validity and predictive validity of the unconditioned positive and negative motivational states, and in particular the conditioned positive and negative motivational states associated with drug use and withdrawal. However, the gaps in knowledge may lie more in the human clinical laboratory domain than in the animal models domain. The study of the changes in the central nervous system that are associated with these models is the subject of the chapters that follow and may provide insights into drug addiction and the etiology of psychopathologies associated with addiction, such as anxiety and affective disorders.

REFERENCES

Ahmed, S. H., and Koob, G. F. (1997). Cocaine- but not food-seeking behavior is reinstated by stress after extinction. *Psychopharmacology* **132**, 289–295.

Ahmed, S. H., and Koob, G. F. (1998). Transition from moderate to excessive drug intake: Change in hedonic set point. *Science* **282**, 298–300.

Ahmed, S. H., and Koob, G. F. (1999). Long-lasting increase in the set point for cocaine self-administration after escalation in rats. *Psychopharmacology* **146**, 303–312.

Ahmed, S. H., Kenny, P. J., Koob, G. F., and Markou, A. (2002). Neurobiological evidence for hedonic allostasis associated with escalating cocaine use. *Nature Neuroscience* **5**, 625–626.

Ahmed, S. H., Walker, J. R., and Koob, G. F. (2000). Persistent increase in the motivation to take heroin in rats with a history of drug escalation. *Neuropsychopharmacology* **22**, 413–421.

Ahmed, S. H., Lin D., Koob, G. F., and Parsons, L. H. (2003). Escalation of cocaine self-administration does not depend on altered cocaine-induced nucleus accumbens dopamine levels. *Journal of Neurochemistry* **86**, 102–113.

Alling, C., Balldin, J., Bokstrom, K., Gottfries, C. G., Karlsson, I., and Langstrom, G. (1982). Studies on duration of a late recovery period after chronic abuse of ethanol. A cross-sectional study of biochemical and psychiatric indicators. *Acta Psychiatrica Scandinavica* **66**, 384–397.

Ambrosio, E., Goldberg, S. R., and Elmer, G. I. (1995). Behavior genetic investigation of the relationship between spontaneous locomotor activity and the acquisition of morphine self-administration behavior. *Behavioural Pharmacology* **6**, 229–237.

American Psychiatric Association (1994). *Diagnostic and Statistical Manual of Mental Disorders*, 4th ed., American Psychiatric Press, Washington DC.

Annis, H. M., Sklar, S. M., and Moser, A. E. (1998). Gender in relation to relapse crisis situations, coping, and outcome among treated alcoholics. *Addictive Behaviors* **23**, 127–131.

Anthony, J. C., Warner, L. A., and Kessler, R. C. (1994). Comparative epidemiology of dependence on tobacco, alcohol, controlled substances, and inhalants, Basic findings from the National Comorbidity Survey. *Experimental and Clinical Psychopharmacology* **2**, 244–268.

Arroyo, M., Markou, A., Robbins, T. W., and Everitt, B. J. (1998). Acquisition, maintenance and reinstatement of intravenous cocaine self-administration under a second-order schedule of reinforcement in rats: effects of conditioned cues and continuous access to cocaine. *Psychopharmacology* **140**, 331–344.

Aufrere, G., Le Bourhis, B., and Beauge, F. (1997). Ethanol intake after chronic intoxication by inhalation of ethanol vapour in rat: behavioural dependence. *Alcohol* **14**, 247–253.

Baker, T. B., and Cannon, D. S. (1979). Potentiation of ethanol withdrawal by prior dependence. *Psychopharmacology* **60**, 105–110.

Baldwin, H. A., and Koob, G. F. (1993). Rapid induction of conditioned opiate withdrawal in the rat. *Neuropsychopharmacology* **8**, 15–21.

Balster, R. L., and Schuster, C. R. (1973). Fixed-interval schedule of cocaine reinforcement: effect of dose and infusion duration. *Journal of the Experimental Analysis of Behavior* **20**, 119–129.

Bauco, P., and Wise, R. A. (1997). Synergistic effects of cocaine with lateral hypothalamic brain stimulation reward: lack of tolerance or sensitization. *Journal of Pharmacology and Experimental Therapeutics* **283**, 1160–1167.

Becker, H. C. (1994). Positive relationship between the number of prior ethanol withdrawal episodes and the severity of subsequent withdrawal seizures. *Psychopharmacology* **116**, 26–32.

Becker, H. C., and Hale, R. L. (1989). Ethanol-induced locomotor stimulation in C57BL/6 mice following RO15-4513 administration. *Psychopharmacology* **99**, 333–336.

Begleiter, H. (1975). Ethanol consumption subsequent to physical dependence. *In Alcohol Intoxication and Withdrawal: Experimental Studies II* (series title: *Advances in Experimental Medicine and Biology*, vol. 59), (M. M. Gross, Ed.), pp. 373–378. Plenum Press, New York.

Ben-Shahar, O., Ahmed, S. H., Koob, G. F., and Ettenberg, A. (2004). The transition from controlled to compulsive drug use is associated with a loss of sensitization. *Brain Research* **995**, 46–54.

Bertalmio, A. J., Herling, S., Hampton, R. Y., Winger, G., and Woods, J. H. (1982). A procedure for rapid evaluation of the discriminative stimulus effects of drugs. *Journal of Pharmacological Methods* **7**, 289–299.

Bindra, D. (1976). *A Theory of Intelligent Behavior*. Wiley, New York.

Branchey, M., Rauscher, G., and Kissin, B. (1971). Modifications in the response to alcohol following the establishment of physical dependence. *Psychopharmacologia* **22**, 314–322.

Brown, S. A., Vik, P. W., Patterson, T. L., Grant, I., and Schuckit, M. A. (1995). Stress, vulnerability and adult alcohol relapse. *Journal of Studies on Alcohol* **56**, 538–545.

Burish, T. G., Maisto, S. A., Cooper, A. M., and Sobell, M. B. (1981). Effects of voluntary short-term abstinence from alcohol on subsequent drinking patterns of college students. *Journal of Studies on Alcohol* **42**, 1013–1020.

Caine, S. B., and Koob, G. F. (1993). Modulation of cocaine self-administration in the rat through D-3 dopamine receptors. *Science* **260**, 1814–1816.

Caine, S. B., and Koob, G. F. (1994). Effects of mesolimbic dopamine depletion on responding maintained by cocaine and food. *Journal of the Experimental Analysis of Behavior* **61**, 213–221.

Caine, S. B., Lintz, R., and Koob, G. F. (1993). Intravenous drug self-administration techniques in animals. *In Behavioural Neuroscience: A Practical Approach*, vol. 2, (A. Sahgal, Ed.), pp. 117–143. IRL Press, Oxford.

Campbell, K. A., Evans, G., and Gallistel, C. R. (1985). A microcomputer-based method for physiologically interpretable measurement of the rewarding efficacy of brain stimulation. *Physiology and Behavior* **35**, 395–403.

Cappell, H., and LeBlanc, A. E. (1981). Tolerance and physical dependence: Do they play a role in alcohol and drug self-administration? *In Research Advances in Alcohol and Drug Problems*, vol. 6, (Y. Israel, F. B. Glaser, H. Kalant, R. E. Popham, W. Schmidt and R. G. Smart, Eds.), pp. 159–196. Plenum Press, New York.

Carboni, E., and Vacca, C. (2003). Conditioned place preference: a simple method for investigating reinforcing properties in laboratory animals. *In Drugs of Abuse: Neurological Reviews and Protocols* (series title: *Methods in Molecular Medicine*, vol. 79), (J. Q. Wang, Ed.), pp. 481–498. Humana Press, Totowa NJ.

Carr, G. D., Fibiger, H. C., and Phillips, A. G. (1989). Conditioned place preference as a measure of drug reward. *In The Neuropharmacological Basis of Reward* (series title: *Topics in Experimental Psychopharmacology*, vol. 1), (J. M. Liebman, S. J. Cooper, Eds.), pp. 264–319. Oxford University Press, New York.

Carrera, M. R. A., Schulteis, G., and Koob, G. F. (1999). Heroin self-administration in dependent Wistar rats: increased sensitivity to naloxone. *Psychopharmacology* **144**, 111–120.

Chen, S. A., O'Dell, L., Lerner, K., Hoefer, M., Zorrilla, E. P., and Koob, G. F. (2005). Unlimited access to heroin self-administration: differential measures of dependence support a set point change in heroin reward, submitted.

Childress, A. R., McLellan, A. T., Ehrman, R., and O'Brien, C. P. (1988). Classically conditioned responses in opioid and cocaine dependence: A role in relapse? *In Learning Factors in Substance Abuse* (series title: *NIDA Research Monograph*, vol. 84), (B. A. Ray, Ed.), pp. 25–43. National Institute on Drug Abuse, Rockville MD.

Ciccocioppo, R., Angeletti, S., and Weiss, F. (2001). Long-lasting resistance to extinction of response reinstatement induced by ethanol-related stimuli: Role of genetic ethanol preference, *Alcoholism: Clinical and Experimental Research* **25**, 1414–1419.

Collins, R. J., Weeks, J. R., Cooper, M. M., Good, P. I., and Russell, R. R. (1984). Prediction of abuse liability of drugs using IV self-administration by rats. *Psychopharmacology* **82**, 6–13.

Colombo, G. (1997). ESBRA-Nordmann 1996 Award Lecture: ethanol drinking behaviour in Sardinian alcohol-preferring rats. *Alcohol and Alcoholism* **32**, 443–453.

Colpaert, F. C., Lal H., Niemegeers, C. J., and Janssen, P. A. (1975). Investigations on drug produced and subjectively experienced discriminative stimuli: I. The fentanyl cue, a tool to investigate subjectively experience narcotic drug actions. *Life Sciences* **16**, 705–715.

Crabbe, J. C., Belknap, J. K., and Buck, K. J. (1994). Genetic animal models of alcohol and drug abuse, *Science* **264**, 1715–1723.

Davis, W. M., and Smith, S. G. (1987). Conditioned reinforcement as a measure of the rewarding properties of drugs. *In Methods of Assessing the Reinforcing Properties of Abused Drugs*, (M. A. Bozarth, Ed.), pp. 199–210. Springer-Verlag, New York.

de Wit, H., and Stewart, J. (1981). Reinstatement of cocaine-reinforced responding in the rat. *Psychopharmacology* **75**, 134–143.

de Wit, H., Uhlenhuth, E. H., and Johanson, C. E. (1986). Individual differences in the reinforcing and subjective effects of amphetamine and diazepam. *Drug and Alcohol Dependence* **16**, 341–360.

Denoble, U., and Begleiter, H. (1976). Response suppression on a mixed schedule of reinforcement during alcohol withdrawal. *Pharmacology Biochemistry and Behavior* **5**, 227–229.

Depoortere, R. Y., Li, D. H., Lane, J. D., and Emmett-Oglesby, M. W. (1993). Parameters of self-administration of cocaine in rats under a progressive-ratio schedule. *Pharmacology Biochemistry and Behavior* **45**, 539–548.

Deroche, V., Marinelli, M., Le Moal, M., and Piazza, P. V. (1997). Glucocorticoids and behavioral effects of psychostimulants: II Cocaine intravenous self-administration and reinstatement depend on glucocorticoid levels. *Journal of Pharmacology and Experimental Therapeutics* **281**, 1401–1407.

Deroche, V., Le Moal, M., and Piazza, P. V. (1999). Cocaine self-administration increases the incentive motivational properties of the drug in rats. *European Journal of Neuroscience* **11**, 2731–2736.

Deroche-Gamonet, V., Belin, D., and Piazza, P. V. (2004). Evidence for addiction-like behavior in the rat. *Science* **305**, 1014–1017.

Deroche-Gamonet, V., Martinez, A., Le Moal, M., and Piazza, P. V. (2003). Relationships between individual sensitivity to CS- and cocaine-induced reinstatement in the rat. *Psychopharmacology* **168**, 201–207.

Deutsch, J. A., and Koopmans, H. S. (1973). Preference enhancement for alcohol by passive exposure. *Science* **179**, 1242–1243.

Deutsch, J. A., and Walton, N. Y. (1977). A rat alcoholism model in a free choice situation. *Behavioral Biology* **19**, 349–360.

Dole, V. P. (1986). On the relevance of animal models to alcoholism in humans. *Alcoholism: Clinical and Experimental Research* **10**, 361–363.

Drummond, D. C. (2000). What does cue-reactivity have to offer clinical research?, *Addiction* **95**(Suppl. 2):s129–s144.

Ebel, R. L. (1961). Must all tests be valid? *American Psychologist* **16**, 640–647.

Edmonds, D. E., and Gallistel, C. R. (1974). Parametric analysis of brain stimulation reward in the rat: III. Effect of performance variables on the reward summation function. *Journal of Comparative and Physiological Psychology* **87**, 876–883.

Edwards, G. (1990). Withdrawal symptoms and alcohol dependence: fruitful mysteries. *British Journal of Addiction* **85**, 447–461.

Ehrman, R. N., Robbins, S. J., Childress, A. R., and O'Brien, C. P. (1992). Conditioned responses to cocaine-related stimuli in cocaine abuse patients. *Psychopharmacology* **107**, 523–529.

Elmer, G. I., Pieper, J. O., Goldberg, S. R., and George, F. R. (1995). Opioid operant self-administration, analgesia, stimulation and respiratory depression in mu-deficient mice. *Psychopharmacology* **117**, 23–31.

Emmett-Oglesby, M. W., Mathis, D. A., Moon, R. T., and Lal, H. (1990). Animal models of drug withdrawal symptoms. *Psychopharmacology* **101**, 292–309.

Epping-Jordan, M. P., Watkins, S. S., Koob, G. F., and Markou, A. (1998). Dramatic decreases in brain reward function during nicotine withdrawal. *Nature* **393**, 76–79.

Epstein, D. H., and Preston, K. L. (2003). The reinstatement model and relapse prevention: a clinical perspective. *Psychopharmacology* **168**, 31–41.

Erb, S., Shaham, Y., Stewart, J. (1996). Stress reinstates cocaine-seeking behavior after prolonged extinction and a drug-free period. *Psychopharmacology* **128**, 408–412.

Eriksson, K., (1968). Genetic selection for voluntary alcohol consumption in the Albino rat, *Science* **159**, 739–741.

Fadda, F., Mosca, E., Colombo, G., and Gessa, G. L. (1989). Effect of spontaneous ingestion of ethanol on brain dopamine metabolism. *Life Sciences* **44**, 281–287.

Files, F. J., Lewis, R. S., Samson, H. H. (1994). Effects of continuous versus limited access to ethanol on ethanol self-administration. *Alcohol* **11**, 523–531.

Fouriezos, G., Bielajew, C., and Pagotto, W. (1990). Task difficulty increases thresholds of rewarding brain stimulation. *Behavioural Brain Research* **37**, 1–7.

France, C. P., and Woods J. H., (1989). Discriminative stimulus effects of naltrexone in morphine-treated rhesus monkeys. *Journal of Pharmacology and Experimental Therapeutics* **250**, 937–943.

Gallistel, C. R. (1983). Self-stimulation. *In The Physiological Basis of Memory*, 2nd ed. (J. A. Deutsch Ed.) pp. 73–77. Academic Press, New York.

Gardner, E. L., and Vorel, S. R. (1998). Cannabinoid transmission and reward-related events. *Neurobiology of Disease* **5**, 502–533.

Gatto, G. J., McBride, W. J., Murphy, J. M., Lumeng, L., and Li, T. K. (1994). Ethanol self-infusion into the ventral tegmental area by alcohol-preferring rats. *Alcohol* **11**, 557–564.

Gauvin, D. V., and Holloway, F. A. (1991). Cue dimensionality in the three-choice pentylenetetrazole-saline-chlordiazepoxide discrimination task. *Behavioural Pharmacology* **2**, 417–428.

Gellert, V. F., and Holtzman, S. G. (1978). Development and maintenance of morphine tolerance and dependence in the rat by scheduled access to morphine drinking solutions, *Journal of Pharmacology and Experimental Therapeutics* **205**, 536–546.

Gellert, V. F., and Holtzman, S. G. (1979). Discriminative stimulus effects of naltrexone in the morphine-dependent rat. *Journal of Pharmacology and Experimental Therapeutics* **211**, 596–605.

Gellert, V. F., and Sparber S. B. (1977). A comparison of the effects of naloxone upon body weight loss and suppression of fixed-ratio operant behavior in morphine-dependent rats. *Journal of Pharmacology and Experimental Therapeutics* **201**, 44–54.

Gerber, G. J., and Stretch, R. (1975). Drug-induced reinstatement of extinguished self-administration behavior in monkeys. *Pharmacology Biochemistry and Behavior* **3**, 1055–1061.

Geyer, M. A., and Markou,. (1995). A. Animal models of psychiatric disorders. In *Psychopharmacology: The Fourth Generation of Progress*, (F. E. Bloom, D. J. Kupfer, Eds.), pp. 787–798. Raven Press, New York.

Geyer, M. A., and Markou, A. (2002). The role of preclinical models in the development of psychotropic drugs. *In Neuropsychopharmacology: The Fifth Generation of Progress*, (K. L. Davis, D. Charney, J. T. Coyle, and C. Nemeroff, Eds.), pp. 445–455. Lippincott Williams and Wilkins, New York.

Goldberg, S. R. (1973). Comparable behavior maintained under fixed-ratio and second-order schedules of food presentation, cocaine injection or d-amphetamine injection in the squirrel monkey. *Journal of Pharmacology and Experimental Therapeutics* **186**, 18–30.

Goldberg, S. R. (1975). Stimuli associated with drug injections as events that control behavior. *Pharmacological Reviews* **27**, 325–340.

Goldberg, S. R., and Kelleher, R. T. (1976). Behavior controlled by scheduled injections of cocaine in squirrel and rhesus monkeys. *Journal of the Experimental Analysis of Behavior* **25**, 93–104.

Goldberg, S. R., and Schuster, C. R. (1967). Conditioned suppression by a stimulus associated with nalorphine in morphine-dependent monkeys. *Journal of the Experimental Analysis of Behavior* **10**, 235–242.

Goldberg, S. R., and Schuster, C. R. (1970). Conditioned nalorphine-induced abstinence changes: Persistence in post morphine-dependent monkeys. *Journal of the Experimental Analysis of Behavior* **14**, 33–46.

Goldberg, S. R., Woods, J. H., and Schuster, C. R. (1969). Morphine: Conditioned increases in self-administration in rhesus monkeys. *Science* **166**, 1306–1307.

Goldberg, S. R., Hoffmeister, F., Schlichting, V., and Wolfgang, W. (1971). Aversive properties of nalorphine and naloxone in morphine-dependent rhesus monkeys *Journal of Pharmacology and Experimental Therapeutics* **179**, 268–276.

Goldberg, S. R., Kelleher, R. T., and Morse, W. H. (1975). Second-order schedules of drug injection, *Federation Proceedings* **34**, 1771–1776.

Goldberg, S. R., Morse, W. H., and Goldberg, D. M. (1976). Behavior maintained under a second-order schedule by intramuscular injection of morphine or cocaine in rhesus monkeys. *Journal of Pharmacology and Experimental Therapeutics* **199**, 278–286.

Goldberg, S. R., Kelleher, R. T., and Goldberg, D. M. (1981). Fixed-ratio responding under second-order schedules of food presentation or cocaine injection. *Journal of Pharmacology and Experimental Therapeutics* **218**, 271–281.

Grasing, K., Li, N., He, S., Parrish, C., Delich, J., and Glowa, J. (2003). A new progressive ratio schedule for support of morphine self-administration in opiate dependent rats. *Psychopharmacology* **168**, 387–396.

Griffiths, R. R., Brady J. V., and Snell, J. D. (1978). Progressive-ratio performance maintained by drug infusions: comparison of cocaine, diethylpropion, chlorphentermine, and fenfluramine. *Psychopharmacology* **56**, 5–13.

Hand, T. H., Koob, G. F., Stinus, L., and Le Moal, M. (1988). Aversive properties of opiate receptor blockade: Evidence for exclusively central mediation in naive and morphine-dependent rats. *Brain Research* **474**, 364–368.

Hebb, D. O. (1949). *Organization of Behavior: A Neuropsychological Theory*. Wiley, New York.

Hershon, H. I. (1977). Alcohol withdrawal symptoms and drinking behavior, *Journal of Studies on Alcohol* **38**, 953–971.

Heyser, C. J., Schulteis, G., Durbin, P., and Koob, G. F. (1998). Chronic acamprosate eliminates the alcohol deprivation effect while having limited effects on baseline responding for ethanol in rats. *Neuropsychopharmacology* **18**, 125–133.

Heyser., C. J., Schulteis, G., and Koob G. F. (1997). Increased ethanol self-administration after a period of imposed ethanol deprivation in rats trained in a limited access paradigm *Alcoholism: Clinical and Experimental Research* **21**, 784–791.

Hilgard, E. R., and Marquis, D. G. (1961). *Hilgard and Marquis' Conditioning and Learning*, 2nd ed., Appleton-Century-Crofts, New York.

Holtzman, S. G. (1990). Discriminative stimulus effects of drugs: relationship to potential for abuse. *In Testing and Evaluation of Drugs of Abuse* (series title: *Modern Methods in Pharmacology*, vol. 6), (M. W. Adler, A. Cowan, Eds.), pp. 193–210. Wiley, New York.

Hubner, C. B., Kornetsky, C. (1992). Heroin, 6-acetylmorphine and morphine effects on threshold for rewarding and aversive brain stimulation. *Journal of Pharmacology and Experimental Therapeutics* **260**, 562–567.

Hunt, W. A., Barnett, L. W., Branch, L. G. (1971). Relapse rates in addiction programs. *Journal of Clinical Psychology* **27**, 455–456.

Hunter,B. E., Walker, D. W., and Riley, J. N. (1974). Dissociation between physical dependence and volitional ethanol consumption: role of multiple withdrawal episodes *Pharmacology Biochemistry and Behavior* **2**, 523–529.

Huston-Lyons, D., and Kornetsky, C. (1992). Effects of nicotine on the threshold for rewarding brain stimulation in rats. *Pharmacology Biochemistry and Behavior* **41**, 755–759.

Iacono, W. G. (1998). Identifying psychophysiological risk for psychopathology: examples from substance abuse and schizophrenia research. *Psychophysiology* **35**, 621–637.

Iacono, W. G., Carlson, S. R., and Malone, S. M. (2000). Identifying a multivariate endophenotype for substance use disorders using psychophysiological measures. *International Journal of Psychophysiology* **38**, 81–96.

Izenwasser, S., and Kornetsky, C. (1992). Brain-stimulation reward: a method for assessing the neurochemical bases of drug-induced euphoria. *In Drugs of Abuse and Neurobiology*, (R. R. Watson, Ed.), pp. 1–21. CRC Press, Boca Raton FL.

Jentsch, J. D., Henry, P. J., Mason, P. A., Merritt, J. H., and Ziriax, J. M. (1998). Establishing orally self-administered cocaine as a reinforcer in rats using home-cage pre-exposure. *Progress in Neuropsychopharmacology and Biological Psychiatry* **22**, 229–239.

Katner, S. N., Magalong, J. G., and Weiss, F. (1999). Reinstatement of alcohol-seeking behavior by drug-associated discriminative stimuli after prolonged extinction in the rat. *Neuropsychopharmacology* **20**, 471–479.

Katz, J. L., and Goldberg, S. R. (1987). Second-order schedules of drug injection. *In Methods of Assessing the Reinforcing Properties of Abused Drugs*, (M. A. Bozarth, Ed.), pp. 105–115. Springer-Verlag, New York.

Katz, J. L., and Goldberg, S. R. (1991). Second-order schedules of drug injection: implications for understanding reinforcing effects of abused drugs. *Advances in Substance Abuse* **4**, 205–223.

Katz, J. L., and Higgins, S. T. (2003). The validity of the reinstatement model of craving and relapse to drug use. *Psychopharmacology* **168**, 21–30 [erratum: 168:244].

Kelleher, R. T. (1975). Characteristics of behavior controlled by scheduled injections of drugs. *Pharmacological Reviews* **27**, 307–323.

Kelleher, R. T., and Goldberg, S. R. (1977). Fixed-interval responding under second-order schedules of food presentation or cocaine injection. *Journal of the Experimental Analysis of Behavior* **28**, 221–231.

Kleven, M. S., and Woolverton, W. L. (1990). Effects of continuous infusions of SCH 23390 on cocaine- or food-maintained behavior in rhesus monkeys. *Behavioural Pharmacology* **1**, 365–373.

Kling, J. W., and Riggs, L. A. (1971). *Woodworth and Schlosberg's Experimental Psychology*, 3rd ed., Holt, Rinehart and Winston, New York.

Kokkinidis, L., and McCarter, B. D. (1990). Postcocaine depression and sensitization of brain-stimulation reward: Analysis of reinforcement and performance effects *Pharmacology Biochemistry and Behavior* **36**, 463–471.

Koob, G. F., and Goeders, N. E. (1989). Neuroanatomical substrates of drug self-administration. *In The Neuropharmacological Basis of Reward* (series title: *Topics in Experimental Psychopharmacology*, vol. 1), (J. M. Liebman, and S. J. Cooper, Eds.), pp. 214–263. Clarendon Press, Oxford.

Koob, G. F., and Le Moal, M. (1997). Drug abuse: Hedonic homeostatic dysregulation. *Science* **278**, 52–58.

Koob, G. F., and Le Moal, M. (2001). Drug addiction, dysregulation of reward, and allostasis. *Neuropsychopharmacology* **24**, 97–129.

Koob, G. F., Wall T. L., and Bloom, F. E. (1989). Nucleus accumbens as a substrate for the aversive stimulus effects of opiate withdrawal. *Psychopharmacology* **98**, 530–534.

Koob, G. F., Markou, A., Weiss, F., and Schulteis, G. (1993). Opponent process and drug dependence: Neurobiological mechanisms. *Seminars in the Neurosciences* **5**, 351–358.

Koob, G. F., Sanna, P. P., and Bloom, F. E. (1998). Neuroscience of addiction. *Neuron* **21**, 467–476.

Kornet, M., Goosen, C., and Van Ree, J. M. (1991). Effect of naltrexone on alcohol consumption during chronic alcohol drinking and after a period of imposed abstinence in free-choice drinking rhesus monkeys. *Psychopharmacology* **104**, 367–376.

Kornetsky, C. (1985). Brain-stimulation reward: a model for the neuronal bases for drug-induced euphoria. *In Neuroscence Methods in Drug Abuse Research* (series title: *NIDA Research Monograph*, vol. 62), (R. M. Brown, D. P. Friedman, and Y. Nimit, Eds.), pp. 30–50. National Institute on Drug Abuse, Rockville MD.

Kornetsky, C., and Bain, G. (1990). Brain-stimulation reward: A model for drug induced euphoria, *In Testing and Evaluation of Drugs of Abuse* (series title: *Modern Methods in Pharmacology*, vol. 6), (M. W. Adler, and A. Cowan, Eds.), pp. 211–231. Wiley-Liss, New York.

Kornetsky, C., and Esposito, R. U. (1979). Euphorigenic drugs: Effects on the reward pathways of the brain. *Federation Proceedings* **38**, 2473–2476.

Kosten, T. R., Rounsaville, B. J., and Kleber, H. D. (1986). A 2.5-year follow-up of depression, life crises, and treatment effects on abstinence among opioid addicts. *Archives of General Psychiatry* **43**, 733–738.

Lallemand, F., Soubrie, P. H., and De Witte, P. H. (2001). Effects of CB1 cannabinoid receptor blockade on ethanol preference after chronic ethanol administration. *Alcoholism: Clinical and Experimental Research* **25**, 1317–1323.

Le Bourhis, B. (1975). Alcoolisation du rat par voie pulmonaire [Alcoholization of the rat through the lungs]. *Comptes Rendus des Séances de la Société de Biologie et de ses Filiales* **169**, 898–904.

Le Magnen, J. (1960) Etude de quelques facteurs associé á des modifications de la consommation spontanée alcool ethylique par le rat [Study of some factors associated with modifications of spontaneous ingestion of ethyl alcohol by the rat]. *Journal de Radiologie, d'Electrologie, et de Médecine Nucléaire* **52**, 873–884.

Leith, N. J., and Barrett, R. J. (1976). Amphetamine and the reward system: Evidence for tolerance and post-drug depression. *Psychopharmacologia* **46**, 19–25.

Li, T. K. (2000). Pharmacogenetics of responses to alcohol and genes that influence alcohol drinking. *Journal of Studies on Alcohol* **61**, 5–12.

Li, T. K., Lumeng, L., and Doolittle, D. P. (1993). Selective breeding for alcohol preference and associated responses. *Behavior Genetics* **23**, 163–170.

Li, T. K., and McBride, W. J. (1995). Pharmacogenetic models of alcoholism. *Clinical Neuroscience* **3**, 182–188.

Liu, Y., Roberts, D. C., and Morgan, D. (2005). Effects of extended-access self-administration and deprivation on breakpoints maintained by cocaine in rats. *Psychopharmacology* **179**, 644–651.

Lowman, C., Allen, J., and Stout R. L. (1996). Replication and extension of Marlatt's taxonomy of relapse precipitants: overview of procedures and results. The Relapse Research Group. *Addiction* **91**(Suppl.), s51–s71.

Lumeng, L., Hawkins, T. D., Li, T. K. (1977). New strains of rats with alcohol preference and nonpreference. *In Alcohol and Aldehyde Metabolizing Systems: Volume III. Intermediary Metabolism and Neurochemistry*, (R. G. Thurman, J. R. Williamson, H. R. Drott, and B. Chance Eds.), pp. 537–544. Academic Press, New York.

Macey, D. J., Schulteis, G., Heinrichs, S. C., and Koob, G. F. (1996). Time-dependent quantifiable withdrawal from ethanol in the rat: effect of method of dependence induction. *Alcohol* **13**, 163–170.

Malin, D. H., Lake, J. R., Newlin-Maultsby, P., Roberts L. K., Lanier J. G., Carter, V. A., Cunningham, J. S., and Wilson, O. B. (1992). Rodent model of nicotine abstinence syndrome. *Pharmacology Biochemistry and Behavior* **43**, 779–784.

Mantsch, J. R., Ho, A., Schlussman, S. D., and Kreek, M. J. (2001). Predictable individual differences in the initiation of cocaine self-administration by rats under extended-access conditions are dose-dependent. *Psychopharmacology* **157**, 31–39.

Mantsch, J. R., Yuferov, V., Mathieu-Kia, A. M., Ho, A., and Kreek, M. J. (2003). Neuroendocrine alterations in a high-dose, extended-access rat self-administration model of escalating cocaine use. *Psychoneuroendocrinology* **28**, 836–862.

Mantsch, J. R., Yuferov, V., Mathieu-Kia, A. M., Ho, A., and Kreek, M. J. (2004). Effects of extended access to high versus low cocaine doses on self-administration, cocaine-induced reinstatement and brain mRNA levels in rats. *Psychopharmacology* **175**, 26–36.

Mardones, J., and Segovia-Riquelme, N. (1983). Thirty-two years of selection of rats by ethanol preference: UChA and UChB strains. *Neurobehavioral Toxicology and Teratology* **5**, 171–178.

Markou, A., and Koob, G. F. (1991). Post-cocaine anhedonia: An animal model of cocaine withdrawal. *Neuropsychopharmacology* **4**, 17–26.

Markou, A., and Koob, G. F. (1992). Construct validity of a self-stimulation threshold paradigm: Effects of reward and performance manipulations. *Physiology and Behavior* **51**, 111–119.

Markou, A., Weiss, F., Gold, L. H., Caine, S. B., Schulteis, G., and Koob, G. F. (1993). Animal models of drug craving. *Psychopharmacology* **112**, 163–182.

Marlatt, G. A., and George, W. H. (1984). Relapse prevention: introduction and overview of the model. *British Journal of Addiction* **79**, 261–273.

Mason, B. J., Ritvo, E. C., Morgan, R. O., Salvato, F. R., Goldberg, G., Welch, B., and Mantero-Atienza E. (1994). A double-blind, placebo-controlled pilot study to evaluate the efficacy and safety of oral nalmefene HCl for alcohol dependence *Alcoholism: Clinical and Experimental Research* **18**, 1162–1167.

McDonald, R. J., and White, N. M. (1993). A triple dissociation of memory systems: hippocampus, amygdala, and dorsal striatum. *Behavioral Neuroscience* **107**, 3–22.

McFarland, K., and Ettenberg, A. (1997). Reinstatement of drug-seeking behavior produced by heroin-predictive environmental stimuli. *Psychopharmacology* **131**, 86–92.

McKinney, W. T. (1988). *Models of Mental Disorders: A New Comparative Psychiatry*. Plenum, New York.

McLellan, A. T., Childress, A. R., Ehrman, R., O'Brien, C. P., and Pashko, S. (1986). Extinguishing conditioned responses during opiate dependence treatment turning laboratory findings into clinical procedures. *Journal of Substance Abuse Treatment* **3**, 33–40.

Meisch, R. A., Lemaire, G. A., and Cutrell, E. B. (1992). Oral self-administration of pentobarbital by rhesus monkeys: Relative reinforcing effects under concurrent signalled differential-reinforcement-of-low-rates schedules. *Drug and Alcohol Dependence* **30**, 215–225.

Meyer, R. E., and Mirin, S. M. (1979). *The Heroin Stimulus: Implications for a Theory of Addiction*. Plenum, New York.

Miliaressis, E., Rompre, P. P., Laviolette, P., Philippe, L., and Coulombe D. (1986). The curve-shift paradigm in self-stimulation. *Physiology and Behavior* **37**, 85–91.

Moolten, M., and Kornetsky, C. (1990). Oral self-administration of ethanol and not experimenter-administered ethanol facilitates rewarding electrical brain stimulation. *Alcohol* **7**, 221–225.

Morgan, A. D., Campbell, U. C., Fons, R. D., and Carroll, M. E. (2002). Effects of agmatine on the escalation of intravenous cocaine and fentanyl self-administration in rats. *Pharmacology Biochemistry and Behavior* **72**, 873–880.

Morgan, A. D., Dess, N. K., and Carroll, M. E. (2005). Escalation of intravenous cocaine self-administration, progressive-ratio performance, and reinstatement in rats selectively bred for high (HiS) and low (LoS) saccharin intake. *Psychopharmacology* **178**, 41–51.

Mucha, R. F., van der Kooy, D., O'Shaughnessy, M., and Bucenieks, P. (1982). Drug reinforcement studied by the use of place conditioning in rat. *Brain Research* **243**, 91–105.

Mueller, D., and Stewart, J. (2000). Cocaine-induced conditioned place preference: reinstatement by priming injections of cocaine after extinction. *Behavioural Brain Research* **115**, 39–47.

Murphy, J. M., Gatto, G. J., Waller, M. B., McBride, W. J., Lumeng, L., and Li, T. K. (1986). Effects of scheduled access on ethanol intake by the alcohol-preferring (P) line of rats, *Alcohol* **3**, 331–336.

Murphy, J. M., Stewart, R. B., Bell, R. L., Badia-Elder, N. E., Carr, L. G., McBride, W. J., Lumeng, L., and Li, T. K. (2002). Phenotypic and genotypic characterization of the Indiana University rat lines selectively bred for high and low alcohol preference. *Behavior Genetics* **32**, 363–388.

Myers, R. D., Stoltman, W. P., and Martin, G. E. (1972). Effects of ethanol dependence induced artificially in the rhesus monkey on the subsequent preference for ethyl alcohol. *Physiology and Behavior* **9**, 43–48.

Naassila, M., Beauge, F. J., Sebire, N., and Daoust, M. (2000). Intracerebroventricular injection of antisense oligos to nNOS decreases rat ethanol intake. *Pharmacology Biochemistry and Behavior* **67**, 629–636.

O'Brien, C.P., Testa J, O'Brien, T.J., Brady, J.P., and Wells B. (1977). Conditioned narcotic withdrawal in humans. *Science* **195**, 1000–1002.

O'Brien, C. P., Ehrman, R. N., Ternes, J. M. (1986). Classical conditioning in human opioid dependence. *In Behavioral Analysis of Drug Dependence*, (S. R. Goldberg, and I. P. Stolerman, Eds.), pp. 329–356. Academic Press, Orlando, F.L.

O'Brien, C. P., Childress, A. R., McLellan, A. T., and Ehrman, R. (1992). Classical conditioning in drug-dependent humans. In *The Neurobiology of Drug and Alcohol Addiction* (series title: *Annals of the New York Academy of Sciences*, vol. 654), (P. W. Kalivas, H. H. Samson, Eds.), pp. 400–415. New York Academy of Sciences, New York.

O'Dell, L. E., Roberts, A. J., Smith, R. T., and Koob, G. F. (2004). Enhanced alcohol self-administration after intermittent versus continuous alcohol vapor exposure. *Alcoholism: Clinical and Experimental Research* **28**, 1676–1682.

Olds, J., and Milner, P. (1954). Positive reinforcement produced by electrical stimulation of septal area and other regions of rat brain. *Journal of Comparative and Physiological Psychology* **47**, 419–427.

Overton, D. A. (1974). Experimental methods for the study of state-dependent learning. *Federation Proceedings* **33**, 1800–1813.

Paterson, N. E., and Markou, A. (2003). Increased motivation for self-administered cocaine after escalated cocaine intake. *Neuroreport* **14**, 2229–2232.

Paterson, N. E., Myers, C., and Markou, A. (2000). Effects of repeated withdrawal from continuous amphetamine administration on brain reward function in rats. *Psychopharmacology* **152**, 440–446.

Phillips, A. G., and Fibiger, H. C. (1987). Anatomical and neurochemical substrates of drug reward determined by the conditioned place preference technique. *In Methods of Assessing the Reinforcing Properties of Abused Drugs*, (M. A. Bozarth, Ed.), pp. 275–290. Springer-Verlag, New York.

Piazza, P. V., and Le Moal, M. L., (1996). Pathophysiological basis of vulnerability to drug abuse: Role of an interaction between stress, glucocorticoids, and dopaminergic neurons. *Annual Review of Pharmacology and Toxicology* **36**, 359–378.

Piazza, P. V., Deminiere, J. M., Le Moal, M., and Simon, H. (1989). Factors that predict individual vulnerability to amphetamine self-administration. *Science* **245**, 1511–1513.

Piazza, P. V., Rouge-Pont, F., Deminiere, J. M., Kharoubi, M, Le Moal, M. and Simon H. (1991). Dopaminergic activity is reduced in the prefrontal cortex and increased in the nucleus accumbens of rats predisposed to develop amphetamine self-administration. *Brain Research* **567**, 169–174.

Piazza, P. V., Rouge-Pont, F., Deroche, V., Maccari, S., Simon, H., Le Moal, M. (1996). Glucocorticoids have state-dependent stimulant effects on the mesencephalic dopaminergic transmission. *Proceedings of the National Academy of Sciences, USA* **93**, 8716–8720.

Piazza, P. V., Deroche-Gamonent, V., Rouge-Pont, F., and Le Moal, M. (2000). Vertical shifts in self-administration dose-response functions predict a drug-vulnerable phenotype predisposed to addiction. *Journal of Neuroscience* **20**, 4226–4232.

Richter, C. P. (1996). Animal behavior and internal drives. *Quarterly Review of Biology* 2, 307–343.

Roberts, A. J., Cole, M. and Koob, G. F. (1996) Intra-amygdala muscimol decreases operant ethanol self-administration in dependent rats. *Alcoholism: Clinical and Experimental Research* **20**, 1289–1298.

Roberts, A. J., Heyser, C. J., Cole, M., Griffin, P., and Koob, G. F. (2000). Excessive ethanol drinking following a history of dependence: Animal model of allostasis, *Neuropsychopharmacology* **22**, 581–594.

Roberts, D. C., Loh, E. A., and Vickers, G., (1989). Self-administration of cocaine on a progressive ratio schedule in rats: dose-response relationship and effect of haloperidol pretreatment. *Psychopharmacology* 97, 535–538.

Rodd, Z. A., Bell, R. L., Kuc, K. A., Murphy, J. M., Lumeng, L., Li, T. K., and McBride, W. J., (2003). Effects of repeated alcohol deprivations on operant ethanol self-administration by alcohol-preferring (P) rats. *Neuropsychopharmacology* **28**, 1614–1621.

Rodd, Z. A., Bell, R. L., Sable, H. J., Murphy, J. M., and McBride, W. J., (2004). Recent advances in animal models of alcohol craving and relapse. *Pharmacology Biochemistry and Behavior* **79**, 439–450.

Rodd-Henricks, Z. A., Bell, R. L., Kuc, K. A., Murphy, J. M., McBride, W. J., Lumeng, L., and Li, T. K. (2001). Effects of concurrent access to multiple ethanol concentrations and repeated deprivations on alcohol intake of alcohol-preferring rats. *Alcoholism: Clinical and Experimental Research* **25**, 1140–1150.

Roelofs, S. M. (1985). Hyperventilation, anxiety, craving for alcohol: a subacute alcohol withdrawal syndrome. *Alcohol* **2**, 501–505.

Roth, M. E., and Carroll, M. E. (2004). Sex differences in the escalation of intravenous cocaine intake following long- or short-access to cocaine self-administration. *Pharmacology Biochemistry and Behavior* **78**, 199–207.

Salimov, R. M., and Salimova, N. B. (1993). The alcohol-deprivation effect in hybrid mice. *Drug and Alcohol Dependence* **32**, 187–191.

Samson, H. H. (1986). Initiation of ethanol reinforcement using a sucrose-substitution procedure in food- and water-sated rats. *Alcoholism: Clinical and Experimental Research* **10**, 436–442.

Samson, H. H., and Falk, J. L. (1974). Alteration of fluid preference in ethanol-dependent animals. *Journal of Pharmacology and Experimental Therapeutics* **190**, 365–376.

Sarkar, M., Kornetsky, C. (1995). Methamphetamine's action on brain-stimulation reward threshold and stereotypy. *Experimental and Clinical Psychopharmacology* **3**, 112–117.

Sayette, M. A., Shiffman, S., Tiffany, S. T., Niaura, R. S., Martin, C. S., and Shadel, W. G. (2000). The measurement of drug craving, *Addiction* **95**(Suppl. 2), s189–s210.

Schindler, C. W., Panlilio, L. V., and Goldberg, S. R. (2002). Second-order schedules of drug self-administration in animals. *Psychopharmacology* **163**, 327–344.

Schulteis, G., Hyytia, P., Heinrichs, S. C., and Koob, G. F. (1996). Effects of chronic ethanol exposure on oral self-administration of ethanol or saccharin by Wistar rats. *Alcoholism: Clinical and Experimental Research* **20**, 164–171.

Schulteis, G., Markou, A., Cole, M., and Koob, G. (1995). Decreased brain reward produced by ethanol withdrawal. *Proceedings of the National Academy of Sciences USA* **92**, 5880–5884.

Schulteis, G., Markou, A., Gold, L. H., Stinus, L., and Koob, G. F. (1994). Relative sensitivity to naloxone of multiple indices of opiate withdrawal: A quantitative dose-response analysis. *Journal of Pharmacology and Experimental Therapeutics* **271**, 1391–1398.

Schuster, C. R., and Woods, J. H., (1968). The conditioned reinforcing effects of stimuli associated with morphine reinforcement. *International Journal of the Addictions* **3**, 223–230.

See, R. E., Grimm, J. W., Kruzich, P. J., and Rustay, N. (1999). The importance of a compound stimulus in conditioned drug-seeking behavior following one week of extinction from self-administered cocaine in rats. *Drug and Alcohol Dependence* **57**, 41–49.

Shannon, H. E., and Holtzman, S. G. (1976). Evaluation of the discriminative effects of morphine in the rat. *Journal of Pharmacology and Experimental Therapeutics* **198**, 54–65.

Sinclair, J. D., and Senter, R. J. (1967). Increased preference for ethanol in rats following alcohol deprivation. *Psychonomic Science* **8**, 11–12.

Sinclair, J. D., and Senter, R. J. (1968). Development of an alcohol-deprivation effect in rats. *Quarterly Journal of Studies on Alcohol* **29**, 863–867.

Spanagel, R., Hölter, S. M., Allingham, K., Landgraf, R., Zieglgänsberger, W. (1996). Acamprosate and alcohol: I. Effects on alcohol intake following alcohol deprivation in the rat. *European Journal of Pharmacology* **305**, 39–44.

Sparber, S. B., and Meyer, D. R. (1978). Clonidine antagonizes naloxone-induced suppression of conditioned behavior and body weight loss in morphine-dependent rats *Pharmacology Biochemistry and Behavior* **9**, 319–325.

Spear, D. J., and Katz, J. L. (1991). Cocaine and food as reinforcers: effects of reinforcer magnitude and response requirement under second-order fixed-ratio and progressive-ratio schedules. *Journal of the Experimental Analysis of Behavior* **56**, 261–275.

Stellar, J. R., and Rice, M. B. (1989). Pharmacological basis of intracranial self-stimulation reward. *In The Neuropharmacological Basis of Reward* (series title: *Topics in Experimental Psychopharmacology*, vol. 1), (J. M. Liebman, S. J. Cooper, Eds.), pp. 14–65. Clarendon Press, Oxford.

Stellar, J. R., and Stellar, E. (1985). *The Neurobiology of Motivation and Reward*. Springer-Verlag, New York.

Stewart, B. S., Lemaire, G. A., Roache, J. D., and Meisch, R. A., (1994). Establishing benzodiazepines as oral reinforcers: Midazolam and diazepam self-administration in rhesus monkeys. *Journal of Pharmacology and Experimental Therapeutics* **271**, 200–211.

Stewart, J., and de Wit, H. (1987). Reinstatement of drug-taking behavior as a method of assessing incentive motivational properties of drugs. *In Methods of Assessing the Reinforcing Properties of Abused Drugs*, (M. A. Bozarth, Ed.), pp. 211–227. Springer-Verlag, New York.

Stewart, J., and Eikelboom, R. (1987). Conditioned drug effects. *In New Directions in Behavioral Pharmacology* (series title: *Handbook of Psychopharmacology*, vol. 19), (L. I. Iversen, S. D. Iversen, and S. H. Snyder, Eds.), pp. 1–57. Plenum Press, New York.

Stewart, J., and Wise, R. A. (1992). Reinstatement of heroin self-administration habits: morphine prompts and naltrexone discourages renewed responding after extinction. *Psychopharmacology* **108**, 79–84.

Stewart, R. B., Bass, A. A., Wang, N. S., and Meisch, R. A. (1996). Ethanol as an oral reinforcer in normal weight rhesus monkeys: Dose-response functions. *Alcohol* **13**, 341–346.

Stinus, L., Le Moal, M., and Koob, G. F. (1990). Nucleus accumbens and amygdala are possible substrates for the aversive stimulus effects of opiate withdrawal. *Neuroscience* **37**, 767–773.

Stinus, L., Caille, S., and Koob, G. F. (2000). Opiate withdrawal-induced place aversion lasts for up to 16 weeks. *Psychopharmacology* **149**, 115–120.

Stolerman, I. P., Rasul, F., and Shine, P. J. (1989). Trends in drug discrimination research analysed with a cross-indexed bibliography, 1984-1987, *Psychopharmacology* **98**, 1–19.

Swerdlow, N. R., Gilbert, D., and Koob, G. F. (1989). Conditioned drug effects on spatial preference: critical evaluation. *In Psychopharmacology* (series title: *Neuromethods*, vol. 13), (A. A. Boulton, G. B. Baker, and A. J. Greenshaw, Eds.), pp. 399–446. Humana Press, Clifton, N. J.

Tiffany, S. T., Carter, B. L., and Singleton, E. G. (2000). Challenges in the manipulation, assessment and interpretation of craving relevant variables. *Addiction* **95**(Suppl. 2), s177–s187.

Valdez, G. R., Zorrilla, E. P., Roberts, A. J., and Koob, G. F. (2003). Antagonism of corticotropin-releasing factor attenuates the enhanced responsiveness to stress observed during protracted ethanol abstinence. *Alcohol* **29**, 55–60.

Valdez, G. R., Sabino, V., and Koob, G. F. (2004). Increased anxiety-like behavior and ethanol self-administration in dependent rats: reversal via corticotropin-releasing factor-2 receptor activation. *Alcoholism: Clinical and Experimental Research* **28**, 865–872.

van der Kooy, D. (1987). Place conditioning: A simple and effective method for assessing the motivational properties of drugs. *In Methods of Assessing the Reinforcing Properties of Abused Drugs*, (M. A. Bozarth, Ed.), pp. 229–240. Springer-Verlag, New York.

Vanderschuren, L. J., and Everitt, B. J. (2004). Drug seeking becomes compulsive after prolonged cocaine self-administration. *Science* **305**, 1017–1019.

Veale, W. L., and Myers, R. D. (1969). Increased alcohol preference in rats following repeated exposures to alcohol. *Psychopharmacologia* **15**, 361–372.

Walker, J. R., Chen, S. A., Moffitt, H., Inturrisi, C. E., and Koob, G. F. (2003). Chronic opioid exposure produces increased heroin self-administration in rats. *Pharmacology Biochemistry and Behavior*. **75**, 349–354.

Waller, M. B., McBride, W. J., Gatto, G. J., Lumeng, L., and Li, T. K. (1984). Intragastric self-infusion of ethanol by ethanol-preferring and -nonpreferring lines of rats. *Science* **225**, 78–80.

Ward, A. S., Li, D. H., Luedtke, R. R., and Emmett-Oglesby, M. W. (1996). Variations in cocaine self-administration by inbred rat strains under a progressive-ratio schedule. *Psychopharmacology* **127**, 204–212.

Weiss, F., Mitchiner, M., Bloom, F. E., and Koob, G. F. (1990). Free-choice responding for ethanol versus water in alcohol preferring (P) and unselected Wistar rats is differentially modified by naloxone, bromocriptine, and methysergide. *Psychopharmacology* **101**, 178–186.

Weiss, F., Parsons, L. H., Schulteis, G., Hyytia, P., Lorang, M. T., Bloom, F. E., and Koob, G. F. (1996). Ethanol self-administration restores withdrawal-associated deficiencies in accumbal dopamine and 5-hydroxytryptamine release in dependent rats. *Journal of Neuroscience* **16**, 3474–3485.

Weiss, F., Maldonado-Vlaar, C. S., Parsons, L. H., Kerr, T. M., Smith, D. L., and Ben-Shahar, O. (2000). Control of cocaine-seeking behavior by drug-associated stimuli in rats: Effects on recovery of extinquished operant-responding and extracellular dopamine levels in amygdala and nucleus accumbens. *Proceedings of the National Academy of Sciences USA* **97**, 4321–4326.

Wikler, A. (1973). Dynamics of drug dependence: Implications of a conditioning theory for research and treatment. *Archives of General Psychiatry* **28**, 611–616.

Wikler, A., and Pescor, F. T. (1967). Classical conditioning of a morphine abstinence phenomenon, reinforcement of opioid-drinking behavior and 'relapse' in morphine-addicted rats. *Psychopharmacologia* **10**, 255–284.

Wikler, A., Pescor, F. T., Miller, D., and Norrell, H. (1971). Persistent potency of a secondary (conditioned) reinforcer following withdrawal of morphine from physically dependent rats. *Psychopharmacologia* **20**, 103–117.

Willner, P. (1984). The validity of animal models of depression. *Psychopharmacology* **83**, 1–16.

Winger, G. (1988). Effects of ethanol withdrawl on ethanol-reinforced responding in rhesus monkeys. *Drug and Alcohol Dependence* **22**, 235–240.

Wolffgramm, J., Heyne, (1991). A. Social behavior, dominance, and social deprivation of rats determine drug choice. *Pharmacology Biochemistry and Behavior* **38**, 389–399.

Yanagita, T. (1973). An experimental framework for evaluation of dependence liability of various types of drugs in monkeys. *Bulletin on Narcotics* **25**, 57–64.

Zywiak, W. H., Connors, G. J., Maisto, S. A., and Westerberg, V. S. (1996). Relapse research and the Reasons for Drinking Questionnaire: a factor analysis of Marlatt's relapse taxonomy. *Addiction* **91**(Suppl.):s121–s130.

CHAPTER

3

Psychostimulants

OUTLINE

DEFINITIONS

Stimulant drugs such as cocaine, D-amphetamine, and methamphetamine have medical uses but also considerable abuse potential (Sanchez-Ramos, 1990) (**Fig. 3.1**). There are two major classes of psychomotor stimulants (**Table 3.1**): (1) direct or indirect sympathomimetics, such as cocaine and amphetamine, and (2) nonsympathomimetics. This chapter will focus on indirect sympathomimetics. Indirect sympathomimetic compounds such as cocaine and amphetamines share a common molecular structure—a benzene ring with an ethylamine side chain (**Fig. 3.2**). Amphetamine differs from the parent compound, β-phenethylamine, by the addition of a methyl group, whereas methamphetamine has two additional methyl groups. Psychomotor stimulants are drugs that produce a behavioral activation usually accompanied by increases in arousal, alertness, and motor activity. The term *sympathomimetic* derives originally from the description of the mechanism of action of these drugs. Sympathomimetics mimic the action of the sympathetic nervous system when it is activated. Indeed, the term *sympathin* originally was used to describe the hormone noradrenaline (norepinephrine) found in the central nervous system (Cannon and Rosenblueth, 1933; von Euler, 1947; Vogt, 1954). Thus, sympathomimetic drugs mimic the peripheral actions of norepinephrine in the autonomic system and neuropharmacologically either directly or indirectly activate monoamine receptors (see below). Indirect sympathomimetics mimic this action by acting on

Neurobiology of Addiction, by George F. Koob and Michel Le Moal.
ISBN – 13: 978-0-12-419239-3 ISBN – 10: 0-12-419239-4

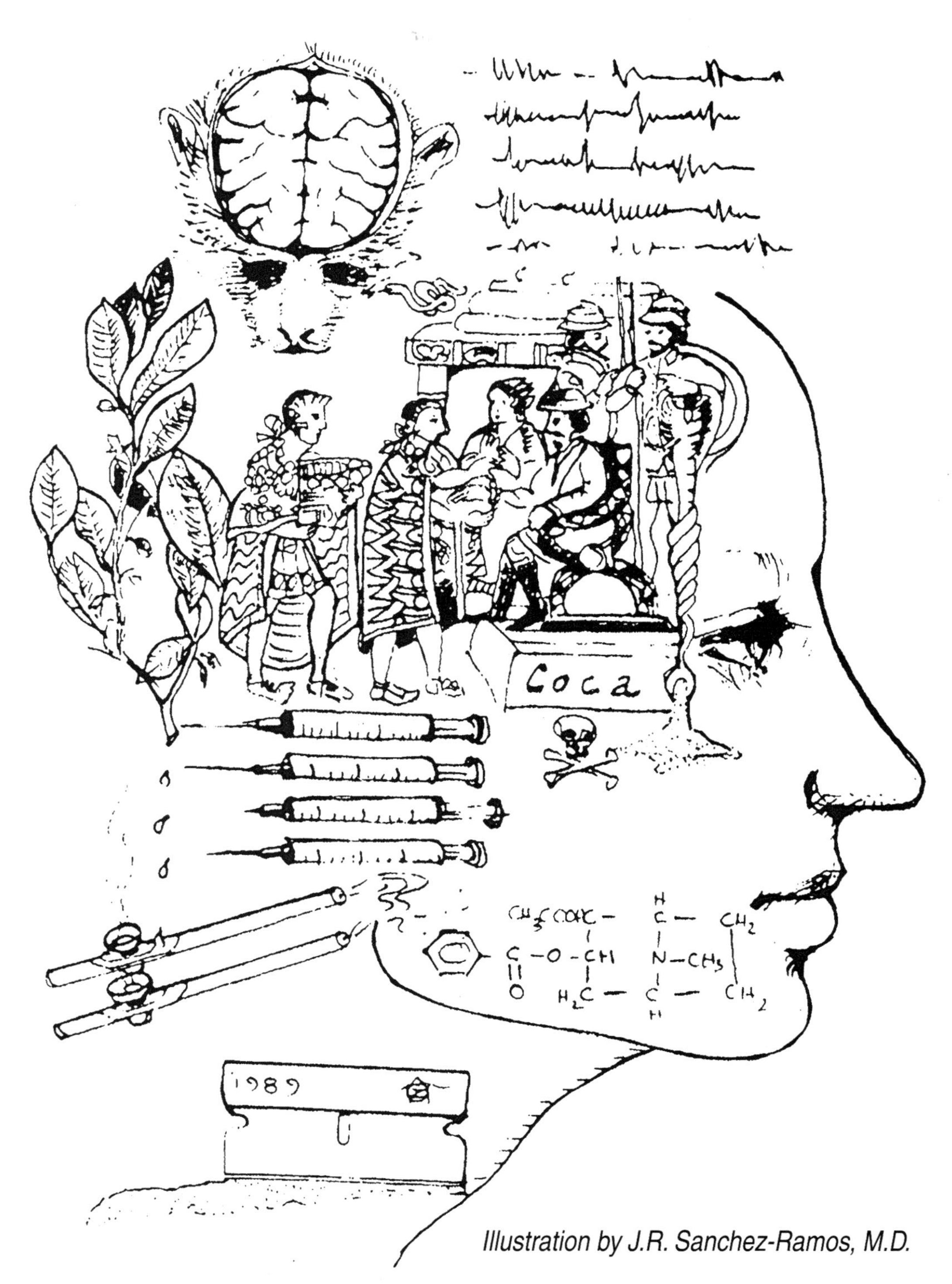

FIGURE 3.1 From non-humans to humans, cocaine and related psychostimulants exhibit myriad effects on brain and behavior. [Reproduced with permission from Sanchez-Ramos, 1990.]

TABLE 3.1 Psychomotor Stimulant Drugs

Direct sympathomimetics	Indirect sympathomimetics	Non-sympathomimetics
Isoproterenol	Amphetamine	Caffeine
Epinephrine	Methamphetamine	Nicotine
Norepinephrine	Cocaine	Scopolamine
Phenylephrine	Methylphenidate	Strychnine
Phenylpropanolamine	Phenmetrazine	Pentylenetetrazol
Apomorphine	Pipradrol	Modafinil
	Tyramine	
	Pemoline	

neuronal mechanisms that do not involve direct activation of postsynaptic receptors, and the present treatise will focus only on the neurobiological mechanisms involved in the addiction liability of indirect sympathomimetics. Nonsympathomimetics act via different neuropharmacological mechanisms altogether. Historically, there have been numerous eras of stimulant addiction often linked to increased availability and distorted or misinformed perceptions of the abuse potential of these drugs.

HISTORY OF PSYCHOSTIMULANT USE, ABUSE, AND ADDICTION

Cocaine is derived from the coca plant *(Erythroxylon coca)* and has a long history as a stimulant. Cocaine has been used for centuries in tonics and other preparations to allay fatigue, sustain performance, and treat a large variety of ailments (Haddad, 1978; Angrist and Sudilovsky, 1978). Cocaine was once a component of Coca Cola®, and its extract (without cocaine) still is used as an ingredient today. In 1886 John Styth Pemberton, a druggist, devised a patent medicine that contained two natural stimulants, cocaine and caffeine, to formulate the syrup base for Coca-Cola®. He blended a whole-leaf extract of coca with an extract from the African kola nut which contains caffeine. Coca-Cola® was initially manufactured and marketed as an 'intellectual beverage' and 'brain tonic,' and until 1903 Coca-Cola® contained approximately 60 mg of cocaine per 8 ounce serving. Later it was touted as a temperance drink because it had no alcohol despite the fact that cocaine was still a key ingredient. The manufacturer believed that their product should not only be strongly associated with cocaine by the product name but also that the product package should stand out as unique. Thus, the unique shape of the Coca-Cola® bottle designed by C.J. Root Co. (Terre Haute, Indiana, USA) was originally intended to resemble the shape of a coca bean. In reality, the bottle shape resembles a cacao (i.e., cocoa) bean because the production artists mistakenly used a *cacao* bean, instead of a *coca* bean, as the model for the bottle design. In 1903, soon after the dangers of cocaine were publicized, the manufacturer of Coca-Cola® removed cocaine from its formulation (Cornish and O'Brien, 1996; Gold and Jacobs, 2005; Louis and Yazijian, 1980).

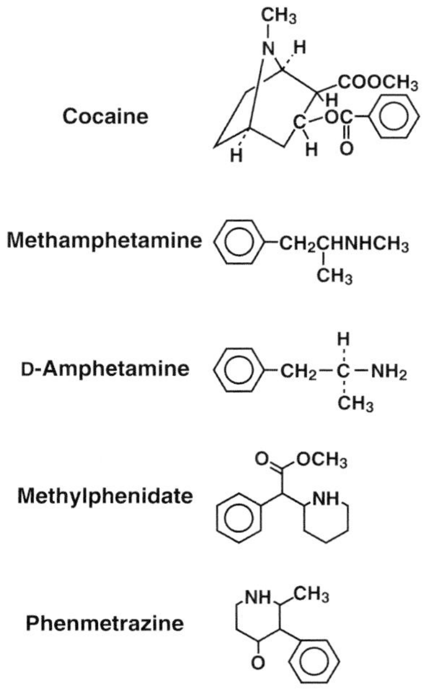

FIGURE 3.2 Chemical structures of various indirect sympathomimetic psychostimulants.

The use of cocaine was even advocated briefly by Dr. Sigmund Freud to treat a variety of disorders, including psychiatric disorders and drug addiction (Freud, 1884), but Freud quickly lost his enthusiasm after observing his first cocaine psychosis (Freud, 1974, for a review of Freud's writings on cocaine), and cocaine and other indirect sympathomimetics have been involved in more than one epidemic of drug abuse in the United States and the world (Byck, 1987) (**Table 3.2**). Presumably because cocaine was used in numerous tonics, there was extensive cultivation of cocaine in South America and exportation to the United States and Europe (Hall, 1884; Chopra and Chopra, 1931; Angrist and Sudilovsky, 1978). Widespread use followed, and in the United States, the first restriction of coca products commenced in 1914 with the Harrison Narcotics Act (United States Treasury Department, Bureau of Internal Revenue, 1915). This act penalized its illicit possession and sale but mislabeled cocaine-containing preparations as narcotics.

In the 1960s, a rise in the smuggling and use of cocaine started on the basis of the high monetary profit involved in its illegal trafficking (Angrist and Sudilovsky, 1978) that culminated in an epidemic of

TABLE 3.2 History of Cocaine Use and Misuse

3000 B.C.	Cocaine is believed to have originated in the subtropical valleys of the eastern slopes of the Andes or Amazonian subtropical valleys (Naranjo, 1981). The earliest archeological evidence from Peru dates coca chewing to 3000 BC (Siegel, 1982).
1493–1527	Coca chewing was restricted to Incan royalty and religious figures (Siegel, 1982). Coca leaves were used as offerings and were used in cultural and religious ceremonies (Naranjo, 1981).
1536	Coca chewing came to the masses following the Spanish conquest (Siegel, 1982). Coca leaves were used by Indian slave laborers in the silver mines to keep themselves alert and working (Naranjo, 1981).
1859	Albert Niemann analyzes a sample of Peruvian coca in the lab of Fredrich Wohler to determine active compound and isolates cocaine (Karch, 1997).
1868–1869	Coca is touted by Angelo Mariani who developed a coca based wine Vin Mariani (Karch, 1997). The wine contained no more than 300 mg of cocaine; the wine was very popular and was marketed as a tonic wine and cure-all. (Karch, 1997; Brain and Coward, 1989).
1884	Karl Koller publishes work on using cocaine as an anesthetic during eye surgery (Karch, 1997).
	Sigmund Freud publishes *On Coca*. Recommends cocaine use for a variety of illnesses, notably for alcoholism and morphine addiction (Grinspoon and Bakalar, 1981).
1886	Albert Erlenmeyer publishes a paper denouncing cocaine use as treatment for opioid addiction. Blames Freud for releasing 'the third scourge of mankind' (Karch, 1997).
1887	Freud publishes *Craving for and Fear of Cocaine*. He admits that cocaine should not be used to treat morphine addiction after his friend Ernst von Fleischl-Marxow experiences severe toxic symptoms of heavy cocaine use (Grinspoon and Bakalar, 1981).
1885	Pemberton, a patent-medicine maker from Atlanta, produced a wine called *Cocaine—Ideal Nerve and Tonic Stimulant*. Because of overriding Prohibition restrictions, he launched a nonalcoholic extract of coca leaves and caffeine-rich African Kola nuts in a sweet carbonated syrup he called *Coca Cola* (Brain and Coward, 1989).
1892	Coca Cola Company is founded. Coca Cola is touted as a medicinal drink (Louis and Yazijian, 1980).
1902	Due to negative public sentiment, Coca Cola 'decocainizes' its preparation, replacing cocaine with caffeine (Louis and Yazijian, 1980).
1910	President Howard Taft presents a State Department report on drug use to Congress. Cocaine officially becomes 'Public Enemy #1.' 'The illicit sales…and the habitual use of it temporarily raises the power of a criminal to a point where in resisting arrest there is no hesitation to murder. It is more appalling in its effects than any other habit-forming drug used in the United States' (Das, 1993).
1914	The Harrison Narcotic Act is passed which tightly regulates the distribution and sale of drugs. Because of public anti-cocaine sentiment the Harrison Act was largely supported and was rather successful (Musto, 1989; Das, 1993).
1970	Cocaine use increases following a backlash against amphetamine use. Stimulant users rediscover cocaine as a 'safe' recreational drug (Gold and Jacobs, 1997). Most used type is for social-recreational reasons among friends or acquaintances (Siegel, 1992).
	Controlled Substances Act passes in United States Congress. Cocaine is made a Schedule II drug by the Drug Enforcement Administration (abusable drugs with officially sanctioned medical uses) (Madge, 2001).
1974	'Free-basing' develops in southern California (Siegel, 1982).
1975	A White Paper issued by the United States government indicated that cocaine is 'not physically addictive' and 'usually does not result in serious social consequences such as crime, hospital emergency room admission, or death" (Madge, 2001).
1980	Approximately 20% of those aged 15–25 admit to using cocaine (Cocaine Anonymous, 2003). Drug abuse treatment facilities and hospitals report dramatic increases in cocaine free base admission (Siegel, 1982).

cocaine use in the 1980s. In the 1970s, cocaine was usually administered intranasally in a powder form (cocaine hydrochloride). The perception among users was that it was safe and non-addictive. In fact, the 1980 edition of the *Comprehensive Textbook of Psychiatry* stated, 'used no more than two–three times a week, cocaine creates no serious problems. In daily and fairly large amounts, it can produce minor psychological disturbances. Chronic cocaine abuse does not appear as a medical problem' (Grinspoon and Bakalar, 1980). During the 1970s, freebase cocaine use developed. Cocaine freebase generally is prepared from its hydrochloride salt by one of two techniques. In one procedure, the hydrochloride salt first is mixed with buffered ammonia, then the alkaloidal cocaine is extracted from the solution using ether, and finally the ether is evaporated to yield cocaine crystals. When heated, the crystals release vaporized cocaine that can be inhaled. The hydrochloride salt of cocaine by itself cannot be smoked because it is quickly destroyed at high temperatures (Cook and Jeffcoat, 1986; Snyder *et al.*, 1988). Cocaine freebase melts at 98°C (Muhtadi and Al-Badr, 1986) and vaporizes at 260°C (Martin *et al.*, 1989). This form of cocaine is very pure and was generally called 'freebase' on the street and began to be seen in the late 1970s and was popular in the

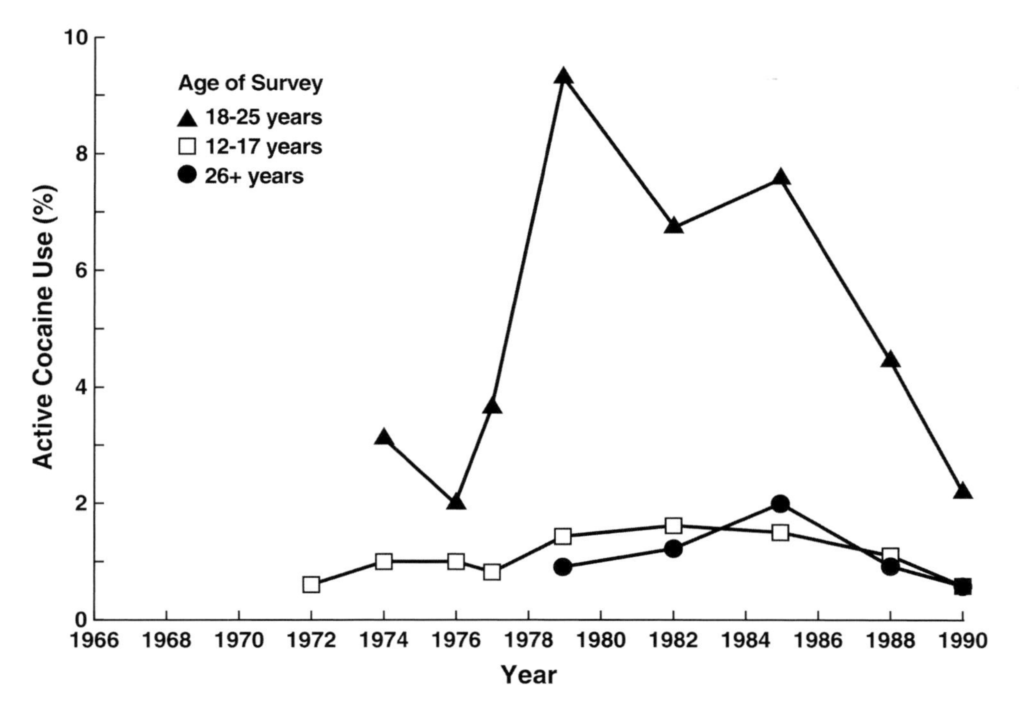

FIGURE 3.3 Estimated prevalence (%) of active cocaine use by year of survey and age of respondent (any cocaine use in the 30 days prior to interview). Data from the National Household Survey on Drug Abuse, 1972–1990. [Reproduced with permission from Anthony, 1992.]

mid-1980s (Gold and Jacobs, 2005). The crystals also make a popping sound when heated; it is this characteristic sound that has been hypothesized to be the origin of the term 'crack' cocaine. Later, crack cocaine was sold in small, ready-to-smoke 'rocks.' Cocaine hydrochloride was combined with baking soda (sodium bicarbonate), and the solution was heated until a solid was formed. Since the mid-1980s, this has been the preferred method of production for smokable cocaine because it is simpler and safer than the ether extraction method, and today most of the available crack cocaine in the United States is produced in this manner (Cornish and O'Brien, 1996; Gold and Jacobs, 2005).

National surveys indicate that prevalence rates reached their peak in 1985 with the *National Household Survey on Drug Use and Health* (formerly the *National Household Survey on Drug Abuse*) reporting 7.1 million individuals in the United States having used cocaine within the past year, with the number steadily declining since that time (Anthony, 1992; Johanson and Schuster, 1995; Substance Abuse and Mental Health Services Administration, 1996) (**Fig. 3.3**). The percentage of young adults aged 18 to 25 who had ever used cocaine was below 1 per cent during the mid-1960s, but rose steadily throughout the 1970s and early 1980s, reaching a high of 17.9 per cent in 1984. Others have argued that because this drug use survey missed many drug-active populations, that these were underestimations and that during this period upwards of 25 per cent of American adults (50 million) had experimented with the drug (Das, 1990). By 1996, the rate had dropped to 10.1 per cent, but had climbed back to 15.4 per cent in 2002 (Substance Abuse and Mental Health Services Administration, 2003). In the European Union, lifetime prevalence ranges from 1 to 10 per cent among 15–34-year-olds. Spain and the United Kingdom report figures over 4 per cent (European Monitoring Centre for Drugs and Drug Addiction, 2004).

Amphetamines had widespread medical use in the treatment of narcolepsy and a variety of other disorders from 1936 to the mid-1940s (Angrist and Sudilovsky, 1978). At the end of World War II (1945–1955), there was an epidemic of methamphetamine use in Japan, and this epidemic was attributed to the 'dumping' on the open market of war stockpiles of stimulant agents (Masaki, 1956). In the 1960s, illegal diversion of amphetamines paralleled the increased use of the drugs resulting in the cyclic pattern of abuse, known as 'speed freaks' (Kramer *et al.*, 1967; Angrist and Sudilovsky, 1978). There was a parallel epidemic of abuse of phenmetrazine, an amphetamine-like stimulant, in Sweden in the 1950s and 1960s (Rylander, 1969).

Methamphetamine was first synthesized in Japan in 1893 (Suwaki *et al.*, 1997) and came into widespread use during World War II for increasing endurance and performance of military personnel. Methamphetamine was sold over-the-counter in Japan as Philopin and Sedrin as a product to fight sleepiness and enhance vitality (Anglin *et al.*, 2000). The most common manufacturing process uses ephedrine or pseudoephedrine in a reduction process with a mixture of iodine and red phosphorous to yield both the dextro (D) and levo (L) isomers of methamphetamine (Cho and Melega, 2002). The D-isomer is 5–10 times more potent than the L-isomer in producing central nervous system effects (Melega *et al.*, 1999). Methamphetamine can still be prescribed in the United States for attention deficit hyperactivity disorder (ADHD) and as a short-term adjunct treatment in a regimen for weight reduction based on caloric restriction (Medical Economics Company, 2004). The L-isomer, called desoxyephedrine (Levmetafetamine), is the active ingredient in nasal decongestant inhalers (Medical Economics Company, 2004).

An epidemic of methamphetamine abuse occurred in Japan after World War II (1945–1957) when military stockpiles flooded the market, with estimates in 1954 of 550 000 methamphetamine abusers in Japan (Anglin *et al.*, 2000). In the 1960s, methamphetamine was synthesized from other precursors and was a mixture of the D- and L-isomers called 'crank' and dominated the 'speed' market (Miller, 1997). Manufacture shifted to the San Diego area in the 1980s with the production of 'crystal meth' also from ephedrine (Morgan and Beck, 1997). Crystal methamphetamine, which is the D-isomer, is called 'ice', allegedly because of its resemblance to ice crystals (Cho and Melega, 2002) and can be smoked or snorted. Smokable methamphetamine became a popular drug of abuse in the 1980s in Hawaii, the Pacific Coast of the United States, and Southern California and has subsequently spread to the rest of the United States (Cho, 1990; Derlet and Heischober, 1990).

In 2002 it was estimated by the United States *National Survey on Drug Use and Health* that there were nearly 34 million individuals who ever used cocaine and over 12 million individuals who ever used methamphetamine (Substance Abuse and Mental Health Services Administration, 2003). According to this survey, nearly 6 million individuals reported using cocaine within the past year. Of these, 25 per cent (1.5 million) could be classified as meeting the criteria for Substance Dependence as defined by the *Diagnostic and Statistical Manual of Mental Disorders*, 4th edition (American Psychiatric Association, 1994; Substance Abuse and Mental Health Services Administration, 2003; see also O'Malley and Gawin, 1990). In the European Union, lifetime prevalence of amphetamine use among the general adult population 15–63 years of age varied from 0.5 to 6 per cent, except in the United Kingdom where amphetamine use is as high as 12 per cent (European Monitoring Centre for Drugs and Drug Addiction, 2004).

BEHAVIORAL EFFECTS AND MEDICAL USES

Behavioral Effects

Cocaine administered intranasally produces stimulant effects similar to those of amphetamines, but with much shorter durations (20–45 min) that include feelings of having much energy, fatigue reduction, a sense of well being, increased confidence, and increased talkativeness. Intoxication includes a euphoric effect that has been described as exhilarating, with a kind of rush that goes straight to one's brain, mild elation, and an enhanced ability to concentrate. Sigmund Freud wrote, 'The psychic effect of cocaine in doses of 50–100 mg consists of exhilaration and lasting euphoria, which does not differ in any way from the normal euphoria of a healthy person'. In a letter to his fiancée Martha dated June 2, 1884, he wrote how 'a small dose lifted me to heights in a wonderful fashion' (Jones, 1961). When taken intravenously or smoked, cocaine produces an intense euphoria, sometimes followed by a crash. William Burroughs wrote in 1959, 'When you shoot coke in the mainline there is a rush of pure pleasure to the head...Ten minutes later you want another shot...intravenous C is electricity through the brain, activating cocaine pleasure connections' (Burroughs, 1959). Cocaine has many of the same stimulant effects of amphetamines, including sustained performance in situations of fatigue (Fischman and Schuster, 1982). Normal healthy volunteers tested with a wide range of intravenous doses of cocaine or D-amphetamine showed that cocaine produced much the same subjective and physiological effects as D-amphetamine, although D-amphetamine was more potent (Fischman and Schuster, 1982) (**Fig. 3.4**).

Amphetamines in recreational dose ranges (see *Pharmacokinetics* below) produce stimulant effects, but the most dramatic effects are observed in situations of fatigue and boredom (Eysenck *et al.*, 1957). Beneficial effects include increased stimulation, improved co-ordination, increased strength and endurance, and increased mental and physical activation, with mood changes of boldness, elation and friendliness (Smith and Beecher, 1960). These drugs can both reduce the subjective sensation of fatigue and can prolong physical performance for long periods (Heyrodt and Weissenstein, 1940; Cuthbertson and Knox, 1947; Kornetsky *et al.*, 1959; Laties and Weiss, 1981). Amphetamines enhance performance in simple motor and cognitive tasks, including measures of reaction time, speed, attention and performance (Heyrodt and Weissenstein, 1940; Cuthbertson and Knox, 1947; Smith

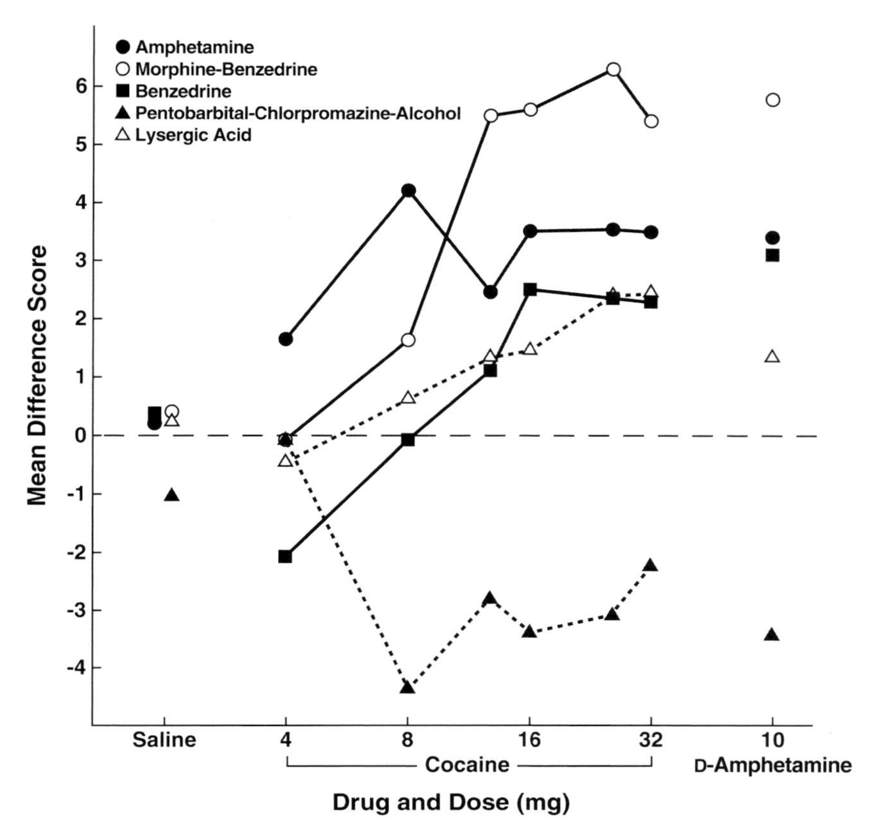

FIGURE 3.4 Mean change in scores of five separate scales of the Addiction Research Center Inventory. This inventory assesses drug-related changes in verbal reports of their effects. A short form of the inventory was answered before drug or saline injection and again 15 min after injection. Subjects received an intravenous dose of 4, 8, 16, and 32 mg of cocaine and 10 mg of D-amphetamine. [Reproduced with permission from Fischman and Schuster, 1982.]

and Beecher, 1959; Weiss and Laties, 1962; Caldwell *et al.*, 1995, 2000; Wiegmann*et al.*, 1996) (**Fig. 3.5**). Amphetamines also can significantly improve athletic performance by slight amounts (0.5–4 per cent), and these small percentage improvements may be sufficient to be significant in competitive situations (Smith and Beecher, 1959; Laties and Weiss, 1981) (**Table 3.3**).

Nevertheless, stimulants such as the amphetamines and cocaine can fail to improve performance of the well-functioning, motivated subject (Balloch *et al.*, 1952; Hauty and Payne, 1957; Kornetsky, 1958), and there is little evidence to suggest that amphetamines can enhance intellectual functioning in complex tasks or tests of intelligence (Weiss and Laties, 1962). In fact, methamphetamine failed to improve performance on a complex attention task, even though methamphetamine did increase the rate at which a visual display was scanned (Mohs *et al.*, 1978). An inverted U-shaped relationship between performance and dose of stimulant that is related to the complexity of the task has been hypothesized (Lyon and Randrup, 1972; Branch and Gollub, 1974; Branch, 1975; Lyon and Robbins, 1975; Skjoldager *et al.*, 1991; Cools and Robbins, 2004) (**Fig. 3.6**). This inverted U-shaped function may reflect the observation that as the dose of stimulant increases (e.g., a single increasing stimulatory effect can explain both sides of the U-shaped function), behavior becomes progressively more constricted and repetitive, resulting in both cognitive and behavioral persever-ation (Lyon and Robbins, 1975) (see below). Consistent with this conceptualization, cognitive deficits have been observed with direct dopamine agonists (Cools *et al.*, 2001; Cools and Robbins, 2004). Drug effects on cognitive performance also depend on initial conditions. The memory effects of methyphenidate in healthy volunteers showed greater involvement in subjects with lower baseline memory capacity (Mehta *et al.*, 2000).

Other acute actions of amphetamines and cocaine include a decrease in appetite for which these drugs have been used therapeutically and to which tolerance develops (Poindexter, 1960; Simkin and Wallace, 1961; Penick, 1969) (**Table 3.4**). Trials over 4 weeks reported significant weight loss; trials over 6 months reported no significant differences. Amphetamines also produce decreases in sleepiness, increased latency to fall asleep, increased latency to the onset of REM (rapid eye movement) sleep, and a reduction in the proportion of REM sleep (Rechtschaffen and Maron, 1964; Baekeland, 1966; Oswald, 1968).

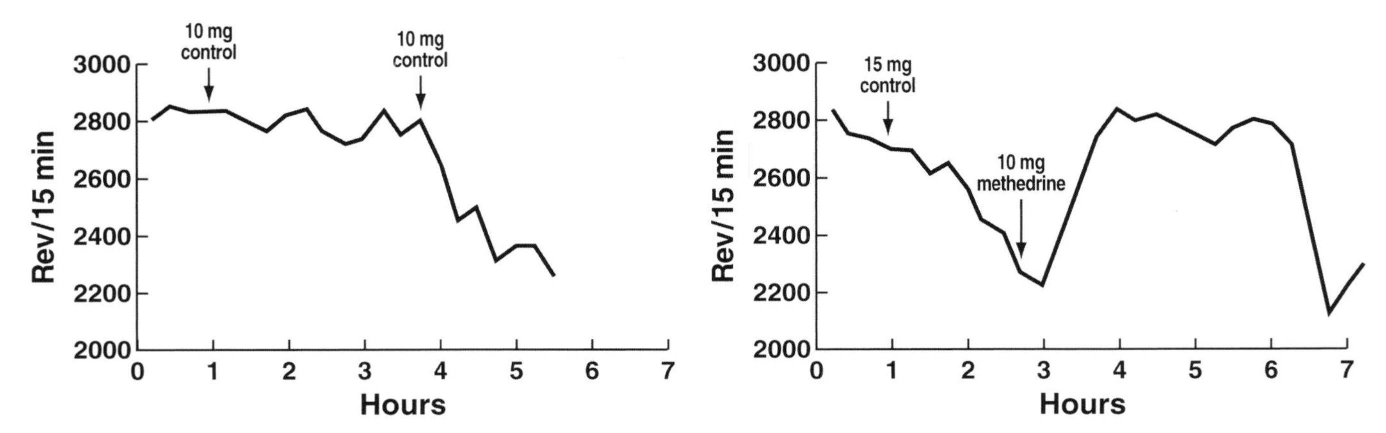

FIGURE 3.5 Performance on a cycle ergometer in humans after treatment with the psychostimulant methedrine. [Reproduced with permission from Cuthbertson and Knox, 1947.]

Finally, amphetamines and cocaine have long been reported to heighten sexual interest and prolong orgasm. In some instances, such delays in ejaculation have led to 'marathon' bouts of intercourse lasting for hours and probably reflect some of the behavioral psychopathology produced by these drugs (Angrist and Gershon, 1969; see next section). However, systematic studies of the effects of amphetamines on sexual behavior show that in amphetamine users, amphetamine use can lead to significant decreases in sexual performance with prolonged use of the drugs (Angrist and Sudilovsky, 1978).

TABLE 3.3 Comparative Effects of Amphetamine Sulfate and Placebo on Swimming Performance Times in Subjects* Under Rested and Fatigued Conditions

	Swim time (s)		
Swim style	*Placebo*	*Amphetamine (0.2 mg/kg)*	Improvement
Rested			
Freestyle (100 yards)	57.47	56.87	1.04%
Butterfly (100 yards)	70.96	69.36	2.25%
Freestyle (200 yards)	136.88	135.94	0.69%
Backstroke (200 yards)	159.80	158.32	0.93%
Breaststroke (200 yards)	171.87	170.22	0.96%
Fatigued			
Freestyle (100 yards)	59.31	58.53	1.32%
Butterfly (100 yards)	76.06	74.80	1.66%
Freestyle (200 yards)	144.24	142.38	1.29%
Backstroke (200 yards)	166.48	167.19	—
Breaststroke (200 yards)	175.14	176.87	—

*Three subjects performed each of the swim tasks specified, under both the rested (1st swim) and fatigued (2nd swim) conditions. The 2nd swim occurred 15 min after the 1st swim.
[Data from Smith and Beecher, 1959.]

The central nervous system effects of amphetamines are more pronounced by 3- to 4-fold with the D-isomer. With methamphetamine at low doses, these central nervous system effects are more pronounced than the autonomic effects, presumably due to its increased lipophilicity which allows it to readily cross the blood–brain barrier (United States Pharmacopeial Convention, 1998).

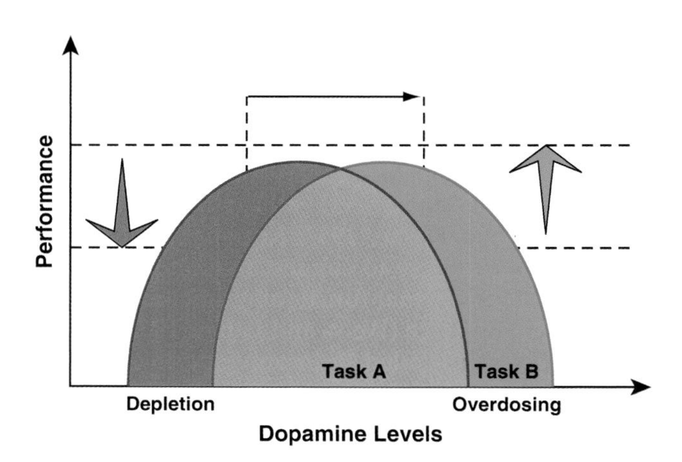

FIGURE 3.6 The relationship between cognitive performance and dopamine levels follows an inverted U-shaped function, where both too little and too much dopamine impairs performance. How likely it is that a drug will cause beneficial or detrimental effects depends partly on basal dopamine levels. A single inverted U-shaped curve is insufficient to predict performance. Some tasks benefit from extra dopamine (green), while some tasks are disrupted by extra dopamine (red). The black arrows represent the dopamine-enhancing effect of a hypothetical drug, leading to a beneficial effect on task B (green), but a detrimental effect on task A (red). [Reproduced with permission from Cools and Robbins, 2004.]

TABLE 3.4 Weight Loss (in Pounds) Induced by Benzphetamine and D-Amphetamine

Medication	Week 0	Week 1	Week 2	Week 3	Week 4
Benzphetamine	1.84 ± 0.23 (n = 20)	0.86 ± 0.1 (n = 19)	0.70 ± 0.24 (n = 16)	0.49 ± 0.20 (n = 14)	0.53 ± 0.23 (n = 9)
D-Amphetamine	1.53 ± 0.16 (n = 19)	0.93 ± 0.16 (n = 17)	0.56 ± 0.14 (n = 12)	0.54 ± 0.24 (n = 11)	0.29 ± 0.25 (n = 8)

Taken with permission from Simkin and Wallace, 1961.

Medical Uses

Cocaine was recognized early as a local anesthetic for ophthalmological work (Knapp, 1884), and the only well-accepted current medical uses are mucous membrane anesthesia and vasoconstriction (Medical Economics Company, 2004). This use ultimately led to the discovery of procaine. Amphetamines were synthesized originally as possible alternative drugs for the treatment of asthma and were the principal component of the original benzedrine inhaler (Benzedrine, 1933) (**Fig. 3.7**). They were used (and still are used) by the United States military for antifatigue indications (Caldwell *et al.*, 2000), and they are currently legally available for medical use as adjuncts for short-term weight control. Amphetamines are effective treatments for narcolepsy (Prinzmetal and Bloomberg, 1935; Mitler *et al.*, 1990). Amphetamines also improve ADHD, and decrease the hyperactivity observed in these children (Bradley, 1937; Lambert *et al.*, 1976; Huey, 1986). A recent reformulation of amphetamine isomers (e.g., Adderall) (for review see McKeage and Scott, 2003) is the best-selling medication for ADHD with a 23 per cent market share (IMS Health data, week ending July 30, 2004; see Kessler *et al.*, 2004; Roper, 2004).

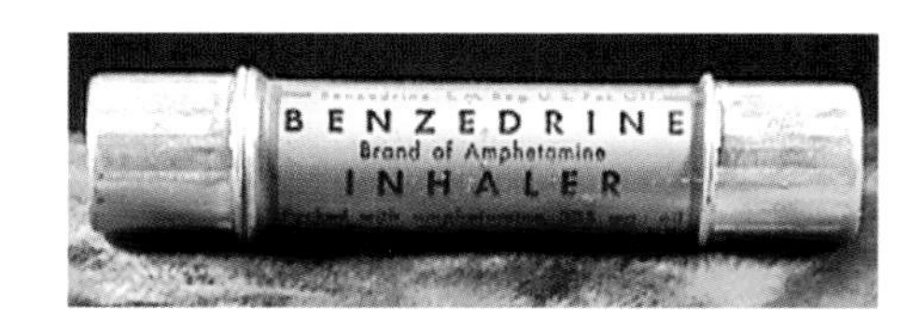

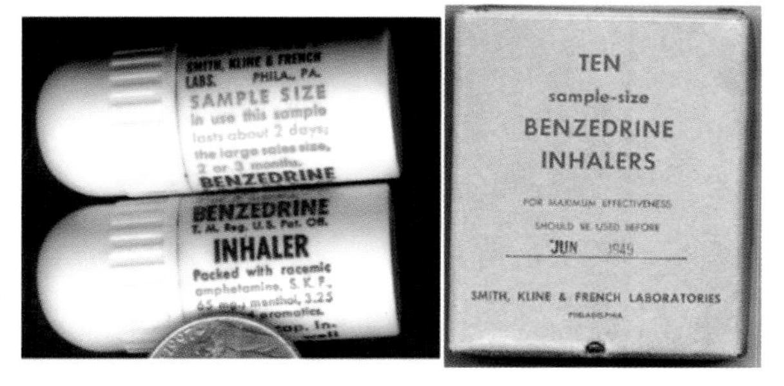

FIGURE 3.7 Photographs of two types of benzedrine inhaler, first introduced to the market in 1932 by Smith, Kline & French Co.

Physiological Actions

Both cocaine and amphetamine produce increases in systolic and diastolic blood pressure. In humans, a dose of 10 mg of D-amphetamine administered

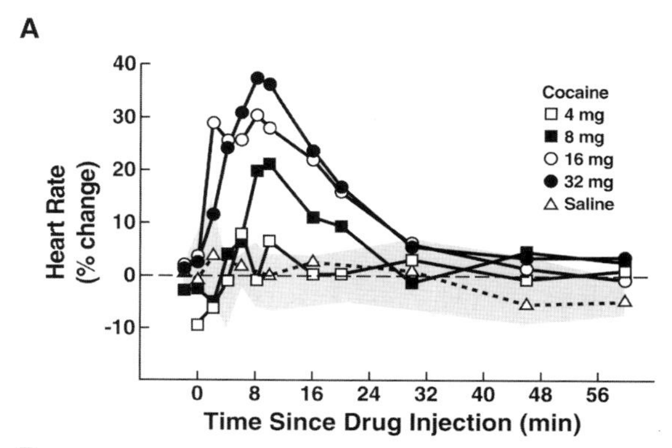

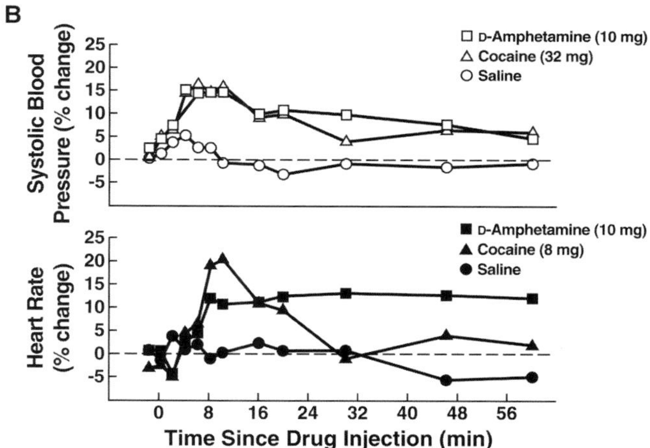

FIGURE 3.8 **(A)** Median percentage change in heart rate in humans 1 h after intravenous injection of saline or 4–32 mg cocaine. Percentage change was calculated for each dose of cocaine with reference to its own 30-min pre-drug baseline. The saline function represents data collected on Day 8 of the experimental series. The shaded region indicates the semi-interquartile range of those data. Mean pre-drug heart rate was 74 beats/min. **(B)** A comparison of the effects of intravenous cocaine and D-amphetamine on heart rate and systolic blood pressure. Median percentage change is shown after 10 mg D-amphetamine compared to 8 mg cocaine for heart rate effects and 32 mg cocaine for blood pressure effects. [Reproduced with permission from Fischman and Schuster, 1982.]

intravenously produces an increase in blood pressure equal to that produced with a dose of 32 mg of intravenous cocaine (Fischman and Schuster, 1982) (**Fig. 3.8**). Cocaine and amphetamines also stimulate heart rate, but amphetamines may cause less of an effect than one would expect based on other physiological measures because of a reflex slowing of heart rate. These drugs also produce bronchial dilation and pupillary dilation as well as decreases in glandular secretions—effects observed after activation of the sympathetic nervous system.

The mechanism of action for the autonomic effects of indirect sympathomimetics such as amphetamines and cocaine has long been known. Both drugs indirectly cause the release of norepinephrine and epinephrine by blocking both reuptake and potentiating release (Ferris *et al.*, 1972; Iversen, 1973). Chronic postganglionic adrenergic denervation, or treatment with reserpine which dramatically depletes tissue stores of catecholamines, abolishes the autonomic effects of amphetamine, establishing that these effects are due to an indirect sympathomimetic action (Trendelenburg *et al.*, 1962).

PHARMACOKINETICS

The nature of stimulant effects of cocaine and amphetamines depends on the route of administration. As noted above, intravenous or inhaled freebase preparations produce marked, intense, pleasurable sensations characterized as a 'rush' that has been likened to sexual orgasm and is thought to be a powerful motivation for the abuse of these drugs. Intravenous doses producing these subjective effects are approximately 8–16 mg of cocaine and 10 mg of D-amphetamine (Fischman and Schuster, 1982). Smoked cocaine in the freebase form has absorption characteristics similar to intravenous administration, and 50 mg of freebase produces cardiovascular effects approximately equivalent to 32 mg of cocaine administered intravenously (Foltin *et al.*, 1990). Intranasal doses of 20–30 mg of cocaine also produce euphoric and stimulant effects that last for approximately 30 min. Cocaine has less powerful effects administered orally, presumably due to a markedly slower absorption rate. Indeed, South American Indians for centuries have used an oral coca leaf preparation combined with ash to promote absorption. This use was effective as a stimulant to reduce fatigue and hunger and is not characterized by any obvious untoward physical or psychic effects (Siegel, 1985). Intranasal or oral administration of D-amphetamine in the dose range of 2.5–15 mg produces stimulant effects similar to cocaine. Subjects report feelings of alertness, energetic vitality, confident assertiveness, and a decrease in appetite and fatigue. Intranasal absorption is faster with more intense effects than oral administration, and the stimulant effects of amphetamines last considerably longer (up to 4–6 h) (Verebey and Gold, 1988) (**Table 3.5**).

Amphetamine is metabolized in the liver via deamination to phenylacetone and ultimately oxidized to benzoic acid, then excreted as glucoronide or glycine conjugates (Dring *et al.*, 1966). However, approximately 30 per cent is excreted unchanged at normal pH, accounting for a significant portion of its removal. Amphetamine has a relatively long half-life of approximately 12 h, but since it has a pKa of 9.9, that half-life can be extended with an alkaline urine to over 16 h and shortened to 8 h with acid urine (Davis *et al.*, 1971). Methamphetamine has a pKa and renal excretion similar to amphetamine. Cocaine is rapidly metabolized to benzoylecgonine and ecgonine methyl ester, both of

TABLE 3.5 Differential Effects Dependent on Routes of Cocaine Administration

Administration: Route	Administration: Mode	Initial onset of action	Duration of 'high'	Average acute dose	Peak plasma levels	Purity	Bioavailability (% absorbed)
Oral	Coca leaf chewing	5–10 mins	45–90 min	20–50 mg	150 ng/ml	0.5–1%	—
Oral	Cocaine HCl	10–30 mins	—	100–200 mg	150–200 ng/ml	20–80%	20–30%
Intranasal	Cocaine HCl	2–3 mins	30–45 min	5 × 30 mg	150 ng/ml	20–80%	20–30%
Intravenous	Cocaine HCl	30–45 s	10–20 min	25–50 mg >200 mg	300–400 ng/ml 1000–1500 ng/ml	7–100 × 58%	100%
Smoking	Coca paste	8–10 s	5–10 min	60–250 mg	300–800 ng/ml	40–85%	6–32%
	Freebase			250–1000 mg	800–900 ng/ml	90–100%	
	'Crack' cocaine			—	—	50–95%	

—, not tested. [Reproduced with permission from Verebey and Gold, 1988.]

which are pharmacologically inactive (Morishima *et al.*, 1999); less than 10 per cent is excreted unchanged in the urine (Schwartz and Oderda, 1980) (**Fig. 3.9**). The half-life of cocaine ranges from 48 to 75 min (Wilkinson *et al.*, 1980).

Methamphetamine has a mean plasma half-life of 11.1 h for the smoked route and 12.2 h for the intravenous route (Cook *et al.*, 1993). The subjective and physiological effects of methamphetamine last less than 1 h, well before marked decreases in plasma levels are observed, suggesting acute tolerance (Cook *et al.*, 1993).

ABUSE AND ADDICTION POTENTIAL

Psychostimulant Abuse Cycle

Amphetamine and cocaine have high abuse potential and now are well documented to produce addiction (Substance Dependence) by most modern definitions (Gawin and Ellinwood, 1988). While most users (85 per cent) do not become addicted to the drug (Anthony *et al.*, 1994), clinical observations indicate that controlled use often shifts to more compulsive use, either when there is increased access to the drug or when a more rapid route of administration is employed.

The natural history of a cocaine abuse pattern that includes dependence follows a trajectory that loops back on itself (Siegel, 1982) (**Fig. 3.10**). First, there is an intense *euphoria* that is enhanced by increased speed of access to the brain with routes of administration such as intravenous or smoked cocaine, followed by *dysphoria* immediately after the euphoric state (Van Dyke and Byck, 1982). The onset and intensity of the 'high' and the subsequent dysphoria are dependent on the route of administration, with a more rapid and intense high and more rapid onset of dysphoria from smoked cocaine than from either the intranasal or oral routes (Van Dyck *et al.*, 1976, 1978; Paly *et al.*, 1980). In the laboratory setting, smoking 50 mg of cocaine base in a session resulted in large transient increases in heart rate, blood pressure, and self-reported 'stimulated' subjective scores (Foltin *et al.*, 1990). Both cardiovascular and subjective effects were greater on the ascending limb than on the descending limb of the cocaine blood level function, suggesting acute tolerance (Fischman *et al.*, 1985; Foltin *et al.*, 1990; Breiter *et al.*, 1997). With chronic use, the dose required to produce euphoria increases, and the subjective high decreases. As cocaine use and duration increases, the positive reinforcing effects are diminished while the resulting dysphoria increases. In a laboratory setting, pretreatment of subjects with a large intranasal dose of cocaine blocked the subsequent arousal and mood responses

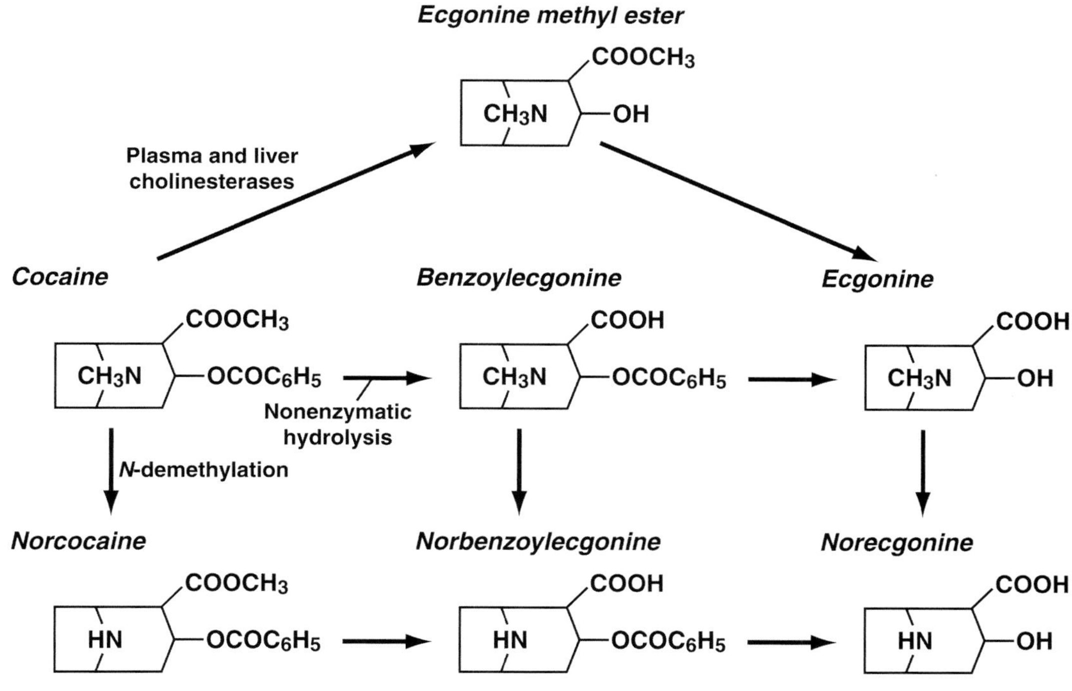

FIGURE 3.9 Cocaine metabolism. [Adapted with permission from Schwartz and Oderda, 1980.]

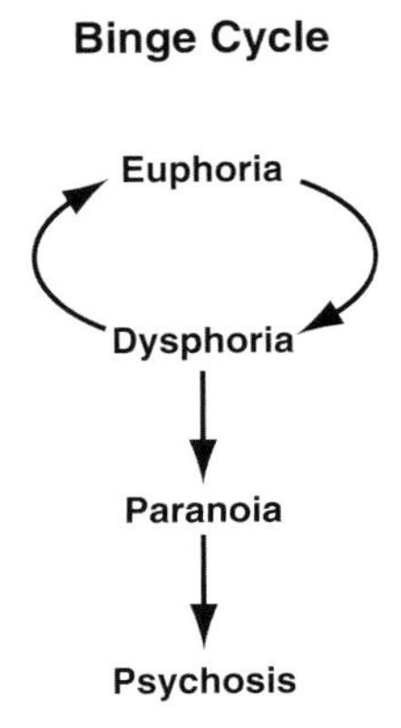

FIGURE 3.10 Different stages of the binge cycle associated with cocaine. Some individuals experience these stages in a single cocaine smoking episode with high doses; others experience them with low dosage, chronic use patterns. Stage 1 (*Euphoria*) is marked by euphoria, affective lability, increased cognitive and motor performance, hyperalertness, hyperactivity, anorexia, and insomnia. Stage 2 (*Dysphoria*) is marked by sadness, melancholia, apathy, difficulty concentrating, anorexia, and insomnia. Stage 3 (*Paranoia*) is marked by suspiciousness, paranoia (both grandiosity and persecutory), hallucinations, and insomnia. Stage 4 (*Psychosis*) is marked by anhedonia, hallucinations, stereotyped behavior, paranoid delusions, insomnia, loss of impulse control, and disorientation (see Siegel, 1982).

to intravenous cocaine (Fischman *et al.*, 1985) (**Fig. 3.11**), suggesting acute tolerance to the arousing and positive mood effects of cocaine. Compulsive use results in an exaggeration of the *binge* stage where a user characteristically re-administers the drug every 10 min for up to 7 days but usually averaging 12 h. Euphoria is replaced by dysphoria, including agitation, anxiety, and even panic attacks. Stereotyped movements such as teeth grinding and pacing may appear, as well as hyperactivity, with pressure speech and labile emotions. High doses may cause paranoia and hallucinations in some users, and some heavy users may present with psychotic symptoms similar to those of acute psychoses (e.g., paranoid schizophrenia) (Withers *et al.*, 1995). Within a binge, euphoria produces dysphoria and then more drug-taking. As this cycle continues, paranoia and psychosis ultimately develop as the dose increases or the binge lengthens (see below).

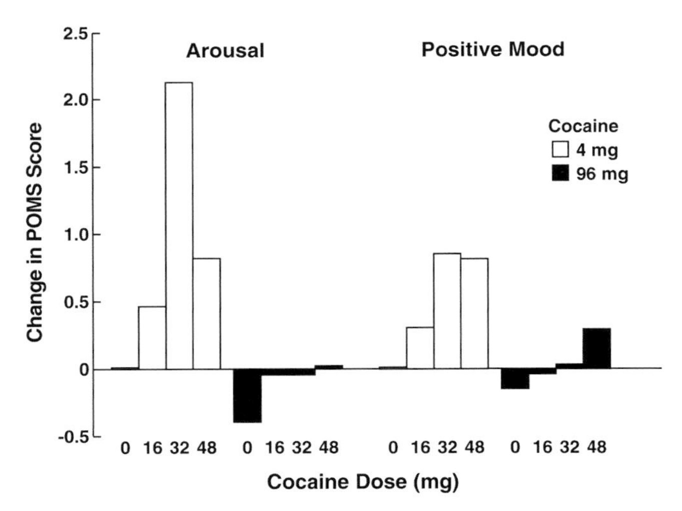

FIGURE 3.11 Mean change in scores on the arousal and positive mood factors of the Profile of Mood States questionnaire. Intravenous injection of saline or 16, 32, or 48 mg of cocaine occurred 1 h after intranasal pretreatment inhalation of 4 or 96 mg of cocaine (saline, 16 mg and 32 mg cocaine: n = 8; 48 mg cocaine: n = 4). [Reproduced with permission from Fischman *et al.*, 1985.]

Withdrawal

Withdrawal from chronic or high dose cocaine use in humans is associated with relatively few overt physical signs but a number of motivationally relevant symptoms such as dysphoria and depression, anxiety, anergia, insomnia, and craving for the drug (Gawin and Kleber, 1986; Weddington *et al.*, 1990; Satel *et al.*, 1991; Miller *et al.*, 1993). Several phases have been identified in outpatient studies of compulsive users (Gawin and Kleber, 1986; Gawin, 1991) (**Fig. 3.12**). *Phase 1* consists of a 'crash' phase which lasts up to 4 days where there is a rapid lowering of mood and energy and acute onset of both an agitated and retarded depression. Craving for the drug, anxiety and paranoia peak and then are replaced by hyperphagia and insomnia. *Phase 2* has been described as a period of prolonged dysphoria, anhedonia and lack of motivation, and increased craving that can last up to 10 weeks; relapse is highly likely during this phase. *Phase 3* is characterized by episodic craving and lasts indefinitely (Gawin, 1991). The withdrawal syndrome contributes to a vicious cycle where cessation of cocaine use leads to withdrawal symptoms, then the associated dysphoria combined with craving leads to relapse. Inpatient studies, however, do not show the three-phase patterns; instead, subjects begin with high mood-distress/craving scores which show decreases gradually and steadily over several weeks (Weddington *et al.*, 1990; Satel *et al.*, 1991) (**Fig. 3.13**) suggesting an important role for the environment in eliciting the cocaine withdrawal syndrome.

Tolerance can be defined as a given drug producing a decreasing effect with repeated dosing or when larger doses must be administered to produce the same effect (Jaffe, 1985, 1990). There is differential tolerance to psychomotor stimulants that depends on the route and frequency of administration. In humans, rapid tolerance develops to the anorexic effects and the lethal effects of amphetamine and cocaine

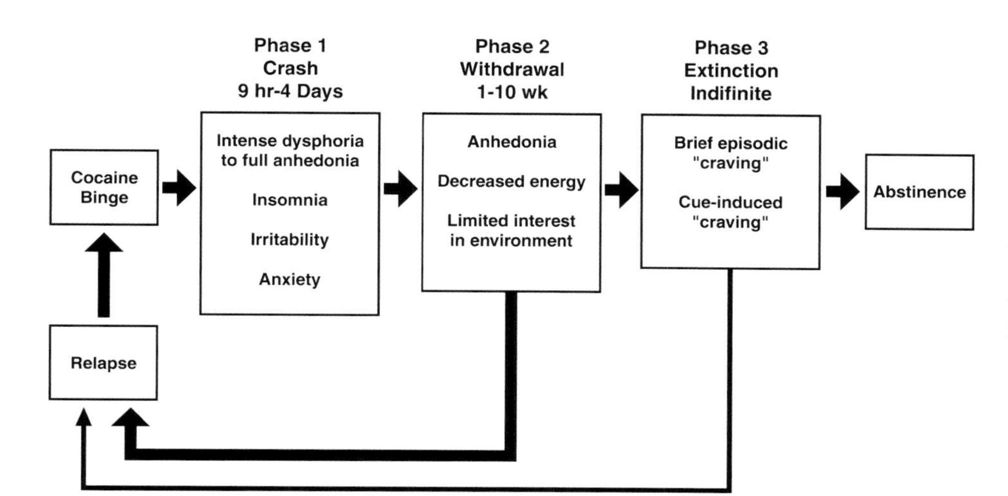

FIGURE 3.12 Phases of cocaine withdrawal following a binge. The duration and intensity of symptoms vary based on binge characteristics and diagnosis. Binges can range from under 4 h to six or more days. [Modified with permission from Gawin and Kleber, 1986.]

(Angrist and Sudilovsky, 1978; Hoffman and Lefkowitz, 1990). No tolerance or changes in sensitivity of behavioral responses was observed after repeated daily oral doses of D-amphetamine (10 mg) (Johanson *et al.*, 1983). Similarly, no tolerance developed to the subjective

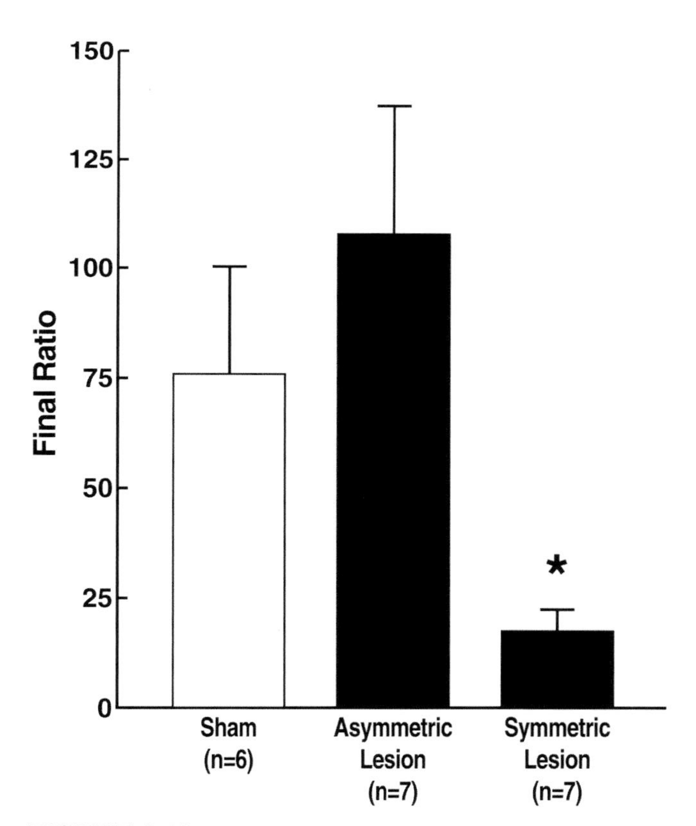

FIGURE 3.13 Mean (± SEM) scores of mood over time using the Profile of Mood States and the Beck Depression Inventory. Day 1 is the day of admission. Addicted subjects demonstrated significantly elevated scores of mood disturbances and rates of mood change over time. [Taken with permission from Weddington *et al.*, 1990.]

'high' after 10 mg of methamphetamine, but tolerance did develop to the cardiovascular effects with repeated daily oral dosing (Perez-Reyes *et al.*, 1991). Some acute tolerance appears to develop to the cardiovascular effects of cocaine even over a 4-h infusion period (Ambre *et al.*, 1988). Subjective, behavioral, and cardiovascular effects do decline after sequential oral doses of D-amphetamine despite substantial plasma levels, suggesting acute tolerance (Angrist *et al.*, 1987). Oral administration in spaced doses is thus less likely to produce tolerance to the subjective effects of psychostimulants, but intravenous or smoked administration can produce rapid acute tolerance (Van Dyke and Byck, 1982; Fischman *et al.*, 1985; Foltin *et al.*, 1990; Breiter *et al.*, 1997) (see **Fig. 3.11**). Tolerance does not develop to the stereotyped behavior and psychosis induced by stimulants, and in fact these behavioral effects appear to show a sensitization (that is, an increase with repeated administration) (Post *et al.*, 1992) (see *What is Addiction?* chapter for details on behavioral sensitization). Similar results have been observed in animal studies, with tolerance developing to the anorexic and lethal effects of amphetamine but not to stereotyped behavior (Lewander, 1974).

Tolerance, or 'apparent tolerance' (Colpaert, 1996), to the euphoric effects also can occur in subjects with one bout of cocaine administration. Human subjects show increases in the subjective sensation of intoxication during the rising phase of plasma cocaine levels after smoking coca paste, but the mood state shifts to a dramatic negative state rapidly thereafter (Van Dyke and Byck, 1982 see **Fig. 1.11**). Indeed, the mood state falls into the dysphoria zone while plasma cocaine levels are still quite high. Similar results have been seen in animal studies where a bout of 10 intravenously administered injections of cocaine results in a lowering of brain reward thresholds (reflecting increased brain reward function) immediately post-bout, but 80 injections only result in a raising of brain reward thresholds (decreased reward) immediately post-bout (Kenny *et al.*, 2003) (see *What is Addiction?* chapter).

BEHAVIORAL MECHANISMS

The behavioral mechanism of action for this book refers to a unifying principle of order and predictability at the behavioral level. Each drug class has different behavioral effects that define its phenotype or appearance. This behavioral mechanism may derive from medical use or behavioral pathology that informs medical use. High doses of amphetamines and cocaine or prolonged use/abuse can lead to significant behavioral pathology. Amphetamine abusers persist in repetitive thoughts or acts for hours. These behaviors can include repetitively cleaning the home or an item such as a car, bathing in a tub all day, elaborately sorting small objects, or endlessly dismantling or putting back together items such as clocks or radios. Termed 'punding' by Rylander (1971), this behavior was described as 'organized, goal-directed, but meaningless activity'. Also, such repetitive behavior under the influence of amphetamines and cocaine is called 'stereotyped behavior' and can be defined as 'integrated behavioral sequences that acquire a stereotyped character, being performed at an increasing rate in a repetitive manner' (Randrup and Munkvad, 1970). Stereotyped behavior is observed in many animal species (Randrup and Munkvad, 1967; Ellinwood *et al.*, 1973). For instance, monkeys will pick at their skin, exhibit mouth and tongue movements, and stare. Rats will sniff intensely in one location. Pigeons will repetitively peck at one location on a stimulus display.

Insights into the nature and behavioral mechanism of action of amphetamine-like drugs were derived from further experimental and theoretical analysis of stereotyped behavior (Lyon and Robbins, 1975) (**Fig. 3.14**). Lyon and Robbins (1975) hypothesized that as the dose of amphetamine increases, the repetition rate of all motor activities increases with the result that the organism will exhibit 'increases in response rates within a decreasing number of response categories'. This type of analysis makes a number of predictions. Complex behavioral chains or behaviors are the first to be eliminated as the response categories decrease. Behaviors capable of repetition without long pauses then dominate, and shorter and shorter response sequences result. As a result, high rates of responding in operant situations decrease and locomotor activity decreases (Segal, 1975; Rapoport *et al.*, 1980). Thus, the classic inverted U-shaped dose-response function relating amphetamines and locomotor activity (or any other high rate behavior) may reflect the competitive nature of that activity and stereotyped behavior (Robbins and Sahakian, 1979).

The inverted U-shaped function relating psychostimulant dose to performance also may be reflected in the famous behavioral pharmacological principle of 'rate dependency' (Dews, 1958; Dews and Wenger, 1977) (**Fig. 3.15**). Clear evidence in behavioral pharmacology established that the effects of a drug changes the rate of responding in a given response situation differently, depending on what the rate of responding would have been without the drug, and this relationship has come to be known as 'rate-dependency' (Dews and Wenger, 1977). One of the strong

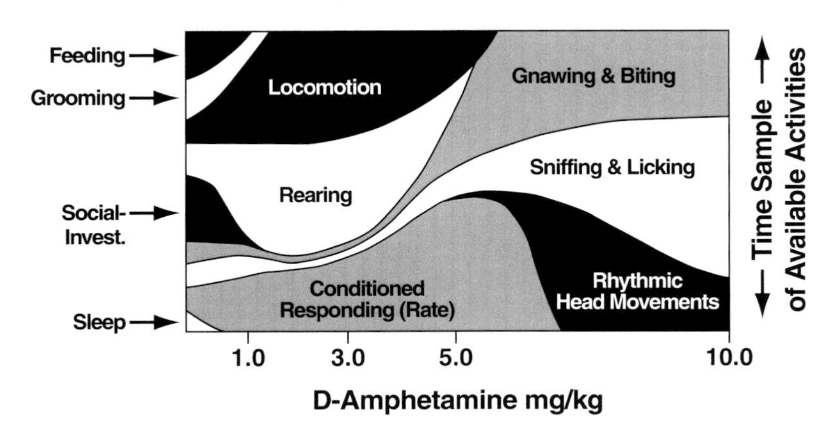

FIGURE 3.14 Schematic drawing depicting the relative distribution of varying behavioral activities within a given time sample relative to increasing doses of D-amphetamine. Note that as the dose increases, the number of activities decreases, but the rate of behavior within a given behavioral activity increases. [Reproduced with permission from Lyon and Robbins, 1975.]

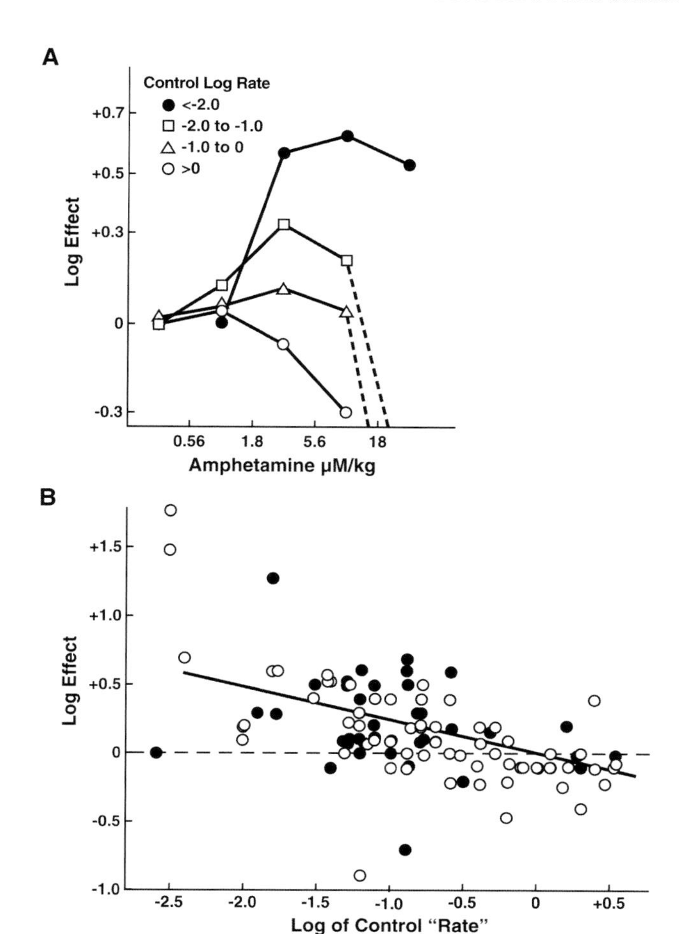

FIGURE 3.15 **(A)** Dose-effect curves for amphetamine in rats on different control rates of responding. ●, Control log rate < −2.0 (less than 0.01 responses/s). □, Control log rate between −2.0 and −1.0 (between 0.01 and 0.1 responses/s). △, Control log rate between −1.0 and 0 (between 0.1 and 1.0 responses/s). ○, Control log rate more than 0 (more than 1 response/s). Note that at control rates of less than 0.1 responses/s, large increases occurred following amphetamine and continued up to doses greater than 18 μM/kg (about 3 mg/kg), while at control rates greater than 1.0 responses/s, essentially no increase was seen and a clear decline was seen at doses in the range 1.8–5.6 μM/kg (0.3–1.0 mg/kg). **(B)** Scatterplot showing the relation between control log rate and log effect for doses from 1.8–5.59 μM/kg (0.3–1.0 mg/kg). ○, Schedule-controlled key operations. ●, All other behavioral situations. Weighted regression line is $y = -0.22x + 0.03$. [Reproduced with permission from Dews and Wenger, 1977.]

propositions associated with rate-dependency is that general differences in the rate of responding will determine differences in the effects of a drug. A second proposition is that the log rate of responding under the influence of a drug is a linear function of the control rate (Ritz *et al.*, 1987). While a more radical view of rate dependency is that the control rate of responding may be the sole determinant of the behavioral effects of any drug (Dews and Wenger, 1977), it is clear that with stimulant drugs high rates of responding are decreased and low rates of responding are increased with administration of psychostimulants, and this effect generalizes to a broad number of behavioral situations. Indeed, some aspects of the behavioral principle for stimulants outlined above, where the increasing rates of behavior combine with a decreasing number of response categories as outlined by Lyon and Robbins (1975), can be considered a form of rate dependency. Clearly, in rodents with psychostimulants, as the dose increases, high rates of behavior in operant situations decrease and locomotor activity decreases, and head bobbing and other forms of stereotyped behavior that have an initial low frequency of behavior increase (Lyon and Robbins, 1975).

A possible understanding of how amphetamines produce paranoid ideation and psychosis extends from this analysis of actual overt motor effects to the actions of these drugs on cognitive function. Amphetamines are well documented to produce paranoid psychotic episodes in individuals abusing chronically, or even taking large doses acutely (Chapman, 1954; Askevold, 1959; Johnson and Milner, 1966; Kramer *et al.*, 1967; Angrist and Sudilovsky, 1978). In a study of nine physically healthy volunteers who had been previously administered large doses of amphetamine (and three of whom had previous amphetamine psychoses), repetitive oral administration of 5–10 mg of D-amphetamine produced paranoid delusions, often with blunted affect in all subjects when a cumulative dose range of 55–75 mg was reached (Griffith *et al.*, 1972) (**Fig. 3.16**). The psychoses also can develop during withdrawal from an abuse cycle of amphetamine (Askevold, 1959) (**Tables 3.6** and **3.7**). This paranoid psychosis induced by stimulants in its most severe form can produce actual physical toxicity where subjects believe that bugs under their skin need to be gouged out (also referred to as 'crank bugs'). Such stereotyped behavior and psychosis associated with high-dose use of stimulants also may contribute to the cycle of abuse associated with compulsive use of these drugs.

NEUROBIOLOGICAL MECHANISM—NEUROCIRCUITRY

Acute Reinforcing and Stimulant Effects of Psychostimulants

Indirect sympathomimetics, such as amphetamine and cocaine, are known to act neuropharmacologically to

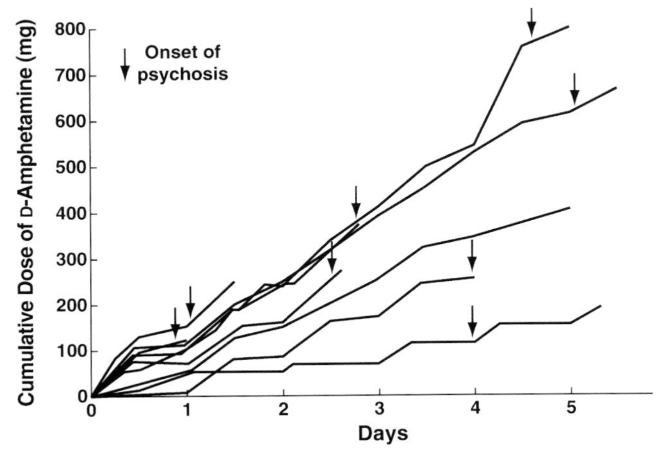

FIGURE 3.16 Onset of psychosis after cumulative administration of D-amphetamine. Individuals were utilized for this study who had previously self-administered large doses of amphetamine without lasting sequelae. Drug-naive subjects were not used. Seven subjects were selected, aged 21 to 37. All subjects were hospitalized before the study on a psychiatric ward for a minimum of six weeks to ensure a drug-free and alcohol-free interval. At the end of this control period, each subject received 10 mg of D-amphetamine intravenously. Subsequent doses of 5–10 mg D-amphetamine were administered orally every hour if it could be tolerated by the subject. This procedure was designed to effect accumulation of the drug within a period of days. The effects of D-amphetamine on the psychological states of the volunteers were determined by (1) tape-recorded interviews rated for depression and paranoid symptoms, (2) narrative descriptions by psychiatrists maintaining continuous observation of the patient, (3) evaluation by one psychiatrist who interviewed each patient four times per day but was not blind to the drug dosage, (4) a symptom checklist of questions designed to quantify changes in affect and presence of paranoid thinking, (5) retrospective descriptions of the psychosis by the subjects, (6) clinical psychological tests which included the Holtzman Inkblot Technique, Ravens Progressive Matrices, House-Tree-Person Drawings, Bender Gestalt, and Tien Organic Integrity Test. [Reproduced with permission from Griffith *et al.*, 1972.]

enhance the amount of monoamines available within the synaptic cleft of monoamine synapses in the central nervous system (Glowinski and Axelrod, 1965; Ferris *et al.*, 1972; Iversen, 1973; Raiteri *et al.*, 1975; Taylor and Ho, 1978; Iversen and Fray, 1982). Amphetamine and cocaine block the reuptake of norepinephrine, dopamine, and serotonin in the central nervous system (Glowinski and Axelrod, 1965; Ferris *et al.*, 1972; Iversen, 1973; Ritz *et al.*, 1987; Matecka *et al.*, 1996; Rothman *et al.*, 2001) (**Table 3.8**). Cocaine has a rank order of potency in blocking monoamine reuptake of *serotonin > dopamine > norepinephrine*, whereas methamphetamine and D-amphetamine have a rank order of potency in blocking monoamine reuptake of *norepinephrine ≥ dopamine > serotonin* (Rothman *et al.*, 2001). Amphetamine also enhances release of norepinephrine, dopamine and serotonin (Glowinski and Axelrod, 1965; Raiteri *et al.*, 1975). Cocaine is thought to enhance release secondary to a blockade of reuptake (Moore *et al.*, 1977). Amphetamine also is a weak monoamine oxidase inhibitor (Robinson, 1985). However, the primary neuropharmacological action responsible for their psychomotor stimulant and reinforcing effects appears to be on the dopamine systems in the central nervous system.

Brain dopamine neurons are organized into major pathways that originate in the midbrain and project to numerous forebrain and cortical regions. Some projections in particular appear to be responsible for different aspects of psychomotor stimulant actions. The mesocorticolimbic dopamine system originates in the ventral tegmental area and projects to the ventral forebrain, including the nucleus accumbens, olfactory tubercle, septum, and frontal cortex. The nigrostriatal dopamine system arises primarily in the substantia nigra and projects to the corpus striatum.

The midbrain dopamine systems long have been associated with motor function and response initiation and also are responsible for the psychostimulant actions of cocaine and amphetamine (**Fig. 3.17**). Degeneration or destruction of the nigrostriatal and mesolimbic dopamine systems together result in the severe motor disturbances of Parkinson's disease, including tremor, dystonic involuntary movements, and akinesia (De Long, 1990). Large bilateral lesions of the midbrain dopamine system using a selective neurotoxin for dopamine, 6-hydroxydopamine, can reproduce many of these deficits. Rats become akinetic to the point of aphagia and adipsia and will die unless intubated (Ungerstedt, 1971). These rats also have severe deficits in learning a conditioned avoidance task, and these deficits can be reversed with L-DOPA treatment (Zis *et al.*, 1974).

Destruction of the mesocorticolimbic dopamine system separately with 6-hydroxydopamine blocks amphetamine- and cocaine-stimulated locomotor activity (Kelly *et al.*, 1975; Kelly and Iversen, 1976; Roberts *et al.*, 1980; Joyce and Koob, 1981). Similar effects have been observed following microinjection of selective dopamine antagonists into the region of the nucleus accumbens (Pijnenburg *et al.*, 1975).

In contrast, disruption of function in the nigrostriatal system blocks the stereotyped behavior associated with administration of high doses of D-amphetamine (Creese and Iversen, 1974; Kelly and Iversen, 1976; Iversen, 1977). When 6-hydroxydopamine lesions are restricted to the striatum itself (Koob *et al.*, 1984), such lesions block the intense, restricted, repetitive behavior produced by high doses of amphetamine, and this results in intense locomotor activity (Koob *et al.*, 1984)

TABLE 3.6 Symptoms of Abstinence Psychoses in Abusers of Barbiturates, Alcohol, and Morphine Compared to Amphetamine

	Barbiturates	Alcohol	Morphine	Amphetamine (Case No.)			
				#2	#6	#4	#9
Vegetative symptoms							
Tremors	+	++	++	0	0	0	++
Perspiration	++	+	++	0	+	0	+
Nausea	+	+	++	0	0	0	0
Retching	+	+	+	0	0	0	0
Diarrhea	+	+	++	0	+	0	+
Sleeplessness	+	+	++	++	++	++	++
Convulsions	++	+	+	0	+	0	+
Psychomotor symptoms							
Motor hyperactivity	+	+	+	++	++	++	++
Motor weakness	++	+	++	0	0	0	+
Unrest	+	+	+	++	++	+	++
Aggression	+	+	+	++	++	++	++
Restlessness	+	+	++	++	++	++	++
Psychic symptoms							
Visual hallucinations	++	++	+	++	++	+	++
Auditory hallucinations	+	+	+	++	++	+	++
Confusion	++	++	+	++	++	++	++
Suspicion	+	+	+	++	++	+	++
Persecutory delusions	+	+	+	+	++	+	++
Anxiety	++	++	++	0	0	0	0

0, not present
+, usually present, slightly pronounced
++, present, strongly pronounced
[Reproduced with permission from Askevold, 1959.]

(**Fig. 3.18**). Such striatal lesions do not block the reinforcing effects of cocaine. Subregions of the corpus striatum have been implicated in the stereotyped behavior produced by amphetamine (Kelley *et al.*, 1989). Amphetamine injected into the ventrolateral striatum of rats produced licking, biting and self-gnawing to the exclusion of other psychomotor behaviors.

TABLE 3.7 Characteristic Syndrome for the Abstinence Delirium of Amphetamine Addicts

1. Delirium is characterized by confusion and hallucination.
2. There is pronounced motor activity with an increase in the quantity of movement but not in the speed. The muscular strength is reduced only slightly or not at all.
3. The activity is unceasing, day and night.
4. There is no sign of open anxiety.
5. There are few or no vegetative symptoms apart from sleeplessness.
6. There is a comparatively lengthy period of latency (3–10 days) between the beginning of the withdrawal and the development of the delirium.
7. The abstinence psychosis seems to have a longer duration.

[Reproduced with permission from Askevold, 1959.]

Thus, the terminal regions of the nigrostriatal and mesocorticolimbic dopamine systems appear to mediate different aspects of psychomotor stimulant actions that can have significant implications for the behavioral effects and the psychopathology associated with stimulant abuse.

Neurotoxin-selective lesions of the mesocorticolimbic dopamine system block the reinforcing effects of cocaine (Roberts *et al.*, 1977, 1980; Koob *et al.*, 1987b) (**Fig. 3.19**). Rats trained to self-administer cocaine or amphetamine intravenously and subjected to a 6-hydroxydopamine lesion of the nucleus accumbens show an extinction-like response pattern (high levels of responding at the beginning of each session and a gradual decline in responding over sessions) and a long-lasting decrease in responding. Neurotoxin-selective lesions of the mesocorticolimbic dopamine system in the nucleus accumbens also block the reinforcing effects of D-amphetamine (Lyness *et al.*, 1979).

Changes in cocaine self-administration also have been observed with serotonin antagonists, reuptake blockers, and lesions, with evidence for both facilitation and inhibition of psychostimulant reinforcement

TABLE 3.8 Pharmacological Profile of (+)Amphetamine, (+)Methamphetamine and Cocaine in Norepinephrine, Serotonin, and Dopamine Release and Reuptake Assays

	Norepinephrine	Serotonin	Dopamine
Release—IC_{50} (nM ± SD)			
(+)Amphetamine	7.07 ± 0.95	1765 ± 94	24.8 ± 3.5
(+)Methamphetamine	12.3 ± 0.7	4640 ± 243	24.5 ± 2.1
Cocaine	>10 000	>10 000	>10 000
Reuptake—K_i (nM ± SD)			
(+)Amphetamine	38.9 ± 1.8	3830 ± 170	34 ± 6
(+)Methamphetamine	48.0 ± 5.1	2137 ± 98	114 ± 11
Cocaine	779 ± 30	304 ± 100*	478 ± 250*

*Data from Matecka *et al.*, 1996.
[Data taken with permission from Rothman *et al.*, 2001.]

(Porrino *et al.*, 1989; Carroll *et al.*, 1990; Loh and Roberts, 1990). Depletion of tryptophan, the serotonin precursor in humans, attenuates the euphorigenic effects of intranasal cocaine (Aronson *et al.*, 1995). Receptor subtypes in specific brain sites may provide clearer insight into serotonergic participation in psychostimulant reinforcement (see below).

The enhancing effects of stress on amphetamine and cocaine self-administration have been described for decades, and many of these effects are related to

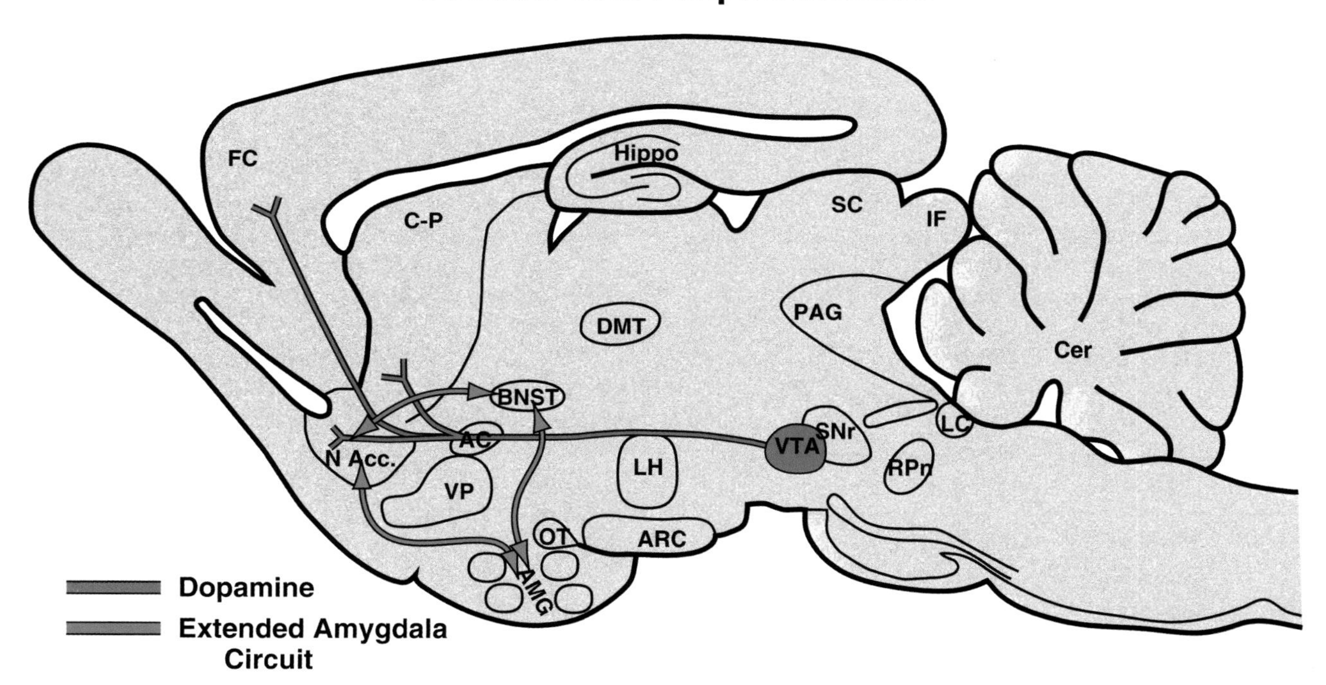

FIGURE 3.17. Sagittal section through a representative rodent brain illustrating the pathways and receptor systems implicated in the acute reinforcing actions of cocaine and amphetamines. Cocaine and amphetamines activate release of dopamine in the nucleus accumbens and amygdala via direct actions on dopamine terminals. The blue arrows represent the interactions within the extended amygdala system hypothesized to have a key role in psychostimulant reinforcement. AC, anterior commissure; AMG, amygdala; ARC, arcuate nucleus; BNST, bed nucleus of the stria terminalis; Cer, cerebellum; C-P, caudate-putamen; DMT, dorsomedial thalamus; FC, frontal cortex; Hippo, hippocampus; IF, inferior colliculus; LC, locus coeruleus; LH, lateral hypothalamus; N Acc., nucleus accumbens; OT, olfactory tract; PAG, periaqueductal gray; RPn, reticular pontine nucleus; SC, superior colliculus; SNr, substantia nigra pars reticulata; VP, ventral pallidum; VTA, ventral tegmental area.

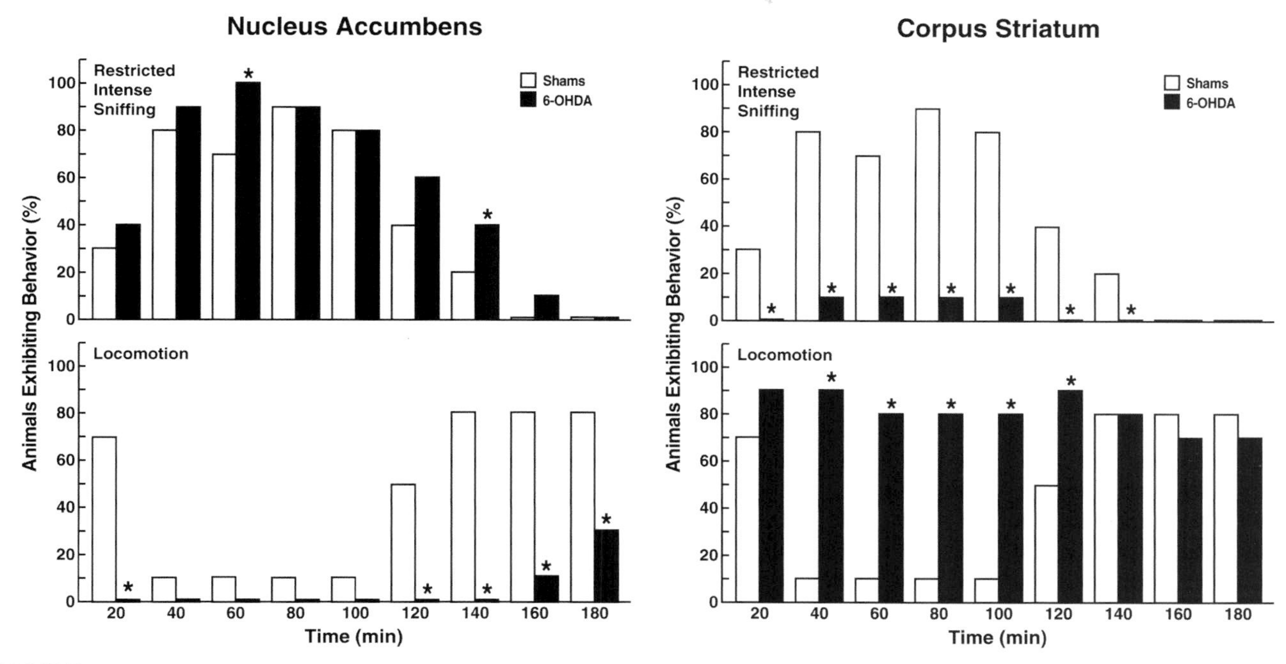

FIGURE 3.18 Rats receiving 6-hydroxydopamine (6-OHDA) lesions to the nucleus accumbens and corpus striatum showed an altered amphetamine response that depended on the motor behavior measured and the dose of D-amphetamine tested. Rats showed a significant blockade of the stimulation of locomotion following treatment with 1.0 mg/kg of D-amphetamine. However, the intense sniffing in one place produced by the 4 mg/kg dose was not significantly decreased, but was actually increased at two time points. In contrast, the rats receiving 6-OHDA lesions to the corpus striatum showed exactly the opposite result. Locomoter activity at the low dose of amphetamine was unaltered, but dramatically increased at the high dose compared to shams. This increased locomotion was a result of virtually complete attenuation of the high dose stereotyped behavior of "Restricted Intense Sniffing." Asterisks (*) indicate statistically significant from sham group ($p < 0.05$, Information statistic). [Data from Koob *et al.*, 1984.]

glucocorticoid release. Further, these effects have been observed for different doses of drugs, during the acquisition phase, during reinstatement, and in motivational measures such as progressive-ratio schedules of reinforcement. Stress, through activation of the hypothalamic-pituitary-adrenal (HPA) axis and the release of glucocorticoids, influences various regions of the brain including dopamine neurons (Piazza and Le Moal, 1996, 1997, 1998) that express corticosteroid receptors (Harfstrand *et al.*, 1986). In normal situations, glucocorticoids state-dependently increase dopaminergic function, especially in mesolimbic regions, during various consummatory behaviors exhibited in the rodents' active period of the light/dark cycle and also in animals shown to self-administer stimulants (Piazza *et al.*, 1996). The interaction of glucocorticoids with the mesolimbic dopamine system may have a significant impact on vulnerability to self-administer psychostimulant drugs. Rats with an initial high sensitivity for exploration in a novel environment with a high initial corticosterone response are much more likely to self-administer psychostimulant drugs (Piazza and Le Moal, 1996, 1997). In addition, rats receiving repeated injections of corticosterone acquire cocaine selfadministration at a lower dose of cocaine than do rats that are administered vehicle (Mantsch *et al.*, 1998), and corticosterone administration causes rats that would not self-administer amphetamine at low doses to self-administer amphetamine (Piazza *et al.*, 1991).

Conversely, adrenalectomy tends to suppress cocaine self-administration in rats (Piazza *et al.*, 1996) (**Fig. 3.20**). Glucocorticoid hormones and stimulants interact in part at the same cellular levels, particularly the shell of the nucleus accumbens (Barrot *et al.*, 2000), and animals, especially those that react more to stimulants, self-administer glucocorticoids in the same way as for cocaine and amphetamine (Piazza *et al.*, 1993). These results suggest that glucocorticoids may be one of the biological factors determining vulnerability to substance use (Deroche *et al.*, 1997).

Functional studies support the hypothesis that a medial part of the ventral forebrain, described as the extended amygdala, may further delineate the neurobiological substrates of psychostimulant reinforcement. The extended amygdala (Alheid and Heimer, 1988) has been conceptualized to be composed of several basal forebrain structures: the bed nucleus of the stria terminalis, centromedial

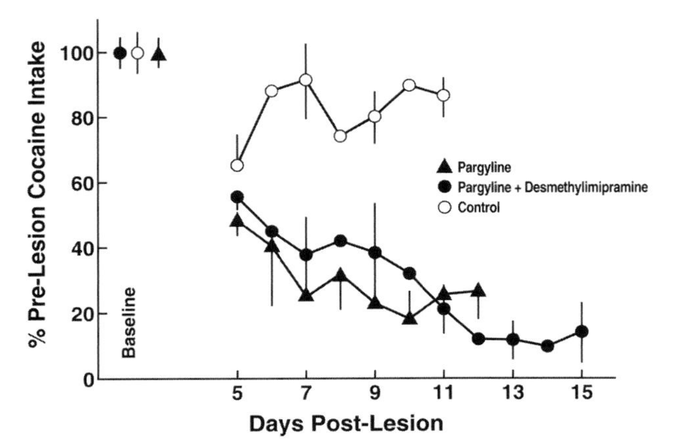

FIGURE 3.19 Effect of 6-hydroxydopamine injections into the nucleus accumbens on cocaine self-administration in rats. Points represent mean (± SEM) daily intake for each group. One group received the monoamine oxidase inhibitor pargyline (50 mg/kg) prior to 6-hydroxydopamine treatment (filled triangles). A second group received both pargyline and the tricyclic antidepressant desmethylimipramine (25 mg/kg) prior to 6-hydroxydopamine (filled circles). The control group received pargyline and desmethylimipramine prior to vehicle infusions into the nucleus accumbens. Cocaine was not available for self-administration under Day 5 post-lesion. Repeated measures analysis of variance revealed a significant difference between the two lesion groups and the control group ($p < 0.01$). No difference was observed between the 6-hydroxydopamine groups. [Reproduced with permission from Roberts *et al.*, 1980.]

amygdala, medial part of the nucleus accumbens (e.g., shell) (Heimer and Alheid, 1991), and sublenticular substantia innominata. Evidence for the extended amygdala construct includes similarities between these structures in morphology, immunohistochemistry, connectivity, and functionality (Alheid and Heimer, 1988). Afferent connections to the extended amygdala complex include mainly limbic regions, while efferent connections from this complex include medial aspects of the ventral pallidum and substantia innominata, as well as a considerable projection to the lateral hypothalamus (Heimer *et al.*, 1991) (see *Neurobiological Theories of Addiction* chapter).

In a series of studies using intravenous self-administration of cocaine under baseline conditions and with progressive-ratio schedules, differential effects were observed using neurotoxin-specific lesions of terminal areas versus microinjection of a dopamine D_1 receptor antagonist (McGregor and Roberts, 1993, 1995; McGregor *et al.*, 1994, 1996). Neurotoxin lesions of the central nucleus of the amygdala and the medial prefrontal cortex facilitated responding on the progressive-ratio schedule (e.g., increased cocaine reinforcing action). In contrast, local intracerebral injections of the D_1 receptor antagonist had opposite effects, decreasing progressive-ratio performance when administered into the central nucleus of the amygdala, medial prefrontal cortex, and nucleus accumbens, with the greatest effects in the nucleus accumbens, followed by the medial prefrontal cortex and the amygdala (McGregor and Roberts, 1993) (**Fig. 3.21**). Similar results were observed on baseline self-administration with microinjections of the D_1 antagonist into the central nucleus of the amygdala, bed nucleus of the stria terminalis and nucleus accumbens (Caine *et al.*, 1995; Epping-Jordan *et al.*, 1998). In addition, lesions of the posterior medial ventral pallidum are particularly

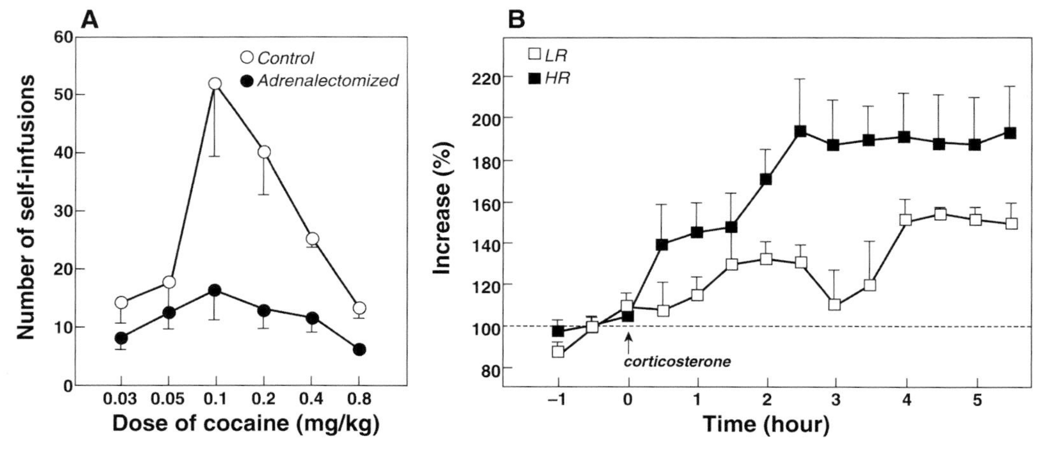

FIGURE 3.20 (A) The effects of adrenalectomy on cocaine self-administration in rats. Animals were trained to self-administer cocaine by nose-poking and subjected to a dose-effect function. Adrenalectomy produced a flattening of the dose-effect function, with decreases of cocaine intake at all doses. [Modified with permission from Deroche *et al.*, 1997.] (B) Corticosterone-induced changes in extracellular concentrations of dopamine in high-responding (HR) and low-responding (LR) animals. HR animals that drank the corticosterone solution (100 mg/ml) in the dark period showed a faster and higher increase in nucleus accumbens dopamine than LR animals. [Reproduced with permission from Piazza *et al.*, 1996.]

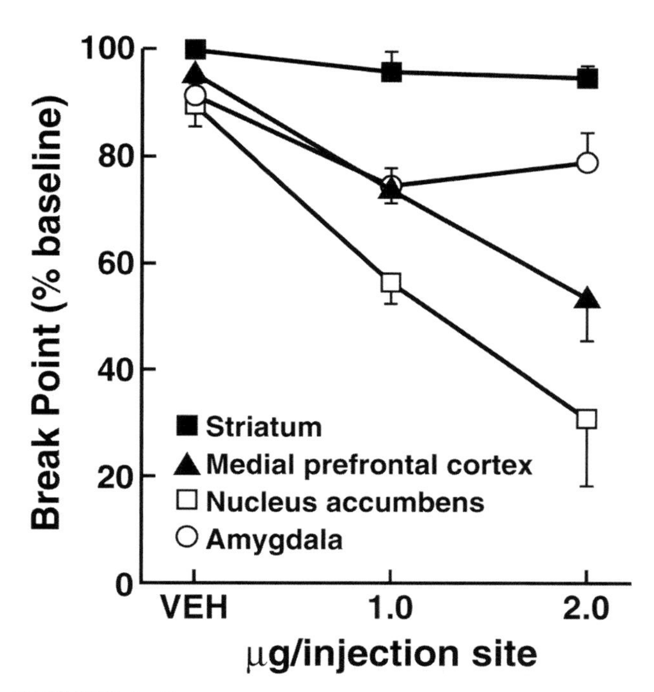

FIGURE 3.21 Mean (± SEM) percent baseline break point in a progressive-ratio schedule of reinforcement in rats following injection of SCH 23390 (1.0 and 2.0 μg/side) or vehicle into the striatum, medial prefrontal cortex, nucleus accumbens, and amygdala (n = 8 per site). [Redrawn with permission. Striatum and medial prefrontal cortex data from McGregor and Roberts, 1995. Nucleus accumbens and amygdala data from McGregor and Roberts, 1993.]

effective in blocking the motivation to work for intravenous cocaine (Hubner and Koob, 1990; Robledo and Koob, 1993). Also, data with *in vivo* microdialysis show that acute administration of psychostimulants preferentially enhance dopamine release in the shell of the nucleus accumbens (Pontieri *et al.*, 1995).

These results suggest that certain neurochemical elements of the projections to the extended amygdala may be important for the acute positive reinforcing effects of psychostimulants. Clearly, the mesocorticolimbic dopamine system is critical for psychomotor stimulant activation and psychomotor stimulant reinforcement and plays a role in the reinforcing actions of other drugs. However, as has been discussed by others (Le Moal and Simon, 1991), the functions of the mesocorticolimbic dopamine system may be determined largely by its specific innervations and not by any intrinsic functional attributes. The heterogeneity of connectivity within subregions of the nucleus accumbens suggests that the medial nucleus accumbens (shell), together with the rest of the extended amygdala, may provide a critical link between the terminals of the mesocortical dopamine system and other forebrain circuitry involved in psychostimulant reinforcement.

Psychostimulant Withdrawal and Dependence

Cocaine withdrawal in animal studies has shown significant behavioral changes but few 'physical' symptoms. Unlimited access to intravenous self-administration of cocaine produced decreases in ingestion of a glucose (3% w/v) + saccharin (0.125% w/v) solution during withdrawal (Carroll and Lac, 1987) (**Fig. 3.22**). Rebound, prolonged hyperthermia and hypoactivity have been observed after administration of 32 mg/kg of cocaine either intraperitoneally or subcutaneously (Gauvin

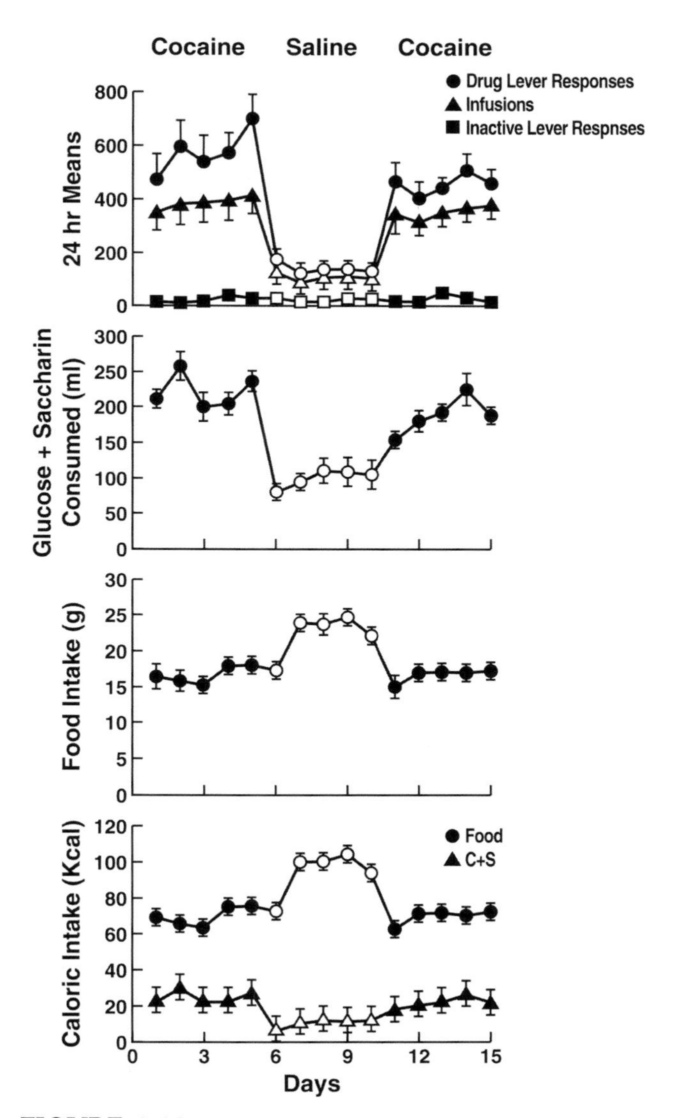

FIGURE 3.22 Mean (± SEM) lever responses and infusions, glucose and saccharin consumed, food intake, and caloric intake over 15 successive days for five rats when cocaine infusions (filled symbols; 0.2 mg/kg) or saline infusions (open symbols; 0.15 ml) were available for intravenous self-administration. [Reproduced with permission from Carroll and Lac, 1987.]

et al., 1997). Others have observed decreases in locomotor activity, decreases in conditioned avoidance, and increases in anxiety-like responses during cocaine withdrawal (Harris and Aston-Jones, 1993; Fung and Richard, 1994; Sarynai *et al.*, 1995; Basso *et al.*, 1999).

Studies of repeated daily administration of cocaine (40 mg/kg, i.p., for 7 days) showed increased brain reward thresholds for up to 7 days post-cocaine (Kokkinidis and McCarter, 1990). Curiously, decreased reward thresholds were observed 20 h post-cocaine using a different chronic cocaine schedule (30 mg/kg, i.p.) twice daily (Kokkinidis and McCarter, 1990). During an acute binge of intravenous self-administration in rats with electrodes in the medial forebrain bundle and tested for brain reward thresholds, there was a dose-dependent elevation in reward thresholds when the self-administration session was terminated (Markou and Koob, 1992) (**Fig. 3.23**). Animals were allowed to self-administer cocaine intravenously for different periods of time, and when the session was terminated brain reward thresholds were measured for up to 144 h post-cocaine. Animals that had access to cocaine for 48 h showed elevations in thresholds that lasted for 5 days post-cocaine. Similar results have been observed with D-amphetamine (Paterson *et al.*, 2000). While these elevations in threshold were robust, it was not clear if they have motivational significance. To test this hypothesis, brain reward thresholds were measured during the development of escalation in drug intake observed previously with extended access to the drug (Ahmed and Koob, 1998). Two groups of rats were differentially exposed to cocaine self-administration (i.e., 1 h short-access, ShA; 6 h long-access, LgA), and brain reward thresholds were measured just prior and just after each self-administration session.

Historically in animal models of stimulant self-administration, the focus was restricted to stable behavior from day to day to reliably interpret within-subject designs aimed at exploring the pharmacological and neuropharmacological basis for the acute reinforcing effects of cocaine. Typically, rats allowed access to less than 3 h of cocaine per day, establish highly stable levels of intake and patterns of responding between daily sessions after acquisition of self-administration. To explore the possibility that differential access to intravenous cocaine self-administration in rats may produce different patterns of drug intake, rats were allowed access to intravenous self-administration of cocaine for 1 h (ShA) and 6 h (LgA) per day. In ShA rats, drug intake remained low and stable, not changing from day to day as observed previously. In contrast, in LgA rats, drug intake gradually escalated over days (Ahmed and Koob, 1998) (see *Animal models of Addiction* chapter). In the escalation group, there was

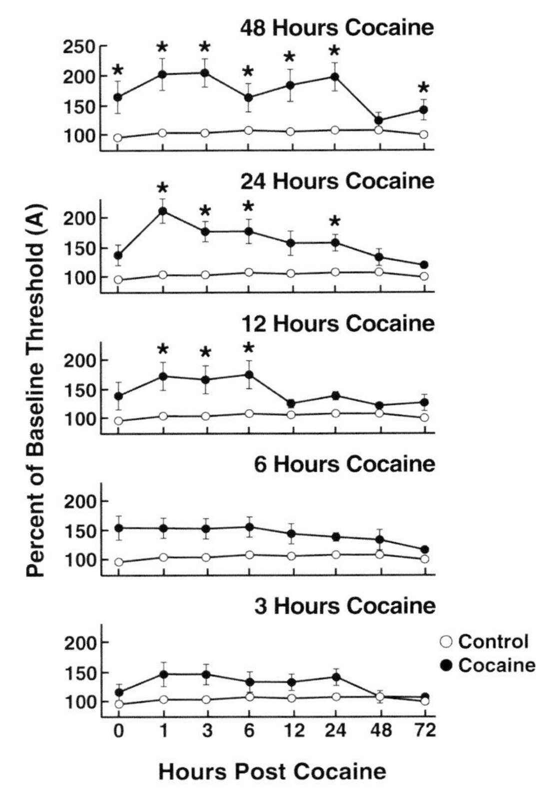

FIGURE 3.23 Intracranial self-stimulation thresholds in rats following 3–48 h of cocaine self-administration at several time points post-cocaine (0, 1, 3, 6, 12, 24, 48, and 72 h). Data are expressed as percent change from baseline threshold levels. The mean ± SEM baseline threshold for the experimental group was 37.414 ± 2.516 μA and for the control group 35.853 ± 3.078 μA. Asterisks (*) indicate statistically significant differences between control and experimental groups with Dunnett's tests, following a significant Group × Hours interaction in the analysis of variance ($p < 0.05$). [Reproduced with permission from Markou and Koob, 1991].

an increased early intake as well as sustained intake over the session and an upward shift in the dose-response function, suggesting an increase in hedonic set point (Ahmed and Koob, 2005). When animals were allowed different doses of cocaine during self-administration, the LgA rats titrated cocaine effects as well as the ShA rats, but the LgA rats consistently self-administered almost twice as much cocaine at any dose tested, further suggesting an upward shift in the set point for cocaine reward in the escalated animals (Ahmed and Koob, 1999).

Elevation in baseline intracranial self-stimulation (ICSS) reward thresholds temporally preceded and was highly correlated with escalation in cocaine intake (Ahmed *et al.*, 2002) (**Fig. 3.24**). Further observation revealed that post-session elevations in ICSS reward thresholds failed to return to baseline levels before the

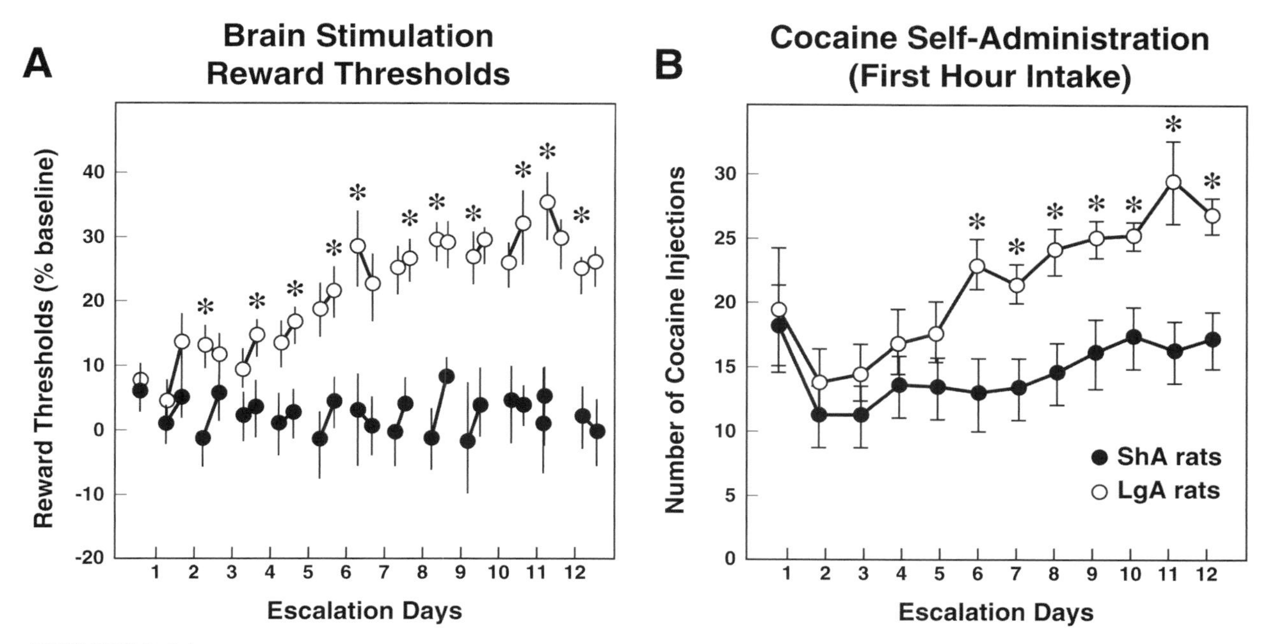

FIGURE 3.24 Relationship between elevation in intracranial self-stimulation (ICSS) reward thresholds and cocaine intake escalation. **(A)** Percentage change from baseline ICSS thresholds. **(B)** Number of cocaine injections earned during the first hour of each session. Rats first were prepared with bipolar electrodes in either the right or left posterior lateral hypothalamus. One week post-surgery, they were trained to respond for electrical brain stimulation. ICSS thresholds measured in microamperes were assessed according to a modified discrete-trial current-threshold procedure (Markou and Koob, 1993). During the screening phase, the 22 rats tested for self-administration were allowed to self-administer cocaine during only 1 h on a fixed-ratio 1 schedule, after which two balanced groups with the same weight, cocaine intake, and ICSS reward thresholds were formed. During the escalation phase, one group had access to cocaine self-administration for only 1 h per day (short-access or ShA rats) and the other group for 6 h per day (long-access or LgA rats). The remaining 8 rats were exposed to the same experimental manipulations as the other rats, except that they were not exposed to cocaine (not shown). ICSS reward thresholds were measured in all rats two times a day, 3 h and 17–22 h after each daily self-administration session (ShA and LgA rats) or the control procedure (drug-naive rats; not shown). Each ICSS session lasted about 30 min. *$p < 0.05$ compared to drug-naive and/or ShA rats, tests of simple main effects. [Taken with permission from Ahmed *et al.*, 2002.]

onset of each subsequent self-administration session, thereby deviating more and more from control levels. The progressive elevation in reward thresholds was associated with a dramatic escalation in cocaine consumption in LgA rats as previously observed. After escalation had occurred, an acute cocaine challenge failed to facilitate brain reward responsiveness to the same degree as before escalation. These results show that the elevation in brain reward thresholds following prolonged access to cocaine failed to return to baseline levels between repeated, prolonged exposure to cocaine self-administration (i.e., residual hysteresis), thus creating a greater and greater elevation in baseline ICSS thresholds. These data provide compelling evidence for brain reward dysfunction in escalated cocaine self-administration.

The neurochemical basis for the motivational withdrawal associated with cocaine and amphetamines involves not only a decrease in dopamine function but also recruitment of anti-reward and brain stress neurochemical function. Dopamine release in the basal forebrain (e.g., nucleus accumbens), as measured by *in vivo* microdialysis, is decreased during withdrawal (Weiss *et al.*, 1992; Maisonneuve *et al.*, 1995; Parsons *et al.*, 1995) (**Figs. 3.25** and **3.26**). Studies using treatments of daily bolus injections of cocaine over periods typically ranging from 1–3 weeks are consistent with an impairment of dopamine function, including decreases in dopamine synthesis (Trulson and Ulissey, 1987). Serotonin release in the nucleus accumbens also is decreased during withdrawal from a 'binge' of self-administration of cocaine (Weiss *et al.*, 1995; Parsons *et al.*, 1995).

Intermittent cocaine exposure (limited access of daily injections of cocaine) with *in vivo* microdialysis studies have revealed a more complex interaction with nucleus accumbens dopamine that depends on dose, duration, and frequency of cocaine pretreatment, as well as the abstinence interval before measurement with intermittent exposure. Extracellular levels of dopamine in the nucleus accumbens appear to be increased during the early cocaine stage following daily

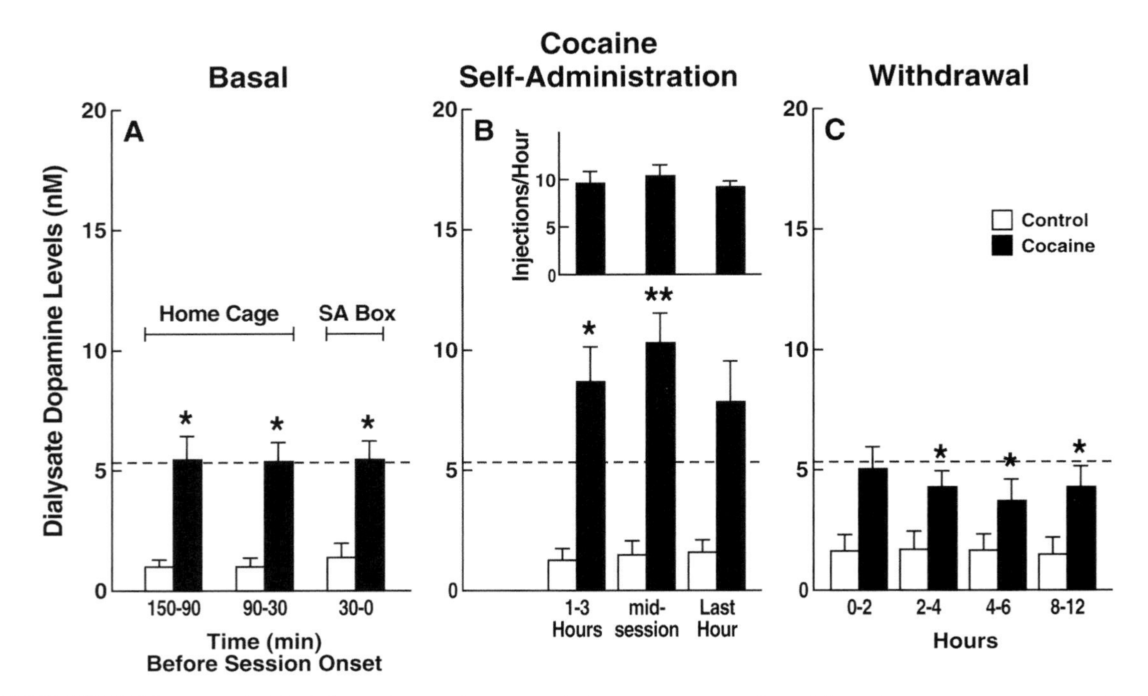

FIGURE 3.25 Mean (+ SEM) dopamine levels in microdialysate fractions collected from the nucleus accumbens of rats ($n = 5$) during unlimited-access cocaine self-administration (0.75 mg/kg/injection) and cocaine withdrawal. Control rats ($n = 3$) were drug-naive animals placed into the self-administration chambers for 30 h without access to cocaine. **(A)** Basal dopamine levels during two 1 h periods in the home cage and 30 min in the self-administration chamber (SA box) prior to cocaine access. **(B)** Response rates for cocaine (inset) and dopamine levels during cocaine self-administration averaged over the first 3 h, mid-session (total self-administration time minus the first 3 h and last 1 h), and the final 60 min of self-administration. **(C)** Dialysate dopamine concentrations during cocaine withdrawal. Dopamine release was significantly suppressed below basal levels between 2–6 h post-cocaine, although dopamine levels tended to increase between 8 and 12 h after onset of the withdrawal period. Dopamine overflow remained significantly below pre-session basal values. The dotted line represents mean pre-session basal dopamine levels for cocaine self-administering rats. $^*p < 0.05$; $^{**}p < 0.01$, significantly different from pre-session basal levels (Newman-Keuls *post hoc* tests). Control data in B and C are arranged with reference to the mean duration of approximately 14 hr in cocaine self-administering rats. Note also that pre-cocaine basal dopamine levels in trained, self-administering rats were significantly higher than in drug-naive control rats (A): $^*p < 0.02$, significantly different from Control. [Taken with permission from Weiss *et al.*, 1992.]

injections or daily limited access conditions, but decreased at time periods over 10 days post-cocaine (Parsons *et al.*, 1991; Robertson *et al.*, 1991; Imperato *et al.*, 1992; Rossetti *et al.*, 1992; Weiss *et al.*, 1992, 1995). Cocaine-induced dopamine release in caudate putamen slices 7 days after continuous (40 mg/kg, s.c., via minipump) or intermittent (40 mg/kg, s.c., daily) cocaine injections revealed decreased cocaine-induced efflux in continuously exposed animals and increased cocaine-induced efflux in intermittently exposed animals (King *et al.*, 1993). Thus, the initial 'crash' associated with a binge is reflected in a decrease in extracellular dopamine followed by an increase and possibly a subsequent decrease.

While much research has focused on a monoamine deficiency hypothesis to explain the decreased reward (and by extrapolation, the human condition of dysphoria) associated with acute withdrawal from psychostimulants, more recent conceptualizations have invoked an activation of brain systems that may act in opposition to brain reward systems. There is evidence that at least two neuropeptide systems may be recruited during acute cocaine withdrawal to contribute to the negative affective state associated with cocaine withdrawal: dynorphin and corticotropin-releasing factor (CRF).

Stressors and the state of stress also can contribute to the components of addiction involving acute withdrawal, protracted abstinence, and vulnerability to relapse (Erb *et al.*, 1996). Hypothalamic-pituitary-adrenal function in humans is activated during psychostimulant dependence and acute withdrawal, and dysregulation can persist even past acute withdrawal (Kreek *et al.*, 1984; Kreek, 1987). An acute binge of cocaine produces a dramatic increase in release of adrenocorticotropic hormone (ACTH) and corticosterone, accompanied by an increase in CRF mRNA levels in the hypothalamus (Zhou *et al.*, 1996, 2003), and this increase shows tolerance with repeated binges

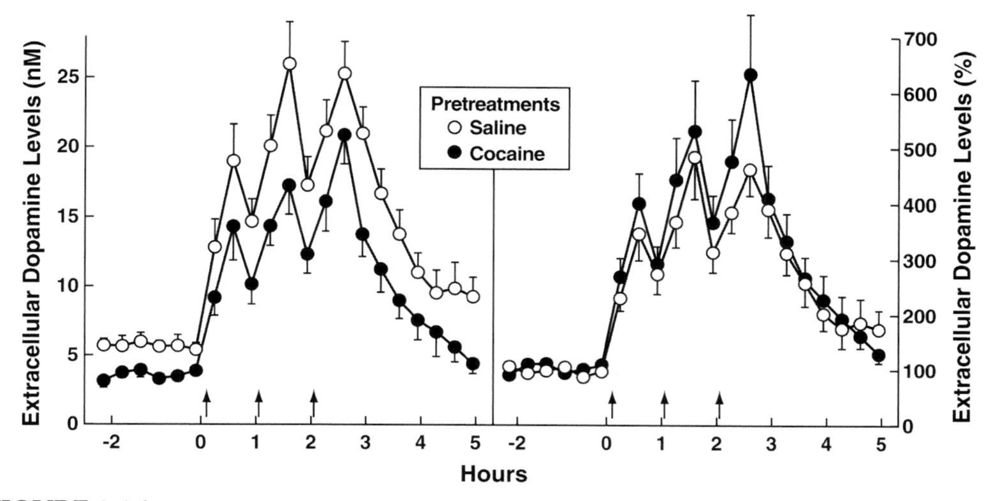

FIGURE 3.26 Estimated extracellular dopamine levels (mean ± SEM) in rats expressed as nM on the left and percentage of their respective basal levels on the right, in the ventromedial striatum before and after cocaine challenge following repeated cocaine administration for 13 days. Animals received injections with saline or cocaine (15 mg/kg, i.p.) at approximately 11:00 am, 12:00 pm and 1:00 pm daily for 13 days. *In vivo* microdialysis was performed on Day 14 when cocaine was administered to all animals. Arrows indicate a cocaine injection (15 mg/kg). The mean estimated extracellular basal dopamine levels and the numbers of animals in each pretreatment group are saline (5.66 ± 0.58 nM, $n = 7$), and cocaine (3.50 ± 0.37 nM, $n = 6$). [Reproduced with permission from Maisonneuve *et al.*, 1995.]

(Mendelson *et al.*, 1998; Sarnyai *et al.*, 1998) and is manifest again during acute withdrawal (Zhou *et al.*, 2003) (**Fig. 3.27**). Following one day of withdrawal from 14 days of binge cocaine, CRF mRNA levels in the hypothalamus returned from a significantly reduced level to normal, and remained normal at 4 and 10 days following cocaine withdrawal (Zhou *et al.*, 2003). During early withdrawal, ACTH and corticosterone levels were again increased, but slowly returned to normal levels by 10 days following withdrawal. Brain CRF function outside of the HPA axis also appeared to be activated during acute withdrawal from cocaine and thus may mediate behavioral aspects of stress associated with abstinence (Richter and Weiss, 1999). Rats treated repeatedly with cocaine showed significant anxiogenic-like responses following cessation of chronic administration. These anxiogenic-like responses were reversed with i.c.v. administration of a CRF receptor antagonist (Rassnick *et al.*, 1993; Sarnyai *et al.*, 1995). Additional evidence supporting a role for activation of brain CRF systems during acute withdrawal were studies showing an increase in extracellular levels of CRF in the region of the central nucleus of the amygdala during acute withdrawal from drugs of abuse. Animals self-administering cocaine intravenously for 12 continuous hours showed a time-related increase in extracellular CRF in the amygdala as measured by *in vivo* microdialysis (Richter and Weiss, 1999) (**Fig. 3.28**).

The opioid peptide dynorphin also has been implicated in the effects of repeated exposure to cocaine. There is evidence that repeated exposure to cocaine increases dynorphin levels in the nucleus accumbens and (Hurd *et al.*, 1992) mRNA expression of prodynorphin (Turchan *et al.*, 1998). κ opioid receptor agonists prevented sensitization to the rewarding effects of cocaine as measured by conditioned place preference for cocaine (Shippenberg *et al.*, 1996), suggesting that recruitment of κ systems may be part of a counteradaptive mechanism associated with cocaine withdrawal.

As described in the *Animal Models of Addiction* chapter, prolonged access to cocaine leads to a time-related escalation in cocaine intake (Ahmed and Koob, 1998). The neuropharmacological basis for the escalation in drug intake associated with more prolonged access is under intense investigation. Preliminary data suggest that the amount of dopamine released by a given amount of cocaine did not differ in escalated versus non-escalated animals, suggesting that escalation is not due to a simple presynaptic sensitization of dopamine neurons (Ahmed *et al.*, 2003; Ahmed and Koob, 2004). However, escalated animals showed an enhanced sensitivity to the mixed D_1/D_2 receptor antagonist *cis*-flupenthixol with a shift of the dose-effect

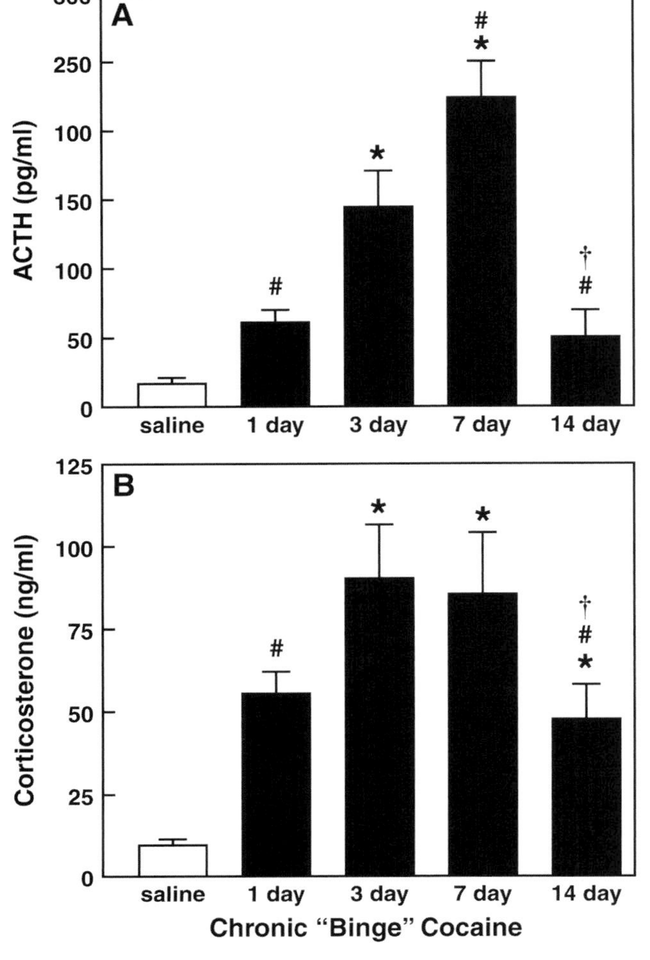

FIGURE 3.27 Effects of 14 days of chronic binge cocaine administration (15 mg/kg, i.p., three times per day) on plasma **(A)** adrenocorticotropic hormone (ACTH) and **(B)** corticosterone levels in rats. Data shown are mean±SEM. Groups: saline (saline for 14 days, $n = 8$), 1 day cocaine (saline for 13 days followed by cocaine for 1 day, $n = 8$), 3 day cocaine (saline for 11 days followed by cocaine for 3 days, $n = 8$), 7 day cocaine (saline for 7 days followed by cocaine for 7 days, $n = 8$), and 14 day cocaine (cocaine for 14 days). $^{*}p < 0.05$ vs saline; $\#p < 0.05$ versus 3 day cocaine; $\dagger p < 0.05$ vs 7 day cocaine. [Taken with permission from Zhou *et al.*, 2003.]

function to the left, suggesting that escalation in cocaine self-administration may be mediated, at least in part, by decreased function of dopamine receptors or changes in signal transduction mechanisms (Ahmed and Koob, 2004).

Reinstatement of Cocaine-Seeking Behavior

Animal models for the preoccupation/anticipation (craving) stage have been extensively characterized using drug-, cue-, and stress-induced reinstatement (see *Animal Models of Addiction* chapter). In animal models of cocaine-induced reinstatement, the neuropharmacological substrates for drug-priming reinstatement with cocaine have focused largely on dopaminergic and glutamatergic systems (Cornish and Kalivas, 2001; McFarland and Kalivas, 2001; Shalev *et al.*, 2002) (**Fig. 3.29**) (**Table 3.9**). Mesolimbic dopamine system activation clearly is implicated in cocaine priming of reinstatement both from studies of agonists mimicking the effects of cocaine and antagonists blocking the effects of cocaine (de Wit and Stewart, 1981; Shalev *et al.*, 2002). Evidence for a role for glutamatergic systems in cocaine reinstatement suggests a facilitation with glutamatergic agonists and inhibition with glutamatergic antagonists at various levels of the mesocorticolimbic dopamine system, including the nucleus accumbens, prefrontal cortex, and the ventral tegmental area (see *Neurobiological Mechanism of Action* below). Particularly striking is the ability of α-amino-3-hydroxy-5-methyl-4-isoxale propionic acid (AMPA) glutamate antagonists, but not dopamine antagonists, when microinjected into the nucleus accumbens, to block cocaine-induced reinstatement (Cornish and Kalivas, 2000) (**Fig. 3.30**).

Morphine infused directly into the ventral tegmental area also reinstates cocaine-seeking behavior and suggests potential 'cross-talk' with opioid peptide systems (Stewart, 1984). A synthetic cannabinoid agonist can induce cocaine reinstatement, and a cannabinoid antagonist can block cocaine-induced reinstatement, suggesting an interaction between cannabinoids similar to that observed with opiates and cocaine (De Vries *et al.*, 2001).

At an anatomical level, there is much evidence for critical roles of the medial prefrontal cortex and nucleus accumbens in cocaine reinstatement of cocaine-seeking behavior (Everitt and Wolf, 2002; Kalivas and McFarland, 2003; See *et al.*, 2003) (see *Neurobiological Theories of Addiction* chapter). Unilateral microinjection of γ-aminobutyric acid agonists into the dorsal prefrontal cortex in one hemisphere and into the ventral pallidum in the other hemisphere provided evidence for a dorsal prefrontal cortex—nucleus accumbens core—ventral pallidum series circuit that mediates cocaine-induced reinstatement (McFarland and Kalivas, 2001). While previous studies have shown that dopamine microinjected into the nucleus accumbens produced drug-related reinstatement, similar effects were observed for dopamine in the dorsal prefrontal cortex. However, in a series of studies, dopamine antagonism in the dorsal prefrontal cortex, but again not in the nucleus accumbens, blocked cocaine-induced reinstatement (Cornish *et al.*, 1999;

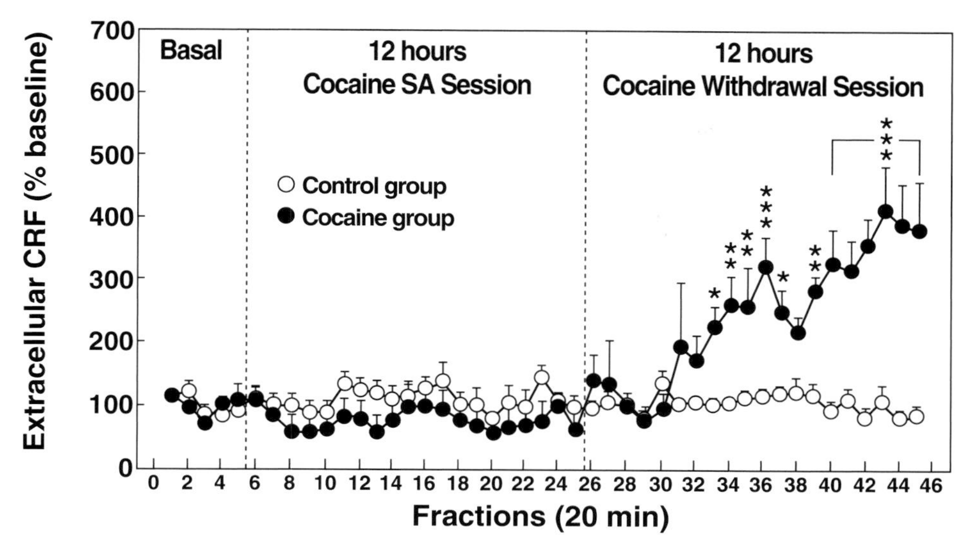

FIGURE 3.28 Mean (± SEM) dialysate corticotropin-releasing factor (CRF) concentrations collected from the central nucleus of the amygdala of rats during baseline, a 12-h cocaine self-administration session, and a subsequent 12-h withdrawal period (Cocaine Group, $n = 5$). CRF levels in animals with the same history of cocaine self-administration training and drug exposure, but not given access to cocaine on the test day, are shown for comparison (Control Group, $n = 6$). The data are expressed as percentages of basal CRF concentrations. Dialysates were collected over 2-h periods alternating with 1-h non-sampling periods. During cocaine self-administration, dialysate CRF concentrations in the cocaine group were decreased by about 25 per cent relative to control animals. In contrast, termination of access to cocaine resulted in a significant increase in CRF efflux, which began approximately 5-h post-cocaine and reached about 400 per cent of pre-session baseline levels at the end of the withdrawal session. *$p < 0.05$; **$p < 0.01$; ***$p < 0.001$; Simple Effects after overall mixed factorial ANOVA. [Reproduced with permission from Richter and Weiss, 1999.]

Cornish and Kalivas, 2000; McFarland and Kalivas, 2001) (**Fig. 3.31**).

Neurobiological studies have focused on basal forebrain projections and connections with the mesolimbic dopamine system as possible substrates for the motivating effects of stimuli that acquire motivational significance in the context of association with cocaine reinforcement (cue-induced responding). Work with animal models of cue-induced reinstatement of self-administration, conditioned place preference, and responding on a second-order schedule of reinforcement show that there is strong evidence of critical roles for the basolateral amygdala, medial prefrontal cortex and nucleus accumbens—ventral pallidum in cue-induced responding (Whitelaw *et al.*, 1996; Meil and See, 1997; Everitt and Robbins, 2000; Everitt and Wolf, 2002; Fuchs *et al.*, 2002; Shalev *et al.*, 2002; Kalivas and McFarland, 2003; See *et al.*, 2003) (**Figs. 3.32** and **3.33**) (see *Neurobiological Theories of Addiction* chapter). Cocaine-predictive contextual stimuli increased Fos protein expression in the basolateral amygdala and medial prefrontal cortex (Weiss *et al.*, 2000; Ciccocioppo *et al.*, 2001). Fos expression within the amygdala in rats parallels the findings in humans of neural activation within the amygdala and anterior cingulate during cue-induced cocaine craving (Maas *et al.*, 1998; Childress *et al.*, 1999). These results are consistent with the importance of the basolateral amygdala in conditioned reinforcement (Cador *et al.*, 1989; Robbins *et al.*, 1989; Burns *et al.*, 1993; Everitt *et al.*, 1999). As with cocaine-primed reinstatement, the medial prefrontal cortex—nucleus accumbens glutamate connection appears critical for cue-induced reinstatement. Injection of a dopamine D_1, but not a D_2 receptor antagonist into the basolateral amygdala blocked cue-induced reinstatement (See *et al.*, 2001). Blockade of glutamate receptors in the nucleus accumbens core, but not the basolateral amygdala (See *et al.*, 2001), also blocked drug-seeking behavior induced by cues (Di Ciano and Everitt, 2001).

There is significant evidence for a critical role of the brain stress neurotransmitter CRF in stress-induced reinstatement. CRF antagonists blocked footshock-induced reinstatement of cocaine-seeking behavior in rats (Erb *et al.*, 1998; Shaham *et al.*, 1998; Lu *et al.*, 2001). Although corticosterone induced reinstatement (Deroche *et al.*, 1997), there is little evidence from hormonal manipulations for a role of corticosterone in stress-induced reinstatement (Shaham *et al.*, 1997). There also is little evidence for either a dopamine or

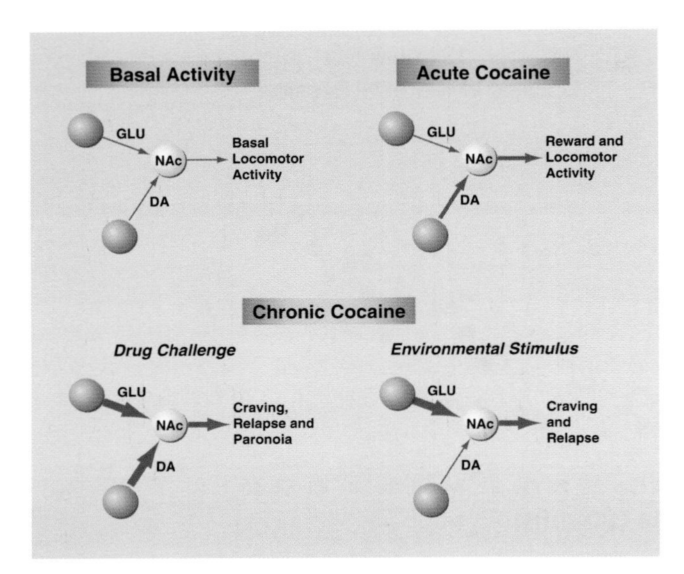

FIGURE 3.29 The role of glutamate (GLU) and dopamine (DA) transmission in the relapse to drug-seeking behavior. During baseline transmission, tonic dopamine and glutamate transmission equally modulate the output of the nucleus accumbens (NAc) to allow normal motor activity. Following the acute administration of cocaine, dopamine levels of the nucleus accumbens are elevated, with little effect on glutamatergic tone, to increase locomotor activity, and stimulate rewarding processes. After withdrawal from chronic intake of the drug, a single administration of cocaine may induce relapse to drug-taking or paranoia through the augmented dopamine release which is associated with and dependent on increased glutamate transmission which may be a consequence of interoceptive cues associated with drug-taking. However, in the absence of cocaine administration, an environmental cue may induce craving and relapse through enhanced glutamate transmission with little dopamine involvement. [Reproduced with permission from Cornish and Kalivas, 2001.]

opioid peptide role in stress-induced reinstatement of drug-related responding (Shalev *et al.*, 2002).

The brain site critical for the CRF role in footshock-induced reinstatement appears to be the bed nucleus of the stria terminalis (BNST), an area rich in CRF receptors, terminals, and cell bodies (Erb and Stewart, 1999) (**Fig. 3.34**). The use of an asymmetric lesion technique to functionally dissect the role of the central nucleus of the amygdala and BNST revealed a critical, but not exclusive, role for the CRF pathway from the central nucleus of the amygdala to the BNST in footshock-induced reinstatement (Erb *et al.*, 2001). Similarly, microinjection studies have shown a role for norepinephrine originating in the ventral noradrenergic pathway and projecting to the BNST, and the central nucleus of the amygdala also has a role in footshock-induced cocaine-related reinstatement (Leri *et al.*, 2002; Shalev *et al.*, 2002) (**Fig. 3.35**), an observation consistent with major reciprocal connections of CRF and norepinephrine in the basal forebrain and brainstem (for reviews see Koob, 1999; Shaham *et al.*, 2003; Aston-Jones and Harris, 2004; Koob and Kreek, 2005) (see *Neurobiological Theories of Addiction* chapter).

NEUROBIOLOGICAL MECHANISM—CELLULAR

Acute Reinforcing and Stimulant Effects of Psychostimulants

The demonstration of synaptic plasticity at excitatory synapses in the source and terminal areas of the mesolimbic dopamine system are argued to be evidence for a role of synaptic plasticity in addiction. This synaptic plasticity derives from two sources: the ability to produce long-term potentiation and long-term depression in both the ventral tegmental area and the nucleus accumbens, and the changes in synaptic strength observed in whole-cell recordings of ventral tegmental area and nucleus accumbens slices. High frequency tetanus-like stimulation of presynaptic fibers to the nucleus accumbens induced long-term potentiation (LTP), and low-frequency stimulation during modest depolarization of the postsynaptic cell-induced long-term depression (LTD), both in the nucleus accumbens and ventral tegmental area. Both LTP and LTD in the nucleus accumbens require activation of *N*-methyl-D-aspartate (NMDA) glutamate

TABLE 3.9 Animal Models of Reinstatement for Cocaine-Seeking ('craving')—Neuropharmacological Interactions

Priming-induced reinstatement		
	Agonists	*Antagonists*
Dopamine	↑ (Nac, FC)	↓ (FC)
Glutamate	↑ (NAc, FC, VTA)	↓ (NAc, FC, VTA)
Opioid	↑ (VTA)	—
Cue-induced reinstatement		
	Agonists	*Antagonists*
Dopamine	—	↓
Glutamate	nt	—
Opioid	nt	nt
Stress-induced reinstatement		
	Agonists	*Antagonists*
Dopamine	nt	nt
CRF	↑ (BNST)	↓ (BNST)
Norepinephrine	nt	↓ (BNST)

nt, not tested; —, no effect; NAc, nucleus accumbens; FC, frontal cortex; VTA, ventral tegmental area; BNST, bed nucleus of the stria terminalis. [Based on Shaham *et al.*, 2003.]

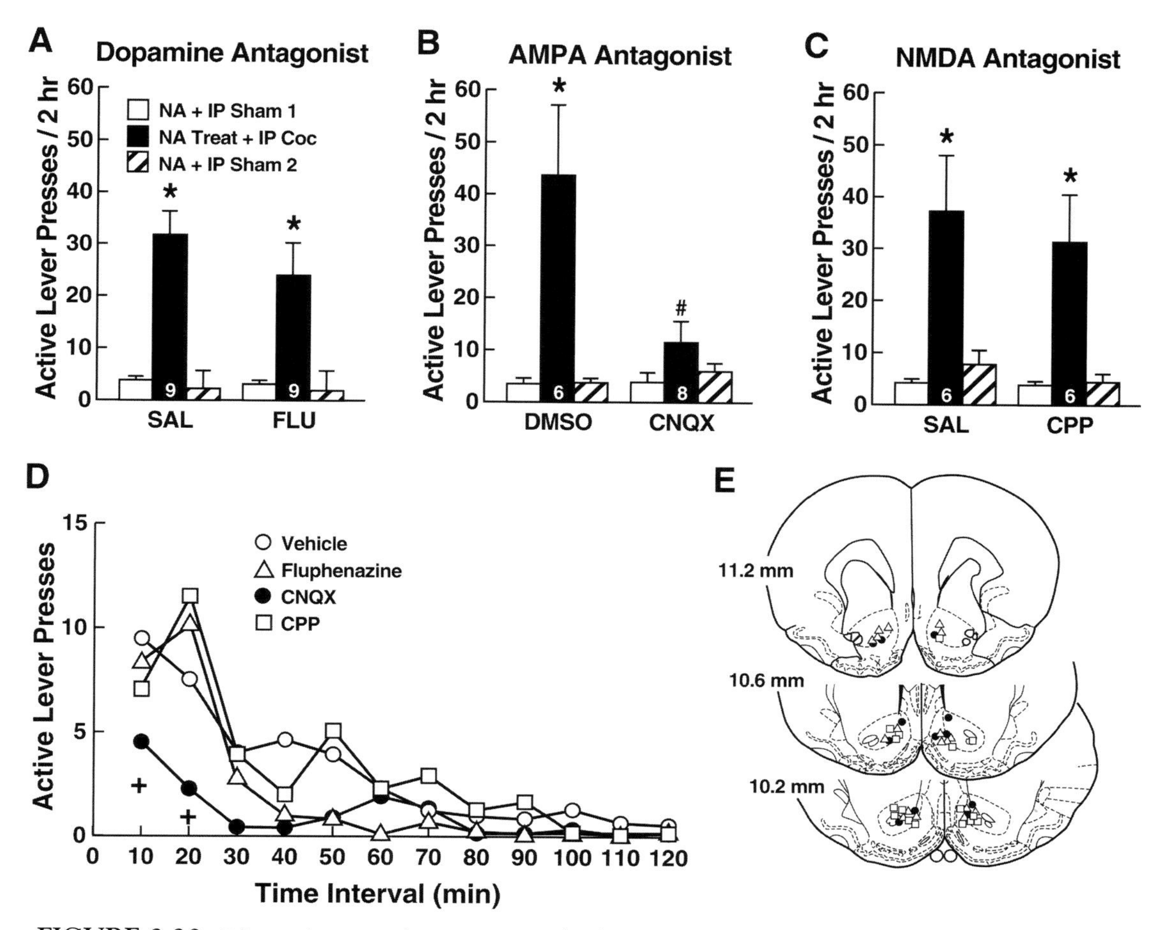

FIGURE 3.30 Effects of intra-nucleus accumbens (NA) treatment with the mixed dopamine receptor antagonist fluphenazine (FLU; 10 nmol/side), the α-amino-3-hydroxy-5-methyl-4-isoxale propionic acid (AMPA) receptor antagonist 6-Cyano-7-nitroquinoxaline-2,3-dione (CNQX) (1.0 nmol/side), and the *N*-methyl-D-aspartate (NMDA)/kainate receptor antagonist 3-(2-carboxypiperazin-4-yl)propyl-1-phosphonic acid (CPP) (0.1 nmol/side) on the drug-paired lever responses produced by a cocaine-priming injection (10 mg/kg, i.p.) in rats. **(A)** The systemic administration of cocaine reinstated drug-seeking behavior that was not affected by dopamine receptor antagonism in the nucleus accumbens. **(B)** The intra-accumbens administration of CNQX completely blocked cocaine-induced drug-seeking behavior compared to dimethyl sulfoxide (DMSO) treatment. **(C)** Pretreating the nucleus accumbens with CPP did not alter the ability for cocaine to induce drug-seeking behavior. **(D)** The time course for the effect of intra-accumbens treatment with vehicle, fluphenazine (FLU), CNQX, and CPP on the number of responses produced by the systemic prime of cocaine. (E) The location of injection sites in the nucleus accumbens corresponding to each of the antagonists used. Average baseline responding before extinction for all animals was 44.2 ± 4.2 cocaine infusions. The number of determinations for each group is written in the solid bar of each treatment. The total number of animals used was 19. Data in A-C are expressed as mean ± SEM and were analyzed by a two-way repeated measures ANOVA (treatment group × treatment day). Each time point in D represents the data expressed as mean number of active lever presses for successive 10 min intervals with the SEM bars eliminated for illustrative clarity, and the data were statistically evaluated using a two-way ANOVA with repeated measures over time. *Post hoc* comparisons were performed in each analysis using a least significant difference test. $^*p < 0.05$ for an increase in drug-paired lever responding compared to sham treatment day. $\#p < 0.05$ for difference compared to vehicle-treated group. $+p < 0.05$ for a difference with the vehicle treatment group. [Taken with permission from Cornish and Kalivas, 2000.]

receptors, and the excitatory input appears to derive from prelimbic cortical afferents (Pennartz *et al.*, 1993; Kombian and Malenka, 1994; Thomas *et al.*, 2000). LTP in the ventral tegmental area also is NMDA receptor dependent (Bonci and Malenka, 1999), but not dopamine receptor dependent (Bonci and Malenka, 1999; Jones *et al.*, 2000). However, LTD is blocked by dopamine or amphetamine acting through dopamine D_2 receptors (Jones *et al.*, 2000).

Further synaptic plasticity has been observed in ventral tegmental area dopamine cells using whole-cell recording techniques in midbrain slices from mice with a single *in vivo* administration of cocaine (15 mg/kg the previous day). There was a marked

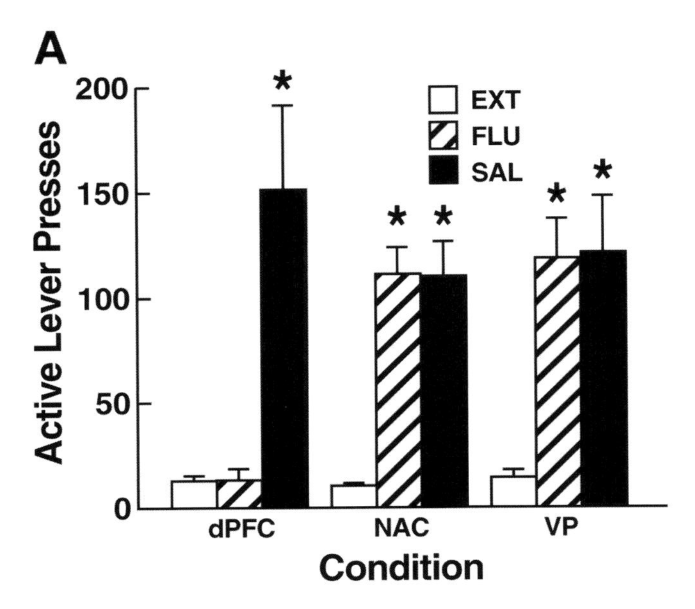

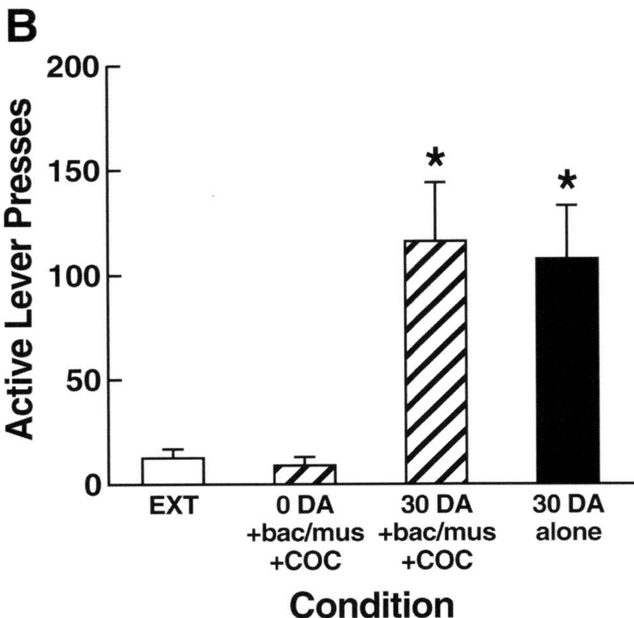

FIGURE 3.31 Role of dopamine in the dorsal prefrontal cortex in cocaine reinstatement in rats. **(A)** Fluphenazine (FLU) infusion into the dorsal prefrontal cortex, but not the nucleus accumbens (NAc) core or ventral pallidum ($n = 6$ in each condition) before reinstatement testing abolished the increase in active lever pressing observed after pretreatment with saline vehicle (SAL). **(B)** After baclofen/muscimol (bac/mus) (0.3 and 0.03 nmol/side, respectively) activation of the ventral tegmental area, subjects received either saline (0 nmol; $n = 5$) or dopamine (30 nmol/side; $n = 7$) infusions into the dorsal prefrontal cortex (dPFC) before a cocaine (COC) reinstatement challenge. When no dopamine was infused into the dorsal prefrontal cortex, subjects exhibited the expected blockade of cocaine-induced reinstatement. However, dopamine replacement into the dorsal prefrontal cortex resulted in a highly significant reinstatement of self-administration behavior. Similarly, dopamine alone into the dorsal prefrontal cortex was sufficient to induce robust responding. $^*p < 0.001$, extinction (EXT) responding compared to other treatments. [Reproduced with permission from McFarland and Kalivas, 2001.]

potentiation in synaptic strength in dopamine cells of animals that received cocaine compared to saline that was due to an upregulation of AMPA glutamate receptors which lasted up to 5 days post-cocaine (Ungless *et al.*, 2001). In contrast, pre-exposure to cocaine *in vivo* produced decreases in LTD and changes in neurons from slices in the nucleus accumbens shell, but not the core, that were decreased in synaptic strength at excitatory synapses from prelimbic cortical afferents (Thomas *et al.*, 2001). These results suggest that even a single administration of cocaine can induce changes in synaptic weights in crucial brain circuits, notably the mesolimbic dopamine circuit that may be of motivational significance (Hyman and Malenka, 2001; Thomas and Malenka, 2003) (see *Neurobiological Theories of Addiction* chapter).

At the *in vivo* cellular level, electrophysiological recordings in animals during intravenous cocaine self-administration have identified several types of neurons in the nucleus accumbens that respond in a manner time-locked to drug infusion and reinforcement. The neurons responsive to cocaine overlap significantly with 'natural' reinforcers (e.g., water and food) (Carelli, 2002). However, others have shown with *in vivo* recordings in animals self-administering cocaine and heroin consecutively within a given session that the majority of neurons in the nucleus accumbens and prefrontal cortex respond differently to cocaine and heroin self-administration as measured by both pre- and post-drug infusion responses0 (Chang *et al.*, 1998). These results suggest that at the cellular level the integration of reinforcement and motivation for cocaine are indeed occurring at the level of the nucleus accumbens (Carelli *et al.*, 1993; Carelli and Deadwyler, 1994; Chang *et al.*, 1994; Peoples and West, 1996). One group of neurons in the nucleus accumbens showed anticipatory neuronal responses, either increased or decreased firing, seconds prior to lever pressing for cocaine, and may be part of an initiation or trigger mechanism. Another group appeared to be inhibited for a few minutes post-cocaine and may represent a direct reward effect. Analysis of videotaped cocaine self-administration behaviors showed that anticipatory responses were specifically associated either with the animal orienting toward and pressing the lever, or with movements directly related to lever pressing (Chang *et al.*, 1994). Dopamine D_1 and D_2 receptor antagonists blocked post-cocaine inhibitory responses but not anticipatory responses (Chang *et al.*, 1994) (**Fig. 3.36**). A third group of neurons appears to have its firing positively correlated with the interinfusion interval of cocaine, linking it to the initiation of the next response (Peoples and West, 1996). Here, the time course of the completion of the progressive reversal from the cocaine-induced

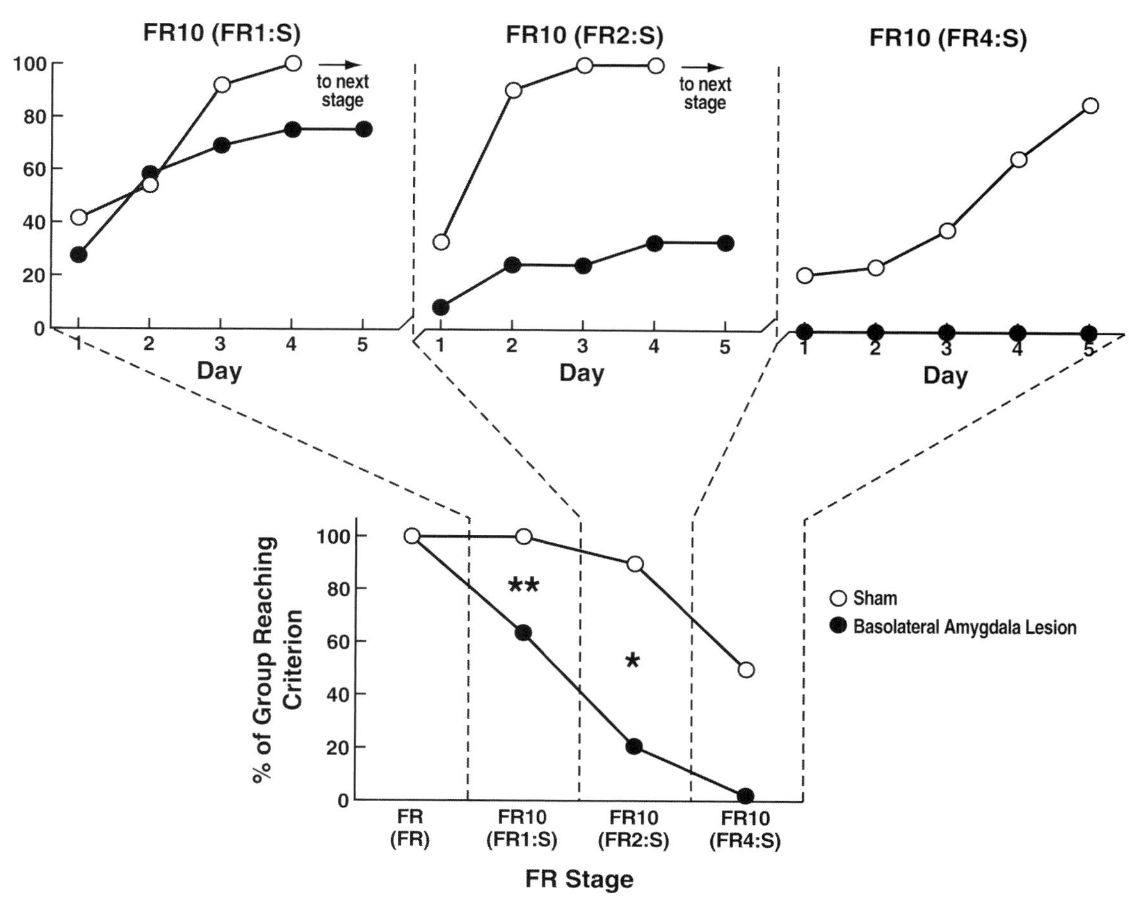

FIGURE 3.32 Acquisition of a second-order schedule of intravenous cocaine self-administration in basolateral amygdala-lesioned and sham-operated rats. The lower panel shows the overall summary of the results. The proportion of rats attaining criterion at successive stages of acquisition are shown. In the upper three panels, the performance of lesioned and control rats during each day of acquisition of each stage is shown in more detail. It can be seen that rats with lesions of the basolateral amygdala repeated more sessions at each stage of acquisition and that a progressively smaller group of basolateral amygdala-lesioned rats moved onto the next stage compared to control rats. The rats were deemed to have failed a stage if they did not reach the criterion response requirement after five repetitions. $^{*}p < 0.05$, $^{**}p < 0.01$. [Reproduced with permission from Whitelaw *et al.*, 1996.]

change in firing approximated that of the interinfusion interval (Peoples and West, 1996) (**Fig. 3.37**).

Much data have been generated with *in vivo* electrophysiology to show that firing of neurons in the nucleus accumbens also can follow the acquisition, extinction, and maintenance of cocaine self-administration. Rats previously trained to self-administer cocaine and allowed to lever press in a drug-free period had a significant number of neurons in the nucleus accumbens that showed an excitatory response during the drug-free period, and these same neurons maintained a higher rate of responding during the actual subsequent self-administration session, suggesting a continuity between neurons firing under drug-free and drug-exposed conditions (Peoples *et al.*, 2004). Similarly, neurons that showed post-response activity during the maintenance phase of cocaine self-administration showed a significant decrease in phasic firing during extinction, but there was no change in pre-response neurons between maintenance and extinction (Carelli and Ijames, 2001). Thus, different neurons may be involved in acquisition and extinction. Even within a given self-administration session, there is some evidence of differential neurochemical mediation with a D_1 antagonist increasing the number of 'load-up' phase responses and both D_1 and D_2 antagonists increased the number of responses following the 'load-up' phase. The transition in firing of neurons in the nucleus accumbens during the session that corresponded with the shift from 'load-up' to maintenance was delayed by administration of a D_1 antagonist (Carelli *et al.*, 1999).

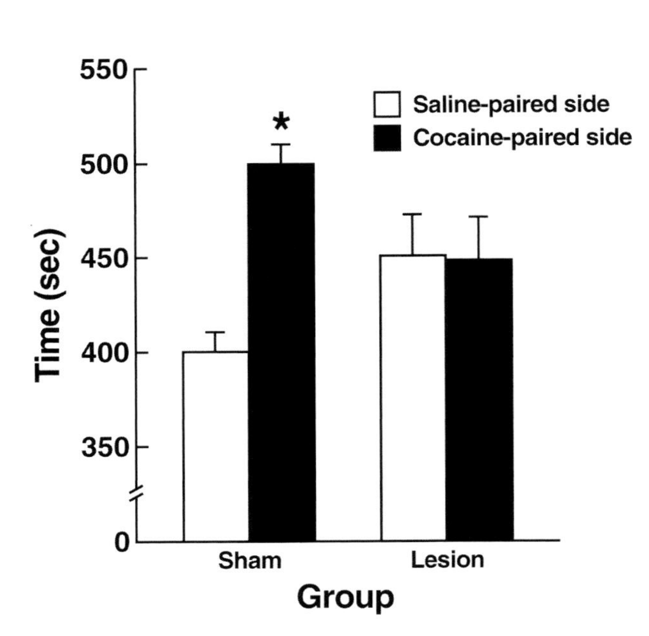

FIGURE 3.33 Effects of pretraining basolateral amygdala complex lesions on acquisition of cocaine conditioned place preference in sham ($n = 9$) and lesion ($n = 9$) groups of rats. The time spent on the cocaine-paired and saline-paired sides of the place conditioning apparatus (± SEM) was measured during a 15 min place conditioning test on post-lesion Day 46–53. Place conditioning testing occurred 24 h after a 2 day conditioning procedure during which animals received one cocaine—environment and one saline—environment pairing. *$p < 0.05$, significant difference from the saline-paired side (Wilcoxon signed ranks test). [Reproduced with permission from Fuchs *et al.*, 2002.]

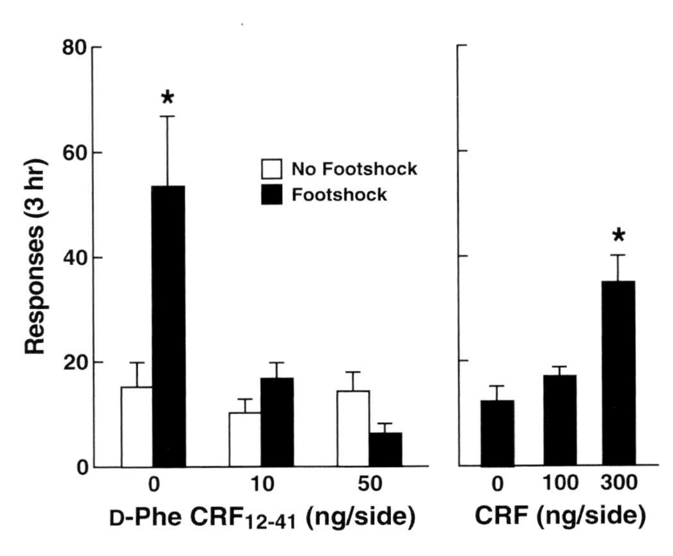

FIGURE 3.34 Mean (± S.E.M.) number of nonreinforced responses on a previously active (cocaine-paired) lever after exposure to 15 min of intermittent footshock in rats pretreated by intra-bed nucleus of the stria terminalis (BNST) injections of the corticotropin-releasing factor (CRF) receptor antagonist D-Phe-CRF$1_{12\text{-}41}$ (left panel), and after intra-BNST injections of CRF itself; no footshock was given (right panel). *$p < 0.05$, significantly different from the other conditions. Similar manipulations in the amygdala had no effects. [Taken with permission from Erb and Stewart, 1999.]

Psychostimulant Withdrawal and Dependence

Early work on cocaine withdrawal revealed that following withdrawal from repeated treatment (10 mg/kg twice daily for 10 days), there was a subsensitivity of the somatodendritic autoreceptor in ventral tegmental area dopamine neurons for 1–4 days post-withdrawal (Ackerman and White, 1990), and during this period

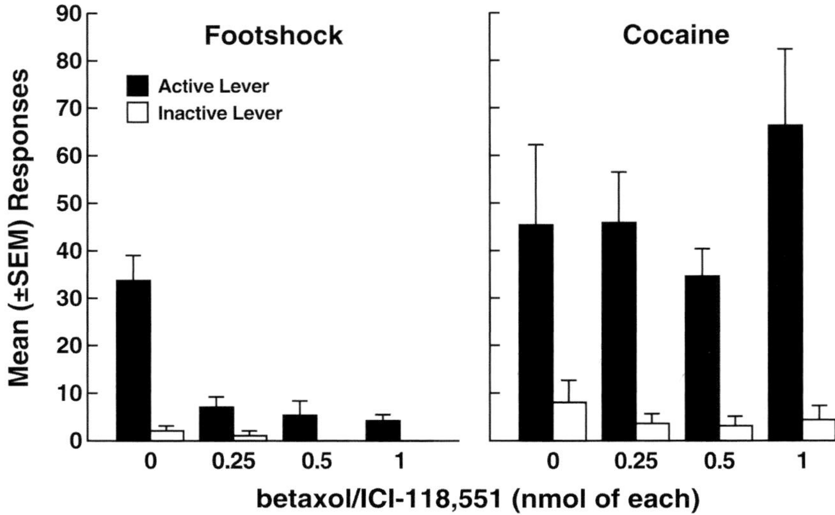

FIGURE 3.35 Effects of different doses of a mixture of the β_1 adrenergic antagonist betaxolol and the β_2 adrenergic antagonist ICI 118 551 infused into the central nucleus of the amygdala on reinstatement of responding induced by intermittent footshock stress (15 min, 0.8 mA) or cocaine (20 mg/kg, i.p.) in rats. All doses of the mixture blocked reinstatement compared to vehicle, apparently in a dose-independent manner. There was a significant Dose × Lever interaction (two-factor ANOVA: $F_{3,12} = 16.7$; $p < 0.001$), and multiple comparisons confirmed that footshock reinstated responding selectively on the active lever only in the animals that received vehicle infusions in the central nucleus of the amygdala. [Reproduced with permission from Leri *et al.*, 2002.]

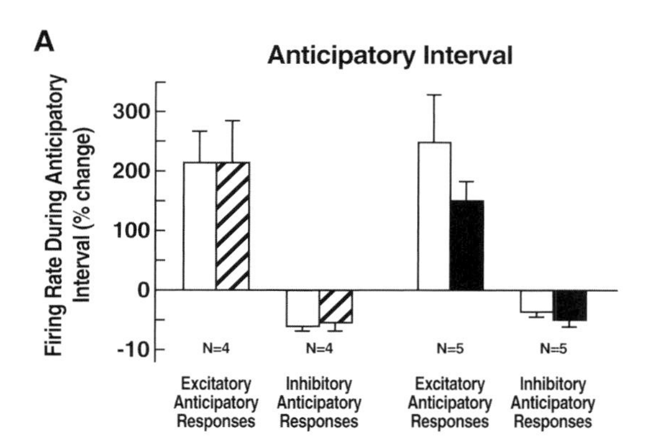

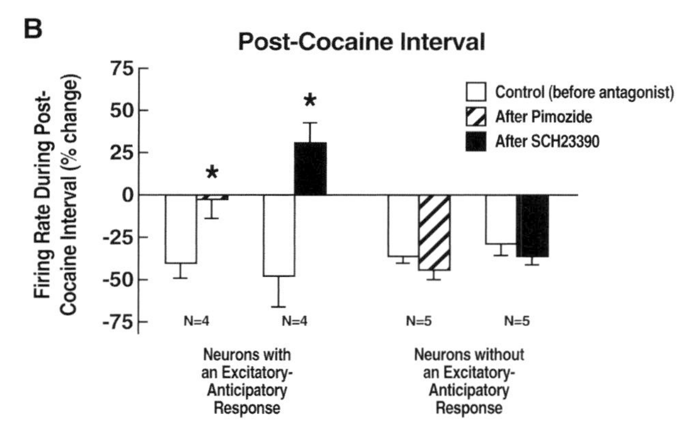

FIGURE 3.36 The effects of the dopamine D_1 receptor antagonist SCH 23390 (10–20 μg/kg, s.c.) and the D_2 receptor antagonist pimozide (0.25 mg/kg, i.p.) on the response properties of nucleus accumbens neurons in rats during cocaine self-administration sessions. **(A)** Effect of the dopamine antagonists on excitatory- and inhibitory-anticipatory responses. Neither pimozide nor SCH 23390 altered excitatory- or inhibitory-anticipatory response. There were no significant differences in pairwise comparisons ($p > 0.05$, *t*-test). **(B)** Effects of SCH 23390 and pimozide on post-cocaine inhibitory responses in two groups of neurons: those with an excitatory-anticipatory and post-cocaine inhibitory response, and those with a post-cocaine inhibitory response only. Pimozide effectively blocked the post-cocaine inhibitory response of the excitatory-anticipatory group, and SCH 23390 reversed the post-cocaine inhibitory response. However, neither antagonist affected the magnitude of post-cocaine inhibition for neurons that lacked an excitatory-anticipatory response. Data presented as mean±SEM; N is the number of neurons tested for each condition. $*p < 0.05$, statistically significant differences in firing rates compared to corresponding control condition (*t*-test). [Taken with permission from Chang *et al.*, 1994.]

dopamine cells were hyperactive but hyporesponsive to cocaine (Henry *et al.*, 1989; Gao *et al.*, 1998). However, at later time points there was a decreased number of spontaneously active cells (Ackerman and White, 1992) but a hyperresponsiveness to cocaine (Akimoto *et al.*, 1989; Kalivas and Stewart, 1991). The response of dopamine neurons also depended on the pattern of exposure. Withdrawal from nonescalating limited access self-administration produced a transient increase in firing rate and bursting activity of midbrain dopamine cells (Marinelli *et al.*, 2003). However, withdrawal from an escalating dose, binge-like regimen of cocaine administration resulted in significantly fewer spontaneously active ventral tegmental area dopamine neurons, and these changes correlated with locomotor depression. These animals also showed a sensitized locomotor response to cocaine (Koeltzow and White, 2003). Withdrawal after continuous infusion of cocaine (40 mg/kg, s.c., via minipump) also reduced the bursting activity of ventral tegmental area neurons and resulted in supersensitive response of impulse-regulating dopamine receptors (Gao *et al.*, 1998).

These results suggested during acute withdrawal a decreased transmission in the mesolimbic dopamine system consistent with results from microdialysis studies and possibly increased transmission during more protracted withdrawal (see above). These results also suggest that the change in dopamine transmission is dose- and time-dependent, with continuous access more likely to produce decreases in firing and transmission, and limited or intermittent access more likely to produce later increases in firing and transmission. For example, it has been argued that the diminished sensitivity of dopamine autoreceptors is a necessary condition for locomotor sensitization to develop to repeated cocaine self-administration (Kalivas and Stewart, 1991; White *et al.*, 1995), and that continuous cocaine administration promotes behavioral tolerance, not sensitization (King *et al.*, 1992). A binge of cocaine actually may produce both effects, decreased firing of ventral tegmental area neurons and later locomotor sensitization, depending on the duration of the binge (Koeltzow and White, 2003).

The excitatory and non-excitatory neurons in the nucleus accumbens show differential changes in firing with repeated administration of cocaine, with decreases in neuronal excitability, and firing rates predominating in repeated cocaine self-administration (Peoples *et al.*, 1999). Studies of repeated administration of cocaine are providing some insight into the cellular neuroadaptations that might be associated with the development of dependence (see Peoples and Cavanaugh, 2003) (see *Neurobiological Theories of Addiction* chapter). The nucleus accumbens neurons that have a phasic excitatory response time-locked to cocaine-reinforced lever presses during intravenous cocaine self-administration showed a resistance to a decline in firing observed during a prolonged self-administration session. In contrast, for these same nucleus accumbens neurons, background firing fell below average pre-drug firing (Peoples and Cavanaugh, 2003). The authors hypothesized that a differential

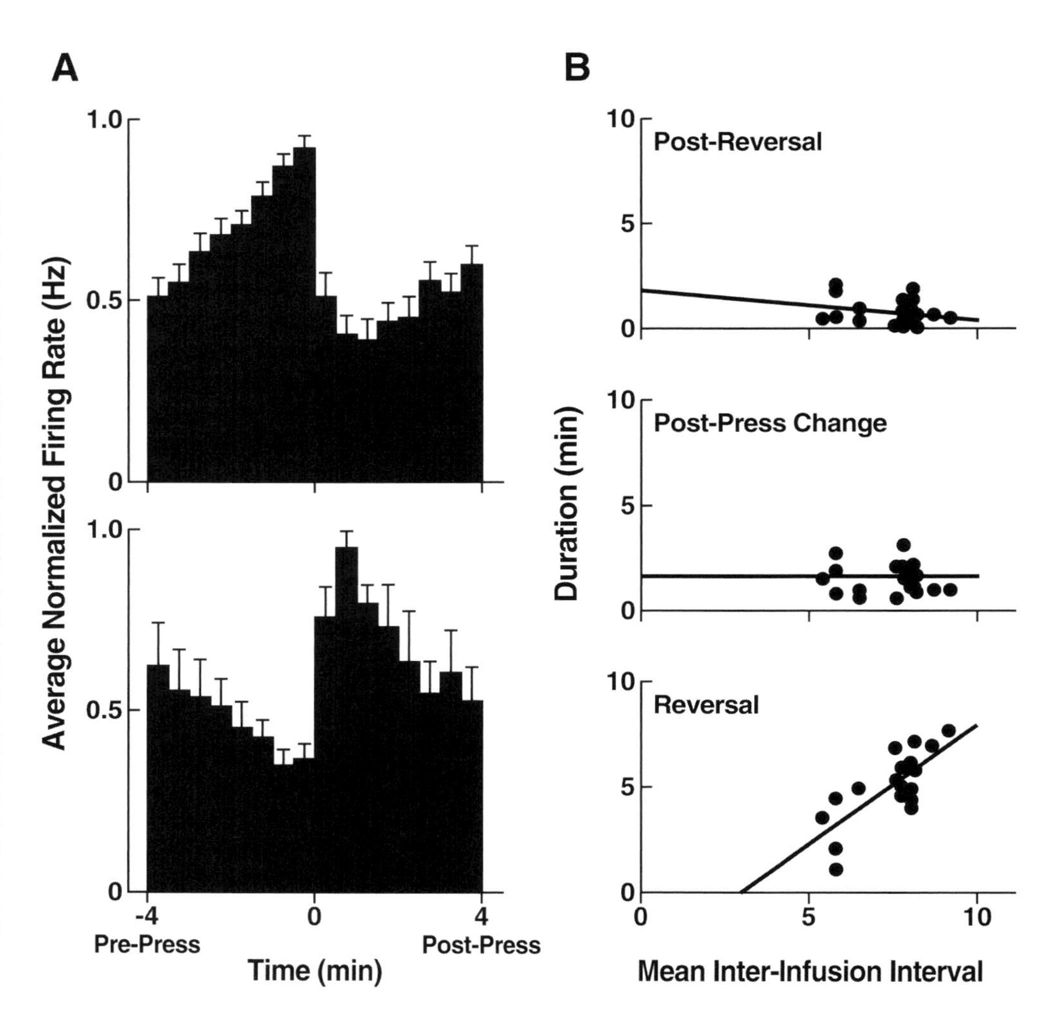

FIGURE 3.37 (A) Summary of the most commonly observed firing patterns time locked to a lever press triggering an intravenous self-administration of cocaine (0.2 ml). The top graph is the average lever-press histogram of all 20 *decrease + progressive reversal* neurons in rats. The bottom graph is the average lever-press histogram of 6 *increase + progressive reversal* neurons. Time 0 represents completion of the lever press. Each neuron that contributed to a summary histogram contributed to each of the 16 bins in that histogram. (B) Relationship between the duration of the complete cycle of self-administration (i.e., interinfusion interval) and the durations of three phases of the *post-press change + progressive reversal* firing pattern. The top graph displays the duration of the post-reversal period that elapsed between reversal culmination and the completed lever press. The middle graph displays the duration of the post-press change. The bottom graph displays the duration of the progressive reversal. [Reproduced with permission from Peoples and West, 1996.]

inhibition in signal and background firing might be expected to increase the relative influence of drug reward-related signals on nucleus accumbens circuits (Peoples and Cavanaugh, 2003) (see Neurobiological *Theories of Addiction* chapter).

Reinstatement of Cocaine-seeking Behavior

Neurons in the nucleus accumbens also come to respond to stimuli paired with cocaine delivery. Cells in the nucleus accumbens that exhibited a post-response change in firing rate within seconds of the reinforced response were controlled, in part, by the stimulus that was paired with cocaine delivery (Carelli, 2000; Carelli and Ijames, 2001). Similarly, neurons in the basolateral amygdala that exhibited increased firing immediately after the response for cocaine were activated by an audio/visual cue paired with cocaine (Carelli *et al.*, 2003). In a particularly intriguing study, neurons in both the nucleus accumbens and medial prefrontal cortex were recorded simultaneously in rats self-administering cocaine using a multi-channel, single-unit recording technique with a focus on the neurons that 'anticipated' a cocaine infusion and increased their firing a few seconds before the lever press (Chang *et al.*, 2000). Cross-correlational analyses revealed inter- and intra-regional-correlated firing patterns between pairs of simultaneously recorded medial prefrontal cortex and nucleus accumbens neurons. The correlations were much higher in neurons that anticipated a lever press, and the temporal correlation revealed that there were many more cases in which nucleus accumbens neurons fired before medial prefrontal cortex neurons. The authors suggested that these data provide evidence that correlated firing between the medial prefrontal cortex and nucleus accumbens may participate in the control of cocaine self-administration (Chang *et al.*, 2000) (see *Neurobiological Theories of Addiction* chapter).

NEUROBIOLOGICAL MECHANISM—MOLECULAR

Acute Reinforcing and Stimulant Effects of Psychostimulants

At the molecular level, several different dopamine and serotonin receptors have been identified, both by pharmacological and molecular biological techniques, and may be involved in the acute reinforcing actions of psychostimulant drugs (Kebabian and Calne, 1979). Five dopamine receptor subtypes have been cloned (D_1–D_5) (Monsma *et al.*, 1990; Sokoloff *et al.*, 1990; Sunahara *et al.*, 1991; Van Tol *et al.*, 1991), and selective ligands exist for three of the subtypes (D_1, D_2, and D_3). There is also some evidence to support the hypothesis of differential functional actions on psychostimulant effects for the D_1 and D_2 receptors at the behavioral level (Bergman *et al.*, 1990; Caine and Koob, 1995). Low doses of the selective D_1 antagonist SCH 23390 potently block amphetamine-induced locomotion (Amalric and Koob, 1993) and block the reinforcing effects of intravenous cocaine self-administration (Koob *et al.*, 1987a; Bergman *et al.*, 1990; Maldonado *et al.*, 1993; Caine and Koob, 1994; Caine *et al.*, 1995) (**Fig. 3.38**).

Similar effects were observed with low doses of D_2 antagonists (Woolverton, 1986; Bergman *et al.*, 1990; Britton *et al.*, 1991; Corrigall and Coen, 1991; Hubner and Moreton, 1991; Caine and Koob, 1994; Haile and Kosten, 2001; Caine *et al.*, 2002b). However, low doses of D_2 antagonists and not D_1 antagonists are effective in impairing responding in a reaction time task particularly sensitive to disruption of nigrostriatal function (Amalric *et al.*, 1993), suggesting a motor component to these antagonists.

The D_3 receptor subtype appears to be restricted in its distribution to the terminal projections of the shell part of the nucleus accumbens mesocorticolimbic dopamine

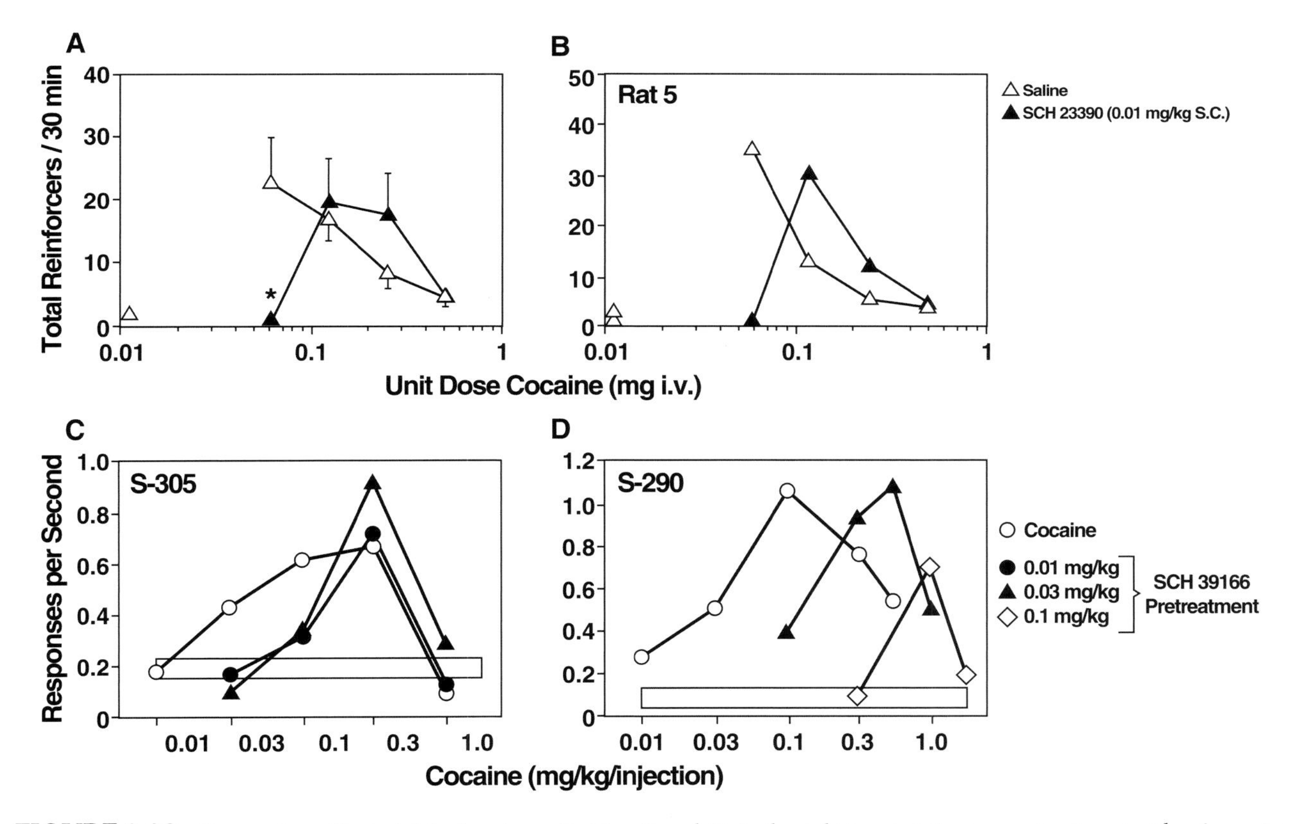

FIGURE 3.38 The cocaine self-administration dose-effect function shifts to the right in rats following pretreatment with a dopamine receptor antagonist. **(A)** The antagonist drug (SCH 23390; 0.01 mg/kg, s.c.) shifts responding maintained by various doses of cocaine (0.06–0.5 mg/injection). Points represent the mean for 4 subjects. **(B)** Same as in A but for an individual rat. It is important to show data from individual subjects because averaging data from multiple subjects can potentially distort overall drug effects. In this case, individual subjects demonstrate the same shifts. [Reproduced with permission from Caine and Koob, 1995.] (C and D) SCH 39166 (0.01, 0.03, 0.1 mg/kg) shifts responding in two individual squirrel monkeys. [Reproduced with permission from Bergman *et al.*, 1990.]

system (Levesque *et al.*, 1992). D_3 agonists dose-dependently facilitate cocaine self-administration, and their potency correlates highly with their potency to activate D_3 receptor transduction mechanisms (Caine and Koob, 1993; Caine *et al.*, 1997). The same D_3 antagonist that is 100 times more selective for D_3 receptors over D_2 receptors dose-dependently decreased cocaine-seeking behavior maintained by a cocaine-associated conditioned reinforcer in a second-order schedule of reinforcement, but had no effect on baseline cocaine self-administration (Reavill *et al.*, 2000; Di Ciano *et al.*, 2003). Recent studies indicate that D_3 antagonists do block cocaine self-administration when response requirements are increased, such as with progressive-ratio schedules (Xi *et al.*, 2004). These results suggest that D_3 receptors may be critically involved in drug-seeking behavior associated with the motivational effects of cocaine.

Molecular biological techniques combined with a molecular genetic approach have provided a selective deletion of the genes for expression of different dopamine receptor subtypes and the dopamine transporter (Xu *et al.*, 1994a; Giros *et al.*, 1996). To date, D_1, D_2, D_3, D_4, and dopamine transporter knockout mice exist and have been subjected to challenges with psychostimulants (Xu *et al.*, 1994a,b; Giros *et al.*, 1996). D_1 receptor knockout mice show no behavioral response to administration of D_1 agonists or antagonists and show a blunted response to the locomotor-activating effects of cocaine and amphetamine (Xu *et al.*, 1994b). D_1 knockout mice do not show a deficit in acquisition of conditioned place preference for cocaine (Miner *et al.*, 1995) but are impaired in their acquisition of intravenous cocaine self-administration compared to wildtype mice (Caine *et al.*, 2002a).

D_2 knockout mice have severe motor deficits and blunted responses to psychostimulants and opioids (Baik *et al.*, 1995; Maldonado *et al.*, 1997). Preliminary results show that D_2 knockout mice actually self-administer more cocaine than their wildtype littermates, an effect similar to that observed from pharmacological blockade of D_2 receptors in intact mice (Caine *et al.*, 2002b). D_3 and D_4 knockout mice show hyperactivity but no blunting of psychostimulant activity, and in some cases supersensitivity to psychostimulants (Rubinstein *et al.*, 1997). Dopamine transporter knockout mice are dramatically hyperactive but also show a blunted response to psychostimulants (Giros *et al.*, 1996).

In summary, pharmacological studies with selective D_1, D_2, and D_3 antagonists and knockout studies have shown that all three receptor subtypes appear to mediate the reinforcing effects of cocaine, albeit possibly different components of the response. Low doses of D_1 antagonists appear in rodents almost as competitive antagonists to cocaine. D_2 antagonists block responding for cocaine but also have pronounced motor response inhibitory actions. D_3 antagonists block drug-seeking behavior associated with cocaine in second-order and progressive-ratio schedules. It is clear that D_1, D_2, and D_3 receptors, and the dopamine transporter play an important role in the actions of psychostimulants, although developmental factors must be taken into account for the compensation or overcompensation. For example, recent evidence suggests that the norepinephrine transporter mediates the self-administration of cocaine in dopamine transporter knockout mice (Carboni *et al.*, 2001). Conditional knockouts, where developmental factors are eliminated, will be required to more precisely delineate the functional roles of the molecular entities within the dopamine system in vulnerability at different stages of the addiction cycle.

In addition, several serotonin receptor subtypes may interact with the mesolimbic dopamine system that may have a role in cocaine-induced behaviors. Serotonin-1B (5-hydroxytryptamine, 5-HT_{1B}) receptor stimulation appears to play a role in facilitating the reinforcing (Neumaier *et al.*, 2002; Parsons *et al.*, 1996) and discriminative properties of cocaine (Callahan and Cunningham, 1995). 5-HT_2 receptor blockade attenuates cocaine-induced hyperactivity in mice (O'Neill *et al.*, 1999). 5-HT_3 receptor antagonists have been shown to block the behavioral expression of cocaine sensitization in rats (Kankaanpaa *et al.*, 1996; King *et al.*, 2000). These results are consistent with pharmacological studies showing 5-HT_3 receptor activation enhances dopamine release in the nucleus accumbens (Costall *et al.*, 1987; Blandina *et al.*, 1989; Jiang *et al.*, 1990; Chen *et al.*, 1991, 1992; McNeish *et al.*, 1993; Matell and King, 1997).

Psychostimulant Withdrawal and Dependence

Repeated administration of cocaine produces changes in dopamine receptors consistent with changes in dopamine release. An initial stage is characterized by increased extracellular dopamine or increased turnover and a supersensitivity of dopamine D_1 receptors (Kalivas *et al.*, 1988; Henry and White, 1991; Weiss *et al.*, 1992), and a subsequent dopamine deficiency stage that is facilitated by high-dose cocaine administration, increased frequency of dosing, or longer abstinence periods (Weiss *et al.*, 1995). Decreases in both D_1-like and D_2-like receptors have been observed after long-term, heavy exposure to passive administration or self-administration of cocaine or other psychostimulant drugs in rats (Tsukada *et al.*, 1996; Graziella de Montis *et al.*, 1998; Maggos *et al.*,

1998), nonhuman primates (Moore *et al.*, 1998a,b; Nader *et al.*, 2002), and humans (Volkow *et al.*, 1993, 2001). A chronic reduction in the number and/or function of mesolimbic dopamine receptors also may explain the chronic shift in baseline brain reward thresholds observed in animal studies of cocaine escalation and the increased sensitivity of these animals to the dopamine antagonist *cis*-flupenthixol (Ahmed and Koob, 2003).

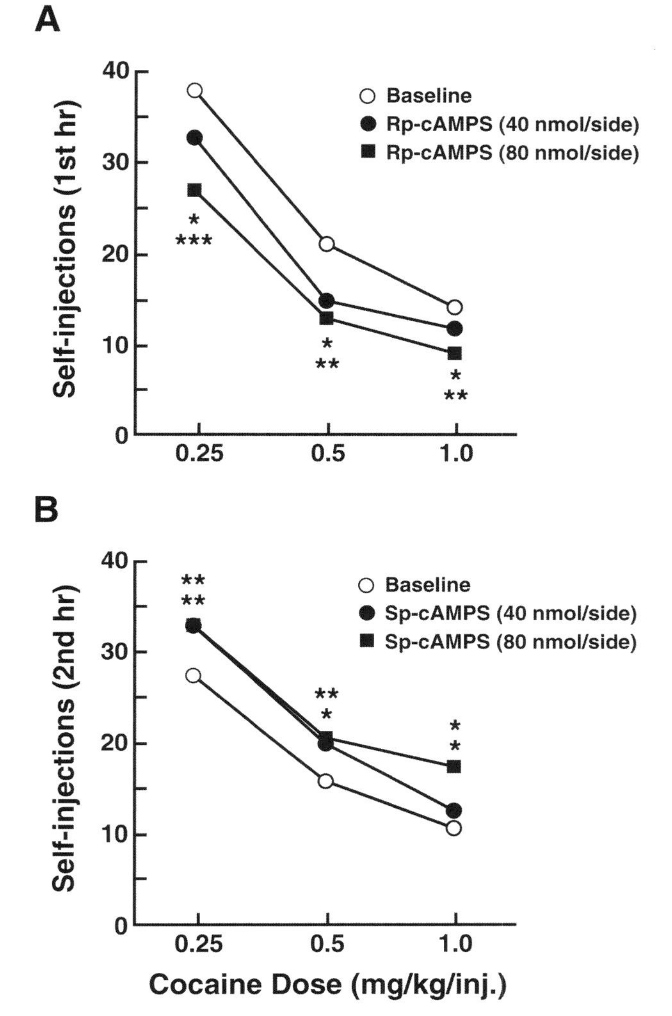

FIGURE 3.39 Effects of bilateral nucleus accumbens infusions of the protein kinase A inhibitor R_p-cAMPS (R_p-adenosine 3′,5′-cyclic monophosphorothioate triethylammonium salt). **(A)** or activator S_p-cAMPS (S_p-adenosine 3′,5′-cyclic monophosphorothioate triethylammonium salt). **(B)** on the dose-response relationship of cocaine self-administration in rats. Self-administration rates are shown for the first hour of the test session in experiments with R_p-cAMPS and during the second hour in experiments with S_p-cAMPS, when the cyclic adenosine monophosphate (cAMP) analogs produced their maximal behavioral effects. The data are expressed as the mean number of self-injections (n = 10–12). Baseline values represent pooled data from tests preceding both the 40 and 80 nmol/1.0 μl per side doses of each cAMP analog. Asterisks (*) indicate that values differ from baseline values by paired t test for the 40 or 80 nmol/side dose (*$p < 0.05$; **$p < 0.01$; ***$p = 0.001$). [Reproduced with permission from Self *et al.*, 1998.]

An early biochemical finding with chronic administration of cocaine (15 mg/kg, i.p., twice daily for 14 days) was an upregulation 16 h after the last injection of the cyclic adenosine monophosphate (cAMP) pathway in the nucleus accumbens (Terwilliger *et al.*, 1991). Stimulation of cAMP-dependent protein kinase A (PKA) which activates cAMP response element binding protein (CREB) in the nucleus accumbens decreased cocaine self-administration (Self *et al.*, 1998) (**Fig. 3.39**). Similarly, elevation of CREB expression in the nucleus accumbens decreased cocaine self-administration (Carlezon *et al.*, 1998). In contrast, blockade of PKA activity or overexpression of a dominant-negative CREB in the form of a mutant CREB which acts as a CREB antagonist in the nucleus accumbens increased cocaine reward (Carlezon *et al.*, 1998; Self *et al.*, 1998).

The effects of CREB in the nucleus accumbens have been hypothesized to involve increased expression of dynorphin, a neuropeptide associated with decreased function of the mesolimbic dopamine system, and dysphoria-like responses (Bals-Kubik *et al.*, 1993). Psychostimulants increase dynorphin expression in the nucleus accumbens (Hurd *et al.*, 1992; Spangler *et al.*, 1993). CREB regulates dynorphin gene expression *in vitro* (Cole *et al.*, 1995) and *in vivo* (Sakai *et al.*, 2002). Using a herpes simplex viral vector to elevate CREB in the nucleus accumbens resulted in increased aversion to cocaine in a conditioned place preference test and a depression-like profile in the forced swim test that was reversed by a κ antagonist, supporting the hypothesis that elevations in CREB and dynorphin in the nucleus accumbens contribute to the dysphoria associated with withdrawal from chronic cocaine (Pliakas *et al.*, 2001). These data have led Nestler to argue:

> 'There is now compelling evidence that upregulation of the cAMP pathway and CREB in the nucleus accumbens represents a mechanism of ''motivational tolerance and dependence''...These molecular adaptations decrease an individual's sensitivity to the rewarding effects of subsequent drug exposures (tolerance) and impair the reward pathway (dependence) so that after removal of the drug the individual is left in an amotivational, depressed-like state' (Nestler, 2004).

Other molecular adaptations in the ventral tegmental area—nucleus accumbens system associated with chronic cocaine exposure include changes in tyrosine hydroxylase, G-protein subunit expression, neurofilament proteins, and glutamate receptors. Chronic but not acute cocaine decreased levels of $G_{i\alpha}$ and $G_{o\alpha}$ in the ventral tegmental area and nucleus accumbens (Nestler *et al.*, 1990), and inactivation of G_i and G_o proteins in the nucleus accumbens decreased cocaine reinforcement, shifting the dose-response function to the right (Self *et al.*, 1994). Chronic but not acute cocaine administration increased extracellular signal-regulated kinase (ERK) catalytic activity in the ventral tegmental area (Berhow *et al.*, 1996b), and chronic but not acute cocaine increased immunoreactivity of Janus kinase, a ciliary neurotrophic factor-regulated protein tyrosine kinase, in the ventral tegmental area (Berhow *et al.*, 1996a). Chronic but not acute cocaine also decreased levels of neurofilament proteins in the ventral tegmental area (Beitner-Johnson *et al.*, 1992). Chronic but not acute cocaine also increased expression of glutamate receptor-1 (GluR1; an AMPA receptor subunit) and NMDAR1 (an NMDA receptor subunit) in the ventral tegmental area (Fitzgerald *et al.*, 1996). These results together suggest that the cAMP system is only one of several intracellular signaling pathways that are altered by cocaine exposure, and these may ultimately lead to long-term changes in structure and function (Koob *et al.*, 1998) (**Fig. 3.40**).

One mechanism for long-term molecular changes in cellular function is the induction of transcription factors. Acute administration of cocaine has long been known to induce c-Fos expression (Graybiel *et al.*, 1990; Young *et al.*, 1991), but c-Fos expression is short-lived in its activation, returning to normal within 12 h of drug exposure. Furthermore, chronic administration of cocaine reduced the ability of cocaine to induce c-Fos expression. However, chronic cocaine did cause the accumulation of activator protein 1 complexes and the Fos proteins responsible for the activator protein-1 complex was ΔFosB, a truncated splice variant of the *FosB* gene (Hope *et al.*, 1994; Nye *et al.*, 1995). Induction of ΔFosB in the nucleus accumbens is long-lived after chronic cocaine exposure, and studies of inducible transgenic mice have shown that ΔFosB overexpression increases the sensitivity to the locomotor-activating and rewarding effects of cocaine, increases cocaine self-administration, and increases progressive-ratio responding for cocaine (Hiroi *et al.*, 1997; Kelz *et al.*, 1999; Colby *et al.*, 2003; Nestler, 2004). From these results, ΔFosB has been hypothesized to have a potential role in initiating and maintaining an addictive state via increasing the drive for drug reward weeks and months after the last drug exposure (Nestler, 2004) (see *Neurobiological Theories of Addiction* chapter).

When the regulation of gene expression in the nucleus accumbens was explored following different cocaine exposure periods in mice combined with mice overexpressing CREB or ΔFosB for different amounts of time, the subset of genes upregulated by CREB also were upregulated by short-term exposure to cocaine for 5 days. In contrast, 4 weeks of cocaine treatment produced an upregulation of genes more related to ΔFosB and few genes related to CREB (McClung and Nestler, 2003) (**Fig. 3.41**). Gene expression induced by short-term ΔFosB and CREB were very similar, and both reduced the rewarding effects of cocaine as measured by conditioned place preference. In contrast, prolonged ΔFosB expression increased cocaine reward as measured by conditioned place preference (McClung and Nestler, 2003). Transient overexpression of ΔFosB in striatal cell types in the mouse facilitated acquisition of cocaine self-administration and increased progressive-ratio performance for cocaine (Colby *et al.*, 2003). Altogether, these results suggest that short-term regulation of gene expression by CREB and ΔFosB reduces cocaine reward during early phases of cocaine exposure, but accumulation of ΔFosB during more chronic treatment mediates increased responsiveness to cocaine.

Reinstatement of Cocaine-Seeking Behavior

Reinstatement of cocaine-seeking behavior in craving models of cocaine-primed reinstatement shows a role for dopaminergic systems, and the specific molecular targets now are under investigation. Most evidence suggests that dopamine agonist-induced reinstatement is dopamine D_2 receptor-mediated, as opposed to D_1 receptor-mediated, although D_1 receptor antagonists do block cocaine-primed reinstatement (Norman *et al.*, 1999). There is also evidence of a glutamatergic involvement in drug-priming reinstatement. Systemic or intracranial injections of agonists and antagonists of ionotropic glutamate receptors also alter cocaine-priming reinstatement. Intra-nucleus accumbens infusions of AMPA selectively reinstated cocaine self-administration, and the AMPA receptor antagonist 6-cyano-7-nitroquinoxaline-2,3-dione (CNQX) injected into the nucleus accumbens blocked the reinstatement produced by cocaine administered systemically. However, NMDA receptor antagonists injected into the nucleus accumbens were without effect on reinstatement produced by cocaine (Cornish *et al.*, 1999; Cornish and Kalivas, 2000). Kalivas and colleagues have identified at least four molecular sites of adaptational change in the medial prefrontal cortex—nucleus accumbens glutamate pathway that are hypothesized to facilitate glutamate release in response to cocaine administration, including increased G-protein signaling through the upregulation of AGS3 (activator of G-protein signaling-3), reduced activity of the cystine-glutamate exchanger, diminished presynaptic metabotropic glutamate function, and decreased Homer-1 protein in the nucleus accumbens (Kalivas

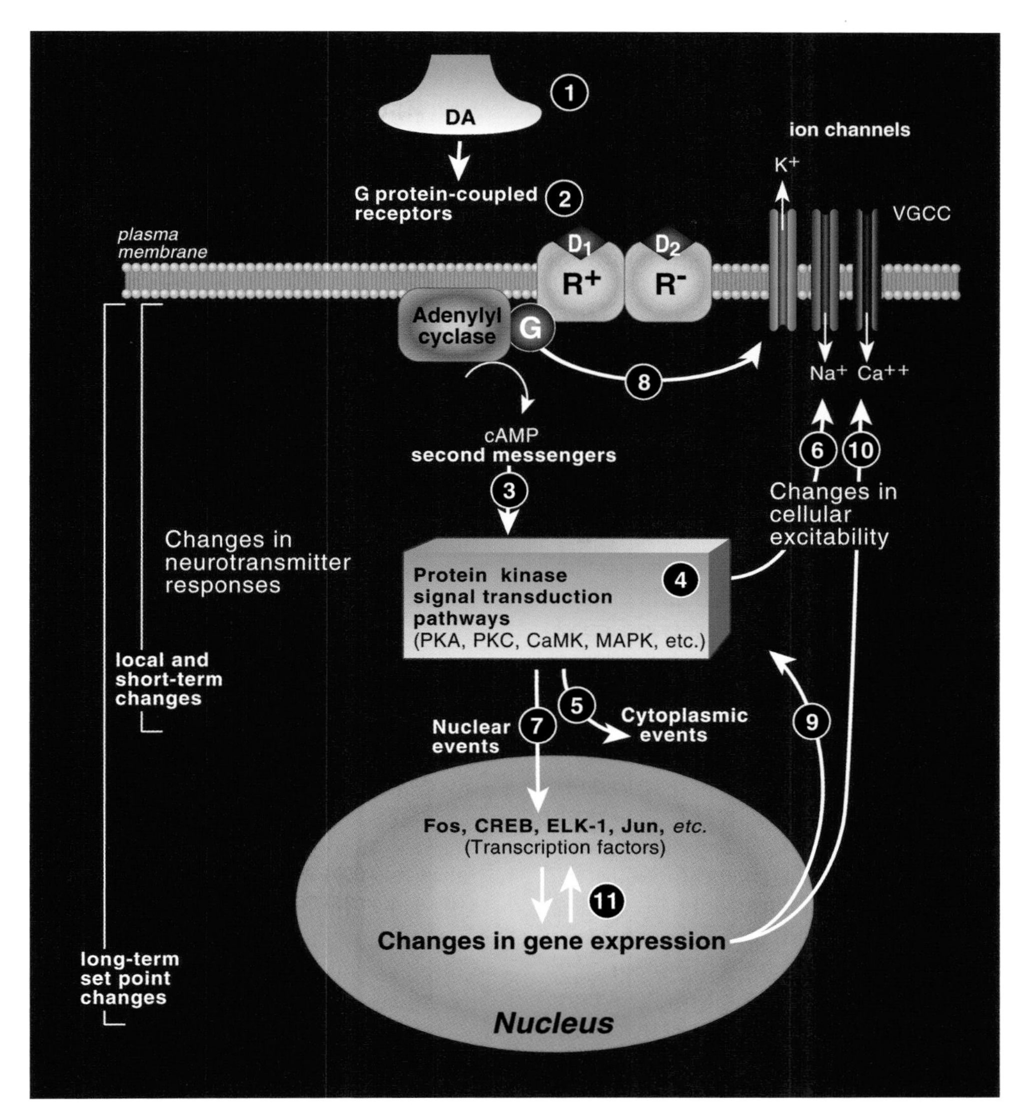

FIGURE 3.40 Molecular mechanisms of neuroadaptation—psychostimulants. Cocaine and amphetamines, as indirect sympathomimetics, stimulate the release of dopamine (①) which acts at G protein-coupled receptors (R), specifically D_1, D_2, D_3, D_4, and D_5 (②). These receptors modulate the levels of second-messengers like cyclic adenosine monophosphate (cAMP) and Ca^{2+} (③), which in turn regulate the activity of protein kinase transducers (④). Such protein kinases affect the functions of proteins located in the cytoplasm, plasma membrane, and nucleus (⑤ ⑥ ⑦). Among membrane proteins affected are ligand-gated and voltage-gated ion channels (⑥). G_i and G_o proteins also can regulate potassium and calcium channels directly through their $\beta\gamma$ subunits (⑧). Protein kinase transduction pathways also affect the activities of transcription factors (⑦). Some of these factors, like cAMP response element binding protein (CREB), are regulated posttranslationally by phosphorylation; others, like Fos, are regulated transcriptionally; still others, like Jun, are regulated both posttranslationally and/or transcriptionally. While membrane and cytoplasmic changes may be only local (e.g., dendritic domains or synaptic boutons), changes in the activity of transcription factors may result in long-term functional changes. These may include changes in gene expression of proteins involved in signal transduction (⑨) and/or neurotransmission (⑩), resulting in altered neuronal responses. For example, chronic exposure to psychostimulants has been reported to increase levels of protein kinase A (⑨) and adenylyl cyclase in the nucleus accumbens and to decrease levels of $G_{i\alpha}$. Chronic exposure to psychostimulants also alters the expression of transcription factors themselves (⑪). CREB expression, for instance, is depressed in the nucleus accumbens by chronic cocaine treatment. Chronic cocaine induces a transition from Fos induction to the induction of the much longer-lasting Fos-related antigens (Fras) such as ΔFosB. The receptor systems depicted in the figure may not coexist in the same cells. [Modified with permission from Koob *et al.*, 1998.]

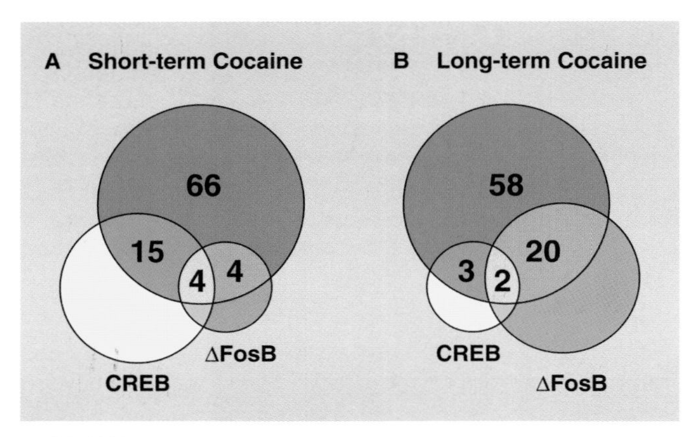

FIGURE 3.41 Regulation of gene expression by cocaine and comparison to effects of cyclic adenosine monophosphate response element binding protein (CREB) and ΔFosB. Mice carrying the *NSE-tTA* transgene alone were given daily i.p. injections of cocaine (10 mg/kg) or saline for 5 days (A), or cocaine (15 mg/kg) or saline for 5 days/week for four consecutive weeks (B), and were used 24 h after the last injection. RNA was isolated from the nucleus accumbens and subjected to microarray analysis. **(A)** Venn diagram showing the number of genes with shared upregulation among short-term cocaine treatment, short-term ΔFosB expression (1–2 weeks), and CREB expression (8 weeks). **(B)** Venn diagram showing the number of genes with shared upregulation among long-term cocaine treatment, long-term ΔFosB (4–8 weeks), and CREB expression. [Reproduced with permission from McClung and Nestler, 2003.]

and McFarland, 2003) (see *Neurobiological Theories of Addiction* chapter). Further evidence supporting a glutamatergic role in cocaine reinstatement was the observation that stimulation of the hippocampal-containing neurons in the ventral subiculum reinstates cocaine-seeking behavior and is blocked by intra-ventral tegmental area infusion of kynurenic acid, a nonselective ionotropic glutamate antagonist (Vorel *et al.*, 2001). These results are consistent with evidence at the cellular level of long-term potentiation in dopamine neurons produced by cocaine that is blocked by co-administration of an NMDA antagonist (Ungless *et al.*, 2001). Interestingly, chronic cocaine treatment resulted in decreased glutamate terminal immunolabeling within the ventral tegmental area, but not within the nucleus accumbens (Kozell and Meshul, 2003).

Serotonin, possibly through an action at the 5-HT_{2C} receptor, also may modulate reinstatement of cocaine-seeking behavior (Grottick *et al.*, 2000). Manipulations of intracellular signaling have shown that activators of PKA injected into the nucleus accumbens also can mimic cocaine reinstatement similar to D_2 receptor and AMPA receptor activation (Self *et al.*, 1998). In conditioned reinstatement of cocaine-seeking, both Fos expression and increased responding for cocaine-predictive cues were reversed by pretreatment with selective D_1 antagonists (Ciccocioppo *et al.*, 2001). Other studies using a conceptually similar model with a second-order schedule of reinforcement have shown that a partial D_3 antagonist and a D_3 antagonist can block the cocaine-seeking component of a second-order schedule without affecting cocaine self-administration *per se* (Pilla *et al.*, 1999). Recent data also show that a cannabinoid CB_1 receptor antagonist can attenuate cue-induced reinstatement for cocaine (De Vries *et al.*, 2001).

SUMMARY

Psychostimulant drugs such as cocaine and amphetamines of the indirect sympathomimetic class have a long history associated with tonics and other preparations to allay fatigue and sustain performance. These drugs also have a long history of abuse and dependence with episodic collective amnesia through history about the behavioral toxicity associated with excessive use. Abuse potential varies with the availability of the drug, both environmentally and physiologically, with intravenous and smoked forms of both cocaine and amphetamines producing much more severe addiction. Cocaine and amphetamines produce euphoria, increase activity, facilitate performance (particularly in situations of fatigue), and decrease appetite. Amphetamines have medical uses as adjuncts for short-term weight control, treatment of ADHD and narcolepsy. Cocaine is available as a local anesthetic for mucous membrane anesthesia and vasoconstriction. An inverted U-shaped dose-response function has been demonstrated relating the performance-enhancing effects of psychostimulants to dose. Cocaine and amphetamines have a characteristic abuse cycle involving binge administration, withdrawal dysphoria, paranoia, and psychosis-like symptoms as the cycle continues or intensifies. The behavioral mechanism of action is hypothesized to reflect a behavioral principle where increases in response rates occur for a given behavior, but there is a decrease in the number of response categories of behavior manifested as the dose increases. This principle has significant explanatory power not only for the acute stimulant effects of the drugs but also for their pathophysiology.

Significant advances have been made in our understanding of the mechanism of action of psychomotor stimulant drugs at the behavioral, neuropharmacological, and molecular levels that have important implications for understanding the neurobiology of addiction to psychostimulants. There is evidence for a critical role for the mesolimbic dopamine system in the acute reinforcing effects of cocaine and D-amphetamine.

Dopamine release in the region of the nucleus accumbens is necessary for intravenous self-administration of cocaine, and microinjection studies have shown important contributions from dopamine receptors in other related structures in the extended amygdala, including the central nucleus of the amygdala and the BNST. These studies suggest that the circuitry for the acute reinforcing effects of cocaine and amphetamines involves the extended amygdala (which includes the BNST and the central nucleus of the amygdala and a transition area in the shell of the nucleus accumbens) through activation of a ventral striatal—ventral pallidal—thalamic circuit. At the cellular level, there are neural elements in the nucleus accumbens that respond both to the self-administration of the drug and the anticipation of drug self-administration. Molecular neuropharmacological studies have revealed important roles for dopamine D_1, D_2, and D_3 receptors, and studies in knockout mice have shown that D_1 receptor knockout mice do not readily self-administer cocaine. Acute withdrawal from psychostimulants produces major elevations in brain reward thresholds that may have a critical role in driving escalation to dependence. At the neurochemical level, decreases in mesolimbic dopaminergic and serotonergic function and recruitment of the brain stress neurotransmitter CRF appear to have roles in this acute motivational withdrawal syndrome. Stimulation of cAMP-dependent protein kinase, and activation of CREB with concomitant expression of dynorphin in the nucleus accumbens all have been hypothesized to provide at least one mechanism for motivational tolerance and dependence. Animal models of relapse have revealed important roles for the basolateral amygdala and medial prefrontal cortex connections to the nucleus accumbens and extended amygdala, with the neurotransmitters dopamine, glutamate and CRF having important roles. How these circuits are altered by molecular and cellular events following the chronic administration of cocaine and amphetamines to produce the neuroadaptive changes associated with addiction are the subject of intense investigation. Several molecular sites in the medial prefrontal cortex—nucleus accumbens glutamate projection have been proposed to mediate the enhanced glutamatergic signal importantly involved in cocaine-induced reinstatement. Another long-term molecular change that may weight circuitry to an increased sensitivity to cocaine long after abstinence, is the recruitment of the transcription factor ΔFosB. Such studies provide key insights into our understanding of the vulnerability to develop psychostimulant addiction and vulnerability to relapse, and will provide novel approaches to prevention and treatment in the human population.

REFERENCES

Ackerman, J. M., and White, F. J. (1990). A10 somatodendritic dopamine autoreceptor sensitivity following withdrawal from repeated cocaine treatment. *Neuroscience Letters* **117**, 181–187.

Ackerman, J. M., and White, F. J. (1992). Decreased activity of rat A10 dopamine neuronsfollowing withdrawal from repeated cocaine. *European Journal of Pharmacology* **218**, 171–173.

Ahmed, S. H., and Koob, G. F. (1998). Transition from moderate to excessive drug intake: Change in hedonic set point. *Science* **282**, 298–300.

Ahmed, S. H., and Koob, G. F. (1999). Long-lasting increase in the set point for cocaine self-administration after escalation in rats. *Psychopharmacology* **146**, 303–312.

Ahmed, S. H., Kenny, P. J., Koob, G. F., and Markou, A. (2002). Neurobiological evidence for hedonic allostasis associated with escalating cocaine use. *Nature Neuroscience* **5**, 625–626.

Ahmed, S. H., Lin, D., Koob, G. F., and Parsons, L. H. (2003). Escalation of cocaine self-administration does not depend on altered cocaine-induced nucleus accumbens dopamine levels. *Journal of Neurochemistry* **86**, 102–113.

Ahmed, S. H., and Koob, G. F. (2004). Changes in response to a dopamine antagonist in rats with escalating cocaine intake. *Psychopharmacology* **172**, 450–454.

Ahmed, S. H., and Koob, G. F. (2005). The transition to drug addiction: A negative reinforcement model based on an allostatic decrease in reward function. *Psychopharmacology* **180**, 473–490.

Akimoto, K., Hamamura, T., and Otsuki, S. (1989). Subchronic cocaine treatment enhances cocaine-induced dopamine efflux, studied by in vivo intracerebral dialysis, *Brain Research* **490**, 339–344 [erratum: 495:203].

Alheid, G. F., and Heimer, L. (1988). New perspectives in basal forebrain organization of special relevance for neuropsychiatric disorders: The striatopallidal, amygdaloid, and corticopetal components of substantia innominata. *Neuroscience* **27**, 1–39.

Amalric, M. A., Berhow, M., Polis, I., and Koob, G. F. (1993). Selective effects of low dose D2 dopamine receptor antagonism in a reaction-time task in rats. *Neuropsychopharmacology* **8**, 195–200.

Amalric, M., and Koob, G. F. (1993). Functionally selective neurochemical afferents and efferents of the mesocorticolimbic and nigrostriatal dopamine system. *In Chemical Signalling in the Basal Ganglia* (series title: Progress in Brain Research, vol. 99) (G. W. Arbuthnott, and P. C. Emson, Eds.), pp. 209–226. Elsevier, New York.

Ambre, J. J., Belknap, S. M., Nelson, J., Ruo, T. I., Shin S-G., and Atkinson, A. J. Jr. (1988). Acute tolerance to cocaine in humans. *Clinical Pharmacology and Therapeutics* **44**, 1–8.

American Psychiatric Association. (1994). *Diagnostic and Statistical Manual of Mental Disorders*, 4th ed., American Psychiatric Press, Washington DC.

Anglin, M. D., Burke, C., Perrochet, B., Stamper, E., and Dawud-Noursi, S. (2000). History of the methamphetamine problem. *Journal of Psychoactive Drugs* **32**, 137–141.

Angrist, B. M., Gershon, S. (1969). Amphetamine abuse in New York City: 1966–1968. *Seminars in Psychiatry* **1**, 195–207.

Angrist, B., and Sudilovsky, A. (1978). Central nervous system stimulants: Historical aspects and clinical effects. *In Stimulants* (series title: *Handbook of Psychopharmacology*, vol. 11), (L. L. Iversen, S. D. Iversen, S. H. Snyder, Eds.), pp. 99–165. Plenum Press, New York.

Angrist, B., Corwin, J., Bartlik, B., and Cooper, T. (1987). Early pharmacokinetics and clinical effects of oral d-amphetamine in normal subjects. *Biological Psychiatry* **22**, 1357–1368.

Anthony, J. C. (1992). Epidemiological research on cocaine use in the USA. In: Bock GR, *Cocaine: Scientific and Social Dimensions* (series title: *Ciba Foundation Symposium*, vol. 166), (J. Whelan, Eds.), pp. 20–39 John Wiley, Chichester.

Anthony, J. C., Warner, L. A., and Kessler, R. C. (1994). Comparative epidemiology of dependence on tobacco, alcohol, controlled substances, and inhalants: Basic findings from the National Comorbidity Survey. *Experimental and Clinical Psychopharmacology* **2**, 244–268.

Aronson, S. C., Black, J. E., McDougle, C. J., Scanley, B. E., Jatlow, P., Kosten, T. R., Heninger, G. R., and Price, L. H. (1995). Serotonergic mechanisms of cocaine effects in humans. *Psychopharmacology* **119**, 179–185.

Askevold, F. (1959). The occurrence of paranoid incidents and abstinence delirium in abusers of amphetamine. *Acta Psychiatrica et Neurologica Scandinavica* **34**, 145–164.

Aston-Jones, G., and Harris, G. C. (2004). Brain substrates for increased drug seeking during protracted withdrawal, Neuropharmacology **47**(Suppl 1.):167–179.

Baekeland, F. (1966). The effect of methyl phenidate on the sleep cycle in man. *Psychopharmacologia* **10**, 179–183.

Baik, J. H., Picetti, R., Saiardi, A., Thiriet, G., Dierich, A., Depaulis, A., Le Meur, M., and Borrelli, E. (1995). Parkinsonian-like locomotor impairment in mice lacking dopamine D2 receptors. *Nature* **377**, 424–428.

Balloch, J. C., LaSaine, T. A., and Robinson, J. M. (1952). An experimental study of the effects of dexedrine (D-amphetamine sulfate) upon motor and mental performance and some factors in mood. *Journal of the Tennessee Academy of Science* **27**, 296–303.

Bals-Kubik, R., Ableitner, A., Herz, A., and Shippenberg, T. S. (1993). Neuroanatomical sites mediating the motivational effects of opioids as mapped by the conditioned place preference paradigm in rats. *Journal of Pharmacology and Experimental Therapeutics* **264**, 489–495.

Barrot, M., Marinelli, M., Abrous, D. N., Rouge-Pont, F., Le Moal, M., and Piazza, P. V. (2000). The dopaminergic hyper-responsiveness of the shell of the nucleus accumbens is hormone-dependent. *European Journal of Neuroscience* **12**, 973–979.

Basso, A. M., Spina, M., Rivier, J., Vale, W., and Koob, G. F. (1999). Corticotropin-releasing factor antagonist attenuates the 'anxiogenic-like' effect in the defensive burying paradigm but not in the elevated plus-maze following chronic cocaine in rats *Psychopharmacology* **145**, 21–30.

Beitner-Johnson, D., Guitart, X., and Nestler, E. J. (1992). Neurofilament proteins and the mesolimbic dopamine system: common regulation by chronic morphine and chronic cocaine in the rat ventral tegmental area. *Journal of Neuroscience* **12**, 2165–2176.

Benzedrine: report of the council on pharmacy and chemistry. (1933). *Journal of the American Medical Association* **101**, 1315.

Bergman, J., Kamien, J. B., and Spealman, R. D. (1990). Antagonism of cocaine self-administration by selective dopamine D1 and D2 antagonists. *Behavioural Pharmacology* **1**, 355–363.

Berhow, M. T., Hiroi, N., Kobierski, L. A., Hyman, S. E., and Nestler, E. J. (1996a). Influence of cocaine on the JAK-STAT pathway in the mesolimbic dopamine system. *Journal of Neuroscience* **16**, 8019–8026.

Berhow, M. T., Hiroi, N., and Nestler, E. J. (1996b). Regulation of ERK (extracellular signal regulated kinase), part of the neurotrophin signal transduction cascade, in the rat mesolimbic dopamine system by chronic exposure to morphine or cocaine, *Journal of Neuroscience* **16**, 4707–4715.

Blandina, P., Goldfarb, J., Craddock-Royal, B., and Green, J. P. (1989). Release of endogenous dopamine by stimulation of 5-hydroxytryptamine3 receptors in rat striatum. *Journal of Pharmacology and Experimental Therapeutics* **251**, 803–809.

Bonci, A., and Malenka, R. C. (1999). Properties and plasticity of excitatory synapses on dopaminergic and GABAergic cells in the ventral tegmental area. *Journal of Neuroscience* **19**, 3723–3730.

Bradley, C. (1937). The behavior of children receiving benzedrine. *American Journal of Psychiatry* **94**, 577–585.

Brain, P. F., and Coward, G. A. (1989). A Review of the history, actions and legitimate uses of cocaine. *Journal of Substance Abuse* **1**, 431–451.

Branch, M. N. (1975). Effects of chlorpromazine and d-amphetamine on observing responses during a fixed-interval schedule. *Psychopharmacologia* **42**, 87–93.

Branch, M. N., and Gollub, L. R. (1974). A detailed analysis of the effects of d-amphetamine on behavior under fixed-interval schedules. *Journal of the Experimental Analysis of Behavior* **21**, 519–539.

Breiter, H. C., Gollub, R. L., Weisskoff, R. M., Kennedy, D. N., Makris, N., Berke, J. D., Goodman, J. M., Kantor, H. L., Gastfriend, D. R., Riorden, J. P., Mathew, R. T., Rosen, B. R., and Hyman, S. E. (1997). Acute effects of cocaine on human brain activity and emotion. *Neuron* **19**, 591–611.

Britton, D. R., Curzon, P., Mackenzie, R. G., Kebabian, J. W., Williams, J. E., and Kerkman, D. (1991). Evidence for involvement of both D1 and D2 receptors in maintaining cocaine self-administration. *Pharmacology Biochemistry and Behavior* **39**, 911–915.

Burns, L. H., and Robbins, T. W. (1993). Everitt BJ, Differential effects of excitotoxic lesions of the basolateral amygdala, ventral subiculum and medial prefrontal cortex on responding with conditioned reinforcement and locomotor activity potentiated by intra-accumbens infusions of D-amphetamine. *Behavioural Brain Research* **55**, 167–183.

Burroughs, W. S. (1959). *The Naked Lunch*, Grove Press, New York.

Byck, R. (1987). Cocaine use and research: three histories. In: Fisher S, Raskin A, *Cocaine: Clinical and Biobehavioral Aspects*, (E. H. Uhlenhuth Ed.), pp. 3–20 Oxford University Press, New York.

Cador, M., Robbins, T. W., and Everitt, B. J. (1989). Involvement of the amygdala in stimulus-reward associations: Interaction with the ventral striatum. *Neuroscience* **30**, 77–86.

Caine, S. B., and Koob, G. F. (1993). Modulation of cocaine self-administration in the rat through D-3 dopamine receptors. *Science* **260**, 1814–1816.

Caine, S. B., and Koob, G. F. (1994). Effects of dopamine D1 and D2 antagonists on cocaine self-administration under different schedules of reinforcement in the rat. *Journal of Pharmacology and Experimental Therapeutics* **270**, 209–218.

Caine, S. B., and Koob, G. F. (1995). Pretreatment with the dopamine agonist 7-OH-DPAT shifts the cocaine self-administration dose-effect function to the left under different schedules in the rat. *Behavioural Pharmacology* **6**, 333–347.

Caine, S. B., Heinrichs, S. C., Coffin, V. L., and Koob, G. F. (1995). Effects of the dopamine D1 antagonist SCH 23390 microinjected into the accumbens, amygdala or striatum on cocaine self-administration in the rat. *Brain Research* **692**, 47–56.

Caine, S. B., Koob, G. F., Parsons, L. H., Everitt, B. J., Schwartz, J-C., Sokoloff, P. (1997). D3 receptor test in vitro predicts decreased cocaine self-administration in rats. *Neuroreport* **8**, 2373–2377.

Caine, S. B., Gabriel, K. I., Berkowitz, J. S., Zhang, J., and Xu, M. (2002a). Decreased cocaine self-administration in dopamine D-1 receptor knockout mice. *Drug and Alcohol Dependence* **66**, s25.

Caine, S. B., Negus, S. S., Mello, N. K., Patel, S., Bristow, L., Kulagowski, J., Vallone, D., Saiardi, A., and Borrelli, E. (2002b). Role of dopamine D2-like receptors in cocaine self-administration: studies with D2 receptor mutant mice and novel D2 receptor antagonists. *Journal of Neuroscience* **22**, 2977–2988.

Caldwell, J. A., Caldwell, J. L., Crowley, J. S., and Jones, H. D. (1995). Sustaining helicopter pilot performance with Dexedrine during periods of sleep deprivation. *Aviation, Space, and Environmental Medicine* **66**, 930–937.

Caldwell, J. A., Smythe, N. K., Leduc, P. A., and Caldwell, J. L. (2000). Efficacy of Dexedrine for maintaining aviator performance during 64 hours of sustained wakefulness: a simulator study. *Aviation, Space, and Environmental Medicine* **71**, 7–18.

Callahan, P. M., and Cunningham, K. A. (1995). Modulation of the discriminative stimulus properties of cocaine by 5-HT1B and 5-HT2C receptors. *Journal of Pharmacology and Experimental Therapeutics* **274**, 1414–1424.

Cannon, W. B., and Rosenblueth, A. (1933). Studies on conditions of activity in endocrine organs: XXIX. Sympathin E and sympathin I. *American Journal of Physiology* **104**, 557–574.

Carboni, E., Spielewoy, C., Vacca, C., Nosten-Bertrand, M., Giros, B., and Di Chiara, G. (2001). Cocaine and amphetamine increase extracellular dopamine in the nucleus accumbens of mice lacking the dopamine transporter gene. *Journal of Neuroscience* **21**, RC141 (1-4).

Carelli, R. M. (1994). Deadwyler SA, A comparison of nucleus accumbens neuronal firing patterns during cocaine self-administration and water reinforcement in rats. *Journal of Neuroscience* **14**, 7735–7746.

Carelli, R. M. (2000). Activation of accumbens cell firing by stimuli associated with cocaine delivery during self-administration. *Synapse* **35**, 238–242.

Carelli, R. M. (2002). Nucleus accumbens cell firing during goal-directed behaviors for cocaine vs. 'natural' reinforcement. *Physiology and Behavior* **76**, 379–387.

Carelli, R. M., King, V. C., Hampson, R. E., and Deadwyler, S. A. (1993). Firing patterns of nucleus accumbens neurons during cocaine self-administration in rats. *Brain Research* **626**, 14–22.

Carelli, R. M., Ijames, S., Konstantopoulos, J., and Deadwyler, S. A. (1999). Examination of factors mediating the transition to behaviorally correlated nucleus accumbens cell firing during cocaine self-administration sessions in rats. *Behavioural Brain Research* **104**, 127–139.

Carelli, R. M., and Ijames, S. G. (2001). Selective activation of accumbens neurons by cocaine-associated stimuli during a water/cocaine multiple schedule. *Brain Research* **907**, 156–161.

Carelli, R. M., Williams, J. G., and Hollander, J. A. (2003). Basolateral amygdala neurons encode cocaine self-administration and cocaine-associated cues. *Journal of Neuroscience* **23**, 8204–8211.

Carlezon, W. A. Jr, Thome, J., Olson, V. G., Lane-Ladd, S. B., Brodkin, E. S., Hiroi, N., Duman, R. S., Neve, R. L., and Nestler, E. J. (1998). Regulation of cocaine reward by CREB. *Science* **282**, 2272–2275.

Carroll, M. E., and Lac, S. T. (1987). Cocaine withdrawal produces behavioral disruptions in rats. *Life Sciences* **40**, 2183–2190.

Carroll, M. E., Lac, S. T., Asencio, M., and Kragh, R. (1990). Intravenous cocaine self-administration in rats is reduced by dietary L-tryptophan. *Psychopharmacology* **100**, 293–300.

Chang, J. Y., Janak, P. H., and Woodward, D. J. (1998). Comparison of mesocorticolimbic neuronal responses during cocaine and heroin self-administration in freely moving rats. *Journal of Neuroscience* **18**, 3098–3115.

Chang, J. Y., Janak, P. H., and Woodward, D. J. (2000). Neuronal and behavioral correlations in the medial prefrontal cortex and nucleus accumbens during cocaine self-administration by rats, *Neuroscience* **99**, 433–443.

Chang, J. Y., Sawyer, S. F., Lee, R-S., and Woodward, D. J. (1994). Electrophysiological and pharmacological evidence for the role of the nucleus accumbens in cocaine self-administration in freely moving rats. *Journal of Neuroscience* **14**, 1224–1244.

Chapman, A. H. (1954). Paranoid psychoses associated with amphetamine usage; a clinical note. *American Journal of Psychiatry* **111**, 43–45.

Chen, J. P., van Praag, H. M., and Gardner, E. L. (1991). Activation of 5-HT3 receptor by 1-phenylbiguanide increases dopamine release in the rat nucleus accumbens. *Brain Research* **543**, 354–357.

Chen, J., Paredes, W., Van Praag, H. M., Lowinson, J. H., and Gardner, E. L. (1992). Presynaptic dopamine release is enhanced by 5-HT3 receptor activation in medial prefrontal cortex of freely moving rats. *Synapse* **10**, 264–266.

Childress, A. R., Mozley, P. D., McElgin, W., Fitzgerald, J., Reivich, M., O'Brien, C. P. (1999). Limbic activation during cue-induced cocaine craving. *American Journal of Psychiatry* **156**, 11–18.

Cho, A. K. (1990). Ice: a new dosage form of an old drug. *Science* **249**, 631–634.

Cho, A. K., and Melega, W. P. (2002). Patterns of methamphetamine abuse and their consequences. *Journal of Addictive Diseases* **21**, 21–34.

Chopra, R. N., and Chopra, G. S. (1931). Cocaine habit in India. *Indian Journal of Medical Research* **18**, 1013–1046.

Ciccocioppo, R., Sanna, P. P., and Weiss, F. (2001). Cocaine-predictive stimulus induces drug-seeking behavior and neural activation in limbic brain regions after multiple months of abstinence: reversal by D(1) antagonists. *Proceedings of the National Academy of Sciences USA* **98**, 1976–1981.

Cocaine Anonymous: Public Information Fact File, (2003). Cocaine Anonymous World Service Office, Los Angeles.

Colby, C. R., Whisler, K., Steffen, C., Nestler, E. J., and Self, D. W. (2003). Striatal cell type-specific overexpression of DeltaFosB enhances incentive for cocaine. *Journal of Neuroscience* **23**, 2488–2493.

Cole, R. L., Konradi, C., Douglass, J., and Hyman, S. E. (1995). Neuronal adaptation to amphetamine and dopamine: molecular mechanisms of prodynorphin gene regulation in rat striatum. *Neuron* **14**, 813–823.

Colpaert, F. C. (1996). System theory of pain and of opiate analgesia: No tolerance to opiates. *Pharmacological Reviews* **48**, 355–402.

Cook, C. E., and Jeffcoat, A. R. (1986). Cocaine thermal degradation, *Chemical and Engineering News* **29**, 4.

Cook, C. E., Jeffcoat, A. R., Hill, J. M., Pugh, D. E., Patetta, P. K., Sadler, B. M., White, W. R., and Perez-Reyes, M. (1993). Pharmacokinetics of methamphetamine self-administered to human subjects by smoking S-(+)-methamphetamine hydrochloride. *Drug Metabolism and Disposition* **21**, 717–723.

Cools, R., Barker, R. A., Sahakian, B. J., and Robbins, T. W. (2001). Enhanced or impaired cognitive function in Parkinson's disease as a function of dopaminergic medication and task demands. *Cerebral Cortex* **11**, 1136–1143.

Cools, R., and Robbins, T. W. (2004). Chemistry of the adaptive mind, *Philosophical Transactions: Series A. Mathematical, Physical, and Engineering Sciences* **362**, 2871–2888.

Cornish, J. W., and O'Brien, C. P. (1996). Crack cocaine abuse: an epidemic with many public health consequences. *Annual Review of Public Health* **17**, 259–273.

Cornish, J. L., Duffy, P., and Kalivas, P. W. (1999). A role for nucleus accumbens glutamate transmission in the relapse to cocaine-seeking behavior. *Neuroscience* **93**, 1359–1367.

Cornish, J. L., and Kalivas, P. W. (2000). Glutamate transmission in the nucleus accumbens mediates relapse in cocaine addiction. *Journal of Neuroscience* **20**, RC89.

Cornish, J. L., and Kalivas, P. W. (2001). Cocaine sensitization and craving: differing roles for dopamine and glutamate in the nucleus accumbens. *Journal of Addictive Diseases* **20**, 43–54.

Corrigall, W. A., and Coen, K. M. (1991). Cocaine self-administration is increased by both D1 and D2 dopamine antagonists. *Pharmacology Biochemistry and Behavior* **39**, 799–802.

Costall, B., Domeney, A. M., Naylor, R. J., and Tyers, M. B. (1987). Effects of the 5-HT3 receptor antagonist, GR38032F, on raised dopaminergic activity in the mesolimbic system of the rat and marmoset brain, *British Journal of Pharmacology* **92**, 881–894.

Creese, I., and Iversen, S. D. (1974). The role of forebrain dopamine systems in amphetamine-induced stereotyped behavior in the rat. *Psychopharmacology* **39**, 345–357.

Cuthbertson, D. P., and Knox, J. A. C. (1947). The effects of analeptics on the fatigued subject. *Journal of Physiology* **106**, 42–58.

Das, G. (1990). Cocaine and the cardiovascular system. *Canadian Journal of Cardiology* **6**, 411–415.

Das, G. (1993). Cocaine abuse in North America: a milestone in history, *Journal of Clinical Pharmacology* **33**, 296–310.

Davis, J. M., Kopin, I. J., Lemberger, L., and Axelrod, J. (1971). Effects of urinary pH on amphetamine metabolism. *In Drug Metabolism in Man* (series title: *Annals of the New York Academy of Sciences*, vol. 179), (E. S. Vessell, Ed.), pp. 493–501. New York Academy of Sciences, New York.

De Long, M. R. (1990). Primate models of movement disorders of basal ganglia origin, *Trends in Neurosciences* **13**, 281–285.

De Vries, T. J., Shaham, Y., Homberg, J. R., Crombag, H., Schuurman, K., Dieben, J., Vanderschuren, L. J., and Schoffelmeer, A. N. (2001). A cannabinoid mechanism in relapse to cocaine seeking. *Nature Medicine* **7**, 1151–1154.

de Wit, H., and Stewart, J. (1981). Reinstatement of cocaine-reinforced responding in the rat. *Psychopharmacology* **75**, 134–143.

Derlet, R. W., and Heischober, B. (1990). Methamphetamine. Stimulant of the 1990s? *Western Journal of Medicine* **153**, 625–628.

Deroche, V., Marinelli, M., Le Moal, M., and Piazza, P. V. (1997). Glucocorticoids and behavioral effects of psychostimulants: II Cocaine intravenous self-administration and reinstatement depend on glucocorticoid levels. *Journal of Pharmacology and Experimental Therapeutics* **281**, 1401–1407.

Dews, P. B. (1958). Studies on behavior. IV. Stimulant actions of methamphetamine. *Journal of Pharmacology and Experimental Therapeutics* **122**, 137–147.

Dews, P. B., and Wenger, G. R. (1977). Rate-dependent effects of amphetamine. *In Advances in Behavioral Pharmacology*, vol. 1, (T. Thompson., P. B. Dews, Eds.), pp. 167–227. Academic Press, New York.

Di Ciano, P., and Everitt, B. J. (2001). Dissociable effects of antagonism of NMDA and AMPA/KA receptors in the nucleus accumbens core and shell on cocaine-seeking behavior. *Neuropsychopharmacology* **25**, 341–360.

Di Ciano, P., Underwood, R. J., Hagan, J. J., and Everitt, B. J. (2003). Attenuation of cue-controlled cocaine-seeking by a selective D3 dopamine receptor antagonist SB-277011-A. *Neuropsychopharmacology* **28**, 329–338.

Dring, L. G., Smith, R. L., and Williams, R. T. (1966). The fate of amphetamine in man and other mammals. *Journal of Pharmacy and Pharmacology* **18**, 402–404.

Ellinwood, E. H. Jr., Sudilovsky, A., and Nelson, L. M. (1973). Evolving behavior in the clinical and experimental amphetamine (model) psychosis. *American Journal of Psychiatry* **130**, 1088–1093.

Epping-Jordan, M. P., Markou, A., and Koob, G. F. (1998). The dopamine D-1 receptor antagonist SCH 23390 injected into the dorsolateral bed nucleus of the stria terminalis decreased cocaine reinforcement in the rat. *Brain Research* **784**, 105–115.

Erb, S., Shaham, Y., and Stewart, J. (1996). Stress reinstates cocaine-seeking behavior after prolonged extinction and a drug-free period. *Psychopharmacology* **128**, 408–412.

Erb, S., Shaham, Y., and Stewart, J. (1998). The role of corticotropin-releasing factor and corticosterone in stress- and cocaine-induced relapse to cocaine seeking in rats. *Journal of Neuroscience* **18**, 5529–5536.

Erb, S., and Stewart, J. (1999). A role for the bed nucleus of the stria terminalis, but not the amygdala, in the effects of corticotropin-releasing factor on stress-induced reinstatement of cocaine seeking. *Journal of Neuroscience* **19**, RC35.

Erb, S., Salmaso, N., Rodaros, D., and Stewart, J. (2001). A role for the CRF-containing pathway from central nucleus of the amygdala to bed nucleus of the stria terminalis in the stress-induced reinstatement of cocaine seeking in rats. *Psychopharmacology* **158**, 360–365.

European Monitoring Centre for Drugs and Drug Addiction, (2004). *Annual Report 2004: The State of the Drugs Problem in the European Union and Norway*, Office for Official Publications of the European Communities, Luxembourg.

Everitt, B. J., Parkinson, J. A., Olmstead, M. C., Arroyo, M., Robledo, P., and Robbins, T. W. (1999). Associative processes in addiction and reward: The role of amygdala-ventral striatal subsystems. *In Advancing from the Ventral Striatum to the Extended Amygdala: Implications for Neuropsychiatry and Drug Abuse* (series title: *Annals of the New York Academy of Sciences*, vol. 877), (J. F. McGinty, Ed.), pp. 412–438. New York Academy of Sciences, New York.

Everitt, B. J., and Robbins, T. W. (2000). Second-order schedules of drug reinforcement in rats and monkeys: measurement of reinforcing efficacy and drug-seeking behaviour. *Psychopharmacology* **153**, 17–30.

Everitt, B. J., and Wolf, M. E. (2002). Psychomotor stimulant addiction: a neural systems perspective. *Journal of Neuroscience* **22**, 3312–3320 [erratum: 22(16):1a].

Eysenck, H. J., Casey, S., and Trouton, D. S. (1957). Drugs and personality: II. The effect of stimulant and depressant drugs on continuous work. *Journal of Mental Science* **103**, 645–649.

Ferris, R. M., Tang, F. L., and Maxwell, R. A. (1972). A comparison of the capacities of isomers of amphetamine, deoxypipradrol and methylphenidate to inhibit the uptake of tritiated catecholamines into rat cerebral cortex slices, synaptosomal preparations of rat cerebral cortex, hypothalamus and striatum and into adrenergic nerves of rabbit aorta. *Journal of Pharmacology and Experimental Therapeutics* **181**, 407–416.

Fischman, M. W., and Schuster, C. R. (1982). Cocaine self-administration in humans. *Federation Proceedings* **41**, 241–246.

Fischman, M. W., Schuster, C. R., Javaid, J., Hatano, Y., and Davis, J. (1985). Acute tolerance development to the cardiovascular and subjective effects of cocaine. *Journal of Pharmacology and Experimental Therapeutics* **235**, 677–682.

Fitzgerald, L. W., Ortiz, J., Hamedani, A. G., and Nestler, E. J. (1996). Drugs of abuse and stress increase the expression of GluR1 and NMDAR1 glutamate receptor subunits in the rat ventral tegmental area: common adaptations among cross-sensitizing agents. *Journal of Neuroscience* **16**, 274–282.

Foltin, R. W., Fischman, M. W., Nestadt, G., Stromberger, H., Cornell, E. E., and Pearlson, G. D. (1990). Demonstration of naturalistic methods for cocaine smoking by human volunteers. *Drug and Alcohol Dependence* **26**, 145–154.

Freud, S. (1974). *Cocaine Papers*, New American Library, New York.

Freud, S. Uber koca, (1884). *Wien Centralblatt fur die Gesamte Therapie* **2**, 289–314.

Fuchs, R. A., Weber, S. M., Rice, H. J., and Neisewander, J. L. (2002). Effects of excitotoxic lesions of the basolateral amygdala on cocaine-seeking behavior and cocaine conditioned place preference in rats. *Brain Research* **929**, 15–25.

Fung, Y. K., and Richard, L. A. (1994). Behavioural consequences of cocaine withdrawal in rats. *Journal of Pharmacy and Pharmacology* **46**, 150–152.

Gao, W. Y., Lee, T. H., King, G. R., and Ellinwood, E. H. (1998). Alterations in baseline activity and quinpirole sensitivity in putative dopamine neurons in the substantia nigra and ventral tegmental area after withdrawal from cocaine pretreatment. *Neuropsychopharmacology* **18**, 222–232.

Gauvin, D. V., Briscoe, R. J., Baird, T. J., Vallett, M., Carl, K. L., and Holloway, F. A. (1997). Physiological and subjective effects of acute cocaine withdrawal (crash) in rats. *Pharmacology Biochemistry and Behavior* **57**, 923–934.

Gawin, F. H. (1991). Cocaine addiction: psychology and neurophysiology. *Science* **251**, 1580–1586.

Gawin, F. H., and Kleber, H. D. (1986). Abstinence symptomatology and psychiatric diagnosis in cocaine abusers: Clinical observations. *Archives of General Psychiatry* **43**, 107–113.

Gawin, F. H., and Ellinwood, E. H. Jr. (1988). Cocaine and other stimulants: actions, abuse, and treatment. *New England Journal of Medicine* **318**, 1173–1182.

Giros, B., Jaber, M., Jones, S. R., Wightman, R. M., and Caron, M. G. (1996). Hyperlocomotion and indifference to cocaine and amphetamine in mice lacking the dopamine transporter. *Nature* **379**, 606–612.

Glowinski, J., and Axelrod, J. (1965). Effect of drugs on the uptake, release, and metabolism of 3H-norepinephrine in the rat brain. *Journal of Pharmacology and Experimental Therapeutics* **149**, 43–49.

Gold, M. S., and Jacobs, W. S. (2005). Cocaine and crack: clinical aspects. *In Substance Abuse: A Comprehensive Textbook*, 4th ed., (J. H. Lowinson, P. Ruiz, R. B. Millman, and J. G. Langrod, Eds.), pp. 218–251. Lippincott Williams and Wilkins, Philadelphia.

Graybiel, A. M., Moratalla, R., and Robertson, H. A. (1990). Amphetamine and cocaine induce drug-specific activation of the c-fos gene in striosome-matrix compartments and limbic subdivisions of the striatum. *Proceedings of the National Academy of Sciences USA* **87**, 6912–6916.

Graziella de Montis, M., Co, C., Dworkin, S. I., and Smith, J. E. (1998). Modifications of dopamine D1 receptor complex in rats self-administering cocaine. *European Journal of Pharmacology* **362**, 9–15.

Griffith, J. D., Cavanaugh, J., Held, J., and Oates, J. A. (1972). Dextroamphetamine: evaluation of psychomimetic properties in man. *Archives of General Psychiatry* **26**, 97–100.

Grinspoon, L., and Bakalar, J. B. (1980). Drug dependence: non-narcotic agents. *In Comprehensive Textbook of Psychiatry: Volume 2*, 3rd ed., (H. I. Kaplan, A. M. Freedman, and B. J. Sadock, Eds.), pp. 1614–1629 Williams and Wilkins, Baltimore.

Grinspoon, L., and Bakalar, J. B. (1981). Coca and cocaine as medicines: an historical. *Journal of Ethnopharmacology* **3**, 161–172.

Grottick, A. J., Fletcher, P. J., and Higgins, G. A. (2000). Studies to investigate the role of 5-HT(2C) receptors on cocaine- and food-maintained behavior. *Journal of Pharmacology and Experimental Therapeutics* **295**, 1183–1191.

Habert, E., Graham, D., Tahraoui, L., Claustre, Y., and Langer, S. Z. (1985). Characterization of [3H]paroxetine binding to rat cortical membranes, *European Journal of Pharmacology* **118**, 107–114.

Haddad, L.M. (1979). 1978: Cocaine in perspective, *Journal of the American College of Emergency Physicians* **8**, 374–376.

Haile, C. N., and Kosten, T. A. (2001). Differential effects of D1- and D2-like compounds on cocaine self-administration in Lewis and Fischer 344 inbred rats. *Journal of Pharmacology and Experimental Therapeutics* **299**, 509–518.

Hall, R. J. (1884). Hydrochlorate of cocaine. *New York Medical Journal* **40**, 643–644.

Härfstrand, A., Fuxe, K., Cintra, A., Agnati, L. F., Zini, I., Wikstrom, A. C., Okret, S., Yu, Z. Y., Goldstein, M., Steinbusch, H., Verhofstad A., and Gustafsson, J-A. (1986). Glucocorticoid receptor immunoreactivity in monoaminergic neurons of rat brain. *Proceedings of the National Academy of Sciences USA* **83**, 9779–9783.

Harris, G. C., and Aston-Jones, G. (1993). Beta-adrenergic antagonists attenuate withdrawal anxiety in cocaine- and morphine-dependent rats. *Psychopharmacology* **113**, 131–136.

Hauty, G. T., and Payne, R. B. (1957). Effects of dextro-amphetamine upon judgment. *Journal of Pharmacology and Experimental Therapeutics* **120**, 33–37.

Heimer, L., and Alheid, G. (1991). Piecing together the puzzle of basal forebrain anatomy. *In The Basal Forebrain: Anatomy to Function* (series title: *Advances in Experimental Medicine and Biology*, vol. 295), (T. C. Napier, P. W. Kalivas, and I. Hanin, Eds.), pp. 1–42. Plenum Press, New York.

Heimer, L., Zahm, D. S., Churchill, L., Kalivas, P. W., and Wohltmann, C. (1991). Specificity in the projection patterns of accumbal core and shell in the rat. *Neuroscience* **41**, 89–125.

Henry, D. J., Greene, M. A., and White, F. J. (1989). Electrophysiological effects of cocaine in the mesoaccumbens dopamine system: repeated administration. *Journal of Pharmacology and Experimental Therapeutics* **251**, 833–839.

Henry, D. J., and White, F. J. (1991). Repeated cocaine administration causes persistent enhancement of D1 dopamine receptor sensitivity within the rat nucleus accumbens. *Journal of Pharmacology and Experimental Therapeutics* **258**, 882–890.

Heyrodt, H., and Weissenstein, H. (1940). Uber steigerung korperlicher leistungfahigkeit durch pervitin [Excessive enhancement of physical performance by pervitin]. *Archiv fur Experimentelle Pathologie und Pharmakologie* **195**, 273–275.

Hiroi, N., Brown, J. R., Haile, C. N., Ye, H., Greenberg, M. E., and Nestler, E. J. (1997). FosB mutant mice: Loss of chronic cocaine induction of Fos-related proteins and heightened sensitivity to cocaine's psychomotor and rewarding effects. *Proceedings of the National Academy of Sciences USA* **94**, 10397–10402.

Hoffman, B. B., and Lefkowitz, R. J. (1990). Catecholamines and sympathomimetic drugs. *In Goodman and Gilman's The Pharmacological Basis of Therapeutics*, 8th ed., (A. G. Gilman, T. W. Rall, A. S. Nies, P. Taylor, Eds.), pp. 187–220. Pergamon Press, New York.

Hope, B. T., Nye, H. E., Kelz, M. B., Self, D. W., Iadarola, M. J., Nakabeppu, Y., Duman, R. S., and Nestler, E. J. (1994). Induction of a long-lasting AP-1 complex composed of altered Fos-like proteins in brain by chronic cocaine and other chronic treatments. *Neuron* **13**, 1235–1244.

Hubner, C. B., and Koob, G. F. (1990). The ventral pallidum plays a role in mediating cocaine and heroin self-administration in the rat. *Brain Research* **508**, 20–29.

Hubner, C. B., and Moreton, J. E. (1991). Effects of selective D1 and D2 dopamine antagonists on cocaine self-administration in the rat. *Psychopharmacology* **105**, 151–156.

Huey, L. Y. (1986). Attention deficit disorders. *In Psychological Foundations of Clinical Psychiatry* (series title: *Psychiatry Series*, vol. 4), (L. L. Judd, and P. M. Groves, Eds.), pp. 1–31. Basic Books, New York.

Hurd, Y. L., Brown, E. E., Finlay, J. M., Fibiger, H. C., and Gerfen, C. R. (1992). Cocaine self-administration differentially alters mRNA expression of striatal peptides. *Molecular Brain Research* **13**, 165–170.

Hyman, S. E., and Malenka, R. C. (2001). Addiction and the brain: the neurobiology of compulsion and its persistence. *Nature Reviews Neuroscience* **2**, 695–703.

Imperato, A., Mele, A., Scrocco, M. G., and Puglisi-Allegra, S. (1992). Chronic cocaine alters limbic extracellular dopamine. Neurochemical basis for addiction. *European Journal of Pharmacology* **212**, 299–300.

Iversen, L. L. (1973). Catecholamine uptake processes. *British Medical Bulletin* **29**, 130–135.

Iversen, S. D. (1977). Brain dopamine system and behavior. *In Drugs, Neurotransmitters, and Behavior* (series title: *Handbook of*

Psychopharmacology, vol. 8), (L. L. Iversen, S. D. Iversen, and S. H. Snyder, Eds.), pp. 334–384. Plenum Press, New York.

Iversen, S. D., and Fray, P. J. (1982). Brain catecholamines in relation to affect. *In Neural Basis of Behavior* (A. L. Beckman, Eds.), pp. 229–269. SP Medical and Scientific Books , New York.

Jaffe, J. H. (1985). Drug addiction and drug abuse. *In Goodman and Gilman's The Pharmacological Basis of Therapeutics*, 7th edition, (A. G. Gilman, L. S. Goodman, and T. W. Rall, Eds.), pp. 522–573. MacMillan, New York.

Jaffe, J. H. (1990). Drug addiction and drug use. *In Goodman and Gilman's The Pharmacological Basis of Therapeutics*, 8th edition, (A. G. Goodman, T. W. Rall, A. S. Nies, and P. Taylor, Eds.), pp. 522–573. Pergamon Press, New York.

Javitch, J. A., Blaustein, R. O., and Snyder, S. H. (1984). [3H]mazindol binding associated with neuronal dopamine and norepinephrine uptake sites. *Molecular Pharmacology* **26**, 35–44.

Jiang, L. H., Ashby, C. R. Jr., Kasser, R. J., and Wang, R. Y. (1990). The effect of intraventricular administration of the 5-HT3 receptor agonist 2-methylserotonin on the release of dopamine in the nucleus accumbens: an in vivo chronocoulometric study. *Brain Research* **513**, 156–160.

Johanson, C. E., Kilgore, K., and Uhlenhuth, E. H. (1983). Assessment of dependence potential of drugs in humans using multiple indices. *Psychopharmacology* **81**, 144–149.

Johanson, C. E., and Schuster, C. R. (1995). Cocaine. *In Psychopharmacology: The Fourth Generation of Progress*, Raven Press, (F. E. Bloom, and D. J. Kupfer, Eds.), pp. 1685–1697. New York.

Johnson, J., and Milner, G. (1966). Psychiatric complications of amphetamine substances *Acta Psychiatrica Scandinavica* **42**, 252–263.

Jones, E. (1961). *The Life and Work of Sigmund Freud*, Basic Books, New York.

Jones, S., Kornblum, J. L., and Kauer, J. A. (2000). Amphetamine blocks long-term synaptic depression in the ventral tegmental area. *Journal of Neuroscience* **20**, 5575–5580.

Joyce, E. M., and Koob, G. F. (1981). Amphetamine-, scopolamine- and caffeine-induced locomotor activity following 6-hydroxydopamine lesions of the mesolimbic dopamine system. *Psychopharmacology* **73**, 311–313.

Kalivas, P. W., Duffy, P., DuMars, L. A., and Skinner, C. (1988). Behavioral and neurochemical effects of acute and daily cocaine administration in rats. *Journal of Pharmacology and Experimental Therapeutics* **245**, 485–492.

Kalivas, P. W., and Stewart, J. (1991). Dopamine transmission in the initiation and expression of drug- and stress-induced sensitization of motor activity. *Brain Research Reviews* **16**, 223–244.

Kalivas, P. W., and McFarland, K. (2003). Brain circuitry and the reinstatement of cocaine-seeking behavior. *Psychopharmacology* **168**, 44–56.

Kankaanpaa, A., Lillsunde, P., Ruotsalainen, M., Ahtee, L., and Seppala, T. (1996). 5-HT3 receptor antagonist MDL 72222 dose-dependently attenuates cocaine- and amphetamine-induced elevations of extracellular dopamine in the nucleus accumbens and the dorsal striatum. *Pharmacology and Toxicology* **78**, 317–321.

Karch, S. (1997). *A Brief History of Cocaine*, CRC Press, Boca Raton.

Kebabian, J. W., and Calne, D. B. (1979). Multiple receptors for dopamine. *Nature* **277**, 93–96.

Kelley, A. E., Gauthier, A. M., and Lang, C. G. (1989). Amphetamine microinjections into distinct striatal subregions cause dissociable effects on motor and ingestive behavior. *Behavioural Brain Research* **35**, 27–39.

Kelly, P. H., Seviour, P. W., and Iversen, S. D. (1975). Amphetamine and apomorphine responses in the rat following 6-OHDA lesions of the nucleus accumbens septi and corpus striatum. *Brain Research* **94**, 507–522.

Kelly, P. H., and Iversen, S. D. (1976). Selective 6-OHDA-induced destruction of mesolimbic dopamine neurons: Abolition of psychostimulant-induced locomotor activity in rats. *European Journal of Pharmacology* **40**, 45–56.

Kelz, M. B., Chen, J., Carlezon, W. A. Jr., Whisler, K., Gilden, L., Beckmann, A. M., Steffen, C., Zhang, Y. J., Marotti, L., Self, D. W., Tkatch, T., Baranauskas, G., Surmeier, D. J., Neve, R. L., Duman, R. S., Picciotto, M. R., and Nestler, E. J. (1999). Expression of the transcription factor deltaFosB in the brain controls sensitivity to cocaine. *Nature* **401**, 272–276.

Kenny, P. J., Koob, G. F., and Markou, A. (2003). Conditioned facilitation of brain reward function after repeated cocaine administration. *Behavioral Neuroscience* **117**, 1103–1107.

Kessler, R. C. (2004). Prevalence of adult ADHD in the United States: results from the National Comorbidity Survey Replication (NCS-R) [abstract]. American Psychiatric Association, 157th Annual Meeting, New York, May 1.

King, G. R., Joyner, C., Lee, T., Kuhn, C., and Ellinwood, E. H. Jr. (1992). Intermittent and continuous cocaine administration: residual behavioral states during withdrawal. *Pharmacology Biochemistry and Behavior* **43**, 243–248.

King, G. R., Kuhn, C., and Ellinwood, E. H. Jr. (1993). Dopamine efflux during withdrawal from continuous or intermittent cocaine. *Psychopharmacology* **111**, 179–184.

King, G. R., Xiong, Z., Douglass, S., and Ellinwood, E. H. (2000). Long-term blockade of the expression of cocaine sensitization by ondansetron, a 5-HT(3) receptor antagonist. *European Journal of Pharmacology* **394**, 97–101.

Knapp, J. (1884). On cocaine and its use in ophthalmic and general surgery. *Archives of Ophthalmology* **13**, 402–448.

Koeltzow, T. E., and White, F. J. (2003). Behavioral depression during cocaine withdrawal is associated with decreased spontaneous activity of ventral tegmental area dopamine neurons. *Behavioral Neuroscience* **117**, 860–865.

Kokkinidis, L., and McCarter, B. D. (1990). Postcocaine depression and sensitization of brain-stimulation reward: Analysis of reinforcement and performance effects. *Pharmacology Biochemistry and Behavior* **36**, 463–471.

Kombian, S. B., and Malenka, R. C. (1994). Simultaneous LTP of non-NMDA- and LTD of NMDA-receptor-mediated responses in the nucleus accumbens, *Nature* **368**, 242–246.

Koob, G. F. (1999). Corticotropin-releasing factor, norepinephrine and stress. *Biological Psychiatry* **46**, 1167–1180.

Koob, G. F., Simon, H., Herman, J. P., and Le Moal, M. (1984). Neuroleptic-like disruption of the conditioned avoidance response requires destruction of both the mesolimbic and nigrostriatal dopamine systems. *Brain Research* **303**, 319–329.

Koob, G. F., Le, H. T., and Creese I. (1987a). The D-1 dopamine receptor antagonist SCH 23390 increases cocaine self-administration in the rat. *Neuroscience Letters* **79**, 315–320.

Koob, G. F., Vaccarino, F., Amalric, M., and Bloom, F. E. (1987b). Positive reinforcement properties of drugs: search for neural substrates. *In Brain Reward Systems and Abuse*, Raven Press, (J. Engel, L. Oreland, D. H. Ingvar, B. Pernow, S. Rossner, L. A. Pellborn, Eds.), pp. 35–50, New York.

Koob, G. F., and Kreek, M. J. (2005). Stress dysregulation of drug reward pathways, and drug dependence: an updated perspective commemorating the 30th anniversary of the National Institute on Drug Abuse. *American Journal of Psychiatry*, in press.

Koob, G. F., Sanna, P. P., and Bloom, F. E. (1998). Neuroscience of addiction. *Neuron* **21**, 467–476.

Kornetsky, C. (1958). Effects of meprobamate, phenobarbital, and dextroamphetamine on reaction time and learning in man.

Journal of Pharmacology and Experimental Therapeutics **123**, 216–219.

Kornetsky, C., Mirsky, A. F., Kessler, E. K., and Dorff, J. E. (1959). The effects of dextro-amphetamine on behavioral deficits produced by sleep loss in humans. *Journal of Pharmacology and Experimental Therapeutics* **127**, 46–50.

Kozell, B., and Meshul, K. (2003). Alterations in verve terminal glutamate immunoreactivity in the nucleus accumbens and the ventral tegmental area following single and repeated doses of cocaine. *Psychopharmacology* **165**, 337–345.

Kramer, J. C., Fischman, V. S., and Littlefield, D. C. (1967). Amphetamine abuse: pattern and effects of high doses taken intravenously. *Journal of the American Medical Association* **201**, 305–309.

Kreek, M. J. (1987). Multiple drug abuse patterns and medical consequences. *In Psychopharmacology: The Third Generation of Progress*, Raven Press, (H. Y. Meltzer, Eds.), pp. 1597–1604. New York.

Kreek, M. J., Ragunath, J., Plevy, S., Hamer, D., Schneider, B., and Hartman, N. (1984). ACTH, cortisol and beta-endorphin response to metyrapone testing during chronic methadone maintenance treatment in humans. *Neuropeptides* **5**, 277–278.

Lambert, N. M., Windmiller, M., Sandoval, J., and Moore, B. (1976). Hyperactive children and the efficacy of psychoactive drugs as a treatment intervention. *American Journal of Orthopsychiatry* **46**, 335–352.

Laties, V. G., and Weiss, B. (1981). The amphetamine margin in sports. *Federation Proceedings* **40**, 2689–2692.

Le Moal, M., and Simon, H. (1991). Mesocorticolimbic dopaminergic network: functional and regulatory roles. *Physiological Reviews* **71**, 155–234.

Leri, F., Flores, J., Rodaros, D., and Stewart, J. (2002). Blockade of stress-induced but not cocaine-induced reinstatement by infusion of noradrenergic antagonists into the bed nucleus of the stria terminalis or the central nucleus of the amygdala. *Journal of Neuroscience* **22**, 5713–5718.

Levesque, D., Diaz, J., Pilon, C., Martres, M-P., Giros, B., Souil, E., Schott, D., Morgat, J-L., Schwartz, J-C., and Sokoloff, P. (1992). Identification, characterization, and localization of the dopamine D3 receptor in rat brain using 7-[3H] hydroxy-N,N,-di-N-propyl-2-aminotetralin. *Proceedings of the National Academy of Sciences USA* **89**, 8155–8159.

Lewander, T. (1974). Effect of chronic treatment with central stimulants on brain monoamines and some behavioral and physiological functions in rats, guinea pigs, and rabbits. *In Neuropsychopharmacology of Monoamines and Their Regulatory Enzymes* (series title: *Advances in Biochemical Psychopharmacology*, vol. 12), (E. Usdin, Ed.), pp. 221–239. Raven Press, New York.

Loh, E. A., and Roberts, D. C. (1990). Break-points on a progressive ratio schedule reinforced by intravenous cocaine increase following depletion of forebrain serotonin. *Psychopharmacology* **101**, 262–266.

Louis, J. C., and Yazijian, H. Z. (1980). *The Cola Wars*, Everest House, New York.

Lu, L., Liu, D., and Ceng, X. (2001). Corticotropin-releasing factor receptor type 1 mediates stress-induced relapse to cocaine-conditioned place preference in rats. *European Journal of Pharmacology* **415**, 203–208.

Lyness, W. H., Friedle, N. M., and Moore, K. E. (1979). Destruction of dopaminergic nerve terminals in nucleus accumbens: Effect on d-amphetamine self-administration. *Pharmacology Biochemistry and Behavior* **11**, 553–556.

Lyon, M., and Randrup, A. (1972). The dose-response effect of amphetamine upon avoidance behaviour in the rat seen as a function of increasing stereotypy. *Psychopharmacologia* **23**, 334–347.

Lyon, M., and Robbins, T. W. (1975). The action of central nervous system stimulant drugs: a general theory concerning amphetamine effects. *In Current Developments in Psychopharmacology*, vol. 2, (W. B. Essman, and L. Valzelli, Eds.), pp. 79–163. Spectrum Publications, New York.

Maas, L. C., Lukas, S. E., Kaufman, M. J., Weiss, R. D., Daniels, S. L., Rogers, V. W., Kukes, T. J., and Renshaw, P. F. (1998). Functional magnetic resonance imaging of human brain activation during cue-induced cocaine craving. *American Journal of Psychiatry* **155**, 124–126.

Madge, T. (2001). *White Mischief: A Cultural History of Cocaine*. Mainstream Publishing, London.

Maggos, C. E., Tsukada, H., Kakiuchi, T., Nishiyama, S., Myers, J. E., Kreuter, J., Schlussman, S. D., Unterwald, E. M., Ho, A., and Kreek, M. J. (1998). Sustained withdrawal allows normalization of in vivo [11C]N-methylspiperone dopamine D2 receptor binding after chronic binge cocaine: a positron emission tomography study in rats. *Neuropsychopharmacology* **19**, 146–153.

Maisonneuve, I. M., Ho, A., and Kreek, M. J. (1995). Chronic administration of a cocaine 'binge' alters basal extracellular levels in male rats: an in vivo microdialysis study. *Journal of Pharmacology and Experimental Therapeutics* **272**, 652–657.

Maldonado, R., Robledo, P., Chover, A. J., Caine, S. B., and Koob, G. F. (1993). D-1 dopamine receptors in the nucleus accumbens modulate cocaine self-administration in the rat. *Pharmacology, Biochemistry, and Behavior* **45**, 239–242.

Maldonado, R., Saiardi, A., Valverde, O., Samad, T. A., Roques, B. P., and Borrelli, E. (1997). Absence of opiate rewarding effects in mice lacking dopamine D-2 receptors. *Nature* **388**, 586–589.

Mantsch, J. R., Saphier, D., and Goeders, N. E. (1998). Corticosterone facilitates the acquisition of cocaine self-administration in rats: opposite effects of the type II glucocorticoid receptor agonist dexamethasone. *Journal of Pharmacology and Experimental Therapeutics* **287**, 72–80.

Marinelli, M., Cooper, D. C., Baker, L. K., and White, F. J. (2003). Impulse activity of midbrain dopamine neurons modulates drug-seeking behavior. *Psychopharmacology* **168**, 84–98.

Markou, A., and Koob, G. F. (1991). Post-cocaine anhedonia: An animal model of cocaine withdrawal. *Neuropsychopharmacology* **4**, 17–26.

Markou, A., and Koob, G. F. (1992). Construct validity of a self-stimulation threshold paradigm: effects of reward and performance manipulations. *Physiology and Behavior* **51**, 111–119.

Markou, A., and Koob, G. F. (1993). Intracranial self-stimulation thresholds as a measure of reward. *In Behavioural Neuroscience: A Practical Approach*, vol. 2, (A. Sahgal, Ed.), pp. 93–115. IRL Press, Oxford.

Martin, B. R., Lue, L. P., and Boni, J. P. (1989). Pyrolysis and volatilization of cocaine. *Journal of Analytical Toxicology* **13**, 158–162.

Masaki, T. (1956). The amphetamine problem in Japan. *In Expert Committee on Addiction-Producing Drugs, Sixth Report: Expert Committee on Drugs Liable to Produce Addiction* (series title: *World Health Organization Technical Report Series*, vol. 102), World Health Organization, Geneva, pp. 14–21.

Matecka, D., Rothman, R. B., Radesca, L., de Costa, B. R., Dersch, C. M., Partilla, J. S., Pert, A., Glowa, J. R., Wojnicki, F. H., and Rice, K. C. (1996). Development of novel, potent, and selective dopamine reuptake inhibitors through alteration of the piperazine ring of 1-[2-(diphenylmethoxy)ethyl]-and 1-[2-[bis(4-fluorophenyl)methoxy]ethyl]-4-(3-phenylpropyl)piperazines *(GBR 12935 and GBR 12909)*. *Journal of Medicinal Chemistry* **39**, 4704–4716.

Matell, M. S., and King, G. R. (1997). 5-HT3 receptor mediated dopamine release in the nucleus accumbens during withdrawal from continuous cocaine. *Psychopharmacology* **130**, 242–248.

McClung, C. A., and Nestler, E. J. (2003). Regulation of gene expression and cocaine reward by CREB and DeltaFosB. *Nature Neuroscience* **6**, 1208–1215.

McFarland, K., and Kalivas, P. W. (2001). The circuitry mediating cocaine-induced reinstatement of drug-seeking behavior. *Journal of Neuroscience* **21**, 8655–8663.

McGregor, A., and Roberts, D. C. (1993). Dopaminergic antagonism within the nucleus accumbens or the amygdala produces differential effects on intravenous cocaine self-administration under fixed and progressive ratio schedules of reinforcement. *Brain Research* **624**, 245–252.

McGregor, A., Baker, G., and Roberts, D. C. (1994). Effect of 6-hydroxydopamine lesions of the amygdala on intravenous cocaine self-administration under a progressive ratio schedule of reinforcement. *Brain Research* **646**, 273–278.

McGregor, A., and Roberts, D. C. S. (1995). Effect of medial prefrontal cortex injections of SCH23390 on intravenous cocaine self-administration under both a fixed and progressive ratio schedule of reinforcement. *Behavioural Brain Research* **67**, 75–80.

McGregor, A., Baker, G., and Roberts, D. C. (1996). Effect of 6-hydroxydopamine lesions of the medial prefrontal cortex on intravenous cocaine self-administration under a progressive ratio schedule of reinforcement. *Pharmacology Biochemistry and Behavior* **53**, 5–9.

McKeage, K., and Scott, L. J. (2003). SLI-381 (Adderall XR), *CNS Drugs* **17**, 669–675.

McNeish, C. S., Svingos, A. L., Hitzemann R., and Strecker, R. E. (1993). The 5-HT3 antagonist zacopride attenuates cocaine-induced increases in extracellular dopamine in rat nucleus accumbens. *Pharmacology Biochemistry and Behavior* **45**, 759–763.

Medical Economics Company, (2004). *Physicians' Desk Reference for Ophthalmology*, 28th edition, Medical Economics Company, Oradell NJ.

Mehta, M. A., Owen, A. M., Sahakian, B. J., Mavaddat, N., Pickard, J. D., and Robbins, T. W. (2000). Methylphenidate enhances working memory by modulating discrete frontal and parietal lobe regions in the human brain. *Journal of Neuroscience* **20,** RC65.

Meil, W. M., and See, R. E. (1997). Lesions of the basolateral amygdala abolish the ability of drug associated cues to reinstate responding during withdrawal from self-administered cocaine. *Behavioural Brain Research* **87**, 139–148.

Melega, W. P., Cho, A. K., Schmitz, D., Kuczenski, R., and Segal, D. S. (1999). l-methamphetamine pharmacokinetics and pharmacodynamics for assessment of in vivo deprenyl-derived l-methamphetamine. *Journal of Pharmacology and Experimental Therapeutics* **288**, 752–758.

Mendelson, J. H., Sholar, M., Mello, N. K., Teoh, S. K., and Sholar, J. W. (1998). Cocaine tolerance: behavioral, cardiovascular, and neuroendocrine function in men. *Neuropsychopharmacology* **18**, 263–271.

Miller, M. A. (1997). History and epidemiology of amphetamine abuse in the United States. *In Amphetamine Misuse: International Perspectives on Current Trends* (H. Klee, Ed.), pp. 113–133. Harwood, Amsterdam,.

Miller, N. S., Summers, G. L., and Gold, M. S. (1993). Cocaine dependence: alcohol and other drug dependence and withdrawal characteristics. *Journal of Addictive Diseases* **12**, 25–35.

Miner, L. L., Drago, J., Chamberlain, P. M., Donovan, D., and Uhl, G. R. (1995). Retained cocaine conditioned place preference in D1 receptor deficient mice. *Neuroreport* **6**, 2314–2316.

Mitler, M. M., Hajdukovic, R., Erman, M., and Koziol, J. A. (1990). Narcolepsy. *Journal of Clinical Neurophysiology* **7**, 93– 118.

Mohs, R. C., Tinklenberg, J. R., Roth, W. T., and Kopell, B. S. (1978). Methamphetamine and diphenhydramine effects on the rate of cognitive processing. *Psychopharmacology* **59**, 13–19.

Monsma, F. J. Jr., Mahan, L. C., McVittie, L. D., Gerfen, C. R., and Sibley, D. R. (1990). Molecular cloning and expression of a D1 dopamine receptor linked to adenylyl cyclase activation. *Proceedings of the National Academy of Sciences USA* **87**, 6723–6727.

Moore, K. E., Chiueh, C. C., and Zeldes, G. (1977). Release of neurotransmitters from the brain in vivo by amphetamine, methlyphenidate and cocaine. *In Cocaine and Other Stimulants* (series title: *Advances in Behavioral Biology*, vol. 21), (E. H. Ellinwood Jr. and M. M. Kilbey, Eds), pp. 143–160, Plenum Press, New York.

Moore, R. J., Vinsant, S. L., Nader, M. A., Porrino, L. J., and Friedman, D. P., (1998a). Effect of cocaine self-administration on striatal dopamine D1 receptors in rhesus monkeys. *Synapse* **28**, 1–9.

Moore, R. J., Vinsant, S. L., Nader, M. A., Porrino, L. J., and Friedman, D. P. (1998b). Effect of cocaine self-administration on dopamine D2 receptors in rhesus monkeys. *Synapse* **30**, 88–96.

Morgan, P., and Beck, J. E. (1997). The legacy and the paradox: hidden contexts of methamphetamine use in the United States. *In Amphetamine Misuse: International Perspectives on Current Trends* (H. Klee, Ed.), pp. 135–162. Harwood, Amsterdam.

Morishima, H. O., Whittington, R. A., Iso, A., and Cooper, T. B. (1999). The comparative toxicity of cocaine and its metabolites in conscious rats. *Anesthesiology* **90**, 1684–1690.

Muhtadi, F. J., and Al-Badr, A. A. (1986). Cocaine hydrochloride. *In Analytical Profiles of Drug Substances*, vol. 15, (K. Florey, Ed.), pp. 151–231. Academic Press, Orlando

Musto, D. F. (1989). America's first cocaine epidemic. *Wilson Quarterly* **13**, 59–64.

Nader, M. A., Daunais, J. B., Moore, T., Nader, S. H., Moore, R. J., Smith, H. R., Friedman, D. P., and Porrino, L. J. (2002). Effects of cocaine self-administration on striatal dopamine systems in rhesus monkeys: initial and chronic exposure. *Neuropsychopharmacology* **27**, 35–46.

Naranjo, P. (1981). Social Function of Coca in Pre-Columbian America. *Journal of Ethnopharmacology* **3**, 161–172.

Nestler, E. J. (2004). Historical review: Molecular and cellular mechanisms of opiate and cocaine addiction. *Trends in Pharmacological Sciences* **25**, 210–218.

Nestler, E. J., Terwilliger, R. Z., Walker, J. R., Sevarino, K. A., and Duman, R. S. (1990). Chronic cocaine treatment decreases levels of the G protein subunits Gi alpha and Go alpha in discrete regions of rat brain. *Journal of Neurochemistry* **55**, 1079–1082.

Neumaier, J. F., Vincow, E. S., Arvanitogiannis, A., Wise, R. A., and Carlezon, W. A. Jr, (2002). Elevated expression of 5-HT1B receptors in nucleus accumbens efferents sensitizes animals to cocaine. *Journal of Neuroscience* **22**, 10856–10863.

Norman, A. B., Norman, M. K., Hall, J. F., and Tsibulsky, V. L. (1999). Priming threshold: a novel quantitative measure of the reinstatement of cocaine self-administration. *Brain Research* **831**, 165–174.

Nye, H. E., Hope, B. T., Kelz, M. B., Iadarola, M., and Nestler, E. J. (1995). Pharmacological studies of the regulation of chronic FOS-related antigen induction by cocaine in the striatum and nucleus accumbens. *Journal of Pharmacology and Experimental Therapeutics* **275**, 1671–1680.

O'Malley, S. S., and Gawin, F. H. (1990). Abstinence symptomatology and neuropsychological impairment in chronic cocaine abusers. *In Residual Effects of Abused Drugs on Behavior* (series title: *NIDA Research Monograph*, vol. 101), (J. W. Spenjer, and J. J. Boren, Eds.), pp. 179–190. National Institute on Drug Abuse, Rockville MD.

O'Neill, M. F., Heron-Maxwell, C. L., and Shaw, G. (1999). 5-HT2 receptor antagonism reduces hyperactivity induced by amphetamine, cocaine, and MK-801 but not D1 agonist C-APB, *Pharmacology Biochemistry and Behavior* **63**, 237–243.

Oswald, I. (1968). Drugs and sleep, *Pharmacological Reviews* **20**, 273–303.

Paly, D., Van Dyke, C., Jatlow, P., Jeri, F. R., and Byck, R. (1980). Cocaine: plasma levels after cocaine paste smoking. In Jeri FR (Ed.), *Cocaine 1980: Proceedings of the Interamerican Seminar on Medical and Sociological Aspects of Coca and Cocaine*, Pacific Press, Lima, pp. 106–110.

Parsons, L. H., Smith, A. D., and Justice, J. B. Jr. (1991). Basal extracellular dopamine is decreased in the rat nucleus accumbens during abstinence from chronic cocaine. *Synapse* **9**, 60–65.

Parsons, L. H., Koob, G. F., and Weiss, F. (1995). Serotonin dysfunction in the nucleus accumbens of rats during withdrawal after unlimited access to intravenous cocaine. *Journal of Pharmacology and Experimental Therapeutics* **274**, 1182–1191.

Parsons, L. H., Weiss, F., and Koob, G. F. (1996). Serotonin 1b receptor stimulation enhances dopamine-mediated reinforcement. *Psychopharmacology* **128**, 150–160.

Paterson, N. E., Myers, C., and Markou, A. (2000). Effects of repeated withdrawal from continuous amphetamine administration on brain reward function in rats. *Psychopharmacology* **152**, 440–446.

Pennartz, C. M., Ameerun, R. F., Groenewegen, H. J., and Lopes da Silva, F. H. (1993). Synaptic plasticity in an in vitro slice preparation of the rat nucleus accumbens. *European Journal of Neuroscience* **5**, 107–117.

Penick, S. B. (1969). Amphetamines on obesity. *Seminars in Psychiatry* **1**, 144–162.

Peoples, L. L., and West, M. O. (1996). Phasic firing of single neurons in the rat nucleus accumbens correlated with the timing of intravenous cocaine self-administration. *Journal of Neuroscience* **16**, 3459–3473.

Peoples, L. L., Uzwiak, A. J., Gee, F., and West, M. O. (1999). Tonic firing of rat nucleus accumbens neurons: changes during the first 2 weeks of daily cocaine self-administration sessions. *Brain Research* **822**, 231–236.

Peoples, L. L., and Cavanaugh, D. (2003). Differential changes in signal and background firing of accumbal neurons during cocaine self-administration. *Journal of Neurophysiology* **90**, 993–1010.

Peoples, L. L., Lynch, K. G., Lesnock, J., and Gangadhar, N. (2004). Accumbal neural responses during the initiation and maintenance of intravenous cocaine self-administration. *Journal of Neurophysiology* **91**, 314–323.

Perez-Reyes, M., White, W. R., McDonald, S. A., Hicks, R. E., Jeffcoat, A. R., Hill, J. M., and Cook, C. E. (1991). Clinical effects of daily methamphetamine administration. *Clinical Neuropharmacology* **14**, 352–358.

Piazza, P. V., Maccari, S., Deminiere, J. M., Le Moal, M., Mormede, P., and Simon, H. (1991). Corticosterone levels determine individual vulnerability to amphetamine self-administration. *Proceedings of the National Academy of Sciences USA* **88**, 2088–2092.

Piazza, P. V., Deroche, V., Deminiere, J. M., Maccari, S., Le Moal, M., and Simon, H. (1993). Corticosterone in the range of stress-induced levels possesses reinforcing properties: Implications for sensation-seeking behaviors. *Proceedings of the National Academy of Sciences USA* **90**, 11738–11742.

Piazza, P. V., and Le Moal, M. L. (1996). Pathophysiological basis of vulnerability to drug abuse: Role of an interaction between stress, glucocorticoids, and dopaminergic neurons. *Annual Review of Pharmacology and Toxicology* **36**, 359–378.

Piazza, P. V., Rouge-Pont, F., Deroche, V., Maccari, S., Simon, H., and Le Moal, M. (1996). Glucocorticoids have state-dependent stimulant effects on the mesencephalic dopaminergic transmission. *Proceedings of the National Academy of Sciences USA* **93**, 8716–8720.

Piazza, P. V., and Le Moal, M. (1997). Glucocorticoids as a biological substrate of reward: Physiological and pathophysiological implications. *Brain Research Reviews* **25**, 359–372.

Piazza, P. V., and Le Moal, M. (1998). The role of stress in drug self-administration. *Trends in Pharmacological Sciences* **19**, 67–74.

Pijnenburg, A. J., Honig, W. M., and Rossum, J. M. (1975). Antagonism of apomorphine and d- amphetamine-induced stereotyped behavior by injection of low doses of haloperidol into the caudate nucleus and the nucleus accumbens. *Psychopharmacologia* **45**, 65–71.

Pilla, M., Perachon, S., Sautel, F., Garrido, F., Mann, A., Wermuth, C. G., Schwartz, J. C., Everitt, B.J., and Sokoloff P. (1999). Selective inhibition of cocaine-seeking behaviour by a partial dopamine D3 receptor agonist. *Nature* **400**, 371–375.

Pliakas, A. M., Carlson, R. R., Neve, R. L., Konradi C., Nestler, E. J., and Carlezon, W. A. Jr. (2001). Altered responsiveness to cocaine and increased immobility in the forced swim test associated with elevated cAMP response element-binding protein expression in nucleus accumbens. *Journal of Neuroscience* **21**, 7397–7403.

Poindexter, A. (1960). Appetite suppressant drugs: a controlled clinical comparison of benzphetamine, phenmetrazine, d-amphetamine and placebo. *Current Therapeutic Research: Clinical and Experimental* **2**, 354–363.

Pontieri, F. E., Tanda, G., and Di Chiara, G. (1995). Intravenous cocaine, morphine, and amphetamine preferentially increase extracellular dopamine in the "shell" as compared with the 'core' of the rat nucleus accumbens. *Proceedings of the National Academy of Sciences USA* **92**, 12304–12308.

Porrino, L. J., Ritz, M. C., Goodman, N. L., Sharpe, L. G., Kuhar, M. J., and Goldberg, S. R. (1989). Differential effects of the pharmacological manipulation of serotonin systems on cocaine and amphetamine self-administration in rats. *Life Sciences* **45**, 1529–1535.

Post, R. M., Weiss, S. R. B., Fontana, D., and Pert, A. (1992). Conditioned sensitization to the psychomotor stimulant cocaine. *In The Neurobiology of Drug and Alcohol Addiction* (series title: *Annals of the New York Academy of Sciences*, vol. 654), (P. W. Kalivas, H. H. Samson, Eds), pp. 386–399. New York Academy of Sciences, New York.

Prinzmetal, M., and Bloomberg, W. (1935). Use of benzedrine for the treatment of narcolepsy. *Journal of the American Medical Association* **105**, 2051–2054.

Raiteri, M., Bertollini, A., Angelini, F., and Levi, G. (1975). d-Amphetamaine as a releaser or reuptake inhibitor of biogenic amines in synaptosomes. *European Journal of Pharmacology* **34**, 189–195.

Randrup, A., and Munkvad, I. (1967). Stereotyped activities produced by amphetamine in several animal species and man. *Psychopharmacologia* **11**, 300–310.

Randrup, A., and Munkvad, I. (1970). Biochemical, anatomical and psychological investigations of stereotyped behavior induced by amphetamines. *In International Symposium on Amphetamines and Related Compounds*, (E. Costa, and S. Garattini, Eds.), pp. 695– 713, Raven Press, New York.

Rapoport, J. L., Buchsbaum, M. S., Weingartner H., Zahn, T. P., Ludlow C., and Mikkelsen, E. J. (1980). Dextroamphetamine: its cognitive and behavioral effects in normal and hyperactive boys and normal men. *Archives of General Psychiatry* **37**, 933–943.

Rassnick, S., Heinrichs, S. C., Britton, K. T., and Koob, G. F. (1993). Microinjection of a corticotropin-releasing factor antagonist into the central nucleus of the amygdala reverses anxiogenic-like effects of ethanol withdrawal. *Brain Research* **605**, 25–32.

Reavill C., Taylor, S. G., Wood, M. D., Ashmeade, T., Austin, N. E., Avenell, K. Y., Boyfield, I., Branch, C. L., Cilia, J., Coldwell, M. C., Hadley, M. S., Hunter, A. J., Jeffrey, P., Jewitt, F., Johnson, C. N., Jones, D. N., Medhurst, A. D., Middlemiss, D. N., Nash, D. J., Riley, G. J., Routledge, C., Stemp, G., Thewlis, K. M., Trail, B., Vong, A. K., and Hagan, J. J. (2000). Pharmacological actions of a novel, high-affinity, and selective human dopamine D(3) receptor antagonist, SB-277011-A. *Journal of Pharmacology and Experimental Therapeutics* **294**, 1154–1165.

Rechtschaffen, A., and Maron, L. (1964). The effect of amphetamine on the sleep cycle. *Electroencephalography and Clinical Neurophysiology* **16**, 438–445.

Richter, R. M., and Weiss, F. (1999). In vivo CRF release in rat amygdala is increased during cocaine withdrawal in self-administering rats. *Synapse* **32**, 254–261.

Ritz, M. C., Lamb, R. J., Goldberg, S. R., and Kuhar, M. J. (1987). Cocaine receptors on dopamine transporters are related to self-administration of cocaine. *Science* **237**, 1219–1223.

Robbins, T. W., Cador, M., Taylor, J. R., and Everitt, B. J. (1989). Limbic-striatal interactions in reward-related processes. *Neuroscience and Biobehavioral Reviews* **13**, 155–162.

Robbins, T. W., and Sahakian, B. J. (1979). 'Paradoxical' effects of psychomotor stimulant drugs in hyperactive children from the standpoint of behavioural pharmacology. *Neuropharmacology* **18**, 931–950.

Roberts, D. C., Corcoran, M. E., and Fibiger, H. C. (1977). On the role of ascending catecholaminergic systems in intravenous self-administration of cocaine. *Pharmacology Biochemistry and Behavior* **6**, 615–620.

Roberts D. C. S., Koob, G. F., Klonoff, P., and Fibiger, H. C. (1980). Extinction and recovery of cocaine self-administration following 6-hydroxydopamine lesions of the nucleus accumbens. *Pharmacology Biochemistry and Behavior* **12**, 781–787.

Robertson, M. W., Leslie, C. A., and Bennett, J. P. Jr. (1991). Apparent synaptic dopamine deficiency induced by withdrawal from chronic cocaine treatment. *Brain Research* **538**, 337–339.

Robinson, J. B. (1985). Stereoselectivity and isoenzyme selectivity of monoamine oxidase inhibitors: enantiomers of amphetamine, N-methylamphetamine and deprenyl. *Biochemical Pharmacology* **34**, 4105–4108.

Robledo, P., and Koob, G. F. (1993). Two discrete nucleus accumbens projection areas differentially mediate cocaine self-administration in the rat. *Behavioural Brain Research* **55**, 159–166.

Roper, A. S. W. (2004). Living with ADHD: A national survey of people living with attention deficit hyperactivity disorder [abstract]. *American Psychiatric Association*, 157th Annual Meeting. New York, May 6, 2004.

Rossetti, Z. L., Hmaidan, Y., and Gessa, G. L. (1992). Marked inhibition of mesolimbic dopamine release: A common feature of ethanol, morphine, cocaine and amphetamine abstinence in rats. *European Journal of Pharmacology* **221**, 227–234.

Rothman, R. B., Baumann, M. H., Dersch, C. M., Romero, D. V., Rice, K. C., Carroll, F. I., and Partilla, J. S. (2001). Amphetamine-type central nervous system stimulants release norepinephrine more potently than they release dopamine and serotonin. *Synapse* **39**, 32–41.

Rubinstein, M., Phillips, T. J., Bunzow, J. R., Falzone, T. L., Dziewczapolski, G., Zhang, G., Fang, Y., Larson, J. L., McDougall, J. A., Chester, J. A., Saez, C., Pugsley, T. A., Gershanik, O., Low, M. J., and Grandy, D. K. (1997). Mice lacking dopamine D4 receptors are supersensitive to ethanol, cocaine, and methamphetamine. *Cell* **90**, 991–1001.

Rylander, G. (1969). Clinical and medical-criminological aspects of addiction to central stimulating drugs. *In Abuse of Central Stimulants*, (F. Sjoqvist, and M. Tottie, Eds), pp. 251–273. Almqvist and Wiksell, Stockholm.

Rylander G. (1971). Stereotype behaviour in man following amphetamine abuse. *In The Correlation of Adverse Effects in Man with Observations in Animals* (series title: *International Congress Series*, vol. 220), (S. B. C. Baker, Ed.), pp. 28–31. Excerpta Medica, Amsterdam.

Sakai, N., Thome, J., Newton, S. S., Chen, J., Kelz, M. B., Steffen, C., Nestler, E. J., and Duman, R. S. (2002). Inducible and brain region-specific CREB transgenic mice. *Molecular Pharmacology* **61**, 1453–1464.

Sanchez-Ramos, J. R. (1990). Neurological complications of cocaine abuse include seizure and strokes. *The Psychiatric Times* February, pp. 20, 22.

Sarnyai, Z., Biro, E., Gardi, J., Vecsernyes, M., Julesz, J., and Telegdy, G. (1995). Brain corticotropin-releasing factor mediates 'anxiety-like' behavior induced by cocaine withdrawal in rats. *Brain Research* **675**, 89–97.

Sarnyai, Z., Dhabhar, F. S., McEwen, B. S., and Kreek, M. J. (1998). Neuroendocrine-related effects of long-term, 'binge' cocaine administration: diminished individual differences in stress-induced corticosterone response. *Neuroendocrinology* **68**, 334–344.

Satel, S. L., Price, L. H., Palumbo, J. M., McDougle, C. J., Krystal, J. H., Gawin F., Charney, D. S., Heninger, G. R., and Kleber, H. D. (1991). Clinical phenomenology and neurobiology of cocaine abstinence: a prospective inpatient study. *American Journal of Psychiatry* **148**, 1712–1716.

Schwartz, W. K., and Oderda, G. M. (1980). Management of cocaine intoxications. *Clinical Toxicology Consultant* **2**, 45–58.

See, R. E., Kruzich, P. J., and Grimm, J. W. (2001). Dopamine, but not glutamate, receptor blockade in the basolateral amygdala attenuates conditioned reward in a rat model of relapse to cocaine-seeking behavior. *Psychopharmacology* **154**, 301–310.

See, R. E., Fuchs, R. A., Ledford, C. C., and McLaughlin J. (2003). Drug addiction, relapse, and the amygdala. *In The Amygdala in Brain Function: Basic and Clinical Approaches* (series title: *Annals of the New York Academy of Sciences*, vol. 985), (P. Shinnick-Gallagher, A. Pitkanen, A. Shekhar, and L. Cahill, Eds.), pp. 294–307. New York Academy of Sciences, New York.

Segal, D. S. (1975). Behavioral characterization of d- and 1-amphetamine: neurochemical implications. *Science* **190**, 475–477.

Self, D. W., Terwilliger, R. Z., Nestler, E. J., and Stein, L. (1994). Inactivation of Gi and Go proteins in nucleus accumbens reduces both cocaine and heroin reinforcement. *Journal of Neuroscience* **14**, 6239–6247.

Self, D. W., Genova, L. M., Hope, B. T., Barnhart, W. J., Spencer, J. J., and Nestler, E. J. (1998). Involvement of cAMP-dependent protein kinase in the nucleus accumbens in cocaine self-administration and relapse of cocaine-seeking behavior. *Journal of Neuroscience* **18**, 1848–1859.

Shaham, Y., Funk, D., Erb, S., Brown, T. J., Walker, C. D., and Stewart, J. (1997). Corticotropin-releasing factor, but not corticosterone, is involved in stress-induced relapse to heroin-seeking in rats. *Journal of Neuroscience* **17**, 2605–2614.

Shaham, Y., Erb, S., Leung, S., Buczek, Y., and Stewart, J. (1998). CP-154,526, a selective, non-peptide antagonist of the corticotropin-releasing factor1 receptor attenuates stress-induced relapse to drug seeking in cocaine- and heroin-trained rats. *Psychopharmacology* **137**, 184–190.

Shaham, Y., Shalev, U., Lu L., De Wit, H., and Stewart, J. (2003). The reinstatement model of drug relapse: history, methodology and major findings. *Psychopharmacology* **168**, 3–20.

Shalev, U., Grimm, J. W., and Shaham, Y. (2002). Neurobiology of relapse to heroin and cocaine seeking: a review. *Pharmacological Reviews* **54**, 1–42.

Shippenberg, T. S., LeFevour, A., and Heidbreder, C. (1996). kappa-Opioid receptor agonists prevent sensitization to the conditioned rewarding effects of cocaine. *Journal of Pharmacology and Experimental Therapeutics* **276**, 545–554.

Siegel, R. K. (1982). Cocaine smoking. *Journal of Psychoactive Drugs* **14**, 271–359.

Siegel, R. K. (1985). New patterns of cocaine use: changing doses and routes. *In Cocaine Use in America: Epidemiologic and Clinical Perspectives* (series title: *NIDA Research Monograph*, vol. 61), (N. L. Kozel, E. H. Adams, Eds), pp. 204–220. National Institute on Drug Abuse, Rockville MD.

Siegel, R.K. (1992). Repeating cycles of cocaine use and abuse. In Gerstein D., Harwood H (Eds.), *Treating Drug Problems*, vol. 2, pp. 289–313. Institute of Medicine, Washington DC.

Simkin, B., and Wallace, L. (1961). A controlled clinical comparison of benzphetamine and D-amphetamine in the management of obesity. *American Journal of Clinical Nutrition* **9**, 632–637.

Skjoldager, P., Winger, G., and Woods, J. H. (1991), Analysis of fixed-ratio behavior maintained by drug reinforcers. *Journal of the Experimental Analysis of Behavior* **56**, 331–343.

Smith, G. M., and Beecher, H. K. (1959). Amphetamine sulfate and athletic performance: I. Objective effects. *Journal of the American Medical Association* **170**, 542–557.

Smith, G. M., and Beecher, H. K. (1960). Amphetamine, secobarbital, and athletic performance: II. Subjective evaluations of performance, mood states and physical states. *Journal of the American Medical Association* **172**, 1502–1514.

Snyder, C. A., Wood, R. W., Graefe, J. F., Bowers, A., and Magar, K. (1988). 'Crack smoke' is a respirable aerosol of cocaine base. *Pharmacology Biochemistry and Behavior* **29**, 93–95 [erratum: 29:835].

Sokoloff, P., Giros, B., Martres, M-P, Bouthenet, M. L., and Schwartz, J-C. (1990). Molecular cloning and characterization of a novel dopamine receptor (D3) as a target for neuroleptics. *Nature* **347**, 146–151.

Spangler, R., Unterwald, E. M., Kreek, M. J. (1993). 'Binge' cocaine administration induces a sustained increase of prodynorphin mRNA in rat caudate-putamen. *Molecular Brain Research* **19,** 323–327.

Stewart, J. (1984). Reinstatement of heroin and cocaine self-administration behavior in the rat by intracerebral application of morphine in the ventral tegmental area. *Pharmacology Biochemistry and Behavior* **20**, 917–923.

Substance Abuse and Mental Health Services Administration, (1996). *National Household Survey on Drug Abuse: Population Estimates 1995* (DHHS publication no. [SMA] 96–3095), U.S. Department of Health and Human Services, Rockville MD.

Substance Abuse and Mental Health Services Administration. (2003). *Results from the 2002 National Survey on Drug Use and Health: National Findings* (Office of Applied Studies, NHSDA Series H-22, DHHS Publication No. SMA 03–3836), Rockville MD.

Sunahara, R. K., Guan, H. C., O'Dowd, B. F., Seeman P., Laurier, L. G., Ng, G., George, S. R., Torchia, J., Van Tol, H. H. M., and Niznik, H. B. (1991). Cloning of the gene for a human dopamine D5 receptor with higher affinity for dopamine than D1. *Nature* **350**, 614–619.

Suwaki, H., Fukui, S., and Konuma, K. (1997). Methamphetamine abuse in Japan: its 45 year history and the current situation. *In Amphetamine Misuse: International Perspectives on Current Trends* (H. Klee, Ed.), pp. 199–214. Harwood, Amsterdam.

Taylor, D., and Ho, B.T. (1978). Comparison of inhibition of monoamine uptake by cocaine, methylphenidate and amphetamine. *Research Communications in Chemical Pathology and Pharmacology* **21**, 67–75.

Terwilliger, R. Z., Beitner-Johnson, D., Sevarino, K. A., Crain, S. M., and Nestler, E. J. (1991). A general role for adaptations in G-proteins and the cyclic AMP system in mediating the chronic actions of morphine and cocaine on neuronal function. *Brain Research* **548**, 100–110.

Thomas, M. J., Beurrier, C., Bonci, A., and Malenka, R. C. (2001). Long-term depression in the nucleus accumbens: a neural correlate of behavioral sensitization to cocaine. *Nature Neuroscience* **4**, 1217–1223.

Thomas, M. J., Malenka, R. C., and Bonci A. (2000). Modulation of long-term depression by dopamine in the mesolimbic system. *Journal of Neuroscience* **20**, 5581–5586.

Thomas, M. J., and Malenka, R. C. (2003). Synaptic plasticity in the mesolimbic dopamine system. *Transactions of the Royal Society of London B Biological Sciences* **358**, 815–819.

Trendelenburg, U., Muskus, A., and Fleming, W. W. (1962). Gomez-Alonso de la Sierra B, Modification by reserpine of the action of sympathomimetic amines in spinal cats: a classification of sympathomimetic amines. *Journal of Pharmacology and Experimental Therapeutics* **138**, 170–180.

Trulson, M. E., and Ulissey, M. J. (1987). Chronic cocaine administration decreases dopamine synthesis rate and increases [3H] spiroperidol binding in rat brain. *Brain Research Bulletin* **19**, 35–38.

Tsukada, H., Kreuter, J., Maggos, C. E., Unterwald, E. M., Kakiuchi, T., Nishiyama, S., Futatsubashi, M., and Kreek, M. J. (1996). Effects of binge pattern cocaine administration on dopamine D1 and D2 receptors in the rat brain: an in vivo study using positron emission tomography. *Journal of Neuroscience* **16**, 7670–7677.

Turchan, J., Przewlocka, B., Lason, W., and Przewlocki, R. (1998). Effects of repeated psychostimulant administration on the prodynorphin system activity and kappa opioid receptor density in the rat brain. *Neuroscience* **85**, 1051–1059.

Ungerstedt, U. (1971). Adipsia and aphagia after 6-hydroxy-dopamine induced degeneration of the nigro-striatal dopamine system. *Acta Physiologica Scandinavica Supplementum* **367**, 95–122.

Ungless, M. A., Whistler, J. L., Malenka, R. C., and Bonci A. (2001). Single cocaine exposure in vivo induces long-term potentiation in dopamine neurons. *Nature* **411**, 583–387.

United States Treasury Department, Bureau of Internal Revenue. (1915). *Compilation of Treasury Decisions Relating to the Act of Dec. 17, 1914, known as the Harrison Narcotic Law*, U.S. Government Printing Office, Washington, D.C.

United States Pharmacopeial Convention. (1998). *USP Dictionary of USAN and International Drug Names: A Compilation of the United States Adopted Names (USAN) Selected and Released from June 15, 1961–1998,* United States Pharmacopeial Convention, Rockville MD.

Van Dyke, C., Barash, P. G., Jatlow, P., and Byck, R. (1976). Cocaine: plasma concentrations after intranasal application in man. *Science* **191**, 859–61.

Van Dyke, C., Jatlow, P., Ungerer, J., Barash, P. G., and Byck R. (1978). Oral cocaine: plasma concentrations and central effects. *Science* **200**, 211–3.

Van Dyke, C., and Byck, R. (1982). Cocaine, *Scientific American* **246**, 128–141.

Van Tol H. H. M., Bunzow, J. R., Guan, H. C., Sunahara, R. K., Seeman, P., Niznik, H. B., and Civelli O. (1991). Cloning of the gene for a human dopamine D4 receptor with high affinity for the antipsychotic clozapine. *Nature* **350**, 610–614.

Verebey, K., and Gold, M. S. (1988). From coca leaves to crack: the effects of dose and routes of administration in abuse liability. *Psychiatric Annals* **18**, 513–520.

Vogt, M. (1954). The concentration of sympathin in different parts of the central nervous system under normal conditions and after the administration of drugs. *Journal of Physiology (London)* **123**, 451–81.

Volkow, N. D., Fowler, J. S., Wang, G. J., Hitzemann, R., Logan, J., Schlyer, D. J., Dewey, S. L., and Wolf, A. P. (1993). Decreased dopamine D2 receptor availability is associated with reduced frontal metabolism in cocaine abusers. *Synapse* **14**, 169–177.

Volkow, N. D., Chang, L., Wang, G. J., Fowler, J. S., Ding, Y.S., Sedler, M., Logan, J., Franceschi, D., Gatley, J., Hitzemann, R., Gifford, A., Wong, C., and Pappas N. (2001). Low level of brain dopamine D2 receptors in methamphetamine abusers: association with metabolism in the orbitofrontal cortex. *American Journal of Psychiatry* **158**, 2015–2021.

von Euler, U.S. (1947). A specific sympathomimetic ergone in adrenergic nerve fibres (sympathin) and its relations to adrenaline and noradrenaline. *Acta Physiologica Scandinavica* **12**, 73–97.

Vorel, S. R., Liu, X., Hayes, R. J., Spector, J. A., and Gardner, E. L. (2001). Relapse to cocaine-seeking after hippocampal theta burst stimulation. *Science* **292**, 1175–1178.

Weddington, W. W., Brown, B. S., Haertzen, C. A., Cone, E. J., Dax, E. M., Herning, R. I., and Michaelson, B. S. (1990). Changes in mood, craving, and sleep during short-term abstinence reported by male cocaine addicts: a controlled, residential study. *Archives of General Psychiatry* **47**, 861–868.

Weiss, B., and Laties, V. G. (1962). Enhancement of human performance by caffeine and the amphetamines. *Pharmacological Reviews* **14**, 1–36.

Weiss, F., Markou, A., Lorang, M. T., and Koob, G. F. (1992). Basal extracellular dopamine levels in the nucleus accumbens are decreased during cocaine withdrawal after unlimited-access self-administration. *Brain Research* **593**, 314–318.

Weiss, F., Parsons, L. H., and Markou, A. (1995). Neurochemistry of cocaine withdrawal. *In The Neurobiology of Cocaine: Cellular and Molecular Mechanisms*, (R. P. Hammer, Jr. Ed.), pp. 163–180, CRC Press, New York.

Weiss, F., Maldonado-Vlaar, C. S., Parsons, L. H., Kerr, T. M., and Smith, D. L. (2000). Ben-Shahar O, Control of cocaine-seeking behavior by drug-associated stimuli in rats: Effects on recovery of extinquished operant-responding and extracellular dopamine levels in amygdala and nucleus accumbens. *Proceedings of the National Academy of Sciences USA* **97**, 4321–4326.

White, F. J., Hu, X. T., Henry, D. J., and Zhang, X. F. (1995). Neurophysiological alterations in the mesocorticolimbic dopamine system with repeated cocaine administration. *In The Neurobiology of Cocaine: Cellular and Molecular Mechanisms*, (R. P. Hammer, Jr Ed.), pp. 99–119. CRC Press, Boca Raton FL.

Whitelaw, R. B., Markou A., Robbins, T. W., and Everitt, B. J. (1996). Excitotoxic lesions of the basolateral amygdala impair the acquisition of cocaine-seeking behaviour under a second-order schedule of reinforcement. *Psychopharmacology* **127**, 213–224.

Wiegmann, D. A., Stanny, R. R., McKay, D. L., Neri, D. F., and McCardie, A. H. (1996). Methamphetamine effects on cognitive processing during extended wakefulness. *International Journal of Aviation Psychology* **6**, 379–397.

Wilkinson, P., Van Dyke, C., Jatlow, P., Barash, P., and Byck, R. (1980). Intranasal and oral cocaine kinetics. *Clinical Pharmacology and Therapeutics* **27**, 386–394.

Withers, N., Pulvirenti, L., Koob, G. F., and Gillin, J. C. (1995). Cocaine abuse and dependence. *Journal of Clinical Psychopharmacology* **15**, 63–78.

Woolverton, W. L. (1986). Effects of a D1 and a D2 dopamine antagonist on the self-administration of cocaine and piribedil by rhesus monkeys. *Pharmacology Biochemistry and Behavior* **24**, 531–535.

Xi, Z. X., Gilbert, J., Campos, A. C., Kline, N., Ashby, C. R. Jr., Hagan, J. J., Heidbreder, C. A., and Gardner, E. L. (2004). Blockade of mesolimbic dopamine D3 receptors inhibits stress-induced reinstatement of cocaine-seeking in rats. *Psychopharmacology* **176**, *f* 57–65.

Xu, M., Hu, X-T., Cooper, D. C., Moratalla, R., Graybiel, A. M., White, F. J., and Tonegawa, S. (1994a). Elimination of cocaine-induced hyperactivity and dopamine-mediated neurophysiological effects in dopamine D1 receptor mutant mice. *Cell* **79**, 945–955.

Xu, M., Moratalla R., Gold, L. H., Hiroi, N., Koob, G. F., Graybiel, A. M., Tonegawa, S., and Dopamine. (1994b). D.1 receptor mutant mice are deficient in striatal expression of dynorphin and in dopamine-mediated behavioral responses. *Cell* **79**, 729–742.

Young, S. T., Porrino, L. J., and Iadarola, M. J. (1991). Cocaine induces striatal c-fos-immunoreactive proteins via dopaminergic D1 receptors. *Proceedings of the National Academy of Sciences USA* **88**, 1291–1295.

Zhou, Y., Spangler R., LaForge, K. S., Maggos, C. E., Ho, A., and Kreek, M. J. (1996). Corticotropin-releasing factor and type 1 corticotropin-releasing factor receptor messenger RNAs in rat brain and pituitary during 'binge'-pattern cocaine administration and chronic withdrawal. *Journal of Pharmacology and Experimental Therapeutics* **279**, 351–358.

Zhou, Y., Spangler, R., Schlussman, S. D., Ho, A., and Kreek, M. J. (2003). Alterations in hypothalamic-pituitary-adrenal axis activity and in levels of proopiomelanocortin and corticotropin-releasing hormone-receptor 1 mRNAs in the pituitary and hypothalamus of the rat during chronic 'binge' cocaine and withdrawal. *Brain Research* **964**, 187–199.

Zis, A. P., Fibiger, H. C., and Phillips, A. G. (1974). Reversal by L-DOPA of impaired learning due to destruction of the dopaminergic nigro-neostriatal projection. *Science* **185**, 960–962.

CHAPTER

4

Opioids

OUTLINE

DEFINITIONS

Opiates were originally derived from extracts of the juice of the opium poppy (*Papaver somniferum*) and were defined as any preparation or semisynthetic derivative of opium. Because of the development of synthetic drugs with morphine-like action, the term *opioid* came into use and can be defined as all drugs, natural and synthetic, with morphine-like actions. Opioids also include endogenous 'morphine-like' substances that bind to the same brain sites (receptors) as opioids, and antagonists of opioid drugs (Martin, 1967; Jaffe and Martin, 1990). Opioids are drugs with major medical uses that have been used for thousands of years and have relieved much human suffering. The two major uses historically have been in the treatment of diarrhea and the treatment of pain.

HISTORY OF OPIOID USE, ABUSE, AND ADDICTION

One of the first references to the medical use of opium was by the Greek philosopher Theophrastus who, at the beginning of the third century B.C., spoke of meconium (Macht, 1915) which was composed of extracts of stems, leaves and fruit of *Papaver somniferum*. Parcelus (1490–1540 A.D.), a famous physician of the middle ages, used opium often, and his followers

Neurobiology of Addiction, by George F. Koob and Michel Le Moal.
ISBN – 13: 978-0-12-419239-3 ISBN – 10: 0-12-419239-4

were equally enthusiastic (Macht, 1915). Thomas Sydenham, one of the great physicians of the seventeenth century, wrote in describing the treatment of a series of dysentery epidemics in 1669–1672:

> 'And here I cannot but break out in praise of the great God, the giver of all good things, who hath granted to the human race, as a comfort in their affliction, no medicine of the value of opium, either in regard to the number of diseases it can control, or its efficiency in extirpating them' (Latham, 1848).

However, equally early along with the description of beneficial opioid medical actions, withdrawal from opioids was described. An early description of opioid withdrawal in 1700 A.D. was written by Dr. John Jones who detailed the withdrawal associated with cessation of chronic use of opium. In Chapter 23 of *The Mysteries of Opium Revealed* (Jones, 1700) he wrote, 'A return of all diseases, pains and disasters, must happen generally, because the opium takes them off by a bare diversion of the sense thereof by pleasure'. Such a description presaged the ultimate dilemma with the opioids. Their beneficial medical effects are accompanied by significant side effects, the most devastating being opioid addiction with chronic uncontrolled use.

The history of opioid abuse in the Western world began with the spread from the Middle East of opium to Europe and also to the Orient. Europeans traded opium for tea from China through the British East India Company. More specifically, British merchants smuggled opium into China to balance their purchases of tea for export to Britain. The Chinese ultimately realized the addictive properties of opium and attempted to stop the trade, resulting in the Opium Wars of the 1840s. The result of the British victories was the opening of ports for British trade, the ceding of Hong Kong to the British, and ultimately the legalization of opium importation to China (Fay, 1975; Wakeman, 1978). Subsequently, opium use spread to the United States with the immigration of Chinese laborers. Unlimited opium use in the United States helped contribute to the Harrison Act of 1914 and paradoxically, as a result, to the social marginalization of opioid use and the development of heroin addiction (Acker, 2002).

Heroin addiction remains a major medical problem in the United States, with estimates of 200 000 heroin-addicted subjects according to the National Survey on Drug Use and Health (Substance Abuse and Mental Health Services Administration, 2003). The rate of heroin abuse and addiction over the past 40 years has fluctuated. There were a total of 2.3 million people who had ever used heroin in 1979, dropping to 1.5 million in 1990, then rising again to 2.4 million in 1996. In 2002, there were an estimated 166 000 current heroin users, defined as having used heroin within the past month. The total number of past-month users in 1996 was 216 000. The percentage of young adults aged 18–25 who had ever used heroin in the late 1960s was about 0.2 per cent, then rose to a peak of 2.3 per cent in 1977, and then declined steadily into the mid-1990s. The rate then rose from 0.7 per cent in 1996 to 1.8 per cent in 2002. The rate for 12–17 year olds hovered between 0.1 and 0.2 per cent up through 1995, but from 1996–2002 the rate rose from 0.2 to 0.4 per cent (Substance Abuse and Mental Health Services Administration, 1996, 2002, 2003). In the European Union, problem drug use is defined as 'injecting drug use or long duration/regular use of opiates, cocaine and/or amphetamines', and estimates of problem drug use in 2003 varied between 0.2 and 1 per cent of the adult population (1–1.5 million). It is estimated that half of the 'problem drug users' in the European Union are drug injectors (approximately 500 000–750 000) (European Monitoring Centre for Drugs and Drug Addiction, 2003).

More recently, there has been an epidemic of oxycodone (Oxycontin) abuse with the diversion and overuse of the opioid pain medication Oxycontin (Substance Abuse and Mental Health Services Administration, 2002). In 2002, 1.9 million people aged 12 and over had used Oxycontin nonmedically (diverted from medical use) at least once (Substance Abuse and Mental Health Services Administration, 2003).

Opium contains 10 per cent morphine but also thebaine and codeine. Morphine was first isolated from opium by Serturner in 1804 (Macht, 1915), and it was named after Morpheus, the God of Dreams, or Morphina, the God of Sleep. Codeine was first isolated from opium in 1832 by Robiquet (Macht, 1915) and was used in the United States as a tonic for a wide variety of problems and ailments. It is still the most widely prescribed legal opioid (Foley, 1993). Heroin (diacetylmorphine) was developed by the Bayer company in 1898 as a cough suppressant with an alleged stimulant action on the respiratory system (the latter later proven false) (Sneader, 1998) (**Fig. 4.1**).

MEDICAL USE AND BEHAVIORAL EFFECTS

Medical Uses

Opioids are the most powerful and effective drugs for the relief of pain known to humans. Pain relief from morphine at a standard dose of 10 mg administered intramuscularly or subcutaneously lasts up to 3–4 h. Opioid analgesia has been described as the selective suppression of pain without effects on other sensations at reasonable analgesic doses (Gutstein and Akil, 2001). However, opioid analgesia also distinguishes

Am. J. Ph.] 7 [December, 1901

BAYER Pharmaceutical Products

HEROIN–HYDROCHLORIDE

is pre-eminently adapted for the manufacture of cough elixirs, cough balsams, cough drops, cough lozenges, and cough medicines of any kind. Price in 1 oz. packages, $4.85 per ounce; less in larger quantities. The efficient dose being very small (1-48 to 1-24 gr.), it is

The Cheapest Specific for the Relief of Coughs

(In bronchitis, phthisis, whooping cough, etc., etc.)

WRITE FOR LITERATURE TO

FARBENFABRIKEN OF ELBERFELD COMPANY

SELLING AGENTS

P. O. Box 2160 40 Stone Street, NEW YORK

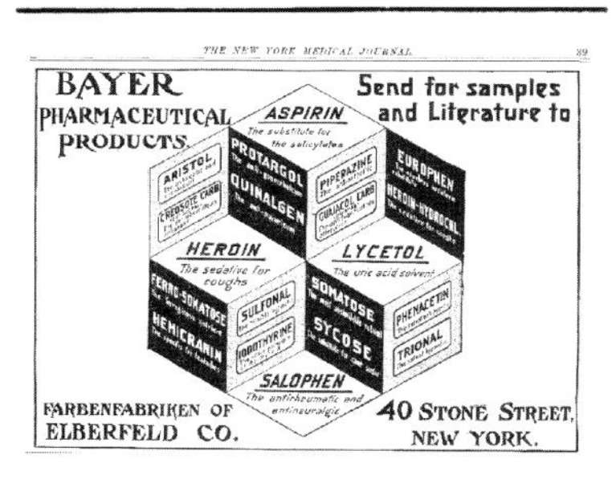

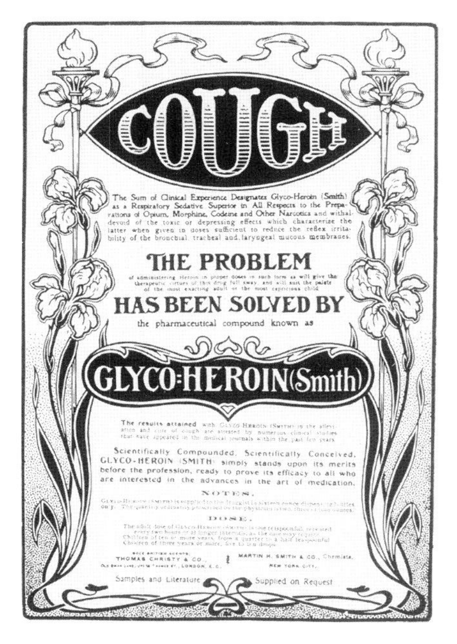

FIGURE 4.1 Advertisements from the Bayer Company and Martin M. Smith & Co. Ltd. for the use of heroin for the treatment of cough, circa early 1900s.

between different types of pain. Opioids have minimal effects on 'first pain', the initial sharp sensation produced by a noxious stimulus, but are very effective against what has been called 'second pain', the dull continuous ache that continues after a noxious stimulus (e.g., the reaction to the specific sensation) (Cooper *et al.*, 1986). Presumably, this distinction has survival value, as selective suppression of second pain by endogenous opioids allows relief from the pain of a previous injury, but does not eliminate awareness of the immediate danger of a new injury. Opioids are less effective in reducing neuropathic pain, requiring higher doses (McQuay, 1988). Perhaps more importantly for the neurobiology of addiction, opioids also are effective in reducing 'emotional pain' (Stewart, 1987).

Opioids at higher doses, however, can produce local anesthetic effects, and these effects have been hypothesized to be mediated by an action on the dorsal root entry zone in the spinal cord (Jaffe and Rowe, 1996). Such effects may be the basis for the potent and effective use of opioids epidurally and intrathecally. Opioids also produce analgesia when administered in the periphery or at local sites, and opioid receptors are present on peripheral nerves (Fields *et al.*, 1980).

Opioids are administered epidurally for the management of obstetric pain, and such use is based on the concept of selective sites of analgesia action to a specific segmental level of the spinal cord. Both epidural and intrathecal administration have been used for chronic pain states, such as lower back pain, neuralgia, and limb pain (Arner *et al.*, 1988). In the case of epidural administration, opioids can bind to opioid receptors in the spinal cord and produce analgesia without motor or sensory dysfunction (Yaksh, 1981). Drugs with high selectivity for the μ opioid receptor and high lipid solubility, such as fentanyl, are taken up rapidly into the spinal cord and show a fast onset of action (Littrell, 1991).

Opioids are often used to relieve pain during general anesthesia. However, they are also a component of 'balanced anesthesia' where a balance of agents is used to produce the different components of anesthesia (analgesia, amnesia, muscle relaxation, and abolition of autonomic reflexes with maintenance of homeostasis) (Woodbridge, 1957). Currently, fentanyl and its congeners alfentanil and remifentanil, because of their rapid onset, are used successfully as components of balanced anesthesia. The inclusion of an opioid can reduce preoperative pain and anxiety, decrease adverse responses to manipulations of the airways, lower the requirements for inhaled anesthetics, and provide immediate postoperative analgesia.

Patient-controlled analgesia is a method of opioid administration where the patient can carefully titrate the rate of opioid administration to meet individual pain relief needs (Hill *et al.*, 1991). While this can be done with oral dosing, there are now specifically designed infusion pumps that can deliver a continuous infusion with bolus doses by the intravenous, subcutaneous, or epidural routes. Pumps are programmed to the needs of the patient, with limits set to prevent overdose. Evidence suggests that patient-controlled analgesia is as safe and effective as other methods, and may be more effective under certain conditions. Patient-controlled analgesia provides a more consistent level of analgesia and is associated with greater patient compliance (Macintyre, 2001).

Table 4.1 provides the equivalent doses of commonly prescribed opioids required to produce analgesia

TABLE 4.1 Dosing Data for Opioid Analgesics

Drug	Approximate equianalgesic oral dose	Approximate equianalgesic parenteral dose
Morphine[1]	30 mg q 3–4 h (Around-the-clock dosing) 60 mg every 3-4 h (Single or intermittent dose)	10 mg q 3–4 h
Codeine[2]	130 mg every 3–4 h	75 mg every 3–4 h
Hydromophone[1] (Dilaudid)	7.5 mg every 3–4 h	1.5 mg every 3–4 h
Hydrocodone (Lorcet, Lortab, Vicodin, others)	30 mg every 3–4 h	not available
Levorphanol (Levo-Dromoran)	4 mg every 6–8 h	2 mg every 6–8 h
Meperidine (Demerol)	300 mg every 2–3 h	100 mg every 3 h
Methadone (Dolophine, others)	20 mg every 6–8 h	10 mg every 6–8 h
Oxycodone (Roxicodone, Percocet, Percodan, Tylox, others)[3]	30 mg every 3–4 h	not available
Oxymorphine[1]	not available	1 mg every 3–4 h
Propoxyphene (Darvon)	130 mg[4]	not available
Tramadol[5] (Ultram)	100 mg[4]	100 mg

[1]For morphine, hydromorphone, and oxymorphone, rectal administration is an alternate route for patients unable to take oral medications, but equianalgesic doses may differ from oral and parenteral doses because of pharmacokinetic differences.
[2]*Caution*: Codeine doses above 65 mg often are not appropriate due to diminishing incremental analgesia with increasing doses but continually increasing constipation and other side effects.
[3]Oxycontin is an extended-release preparation containing up to 160 mg of oxycodone per tablet and is recommended for use every 12 h.
[4]Doses for moderate pain not necessarily equivalent to 30 mg oral or 10 mg parenteral morphine.
[5]Risk of seizures; parenteral formulation not available in the United States.
[Reproduced with permission from Gutstein and Akil, 2001.]

equivalent to that of a standard dose of morphine (10 mg) (Gutstein and Akil, 2001). There are numerous natural and synthetic opioids available for mild to moderate pain relief, many of which have a high oral-to-parenteral ratio (Foley, 1993). Codeine, a natural component of opium, is the most commonly used opioid analgesic for the management of mild to moderate pain, and is often combined with aspirin or acetaminophen. It is significantly less potent than morphine. Oxycodone (the long acting preparation Oxycontin) is a synthetic derivative of morphine that is used in the management of mild to moderate pain and is nearly equipotent with morphine. It is 20–30 times more potent than morphine via the parenteral route. It has a relatively short half-life of 2–3 h and is excreted mainly via the kidney. Other medical uses of opioids include treatment of severe diarrhea and treatment of cough.

Peripheral Physiological Actions

In nontolerant adults, morphine can produce coma, constricted (pinpoint) pupils, and respiratory depression at toxic doses (Ellenhorn and Barceloux, 1988). At analgesic doses, morphine produces decreased body temperature, decreased pituitary function (as reflected in decreased luteinizing hormone, follicle-stimulating hormone, and adrenocorticotropic hormone), decreased respiration, suppressed cough reflex, and nausea (Gutstein and Akil, 2001). Opioids at therapeutic doses have no effects on blood pressure or heart rate but can cause orthostatic hypotension, particularly in the elderly (Hugues *et al.*, 1992; Medical Economics Company, 2004). Opioids decrease gastrointestinal secretions and decrease gastrointestinal motility (Manara and Bianchetti, 1985). Opioids relieve diarrhea by an action on the intestine to slow gastrointestinal motility and delay transit of intraluminal contents (Galligan and Burks, 1982), but also by a specific transport anti-secretory action (Sandhu *et al.*, 1983; Schiller, 1995). Opioids produce pruritis when administered either systemically or intraspinally (Ballantyne *et al.*, 1988) and suppress the immune system (Gutstein and Akil, 2001).

PHARMACOKINETICS

Heroin injected intravenously enters rapidly in the blood, and after smoking, peaks in 1–5 min. Heroin levels then decrease precipitously, reaching the limits of detection in 30 min. Heroin (3-6-diacetylmorphine) is rapidly converted to 6-monacetylmorphine and then to morphine (by removal of the 3 and then the 6 acetyl group) after systemic administration (Inturrisi *et al.*, 1984; Pichini *et al.*, 1999) (**Fig. 4.2**). Intravenous heroin has a half-life of only 3 min and is rapidly converted to 6-monacetylmorphine and then more slowly to morphine (Inturrisi *et al.*, 1984). The elimination half-life by the smoked route has been shown to be 3.3 min for heroin, 5.4 min for 6-acetylmorphine, and 18.8 min

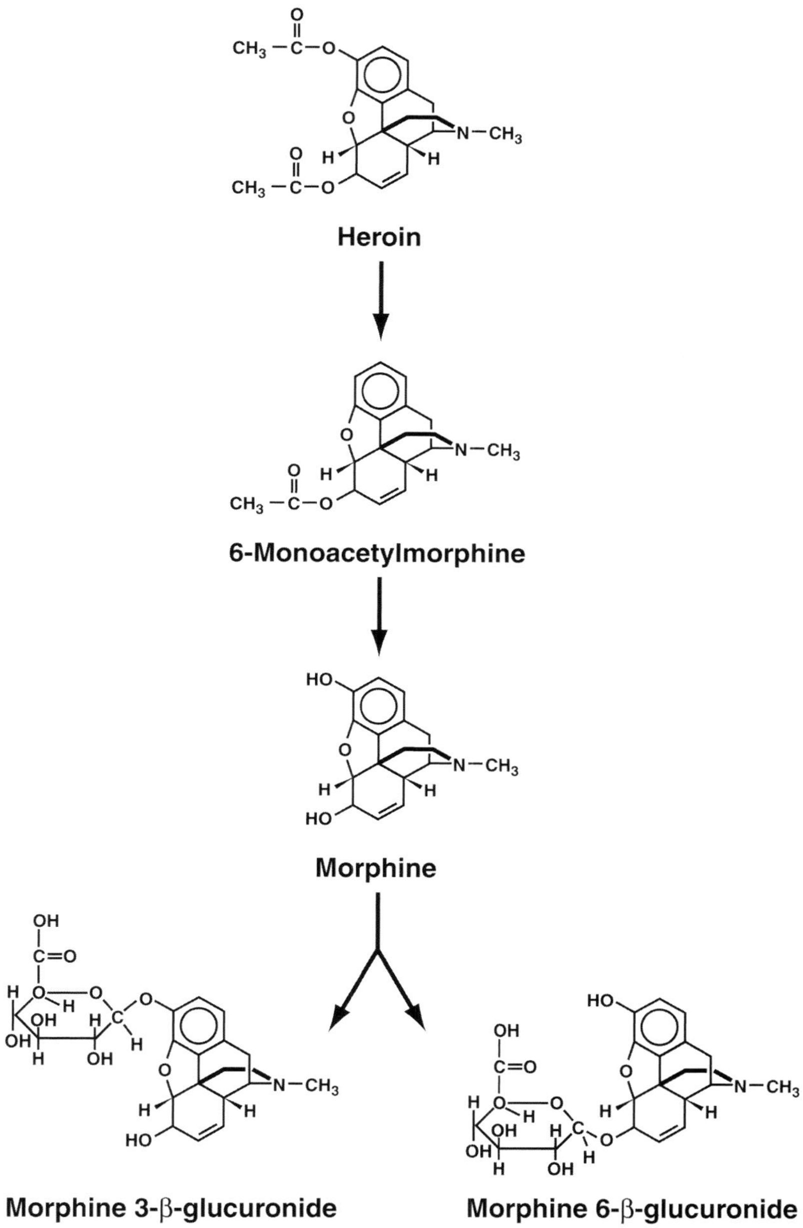

Heroin

6-Monoacetylmorphine

Morphine

Morphine 3-β-glucuronide

Morphine 6-β-glucuronide

FIGURE 4.2 Biotransformation pathway for heroin in humans. [Reproduced with permission from Pichini *et al.*, 1999.]

for morphine (Jenkins *et al.*, 1994). Morphine then is largely metabolized to morphine 3-β-glucuronide and morphine 6-β-glucuronide (Osborne *et al.*, 1990).

Systemic 6-monacetylmorphine is 3–10 times more potent than morphine (depending on the nociceptive test) (Umans and Inturrisi, 1981). Evidence from opioid binding studies shows that heroin acts through its metabolites because it does not bind to opioid receptors in brain homogenates (Inturrisi *et al.*, 1983). Both 6-monacetylmorphine and morphine bind with near equal affinity to opioid binding sites.

Thus, heroin is basically a prodrug that acts via its conversion to 6-monoacetylmorphine and then to morphine. The conversion to 6-monacetylmorphine is very rapid, occurring via esterase enzymes in the brain and the blood as well as in virtually every tissue, including the liver, and thus the 6-monacetylmorphine conversion accounts for the rapid onset and greater potency of heroin than morphine. Eventually the 6-monacetylmorphine gets converted to morphine, so morphine also contributes to the duration of heroin's effect.

Morphine 6-β-glucuronide is produced following morphine administration in humans (Osborne *et al.*, 1988). It is a pharmacologically active metabolite, and has a potency in humans and animals similar to that of morphine, depending on the route of administration (Shimora *et al.*, 1971; Paul *et al.*, 1989; Frances *et al.*, 1992; Hanna *et al.*, 2005). Morphine 3-β-glucuronide does not bind to the opioid receptor but has some excitatory effects if injected into the brain, consistent with convulsant effects of some opioids in rodents. In a controlled clinical trial, morphine 3-β-glucuronide was shown to be devoid of activity and also had no antimorphine effects (Penson *et al.*, 2000).

ABUSE AND ADDICTION POTENTIAL

A popular, virtually universal misconception of opioid use is that any opioid use within or outside of medical situations leads to intractable physiological dependence and addiction. However, extensive work has established that there is wide variety of nonmedical drug consumption, including opioids, ranging from nonproblematic to abusive. Three large groups of opioid use have been elaborated: controlled subjects or 'chippers', marginal subjects or abusers, and compulsive subjects or addicts. Controlled use is generally recognized as occasional use or 'chipping' and most often indicates a nonaddictive pattern of opioid use (Harding and Zinberg, 1983). In contrast, an addict, or 'junkie' would meet the criteria for Substance Dependence on opioids as defined by the *Diagnostic and Statistical Manual of Mental Disorders*, 4th edition (DSM-IV) (American Psychiatric Association, 1994). A marginal user would probably meet the criteria for Substance Abuse as defined by the DSM-IV (Harding and Zinberg, 1983). Controlled opioid use, or 'chipping', has several characteristics that differ from the other extreme of a compulsive user or someone considered suffering from addiction. First, controlled substance users limit their use of the drug, often to amounts or periods that do not interfere with social and occupational functioning:

> 'Arthur, a "controlled user", was a forty-year-old white male who had been married for sixteen years and was the father of three children. He had lived in his own home in a middle-class suburb for twelve years. He had been steadily employed as a union carpenter, with the same construction firm for five years. During the ten years prior to the first interview, Arthur used heroin on weekends, occasionally injecting during the week, but during the previous five years mid-week use had not occurred' (Harding and Zinberg, 1983).

> 'Bob, a "compulsive user", was a twenty-six-year-old white male who lived alone, a college graduate with a degree in psychology. Following separation from his wife and child three years before interview, he had worked sporadically at part-time jobs. Dealing drugs had become his major source of income. He had used heroin at least three to four times per week since beginning use thirty months before interview and had had many periods of daily use lasting for as long as two weeks' (Harding and Zinberg, 1983).

Second, patterns of controlled use are stable and can last as long as 15 years (Harding and Zinberg, 1983) (**Table 4.2**). In a sample of 61 controlled users and 30 compulsive users, the average period of controlled use was 4.5 years, similar to that of compulsive users. Because of this, and because compulsive users had few years of controlled use prior to compulsive use, it was concluded that a few long-term compulsive users actually pass through an early stage of controlled use, and if they do pass through a period

TABLE 4.2 Characteristics of Controlled Versus Compulsive Chippers

	Controlled Subjects ($n = 61$)	Compulsive Subjects ($n = 30$)
Age (mean)	25.9 years	25.9 years
Male	77%	67%
Female	23%	33%
Duration of current style of use (months)	53.4 months	59.5 months
Average frequency of use		
(last 12 months):	—	23%
> 2 times daily	—	23%
once every 1–2 days	41%	47%
2 times weekly	36%	3%
1–3 times monthly	23%	3%
< once monthly		
Length of use (mean)	7.2 years	6.8 years
Peak frequency of use:	23%	87%
> 2 times daily	54%	13%
once every 1–2 days	16%	—
2 times weekly	5%	—
1–3 times monthly	2%	—
< once monthly		
Percentage who selected rules for use:		
Never use in strange place	20%	13%
Never use with strangers	31%	28%
Never inject	13%	—
Special schedule for use	34%	20%
Plan for use	30%	—
Never share needles	28%	31%
Never use alone	26%	10%
Caution in 'copping'	53%	20%
Budgets for use	48%	23%
History of adverse reactions	36%	45%

[Data from Harding and Zinberg, 1983.]

of controlled use, they do not remain in that style of use for as long as controlled users.

Third, controlled users are also much more cautious about the manner in which they use opioids (Harding and Zinberg, 1977; Zinberg *et al.*, 1977). Controlled users were much more likely than compulsives to adopt a variety of rules to minimize the risks of opioid use. Four prominent rules were used: (1) refusing to inject opioids, (2) planning for their use, (3) exercising caution when 'copping', and (4) budgeting money to be spent on opioids. As a result, controlled users suffered fewer adverse negative consequences of opioid use.

The question of prevalence of controlled opioid use versus compulsive use is difficult to address, but evidence from a variety of sources suggests that there are approximately as many controlled opioid users as compulsive opioid users at any given time (Harding and Zinberg, 1983). Consistent with this observation is that when naloxone was administered to applicants in a methadone maintenance program, 55 per cent were found to be positive for opioid dependence, 18 per cent were weakly positive, and 15 per cent negative (O'Brien, 1975). Others have estimated that 45 per cent of the applicants were not addicted (Glaser, 1974). In a random sample of clients visiting a Los Angeles Health Department youth clinic, there were more than twice as many occasional heroin users as daily users (Hunt, 1979). Similar significant nondaily percentages of opioid use have been reported by numerous sources (Harding and Zinberg, 1983).

The natural history of opioid addiction is that of a disorder that is remarkably stable over time. While repeated cycles of remission and resumption of use occur, these patterns extend over long periods. Longitudinal studies have shown that heroin addiction, at least for some individuals, is a lifelong condition. In a 20 year follow up of 100 heroin addicts in New York, 23 per cent died, 25–35 per cent were still known to be using drugs, and 35–42 per cent achieved stable abstinence (Vaillant, 1973). Addicts that had achieved more than three years of abstinence appeared to maintain their abstinence indefinitely (Vaillant, 1973). In a longitudinal study of 581 male heroin addicts admitted to the California Civil Addict Program during the years 1962–1964 and followed for 33 years, there were similar results as obtained with the New York heroin addicts (Hser *et al.*, 2001). In 1995–1997, 33 years later, 21 per cent tested positive for heroin, 10 per cent refused urine analysis, and 14 per cent were incarcerated. This was very similar to the data from 1974–1975 where 23 per cent reported positive urines, 6 per cent refused urine analysis, and 18 per cent were incarcerated (Hser *et al.*, 2001) (**Table 4.3** and **Fig. 4.3**). Again, the rate of abstinence in 1996–1997 was related to the amount of time previously abstinent, and most subjects (75 per cent) who reported abstinence greater than 5 years were still abstinent in 1996–1997 (Hser *et al.*, 2001). The number of negative urines also was relatively stable, but the number of deaths progressively increased (Hser *et al.*, 2001). The most

TABLE 4.3 Longitudinal Patterns of Heroin use, other Substance use, Criminal Involvement, and Mortality among Heroin Addicts

	1974–1975		1985–1986		1996–1997	
Status	*Total (%)* (*n* = 581)	*Living and interviewed (%)* (*n* = 439)	*Total (%)* (*n* = 581)	*Living and interviewed (%)* (*n* = 354)	*Total (%)* (*n* = 581)	*Living and interviewed (%)* (*n* = 242)
Inactive use Urine negative for opiates	28.6	37.8	25.0	41.0	23.2	55.8
Active use Urine positive for opiates	23.1	30.5	19.4	31.9	8.6	20.7
Refused to provide urine	6.2	8.2	4.8	7.9	4.0	9.5
Incarcerated	17.7	23.5	11.7	19.2	5.8	14.0
Deceased	13.8	na	27.7	na	48.9	na
Not interviewed	10.7	na	11.4	na	9.5	na
Total	**100.0**	**100.0**	**100.0**	**100.0**	**100.0**	**100.0**

na, not applicable
[Reproduced with permission from Hser *et al.*, 2001.]

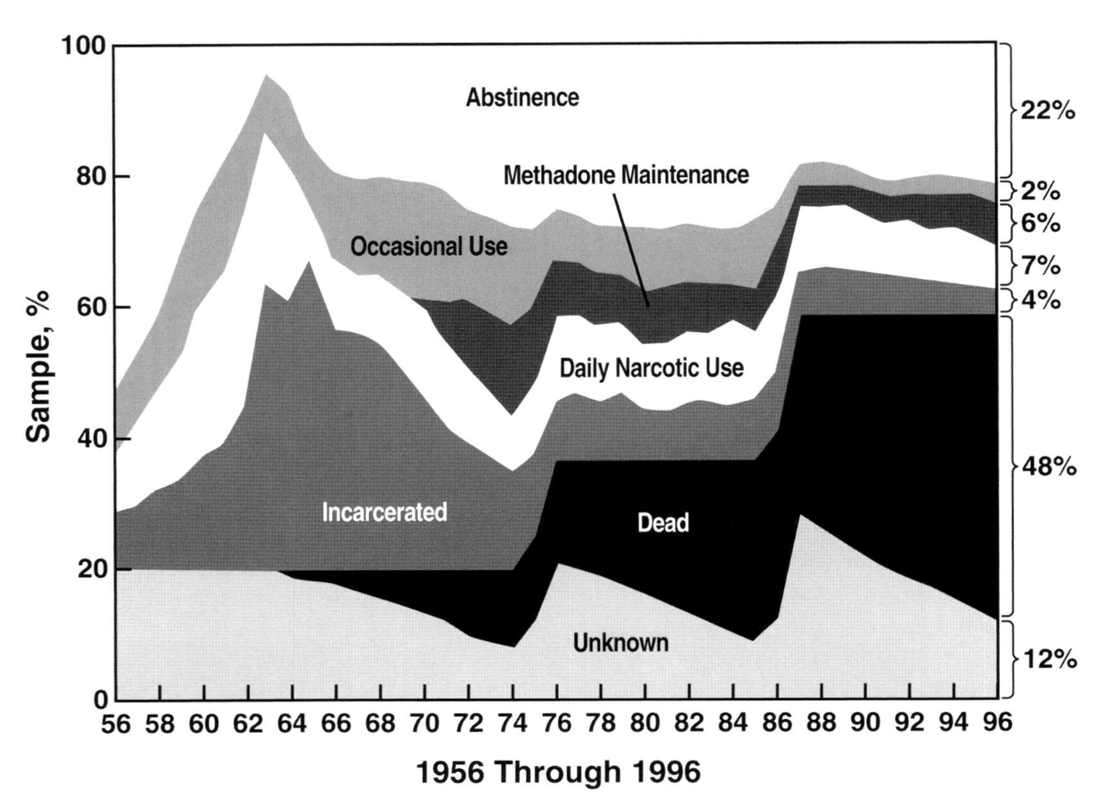

FIGURE 4.3 The natural history of narcotics addiction among a male sample of heroin addicts ($n = 581$) admitted to the California Civil Addict Program, a compulsory drug treatment program for heroin-dependent criminal offenders, during the years 1962 through 1964. This 40-year follow-up study updated information previously obtained from admission records and two face-to-face interviews conducted in 1974–1975 and 1985–1986. In 1996–1997, at the latest follow-up, 284 were dead and 242 were interviewed. The mean age of the 242 interviewed subjects was 57.4 years. Age, disability, years since first heroin use, and heavy alcohol use were significant correlates of mortality. Of the 242 interviewed subjects, 20.7 per cent tested positive for heroin. The group also reported high rates of health problems, mental health problems, and criminal justice system involvement. Long-term heroin abstinence was associated with less criminality, morbidity, psychological distress, and higher employment. While the number of deaths increased steadily over time, heroin use patterns were remarkably stable for the group as a whole. For some, heroin addiction had been a lifelong condition associated with severe health and social consequences. [Reproduced with permission from Hser *et al.*, 2001.]

common cause of death was overdose, chronic liver disease, cancer, and cardiovascular disease. It should be noted that by age 50–60, only about half of the interviewed subjects tested negative for heroin, which argues against the hypothesis that drug addiction fades with age.

Opioid Intoxication

Intoxication for an opioid addict following an intravenous self-injection or smoking has been described as four different components which can overlap in time (Agar, 1973, p. 55). First, there is a profound euphoria termed the *rush* which has been described as occurring about 10 seconds after the beginning of the injection and includes a wave of euphoric feelings, frequently characterized in sexual terms (Dole, 1980):

> 'So I snort again and *holy f—ing s—t!* I felt like I died and went to heaven. My whole body was like one giant f—ing incredible orgasm' (Inciardi, 1986, p. 61).

> 'After a while she asks me if I want to try the needle and I say no, but then I decide to go halfway and *skin-pop* [injecting into the muscle just beneath the skin]. Well, man, it was wonderful. Popping was just like snorting, only stronger, finer, better, and faster' (Inciardi, 1986, p. 62).

> 'Travelin' along the mainline was like a grand slam home run f___, like getting a blow job from Miss America. The rush hits you instantly, and all of

a sudden you're up there on Mount Olympus talking to Zeus' (Inciardi, 1986, p. 62).

In this first state, there also are visceral sensations, a facial flush, and deepening of the voice. While other effects show tolerance with chronic use, the *rush* is resistant to tolerance. Second, the *high* follows and is a general feeling of well-being and can extend several hours beyond the rush that does show tolerance. Third, the *nod* is a state of escape from reality that can range from sleepiness to virtual unconsciousness. Addicts are described as calm and detached and very uninterested in external events (Dole, 1980). Fourth, *being straight* is the state where the user is no longer experiencing the rush or nod or high but also is not experiencing withdrawal. This state can last up to 8 h following an injection or smoking of heroin.

The route of administration and the infusion rate of that administration have a profound effect on the subjective and physiological effects of opioid administration (Marsch *et al.*, 2001). Healthy volunteers received intravenous injections of two doses of morphine at three different infusion rates. Faster infusions produced greater positive subjective effects than slower infusions on measures of *good drug effect, drug liking,* and *high*. Faster infusions also resulted in greater opioid agonist effects (Marsch *et al.*, 2001) (**Fig. 4.4**). In experimental studies of heroin addiction, detoxified opioid addicts were allowed to intravenously self-administer heroin with self-regulated

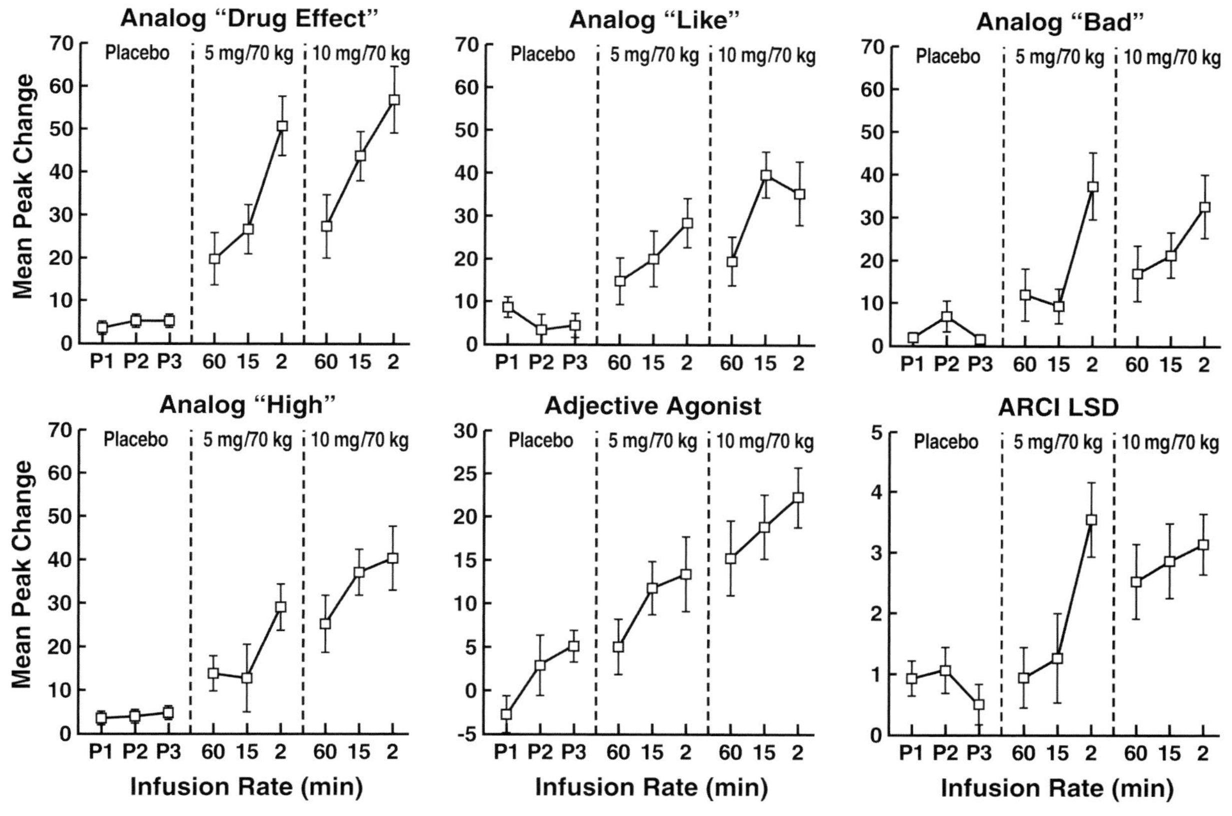

FIGURE 4.4 Mean peak change scores for the drug effect, drug liking, bad, and high effects of the Visual Analog Scale (VAS); the adjective agonist measure of the Adjective Rating Scale (ARS); and the Lysergic Acid Diethylamide (LSD) scale of the Addiction Research Center Inventory (ARCI) for each infusion rate and dosing condition. Morphine Sulfate (5 or 10 mg/70 kg body weight, i.v.) or placebo (0 mg on saline) were administered at one of three infusion rates (2, 15, and 60 min). *VAS*: On this measure (Preston *et al.*, 1988), subjects rated the extent to which they experienced six effects (drug effect, drug liking, good drug effects, bad drug effects, drug-induced high, and sick). The analog scales consisted of a line approximately 100 mm in length, anchored at each end by 'not at all' and 'severe.' Responses were recorded as a score ranging from 0 to 100. *ARS*: Self-reports of drug effects were rated on a modified version of an adjective rating scale (Bickel *et al.*, 1988a) listing 32 items describing typical opioid drug effects and withdrawal effects. Responses were recorded as a score ranging from 0 to 9. Opioid drug effects included nodding, rush, loaded/high, coasting, itchy skin, etc., and withdrawal effects included such items as irritability, chills/gooseflesh, runny nose, yawning, etc. *ARCI short form*: Subjects completed this 49-item true–false questionnaire derived from a 500-item questionnaire containing numerous empirically derived drug-sensitive scales (Haertzen, 1970; Martin *et al.*, 1971). The LSD scale measures dysphoria and psychotomimetic effects. [Reproduced with permission from Marsch *et al.*, 2001.]

access to increasing doses over a 10-day period in a residential laboratory setting in a locked unit of a large psychiatric hospital. (Babor *et al.*, 1976; Mirin *et al.*, 1976). The early phase of acquisition of heroin self-administration was accompanied by an increase in elated mood and a decrease in 'somatic concern'. The later stages of acquisition were characterized by a profound shift toward increasing dysphoria, a notable rise in somatic concern, anxiety, depression, social isolation, and motor retardation. Initially, the reinforcing properties of heroin stemmed primarily from its ability to relieve tension and produce euphoria. However, as the frequency of drug self-administration increased, tolerance quickly developed to the euphorigenic effects, although single injections remained capable of producing brief periods (30–60 min) of positive mood (Mirin *et al.*, 1976). However, this tolerance was accompanied by a distinct shift in the direction of increasing psychopathology and dysphoria. Symptoms included sleep disturbances, social isolation, belligerence, irritability, less motivation for sexual activity, and motor retardation (Haertzen and Hooks, 1969; Babor *et al.*, 1976; Mirin *et al.*, 1976).

Opioid Withdrawal

The characteristic withdrawal syndrome associated with withholding derivatives of opium from chronic users was described well early on by C.K. Himmelsbach, Director of Research, U.S. Public Health Service Hospital at Lexington, Kentucky (Himmelsbach, 1943). The symptoms included yawning, lacrimation, rhinorrhea, perspiration, gooseflesh, tremor, dilated pupils, anorexia, nausea, emesis, diarrhea, restlessness, insomnia, weight loss, dehydration, hyperglycemia, elevations of temperature and blood pressure, and alterations of pulse rate. While many of these symptoms were recognized at that time as manifestations of disturbances in the function of the autonomic nervous system (Himmelsbach, 1943), it was also recognized at this time that a negative affective state could accompany these physical signs of opioid withdrawal. A negative affective state is defined as a dysphoric state accompanied by depressive-like and anxiety-like symptoms that do not fully meet the criteria of a major mental disorder such as a major depressive episode or generalized anxiety disorder. Addicts were described as attempting to obtain sufficient drug to 'prevent the dysphoria associated with the [opioid] withdrawal syndrome' (Reichard, 1943).

Subsequent descriptions of acute opioid withdrawal included two types of symptoms: purposive symptoms and nonpurposive symptoms (Council Reports, 1972). Purposive symptoms were defined as symptoms which were goal-oriented (i.e., directed at getting more drug). The purposive symptoms include complaints, pleas, demands, and manipulations, and were significantly decreased in a hospital setting where such efforts to get more drug had no consequences. In contrast, nonpurposive symptoms included those which were not goal-oriented and were relatively independent of the observer, the patient's will, and the environment. A more modern framework probably would consider purposive symptoms as the 'craving' or 'motivational' domains and nonpurposive as the physical or somatic domains.

The symptoms of opioid withdrawal also change significantly over time since the last dose. In the early stages of withdrawal from heroin (6–8 h after the last dose), the more purposive behavior, goal-oriented, is prominent and peaks at 36–72 h after the last dose of morphine or heroin. The more nonpurposive, autonomic signs appear at about 8–12 h after the last dose and include yawning and sweating, runny nose, and watery eyes (Blachly, 1966; Bewley, 1968) (**Table 4.4**). These mild autonomic signs continue to increase in intensity during the first 24 h and level off. Subsequently, additional nonpurposive, physical symptoms appear which peak at 36–48 h and continue up to 72 h. There is pupillary dilation, gooseflesh, hot and cold flashes, loss of appetite, muscle cramps, tremor, and insomnia. Other autonomic signs include elevations in blood pressure, heart rate, respiratory rate, body temperature, nausea, and vomiting.

Withdrawing individuals have complaints of feeling chilled, alternating with a flushing sensation, and excessive sweating. Waves of gooseflesh are prominent, resulting in skin that looks like a 'plucked turkey,' the basis of the expression 'cold turkey.' Accompanying these symptoms are also the subjective symptoms of aches and pains and general misery such as that associated with a flu-like state. Muscle spasms, uncontrollable muscle twitching, and kicking movements have been hypothesized to represent the basis for the expression 'kicking the habit'. At 24–36 h, there may be diarrhea and dehydration. The peak of the physical withdrawal syndrome appears to be approximately 48–72 h after the last dose. Without treatment, the syndrome completes its course in 7–10 days. However, residual, subclinical signs may persist for many weeks after withdrawal (Council Reports, 1972).

The persistent signs of abstinence in detoxified subjects, including hyperthermia, mydriasis, increased blood pressure, and increased respiratory rate, have been observed to continue for months after opioid withdrawal and have been termed 'protracted abstinence' (Himmelsbach, 1942, 1943). Protracted abstinence

TABLE 4.4 Abstinence Signs in Sequential Appearance After Last Dose of Narcotic

Grade of Abstinence	Withdrawal signs	Hours after last dose		
		Methadone	*Morphine*	*Heroin*
Grade 0	Craving for drug Anxiety	12	6	4
Grade 1	Yawning Perspiration Runny nose Lacrimation	34–48	14	8
Grade 2	Increase in above signs Mydriasis Gooseflesh Piloerection Tremors Muscle twitches Hot and cold flashes Aching bones and muscles Anorexia	48–72	16	12
Grade 3	Increased intensity of above Insomnia Increased blood pressure Increased temperature Increased respiratory rate and depth Increased pulse rate Restlessness Nausea	na	24–36	18–24
Grade 4	Increased intensity of above Febrile facies Curled up position on hard surface Vomiting Diarrhea Weight loss Spontaneous ejaculation or orgasm Hemoconcentration leucocytosis Eosinopenia Increased blood sugar	na	36–48	24–36

na, not applicable. [Modified with permission from Blachly, 1966.]

can be defined as signs of drug abstinence that persist after an acute withdrawal syndrome (usually one week postacute withdrawal). Metabolic changes have been reported in an even later stage of protracted abstinence where the direction of the changes is the opposite of the acute signs of abstinence (e.g., hypothermia, miosis, decreased blood pressure, etc.) (Martin and Jasinsky, 1969). Perhaps more importantly, this protracted abstinence state also has a motivational component, with individuals reporting a 'gray-like' mood state where few stimuli or activities produce pleasure:

> 'It's staying off that is the hard part. It takes a lot of willpower. But seeing smack eats away at your willpower; it makes it very hard. When I stop I just feel vacant with no direction or energy and that lasts for months' (Stewart, 1987, p. 166).

Opioid drugs show qualitatively similar opioid withdrawal effects that vary in duration and intensity. Abrupt withdrawal from methadone, even after large doses, is slower to develop than that of heroin, and is less intense and more prolonged. Few or no symptoms are observed for almost 2 days, and peak intensity is reached on the sixth day. In contrast, the withdrawal syndrome from meperidine (Demerol®) usually develops within 3 h of the last dose, reaches a peak within 8–12 h and then decreases (Council Reports, 1972). These differences in time course and intensity of opioid withdrawal with different opioids speak to the general principle that long-acting drugs produce longer onset,

TABLE 4.5 Time-Course of Withdrawal from Various Narcotic Agents

	Nonpurposive withdrawal symptoms (hours)	Peak (hours)	Time in which majority of symptoms terminate (days)
Morphine	14–20	36–48	5–10
Heroin	8–12	48–72	5–10
Methadone	36–72 (Second day)	72–96 (Sixth day)	14–21
Codeine	24		
Dilaudid	4–5		
Meperidine	4–6	8–12	4–5

Purposive symptoms are goal-oriented, highly dependent on the observer and environment, and directed at getting more drugs.
Nonpurposive symptoms are not goal-oriented, are relatively independent of the observer and of the patient's will and the environment.
The purposive phenomena, including complaints, pleas, demands and manipulations, and symptom mimicking are as varied as the psychodynamics, psychopathology, and imagination of the drug-dependent person. In a hospital setting, these phenomena are considerably less pronounced when the patient becomes aware that this behavior will not affect the decision to give him a drug.
[Reproduced with permission from Kleber, 1981.]

longer duration, and less intense withdrawal than short-acting drugs (Kleber, 1981) (**Table 4.5**).

Environmental stimuli can be conditioned both to the acute reinforcing effects of opioids and the withdrawal associated with opioids, and both have been hypothesized to contribute to craving. In humans, stimuli paired with morphine injection also have been shown to alleviate withdrawal. The following quotation describes the signs and symptoms of opioid withdrawal after 48 h of abstinence and the conditioned injection effect:

> 'Further evidence that the picture of withdrawal symptoms has its basis in an emotional state is the response on the part of one of our addicts at the end of a 36-hour withdrawal period to the hypodermic injection of sterile water. Despite his obvious suffering, he immediately went to sleep and slept for eight hours. Addicts frequently speak about the 'needle habit,' in which the single prick of the needle brings about relief. It is not uncommon for one addict to give another a hypodermic injection of sterile water and the recipient to derive a 'kick' and become quiet. On the other hand, it has been our experience just as frequently to have the addict know that he was given a hypodermic injection of sterile water and to have him fail to respond to its effect. Paradoxical as it may seem, we believe that the greater the craving of the addict and the severity of the withdrawal symptoms, the better are the chances of substituting a hypodermic injection of sterile water to obtain temporary relief' (Light and Torrance, 1929).

Subsequent studies of this phenomenon described these individuals as 'needle freaks,' where at least part of the relief and pleasure they obtain from injecting the drug was hypothesized to be the result of a conditioned positive response to the injection procedure associated with heroin use (Levine, 1974; O'Brien, 1974; Meyer and Mirin, 1979). Indeed, in an experimental laboratory situation under double-blind conditions, subjects were administered an opioid antagonist and then allowed to self-administer vehicle or opioid. All of the self-injections were rated as pleasurable at first, and after 3–5 injections, the subjects reported neutral effects (O'Brien, 1974; Meyer and Mirin, 1979). Conditioned withdrawal can be observed in the human situation and can set up a condition of negative reinforcement. O'Brien described the experience of an opioid addict upon returning to an environment where he had previously experienced opioid withdrawal:

> 'For example, one patient who was slowly detoxified after methadone maintenance went to visit relatives in Los Angeles after receiving his last dose. Since he knew that he would be away from the clinic in Philadelphia for three weeks, he saved one take-home bottle of methadone in case he got sick while in California. To his surprise, he felt no sickness while in this new environment and never even thought about the bottle of methadone in his suitcase. He felt healthy over the three-week, drug-free period, but as soon as he arrived in the Philadelphia airport, he began to experience craving. By the time he reached his home, there was yawning and tearing. He immediately took the methadone he had been saving and felt relieved, but the symptoms re-occurred the next day. After three weeks of being symptom-free in Los Angeles, he experienced regular withdrawal in Philadelphia' (O'Brien *et al.*, 1986).

This conditioned withdrawal has been experimentally induced in the laboratory. Methadone-maintained volunteers were subjected to repeated episodes of precipitated opioid withdrawal by a very small dose of naloxone (0.1 mg, i.m.) in the context of a tone and peppermint smell in a particular environment. This dose of naloxone elicited tearing, rhinorrhea, yawning, decreased skin temperature, increased respiratory rate, and subjective feelings of drug sickness and craving. Subsequently, injection of vehicle (physiological saline) accompanied by the peppermint smell and tone elicited reliable signs and symptoms of opioid withdrawal that were similar to the precipitated withdrawal, though less severe (O'Brien *et al.*, 1977). Thus, stimuli paired with opioid withdrawal or with opioids themselves can be of motivational significance, both eliciting drug taking or alleviating withdrawal, respectively, and contributing to the maintenance and relapse associated with opioid addiction.

Opioid Tolerance

Tolerance can be defined as a decreased response to a drug with repeated administration or the requirement for larger doses of a drug to produce the same effect. Tolerance develops to the analgesic, euphorigenic, sedative, and other central nervous system depressant effects of opioids. Tolerance also develops to the lethal effects of opioids. Opioid addicts can increase intake to enormous doses, such as 2 g of morphine intravenously over a period of 2–3 h without significant changes in blood pressure or heart rate. The lethal dose of morphine in a nontolerant individual is approximately 30 mg parenterally and 120 mg orally (Ellenhorn and Barceloux, 1988).

Tolerance to opioids is also characterized by a shortening of the duration of action. The rate of tolerance development is dependent on not only the dose of the drug, but also the pattern of use and the context of use (see below). With doses in the therapeutic range and appropriate intermittent use, it may be possible to obtain the desired analgesic effect for an indefinite period. When the opioid is taken in a continuous pattern, significant tolerance develops rapidly. Cross-tolerance develops to a high degree with opioids as long as the opioids are acting through the same receptor subtype (Moulin *et al.*, 1988).

Tolerance can be not only dramatic but also differential. Subjects may become very tolerant to the lethal or respiratory depressant effects of opioids but still continue to show sedation, miosis, and constipation (Martin *et al.*, 1973). For example, constipation continues in a large number of methadone-maintained individuals for up to 8 months of daily use. Insomnia is observed in 10–20 per cent of patients and excessive sweating in 50 per cent of patients.

Tolerance to many effects of opioids also develops rapidly in laboratory animals and is usually reflected in a shift of the dose-effect function to the right. Morphine treatment in mice over a period of 20 days produced parallel shifts of the analgesia dose-effect function to the right using two different measures of analgesia (Fernandes *et al.*, 1977) (**Fig. 4.5**). However, there was no tolerance to the motor-activating effects of morphine, but in fact, some evidence of sensitization (Fernandes *et al.*, 1977). Robust tolerance has been observed for the discriminative stimulus effects of opioids in animals, suggesting parallels with the subjective reports of tolerance in humans (Young *et al.*, 1990, 1992; Young and Goudie, 1995). Tolerance to the rewarding effects of opioids with continuous administration, as measured by conditioned place preference, also has been observed (Shippenberg *et al.*, 1988).

Two types of tolerance exist: dispositional and pharmacodynamic. Dispositional tolerance is the decreased response to a drug with repeated administration due to a reduction in the amount of drug at its site of action. Pharmacodynamic tolerance refers to changes in response to a drug that result from adaptive changes excluding changes in the disposition of the drug. Although there is some evidence that metabolism to opioid drugs can be enhanced in tolerant animals, most opioid tolerance is thought to be pharmacodynamic (Jaffe, 1990; Jaffe and Martin, 1990).

Acute and chronic tolerance, and acute and chronic withdrawal, also have an important associative basis (Siegel *et al.*, 2000). Drug administration produces disturbances in the body (presumably largely via the brain for psychotropic effects), and these disturbances come to elicit unconditioned responses that compensate for the perturbation produced by the drug. Thus, the unconditioned stimulus is the drug effect, and the unconditioned responses are responses that compensate for the drug effect. These compensatory responses ultimately come to be elicited by drug-associated cues in a Pavlovian conditioning process. Such conditioned compensatory responses often are opposite in direction to the unconditioned effects of a drug (Siegel, 1975) and can mediate tolerance in the presence of drug and withdrawal in the absence of drug (Ramsay and Woods, 1997). Pavlovian conditioning, and learning in general, have long been hypothesized to contribute to the development of tolerance (Cochin and Kornetsky, 1964; Cohen *et al.*, 1965; Cochin, 1970).

Conditioned compensatory responses are most clearly observed by presenting the predrug cues that have been paired with drug effects in the absence of the drug. Under such conditions, opioids produce conditioned compensatory responses (Krank *et al.*, 1981). In such an experiment, rats are assigned to a paired or unpaired condition. For the paired animals, morphine injections were signaled by an audiovisual cue. Unpaired animals received the same cues and injections but in an unpaired manner. When subsequently tested in the presence of the cue, paired animals were more tolerant than unpaired animals (Siegel, 1978). There are many parallels between the Pavlovian conditioning model of Siegel and colleagues and tolerance, including extinction, retardation of the development of tolerance, and other predrug cue manipulations such as inhibitory learning and stimulus generalization (Siegel *et al.*, 2000). Conditioned compensatory responses produced in the absence of

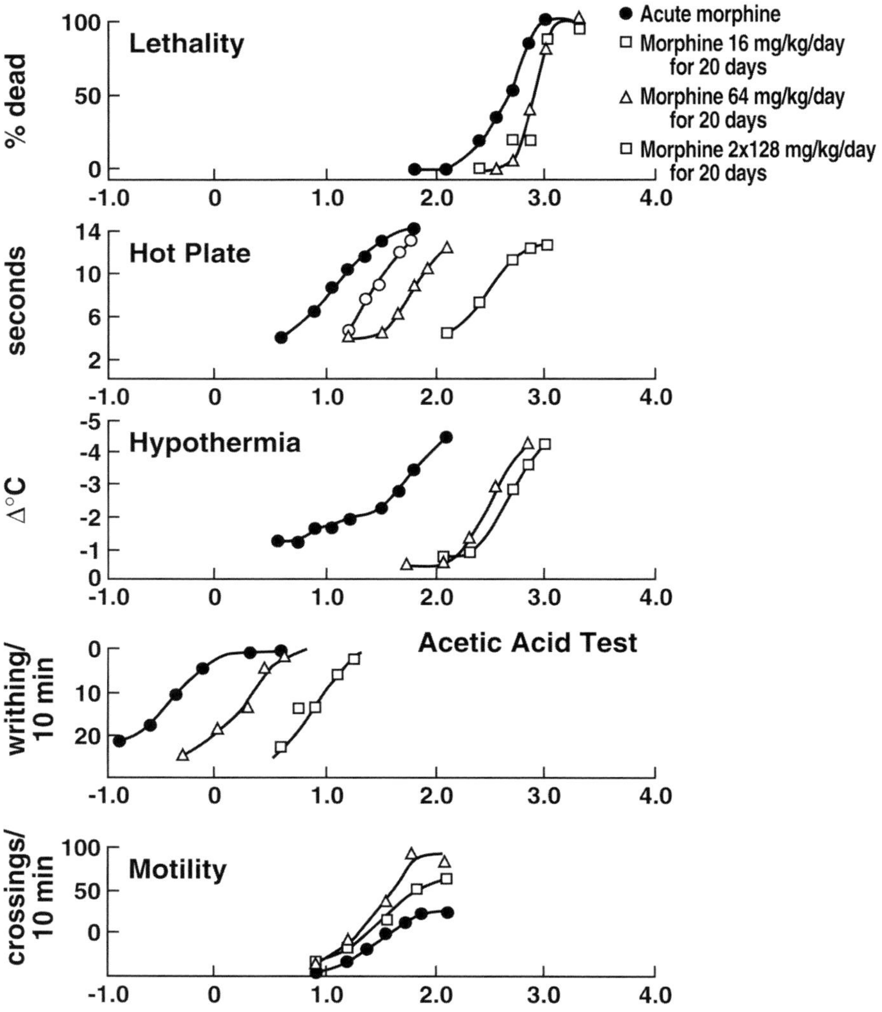

FIGURE 4.5 Qualification of tolerance in mice. Effects of morphine pretreatment on the dose-response functions for morphine in five different measures in mice. Values represent dose-response curves for morphine hydrochloride in different test situations. Solid circles indicate acute morphine. Open circles indicate 1 × 16 mg/kg/day for 20 days. Open triangles indicate 1 × 64 mg/kg/day for 20 days. Open squares indicate 2 × 128 g/kg/day for 20 days. Each point represents the mean results from at least 15 animals. Horizontal axis represents log mg/kg morphine hydrochloride. [Reproduced with permission from Fernandes *et al.*, 1977.]

drugs also would be reflected in conditioned withdrawal-like effects (Krank *et al.*, 1981; Grisel *et al.*, 1994).

The role of associative mechanisms in opioid tolerance and withdrawal not only impacts the neurobiological substrates and adds an important layer of neurocircuitry in relapse that must be considered, but also may have practical implications for opioid treatment and addiction. Patients or addicts receiving drugs in a different context, even a different route of administration, may be at risk for overdose because the conditioned compensatory responses will not be elicited (Siegel *et al.*, 1982; Siegel and Kim, 2000; Siegel and Ramos, 2002).

BEHAVIORAL MECHANISM OF ACTION

The behavioral mechanism most associated with opioids is their effects on pain processing and relief of pain and suffering. Subsequently, opioids can induce pain as a result of counteradaptive processes (Larcher *et al.*, 1998; Laulin *et al.*, 1998, 1999; Celerier *et al.*, 2001; Ossipov *et al.*, 2003). The pain relief takes on many forms in several parts of the central nervous system and, in fact, outside of the central nervous system. Within this book, the behavioral mechanism of action refers to a unifying principle of order and predictability at the behavioral level. Each drug class has behavioral effects that define its phenotype. This behavioral mechanism may derive from medical use, behavioral pathology, or some combination of both.

Opioids also relieve emotional pain, and this is one of the behavioral mechanisms implicated in the addiction cycle (Khantzian, 1985, 1990, 1997) (see *What is Addiction?* chapter). A unique aspect of heroin addiction has been described as:

> 'the special role the drug comes to play in the personality organization of these patients. They have not successfully established familiar defensive, neurotic, characterological or other common adaptive mechanisms as a way of dealing with their distress. Instead, they have resorted to the use of opioids as a way of coping with a range of problems including ordinary human pain, disappointment, anxiety, loss, anguish, sexual frustration, and other suffering' (Khantzian *et al.*, 1974, p. 162).

This aspect of drug addiction has been extended to an overall hypothesis of self-medication where patients are hypothesized to experiment with various classes of drugs to discover one that is particularly compelling because it changes affective states that the patient finds particularly problematic or painful. In the case of opioid drugs, opioids are hypothesized to be preferred because of their powerful actions in diminishing the disorganizing and threatening effects of rage and aggression (Khantzian, 1985, 1997). Subjects who experienced or expressed physical abuse and violent behavior described how opioids helped them feel normal, calm, mellow, soothed, and relaxed (Khantzian, 1985).

For clinicians, it is now becoming clearer that patients receiving long-term opioid therapy can develop unexpected abnormal pain as well as hyperalgesia upon withdrawal (Ossipov *et al.*, 2003). Importantly, pain also is a component of the human opioid withdrawal syndrome. In a review of the clinical experiences of over 750 patients receiving epidural or spinal morphine over a mean period of 124 days, many patients developed an increased sensitivity to sensory stimuli (i.e., hyperesthesias) and pain elicited by normally innocuous sensory stimulation (i.e., allodynia). In an extreme case, a patient developed severe allodynia after receiving only one 30 mg/ml infusion of morphine (Arner *et al.*, 1988). The same observations have been reported during the last two decades in various clinical situations (Ali, 1986; de Conno *et al.*, 1991). In one case of lower lumbar arachoiditis treated with intrathecal morphine, the abnormal pain (hyperalgesia) was manifested differently than the original pain complaint in the legs and lower extremities (Devulder, 1997).

Remifentanil is one of the most widely used opioids for surgery. It is a powerful, quick, and short-lasting drug acting at a dose of 0.1–0.2 μg/kg/min, or 1 ng/ml. It has been demonstrated that intraoperative remifentanil increased postoperative pain and morphine consumption, suggesting the development of opioid tolerance and hyperalgesia (Guignard *et al.*, 2000). However, while well documented at the bedside and practice level by physicians and anesthesiologists, the problem had not raised more direct rigorous investigation in humans until recently.

In a series of randomized, double-blind, placebo-controlled crossover studies using remifentanil and transcutaneous electrical stimulation at high electrical current densities in human volunteers, postinfusion pain and hyperalgesia were clearly shown, and thus, apparent tolerance to opioids (Angst *et al.*, 2003; Hood *et al.*, 2003; Koppert *et al.*, 2003; Luginbühl *et al.*, 2003) (**Fig. 4.6**). These effects appeared after one administration and lasted for hours after exposure and were suppressed by the *N*-methyl-D-aspartate (NMDA)-receptor antagonist S-ketamine, pointing to a potential role for the NMDA receptor system in mediating such a pain sensitization response, an effect demonstrated earlier in animal studies (Laulin *et al.*, 1998) (see below). Clonidine, an α_2 adrenergic receptor agonist, when given in combination with remifentanil, also attenuated the post-infusion increase of pain rating (Koppert *et al.*, 2003). The hyperalgesia to touch did not extend to the heat modality, and it is possible that there are differential susceptibilities of various pain modalities for expressing opioid-associated hyperalgesia (Angst *et al.*, 2003). Data from laboratory studies in rodents confirmed that opioids, in doses reflecting those prescribed in the clinic, induced 'paradoxical' opioid-induced pain, pain 'sensitization,' or hyperalgesia, even after the first administration (Laulin *et al.*, 1998).

NEUROBIOLOGICAL MECHANISM—NEUROCIRCUITRY

Neurobiology of the Acute Reinforcing Effects of Opioids

The analgesic effects of opioids are due to direct inhibition of nociceptive activity ascending from the dorsal horn of the spinal cord and activation of pain control circuits that descend from the midbrain via the rostral

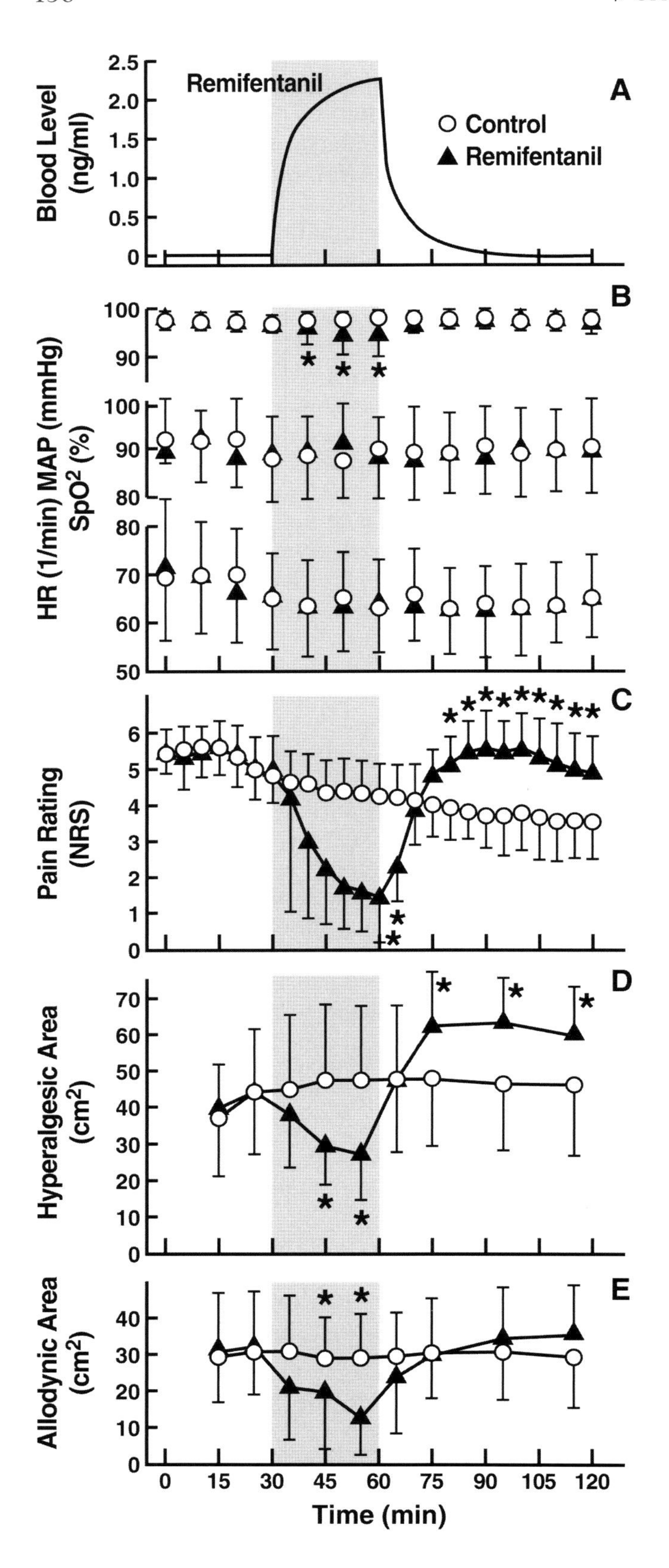

FIGURE 4.6 Analgesia and hyperalgesia produced by an opiod in humans. **(A)** Time course of calculated remifentanil plasma concentrations in human volunteers after a constant-rate infusion of 0.1 μg/kg/min (shaded area). (B) Infusion of remifentanil resulted in a significant decrease in oxygen saturation measured by pulse oximetry (SpO_2) ($p < 0.001$ by ANOVA), whereas mean arterial pressure (MAP) and heart rate (HR) remained unchanged. Pain ratings (C) and areas of punctate hyperalgesia (D) and touch-evoked allodynia (E) were reduced significantly during infusion of remifentanil ($p < 0.05$ by ANOVA). However, shortly after cessation of the infusion, pain ratings and hyperalgesic areas increased and exceeded control values ($p < 0.01$ by ANOVA). Data are expressed as mean ± SD ($n = 13$), $^*p < 0.05$, planned comparisons corrected with the Bonferroni procedure. NRS = numerical rating scale. [Reproduced with permission from Koppert *et al.*, 2003.]

ventromedial medulla to the dorsal horn of the spinal cord (Basbaum and Fields, 1984; Mansour *et al.*, 1995). Opioid peptides and their receptors are highly localized to these descending pain control circuits. The selectivity of opioids has a cellular basis in that it appears to be due to presynaptic inhibition of slow-conducting, small nociceptors via inhibition of calcium channels (Taddese *et al.*, 1995).

Opioid drugs, such as heroin, are readily self-administered intravenously by mice, rats, and monkeys. If provided limited access, rats will maintain stable levels of drug intake on a daily basis without any major signs of physical dependence (Koob *et al.*, 1984). This heroin self-administration typically involves some rapid responding at the beginning of the session (loading phase), followed by regular interinjection intervals (Koob, 1987) (**Fig. 4.7**). This model probably best reflects the human equivalent of 'chipping' (Zinberg

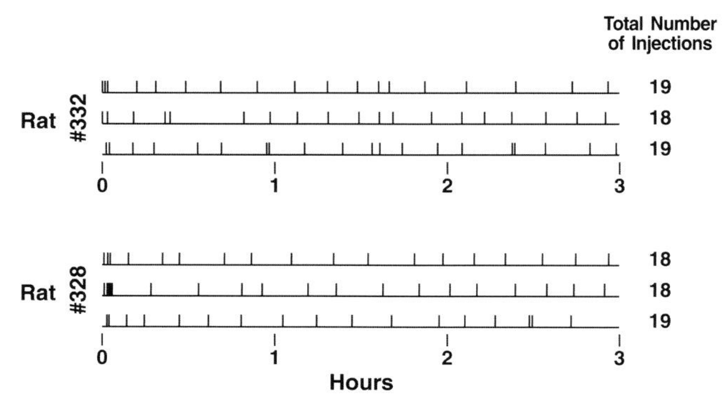

FIGURE 4.7 Representative response records for two rats self-administering heroin. Test sessions were 3 h in duration. Each mark represents a response/infusion of intravenous heroin (0.06 mg/kg/injection). [Reproduced with permission from Koob, 1987.]

and Jacobson, 1976) and has been used to study the neurobiologic basis of heroin reward independent of confounds of physical dependence (see *What is Addiction?* chapter).

Using this model, decreases in the dose of heroin available to the animal change the pattern of self-administration such that the interinjection interval decreases, and the number of injections increases (Koob, 1987). Similar results have been obtained by both systemic and central administration of competitive opioid antagonists (Goldberg *et al.*, 1971; Weeks and Collins, 1976; Ettenberg *et al.*, 1982; Koob *et al.*, 1984; Vaccarino *et al.*, 1985b), suggesting that the animals attempt to compensate for the opioid antagonism by increasing the amount of drug injected. Opioid drugs also support place preference in place conditioning in nondependent animals (Mucha *et al.*, 1982; Bals-Kubik *et al.*, 1993).

Much data suggest that neural elements in the region of the ventral tegmental area and the nucleus accumbens are responsible for the activational and reinforcing properties of opioids and that there are both dopamine-dependent and dopamine-independent mechanisms of opioid action (Stinus *et al.*, 1980, 1989; Spyraki *et al.*, 1983; Pettit *et al.*, 1984; Shippenberg *et al.*, 1992; van Ree *et al.*, 1999) (**Fig. 4.8**). Direct intracerebral administration of quaternary derivatives of naloxone have been used to identify the sites of opioid action in the brain because these quaternary derivatives do not spread readily from the injection site (Schroeder *et al.*,

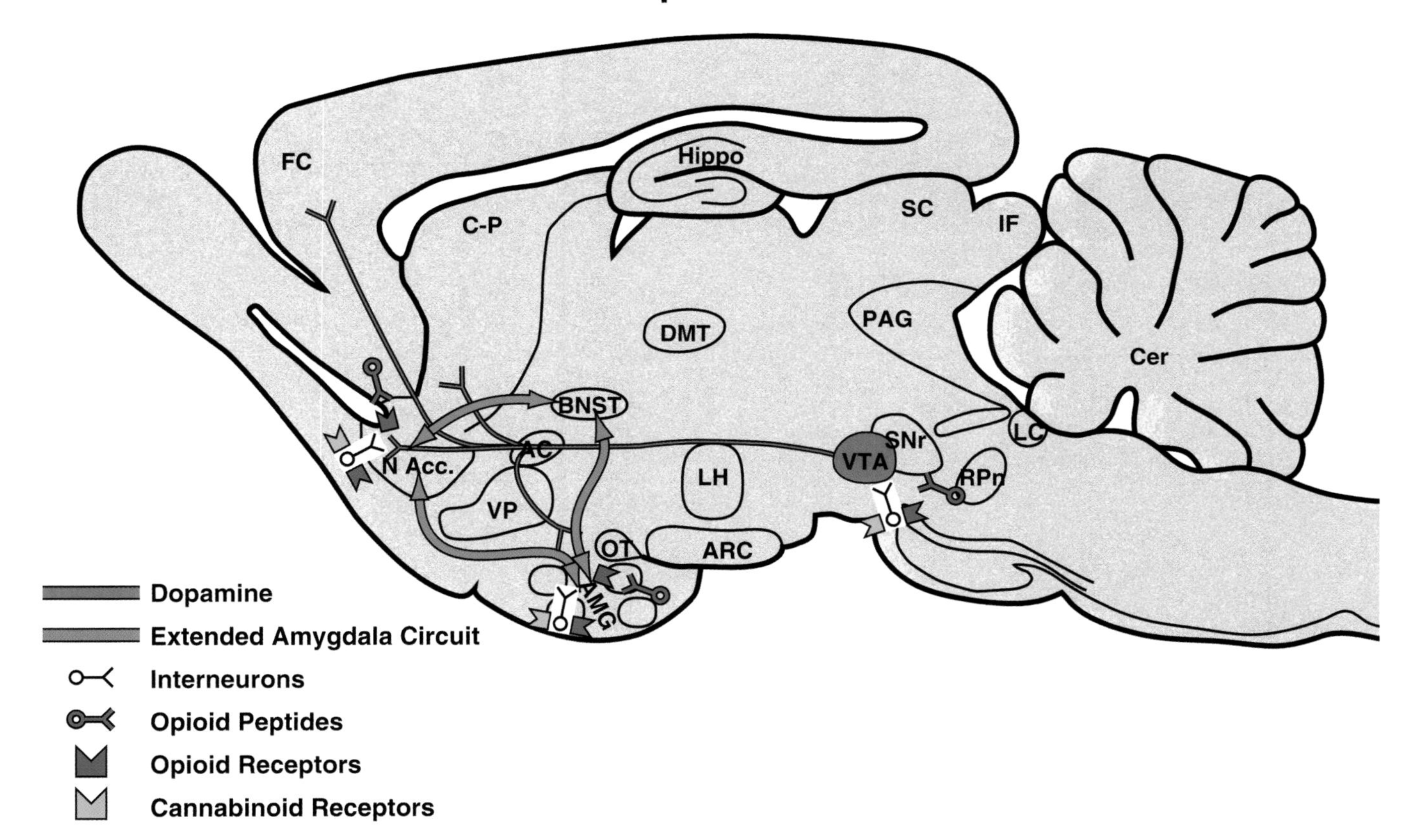

FIGURE 4.8 Sagittal section through a representative rodent brain illustrating the pathways and receptor systems implicated in the acute reinforcing actions of opioids. Opioids activate opioid receptors in the ventral tegmental area, nucleus accumbens, and amygdala via direct actions on interneurons. Opioids facilitate the release of dopamine (red) in the nucleus accumbens via an action either in the ventral tegmental area or the nucleus accumbens, but also are hypothesized to activate elements independent of the dopamine system. Endogenous cannabinoids may interact with postsynaptic elements in the nucleus accumbens involving dopamine and/or opioid peptide systems. The blue arrows represent the interactions within the extended amygdala system hypothesized to have a key role in opioid reinforcement. AC, anterior commissure; AMG, amygdala; ARC, arcuate nucleus; BNST, bed nucleus of the stria terminalis; Cer, cerebellum; C-P, caudate-putamen; DMT, dorsomedial thalamus; FC, frontal cortex; Hippo, hippocampus; IF, inferior colliculus; LC, locus coeruleus; LH, lateral hypothalamus; N Acc., nucleus accumbens; OT, olfactory tract; PAG, periaqueductal gray; RPn, reticular pontine nucleus; SC, superior colliculus; SNr, substantia nigra pars reticulata; VP, ventral pallidum; VTA, ventral tegmental area.

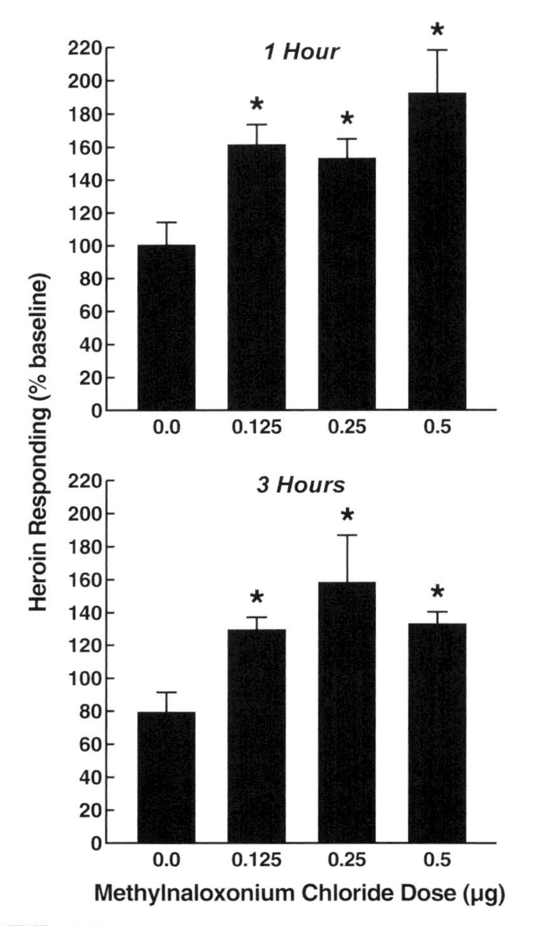

FIGURE 4.9 Percentage predrug day responding for intravenous heroin during the first hour and for the total 3-h session of heroin self-administration (0.06 mg/kg/injection) following methylnaloxonium injections into the nucleus accumbens of seven rats. *$p < 0.05$, compared to saline vehicle, Duncan Multiple Range *a posteriori* test. [Reproduced with permission from Vaccarino *et al.*, 1985a.]

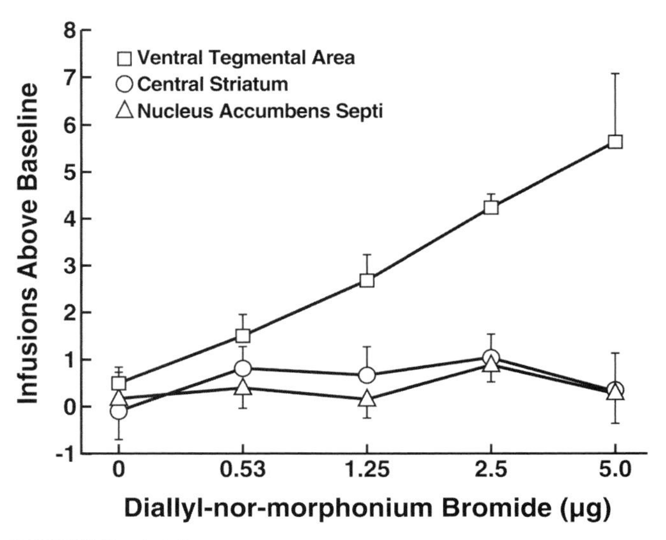

FIGURE 4.10 Mean increases in intravenous heroin self-administration (0.05 mg/kg/injection) in the hour following various doses of intracranial diallyl-nor-morphinium bromide in rats. [Reproduced with permission from Britt and Wise, 1983.]

1991). The hydrophilic quarternary opioid antagonist methylnaloxonium injected into the lateral ventricle, ventral tegmental area, and nucleus accumbens increased heroin self-administration (decreased the interinjection interval) with the most sensitive site being the nucleus accumbens (Vaccarino *et al.*, 1985a,b) (**Fig. 4.9**). Others have seen similar effects in both the nucleus accumbens and ventral tegmental area with other quaternary derivatives of opioid antagonists (Britt and Wise, 1983; Corrigall and Vaccarino, 1988) (**Fig. 4.10**). Similarly, injections of methylnaloxonium into the nucleus accumbens were most potent in reversing the locomotor stimulation produced by heroin (Amalric and Koob, 1985). These results suggested that opioid receptors in both the ventral tegmental area and the nucleus accumbens may be important for the acute reinforcing actions of opioids (Bozarth and Wise, 1983). Strong evidence for a role of the ventral tegmental area in the acute reinforcing effects of opioids in nondependent rats also comes from place conditioning studies (Bals-Kubik *et al.*, 1993) (but see below).

Intracranial self-administration studies have established that the lateral hypothalamus, nucleus accumbens, amydala, periaqueductal gray, and ventral tegmental area all support morphine self-administration (Olds, 1979, 1982; Bozarth and Wise, 1981; Cazala *et al.*, 1987; Welzl *et al.*, 1989; Cazala, 1990; Devine and Wise, 1994; David and Cazala, 1994, 1998). Intracranial self-administration was established in the lateral hypothalamus and nucleus accumbens of the rat (Olds, 1979, 1982) and the amygdala, lateral hypothalamus, and periaqueductal gray in the mouse (Cazala *et al.*, 1987; Cazala, 1990; David and Cazala, 1994, 1998), and the ventral tegmental area in both rats and mice (Bozarth and Wise, 1981; Welzl *et al.*, 1989; Devine and Wise, 1994; David and Cazala, 1994, 1998).

In opioid reinforcement, a role for neural elements in the nucleus accumbens postsynaptic to dopamine afferents became more important with the observation that dopamine receptor blockade and dopamine denervation of the nucleus accumbens (with the neurotoxin 6-hydroxydopamine), can eliminate cocaine and amphetamine self-administration (Roberts *et al.*, 1977, 1980; Lyness *et al.*, 1979), but can spare heroin and morphine self-administration (Ettenberg *et al.*, 1982; Pettit *et al.*, 1984; Smith *et al.*, 1985; Dworkin *et al.*, 1988b) (**Fig. 4.11**). Similar dopamine-independent effects on heroin self-administration have been seen with pharmacological studies using dopamine antagonists

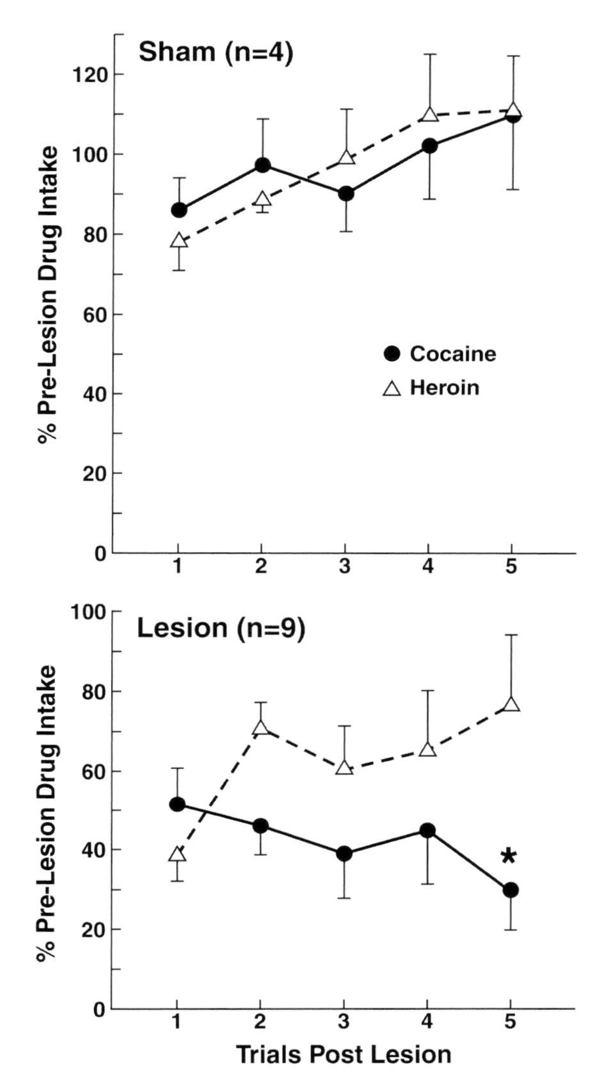

FIGURE 4.11 Top: Percentage of presurgery cocaine and heroin intake over the first five trials postlesion for rats receiving sham lesions. Both cocaine and heroin responding increased significantly over time. Bottom: Percentage of prelesion cocaine and heroin intake over the first five trials postlesion for rats receiving 6-hydroxydopamine lesions to the nucleus accumbens. Cocaine responding showed a trial-dependent decrease, whereas heroin self-administration showed a trial-dependent recovery. The asterisk (*) indicates significant difference between the two drugs ($p < 0.05$, paired t-test). [Reproduced with permission from Pettit *et al.*, 1984.]

(Ettenberg *et al.*, 1982; Gerrits *et al.*, 1994; Hemby *et al.*, 1996). While systemic administration of dopamine antagonists has been shown to attenuate opioid self-administration (Glick and Cox, 1975; van Ree and Ramsey, 1987; Gerrits *et al.*, 1994; Hemby *et al.*, 1996; Awasaki *et al.*, 1997), most of these effects were observed only at doses which affect motor effects or rate of responding (van Ree *et al.*, 1999). Initiation of heroin self-administration is not altered by injection of large doses of haloperidol into the terminals of the midbrain dopamine system, including the nucleus accumbens, prefrontal cortex, caudate putamen, and amygdala (van Ree and Ramsey, 1987).

In contrast, cell body lesions (kainic acid) of the nucleus accumbens block cocaine, heroin, and morphine self-administration (Zito *et al.*, 1985; Dworkin *et al.*, 1988a). Selective excitotoxic lesions of the shell and core subregions of the nucleus accumbens did not produce long-term deficits in self-administration of heroin on a continuous reinforcement schedule (Hutcheson *et al.*, 2001). However, core lesions but not shell lesions severely impaired acquisition of a second-order schedule of reinforcement (Hutcheson *et al.*, 2001), suggesting an important role of the core of the nucleus accumbens in the acquisition of heroin-seeking behavior under the control of drug-associated stimuli. Ventral pallidal cell body lesions (ibotenic acid) block both heroin and cocaine self-administration (Hubner and Koob, 1990) (**Fig. 4.12**). Locomotor activation to opioids also can be independent of dopamine release in the nucleus accumbens and is augmented following dopamine depletion (Churchill *et al.*, 1995). Other work has shown the ventral pallidum (region of the substantia innominata) to be a critical connection in the expression of nucleus accumbens behavioral stimulation (Swerdlow and Koob, 1984; Swerdlow *et al.*, 1984a,b; Churchill *et al.*, 1998). Lesions of the pedunculopontine tegmental nucleus also block heroin reinforcement as measured by intravenous self-administration (Olmstead *et al.*, 1998) (**Fig. 4.13**). Together, these studies suggest that neurons in the region of the nucleus accumbens may be critical for mediating the acute reinforcing properties of opioids and psychomotor stimulants. They also indicate that the nucleus accumbens–ventral pallidum–pedunculopontine nucleus circuit may be a common second-order link for both stimulant and opioid reinforcement.

Using place conditioning, a key role for the mesocorticolimbic dopamine system in opioid-induced reward has been hypothesized, but some have argued that this role is only observed in dependent animals. Dopaminergic lesions of the nucleus accumbens block the acquisition of opioid place preference (Spyraki *et al.*, 1983), and systemic treatment as well as intra-nucleus accumbens treatment with dopamine antagonists also blocked the development of opioid place preference (Leone and Di Chiara, 1987; Shippenberg and Herz, 1987, 1988; Acquas *et al.*, 1989; Shippenberg *et al.*, 1993; Longoni *et al.*, 1998). Also, morphine place conditioning was blocked in knockout mice lacking the dopamine D_2 receptor (Maldonado *et al.*, 1997).

However, in studies of place conditioning produced by opioids in both nondependent and dependent

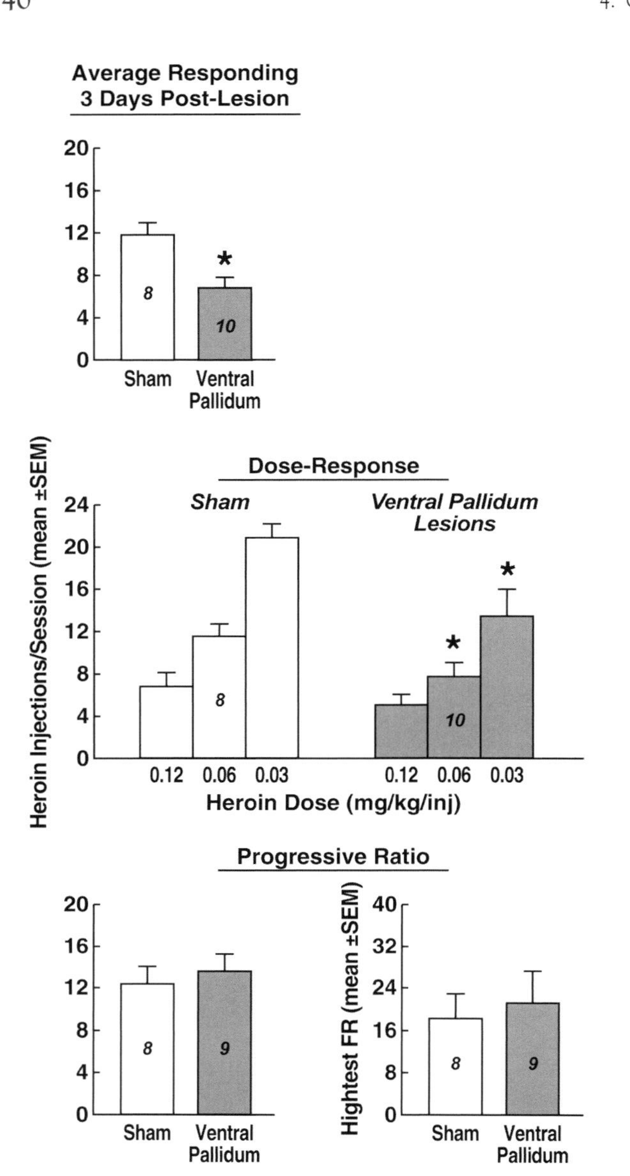

FIGURE 4.12 Effects of bilateral ibotenic acid lesions of the ventral pallidum on responding in rats self-administering heroin. Sham: vehicle (pH 7.4 phosphate buffered saline) injected controls. Ventral Pallidum: Ibotenic acid (5 μg/0.5 μl; expressed as salt) injected into the ventral pallidum. The top panel shows the mean number of injections maintained on a fixed-ratio 5 schedule of reinforcement averaged over the first 3 days postlesion. The middle panel shows the dose-effect function for heroin, expressed as mean number of injections, maintained on a fixed-ratio 5 schedule. The bottom panel shows the number of injections and the mean highest ratio completed on a progressive-ratio schedule.
*Significantly different from the Sham group ($p < 0.05$, *t*-test following significant ANOVA main effect.) [Reproduced with permission from Hubner and Koob, 1990.]

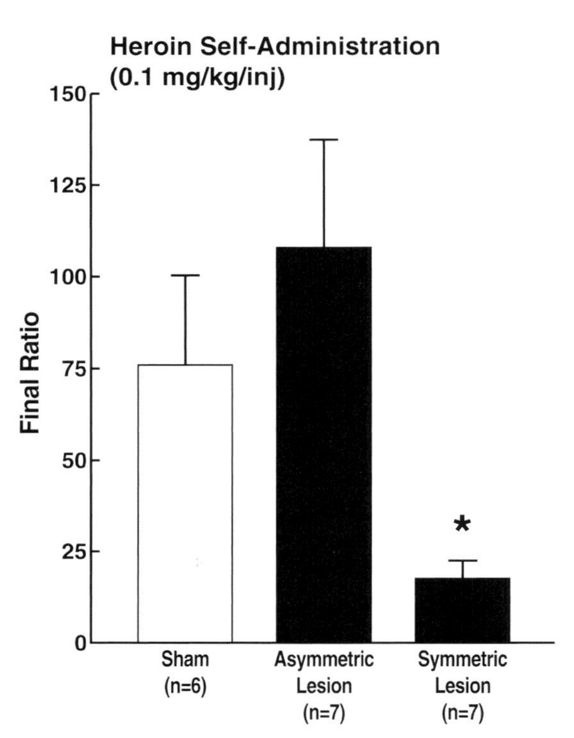

FIGURE 4.13 Effects of pedunculopontine tegmental nucleus lesions on intravenous heroin self-administration in rats. Twenty rats that had learned to self-administer intravenous heroin (0.1 mg/kg/inj) on a fixed-ratio 1 schedule of reinforcement during an initial 15 acquisition sessions were tested on a progressive-ratio schedule of reinforcement. Break points of all animals stabilized within 8.05 ± 0.71 days (sham lesions: 7.0 ± 1.63 days; symmetric lesions: 7.14 ± 0.34 days; asymmetric lesions: 9.29 ± 1.63 days). The final ratios (number of bar presses made for the last infusion of heroin) and the ordinal values of these final ratios (progressive-ratio step number) are shown. Rats with pedunculopontine tegmental nucleus lesions stopped responding at a lower step in the progressive-ratio schedule than sham-lesioned rats (i.e., lesioned animals made fewer lever presses for their final infusion of heroin) ($^*p = 0.0006$). Break points of rats with sham and asymmetric lesions were not significantly different. [Reproduced with permission from Olmstead *et al.*, 1998.]

rats, van der Kooy and associates have argued for a hypothesis of two separate motivational systems critical for opioid addiction that involves the mesocorticolimbic dopamine system and the pedunculopontine tegmental nucleus. Systemic administration of dopamine antagonists, even at very high doses, did not block the acquisition of place preference in nondependent animals but blocked the acquisition of opioid place preference in *dependent* rats induced by systemic injection of morphine or by injection of morphine directly into the ventral tegmental area (Bechara *et al.*, 1992, 1995; Nader and van der Kooy, 1997). The aversive effects of spontaneous opioid withdrawal in dependent rats as measured by place aversion also were blocked by dopamine antagonists (Bechara *et al.*, 1995). In contrast, the acute rewarding effects of opioids in nondependent rats were blocked by lesions of the pedunculopontine tegmental nucleus of the midbrain (Bechara and van der Kooy, 1989; Olmstead and Franklin, 1993; Nader and van der Kooy, 1997), but lesions of the pedunculopontine tegmental nucleus did not alter the rewarding effects of opioids in dependent animals (Bechara and van der Kooy, 1992). To account for the different neural substrates for opioid reward, the existence of a two-motivational-systems hypothesis for opioid addiction was proposed, with the critical substrate in the nondependent state being the pedunculopontine tegmental nucleus of the brain stem, and a second reward system in the dependent state being mediated by dopaminergic systems. A major limitation of this hypothesis is that, to date, there is no confirmation of these effects with other dependent variables such as self-administration, which will be required to test the generalizability of the place preference results. Also, much data from other laboratories showed blockade of opioid place preference by dopamine antagonists or dopamine-specific lesions in nondependent rats (see above), and rats showed a place preference when microinjected with opioids only in the ventral tegmental area and not the nucleus accumbens or frontal cortex (Bals-Kubik *et al.*, 1993). Nevertheless, a potential key role for the pedunculopontine tegmental nucleus of the midbrain in opioid reward, as well as in nicotine reward (see *Nicotine* chapter), is supported by *both* self-administration and place preference studies.

Neurobiology of Tolerance

Despite much research, it has not been conclusively demonstrated *in vivo* that functional changes observed at the opioid receptor level in response to opioids account for development of tolerance (Simonnet and Rivat, 2003). In brief, two types of biological processes for tolerance have been proposed: a within-system adaptation and a between-system adaptation (Koob and Bloom, 1988). In the within-system process, the drug elicits an opposing, neutralizing reaction within the same system in which the drug elicits its primary and unconditioned reinforcing actions. Early *in vitro* studies focused on the obvious mechanisms of opioid receptor downregulation and desensitization leading to tolerance after chronic morphine administration (Chang *et al.*, 1981; Law *et al.*, 1983; Puttfarcken *et al.*, 1988; Zadina *et al.*, 1994; Liu *et al.*, 1999a), a position not generally accepted (Smith *et al.*, 1988; Cox, 1991). In particular, *in vivo* investigation using current pharmacological doses of morphine have failed to show downregulation, and in fact showed upregulation (Rothman *et al.*, 1989; Harrison *et al.*, 1998). A more accepted view holds that tolerance results from a decreased coupling of the receptor with an inhibitory G protein (Christie *et al.*, 1987; Cox, 1991). The ability of opioids to acutely inhibit adenylate cyclase appears to be lost during chronic treatment and is associated with the phosphorylation and altered association of multiple signaling proteins such as G protein receptor kinase and β-arrestin (Whistler and von Zastrow, 1998; Zhang *et al.*, 1998; Bohn *et al.*, 2000). A limitation of the decreased coupling hypothesis is the increased ability of the opioid receptor antagonist naloxone, which is ineffective in nontolerant animals, to induce hyperalgesia in tolerant animals (Laulin *et al.*, 1998). This implies that the receptors were still functional and actually were activated by opioids. Moreover, rats tested in the context of the customary predrug cues are more tolerant to the analgesic action of morphine than equally drug-experienced rats tested in the context of alternative cues, suggesting that biological changes underlying tolerance might not be limited to changes at the receptor level and that tolerance is associated with learning processes (Siegel, 1975; Poulos and Cappell, 1991; Young and Goudie, 1995).

Animal studies have documented hyperalgesia during spontaneous or precipitated opioid withdrawal following acute or chronic opioid exposure (Kayan *et al.*, 1971; Mao *et al.*, 1994; Larcher *et al.*, 1998; Laulin *et al.*, 1999; Celerier *et al.*, 2001; Li *et al.*, 2001; Vanderah *et al.*, 2001). Some studies suggest that pain sensitization might develop while subjects are still exposed to the opioid (Mao *et al.*, 1994, 2002; Vanderah *et al.*, 2001). Heroin was effective at inducing pain facilitation in rats after one exposure to the drug (Laulin *et al.*, 1998) (**Fig. 4.14**). The analgesic action of one injection of heroin (2.5 mg/kg, s.c.) lasts for approximately 2 h (vocalization threshold to paw pressure in rats) and declines to the predrug threshold in 4 h. Afterward, a significant lowering of nociceptive threshold appears when rats are tested several days following the injection, as demonstrated by a dose-dependent increase in pain reactivity (i.e., allodynia). The maximum decrease in nociceptive threshold was observed 24 h after the injection and lasted 3 days for the higher doses (1.25 and 2.5 mg/kg). It was suggested that allodynia was an early sign reflecting neural plasticity associated with the development of dependence (Laulin, 1998), but the same neuroplasticity also may be involved in tolerance.

This phenomenon is hypothesized to be an actual sensitization of pain facilitatory systems because both magnitude and duration of the hyperalgesia increase as a function of opioid administration (i.e., a first heroin administration induces a moderate hyperalgesia for two days, whereas the fifth injection of the same dose is followed by hyperalgesia for six days) (Celerier *et al.*, 2001). When the opioid is administered repeatedly (once daily for two weeks), a gradual and dose-dependent lowering of the nociceptive threshold is observed, lasting several weeks after

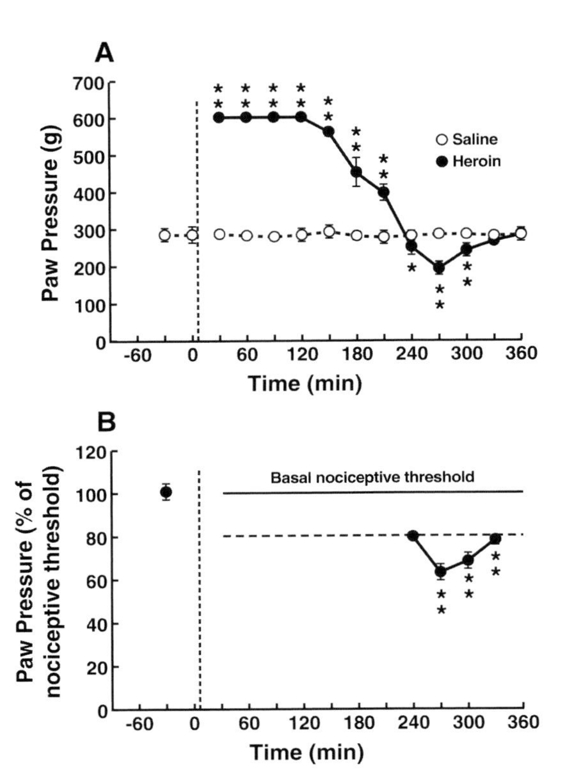

FIGURE 4.14 Heroin produced analgesia and hyperalgesia in rats as measured by vocalization threshold to paw pressure. **(A)** Rats received a subcutaneous injection of 2.5 mg/kg of heroin or saline after determination of basal paw pressure inducing vocalization, and then values were determined every 30 min for 6 h. Each point represents the mean paw pressure ± SEM of 8–10 animals. **(B)** After a preliminary determination of the value of the paw pressure which induced a vocalization (basal nociceptive threshold), animals ($n = 10$) were subjected every 30 min to increasing paw pressure, limited to 80 per cent of the basal value. Vocalization was not obtained until 210 min, but was observed in 3, 10, 10, and 10 out of the 10 tested rats at 240, 270, 300, and 330 min, respectively. $^*p < 0.05$, $^{**}p < 0.01$ (Dunnett's test subsequent to ANOVA for A; Student's t-test for B). [Reproduced with permission from Laulin *et al.*, 1998.]

drug administration (Celerier *et al.*, 2001) (**Fig. 4.15**). A small dose of heroin, which is ineffective at triggering a delayed hyperalgesia in nonheroin-treated rats, induced an enhancement in pain sensitivity for several days after a series of heroin administrations, suggesting a sensitization phenomenon (Celerier *et al.*, 2001). The effectiveness of the opioid receptor antagonist naloxone in precipitating hyperalgesia in rats that had recovered their predrug nociceptive value after single or repeated heroin administrations indicates that heroin-deprived rats were in a new biological state associated with a functional balance between opioid-dependent analgesic systems and pronociceptive systems.

Trujillo and Akil (1991) found that tolerance and physical dependence were dramatically reduced when the noncompetitive glutamate receptor antagonist MK-801 was administered to rats along with chronic morphine. MK-801 also prevented both heroin-induced, long-lasting enhancement in pain sensitivity and naloxone-precipitated hyperalgesia. Mao and colleagues (1994) reported a parallel between the development of thermal hyperalgesia in rats and tolerance to the analgesic effects of morphine administered intrathecally for eight consecutive days. The concurrent development of tolerance was prevented by intrathecal co-administration of morphine with MK-801 and a protein kinase C inhibitor (Mao *et al.*, 1994).

Consistent with this hypothesis is the observation that intermittent systemic opioid administration induces a state of intermittent withdrawal associated with a prolonged increased excitability in nociceptive pathways which increasingly opposes analgesia induced by repeated opioid administration, giving rise to tolerance. Repeated, once-daily heroin injections produce a gradual lowering of the nociceptive threshold which progressively masked a sustained heroin analgesic functional effect. Again, MK-801 prevented such opioid-induced allodynia, thereby preventing the development of an apparent decrease in the effectiveness of heroin (Laulin *et al.*, 1999). The persistent lowering of the basal nociceptive threshold has led in the past to an erroneous conclusion of an absolute decrease in the effectiveness of heroin. The data clearly show that allodynia increases with the passage of time and with repeated opioid administration. Moreover, hyperalgesia and opioid analgesic tolerance are partly mediated by the same pain facilitatory systems in such a way that rats tolerant to opioids must be hyperalgesic (McNally, 1999; Simonnet and Rivat, 2003). Thus, pain sensitization following opioid administration is hypothesized to take place in the excitatory amino acid neuronal systems and not in the neurons where the opioids' primary effects take place. Apparent tolerance is hypothesized to result from an algebraic sum between a still present primary unconditional action and the increasing counteradaptive hyperalgesia. The physiological basis of conditioned tolerance to opioids appears to involve cholecystokinin (Hoffman and Wiesenfeld-Hallin, 1994; Kim and Siegel, 2001) and possibly some of the same transduction factors associated with chronic administration of opioids (Baptista *et al.*, 1998). One can speculate that a sort of neuronal 'memory', reflected as a vulnerable state, remains long after the complete wash-out of the drug and when an apparent equilibrium near the predrug state has been re-established. These long-term effects may involve other pathological expressions such as negative affective states observed in humans even long after drug withdrawal.

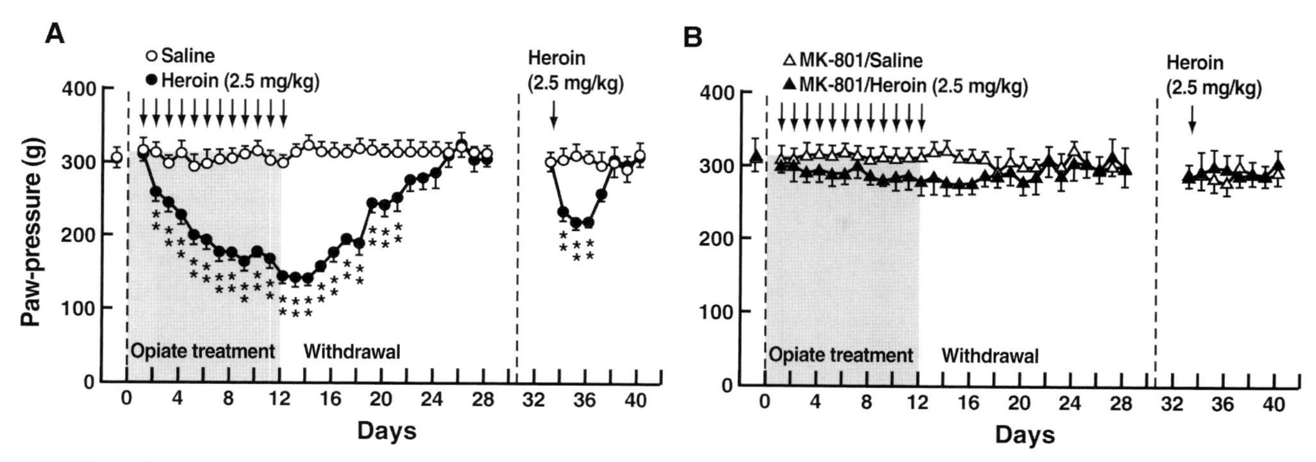

FIGURE 4.15 **(A)** Delayed effects of 12 once-daily heroin (2.5 mg/kg, s.c.) or saline administrations on basal nociceptive threshold in rats (black arrows; $n = 9$–10 rats per group). The basal nociceptive threshold was determined daily before each heroin or saline administration and after the heroin treatment was stopped. Delayed effects of a low heroin dose (0.2 mg/kg, s.c.; white arrow) on Day 33 when animals had recovered their predrug nociceptive threshold after the heroin treatment. Mean pressure values for triggering vocalization (± SEM) were expressed in grams. $^{**}p < 0.01$ with Dunnett's test as compared to basal nociceptive value on Day 1. **(B)** Results obtained in experiment similar to A in rats receiving 12 coadministrations of the noncompetitive glutamate receptor antagonist MK-801 (0.15 mg/kg, s.c.) and heroin (or saline). MK-801 was administered 30 min before each 2.5 mg/kg heroin administration. [Reproduced with permission from Celerier *et al.*, 2001.]

Neurobiology of Withdrawal

Dependence (with a 'little d'; see *What is Addiction?* chapter) on opioid drugs is defined by a characteristic withdrawal syndrome that appears with the abrupt termination of opioid administration or with administration of competitive opioid antagonists. One important observation is that the neural substrates for the physical signs of opioid withdrawal and the motivational ('affective') signs of opioid withdrawal can be dissociated (Bozarth and Wise, 1984). In rats, physical dependence has been characterized by an abstinence syndrome that includes the appearance of ptosis, teeth chattering, wet dog shakes, and diarrhea (Way *et al.*, 1969). This syndrome can be dramatically precipitated in dependent animals by systemic injections of opioid antagonists (Way *et al.*, 1969). Motivational withdrawal includes changes in brain reward thresholds, the disruption of trained operant behavior for food reward following precipitated withdrawal with systemic opioid antagonist administration (Gellert and Sparber, 1977), and place aversions produced by precipitated opioid withdrawal (Hand *et al.*, 1988; Stinus *et al.*, 2000).

Opioid withdrawal, as with other drugs of abuse, is associated with decreases in brain stimulation reward (Schaefer and Michael, 1983) and elevations in brain reward thresholds (Schaefer and Michael, 1986). Precipitated opioid withdrawal is also associated with increased brain reward thresholds at doses of naloxone that do not elicit physical signs of withdrawal, suggesting a dissociation of the mechanisms for the physical and motivational components of opioid withdrawal (Schulteis *et al.*, 1994) (**Fig. 4.16**).

Opioid dependence, as well as opioid tolerance, may begin with a single administration of the drug (Schulteis and Koob, 1996; Schulteis *et al.*, 1997). Single injections of opioids can induce a state of acute opioid dependence in humans and animals (Bickel *et al.*, 1988b; Heishman *et al.*, 1989; Azolosa *et al.*, 1994) as measured by precipitated withdrawal when a competitive opioid antagonist such as naloxone is administered 4–24 h after the initial dosing. Acute dependence is particularly manifest using 'motivational' measures of withdrawal, such as suppression of operant responding for

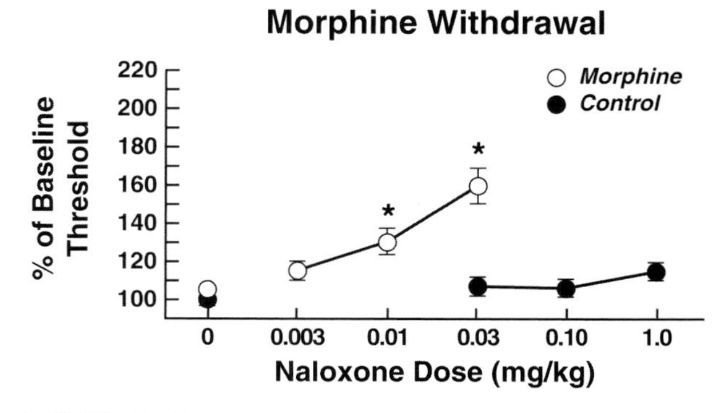

FIGURE 4.16 Naloxone dose-dependently elevated intracranial self-stimulation reward thresholds in rats (reflecting a decrease in reward function) in the morphine group but not in the placebo group. The minimum dose of naloxone that elevated ICSS thresholds in the morphine group was 0.01 mg/kg ($^{*}p < 0.05$, compared to vehicle). [Reproduced with permission from Schulteis *et al.*, 1994.]

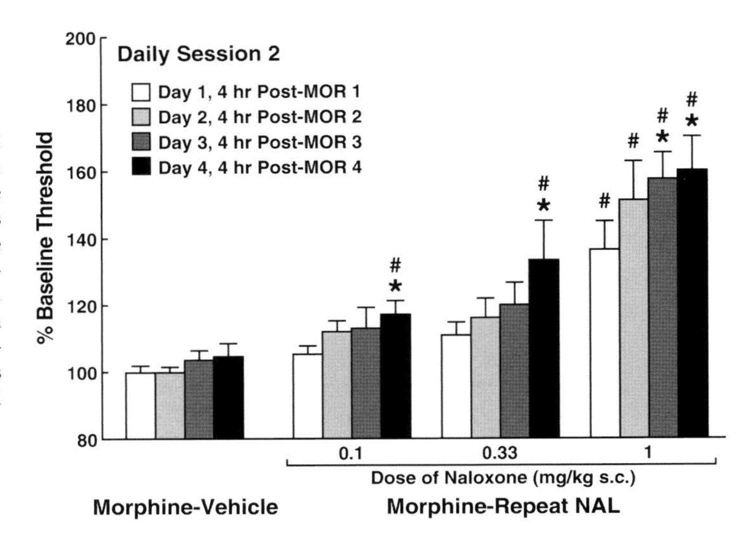

FIGURE 4.17 Effects of acute and repeated morphine and naloxone administration on brain reward thresholds in rats. Data reflect reward thresholds measured 4 h after that day's morphine injection and 5 min after that day's naloxone injection. All groups of rats were treated on four consecutive days with morphine (5.6 mg/kg), but differed with respect to the treatment administered 4 h later (vehicle [Morphine–Vehicle] or naloxone 0.10, 0.33, 1.0 mg/kg [Morphine–Repeat NAL groups]). Data represent mean (± SEM) percentage of baseline threshold. *$p < 0.05$ versus threshold measured in same treatment group on Day 1; #$p < 0.05$ versus Morphine–Vehicle group on the same treatment day. [Reproduced with permission from Liu and Schulteis, 2004.]

food reward (Schulteis *et al.*, 1999), place aversion (Azar *et al.*, 2003), and brain stimulation reward (Liu and Schulteis, 2004) (**Figs. 4.17** and **4.18**).

Repeated administration of morphine results in a progressive shift to the left of the potency for naloxone to elicit precipitated withdrawal, regardless of whether naloxone is repeatedly paired with the repeated morphine administration (Schulteis *et al.*, 2003). However, greater shifts in potency were observed when naloxone was administered on all treatment days, but only in the testing situation and not in the home cage, suggesting an important conditioning component to acute dependence (Schulteis *et al.*, 2003).

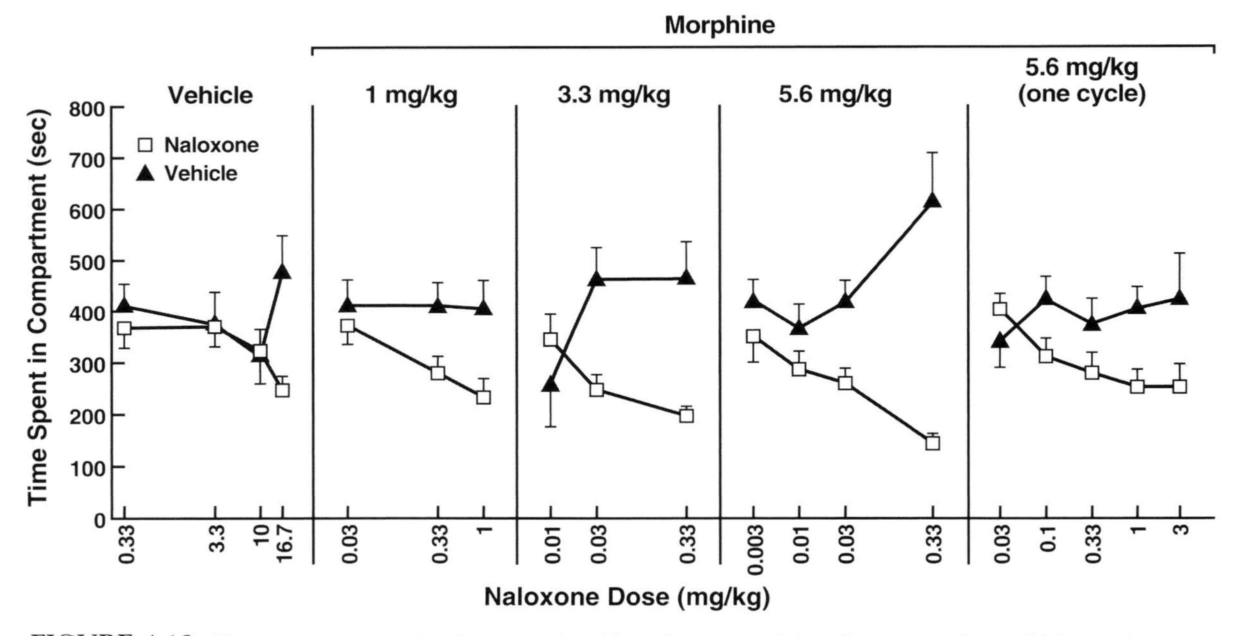

FIGURE 4.18 The aversive motivational state produced by naloxone-precipitated acute morphine withdrawal in rats as measured by conditioned place aversion. Doses of morphine and naloxone, as well as number of conditioning trials, were systematically varied to determine the minimum conditions that would result in a detectable place aversion. Naloxone (0.003–16.7 mg/kg) was administered 4 h after an injection of vehicle or morphine (1.0, 3.3, or 5.6 mg/kg) and immediately prior to confinement to one compartment of the conditioning apparatus. Rats received either one or two such naloxone-conditioning trials, separated by 48 h. Mean (± SEM) time spent in each of three compartments of a place conditioning apparatus. All morphine/naloxone treatment conditions represent two conditioning cycles, except where noted. As time spent in the naloxone-paired compartment declines as a function of naloxone dose, this extra time appears to be randomly spent in either the vehicle-paired or neutral compartment. [Reproduced with permission from Azar *et al.*, 2003.]

Studies of the neurobiological substrates of physical dependence on opioids have revealed multiple sites of action, including the periaqueductal gray (Wei *et al.*, 1972, 1973), dorsal thalamus (Bozarth and Wise, 1984), and locus coeruleus (Maldonado *et al.*, 1992). However, the sites in the brain responsible for more motivational or emotional changes associated with opioid withdrawal have been localized to structures in the basal forebrain.

In support of this hypothesis, injection of hydrophilic opioid antagonists into certain sites such as the nucleus accumbens and the amygdala were largely negative for eliciting physical signs of opioid withdrawal but produced motivational signs of withdrawal as measured by both a suppression of operant responding in dependent rats (Koob *et al.*, 1989b) and the establishment of conditioned place aversions (Stinus *et al.*, 1990). The aversive stimulus effects of opioid withdrawal that form the basis for the motivational effects of opioid withdrawal and conditioned withdrawal can be measured readily using place aversion, and such place aversions last up to 16 weeks (Stinus *et al.*, 1990, 2000). These results suggest that there is a lasting association of a specific environment with the aversive stimulus effects of opioid withdrawal that persists even in the absence of opioids and a long time after the acute withdrawal state has dissipated (Wikler, 1973).

Local intracerebral administration of a hydrophilic opioid antagonist in rats dependent on morphine revealed that very low doses of methylnaloxonium (4–64 mg) injected into the nucleus accumbens produced a disruption of food-motivated operant responding (Koob *et al.*, 1989b) (**Fig. 4.19**). Injections of methylnaloxonium into the periaqueductal gray or the dorsal thalamus produced a dose-effect function similar to that following intracerebroventricular injection (Koob *et al.*, 1989b). These results show that during chronic morphine exposure, neural elements in the region of the nucleus accumbens may have become sensitized to opioid antagonists and may be responsible for the negative stimulus effects of opioid withdrawal. Confirmation of an aversive stimulus effect for intracerebral methylnaloxonium was shown using the place aversion test (Hand *et al.*, 1988), and the nucleus accumbens proved to be the most sensitive site for intracerebral injections of methylnaloxonium to produce place aversions (Stinus *et al.*, 1990) (**Fig. 4.20**).

Data from both the conditioned place aversion measures and unlimited access to intravenous heroin self-administration have converged to provide a neuropharmacological framework for the neurocircuitry associated with opioid addiction. Precipitated opioid withdrawal is associated with decreases in extracellular dopamine in the nucleus accumbens (Pothos *et al.*, 1991), but increases in extracellular dopamine in the medial prefrontal cortex (Cadoni and Di Chiara, 1999), possibly due to dopamine release in noradrenergic terminals in the medial prefrontal cortex (Devoto *et al.*, 2002). Consistent with a functional role of dopamine in opioid withdrawal, activation of dopamine D_2 receptors in the nucleus accumbens prevents the somatic signs of precipitated opioid withdrawal (Harris and Aston-Jones, 1994). However, others have failed to alter precipitated opioid somatic or motivational withdrawal with dopamine-specific lesions of the nucleus accumbens (Caille *et al.*, 2003).

While opioids and corticotropin-releasing factor (CRF) have opposing actions on the activity of the noradrenergic cells of the locus coeruleus (Valentino and Van Bockstaele, 2001), virtually complete lesions of the dorsal noradrenergic bundle using selective injections of the neurotoxin 6-hydroxydopamine failed to alter the place aversion produced by precipitated opioid withdrawal (Caille *et al.*, 1999). However, ventral noradrenergic bundle lesions did attenuate opioid withdrawal-induced place aversions (Delfs *et al.*, 2000). Global increases or decreases in serotonergic function failed to alter the aversive stimulus effects of

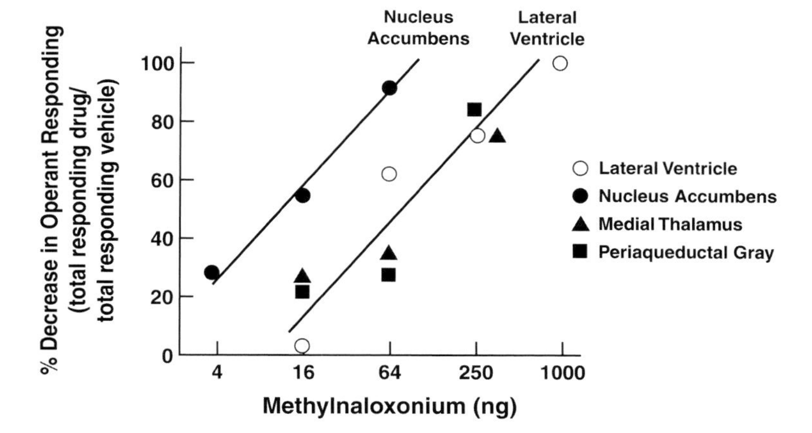

FIGURE 4.19 Summary graph depicting the percentage decrease in operant responding in rats produced by different doses of methylnaloxonium injected either intracerebroventricularly or intracerebrally into the nucleus accumbens, medial thalamus, or periaqueductal gray. Total values for the 20 min sessions were expressed as a percentage decrease relative to values for controls intracerebroventricularly injected with saline. [Reproduced with permission from Koob *et al.*, 1989a.]

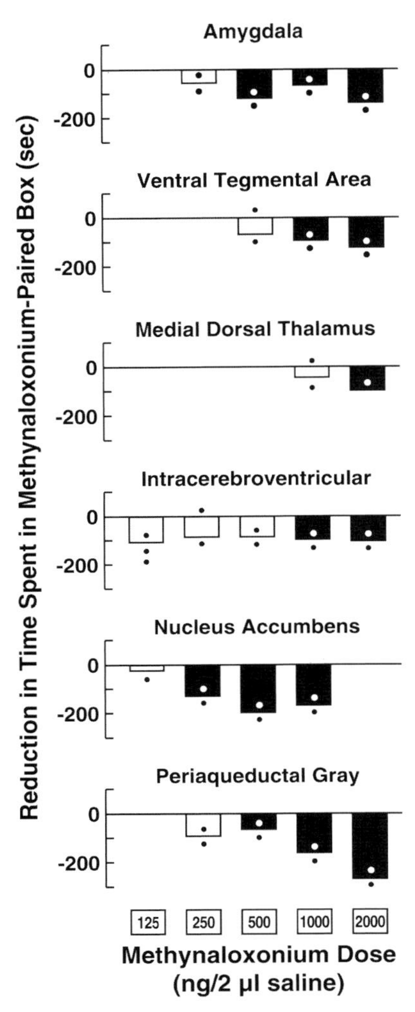

FIGURE 4.20 The effect of intracerebral methylnaloxonium paired with a particular environment on the amount of time spent by rats in that environment during an injection-free test session. Values represent the median difference between the postconditioning score and the preconditioning score. Dots refer to the interquartile range of this distribution. Darkened bars represent those doses where the conditioning scores were significantly different from the preconditioniong scores using the nonparametric Wilcoxon matched pairs signed ranks test. Significance was set at $*p < 0.02$ to control for multiple comparisons. [Reproduced with permission from Stinus *et al.*, 1990.]

opioid withdrawal, with no effects of total forebrain depletion of serotonin (Caille *et al.*, 2002). No effects on opioid withdrawal-induced place aversion were observed with a dopamine partial agonist or with the NMDA partial agonist acamprosate (Stinus *et al.*, 2005). However, buprenorphine, an opioid partial agonist used for detoxification and treatment of opioid dependence, dose-dependently decreased opioid withdrawal-induced place aversion (Stinus *et al.*, 2005). Even more interesting, a CRF_1 receptor antagonist also blocked opioid withdrawal-induced place aversion (Stinus *et al.*, 2005) (**Fig. 4.21**).

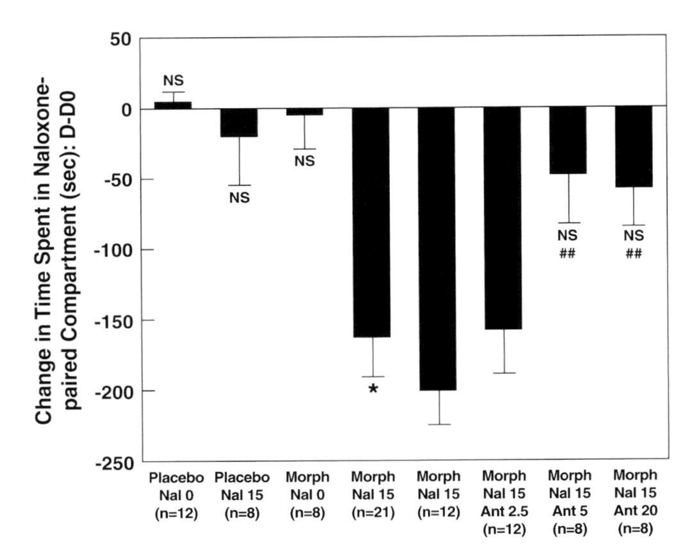

FIGURE 4.21 Antalarmin reduced naloxone-precipitated place aversion conditioning in morphine-dependent rats. Within each dose group treatment, Wilcoxon signed ranks test (D versus D_0), $*p < 0.05$; NS refers to no significant place preference or place aversion with the Wilcoxon signed ranks test; between-group comparison, Mann-Whitney test (ΔD), $\#\#p < 0.01$, compared to Morph-Nal 15 group. [Reproduced with permission from Stinus *et al.*, 2005.]

Recent neuroanatomical data and new functional observations have provided support for the hypothesis that the neuroanatomical substrates for many of the motivational effects of drugs of abuse may involve a common neural circuitry that forms a separate entity within the basal forebrain, termed the 'extended amygdala' (Alheid and Heimer, 1988) (see *Neurobiological Theories of Addiction* chapter). Opioid withdrawal produced an increase in expression of the transcription factor c-Fos protein, a marker of neural activation with particular sensitivity in the extended amygdala (Stornetta *et al.*, 1993; Gracy *et al.*, 2001; Le Guen *et al.*, 2001; Frenois *et al.*, 2002; Veinante *et al.*, 2003). The c-Fos changes correlated with the motivational effects rather than the somatic effects of opioid withdrawal as measured by place aversion (Gracy *et al.*, 2001; Frenois *et al.*, 2002). Further support for the role of the extended amygdala in the aversive stimulus effects of opioid withdrawal comes from studies of lesions of the extended amygdala and precipitated withdrawal-induced place aversions (Watanabe *et al.*, 2002a,b, 2003). Excitotoxic lesions of the central nucleus of the amygdala blocked the development of morphine withdrawal-induced conditioned place aversion, but with much less effects on the somatic signs of withdrawal (Kelsey and Arnold, 1994; Watanabe *et al.*, 2002b).

Neurochemical elements within the extended amygdala involved in the aversive stimulus effects of opioid withdrawal include CRF and norepinephrine systems. Blockade of CRF receptors in the central nucleus of the amygdala blocked conditioned place aversion produced by opioid withdrawal (Heinrichs *et al.*, 1995). Inactivation of noradrenergic function in the bed nucleus of the stria terminalis (BNST) has been shown to block the aversive motivational effects of opioid withdrawal (Delfs *et al.*, 2000) (**Fig. 4.22**). Microinjection of a β-adrenergic antagonist or an α_2 agonist into the lateral BNST blocked the place aversion associated with precipitated opioid withdrawal (Delfs *et al.*, 2000). Noradrenergic terminals in the BNST represent the highest concentration of norepinephrine in the brain (Brownstein and Palkovits, 1984). Blockade of norepinephrine receptors in the central nucleus of the amygdala with microinjection of β-adrenoceptor antagonists also attenuated morphine withdrawal-induced conditioned place aversion (Watanabe *et al.*, 2003).

Neuropeptide Y (NPY) administered intracerebroventricularly attenuates the somatic signs of opioid withdrawal (Woldbye *et al.*, 1998; Clausen *et al.*, 2001). NPY has a wide distribution in the forebrain with significant effects in the amygdala in mediating behavioral responses to stressors (Heilig *et al.*, 1993). Together, these results suggest that CRF, NPY, and noradrenergic systems within the extended amygdala may have a role in the motivational (aversive stimulus) effects of opioid withdrawal. Changes in sensitivity to opioid antagonists in opioid-dependent rats have been observed in the nucleus accumbens, BNST, and central nucleus of the amygdala (Koob *et al.*, 1989b; Stinus *et al.*, 1990), and opioid-dependent rats self-administering heroin are particularly sensitive to opioid antagonism in the BNST (Walker *et al.*, 2000).

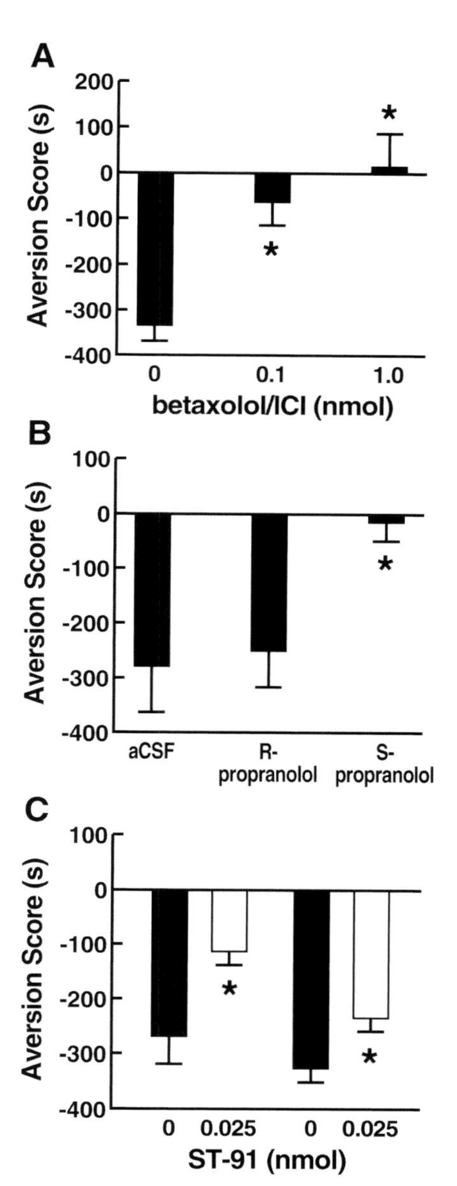

FIGURE 4.22 Effects of infusing noradrenergic drugs into the bed nucleus of the stria terminalis on naloxone-induced place aversion in morphine-dependent rats. **(A)** Effects of a combination of the β_1-adrenergic antagonist betaxolol and the β_2-adrenergic antagonist ICI 118,551 (*$p < 0.05$, ANOVA followed by Fisher's PLSD for multiple comparisons). **(B)** Effects of the propranolol isomers *R*-propranolol and *S*-propranolol (*$p < 0.05$, ANOVA followed by Fisher's PLSD for multiple comparisons). **(C)** Effects of ST-91, a polar analog of the α_2-adrenergic agonist clonidine (*$p < 0.05$, Student's *t*-test). [Reproduced with permission from Delfs *et al.*, 2000.]

Neurobiology of Relapse

In animal models related to the preoccupation/anticipation ('craving') stage, there is evidence for roles of μ opioid receptors in the ventral tegmental area in opioid-induced reinstatement, and CRF and norepinephrine in the extended amygdala in stress-induced reinstatement of heroin-seeking. Heroin priming of reinstatement of heroin self-administration after extinction is dependent on activation of μ opioid receptors. μ agonists reinstate responding when injected systemically or into the ventral tegmental area, and these effects are blocked by naloxone (Stewart and Wise, 1992).

Using a runway model of cue-induced reinstatement with heroin-trained rats, there is evidence of a resistance to dopamine and opioid blockade of reinstatement to an olfactory cue that induced reinstatement (McFarland and Ettenberg, 1995, 1997, 1998; Ettenberg and McFarland, 2003). Similarly, haloperidol does not block conditioned place preference produced by heroin cues (McFarland and Ettenberg, 1999).

Intermittent footshock produced reinstatement of heroin self-administration in nondependent rats after extinction (Shaham and Stewart, 1995; Ahmed *et al.*,

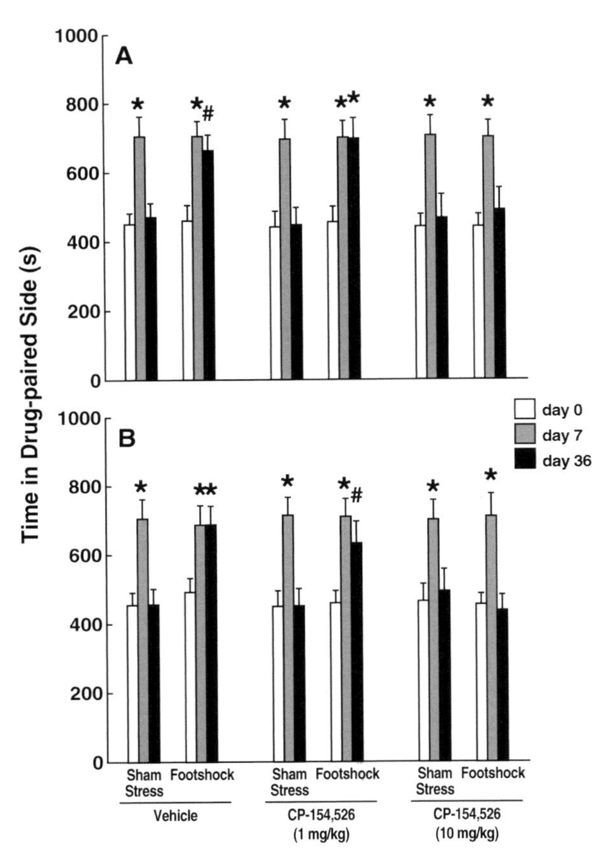

FIGURE 4.23 Effect of pretreatment with the corticotropin-releasing factor antagonists α-helical-CRF_{9-41} and CP-154,526 on footshock-induced maintenance of morphine conditioned place preference in rats. α-helical-CRF_{9-41} was tested on Day 0. CP-154,526 was tested on Day 7. The time each rat spent in the drug-paired side was measured for 15 min on the test day. Each column represents the mean ± SEM of 10 animals. #$p < 0.05$; *$p < 0.01$ versus time spent in the drug-paired side on Day 0. [Reproduced with permission from Lu *et al.*, 2000.]

2000), and reactivated opioid-induced place preferences after drug-free periods without extinction (Lu *et al.*, 2000). These effects were not readily blocked by opioid or dopamine antagonists (Shaham and Stewart, 1996; McFarland and Ettenberg, 1998). However, CRF reinstated heroin-seeking (Shaham *et al.*, 1997), and stress-induced reinstatement was blocked with peptide CRF_1/CRF_2 antagonists and a selective CRF_1 receptor-selective antagonist (Shaham *et al.*, 1998; Lu *et al.*, 2000) (**Fig. 4.23**). A CRF_2 antagonist was ineffective (Lu *et al.*, 2000). Inactivation of the medial septum, which produces a 'stress-like' effect, also reinstated heroin-seeking behavior (Highfield *et al.*, 2000). The brain sites critical for the CRF role in footshock-induced reinstatement appear to be the BNST and central nucleus of the amygdala because reversible inactivation of these structures blocked footshock-induced reinstatement of heroin-seeking (Shaham *et al.*, 2000a).

TABLE 4.6 Animal Models of Heroin Craving—Neuropharmacological Interactions

Heroin-primed reinstatement	*Agonists*	*Antagonists*
Dopamine		↓[a]
Glutamate		
Opioid	↑[a]	↓[a]
Cue-induced reinstatement	***Agonists***	***Antagonists***
Dopamine		—[b]
Glutamate		
Opioid		—[a]
Stress-induced reinstatement	***Agonists***	***Antagonists***
Dopamine		—[a]
CRF	↑[c]	↓[c, e]
Norepinephrine		↓[d]
Opioid		—[a]

—, no effect
[a]Shaham and Stewart, 1996; [b]McFarland and Ettenberg, 1997; [c]Shaham *et al.*, 1997; [b]Shaham *et al.*, 2000b; [e]Lu *et al.*, 2000

Similarly, norepinephrine originating in the ventral noradrenergic pathway and projecting to the BNST and the central nucleus of the amygdala also has a role in footshock-induced opioid-related reinstatement. α_2 adrenergic agonists, drugs known to inhibit norepinephrine release, blocked footshock-induced reinstatement of heroin-seeking (Erb *et al.*, 2000; Shaham *et al.*, 2000b; Highfield *et al.*, 2001). The brain sites for these effects appear to be the ventral noradrenergic bundle from the lateral tegmental nucleus because neurotoxin-specific lesions of this pathway attenuated footshock-induced reinstatement of heroin-seeking (Shaham *et al.*, 2000b; see also Shaham and Stewart, 1996; McFarland and Ettenberg, 1997; Shaham *et al.*, 1997; Lu *et al.*, 2000) (**Table 4.6**).

Neurobiology of Protracted Abstinence–Stress and Dependence

Opioid addicts are well known to have dysregulated stress axis activity that persists during cycles of addiction. These persistent physiological abnormalities stabilize during methadone maintenance (Kreek *et al.*, 1983, 1984). Using the metyrapone test to block cortisol synthesis and this reduced feedback inhibition, heroin addicts showed that the reduced hypothalamic–pituitary–adrenal (HPA) axis reserve returned to normal then stabilized during methadone maintenance. A reactivation of a hyperactive HPA axis occurred during opioid withdrawal (Rosen *et al.*, 1995; Kreek, 1996a), and during protracted abstinence there

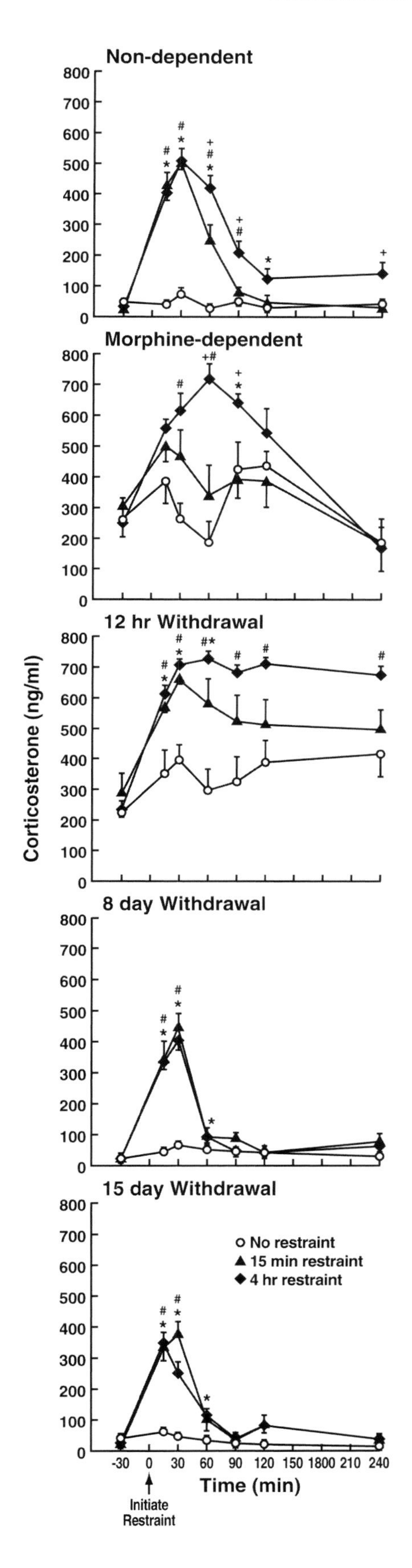

FIGURE 4.24 Plasma corticosterone responses to 15 min and 4 h restraint stress in vehicle- and morphine-treated rats. The arrow indicates the time of injection or the initiation of restraint. Data are mean ± SEM (n = 5–9). *$p < 0.05$, 15 min restraint versus no restraint groups. #$p < 0.05$, 4 h restraint versus no restraint groups. +$p < 0.05$, 15 min restraint versus 4 h restraint groups. [Reproduced with permission from Houshyar *et al.*, 2001a.]

is a hyperresponsiveness of the cortisol negative feedback system that regulates the HPA axis (Kreek, 1996b). During cycles of heroin addiction, there was a hyporesponsivity of the temporary shutoff of the negative feedback on the HPA axis produced by blockade of cortisol synthesis (Cushman and Kreek, 1974; Kreek, 1996a).

Rats differentiated on their individual reactivity to novelty, a mild stress, display different locomotor increases after morphine administration, and stress-induced corticosterone secretion plays a role in the determination of differential sensitivity to opioids (Deroche *et al.*, 1993). In animals stressed by food restriction, the locomotor response induced by morphine infused into the ventral tegmental area (1 μg/side) was increased, but this effect was abolished by suppressing corticosterone secretion by adrenalectomy, and reinstated by administering corticosterone (Deroche *et al.*, 1995). The locomotor response due to morphine administration was dose-dependently reduced by selective glucocorticoid receptor antagonists (but not mineralocorticoid). Glucocorticoid antagonists also reduced the basal and morphine-induced increase in nucleus accumbens dopamine (Marinelli *et al.*, 1998). Adrenalectomy profoundly decreased basal and morphine-induced nucleus accumbens dopamine release.

Glucocorticoids have a profound heterogenous influence within dopaminergic projections. While glucocorticoids appear to have little influence on morphine-induced dopamine release in the dorsolateral striatum, after morphine, as well as after mild stress, the shell region of the nucleus accumbens exhibited the largest increase in dopamine and Fos-like immunoreactivity (Barrot *et al.*, 2001). The changes are far less noticeable in the core of the nucleus accumbens, a region more closely related anatomically and functionally to the dorsal striatum than with the shell of the nucleus accumbens (which forms a transition zone between the extended amygdala and the ventral striatum) (Zahm and Brog, 1992). In morphine-dependent rats, during opioid dependence and acute withdrawal, there is a marked elevation of basal corticosterone concentrations and an exaggerated response to a stressor (Houshyar *et al.*, 2001a,b) (**Fig. 4.24**). Rats that had undergone 12 h withdrawal displayed increased basal corticosterone and potentiated and prolonged the corticosterone response to restraint stress, but after 8- and 16-day withdrawal they recovered normal baseline HPA activity and showed a blunted response to a stressor. This reduced response to a stressor was hypothesized to be due to increased sensitivity of the negative-feedback systems to glucocorticoids and reduced CRF function (Houshyar *et al.*, 2001b) similar to that seen in heroin addicts (Kreek *et al.*, 1984). Thus, acute and chronic stress exposure and acute and chronic morphine exposure have similar effects in rats, and hormonal stress responses may play a role in the maintenance and relapse to opioid use long after acute withdrawal from opioids (Houshyar *et al.*, 2001a,b).

The protracted abstinence part of the preoccupation/anticipation (craving) stage of drug addiction also is accompanied by an attentional bias for drugs and drug-related stimuli that helps channel the behavior toward drug-seeking. Heroin-dependent subjects showed larger slow wave components of event-related potentials to heroin-related stimuli during abstinence which correlated with postexperiment craving (Franken *et al.*, 2003). In animal studies, hippocampal long-term potentiation was reduced by chronic opioid treatment, and the chronically treated animals showed partial learning deficits in the Morris water maze (Pu *et al.*, 2002). Together, these results suggest that opioids can impair cognitive processing, and this may contribute to the phenomenon of craving.

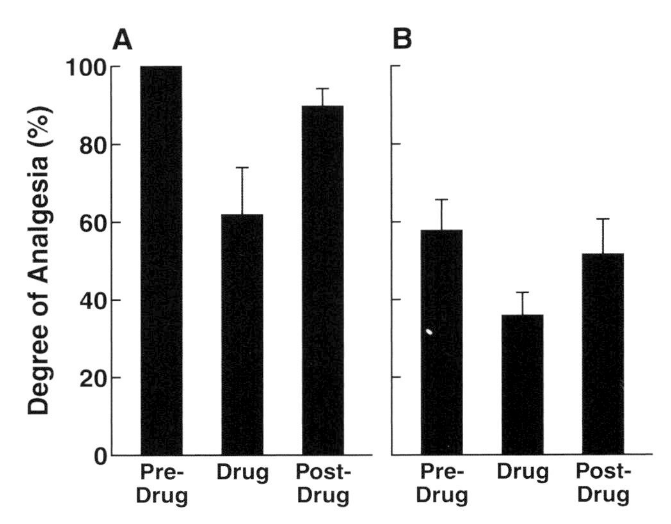

FIGURE 4.25 The effects of naloxone (1 mg/kg) on stimulation-produced analgesia in **(A)** rats showing an initial degree of analgesia of 100 per cent and in **(B)** rats showing a mean initial degree of analgesia of 58 per cent. [Reproduced with permission from Akil *et al.*, 1976.]

NEUROBIOLOGICAL MECHANISM—CELLULAR

Endogenous Opioid Peptides

Great excitement heralded the discovery of endogenous opioid-like substances in the brain that bound to the same receptors as morphine but were peptides, not alkaloids. The discovery of opioid peptides was preceded by physiological studies which showed that animals stimulated through electrodes in the central gray showed a stimulation-induced analgesia that was reversed by the opioid antagonist naloxone (Mayer *et al.*, 1971). Reported initially at the Society for Neuroscience meeting in 1971 by the laboratory of John Liebeskind (Mayer *et al.*, 1971), this study was ultimately published in 1976 (Akil *et al.*, 1976) (**Fig. 4.25**). During the intervening period, Hughes and Kosterlitz discovered and sequenced two 5-amino acid peptides, methionine and leucine enkephalin, that bound to the opioid receptor and had opioid activity *in vivo* (Waterfield *et al.*, 1976). Subsequently, β-endorphin (Li *et al.*, 1976) and dynorphin (Goldstein *et al.*, 1979) were isolated, and the precursor molecules for enkephalins and endorphins were identified (Crine *et al.*, 1978; Legon *et al.*, 1982; Noda *et al.*, 1982). Ultimately, three distinct families of peptides have been identified: enkephalins, endorphins, and dynorphins. There is a distinct polypeptide precursor for each peptide and a distinct, but overlapping neuroanatomical distribution of these precursors as measured by opioid mRNA expression (Mansour *et al.*, 1993, 1994a,b,c, 1995) (**Fig. 4.26**).

Opioid peptides produce self-administration, analgesia, locomotor activation, and place preference when administered directly into the brain. Opioid peptides injected into the brain produce analgesia (Belluzzi *et al.*, 1976; Malick and Goldstein, 1977; Nemeroff *et al.*, 1979), and with intracerebral injections, sensitive sites include the raphe (Dickenson *et al.*, 1979), periaqueductal gray (Satoh *et al.*, 1983), nucleus reticularis gigantocellularis of the medulla (Takagi *et al.*, 1978; Kaneko *et al.*, 1983), and spinal cord (Satoh *et al.*, 1979), among other sites. Microinjection of an enkephalinase inhibitor into the central nucleus of the amygdala, periaqueductal gray, and ventral medulla also produced analgesia in the hot plate test (al Rodhan *et al.*, 1990).

Opioid peptides injected directly into the nucleus accumbens (Kalivas *et al.*, 1983) and ventral tegmental area produce locomotor activation (Stinus *et al.*, 1980). Opioid peptides are self-administered intracerebroventricularly (Belluzzi and Stein, 1977) and directly into several brain sites, notably the lateral hypothalamus (Olds, 1979) and nucleus accumbens (Goeders *et al.*, 1984). Numerous studies have demonstrated intracranial self-administration of opioids other than endogenous opioid peptides into the ventral tegmental area (McBride *et al.*, 1999). β-Endorphin injected intracerebroventricularly produced place preference in rats (Amalric *et al.*, 1987), and D-ala2-met5-enkephalin injected into the ventral tegmental area produced place preference (Phillips and LePiane, 1982). The recently discovered opioid peptide endomorphin-1, that has high selectivity for the μ opioid receptor, produced locomotor activity when injected into the ventral tegmental area but not the nucleus accumbens (Zangen *et al.*, 2002). Opioid peptides injected into the brain can produce tolerance (Tseng *et al.*, 1977) and dependence (Wei and Loh, 1976). Thus, opioid peptides have much the same actions in the brain in the domain of behavioral effects relevant to addiction as systemically administered nonpeptide opioids.

Little work has been done linking endogenous opioid peptides directly to the reinforcement associated with opioids that are abused. However, in a recent study, enkephalin levels as measured by *in vivo* microdialysis are increased in the nucleus accumbens in a place conditioning situation when the animals were placed in the drug-associated side, but decreased when the animals were placed in the saline-paired side (Nieto *et al.*, 2002) (**Fig. 4.27**). Together these results suggest that opioid peptides might be activated in hedonically positive situations and may be involved in reward processing, possibly independent of the dopamine system (Koob, 1992).

Electrophysiological Studies

The pioneering cellular (electrophysiological) work on opioid peptides utilized the hippocampus as a model system. These studies have taken on added significance for understanding the cellular mechanisms of neuroplasticity in drug dependence with the further refinement of neuroadaptive models of dependence (see *Neurobiological Theories of Addiction* chapter). Opioids produce an excitation of neurons in the CA1 and CA3 regions of the hippocampus, and these effects are mediated by an indirect inhibitory action on interneurons or presynaptic terminals (Zieglgansberger *et al.*, 1979) leading to the disinhibition hypothesis of μ and δ opioid peptide action at the cellular level (Siggins and Zieglgansberger, 1981). Subsequent studies showed that opioid peptides hyperpolarize interneurons in the hippocampus (Madison and Nicoll, 1988) and activate voltage-gated potassium conductance (Wimpey and Chavkin, 1991). Dynorphin, which has mainly a κ agonist action, also has a strong inhibitory effect in the hippocampus, and this inhibition was hypothesized to be mediated via presynaptic inhibition of glutamate release (Wagner *et al.*, 1993; Weisskopf *et al.*, 1993).

Using the hippocampal model, subsequent cellular studies have been focused on elements of the brain reward pathways as defined by neuropharmacological studies of drug reward. In a slice preparation of the nucleus accumbens with intracellular recording, μ opioid agonists produce a reduction in presynaptic release of glutamate and an enhancement of postsynaptic NMDA glutamate effects (Siggins *et al.*, 1995; Martin *et al.*, 1997). The μ opioid agonists also hyperpolarized (inhibited firing) a subpopulation of neurons [probably γ-aminobutyric acid (GABA)] in the central nucleus of the amygdala mediated by the opening of inwardly rectifying potassium channels (Zhu and Pan, 2004).

In the amygdala, orphanin FQ produced marked inhibitory effects postsynaptically and inhibition of GABA and glutamate release (Meis and Pape, 1998; Meis and Pape, 2001). Orphanin FQ (also known as nociceptin) is an endogenous agonist for the orphan receptor that has a close homology with classical opioid receptors (Mollereau *et al.*, 1994; Meunier *et al.*, 1995; Reinscheid *et al.*, 1995). Orphanin FQ shares many of the cellular actions of classical opioids (Grudt and Williams, 1995). Orphanin FQ also decreases dopamine release in the nucleus accumbens (Murphy *et al.*, 1996) and produces anti-stress-like actions (Ciccocioppo *et al.*, 1994; Koster *et al.*, 1999; Jenck *et al.*, 2000; Mogil and Pasternak, 2001) and also produces inhibitory effects on ventral tegmental neurons *in vitro* (Zheng *et al.*, 2002).

Perhaps even more interesting are studies showing that nucleus accumbens and amygdala neuronal plasticity are associated with the development of morphine dependence. Chronic morphine produces a shift in the nucleus accumbens to the subunit expression of NMDA glutamate receptors to the NMDA NR2 receptor subunits in the nucleus accumbens (Martin *et al.*, 2004). Chronic morphine also alters gating of potassium channels that are modulated by opioid receptors (Chen *et al.*, 2000, 2001).

Activation of dopamine cells in the ventral tegmental area by opioids also have been hypothesized to contribute significantly to their reinforcing effects. Intracellular recording of slices from the ventral tegmental area have revealed that opioids do not directly affect the dopamine-containing neurons but rather hyperpolarize or inhibit GABA-containing interneurons which increases firing (Johnson and North,

Mu

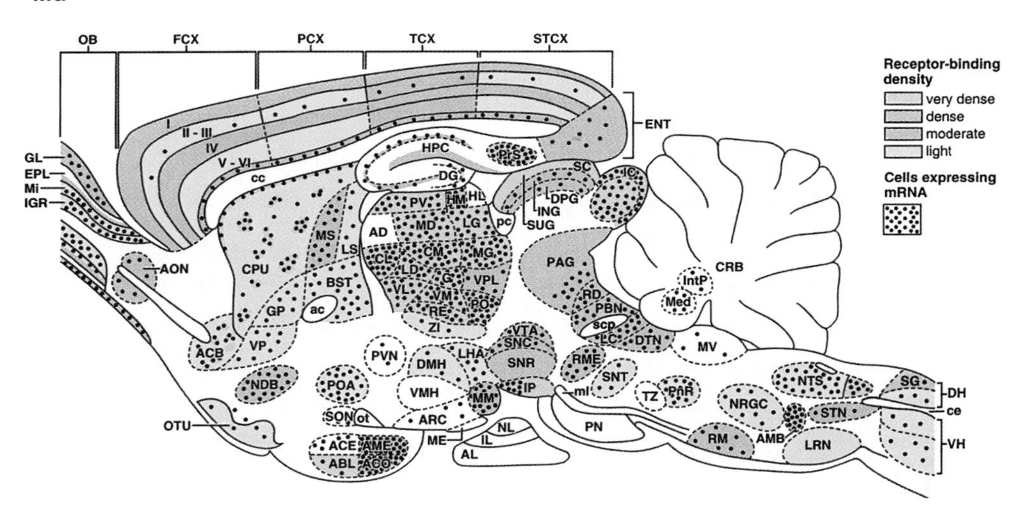
OB
FCX
PCX
TCX
STCX
ENT
I
II - III
IV
V - VI
cc
GL
EPL
Mi
IGR
AON
CPU
MS
LS
BST
GP
ac
ACB
VP
HPC
DG
PrS
SC
IC
DPG
ING
SUG
pc
PV
HM
HL
MD
LG
AD
CL
CM
MG
LD
G
VPL
VL
VM
PO
RE
ZI
PAG
CRB
IntP
Med
RD
PBN
scp
LC
DTN
MV
VTA
SNC
SNR
LHA
PVN
DMH
RME
SNT
NDB
POA
VMH
MM
IP
ml
TZ
PnR
NTS
SG
DH
ce
NRGC
STN
SON
ot
ARC
NL
IL
PN
OTU
ACE
AME
ABL
ACO
ME
AL
RM
AMB
LRN
VH
Receptor-binding density
very dense
dense
moderate
light
Cells expressing mRNA

Delta

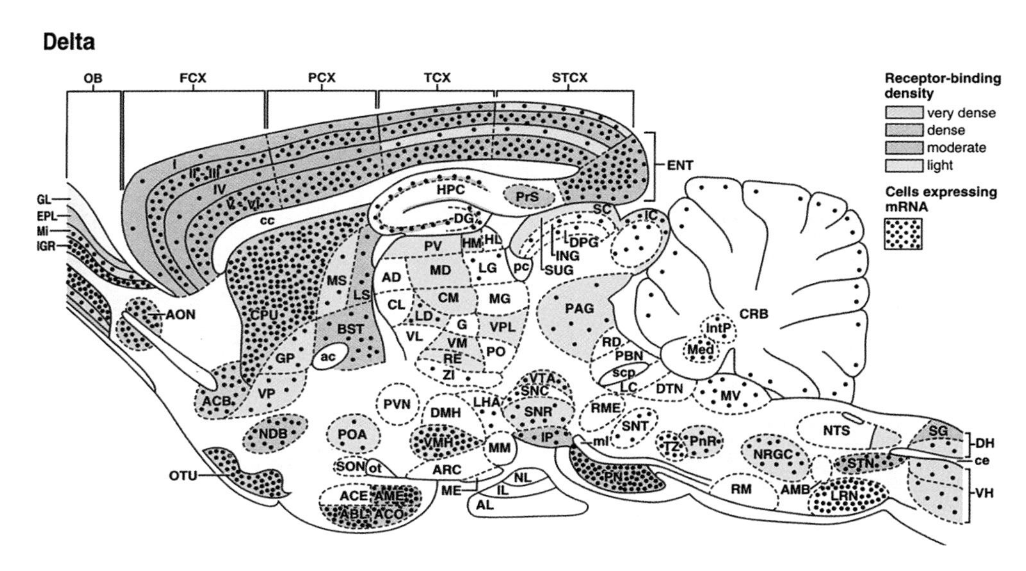
OB
FCX
PCX
TCX
STCX
ENT
I
II - III
IV
V - VI
cc
GL
EPL
Mi
IGR
AON
CPU
MS
LS
BST
GP
ac
ACB
VP
HPC
DG
PrS
SC
IC
DPG
ING
SUG
pc
PV
HM
HL
MD
LG
AD
CL
CM
MG
LD
G
VPL
VL
VM
PO
RE
ZI
PAG
CRB
IntP
Med
RD
PBN
scp
LC
DTN
MV
VTA
SNC
SNR
LHA
PVN
DMH
RME
SNT
NDB
POA
VMH
MM
IP
ml
TZ
PnR
NTS
SG
DH
ce
NRGC
STN
SON
ot
ARC
NL
IL
PN
OTU
ACE
AME
ABL
ACO
ME
AL
RM
AMB
LRN
VH
Receptor-binding density
very dense
dense
moderate
light
Cells expressing mRNA

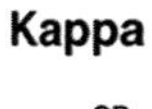
Kappa

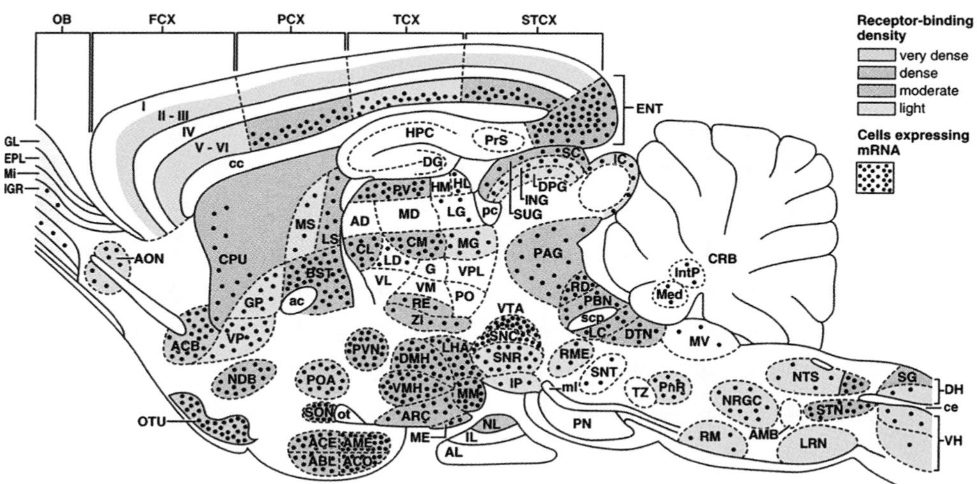
OB
FCX
PCX
TCX
STCX
ENT
I
II - III
IV
V - VI
cc
GL
EPL
Mi
IGR
AON
CPU
MS
LS
BST
GP
ac
ACB
VP
HPC
DG
PrS
SC
IC
DPG
ING
SUG
pc
PV
HM
HL
MD
LG
AD
CL
CM
MG
LD
G
VPL
VL
VM
PO
RE
ZI
PAG
CRB
IntP
Med
RD
PBN
scp
LC
DTN
MV
VTA
SNC
SNR
LHA
PVN
DMH
RME
SNT
NDB
POA
VMH
MM
IP
ml
TZ
PnR
NTS
SG
DH
ce
NRGC
STN
SON
ot
ARC
NL
IL
PN
OTU
ACE
AME
ABL
ACO
ME
AL
RM
AMB
LRN
VH
Receptor-binding density
very dense
dense
moderate
light
Cells expressing mRNA

FIGURE 4.26 Schematic representation of μ, δ, and κ_1 receptor mRNA expression and binding in the rat central nervous system. μ receptor binding sites are labeled with [^{3}H]DAMGO (D-Ala^2-MePhe^4-Gly-ol^5-enkephalin). δ receptors are labeled with [^{3}H]DPDPE (D-Pen2-D-Pen5-enkephalin). κ_1 receptors are labeled with [^{3}H]U69593 {5α, 7α. 8β-(–)-N-methyl-N-[7-(1-pyrrolidinyl)-1-oxaspiro(4,5)dec-8-yl]-benzene acetamide}. The receptor densities, shown as different color hues, are within a receptor type and do not represent absolute receptor capacity. Therefore, dense binding of one receptor type is not equivalent in terms of receptor number to dense binding of another receptor type. This is particularly true for the κ_1 receptor binding sites, which represent only 10 per cent of the total opioid-receptor binding sites in the rat brain. These rat parasagittal sections are designed, however, to transmit qualitatively the correspondence between receptor mRNA expression and receptor binding distribution. This figure demonstrates that each opioid receptor mRNA has a unique distribution which correlates well with the known μ, δ, and κ_1 receptor binding distributions (Mansour *et al.*, 1993, 1994a,b,c).

μ Receptor Distribution. A high correlation between μ receptor mRNA expression and binding is observed in the striatal clusters and patches of the nucleus accumbens and caudate-putamen, diagonal band of Broca, globus pallidus and ventral pallidum, bed nucleus of the stria terminalis, most thalamic nuclei, medial and cortical amygdala, mammillary nuclei, presubiculum, interpeduncular nucleus, median raphe, raphe magnus, parabrachial nucleus, locus coeruleus, nucleus ambiguus and nucleus of the solitary tract. Differences in μ receptor mRNA and binding distributions are observed in regions such as the neocortex, olfactory bulb, superior colliculus, spinal trigeminal nucleus, and spinal cord which might be a consequence of receptor transport to presynaptic terminals.

δ Receptor Distribution. A high correlation between δ receptor mRNA expression and binding is observed in such regions as the anterior olfactory nucleus, neocortex, caudate-putamen, nucleus accumbens, olfactory tubercle, diagonal band of Broca, globus pallidus and ventral pallidum, septal nuclei, amygdala, and pontine nuclei, suggesting local receptor synthesis. Regions of high δ receptor mRNA expression and comparatively low receptor binding include the internal granular layer of the olfactory bulb and ventromedial nucleus of the hypothalamus. The apparent discrepancy between δ receptor mRNA expression and δ receptor binding in several brainstem nuclei and in the cerebellum is, in part, due to the increased sensitivity of *in situ* hybridization methods and high levels of nonspecific binding observed with δ-selective ligands. Differences in distributions indicative of receptor transport are observed in the substantia gelatinosa of the spinal cord, external plexiform layer of the olfactory bulb, and the superficial layer of the superior colliculus.

κ Receptor Distribution. A high degree of correlation between κ_1 receptor mRNA expression and binding is observed in regions such as the nucleus accumbens, caudate-putamen, olfactory tubercle, bed nucleus of the stria terminalis, medial preoptic area, paraventricular nucleus, supraoptic nucleus, dorsomedial and ventromedial hypothalamus, amygdala, midline thalamic nuclei, periaqueductal gray, raphe nuclei, parabrachial nucleus, locus coeruleus, spinal trigeminal nucleus, and the nucleus of the solitary tract. Differences in κ_1 receptor binding and mRNA distribution in the substantia nigra pars compacta, ventral tegmental area, and neural lobe of the pituitary might be due to receptor transport.

Abbreviations
ABL, basolateral amygdaloid nucleus; ac, anterior commissure; ACB, nucleus accumbens; ACE, central amygdaloid nucleus; ACO, cortical amygdaloid nucleus; AD, anteriodorsal thalamus; AL, anterior lobe, pituitary; AMB, nucleus ambiguus; AME, medial amygdaloid nucleus; AON, anterior olfactory nucleus; ARC, arcuate nucleus, hypothalamus; BST, bed nucleus of the stria terminalis; cc, corpus callosum; ce, central canal; CL, centrolateral thalamus; CM, centromedial thalamus; CPU, caudate-putamen; CRB, cerebellum; DG, dentate gyrus; DH, dorsal horn, spinal cord; DMH, dorsomedial hypothalamus; DPG, deep gray layer, superior colliculus; DTN, dorsal tegmental nucleus; ENT, entorhinal cortex; EPL, external plexiform layer, olfactory bulb; FCX, frontal cortex; G, nucleus gelatinosus thalamus; GL, glomerular layer, olfactory bulb; GP, globus pallidus; HL, lateral habenula; HM, medial habenula; HPC, hippocampus; IC, inferior colliculus; IGR, intermediate granular layer, olfactory bulb; IL, intermediate lobe, pituitary; ING, intermediate gray layer, superior colliculus; IntP, interposed cerebellar nucleus; IP, interpeduncular nucleus; LC, locus coeruleus; LD, laterodorsal thalamus; LG, lateral geniculate thalamus; LHA, lateral hypothalamic area; LRN, lateral reticular nucleus; LS, lateral septum; MD, dorsomedial thalamus; ME, median eminence; Med, medial cerebellar nucleus; MG, medial geniculate; Mi, mitral cell layer, olfactory bulb; ml, medial leminscus; MM, medial mammillary nucleus; MS, medial septum; MV, medial vestibular nucleus; NDB, nucleus diagonal band; NL, neural lobe, pituitary; NRGC, nucleus reticularis gigantocellularis; NTS, nucleus tractus solitarius; OB, olfactory bulb; ot, optic tract; OTU, olfactory tubercle; PaG, periaqueductal gray; PBN, parabrachial nucleus; pc, posterior commissure; PCX, parietal cortex; PN, pons; PnR, pontine reticular; PO, posterior nucleus thalamus; POA, preoptic area; PrS, presubiculum; PV, paraventricular thalamus; PVN, paraventricular hypothalamus; RD, dorsal raphe; RE, reuniens thalamus; RM, raphe magnus; RME, median raphe; SC, superior colliculus; scp, superior cerebellar peduncle; SG, substantia gelatinosa; SNC, substantia nigra pars compacta; SNR, substantia nigra pars reticulata; SNT, sensory nucleus trigeminal; SON, supraoptic nucleus; STCX, striate cortex; STN, spinal trigeminal nucleus; SUG, superficial gray layer, superior colliculus; TCX, temporal cortex; TZ, trapezoid nucleus; VH, ventral horn, spinal cord; VL, ventrolateral thalamus; VM, ventromedial thalamus; VMH, ventromedial hypothalamus; VP, ventral pallidus; VPL, ventroposteriolateral thalamus; VTA, ventral tegmental area; ZI, zona incerta.
[Reproduced with permission from Mansour *et al.*, 1995.]

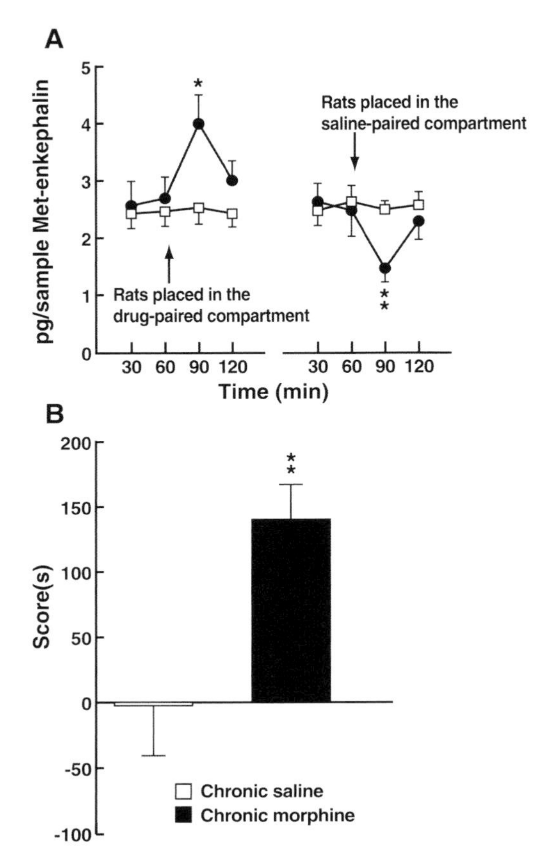

FIGURE 4.27 **(A)** Determination of the extracellular levels of enkephalin in the nucleus accumbens of rats chronically treated with saline or morphine to induce a positive place conditioning. The dialysis samples were collected every 30 min. Results are expressed as mean ± SEM of enkephalin in picograms per sample ($n = 6$–8 rats). $^{*}p < 0.05$ and $^{**}p < 0.01$ compared to microdialysis samples collected from control rats placed in the same conditions (Newman–Keuls test). **(B)** Conditioned place preference induced by morphine (5 mg/kg, i.p.). Data are expressed as scores (mean ± SEM) calculated as the difference between postconditioning and preconditioning time spent in the compartment associated with the drug ($n = 8$ rats per group). $^{**}p < 0.01$ versus the saline group (Newman–Keuls test). [Reproduced with permission from Nieto *et al.*, 2002.]

1992). Subsequent studies have shown that dopamine neuronal firing is decreased during opioid withdrawal, and concomitant decreases in extracellular dopamine as measured by microdialysis have been observed (Rosetti *et al.*, 1992). Also, chronic morphine is associated with a decrease in the size of ventral tegmental area dopamine neurons (Sklair-Tavron *et al.*, 1996). These neurochemical changes during withdrawal from opioids are accompanied by increased presynaptic inhibition by GABA (more GABA activity) and increased sensitivity of metabotropic glutamate receptors, both causing decreased release of glutamate in the ventral tegmental area leading to decreased dopamine cell firing (Bonci and Williams, 1997; Manzoni and Williams, 1999). Thus, chronic administration of opioids has multiple effects, presumably via opioid receptors, on the neuropharmacological systems that interface with the ventral tegmental area and the extended amygdala that may convey the cellular basis for the neuroadaptation associated with the development of dependence. Opioids act indirectly via GABA or glutamate systems to activate reward pathways, and these same systems show neuroplasticity with chronic opioid exposure associated with dependence.

NEUROBIOLOGICAL MECHANISM—MOLECULAR

Receptor Mechanisms

Pharmacologists initially identified three putative opioid receptors, μ, δ, and κ, based on differential pharmacological actions and differential antagonism (Martin, 1983). Subsequently, opioid receptors were demonstrated using radioligand binding techniques facilitated by the availability of radiolabeled ligands with high specific radioactivity, extensive washing techniques to remove nonspecific binding (Pert and Snyder, 1973), and the use of stereospecificity as a criterion to identify opioid receptors (Goldstein *et al.*, 1971). Using this approach, saturable, stereospecific binding with high affinity for known opioids was established (Pert and Snyder, 1973; Snyder and Pasternak, 2003) (**Table 4.7**). The opioid receptors were localized in differential density to different brain tissues using autoradiography with high concentrations in pain areas such as the periaqueductal gray, medial thalamus, and amygdala (Hiller *et al.*, 1973; Kuhar *et al.*, 1973; Pert *et al.*, 1976). Binding studies combined with the development of highly specific ligands eventually supported the existence of the three originally proposed opioid receptors—μ, δ, and κ (Martin, 1967; Lord *et al.*, 1977).

Opioids produce their effects in target neurons via interactions with μ, δ, and κ classes of receptors. At the receptor level, two mechanisms in *in vitro* preparations have been implicated in 'within-system' changes that may contribute to acute tolerance: receptor desensitization and internalization (Harris and Williams, 1991; Liu *et al.*, 1999b; Noble and Cox, 1996; Trapaidze *et al.*, 1996; Liu and Anand, 2001), but the role of desensitization in opioid tolerance remains

TABLE 4.7 Drug competition for [^{3}H]naloxone binding*

Drug	ED_{50} (nM)
(−)Naloxone	10
(−)3-Hydroxy-*N*-allylmorphinan (levallorphan)	1
Levorphanol	2
(−)Nalorphine	2
(−)Morphine	6
(−)Methadone	20
(±)Pentazocine	50
(+)Methadone	200
(±)Propoxyphene	1000
(+)3-Hydroxy-*N*-allylmorphinan	5000
Dextrorphan	8000
(−)Codeine	20 000

*Drug influences on opioid receptor binding illustrates the power of reversible ligand binding to characterize pharmacological actions, which was a novel approach in the early 1970s. Stereospecificity of the receptor is evident by the 4000-fold greater potency of levorphanol than its enantiomer dextrorphan. The inactivity of codeine (3-O-methylmorphine) indicates that the drug acts only following its O-demethylation by hepatic cytochrome P450 enzymes to morphine, providing more gradual access to the brain and hence lesser euphoria. The similar potencies of the agonist morphine and the corresponding antagonist nalorphine (which differs from morphine by transformation of an *N*-methyl to an *N*-allyl substituent) establishes the challenge of differentiating agonists and antagonists. Phenobarbital, serotonin, norepinephrine, carbamylcholine, choline, atropine, histamine and colchicine have no effect at 100 000 nM. [Modified with permission from Pert and Snyder, 1973.]

controversial because tolerance develops at doses that do not produce desensitization, but in fact activate second and third messenger transduction mechanisms (see below).

μ, δ, and κ opioid receptors are coupled via pertussis toxin-sensitive G-protein binding proteins to inhibit adenylyl cyclase (Law *et al.*, 2000). However, opioid receptors may couple to a number of other second messenger systems, including activation of mitogen-activated protein kinases and activation of phospholipase C (Liu and Anand, 2001). μ opioid receptors show short-term desensitization via a protein kinase mechanism, probably protein kinase C, which has been hypothesized to be involved in acute tolerance (Mestek *et al.*, 1995; Narita *et al.*, 1995).

μ and δ receptors, but not κ receptors, also can undergo rapid agonist-mediated internalization via endocytosis (Gaudriault *et al.*, 1997). Rapid internalization may have a role in functional recovery by promoting receptor ligand dissociation (Yu *et al.*, 1993). Some ligands cause rapid internalization while others such as morphine do not show internalization (Keith *et al.*, 1996), suggesting a role for such intracellular events in the relative efficacy of opioids. Consistent with a differential role for internalization in the efficacy of opioids, promotion of receptor internalization by deleting the β-arrestin-2 gene in mice potentiated and prolonged the analgesic action of morphine, suggesting an impairment of receptor desensitization (Bohn *et al.*, 1999). The link of receptor internalization to acute tolerance is elaborated by the observation that ligands that produce rapid acute tolerance (e.g., morphine) do not show receptor internalization, whereas ligands that do not show rapid tolerance (e.g., dihydroetorphine) do show acute internalization (Duttaroy and Yoburn, 1995; Aceto *et al.*, 1997). Dihydroetorphine is a very potent opioid agonist described as 1000 (Tokuyama *et al.*, 1996) to 1200 (Bentley and Hardy, 1967) times more potent than morphine but with a shorter duration of action (see Ohmori and Morimoto, 2002). The tendency for such rapid internalization to produce less tolerance and dependence raises the hypothesis that activation of opioid receptors without internalization may cause changes in 'between-system' neuronal homeostasis that contribute to the development of addiction (Huang *et al.*, 1994; Qin *et al.*, 1994; Keith *et al.*, 1998).

More specifically, activation of opioid receptors leads to recruitment of G_i and related G proteins. Recruitment of G_i also leads to the inhibition of adenylyl cyclase and of the cyclic adenosine monophosphate (cAMP) protein phosphorylation cascade (Koob *et al.*, 1998; Gutstein and Akil, 2001; Liu and Anand, 2001) (**Fig. 4.28**). Recruitment of G_i also leads to activation of certain K^+ channels and inhibition of voltage-gated Ca^{2+} channels, although the two actions occur to varying extents in different neuronal cell types. Both are inhibitory actions—more K^+ flows out of the cell and less Ca^{2+} flows into the cell—that mediate some of the relatively rapid inhibitory effects of opioids on the electrical properties of their target neurons. Similarly, reductions in cellular Ca^{2+} levels alter Ca^{2+}-dependent protein phosphorylation cascades. Altered activity of these protein phosphorylation cascades, which also can vary among different cell types, leads to the regulation of still additional ion channels which contribute further to the acute effects of the drug (Nestler, 2004; Nestler and Malenka, 2004).

Chronic exposure to opioids increases adenylyl cyclase (Collier and Francis, 1975; Sharma *et al.*, 1975), leading to other perturbations in intracellular messenger pathways, such as in protein phosphorylation mechanisms. These perturbations eventually elicit long-term adaptations that can lead to changes in many other neural processes within target neurons. These target neurons include those that trigger the long-term effects of the drugs that lead eventually to tolerance, dependence, withdrawal, sensitization,

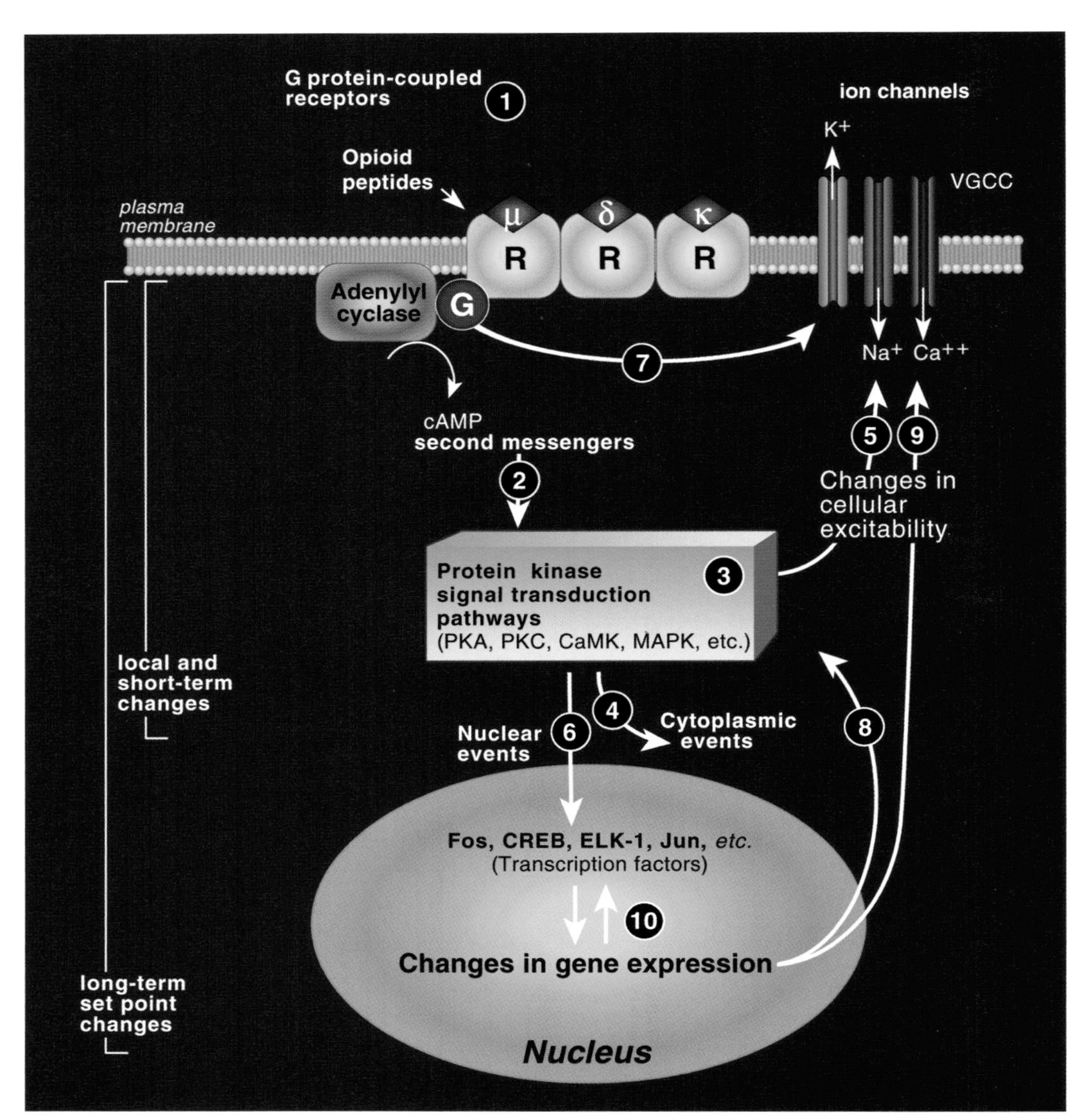

FIGURE 4.28 Molecular mechanisms of neuroadaptation—opioids. Opioids, by acting on neurotransmitter systems, affect the phenotypic and functional properties of neurons through the general mechanisms outlined in the diagram. Shown are examples of G protein-coupled receptors (R) such as the opioid receptors (①). These receptors modulate the levels of second messengers like cyclic adenosine monophosphate (cAMP) and Ca^{2+} (②), which in turn regulate the activity of protein kinase transducers (③). Such protein kinases affect the functions of proteins located in the cytoplasm, plasma membrane, and nucleus (④ ⑤ ⑥). Among membrane proteins affected are ligand-gated and voltage-gated ion channels (⑤). G_i and G_0 proteins also can regulate potassium and calcium channels directly through their βγ subunits (⑦). Protein kinase transduction pathways also affect the activities of transcription factors (⑥). Some of these factors, like cAMP response element binding protein (CREB), are regulated posttranslationally by phosphorylation; others, like Fos, are regulated transcriptionally; still others, like Jun, are regulated both posttranslationally and/or transcriptionally. While membrane and cytoplasmic changes may be only local (e.g., dendritic domains or synaptic boutons), changes in the activity of transcription factors may result in long-term functional changes. These may include changes in gene expression of proteins involved in signal transduction (⑧) and/or neurotransmission (⑨), resulting in altered neuronal responses. For example, chronic exposure to opioids has been reported to increase levels of protein kinase A (⑧) and adenylyl cyclase in the nucleus accumbens and to decrease levels of $G_{i\alpha}$ (Self and Nestler, 1995; Nestler, 1996). Chronic exposure to opioids also alters the expression of transcription factors themselves (⑩). CREB expression, for instance, is depressed in the nucleus accumbens by chronic morphine treatment and increased during withdrawal (Nestler, 1996; Widnell *et al.*, 1996), while chronic opioid exposure activates Fos-related antigens (FRAs) such as ΔFosB (Nestler, 2004). The receptor systems depicted in the figure may not coexist in the same cells. [Modified from Koob *et al.*, 1998.]

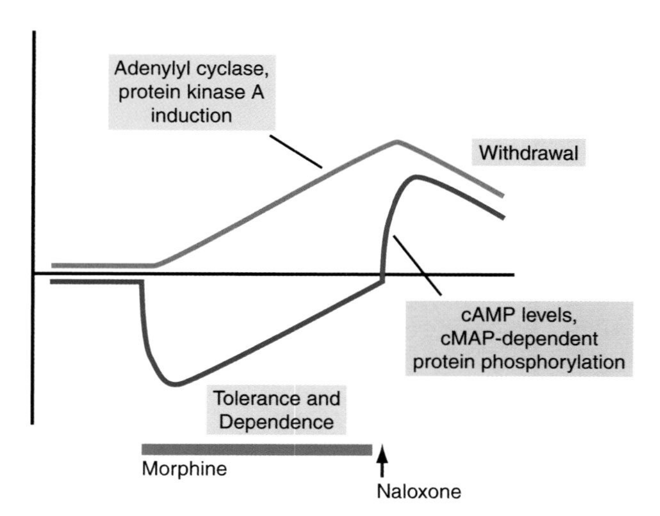

FIGURE 4.29 Upregulation of the cyclic adenosine monophosphate (cAMP) pathway as a mechanism of opioid tolerance and dependence. Opioids acutely inhibit the functional activity of the cAMP pathway (indicated by cellular levels of cAMP and cAMP-dependent protein phosphorylation). With continued opioid exposure, functional activity of the cAMP pathway gradually recovers, and then increases far above control levels following removal of the opioid (e.g., by administration of the opioid receptor antagonist naloxone). These changes in the functional state of the cAMP pathway are mediated via the induction of adenylyl cyclase and protein kinase A in response to chronic administration of opioids. Induction of these enzymes accounts for the gradual recovery in the functional activity of the cAMP pathway that occurs during chronic opioid exposure (tolerance and dependence) and activation of the cAMP pathway observed upon removal of opioid (withdrawal). [Reproduced with permission from Nestler, 2004.]

and, ultimately, addiction (Nestler, 2004; Nestler and Malenka, 2004) (**Figs. 4.29** and **4.30**).

Early studies involving the locus coeruleus showed regulation of the cAMP pathway in locus coeruleus neurons (Nestler *et al.*, 1993) and formed a molecular model of opioid neuroadaptation in the brain (Sharma *et al.*, 1975; Collier, 1980; Nestler, 1996). Here acute administration of opioids inhibited the firing of the neurons via regulation of two ion channels: activation of K^+ channels via direct G_i protein coupling and inhibition of a Na^+ current indirectly via inhibition of adenylyl cyclase and cAMP-dependent protein kinase (Alreja and Aghajanian, 1993). In contrast, chronic opioids increased adenylyl cyclase and cAMP-dependent protein kinase expressed in locus coeruleus neurons (Duman *et al.*, 1988; Nestler and Tallman, 1988). This upregulated cAMP pathway was hypothesized to contribute to the increase in the intrinsic electrical excitability of locus coeruleus neurons that underlies the cellular forms of tolerance and dependence exhibited by these neurons (Rassmussen *et al.*, 1990; Nestler, 1996). Several observations show that the transcription factor cAMP response element binding protein (CREB), has a role in the upregulation of adenylyl cyclase and protein kinase A in the locus coeruleus (Widnell *et al.*, 1994; Maldonado *et al.*, 1996). CREB is acutely inhibited by opioids and increased during opioid withdrawal (Guitart *et al.*, 1992; Widnell *et al.*, 1994).

The studies on opioid regulation of second messenger systems and transcription factors in the locus coeruleus led to a focus on similar molecular elements of the brain motivational circuits, notably the nucleus accumbens, and have subsequently been expanded to incorporate other transcription factors that are activated for long periods following drug withdrawal in chronically treated animals. Tyrosine hydroxylase activity was increased in the ventral tegmental area and decreased in the nucleus accumbens of rats treated chronically with morphine and in rats self-administering heroin (Self *et al.*, 1995). Heroin exposure also increased protein kinase A activity in the nucleus accumbens (Self *et al.*, 1995). CREB in the nucleus accumbens was decreased during chronic morphine exposure (Widnell *et al.*, 1996), whereas upregulation of cAMP response element (CRE) transcription also has been observed in the nucleus accumbens during opioid withdrawal (Shaw-Lutchman *et al.*, 2002). Another transcription factor activated by

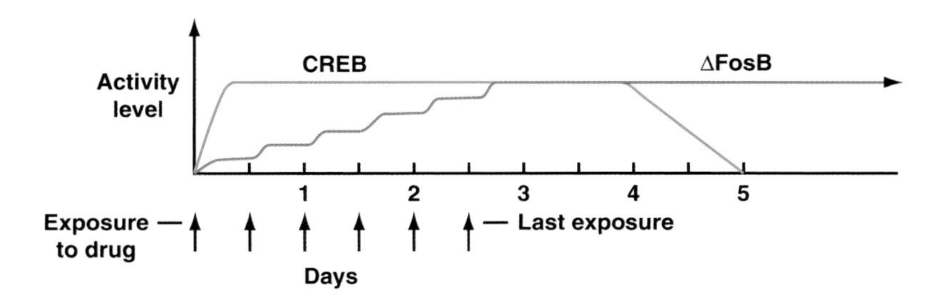

FIGURE 4.30 Whether a user is tolerant to a drug or, conversely, sensitized to it, depends in part on the levels of active cyclic adenosine monophosphate response-element binding protein (CREB) and ΔFosB in nucleus accumbens cells. Initially, CREB dominates, leading to tolerance and, in the drug's absence, discomfort that only more drug can alleviate. But CREB activity decreases within days when not boosted by repeated administrations of opioids. In contrast, ΔFosB concentrations stay elevated for weeks after the last drug exposure. As CREB activity declines, the dangerous long-term sensitizing effects of ΔFosB come to dominate. [Reproduced with permission from Nestler and Malenka, 2004.]

chronic administration of opioids and other drugs of abuse is ΔFosB (Nye and Nestler, 1996). ΔFosB is responsible for persistent high levels of the activator protein 1 (AP-1) complexes that are transcriptionally active dimers of Fos and related Jun-family proteins (Nye and Nestler, 1996). Overexpression of ΔFosB increases sensitivity to the rewarding effects of morphine and expression of the dominant negative ΔFosB decreases sensitivity to morphine (Nestler, 2004). Because the effects of drugs of abuse on ΔFosB are stable for weeks and months after the last drug exposure, Nestler has hypothesized that ΔFosB 'could be a sustained molecular switch that helps to initiate and maintain a state of addiction' (Nestler *et al.*, 2001) (see *Neurobiological Theories of Addiction* chapter).

Opioid Receptor Knockouts

Progress towards understanding the molecular basis of opioid action and addiction has been aided by the cloning and characterization of genes encoding three families of opioid peptides and their receptors: μ opioid receptor, δ opioid receptor, and κ opioid receptor (Akil *et al.*, 1984; Evans *et al.*, 1992; Kieffer *et al.*, 1992; Chen *et al.*, 1993; Yasuda *et al.*, 1993). The three opioid receptor genes correspond to the μ, δ, and κ receptors originally described by *in vivo* and *in vitro* pharmacology (Martin, 1967). The amino acid sequences of the three opioid receptors are highly homologous (approximately 70 per cent) but derive from different chromosomes, suggesting a high conservation and lack of divergence in recent evolutionary history (Zaki *et al.*, 1996). A subsequent orphanin FQ/nociceptin receptor was discovered later (Meunier *et al.*, 1995), making a four-member gene subfamily of the G protein-coupled receptor family.

Mutant mouse strains lacking the genes of the opioid system have been generated by utilizing homologous recombination technology. Functional studies have provided important confirmation and extension of pharmacological studies of the molecular basis of opioid effects in general, and of the reinforcing and dependence-inducing effects of opioids in particular (Gaveriaux-Ruff and Kieffer, 2002; Kieffer and Gaveriaux-Ruff, 2002). Opioid (morphine) reinforcement as measured by place preference or self-administration is absent in μ knockout mice (Matthes *et al.*, 1996; Becker *et al.*, 2000; Sora *et al.*, 2001), and there is no development of somatic signs of dependence to morphine in these mice (Matthes *et al.*, 1996; Sora *et al.*, 2001). Indeed, to date all morphine effects tested, including analgesia, hyperlocomotion, respiratory depression, and inhibition of gastrointestinal transit, are abolished in μ knockout mice (Gaveriaux-Ruff and Kieffer, 2002) (**Fig. 4.31**). Opioid reinforcement as measured by place preference also is blocked in D_2 knockout mice, although these mice had a blunted locomotor response to morphine (Maldonado *et al.*, 1997).

κ knockout mice fail to show the aversive stimulus effects of κ agonists as measured by conditioned place aversion to the κ agonist U50,288 (Simonin *et al.*, 1998). κ knockout mice also show an attenuation of naloxone-precipitated withdrawal following chronic morphine treatment.

In contrast, studies of δ receptor knockout mice have revealed a phenotype of increased levels of anxiety-like behavior in a number of behavioral tests. δ knockout mice also show a blockade of tolerance to the analgesic effects of morphine suggesting that δ receptors may contribute to this adaptive aspect of chronic opioid exposure (Zhu *et al.*, 1999). Mice lacking β-endorphin and/or enkephalins reduced

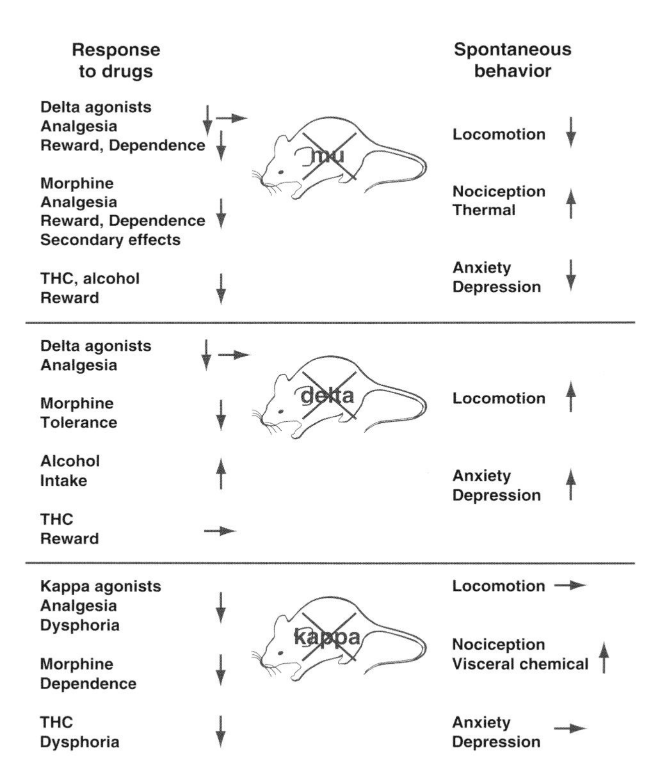

FIGURE 4.31 Summary of responses to drugs and spontaneous behaviors in μ, δ, and κ opioid receptor knockout mice. Responses to drugs are shown on the left with drugs indicated; behaviors in the absence of drugs are shown on the right. ↓, strongly reduced or abolished. ↑, increased. →, unchanged. ↓→, unchanged or decreased depending on the experimental conditions. Abbreviation: THC, Δ^9-tetrahydrocannabinol. [Modified with permission from Gaveriaux-Ruff and Kieffer, 2002.]

responding for food in a progressive-ratio schedule of reinforcement, but only in a nondeprivation condition, suggesting that the endogenous opioids have a role in the hedonics of feeding (Hayward *et al.*, 2002).

SUMMARY

Opioids are defined as all drugs, natural and synthetic, with morphine-like action. Opium, which contains morphine, and the other opioid alkaloids such as thebaine and codeine has been used for centuries to relieve pain and treat diarrhea. As with other drugs of abuse, self-limited, controlled use of opioids is accompanied by a lack of withdrawal symptoms and limited pathology. However, opioid addiction remains a chronic problem in the world, and the diversion of opioid pain medications continues to evolve as an addiction problem. Opioids have powerful effects to relieve not only physical but also emotional pain, and this forms the main behavioral mechanism of action. Opioid addiction is characterized by different stages that follow the regular intravenous use of opioids with self-administration 1–3 times per day, and includes the rush, nod, high, and being straight. Opioid withdrawal is characterized by severe motivational (purposive) and physical (nonpurposive) signs, and the motivational signs, in particular, begin with a single administration. Protracted abstinence features powerful craving, derived from both conditioned positive and conditioned negative reinforcement. Much is known about the neurobiological substrates for the acute reinforcing effects of opioid drugs. At a neurocircuitry level, neural elements in the region of the ventral tegmental area and the nucleus accumbens appear to be important. Heroin-primed reinstatement is attenuated by blockade of dopamine D_1 and D_2 receptor antagonists and opioid antagonists. CRF antagonists and the α_2 adrenergic agonist clonidine block stress-induced reinstatement of heroin self-administration, suggesting a role for opioid and dopaminergic mechanisms in opioid-induced reinstatement and a role for brain stress systems in stress-induced reinstatement. At the cellular level, evidence exists for a disinhibitory effect of opioids via glutamatergic or GABAergic interneurons both in the extended amygdala and the ventral tegmental area. At the molecular level, the μ opioid receptor is most critical for the acute reinforcing effects of opioids as defined by neuropharmacological and knockout approaches. During the development of dependence on opioids, there is evidence from studies at the neurocircuitry and cellular levels that both within-system and between-system changes occur at the level of the ventral tegmental area and the extended amygdala which contribute to the motivational effects of withdrawal. Chronic morphine produces a sensitized aversive response to opioid antagonists in the extended amygdala, a recruitment of norepinephrine and CRF activity in the amygdala and BNST, and a shift in NMDA glutamate activity in the amygdala to the NMDA NR2 receptor subunits. In the ventral tegmental area, there is a decrease in neuronal firing of dopamine neurons during withdrawal that may be mediated by increased release of GABA and decreased release of glutamate. μ opioid receptor knockout mice do not show opioid dependence or withdrawal. All of these changes may contribute to the decreased reward function associated with opioid dependence and abstinence.

REFERENCES

Aceto, M. D., Harris, L. S., and Bowman, E. R. (1997). Etorphines: mu-opioid receptor-selective antinociception and low physical dependence capacity, *European Journal of Pharmacology* **338**, 215–223.

Acker, C. J. (2002). *Creating the, American, Junkie: Addiction, Research in the, Classic, Era of, Narcotic, Control*, Johns, Hopkins, University, Press, Baltimore.

Acquas, E., Carboni, E., Leone, P., and Di Chiara, G. (1989). SCH 23390 blocks drug-conditioned place-preference and place-aversion: anhedonia (lack of reward) or apathy (lack of motivation) after dopamine-receptor blockade?. *Psychopharmacology* **99**, 151–155.

Agar, M. (1973). *Ripping and Running: A Formal Ethnography of Urban Heroin Addicts*, Seminar Press, New York.

Ahmed, S. H., Walker, J. R., and Koob, G. F. (2000). Persistent increase in the motivation to take heroin in rats with a history of drug escalation. *Neuropsychopharmacology* **22**, 413–421.

Akil, H., Mayer, D. J., and Liebeskind, J. C. (1976). Antagonism of stimulation-produced analgesia by naloxone, a narcotic antagonist. *Science* **191**, 961–962.

Akil, H., Watson, S. J., Young, E., Lewis, M. E., Khachaturian H., and Walker, J.M. (1984). Endogenous opioids: biology and function. *Annual Review of Neuroscience* **7**, 223–255.

al Rodhan, N., Chipkin, R., and Yaksh, T. L. (1990). The antinociceptive effects of SCH-32615, a neutral endopeptidase (enkephalinase) inhibitor, microinjected into the periaqueductal, ventral medulla and amygdala. *Brain Research* **520**, 123–130.

Alheid, G. F., and Heimer, L. (1988). New perspectives in basal forebrain organization of special relevance for neuropsychiatric disorders: The striatopallidal, amygdaloid, and corticopetal components of substantia innominata. *Neuroscience* **27**, 1–39.

Ali, N. M. (1986). Hyperalgesic response in a patient receiving high concentrations of spinal morphine. *Anesthesiology* **65**, 449.

Alreja, M., and Aghajanian, G. K. (1993). Opiates suppress a resting sodium-dependent inward current and activate an outward potassium current in locus coeruleus neurons. *Journal of Neuroscience* **13**, 3525–3532.

Amalric, M., and Koob, G. F. (1985). Low doses of methylnaloxonium in the nucleus accumbens antagonize hyperactivity induced by heroin in the rat. *Pharmacology Biochemistry and Behavior* **23**, 411–415.

Amalric, M., Cline, E. J., Martinez, J. L. Jr., Bloom, F. E., and Koob, G. F. (1987). Rewarding properties of ß-endorphin as measured by conditioned place preference. *Psychopharmacology* **91**, 14–19.

American Psychiatric Association. (1994). *Diagnostic and Statistical Manual of Mental Disorders*, 4th edition, American Psychiatric Press, Washington, D.C.

Angst, M. S., Koppert, W., Pahl, I., Clark, D. J., and Schmelz, M. (2003). Short-term infusion of the mu-opioid agonist remifentanil in humans causes hyperalgesia during withdrawal. *Pain* **106**, 49–57.

Arner, S., Rawal, N., and Gustafsson, L. L. (1988). Clinical experience of long-term treatment with epidural and intrathecal opioids: a nationwide survey. *Acta Anaesthesiologica Scandinavica* **32**, 253–259.

Awasaki, Y., Nishida, N., Sasaki, S., and Sato, S. (1997). Dopamine D-1 antagonist SCH 23390 attenuates self-administration of both cocaine and fentanyl in rats. *Environmental Toxicology and Pharmacology* **3**, 115–122.

Azar, M. R., Jones, B. C., and Schulteis, G. (2003). Conditioned place aversion is a highly sensitive index of acute opioid dependence and withdrawal. *Psychopharmacology* **170**, 42–50.

Azolosa, J. L., Stitzer, M. L., and Greenwald, M. K. (1994). Opioid physical dependence development: effects of single versus repeated morphine pretreatments and of subjects' opioid exposure history. *Psychopharmacology* **114**, 71–80.

Babor, T. F., Meyer, R. E., Mirin, S. M., McNamee, H. B., and Davies, M. (1976). Behavioral and social effects of heroin self-administration and withdrawal. *Archives of General Psychiatry* **33**, 363–367.

Ballantyne, J. C., Loach, A. B., and Carr, D. B. (1988). Itching after epidural and spinal opiates. *Pain* **33**, 149–160.

Bals-Kubik, R., Ableitner, A., Herz, A., and Shippenberg, T. S. (1993). Neuroanatomical sites mediating the motivational effects of opioids as mapped by the conditioned place preference paradigm in rats. *Journal of Pharmacology and Experimental Therapeutics* **264**, 489–495.

Baptista, M. A., Siegel, S., MacQueen, G., and Young, L. T. (1998). Predrug cues modulate morphine tolerance, striatal c-Fos, and AP-1 DNA binding. *Neuroreport* **9**, 3387–3390.

Barrot, M., Abrous, D. N., Marinelli, M., Rouge-Pont F., Le Moal, M., and Piazza, P. V. (2001). Influence of glucocorticoids on dopaminergic transmission in the rat dorsolateral striatum. *European Journal of Neuroscience* **13**, 812–818 [erratum: 13:2013].

Basbaum, A. I., and Fields, H. L. (1984). Endogenous pain control systems: brainstem spinal pathways and endorphin circuitry. *Annual Review of Neuroscience* **7**, 309–338.

Bechara, A., and van der Kooy, D. (1989). The tegmental pedunculopontine nucleus: a brain-stem output of the limbic system critical for the conditioned place preferences produced by morphine and amphetamine. *Journal of Neuroscience* **9**, 3400–3409.

Bechara, A., and van der Kooy, D. (1992). Lesions of the tegmental pedunculopontine nucleus: effects on the locomotor activity induced by morphine and amphetamine. *Pharmacology Biochemistry and Behavior* **42**, 9–18.

Bechara, A., Harrington, F., Nader, K., and van der Kooy, D. (1992). Neurobiology of motivation: double dissociation of two motivational mechanisms mediating opiate reward in drug-naive versus drug-dependent animals. *Behavioral Neuroscience* **106**, 798–807.

Bechara, A., Nader K., and van der Kooy, D. (1995). Neurobiology of withdrawal motivation: Evidence for two separate aversive effects produced in morphine-naive versus morphine-dependent rats by both naloxone and spontaneous withdrawal. *Behavioral Neuroscience* **109**, 91–105

Becker, A., Grecksch, G., Brodemann, R., Kraus, J., Peters, B., Schroeder, H., Thiemann, W., and Loh, H. H. (2000). Hollt V., Morphine self-administration in mu-opioid receptor-deficient mice. *Naunyn Schmiedebergs Archives of Pharmacology* **361**, 584–589.

Belluzzi, J. D., Grant, N., Garsky, V., Sarantakis, D., Wise, C. D., and Stein, L. (1976). Analgesia induced in vivo by central administration of enkephalin in rat. *Nature* **260**, 625–626.

Belluzzi, J. D., and Stein, L. (1977). Enkephaline may mediate euphoria and drive-reduction reward. *Nature* **266**, 556–558.

Bentley, K. W., and Hardy, D. G. (1967). Novel analgesics and molecular rearrangements in the morphine-thebaine group: 3. Alcohols of the 6,14-endo-ethenotetrahydrooripavine series and derived analogs of N-allylnormorphine and –norcodeine. *Journal of the American Chemical Society* **89**, 3281–3292.

Bewley, T. H. (1968). The diagnosis and management of heroin addiction. *Practitioner* **200**, 215–219.

Bickel, W. K., Stitzer, M. L., Bigelow, G. E., Liebson, I. A., Jasinski, D. R., and Johnson, R. E. (1988a). Buprenorphine: dose-related blockade of opioid challenge effects in opioid dependent humans, *Journal of Pharmacology and Experimental Therapeutics* **247**, 47–53.

Bickel, W. K., Stitzer, M. L., Liebson, I. A., and Bigelow, G. E. (1988b). Acute physical dependence in man: effects of naloxone after brief morphine exposure. *Journal of Pharmacology and Experimental Therapeutics* **244**, 126–132.

Blachly, P. H. (1966). Management of the opiate abstinence syndrome. *American Journal of Psychiatry* **122**, 742–744.

Bohn, L. M., Lefkowitz, R. J., Gainetdinov, R. R., Peppel, K., Caron, M. G., and Lin, F. T. (1999). Enhanced morphine analgesia in mice lacking beta-arrestin 2. *Science* **286**, 2495–2498.

Bohn, L. M., Gainetdinov, R. R., Lin, F. T., Lefkowitz, R. J., and Caron, M. G. (2000). Mu-opioid receptor desensitization by beta-arrestin-2 determines morphine tolerance but not dependence. *Nature* **408**, 720–723.

Bonci, A., and Williams, J. T. (1997). Increased probability of GABA release during withdrawal from morphine. *Journal of Neuroscience* **17**, 796–803.

Bozarth, M. A., and Wise, R. A. (1981). Intracranial self-administration of morphine into the ventral tegmental area in rats. *Life Sciences* **28**, 551–555.

Bozarth, M. A., and Wise, R. A. (1983). Neural substrates of opiate reinforcement. *Progress in Neuropsychopharmacology and Biological Psychiatry* **7**, 569–575.

Bozarth, M. A., and Wise, R. A. (1984). Anatomically distinct opiate receptor fields mediate reward and physical dependence. *Science* **224**, 516–517.

Britt, M. D., and Wise, R. A. (1983). Ventral tegmental site of opiate reward: Antagonism by a hydrophilic opiate receptor blocker. *Brain Research* **258**, 105–108.

Brownstein, M. J., and Palkovits, M. (1984). Catecholamines, serotonin, acetylcholine, and gamma-aminobutyric acid in the rat brain: biochemical studies. *In Classical Transmitters in the CNS: Part I* (series title: *Handbook of Chemical Neuroanatomy*, vol. 2), (A. Bjorklund, A. T. Hokfelt, and M. J. Kuhar, Eds.), pp. 23–54, Elsevier, Amsterdam.

Cadoni, C., and Di Chiara, G. (1999). Reciprocal changes in dopamine responsiveness in the nucleus accumbens shell and core and in the dorsal caudate-putamen in rats sensitized to morphine. *Neuroscience* **90**, 447–455.

Caille, S., Espejo, E. F., Reneric, J-P., Cador, M., Koob, G. F., and Stinus, L. (1999). Total neurochemical lesion of noradrenergic neurons of the locus coeruleus does not alter either naloxone-precipitated or spontaneous opiate withdrawal nor does it influence ability of clonidine to reverse opiate withdrawal. *Journal of Pharmacology and Experimental Therapeutics* **290**, 881–892.

Caille, S., Espejo, E. F., Koob, G. F., and Stinus, L. (2002). Dorsal and median raphe serotonergic system lesion does not alter the opiate withdrawal syndrome. *Pharmacology Biochemistry and Behavior* **72**, 979–986.

Caille, S., Rodriguez-Arias, M., Minarro, J., Espejo, E. F., Cador, M., and Stinus, L. (2003). Changes in dopaminergic neurotransmission do not alter somatic or motivational opiate withdrawal-induced symptoms in rats. *Behavioral Neuroscience* **117**, 995–1005.

Cazala, P. (1990). Dose-dependent effects of morphine differentiate self-administration elicited from lateral hypothalamus and mesencephalic central gray area in mice. *Brain Research* **527**. 280–285.

Cazala, P., and Darracq, C. (1987). Saint-Marc M., Self-administration of morphine into the lateral hypothalamus in the mouse. *Brain Research* **416**, 283–288.

Celerier, E., Laulin, J-P., Corcuff, J-B., Le Moal, M., and Simonnet, G. (2001). Progressive enhancement of delayed hyperalgesia induced by repeated heroin administration: a sensitization process. *Journal of Neuroscience* **21**, 4074–4080.

Chang, K. J., Hazum, E., and Cuatrecasas, P. (1981). Novel opiate binding sites selective for benzomorphan drugs. *Proceedings of the National Academy of Sciences USA* **78**, 4141–4145.

Chen, X., Marrero, H. G., Murphy, R., Lin, Y. J., and Freedman, J. E. (2000). Altered gating of opiate receptor-modulated K+ channels on amygdala neurons of morphine-dependent rats. *Proceedings of the National Academy of Sciences USA* **97**, 14692–14696.

Chen, X., Marrero, H. G., and Freedman, J. E. (2001). Opioid receptor modulation of a metabolically sensitive ion channel in rat amygdala neurons. *Journal of Neuroscience* **21**, 9092–9100.

Chen, Y., Mestek, A., Liu, J., Hurley, J. A., and Yu, L. (1993). Molecular cloning and functional expression of a mu-opioid receptor from rat brain. *Molecular Pharmacology* **44**, 8–12.

Christie, M. J., Williams, J. T., and North, R. A. (1987). Cellular mechanisms of opioid tolerance: studies in single brain neurons. *Molecular Pharmacology* **32**, 633–638.

Churchill, L., Roques, B. P., and Kalivas, P. W. (1995). Dopamine depletion augments endogenous opioid-induced locomotion in the nucleus accumbens using both mu 1 and delta opioid receptors. *Psychopharmacology* **120**, 347–355.

Churchill, L., Klitenick, M. A., and Kalivas, P. W. (1998). Dopamine depletion reorganizes projections from the nucleus accumbens and ventral pallidum that mediate opioid-induced motor activity. *Journal of Neuroscience* **18**, 8074–8085.

Ciccocioppo, R., Cippitelli, A., Economidou, D., Fedeli, A., and Massi, M. (2004). Nociceptin/orphanin FQ acts as a functional antagonist of corticotropin-releasing factor to inhibit its anorectic effect. *Physiology and Behavior* **82**, 63–68.

Clausen, T. R., Moller, M., and Woldbye, D. P. (2001). Inhibitory effect of neuropeptide Y on morphine withdrawal is accompanied by reduced c-fos expression in specific brain regions. *Journal of Neuroscience Research* **64**, 410–417.

Cochin, J. (1970). Possible mechanisms in development of tolerance. *Federation Proceedings* **29**, 19–27.

Cochin, J., and Kornetsky, C. (1964). Development and loss of tolerance to morphine in the rat after single and multiple injections. *Journal of Pharmacology and Experimental Therapeutics* **145**, 1–10.

Cohen, M., Keats, A. S., Krivoy, W., and Ungar, G. (1965). Effect of actinomycin D on morphine tolerance. *Proceedings of the Society for Experimental Biology and Medicine* **119**, 381–384.

Collier, H. O. (1980). Cellular site of opiate dependence. *Nature* **283**, 625–629.

Collier, H. O., and Francis, D. L. (1975). Morphine abstinence is associated with increased brain cyclic AMP. *Nature* **255**, 159–162.

Cooper, B. Y., Vierck, C. J. Jr., and Yeomans, D. C. (1986). Selective reduction of second pain sensations by systemic morphine in humans. *Pain* **24**, 93–116.

Corrigall, W. A., and Vaccarino, F. J. (1988). Antagonist treatment in nucleus accumbens or periaqueductal grey affects heroin self-administration. *Pharmacology Biochemistry and Behavior* **30**, 443–450.

Council Reports. (1972). Treatment of morphine-type dependence by withdrawal methods. *Journal of the American Medical Association* **219**, 1611–1615.

Cox, B. M. (1991). Molecular and Cellular Mechanims of Opioid Tolerance. *In Towards a New Pharmacotherapy of Pain* (series title: *Life Sciences Research Report*, vol. 49), (A. I. Basbaum, J. M. Besson, Eds.), pp. 137–156, Wiley, Chichester.

Crine, P., Gianoulakis, C., Seidah, N. G., Gossard, F., Pezalla, P. D., Lis, M., and Chretien, M. (1978). Biosynthesis of beta-endorphin from beta-lipotropin and a larger molecular weight precursor in rat pars intermedia. *Proceedings of the National Academy of Sciences USA* **75**, 4719–4723.

Cushman, P., and Kreek, M. J. (1974). Some endocrinological observations in narcotic addicts. *In Narcotics and the Hypothalamus* (series title: *Kroc Foundation Symposia*, vol. 2), (E. Zimmerman, F. George, Eds.), pp. 161–173, Raven Press, New York.

David, V., and Cazala, P. (1994). A comparative study of self-administration of morphine into the amygdala and the ventral tegmental area in mice. *Behavioural Brain Research* **65**, 205–211.

David, V., and Cazala, P. (1998). Preference for self-administration of a low dose of morphine into the ventral tegmental area rather than into the amygdala in mice. *Psychobiology* **24**, 211–218.

Davis, M. (1997). Neurobiology of fear responses: the role of the amygdala. *Journal of Neuropsychiatry and Clinical Neurosciences* **9**, 382–402.

de Conno, F., Caraceni, A., Martini, C., Spoldi, E., Salvetti, M., and Ventafridda, V. (1991). Hyperalgesia and myoclonus with intrathecal infusion of high-dose morphine. *Pain* **47**, 337–339.

Delfs, J. M., Zhu, Y., Druhan, J. P., and Aston-Jones, G. (2000). Noradrenaline in the ventral forebrain is critical for opiate withdrawal-induced aversion. *Nature* **403**, 430–434.

Deroche, V., Piazza, P. V., Le Moal, M., and Simon, H. (1993). Individual differences in the psychomotor effects of morphine are predicted by reactivity to novelty and influenced by corticosterone secretion. *Brain Research* **623**, 341–344.

Deroche, V., Marinelli, M., Maccari, S., Le Moal, M., Simon, H., and Piazza, P. V. (1995). Stress-induced sensitization and glucocorticoids. I. Sensitization of dopamine-dependent locomotor effects of amphetamine and morphine depends on stress-induced corticosterone secretion. *Journal of Neuroscience* **15**, 7181–7188.

Devine, D. P., and Wise, R. A. (1994). Self-administration of morphine, DAMGO, and DPDPE into the ventral tegmental area of rats. *Journal of Neuroscience* **14**, 1978–1984.

Devoto, P., Flore, G., Pira, L., Diana, M., and Gessa, G. L. (2002). Co-release of noradrenaline and dopamine in the prefrontal cortex after acute morphine and during morphine withdrawal. *Psychopharmacology* **160**, 220–224.

Devulder, J. (1997). Hyperalgesia induced by high-dose intrathecal sufentanil in neuropathic pain. *Journal of Neurosurgery and Anesthesiology* **9**, 146–148.

Dickenson, A. H., Fardin, V., Le Bars, D., and Besson, J. M. (1979). Antinociceptive action following microinjection of methionine-enkephalin in the nucleus raphe magnus of the rat. *Neuroscience Letters* **15**, 265–270.

Dole, V. P. (1980). Addictive behavior. *Scientific American* **243**, 138–140, 142, 144 passim.

Duman, R. S., Tallman, J. F., and Nestler, E. J. (1988). Acute and chronic opiate-regulation of adenylate cyclase in brain: specific effects in locus coeruleus. *Journal of Pharmacology and Experimental Therapeutics* **246**, 1033–1039.

Duttaroy, A., and Yoburn, B. C. (1995). The effect of intrinsic efficacy on opioid tolerance. *Anesthesiology* **82**, 1226–1236.

Dworkin, S. I., Guerin, G. F., Goeders, N. E., and Smith, J. E. (1988a). Kainic acid lesions of the nucleus accumbens selectively attenuate morphine self-administration. *Pharmacology Biochemistry and Behavior* **29**, 175–181.

Dworkin, S. I., Guerin, G. F., Co, C., Goeders, N. E., and Smith, J. E. (1988b). Lack of an effect of 6-hydroxydopamine lesions of the nucleus accumbens on intravenous morphine self-administration. *Pharmacology Biochemistry and Behavior* **30**, 1051–1057.

Ellenhorn, M. J., and Barceloux, D. G. (1988). *Medical Toxicology: Diagnosis and Treatment of Human Poisoning*, Elsevier, New York.

Erb, S., Hitchcott, P. K., Rajabi, H., Mueller, D., Shaham, Y., and Stewart, J. (2000). Alpha-2 adrenergic receptor agonists block stress-induced reinstatement of cocaine seeking. *Neuropsychopharmacology* **23**, 138–150.

Ettenberg, A., and McFarland, K. (2003). Effects of haloperidol on cue-induced autonomic and behavioral indices of heroin reward and motivation. *Psychopharmacology* **168**, 139–145.

Ettenberg, A., Pettit, H. O., Bloom, F. E., and Koob, G. F. (1982). Heroin and cocaine intravenous self-administration in rats: Mediation by separate neural systems. *Psychopharmacology* **78**, 204–209.

European Monitoring Centre for Drugs and Drug Addiction. (2003). *Annual Report 2004: The State of the Drugs Problem in the European Union and Norway*, Office for Official Publications of the European Communities, Luxembourg.

Evans, C. J., Keith, D. E. Jr., Morrison, H., Magendzo, K., and Edwards, R. H. (1992). Cloning of a delta opioid receptor by functional expression. *Science* **258**, 1952–1955.

Fay, W. P. (1975). *The Opium War, 1840-1842: Barbarians in the Celestial Empire in the Early Part of the Nineteenth Century and the War by which They Forced Her Gates Ajar*, University of North Carolina Press, Chapel Hill NC.

Fernandes, M., Kluwe, S., and Coper, H. (1977). Quantitative assessment of tolerance to and dependence on morphine in mice, *Naunyn Schmiedebergs Archives of Pharmacology* **297**, 53–60.

Fields, H. L., Emson, P. C., Leigh, B. K., Gilbert, R. F., and Iversen, L. L. (1980). Multiple opiate receptor sites on primary afferent fibres. *Nature* **284**, 351–353.

Foley, K. M. (1993). Opioid analgesics in clinical pain management. *In Opioids II* (series title: *Handbook of Experimental Pharmacology*, vol. 104/2), (M. W. Adler, A. Herz, H. Akil, E. J. Simon, Eds.), pp. 697–743, Springer-Verlag, Berlin.

Frances, B., Gout, R., Monsarrat, B., Cros, J., and Zajac, J. M. (1992). Further evidence that morphine-6 beta-glucuronide is a more potent opioid agonist than morphine. *Journal of Pharmacology and Experimental Therapeutics* **262**, 25–31.

Franken, I. H., Stam, C. J., Hendriks, V. M., and van den Brink, W. (2003). Neurophysiological evidence for abnormal cognitive processing of drug cues in heroin dependence. *Psychopharmacology* **170**, 205–212.

Frenois, F., Cador, M., Caille, S., Stinus, L., and Le Moine, C. (2002). Neural correlates of the motivational and somatic components of naloxone-precipitated morphine withdrawal. *European Journal of Neuroscience* **16**, 1377–1389.

Galligan, J. J., and Burks, T. F. (1982). Opioid peptides inhibit intestinal transit in the rat by a central mechanism. *European Journal of Pharmacology* **85**, 61–68.

Gaudriault, G., Nouel, D., Dal Farra, C., Beaudet, A., and Vincent, J. P. (1997). Receptor-induced internalization of selective peptidic mu and delta opioid ligands. *Journal of Biological Chemistry* **272**, 2880–2888.

Gaveriaux-Ruff, C., and Kieffer, B. L. (2002). Opioid receptor genes inactivated in mice: the highlights. *Neuropeptides* **36**, 62–71.

Gellert, V. F., and Sparber, S. B. (1977). A comparison of the effects of naloxone upon body weight loss and suppression of fixed-ratio operant behavior in morphine-dependent rats. *Journal of Pharmacology and Experimental Therapeutics* **201**, 44–54.

Gerrits, M. A., Ramsey, N. F., Wolterink, G., and van Ree, J. M. (1994). Lack of evidence for an involvement of nucleus accumbens dopamine D1 receptors in the initiation of heroin self-administration in the rat. *Psychopharmacology* **114**, 486–494.

Glaser, F. B. (1974). Psychologic vs. pharmacologic heroin dependence. *New England Journal of Medicine* **290**, 231.

Glick, S. D., and Cox, R. D. (1975). Dopaminergic and cholinergic influences on morphine self-administration in rats. *Research Communications in Chemical Pathology and Pharmacology* **12**, 17–24.

Goeders, N. E., Lane, J. D., and Smith, J. E. (1984). Self-administration of methionine enkephalin into the nucleus accumbens. *Pharmacology Biochemistry and Behavior* **20**, 451–455.

Goldberg, S. R., Woods, J. H., and Schuster, C. R. (1971). Nalorphine-induced changes in morphine self-administration in rhesus monkeys. *Journal of Pharmacology and Experimental Therapeutics* **176**, 464–471.

Goldstein, A., Lowney, L. I., and Pal, B. K. (1971). Stereospecific and nonspecific interactions of the morphine congener levorphanol in subcellular fractions of mouse brain. *Proceedings of the National Academy of Sciences USA* **68**, 1742–1747.

Goldstein, A., Tachibana, S., Lowney, L. I., Hunkapiller, M., and Hood, L. (1979). Dynorphin-(1-13), an extraordinarily potent opioid peptide. *Proceedings of the National Academy of Sciences USA* **76**, 6666–6670.

Gracy, K. N., Dankiewicz, L. A., and Koob, G. F. (2001). Opiate withdrawal-induced Fos immunoreactivity in the rat extended amygdala parallels the development of conditioned place aversion. *Neuropsychopharmacology* **24**, 152–160.

Grisel, J. E., Wiertelak, E. P., Watkins, L. R., and Maier, S. F. (1994). Route of morphine administration modulates conditioned analgesic tolerance and hyperalgesia. *Pharmacology Biochemistry and Behavior* **49**, 1029–1035.

Grudt, T. J., and Williams, J. T. (1995). Opioid receptors and the regulation of ion conductances. *Reviews in the Neurosciences* **6**, 279–286.

Guignard, B., Bossard, A. E., Coste, C., Sessler, D. I., Lebrault, C., Alfonsi, P., Fletcher, D., and Chauvin, M. (2000). Acute opioid tolerance: intraoperative remifentanil increases postoperative pain and morphine requirement. *Anesthesiology* **93**, 409–417.

Guitart, X., Thompson, M. A., Mirante, C. K., Greenberg, M. E., and Nestler, E. J. (1992). Regulation of cyclic AMP response element-binding protein (CREB) phosphorylation by acute and chronic morphine in the rat locus coeruleus. *Journal of Neurochemistry* **58**, 1168–1171.

Gutstein, H. B., Akil, H. (2001). Opioid analgesics. *In Goodman and Gilman's The Pharmacological Basis of Therapeutics*, 10th ed., (J. G. Hardman, L. E. Limbird, and A. Goodman-Gilman, Eds.), pp. 569–619, McGraw-Hill, New York.

Haertzen, C. A. (1970). Subjective effects of narcotic antagonists cyclazocine and nalorphine on the Addiction Research Center Inventory (ARCI). *Psychopharmacologia* **18**, 366–377.

Haertzen, C. A., and Hooks, N. T. Jr. (1969). Changes in personality and subjective experience associated with the chronic administration and withdrawal of opiates. *Journal of Nervous and Mental Disease* **148**, 606–614.

Hanna, M. H., Elliott, K. M., and Fung, M. (2005). Randomized, double-blind study of the analgesic efficacy of morphine-6-glucuronide versus morphine sulfate for postoperative pain in major surgery. *Anesthesiology* **102**, 815–821.

Hand, T. H., Koob, G. F., Stinus, L., and Le Moal, M. (1988). Aversive properties of opiate receptor blockade: Evidence for exclusively central mediation in naive and morphine-dependent rats. *Brain Research* **474**, 364–368.

Harding, W. M., and Zinberg, N. E. (1977). The effectiveness of the subculture in developing rituals and social sanctions for controlled use. *In Drugs, Rituals and Altered States of Consciousness*, (B. M. Du Toit, Ed.), pp. 111–133. A.A. Balkema, Rotterdam.

Harding, W. M., and Zinberg, N. E. (1983). Occasional opiate use. *Advances in Substance Abuse* **3**, 27–61.

Harris, G. C., and Williams, J. T. (1991). Transient homologous mu-opioid receptor desensitization in rat locus coeruleus neurons. *Journal of Neuroscience* **11**, 2574–2581.

Harris, G. C., and Aston-Jones, G. (1994). Involvement of D2 dopamine receptors in the nucleus accumbens in the opiate withdrawal syndrome. *Nature* **371**, 155–157.

Harrison, L. M., Kastin, A. J., and Zadina, J. E. (1998). Opiate tolerance and dependence: receptors, G-proteins, and antiopiates. *Peptides* **19**, 1603–1630.

Hayward, M. D., Pintar, J. E., and Low, M. J. (2002). Selective reward deficit in mice lacking beta-endorphin and enkephalin. *Journal of Neuroscience* **22**, 8251–8258.

Heilig, M., McLeod, S., Brot, M., Heinrichs, S. C., Menzaghi, F., Koob, G. F., and Britton, K. T. (1993). Anxiolytic-like action of neuropeptide Y: mediation by Y1 receptors in amygdala, and dissociation from food intake effects. *Neuropsychopharmacology* **8**, 357–363.

Heinrichs, S. C., Menzaghi, F., Schulteis, G., Koob, G. F., and Stinus, L. (1995). Suppression of corticotropin-releasing factor in the amygdala attenuates aversive consequences of morphine withdrawal. *Behavioural Pharmacology* **6**, 74–80.

Heishman, S. J., Stitzer, M. L., Bigelow, G. E., and Liebson, I. A. (1989). Acute opioid physical dependence in humans: effect of varying the morphine-naloxone interval: I., *Journal of Pharmacology and Experimental Therapeutics* **250**, 485–491.

Hemby, S. E., Smith, J. E., and Dworkin, S. I. (1996). The effects of eticlopride and naltrexone on responding maintained by food, cocaine, heroin and cocaine/heroin combinations in rats. *Journal of Pharmacology and Experimental Therapeutics* **277**, 1247–1258 [erratum in: *Journal of Pharmacology and Experimental Therapeutics*, 1996, **279**, 442].

Highfield, D., Clements, A., Shalev, U., McDonald, R., Featherstone, R., Stewart, J., and Shaham, Y. (2000). Involvement of the medial septum in stress-induced relapse to heroin seeking in rats. *European Journal of Neuroscience* **12**, 1705–1713.

Highfield, D., Yap, J., Grimm, J. W., Shalev, U., and Shaham, Y. (2001). Repeated lofexidine treatment attenuates stress-induced, but not drug cues-induced reinstatement of a heroin-cocaine mixture (speedball) seeking in rats. *Neuropsychopharmacology* **25**, 320–331.

Hill, H. F., Mackie, A. M., Coda, B. A., Iverson, K., and Chapman, C. R. (1991). Patient-controlled analgesic administration. A comparison of steady-state morphine infusions with bolus doses. *Cancer* **67**, 873–882.

Hiller, J. M., Pearson, J., and Simon, E. J. (1973). Distribution of stereospecific binding of the potent narcotic analgesic etorphine in the human brain: predominance in the limbic system. *Research Communications in Chemical Pathology and Pharmacology* **6**, 1052–1062.

Himmelsbach, C. K. (1942). Clinical studies of drug addiction: Physical dependence, withdrawal and recovery. *Archives of Internal Medicine* **69**, 766–772.

Himmelsbach, C. K. (1943). Can the euphoric, analgetic, and physical dependence effects of drugs be separated? IV With reference to physical dependence. *Federation Proceedings* **2**, 201–203.

Hoffmann, O., and Wiesenfeld-Hallin, Z. (1994). The CCK-B receptor antagonist Cl 988 reverses tolerance to morphine in rats. *Neuroreport* **5**, 2565–2568.

Hood, D. D., Curry, R., and Eisenach, J. C. (2003). Intravenous remifentanil produces withdrawal hyperalgesia in volunteers with capsaicin-induced hyperalgesia. *Anesthesia and Analgesia* **97**, 810–815.

Houshyar, H., Cooper, Z. D., and Woods, J. H. (2001a) Paradoxical effects of chronic morphine treatment on the temperature and pituitary-adrenal responses to acute restraint stress: a chronic stress paradigm. *Journal of Neuroendocrinology* **13**, 862–874.

Houshyar, H., Galigniana, M. D., Pratt, W. B., and Woods, J. H. (2001b). Differential responsivity of the hypothalamic-pituitary-adrenal axis to glucocorticoid negative-feedback and corticotropin releasing hormone in rats undergoing morphine withdrawal: possible mechanisms involved in facilitated and attenuated stress responses. *Journal of Neuroendocrinology* **13**, 875–886.

Hser, Y. I., Hoffman, V., Grella, C. E., and Anglin, M. D. (2001). A 33-year follow-up of narcotics addicts. *Archives of General Psychiatry* **58**, 503–508.

Huang, M., Wang, D. X., and Qin, B. Y. (1994). Dihydroetorphine, a potent opioid with low dependent potential, *Regulatory Peptides* **53**(Suppl. 1), s81–s82.

Hubner, C. B., and Koob, G. F. (1990). The ventral pallidum plays a role in mediating cocaine and heroin self-administration in the rat. *Brain Research* **508**, 20–29.

Hugues, F. C., Munera, Y., and Le Jeunne, C. (1992). Drug induced orthostatic hypotension [French], *Revue de Medecine Interne* **13**, 465–470.

Hunt, L. G. (1979). Growth of substance use and misuse: some speculations and data. *Journal of Drug Issues* **9**, 257–265.

Hutcheson, D. M., Parkinson, J. A., Robbins, T. W., and Everitt, B. J. (2001). The effects of nucleus accumbens core and shell lesions on intravenous heroin self-administration and the acquisition of drug-seeking behaviour under a second-order schedule of heroin reinforcement. *Psychopharmacology* **153**, 464–472.

Inciardi, J. A. (1986). *The War on Drugs: Heroin, Cocaine, Crime, and Public Policy*, Mayfield, Palo Alto CA.

Inturrisi, C. E., Max, M. B., Foley, K. M., Schultz, M., Shin, S. U., and Houde, R. W. (1984). The pharmacokinetics of heroin in patients with chronic pain. *New England Journal of Medicine* **310**, 1213–1217.

Inturrisi, C. E., Schultz, M., Shin, S., Umans, J. G., Angel, L., and Simon, E. J. (1983). Evidence from opiate binding studies that heroin acts through its metabolites. *Life Sciences* **33**(Suppl. 1), 773–776.

Jaffe, J. H. (1990). Drug addiction and drug abuse. *In Goodman and Gilman's The Pharmacological Basis of Therapeutics*, 8th ed., (A. Goodman Gilman, L. S. Goodman, and T. W. Rall, Eds.), pp. 522–573. MacMillan, New York.

Jaffe, J. H., and Martin, W. R. (1990). Opioid analgesics and antagonists. *In Goodman and Gilman's The Pharmacological Basis of Therapeutics*, 8th ed., (A. Goodman Gilman, T. W. Rall, A. S. Nies, and P. Taylor P, Eds.), pp. 485–521. Pergmon Press, New York.

Jaffe, R. A., and Rowe, M. A. (1996). A comparison of the local anesthetic effects of meperidine, fentanyl, and sufentanil on dorsal root axons. *Anesthesia and Analgesia* **83**, 776–781.

Jenck, F., Ouagazzal, A. M., Pauly-Evers, M., and Moreau, J. L. (2000). OrphaninFQ: role in behavioral fear responses and vulnerability to stress?. *Molecular Psychiatry* **5**, 572–574.

Jenkins, A. J., Keenan, R. M., Henningfield, J. E., and Cone, E. J. (1994). Pharmacokinetics and pharmacodynamics of smoked heroin. *Journal of Analytical Toxicology* **18**, 317–330.

Johnson, S. W., and North, R. A. (1992). Opioids excite dopamine neurons by hyperpolarization of local interneurons. *Journal of Neuroscience* **12**, 483–488.

Jones, J. (1700). The Mysteries of Opium Reveal'd, Printed for Richard Smith, London.

Kalivas, P. W., Widerlov, E., Stanley, D., Breese, G., and Prange, A. J. Jr. (1983). Enkephalin action on the mesolimbic system: a dopamine-dependent and a dopamine-independent increase in locomotor activity. *Journal of Pharmacology and Experimental Therapeutics* **227**, 229–237.

Kaneko, T., Nakazawa, T., Ikeda, M., Yamatsu, K., Iwama, T., Wada, T., Satoh, M., and Takagi, H. (1983). Sites of analgesic action of dynorphin. *Life Sciences* **33** (Suppl. 1), 661–664.

Kayan, S., Woods, L. A., and Mitchell, C. L. (1971). Morphine-induced hyperalgesia in rats tested on the hot plate. *Journal of Pharmacology and Experimental Therapeutics* **177**, 509–513.

Keith, D. E., Murray, S. R., Zaki, P. A., Chu, P. C., Lissin, D. V., Kang L., Evans, C. J., and von Zastrow M. (1996). Morphine activates opioid receptors without causing their rapid internalization. *Journal of Biological Chemistry* **271**, 19021–19024.

Keith, D. E., Anton, B., Murray, S. R., Zaki, P. A., Chu, P. C., Lissin, D. V., Monteillet-Agius, G., Stewart, P. L., Evans, C. J., and von Zastrow, M. (1998). mu-Opioid receptor internalization: opiate drugs have differential effects on a conserved endocytic mechanism in vitro and in the mammalian brain. *Molecular Pharmacology* **53**, 377–384.

Kelsey, J. E., and Arnold, S. R. (1994). Lesions of the dorsomedial amygdala, but not the nucleus accumbens, reduce the aversiveness of morphine withdrawal in rats. *Behavioral Neuroscience* **108**, 1119–1127 [erratum: **109**, 203].

Khantzian, E. J. (1985). The self-medication hypothesis of affective disorders: focus on heroin and cocaine dependence. *American Journal of Psychiatry* **142**, 1259–1264.

Khantzian, E. J. (1990). Self-regulation and self-medication factors in alcoholism and the addictions: similarities and differences. *In Combined Alcohol and Other Drug Dependence* (series title: *Recent Developments in Alcoholism*, vol. 8), (M. Galanter, Ed.), pp. 255–271. Plenum Press, New York.

Khantzian, E. J. (1997). The self-medication hypothesis of substance use disorders: a reconsideration and recent applications. *Harvard Review of Psychiatry* **4**, 231–244.

Khantzian, E. J., Mack, J. E., and Schatzberg, A. F. (1974). Heroin use as an attempt to cope: clinical observations. *American Journal of Psychiatry* **131**, 160–164.

Kieffer, B. L., Befort, K., Gaveriaux-Ruff, C., and Hirth, C. G. (1992). The delta-opioid receptor: isolation of a cDNA by expression cloning and pharmacological characterization. *Proceedings of the National Academy of Sciences USA* **89**, 12048–12052 [erratum in: *Proceedings of the National Academy of Sciences USA*, 1994, **91**, 1193].

Kieffer, B. L., Gaveriaux-Ruff, C. (2002). Exploring the opioid system by gene knockout. *Progress in Neurobiology* **66**, 285–306.

Kim, J. A., and Siegel, S. (2001). The role of cholecystokinin in conditional compensatory responding and morphine tolerance in rats. *Behavioral Neuroscience* **115**, 704–709.

Kleber, H. (1981). Detoxification from narcotics. *In Substance Abuse: Clinical Problems and Perspectives*, (J. H. Lowinson, P. Ruiz, Eds.), Williams and Wilkins, Baltimore 317–338.

Koob, G. F., (1987). Neural substrates of opioid tolerance and dependence. *In: Problems of Drug Dependence 1986: Proceedings of the 48th Annual Scientific Meeting, The Committee on Problems of Drug Dependence, Inc.* (series title: *NIDA Research Monograph*, vol. 76), (L. S. Harris, Ed.) pp. 46–52. National Institute on Drug Abuse, Rockville, MD.

Koob, G. F. (1992). Drugs of abuse: anatomy, pharmacology, and function of reward pathways. *Trends in Pharmacological Sciences* **13**, 177–184.

Koob, G. F., and Bloom, F. E. (1988). Cellular and molecular mechanisms of drug dependence. *Science* **242**, 715–723.

Koob, G. F., Pettit, H. O., Ettenberg, A., and Bloom, F. E. (1984). Effects of opiate antagonists and their quaternary derivatives on heroin self-administration in the rat. *Journal of Pharmacology and Experimental Therapeutics* **229**, 481–486.

Koob, G. F., Stinus, L., Le Moal, M., and Bloom, F. E. (1989a). Opponent process theory of motivation: neurobiological evidence from studies of opiate dependence. *Neuroscience and Biobehavioral Reviews* **13**, 135–140.

Koob, G. F., Wall, T. L., and Bloom, F. E. (1989b). Nucleus accumbens as a substrate for the aversive stimulus effects of opiate withdrawal. *Psychopharmacology* **98**, 530–534.

Koob, G. F., Sanna, P. P., and Bloom, F. E. (1998). Neuroscience of addiction. *Neuron* **21**, 467–476.

Koppert, W., Sittl, R., Scheuber, K., Alsheimer, M., Schmelz, M., and Schuttler, J. (2003). Differential modulation of remifentanil-induced analgesia and postinfusion hyperalgesia by S-ketamine and clonidine in humans. *Anesthesiology* **99**, 152–159.

Koster, A., Montkowski, A., Schulz, S., Stube, E. M., Knaudt, K., Jenck, F., Moreau, J. L., Nothacker, H. P., Civelli, O., and Reinscheid, R. K. (1999). Targeted disruption of the orphanin FQ/nociceptin gene increases stress susceptibility and impairs stress adaptation in mice. *Proceedings of the National Academy of Sciences USA* **96**, 10444–10449.

Krank, M. D., Hinson, R. E., and Siegel, S. (1981). Conditional hyperalgesia is elicited by environmental signals of morphine. *Behavioral and Neural Biology* **32**, 148–157.

Kreek, M. J. (1996a). Opiates, opioids and addiction. *Molecular Psychiatry* 1, 232–254.

Kreek, M. J. (1996b). Opioid receptors: some perspectives from early studies of their role in normal physiology, stress responsivity, and in specific addictive diseases. *Neurochemical Research* **21**, 1469–1488.

Kreek, M. J., Wardlaw, S. L., Hartman, N., Raghunath, J., Friedman, J., Schneider, B., and Frantz, A. G. (1983). Circadian rhythms and levels of beta-endorphin, ACTH, and cortisol during chronic methadone maintenance treatment in humans. *Life Sciences* **33** (Suppl. 1), 409–411.

Kreek, M. J., Ragunath, J., Plevy, S., Hamer, D., Schneider, B., and Hartman, N. (1984). ACTH, cortisol and beta-endorphin response to metyrapone testing during chronic methadone maintenance treatment in humans. *Neuropeptides* **5**, 277–278.

Kuhar, M. J., Pert, C. B., and Snyder, S. H. (1973). Regional distribution of opiate receptor binding in monkey and human brain. *Nature* **245**, 447–450.

Larcher, A., Laulin, J. P., Celerier, E., Le Moal, M., and Simonnet, G. (1998). Acute tolerance associated with a single opiate administration: Involvement of N-methyl-D-aspartate-dependent pain facilitatory systems. *Neuroscience* **84**, 583–589.

Latham, R. G. (1848). *The Works of Thomas Sydenham*, Syndenham Society, London.

Laulin, J. P., Larcher, A., Celerier, E., Le Moal, M., and Simonnet, G. (1998). Long-lasting increased pain sensitivity in rat following exposure to heroin for the first time. *European Journal of Neuroscience* **10**, 782–785.

Laulin, J. P., Celerier, E., Larcher, A., Le Moal, M., and Simonnet, G. (1999). Opiate tolerance to daily heroin administration: An apparent phenomenon associated with enhanced pain sensitivity. *Neuroscience* **89**, 631–636.

Law, P. Y., Hom, D. S., and Loh, H. H. (1983). Opiate receptor down-regulation and desensitization in neuroblastoma × glioma NG108-15 hybrid cells are two separate cellular adaptation processes. *Molecular Pharmacology* **24**, 413–424.

Law, P. Y., Wong, Y. H., and Loh, H. H. (2000). Molecular mechanisms and regulation of opioid receptor signaling. *Annual Review of Pharmacology and Toxicology* **40**, 389–430.

Le Guen, S., and Gestreau, C. (2001). Besson, J. M., Sensitivity to naloxone of the behavioral signs of morphine withdrawal and c-Fos expression in the rat CNS: a quantitative dose-response analysis. *Journal of Comparative Neurology* **433**, 272–296.

Legon, S., Glover, D. M., Hughes, J., Lowry, P. J., Rigby, P. W., and Watson, C. J. (1982). The structure and expression of the preproenkephalin gene. *Nucleic Acids Research* **10**, 7905–7918.

Leone, P., and Di Chiara, G. (1987). Blockade of D-1 receptors by SCH 23390 antagonizes morphine- and amphetamine-induced place preference conditioning. *European Journal of Pharmacology* **135**, 251–254.

Levine, D. G. (1974). 'Needle freaks': compulsive self-injection by drug users. *American Journal of Psychiatry* **131**, 297–300.

Li, C. H., Lemaire, S., Yamashiro, D., and Doneen, B. A. (1976). The synthesis and opiate activity of beta-endorphin. *Biochemical and Biophysical Research Communications* **71**, 19–25.

Li, X., Angst, M. S., and Clark, J. D. (2001). Opioid-induced hyperalgesia and incisional pain. *Anesthesia and Analgesia* **93**, 204–209.

Light, A. B., and Torrance, E. G. (1929). Opium addiction: VI. The effects of abrupt withdrawal followed by readministration of morphine in human addicts, with special reference to the composition of the blood, the circulation and the metabolism. *Archives of Internal Medicine* **44**, 1–16.

Littrell, R. A. (1991). Epidural analgesia. *American Journal of Hospital Pharmacy* **48**, 2460–2474.

Liu, J. G., and Anand, K. J. (2001). Protein kinases modulate the cellular adaptations associated with opioid tolerance and dependence. *Brain Research Reviews* **38**, 1–19.

Liu, J., and Schulteis, G. (2004). Brain reward deficits accompany naloxone-precipitated withdrawal from acute opioid dependence. *Pharmacology Biochemistry and Behavior* **79**, 101–108.

Liu, J. G., Liao, X. P., Gong, Z. H., and Qin, B. Y. (1999a). The difference between methadone and morphine in regulation of delta-opioid receptors underlies the antagonistic effect of methadone on morphine-mediated cellular actions. *European Journal of Pharmacology* **373**, 233–239.

Liu, J. G., Liao, X. P., Gong, Z. H., and Qin, B. Y.(1999b). Methadone-induced desensitization of the delta-opioid receptor is mediated by uncoupling of receptor from G protein. *European Journal of Pharmacology* **374**, 301–308.

Longoni, R., Cadoni, C., Mulas, A., Di Chiara, G., and Spina, L. (1998). Dopamine-dependent behavioural stimulation by nonpeptide delta opioids BW373U86 and SNC 80, 2. Place-preference and brain microdialysis studies in rats. *Behavioural Pharmacology* **9**, 9–14.

Lord, J. A., Waterfield, A. A., Hughes, J., and Kosterlitz, H. W. (1977). Endogenous opioid peptides: multiple agonists and receptors. *Nature* **267**, 495–499.

Lu, L., Ceng, X., and Huang, M. (2000). Corticotropin-releasing factor receptor type 1 mediates stress-induced relapse to opiate dependence in rats. *Neuroreport* **11**, 2373–2378.

Luginbuhl, M., Gerber, A., Schnider, T. W., Petersen-Felix, S., Arendt-Nielsen, L., and Curatolo, M. (2003). Modulation of remifentanil-induced analgesia, hyperalgesia, and tolerance by small-dose ketamine in humans. *Anesthesia and Analgesia* **96**, 726–732.

Lyness, W. H., Friedle, N. M., and Moore, K. E. (1979). Destruction of dopaminergic nerve terminals in nucleus accumbens: Effect on d-amphetamine self-administration. *Pharmacology Biochemistry and Behavior* **11**, 553–556.

Macht, D. I. (1915). The history of opium and some of its preparations and alkaloids. *Journal of the American Medical Association* **64**, 477–481.

Macintyre, P. E. (2001). Safety and efficacy of patient-controlled analgesia. *British Journal of Anaesthesia* **87**, 36–46.

Madison, D. V., and Nicoll, R. A. (1988). Enkephalin hyperpolarizes interneurones in the rat hippocampus *Journal of Physiology* **398**, 123–130.

Maldonado, R., Stinus, L., Gold, L. H., and Koob, G. F. (1992). Role of different brain structures in the expression of the physical morphine withdrawal syndrome. *Journal of Pharmacology and Experimental Therapeutics*. **261**, 669–677.

Maldonado, R., Blendy, J. A., Tzavara, E., Gass, P., Roques, B. P., Hanoune, J., and Schutz, G. (1996). Reduction of morphine abstinence in mice with a mutation in the gene encoding CREB. *Science* **273**, 657–659.

Maldonado, R., Saiardi, A., Valverde, O., Samad, T. A., Roques, B. P., and Borrelli, E. (1997). Absence of opiate rewarding effects in mice lacking dopamine D-2 receptors. *Nature* **388**, 586–589.

Malick, J. B., and Goldstein, J. M. (1977). Analgesic activity of enkephalins following intracerebral administration in the rat. *Life Sciences* **20**, 827–832.

Manara, L., and Bianchetti, A. (1985). The central and peripheral influences of opioids on gastrointestinal propulsion. *Annual Review of Pharmacology and Toxicology* **25**, 249–273.

Mansour, A., Khachaturian, H., Lewis, M. E., Akil, H., and Watson, S. J., (1988). Anatomy of CNS opioid receptors. *Trends in Neurosciences* **11**, 308–314.

Mansour, A., Thompson, R. C., Akil, H., and Watson, S. J. (1993). Delta opioid receptor mRNA distribution in the brain: comparison to delta receptor binding and proenkephalin mRNA. *Journal of Chemical Neuroanatomy* **6**, 351–362.

Mansour, A., Fox, C. A., Burke, S., Meng, F., Thompson, R. C., Akil, H., and Watson, S. J. (1994a). Mu, delta, and kappa opioid receptor mRNA expression in the rat CNS: an in situ hybridization study. *Journal of Comparative Neurology* **350**, 412–438.

Mansour, A., Fox, C. A., Meng, F., Akil, H., and Watson, S. J. (1994b). Kappa 1 receptor mRNA distribution in the rat CNS: comparison to kappa receptor binding and prodynorphin mRNA. *Molecular and Cellular Neuroscience* **5**, 124–144.

Mansour, A., Fox, C. A., Thompson, R. C., Akil, H., and Watson, S. J. (1994c). mu-Opioid receptor mRNA expression in the rat CNS: comparison to mu-receptor binding. *Brain Research* **643**, 245–265.

Mansour, A., Fox, C. A., Akil, H., and Watson, S. J. (1995). Opioid-receptor mRNA expression in the rat CNS: anatomical and functional implications. *Trends in Neurosciences* **18**, 22–29.

Manzoni, O. J., and Williams, J. T. (1999). Presynaptic regulation of glutamate release in the ventral tegmental area during morphine withdrawal. *Journal of Neuroscience* **19**, 6629–6636.

Mao, J., Price, D. D., and Mayer, D. J. (1994). Thermal hyperalgesia in association with the development of morphine tolerance in rats: roles of excitatory amino acid receptors and protein kinase C. *Journal of Neuroscience* **14**, 2301–2312.

Mao, J., Sung, B., Ji, R. R., and Lim, G. (2002). Chronic morphine induces downregulation of spinal glutamate transporters: implications in morphine tolerance and abnormal pain sensitivity. *Journal of Neuroscience* **22**, 8312–8323.

Marinelli, M., Aouizerate, B., Barrot, M., Le Moal, M., and Piazza, P. V. (1998). Dopamine-dependent responses to morphine depend on glucocorticoid receptors. *Proceedings of the National Academy of Sciences USA* **95**, 7742–7747.

Marsch, L. A., Bickel, W. K., Badger, G. J., Rathmell, J. P., Swedberg, M. D., Jonzon, B., and Norsten-Hoog C. (2001). Effects of infusion rate of intravenously administered morphine on physiological, psychomotor, and self-reported measures in humans. *Journal of Pharmacology and Experimental Therapeutics* **299**, 1056–1065.

Martin, G., Nie, Z., and Siggins, G. R. (1997). mu-Opioid receptors modulate NMDA receptor-mediated responses in nucleus accumbens neurons *Journal of Neuroscience* **17**, 11–22.

Martin, G., Guadano-Ferraz, A., Morte, B., Ahmed, S., Koob, G. F., de Lecea, L., and Siggins, G. R. (2004). Chronic morphine treatment alters *N*-methyl-D-aspartate receptors in freshly isolated neurons from nucleus accumbens **311**, 265–273.

Martin, W. R. (1967). Opioid antagonists. *Pharmacological Reviews* **19**, 463–521.

Martin, W. R. (1983). Pharmacology of opioids. *Pharmacological Reviews* **35**, 283–323.

Martin, W. R., and Jasinski, D. R. (1969). Physiological parameters of morphine dependence in man—tolerance, early abstinence, protracted abstinence. *Journal of Psychiatric Research* **7**, 9–17.

Martin, W. R., Sloan, J. W., Sapira, J. D., and Jasinski, D. R., (1971). Physiologic, subjective, and behavioral effects of amphetamine, methamphetamine, ephedrine, phenmetrazine, and methylphenidate in man. *Clinical Pharmacology and Therapeutics* **12**, 245–258.

Martin, W. R., Jasinski, D. R., Haertzen, C. A., Kay, D. C., Jones, B. E., Mansky, P. A., and Carpenter, R. W. (1973). Methadone: a reevaluation. *Archives of General Psychiatry* **28**, 286–295.

Matthes, H. W. D., Maldonado, R., Simonin, F., Valverde, O., Slowe, S., Kitchen, I., Befort, K., Dierich, A., Le Meur, M. O., Dolle, P., Tzavara, E., Hanoune, J., Roques, B. P., and Kieffer, B. L. (1996). Loss of morphine-induced analgesia, reward effect and withdrawal symptoms in mice lacking the mu-opioid-receptor gene. *Nature* **383**, 819–823.

Mayer, D. J., Akil, H., and Liebeskind, J. C. (1971). Behavioral studies of analgesia resulting from electrical stimulation of the brain. *Society for Neuroscience Abstracts* **1**, 139.

McBride, W. J., Murphy, J. M., and Ikemoto, S. (1999). Localization of brain reinforcement mechanisms: intracranial self-administration and intracranial place-conditioning studies. *Behavioural Brain Research* **101**, 129–152.

McFarland, K., and Ettenberg, A. (1995). Haloperidol differentially affects reinforcement and motivational processes in rats running an alley for intravenous heroin. *Psychopharmacology* **122**, 346–350.

McFarland, K., and Ettenberg, A. (1997). Reinstatement of drug-seeking behavior produced by heroin-predictive environmental stimuli. *Psychopharmacology* **131**, 86–92.

McFarland, K., and Ettenberg, A. (1998). Naloxone blocks reinforcement but not motivation in an operant runway model of heroin-seeking behavior. *Experimental and Clinical Psychopharmacology* **6**, 353–359.

McFarland, K., and Ettenberg, A. (1999). Haloperidol does not attenuate conditioned place preferences or locomotor activation produced by food- or heroin-predictive discriminative cues. *Pharmacology Biochemistry and Behavior* **62**, 631–641.

McNally, G. P. (1999). Pain facilitatory circuits in the mammalian central nervous system: their behavioral significance and role in morphine analgesic tolerance. *Neuroscience and Biobehavioral Reviews* **23**, 1059–1078.

McQuay, H. J. (1988). Pharmacological treatment of neuralgic and neuropathic pain. *Cancer Surveys* **7**, 141–159.

Medical Economics Company. (2004). *Physicians' Desk Reference*, 58th ed., Medical Economics Company, Oradell NJ.

Meis, S., and Pape, H. C. (1998). Postsynaptic mechanisms underlying responsiveness of amygdaloid neurons to nociceptin/orphanin FQ. *Journal of Neuroscience* **18**, 8133–8144.

Meis, S., and Pape, H. C. (2001). Control of glutamate and GABA release by nociceptin/orphanin FQ in the rat lateral amygdala. *Journal of Physiology* **532**, 701–712.

Mestek, A., Hurley, J. H., Bye, L. S., Campbell, A. D., Chen, Y., Tian, M., Liu, J., Schulman, H., and Yu, L. (1995). The human mu opioid receptor: modulation of functional desensitization by calcium/calmodulin-dependent protein kinase and protein kinase C. *Journal of Neuroscience* **15**, 2396–2406.

Meunier, J. C., Mollereau, C., Toll, L., Suaudeau, C., Moisand, C., Alvinerie, P., Butour, J. L., Guillemot, J. C., Ferrara, P., Monsarrat, B., Mazarguil, H., Vassart, G., Parmentier, M., and Costentin, J. (1995). Isolation and structure of the endogenous agonist of opioid receptor-like ORL1 receptor. *Nature* **377**, 532–535.

Meyer, R. E., and Mirin, S. M. (1979). *The Heroin Stimulus: Implications for a Theory of Addiction*, Plenum, New York.

Mirin, S. M., Meyer, R. E., and McNamee, B. (1976). Psychopathology and mood during heroin use *Archives of General Psychiatry* **33**, 1503–1508.

Mogil, J. S., and Pasternak, G. W. (2001). The molecular and behavioral pharmacology of the orphanin FQ/nociceptin peptide and receptor family. *Pharmacological Reviews* **53**, 381–415.

Mollereau, C., Parmentier, M., Mailleux, P., Butour, J. L., Moisand, C., Chalon, P., Caput, D., Vassart, G., and Meunier, J. C. (1994). ORL1, a novel member of the opioid receptor family. Cloning, functional expression and localization. *FEBS Letters* **341**, 33–38.

Moulin, D. E., Ling, G. S., and Pasternak, G. W. (1988). Unidirectional analgesic cross-tolerance between morphine and levorphanol in the rat. *Pain* **33**, 233–239.

Mucha, R. F., van der Kooy, D., O'Shaughnessy, M., and Bucenieks, P. (1982). Drug reinforcement studied by the use of place conditioning in rat. *Brain Research* **243**, 91–105.

Murphy, N. P., and Ly, H. T. (1996). Maidment, N. T., Intracerebroventricular orphanin FQ/nociceptin suppresses dopamine release in the nucleus accumbens of anaesthetized rats. *Neuroscience* **75**, 1–4.

Nader, K., and van der Kooy, D. (1997). Deprivation state switches the neurobiological substrates mediating opiate reward in the ventral tegmental area. *Journal of Neuroscience* **17**, 383–390.

Narita, M., Narita, M., Mizoguchi, H., and Tseng, L. F. (1995). Inhibition of protein kinase C, but not of protein kinase A, blocks the development of acute antinociceptive tolerance to an intrathecally administered mu-opioid receptor agonist in the mouse. *European Journal of Pharmacology* **280**, R1–R3.

Nemeroff, C. B., Osbahr, A. J. 3rd, Manberg, P. J., Ervin, G. N., and Prange, A. J. Jr. (1979). Alterations in nociception and body temperature after intracisternal administration of neurotensin, beta-endorphin, other endogenous peptides, and morphine. *Proceedings of the National Academy of Sciences USA* **76**, 5368–5371.

Nestler, E. J. (1996). Under siege: The brain on opiates. *Neuron* **16**, 897–900.

Nestler, E. J. (2004). Historical review: Molecular and cellular mechanisms of opiate and cocaine addiction. *Trends in Pharmacological Sciences* **25**, 210–218.

Nestler, E. J., and Tallman, J. F. (1988). Chronic morphine treatment increases cyclic AMP-dependent protein kinase activity in the rat locus coeruleus. *Molecular Pharmacology* **33**, 127–132.

Nestler, E. J., Hope, B. T., and Widnell, K. L. (1993). Drug addiction: A model for the molecular basis of neural plasticity. *Neuron* **11**, 995–1006.

Nestler, E. J., Barrot, M., and Self, D. W. (2001). DeltaFosB: a sustained molecular switch for addiction. *Proceedings of the National Academy of Sciences USA* **98**, 11042–11046.

Nestler, E. J., and Malenka, R. C. (2004). The addicted brain, *Scientific American* **290**, 78–85.

Nieto, M. M., Wilson, J., Cupo, A., Roques, B. P., and Noble, F. (2002). Chronic morphine treatment modulates the extracellular levels of endogenous enkephalins in rat brain structures involved in opiate dependence: a microdialysis study. *Journal of Neuroscience* **22**, 1034–1041.

Noble, F., and Cox, B. M. (1996). Differential desensitization of mu- and delta- opioid receptors in selected neural pathways following chronic morphine treatment. *British Journal of Pharmacology* **117**, 161–169.

Noda, M., Furutani, Y., Takahashi, H., Toyosato, M., Hirose, T., Inayama, S., Nakanishi S., and Numa, S. (1982). Cloning and sequence analysis of cDNA for bovine adrenal preproenkephalin. *Nature* **295**, 202–206.

Nye, H. E., and Nestler, E. J. (1996). Induction of chronic Fos-related antigens in rat brain by chronic morphine administration. *Molecular Pharmacology* **49**, 636–645.

O'Brien, C. P. (1974). 'Needle freaks': Psychological dependence on shooting up. *In Medical World News, Psychiatry Annual*. McGraw Hill, New York.

O'Brien, C. P. (1975). Experimental analysis of conditioning factors in human narcotic addiction. *Pharmacological Review* **27**, 533–543.

O'Brien, C. P., Testa, J., O'Brien, T. J., Brady, J. P., and Wells, B. (1977). conditioned narcotic withdrawal in humans. *Science* **195,** 1000–1002.

O'Brien, C. P., Ehrman, R. N., and Ternes, J. M. (1986). Classical conditioning in human opioid dependence. *In Behavioral Analysis of Drug Dependence*, (S. R. Goldberg, and I. P. Stolerman, Eds.), pp. 329–356. Academic Press, Orlando FL.

Ohmori, S., and Morimoto, Y. (2002). Dihydroetorphine: a potent analgesic: Pharmacology, toxicology, pharmacokinetics, and clinical effects. *CNS Drug Reviews* **8**, 391–404.

Olds, M. E. (1979). Hypothalamic substrate for the positive reinforcing properties of morphine in the rat. *Brain Research* **168**, 351–360.

Olds, M. E. (1982). Reinforcing effects of morphine in the nucleus accumbens. *Brain Research* **237**, 429–440.

Olmstead, M. C., Munn, E. M., Franklin, K. B., and Wise, R. A. (1998). Effects of pedunculopontine tegmental nucleus lesions on responding for intravenous heroin under different schedules of reinforcement. *Journal of Neuroscience* **18**, 5035–5044.

Olmstead, M. C., and Franklin, K. B. (1993). Effects of pedunculopontine tegmental nucleus lesions on morphine-induced conditioned place preference and analgesia in the formalin test. *Neuroscience* **57**, 411–418.

Osborne, R., Joel, S., Trew, D., and Slevin, M. (1988). Analgesic activity of morphine-6-glucuronide. *Lancet* **331**(8589), 828.

Osborne, R., Joel, S., Trew, D., and Slevin, M. (1990). Morphine and metabolite behavior after different routes of morphine administration: demonstration of the importance of the active metabolite morphine-6-glucuronide. *Clinical Pharmacology and Therapeutics* **47**, 12–19.

Ossipov, M. H., Lai, J., Vanderah, T. W., and Porreca, F. (2003). Induction of pain facilitation by sustained opioid exposure: relationship to opioid antinociceptive tolerance. *Life Sciences* **73**, 783–800.

Paul, D., Standifer, K. M., Inturrisi, C. E., and Pasternak, G. W. (1989). Pharmacological characterization of morphine-6 beta-glucuronide, a very potent morphine metabolite. *Journal of Pharmacology and Experimental Therapeutics* **251**, 477–483.

Penson, R. T., Joel, S. P., Bakhshi, K., Clark, S. J., Langford, R. M., and Slevin, M. L. (2000). Randomized placebo-controlled trial of the activity of the morphine glucuronides. *Clinical Pharmacology and Therapeutics* **68**, 667–676.

Pert, C. B., and Snyder, S. H. (1973). Opiate receptor: demonstration in nervous tissue. *Science* **179**, 1011–1014.

Pert, C. B., Kuhar, M. J., and Snyder, S. H. (1976). Opiate receptor: autoradiographic localization in rat brain. *Proceedings of the National Academy of Sciences USA* **73**, 3729–3733.

Pettit, H. O., Ettenberg, A., Bloom, F. E., and Koob, G. F. (1984). Destruction of dopamine in the nucleus accumbens selectively attenuates cocaine but not heroin self-administration in rats. *Psychopharmacology* **84**, 167–173.

Phillips, A. G., and LePiane, F. G. (1982). Reward produced by microinjection of (D-Ala2), Met5-enkephalinamide into the ventral tegmental area. *Behavioural Brain Research* **5**, 225–229.

Pichini, S., Altieri, I., Pellegrini, M., Zuccaro, P., and Pacifici, R. (1999). The role of liquid chromatography-mass spectrometry in the determination of heroin and related opioids in biological fluids. *Mass Spectrometry Reviews* **18**, 119–130.

Pothos, E., Rada, P., Mark, G. P., and Hoebel, B. G. (1991). Dopamine microdialysis in the nucleus accumbens during acute and chronic morphine, naloxone-precipitated withdrawal and clonidine treatment. *Brain Research* **566**, 348–350.

Poulos, C. X., and Cappell, H. (1991). Homeostatic theory of drug tolerance: A general model of physiological adaptation. *Psychological Reviews* **98**, 390–408.

Preston, K. L., Bigelow, G. E., and Liebson, I. A. (1988). Buprenorphine and naloxone alone and in combination in opioid-dependent humans. *Psychopharmacology* **94**, 484–490.

Pu, L., Bao, G. B., Xu, N. J., Ma, L., and Pei, G. (2002). Hippocampal long-term potentiation is reduced by chronic opiate treatment and can be restored by re-exposure to opiates. *Journal of Neuroscience* **22**, 1914–1921.

Puttfarcken, P. S., Werling, L. L., and Cox, B. M. (1988). Effects of chronic morphine exposure on opioid inhibition of adenylyl cyclase in 7315c cell membranes: a useful model for the study of tolerance at mu opioid receptors. *Molecular Pharmacology* **33**, 520–527.

Qin, B. Y., Wang, D. X., and Huang, M. (1994). The application of dihydroetorphine to detoxification of heroin addicts. *Regulatory Peptides* **53**(Suppl. 1), s293–s294.

Ramsay, D. S., and Woods, S. C. (1997). Biological consequences of drug administration: implications for acute and chronic tolerance. *Psychological Review* **104**, 170–193.

Rasmussen, K., Beitner-Johnson, D. B., Krystal, J. H., Aghajanian, G. K., and Nestler, E. J. (1990). Opiate withdrawal and the rat locus coeruleus: behavioral, electrophysiological, and biochemical correlates. *Journal of Neuroscience* **10**, 2308–2317.

Reichard, J. D. (1943). Can the euphoric, analgetic and physical dependence effects of drugs be separated? I. With reference to euphoria. *Federation Proceedings* **2**, 188–191.

Reinscheid, R. K., Nothacker, H. P., Bourson, A., Ardati, A., Henningsen, R. A., Bunzow, J. R., Grandy, D. K., Langen, H., Monsma, F. J., Jr., Civelli, O., and Orphanin, F. Q. (1995). a neuropeptide that activates an opioidlike G protein-coupled receptor. *Science* **270**, 792–794.

Roberts, D. C., Corcoran, M. E., and Fibiger, H. C. (1977). On the role of ascending catecholaminergic systems in intravenous self-administration of cocaine. *Pharmacology Biochemistry and Behavior* **6**, 615–620.

Roberts, D. C. S., Koob, G. F., Klonoff, P., and Fibiger, H. C. (1980). Extinction and recovery of cocaine self-administration following 6-hydroxydopamine lesions of the nucleus accumbens. *Pharmacology Biochemistry and Behavior* **12**, 781–787.

Rosen, M. I., McMahon, T. J., Margolin, A., Gill, T. S., Woods, S. W., Pearsall, H. R., Kreek, M. J., and Kosten, T. R. (1995). Reliability of sequential naloxone challenge tests. *American Journal of Drug and Alcohol Abuse* **21**, 453–467.

Rossetti, Z. L., Hmaidan, Y., and Gessa, G. L. (1992). Marked inhibition of mesolimbic dopamine release: A common feature of ethanol, morphine, cocaine and amphetamine abstinence in rats. *European Journal of Pharmacology* **221**, 227–234.

Rothman, R. B., Bykov, V., Long, J. B., Brady, L. S., Jacobson, A. E., Rice, K. C., and Holaday, J. W. (1989). Chronic administration of morphine and naltrexone up-regulate mu-opioid binding sites labeled by [3H][D-Ala2,MePhe4,Gly-ol5]enkephalin: further evidence for two mu-binding sites. *European Journal of Pharmacology* **160**, 71–82.

Sandhu, B. K., Milla, P. J., and Harries, J. T. (1983). Mechanisms of action of loperamide. *Scandinavian Journal of Gastroenterology Supplement* **84**, 85–92.

Satoh, M., Kawajiri, S. I., Ukai, Y., and Yamamoto, M. (1979). Selective and nonselective inhibition by enkephalins and noradrenaline of nociceptive response of lamina V type neurons in the spinal dorsal horn of the rabbit. *Brain Research* **177**, 384–387.

Satoh, M., Kubota, A., Iwama, T., Wada, T., Yasui, M., Fujibayashi, K., and Takagi, H. (1983). Comparison of analgesic potencies of mu, delta and kappa agonists locally applied to various CNS regions relevant to analgesia in rats. *Life Sciences* **33**(Suppl. 1), 689–692.

Schaefer, G. J., and Michael, R. P. (1983). Morphine withdrawal produces differential effects on the rate of lever-pressing for brain self-stimulation in the hypothalamus and midbrain in rats. *Pharmacology Biochemistry and Behavior* **18**, 571–577.

Schaefer, G. J., and Michael, R. P. (1986). Changes in response rates and reinforcement thresholds for intracranial self-stimulation during morphine withdrawal. *Pharmacology Biochemistry and Behavior* **25**, 1263–1269.

Schiller, L. R. (1995). Review article: anti-diarrhoeal pharmacology and therapeutics. *Alimentary Pharmacology and Therapeutics* **9**, 87–106.

Schroeder, R. L., Weinger, M. B., Vakassian, L., and Koob, G. F. (1991). Methylnaloxonium diffuses out of the rat brain more slowly than naloxone after direct intracerebral injection. *Neuroscience Letters* **121**, 173–177.

Schulteis, G., Markou, A., Gold, L. H., Stinus, L., and Koob, G. F. (1994). Relative sensitivity to naloxone of multiple indices of opiate withdrawal: A quantitative dose-response analysis. *Journal of Pharmacology and Experimental Therapeutics* **271**, 1391–1398.

Schulteis, G., and Koob, G. F. (1996). Reinforcement processes in opiate addiction: a homeostatic model. *Neurochemical Research* **21**, 1437–1454.

Schulteis, G., Heyser, C. J., and Koob, G. F. (1997). Opiate withdrawal signs precipitated by naloxone following a single exposure to morphine: Potentiation with a second morphine treatment. *Psychopharmacology* **129**, 56–65.

Schulteis, G., Heyser, C. J., and Koob, G. F. (1999). Differential expression of response-disruptive and somatic indices of opiate withdrawal during the initiation and development of opiate dependence. *Behavioural Pharmacology* **10**, 235–242.

Schulteis, G., Morse, A. C., and Liu, J. (2003). Repeated experience with naloxone facilitates acute morphine withdrawal: potential role for conditioning processes in acute opioid dependence. *Pharmacology Biochemistry and Behavior* **76**, 493–503.

Self, D. W., McClenahan, A. W., Beitner-Johnson, D., Terwilliger, R. Z., and Nestler, E. J. (1995). Biochemical adaptations in the mesolimbic dopamine system in response to heroin self-administration. *Synapse* **21**, 312–318.

Self, D. W., and Nestler, E. J. (1995). Molecular mechanisms of drug reinforcement and addiction. *Annual Review of Neuroscience* **18**, 463–495.

Shaham, Y., and Stewart, J. (1995). Effects of restraint stress and intra-ventral tegmental area injections of morphine and methyl naltrexone on the discriminative stimulus effects of heroin in the rat. *Pharmacology Biochemistry and Behavior* **51**, 491–498.

Shaham, Y., and Stewart, J. (1996). Effects of opioid and dopamine receptor antagonists on relapse induced by stress and re-exposure to heroin in rats, *Psychopharmacology* **125**, 385–391.

Shaham, Y., Funk, D., Erb, S., Brown, T. J., Walker, C. D., and Stewart, J. (1997). Corticotropin-releasing factor, but not corticosterone, is involved in stress-induced relapse to heroin-seeking in rats. *Journal of Neuroscience* **17**, 2605–2614.

Shaham, Y., Erb, S., Leung, S., Buczek, Y., and Stewart, J. (1998). CP-154,526, a selective, nonpeptide antagonist of the corticotropin-releasing factor1 receptor attenuates stress-induced relapse to drug seeking in cocaine- and heroin-trained rats. *Psychopharmacology* **137**, 184–190.

Shaham, Y., Erb, S., and Stewart, J. (2000a). Stress-induced relapse to heroin and cocaine seeking in rats: a review. *Brain Research Reviews* **33**, 13–33.

Shaham, Y., Highfield, D., Delfs, J., Leung, S., and Stewart, J. (2000b). Clonidine blocks stress-induced reinstatement of heroin seeking in rats: an effect independent of locus coeruleus noradrenergic neurons. *European Journal of Neuroscience* **12**, 292–302.

Sharma, S. K., Klee, W. A., and Nirenberg, M. (1975). Dual regulation of adenylate cyclase accounts for narcotic dependence and tolerance. *Proceedings of the National Academy of Sciences USA* **72**, 3092–3096.

Shaw-Lutchman, T. Z., Barrot, M., Wallace, T., Gilden, L., Zachariou, V., Impey, S., Duman, R. S., Storm, D., and Nestler, E. J. (2002). Regional and cellular mapping of cAMP response element-mediated transcription during naltrexone-precipitated morphine withdrawal. *Journal of Neuroscience* **22**, 3663–3672.

Shimora, K., Kamata, O., Ueki, S., Ida, S., Oguri, K., Yoshimora, H., and Tsukamoto, H. (1971). Analgesic effects of morphine glucuronides. *Tohoku Journal of Experimental Medicine* **105**, 45–52.

Shippenberg, T. S., and Herz, A. (1987). Place preference conditioning reveals the involvement of D1-dopamine receptors in the motivational properties of mu- and kappa-opioid agonists. *Brain Research* **436**, 169–172.

Shippenberg, T. S., and Herz, A. (1988). Motivational effects of opioids: influence of D-1 versus D-2 receptor antagonists. *European Journal of Pharmacology* **151**, 233–242.

Shippenberg, T. S., Emmett-Oglesby, M. W., Ayesta, F. J., and Herz, A. (1988). Tolerance and selective cross-tolerance to the motivational effects of opioids. *Psychopharmacology* **96**, 110–115.

Shippenberg, T. S., Herz, A., Spanagel, R., Bals-Kubik, R., and Stein, C. (1992). Conditioning of opioid reinforcement: Neuroanatomical and neurochemical substrates. *In The Neurobiology of Drug and Alcohol Addiction* (series title: *Annals of the New York Academy of Sciences*, vol. 654), (P. W. Kalivas, H. H. Samson, Eds.), pp. 347–356. New York Academy of Sciences, New York.

Shippenberg, T. S., Bals-Kubik, R., and Herz, A. (1993). Examination of the neurochemical substrates mediating the motivational effects of opioids: role of the mesolimbic dopamine system and D-1 vs. D-2 dopamine receptors. *Journal of Pharmacology and Experimental Therapeutics* **265**, 53–59.

Siegel, S. (1975). Evidence from rats that morphine tolerance is a learned response *Journal of Comparative and Physiological Psychology* **89**, 498–506.

Siegel, S. (1978). Response to: Hayes, R. L., Mayer, D. J., Morphine tolerance: is there evidence for a conditioning model? *Science* **200**, 344–345.

Siegel, S., and Kim, J. A. (2000). Absence of cross-tolerance and the situational specificity of tolerance, *Palliative Medicine* **14**, 75–77.

Siegel, S., and Ramos, B. M. (2002). Applying laboratory research: drug anticipation and the treatment of drug addiction. *Experimental and Clinical Psychopharmacology* **10**, 162–183.

Siegel, S., Hinson, R. E., Krank, M. D., and McCully, J. (1982). Heroin 'overdose' death: contribution of drug-associated environmental cues. *Science* **216**, 436–437.

Siegel, S., Baptista, M. A., Kim, J. A., McDonald, R. V., and Weise-Kelly, L. (2000). Pavlovian psychopharmacology: the associative basis of tolerance. *Experimental and Clinical Psychopharmacology* **8**, 276–293.

Siggins, G. R., and Zieglgansberger, W. (1981). Morphine and opioid peptides reduce inhibitory synaptic potentials in hippocampal pyramidal cells in vitro without alteration of membrane potential. *Proceedings of the National Academy of Sciences USA* **78**, 5235–5239.

Siggins, G. R., Martin, G., Yuan, X., Nie, Z., and Madamba, S. (1995). Opiate modulation of glutamatergic transmission in nucleus accumbens neurons in vitro. *Analgesia* **1**, 728–733.

Simonin, F., Valverde, O., Smadja, C., Slowe, S., Kitchen, I., Dierich, A., Le Meur, M., Roques, B. P., Maldonado, R., and Kieffer, B. L. (1998). Disruption of the kappa-opioid receptor gene in mice enhances sensitivity to chemical visceral pain, impairs pharmacological actions of the selective kappa-agonist U-50,488H and attenuates morphine withdrawal. *EMBO Journal* **17**, 886–897.

Simonnet, G., and Rivat, C. (2003). Opioid-induced hyperalgesia: abnormal or normal pain? *Neuroreport* **14**, 1–7.

Sklair-Tavron, L., Shi, W. X., Lane, S. B., Harris, H. W., Bunney, B. S., and Nestler, E. J. (1996). Chronic morphine induces visible changes in the morphology of mesolimbic dopamine neurons. *Proceedings of the National Academy of Sciences USA* **93**, 11202–11207.

Smith, A. P., Law, P. Y., and Loh, H. H. (1988). Role of opioid receptors in narcotic tolerance/dependence. *In The Opiate Receptors* (G. W. Pasternak, Ed.), pp. 441–485. Humana Press, Clifton NJ.

Smith, J. E., Guerin, G. F., Co, C., Barr, T. S., and Lane, J. D. (1985). Effects of 6-OHDA lesions of the central medial nucleus accumbens on rat intravenous morphine self-administration. *Pharmacology Biochemistry and Behavior* **23**, 843–849.

Sneader, W. (1998). The discovery of heroin. *Lancet* **352**, 1697–1699.

Snyder, S. H., and Pasternak, G. W. (2003). Historical review: opioid receptors. *Trends in Pharmacological Sciences* **24**, 198–205.

Sora, I., Elmer, G., Funada, M., Pieper, J., Li, X. F., Hall, F. S., and Uhl, G. R. (2001). Mu opiate receptor gene dose effects on different morphine actions: evidence for differential in vivo mu receptor reserve. *Neuropsychopharmacology* **25**, 41–54.

Spyraki, C., Fibiger, H. C., and Phillips, A. G. (1983). Attenuation of heroin reward in rats by disruption of the mesolimbic dopamine system. *Psychopharmacology* **79**, 278–283.

Stewart, J., and Wise, R. A. (1992). Reinstatement of heroin self-administration habits, morphine prompts and naltrexone discourages renewed responding after extinction. *Psychopharmacology* **108**, 79–84.

Stewart, T. (1987). *The Heroin Users*. Pandora, London.

Stinus, L., Koob, G. F., Ling, N., Bloom, F. E., and Le Moal, M. (1980). Locomotor activation induced by infusion of endorphins into the ventral tegmental area: evidence for opiate-dopamine interactions. *Proceedings of the National Academy of Science USA* **77**, 2323–2327.

Stinus, L., Nadaud, D., Deminiere, J. M., Jauregui, J., Hand, T. T., and Le Moal, M. (1989). Chronic flupentixol treatment potentiates the reinforcing properties of systemic heroin administration. *Biological Psychiatry* **26**, 363–371.

Stinus, L., Le Moal, M., and Koob, G. F. (1990). Nucleus accumbens and amygdala are possible substrates for the aversive stimulus effects of opiate withdrawal. *Neuroscience* **37**, 767–773.

Stinus, L., Caille, S., and Koob, G. F. (2000). Opiate withdrawal-induced place aversion lasts for up to 16 weeks. *Psychopharmacology* **149**, 115–120.

Stinus, L., Cador, M., Zorrilla, E. P., and Koob, G. F. (2005). Buprenorphine and a CRF1 antagonist block the acquisition of opiate withdrawal-induced conditioned place aversion in rats. *Neuropsychopharmacology* **30**, 90–98.

Stornetta, R. L., Norton, F. E., and Guyenet, P. G. (1993). Autonomic areas of rat brain exhibit increased Fos-like immunoreactivity during opiate withdrawal in rats. *Brain Research* **624**, 19–28.

Substance Abuse and Mental Health Services Administration. (1996). *National Household Survey on Drug Abuse: Population Estimates 1995* (DHHS publication no. [SMA] 96–3095), U.S. Department of Health and Human Services, Rockville MD.

Substance Abuse and Mental Health Services Administration. (2002). *Emergency Department Trends from the Drug Abuse Warning Network, Final Estimates 1994–2001* (Office of Applied Studies, DAWN Series D-21, DHHS Publication No. SMA 02–3635), Rockville, MD.

Substance Abuse and Mental Health Services Administration. (2003). *Results from the 2002 National Survey on Drug Use and Health: National Findings* (Office of Applied Studies, NHSDA Series H-22, DHHS Publication No. SMA 03–3836), Rockville MD.

Swerdlow, N. R., and Koob, G. F. (1984). The neural substrates of apomorphine-stimulated locomotor activity following denervation of the nucleus accumbens. *Life Sciences* **35**, 2537–2544.

Swerdlow, N. R., Swanson, L. W., and Koob, G. F. (1984a). Electrolytic lesions of the substantia innominata and lateral preoptic area attenuate the 'supersensitive' locomotor response to apomorphine resulting from denervation of the nucleus accumbens. *Brain Research* **306**, 141–148.

Swerdlow, N. R., Swanson, L. W., and Koob, G. F. (1984b). Substantia innominata: critical link in the behavioral expression of mesolimbic dopamine stimulation in the rat. *Neuroscience Letters* **50**, 19–24.

Taddese, A., Nah, S. Y., and McCleskey, E. W. (1995). Selective opioid inhibition of small nociceptive neurons. *Science* **270**, 1366–1369.

Takagi, H., Satoh, M., Akaike, A., Shibata, T., Yajima, H., and Ogawa, H. (1978). Analgesia by enkephalins injected into the nucleus reticularis gigantocellularis of rat medulla oblongata. *European Journal of Pharmacology* **49**, 113–116.

Tokuyama, S., Nakamura, F., Takahashi, M., and Kaneto, H. (1996). Antinociceptive effect of dihydroetorphine following various routes of administration: a comparative study with morphine. *Biological and Pharmaceutical Bulletin* **19**, 477–479.

Trapaidze, N., Keith, D. E., Cvejic, S., Evans, C. J., and Devi, L. A. (1996). Sequestration of the delta opioid receptor. Role of the C terminus in agonist-mediated internalization. *Journal of Biological Chemistry* **271**, 29279–29285.

Trujillo, K. A., and Akil, H. (1991). Inhibition of morphine tolerance and dependence by the NMDA receptor antagonist MK-801. *Science* **251**, 85–87.

Tseng, L. F., Loh, H. H., and Li, C. H. (1977). Human beta-endorphin: development of tolerance and behavioral activity in rats. *Biochemical and Biophysical Research Communications* **74**, 390–396.

Umans, J. G., and Inturrisi, C. E. (1981). Pharmacodynamics of subcutaneously administered diacetylmorphine, 6-acetylmorphine and morphine in mice. *Journal of Pharmacology and Experimental Therapeutics* **218**, 409–415.

Vaccarino, F. J., Bloom, F. E., and Koob, G. F. (1985a). Blockade of nucleus accumbens opiate receptors attenuates intravenous heroin reward in the rat. *Psychopharmacology* **86**, 37–42.

Vaccarino, F. J., Pettit, H. O., Bloom, F. E., and Koob, G. F. (1985b). Effects of intracerebroventricular administration of methyl naloxonium chloride on heroin self-administration in the rat. *Pharmacology Biochemistry and Behavior* **23**, 495–498.

Vaillant, G. E. (1973). A 20-year follow-up of New York narcotic addicts. *Archives of General Psychiatry* **29**, 237–241.

Valentino, R. J., and Van Bockstaele, E. (2001). Opposing regulation of the locus coeruleus by corticotropin-releasing factor and opioids: potential for reciprocal interactions between stress and opioid sensitivity. *Psychopharmacology* **158**, 331–342.

van Ree, J. M., and Ramsey, N. (1987). The dopamine hypothesis of opiate reward challenged. *European Journal of Pharmacology* **134**, 239–243.

van Ree, J. M., Gerrits, M. A., and Vanderschuren, L. J. (1999). Opioids, reward and addiction: An encounter of biology, psychology, and medicine. *Pharmacological Reviews* **51**, 341–396.

Vanderah, T. W., Suenaga, N. M., Ossipov, M. H., Malan, T. P. Jr., Lai, J., and Porreca, F. (2001). Tonic descending facilitation from the rostral ventromedial medulla mediates opioid-induced abnormal pain and antinociceptive tolerance. *Journal of Neuroscience* **21**, 279–286.

Veinante, P., Stoeckel, M. E., Lasbennes, F., and Freund-Mercier, M. J. (2003). c-Fos and peptide immunoreactivities in the central extended amygdala of morphine-dependent rats after naloxone-precipitated withdrawal. *European Journal of Neuroscience* **18**, 1295–1305.

Wagner, J. J., Terman, G. W., and Chavkin, C. (1993). Endogenous dynorphins inhibit excitatory neurotransmission and block LTP induction in the hippocampus. *Nature* **363**, 451–454.

Wakeman, F. Jr. (1978). The Canton Trade and the Opium War. *In The Cambridge History of China: Volume 10.* (J. K. Fairbank, Ed.), pp. 163–212. *Late Ch-ing, 1800–1911, Part I*, Cambridge University Press, Cambridge.

Walker, J. R., Ahmed, S. H., Gracy, K. N., and Koob, G. F. (2000). Microinjections of an opiate receptor antagonist into the bed nucleus of the stria terminalis suppress heroin self-administration in dependent rats. *Brain Research* **854**, 85–92.

Watanabe, T., Nakagawa, T., Yamamoto, R., Maeda, A., Minami, M., and Satoh, M. (2002a). Involvement of glutamate receptors within the central nucleus of the amygdala in naloxone-precipitated morphine withdrawal-induced conditioned place aversion in rats. *Japanese Journal of Pharmacology* **88**, 399–406.

Watanabe, T., Yamamoto, R., Maeda, A., Nakagawa, T., Minami, M., and Satoh, M. (2002b). Effects of excitotoxic lesions of the central or basolateral nucleus of the amygdala on naloxone-precipitated withdrawal-induced conditioned place aversion in morphine-dependent rats. *Brain Research* **958**, 423–428.

Watanabe, T., Nakagawa, T., Yamamoto, R., Maeda, A., Minami, M., and Satoh, M. (2003). Involvement of noradrenergic system within the central nucleus of the amygdala in naloxone-precipitated morphine withdrawal-induced conditioned place aversion in rats. *Psychopharmacology* **170**, 80–88.

Waterfield, A. A., Hughes, J., and Kosterlitz, H. W. (1976). Cross tolerance between morphine and methionine-enkephalin. *Nature* **260**, 624–625.

Way, E. L., Loh, H. H., and Shen, F. H. (1969). Simultaneous quantitative assessment of morphine tolerance and physical dependence. *Journal of Pharmacology and Experimental Therapeutics* **167**, 1–8.

Weeks, J. R., and Collins, R. J. (1976). Changes in morphine self-administration in rats induced by prostaglandin E1 and naloxone. *Prostaglandins* **12**, 11–19.

Wei, E., and Loh, H. (1976). Physical dependence of opiate–like peptides. *Science*. **193**, 1262–1263.

Wei, E., Loh, H. H., and Way, E. L. (1972). Neuroanatomical correlates of morphine dependence. *Science* **177**, 616–617.

Wei, E., Loh, H. H., and Way, E. L. (1973). Brain sites of precipitated abstinence in morphine-dependent rats. *Journal of Pharmacology and Experimental Therapeutics* **185**, 108–115.

Weisskopf, M. G., Zalutsky, R. A., and Nicoll, R. A. (1993). The opioid peptide dynorphin mediates heterosynaptic depression of hippocampal mossy fibre synapses and modulates long-term potentiation. *Nature* **362**, 423–427.

Welzl, H., Kuhn, G., and Huston, J. P. (1989). Self-administration of small amounts of morphine through glass micropipettes into the ventral tegmental area of the rat. *Neuropharmacology*. **28**, 1017–1023.

Whistler, J. L., and von Zastrow, M. (1998). Morphine-activated opioid receptors elude desensitization by beta-arrestin. *Proceedings of the National Academy of Sciences USA* **95**, 9914–9919.

Widnell, K. L., Russell, D. S., and Nestler, E. J. (1994). Regulation of expression of cAMP response element-binding protein in the locus coeruleus in vivo and in a locus coeruleus-like cell line in vitro. *Proceedings of the National Academy of Sciences USA* **91**, 10947–10951.

Widnell, K. L., Self, D. W., Lane, S. B., Russell, D. S., Vaidya, V. A., Miserendino, M. J., Rubin, C. S., Duman, R. S., and Nestler, E. J. (1996). Regulation of CREB expression: in vivo evidence for a functional role in morphine action in the nucleus accumbens. *Journal of Pharmacology and Experimental Therapeutics* **276**, 306–315.

Wikler, A. (1973). Dynamics of drug dependence: Implications of a conditioning theory for research and treatment. *Archives of General Psychiatry* **28**, 611–616.

Wimpey, T. L., and Chavkin, C. (1991). Opioids activate both an inward rectifier and a novel voltage-gated potassium conductance in the hippocampal formation. *Neuron* **6**, 281–289.

Woldbye, D. P., Klemp, K., and Madsen, T. M. (1998). Neuropeptide Y attenuates naloxone-precipitated morphine withdrawal via Y5-like receptors. *Journal of Pharmacology and Experimental Therapeutics* **284**, 633–636.

Woodbridge, P. D. (1957). Changing concepts concerning depth of anesthesia. *Anesthesiology* **18**, 536–550.

Yaksh, T. L. (1981). Spinal opiate analgesia: characteristics and principles of action. *Pain* **11**, 293–346.

Yasuda, K., Raynor, K., Kong, H., Breder, C. D., Takeda, J., Reisine, T., and Bell, G. I. (1993). Cloning and functional comparison of kappa and delta opioid receptors from mouse brain. *Proceedings of the National Academy of Sciences USA*, **90**, 6736–6740.

Young, A. M., and Goudie, A. J. (1995). Adaptive processes regulating tolerance to the behavioral effects of drugs. *In Psychopharmacology: The Fourth Generation of Progress* (F. E. Bloom, D. J. Kupfer, Eds.), pp. 733–742. Raven Press, New York.

Young, A. M., Sannerud, C. A., Steigerwald, E. S., Doty, M. D., Lipinski, W. J., and Tetrick, L. E. (1990). Tolerance to morphine stimulus control: role of morphine maintenance dose. *Psychopharmacology* **102**, 59–67.

Young, A. M., Walton, M. A., and Carter, T. L. (1992). Selective tolerance to discriminative stimulus effects of morphine or d-amphetamine. *Behavioural Pharmacology* **3**, 201–209.

Yu, S. S., Lefkowitz, R. J., and Hausdorff, W. P. (1993). Beta-adrenergic receptor sequestration. A potential mechanism of receptor resensitization. *Journal of Biological Chemistry* **268**, 337–341.

Zadina, J. E., Harrison, L. M., Ge, L. J., Kastin, A. J., and Chang, S. L. (1994). Differential regulation of mu and delta opiate receptors by morphine, selective agonists and antagonists and differentiating agents in, S. H-SY5Y human neuroblastoma cells. *Journal of Pharmacology and Experimental Therapeutics* **270**, 1086–1096.

Zahm, D. S., and Brog, J. S. (1992). On the significance of subterritories in the accumbens' part of the rat ventral striatum. *Neuroscience* **50**, 751–767.

Zaki, P. A., Bilsky, E. J., Vanderah, T. W., Lai, J., Evans, C. J., and Porreca, F. (1996). Opioid receptor types and subtypes: the delta receptor as a model. *Annual Review of Pharmacology and Toxicology* **36**, 379–401.

Zangen, A., Ikemoto, S., Zadina, J. E., and Wise, R. A. (2002). Rewarding and psychomotor stimulant effects of endomorphin-1: anteroposterior differences within the ventral tegmental area and lack of effect in nucleus accumbens. *Journal of Neuroscience* **22**, 7225–7233.

Zhang, J., Ferguson, S. S., Barak, L. S., Bodduluri, S. R., Laporte, S. A., Law, P. Y., and Caron, M. G. (1998). Role for G protein-coupled receptor kinase in agonist-specific regulation of mu-opioid receptor responsiveness. *Proceedings of the National Academy of Sciences USA* **95**, 7157–7162.

Zheng, F., Grandy, D. K., and Johnson, S. W. (2002). Actions of orphanin FQ/nociceptin on rat ventral tegmental area neurons in vitro. *British Journal of Pharmacology* **136**, 1065–1071.

Zhu, W., and Pan, Z. Z. (2004). Synaptic properties and postsynaptic opioid effects in rat central amygdala neurons. *Neuroscience* **127**, 871–879.

Zhu, Y., King, M. A., Schuller, A. G., Nitsche, J. F., Reidl, M., Elde, R. P., Unterwald, E., Pasternak, G. W., and Pintar, J. E. (1999). Retention of supraspinal delta-like analgesia and loss of morphine tolerance in delta opioid receptor knockout mice. *Neuron* **24**, 243–252.

Zieglgansberger, W., French, E. D., Siggins, G. R., and Bloom, F. E. (1979). Opioid peptides may excite hippocampal pyramidal neurons by inhibiting adjacent inhibitory interneurons. *Science* **205**, 415–417.

Zinberg, N. E., and Jacobson, R. C. (1976). The natural history of 'chipping.' *American Journal of Psychiatry* **133**, 37–40.

Zinberg, N. E., Harding, W. M., and Winkeller, M. (1977). A study of social regulatory mechanisms in controlled illicit drug users. *Journal of Drug Issues* **7**, 117–133.

Zito, K. A., Vickers, G., and Roberts, D. C. (1985). Disruption of cocaine and heroin self-administration following kainic acid lesions of the nucleus accumbens. *Pharmacology Biochemistry and Behavior* **23**, 1029–1036.

C H A P T E R

5

Alcohol

OUTLINE

DEFINITIONS

> 'Alcohol is the king of liquids. It excites the taste to the highest degree, its various preparations have opened up to mankind new sources of enjoyment. It supplies to certain medicines an energy which they could not have without it' (Brillat-Savarin, 1826; see also De Rasor and Youra, 1980).

The word 'alcohol', according to Merriam-Webster's dictionary, finds its roots in the Arabic *al-kuhul* (or *kohl, cohol*, or *kohol*), to mean a powder of antimony or galena, used by women to darken the eyebrows. The name *alcohol* was derived through Medieval Latin from Arabic and was afterward applied, on account of the fineness of this powder, to highly rectified spirits, a signification unknown in Arabia (from *Webster's Unabridged Dictionary*, 1996, 1998, 2002). Alcohol represents a whole series of compounds, but the alcohol suitable for drinking is ethanol (see **Fig. 5.1**). Alcohol is found in all substances that contain glucose. It is the product of the saccharine principle which takes place during alcoholic fermentation (De Rasor and Youra, 1980). Fermentation is the conversion of glucose and water in the presence of yeast to produce ethanol and carbon dioxide, and this is a common biological reaction in nature. Five agents are required for alcoholic fermentation: sugar (or starch to form glucose), water, heat, ferment (usually yeast – *Saccharomyces cerevisiae*), and air. Yeast will convert glucose into alcohol up to about the 12 per cent level until the alcohol level rises to a concentration which is toxic to the yeast and the yeast dies. Such a process occurs in nature in seed

Neurobiology of Addiction, by George F. Koob and Michel Le Moal.
ISBN – 13: 978-0-12-419239-3 ISBN – 10: 0-12-419239-4

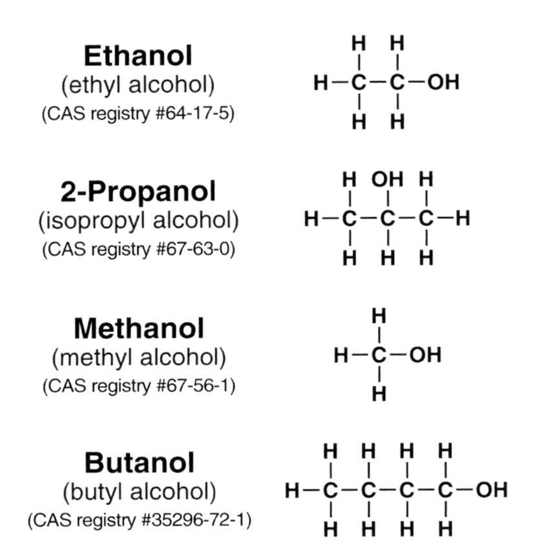

FIGURE 5.1 Chemical structures and registry numbers for various alcohols from the Chemical Abstracts database.

germination and ripening of fruit. Any source of glucose is sufficient to produce alcohol through the process of fermentation, and thus forms the basis of numerous alcoholic beverages worldwide (**Table 5.1**). In order to raise alcohol (ethanol) concentrations above 12 per cent, the yeast-converted fermentation mixture must be distilled. The fermentation mixture is heated, and the ethanol vaporizes at a lower temperature than water. When cooled in some form of condensation device (often a cool metal or glass apparatus) the ethanol can be captured as a liquid again.

HISTORY OF ALCOHOL USE, ABUSE, AND ALCOHOLISM

Alcohol is a ubiquitous substance in our society and is widely used in moderate doses in the form of alcoholic beverages for beneficial effects, both social and medical (see National Institute on Alcohol Abuse and Alcoholism, 2005). Alcoholic beverages are considered to have both nutritional and drug effects and as such, alcohol is unique among drug preparations. Alcohol ingestion per capita has been steady since 1850 in the United States with the exception of the period of 'Prohibition,' from 1919 to 1933, when alcohol was prohibited from sale (National Institute on Alcohol Abuse and Alcoholism, 1999). A large proportion of persons in the United States population have used alcohol at least once in their lifetime (88 per cent) (Substance Abuse and Mental Health Services Administration, 2000). Alcohol is a well-known 'social lubricant' used to produce disinhibition in social

TABLE 5.1 Common Alcoholic Beverages

Beverage	Source of glucose	Fermentation	Distillation	Alcohol by volume
Beer (ale, lager)	Barley, hops	X		3–6%
Beer (wheat)	Wheat, hops	X		3–6%
Cider (alcohol)	Apples	X		3–6%
Gin	Juniper berries		X	40%
Vodka	Rye, wheat, potatoes	X	X	35–60%
Schnapps	Potatoes (flavored with various fruit juices)		X	20–45%
Rum	Sugar cane molasses	X	X	35–40%
Tequila	Agave cactus, corn, sugar cane	X	X	40–70%
Mead	Honey	X		11–17%
White wine	White grapes and other grapes without skins	X		13%
Red wine	Red or black grapes with skins	X		12.5%
Rose wine	Red or black grapes with skins for short time	X		13%
Port wine	Grapes (fermentation halted by addition of distilled grape spirits)	X	X	20%
Champagne	Grapes	X		12%
Sherry	Grapes (fortified with neutral spirits)	X		14–20%
Sake	Rice	X		10–20%
Brandy (cognac, armagnac)	Grapes, plums		X	40–60%
Liqueur	Grapes (flavored with fruit, herbs, spices, flowers, wood)		X	14–40%
Drambuie	Scotch, heather honey (flavored with herbs)		X	40%
Whisky	Corn, rye, barley		X	40–60%
Scotch	Barley, wheat, maize (flavored with peat)		X	40–60%

For more information, visit *http://en.wikipedia.org/wiki/Alcoholic_beverage#Types_of_alcoholic_beverages.*

situations, but at the same time, excessive use produces the most harm to society of all drugs of abuse (see below). The following address to the legislature by Mississippi State Senator Judge Noah S. Sweat in 1952 represents the dilemma faced by society in addressing the various aspects of the impact of alcohol on society:

'You have asked me how I feel about whisky. All right, here is just how I stand on this question: If when you say whisky, you mean the devil's brew, the poison scourge; the bloody monster that defiles innocence, yea, literally takes the bread from the mouths of little children; if you mean the evil drink that topples the Christian man and woman from the pinnacles of righteous, gracious living into the bottomless pit of degradation and despair, shame and helplessness and hopelessness, then certainly I am against it with all of my power...But, if when you say whisky, you mean the oil of conversation, the philosophic wine, the stuff that is consumed when good fellows get together, that puts a song in their hearts and laughter on their lips and the warm glow of contentment in their eyes, if you mean holiday cheer; if you mean the stimulating drink that puts the spring in the old gentlemen's step on a frosty morning; if you mean the drink that enables a man to magnify his joy, and his happiness and to forget, if only for a little while, life's great tragedies and heartbreaks and sorrows, if you mean that drink, the sale of which pours into our treasuries untold millions of dollars, which are used to provide tender care for our little crippled children, our blind, our deaf, our dumb, our pitiful aged and infirmed, to build highways, hospitals and schools, then certainly I am in favor of it. This is my stand. I will not retreat from it; I will not compromise' (Associated Press, 1996).

In 2003 in the United States, approximately 51 per cent (120 million) of persons over the age of 12 years were current alcohol users, 23 per cent (54 million) engaged in binge drinking, and about 7 per cent (16 million) were defined as heavy drinkers (Substance Abuse and Mental Health Services Administration, 2003b) (**Table 5.2**). Of these current users, 7.7 per cent (18 million) met the criteria for Substance Abuse or Dependence on alcohol (Grant *et al.*, 2004). In Europe, approximately 53 per cent of the population are current drinkers (drinking at least once per week), and 13 per cent engage in binge drinking (defined as 5 or more standard drinks in a single occasion at least once per week). Approximately 27 per cent met the criteria for alcohol-related social harm (at least one social harm incident in the past 12 months). Regular daily drinking is more common in Southern Europe, and binge drinking is more common in the United Kingdom, Ireland, and northern Europe (Norstrom, 2002).

TABLE 5.2 Alcohol Use Definitions

Term	Definition
Current use	At least one drink in the past 30 days (includes binge and heavy use).
Binge use	Five or more drinks on the same occasion at least once in the past 30 days. A pattern of drinking that brings blood alcohol concentration to 0.08 g% or above. For the typical adult, this pattern corresponds to consuming five or more drinks (male), or four or more drinks (female), in about two hours.
Heavy use	Five or more drinks on the same occasion on at least five different days in the past 30 days.
Drink	A 12-ounce can or bottle of beer or wine cooler, a 5-ounce glass of wine, or 1.5-ounces of 80 proof (40% alcohol) distilled spirits.

Source: Substance Abuse and Mental Health Services Administration, 2003b; National Institute on Alcohol Abuse and Alcoholism, 2005.

Alcohol abuse and alcoholism can lead to numerous medical conditions ranging from cirrhosis of the liver and heart disease to pancreatitis, Korsakoff's dementia, and fetal alcohol syndrome. Alcohol was implicated in 32 per cent of all deaths in the United States in 1996 (National Institute on Alcohol Abuse and Alcoholism, 1996), and this includes fatal automobile crashes, other accidents, liver disease, heart disease, neurological diseases, and cancer. Although steadily declining since 1980, in 2002, 17 419 individuals in the United States died in alcohol-related traffic crashes, and more than 15 000 of these involved a driver or pedestrian with a blood alcohol level of 0.08 g% or higher (Grant *et al.*, 2004). Alcohol abuse and alcoholism accounted for over $180 billion in costs to society in 1998, including health care expenditures, impacts on productivity, and social costs such as motor vehicle crashes, crime, and social welfare (Yi *et al.*, 2000).

BEHAVIORAL EFFECTS OF ALCOHOL

Alcohol has a range of behavioral effects, and for centuries has been widely regarded as both a sedative and a stimulant. To quote Shakespeare from *Macbeth* Act II, Scene III:

Porter. Faith, sir, we were carousing till the second cock; and drink, sir, is a great provoker of three things.

Macduff. What three things does drink especially provoke?

Porter. Merry, sir, nose-painting, sleep, and urine. Lechery, sir, it provokes, and unprovokes; it provokes

> the desire, but it takes away the performance; therefore, much drink may be said to be an equivocator with lechery; it makes him, and it mars him; it sets him on, and it takes him off; it persuades him, and disheartens him; makes him stand to, and not stand to; in conclusion, equivocates him in a sleep, and, giving him the lie, leaves him.

Alcohol, a sedative hypnotic, produces dose-dependent behavioral effects in humans such as sedation (decreases in activity) and hypnosis (sleep induction). At low blood alcohol levels of 0.01–0.05 g%, alcohol produces personality changes, including increased sociability, increased talkativeness, and a more expansive personality (**Fig. 5.2**). There is a mild euphoria with increased mood, good feelings, increased confidence, and increased assertiveness. There is also some release of inhibitions, tension reduction, and increased responsiveness in conflict situations. As the level increases from 0.08 to 0.10 g%, mood swings become more pronounced, with euphoria, emotional outbursts, and release of inhibitions. Blood alcohol levels of 0.08 g% produce distinct impairment in judgment and motor function. At levels of 0.15–0.20 g% there is marked ataxia, major motor impairment, staggering, slurred speech, muscular incoordination, and impairment in reaction time. Sensory responses are also impaired, including a loss of vestibular sense and decreased pain sensation. There is also a dulling of concentration and insight, impairments in discrimination and memory, and significant impairment in judgment, and even greater emotional instability and release of inhibitions. At this level, 'blackouts' can occur where a person, post-intoxication, will have no memory of events that transpired while intoxicated. At levels of 0.30 g%, subjects have been described as stuporous but conscious. Here, one has reached the anesthetic level, with marked decreases in responsivity to environmental stimuli, severely impaired motor function, and rapid and dramatic changes in mood. Vomiting can occur at this level. The lethal dose 50 per cent (LD50) for alcohol is considered to be approximately 0.40–0.50 g% in nondependent individuals.

In animals, alcohol has similar behavioral effects. Early paradigms which assessed the reinforcing effects of alcohol typically used an oral preference paradigm where animals were allowed to drink alcohol or water. A major breakthrough in this area was the development of a training procedure involving access to a sweetened solution and a subsequent fading in of alcohol to avoid the aversiveness of the alcohol taste (for review, see Samson, 1987) (see *Animal Models of Addiction* chapter). Following a sweet solution fadeout, rats will readily self-administer doses of alcohol in limited access situations that result in blood alcohol levels that range from 0.04 to 0.08 g% (Weiss *et al.*, 1990; Rassnick *et al.*, 1993c) (**Fig. 5.3**) (see *Animal Models of Addiction* chapter).

The antianxiety or tension-reduction properties of alcohol have been demonstrated in a variety of behavioral situations in both animals and humans. An early observation involved the study of the effects of alcohol on 'neurotic' behavior induced in cats by an

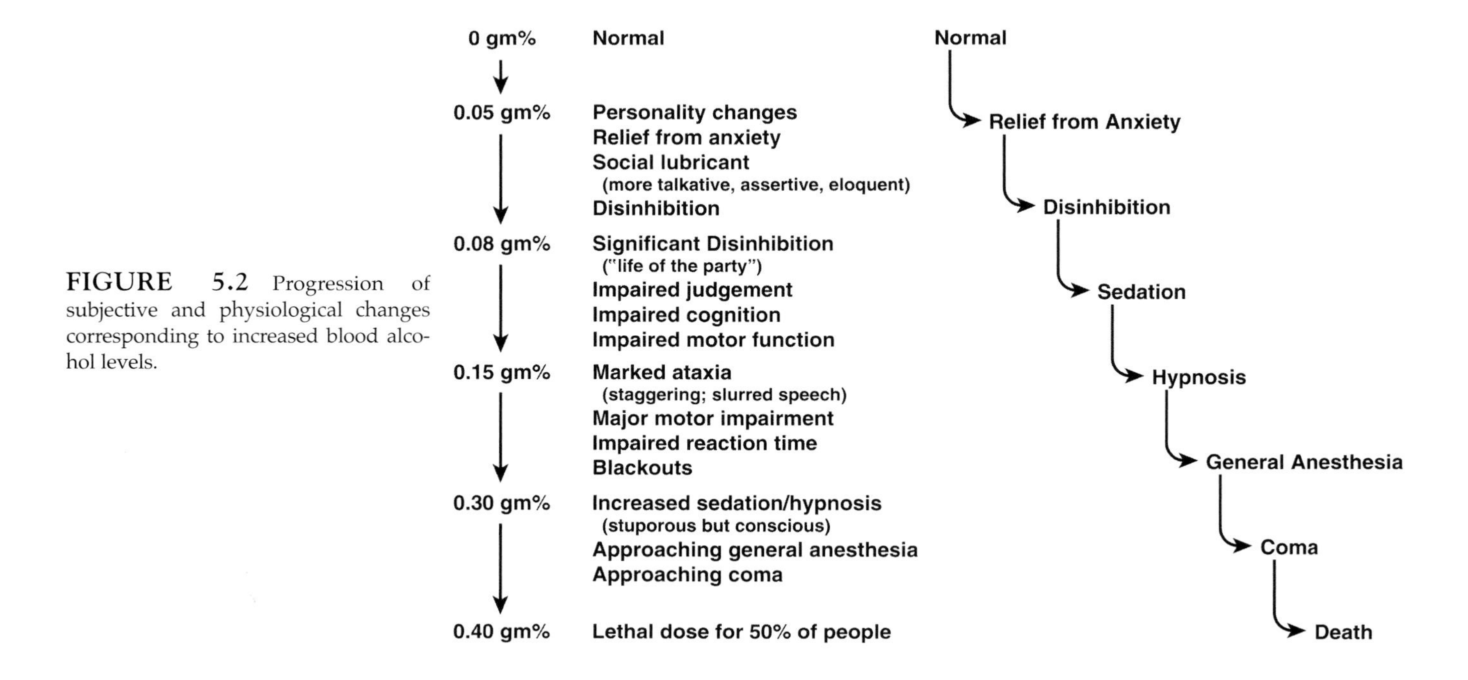

FIGURE 5.2 Progression of subjective and physiological changes corresponding to increased blood alcohol levels.

Training	Saccharin (w/v)	Ethanol (w/v)	
Days 1-3	0.2%	0%	* Rats trained to lever press on a FR-1 schedule
Days 4-9	0.2%	5%	* Ethanol added to the saccharin solution
Day 10	–	5%	* Access to ethanol and water or ethanol + saccharin and water
Days 11-12	0.2%	5%	
Day 13	–	5%	
Day 14	0.2%	8%	
Days 15-16	–	8%	
Day 17	0.2%	10%	
Day 18+	–	10%	Initiation of the free-choice operant task: ethanol (10%) and water

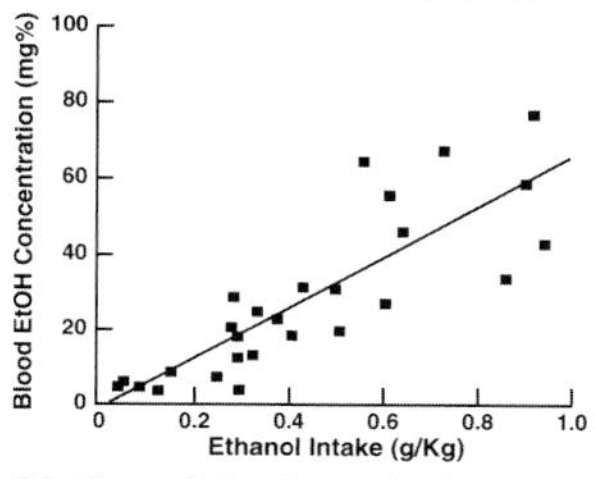

FIGURE 5.3 (Left) A saccharin fadeout protocol for alcohol dependence induction. Alcohol concentrations progressively increase, while saccharin concentrations are gradually decreased to zero (see text). (Right) Blood ethanol concentrations as a function of ethanol intake during a baseline session in the two-lever, free-choice operant task in rats. The variation in these measures in this distribution reflects the nature of responding that typically is observed in a group of nonselected, heterogeneous Wistar rats. [Reproduced with permission from Rassnick *et al.*, 1993b.]

approach-avoidance (conflict) situation (Masserman and Yum, 1946). Cats were trained to run down a runway to open a box for food when signaled by a bell-light conditioned stimulus, but a blast of air accompanied the food at irregular intervals. The animals developed bizarre, 'neurotic' behaviors such as inhibition of feeding, startle, phobic responses, and aversive behavior to stimuli associated with the food/air blast conflict. Alcohol in doses of approximately 1.0–1.5 g/kg significantly reduced the 'neurotic' behavior (Masserman and Yum, 1946). The cats showed a restoration of the simpler switch-and-signal responses and an attenuation of the phobias, motor disturbances, and other abnormal behaviors. Interestingly, alcohol disrupted, although to a lesser extent, the 'timing, spatial orientation, sequence and efficiency of "normal" goal-oriented responses' (Masserman and Yum, 1946). This early study elaborated many of the features of alcohol's effect on conflict behavior, notably the ability of alcohol to block or reduce the suppression of behavior induced by punishment but also to decrease unpunished behavior, often at the same time or at the same doses.

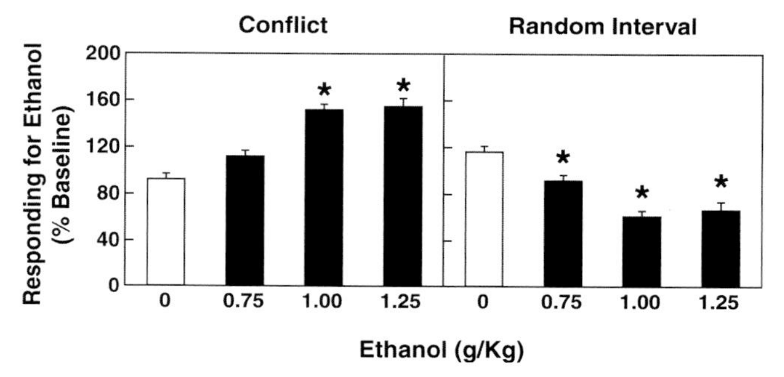

FIGURE 5.4 Dose-response curves expressed as percentage of baseline (i.e., uninjected response rates) for ethanol alone during conflict and random interval schedule components in a multiple-schedule conflict procedure in rats ($n = 18$ for each component). Baseline data were obtained the day prior to drug test days. There was a significant effect of overall ethanol treatment on both conflict and random interval response rates. Comparisons of individual means revealed that at least a 1 g/kg dose of ethanol was required to significantly increase punished responding during the conflict component ($^*p < 0.05$, Newman-Keuls test) and a 0.75 g/kg dose was required to significantly decrease responding during the food-alone, random interval component ($^*p < 0.05$, Newman-Keuls test). [Reproduced with permission from Aston-Jones *et al.*, 1984.]

Subsequently, the tension-reducing properties of alcohol have been demonstrated in a variety of behavioral situations (Cappell and Herman, 1972; Sepinwall and Cook, 1978; Pohorecky, 1981; Liljequist and Engel, 1984). Alcohol produces anticonflict effects in the social interaction test (File and Hyde, 1978; File, 1980; Lister and Hilakivi, 1988) and the elevated plus maze (Pellow and File, 1986; Lister, 1987), and alcohol reduces the negative contrast associated with dramatic shifts in reinforcer value (Becker and Flaherty, 1982). Alcohol also produces an anticonflict effect in the lick-suppression test (Vogel *et al.*, 1971) and in a modification of the Geller–Seifter procedure (Geller and Seifter, 1960) using incremental shock (Pollard and Howard, 1979; Aston-Jones *et al.*, 1984; Koob *et al.*, 1984, 1986, 1989; Liljequist and Engel, 1984; Thatcher-Britton and Koob, 1986) (**Fig. 5.4**). Effective doses in rats range between 0.5 and 1.0 g/kg intraperitoneally which produce blood alcohol levels up to 0.077 g% 1 h postinjection (Morse *et al.*, 2000). These same doses of alcohol produce dose-dependent decreases in responding during the unpunished component which presumably reflects the acute motor-debilitating effects of alcohol (Aston-Jones *et al.*, 1984; Koob *et al.*, 1984). These anticonflict effects of alcohol also show rapid tolerance with repeated administration every 8 h, an effect not observed with benzodiazepines (Koob *et al.*, 1987).

PHARMACOKINETICS

Ethanol or ethyl alcohol or alcohol has been described as the 'universal solvent'. It is readily miscible in water and has low lipid solubility ($P_{oil/water}$ = 0.035; $P_{membrane/buffer}$ = 0.096) (Lindenberg, 1951; Leo *et al.*, 1971; McCreery and Hunt, 1978). As such, alcohol readily crosses cell membranes, although at equilibrium the membrane concentration of alcohol is only 3–9 per cent as high as in an aqueous medium (Kalant, 1996). Alcohol is absorbed in the stomach (20 per cent) and small intestine (80 per cent). In normal adults, 80–90 per cent of absorption occurs in 30–60 min. It is well documented that absorption of alcohol in lipid mediums is delayed if food is present in the stomach and the total amount of alcohol ingested is reduced, up to 4–6 h (Goldberg, 1943) (**Fig. 5.5**). By definition, absorption of a drug is the movement of the drug into the blood stream. Blood alcohol levels have become the standard by which to assess not only the absorption of alcohol but also the dosing between animals and humans and between experimental conditions. Blood alcohol levels are measured in g% where *0.08 g alcohol/100 ml = 0.08 g% = 17 mM*. This is the legal limit of intoxication throughout the United States. The amount one can drink to obtain a given blood alcohol level is illustrated in **Table 5.3,** but in general for a male weighing 150 lbs, 4 ounces of spirits (100 proof or 50 per cent alcohol), 4 glasses of wine, or 4 beers will result in a blood alcohol level of 0.10 g%. For a female weighing 150 lbs, the same amounts would result in a blood alcohol level of 0.12 g%. The difference in blood alcohol levels in males and females has been attributed both to the differences in distribution of body fat mass, with more fat per kilogram (thus less water) for females, and to lower gastric levels of the alcohol metabolizing enzyme alcohol dehydrogenase in females (Frezza *et al.*, 1990; Baraona *et al.*, 2001). Given that there are no blood alcohol differences observed following intravenous alcohol administration, the most critical factor appears to be the difference in gastric metabolism (Baraona *et al.*, 2001).

Alcohol is largely eliminated by its catabolism in the liver by three enzymatic pathways (Riveros-Rosas *et al.*, 1997; Nagy, 2004) (**Fig. 5.6**). Only 5–10 per cent of alcohol is excreted unchanged from the lungs or in the form of urine. A factor of 2100 will convert breath alcohol to arterial blood concentration, and the concentration in breath alcohol is linear with the blood alcohol concentration which is the basis for the 'breathalyzer' test (Pennington and Brien, 1964; Preston, 1969). Alcohol is broken down to acetaldehyde in the liver by the rate-limiting enzyme alcohol dehydrogenase which is responsible for a large part of

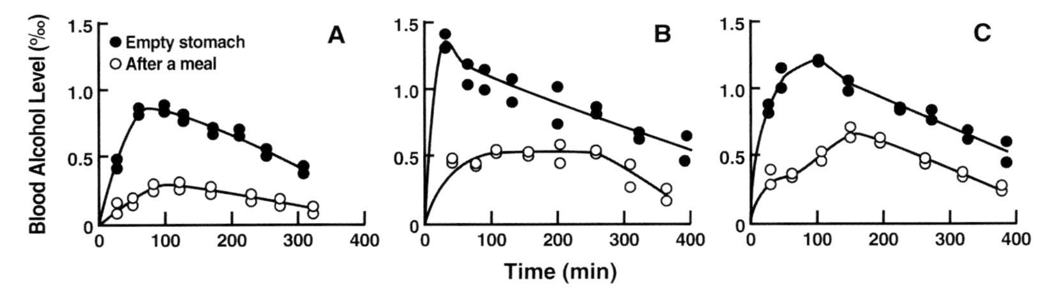

FIGURE 5.5 Influence of a meal, taken with alcohol, on blood alcohol levels. Closed circles correspond to the blood alcohol level when the alcohol was taken on an empty stomach, and open circles correspond to the alcohol level after consuming the same quantity of alcohol with a meal. Units are expressed as ‰, which refers to *Δ/baseline per mil*. In an example of such annotation, 1.5‰ would be equivalent to 150 mg%. **(A)** This 'food curve' shows a large depression in relation to the fasting curve with a low maximum and a slowly descending limb. This finding is interpreted as the result of a great delay in absorption, both with respect to rate and time, with absorption continuing for 4–6 h. **(B)** A plateau was maintained for the 2–3 h interval after a delayed absorption period. The horizontal course was followed by a rather rapid fall, implying that absorption, distribution, and combustion balance each other. When absorption is completed, distribution is maintained which caused the rapid fall of the postabsorptive phase. **(C)** The postabsorptive period had a parallel course to that of the fasting curve. Absorption in this case was delayed to such an extent that the maximum appeared later, and the normal over-shooting of alcohol was prevented. The distribution phase was not reflected in the blood alcohol level as under fasting conditions, and the rate of disappearance of alcohol from the blood stream during the postabsorptive period was the same as in fasting. [Reproduced with permission from Goldberg, 1943.]

TABLE 5.3 Blood Alcohol Level (g%) Estimations for Men and Women

BAL for men								
	Body weight (lb)							
Drinks	100	120	140	160	180	200	220	240
1	0.04	0.03	0.03	0.02	0.02	0.02	0.02	0.02
2	0.08	0.06	0.05	0.05	0.03	0.03	0.03	0.03
3	0.11	0.09	0.08	0.07	0.06	0.06	0.05	0.05
4	0.15	0.12	0.11	0.09	0.08	0.08	0.07	0.06
5	0.19	0.16	0.13	0.12	0.11	0.09	0.09	0.08
6	0.23	0.19	0.16	0.14	0.13	0.11	0.10	0.09
7	0.26	0.22	0.19	0.16	0.15	0.13	0.12	0.11
8	0.30	0.25	0.21	0.19	0.17	0.15	0.14	0.13
9	0.38	0.31	0.27	0.23	0.21	0.19	0.17	0.16
BAL for women								
	Body weight (lb)							
Drinks	100	120	140	160	180	200	220	240
1	0.05	0.04	0.03	0.03	0.03	0.02	0.02	0.02
2	0.09	0.08	0.07	0.06	0.05	0.05	0.04	0.04
3	0.14	0.11	0.10	0.09	0.08	0.07	0.06	0.06
4	0.18	0.15	0.13	0.11	0.10	0.09	0.08	0.08
5	0.23	0.19	0.16	0.14	0.13	0.11	0.10	0.09
6	0.27	0.23	0.19	0.17	0.15	0.14	0.12	0.11
7	0.32	0.27	0.23	0.20	0.18	0.16	0.14	0.13
8	0.36	0.30	0.26	0.23	0.20	0.18	0.17	0.15
9	0.41	0.34	0.29	0.26	0.23	0.20	0.19	0.17
10	0.45	0.38	0.32	0.28	0.25	0.23	0.21	0.19

Subtract 0.01 g% for each 40 min of drinking.
1 drink = 1.25 ounces 80 Proof liquor, 12 ounces beer, or 5 ounces wine.
[http://www.alcohol.vt.edu/Students/alcoholEffects/estimatingBAC/]

the oxidation of alcohol. Acetaldehyde then is broken down to acetic acid, water, and carbon dioxide by acetaldehyde dehydrogenase. Allelic variation in this pathway is responsible for differential alcohol elimination in the human population and individual differences in elimination. Inactivation of acetaldehyde dehydrogenase-2 (ALDH2; the major liver mitochondrial ALDH which has a very low K_m for acetaldehyde) provides the basis for the alcohol-induced flush reaction seen in Asian populations, similar to the effects of the antirelapse prevention medication disulfiram (Antabuse) if one drinks alcohol (see below). The flush reaction is characterized by marked facial flushing, tachycardia, hypotension, elevation of skin temperature, increased body sway, nystagmus, and more intense feelings of intoxication (Wall *et al.*, 1993).

In certain ethnic groups, a mutation in two nucleotides of the ALDH2 gene produces a complete inactivity of acetaldehyde dehydrogenase if one is homozygous for both alleles and a partial blockade if one is heterozygous. The Japanese, Chinese, and Vietnamese show approximately 40–50 per cent of the population heterozygous for the ALDH2 gene mutation (Goedde *et al.*, 1983a,b; Mizoi *et al.*, 1983). The Japanese are approximately 7 per cent homozygous for the mutation (Higuchi *et al.*, 1995). Koreans show 30 per cent heterozygous and approximately 3 per cent homozygous for the mutation (Goedde *et al.*, 1989; Agarwal and Goedde, 1992). Native North Americans also show a very low incidence of the ALDH2 mutation. In contrast, native South Americans show a high incidence of the mutation (Goedde *et al.*, 1986). Europeans show virtually 0 per cent of the population with the mutation (Goedde *et al.*, 1983a,b; Mizoi *et al.*, 1983) (**Table 5.4**).

Inactivation of ALDH2 also provides the basis for the pharmacological effects of the antirelapse medication disulfiram (Antabuse). Disulfiram in average therapeutic doses of 250 mg/day (not to exceed 500 mg/day) produces a blockade of acetaldehyde

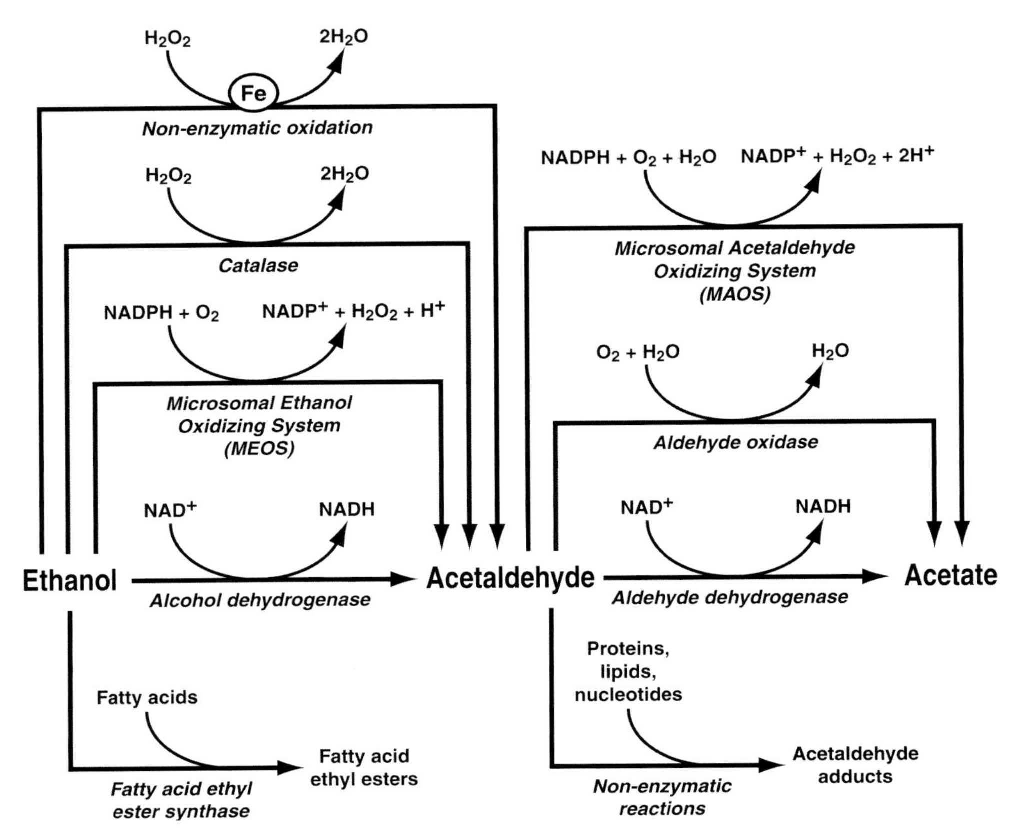

FIGURE 5.6 Principal metabolic pathways related to ethanol and acetaldehyde metabolism. In the liver, there are three metabolic systems capable of carrying out ethanol oxidation. (1) Alcohol dehydrogenases are found in the cytosol of different tissues which make up the human body, principally hepatic. These enzymes promote the oxidation of ethanol into acetaldehyde, coupling this oxidation with the reduction of a nicotinamide adenine dinucleotide (NAD^+). (2) Hepatic cell peroxisome oxidation couples ethanol oxidation into acetaldehyde with the simultaneous decomposition of hydrogen peroxide in a reaction catalyzed by the catalase enzyme. (3) The microsomal ethanol oxidizing system (MEOS) requires cytochrome P-450 participation by coupling ethanol and nicotinamide adenine dinucleotide phosphate oxidation (NADPH) to the reduction of an oxygen molecule to form hydrogen peroxide. Note that excessive nicotinamide adenine dinucleotide reductase (NADH) levels can inhibit glucose production (i.e., gluconeogenesis) and breakdown (i.e., oxidation) of fat molecules as well as stimulate production of fat molecules (Weathermon and Crabb, 1999). [Reproduced with permission from Riveros-Rosas *et al.*, 1997.]

dehydrogenase (Medical Economics Company, 2004). A disulfiram flush reaction after an alcohol challenge was observed in 48 of 52 volunteers after 100–200 mg doses of disulfiram (Christensen *et al.*, 1991; Johansson *et al.*, 1991).

A second major pathway for alcohol elimination is the microsomal ethanol oxidizing system (MEOS) catalyzed by cytochrome P4502E1-CYP2E1. Activity of this system is responsible for metabolic tolerance and can contribute to the production of highly toxic metabolites in the liver of alcoholics (Lieber, 1997). In a 150 lb person, the average rate of metabolism of pure alcohol is approximately 7–9 ml/h (0.3 ounces/hour). This converts to approximately 0.01–0.015 per cent per hour or approximately half of a standard drink per hour (Goldstein, 1983; Jones *et al.*, 1993). Alcoholics can double this rate of metabolism by the phenomenon of metabolic tolerance, and this is hypothesized to be due largely to the induction of the liver enzyme CYP2E1 although some have argued for increased activity of alcohol dehydrogenase as well (Buris *et al.*, 1985; Lieber, 1997). P450IIE1 is increased 4 times in liver biopsies of alcoholics (1 day of abstinence) (Tsutsumi *et al.*, 1989). This type of tolerance results in lower blood alcohol levels for the tolerant individual than the nontolerant individual for the same dose (Mendelson *et al.*, 1965; Buris *et al.*, 1985).

A third pathway for alcohol metabolism is a nonoxidative pathway catalyzed by fatty acid ethyl ester (FAEE) synthase which leads to the formation of fatty

TABLE 5.4 Frequency of Acetaldehyde Dehydrogenase Isozyme Deficiency in American Indians, Asian Mongoloids, and other Populations

Population	Sample size	% Deficient	Reference
South American Indians			
Atacameños (Chile)	133	43	Goedde *et al.*, 1984b
Mapuche (Chile)	64	41	Goedde *et al.*, 1986
Shuara (Ecuador)	99	42	Goedde *et al.*, 1983b
North American Indians			
Sioux (North Dakota)	90	5	Goedde *et al.*, 1986
Navajo (New Mexico)	56	2	Goedde *et al.*, 1986
Mexican Indians			
Mestizo (Mexico City)	43	4	Goedde *et al.*, 1986
Asian Mongoloids			
Japanese	184	44	Goedde *et al.*, 1979, 1983a Agarwal *et al.*, 1981
Chinese			
Mongolian	198	30	Goedde *et al.*, 1984a
Zhuang	106	45	Goedde *et al.*, 1984a
Han	120	50	Goedde *et al.*, 1984a
Korean (Mandschu)	209	25	Goedde *et al.*, 1984a
Chinese (living abroad)	196	35	Goedde *et al.*, 1983b
South Korean	75	27	Goedde *et al.*, 1983c
Vietnamese	138	53	Goedde *et al.*, 1983c
Indonesian	30	39	Goedde *et al.*, 1983c
Thai (North)	110	8	Goedde *et al.*, 1983c
Filipino	110	13	Goedde *et al.*, 1986
Ainu	80	20	Goedde *et al.*, 1986
Other populations			
German	300	0	Goedde *et al.*, 1983c
Egyptian	260	0	Goedde *et al.*, 1983c
Sudanese	40	0	Goedde *et al.*, 1983c
Kenyan	23	0	Goedde *et al.*, 1983c
Liberian	184	0	Goedde *et al.*, 1983c
Turk	65	0	Goedde *et al.*, 1986
Fang	37	0	Goedde *et al.*, 1986
Israeli	77	0	Goedde *et al.*, 1986
Asian Indian	50	0	Goedde *et al.*, 1986

[Reproduced with permission from Goedde *et al.*, 1986.]

acid esters. FAEEs are found in the highest concentrations in organs that are susceptible to the toxic effects of alcohol, including the liver (Best and Laposata, 2003).

ABUSE AND ADDICTION POTENTIAL

Alcoholism

Alcoholism is equivalent to Substance Dependence on Alcohol (with a 'big D'; see *What is Addiction?* chapter) as defined by *Diagnostic and Statistical Manual of Mental Disorders*, 4th edition (DSM-IV) (American Psychiatric Association, 1994). Alcoholism is a chronic relapsing disorder and is characterized by a preoccupation with obtaining alcohol, loss of control over its consumption, development of tolerance, dependence ('little d'), and impairment in social and occupational functioning. The stages of alcoholism vary significantly and can be manifest at any time of life and take on many forms. A representative time course associated with a person in a work situation is shown in **Table 5.5** (Cline, 1975).

Patterns of alcohol use vary with a given culture (**Table 5.6**), and per capita alcohol consumption varies across countries from a low of 1 liter per capita in India to a high of over 11 liters per capita in France (**Table 5.7**). France, Spain, UK, Greece, and Italy have

TABLE 5.5 Progressively Worsening Alcoholic Employee Behavior Patterns

	Observable signs		
Employee behavior pattern	***Absenteeism***	***General behavior***	***Performance on the job***
Early stage Drinking to relieve tension Increase in tolerance Memory lapses	Tardiness at lunchtime Early departure	Overreacts to real or imagined criticism Complains of not feeling well	Misses deadlines Lowered job efficiency
Middle stage Sneaking drinks Feeling guilty Tremors Loss of interest	Frequent days off for vague ailments or implausible reasons	Marked changes (e.g., statements are not dependable, begins avoiding associates, repeated minor injuries on and off job)	Spasmodic work pace Lapses of attention Cannot concentrate
Late Middle stage Unable to discuss problems Efforts for control fail Neglect of food Drinking alone	Frequent time off (perhaps several days) Does not return from lunch	Domestic problems interfere with work Financial problems More frequent hospitalization Will not discuss problems	Far below what is expected
Terminal stage Now thinks, 'My job interferes with my drinking'	Prolonged unpredictable absences	Completely undependable Repeated hospitalization Physical deterioration visible Serious financial problems Serious family problems	Uneven Generally incompetent

Source: Cline, 1975.

TABLE 5.6 Cultural Differences in Patterns of Alcohol Consumption

France. As shown in **Table 5.7**, France has the highest per capita alcohol consumption (approximately 12 liters per person per year) of any country. Two-thirds of this total consumption is in the form of wine drinking. For men, and to a lesser extent women, wine is drunk at meals as part of the diet. This has declined, however, in the last 50 years. Wine also is consumed as part of social and festive occasions, and as part of an entirely aesthetic experience (i.e., for oenological reasons). There has existed a popular French belief that 'wine is not alcohol' (which stems from the mid-1800s when Pasteur deemed wine as beneficial and 'hygienic' because it could kill microbes) which culturally persists today. The word *alcohol* had two meanings, one in the singular referring to a chemical element in fermented beverages, and the other referring to distilled drinks. The majority of habitual or excessive drinkers are men—although drinking has increased in women—and alcohol consumption remains an expression of the masculine identity (Nahoum-Grappe, 1995). Culturally, there is debate in France whether wine is considered 'alcohol'.

Italy. Similar to France, wine pervades most public and private sectors of life in Italy, with wine drinking the most prevalent. Italians rank 5th in alcohol consumption in **Table 5.7**. Abstainers are viewed with curiosity, as 'odd persons' who should explain why they do not drink. In young people, males drink more than females, more wine in northeastern Italy in the home, and more beer in southern Italy outside the home. Though Italians do not stigmatize heavy consumption, they do tend to blame problem drinkers who lose control. Italian physicians are known to suggest moderate wine consumption to prevent cardiovascular disease, an approach recently receiving scientific justification, but both media and medical lobbies of late are increasingly emphasizing the breadth of alcohol-related problems, particularly drunk driving and cirrhosis of the liver (Cottino, 1995).

Sweden. Sweden has half the alcohol consumption of France, but Swedes often are characterized as binge-drinkers, as one of the countries of the so-called 'vodka belt' across northeastern Europe, a long-standing drinking pattern that has ancient cultural roots. The most prevalent beverage is beer being drunk on rare occasions, usually weekends; therefore, a small part of the population is responsible for a large part of total consumption. For men, getting deliberately intoxicated is a sign of masculinity, but for women, the act can bring condemnation. Swedes also look down on those with persistent alcohol problems, a blame directed toward the individual rather than on the family or workplace. Even though the Swedes are labeled as binge-drinkers, they have a relatively low incidence of alcohol-related problems compared to other Western countries and a significantly lower overall consumption (Nyberg and Allebeck, 1995) (**Table 5.7**).

China. China ranks low on total alcohol consumption, consuming significantly less than half of France, with the most being in the form of spirits (liquors). The Chinese stress the role of alcohol in all aspects of life, affecting structures of belief, behavior, values, attitudes, and religion. It is an important adjunct to hospitality, with both the host and guest honoring each other with drink as a combined expression of welcome and appreciation. Everyone toasts for the Chinese New Year, for good health, and for long-lasting prosperity for their elders. Alcohol still plays a role in Chinese medicine (in fact, the Chinese written characters for 'alcohol' and 'medicine' share the same root). The majority of men (64%) are moderate drinkers, while the majority of women (51%) abstain. A very low percentage of the population (10%, all male), however, are heavy drinkers. Overall, approximately 30% abstain (Chi *et al.*, 1989). There is a high proportion of individuals (50%) with the inactive allele for acetaldehyde dehydrogenase which produces a flushing reaction which is hypothesized to form a protective factor against excessive drinking and subsequent alcoholism. However, socially, the Chinese people seem generally unaware of the flushing reaction, and if they are, it has little or no impact on their attitudes in relation to alcohol despite the high abstinence rate and very low heavy drinking (Xiao, 1995).

TABLE 5.7 Representative Listing by the World Health Organization of Per Capita Alcohol Consumption (in liters) in 2000

Country	Total	Beer	Wine	Spirits
France	11.7**	2.0**	7.1**	2.6**
Spain	11.17	3.76	4.61	2.79
United Kingdom*	10.39	4.86	2.10	1.50
Greece	9.16	2.10	4.82	2.24
Italy	9.16	1.47	7.11	0.58
United States	9.08	5.11	1.67	2.29
Venezuela	7.28	4.86	0.07	2.34
Sweden*	6.86	2.77	1.86	1.00
Japan	6.26	2.30	0.41	2.81
Chile	6.05	1.18	4.25	0.63
China	5.17	1.03	0.09	4.04
Brazil	4.79	2.40	0.36	2.03
South Korea	4.13	0.25	0	3.88
Mexico	4.01	2.92	0.30	0.79
Israel	2.06	0.93	0.25	0.87
India	1.01	0.03	0	0.98

These values estimate the level of alcohol consumption per adult (15 years or age or older) of pure alcohol during a calendar year as calculated from official statistics on production, sales, import and export, taking into account stocks whenever possible. Conversion factors used to estimate amount of pure alcohol in (barley) beer is 5%, wine 12%, and spirits 40% of alcohol (other conversion factors were used for some types of beer and other beverages). Data were collected, and calculations were made mainly using three sources: FAOSTAT - United Nations Food and Agriculture Organization's Statistical Database, World Drink Trends, regularly published by Produktschap voor Gedistilleerde Dranken (Netherlands), and in some cases direct government data. Data are available from the World Health Organization 'Adult Per Capita Alcohol Consumption' database for most countries of the world since 1961 onward. Data are presented for the groups of total, and also beer, wine, and spirits separately. The category *Beer* includes data on barley, maize, millet, and sorghum beer combined. The amounts from beer, wine, and spirits do not necessarily add up to the presented total, as in some cases the total includes other beverage categories, including palm wine, vermouths, cider, fruit wines, and so on. It is important to note that these figures comprise in most cases the recorded alcohol consumption only and have some inherent problems. Factors which influence the accuracy of per capita data are: informal production, tourists, overseas consumption, stockpiling, waste, spillage, smuggling, duty-free sales, variation in beverage strength, and the quality of the data upon which the this table is based. In some countries there exists a significant unrecorded alcohol consumption that should be added for a comprehensive picture of total alcohol consumption.
*Data from 2001.
**Data from World Drink Trends, 2002.

the highest per capita consumption which corresponds to a pattern of high wine drinking pervading all facets of daily life, including a high incidence of emergency room visits. However, the amount of wine consumption in France has been steadily declining since 1970, dropping from 12 liters per capita to 7 liters per capita in 2000 (World Drink Trends, 2002).

A strong genetic component is well established in alcoholism. Twin registry and adoption studies have shown that the heritability of alcoholism may be as high as 50–60 per cent (Prescott and Kendler, 1999) (see *What is Addiction?* chapter). The molecular genetic basis for this heritability is largely unknown, but several lines of research have provided some insights into possible genetic factors.

Early studies on alcoholism identified that alcoholics or certain subgroups of alcoholics have reduced cerebrospinal fluid level of 5-hydroxyindoleacetic acid (5- HIAA), the major metabolite of serotonin (Ballenger *et al.*, 1979; Linnoila *et al.*, 1994). A series of studies has highlighted the role of serotonin in impulse control (Virkkunen and Linnoila, 1993), and low cerebrospinal 5-HIAA was primarily associated with increased irritability and impaired impulse control in violent alcoholics with antisocial personality disorder (Virkkunen *et al.*, 1994). These characteristics are hallmarks of the young male alcoholic, or Type II alcoholic as described by Cloninger (Cloninger *et al.*, 1981).

Alcohol reinforcement has long been associated with activation of opioid peptides in animal studies since opioid receptor antagonists block alcohol self-administration (Altshuler *et al.*, 1980), and opioid antagonists have been shown to have efficacy in treating alcoholism in humans (O'Malley *et al.*, 1992; Volpicelli *et al.*, 1992). A recent study has identified a functional polymorphism of the μ opioid receptor gene (A118G single nucleotide polymorphism in exon I) that was over-represented in a population of Swedish alcoholics (11 per cent), and there was a significant overall association between genotypes with the 118G allele and Substance Dependence on Alcohol (Bart *et al.*, 2005). Other groups have reported that the presence of this allele variant increases the stress response and therapeutic response to naltrexone in the treatment of alcoholism (O'Malley *et al.*, 2002; Wand *et al.*, 2002). Receptors encoded by this A118G variant bind the endogenous opioid peptide β-endorphin with a three-fold greater affinity (Bond *et al.*, 1998). Other studies have not found an association between the A118 allele or the 118G allele and alcoholism (Kranzler *et al.*, 1998; Schinka *et al.*, 2002). Together these results point to two neurochemical systems that may contribute to the genetic vulnerability for alcoholism.

Alcohol Withdrawal

Alcohol withdrawal has long been characterized as a latent state of hyperexcitability, representing a 'rebound' phenomenon from the previously chronically depressed central nervous system. This is a reasonably accurate description of the somatic signs of alcohol withdrawal but does not address the motivational effects of alcohol withdrawal. In humans, alcohol withdrawal from ingestion of 10–15 drinks per day over a ten-day period leads to time-dependent effects that begin within a few hours of abstinence depending

TABLE 5.8 Stages of Alcohol Withdrawal in Humans Following Ingestion of 10–15 Drinks Per Day Over a Ten-Day Period

Stage	Period of withdrawal	Signs	Comments
Early Stage	From a few hours to 36–48 h	Anxiety	
		Anorexia	
		Insomnia	
		Tremor	This tremulous phase (shakes) often is observed on admission and can be severe enough to prevent a patient from feeding himself.
		Mild disorientation	
		Convulsions	Between 24 and 48 h; similar to tonic-clonic seizures of grand mal epilepsy
		Sympathetic response	Elevated blood pressure and body temperature; increased heart rate (heart rate is a useful index of continuing toxicity and may warn of impending delirium tremens)
Late Stage	Begins after 2–4 days and can last 2–3 days	Delirium tremens	The most severe withdrawal states are: • marked tremor, anxiety, and insomnia • marked paranoia and disorientation • severe autonomic overactivity, including sweating, nausea, vomiting, diarrhea, and fever • agitation • vivid hallucinations • reality testing fails, and patient must be protected from self-harm during outbursts of irrational behavior • seizures are rare in this stage • *high* fever, in which *prognosis is poor* and can be life-threatening • death usually occurs secondary to complicating illness, infection, or injury • shock and hyperthermia can be fatal • fatality rate used to be upward of 15% but now has decreased to 1–2%

Source: Goldstein, 1983.

on the blood alcohol level obtained (Goldstein, 1983) (**Table 5.8**). Early stages (up to 36 h) are characterized by tremor, elevated sympathetic responses, including increase in heart rate, blood pressure, and body temperature. Such physical signs are accompanied by insomnia, anxiety, anorexia, and dysphoria. Late stages of withdrawal, if left untreated (which is now rare), can include more severe tremor, sympathetic responses, anxiety, and delirium tremens. Delirium tremens is associated with vivid hallucinations and psychotic-like behavior. High fever can result without treatment and could be life-threatening (Emerson, 1932, p. 295), but again is rare in modern times because of the availability of benzodiazepines for treatment (Sereny and Kalant, 1965; Golbert *et al.*, 1967).

In rodents, alcohol withdrawal is characterized by irritability, hyperresponsiveness, abnormal motor responses, anxiety-like responses, and decreased reward sensitivity. Seizures, or auditory-induced seizures, can occur at high doses (Majchrowicz, 1975) (**Fig. 5.7**). With high doses (9.15 g/kg/day), alcohol withdrawal signs in rats become maximal after 3–4 days (Majchrowicz and Hunt, 1976). The anxiety-like responses and seizures can be 'kindled,' that is they increase with repeated withdrawal (Becker and Hale, 1993) (**Fig. 5.8**). As with all other major drugs of abuse, chronic administration of alcohol produces elevations in brain reward thresholds during withdrawal (Schulteis *et al.*, 1995) (**Fig. 5.9**).

Alcohol Tolerance

Multiple forms of tolerance develop to the effects of alcohol but can be divided into two major types: dispositional and functional. With dispositional tolerance, chronic use of alcohol can lead to an increased capacity to breakdown (metabolize) alcohol by the induction of P4502E1 (Pringsheim, 1908; Hawkins *et al.*, 1966) (see above). This increase can double, reaching a level of 30–40 mg/dl/h compared to a nontolerant average adult who will metabolize alcohol at a rate of

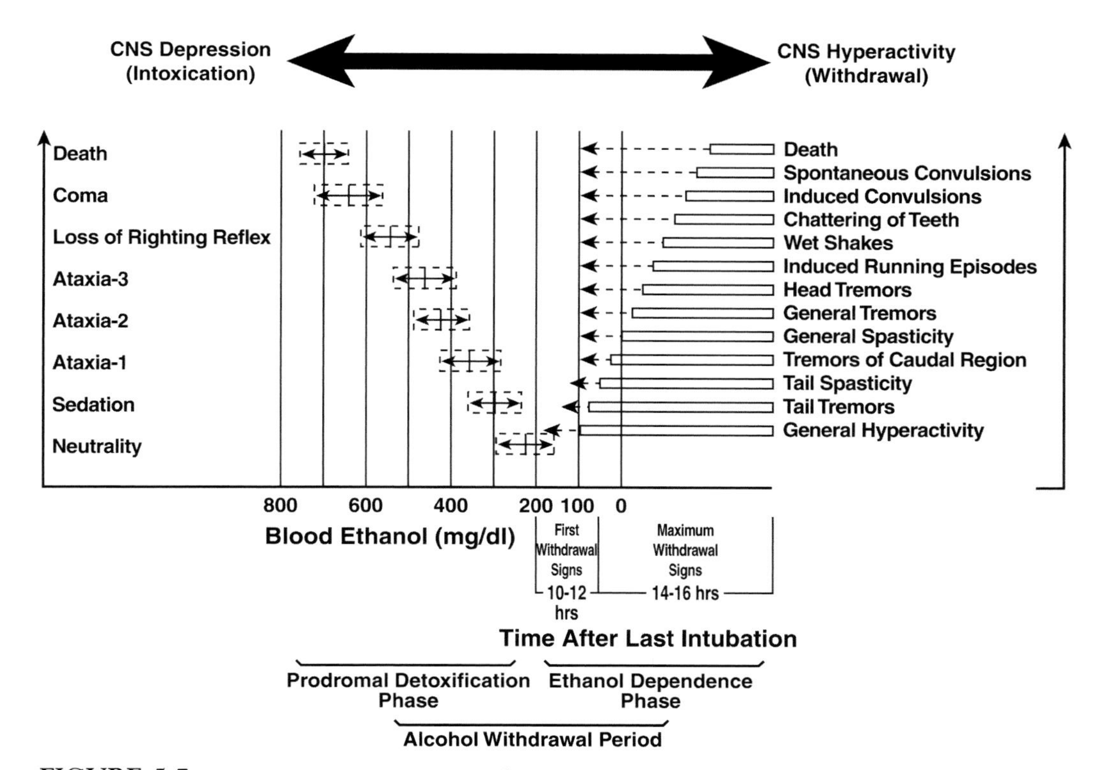

FIGURE 5.7 Spectrum and continuum of ethanol intoxication and withdrawal in rats. The entire withdrawal period is divided into two phases: prodromal detoxification and ethanol dependence. The extreme ethanol intoxication may terminate in death due to central nervous system depression, whereas the ethanol withdrawal phase may terminate in death due to central nervous system hyperactivity (as shown by the wide arrow at the top of the diagram). The signs and reactions of both phases increase in severity from the bottom to the top (as indicated by the vertical arrows on either side). The signs of the detoxification phase are directly related to blood alcohol level (heavy vertical lines represent the means, and dotted-box widths represent the standard deviations). Signs and reactions of the ethanol dependence phase are arranged according to subjective evaluation relative to order of appearance, specificity, frequency of occurrence, and degrees of severity. The earliest, least specific, most frequently observed, and least severe signs and reactions are at the bottom of the diagram. [Reproduced with permission from Majchrowicz, 1975.]

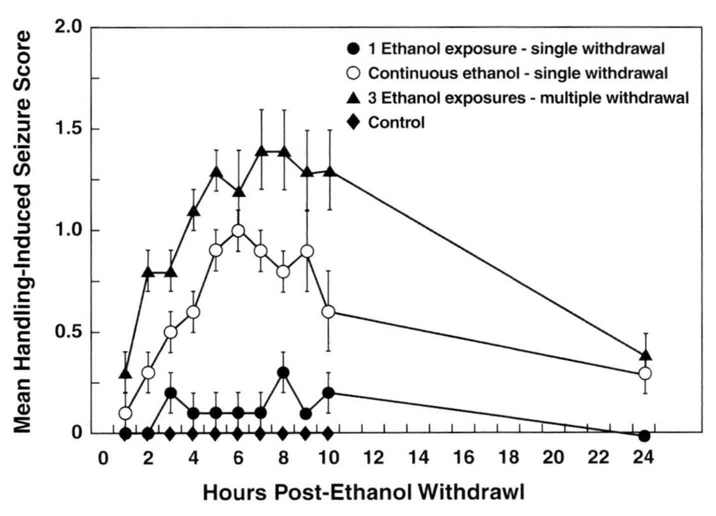

FIGURE 5.8 The progressive development of handling-induced seizures during multiple alcohol withdrawal episodes in male C3H mice. The multiple withdrawal group received three cycles of 16 h ethanol vapor separated by 8 h periods of abstinence. A single withdrawal group received a single bout of ethanol exposure (16 h). A second group experienced a single withdrawal episode after receiving the equivalent amount of ethanol intoxication as the multiple withdrawal group ($16 \times 3 = 48$ h), but received it continuously (uninterrupted). The control group did not receive ethanol. The severity of handling-induced seizures was greatest for the multiple withdrawal group, intermediate for the continuous ethanol—single withdrawal group, and minimal for the single ethanol exposure—single withdrawal group. The incidence of spontaneous handling-induced seizures was virtually negligible (Control group). Peak seizure intensity was reached approximately 6–8 h after withdrawal for all ethanol-exposed groups and generally subsided by 24 h, although withdrawal signs still were evident in the multiple withdrawal and continuous ethanol—single withdrawal groups at this time. [Reproduced with permission from Becker and Hale, 1993.]

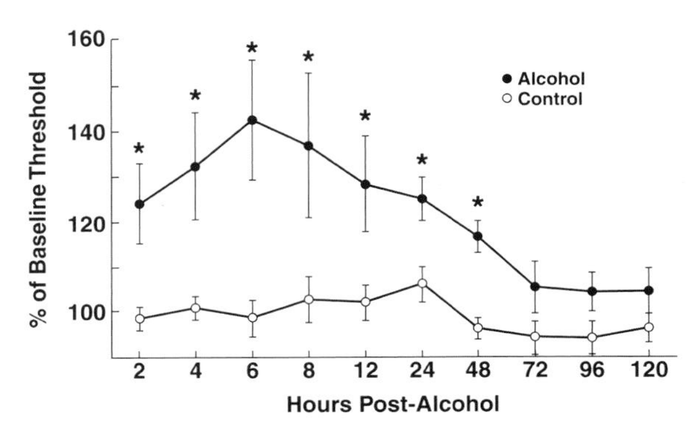

FIGURE 5.9 Time-dependent elevation of intracranial self-stimulation thresholds in rats during alcohol withdrawal. Alcohol dependence was induced by means of alcohol vapor chambers (22 mg/liter). Mean blood alcohol limits of 197.29 mg% were achieved after 10 days and were maintained for at least 7 days prior to the onset of withdrawal. Data are expressed as mean ± SEM percentage of baseline threshold. Asterisks (*) indicate thresholds that were significantly elevated above control levels at 2–48 h postethanol ($p < 0.05$). Open circles indicate the control condition. Closed circles indicate the ethanol withdrawal condition. [Reproduced with permission from Schulteis *et al.*, 1995.]

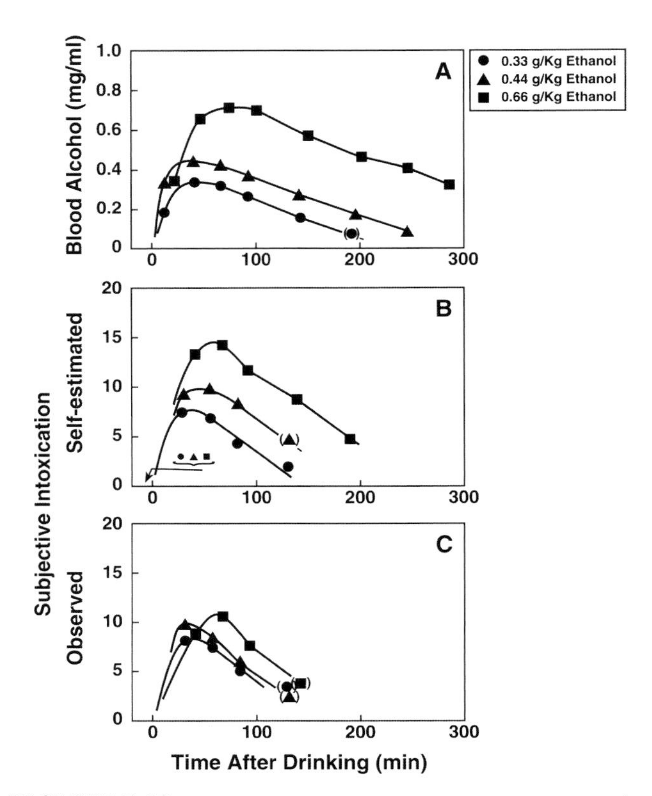

FIGURE 5.10 Comparison of effects of 0.33, 0.44, and 0.66 g/kg ethanol on blood alcohol levels, self-estimated subjective intoxication, and observed degree of intoxication. Alcohol was taken as whisky on an empty stomach, and the time allowed for drinking was 10–15 min. Means within parentheses include one zero and thus may be too high. [Modified with permission from Ekman *et al.*, 1964.]

15–20 mg/dl/h (Mendelson *et al.*, 1965; Ellenhorn and Barceloux, 1988). Such dispositional tolerance takes 3–14 days of drinking and can take several weeks of abstinence to return to normal levels (e.g., a group of detoxified alcoholics drinking a 43% alcohol beverage every 4 h for a total maximum of 26 ounces per day, reaching blood alcohol levels of 0.05–0.19 g%/day) (Mendelson *et al.*, 1965).

Functional tolerance requires a higher blood level to achieve intoxication and can be much more dramatic. Functional tolerance can be divided into acute, chronic, and behavioral categories. Functional tolerance has been hypothesized to be mediated by neuroadaptational changes in the central nervous system (Tabakoff, 1980; Tabakoff *et al.*, 1982; Tabakoff and Hoffman, 1992). Acute tolerance was defined originally as the 'Mellanby effect,' where a greater degree of impairment was observed at a given concentration of alcohol during the ascending portion of the blood alcohol curve than at the same given concentration of alcohol during the descending portion of the blood alcohol curve (Mellanby, 1919). Thus, as with other drugs of abuse there is a rapid initial tolerance within one drinking session where the subjective sense of intoxication follows the ascending limb of the blood alcohol concentration time course but not the descending limb (Ekman *et al.*, 1964) (**Fig. 5.10**). Such tolerance to the intoxicating effects of alcohol is not accompanied by a marked elevation of the lethal dose. Similar effects have been seen in tolerance to the acute motor-impairing effects of alcohol in animals (LeBlanc *et al.*, 1975; Goldstein, 1983) (**Fig. 5.11**).

Chronic functional tolerance in man is reflected in increased intake of alcohol to produce intoxication or a shift to the right in the concentration-response function. In humans, heavy drinkers were less sensitive to alcohol challenge than moderate drinkers or abstainers using the 'finger–finger' motor coordination test (Goldberg, 1943) (**Fig. 5.12**). In rodents, this chronic tolerance has largely been measured by changes in the sedative effects of alcohol usually reflected in the ability to perform a motor task, such as a treadmill or rotarod, following injection of alcohol (Jones and Roberts, 1968). Such motor effects are significantly influenced by environment and conditioning and led to the construct of behavioral tolerance. Here, animals (and humans) can learn either by Pavlovian or operant conditioning to counteract the effects of alcohol (Wenger *et al.*, 1981; Le and Kalant, 1990). Tolerance has been shown to the discriminative stimulus effects of

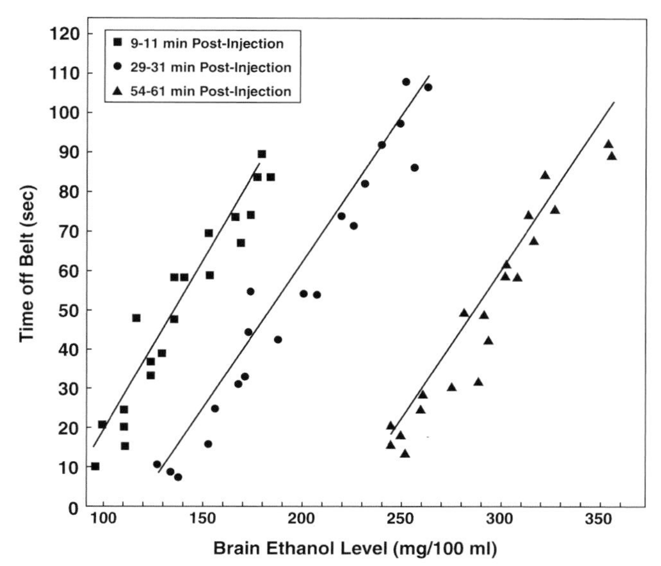

FIGURE 5.11 Error scores on the moving belt test in rats as a function of brain ethanol concentration at approximately 10, 30, and 60 min after administration of 1.0–2.8 g/kg ethanol. Measurements were made 9–11, 29–31, or 59–61 min after interperitoneal injection of ethanol. The slopes of the calculated regression lines are 0.805, 0.741, and 0.778, respectively. [Reproduced with permission from LeBlanc *et al.*, 1975.]

alcohol (Emmett-Oglesby, 1990) and conditioned place preference in rats (Gauvin *et al.*, 2000). However, conditioned place preference relies on drug/environment conditioning and thus could be attributed to general behavioral tolerance rather than tolerance to the positive reinforcing effects of alcohol.

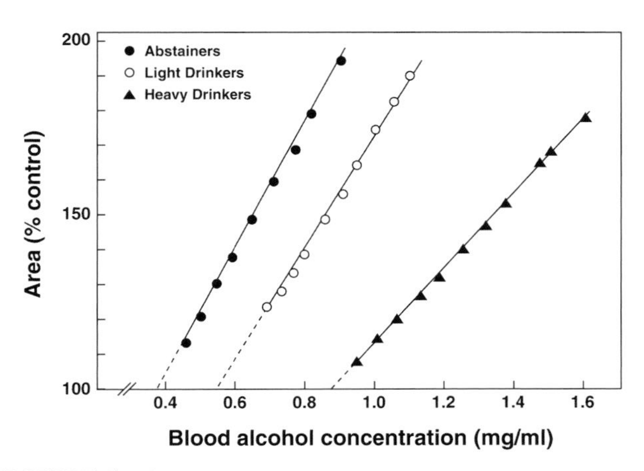

FIGURE 5.12 Performance on the 'finger–finger' test (in which subjects must touch the tip of the left index finger to the tip of the right index finger at arm's length) as a function of blood alcohol concentration in abstainers, light drinkers, and heavy drinkers administered 0.63–0.80, 1.0–1.42, and 1.2–1.35 g/kg alcohol, respectively. The nonoverlapping lines were described as evidence of tolerance to the motor-incoordinating effects of alcohol. [Reproduced with permission from Goldberg, 1943.]

Alcohol Toxicity

Alcohol can be acutely toxic, and overdose can occur under situations of binge drinking or situations of accidents fostered by ignorance of the potential toxicity. Numerous examples of young people overdosing on alcohol at parties, especially those involving challenges to drink excessively, have been covered extensively in the press. In 2001, the National Household Survey on Drug Abuse estimated approximately 6 per 10 000 first-listed alcohol-related diagnoses for discharge episodes from short-stay hospitals for persons aged 15–24 (a decline from 12 per 10000 in 1981) (Chen *et al.*, 2003). In contrast, direct deaths attributed to alcohol have remained stable at approximately 20 000 per year from the mid-1970s to the mid-1990s (Hoyert *et al.*, 1999), though that number since 1997 has decreased slightly (Kochanek and Smith, 2004). The incidence of emergency room visits caused by alcohol in combination with other drugs has increased from 167 000 in 1995 to over 207 000 in 2002, a 24 per cent increase (Substance Abuse and Mental Health Services Administration, 2003a). Others have estimated that over 1400 students aged 18–24 and enrolled in 2–4 year colleges died in 1998 due to alcohol-related unintentional injuries (Hingson *et al.*, 2002). 25 per cent of the 8 million college students in the United States drove under the influence of alcohol; 37 per cent of the 8 million rode with a drinking driver. Alcohol misuse-related damage is not limited to death and emergency room visits, but extends also to a wide range of individual and social consequences (Perkins, 2002) (**Table 5.9**). Half a million college students were unintentionally injured under the influence of alcohol, and an additional 100 000 were assaulted by another student who had been drinking.

As described by Shakespeare above, alcohol has long been associated with facilitation of sexual desire and behavior but at the same time, with impaired performance. Indeed, alcohol has been argued to either directly or indirectly contribute to decreased sexual response, erectile dysfunction, and other sexual dysfunctions (Miller and Gold, 1988).

In males, acute small amounts of alcohol have been associated with an increase in self-reported sexual arousal and a slight increase or decrease in sexual response as measured by penile vasocongestion (Briddell and Wilson, 1976; Farkas and Rosen, 1976; Rubin and Henson, 1976). However, higher doses of alcohol led to substantial reductions in sexual arousal as measured by self-report and penile vasocongestion, and an impaired ability to ejaculate (Briddell and Wilson, 1976; Malatesta *et al.*, 1979). Similar results for acute administration of alcohol to females have been

TABLE 5.9 Potential Negative Consequences of College Student Drinking

Damage to self	Academic impairment Blackouts Personal injury or death Short-term and long-term physical illness Unintended and unprotected sexual activity Suicide Sexual coercion/rape victimization Impaired driving Legal repercussions Impaired athletic performance
Damage to other people	Property damage and vandalism Fights and interpersonal violence Sexual violence Hate-related incidents Noise disturbances
Institutional costs	Property damage Student attrition Loss of perceived academic rigor Poor town relations Added time demands and emotional strain on staff Legal costs

From: Perkins, 2002.

reported (Malatesta *et al.*, 1982). While many women in self-report studies reported that low-dose alcohol use increases sexual pleasure (Beckman, 1979; Wilsnack, 1984; Klassen and Wilsnack, 1986), the sexual arousal and ability to achieve orgasm showed a dose-dependent decrease with increases in blood alcohol levels (Wilson and Lawson, 1976; Malatesta *et al.*, 1982).

Chronic alcohol use and alcoholism in males is associated with a higher frequency of sexual dysfunction (Mandell and Miller, 1983; Peugh and Belenko, 2001). Alcoholics in outpatient treatment reported 3 times the prevalence of serious erectile dysfunction compared to nonalcoholic-matched controls (O'Farrell *et al.*, 1998). Chronic alcohol use is associated with low testosterone and sperm count, smaller testes (Wright *et al.*, 1991), feminization, and gynecomastia (Gordon *et al.*, 1979). Some of these effects may persist into protracted abstinence (Lemere and Smith, 1973) but ultimately are reversible with abstinence (O'Farrell *et al.*, 1998). Alcoholic women were more likely to report increased sexual desire and enjoyment after drinking than nonalcoholic women (Beckman, 1979), and that alcohol relieved their sexual problems by relieving sexual inhibitions (Jensen, 1984; Klassen and Wilsnack, 1986). Expectations of disinhibition of sexuality may serve as a motive for drinking in women (Beckman and Ackerman, 1995). Alcoholic women also reported more guilt surrounding sexual behavior and sexuality than nonalcoholic women (Pinhas, 1980). Alcoholism in women is associated with a number of sexual dysfunction problems, including low sexual desire, lack of orgasm, contractions of the vagina that interfere with intercourse, and painful intercourse (Covington and Kohen, 1984; Beckman and Ackerman, 1995). Thus, sexual dysfunction may precede or result from high alcohol use in women (Beckman and Ackerman, 1995).

Heavy alcohol consumption in women is also associated with menstrual dysfunctions, including anovulation, recurrent amenorrhea, early menopause (Hugues *et al.*, 1980), and a higher incidence of spontaneous abortions (Harlap and Shiono, 1980). These dysfunctions may be related to the hormonal changes in females associated with acute and chronic alcohol ingestion. Acute alcohol intoxication elevated estradiol in females in both pre- and post-menopausal women (Mendelson *et al.*, 1981; Mendelson *et al.*, 1988; Ginsburg *et al.*, 1996). Low doses of alcohol elevated testosterone and estradiol in women on oral contraceptives, but there was no increase in estradiol and there was less of an increase in testosterone in women who were not currently using oral contraceptives (Sarkola *et al.*, 1999, 2000). The authors hypothesized that the alcohol-induced increases in estradiol and testosterone may be related to a decrease in steroid catabolism (Sarkola *et al.*, 1999). There is also a significant relationship between overall alcohol use and engaging in risky sexual behaviors, including more sexual partners and less frequent use of contraception (Leigh and Stall, 1993) which presumably contributes to the higher likelihood of contracting sexually transmitted diseases with alcohol use (Ericksen and Trocki, 1992).

Alcohol also contributes to several major medical diseases, including liver damage, heart disease, and neurological disorders. Chronic, high-dose alcohol can lead to fatty liver, alcoholic hepatitis, and ultimately cirrhosis of the liver. Here, alcohol is used preferentially for fuel in the liver, displacing up to 90 per cent of the liver's normal metabolic substrates. Acetaldehyde accumulates, microsomal alcohol-oxidizing activity (CYP2E1) increases, and alcohol dehydrogenase-mediated nicotinamide adenine dinucleotide reductase (NADH) accumulates (Lieber, 1997). All of these metabolic effects can lead to hepatotoxicity (Lieber, 1997). Alcoholic cirrhosis usually develops after more than a decade of heavy drinking. The amount of alcohol that can injure the liver varies greatly from person to person. In women, as few as two to three drinks per day have been linked with cirrhosis, and in men, as few as three to four drinks per day (National Institute of Diabetes and Digestive and Kidney Diseases, 2004). In 2000, liver cirrhosis was the 12th leading cause of death in the United States, ranking above hypertension

(13th), homicide (14th), and pneumonitis (15th) (Minino *et al.*, 2002). A nearly 10 per cent mortality rate exists for those afflicted with the disease (numbering over 27 000 in 2000—over 12 000 of which are alcohol-related) (Yoon *et al.*, 2003).

Heavy drinking also is associated with increased mortality from heart disease. In effect, alcohol causes cardiomyopathy with damage to the heart muscle (Piano, 2002). Heavy alcohol intake also is associated with cancers of the mouth, tongue, esophagus, liver, pancreas, pharynx, larynx, stomach, lung, colon, and rectum (Seitz *et al.*, 1998; Bandera *et al.*, 2001; Dennis and Hayes, 2001; Singletary and Gapstur, 2001; Salaspuro, 2003; Figuero Ruiz *et al.*, 2004). It should be noted, however, that moderate drinking is associated with beneficial effects to the cardiovascular system hypothesized to be due to increases in high density lipoproteins (i.e., 'good cholesterol') (Redmond *et al.*, 2000).

Alcohol and cancer vulnerability involves a powerful interaction with smoking which historically is high among alcoholics. According to a recent survey (Dawson, 2000), smoking prevalence was 28 per cent among lifetime abstainers from alcohol, 49 per cent among light drinkers, and 73 per cent among heavy drinkers. 68 per cent of people with lifetime alcohol abuse or dependence ever smoked compared to 46 per cent of those without these disorders. Over the decades, the association between alcohol drinking and smoking has been high (Bien and Burge, 1990; Dawson, 2000), as has been the negative association between alcohol consumption and the likelihood to quit smoking (Kaprio and Koskenvuo, 1988; Zimmerman *et al.*, 1990; Carmelli *et al.*, 1993).

Heavy drinking associated with thiamine deficiency can lead to two neuropsychiatric disorders, one of which is reversible—Wernicke's encephalopathy—and the other which is not—Korsakoff's psychosis. These syndromes now are considered a unitary disorder termed Wernicke's Korsakoff Syndrome (Day *et al.*, 2004). Wernicke's encephalopathy is characterized by a confusional state, ataxia, abnormal eye movements, blurred vision, double vision, nystagmus, and tremor. It is associated with a prolonged history of alcoholism with steady drinking and an inadequate nutritional state. The neurological syndrome (ataxia, opthalmoplegia, and nystagmus) can be reversed in early stages by treatment with thiamine in the diet, but the impairments of memory and learning respond more slowly and incompletely (Thomson, 2000; Day *et al.*, 2004). Untreated Wernicke's leads to death in up to 20 per cent of the cases (Harper *et al.*, 1986), or a larger percentage (85 per cent) of surviving Wernicke's go on to develop Korsakoff's psychosis (Victor *et al.*, 1971, 1989; Day *et al.*, 2004). This disease is characterized by severe anterograde amnesia with intact retrograde memory (i.e., memories prior to onset of Korsakoff's remain intact). Other cognitive functions may be spared, but the amnesia is irreversible and is associated with loss of neurons in areas such as anterior portions of the diencephalon, including the paratenial nucleus (Mair *et al.*, 1979; Mayes *et al.*, 1988), mesial temporal lobe structures, orbitofrontal cortices (Jernigan *et al.*, 1991), nucleus basalis (Cullen *et al.*, 1997), basal forebrain structures, and the hippocampus (Mayes *et al.*, 1988).

Alcohol has significant teratogenic effects on the developing embryo imparted by drinking during pregnancy. Originally described as fetal alcohol syndrome (Jones and Smith, 1973), the syndrome now is recognized as one of a spectrum of outcomes of the teratogenic effects of alcohol termed fetal alcohol spectrum disorder (Sokol *et al.*, 2003). Fetal alcohol syndrome is diagnosed when the following are present: characteristic facial dysmorphology, growth restriction, and central nervous system/neurodevelopmental abnormalities. These include for growth: prenatal growth deficiency, postnatal growth deficiency, low weight-to-height ratio; for characteristic facial features: short palpebral tissues, maxillary hypoplasia, epicanthal folds, thin upper lip, and flattened philtrum; for central nervous system anomalies or dysfunction: microcephaly, developmental delays, intellectual disability, and neonatal problems. Fetal alcohol effects or partial fetal alcohol syndrome refer to a complex pattern of behavioral or cognitive dysfunction resulting from fetal exposure to alcohol that includes difficulties in learning, poor school performance, poor impulse control, problems in relating to others, and deficits in language, abstract thinking, memory, attention, and judgment (O'Leary, 2004). Other disorders within the spectrum include alcohol-related neurodevelopmental disorders with only central nervous system abnormalities and alcohol-related birth defects with physical abnormalities (O'Leary, 2004). Risk-drinking during pregnancy has been defined as an average of more than one drink per day, or less if massed as a binge (Hankin and Sokol, 1995), and there is evidence of deleterious effects with even small amounts of alcohol (Sood *et al.*, 2001). Factors such as the pattern and quantity of maternal drinking, stage of development of the fetus at the time of exposure, use of other drugs, and socio-behavioral risk factors strongly influence the probability of developing fetal alcohol effects (O'Leary, 2004). The prevalence of fetal alcohol syndrome in the United States has been estimated to be an average of 1.95 per 1000 births (Abel, 1995). Others have estimated the combined prevalence of fetal alcohol syndrome and fetal alcohol effects in the United States to be approximately 9 per 1000 births

(Sampson *et al.*, 1997). The highest documented prevalence was 39.2 per 1000 births in the Western Cape Province of South Africa (May *et al.*, 2000).

BEHAVIORAL MECHANISM OF ACTION

Sedative hypnotic drugs in general, and alcohol in particular, produce a common behavioral action—disinhibition—that can explain many of their behavioral effects, from 'social lubricant' to paradoxical rage reactions. At low to moderate doses (blood alcohol levels of 0.05–0.10 g%) alcohol disinhibits behavior (i.e., individuals show a release of inhibitions in situations where normally they might be socially constrained). Alcohol is widely used as a 'social lubricant' at social events to promote conversation and social interaction. This disinhibition often is mistaken for a psychostimulant effect, and as a result, alcohol at lower doses often is labeled as a stimulant. Benzodiazepines have long been recognized to produce a paradoxical rage reaction where a sedative hypnotic drug produces aggressive and impulsive behavior, possibly due to the release of long-suppressed anger that only surfaces when the repression is lifted by the drug treatment (Miczek *et al.*, 1997). Presumably, alcohol's well-documented aggression-promoting actions have a similar behavioral basis (Chermack and Giancola, 1997; Miczek *et al.*, 1997). The disinhibition produced by alcohol is part of the classic continuum of behavioral effects of sedative hypnotic drugs that relates their behavioral action to dose. In addition, such disinhibition may contribute significantly to the use of alcohol as a form of self-medication (see *What is Addiction?* chapter).

Suffering in alcoholics has been hypothesized to be deeply rooted in disordered emotions characterized at either extreme by unbearable painful affect or by a painful sense of emptiness, an inability to express personal feelings, and as a result, self-humiliation and frustration in interrelationships (Khantzian and Wilson, 1993; Khantzian, 1995, 1997). Individuals cut off from their emotions welcome repeated moderate doses of alcohol or depressants as medicine to express feelings they are not able to communicate (referred to as alexithymia) (see *What is Addiction?* chapter). Individuals thus are intrinsically vulnerable and present self-care deficits. They cannot control their behaviors and cannot contain what is repeatedly endangering well-being, family, social relations, and survival. Fundamentally, they cannot adequately evaluate the consequences of the situation, or the dangers of the situation, rationally or emotionally. Thus, in some cases, the subjects seek to relieve painful feelings or to control feelings. The paradox is that using alcohol to self-medicate emotional pain or a life out of control, will later perpetuate a life revolving around alcohol.

The disinhibition observed in humans with alcohol is reflected in animal models related to the antistress or antianxiety effects of alcohol. Alcohol produces a dose-dependent release of punished responding in the Geller–Seifter conflict test (see **Fig. 5.4**) similar to that of benzodiazepines and barbiturates (Cook and Sepinwall, 1975) (**Fig. 5.13**). As noted above, alcohol blocked the 'neurotic-like' behavior of cats in a conflict situation (Masserman and Yum, 1946) and subsequently showed antianxiety-like effects in a number of animal models of anxiety including the Geller–Seifter test (Liljequist and Engel, 1984), lick-suppression test (Vogel *et al.*, 1971), social interaction test (File, 1980), elevated plus maze (Pellow and File, 1986), and behavioral contrast test (Becker and Flaherty, 1982). Alcohol, like benzodiazepines and barbiturates, has anticonflict effects that generalize across a variety of conflict/anxiety-like situations, and this generalization contrasts with the actions of other anxiolytic-like substances which act more selectively, such as serotonin drugs and corticotropin-releasing factor antagonists.

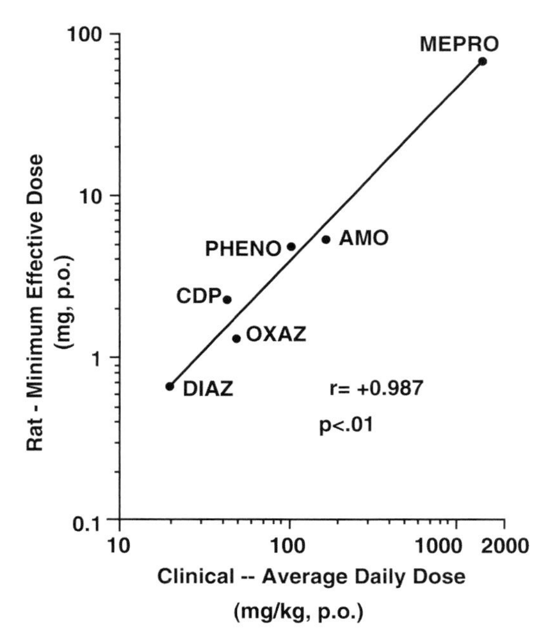

FIGURE 5.13 Correlation of rat punishment potency with clinical potency. Rat minimum effective dose represents the lowest dose tested which significantly attenuated the effects of punishment. Average clinical daily dose represents the dose found to be effective in treating psychoneurotic patients in 74 published clinical reports. CDP, chlordiazepoxide; OXAZ, oxazepam; PHENO, phenobarbital; AMO, amobarbital, MEPRO, meprobamate, DIAZ, diazepam. [Data from Cook and Davidson, 1973. Graphical representation from Cook and Sepinwall, 1975.]

NEUROBIOLOGICAL MECHANISM—NEUROCIRCUITRY

Acute Reinforcing and Anxiolytic-like Effects of Alcohol

Animal models for the acute reinforcing effects of alcohol (see *Animal Models of Addiction* chapter), and the use of selective antagonists for specific neurochemical systems have supported the hypothesis that alcohol at intoxicating doses has a wide, but selective, action on neurotransmitter systems in the brain reward systems. Neuropharmacological studies have implicated multiple neurochemical systems in the acute reinforcing effects of alcohol, including γ-aminobutyric acid (GABA), opioid peptides, dopamine, serotonin, and glutamate (**Fig. 5.14**). How exactly these neurochemical systems are engaged at the cellular and molecular levels will be discussed below.

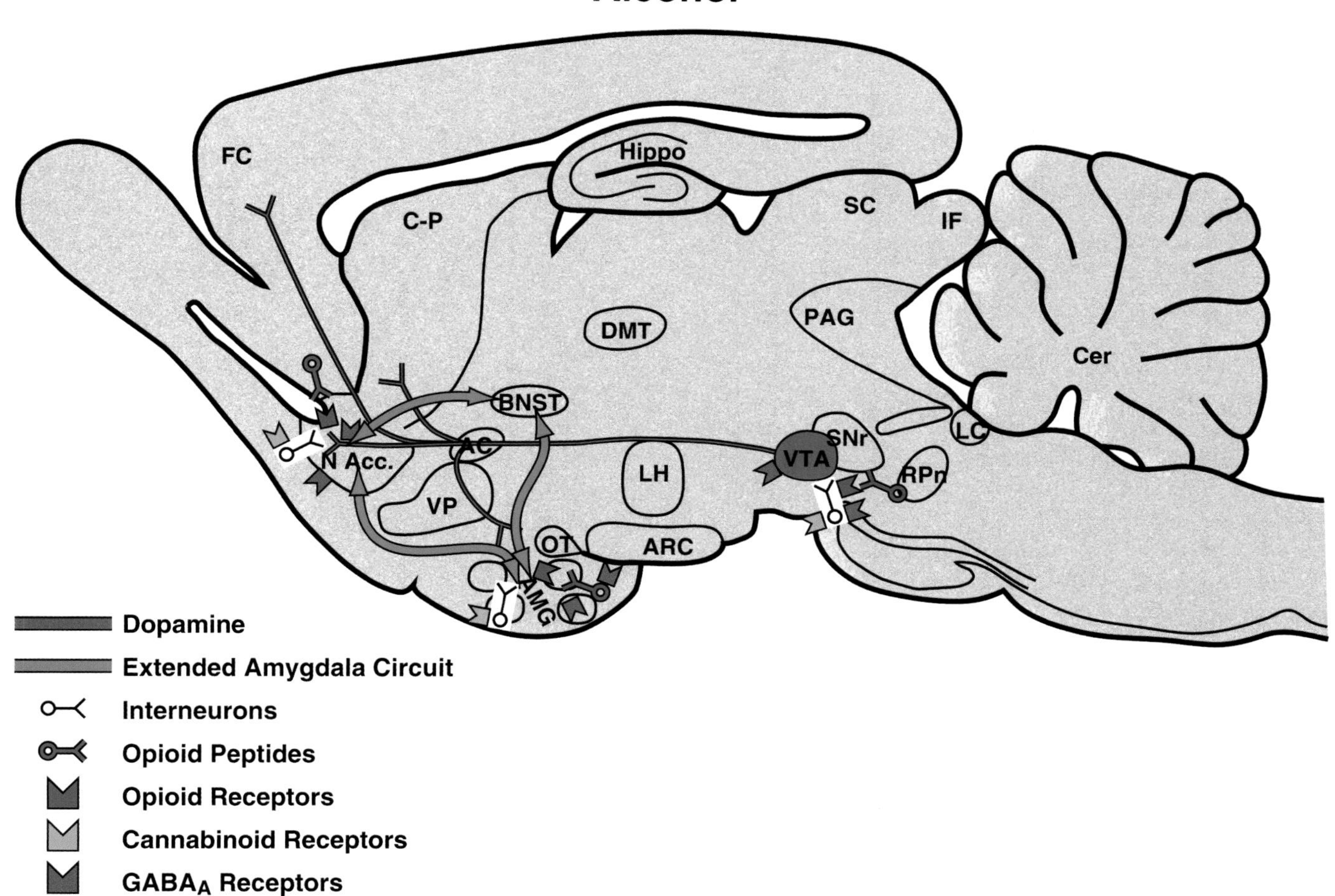

FIGURE 5.14 Sagittal section through a representative rodent brain illustrating the pathways and receptor systems implicated in the acute reinforcing actions of alcohol. Alcohol activates γ-aminobutyric acid-A ($GABA_A$) receptors in the ventral tegmental area, nucleus accumbens, and amygdala via either direct actions at the $GABA_A$ receptor or through indirect release of GABA. Alcohol is hypothesized to facilitate the release of opioid peptides in the ventral tegmental area, nucleus accumbens, and central nucleus of the amygdala. Alcohol facilitates the release of dopamine (*red*) in the nucleus accumbens via an action either in the ventral tegmental area or the nucleus accumbens. Endogenous cannabinoids and adenosine may interact with postsynaptic elements in the nucleus accumbens involving dopamine and/or opioid peptide systems. The blue arrows represent the interactions within the extended amygdala system hypothesized to have a key role in alcohol reinforcement. AC, anterior commissure; AMG, amygdala; ARC, arcuate nucleus; BNST, bed nucleus of the stria terminalis; Cer, cerebellum; C-P, caudate-putamen; DMT, dorsomedial thalamus; FC, frontal cortex; Hippo, hippocampus; IF, inferior colliculus; LC, locus coeruleus; LH, lateral hypothalamus; N Acc., nucleus accumbens; OT, olfactory tract; PAG, periaqueductal gray; RPn, reticular pontine nucleus; SC, superior colliculus; SNr, substantia nigra pars reticulata; VP, ventral pallidum; VTA, ventral tegmental area.

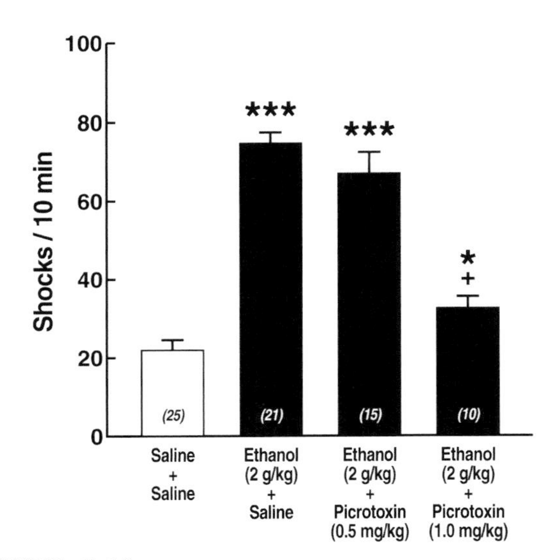

FIGURE 5.15 Effect of the γ-aminobutyric acid-A receptor antagonist picrotoxin (0.5 and 1.0 mg/kg, i.p.) on the antipunishment effect of ethanol (2.0 g/kg, i.p.) in rats. Ethanol and picrotoxin were administered 50 min and 30 min prior to the start of the recordings, respectively. The number of electric shocks accepted was recorded for 10 min. Shown are the means ± SEM. Numbers within the bars are the number of animals per group (*$p < 0.05$, ***$p < 0.001$, compared to saline-pretreated controls; +p < 0.001, compared to ethanol-pretreated animals). [Reproduced with permission from Liljequist and Engel, 1984.]

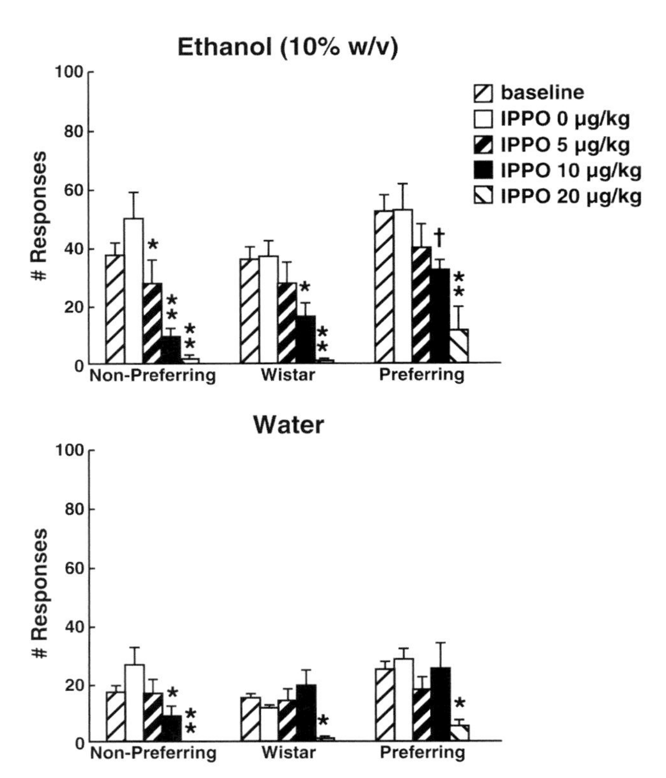

FIGURE 5.16 Effects of the γ-aminobutyric acid-A receptor inverse agonist isopropylbicyclophosphate (IPPO) on responding for ethanol or water in a free-choice operant task in rats. Responses on a fixed-ratio 1 schedule resulted in the delivery of response-contingent ethanol (10% w/v) or water reinforcement. Values shown represent mean ± SEM number of lever presses for ethanol and for water during 30 min sessions (*$p < 0.05$, **$p < 0.01$, compared to vehicle, Newman-Keuls test; †p < 0.05, compared to baseline). [Reproduced with permission from Rassnick *et al.*, 1993a.]

There has been significant work showing, at least at the pharmacological level, a role for GABA in the intoxicating effects of alcohol (Martz *et al.*, 1983). Systemic injections of $GABA_A$ antagonists or inverse agonists reverse the motor-impairing effects of alcohol Hellevuo *et al.*, 1989), the anxiolytic-like effects of alcohol (Liljequist and Engel, 1984) (**Fig. 5.15**), and alcohol drinking (McBride *et al.*, 1988; Balakleevsky *et al.*, 1990; June *et al.*, 1991; Wegelius *et al.*, 1994). $GABA_A$ antagonists or inverse agonists also decrease operant alcohol self-administration (Samson *et al.*, 1987; Rassnick *et al.*, 1993a; Samson and Chappell, 2001) (**Fig. 5.16**) and block alcohol stimulus effects in drug discrimination (Gatto and Grant, 1997; Grant *et al.*, 2000).

Other modulators of the GABA system may act not only through modulation of the $GABA_A$ receptor, but also may modulate GABA release by interactions with the $GABA_B$ receptor. The selective $GABA_B$ receptor agonist baclofen decreases alcohol self-administration in nondependent rats (Janak and Gill, 2003) and decreases the alcohol deprivation effect in alcohol-preferring rats (Colombo *et al.*, 2003a,b). Several clinical studies also have shown potential efficacy of baclofen in reducing alcohol craving and alcohol withdrawal (Addolorato *et al.*, 2002a,b).

The endogenous opioid peptide systems have long been hypothesized to have a role in the reinforcing effects of alcohol (Reid and Hunter, 1984; Herz, 1997). Naltrexone decreases alcohol drinking and self-administration in a variety of animal models, and these results led to the clinical use of naltrexone in reducing consumption of alcohol and in preventing relapse (Altshuler *et al.*, 1980; Reid and Hunter, 1984; Samson and Doyle, 1985: Froehlich *et al.*, 1990; Weiss *et al.*, 1990; Hubbell *et al.*, 1991; Kornet *et al.*, 1991; O'Malley *et al.*, 1992; Volpicelli *et al.*, 1992; Hyytia and Sinclair, 1993; Davidson and Amit, 1997) (**Fig. 5.17**).

Consistent with the overlap in the neuropharmacological actions of cannabinoid CB_1 receptors and µ opioid receptors (see *Cannabinoids* chapter), CB_1 receptors modulate alcohol preference in mice. The high alcohol preference of young mice is reduced by a competitive CB_1 receptor antagonist, and young CB_1 receptor knockout mice show a decrease in alcohol preference (Wang *et al.*, 2003). Older mice show less of an alcohol preference and no effect of the cannabinoid antagonist and no effects of the knockout (Wang *et al.*, 2003).

Significant evidence also supports a role for the mesolimbic dopamine system in alcohol reinforcement.

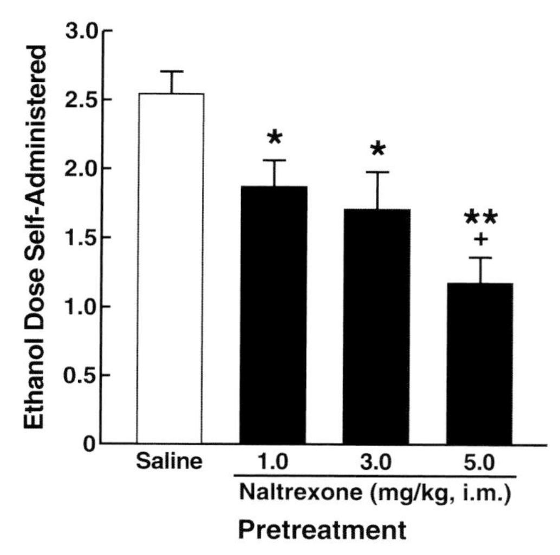

FIGURE 5.17 Effects of saline (0.9% NaCl, 1.0 ml, i.m.) or naltrexone (1.0, 3.0, or 5.0 mg/kg, i.m.) on intravenous ethanol self-administration over a 4-h session in rhesus monkeys. Each bar represents the mean ethanol intake for all animals and all days of pretreatment with a given naltrexone dose or saline. All doses of naltrexone were associated with ethanol intake below saline levels (* $p < 0.05$, compared to saline; ** $p < 0.01$, compared to saline). The differences between the effects of 1.0 and 3.0 mg/kg naltrexone were not significant, but the effects of 5.0 mg/kg differed significantly from 1.0 and 3.0 mg/kg doses (+ $p < 0.05$). [Reproduced with permission from Altschuler *et al.*, 1980.]

Systemic injections of dopamine receptor antagonists decrease responding for alcohol (Pfeffer and Samson, 1986; Rassnick *et al.*, 1992). Alcohol self-administration increases extracellular levels of dopamine in the nucleus accumbens in nondependent rats (Weiss *et al.*, 1993) (**Fig. 5.18**). Such increases occur not only during the actual self-administration session but also precede the self-administration session, possibly reflecting the incentive motivational properties of cues associated with alcohol (Weiss *et al.*, 1993). Consistent with this view, mesolimbic dopamine does not appear to be critical for the acute reinforcing effects of alcohol in that lesions of the mesolimbic dopamine system fail to block operant self-administration of alcohol (Myers, 1990; Lyness and Smith, 1992; Rassnick *et al.*, 1993d).

Modulation of various aspects of serotonergic transmission, including increases in the synaptic availability of serotonin with precursor loading and blockade of serotonin reuptake, can decrease alcohol intake (Sellers *et al.*, 1992a) (**Table 5.10**). Antagonists of several serotonin receptor subtypes can decrease alcohol self-administration. Serotonin-3 ($5\text{-}HT_3$) receptor antagonists decrease alcohol self-administration under 24-h access conditions but are less effective

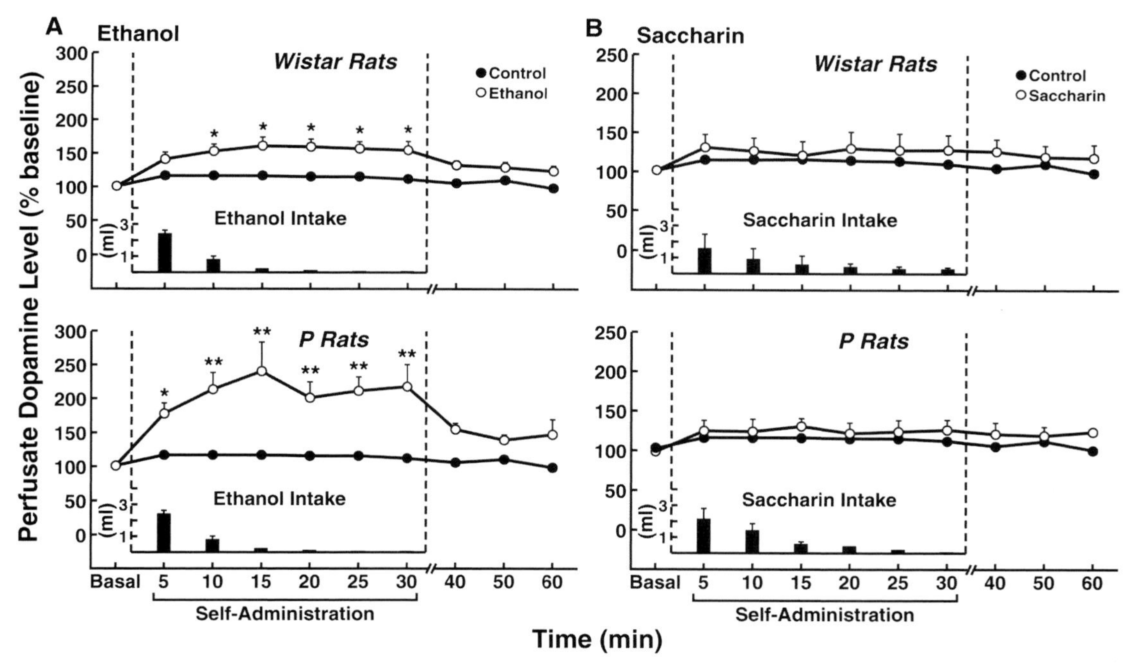

FIGURE 5.18 **(A)** Effects of oral ethanol self-administration on dopamine release in the nucleus accumbens in genetically heterogeneous Wistar ($n = 11$) and alcohol-preferring (P) rats ($n = 9$). Data from both strains are contrasted against the same control group ($n = 9$) consisting of ethanol-naive Wistar and P rats trained to respond for water only. Ethanol produced significant increases in dopamine release in both groups of rats (Wistar: * $p < 0.01$; P rats: * $p < 0.05$, ** $p < 0.01$; simple effects ANOVA). Insets show the mean ± SEM ethanol intake per 5 min interval. **(B)** Dopamine release in the nucleus accumbens of Wistar ($n = 4$) and P rats ($n = 4$) responding for saccharin reinforcement. The data of both strains are contrasted against the same control group consisting of saccharin-naive Wistar and P rats trained to respond for water. Saccharin produced only negligible increases in dopamine efflux compared to ethanol. [Reproduced with permission from Weiss *et al.*, 1993.]

TABLE 5.10 Effect of Serotonin Manipulations on Ethanol and Food Intake in Rats

Manipulation	Effect on ethanol intake	Reference
Increasing serotonin function		
Precursor loading	↓	Zabik and Roach, 1983
Serotonin releasers	↓	Sellers *et al.*, 1992b
Serotonin agonists		
5–HT_{1A}	↑	Tomkins *et al.*, 1994
	↓	Svensson *et al.*, 1989; McBride *et al.*, 1990
$5\text{–HT}_{1A/1B}$	↓	Tomkins and O'Neill, 2000
$5\text{–HT}_{1B/1C}$	↓	McBride *et al.*, 1990
$5\text{–HT}_{1C/2}$	↓	McBride *et al.*, 1990
5–HT_{2C}	↑	Tomkins *et al.*, 2002
Intraventricular serotonin	↓	Hill, 1974
Serotonin uptake blockers	↓	Gill *et al.*, 1988; Murphy *et al.*, 1988
Decreasing serotonin function		
Serotonin antagonists		
5–HT_{1B}	↑	Tomkins and O'Neill, 2000
$5\text{–HT}_{1/2}$	—	Rockman *et al.*, 1982; Weiss *et al.*, 1990
5–HT_{2}	↓	Roberts *et al.*, 1998
5–HT_{3}	↓	Fadda *et al.*, 1991; Sellers *et al.*, 1992b; Hodge *et al.*, 1993; Tomkins *et al.*, 1995
Synthesis inhibition	↓	Parker and Radow, 1976; Walters, 1977
Central 5–HT lesions	↑	Ho *et al.*, 1974
	—	Kiianmaa, 1976

[Table adapted from Sellers *et al.*, 1992a.]

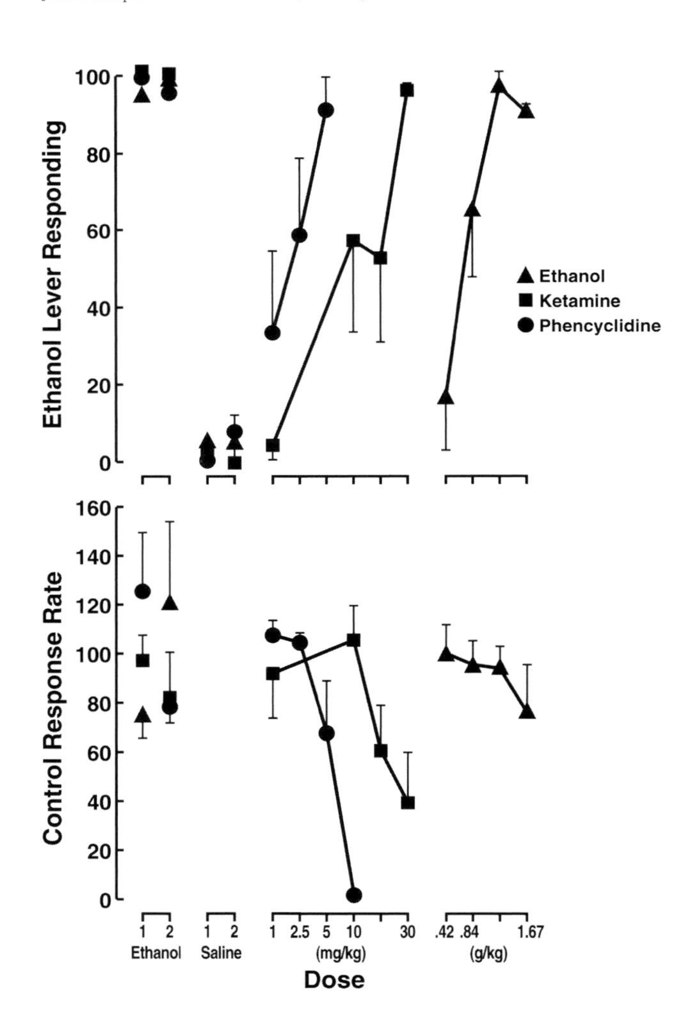

FIGURE 5.19 Mean percentage of ethanol lever-responding and mean percentage of control response rates following administration of ethanol ($n = 9$), ketamine ($n = 5$), or phencyclidine ($n = 6$) in mice trained to discriminate ethanol from saline. At 1.67 g/kg ethanol, 30 mg/kg ketamine, and 5 mg/kg phencyclidine, the n used to calculate the mean percentage of ethanol-lever responding was 7, 3, and 5, respectively. Ethanol (1.25 g/kg) and saline were tested before (1) and after (2) each dose-response curve determination. [Reproduced with permission from Grant *et al.*, 1991.]

in limited-access conditions (Fadda *et al.*, 1991; Knapp and Pohorecky, 1992; Kostowski *et al.*, 1993; Beardsley *et al.*, 1994; McKinzie *et al.*, 1998; but see Hodge *et al.*, 1993) and have been effective at decreasing relapse in clinical trials (Johnson *et al.*, 2000). The 5-HT_3 receptor antagonists block alcohol-stimulated release of dopamine in the mesolimbic dopamine system (Campbell and McBride, 1995). 5-HT_2 receptor antagonists, including some with both 5-HT_{1A} agonist and 5-HT_2 antagonist action, can selectively decrease acute alcohol reinforcement (Roberts *et al.*, 1998).

Another approach used extensively to explore the neuropharmacological basis of alcohol intoxication is the drug discrimination procedure. Animals are trained to discriminate a given response for food in an alcohol-intoxicated state (see *Animal Models of Addiction* chapter). Drug discrimination studies where alcohol injection is discriminated from water have revealed prominent substitution with glutamate antagonists (Grant *et al.*, 1991) (**Fig. 5.19**) (**Table 5.11**), and alcohol at low concentrations modulates the *N*-methyl-D-aspartate (NMDA) receptor (Hoffman *et al.*, 1989; Lovinger *et al.*, 1989). Other drugs that substitute for alcohol include barbiturates and benzodiazepines, 5-HT_1 receptor agonists, and positive GABA-modulating neuroactive steroids such as allopregnanolone and pregnanolone. Drugs that do not substitute for alcohol include morphine, cocaine, phenytoin, 5-HT_2 agonists, nonpositive GABA-modulating neuroactive steroids, taurine, and acetaldehyde (**Table 5.11**).

The neuroanatomical substrates for the reinforcing actions of alcohol also may involve neurocircuitry that forms a separate entity within the basal forebrain, termed the 'extended amygdala' (Alheid and Heimer, 1988) (**Fig. 5.14** above). The extended amygdala has been conceptualized to be composed of several basal forebrain structures: the bed nucleus of the stria terminalis, centromedial amygdala, sublenticular

TABLE 5.11 **Drug Substitution for Ethanol Using Two-Bottle Choice Drug Discrimination Procedures**

	Reference
Drugs substituting for ethanol	
Noncompetitive NMDA antagonists	Grant *et al.*, 1991; Holter *et al.*, 2000; Shelton and Grant, 2002
Ketamine	Grant *et al.*, 1991
Phencyclidine	Grant and Colombo, 1993a
5–HT_1 agonist	Grant and Colombo, 1993b
5–HT_{1B} agonist	Grant *et al.*, 1997
Allotetrahydrodeoxycorticosterone	Ator *et al.*, 1993; Bowen *et al.*, 1999
Allopregnanolone	Ator *et al.*, 1993; Bowen *et al.*, 1999; Shelton and Grant, 2002
Pregnanolone	Bowen *et al.*, 1999
Isradipine	Green and Grant, 1999
Pentobarbital	Grant *et al.*, 2000
Midazolam	Grant *et al.*, 2000; Shelton and Grant, 2002
Drugs not substituting for ethanol	
Taurine	Quertemont and Grant, 2004
L–NAME (nitric oxide synthase inhibitor)	Green *et al.*, 1997
5–HT_{2A} agonist	Szeliga and Grant, 1998
Dehydroepiandrosterone sulfate	Bowen *et al.*, 1999
Pregnenolone sulfate	Bowen *et al.*, 1999
Pregnanolone sulfate	Bowen *et al.*, 1999
Epipregnanolone sulfate	Bowen *et al.*, 1999
Cocaine	Grant *et al.*, 1991
Phenytoin	Grant *et al.*, 1991
Morphine	Grant *et al.*, 1991
Acetaldehyde	Quertemont and Grant, 2002
Adenosine A_1 and A_2 agonists	Michaelis *et al.*, 1987
Drugs blocking discriminative stimulus effects of ethanol	
5–HT_3 antagonist	Grant and Barrett, 1991; Mhatre *et al.*, 2001
Benzodiazepine inverse agonist	Gatto and Grant, 1997
Drugs not blocking discriminative stimulus effects of ethanol	
δ_2 opioid receptor antagonist	Mhatre *et al.*, 2000
Caffeine*	Michaelis *et al.*, 1987

*Except at high doses.

substantia innominata, and a transition zone in the medial part of the nucleus accumbens (e.g., shell) (Heimer and Alheid, 1991). The very potent $GABA_A$ receptor antagonist SR 95531, when microinjected into the basal forebrain, significantly decreased responding for alcohol but not water in rats trained to lever press for alcohol (Hyytia and Koob, 1995). When the GABA antagonist was injected bilaterally into the nucleus accumbens, bed nucleus of the stria terminalis, and central nucleus of the amygdala, the most sensitive site for decreasing alcohol responding was the central nucleus of the amygdala (Hyytia and Koob, 1995) (**Fig. 5.20**). Very low doses of the dopamine antagonist fluphenazine injected into the nucleus accumbens blocked alcohol self-administration at low doses (Rassnick *et al.*, 1992; Samson *et al.*, 1993) (**Fig. 5.21**). Injections of an opiate antagonist into the central nucleus of the amygdala also significantly reduced alcohol consumption at lower doses than for other sites such as the nucleus accumbens or lateral ventricle (Heyser *et al.*, 1999), suggesting a role for opioid peptides in the extended amygdala in the acute reinforcing actions of alcohol (Heyser *et al.*, 1999) (**Fig. 5.22**). Microinjection of $5\text{-}HT_3$ antagonists into the amygdala of nondependent Wistar rats also significantly attenuated alcohol drinking in limited-access paradigms (2-h daily sessions) (Dyr and Kostowski, 1995).

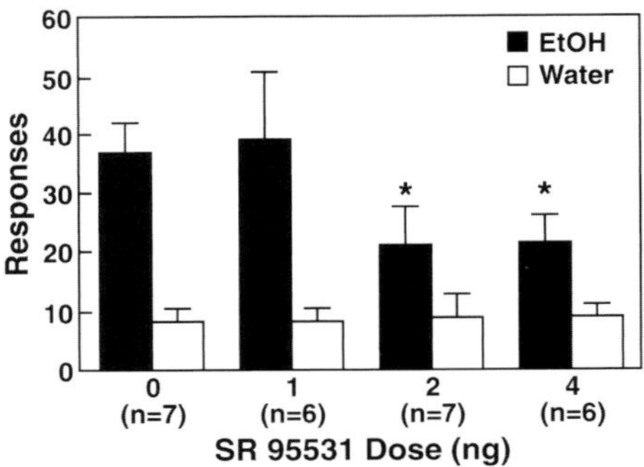

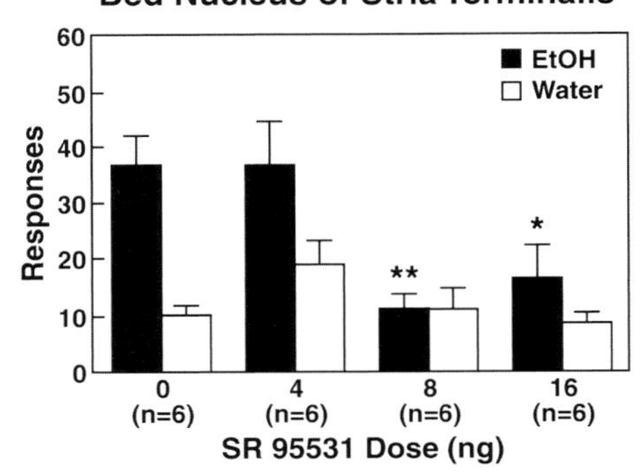

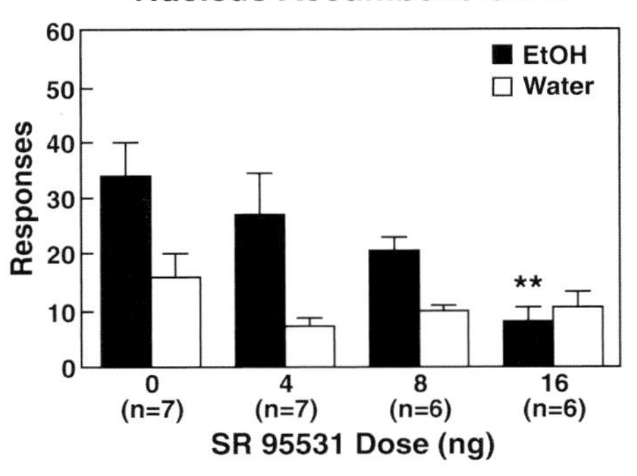

FIGURE 5.20 The effect of the γ-aminobutyric acid-A receptor antagonist SR 95531 injections into the central nucleus of the amygdala, the bed nucleus of the stria terminalis, and the shell of the nucleus accumbens on responding for ethanol (EtOH) and water in nondependent rats. Data are expressed as the mean ± SEM numbers of responses for ethanol and water during 30-min sessions for each injection site. Note the change in the abscissa scale for injections into the bed nucleus of the stria terminalis and the nucleus accumbens shell. Significance of differences from the corresponding saline control values: $^*p < 0.05$, $^{**}p < 0.01$ for ethanol responses; $\#p < 0.05$ for water responses (adjusted means test). [Reproduced with permission from Hyytia and Koob, 1995.]

Alcohol Tolerance

While metabolic tolerance can account for some of the resistance of the body to alcohol, early studies showed that rats treated chronically with alcohol had higher brain levels of alcohol than controls at the time of recovery from the loss of the righting reflex, thus arguing for pharmacodynamic tolerance (Levy, 1935). The neurobiological mechanisms hypothesized for pharmacodynamic tolerance have ranged from an increase in oxidative metabolism, to alterations in membrane lipid composition and fluidization, to neuroadaptive changes (Chin and Goldstein, 1977; Kalant, 1998). Lipid changes were produced by normal treatments (vapor or liquid diet) but were very small. In acute studies, high *in vitro* concentrations of ethanol produced small changes.

Studies of tolerance to alcohol engaged neuroadaptive mechanisms with the observation that behavioral factors (learning) have a prominent role in the development of tolerance. Rats allowed to experience a task in the presence of alcohol intoxication showed much more rapid and complete tolerance to alcohol than rats not allowed to experience the task under the influence of alcohol but receiving the alcohol after task exposure (Chen, 1968). Subsequent studies of behavioral tolerance as a model of the neuroadaptive processes associated with the development of pharmacodynamic tolerance have revealed a number of neuropharmacological substrates (**Table 5.12**). These include serotonin, NMDA glutamate systems, and vasopressin. Blockade of

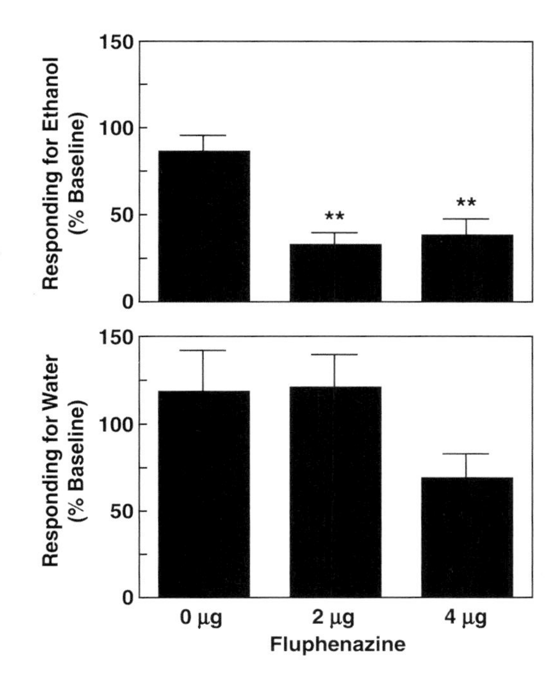

FIGURE 5.21 Effects of microinjection of the dopamine antagonist fluphenazine into the nucleus accumbens in nondependent rats on responding for ethanol (10% w/v) and water. Data are expressed as percentage of baseline responding (mean ± SEM) and are plotted as a function of dose of fluphenazine. Asterisks (**) indicate a significant difference for responding compared to responding after vehicle injection ($p<0.01$, Newman-Keuls *a posteriori* test). [Reproduced with permission from Rassnick *et al.*, 1992.]

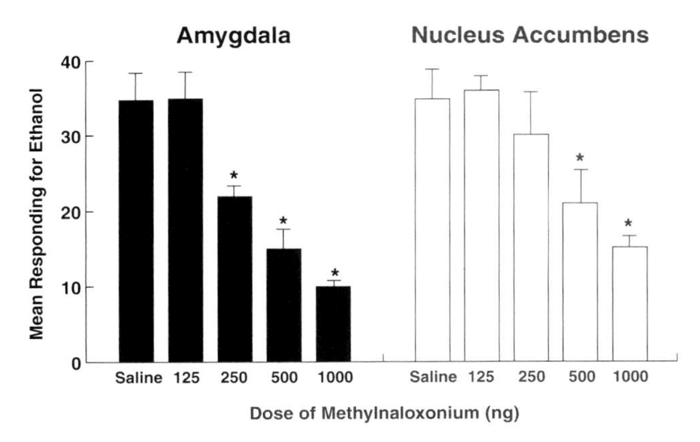

FIGURE 5.22 The effects of intracerebral administration of the opioid receptor antagonist methylnaloxonium into the amygdala and nucleus accumbens on responding for ethanol in rats. Data are presented as mean total responses ± SEM. Asterisks (*) indicate significant differences from saline injections ($p<0.05$). [Modified with permission from Heyser *et al.*, 1999.]

activity in any of the three systems blocks acute and chronic tolerance, and it has been hypothesized that a neural circuit involving these neurotransmitters and connections between the septum and hippocampus plays an essential role in the development and retention of tolerance (Kalant, 1998).

TABLE 5.12 Neuropharmacological Studies of Tolerance to Alcohol

Functional effect	Reference
Serotonin	
↓ Serotonin synthesis	Frankel *et al.*, 1975
↓ Serotonin cell body lesions (raphe)	Campanelli *et al.*, 1988
N-methyl-D-aspartate	
↓ Noncompetitive NMDA antagonist	Khanna *et al.*, 1992, 1993c; Wu *et al.*, 1993
↓ Nitric oxide synthase inhibition	Khanna *et al.*, 1993b
↑ Glycine	Khanna *et al.*, 1993a
Arginine vasopressin	
↑ V_1 Agonists	Hoffman *et al.*, 1978; Le *et al.*, 1982; Speisky and Kalant, 1985
Protein synthesis	
Protein synthesis inhibitors	LeBlanc *et al.*, 1976; Bitran and Kalant, 1993

Alcohol Withdrawal and Dependence

The neurobiological basis for the motivational effect of alcohol withdrawal includes counteradaptive neurochemical events within the brain emotional systems normally used to maintain emotional homeostasis (Lewis, 1996; Koob and Le Moal, 2001). Key to this hypothesis is the observation that during acute withdrawal from alcohol there is a compromised brain reward system as reflected in an increase in brain reward thresholds (Schulteis *et al.*, 1995) which is opposite in direction to the threshold-lowering action of acute alcohol (de Witte and Bada, 1983; Kornetsky *et al.*, 1988; Bespalov *et al.*, 1999). Rats made dependent on alcohol using alcohol vapor that results in blood alcohol levels of 150–200 mg% show elevations in reward thresholds during withdrawal from alcohol that persist up to 72 h postexposure (Schulteis *et al.*, 1995) (**Fig. 5.9**).

Decreases in neurotransmitter function implicated in the acute reinforcing effects of alcohol within the extended amygdala (e.g., GABAergic, opioid peptidergic, dopaminergic, serotonergic, and glutamatergic systems) may represent one of the neurochemical counteradaptations that has motivational significance with the development of alcohol dependence. Rats (Roberts *et al.*, 1996, 2000a) and mice (Becker and Lopez, 2004) increased drinking and working for alcohol during withdrawal after being made dependent (see *Animals Models of Addiction* chapter). Neuropharmacological studies have shown that the enhanced alcohol self-administration during acute

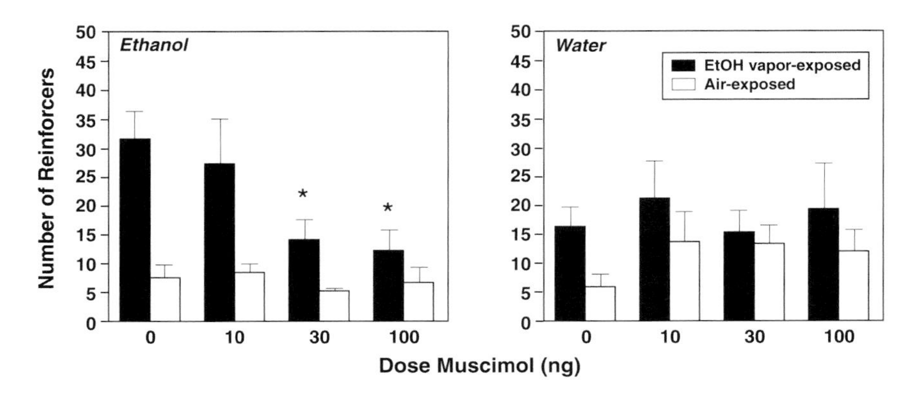

FIGURE 5.23 Operant responding for ethanol and water in ethanol vapor-exposed (dependent) and control air-exposed rats after intra-amygdala muscimol administration. Data are the mean ± SEM of hours 7 and 8 postwithdrawal ($^*p<0.05$). [Adapted with permission from Roberts *et al.*, 1996.]

withdrawal can be reduced dose-dependently by intracerebral pretreatment of the $GABA_A$ agonist muscimol into the central nucleus of the amygdala (Roberts *et al.*, 1996) (**Fig. 5.23**). Acamprosate, a hypothesized partial agonist or functional antagonist of the brain glutamate systems also decreased excessive drinking associated with dependence and abstinence in rats (Boismare *et al.*, 1984; Le Magnen *et al.*, 1987; Gewiss *et al.*, 1991; Spanagel *et al.*, 1996; Holter *et al.*, 1997; Heyser *et al.*, 1998).

Dopaminergic function also is compromised during acute alcohol withdrawal. Animals sustained on a liquid diet showed a decrease in extracellular levels of dopamine in the nucleus accumbens (Weiss *et al.*, 1996) (**Fig. 5.24**). Similar effects have been observed for virtually all major drugs of abuse. Particularly compelling in the above study, however, was the observation that when animals were allowed to self-administer alcohol during acute withdrawal, the animals self-administered just enough alcohol to return extracellular dopamine levels in the nucleus accumbens back to pre-dependence baseline levels (Weiss *et al.*, 1996). Overall, these observations suggest that the classic neurotransmitters associated with regulating the positive reinforcing properties of drugs of abuse, including alcohol, are compromised during alcohol withdrawal.

Alcohol also is a powerful modulator of 'stress' systems, and dysregulation of the brain stress systems has been hypothesized to be another neurochemical counteradaptation that may be of motivational significance during the development of alcohol dependence, an effect that may be crucial in the understanding of dependence and relapse. Both acute and chronic alcohol activate the hypothalamic-pituitary-adrenal (HPA) axis, and this appears to be the result of release of corticotropin-releasing factor (CRF) in the hypothalamus to, in turn, activate the classic neuroendocrine stress response (Rivier *et al.*, 1984; Rasmussen *et al.*, 2000).

Systemic administration of alcohol rapidly increased plasma adrenocorticotropic hormone (ACTH) levels as well as activity in the paraventricular nucleus of the hypothalamus (Lee and Rivier, 1994; Ogilvie *et al.*, 1997, 1998), and immunoneutralization of CRF virtually abolished this response (Rivier *et al.*, 1984). Similar effects were observed following direct injection of alcohol into the brain, adding further support to the hypothesis that the activation of the HPA axis by acute alcohol is mediated via the activation of CRF and vasopressin in the paraventricular nucleus of the hypothalamus (Lee *et al.*, 2004).

Abstinent alcoholics are well documented to have persistent impaired HPA function as reflected in low basal cortisol and blunted ACTH and cortisol responses to CRF (Adinoff *et al.*, 1990; Wand and Dobs, 1991). Similar results have been observed in animal studies where rats fed on a liquid diet and then tested 4 weeks after diet removal, showed disruption of the HPA axis (Rasmussen *et al.*, 2000). Intragastric 'binge' administration of acute alcohol activated the HPA axis, but after 14 days of chronic alcohol, no elevation in plasma ACTH and attenuated elevation of corticosterone were observed. This effect was accompanied by tolerance to the acute alcohol reduction in CRF_1 receptor mRNA levels in the hypothalamus (Zhou *et al.*, 2000). Repeated administration of alcohol by intragastric injection for three days produced a blunted corticosterone, ACTH, and CRF response to subsequent alcohol challenge. The blunted response was hypothesized to be due to decreased neuronal activity of CRF neurons in the paraventricular nucleus of the hypothalamus (Lee *et al.*, 2001). Thus, functional changes in the paraventricular nucleus CRF system may be a mechanism by which the HPA system becomes dysregulated in alcoholism.

Chronic alcohol also may interact with an extensive extrahypothalamic, extra-neuroendocrine CRF system implicated in behavioral responses to stress (for review, see Koob *et al.*, 1994). Chronic alcohol produces

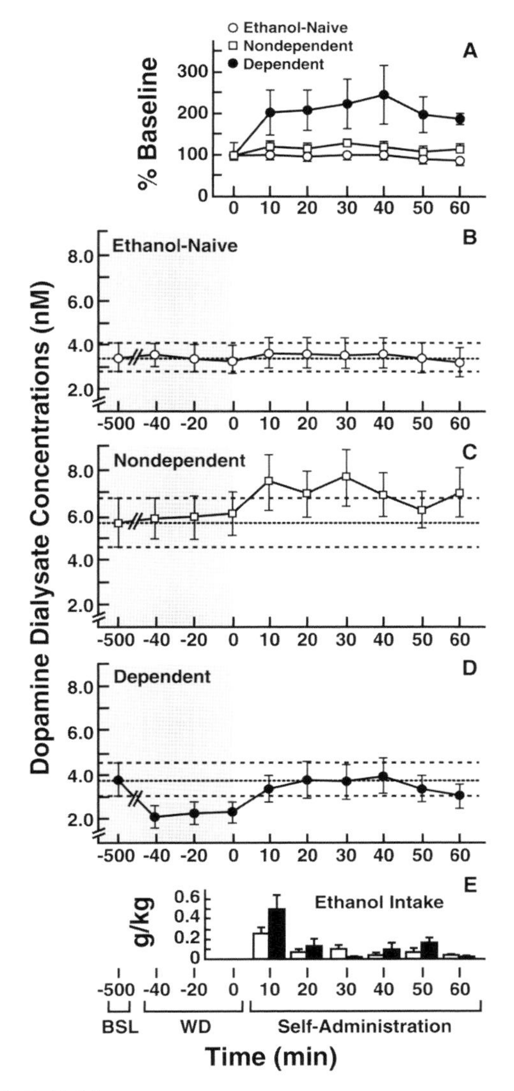

FIGURE 5.24 Effects of operant alcohol self-administration in nondependent and dependent rats undergoing ethanol withdrawal on dopamine efflux in the nucleus accumbens. Dialysate neurotransmitter levels are compared to those in ethanol-naive rats trained to self-administer water. Average water intake in this group was negligible (<0.8 ml) and is not shown. **(A)** Changes in neurotransmitter output from levels recorded during the last hour of withdrawal. Data are expressed as a percentage of baseline values calculated as the average of three 20-min samples collected during hour 8 of withdrawal shown in **B–D**. The corresponding dialysate neurotransmitter concentrations are shown in **B** (Ethanol-Naive), **C** (Nondependent), and **D** (Dependent). To illustrate the changes in neurotransmitter efflux over the various experimental phases, **B–D** also show prewithdrawal (BSL) and withdrawal (WD) dialysate concentrations of dopamine during hour 8 of withdrawal. Dashed lines represent mean ± SEM prewithdrawal dialysate dopamine concentrations. **(E)** Amounts of self-administered ethanol (10% w/v) during 10-min intervals for the dependent (solid bars) and nondependent (open bars) groups. Ethanol self-administration in dependent rats restored dopamine levels to prewithdrawal values. [Modified with permission from Weiss *et al.*, 1996.]

anxiogenic-like responses during acute and protracted withdrawal (Baldwin *et al.*, 1991; Rassnick *et al.*, 1993b; Spanagel *et al.*, 1995; Knapp *et al.*, 1998; Rasmussen *et al.*, 2001; Overstreet *et al.*, 2004b). The anxiogenic-like effect of alcohol withdrawal can be reversed by intracerebral administration of a CRF antagonist into the central nucleus of the amygdala (Rassnick *et al.*, 1993b) and by systemic administration of CRF_1 antagonists (Knapp *et al.*, 2004; Overstreet *et al.*, 2004b; Breese *et al.*, 2005). Increases in extracellular levels of CRF are observed in the amygdala and bed nucleus of the stria terminalis during alcohol withdrawal (Merlo-Pich *et al.*, 1995; Olive *et al.*, 2002) **(Figs. 5.25** and **5.26)**. Even more compelling is the observation that a competitive CRF antagonist that has no effect on alcohol self-administration in nondependent rats effectively eliminates excessive drinking in dependent rats (Valdez *et al.*, 2002) **(Fig. 5.27)**.

Neuropeptide Y (NPY) is a 36 amino acid polypeptide distributed widely throughout the central nervous system but with high concentrations within the extended amygdala (Adrian *et al.*, 1983). Central administration of NPY increases feeding behavior (Clark *et al.*, 1984; Levine and Morley, 1984), reduces anxiety-like behavior (Kask *et al.*, 1998), and potentiates the effects of sedative hypnotic drugs (Heilig and Murison, 1987;

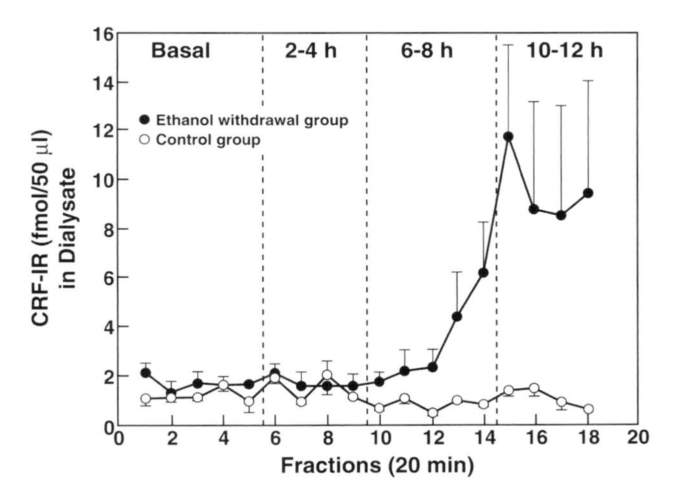

FIGURE 5.25 Effects of ethanol withdrawal on corticotropin-releasing factor immunoreactivity (CRF-IR) levels in the rat amygdala as determined by microdialysis. Rats were maintained on an 8.5% (v/v) ethanol liquid diet for 2–3 weeks which produced blood alcohol levels of approximately 126 mg%. Dialysate was collected over four 2-h periods regularly alternated with nonsampling 2-h periods. The four sampling periods correspond to the basal collection (before removal of ethanol), and 2–4 h, 6–8 h, and 10–12 h after withdrawal. Fractions were collected every 20 min. Data are represented as mean ± SEM ($n = 5$ per group). ANOVA confirmed significant differences between the two groups over time ($p < 0.05$). [Reproduced with permission from Merlo-Pich *et al.*, 1995.]

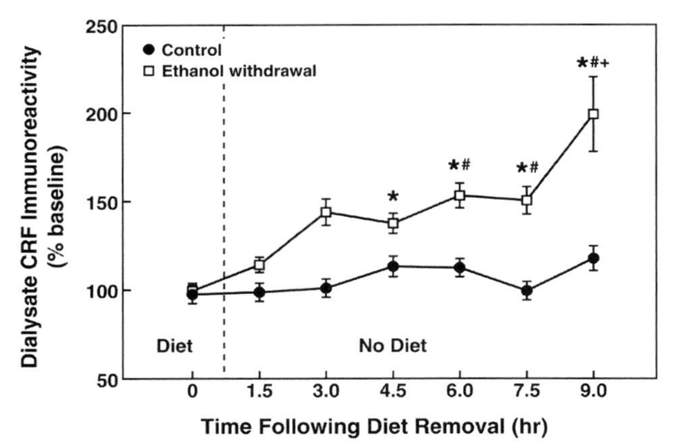

FIGURE 5.26 Effect of acute ethanol withdrawal in rats from an ethanol-containing or control liquid diet on extracellular corticotropin-releasing factor (CRF) levels in the bed nucleus of the stria terminalis. Each data point represents the mean ± SEM dialysate level of CRF (expressed as a percentage of basal levels) in three 30-min microdialysis samples for each animal. Treatment groups are designated as control-fed rats with subsequent access to the control diet (●, $n = 7$) and ethanol-fed rats with subsequent access to the control diet (withdrawal) (□, $n = 7$). *$p < 0.05$ versus baseline. #$p < 0.05$ versus control-fed animals at the same time point. +$p < 0.05$ versus ethanol-fed animals at the same time point. [Reproduced with permission from Olive *et al.*, 2002.]

Heilig *et al.*, 1989). Acute withdrawal from alcohol is associated with decreases in the levels of NPY in the central and medial nuclei of the amygdala and the piriform cortex (Roy and Pandey, 2002). Further, Wistar rats show a blunted electrophysiological response to central injections of NPY in the amygdala following chronic alcohol exposure (Slawecki *et al.*, 1999). Selectively bred alcohol-preferring (P) and high alcohol drinking (HAD) rats also support a role for NPY in excessive drinking in that NPY administered intracerebroventricularly decreases alcohol intake (Badia-Elder *et al.*, 2001). Blockade of NPY_2 receptors with intracerebroventricular administration of an NPY_2 antagonist also reduces alcohol self-administration in rats (Thorsell *et al.*, 2002), and this effect may be due to an increase in endogenous release of NPY with NPY_2 receptors functioning as presynaptic autoreceptors (King *et al.*, 1999).

Other systems potentially involved in the anxiogenic-like effects of alcohol withdrawal include $5\text{–}HT_{1A}$ and $5\text{–}HT_{2C}$ (Overstreet *et al.*, 2003), GABA-benzodiazepine, and glutamate receptor systems (Moy *et al.*, 1997; Knapp *et al.*, 2004). Benzodiazepines and benzodiazepine antagonists reversed the anxiogenic-like effects of alcohol withdrawal in the elevated plus maze and social interaction test (Jung *et al.*, 2000; Knapp *et al.*, 2004; Breese *et al.*, 2005). A $5\text{–}HT_{1A}$ agonist blocked the anxiogenic-like effects of alcohol withdrawal in the social interaction test (Breese *et al.*, 2005). Ritanserin, a $5\text{–}HT_2$ antagonist, blocked the anxiogenic-like effects of alcohol withdrawal in the elevated plus maze, but not in a pentylenetetrazol drug discrimination model (Gatch *et al.*, 2000). Competitive glutamate NMDA antagonists partially reversed the anxiogenic-like effects of alcohol withdrawal (Gatch *et al.*, 1999b), as did an adenosine antagonist (Gatch *et al.*, 1999a). Thus, drugs largely known to have anxiolytic effects in humans or

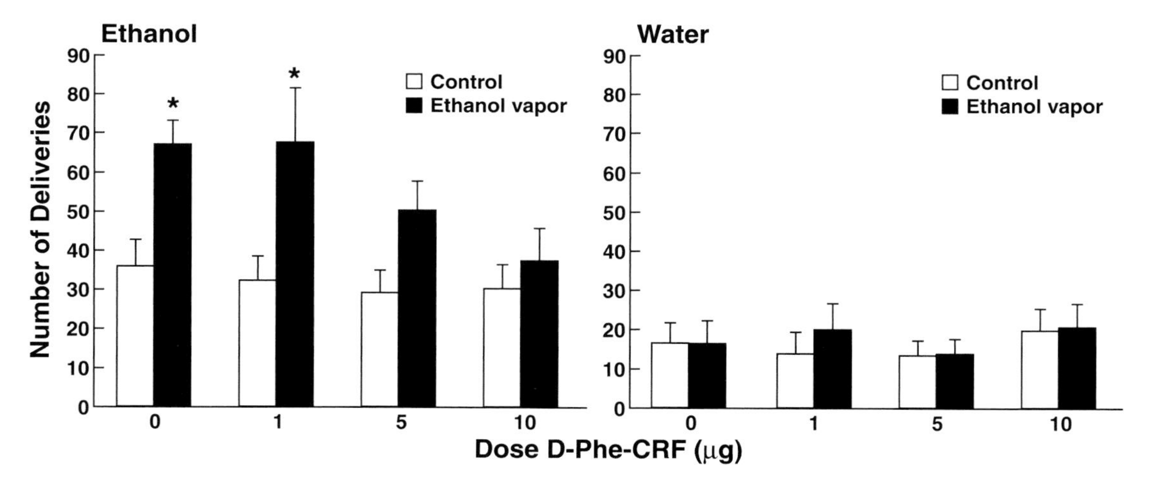

FIGURE 5.27 Effects of the corticotropin-releasing factor antagonist D-Phe-$CRF_{12\text{–}41}$ on responding for ethanol and water in rats 2 h following chronic ethanol vapor exposure. Control rats were exposed to air vapor. Rats were microinjected intracerebroventricularly with 0–10 μg of D-Phe-$CRF_{12\text{–}41}$ ($n = 10\text{–}12$ per group) using a within-subjects Latin square design 2 h after removal from the vapor chambers. The number of lever presses for ethanol and water ± SEM were measured 10 min after injection. Following the initial test session, rats were re-exposed to ethanol vapor or air, and the procedures were repeated until the Latin square design was complete. *$p < 0.05$, Tukey's test, compared to controls. [Reproduced with permission from Valdez *et al.*, 2002.]

TABLE 5.13 Effects of Neuropharmacological Treatments on the Anxiogenic-like Effects of Alcohol Withdrawal

	Elevated plus maze	Social interaction test	Reference
Treatments that block the anxiogenic-like effects of alcohol withdrawal			
Benzodiazepine	X		Jung *et al.*, 2000
Benzodiazepine partial agonist	X		Jung *et al.*, 2000
Benzodiazepine antagonist	X	X	Moy *et al.*, 1997; Knapp *et al.*, 2004; Breese *et al.*, 2005
CRF_1 antagonist	X	X	Knapp *et al.*, 2004; Breese *et al.*, 2005
NMDA competitive antagonist	X		Gatch *et al.*, 1999b
5–HT_2 antagonist	X	X	Gatch *et al.*, 2000; Knapp *et al.*, 2004
5–HT_{1A} agonist		X	Breese *et al.*, 2005
Adenosine antagonist	X		Gatch *et al.*, 1999a
Treatments that do not block the anxiogenic-like effects of alcohol withdrawal			
CRF_2 antagonist		X	Overstreet *et al.*, 2004a
NMDA noncompetitive antagonist	X	X	Gatch *et al.*, 1999a; Knapp *et al.*, 2004
Glycine antagonist	X		Gatch *et al.*, 1999b
5–HT_3 antagonist		X	Knapp *et al.*, 2004
Adenosine agonist	X		Gatch *et al.*, 1999a

anxiolytic-like effects in animal models of anxiety also reversed the anxiogenic-like effects of alcohol withdrawal (**Table 5.13**).

An increase in alcohol consumption has been observed in human social drinkers after a period of abstinence from alcohol (Burish *et al.*, 1981) and nondependent and dependent rats (Le Magnen, 1960; Sinclair and Senter, 1967, 1968; Wolffgramm and Heyne, 1995; Spanagel *et al.*, 1996; Heyser *et al.*, 1997). Termed the 'alcohol deprivation effect' (Sinclair and Senter, 1967), it may be relevant to the abstinence conditions seen in human alcoholics (see *Animal Models of Addiction* chapter). The alcohol deprivation effect can be blocked by a number of neuropharmacological agents that have been shown to block excessive drinking in animal models. Naltrexone at chronic, low doses blocked the alcohol deprivation effect (Holter and Spanagel, 1999; Heyser *et al.*, 2003). Chronic administration of acamprosate also blocked the alcohol deprivation effect (Spanagel *et al.*, 1996; Holter *et al.*, 1997; Heyser *et al.*, 1998) (**Figs. 5.28** and **5.29**), and intracerebral administration of acamprosate showed that the bed nucleus of the stria terminalis is a particularly sensitive site (A. Morse and G.F. Koob, unpublished results). Administration of identical doses of these neuropharmacological agents to nondependent rats had no effect on self-administration of alcohol. The $GABA_B$ agonist baclofen also blocked the alcohol deprivation effect in alcohol-preferring rats (Colombo *et al.*, 2003a,b) (**Fig. 5.30**). Memantine, a metabotropic glutamate 5 receptor antagonist, also blocked the alcohol deprivation effect (Holter *et al.*, 1996; Backstrom *et al.*, 2004). Both naltrexone and acamprosate have efficacy in clinical studies to promote abstinence and prevent relapse (Mason, 2003; O'Malley and Froelich, 2003). Naltrexone (Revia) was approved for the treatment of alcoholism by the United States Food and Drug Administration in 1994, and acamprosate (Campral™) was approved in 2004.

A separate but parallel approach to the study of the neuropharmacological basis for excessive drinking has been a genetic approach. Investigators capitalized on well-known large within-species preference in alcohol

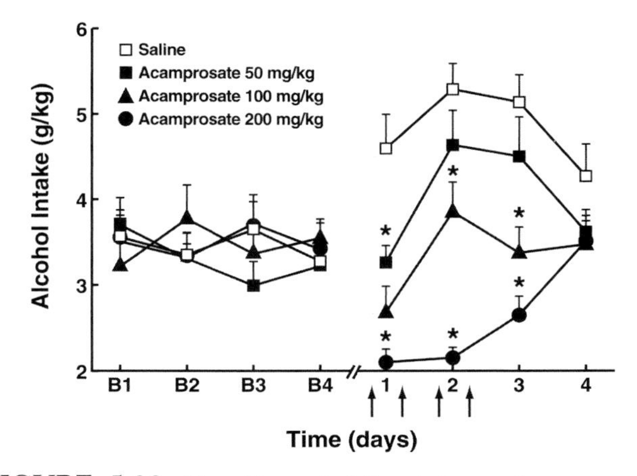

FIGURE 5.28 The effects of different doses of acamprosate administered twice daily on the first two days following alcohol abstinence (indicated by arrows) in rats intermittently exposed to a free choice of water and three different alcohol solutions (5, 10, and 20% v/v) for eight months. *$p < 0.01$, compared to saline injections. [Reproduced with permission from Spanagel *et al.*, 1996.]

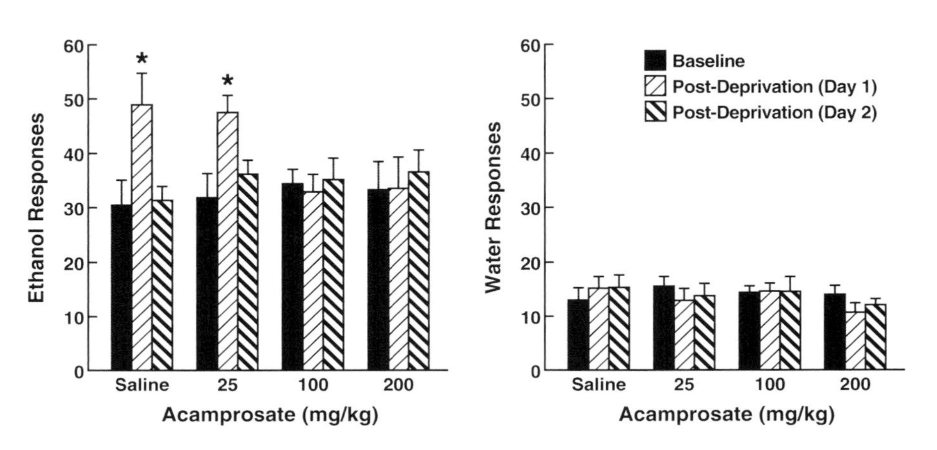

FIGURE 5.29 The effect of chronic administration of saline or acamprosate on responding for ethanol and water in rats after 5 days of ethanol deprivation. Ethanol responding increased above baseline after 5 days of ethanol deprivation in animals treated with saline or 25 mg/kg acamprosate. Twice daily administration of acamprosate during the deprivation period eliminated the alcohol deprivation effect without affecting water responding. Data are expressed as mean total response + SEM. $*p<0.05$, compared to baseline. [Reproduced with permission from Heyser *et al.*, 1998.]

drinking (Richter and Campbell, 1940) to use selective breeding to develop lines of rats that voluntarily consume large amounts of alcohol (Li and McBride, 1995; Li *et al.*, 2001; Murphy *et al.*, 2002) (for details, see *Animal Models of Addiction* chapter). Other phenotypes relevant to alcoholism in alcohol-preferring rats include increased anxiety (Stewart *et al.*, 1993; Salimov *et al.*, 1996), a higher responsivity to the low-dose stimulatory effects of alcohol (Waller *et al.*, 1984), and greater alcohol-induced stimulatory effects as measured by electrophysiological evoked potentials (Ehlers *et al.*, 1991) (see *Animal Models of Addiction* chapter).

There are extensive innate differences in the neurochemical systems implicated in the reinforcing effects of alcohol between the selectively bred high and low drinking rodents that significantly add to the base of information regarding the neuroadaptational changes associated with excessive drinking. Alcohol-preferring rats have lower levels of serotonin and dopamine function (Murphy *et al.*, 2002) (**Table 5.14**) that resemble what is observed during acute alcohol withdrawal in dependent rats. Serotonin and dopamine receptor agonists decrease self-administration of alcohol in alcohol-preferring rats (Murphy *et al.*, 1985; McBride *et al.*, 1990, 1992; Nowak *et al.*, 2000b). Perhaps most intriguing is that dopamine D_2 receptors are decreased in alcohol-preferring rats (Stefanini *et al.*, 1992), similar to what is observed in human alcoholics (McBride *et al.*, 1993) (**Fig. 5.31**) (see *Imaging* chapter).

In contrast, there appears to be an upregulation of the GABA system, and the responses to GABA were studied in alcohol-preferring rats. Higher densities of GABAergic terminals were observed in the nucleus accumbens of alcohol-naive alcohol-preferring compared to alcohol-nonpreferring lines (Hwang *et al.*, 1990). Ventral pallidum injections of a GABA agonist selective for the α_1 receptor subunit of the $GABA_A$-benzodiazepine receptor complex significantly blocked alcohol self-administration in alcohol-preferring rats, suggesting a specific receptor subtype important for the reinforcing effects of alcohol in the alcohol-preferring animals (Harvey *et al.*, 2002; June *et al.*, 2003). Higher levels of opioid receptors have been observed in some lines of alcohol-preferring animals

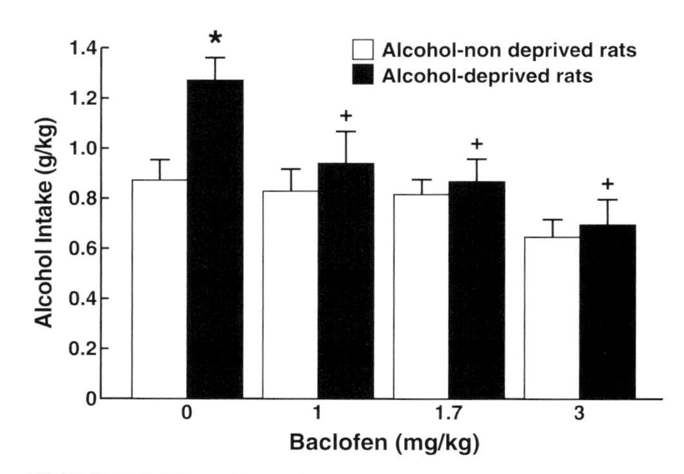

FIGURE 5.30 Effect of the $GABA_B$ receptor agonist baclofen on the alcohol deprivation effect in Sardinian alcohol-preferring (sP) rats. Alcohol-deprived rats were initially allowed to consume alcohol (10%, v/v) and water under the homecage, two-bottle free choice regimen with unlimited access for 8 consecutive weeks and then deprived of alcohol for 7 consecutive days. Conversely, alcohol-nondeprived rats had continuous access to alcohol and water (with the exception of the last 6 h before baclofen injection, when the alcohol bottle was removed to ensure that blood alcohol levels were equal to zero at the time of the test). 30 min before representation of the alcohol bottle (which coincided with lights off), rats of both groups were injected intraperitoneally with 0, 1, 1.7, or 3 mg/kg baclofen. Alcohol intake was registered 60 min after lights off. Each bar is the mean ± SEM of $n = 7–8$. $*p < 0.05$, compared to saline-treated alcohol-nondeprived rats (Newman–Keuls test); $+p < 0.05$ with respect to saline-treated alcohol-deprived rats (Newman–Keuls test). [Reproduced with permission from Colombo *et al.*, 2003a.]

TABLE 5.14 Summary of Major Innate Neurobiological Differences in Limbic Regions Between High Ethanol-Preferring and Low Ethanol-Preferring Rats

Neurotransmitter system/receptor	Differences	Reference
Serotonin		
5–HT content, innervation	NP > P	Murphy *et al.*, 1982, 1987; Zhou *et al.*, 1991a, b
	LAD > HAD	Gongwer *et al.*, 1989
	AA ≥ ANA	Ahtee and Eriksson, 1972, 1973; Korpi *et al.*, 1988
	sNP > sP (frontal cortex)	Devoto *et al.*, 1998
Postsynaptic 5-HT_{1A} receptor	P > NP	McBride *et al.*, 1993, 1997b
	HAD = LAD	McBride *et al.*, 1997a
	AA = ANA	Korpi *et al.*, 1992
5–HT_{1B} receptor	NP > P	McBride *et al.*, 1997b
5–HT_2 receptor	NP > P	McBride *et al.*, 1993b
	HAD = LAD	McBride *et al.*, 1997a
	AA = ANA	Korpi *et al.*, 1992
	sNP > sP	Ciccocioppo *et al.*, 1999
5–HT_{2C} receptor	P > NP	Pandey *et al.*, 1996
5–HT_3 receptor	P = NP	McBride *et al.*, 1997b
	AA = ANA	Ciccocioppo *et al.*, 1998
	NP > P (amygdala)	Ciccocioppo *et al.*, 1998
Dopamine		
Dopamine content, innervation	NP > P	Murphy *et al.*, 1987; Zhou *et al.*, 1995
	LAD > HAD	Gongwer *et al.*, 1989
	AA = ANA	Korpi *et al.*, 1988
	sNP > sP	Casu *et al.*, 2002
D_1 receptor	P = NP	McBride *et al.*, 1997b
	HAD = LAD	McBride *et al.*, 1997a
	sNP > sP	De Montis *et al.*, 1993
D_2 receptor	sNP > sP	Stefanini *et al.*, 1992
	NP > P	McBride *et al.*, 1993a
	HAD = LAD	McBride *et al.*, 1997a
	AA = ANA	Syvalahti *et al.*, 1994
D_3 receptor	P = NP	McBride *et al.*, 1997b
	HAD = LAD	McBride *et al.*, 1997a
GABA		
Innervation	P > NP (nucleus accumbens)	Hwang *et al.*, 1990
	HAD > LAD (nucleus accumbens)	Hwang *et al.*, 1990
$GABA_A$ receptor		
Response to agonist	P = NP	Thielen *et al.*, 1993, 1998
	AA = ANA	Wong *et al.*, 1996
Response to benzodiazepine	P > NP	Thielen *et al.*, 1997
	AA > ANA	Wong *et al.*, 1996
Response to barbiturate	P = NP	Thielen *et al.*, 1998
Opioid		
β-endorphin content	ANA ≥ AA	Gianoulakis *et al.*, 1992; Nylander *et al.*, 1994
Enkephalin mRNA	sP > sNP	Fadda *et al.*, 1999
	P = NP	Li *et al.*, 1998
μ opioid receptor	P > NP (opposite in hippocampus)	McBride *et al.*, 1998
	AA > ANA	De Waele *et al.*, 1995
	sNP > sP	Fadda *et al.*, 1999
	LAD ≥ HAD	Gong *et al.*, 1997
δ opioid receptor	AA ≥ ANA	De Waele *et al.*, 1995; Soini *et al.*, 1998
	NP > P	Strother *et al.*, 2001
Neuropeptides		
Neuropeptide Y content	NP > P	Ehlers *et al.*, 1998
(central amygdala)	NP > P	Hwang *et al.*, 1999
	LAD > HAD	Hwang *et al.*, 1999
(hypothalamus)	P > NP	Hwang *et al.*, 1999
	LAD > HAD	Hwang *et al.*, 1999

Continued

TABLE 5.14 Summary of Major Innate Neurobiological Differences in Limbic Regions Between High Ethanol-Preferring and Low Ethanol-Preferring Rats–Cont'd

Neurotransmitter system/receptor	Differences	Reference
Neuropeptide Y mRNA	ANA > AA	Caberlotto *et al.*, 2001
Arginine vasopressin content	P > NP	Hwang *et al.*, 1998
(hypothalamus)	LAD > HAD	Hwang *et al.*, 1998
CRF content	NP > P	Ehlers *et al.*, 1992
Substance P content	NP > P	Slawecki *et al.*, 2001
Neurokinin content	NP > P	Ehlers *et al.*, 1999; Slawecki *et al.*, 2001
Thyrotropic releasing hormone content	NP > P (septum)	Morzorati and Kubek, 1993

P/NP, alcohol preferring/nonpreferring; HAD/LAD, high/low alcohol drinking; AA/ANA, ALKO alcohol/nonalcohol; sP/sNP, Sardinian alcohol preferring/nonpreferring. [Reproduced with permission from Murphy *et al.*, 2002.]

but not in all. Opioid receptor antagonists block alcohol intake in most of the alcohol-preferring lines (Froehlich *et al.*, 1987; 1990; Sinclair, 1990; Badia-Elder *et al.*, 1999; Overstreet *et al.*, 1999).

An equally exciting observation is the consistent decrease in the activity of the NPY system in the alcohol-preferring animals (Ehlers *et al.*, 1998; Hwang *et al.*, 1999) (**Table 5.14** above). In addition, there is a quantitative trait locus for the NPY precursor in the F2 progeny of an *alcohol-preferring × alcohol-nonpreferring* cross (Carr *et al.*, 1998) (see below). Neuropharmacological studies have shown that NPY administered intracerebroventricularly decreased alcohol intake in alcohol-preferring rats and to a lesser extent in nonpreferring rats (Badia-Elder *et al.*, 2001) (**Fig. 5.32**).

Reinstatement of Alcohol Reinforcement

Protracted abstinence from alcohol, as defined in the rat, spans a period when acute physical withdrawal has disappeared, but changes in behavior persist. Elevations in alcohol intake over baseline and increased stress responsivity persists 2–8 weeks postwithdrawal from chronic alcohol (Roberts *et al.*, 2000a; Valdez *et al.*, 2002). Rats tested on the elevated plus maze 3–5 weeks postwithdrawal did not show an anxiogenic-like response at baseline, but an anxiogenic-like response was provoked by a mild restraint stress only in the rats with a history of alcohol dependence. This stress-induced anxiogenic-like response was reversed by a competitive CRF antagonist (Valdez *et al.*, 2003) (**Fig. 5.33**). The increased self-administration

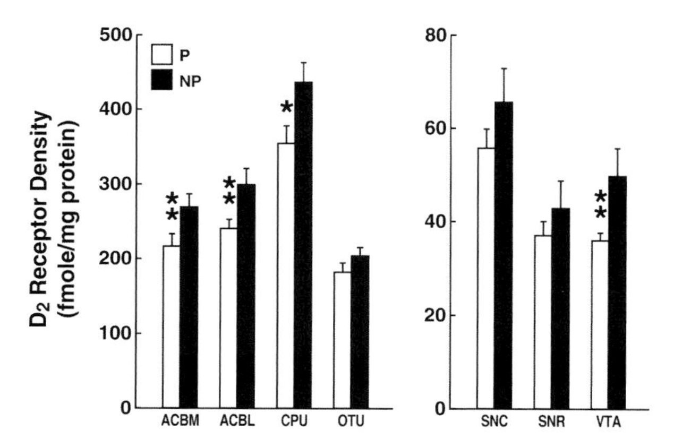

FIGURE 5.31 Densities of dopamine D_2 recognition sites labeled with 20 nM [^{3}H]sulpiride in the medial and lateral nucleus accumbens (ACBM and ACBL), caudate-putamen (CPU), olfactory tubercle (OTU), substantia nigra pars compacta (SNC), substantia nigra pars reticulata (SNR), and ventral tegmental area (VTA) of alcohol-naive P (n = 7) and NP (n = 7) rats. Data are the means + SEM. *$p < 0.025$, **$p < 0.05$, two-tailed Student t-test. [Reproduced with permission from McBride *et al.*, 1993.]

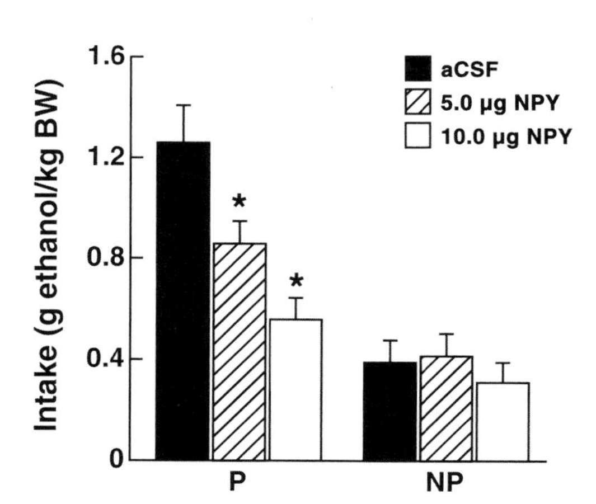

FIGURE 5.32 Mean ± SEM intake (g ethanol/kg body weight) of 8% v/v ethanol solution in alcohol-preferring (P) (n = 8) and nonpreferring (NP) (n = 8) rats after intracerebroventricular administration of neuropeptide Y (5.0 and 10.0 µg/5.0 µl) and artificial cerebrospinal fluid (aCSF) (5.0 µl) during 2-h access to one bottle of solution. [Reproduced with permission from Badia-Elder *et al.*, 2001.]

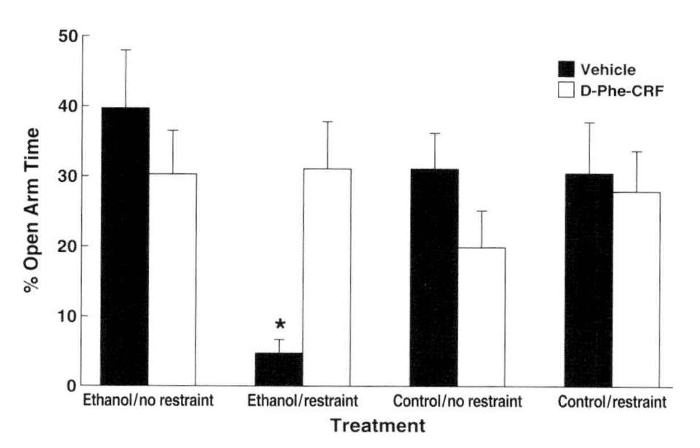

FIGURE 5.33 Effect of restraint stress on exploratory behavior of rats in the elevated plus maze 6 weeks after exposure to an ethanol liquid diet over a 3-week period. Control rats received a sucrose-containing liquid diet. Rats were injected intracerebroventricularly with 10 μg of the corticotropin-releasing factor antagonist D-Phe-CRF_{12-41} (n = 8–11/group) or vehicle (n = 7–8/group) and subsequently placed in restraint tubes or returned to their home cages for 15 min. The mean (± SEM) percentage of time spent in the open arms of the elevated plus maze was measured. $^*p < 0.05$, Tukey test compared to all other groups. [Reproduced with permission from Valdez *et al.*, 2003.]

of alcohol observed during protracted abstinence also was blocked by a competitive CRF antagonist. In rats trained to self-administer alcohol and made dependent and subsequently retested postwithdrawal, a CRF antagonist dose-dependently reduced self-administration only in the rats with a history of dependence (Valdez *et al.*, 2002) (**Fig. 5.34**). Rats similarly made dependent with chronic continuous exposure to ethanol showed anxiety-like behavior on the elevated plus maze at 4 weeks postwithdrawal (Valdez *et al.*, 2002). These results suggest that brain CRF systems remain hyperactive during protracted abstinence, and this hyperactivity is of motivational significance for excessive alcohol drinking.

Behavioral procedures have been developed to reinstate alcohol self-administration with previously neutral stimuli that have been paired with alcohol self-administration or that predict alcohol self-administration (Katner and Weiss, 1999; Katner *et al.*, 1999; Martin-Fardon *et al.*, 2000; Ciccocioppo *et al.*, 2001). Rats come to associate a specific olfactory stimulus with availability of alcohol, and that stimulus then will reinstate responding in animals subjected to extinction. Consistent with the well-established conditioned cue reactivity in human alcoholics, the motivating effects of alcohol-related stimuli are highly resistant to extinction in that they retain their efficacy to elicit alcohol-seeking behavior over more than one month of repeated testing (Katner *et al.*, 1999; Ciccocioppo *et al.*, 2001). This reinstatement was blocked by systemic administration of naltrexone and a selective μ and δ opioid receptor antagonist (Ciccocioppo *et al.*, 2002), results consistent with human studies showing that opioid antagonists may blunt the urge to drink elicited by presentation of alcohol-related cues in alcoholics (Monti *et al.*, 1999). Dopamine D_1 and D_2 receptor antagonists also blocked cue-induced reinstatement (Liu and Weiss, 2002a) (**Fig. 5.35**). Nitric oxide synthesis

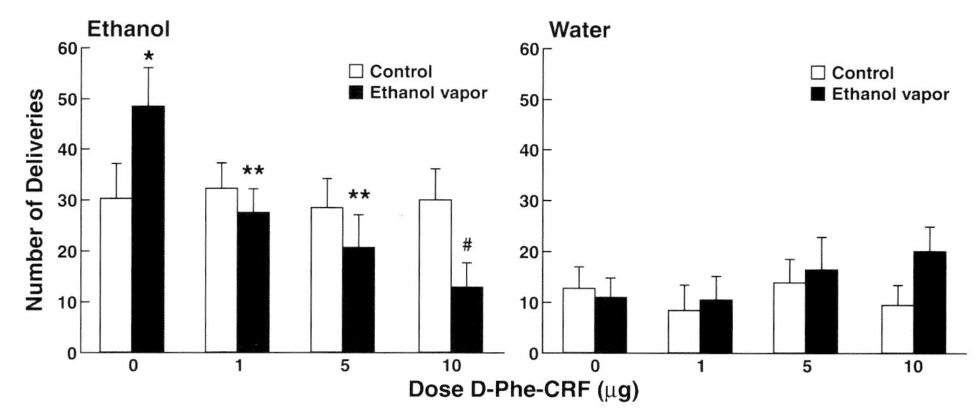

FIGURE 5.34 Effects of the corticotropin-releasing factor antagonist D-Phe-CRF_{12-41} on responding for ethanol and water in rats 2–5 weeks following chronic ethanol vapor exposure. Control rats were exposed to air vapor. Rats were microinjected intracerebroventricularly with 0–10 μg of D-Phe-CRF_{12-41} (n = 8 per group) using a within-subjects Latin square design 2 weeks after removal from the vapor chambers. The number of lever presses for ethanol and water ± SEM were measured 10 min after injection. Following the initial test session, rats were returned to their home cages and left undisturbed. The testing procedures were repeated over the next 3 weeks until the Latin square design was complete. $^*p < 0.05$, Tukey's test, compared to controls; $^{**}p < 0.05$, Tukey's test, compared to ethanol-exposed rats injected with 0 μg D-Phe-CRF_{12-41}; $\#p < 0.05$, Tukey's test, compared to ethanol-exposed rats injected with 0 μg D-Phe-CRF_{12-41} and controls. [Taken with permission from Valdez *et al.*, 2002.]

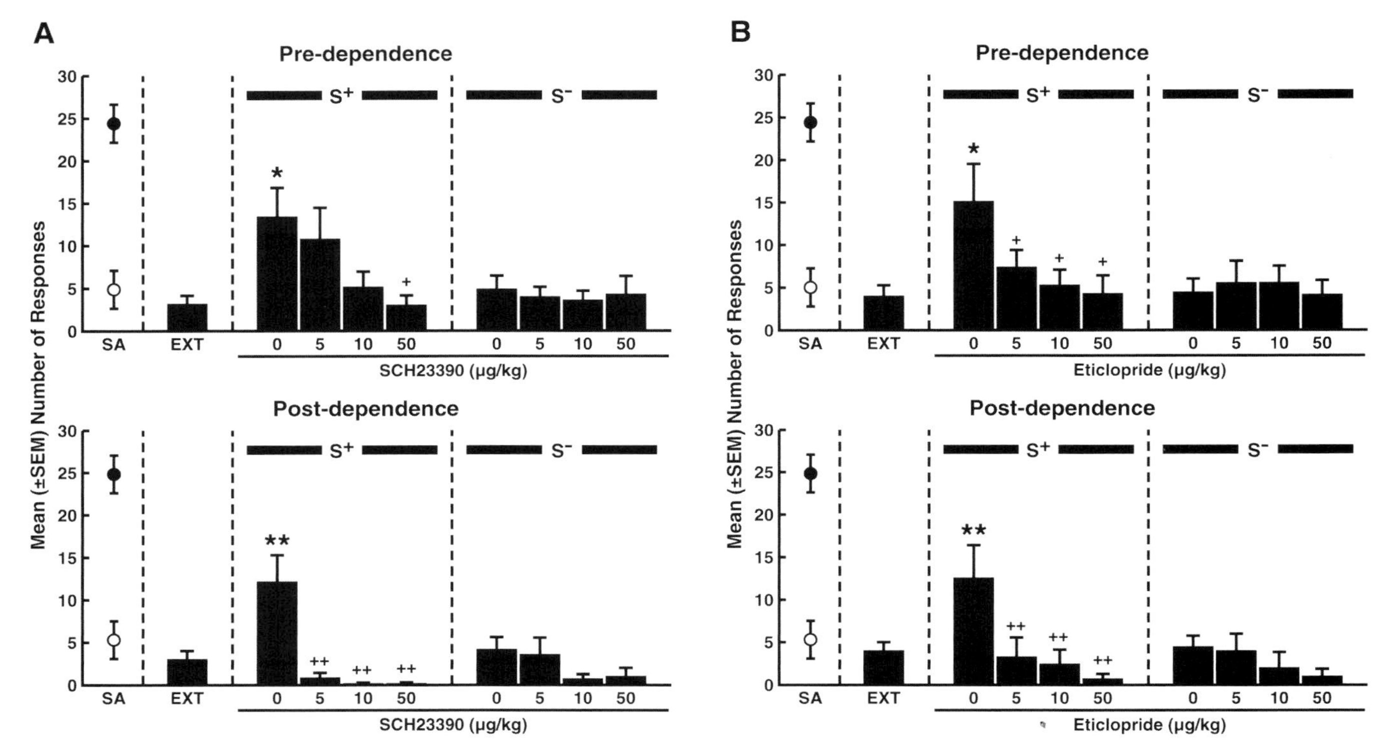

FIGURE 5.35 **(A)** Effects of olfactory discriminative stimuli associated with ethanol availability (S^+) vs. nonreward (S^-) on responding for ethanol in rats, and modification of these effects by the dopamine D_1 receptor antagonist SCH 23390. Tests lasted 30 min and were conducted in nondependent rats before a 12-day ethanol vapor inhalation procedure (Predependence) and again beginning on day 21 after termination of ethanol vapor exposure (Postdependence). Rats were treated with SCH 23390 (0, 5, 10, 50 μg/kg, s.c.) 30 min before reinstatement tests in a counterbalanced order. For comparison, the figure shows also the mean (± SEM) number of responses during ethanol self-administration (•) and nonreward (○) sessions collapsed across the last 3 days of the conditioning phase (SA), as well as the mean (± SEM) number of extinction responses (EXT) at criterion. *$p < 0.05$; **$p < 0.01$ different from extinction baseline; +$p < 0.05$; ++$p < 0.01$ different from vehicle (Newman-Keuls *post hoc* tests after ANOVA for predependence, $F_{4,40} = 3.76$, $p < 0.05$; and postdependence, $F_{4,40} = 13.99$, $p < 0.01$) conditions). **(B)** Effects of olfactory discriminative stimuli associated with ethanol availability (S^+) versus nonreward (S^-) on responding for ethanol, and modification of these effects by the D_2 antagonist eticlopride. Tests lasted 30 min and were conducted in nondependent rats before a 12-day ethanol vapor inhalation procedure (Predependence) and again beginning on day 21 after termination of ethanol vapor exposure (Postdependence). Rats were treated with eticlopride (0, 5, 10, 50 μg/kg, s.c.) 30 min before reinstatement tests in a counterbalanced order. For comparison, the figure shows also the mean (± SEM) number of responses during ethanol self-administration (•) and nonreward (○) sessions collapsed across the last 3 days of the conditioning phase (SA), as well as the mean (± SEM) number of extinction responses (EXT) at criterion. *$p < 0.05$; **$p < 0.01$ different from extinction baseline; +$p < 0.05$; ++$p < 0.01$ different from vehicle (Newman-Keuls *post hoc* tests after ANOVA for predependence, $F_{4,40} = 3.59$, $p < 0.05$; and postdependence $F_{4,40} = 4.93$, $p < 0.01$ conditions). [Reproduced with permission from Liu and Weiss, 2002a.]

inhibition by *N*-ω-nitro-L-arginine methyl ester (L-NAME) also blocked the conditioned reinstatement by alcohol cues, suggesting a glutamatergic component to cue-induced reinstatement to alcohol (Liu and Weiss, 2004).

Stress exposure also can reinstate responding for alcohol in rats previously extinguished, and this stress-induced reinstatement is blocked by the competitive CRF antagonist D-Phe-CRF_{12-41}. In fact, the CRF antagonist blocked stress-induced reinstatement but not cue-induced reinstatement, and naltrexone blocked cue-induced reinstatement but not stress-induced reinstatement, suggesting two independent neuropharmacological routes to reinstatement and by extrapolation relapse (Ciccocioppo *et al.*, 2002, 2003; Liu and Weiss, 2002b) (**Fig. 5.36**).

NEUROBIOLOGICAL MECHANISM—CELLULAR

The action of alcohol at the system level (see above) is based on the premise that alcohol has a specific action on synaptic transmission and more specifically that the synapse might be the most sensitive substrate for alcohol action. This hypothesis originates in part from early electrophysiological findings showing a greater alcohol effect in multisynaptic than monosynaptic pathways (Berry and Pentreath, 1980). Cellular studies on hippocampal and nucleus accumbens brain slices provide support for this hypothesis (Siggins *et al.*, 1987; Nie *et al.*, 1993, 1994). The specific neurotransmitter systems most studied at the cellular level are GABA

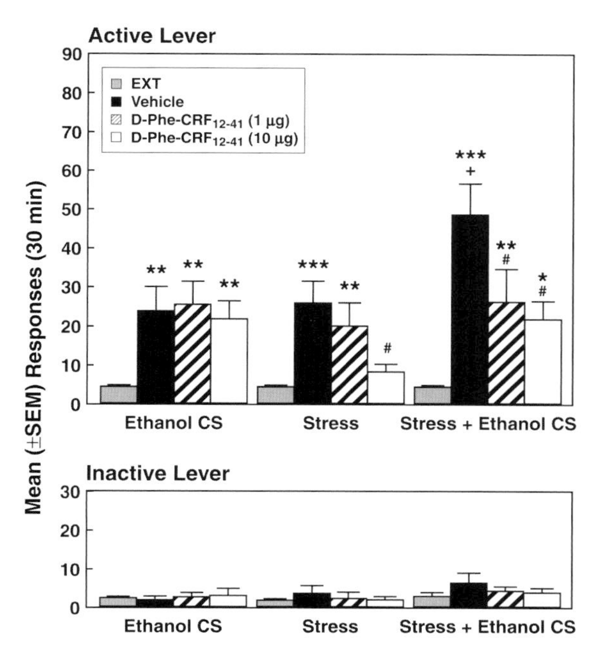

FIGURE 5.36 Effects of the corticotropin-releasing factor antagonist D-Phe-CRF_{12-41} on reinstatement after extinction (EXT) in Stress, Ethanol conditioned stimulus (CS), and Stress + Ethanol CS conditions in rats. D-Phe-CRF_{12-41} antagonized footshock but not Ethanol CS-induced responding and partially reversed reinstatement in the Stress + Ethanol CS condition, but responding remained significantly above extinction levels. For clarity, EXT data have been collapsed across the three groups in each reinstatement condition. Asterisks (*) indicate statistically significant difference from respective baseline performance ($^*p < 0.05$; $^{**}p < 0.01$; $^{***}p < 0.001$). The plus sign (+) indicates statistically significant difference from Stress and Ethanol CS vehicle control conditions. Pound signs (#) indicate statistically significant difference from vehicle. [Reproduced with permission from Liu and Weiss, 2002b.]

and glutamate. Molecular studies have shown that GABA-induced chloride fluxes are increased by alcohol in cultured neurons (Mehta and Ticku, 1988) and synaptoneurosomes (Suzdak *et al.*, 1988; Allan *et al.*, 1991), and alcohol reduced Ca^{2+} influx evoked by NMDA glutamate receptors in cultured cerebellar granule cells (Hoffman *et al.*, 1989) (see below). Electrophysiological studies may bridge the molecular effects of alcohol at the membrane level, and the neurotransmitter actions at the system level, that convey its reinforcing and dependence-inducing effects (Deitrich *et al.*, 1989; Shefner, 1990; Weight, 1992) (**Fig. 5.37**).

Various *in vivo* and *in vitro* electrophysiological studies have since either verified the alcohol–$GABA_A$ receptor interaction (Allan and Harris, 1987; Givens and Breese, 1990; Aguayo, 1991; Soldo *et al.*, 1994, 1998; Aguayo *et al.*, 2002) or have not (Harrison *et al.*, 1987; Siggins *et al.*, 1987; Osmanovic and Shefner, 1990; Marszalec *et al.*, 1998; Davies, 2003) with more reasonable doses of alcohol and in a variety of cell types. These disparate findings demonstrate that alcohol does not increase $GABA_A$-mediated inhibition in all brain regions, all cell types in the same region, nor at all $GABA_A$ sites on the same neuron, nor across species in the same brain region (Criswell *et al.*, 1993, 1995; Mihic, 1999; Siggins *et al.*, 1999; Criswell and Breese, 2005). Substantial controversy on this issue still remains, but the molecular basis for the selectivity of the action of alcohol on the $GABA_A$ receptor has been proposed to depend on the heterogeneity of $GABA_A$ receptor subunit composition and the complex neuropharmacological interactions at multiple sites (Criswell *et al.*, 1993, 1995, 2005; Kumar *et al.*, 2004) (**Table 5.15**).

Multiple mechanisms may influence the sensitivity of $GABA_A$ inhibitory postsynaptic potentials (IPSPs) to alcohol. In some studies, certain experimental conditions have been manipulated to uncover an effect of alcohol (Weiner *et al.*, 1994, 1997; Wan *et al.*, 1996; Ariwodola and Weiner, 2004). For example, in hippocampal slices, alcohol had no effect on compound inhibitory postsynaptic potentials/currents (IPSP/Cs) containing both $GABA_A$ and $GABA_B$ components. However, after pharmacological blockade of $GABA_B$ receptors by selective antagonists, alcohol enhanced the $GABA_A$ component, suggesting that activation of $GABA_B$ receptors by endogenous GABA release prevents alcohol enhancement of $GABA_A$ IPSP/Cs in hippocampus (or in other brain regions showing negative or mixed alcohol-GABA interactions).

In the central nucleus of the amygdala, alcohol produces a more direct facilitation of GABAergic function. Alcohol enhanced GABAergic IPSP amplitudes in the central nucleus of the amygdala in both rats and mice (Roberto *et al.*, 2003; Nie *et al.*, 2004). Alcohol at 44 mM and CRF at 100 nM significantly enhanced GABAergic IPSCs in neurons of the central nucleus of the amygdala of rats and wildtype and control mice, but not in neurons of CRF_1 receptor knockout mice. In addition, several selective CRF_1 antagonists blocked both the CRF and alcohol effects in rats and wildtype mice (Nie *et al.*, 2004; Roberto *et al.*, 2004a) (**Fig. 5.38**). In contrast, central nucleus of the amygdala neurons of C57BL/6J × 129 mice with a null mutation of the μ opioid receptor showed a greater enhancement by acute alcohol (44 mM) of the IPSPs than the wildtype mice (unpublished data). These results suggest that the GABAergic facilitation observed in the central nucleus of the amygdala may depend on an interaction with CRF and opioid neurons. One hypothesis is that the increase in CRF activity observed at the system level during alcohol withdrawal may be an early neuroadaptation of the brain to the effects of alcohol, and these cellular changes represent an early manifestation of the development of dependence.

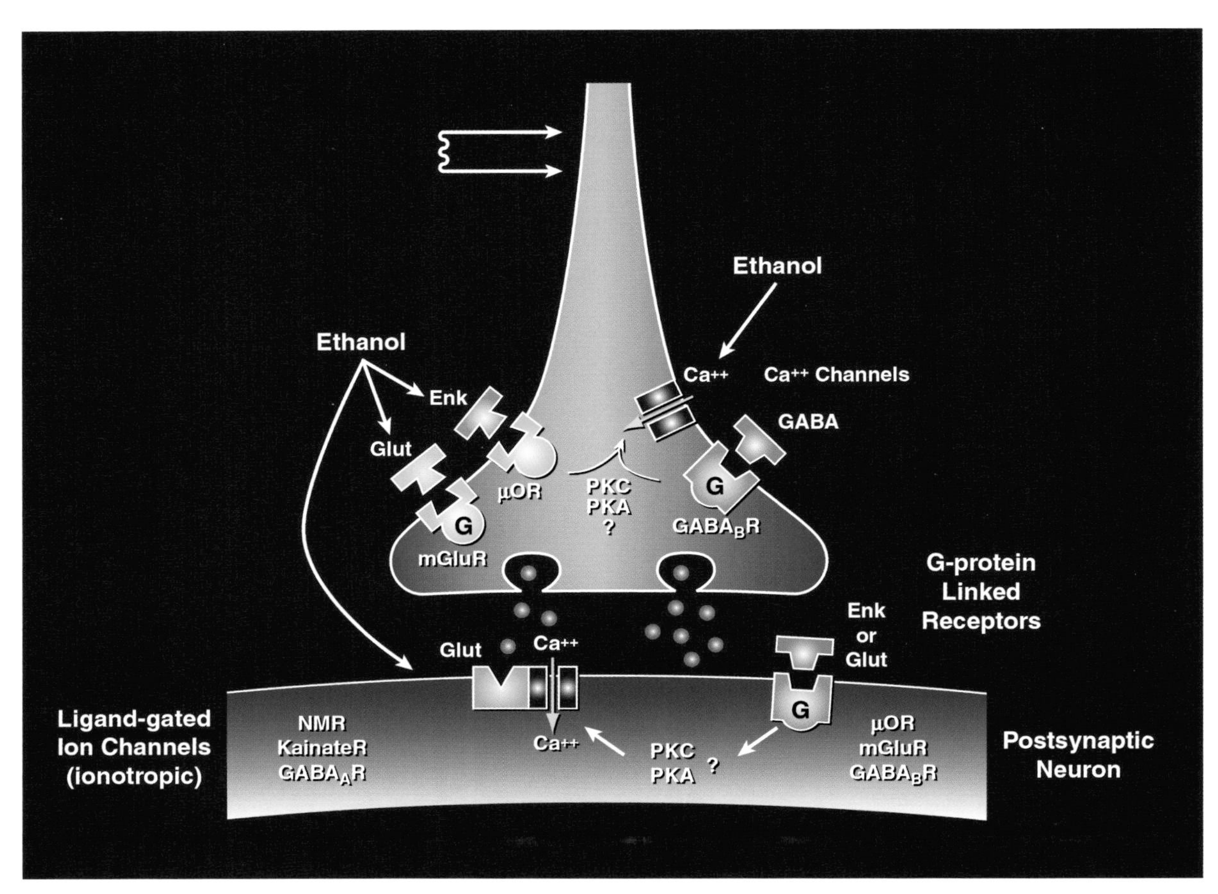

FIGURE 5.37 Probable synaptic sites of ethanol in central neurons. Alcohol facilitates γ-aminobutyric acid (GABA) release in the nucleus accumbens, ventral tegmental area, and central nucleus of the amygdala, and these actions may be mediated in some cases by presynaptic GABA$_B$ receptors. In chronic ethanol-treated animals, there is an increase in release of GABA at some of these sites. Alcohol acutely inhibits the release of glutamate in both the nucleus accumbens and the central nucleus of the amygdala possibly via metabotropic glutamate (Glut) receptors (mGluR), μ opioid receptors (μOR), or calcium (Ca^{++}) channels. In chronic ethanol-treated animals an increase in release is observed (see text for details). Enk, enkephalin; PKA, protein kinase A; PKC, protein kinase C.

TABLE 5.15 Concordance Between the Effects of Zolpidem and Ethanol on γ-Aminobutyric Acid-Induced Inhibition for Individual Neurons in Rats

Brain site	Zolpidem sensitive	Zolpidem insensitive	Concordance of zolpidem and ethanol responses
Globus pallidus	7 of 14	7 of 14	14/14 (100%)[a]
Red nucleus	2 of 5	3 of 5	5/5 (100%)[a]
Ventral pallidum	5 of 9	4 of 9	8/9 (89%)[b]
Medial septum/diagonal band	9 of 13	4 of 13	10/13 (77%)[b]
Summary—all sites	23 of 41	18 of 41	37/41 (90%)[c]

[a]Zolpidem predicted the ethanol enhancement of γ-aminobutyric acid (GABA)-induced inhibition on all neurons tested in this brain area.

[b]There was one neuron in the ventral pallidum and three neurons in the medial septum-diagonal band which provided differing responses for ethanol and zolpidem. Of these four nonconcordant neurons, zolpidem enhanced GABA on three, but ethanol did not, and one neuron responded to ethanol but not zolpidem enhancement of GABA.

[c]Probability of a random distribution < 0.01; $X^2 = 26.77$.

Protocol: Ethanol or zolpidem were applied for 40 s, beginning 30 s before GABA with application terminating at the end of a 10 s application of GABA. Some neurons responded to ethanol or zolpidem by an increase in the effectiveness of GABA (+), whereas other neurons were not affected (–). A response was defined as an increase in GABA-induced inhibition which occurred on at least two different occasions. A negative response was defined as less than 20% enhancement of GABA-induced inhibition at any dose tested. Each neuron was tested at a minimum of three currents or until two positive responses were recorded. The mean ± SEM baseline level of GABA-induced inhibition for all brain areas was 34.1 ± 2.3% for the ethanol-sensitive neurons and 35.2 ± 2.5% for the ethanol-insensitive neurons (i.e., this difference was not significant, $p > 0.1$). [Reproduced with permission from Criswell *et al.*, 1995.]

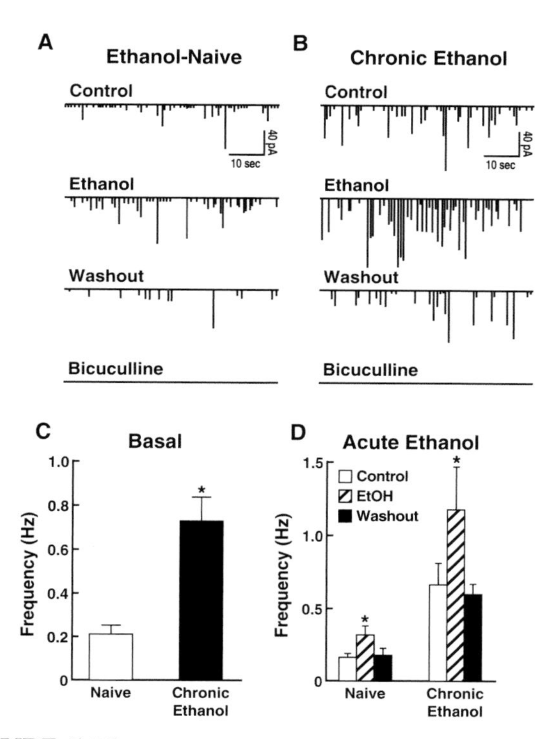

FIGURE 5.38 Effects of acute and chronic ethanol on whole-cell voltage-clamp recordings of mini inhibitory postsynaptic currents (mIPSC) from representative central nucleus of the amygdala neurons in the presence of the following (in μM): 1 TTX, 30 APV, 10 CNQX, and 1 CGP 55845A. **(A)** Current traces from a naive rat. Acute superfusion of 44 mM ethanol onto this cell increased the frequency and amplitude of the mIPSCs, with recovery on washout. Subsequent superfusion of 30 μM bicuculline totally blocked these mIPSCs. **(B)** Traces of mIPSCs from a central nucleus of the amygdala neuron of a chronic ethanol-treated rat. A 44 mM concentration of ethanol increased the frequency and amplitude of the mIPSCs, with recovery on washout. **(C and D)** Pooled data. Chronic ethanol treatment increased the mean frequency of spontaneous mIPSCs, and acute ethanol further increased the mean frequency of mIPSCs. **(C)** Average (mean ± SEM) frequency of mIPSCs (in TTX) for central nucleus of the amygdala neurons from naive rats (n = 9) and chronic ethanol-treated rats (n = 15) ($^*p < 0.001$). **(D)** Acute superfusion of 44 mM ethanol significantly increased the mean frequency of mIPSCs in neurons from both naive (n = 7) and chronic ethanol-treated (n = 9) rats, with recovery on washout. TTX, tetrodotoxin, selective blocker of voltage-gated Na^+ channels used to block action potentials; APV, NMDA antagonist DL-2-amino-5-phosphonovaleric acid; CNQX, α-amino-3-hydroxy-5-methyl-4-isoxale propionic acid (AMPA) receptor antagonist 6-Cyano-7-nitroquinoxaline-2,3-dione; CGP 55845A, γ-aminobutyric acid-B receptor antagonist. [Reproduced with permission from Roberto *et al.*, 2004a.]

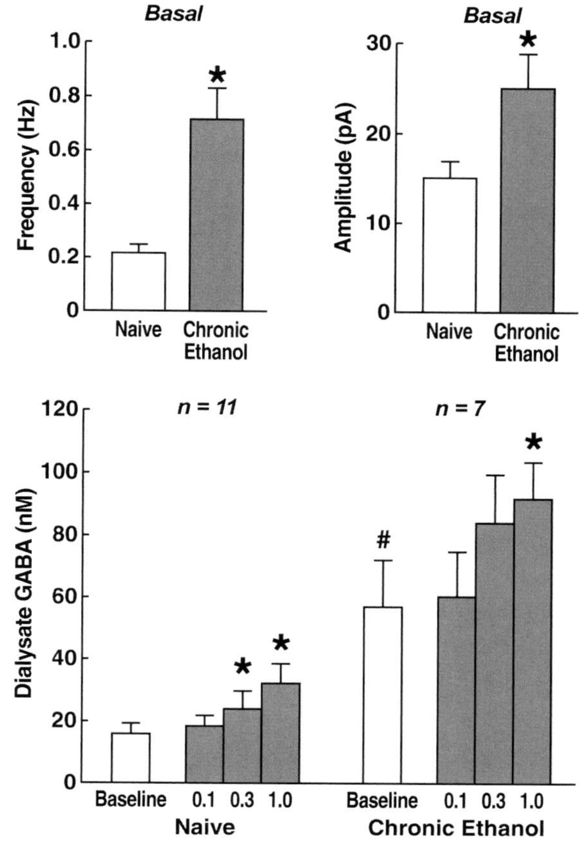

FIGURE 5.39 Chronic ethanol treatment increased the mean frequency and amplitude of spontaneous mini inhibitory postsynaptic currents (mIPSCs), and acute ethanol further increased the mean frequency of mIPSCs. **(A)** Average (mean ± SEM) frequency of mIPSCs (in tetrodotoxin, a selective blocker of voltage-gated Na^+ channels that is used to block action potentials) for central nucleus of the amygdala neurons from naive rats (n = 9) and chronic ethanol-treated rats (n = 15) ($^*p<0.001$). **(B)** The same group of neurons showed an increase in mean amplitude of mIPSCs from chronic ethanol-treated rats ($^*p<0.05$). **(C)** Acute and chronic ethanol increased dialysate levels of γ-aminobutyric acid (GABA) in the central nucleus of the amygdala *in vivo*. In both naive and chronic ethanol-treated rats, ethanol administration into the central nucleus of the amygdala significantly and dose-dependently increased mean local dialysate GABA levels ($^*p < 0.05$). The mean baseline dialysate GABA level was significantly increased in chronic ethanol-treated compared to naive rats ($^*p < 0.001$). [Reproduced with permission from Roberto *et al.*, 2004a.]

Acute alcohol-induced increases in $GABA_A$ receptor-mediated IPSP/Cs in neurons of the central nucleus of the amygdala, at least in part, may be due to an increase in presynaptic GABA release (Roberto *et al.*, 2003; Nie *et al.*, 2004). However, even more compelling is the observation that this presynaptic effect is enhanced in dependent animals (Roberto *et al.*, 2004a). In slices from chronically alcohol-exposed rats (2–8 h after 2 weeks of continuous alcohol vapor exposure with BALs of 150–200 mg%), baseline evoked IPSP/C amplitudes were increased, and paired-pulse facilitation ratios were lower than in naive rats, suggesting increased GABAergic transmission. Microdialysis data confirmed this increase, showing a 400 per cent increase in GABA dialysate in chronic alcohol-exposed animals in the central nucleus of the amygdala (Roberto *et al.*, 2004a) (**Fig. 5.39**).

Alcohol at very low doses blocks responses of the NMDA receptor, seen initially *in vitro* in electrophysiological studies (Lima-Landman and Albuquerque, 1989; Lovinger *et al.*, 1989, 1990; White *et al.*, 1990; Morrisett *et al.*, 1991; Nie *et al.*, 1994) and subsequently *in vivo* (Simson *et al.*, 1991; Roberto *et al.*, 2004a). This interaction is robust and not as dependent upon region or cell type as alcohol–GABA interactions, and the interaction is observed in most neuron types studied, including those in dorsal root ganglia, hippocampus, dentate, cerebellum, medial septum, inferior colliculus, nucleus accumbens (Nie *et al.*, 1994), and amygdala (Calton *et al.*, 1998, 1999; but see Simson *et al.*, 1993; Yang *et al.*, 1996; Roberto *et al.*, 2004a). Acute administration of alcohol also reduced the firing of spontaneous, glutamate-activated, and fimbria-activated neurons in the core of the nucleus accumbens in both anesthetized and unanesthetized, freely moving rats (Criado *et al.*, 1995). A similar interaction of alcohol with non-NMDA glutamate (and especially kainate) receptors has been observed (Dildy-Mayfield and Harris, 1992; Lovinger, 1993; Nie *et al.*, 1994; Martin *et al.*, 1995; Carta *et al.*, 2003; Roberto *et al.*, 2004b).

Electrophysiological studies have provided insights into the molecular changes in glutamate receptors associated with chronic alcohol treatment and withdrawal. Initial studies suggested that the NMDAR2A and 2B subunits were most sensitive to alcohol, while the NMDAR2C and 2D subunits were less sensitive (Masood *et al.*, 1994; Chu *et al.*, 1995; Lovinger, 1995; Mirshahi and Woodward, 1995; Roberto *et al.*, 2004b). However, one or more splice variants of the NMDAR1 subunit also may be alcohol-sensitive (Koltchine *et al.*, 1993). Several behavioral, neurochemical, and electrophysiological findings have suggested that chronic alcohol or withdrawal may upregulate NMDA receptor function in brain (Engberg and Hajos, 1992; Chandler *et al.*, 1993; Davidson *et al.*, 1993; Snell *et al.*, 1993; Ahern *et al.*, 1994; Morrisett, 1994; Davidson *et al.*, 1995; Hu and Ticku, 1995; Ripley and Little, 1995; Roberto *et al.*, 2004b). Upregulation of the NMDAR1, R2A, and 2B subunits with chronic alcohol and withdrawal in some brain regions has been observed (Trevisan *et al.*, 1994; Follesa and Ticku, 1995, 1996). These studies suggested a role for glutamate in the behavioral sensitization elicited by repeated alcohol withdrawal, suggesting that glutamate receptors play a major role in the hyperexcitability following withdrawal from alcohol, and perhaps in alcohol-seeking behavior associated with dependence (Stephens, 1995).

Alcohol in intoxicating doses (20–40 mM) activates neurons in the ventral tegmental area both *in vivo* in unanesthetized rats (Gessa *et al.*, 1985) and *in vitro* using extracellular recordings in brain slice preparations (Brodie *et al.*, 1990). More detailed analysis revealed that at least part of the activation was via an action directly on the dopaminergic neurons in the ventral tegmental area because the alcohol activation did not require calcium (Brodie *et al.*, 1990), and dissociated dopamine neurons from the ventral tegmental area also showed a dose-dependent activation at doses of 20–40 mM (Brodie *et al.*, 1999). This activation appears to be associated with a reduction in the amplitude of the after-spike hyperpolarization that may be related to changes in potassium currents (Brodie and Appel, 1998). Acetaldehyde also activates dopamine neuronal activity in the ventral tegmental area (Foddai *et al.*, 2004) and may contribute in the brain to the reinforcing effects of alcohol as has been suggested by others (Amit *et al.*, 1977).

Other studies have identified nondopaminergic, alcohol-sensitive neurons in the ventral tegmental area that were identified as GABA neurons (Lee *et al.*, 1996; Steffensen *et al.*, 1998). These neurons showed rapid-firing, nonbursting spikes, and other characteristics that distinguished them from dopamine neurons, and these neurons were GABAergic projections to the nucleus accumbens and cortex (Steffensen *et al.*, 1998). Acute administration of alcohol within the intoxicating range of 0.2–2.0 g/kg decreased spontaneous activity of ventral tegmental area GABAergic neurons (Gallegos *et al.*, 1999).

In contrast, during withdrawal from alcohol, there is a decrease in dopaminergic activity in the ventral tegmental area that has been linked to the dysphoria of acute and protracted withdrawal (Diana *et al.*, 1992, 1993, 1996; Shen and Chiodo, 1993). The decrease in dopaminergic activity in the ventral tegmental area is consistent with microdialysis studies showing decreases in dopamine release in the nucleus accumbens during alcohol withdrawal (see above). Reduced dopaminergic neurotransmission is prolonged, outlasting the physical signs of withdrawal (Diana *et al.*, 1996; Bailey *et al.*, 1998, 2000). The decrease in activity of the dopaminergic neurons in the ventral tegmental area also was accompanied by a decrease in the morphology of the tyrosine hydroxylase staining cells in the ventral tegmental area with a reduction in cell size (Diana *et al.*, 2003). However, chronic intermittent alcohol treatment to C57/BL6J mice, not sufficient to produce behavioral signs of withdrawal, produced no differences in the baseline firing rate of dopamine neurons in the ventral tegmental area but did increase the excitatory response produced by alcohol (Brodie, 2002). There was a blunted response to the inhibition produced by GABA in the dopamine neurons, suggesting one possible mechanism for the increased sensitivity to alcohol (Brodie, 2002). In contrast, non-dopamine,

GABAergic neurons in the ventral tegmental area showed enhanced baseline activity following chronic alcohol administration (Gallegos *et al.*, 1999) and thus may contribute to the increased GABAergic activity observed in dependent animals (see above). Together these results suggest that, as with other drugs of abuse (e.g., see *Psychostimulants* chapter), chronic dependence-inducing effects of alcohol can depress dopaminergic function in the ventral tegmental area during withdrawal, but at the same time enhance responsiveness to alcohol under certain conditions. In addition, there may be nondopaminergic GABAergic neurons that are activated during dependence that contribute to the motivating effects of alcohol in the dependent state (see above).

NEUROBIOLOGICAL MECHANISM—MOLECULAR

Ligand-gated Ionotropic Receptors

Alcohol has been hypothesized to interact at the molecular level with specific neuronal elements, termed 'ethanol-receptive elements,' to produce changes in specific synaptic targets at the cellular and system levels to produce its pharmacological effects (Tabakoff and Hoffman, 1992; Deitrich and Erwin, 1996). Low doses of alcohol (10–50 mM) may act directly upon proteins that form ligand-gated receptor channels such as the GABA receptor complex, particularly in specific brain areas, to facilitate GABA transmission (Liljequist and Engel, 1982; Allan and Harris, 1986; Suzdak *et al.*, 1986b; Ticku *et al.*, 1986; Lima-Landman and Albuquerque, 1989; Nie *et al.*, 1994; Mihic and Harris, 1996; Ming *et al.*, 2001). Alcohol also interacts with 5-HT_3 receptors (McBride *et al.*, 2004), glutamate receptors (Lovinger *et al.*, 1989), calcium channels (Leslie *et al.*, 1983), potassium M channels (Moore *et al.*, 1990; Madamba *et al.*, 1995), G-proteins (Diehl *et al.*, 1992), and protein kinases (Olive *et al.*, 2000; Walter *et al.*, 2000) (**Fig. 5.40**).

Alcohol potentiates GABA-gated currents by activating GABA-mediated chloride ion uptake into cerebral microsacs (Allan and Harris, 1985, 1986), synaptoneurosomes (Suzdak *et al.*, 1986a), and cultured spinal cord neurons (Ticku *et al.*, 1986; Grobin *et al.*, 1998) (**Fig. 5.41**). Several hypotheses have been suggested for the mechanism of this action, including a direct action on the GABA–benzodiazepine ionophore complex, an action on the GABA–benzodiazepine ionophore complex only in the presence of exogenous GABA, and a facilitation through coactivation of $GABA_B$ receptors. Also, not all laboratories have reliably observed the chloride flux enhancement (Mihic and Harris, 1996). Similar to chloride influx assays, electrophysiological studies have shown alcohol enhancement of GABAergic currents, but often only in specific brain regions (Givens and Breese, 1990; Proctor *et al.*, 1992; Aguayo *et al.*, 1994; Frye *et al.*, 1994; Soldo *et al.*, 1994; Mihic and Harris, 1996) (see above).

An important question is whether there is subunit dependence for the actions of alcohol on the GABA-benzodiazepine ionophore complex. Molecular studies have revealed multiple $GABA_A$ subunits that can be divided by homology into subunit classes with several members: α1–6, β1–4, γ1–3, δ, ε, and τ (Sieghart and Sperk, 2002). A number of subunits and mechanisms have been implicated in alcohol's enhancement of $GABA_A$ currents at reasonable doses (10–50 mM). The presence of the γ_2 long subunit of the $\alpha1\beta2\gamma2$ GABA receptor (Wafford *et al.*, 1991), the $\alpha4\beta1\delta$ subunit (Sundstrom-Poromaa *et al.*, 2002), and the $\alpha4\beta3\delta$ and $\alpha6\beta3\delta$ subunits (Wallner *et al.*, 2003) have all been implicated in sensitivity to alcohol.

Recent studies have suggested a direct, highly sensitive action of alcohol on $GABA_A$ receptors if the δ subunit is expressed. Recombinant $\alpha4\beta3\delta$ and $\alpha6\beta3\delta$ subunits expressed in *Xenopus* oocytes are particularly sensitive to alcohol with thresholds as low as 0.1 mM (Wallner *et al.*, 2003). Alcohol also was much more effective on $\beta3$ subunits than on $\beta2$ subunits with a 10-fold increase when the $\beta2$ subunit was replaced by the $\beta3$ (Wallner *et al.*, 2003). The authors hypothesized as a result of this increased sensitivity that extrasynaptic receptors in some cells are composed of $\alpha4\beta3\delta$ subunits and may be the primary targets for alcohol in the GABA receptor system, including such functions as sleep, anxiety, memory, and cognition (Hanchar *et al.*, 2004; Olsen *et al.*, 2004). However, total knockout of the $\alpha6\beta3$ and δ GABA receptor subunits does not show dramatic changes in alcohol sensitivity (Homanics *et al.*, 1997; Quinlan *et al.*, 1998; Mihalek *et al.*, 2001; Shannon *et al.*, 2004), although δ knockout mice do drink less alcohol (Mihalek *et al.*, 2001). Nevertheless, it has been argued that point mutations can show more pronounced phenotypes than total knockouts, so the hypothesis of a critical role for the $\beta3$ and δ subunits awaits further study (Jurd *et al.*, 2003; Hanchar *et al.*, 2004). Deletion of the $\alpha1$ subunit and the $\alpha5$ subunit in knockout studies decreased alcohol drinking and increased the locomotor stimulant effects of alcohol (Blednov *et al.*, 2003; Boehm *et al.*, 2004) (**Table 5.16**).

The possibility of an actual alcohol binding site on $GABA_A$ receptors was explored using chimeric receptor constructs where a region of two amino acids in transmembrane domains 2 and 3 were critical for the

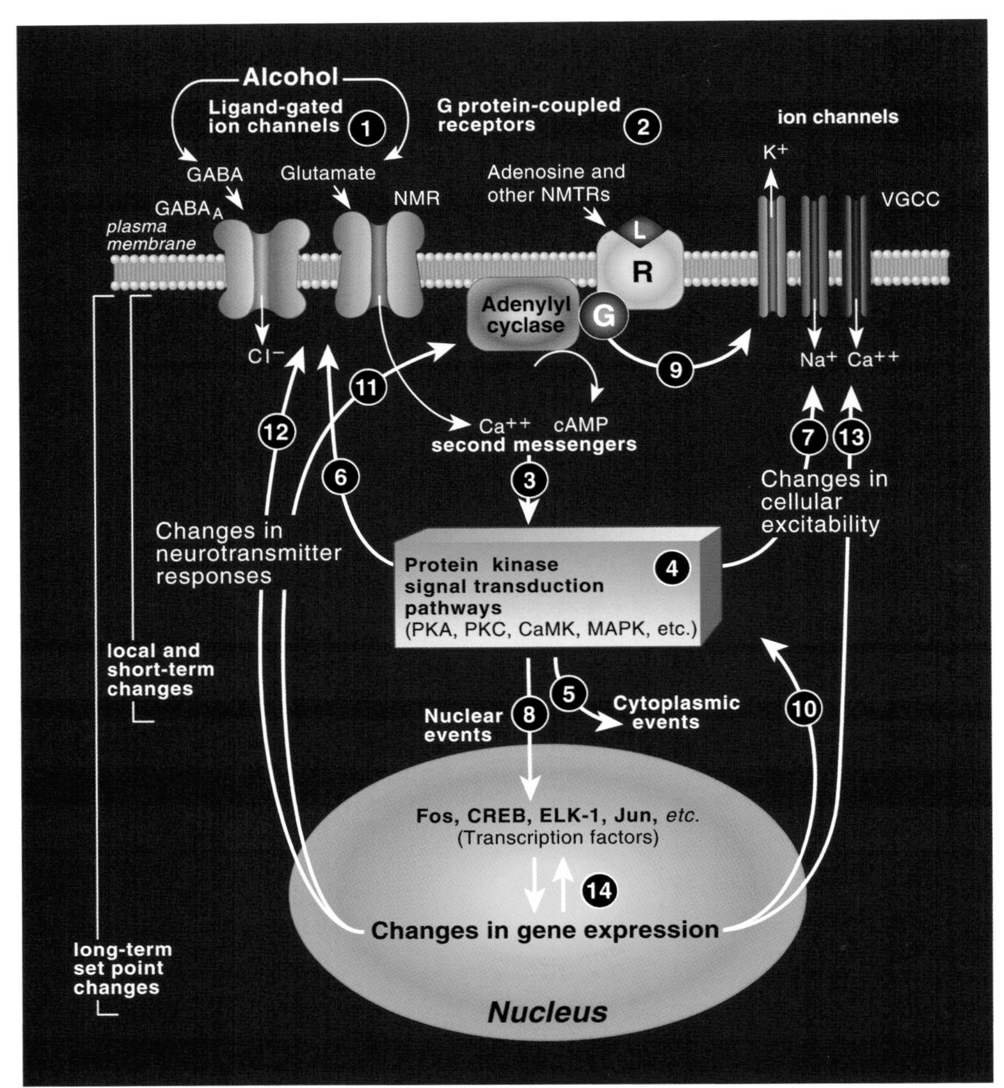

FIGURE 5.40 Alcohol abuse, by acting on neurotransmitter systems, affects the phenotypic and functional properties of neurons through the general mechanisms outlined in the diagram. Shown are examples of ligand-gated ion channels (①) such as the γ-aminobutyric acid-A ($GABA_A$) and the glutamate *N*-methyl-D-aspartate (NMDA) receptor (NMR) and G protein-coupled receptors (R) such as opioid, dopamine (DA), or the cannabinoid CB_1 receptors, among others (②). The latter also is activated by endogenous cannabinoids such as anandamide. These receptors modulate the levels of second messengers such as cyclic adenosine monophosphate (cAMP) and Ca21 (③), which in turn regulate the activity of protein kinase transducers (④). Such protein kinases affect the functions of proteins located in the cytoplasm, plasma membrane, and nucleus (⑤⑥⑦⑧). Among membrane proteins affected are ligand-gated and voltage-gated ion channels (⑥ and ⑦). Alcohol, for instance, has been proposed to affect the $GABA_A$ response via protein kinase C (PKC) phosphorylation. G_i and G_o proteins also can regulate potassium and calcium channels directly through their βγ subunits (⑨). Protein kinase transduction pathways also affect the activities of transcription factors (⑧). Some of these factors, like cAMP response element binding protein (CREB), are regulated posttranslationally by phosphorylation; others, like Fos, are regulated transcriptionally; still others, like Jun, are regulated both posttranslationally and/or transcriptionally. While membrane and cytoplasmic changes may be only local (e.g., dendritic domains or synaptic boutons), changes in the activity of transcription factors may result in long-term functional changes. These may include changes in gene expression of proteins involved in signal transduction (⑩) and/or neurotransmission (⑪⑫⑬), resulting in altered neuronal responses. For example, chronic exposure to alcohol has been reported to increase levels of protein kinase A (PKA) (⑩) and adenylyl cyclase (⑪) in the nucleus accumbens and to decrease levels of $G_{i\alpha}$ (⑪). Moreover, chronic ethanol induces differential changes in subunit composition in the $GABA_A$ and glutamate inotropic receptors (⑫) and increases expression of voltage-gated calcium channels (VGCC) (⑬). Chronic exposure to alcohol also alters the expression of transcription factors themselves (⑭). CREB expression, for instance, is increased in the nucleus accumbens and decreased in the amygdala by chronic alcohol treatment. Chronic alcohol induces a transition from Fos induction to the induction of the longer-lasting Fos-related antigens (FRAs). The receptor systems depicted in this figure may not coexist in the same cells. [Modified with permission from Koob *et al.*, 1998.]

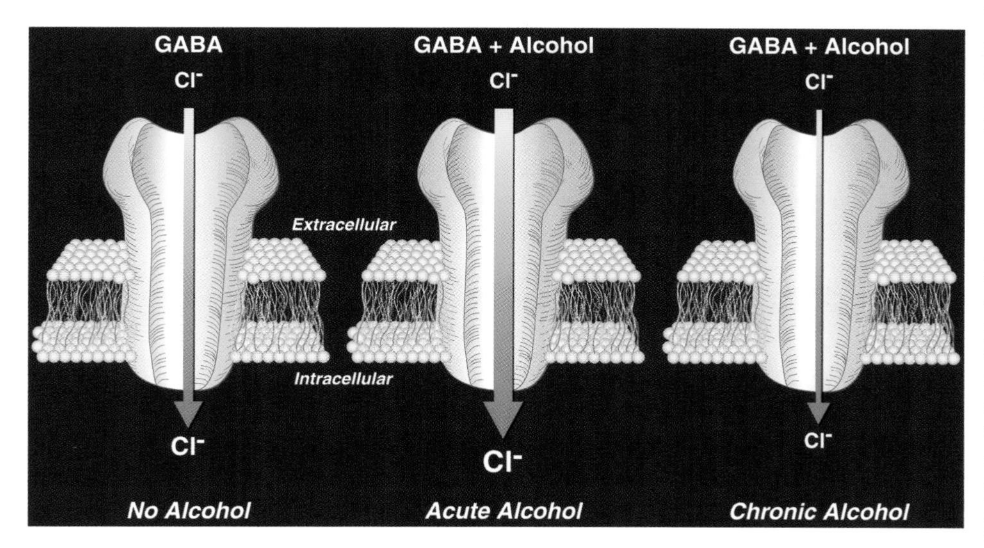

FIGURE 5.41 Representation of γ-aminobutyric acid-A ($GABA_A$) receptor sensitivity during acute and chronic alcohol administration. Normally, $GABA_A$ receptors, in the presence of GABA, will open to allow infux of Cl^- ions along the chloride gradient from outside to inside the cell (leftmost panel), hyperpolarizing the membrane and inhibiting cell firing. Acute alcohol administration increases the effects of GABA, allowing more Cl^- to enter the cell (center). After prolonged ethanol exposure, ethanol and GABA both have reduced effects at the receptor; inhibition is reduced and Cl^- flux is reduced (right panel). Therefore, the inhibitory tone of the neuron is enhanced acutely by ethanol but reduced after prolonged exposure to ethanol. [Reproduced with permission from Grobin *et al.*, 1998.]

allosteric modulation of $GABA_A$ receptors by alcohol (Mihic *et al.*, 1997). The authors hypothesized that one explanation for these results would be the existence of hydrophobic binding sites in pockets formed between at least two transmembrane domains (Mihic *et al.*, 1997).

Alcohol is a small molecule with low binding energy and a high Hill coefficient compared to other drugs of abuse, which is reflected in the requirement for much higher concentrations of alcohol for intoxication (millimolar levels compared to nanomolar). These pharmacological characteristics have been hypothesized to suggest that multiple alcohol binding sites are required to influence the structure and function of proteins (Trudell and Harris, 2004). As noted above, mutation of key amino acids in mammalian ligand-gated ion channel receptors led to the further hypothesis that alcohol may be binding in a water-filled pocket of such receptor proteins (Mihic *et al.*, 1997; Mascia *et al.*, 2000; Harris and Mihic, 2004). Selective expression of specific subunits of the $GABA_A$ receptor shows that coexpression of both the δ and β3 subunits is required for full sensitivity of $GABA_A$ receptors (Wallner *et al.*, 2003) (see above).

How these subunits are linked to specific protein pockets may be modeled by the LUSH protein found in the fruit fly (*Drosophila*), an odorant binding protein that is used by the fly to sense alcohols. Short-chain alcohols bind at only a single site in the LUSH protein, and this binding conveys a dramatic stabilization of the protein (Kruse *et al.*, 2003; Trudell and Harris, 2004) (**Fig. 5.42**). The alcohol binding site in LUSH is between two α-helical segments, an arrangement also suggested for alcohol binding sites in ligand-gated ion channels (Trudell and Harris, 2004). The high resolution studies of the LUSH crystal structure show that when it accommodates alcohols, the binding site makes distinct changes in the orientation of amino acid side chains and the placement of internal water molecules. Unknown at this time is how substitution of a few molecules of water convey the molecular changes in ligand-gated ion channels and other protein functional units in the brain that lead to the cellular and system changes associated with alcohol intoxication, dependence, and addiction.

Alcohol also can interact with neuromodulators such as neuroactive steroids that indirectly act on the $GABA_A$ receptor (Morrow *et al.*, 2001; Partridge and Valenzuela, 2001) as well as other receptors (Simmonds, 1991; Randall *et al.*, 1995; Brot *et al.*, 1997). Systemic administration of alcohol dramatically elevated both plasma and cortical levels of 3α-hydroxy,5α-pregnane-20-one (allopregnanolone) in both male and female rats (Morrow *et al.*, 1998; VanDoren *et al.*, 2000) (**Fig. 5.43**). Such an increase in allopregnanolone is sufficient to potentiate $GABA_A$ receptor-mediated chloride flux which suggested that allopregnanolone may mediate some of the acute actions of alcohol (Morrow *et al.*, 1987). Allopregnanolone increases alcohol self-administration in nondependent rats (Janak *et al.*, 1998), but decreases alcohol self-administration in alcohol-dependent, alcohol-preferring rats (Morrow *et al.*, 2001). Administration of the steroid biosynthetic inhibitor finasteride partially reversed the increase in allopregnanolone observed following alcohol administration (VanDoren *et al.*, 2000). Finasteride blocked the anticonvulsant actions of alcohol but not the hypnotic effects as

TABLE 5.16. Knockout Mice and Alcohol Self-Administration

Knockout	Operant self-administration	Two-bottle choice	Reference
Cannabinoid			
CB_1		↓↑	Hungund *et al.*, 2003; Poncelet *et al.*, 2003; Naassila *et al.*, 2004; Lallemand and De Witte, 2005
Opioid			
μ receptor	↓	↓	Roberts *et al.*, 2000b; Hall *et al.*, 2001; Becker *et al.*, 2002
δ receptor	↑	↑	Roberts *et al.*, 2001
β endorphin		↑	Grisel *et al.*, 1999; Grahame *et al.*, 2000 (postethanol deprivation)
Adenosine			
A_{2A} receptor		↑	Naassila *et al.*, 2002
Dopamine			
D_1 receptor		↓	El Ghundi *et al.*, 1998
D_2 receptor	↑	↑↓	Risinger *et al.*, 2000; Palmer *et al.*, 2003; Thanos *et al.*, 2005
Transporter		↑↓	Savelieva *et al.*, 2002; Hall *et al.*, 2003 (females)
Serotonin			
$5HT_{1B}$ receptor	↑	↑	Crabbe *et al.*, 1996; Risinger *et al.*, 1999
Transporter		↓	Kelai *et al.*, 2003
γ-Aminobutyric acid			
$GABA_{A\alpha1}$ receptor		↓	Blednov *et al.*, 2003
$GABA_{A\delta}$ receptor		↓	Mihalek *et al.*, 2001
Corticotropin-releasing factor			
CRF_1 receptor		↑	Sillaber *et al.*, 2002
Neuropeptide Y			
NPY	↑	↑	Thiele *et al.*, 2000
NPY_1 receptor		↑	Thiele *et al.*, 2002
NPY_2 receptor	↓		Thiele *et al.*, 2004
Other			
Norepinephrine		↓	Weinshenker *et al.*, 2000
Vesicular monoamine transporter		↑	Hall *et al.*, 2003
Protein kinase A		↑	Thiele *et al.*, 2000
PKCε	↓		Olive *et al.*, 2000
		↓	Hodge *et al.*, 1999; Choi *et al.*, 2002; Olive *et al.*, 2005
PKCγ		↑	Bowers and Wehner, 2001
Fyn kinase		↓	Boehm *et al.*, 2003, 2004
nNOS		↑	Spanagel *et al.*, 2002
DARPP-32	↓		Risinger *et al.*, 2001
Homer-2		↓	Szumlinski *et al.*, 2003
Endopeptidase		↓	Siems *et al.*, 2000
Angiotensin II		↓	Maul *et al.*, 2001
ENT–1		↑	Choi *et al.*, 2004
Per2-Brdm1	↑	↑	Spanagel *et al.*, 2005
N-type calcium channel		↓	Newton *et al.*, 2004

measured by alcohol-induced sleep or motor-impairing effects, calling into question the exact physiological relevance of endogenous allopregnanolone in the reinforcing effects of alcohol (Van Doren *et al.*, 2000).

Another mechanism by which $GABA_A$ receptors could be modulated by alcohol is via changes in protein kinase phosphorylation. Protein kinase C (PKC)-mediated phosphorylation (Weiner *et al.*, 1994), and the γ isoform of PKC in particular, may be involved in the actions of alcohol to facilitate GABA receptor activation (Abeliovich *et al.*,1993; Harris *et al.*, 1995). PKCγ knockout mice show reduced sensitivity to alcohol-induced loss of righting reflex and hypothermia (Harris *et al.*, 1995). Perhaps most compelling,

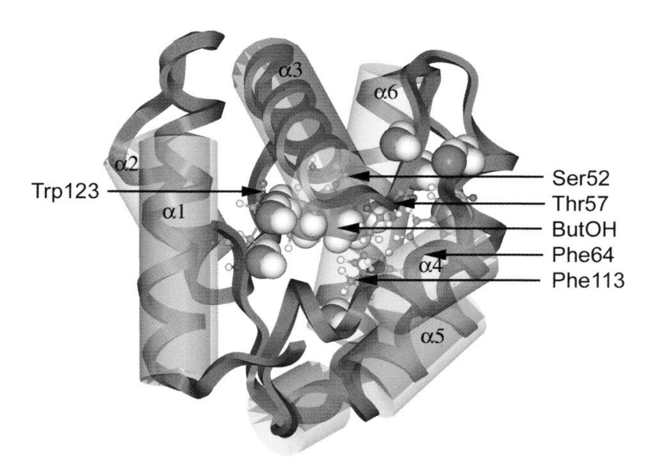

FIGURE 5.42 Butanol (ButOH) at the interhelical binding site in the LUSH protein of *Drosophila*. The protein data bank file of butanol in the binding site of the LUSH protein (100H), complete with 329 water molecules, was provided by Kruse *et al.* (2003). A subset of nine water molecules within 10A of the oxygen atom of butanol was identified with Insight II (Accelrys, San Diego, CA); the subset is highlighted with pink van der Waals surfaces. The five amino acid residues identified in the binding pocket of LUSH by Kruse *et al.* (2003) are rendered with stick surfaces: carbon (*green*), oxygen (*red*), nitrogen (*blue*), and hydrogen (*white*). [Reproduced with permission from Trudell and Harris, 2004.]

in microsacs from PKCγ knockout mice, alcohol no longer enhanced GABA-mediated chloride flux (Harris *et al.*, 1995), and PKCγ knockout mice show enhanced alcohol drinking (Bowers and Wehner, 2001). The observation that mutation of the consensus site for phosphorylation by PKC contained in the γ2 long splice variant in *Xenopus* oocytes blocked alcohol enhancement of GABAergic currents can be integrated with the observation that the γ2 long subunit may be a site of action for alcohol postsynaptic potentiation via a protein kinase interaction (Wafford and Whiting, 1992).

Another PKC isoform, PKCε, also has been hypothesized to mediate several actions of alcohol (Messing *et al.*, 1991). PKCε knockout mice show an enhanced sensitivity to the effects of alcohol on $GABA_A$ receptors and less alcohol self-administration (Hodge *et al.*, 1999; Olive *et al.*, 2000; Choi *et al.*, 2002; Hodge *et al.*, 2002). Confirmation of these results was obtained using a conditional rescue technique where mice lacking the PKCε gene showed a restoration of normal drinking after a conditional expression of PKCε in the basal forebrain, amygdala, and cerebellum (Choi *et al.*, 2002). These results suggest that reduced phosphorylation by PKCε enhances $GABA_A$ function. Thus, PKCγ and PKCε may mediate phosphorylation of $GABA_A$ receptors and regulate $GABA_A$ receptor function and alcohol drinking in opposite ways that may depend on the brain region involved. Importantly, the expression

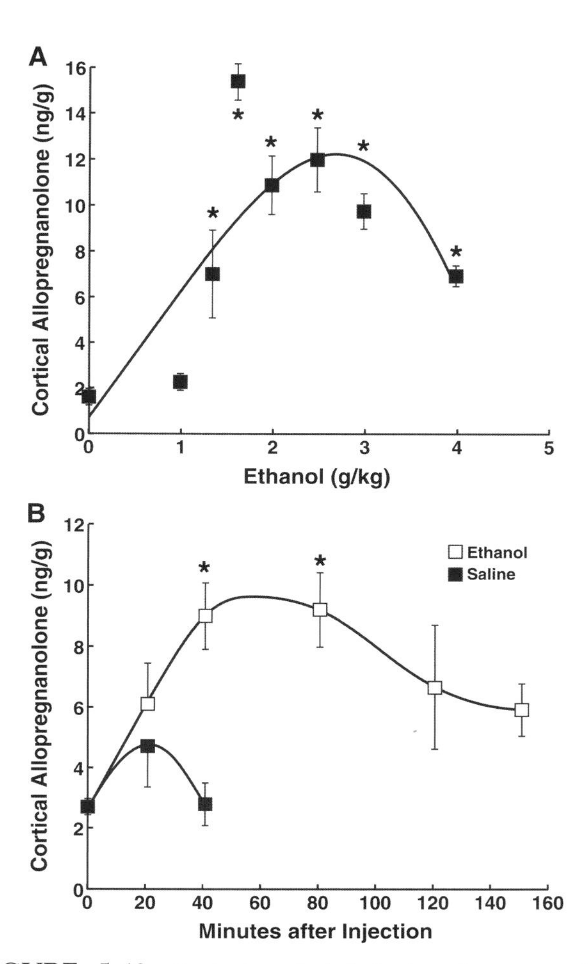

FIGURE 5.43 Cerebral cortical 3α-hydroxy,5α-pregnane-20-one (allopregnanolone) levels are elevated after acute, systemic ethanol administration in rats. **(A)** Cerebral cortical allopregnanolone levels exhibit a biphasic response to increasing ethanol concentrations. Ethanol increased allopregnanolone levels to pharmacologically active concentrations at doses between 1.35 and 4.0 gm/kg (*$p < 0.0001$, ANOVA; $p<0.05$, Dunnett's *post hoc* test) measured 60 min after ethanol administration. Data represent mean ± SEM of duplicate determinations in 6–10 rats per dose from two independent experiments. Further analysis of trend shows a significant match to a quadratic equation ($F_{1,32} = 17.49$; $p < 0.05$) using random samples of five rats per dose. **(B)** Cerebral cortical allopregnanolone levels peak between 40 and 80 min after an injection of ethanol (2 gm/kg, i.p.) followed by a gradual decrease. Data shown are the mean ± SEM of a representative experiment repeated twice with similar results ($n = 6$/time point; *$p<0.05$ Dunnett's *post hoc* test). [Reproduced with permission from VanDoren *et al.*, 2000.]

of PKC isoforms in the brain varies by region (Wetsel *et al.*, 1992).

Alcohol also activates cyclic adenosine monophosphate (cAMP)-dependent protein kinase A (PKA), and this pathway also has been hypothesized to contribute to alcohol potentiation of $GABA_A$ receptors (Kano and Konnerth, 1992; Freund and Palmer, 1997). Knockout mice with a disrupted regulatory unit of PKA

TABLE 5.17 Effect of Chronic Ethanol and its Withdrawal on γ-Aminobutyric Acid-A α1, α2, α3, and α5 Subunit mRNA Levels in Cortex

	mRNA Levels				
			Ethanol-withdrawn		
Subunits	Control	Ethanol-treated	*24 h*	*36 h*	*48 h*
α_1 (3.8, 4.3 kb)	1.12 ± 0.06	0.44 ± 0.06[a] (−61%)	0.33 ± 0.02[a] (−71%)	ND[b]	ND
α_2 (6 kb)	1.61 ± 0.06	0.62 ± 0.08[a] (−61%)	0.63 ± 0.05[a] (−61%)	1.60 ± 0.20[c]	1.87 ± 0.10[c]
α_2 (3 kb)	0.92 ± 0.01	0.51 ± 0.01[a] (−45%)	0.52 ± 0.08[a] (−43%)	1.01 ± 0.10[c]	1.16 ± 0.10[c]
α_3 (3 kb)	0.63 ± 0.02	0.54 ± 0.01 (−14%)	0.62 ± 0.03[c]	ND	ND
α_5 (2.8 kb)	1.27 ± 0.15	0.62 ± 0.10[a] (−51%)	0.70 ± 0.20[a] (−45%)	ND	ND

[a] $p < 0.005$, compared to control.
[b] ND, not determined.
[c] not significant, compared to control.
The values in parentheses represent percentage decreases in the mRNA levels.
[Reproduced with permission from Mhatre and Ticku, 1992.]

are much less sensitive to the sedative effects of alcohol and drink more alcohol than controls (Thiele *et al.*, 2000).

Chronic alcohol, in contrast to the effects of acute alcohol, generally downregulates $GABA_A$ receptor function as measured by chloride ion flux (Mhatre *et al.*, 1993), without a change in receptor density or affinity. Chronic studies of alcohol have shown decreases in GABA-mediated chloride ion flux shortly after withdrawal from chronic exposure in cortex and cerebellum (Morrow *et al.*, 1988; Sanna *et al.*, 1993) **(Fig. 5.41** above). However, such decreases are not sustained with longer withdrawal periods (Frye *et al.*, 1991; Kuriyama and Ueha, 1992). Increased benzodiazepine inverse agonist inhibition of GABA-mediated chloride uptake by brain microsacs has been shown in chronic alcohol-exposed mice, suggesting an additional coupling of inverse agonists that may combine functionally with the decreased coupling of GABA agonists (Buck and Harris, 1990). Potentiation of chloride uptake by the neuroactive steroid allopregnanolone also is enhanced in alcohol-dependent rats (Devaud *et al.*, 1996).

Chronic alcohol has complex effects on $GABA_A$ receptor expression and subunit composition, depending on the brain region and preparation. Chronic alcohol treatment of rats downregulated mRNA and protein for α1, α2, α3, and α5 subunits in cortex and α1 in cerebellum, and upregulated α6 in cerebellum. α2 and α3 mRNA and protein were downregulated with chronic treatment of cortical cultures. The mRNA changes generally returned to control levels at 36 h after withdrawal (Ticku *et al.*, 1986; Mhatre and Ticku, 1992) (**Table 5.17**). However, others have reported that chronic alcohol consumption in rats *increased* mRNAs for α4, γ1, and γ2S (but not γ2L), and as found by Ticku and colleagues (1986), α1 mRNA decreased, but there was no change in mRNAs for α5, β1, β2, β3, γ3 and δ (Mhatre *et al.*, 1993; Devaud *et al.*, 1997). Chronic intermittent alcohol treatment (to model human binge drinking) decreased the γ2L subunit and increased the α4 subunit for many days after withdrawal (Mahmoudi *et al.*, 1997). Thus, the most consistent finding is that chronic alcohol decreases expression of the α1 subunit, but increases the α4 subunit expression, changes that could underlie aspects of alcohol dependence.

Alcohol in concentrations within the intoxication range also potentiates ion currents mediated by the $5\text{–}HT_3$ receptor (Lovinger and White, 1991; Machu and Harris, 1994; Jenkins *et al.*, 1996). The $5\text{–}HT_3$ receptor is a ligand-gated ion channel of the cysteine loop-containing superfamily and closely resembles the $GABA_A$ receptor (Maricq *et al.*, 1991). The potentiation is rapid in onset, rapidly reversible, and hypothesized to be via a direct action on the receptor protein itself (Machu and Harris, 1994), independent of agonist activity (Lovinger *et al.*, 2000; Zhang *et al.*, 2002). In fact, when the receptor is fully occupied by a high efficacy agonist, alcohol is ineffective (Lovinger and White, 1991). Receptor mutation of a specific residue in the $5\text{–}HT_3$ receptor near the end of the TMII domain eliminated alcohol potentiation, suggesting that alcohol may be acting at some element of the receptor channel (Sessoms-Sikes *et al.*, 2003).

Alcohol decreases the function of all three major classes of ionotropic glutamate receptor subtypes: NMDA, kainate, and α-amino-3-hydroxy-5-methyl-4-isoxazolepropionic acid (AMPA) at concentrations within a physiological range (Lovinger *et al.*, 1989; Weiner *et al.*, 1999; Carta *et al.*, 2003). Most studies have

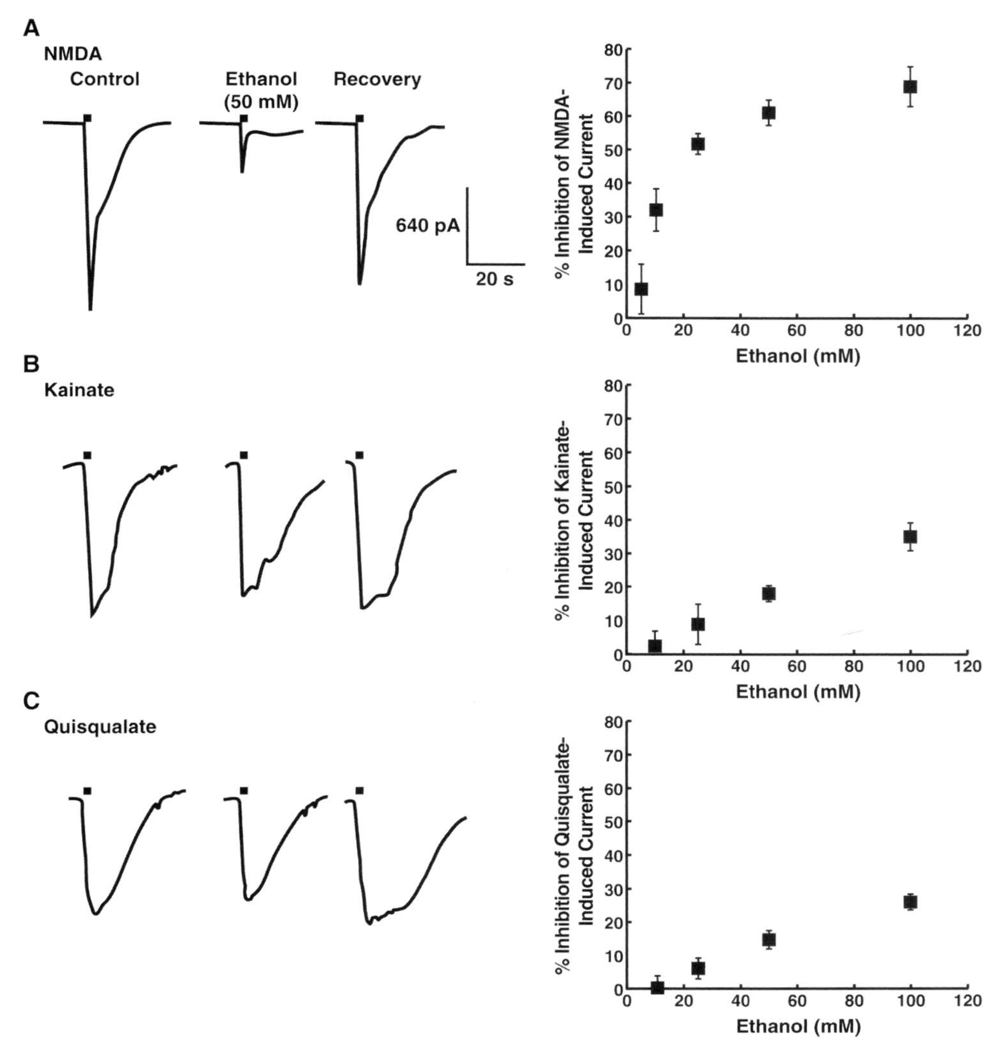

FIGURE 5.44 Ethanol effects on excitatory amino acid-induced ion currents in hippocampal neurons isolated from 16–17 day mouse embryos and grown in culture. Experiments were performed on neurons 2–4 weeks after plating. **(A)** (Left) Effect of 50 mM ethanol on current elicited by application of N-methyl-D-aspartate (NMDA). (Control) Inward current induced by application of 50 μM NMDA. (Ethanol) Response to 50 μM NMDA in the presence of 50 μM ethanol. (Recovery) Current induced by 50 μM NMDA alone (in the absence of ethanol), 2 min after termination of ethanol application. (Right) Average percentage inhibition of NMDA-activated current as a function of ethanol concentration. **(B)** (Left) Effect of 50 mM ethanol on current elicited by application of kainate. The three records show inward current induced by application of 10 μM kainate before, during, and 2 min after the application of 50 mM ethanol. (Right) Average percentage inhibition of kainate-activated current as a function of ethanol concentration. **(C)** (Left) Effect of 50 mM ethanol on current elicited by quisqualate-induced current as a function of ethanol concentration. The concentrations of the different agonists used produced currents of comparable amplitude in the absence of ethanol. Records in **(A)**, **(B)**, and **(C)** are taken from different neurons. Experiments examining NMDA-activated currents in hippocampal neurons were performed in extracellular medium containing no added Mg^{2+}. The records in **(B)** and **(C)** were collected in the presence of concentrations of Mg^{2+} (1 mM) and D,L-2-amino-5-phosphono valeric acid (50 μM) sufficient to prevent any contribution of NMDA receptor-activated currents to the currents induced by kainate and quisqualate. All records were taken at a membrane potential of –50 mV. The solid bar above each record denotes time of drug application. Time and current calibrations in (A) apply to all records. Each point in the graphs represents the mean ± SEM percentage inhibition observed in at least four neurons tested at the indicated ethanol concentration. [Taken with permission from Lovinger *et al.*, 1989.]

focused on NMDA receptors (Lovinger *et al.*, 1989) (**Fig. 5.44**). Single-channel analysis revealed that alcohol acts on an allosteric site on the NMDA receptor and inhibits the channel by reducing agonist efficacy (Lima-Landman and Albuquerque, 1989; Wright *et al.*, 1996). Alcohol also inhibits NMDA-elicited biochemical responses (Dildy and Leslie, 1989; Hoffman *et al.*, 1989). In several neuronal systems, alcohol sensitivity is determined by the relative level of the NMDAR2B subunit as measured by ifenprodil sensitivity (Fink and Gothert, 1996; Yang *et al.*, 1996). However, in primary cultured cerebellar granule cells where the NMDAR2B subunit does not appear to have a role, evidence suggested that alcohol sensitivity involved an interaction with glycine such that a high dose of glycine reversed alcohol inhibition of steady-state NMDA-induced currents (Popp *et al.*, 1999).

In cultured neurons and brain slices from the nucleus accumbens, alcohol inhibition of NMDA receptors is greatly reduced following activation of dopamine D_1 receptors, possibly involving the activation of adenylate cyclase, a cAMP-dependent protein kinase, and the DARPP-32 cascade (Maldve *et al.*, 2002). These results also suggest interactions at the second messenger level whereby disinhibition of NMDA receptors could be effected by activation of D_1 receptors (Lovinger, 2002; Maldve *et al.*, 2002).

Glutamate activation of AMPA receptors is thought to mediate fast synaptic excitatory neurotransmission in the brain, and alcohol does inhibit AMPA receptors in a number of preparations. Alcohol inhibits AMPA receptors expressed in *Xenopus* oocytes (Dildy-Mayfied and Harris, 1992), in cultured rat cortical neurons (Wirkner *et al.*, 2000), and in mouse cortical and hippocampal neurons using whole cell voltage-clamp recordings (Moykkynen *et al.*, 2003). This alcohol inhibition was reduced in a point mutation of the AMPA glutamate receptor that lacks desensitization, suggesting that alcohol inhibits AMPA receptors by stabilizing the desensitized state (Moykkynen *et al.*, 2003).

Acute alcohol administration also inhibits voltage-dependent L-type calcium channels, N channels, and T channels (Dildy-Mayfield *et al.*, 1992). Chronic alcohol exposure has the opposite effect and leads to increased depolarization-stimulated calcium influx (Grant *et al.*, 1993). The action of alcohol on voltage-dependent calcium channels requires PKC activity. The increase in calcium channel activity produced by chronic alcohol peaks during acute withdrawal and has been hypothesized to mediate some of the effects of alcohol withdrawal because calcium channel blockers can reduce physical symptoms of alcohol withdrawal in animals and humans (Koppi *et al.*, 1987; Bone *et al.*, 1989).

Second-messenger Systems

Alcohol also activates various elements of second-messenger systems via direct and indirect mechanisms. Alcohol inhibits adenosine reuptake via inhibition of the equilibrative nucleoside transporter-1 which increases extracellular adenosine and promotes activation of adenosine A_2 receptors (Nagy *et al.*, 1991; Mailliard and Diamond, 2004). Adenosine A_2 receptor activation in turn activates the cAMP response-element binding protein (CREB) system via the cAMP-PKA second-messenger system (Diamond and Gordon, 1997; Asher *et al.*, 2002). This activation ultimately may lead to gene expression changes via stimulation of CREB transcription (Asher *et al.*, 2002) (**Fig. 5.40** above).

More specifically, in NG108-15 neuroblastoma × glioma hybrid cell lines, alcohol increases cellular cAMP levels via activation of adenosine A_2 receptors which leads to the phosphorylation of CREB (Asher *et al.*, 2002). Repeated administration of alcohol increases protein kinase activity and moves the PKA catalytic subunit to the nucleus, suggesting a mechanism by which alcohol may prolong CREB phosphorylation and promote alcohol-induced changes in gene expression (Dohrman *et al.*, 1996; Constantinescu *et al.*, 1999; Asher *et al.*, 2002). Significant gene expression is thought to persist long after cAMP has been degraded and may explain how a single dose of alcohol can change GABA synaptic function for days (Melis *et al.*, 2002; Mailliard and Diamond, 2004). Because there also is synergy at the second-messenger level between alcohol activating adenosine A_2 receptors via blockade of adenosine uptake and dopamine agonists activating dopamine D_2 receptors, it was hypothesized that neurons in the nucleus accumbens may become hypersensitive to alcohol with the simultaneous activation of dopaminergic signaling. Normally, activation of D_2 receptors inhibits adenylate cyclase in PC12 cells stably expressing D_2 receptors, but activation of D_2 receptors can activate adenylate cyclase through release of $\beta\gamma$ dimers from G_i (Yao *et al.*, 2002). Blockade of the synergy between adenosine and D_2 receptors in the nucleus accumbens decreased alcohol drinking (Arolfo *et al.*, 2004).

Chronic alcohol produces tolerance to the alcohol inhibition of adenosine reuptake (Sapru *et al.*, 1994), and chronic exposure to alcohol causes a decrease in receptor-stimulated cAMP and G-protein production in NG108-15 cell cultures. Decreases in G_s with chronic alcohol have been observed with chronic alcohol in the hippocampus (Charness *et al.*, 1988; Mochly-Rosen *et al.*, 1988; Wand and Levine, 1991; Gordon *et al.*, 1992; Rabin, 1993; Davis-Cox *et al.*, 1996). Decreased CREB phosphorylation and decreases in CaMKIV protein levels have been observed in the central nucleus of the amygdala during alcohol withdrawal (Pandey *et al.*, 1999, 2001).

Others have observed increases in protein kinase activity in the nucleus accumbens with chronic administration of drugs of abuse such as morphine and cocaine (Nestler, 2001, 2004). Alcohol also has been shown to increase protein kinase activity in the nucleus accumbens after 12 weeks of chronic exposure but not after 1 or 6 weeks (Ortiz *et al.*, 1995). However, chronic alcohol exposure has been shown to decrease phosphorylated CREB in the nucleus accumbens (Misra *et al.*, 2001; Li *et al.*, 2003) similar to what has been observed in the amygdala (Pandey, 2004). The changes in the amygdala have been linked to the increased anxiety-like responses associated with acute alcohol withdrawal (Pandey *et al.*, 2003).

Preclinical and clinical studies suggest a relationship between high anxiety-like states and greater alcohol intake (see above), and as such, decreased CREB phosphorylation in the central nucleus of the amygdala may be responsible for increased anxiety-like responses in alcohol-dependent animals, and thus increased drinking in dependent animals possibly through actions on the NPY system, which is a CREB target gene (Chance *et al.*, 2000). Consistent with these results is the observation that cAMP-response element modular (CREM) knockout mice have decreased anxiety-like responses (Maldonado *et al.*, 1999). CREM activates the inducible cAMP repressor gene which is a powerful transcriptional repressor.

What is difficult to resolve is how decreased CREB phosphorylation in the central nucleus of the amygdala and the nucleus accumbens yields decreased NPY, which then yields increased anxiety, and then increased alcohol drinking, while increased CREB phosphorylation in the nucleus accumbens has been hypothesized to lead to increased dynorphin gain which leads to dysphoria and increased cocaine intake (Carlezon *et al.*, 1998) (see *Psychostimulants* chapter and *Neurobiological Theories of Addiction* chapter). One possibility is that the anxiety-like responses associated with enhanced alcohol drinking during acute withdrawal reflects decreased CREB function in the central nucleus of the amygdala and nucleus accumbens (Pandey *et al.*, 2003), but with other drugs of abuse, increased CREB function in the nucleus accumbens occurs during active drug taking to trigger the negative affective state of acute withdrawal (Carlezon *et al.*, 1998; Nestler, 2004) (see *Neurobiological Theories of Addiction* chapter).

In summary, cell culture systems show that acute alcohol can stimulate the cAMP-PKA system possibly through an action by adenosine release, and this action may be potentiated by dopamine. This activation can affect CREB phosphorylation and gene expression. However, chronic alcohol is associated with decreased cAMP-PKA activity and decreased CREB phosphorylation that may be related in the amygdala and nucleus accumbens with the negative affective state of alcohol withdrawal.

Molecular Genetic Approaches using Quantitative Trait Loci Analysis

Using inbred strains of mice and rats and selected crosses, the genetic technique of quantitative trait loci (QTL) analysis can be used to isolate candidate chromosomal regions and loci. A QTL is a genetic locus, the alleles of which affect variation in a quantitative trait (measurable phenotypic variation owing to genetic and/or environmental influences) (Abiola *et al.*, 2003). Quantitative traits are generally multifactorial and influenced by several polymorphic genes, so one or many QTLs can influence a trait or phenotype. Using such an approach, a QTL for alcohol preference was provisionally identified on mouse chromosome 9 (syntenic to rat chromosome 8) in C57BL $\times$ DBA recombinant inbred strains (Phillips *et al.*, 1994). Two receptor genes, *Drd2* encoding the dopamine D_2 receptor and *Htr1b* encoding the 5-HT_{1B} receptor, appear to be within or near this region. A QTL for alcohol preference also has been identified on chromosome 2 (syntenic with rat chromosome 3) (Phillips *et al.*, 1994, 1998; Rodriguez *et al.*, 1995; Melo *et al.*, 1996), and appears to be male-specific and contains a gene for the voltage-gated sodium channel.

Using three populations derived from the C57/BL6 and DBA2 strains (BXD RI, F2 cross and selectively bred mice), QTLs for acute and chronic alcohol withdrawal have been mapped on chromosomes 1, 4, 11, 13, and 14 in the mouse (Buck *et al.*, 1997). Candidate genes in proximity to chromosome 11 encode several $GABA_A$ receptor subunits (Crabbe, 1998; Buck and Finn, 2001). The quantitative trait gene for alcohol withdrawal seizures was identified as *Mpdz* (Shirley *et al.*, 2004). *Mpdz* is a cis-regulated gene, and *Mpdz* expression in a panel of standard inbred mouse strains significantly correlated with the severity of withdrawal from alcohol (Shirley *et al.*, 2004). *Mpdz* might interact with $GABA_B$ receptors, and as such may regulate glutamate and/or GABA release (Shirley *et al.*, 2004).

Using alcohol-preferring and alcohol-nonpreferring inbred rats (see *Animal Models of Addiction* chapter) at the 19th generation of inbreeding, reciprocal crosses were made, and 384 F2 progeny were subjected to a genome screen. QTLs were identified on chromosomes 3, 4, and 8, with a logarithm of odds (LOD) score of

8.6 identified on chromosome 4 (Carr *et al.*, 1998). The interval was narrowed to 12.5 cM (Bice *et al.*, 1998). Total gene expression analysis then was used to identify genes that were differentially expressed in brain regions between alcohol-naive inbred alcohol-preferring (iP) and inbred alcohol-nonpreferring (iNP) rats (Liang *et al.*, 2003). α-Synuclein was expressed at 2-fold higher levels in the hippocampus of iP rats compared to iNP rats. Higher levels of α-synuclein protein also were found in the striatum of iP rats. α-Synuclein was located at the region of chromosome 4, representing the peak of the LOD score. α-Synuclein has been implicated in the etiology of several neurodegenerative disorders, including Parkinson's disease (Spillantini *et al.*, 1998), and both alcoholics and alcohol-preferring rats have been shown to have hypodopaminergic function (see above). A prominent gene also within the candidate interval on chromosome 4 is the preproneuropeptide Y gene.

A similar procedure was used to identify QTLs from high alcohol drinking/low alcohol drinking (HAD/LAD) reciprocal crosses where the HAD and LAD founders were crossed to create F1 progeny which then were intercrossed to generate six families (Foroud *et al.*, 2000). Five chromosomal regions were identified with LOD scores greater than 2, and these were chromosomes 1, 5, 10, 12, and 16 (Foroud *et al.*, 2000). Two of the QTLs on chromosomes 5, 10, and 16 were confirmed in a replicate line (HAD2 /LAD2) (Carr *et al.*, 2003; Foroud *et al.*, 2003). Gene candidates located on these chromosomes include proenkephalin (chromosome 5) and neuronal nitric oxide synthase-1 (chromosome 12).

In humans, genotyping of two sets of 105 and 157 alcoholic families in two separate studies has revealed linkage to chromosomes 1, 2, 3, and 7 (Reich *et al.*, 1998; Foroud *et al.*, 2000). Several notable candidate genes are within chromosome 1 and include genes that encode calcium-activated chloride channels and voltage-gated potassium channels, cAMP, and PKC (Faroud *et al.*, 2000). Chromosome 2 contains genes for PKCε. Human chromosome 7 is syntenic to chromosome 4 in rats in which linkage has been reported in alcohol-preferring and alcohol-nonpreferring rats (Carr *et al.*, 1998).

Molecular Genetic Approaches Using Knockout Preparations

A number of molecular genetic knockouts have now been performed, and an alcohol drinking phenotype has been identified using either a two-bottle preference test or operant responding for alcohol (**Table 5.16**) (see

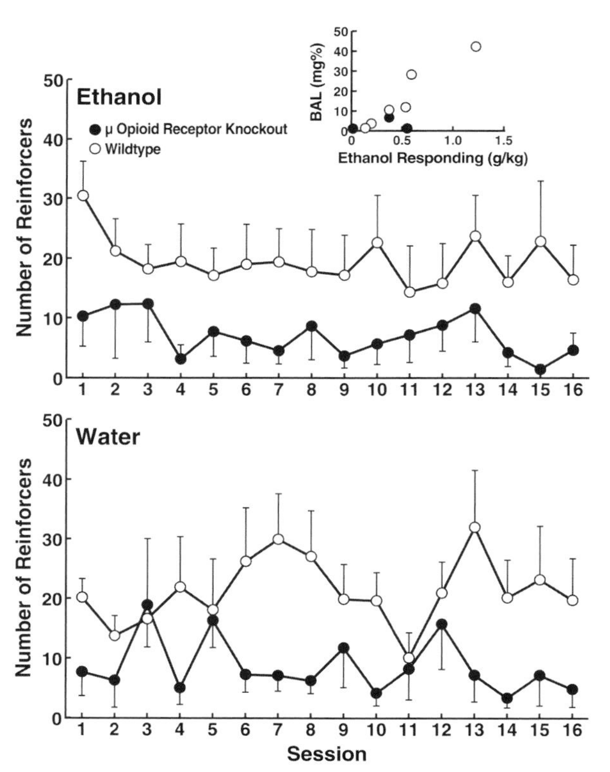

FIGURE 5.45 Operant self-administration (lever pressing) of 10% ethanol (top) versus water (bottom) in knockout and wildtype mice after a saccharin fadeout procedure. The last 16 days of a total of 40 is shown. There was a significant overall strain difference, with higher responding in wildtype relative to knockout mice. The lack of significant interactions involving strain suggests that neither strain responds differentially for ethanol and water. Blood alcohol levels were determined immediately after the final operant session, and the inset to this figure depicts these values (mg/100 ml) against ethanol responding relative to body weight (g/kg). Knockout mice responded for significantly less ethanol relative to body weight than wildtype mice and had lower resulting blood alcohol levels (BAL) [Reproduced with permission from Roberts *et al.*, 2000b.]

Animal Models of Addiction chapter). Knockout of dopamine D_1 and D_2 receptors, μ opioid receptors, NPY_2 receptors, dopamine and serotonin transporters, PKCε, and DARPP-32 decrease drinking. For example, μ opioid receptor knockout mice do not self-administer alcohol (Roberts *et al.*, 2000b) (**Fig. 5.45**). Knockout of cannabinoid CB_1 and δ opioid receptors, NPY and NPY_1 receptors, and PKA, among others, increase alcohol drinking. For example, NPY knockout mice self-administer significantly higher amounts of alcohol compared to wildtype controls (Thiele *et al.*, 1998). These mice are less sensitive to the sedative effects of alcohol and are able to recover from alcohol-induced inactivity faster than wildtype controls with similar blood alcohol levels. NPY-overexpressing mice show a lower preference for alcohol and are more sensitive to the sedative effects of alcohol compared to controls. Extensions of these phenotypes to animal models of excessive drinking and dependence will be of critical interest in future studies.

SUMMARY

Alcohol is a legal drug that is widely used in society for social and medical benefits. Alcohol is readily derived in nature by the process of fermentation. Early in human history, individuals learned to exploit the fermentation process for beverages and tonics. Alcohol taken in excess also is one of the most toxic substances in society from the perspective of both behavioral and physical damage to the body. Alcohol is a sedative hypnotic and as such has euphoric disinhibitory effects and is widely used as a 'social lubricant'. Much of the alleged stimulant effects of alcohol probably result from its disinhibitory effects. As the dose of alcohol increases, disinhibition gives way to motor impairment, muscular incoordination, impairments in reaction time, impairments in judgment, impairments in sensory processing, and impairments in cognitive function—all behavioral effects that contribute to its behavioral toxicity. Chronic use of alcohol can lead to alcoholism (Substance Dependence on Alcohol as defined by the DSM-IV) and/or numerous other medical diseases that range from cirrhosis of the liver to heart disease and pancreatitis. The behavioral mechanism of action of alcohol is its disinhibitory effects which account for a large part of the intoxication and euphoria associated with alcohol and its social use. The neurobiological mechanism of action for the acute reinforcing effects of alcohol involves activation of some of the same reward neurotransmitters implicated in the actions of psychostimulants and opioids, including dopamine and opioid peptides, but with an initial action at ligand-gated ion channel receptors such as GABA and glutamate. At the system level, chronic administration of alcohol, like other drugs of abuse, not only disrupts reward neurotransmitter function, such as dopamine, opioid peptides, and GABA, but also recruits the brain stress CRF system and dysregulates the brain antistress NPY system, all of which appear to contribute to motivational withdrawal and excessive drinking during dependence. Reinstatement studies in animal models have provided evidence for a role of opioid peptides, dopamine, and glutamate in cue-induced reinstatement, and CRF in stress-induced reinstatement. At the cellular level, alcohol has been shown to have specific synaptic effects, with very low doses showing an enhancement of GABAergic neurotransmission and an inhibition of glutamate neurotransmission, actions that show neuroadaptation during alcohol abstinence. These effects have been observed to be particularly localized to elements of the extended amygdala, providing support for the role of the extended amygdala as important for the acute and chronic motivational effects of alcohol in the development of dependence. Alcohol acutely activates dopaminergic cells in the ventral tegmental area, but withdrawal from chronic alcohol is associated with decreased dopaminergic activity in the ventral tegmental area. At the molecular level, while alcohol does not bind to any particular receptor, evidence is accumulating for an action of alcohol at 'ethanol-receptive elements', particularly in ligand-gated ion channel receptors. Studies with key amino acids in mammalian ligand-gated ion channel receptors led to the hypothesis that alcohol may be binding in a water-filled pocket of such receptor proteins. Studies on the $GABA_A$ receptor largely reveal that the presence of certain $GABA_A$ subunits convey extraordinary sensitivity to alcohol, while other receptor elements appear critical for some of the facilitatory actions of alcohol. Alcohol also interacts with second-messenger systems, notably the actions of protein kinases that can change the sensitivity of ligand-gated receptors or enhance gene transcription via adenosine activation of cAMP-PKA and ultimately CREB systems. Chronic alcohol administration is associated with decreases in CREB that may have a role in the motivational effects of alcohol withdrawal. Molecular genetic studies using QTL analysis have identified specific chromosomes and genes that are implicated in alcohol preference and withdrawal, and knockout studies have confirmed some of the more powerful neuropharmacological effects observed, such as in the μ opioid and NPY systems. The challenge for future research is the identification of those receptor proteins, on which neurons, in what circuits, that convey vulnerability to excessive and compulsive drinking associated with alcohol abuse and alcoholism.

REFERENCES

Abel, E. L. (1995). An update on incidence of FAS: FAS is not an equal opportunity birth defect. *Neurotoxicology and Teratology* **17**, 437–443.

Abeliovich, A., Chen, C., Goda, Y., Silva, A. J., Stevens, C. F., and Tonegawa, S. (1993). Modified hippocampal long-term potentiation in PKC gamma-mutant mice. *Cell* **75**, 1253–1262.

Abiola, O., Angel, J. M., Avner, P., Bachmanov, A. A., Belknap, J. K., Bennett, B., Blankenhorn, E. P., Blizard, D. A., Bolivar, V.,

Brockmann, G. A., Buck, K. J., Bureau, J. F., Casley, W. L., Chesler, E. J., Cheverud, J. M., Churchill, G. A., Cook, M., Crabbe, J. C., Crusio, W. E., Darvasi, A., de Haan, G., Dermant, P., Doerge, R. W., Elliot, R. W., Farber, C. R., Flaherty, L., Flint, J., Gershenfeld, H., Gibson, J. P., Gu, J., Gu, W., Himmelbauer, H., Hitzemann, R., Hsu, H. C., Hunter, K., Iraqi, F. F., Jansen, R. C., Johnson, T. E., Jones, B. C., Kempermann, G., Lammert, F., Lu, L., Manly, K. F., Matthews, D. B., Medrano, J. F., Mehrabian, M., Mittlemann, G., Mock, B. A., Mogil, J. S., Montagutelli, X., Morahan, G., Mountz, J. D., Nagase, H., Nowakowski, R. S., O'Hara, B. F., Osadchuk, A. V., Paigen, B., Palmer, A. A., Peirce, J. L., Pomp, D., Rosemann, M., Rosen, G. D., Schalkwyk, L. C., Seltzer, Z., Settle, S., Shimomura, K., Shou, S., Sikela, J. M., Siracusa, L. D., Spearow, J. L., Teuscher, C., Threadgill, D. W., Toth, L. A., Toye, A. A., Vadasz, C., Van Zant, G., Wakeland, E., Williams, R. W., Zhang, H. G., and Zou, F. (2003). The nature and identification of quantitative trait loci: a community's view; Complex Trait Consortium (CTC). *Nature Reviews Genetics* **4**, 911–916.

Addolorato, G., Caputo, F., Capristo, E., Domenicali, M., Bernardi, M., Janiri, L., Agabio, R., Colombo, G., Gessa, G. L., and Gasbarrini, G. (2002a). Baclofen efficacy in reducing alcohol craving and intake: a preliminary double-blind randomized controlled study. *Alcohol and Alcoholism* **37**, 504–508.

Addolorato, G., Caputo, F., Capristo, E., Janiri, L., Bernardi, M., Agabio, R., Colombo, G., Gessa, G. L., and Gasbarrini, G. (2002b). Rapid suppression of alcohol withdrawal syndrome by baclofen. *American Journal of Medicine* **112**, 226–229.

Adinoff, B., Martin, P. R., Bone, G. H., Eckardt, M. J., Roehrich, L., George, D. T., Moss, H. B., Eskay, R., Linnoila, M., and Gold, P. W. (1990). Hypothalamic-pituitary-adrenal axis functioning and cerebrospinal fluid corticotropin releasing hormone and corticotropin levels in alcoholics after recent and long-term abstinence. *Archives of General Psychiatry* **47**, 325–330.

Adrian, T. E., Allen, J. M., Bloom, S. R., Ghatei, M. A., Rossor, M. N., Roberts, G. W., Crow, T. J., Tatemoto, K., and Polak, J. M. (1983). Neuropeptide Y distribution in human brain. *Nature* **306**, 584–586.

Agarwal, D. P., and Goedde, H. W. (1992). Pharmacogenetics of alcohol metabolism and alcoholism. *Pharmacogenetics* **2**, 48–62.

Agarwal, D. P., Harada, S., and Goedde, H. W. (1981). Racial differences in biological sensitivity to ethanol: the role of alcohol dehydrogenase and aldehyde dehydrogenase isozymes. *Alcoholism: Clinical and Experimental Research* **5**, 12–16.

Aguayo, L. G. (1991). Demonstration that ethanol potentiates the GABAA-activated Cl- current in central mammalian neurons. *Alcohol and Alcoholism Supplement* **1,** 187–190.

Aguayo, L. G., Pancetti, F. C., Klein, R. L., and Harris, R. A. (1994). Differential effects of GABAergic ligands in mouse and rat hippocampal neurons. *Brain Research* **647**, 97–105.

Aguayo, L. G., Peoples, R. W., Yeh, H. H., and Yevenes, G. E. (2002). GABA(A) receptors as molecular sites of ethanol action: direct or indirect actions? *Current Topics in Medicinal Chemistry* **2**, 869–885.

Ahern, K. B., Lustig, H. S., and Greenberg, D. A. (1994). Enhancement of NMDA toxicity and calcium responses by chronic exposure of cultured cortical neurons to ethanol. *Neuroscience Letters* **165**, 211–214.

Ahtee, L., and Eriksson K. (1972). 5-Hydroxytryptamine and 5-hydroxyindolylacetic acid content in brain of rat strains selected for their alcohol intake. *Physiology and Behavior* **8**, 123–126.

Ahtee, L., and Eriksson, K. (1973). Regional distribution of brain 5-hydroxytryptamine in rat strains selected for their alcohol intake. *In Alcoholism and the Central Nervous System* (series title: *Annals of the New York Academy of Sciences*, vol. 215), (F. A. Seixas, and S. Eggleston, Eds.), pp. 26–134. New York Academy of Sciences, New York.

Alheid, G. F., and Heimer, L. (1988). New perspectives in basal forebrain organization of special relevance for neuropsychiatric disorders: The striatopallidal, amygdaloid, and corticopetal components of substantia innominata. *Neuroscience* **27**, 1–39.

Allan, A. M., Burnett, D., and Harris, R. A. (1991). Ethanol-induced changes in chloride flux are mediated by both GABA(A) and GABA(B) receptors. *Alcoholism: Clinical and Experimental Research* **15**, 233–237.

Allan, A. M., and Harris, R. A. (1985). Ethanol and barbiturates enhance GABA-stimulated influx of ^{36}Cl in isolated brain membranes. *Pharmacologist* **27**, 125.

Allan, A. M., and Harris, R. A. (1986). Gamma-aminobutyric acid and alcohol actions: neurochemical studies of long sleep and short sleep mice. *Life Sciences* **39**, 2005–2015.

Allan, A. M., and Harris, R. A. (1987). Acute and chronic ethanol treatments alter GABA receptor-operated chloride channels. *Pharmacology Biochemistry and Behavior* **27**, 665–670.

Altshuler, H. L., Phillips, P. E., and Feinhandler, D. A. (1980). Alteration of ethanol self-administration by naltrexone. *Life Sciences* **26**, 679–688.

American Psychiatric Association. (1994). *Diagnostic and Statistical Manual of Mental Disorders*, 4th ed., American Psychiatric Press, Washington DC.

Amit, Z., Brown, Z. W., and Rockman, G. E. (1977). Possible involvement of acetaldehyde, norepinephrine and their tetrahydroisoquinoline derivatives in the regulation of ethanol self-administration. *Drug and Alcohol Dependence* **2**, 495–500.

Ariwodola, O. J., and Weiner, J. L. (2004). Ethanol potentiation of GABAergic synaptic transmission may be self-limiting: role of presynaptic GABAB receptors. *Journal of Neuroscience* **24**, 10679–10686.

Arolfo, M. P., Yao, L., Gordon, A. S., Diamond, I., and Janak, P. H. (2004). Ethanol operant self-administration in rats is regulated by adenosine A2 receptors, *Alcoholism: Clinical and Experimental Research* **28**, 1308–1316.

Asher, O., Cunningham, T. D., Yao, L., Gordon, A. S., and Diamond, I. (2002). Ethanol stimulates cAMP-responsive element (CRE)-mediated transcription via CRE-binding protein and cAMP-dependent protein kinase. *Journal of Pharmacology and Experimental Therapeutics* **301**, 66–70.

Associated Press. (1996). Ex-judge Sweat, of 'Whisky Speech' fame, dies. *The Commercial Appeal*, Memphis TN, Feb 24, p. A4.

Aston-Jones, S., Aston-Jones, G., and Koob, G. F. (1984). Cocaine antagonizes anxiolytic effects of ethanol. *Psychopharmacology* **84**, 28–31.

Ator, N. A., Grant, K. A., Purdy, R. H., Paul, S. M., and Griffiths, R. R. (1993). Drug discrimination analysis of endogenous neuroactive steroids in rats. *European Journal of Pharmacology* **241**, 237–243.

Backstrom, P., Bachteler, D., Koch, S., Hyytia, P.,and Spanagel, R. (2004). mGluR5 antagonist MPEP reduces ethanol-seeking and relapse behavior. *Neuropsychopharmacology* **29**, 921–928.

Badia-Elder, N. E., Mosemiller, A. K., Elder, R. L., and Froehlich, J. C. (1999). Naloxone retards the expression of a genetic predisposition toward alcohol drinking. *Psychopharmacology* **144**, 205–212.

Badia-Elder, N. E., Stewart, R. B., Powrozek, T. A., Roy, K. F., Murphy, J. M., and Li, T. K. (2001). Effect of neuropeptide Y (NPY) on oral ethanol intake in Wistar, alcohol-preferring (P), and -nonpreferring (NP) rats. *Alcoholism: Clinical and Experimental Research* **25**, 386–390.

Bailey, C. P., Andrews, N., McKnight, A. T., Hughes, J., and Little, H. J. (2000). Prolonged changes in neurochemistry of dopamine neurones after chronic ethanol consumption. *Pharmacology Biochemistry and Behavior* **66**, 153–161.

Bailey, C. P., Manley, S. J., Watson, W. P., Wonnacott, S., Molleman, A., and Little, H. J. (1998). Chronic ethanol administration alters activity in ventral tegmental area neurons after cessation of withdrawal hyperexcitability. *Brain Research* **803**, 144–152.

Balakleevsky, A., Colombo, G., Fadda, F., and Gessa, G. L. (1990). Ro 19-4603, a benzodiazepine receptor inverse agonist, attenuates voluntary ethanol consumption in rats selectively bred for high ethanol preference. *Alcohol and Alcoholism* **25**, 449–452.

Baldwin, H. A., Rassnick, S., Rivier, J., Koob, G. F., and Britton, K. T. (1991). CRF antagonist reverses the 'anxiogenic' response to ethanol withdrawal in the rat. *Psychopharmacology* **103**, 227–232.

Ballenger, J. C., Goodwin, F. K., Major, L. F., and Brown, G. L. (1979). Alcohol and central serotonin metabolism in man. *Archives of General Psychiatry* **36**, 224–227.

Bandera, E. V., Freudenheim, J. L., and Vena, J. E. (2001). Alcohol consumption and lung cancer: a review of the epidemiologic evidence. *Cancer Epidemiology, Biomarkers and Prevention* **10**, 813–821.

Baraona, E., Abittan, C. S., Dohmen, K., Moretti, M., Pozzato, G., Chayes, Z. W., Schaefer, C., and Lieber, C. S. (2001). Gender differences in pharmacokinetics of alcohol. *Alcoholism: Clinical and Experimental Research* **25**, 502–507.

Bart, G., Kreek, M. J., Ott, J., LaForge, K. S., Proudnikov, D., Pollak, L., and Heilig, M. (2005). Increased attributable risk related to a functional mu-opioid receptor gene polymorphism in association with alcohol dependence in central Sweden. *Neuropsychopharmacology* **30**, 417–422.

Beardsley, P. M., Lopez, O. T., Gullikson, G., and Flynn, D. (1994). Serotonin 5-HT3 antagonists fail to affect ethanol self-administration of rats. *Alcohol* **11**, 389–395.

Becker, A., Grecksch, G., Kraus, J., Loh, H. H., Schroeder, H., and Hollt, V. (2002). Rewarding effects of ethanol and cocaine in mu opioid receptor-deficient mice. *Naunyn Schmiedebergs Archives of Pharmacology* **365**, 296–302.

Becker, H. C., and Hale, R. L. (1993). Repeated episodes of ethanol withdrawal potentiate the severity of subsequent withdrawal seizures: An animal model of alcohol withdrawal 'kindling,' *Alcoholism: Clinical and Experimental Research* **17**, 94–98.

Becker, H. C., and Flaherty, C. F. (1982). Influence of ethanol on contrast in consummatory behavior. *Psychopharmacology* **77**, 253–258.

Becker, H. C., and Lopez, M. F. (2004). Increased ethanol drinking after repeated chronic ethanol exposure and withdrawal experience in C57BL/6 mice. *Alcoholism: Clinical and Experimental Research* **28**, 1829–1838.

Beckman, L. J. (1979). Reported effects of alcohol on the sexual feelings and behavior of women alcoholics and nonalcoholics. *Journal of Studies on Alcohol* **40**, 272–282.

Beckman, L. J., and Ackerman, K. T. (1995). Women, alcohol, and sexuality. *In Alcoholism and Women* (series title: *Recent Developments in Alcoholism*, vol. 12), (M. Galanter, Ed.), pp. 267–285. Plenum Press, New York.

Berry, M. S., and Pentreath, V. W. (1980). The neurophysiology of alcohol. *In Psychopharmacology of Alcohol*, (M. Sandler, Ed.), pp. 43–72. Raven Press, New York.

Bespalov, A., Lebedev, A., Panchenko, G., and Zvartau, E. (1999). Effects of abused drugs on thresholds and breaking points of intracranial self-stimulation in rats. *European Neuropsychopharmacology* **9**, 377–383.

Best, C. A., and Laposata, M. (2003). Fatty acid ethyl esters: toxic non-oxidative metabolites of ethanol and markers of ethanol intake. *Frontiers in Bioscience* **8**, e202–e217.

Bice, P., Foroud, T., Bo, R., Castelluccio, P., Lumeng, L., Li, T. K., and Carr, L. G. (1998). Genomic screen for QTLs underlying alcohol consumption in the P and NP rat lines. *Mammalian Genome* **9**, 949–955.

Bien, T. H., and Burge, R. (1990). Smoking and drinking: a review of the literature. *International Journal of the Addictions* **25**, 1429–1454.

Bitran, M., and Kalant, H. (1993). Effect of anisomycin on the development of rapid tolerance to ethanol-induced motor impairment. *Pharmacology Biochemistry and Behavior* **45**, 225–228.

Blednov, Y. A., Walker, D., Alva, H., Creech, K., Findlay, G., and Harris, R. A. (2003). GABAA receptor alpha 1 and beta 2 subunit null mutant mice: behavioral responses to ethanol. *Journal of Pharmacology and Experimental Therapeutics* **305**, 854–863.

Boehm, S. L. 2nd, Peden, L., Chang, R., Harris, R. A., and Blednov, Y. A. (2003). Deletion of the fyn-kinase gene alters behavioral sensitivity to ethanol. *Alcoholism: Clinical and Experimental Research* **27**, 1033–1040.

Boehm, S. L. 2nd, Ponomarev, I., Jennings, A. W., Whiting, P. J., Rosahl, T. W., Garrett, E. M., Blednov, Y. A., and Harris, R. A. (2004). Gamma-Aminobutyric acid A receptor subunit mutant mice: new perspectives on alcohol actions. *Biochemical Pharmacology* **68**, 1581–1602.

Boismare, F., Daoust, M., Moore, N., Saligaut, C., Lhuintre, J. P., Chretien, P., and Durlach, J. (1984). A homotaurine derivative reduces the voluntary intake of ethanol by rats: are cerebral GABA receptors involved?. *Pharmacology Biochemistry and Behavior* **21**, 787–789.

Bond, C., LaForge, K. S., Tian, M., Melia, D., Zhang, S., Borg, L., Gong, J., Schluger, J., Strong, J. A., Leal, S. M., Tischfield, J. A., Kreek, M. J., and Yu, L. (1998). Single-nucleotide polymorphism in the human mu opioid receptor gene alters beta-endorphin binding and activity: possible implications for opiate addiction. *Proceedings of the National Academy of Sciences USA* **95**, 9608–9613.

Bone, G. H., Majchrowicz, E., Martin, P. R., Linnoila, M., and Nutt, D. J. (1989). A comparison of calcium antagonists and diazepam in reducing ethanol withdrawal tremors. *Psychopharmacology* **99**, 386–388.

Bowen, C. A., Purdy, R. H., and Grant, K. A. (1999). Ethanol-like discriminative stimulus effects of endogenous neuroactive steroids: effect of ethanol training dose and dosing procedure. *Journal of Pharmacology and Experimental Therapeutics* **289**, 405–411.

Bowers, B. J., and Wehner, J. M. (2001). Ethanol consumption and behavioral impulsivity are increased in protein kinase Cγ null mutant mice. *Journal of Neuroscience* **21**, RC180.

Breese, G. R., Overstreet, D. H., Knapp, D. J., and Navarro, M. (2005). Prior multiple ethanol withdrawals enhance stress-induced anxiety-like behavior: inhibition by CRF(1)- and benzodiazepine-receptor antagonists and a 5-HT(1a)-receptor agonist. *Neuropsychopharmacology*, in press.

Briddell, D. W., and Wilson, G. T. (1976). Effects of alcohol and expectancy set on male sexual arousal. *Journal of Abnormal Psychology* **85**, 225–234.

Brillat-Savarin, J. A. (1826). *Physiologie du Gout*, Feydeau, Paris.

Brodie, M. S. (2002). Increased ethanol excitation of dopaminergic neurons of the ventral tegmental area after chronic ethanol treatment. *Alcoholism: Clinical and Experimental Research* **26**, 1024–1030.

Brodie, M. S., and Appel, S. B. (1998). The effects of ethanol on dopaminergic neurons of the ventral tegmental area studied with intracellular recording in brain slices. *Alcoholism: Clinical and Experimental Research* **22**, 236–244.

Brodie, M. S., Pesold, C., and Appel, S. B. (1999). Ethanol directly excites dopaminergic ventral tegmental area reward neurons. *Alcoholism: Clinical and Experimental Research* **23**, 1848–1852.

Brodie, M. S., Shefner, S. A., and Dunwiddie, T. V. (1990). Ethanol increases the firing rate of dopamine neurons of the rat ventral tegmental area in vitro. *Brain Research* **508**, 65–69.

Brot, M. D., Akwa, Y., Purdy, R. H., Koob, G. F., and Britton, K. T. (1997). The anxiolytic-like effects of the neurosteroid allopregnanolone: interactions with $GABA_A$ receptors. *European Journal of Pharmacology* **325**, 1–7.

Buck, K. J., Finn, and D. A. (2001). Genetic factors in addiction: QTL mapping and candidate gene studies implicate GABAergic genes in alcohol and barbiturate withdrawal in mice. *Addiction* **96**, 139–149.

Buck, K. J., and Harris, R. A. (1990). Benzodiazepine agonist and inverse agonist actions on GABAA receptor-operated chloride channels. II. Chronic effects of ethanol. *Journal of Pharmacology and Experimental Therapeutics* **253**, 713–719.

Buck, K. J., Metten, P., Belknap, J. K., and Crabbe, J. C. (1997). Quantitative trait loci involved in genetic predisposition to acute alcohol withdrawal in mice. *Journal of Neuroscience* **17**, 3946–3955.

Buris, L., Csabai, G., Fodor, M., and Varga, M. (1985). Increase of alcohol dehydrogenase and protein content of liver following chronic ethanol administration. *FEBS Letters* **183**, 143–144.

Burish, T. G., Maisto, S. A., Cooper, A. M., and Sobell, M. B. (1981). Effects of voluntary short-term abstinence from alcohol on subsequent drinking patterns of college students. *Journal of Studies on Alcohol* **42**, 1013–1020.

Caberlotto, L., Thorsell, A., Rimondini, R., Sommer, W., Hyytia, P., and Heilig, M. (2001). Differential expression of NPY and its receptors in alcohol-preferring AA and alcohol-avoiding ANA rats. *Alcoholism: Clinical and Experimental Research* **25**, 1564–1569.

Calton, J. L., Wilson, W. A., and Moore, S. D. (1998). Magnesium-dependent inhibition of N-methyl-D-aspartate receptor-mediated synaptic transmission by ethanol. *Journal of Pharmacology and Experimental Therapeutics* **287**, 1015–1019.

Calton, J. L., Wilson, W. A., and Moore, S. D. (1999). Reduction of voltage-dependent currents by ethanol contributes to inhibition of NMDA receptor-mediated excitatory synaptic transmission. *Brain Research* **816**, 142–148.

Campanelli, C., Le, A. D., Khanna, J. M., and Kalant, H. (1988). Effect of raphe lesions on the development of acute tolerance to ethanol and pentobarbital. *Psychopharmacology* **96**, 454–457.

Campbell, A. D., McBride, and W. J. (1995). Serotonin-3 receptor and ethanol-stimulated dopamine release in the nucleus accumbens. *Pharmacology Biochemistry and Behavior* **51**, 835–842.

Cappell, H., and Herman, C. P. (1995). Alcohol and tension reduction: a review. *Quarterly Journal of Studies on Alcohol* **33**, 33–64.

Carlezon, W. A. Jr., Thome, J., Olson, V. G., Lane-Ladd, S. B., Brodkin, E. S., Hiroi, N., Duman, R. S., Neve, R. L., and Nestler, E. J. (1998). Regulation of cocaine reward by CREB. *Science* **282**, 2272–2275.

Carmelli, D., Swan, G. E., and Robinette, D. (1993). The relationship between quitting smoking and changes in drinking in World War II veteran twins. *Journal of Substance Abuse* **5**, 103–116.

Carr, L. G., Foroud, T., Bice, P., Gobbett, T., Ivashina, J., Edenberg, H., Lumeng, L., and Li, T. K. (1998). A quantitative trait locus for alcohol consumption in selectively bred rat lines. *Alcoholism: Clinical and Experimental Research* **22**, 884–887.

Carr, L. G., Habegger, K., Spence, J., Ritchotte, A., Liu, L., Lumeng, L., Li, T. K., and Foroud, T. (2003). Analyses of quantitative trait loci contributing to alcohol preference in HAD1/LAD1 and HAD2/LAD2 rats. *Alcoholism: Clinical and Experimental Research* **27**, 1710–1717.

Carta, M., Ariwodola, O. J., Weiner, J. L., and Valenzuela, C. F. (2003). Alcohol potently inhibits the kainate receptor-dependent excitatory drive of hippocampal interneurons. *Proceedings of the National Academy of Sciences USA* **100**, 6813–6818.

Casu, M. A., Colombo, G., Gessa, G. L., and Pani, L. (2002). Reduced, T. H.-immunoreactive fibers in the limbic system of Sardinian alcohol-preferring rats. *Brain Research* **24**, 242–251.

Chance, W. T., Sheriff, S., Peng, F., and Balasubramaniam, A. (2000). Antagonism of NPY-induced feeding by pretreatment with cyclic AMP response element binding protein antisense oligonucleotide. *Neuropeptides* **34**, 167–172.

Chandler, L. J., Newsom, H., Sumners, C., and Crews, F. (1993). Chronic ethanol exposure potentiates NMDA excitotoxicity in cerebral cortical neurons. *Journal of Neurochemistry* **60**,1578–1581.

Charness, M. E., Querimit, L. A., and Henteleff, M. (1988). Ethanol differentially regulates G proteins in neural cells. *Biochemical and Biophysical Research Communications* **155**, 138–143.

Chen, C. M., Yi, H.-Y., and Dufour, M. C. (2003). *Trends in Alcohol-Related Morbidity among Short-Stay Community Hospital discharges, United States, 1979–2001* (series title: *Surveillance Report*, vol. 64), National Institute on Alcohol Abuse and Alcoholism, Bethesda MD.

Chen, C. S. (1968). A study of the alcohol-tolerance effect and an introduction of a new behavioral technique. *Psychopharmacologia* **12**, 433–440.

Chermack, S. T., and Giancola, P. R. (1997). The relation between alcohol and aggression: an integrated biopsychosocial conceptualization. *Clinical Psychology Review* **17**, 621–649.

Chi, I., Lubben, J. E., and Kitano, H. H. (1989). Differences in drinking behavior among three Asian–American groups. *Journal of Studies on Alcohol* **50**, 15–23.

Chin, J. H., and Goldstein, D. B. (1977). Drug tolerance in biomembranes: a spin label study of the effects of ethanol. *Science* **196**, 684–685.

Choi, D. S., Cascini, M. G., Mailliard, W., Young, H., Paredes, P., McMahon, T., Diamond, I., Bonci, A., and Messing, R. O. (2004). The type 1 equilibrative nucleoside transporter regulates ethanol intoxication and preference. *Nature Neuroscience* **7**, 855–861.

Choi, D. S., Wang, D., Dadgar, J., Chang, W. S., and Messing, R. O. (2002). Conditional rescue of protein kinase C epsilon regulates ethanol preference and hypnotic sensitivity in adult mice. *Journal of Neuroscience* **22**, 9905–9911.

Christensen, J. K., Moller, I. W., Ronsted, P., Angelo, H. R., and Johansson, B. (1991). Dose-effect relationship of disulfiram in human volunteers. I: Clinical studies. *Pharmacology and Toxicology* **68**, 163–165.

Chu, B., Anantharam, V., and Treistman, S. N. (1995). Ethanol inhibition of recombinant heteromeric NMDA channels in the presence and absence of modulators. *Journal of Neurochemistry* **65**, 140–148.

Ciccocioppo, R., Angeletti, S., Colombo, G., Gessa, G., and Massi, M. (1999). Autoradiographic analysis of 5-HT2A binding sites in the brain of Sardinian alcohol-preferring and nonpreferring rats. *European Journal of Pharmacology* **373**, 3–19.

Ciccocioppo, R., Angeletti, S., and Weiss, F. (2001). Long-lasting resistance to extinction of response reinstatement induced by ethanol-related stimuli: Role of genetic ethanol preference. *Alcoholism: Clinical and Experimental Research* **25**, 1414–1419.

Ciccocioppo, R., Ge, J., Barnes, N. M., and Cooper, S. J. (1998). Central 5-HT3 receptors in P and in AA alcohol-preferring rats: An autoradiographic study. *Brain Research Bulletin* **46**, 311–315.

Ciccocioppo, R., Lin, D., Martin-Fardon, R., and Weiss, F. (2003). Reinstatement of ethanol-seeking behavior by drug cues following single versus multiple ethanol intoxication in the rat: effects of naltrexone. *Psychopharmacology* **168**, 208–215.

Ciccocioppo, R., Martin-Fardon, R., and Weiss, F. (2002). Effect of selective blockade of mu(1) or delta opioid receptors on reinstatement of alcohol-seeking behavior by drug-associated stimuli in rats. *Neuropsychopharmacology* **27**, 391–399.

Clark, J. T., Kalra, P. S., Crowley, W. R., and Kalra, S. P. (1984). Neuropeptide Y and human pancreatic polypeptide stimulate feeding behavior in rats. *Endocrinology* **115**, 427–429.

Cline, S. (1975). *Alcohol and Drugs at Work*, Drug Abuse Council, Chicago.

Cloninger, C. R., Bohman, M., and Sigvardsson, S. (1981). Inheritance of alcohol abuse: Cross-fostering analysis of adopted men. *Archives of General Psychiatry* **38**, 861–868.

Colombo, G., Serra, S., Brunetti, G., Vacca, G., Carai, M. A., and Gessa, G. L. (2003a). Suppression by baclofen of alcohol deprivation effect in Sardinian alcohol-preferring (sP) rats. *Drug and Alcohol Dependence* **70**, 105–108.

Colombo, G., Vacca, G., Serra, S., Brunetti, G., Carai, M. A., and Gessa, G. L. (2003b). Baclofen suppresses motivation to consume alcohol in rats. *Psychopharmacology* **167**, 221–224.

Constantinescu, A., Diamond, I., and Gordon, A. S. (1999). Ethanol-induced translocation of cAMP-dependent protein kinase to the nucleus: mechanism and functional consequences. *Journal of Biological Chemistry* **274**, 26985–26991.

Cook, L., and Davidson, A. B. (1973). Effects of behaviorally active drugs in a conflict-punishment procedure in rats. *In The Benzodiazepines* (S. Garattini, E. Mussini, and L. O. Randall, Eds.), pp. 327–345. Raven Press, New York.

Cook, L., and Sepinwall, J. (1975). Behavioral analysis of the effects and mechanisms of action of benzodiazepines. *In Mechanism of Action of Benzodiazepines* (series title: *Advances in Biochemical Psychopharmacology*, vol. 14), (E. Costa, and P. Greengard, Eds.), pp. 1–28. Raven Press, New York.

Cottino, A. (1995). Italy. *In: International Handbook on Alcohol and Culture*, (D. B. Heath, Ed.), pp. 156–167. Greenwood Press, Westport CT.

Covington, S. S., and Kohen, J. (1984). Women, alcohol, and sexuality. *Advances in Alcohol and Substance Abuse* **4**, 41–56.

Crabbe, J. C. (1998). Provisional mapping of quantitative trait loci for chronic ethanol withdrawal severity in BXD recombinant inbred mice. *Journal of Pharmacology and Experimental Therapeutics* **286**, 263–271.

Crabbe, J. C., Phillips, T. J., Feller, D. J., Hr, Wenger, C. D., Lessov, C. N., and Schafer, G. L. (1996). Elevated alcohol consumption in null mutant mice lacking 5-HT1B serotonin receptors. *Nature Genetics* **14**, 98–101.

Criado, J. R., Lee, R. S., Berg, G. I., and Henriksen, S. J. (1995). Sensitivity of nucleus accumbens neurons in vivo to intoxicating doses of ethanol. *Alcoholism: Clinical and Experimental Research* **19**, 164–169.

Criswell, H. E., and Breese, G. R. (2005). The effect of ethanol on ion channels in the brain: a new look. *In Comprehensive Handbook of Alcohol Related Pathology*, vol. 2, (V. R. Preedy, and R. R. Watson, Eds.), pp. 856–869. Elsevier, London.

Criswell, H. E., Simson, P. E., Duncan, G. E., McCown, T. J., Herbert, J. S., Morrow, A. L., and Breese, G. R. (1993). Molecular basis for regionally specific action of ethanol on gamma-aminobutyric acidA receptors: generalization to other ligand-gated ion channels. *Journal of Pharmacology and Experimental Therapeutics* **267**, 522–537.

Criswell, H. E., Simson, P. E., Knapp, D. J., Devaud, L. L., McCown, T. J., Duncan, G. E., Morrow, A. L., and Breese, G. R. (1995). Effect of zolpidem on gamma-aminobutyric acid (GABA)-induced inhibition predicts the interaction of ethanol with GABA on individual neurons in several rat brain regions. *Journal of Pharmacology and Experimental Therapeutics* **273**, 526–536.

Cullen, K. M., Halliday, G. M., Caine, D., and Kril, J. J. (1997). The nucleus basalis (Ch4) in the alcoholic Wernicke-Korsakoff syndrome: reduced cell number in both amnesic and non-amnesic patients. *Journal of Neurology, Neurosurgery, and Psychiatry* **63**, 315–320.

Davidson, D., and and Amit, Z. (1997). Naltrexone blocks acquisition of voluntary ethanol intake in rats. *Alcoholism: Clinical and Experimental Research* **21**, 677–683.

Davidson, M., Shanley, B., and Wilce, P. (1995). Increased NMDA-induced excitability during ethanol withdrawal: a behavioural and histological study. *Brain Research* **674**, 91–96.

Davidson, M. D., Wilce, P., and Shanley, B. C. (1993). Increased sensitivity of the hippocampus in ethanol-dependent rats to toxic effect of N-methyl-D-aspartic acid in vivo. *Brain Research* **606**, 5–9.

Davies, M. (2003). The role of GABAA receptors in mediating the effects of alcohol in the central nervous system. *Journal of Psychiatry and Neuroscience* **28**, 263–274.

Davis-Cox, M. I., Fletcher, T. L., Turner, J. N., Szarowski, D., and Shain, W. (1996). Three-day exposure to low-dose ethanol alters guanine nucleotide binding protein expression in the developing rat hippocampus. *Journal of Pharmacology and Experimental Therapeutics* **276**, 758–764.

Dawson, D. A. (2000). Drinking as a risk factor for sustained smoking. *Drug and Alcohol Dependence* **59**, 235–249.

Day, E., Bentham, P., Callaghan, R., Kuruvilla, T., and George, S. (2004). Thiamine for Wernicke-Korsakoff Syndrome in people at risk from alcohol abuse. *Cochrane Database of Systematic Reviews* **1**, CD004033.

De Montis, M. G., Gambarana, C., Gessa, G. L., Meloni, D., Tagliamonte, A., and Stefanini, E. (1993). Reduced [3H]SCH 23390 binding and DA-sensitive adenylyl cyclase in the limbic system of ethanol-preferring rats. *Alcohol and Alcoholism* **28**, 397–400.

de Rasor, R., and Youra, D. G. (1980). *Alcohol Distiller's Manual for Gasahol and Spirits*, Dona Carolina Distillers, San Antonio TX.

de Waele, J. P., Kiianmaa, K., and Gianoulakis, C. (1995). Distribution of the mu and delta opioid binding sites in the brain of the alcohol-preferring AA and alcohol-avoiding ANA lines of rats. *Journal of Pharmacology and Experimental Therapeutics* **275**, 518–527.

De Witte, P., and Bada, M. F. (1983). Self-stimulation and alcohol administered orally or intraperitoneally. *Experimental Neurology* **82**, 675–682.

Deitrich, R. A., Dunwiddie, T. V., Harris, R. A., and Erwin, V. G. (1989). Mechanism of action of ethanol: initial central nervous system actions. *Pharmacological Reviews* **41**, 489–537.

Deitrich, R. A., and Erwin, V. G. E. (1996). *Pharmacological Effects of Ethanol on the Nervous System*, CRC Press, Boca Raton.

Dennis, L. K., and Hayes, R. B. (2001). Alcohol and prostate cancer. *Epidemiologic Reviews* **23**, 110–114.

Devaud, L. L., Fritschy, J. M., Sieghart, W., and Morrow, A. L. (1997). Bidirectional alterations of $GABA_A$ receptor subunit peptide levels in rat cortex during chronic ethanol consumption and withdrawal. *Journal of Neurochemistry* **69**, 126–130.

Devaud, L. L., Purdy, R. H., Finn, D. A., and Morrow, A. L. (1996). Sensitization of gamma-aminobutyric acidA receptors to neuroactive steroids in rats during ethanol withdrawal. *Journal of Pharmacology and Experimental Therapeutics* **278**, 510–517.

Devoto, P., Colombo, G., Stefanini, E., and Gessa, G. L. (1998). Serotonin is reduced in the frontal cortex of Sardinian ethanol-preferring rats. *Alcohol and Alcoholism* **33**, 226–229.

Diamond, I., and Gordon, A. S. (1997). Cellular and molecular neuroscience of alcoholism. *Physiological Reviews* **77**, 1–20.

Diana, M., Brodie, M., Muntoni, A., Puddu, M. C., Pillolla, G., Steffensen, S., Spiga, S., and Little, H. J. (2003). Enduring effects of chronic ethanol in the CNS: basis for alcoholism. *Alcoholism: Clinical and Experimental Research* **27**, 354–361.

Diana, M., Pistis, M., Carboni, S., Gessa, G. L., and Rossetti, Z. L. (1993). Profound decrement of mesolimbic dopaminergic neuronal activity during ethanol withdrawal syndrome in rats: Electrophysiological and biochemical evidence. *Proceedings of the National Academy of Sciences USA* **90**, 7966–7969.

Diana, M., Pistis, M., Muntoni, A., and Gessa, G. (1996). Mesolimbic dopaminergic reduction outlasts ethanol withdrawal syndrome: evidence of protracted abstinence. *Neuroscience* **71**, 411–415.

Diana, M., Pistis, M., Muntoni, A., Rossetti, Z. L., and Gessa, G. (1992). Marked decrease of A10 dopamine neuronal firing during ethanol withdrawal syndrome in rats. *European Journal of Pharmacology* **221**, 403–404.

Diehl, A. M., Yang, S. Q., Cote, P., and Wand, G. S. (1992). Chronic ethanol consumption disturbs G-protein expression and inhibits cyclic AMP-dependent signaling in regenerating rat liver. *Hepatology* **16**, 1212–1219.

Dildy, J. E., and Leslie, S. W. (1989). Ethanol inhibits NMDA-induced increases in free intracellular Ca2+ in dissociated brain cells. *Brain Research* **499**, 383–387.

Dildy-Mayfield, J. E., and Harris, R. A. (1992). Comparison of ethanol sensitivity of rat brain kainate, DL-alpha-amino-3-hydroxy-5-methyl-4-isoxalone proprionic acid and N-methyl-D-aspartate receptors expressed in Xenopus oocytes. *Journal of Pharmacology and Experimental Therapeutics* **262**, 487–494.

Dildy-Mayfield, J. E., Machu, T., and Leslie, S. W. (1992). Ethanol and voltage- or receptor-mediated increases in cytosolic Ca2+ in brain cells. *Alcohol* **9**, 63–69.

Dohrman, D. P., Diamond, I., and Gordon, A. S. (1996). Ethanol causes translocation of cAMP-dependent protein kinase catalytic subunit to the nucleus. *Proceedings of the National Academy of Sciences USA* **93**, 10217–10221.

Dyr, W., and Kostowski, W. (1995). Evidence that the amygdala is involved in the inhibitory effects of 5-HT3 receptor antagonists on alcohol drinking in rats. *Alcohol* **12**, 387–391.

Ehlers, C. L., Chaplin, R. I., Lumeng, L., and Li, T. K. (1991). Electrophysiological response to ethanol in P and NP rats. *Alcoholism: Clinical and Experimental Research* **15**, 739–744.

Ehlers, C. L., Chaplin, R. I., Wall, T. L., Lumeng, L., Li, T. K., Owens, M. J., and Nemeroff, C. B. (1992). Corticotropin releasing factor (CRF): studies in alcohol preferring and non-preferring rats. *Psychopharmacology* **106**, 359–364.

Ehlers, C. L., Li, T. K., Lumeng, L., Hwang, B. H., Somes, C., Jimenez, P., and Mathe, A. A. (1998). Neuropeptide Y levels in ethanol-naive alcohol-preferring and nonpreferring rats and in Wistar rats after ethanol exposure *Alcoholism: Clinical and Experimental Research* **22**, 1778–1782.

Ehlers, C. L., Somes, C., Li, T. K., Lumeng, L., Kinkead, B., Owens, M. J., and Nemeroff, C. B. (1999). Neurontensin studies in alcohol naive, preferring and non-preferring rats. *Neuroscience* **93**, 227–236.

Ekman, G., Frankenhaeuser, M., Goldberg, L., Hagdahl, R., and Myrsten A-L. (1964). Subjective and objective effects of alcohol as functions of dosage and time. *Psychopharmacologia* **6**, 399–409.

El Ghundi, M., George, S. R., Drago, J., Fletcher, P. J., Fan, T., Nguyen, T., Liu, C., Sibley, D. R., Westphal, H., and O'Dowd, B. F. (1998). Disruption of dopamine D1 receptor gene expression attenuates alcohol-seeking behavior. *European Journal of Pharmacology* **353**, 149–158.

Ellenhorn, M. J., and Barceloux, D. G. (1988). *Medical Toxicology: Diagnosis and Treatment of Human Poisoning*, Elsevier, New York.

Emerson, H. (1932). *Alcohol and Man: The Effects of Alcohol on Man in Health and Disease*, Macmillan, New York.

Emmett-Oglesby, M. W. (1990). Tolerance to the discriminative stimulus effects of ethanol. *Behavioural Pharmacology* **1**, 497–503.

Engberg, G., and Hajos, M. (1992). Alcohol withdrawal reaction as a result of adaptive changes of excitatory amino acid receptors. *Naunyn Schmiedebergs Archives of Pharmacology* **346**, 437–441.

Ericksen, K. P., and Trocki, K. F. (1992). Behavioral risk factors for sexually transmitted diseases in American households. *Social Science and Medicine* **34**, 843–853.

Fadda, F., Garau, B., Marchei, F., Colombo, G., and Gessa, G. L. (1991). MDL 72222, a selective 5-HT3 receptor antagonist, suppresses voluntary ethanol consumption in alcohol-preferring rats. *Alcohol and Alcoholism* **26**, 107–110.

Fadda, P., Tronci, S., Colombo, G., and Fratta, W. (1999). Differences in the opioid system in selected brain regions of alcohol-preferring and alcohol-nonpreferring rats. *Alcoholism: Clinical and Experimental Research* **23**, 1296–1305.

Farkas, G. M., and Rosen, R. C. (1976). Effect of alcohol on elicited male sexual response. *Journal of Studies on Alcohol* **37**, 265–272.

Figuero Ruiz, E., Carretero Pelaez, M. A., Cerero Lapiedra, R., Esparza Gomez, G., and Moreno Lopez, L. A. (2004). Effects of the consumption of alcohol in the oral cavity: relationship with oral cancer. *Medicina Oral* **9**, 14–23.

File, S. E. (1980). The use of social interaction as a method for detecting anxiolytic activity of chlordiazepoxide-like drugs. *Journal of Neuroscience Methods* **2**, 219–238.

File, S. E., and Hyde, J. R. (1978). Can social interaction be used to measure anxiety? *British Journal of Pharmacology* **62**, 19–24.

Fink, K., and Gothert, M. (1996). Both ethanol and ifenprodil inhibit NMDA-evoked release of various neurotransmitters at different, yet proportional potency: potential relation to NMDA receptor subunit composition. *Naunyn Schmiedebergs Archives of Pharmacology* **354**, 312–319.

Foddai, M., Dosia, G., Spiga, S., and Diana, M. (2004). Acetaldehyde increases dopaminergic neuronal activity in the VTA. *Neuropsychopharmacology* **29**, 530–536.

Follesa, P., and Ticku, M. K. (1995). Chronic ethanol treatment differentially regulates NMDA receptor subunit mRNA expression in rat brain. *Molecular Brain Research* **29**, 99–106.

Follesa, P., and Ticku, M. K. (1996). Chronic ethanol-mediated up-regulation of the N-methyl-D-aspartate receptor polypeptide subunits in mouse cortical neurons in culture. *Journal of Biological Chemistry* **271**, 13297–13299.

Foroud, T., Edenberg, H. J., Goate, A., Rice, J., Flury, L., Koller, D. L., Bierut, L. J., Conneally, P. M., Nurnberger, J. I., Bucholz, K. K., Li, T. K., Hesselbrock, V., Crowe, R., Schuckit, M., Porjesz, B., Begleiter, H., and Reich, T. (2000). Alcoholism susceptibility loci: confirmation studies in a replicate sample and further mapping. *Alcoholism: Clinical and Experimental Research* **24**, 933–945.

Foroud, T., Ritchotte, A., Spence, J., Liu, L., Lumeng, L., Li, T. K., and Carr, L. G. (2003). Confirmation of alcohol preference quantitative trait loci in the replicate high alcohol drinking and low alcohol drinking rat lines. *Psychiatric Genetics* **13**, 155–161.

Frankel, D., Khanna, J. M., LeBlanc, A. E., and Kalant, H. (1975). Effect of p-chlorophenylalanine on the acquisition of tolerance to ethanol and pentobarbital. *Psychopharmacologia* **44**, 247–252.

Freund, R. K., and Palmer, M. R. (1997). Beta adrenergic sensitization of gamma-aminobutyric acid receptors to ethanol involves a cyclic AMP/protein kinase A second-messenger mechanism. *Journal of Pharmacology and Experimental Therapeutics* **280**, 1192–1200.

Frezza, M., di Padova, C., Pozzato, G., Terpin, M., Baraona, E., and Lieber, C. S. (1990). High blood alcohol levels in women. The role of decreased gastric alcohol dehydrogenase activity and first-pass metabolism. *New England Journal of Medicine* **322**, 95–99 [erratum: 323:553; 322:1540].

Froehlich, J. C., Harts, J., Lumeng, L., and Li, T. K. (1987). Naloxone attenuation of voluntary alcohol consumption. *Alcohol and Alcoholism Suppl.* **1**, 333–337.

Froehlich, J. C., Harts, J., Lumeng, L., and Li, T. K. (1990). Naloxone attenuates voluntary ethanol intake in rats selectively bred for high ethanol preference. *Pharmacology Biochemistry and Behavior* **35**, 385–390.

Frye, G. D., Fincher, A. S., Grover, C. A., and Griffith, W. H. (1994). Interaction of ethanol and allosteric modulators with GABAA-activated currents in adult medial septum/diagonal band neurons. *Brain Research* **635**, 283–292.

Frye, G. D., Mathew, J., and Trzeciakowski, J. P. (1991). Effect of ethanol dependence on GABAA antagonist-induced seizures and agonist-stimulated chloride uptake. *Alcohol* **8**, 453–459.

Gallegos, R. A., Lee, R. S., Criado, J. R., Henriksen, S. J., and Steffensen, S. C. (1999). Adaptive responses of gamma-aminobutyric acid neurons in the ventral tegmental area to chronic ethanol. *Journal of Pharmacology and Experimental Therapeutics* **291**, 1045–1053.

Gatch, M. B., Wallis, C. J., and Lal, H. (1999a). The effects of adenosine ligands R-PIA and CPT on ethanol withdrawal. *Alcohol* **19**, 9–14.

Gatch, M. B., Wallis, C. J., and Lal, H. (1999b). Effects of NMDA antagonists on ethanol-withdrawal induced 'anxiety' in the elevated plus maze. *Alcohol* **19**, 207–211.

Gatch, M. B., Wallis, C. J., and Lal, H. (2000). Effects of ritanserin on ethanol withdrawal-induced anxiety in rats. *Alcohol* **21**, 11–17.

Gatto, G. J., and Grant, K. A. (1997). Attenuation of the discriminative stimulus effects of ethanol by the benzodiazepine partial inverse agonist Ro 15-4513. *Behavioural Pharmacology* **8**, 139–146.

Gauvin, D. V., Baird, T. J., and Briscoe, R. J. (2000). Differential development of behavioral tolerance and the subsequent hedonic effects of alcohol in AA and ANA rats. *Psychopharmacology* **151**, 335–343.

Geller, I., and Seifter, J. (1960). The effect of meprobamate, barbiturates, d-amphetamine and promazine on experimentally-induced conflict in the rat. *Psychopharmacologia* **1**, 482–491.

Gessa, G. L., Muntoni, F., Collu, M., Vargiu, L., and Mereu, G. (1985). Low doses of ethanol activate dopaminergic neurons in the ventral tegmental area. *Brain Research* **348**, 201–203.

Gewiss, M., Heidbreder, C., Opsomer, L., Durbin, P., and De Witte, P. (1991). Acamprosate and diazepam differentially modulate alcohol-induced behavioral and cortical alterations in rats following chronic inhalation of ethanol vapour. *Alcohol and Alcoholism* **26**, 129–137.

Gianoulakis, C., de Waele, J. P., and Kiianmaa, K. (1992). Differences in the brain and pituitary beta-endorphin system between the alcohol-preferring AA and alcohol-avoiding ANA rats. *Alcoholism: Clinical and Experimental Research* **16**, 453–459.

Gill, K., Amit, Z., and Koe, B. K. (1988). Treatment with sertraline, a new serotonin uptake inhibitor, reduces voluntary ethanol consumption in rats. *Alcohol* **5**, 349–354.

Ginsburg, E. S., Mello, N. K., Mendelson, J. H., Barbieri, R. L., Teoh, S. K., Rothman, M., Gao, X., and Sholar, J. W. (1996). Effects of alcohol ingestion on estrogens in postmenopausal women. *Journal of the American Medical Association* **276**, 1747–1751.

Givens, B. S., and Breese, G. R. (1990). Site-specific enhancement of gamma-aminobutyric acid-mediated inhibition of neural activity by ethanol in the rat medial septal area. *Journal of Pharmacology and Experimental Therapeutics* **254**, 528–538.

Goedde, H. W., Agarwal, D. P., and Harada, S. (1983a). The role of alcohol dehydrogenase and aldehyde dehydrogenase isozymes in alcohol metabolism, alcohol sensitivity and alcoholism. *In Cellular Localization, Metabolism, and Physiology* (series title: *Isozymes: Current Topcs in Biological and Medical Research*, vol. 8, (M. C. Rattazzi, J. G. Scandalios, and G. S. Whitt, Eds.), pp. 175–193. Alan R. Liss, New York.

Goedde, H. W., Agarwal, D. P., Harada, S., Meier-Tackmann, D., Ruofu, D., Bienzle, U., Kroeger, A., and Hussein, L. (1983b). Population genetic studies on aldehyde dehydrogenase isozyme deficiency and alcohol sensitivity. *American Journal of Human Genetics* **35**, 769–772.

Goedde, H. W., Agarwal, D. P., Harada, S., Rothhammer, F., Whittaker, J. O., and Lisker, R. (1986). Aldehyde dehydrogenase polymorphism in North American, South American, and Mexican Indian populations. *American Journal of Humam Genetics* **38**, 395–399.

Goedde, H. W., Agarwal, D. P., and Paik, Y. K. (1983c). Frequency of aldehyde dehydrogenase I isozyme deficiency in Koreans: a pilot study. *Korean Journal of Genetics* **5**, 88–90.

Goedde, H. W., Benkmann, H. G., Kriese, L., Bogdanski, P., Agarwal, D. P., Du, R. F., Chen, L. Z., Cui, M. Y., Yuan, Y. D., Xu, J. J., Lu, S., and Wu, Y. (1984a). Aldehyde dehydrogenase isozyme deficiency and alcohol sensitivity in four different Chinese populations. *Human Heredity* **34**, 183–186.

Goedde, H. W., Harada, S., and Agarwal, D. P. (1979). Racial differences in alcohol sensitivity: a new hypothesis. *Human Genetics* **51**, 331–334.

Goedde, H. W., Rothhammer, F., Benkmann, H. G., and Bogdanski, P. (1984b). Ecogenetic studies in Atacameno Indians. *Human Genetics* **67**, 343–346.

Goedde, H. W., Singh, S., Agarwal, D. P., Fritze, G., Stapel, K., and Paik, Y. K. (1989). Genotyping of mitochondrial aldehyde dehydrogenase in blood samples using allele-specific oligonucleotides: comparison with phenotyping in hair roots. *Human Genetics* **81**, 305–307.

Goldberg, L. (1943). Quantitative studies on alcohol tolerance in man. *Acta Physiologica Scandinavica Supplementum* **5**, 1–128.

Golbert, T. M., Sanz, C. J., Rose, H. D., and Leitschuh, T. H. (1967). Comparative evaluation of treatments of alcohol withdrawal syndromes. *Journal of the American Medical Association* **201**, 99–102.

Goldstein, D. B. (1983). *Pharmacology of Alcohol*. Oxford University Press, New York.

Gong, J., Li, X. W., Lai, Z., Froehlich, J. C., and Yu, L. (1997). Quantitative comparison of mu opioid receptor mRNA in selected CNS regions of alcohol naive rats selectively bred for high and low alcohol drinking. *Neuroscience Letters* **227**, 9–12.

Gongwer, M. A., Murphy, J. M., McBride, W. J., Lumeng, L., and Li, T. K. (1989). Regional brain contents of serotonin, dopamine and their metabolites in the selectively bred high- and low-alcohol drinking lines of rats. *Alcohol* **6**, 317–320.

Gordon, A. S., Mochly-Rosen, D., and Diamond, I. (1992). Alcoholism: a possible G protein disorder. *In G Proteins, Signal Transduction and Disease*, (G. Milligan, and M. J. O. Wakelam, Eds), pp. 191–216. Academic Press, London.

Gordon, G. G., Southren, A. L., and Lieber, C. S. (1979). Hypogonadism and feminization in the male: a triple effect of alcohol. *Alcoholism: Clinical and Experimental Research* **3**, 210–212.

Grahame, N. J., Mosemiller, A. K., Low, M. J., and Froehlich, J. C. (2000). Naltrexone and alcohol drinking in mice lacking beta-endorphin by site-directed mutagenesis. *Pharmacology Biochemistry and Behavior* **67**, 759–766.

Grant, A. J., Koski, G., and Treistman, S. N. (1993). Effect of chronic ethanol on calcium currents and calcium uptake in undifferentiated PC12 cells. *Brain Research* **600**, 280–284.

Grant, B., Dawson, D., Stinson, F., Chou, P., Dufour, M., and Pickering, R. (2004). The 12-month prevalence and trends in DSM-IV alcohol abuse and dependence: United States, 1991-1992 and 2001-2002. *Drug and Alcohol Dependence*, in press.

Grant, K. A., and Colombo, G. (1993a). Discriminative stimulus effects of ethanol: Effect of training dose on the substitution of N-methyl-D-aspartate antagonists. *Journal of Pharmacology and Experimental Therapeutics* **264**, 1241–1247.

Grant, K. A., and Colombo, G. (1993b). Substitution of the 5-HT1 agonist trifluoromethylphenylpiperazine (TFMPP) for the discriminative stimulus effects of ethanol: effect of training dose. *Psychopharmacology* **113**, 26–30.

Grant, K. A., Colombo, G., and Gatto, G. J. (1997). Characterization of the ethanol-like discriminative stimulus effects of 5-HT receptor agonists as a function of ethanol training dose. *Psychopharmacology* **133**, 133–141.

Grant, K. A., Knisely, J. S., Tabakoff, B., Barrett, J. E., and Balster, R. L. (1991). Ethanol-like discriminative stimulus effects of non-competitive N-methyl-D-aspartate antagonists. *Behavioral Pharmacology* **2**, 87–95.

Grant, K. A., Waters, C. A., Green-Jordan, K., Azarov, A., and Szeliga, K. T. (2000). Characterization of the discriminative stimulus effects of GABA(A) receptor ligands in Macaca fascicularis monkeys under different ethanol training conditions. *Psychopharmacology* **152**, 181–188.

Green, K. L., Gatto, G. J., and Grant, K. A. (1997). The nitric oxide synthase inhibitor L-NAME (N omega-nitro-L-arginine methyl ester) does not produce discriminative stimulus effects similar to ethanol. *Alcoholism: Clinical and Experimental Research* **21**, 483–488.

Green, K. L., and Grant, K. A. (1999). Effects of L-type voltage-sensitive calcium channel modulators on the discriminative stimulus effects of ethanol in rats. *Alcoholism: Clinical and Experimental Research* **23**, 806–814.

Grisel, J. E., Mogil, J. S., Grahame, N. J., Rubinstein, M., Belknap, J. K., Crabbe, J. C., and Low, M. J. (1999). Ethanol oral self-administration is increased in mutant mice with decreased beta-endorphin expression. *Brain Research* **835**, 62–67.

Grobin, A. C., Matthews, D. B., Devaud, L. L., and Morrow, A. L. (1998). The role of GABA(A) receptors in the acute and chronic effects of ethanol, *Psychopharmacology* **139**, 2–19.

Hall, F. S., Sora, I., and Uhl, G. R. (2001). Ethanol consumption and reward are decreased in mu-opiate receptor knockout mice. *Psychopharmacology* **154**, 43–49.

Hall, F. S., Sora, I., and Uhl, G. R. (2003). Sex-dependent modulation of ethanol consumption in vesicular monoamine transporter 2 (VMAT2) and dopamine transporter (DAT) knockout mice. *Neuropsychopharmacology* **28**, 620–628.

Hanchar, H. J., Wallner, M., and Olsen, R. W. (2004). Alcohol effects on gamma-aminobutyric acid type A receptors: are extrasynaptic receptors the answer? *Life Sciences* **76**, 1–8.

Hankin, J. R., and Sokol, R. J. (1995). Identification and care of problems associated with alcohol ingestion in pregnancy. *Seminars in Perinatology* **19**, 286–292.

Harlap, S., and Shiono, P. H. (1980). Alcohol, smoking, and incidence of spontaneous abortions in the first and second trimester. *Lancet* **2**(8187), 173–176.

Harper, C. G., Giles, M., and Finlay-Jones, R. (1986). Clinical signs in the Wernicke-Korsakoff complex: a retrospective analysis of 131 cases diagnosed at necropsy. *Journal of Neurology, Neurosurgery and Psychiatry* **49**, 341–345.

Harris, R. A., McQuilkin, S. J., Paylor, R., Abeliovich, A., Tonegawa, S., and Wehner, J. M. (1995). Mutant mice lacking the gamma isoform of protein kinase C show decreased behavioral actions of ethanol and altered function of gamma-aminobutyrate type A receptors. *Proceedings of the National Academy of Sciences USA* **92**, 3658–3662.

Harris, R. A., and Mihic, S. J. (2004). Alcohol and inhibitory receptors: unexpected specificity from a nonspecific drug. *Proceedings of the National Academy of Sciences USA* **101**, 2–3.

Harrison, N. L., Majewska, M. D., Harrington, J. W., and Barker, J. L. (1987). Structure-activity relationships for steroid interaction with the gamma-aminobutyric acidA receptor complex. *Journal of Pharmacology and Experimental Therapeutics* **241**, 346–353.

Harvey, S. C., Foster, K. L., McKay, P. F., Carroll, M. R., Seyoum, R., Woods, J. E. 2nd, Grey, C., Jones, C. M., McCane, S., Cummings, R., Mason, D., Ma, C., Cook, J. M., and June, H. L. (2002). The GABA(A) receptor alpha1 subtype in the ventral pallidum regulates alcohol-seeking behaviors. *Journal of Neuroscience* **22**, 3765–3775.

Hawkins, R. D., Kalant, H., and Khanna, J. M. (1966). Effects of chronic intake of ethanol on rate of ethanol metabolism. *Canadian Journal of Physiology and Pharmacology* **44**, 241–257.

Heilig, M., and Murison, R. (1987). Intracerebroventricular neuropeptide Y suppresses open field and home cage activity in the rat. *Regulatory Peptides* **19**, 221–231.

Heilig, M., Soderpalm, B., Engel, J. A., and Widerlov, E. (1989). Centrally administered neuropeptide Y (NPY) produces anxiolytic-like effects in animal anxiety models. *Psychopharmacology* **98**, 524–529.

Heimer, L., and Alheid, G. (1991). Piecing together the puzzle of basal forebrain anatomy. *In The Basal Forebrain: Anatomy to Function* (series title: *Advances in Experimental Medicine and Biology*, vol. 295), (T. C. Napier, P. W. Kalivas, and I. Hanin, Eds.), pp. 1–42. Plenum Press, New York.

Hellevuo, K., Kiianmaa, K., and Korpi, E. R. (1989). Effect of GABAergic drugs on motor impairment from ethanol, barbital and lorazepam in rat lines selected for differential sensitivity to ethanol. *Pharmacology Biochemistry and Behavior* **34**, 399–404.

Herz, A. (1997) Endogenous opioid systems and alcohol addiction *Psychopharmacology* **129**, 99–111.

Heyser, C. J., Moc, K., and Koob, G. F. (2003). Effects of naltrexone alone and in combination with acamprosate on the alcohol deprivation effect in rats. *Neuropsychopharmacology* **28**, 1463–1471.

Heyser, C. J., Roberts, A. J., Schulteis, G., and Koob, G. F. (1999). Central administration of an opiate antagonist decreases oral ethanol self-administration in rats. *Alcoholism: Clinical and Experimental Research* **23**, 1468–1476.

Heyser, C. J., Schulteis, G., Durbin, P., and Koob, G. F. (1998). Chronic acamprosate eliminates the alcohol deprivation effect while having limited effects on baseline responding for ethanol in rats. *Neuropsychopharmacology* **18**, 125–133.

Heyser, C. J., Schulteis, G., and Koob, G. F. (1997). Increased ethanol self-administration after a period of imposed ethanol deprivation in rats trained in a limited access paradigm. *Alcoholism: Clinical and Experimental Research* **21**, 784–791.

Higuchi, S., Matsushita, S., Murayama, M., Takagi, S., and Hayashida, M. (1995). Alcohol and aldehyde dehydrogenase polymorphisms and the risk for alcoholism. *American Journal of Psychiatry* **152**, 1219–1221.

Hill, S. Y. (1974). Intraventricular injection of 5-hydroxytryptamine and alcohol consumption in rats. *Biological Psychiatry* **8**, 151–158.

Hingson, R. W., Heeren, T., Zakocs, R. C., Kopstein, A., and Wechsler, H., (2002). Magnitude of alcohol-related mortality and morbidity among U.S. college students ages 18–24. *Journal of Studies on Alcohol* **63**, 136–144.

Ho, A. K., Tsai, C. S., Chen, R. C., Begleiter, H., and Kissin, B. (1974). Experimental studies on alcoholism. I. Increased in alcohol preference by 5.6-dihydroxytryptamine and brain acetylcholine. *Psychopharmacologia* **40**, 101–107.

Hodge, C. W., Mehmert, K. K., Kelley, S. P., McMahon, T., Haywood, A., Olive, M. F., Wang, D., Sanchez-Perez, A. M., and Messing, R. O. (1999). Supersensitivity to allosteric GABA(A) receptor modulators and alcohol in mice lacking PKCepsilon. *Nature Neuroscience* **2**, 997–1002.

Hodge, C. W., Raber, J., McMahon, T., Walter, H., Sanchez-Perez, A. M., Olive, M. F., Mehmert, K., Morrow, A. L., and Messing, R. O. (2002). Decreased anxiety-like behavior, reduced stress hormones, and neurosteroid supersensitivity in mice lacking protein kinase C-epsilon. *Journal of Clinical Investigation* **110**, 1003–1010.

Hodge, C. W., Samson, H. H., Lewis, R. S., and Erickson, H. L. (1993). Specific decreases in ethanol- but not water-reinforced responding produced by the 5-HT3 antagonist ICS 205-930. *Alcohol* **10**, 191–196.

Hoffman, P. L., Rabe, C. S., Moses, F., and Tabakoff, B. (1989). N-methyl-D-aspartate receptors and ethanol: Inhibition of calcium flux and cyclic GMP production. *Journal of Neurochemistry* **52**, 1937–1940.

Hoffman, P. L., Ritzmann, R. F., Walter, R., and Tabakoff, B. (1978). Arginine vasopressin maintains ethanol tolerance. *Nature* **276**, 614–616.

Holter, S. M., Danysz, W., and Spanagel, R. (1996). Evidence for alcohol anti-craving properties of memantine. *European Journal of Pharmacology* **314**, R1–R2.

Holter, S. M., Danysz, W., and Spanagel, R. (2000). Novel uncompetitive N-methyl-D-aspartate (NMDA)-receptor antagonist MRZ 2/579 suppresses ethanol intake in long-term ethanol-experienced rats and generalizes to ethanol cue in drug discrimination procedure. *Journal of Pharmacology and Experimental Therapeutics* **292**, 545–552.

Holter, S. M., Landgraf, R., Zieglgansberger, W., and Spanagel, R. (1997). Time course of acamprosate action on operant ethanol self-administration after ethanol deprivation. *Alcoholism: Clinical and Experimental Research* **21**, 862–868.

Holter, S. M., and Spanagel, R. (1999). Effects of opiate antagonist treatment on the alcohol deprivation effect in long-term ethanol-experienced rats. *Psychopharmacology* **145**, 360–369.

Homanics, G. E., Ferguson, C., Quinlan, J. J., Daggett, J., Snyder, K., Lagenaur, C., Mi, Z. P., Wang, X. H., Grayson, D. R., and Firestone, L. L. (1997). Gene knockout of the alpha6 subunit of the gamma-aminobutyric acid type A receptor: lack of effect on responses to ethanol, pentobarbital, and general anesthetics. *Molecular Pharmacology* **51**, 588–596.

Hoyert, D. L., Kochanek, K. D., and Murphy, S. L. (1999). *Deaths: Final Data for 1997* (series title: *National Vital Statistics Reports*, vol. **47**(19); DHHS publication No. PHS 99-1120), National Center for Health Statistics, Hyattsville MD.

Hu, X. J., and Ticku, M. K. (1995). Chronic ethanol treatment upregulates the NMDA receptor function and binding in mammalian cortical neurons. *Molecular Brain Research* **30**, 347–356.

Hubbell, C. L., Marglin, S. H., Spitalnic, S. J., Abelson, M. L., Wild, K. D., and Reid, L. D. (1991). Opioidergic, serotonergic, and dopaminergic manipulations and rats' intake of a sweetened alcoholic beverage. *Alcohol* **8**, 355–367.

Hugues, J. N., Coste, T., Perret, G., Jayle, M. F., Sebaoun, J., and Modigliani, E. (1980). Hypothalamo-pituitary ovarian function in thirty-one women with chronic alcoholism. *Clinical Endocrinology* **12**, 543–551.

Hungund, B. L., Szakall, I., Adam, A., Basavarajappa, B. S., and Vadasz, C. (2003). Cannabinoid, CB1 receptor knockout mice exhibit markedly reduced voluntary alcohol consumption and lack alcohol-induced dopamine release in the nucleus accumbens. *Journal of Neurochemistry* **84**, 698–704.

Hwang, B. H., Froehlich, J. C., Hwang, W. S., Lumeng L, and Li, T. K. (1998). More vasopressin mRNA in the paraventricular hypothalamic nucleus of alcohol-preferring rats and high alcohol-drinking rats selectively bred for high alcohol preference. *Alcoholism: Clinical and Experimental Research* **22**, 664–669.

Hwang, B. H., Lumeng, L., Wu, J. Y., and Li, T. K. (1990). Increased number of GABAergic terminals in the nucleus accumbens is associated with alcohol preference in rats. *Alcoholism: Clinical and Experimental Research* **14**, 503–507.

Hwang, B. H., Zhang, J. K., Ehlers, C. L., Lumeng, L., and Li, T. K. (1999). Innate differences of neuropeptide Y (NPY) in hypothalamic nuclei and central nucleus of the amygdala between selectively bred rats with high and low alcohol preference. *Alcoholism: Clinical and Experimental Research* **23**, 1023–1030.

Hyytia, P., and Koob, G. F. (1995). GABA-A receptor antagonism in the extended amygdala decreases ethanol self-administration in rats. *European Journal of Pharmacology* **283**, 151–159.

Hyytia, P., and Sinclair, J. D. (1993). Responding for oral ethanol after naloxone treatment by alcohol-preferring AA rats. *Alcoholism: Clinical and Experimental Research* **17**, 631–636.

Janak, P. H., and Gill, M. T. (2003). Comparison of the effects of allopregnanolone with direct GABAergic agonists on ethanol self-administration with and without concurrently available sucrose. *Alcohol* **30**, 1–7.

Janak, P. H., Redfern, J. E., and Samson, H. H. (1998). The reinforcing effects of ethanol are altered by the endogenous neurosteroid, allopregnanolone. *Alcoholism: Clinical and Experimental Research* **22**, 1106–1112.

Jenkins, A., Franks, N. P., and Lieb, W. R. (1996). Actions of general anaesthetics on 5-HT3 receptors in N1E-115 neuroblastoma cells. *British Journal of Pharmacology* **117**, 1507–1515.

Jensen, S. B. (1984). Sexual function and dysfunction in younger married alcoholics: A comparative study. *Acta Psychiatrica Scandinavica* **69**, 543–549.

Jernigan, T. L., Schafer, K., Butters, N., and Cermak, L. S. (1991). Magnetic resonance imaging of alcoholic Korsakoff patients. *Neuropsychopharmacology* **4**, 175–186.

Johansson, B., Angelo, H. R., Christensen, J. K., Moller, I. W., and Ronsted, P. (1991). Dose-effect relationship of disulfiram in human volunteers: II, A study of the relation between the disulfiram-alcohol reaction and plasma concentrations of acetaldehyde, diethyldithiocarbamic acid methyl ester, and erythrocyte aldehyde dehydrogenase activity. *Pharmacology and Toxicology* **68**, 166–170.

Johnson, B. A., Roache, J. D., Javors, M. A., DiClemente, C. C., Cloninger, C. R., Prihoda, T. J., Bordnick, P. S., Ait-Daoud, N., and Hensler, J. (2000). Ondansetron for reduction of drinking among biologically predisposed alcoholic patients: a randomized controlled trial. *Journal of the American Medical Association* **284**, 963–971.

Jones, A. W. (1993). Disappearance rate of ethanol from the blood of human subjects: implications in forensic toxicology. *Journal of Forensic Science* **38**, 104–118 [erratum: 1994, **39**, 591].

Jones, B. J., and Roberts, D. J. (1968). The quantiative measurement of motor inco-ordination in naive mice using an acelerating rotarod. *Journal of Pharmacy and Pharmacology* **20**, 302–304.

Jones, K. L., and Smith, D. W. (1973). Recognition of the fetal alcohol syndrome in early infancy. *Lancet* **2**(7836), 999–1001.

June, H. L., Foster, K. L., McKay, P. F., Seyoum, R., Woods, J. E., Harvey, S. C., Eiler, W. J., Grey, C., Carroll, M. R., McCane, S., Jones, C. M., Yin, W., Mason, D., Cummings, R., Garcia, M., Ma, C., Sarma, P. V., Cook, J. M., and Skolnick, P. (2003). The reinforcing properties of alcohol are mediated by GABA(A1) receptors in the ventral pallidum. *Neuropsychopharmacology* **28**, 2124–2137.

June, H. L., Lummis, G. H., Colker, R. E., Moore, T. O., and Lewis, M. J. (1991). Ro15-4513 attenuates the consumption of ethanol in deprived rats. *Alcoholism: Clinical and Experimental Research* **15**, 406–411.

Jung, M. E., Wallis, C. J., Gatch, M. B., and Lal, H. (2000). Abecarnil and alprazolam reverse anxiety-like behaviors induced by ethanol withdrawal. *Alcohol* **21**, 161–168.

Jurd, R., Arras, M., Lambert, S., Drexler, B., Siegwart, R., Crestani, F., Zaugg, M., Vogt, K. E., Ledermann, B., Antkowiak, B., and Rudolph, U. (2003). General anesthetic actions in vivo strongly attenuated by a point mutation in the GABA(A) receptor beta3 subunit. *FASEB Journal* **17**, 250–252.

Kalant, H. (1996). Pharmacokinetics of ethanol: absorption, distribution, and elimination. *In The Pharmacology of Alcohol and Alcohol Dependence* (series title: *Alcohol and Alcoholism*, vol. 2), (H. Begleiter, and B. Kissin, Eds.), pp. 15–58. Oxford University Press, New York.

Kalant, H. (1998). Research on tolerance: what can we learn from history? *Alcoholism: Clinical and Experimental Research* **22**, 67–76.

Kano, M., and Konnerth, A. (1992). Potentiation of GABA-mediated currents by cAMP-dependent protein kinase. *Neuroreport* **3**, 563–566.

Kaprio, J., and Koskenvuo, M. (1988). A prospective study of psychological and socioeconomic characteristics, health behavior and morbidity in cigarette smokers prior to quitting compared to persistent smokers and non-smokers. *Journal of Clinical Epidemiology* **41**, 139–150.

Kask, A., Rago, L., Harro, J. (1998). Anxiogenic-like effect of the NPY Y1 receptor antagonist BIBP3226 administered into the dorsal periaqueductal gray matter in rats. *Regulatory Peptides* **75–76**, 255–262.

Katner, S. N., Magalong, J. G., and Weiss, F. (1999). Reinstatement of alcohol-seeking behavior by drug-associated discriminative stimuli after prolonged extinction in the rat. *Neuropsychopharmacology* **20**, 471–479.

Katner, S. N., and Weiss, F. (1999). Ethanol-associated olfactory stimuli reinstate ethanol-seeking behavior after extinction and modify extracellular dopamine levels in the nucleus accumbens. *Alcoholism: Clinical and Experimental Research* **23**, 1751–1760.

Kelai, S., Aissi, F., Lesch, K. P., Cohen-Salmon, C., Hamon, M., and Lanfumey, L. (2003). Alcohol intake after serotonin transporter inactivation in mice. *Alcohol and Alcoholism* **38**, 386–389.

Khanna, J. M., Kalant, H., Shah, G., and Chau, A. (1992). Effect of (+)MK-801 and ketamine on rapid tolerance to ethanol. *Brain Research Bulletin* **28**, 311–314.

Khanna, J. M., Kalant, H., Shah, G., and Chau, A. (1993a). Effect of D-cycloserine on rapid tolerance to ethanol. *Pharmacology Biochemistry and Behavior* **45**, 983–986.

Khanna, J. M., Morato, G. S., Shah, G., Chau, A., and Kalant, H. (1993b). Inhibition of nitric oxide synthesis impairs rapid tolerance to ethanol. *Brain Research Bulletin* **32**, 43–47.

Khanna, J. M., Shah, G., Weiner, J., Wu, P. H., and Kalant, H. (1993c). Effect of NMDA receptor antagonists on rapid tolerance to ethanol. *European Journal of Pharmacology* **230**, 23–31.

Khantzian, E. J. (1995). The 1994 distinguished lecturer in substance abuse. *Journal of Substance Abuse Treatment* **12**, 157–165.

Khantzian, E. J. (1997). The self-medication hypothesis of substance use disorders: a reconsideration and recent applications. *Harvard Review of Psychiatry* **4**, 231–244.

Khantzian, E. J., and Wilson, A. (1993). Substance abuse, repetition, and the nature of addictive suffering. *In Hierarchical Concepts in Psychoanalysis: Theory, Research, and Clinical Practice*, (A. Wilson, and J. E. Gedo, Eds.), pp. 263–283. Guilford Press, New York.

Kiianmaa, K. (1976). Alcohol intake in the rat after lowering brain 5-hydroxtryptamine content by electrolytic midbrain raphe lesions, 5, 6-dihydroxytryptamine or p-chlorophenylalanine. *Medical Biology* **54**, 203–209.

King, P. J., Widdowson, P. S., Doods, H. N., and Williams, G. (1999). Regulation of neuropeptide Y release by neuropeptide Y receptor ligands and calcium channel antagonists in hypothalamic slices. *Journal of Neurochemistry* **73**, 641–646.

Klassen, A. D., and Wilsnack, S. C. (1986). Sexual experience and drinking among women in a U.S. national survey. *Archives of Sexual Behavior* **15**, 363–392.

Knapp, D. J., and Pohorecky, L. A. (1992). Zacopride, a 5-HT3 receptor antagonist, reduces voluntary ethanol consumption in rats. *Pharmacology Biochemistry and Behavior* **41**, 847–850.

Knapp, D. J., Duncan, G. E., Crews, F. T., and Breese, G. R. (1998). Induction of Fos-like proteins and ultrasonic vocalizations during ethanol withdrawal: further evidence for withdrawal-induced anxiety. *Alcoholism: Clinical and Experimental Research* **22**, 481–493.

Knapp, D. J., Overstreet, D. H., Moy, S. S., and Breese, G. R. (2004). SB242084, flumazenil, and CRA1000 block ethanol withdrawal-induced anxiety in rats. *Alcohol* **32**, 101–111.

Kochanek, K. D., and Smith, B. L. (2004). *Deaths: Preliminary Data for 2002* (series title: *National Vital Statistics Reports*, vol. 52[13]; DHHS publication No. PHS 2004-1120), National Center for Health Statistics, Hyattsville MD.

Koltchine, V., Anantharam, V., Wilson, A., Bayley, H., and Treistman, S. N. (1993). Homomeric assemblies of NMDAR1 splice variants are sensitive to ethanol. *Neuroscience Letters* **152**, 13–16.

Koob, G. F., and Le Moal, M. (2001). Drug addiction, dysregulation of reward, and allostasis. *Neuropsychopharmacology* **24**, 97–129.

Koob, G. F., and Wall, T. L. (1987). Schafer J, Rapid induction of tolerance to the antipunishment effects of ethanol. *Alcohol* **4**, 481–484.

Koob, G. F., Thatcher-Britton, K., Britton, D. R., Roberts, D. C. S., and Bloom, F. E. (1984). Destruction of the locus coeruleus or the dorsal NE bundle does not alter the release of punished responding by ethanol and chlordiazepoxide. *Physiology and Behavior* **33**, 479–485.

Koob, G. F., Braestrup, C., and Thatcher-Britton, K. (1986). The effects of FG 7142 and RO 15-1788 on the release of punished responding produced by chlordiazepoxide and ethanol in the rat. *Psychopharmacology* **90**, 173–178.

Koob, G. F., Percy, L., and Britton, K. T. (1989). The effects of RO 15-4513 on the behavioral actions of ethanol in an operant reaction time task and a conflict test. *Pharmacology Biochemistry and Behavior* **31**, 757–760.

Koob, G. F., Heinrichs, S. C., Menzaghi, F., Pich, E. M., and Britton, K. T. (1994). Corticotropin releasing factor, stress and behavior. *Seminars in the Neurosciences* **6**, 221–229.

Koob, G. F., Sanna, P. P., and Bloom, F. E. (1998). Neuroscience of addiction. *Neuron* **21**, 467–476.

Koppi, S., Eberhardt, G., Haller, R., Konig, P. (1987). Calcium-channel-blocking agent in the treatment of acute alcohol withdrawal: caroverine versus meprobamate in a randomized double-blind study *Neuropsychobiology* **17**, 49–52.

Kornet, M., Goosen, C., and Van Ree, J. M. (1991). Effect of naltrexone on alcohol consumption during chronic alcohol drinking and after a period of imposed abstinence in free-choice drinking rhesus monkeys. *Psychopharmacology* **104**, 367–376.

Kornetsky, C., Bain, G. T., Unterwald, E. M., and Lewis, M. J. (1988). Brain stimulation reward: effects of ethanol. *Alcoholism: Clinical and Experimental Research* **12**, 609–616.

Korpi, E. R., Sinclair, J. D., Kaheinen, P., Viitamaa, T., Hellevuo, K., and Kiianmaa, K. (1988). Brain regional and adrenal monoamine concentrations and behavioral responses to stress in alcohol-preferring AA and alcohol-avoiding ANA rats. *Alcohol* **5**, 417–425.

Korpi, E. R., Paivarinta, P., Abi-Dargham, A., Honkanen, A., Laruelle, M., Tuominen, K., and Hilakivi, L. A. (1992). Binding of serotonergic ligands to brain membranes of alcohol-preferring AA and alcohol-avoiding ANA rats. *Alcohol* **9**, 369–374.

Kostowski, W., Dyr, W., and Krzascik, P. (1993). The abilities of 5-HT3 receptor antagonist ICS 205-930 to inhibit alcohol preference and withdrawal seizures in rats. *Alcohol* **10**, 369–373.

Kranzler, H. R., Gelernter, J., O'Malley, S., Hernandez-Avila, C. A., and Kaufman, D. (1998). Association of alcohol or other drug dependence with alleles of the mu opioid receptor gene (OPRM1). *Alcoholism: Clinical and Experimental Research* **22**, 1359–1362.

Kruse, S. W., Zhao, R., Smith, D. P., and Jones, D. N. (2003). Structure of a specific alcohol-binding site defined by the odorant binding protein LUSH from *Drosophila melanogaster*. *Nature Structural Biology* **10**, 694–700 [erratum in 11:102].

Kumar, S., Fleming, R. L., and Morrow, A. L. (2004). Ethanol regulation of gamma-aminobutyric acid A receptors: genomic and nongenomic mechanisms. *Pharmacology and Therapeutics* **101**, 211–226.

Kuriyama, K., and Ueha, T. (1992). Functional alterations in cerebral GABAA receptor complex associated with formation of alcohol dependence: analysis using GABA-dependent 36Cl-influx into neuronal membrane vesicles. *Alcohol and Alcoholism* **27**, 335–343.

Lallemand, F., and de Witte, P. (2005). Ethanol induces higher BEC in CB1 cannabinoid receptor knockout mice while decreasing ethanol preference. *Alcohol and Alcoholism* **40**, 54–62.

Le, A. D., Kalant, H., and Khanna, J. M. (1982). Interaction between des-glycinamide9-[Arg8]vasopressin and serotonin on ethanol tolerance. *European Journal of Pharmacology* **80**, 337–345.

Le, A. D., and Kalant, H. (1992). Learning as a factor in ethanol tolerance. *In Neurobiology of Drug Abuse: Learning and Memory* (series title: *NIDA Research Monograph*, vol. 97), (L. Erinoff, Ed.), pp. 193–207. National Institute on Drug Abuse, Rockville, M. D.

Le, M., Agnen, J., Tran, G., Durlach, J., and Martin, C. (1987). Dose-dependent suppression of the high alcohol intake of chronically intoxicated rats by Ca-acetyl homotaurinate. *Alcohol* **4**, 97–102.

Le Magnen, J. (1960). Etude de quelques facteurs associe a des modification de la consommation spontanee d'alcool ethylique par le rat [Study of some factors associated with modifications of spontaneous ingestion of ethyl alcohol by the rat]. *Journal de Physiologie* **52**, 873–884.

LeBlanc, A. E., Kalant, H., and Gibbins, R. J. (1975). Acute tolerance to ethanol in the rat. *Psychopharmacologia* **41**, 43–46.

LeBlanc, A. E., Matsunaga, M., and Kalant, H. (1976). Effects of frontal polar cortical ablation and cycloheximide on ethanol tolerance in rats. *Pharmacology Biochemistry and Behavior* **4**, 175–179.

Lee, S., and Rivier, C. (1994). Hypophysiotropic role and hypothalamic gene expression of corticotropin-releasing factor and vasopressin in rats injected with interleukin-1 beta systemically or into the brain ventricles. *Journal of Neuroendocrinology* **6**, 217–224.

Lee, R. S., Steffensen, S. C., Berg, G. I., and Henriksen, S. J. (1996). Ethanol inhibits firing of ventral tegmental area interneurons and nucleus accumbens neurons in freely behaving rats. *Alcoholism: Clinical and Experimental Research* **20** (2 *Suppl.*), 61A.

Lee, S., Schmidt, E. D., Tilders, F. J., and Rivier, C. (2001). Effect of repeated exposure to alcohol on the response of the hypothalamic-pituitary-adrenal axis of the rat: I. Role of changes in hypothalamic neuronal activity. *Alcoholism: Clinical and Experimental Research* **25**, 98–105.

Lee, S., Selvage, D., Hansen, K., and Rivier, C. (2004). Site of action of acute alcohol administration in stimulating the rat hypothalamic-pituitary-adrenal axis: comparison between the effect of systemic and intracerebroventricular injection of this drug on pituitary and hypothalamic responses. *Endocrinology* **145**, 4470–4479.

Leigh, B. C., and Stall, R. (1993). Substance use and risky sexual behavior for exposure to HIV: Issues in methodology, interpretation, and prevention. *American Psychologist* **48**, 1035–1045.

Lemere, F., and Smith, J. W. (1973). Alcohol-induced sexual impotence. *American Journal of Psychiatry* **130**, 212–213.

Leo, A., Hansch, C., and Elkins, D. (1971). Partition coefficients and their uses. *Chemical Reviews* **71**, 525–615.

Leslie, S. W., Barr, E., Chandler, J., and Farrar, R. P. (1983). Inhibition of fast- and slow-phase depolarization-dependent synaptosomal calcium uptake by ethanol. *Journal of Pharmacology and Experimental Therapeutics* **225**, 571–575.

Levine, A. S., Morley, J. E. (1984). Neuropeptide, Y: a potent inducer of consummatory behavior in rats. *Peptides* **5**, 1025–1029.

Levy, J. (1935). Contribution a l'etude de l'accoutumance experimentale aux poisons: III. Alcoolisme experimentale. L'accoutumance a l'alcool peut-elle etre consideree comme une consequence de l'hyposensibilite cellulaire? [Study of the experimental tolerance to alcohol: III. Can tolerance to alcohol be considered a consequence of cellular hyposensitivity?]. *Bulletin de la Societe de Chimie Biologique* **17**, 47–59.

Lewis, M. J. (1996). Alcohol reinforcement and neuropharmacological therapeutics. *Alcohol and Alcoholism* **31 (Suppl1),** 17–25.

Li, J., Li, Y. H., and Yuan, X. R. (2003). Changes of phosphorylation of cAMP response element binding protein in rat nucleus accumbens after chronic ethanol intake: naloxone reversal. *Acta Pharmacologica Sinica* **24**, 930–936.

Li, T. K., and McBride, W. J. (1995). Pharmacogenetic models of alcoholism. *Clinical Neuroscience* **3**, 182–188.

Li, T. K., Spanagel, R., Colombo, G., McBride, W. J., Porrino, L. J., Suzuki, T., and Rodd-Henricks, Z. A. (2001). Alcohol reinforcement and voluntary ethanol consumption. *Alcoholism: Clinical and Experimental Research* **25**(5 Suppl. ISBRA), 117s–126s.

Li, X. W., Li, T. K., and Froehlich, J. C. (1998). Enhanced sensitivity of the nucleus accumbens proenkephalin system to alcohol in rats selectively bred for alcohol preference. *Brain Research* **794**, 35–47.

Liang, T., Spence, J., Liu, L., Strother, W. N., Chang, H. W., Ellison, J. A., Lumeng, L., Li, T. K., Foroud, T., and Carr, L. G. (2003). Alpha-Synuclein maps to a quantitative trait locus for alcohol preference and is differentially expressed in alcohol-preferring and -nonpreferring rats. *Proceedings of the National Academy of Sciences USA* **100**, 4690–4695.

Lieber, C. S. (1997). Ethanol metabolism, cirrhosis and alcoholism. *Clinica Chimica Acta* **257**, 59–84.

Liljequist, S., and Engel, J. (1982). Effects of GABAergic agonists and antagonists on various ethanol-induced behavioral changes. *Psychopharmacology* **78**, 71–75.

Liljequist, S., and Engel, J. A. (1984). The effects of GABA and benzodiazepine receptor antagonists on the anti-conflict actions of diazepam or ethanol. *Pharmacology Biochemistry and Behavior* **21**, 521–525.

Lima-Landman, M. T. (1989). Albuquerque, E. X., Ethanol potentiates and blocks NMDA-activated single-channel currents in rat hippocampal pyramidal cells. *FEBS Letters* **247**, 61–67.

Lindenberg, B. A. (1951). Sur la solubilite des substances organiques amphipatiques dans les glycerides neutres et hydroxyls. *Journal de Chimie Physique* **48**, 350–355.

Linnoila, M., Virkkunen, M., George, T., Eckardt, M., Higley, J. D., Nielsen, D., and Goldman, D. (1994). Serotonin, violent behavior and alcohol. *In Toward a Molecular Basis of Alcohol Use and Abuse* (series title: *Experientia Supplementum*, vol. 71), (B. Jansson., H. Jornvall, U. Rydberg, L. Terenius, and B. L. Vallee, Eds.), pp. 155–163. Birkhauser Verlag, Boston.

Lister, R. G., and Hilakivi, L. A. (1988). The effects of novelty, isolation, light and ethanol on the social behavior of mice. *Psychopharmacology* **96**, 181–187.

Lister, R. G. (1987). The use of a plus-maze to measure anxiety in the mouse. *Psychopharmacology* **92**, 180–185.

Liu, X., and Weiss, F. (2002a). Reversal of ethanol-seeking behavior by D1 and D2 antagonists in an animal model of relapse: differences in antagonist potency in previously ethanol-dependent versus nondependent rats. *Journal of Pharmacology and Experimental Therapeutics* **300**, 882–889.

Liu, X., and Weiss, F. (2002b). Additive effect of stress and drug cues on reinstatement of ethanol seeking: exacerbation by history of dependence and role of concurrent activation of corticotropin-releasing factor and opioid mechanisms. *Journal of Neuroscience* **22**, 7856–7861.

Liu, X., and Weiss, F. (2004). Nitric oxide synthesis inhibition attenuates conditioned reinstatement of ethanol-seeking, but not the primary reinforcing effects of ethanol. *Alcohol Clin Exp Res.* Aug **28**(8), 1194–1199.

Lovinger, D. M. (1995). Developmental decrease in ethanol inhibition of N-methyl-D-aspartate receptors in rat neocortical neurons: relation to the actions of ifenprodil. *Journal of Pharmacology and Experimental Therapeutics* **274**, 164–172.

Lovinger, D. M. (2002). NMDA receptors lose their inhibitions. *Nature Neuroscience* **5**, 614–616.

Lovinger, D. M. (1993). High ethanol sensitivity of recombinant AMPA-type glutamate receptors expressed in mammalian cells. *Neuroscience Letters* **159**, 83–87.

Lovinger, D. M., Sung, K. W., and Zhou, Q. (2000). Ethanol and trichloroethanol alter gating of 5-HT3 receptor-channels in NCB-20 neuroblastoma cells. *Neuropharmacology* **39**, 561–570.

Lovinger, D. M., and White, G. (1991). Ethanol potentiation of 5-hydroxytryptamine3 receptor-mediated ion current in neuroblastoma cells and isolated adult mammalian neurons. *Molecular Pharmacology* **40**, 263–270.

Lovinger, D. M., White, G., and Weight, F. F. (1989). Ethanol inhibits NMDA-activated ion current in hippocampal neurons. *Science* **243**, 1721–1724.

Lovinger, D. M., White, G., and Weight, F. F. (1990). NMDA receptor-mediated synaptic excitation selectively inhibited by ethanol in hippocampal slice from adult rat. *Journal of Neuroscience* **10**, 1372–1379.

Lyness, W. H., and Smith, F. L. (1992). Influence of dopaminergic and serotonergic neurons on intravenous ethanol self-administration in the rat. *Pharmacology Biochemistry and Behavior* **42**, 187–192.

Machu, T. K., and Harris, R. A. (1994). Alcohols and anesthetics enhance the function of 5-hydroxytryptamine3 receptors expressed in Xenopus laevis oocytes. *Journal of Pharmacology and Experimental Therapeutics* **271**, 898–905.

Madamba, S. G., Hsu, M., Schweitzer, P, and Siggins, G. R. (1995). Ethanol enhances muscarinic cholinergic neurotransmission in rat hippocampus in vitro. *Brain Research* **685,** 21–32.

Mahmoudi, M., Kang, M. H., Tillakaratne, N., Tobin, A. J., and Olsen, R. W. (1997). Chronic intermittent ethanol treatment in rats increases $GABA_A$ receptor $alpha_4$-subunit expression: Possible relevance to alcohol dependence. *Journal of Neurochemistry* **68**, 2485–2492.

Mailliard, W. S. (2004). Diamond I. Recent advances in the neurobiology of alcoholism: the role of adenosine. *Pharmacology and Therapeutics* **101**, 39–46.

Mair, W. G., Warrington, E. K., and Weiskrantz, L. (1979). Memory disorder in Korsakoff's psychosis: a neuropathological and neuropsychological investigation of two cases. *Brain* **102**, 749–783.

Majchrowicz, E. (1975). Induction of physical dependence upon ethanol and the associated behavioral changes in rats. *Psychopharmacologia* **43**, 245–254.

Majchrowicz, E., and Hunt, W. A. (1976). Temporal relationship of the induction of tolerance and physical dependence after continuous intoxication with maximum tolerable doses of ethanol in rats. *Psychopharmacology* **50**, 107–112.

Malatesta, V. J., Pollack, R. H., Crotty, T. D., and Peacock, L. J. (1982). Acute alcohol intoxication and female orgasmic response. *Journal of Sex Research* **18**, 1–17.

Malatesta, V. J., Pollack, R. H., Wilbanks, W. A., and Adams, H. E. (1979). Alcohol effects on the orgasmic-ejaculatory response in human males. *Journal of Sex Res.* May;**15**(2), 101–107.

Maldonado, R., Smadja, C., Mazzucchelli, C., and Sassone-Corsi, P. (1999). Altered emotional and locomotor responses in mice deficient in the transcription factor CREM. *Proceedings of the National Academy of Sciences USA* **96**, 14094–14099 [erratum: 97, 1949].

Maldve, R. E., Zhang, T. A., Ferrani-Kile, K., Schreiber, S. S., Lippmann, M. J., Snyder, G. L., Fienberg, A. A., Leslie, S. W., Gonzales, R. A., and Morrisett, R. A., (2002). DARPP–32 and regulation of the ethanol sensitivity of NMDA receptors in the nucleus accumbens. *Nature Neuroscience* **5**, 641–648.

Mandell, W., and Miller, C. M., (1983). Male sexual dysfunction as related to alcohol consumption: a pilot study. *Alcoholism: Clinical and Experimental Research* **7**, 65–69.

Maricq, A. V., Peterson, A. S., Brake, A. J., Myers, R. M., and Julius, D. (1991). Primary structure and functional expression of the 5HT3 receptor, a serotonin-gated ion channel. *Science* **254**, 432–437.

Marszalec, W., Aistrup, G. L., and Narahashi, T. (1998). Ethanol modulation of excitatory and inhibitory synaptic interactions in cultured cortical neurons. *Alcoholism: Clinical and Experimental Research* **22**, 1516–1524.

Martin, D., Tayyeb, M. I., and Swartzwelder, H. S. (1995). Ethanol inhibition of AMPA and kainate receptor-mediated depolarizations of hippocampal area CA1. *Alcoholism: Clinical and Experimental Research* **19**, 1312–1316.

Martin-Fardon, R., Ciccocioppo, R., Massi, M., and Weiss, F. (2000). Nociceptin prevents stress-induced ethanol- but not cocaine-seeking behavior in rats. *Neuroreport* **11**, 1939–1943.

Martz, A., Deitrich, R. A., and Harris, R. A. (1983). Behavioral evidence for the involvement of gamma-aminobutyric acid in the actions of ethanol. *European Journal of Pharmacology* **89**, 53–62.

Mascia, M. P., Trudell, J. R., and Harris, R. A. (2000). Specific binding sites for alcohols and anesthetics on ligand-gated ion channels. *Proceedings of the National Academy of Sciences USA* **97**, 9305–9310.

Mason, B. J. (2003). Acamprosate and naltrexone treatment for alcohol dependence: an evidence-based risk-benefits assessment. *European Neuropsychopharmacology* **13**, 469–475.

Masood, K., Wu, C., Brauneis, U., and Weight, F. F. (1994). Differential ethanol sensitivity of recombinant N-methyl-D-aspartate receptor subunits. *Molecular Pharmacology* **45**, 324–329.

Masserman, J. H., and Yum, K. S. (1946). An analysis of the influence of alcohol on experimental neuroses in cats. *Psychosomatic Medicine* **8**, 36–52.

Maul, B., Siems, W. E., Hoehe, M. R., Grecksch, G., Bader, M., and Walther, T. (2001). Alcohol consumption is controlled by angiotensin II. *FASEB. Journal* **15**, 1640–1642.

May, P. A., Brooke, L., Gossage, J. P., Croxford, J., Adnams, C., Jones, K. L., Robinson, L., and Viljoen, D. (2000). Epidemiology of fetal alcohol syndrome in a South African community in the Western Cape Province. *American Journal of Public Health* **90**, 1905–1912.

Mayes, A. R., Meudell, P. R., Mann, D., and Pickering, A. (1988). Location of lesions in Korsakoff's syndrome: neuropsychological and neuropathological data on two patients. *Cortex* **24**, 367–388.

McBride, W. J., Murphy, J. M., Lumeng, L., and Li, T. K. (1988). Effects of Ro 15-4513, fluoxetine and desipramine on the intake of ethanol, water and food by the alcohol-preferring (P) and -nonpreferring (NP) lines of rats. *Pharmacology Biochemistry and Behavior* **30**, 1045–1050.

McBride, W. J., Murphy, J. M., Lumeng, L., and Li, T. K. (1990). Serotonin, dopamine and GABA involvement in alcohol drinking of selectively bred rats. *Alcohol* **7**, 199–205.

McBride, W. J., Murphy, J. M., Lumeng, L., and Li, T. K. (1992). Serotonin and alcohol consumption. *In Novel Pharmacological Interventions for Alcoholism*, (C. A. Naranjo, and E. M. Sellers Eds.), pp. 59–67. Springer, New York.

McBride, W. J., Chernet, E., Dyr, W., Lumeng, L., and Li, T. K. (1993a). Densities of dopamine D2 receptors are reduced in CNS regions of alcohol-preferring P rats. *Alcohol* **10**, 387–390.

McBride, W. J., Chernet, E., Rabold, J. A., Lumeng, L., and Li, T. K. (1993b). Serotonin-2 receptors in the CNS of alcohol-preferring and -nonpreferring rats. *Pharmacology Biochemistry and Behavior* **46**, 631–636.

McBride, W. J., Chernet, E., Russell, R. N., Chamberlain, J. K., Lumeng, L., and Li, T. K. (1997a). Regional CNS densities of serotonin and dopamine receptors in high alcohol-drinking (HAD) and low alcohol-drinking (LAD) rats. *Alcohol* **14**, 603–609.

McBride, W. J., Chernet, E., Russell, R. N., Wong, D. T., Guan, X. M., Lumeng, L., and Li, T. K. (1997b). Regional CNS densities of monoamine receptors in alcohol-naive alcohol-preferring P and -nonpreferring NP rats. *Alcohol* **14**, 141–148.

McBride, W. J., Chernet, E., McKinzie, D. L., Lumeng, L., and Li, T. K. (1998). Quantitative autoradiography of mu-opioid receptors in the CNS of alcohol-naive alcohol-preferring P and -nonpreferring NP rats. *Alcohol* **16**, 317–323.

McBride, W. J., Lovinger, D. M., Machu, T., Thielen, R. J., Rodd, Z. A., Murphy, J. M., Roache, J. D., and Johnson, B. A. (2004). Serotonin-3 receptors in the actions of alcohol, alcohol reinforcement, and alcoholism. *Alcoholism: Clinical and Experimental Research* **28**, 257–267.

McCreery, M. J., and Hunt, W. A. (1978). Physico-chemical correlates of alcohol intoxication. *Neuropharmacology* **17**, 451–461.

McKim, W. A. (2003). *Drug and Behavior: An Introduction to Behavioral Pharmacology*, 5th ed., Prentice Hall, Upper Saddle River NJ.

McKinzie, D. L., Eha, R., Cox, R., Stewart, R. B., Dyr, W., Murphy, J. M., McBride, W. J., Lumeng, L., and Li, T. K. (1998). Serotonin-3 receptor antagonism of alcohol intake: effects of drinking conditions. *Alcohol* **15**, 291–298.

Medical Economics Company, (2004). *Physicians' Desk Reference*, 58th ed., Medical Economics Company, Oradell NJ.

Mehta, A. K., and Ticku, M. K. (1988). Ethanol potentiation of GABAergic transmission in cultured spinal cord neurons involves gamma-aminobutyric acidA-gated chloride channels. *Journal of Pharmacology and Experimental Therapeutics* **246**, 558–564.

Melis, M., Camarini, R., Ungless, M. A., and Bonci A. (2002). Long-lasting potentiation of GABAergic synapses in dopamine neurons after a single in vivo ethanol exposure. *Journal of Neuroscience* **22**, 2074–2082.

Mellanby, E. (1919). *Alcohol: Its Absorption into and Disappearance from the Blood Under Different Conditions* (series title: *Medical Research Council Special Report Series*, vol. 31), Her Majesty's Stationery Office, London.

Melo, J. A., Shendure J., Pociask K., and Silver, L. M. (1996). Identification of sex-specific quantitative trait loci controlling alcohol preference in C57BL/6 mice. *Nature Genetics* **13**, 147–153.

Mendelson, J. H., Stein, S., and Mello, N. K. (1965). Effects of experimentally induced intoxication on metabolism of ethanol-1-C-14 in alcoholic subjects. *Metabolism* **14**, 1255–1266.

Mendelson, J. H., Mello, N. K., and Ellingboe, J. (1981). Acute alcohol intake and pituitary gonadal hormones in normal human females. *Journal of Pharmacology and Experimental Therapeutics*. **218**, 23–26.

Mendelson, J. H., Lukas, S. E., Mello, N. K., Amass, L., Ellingboe, J., and Skupny, A. (1988). Acute alcohol effects on plasma estradiol levels in wome. *Psychopharmacology* **94**, 464–467.

Merlo-Pich, E., Lorang, M., Yeganeh, M., Rodriguez de Fonseca, F., Raber, J., Koob, G. F., and Weiss, F. (1995). Increase of extracellular corticotropin-releasing factor-like immunoreactivity levels in the amygdala of awake rats during restraint stress and ethanol withdrawal as measured by microdialysis. *Journal of Neuroscience* **15**, 5439–5447.

Messing, R. O., Petersen, P. J., and Henrich, C. J., (1991). Chronic ethanol exposure increases levels of protein kinase C delta and epsilon and protein kinase C-mediated phosphorylation in cultured neural cells. *Journal of Biological Chemistry* **266**, 23428–23432.

Mhatre, M. C., Carl, K., Garrett, K. M., and Holloway, F. A. (2000). Opiate delta-2-receptor antagonist naltriben does not alter discriminative stimulus effects of ethanol. *Pharmacology Biochemistry and Behavior* **66**, 701–706.

Mhatre, M. C., Garrett, K. M., and Holloway, F. A. (2001). 5-HT 3 receptor antagonist ICS 205-930 alters the discriminative effects of ethanol. *Pharmacology Biochemistry and Behavior* **68**, 163–170.

Mhatre, M. C., Pena, G., Sieghart, W., and Ticku, M. K. (1993). Antibodies specific for GABA-A receptor alpha subunits reveal that chronic alcohol treatment down-regulates alpha-subunit expression in rat brain regions. *Journal of Neurochemistry* **61**, 1620–1625.

Mhatre, M. C., and Ticku, M. K. (1992). Chronic ethanol administration alters gamma-aminobutyric acidA receptor gene expression. *Molecular Pharmacology* **42**, 415–422.

Michaelis, R. C., Holohean, A. M., Hunter, G. A., and Holloway, F. A. (1987). Endogenous adenosine-receptive systems do not mediate the discriminative stimulus properties of ethanol. *Alcohol and Drug Research* **7**, 175–185.

Miczek, K. A., DeBold, J. F., van Erp, A. M., and Tornatzky, W. (1997). Alcohol, GABAA-benzodiazepine receptor complex, and aggression. *In Alcohol and Violence: Epidemiology, Neurobiology, Psychology, Family Issues* (series title: *Recent Developments in Alcoholism*, vol. 13), (M. Galanter, Ed.), pp. 139–171. Plenum Press, New York.

Mihalek, R. M., Bowers, B. J., Wehner, J. M., Kralic, J. E., VanDoren, M. J., Morrow, A. L., and Homanics, G. E. (2001). GABA(A)-receptor delta subunit knockout mice have multiple defects in behavioral responses to ethanol. *Alcoholism: Clinical and Experimental Research* **25**, 1708–1718.

Mihic, S. J. (1999). Acute effects of ethanol on GABAA and glycine receptor function. *Neurochemistry International* **35**, 115–123.

Mihic, S. J., and Harris, R. A. (1996). Alcohol actions at the GABA-A receptor/chloride channel complex. *In Pharmacological Effects of Ethanol on the Nervous System*, (R. A. Deitrich, and V. G. Erwin, Eds.), pp. 51–72. CRC Press, Boca Raton.

Mihic, S. J., Ye, Q., Wick, M. J., Koltchine, V. V., Krasowski, M. D., Finn, S. E., Mascia, M. P., Valenzuela, C. F., Hanson, K. K., Greenblatt, E. P., Harris, R. A., and Harrison, N. L. (1997). Sites of alcohol and volatile anaesthetic action on GABA-A and glycine receptors. *Nature* **389**, 385–389.

Miller, N. S., and Gold, M. S., (1988). The human sexual response and alcohol and drugs. *Journal of Substance Abuse Treatment* **5**, 171–177.

Ming, Z., Knapp, D. J., Mueller, R. A., Breese, G. R., and Criswell, H. E. (2001). Differential modulation of GABA- and NMDA-gated currents by ethanol and isoflurane in cultured rat cerebral cortical neurons. *Brain Research* **920**, 117–124.

Minino, A. M., Arias, E., Kochanek, K. D., Murphy, S. L., and Smith, B. L. (2002). Deaths: final data for 2000. *National Vital Statistics Report* **50**, 1–119.

Mirshahi, T., and Woodward, J. J. (1995). Ethanol sensitivity of heteromeric NMDA receptors: effects of subunit assembly, glycine and NMDAR1 Mg(2+)-insensitive mutants. *Neuropharmacology* **34**, 347–355.

Misra, K., Roy, A., and Pandey, S. C. (2001). Effects of voluntary ethanol intake on the expression of Ca(2+)/calmodulin-dependent protein kinase IV and on CREB expression and phosphorylation in the rat nucleus accumbens. *Neuroreport* **12**, 4133–4137 [erratum: **13**(2):inside back cover].

Mizoi, Y., Tatsuno, Y., Adachi, J., Kogame, M., Fukunaga, T., Fujiwara, S., Hishida, S., and Ijiri, I. (1983). Alcohol sensitivity related to polymorphism of alcohol-metabolizing enzymes in Japanese. *Pharmacology Biochemistry and Behavior* **18**(*Suppl.* 1), 127–133.

Mochly-Rosen, D., Chang, F. H., Cheever, L., Kim, M., Diamon, I., and Gordon, A. S. (1988). Chronic ethanol causes heterologous desensitization of receptors by reducing alpha s messenger RNA. *Nature* **333**, 848–850.

Monti, P. M., Rohsenow, D. J., Hutchison, K. E., Swift, R. M., Mueller, T. I., Colby, S. M., Brown, R. A., Gulliver, S. B., Gordon, A., and Abrams, D. B. (1999). Naltrexone's effect on cue-elicited craving among alcoholics in treatment. *Alcoholism: Clinical and Experimental Research* **23**, 1386–1394.

Moore, S. D., Madamba, S. G., and Siggins, G. R. (1990). Ethanol diminishes a voltage-dependent K+ current, the M-current, in CA1 hippocampal pyramidal neurons in vitro. *Brain Research* **516**, 222–228.

Morrisett, R. A. (1994). Potentiation of N-methyl-D-aspartate receptor-dependent afterdischarges in rat dentate gyrus following in vitro ethanol withdrawal. *Neuroscience Letters* **167**, 175–178.

Morrisett, R. A., Martin, D., Oetting, T. A., Lewis, D. V., Wilson, W. A., and Swartzwelder, H. S., (1991). Ethanol and magnesium ions inhibit N-methyl-D-aspartate-mediated synaptic potentials in an interactive manner. *Neuropharmacology* **30**, 1173–1178.

Morrow, A. L., Suzdak, P. D., Karanian, J. W., and Paul, S. M. (1988). Chronic ethanol administration alters gamma-aminobutyric acid, pentobarbital and ethanol-mediated $^{36}Cl^{-}$ uptake in cerebral cortical synaptoneurosomes. *Journal of Pharmacology and Experimental Therapeutics* **246**, 158–164.

Morrow, A. L., Suzdak, P. D., and Paul, S. M. (1987). Steroid hormone metabolites potentiate GABA receptor-mediated chloride ion flux with nanomolar potency. *European Journal of Pharmacology* **142**, 483–485.

Morrow, A. L., VanDoren, M. J., and Devaud, L. L. (1998). Effects of progesterone or neuroactive steroid? *Nature* **395**, 652–653.

Morrow, A. L., Van Doren, M. J., Penland, S. N., and Matthews, D. B. (2001). The role of GABAergic neuroactive steroids in ethanol action, tolerance and dependence. *Brain Research Reviews* **37**, 98–109.

Morse, A. C., Schulteis, G., Holloway, F. A., and Koob, G. F. (2000). Conditioned place aversion to the 'hangover' phase of acute ethanol administration in the rat. *Alcohol* **22**, 19–24.

Morzorati, S., and Kubek, M. J. (1993). Septal TRH in alcohol-naive P and NP rats and following alcohol challenge. *Brain Research Bulletin* **31**, 301–304.

Moy, S. S., Knapp, D. J., Criswell, H. E., and Breese, G. R. (1997). Flumazenil blockade of anxiety following ethanol withdrawal in rats. *Psychopharmacology* **131**, 354–360.

Moykkynen, T., Korpi, E. R., and Lovinger, D. M. (2003). Ethanol inhibits alpha-amino-3-hydyroxy-5-methyl-4-isoxazolepropionic acid (AMPA) receptor function in central nervous system neurons by stabilizing desensitization. *Journal of Pharmacology and Experimental Therapeutics*. **306**, 546–555.

Murphy, J. M., McBride, W. J., Lumeng, L., and Li, T. K. (1982). Regional brain levels of monoamines in alcohol-preferring and -nonpreferring lines of rats. *Pharmacology Biochemistry and Behavior* **16**, 145–149.

Murphy, J. M., McBride, W. J., Lumeng, L., and Li, T.-K. (1987). Contents of monoamines in forebrain regions of alcohol-preferring (P) and non-preferring (NP) lines of rats. *Pharmacology Biochemistry and Behavior* **26**, 389–392.

Murphy, J. M., Stewart, R. B., Bell, R. L., Badia-Elder, N. E., Carr, L. G., McBride, W. J., Lumeng, L., and Li, T. K. (2002). Phenotypic and genotypic characterization of the Indiana University rat lines selectively bred for high and low alcohol preference. *Behavior Genetics* **32**, 363–388.

Murphy, J. M., Waller, M. B., Gatto, G. J., McBride, W. J., Lumeng, L., and Li, T. K. (1985). Monoamine uptake inhibitors attenuate ethanol intake in alcohol-preferring (P) rats. *Alcohol* **2**, 349–352.

Murphy, J. M., Waller, M. B., Gatto, G. J., McBride, W. J., Lumeng, L., and Li, T. K. (1988). Effects of fluoxetine on the intragastric self-administration of ethanol in the alcohol preferring P line of rats. *Alcohol* **5**, 283–286.

Myers, R. D. (1990). Anatomical 'circuitry' in the brain mediating alcohol drinking revealed by THP-reactive sites in the limbic system. *Alcohol* **7**, 449–459.

Naassila, M., Ledent, C., and Daoust, M. (2002). Low ethanol sensitivity and increased ethanol consumption in mice lacking adenosine A2A receptors. *Journal of Neuroscience* **22**, 10487–10493.

Naassila, M., Pierrefiche, O., Ledent, C., and Daoust, M. (2004). Decreased alcohol self-administration and increased alcohol sensitivity and withdrawal in CB1 receptor knockout mice. *Neuropharmacology* **46**, 243–253.

Nagy, L. E. (2004). Molecular aspects of alcohol metabolism: transcription factors involved in early ethanol-induced liver injury. *Annual Review of Nutrition* **24**, 55–78.

Nagy, L. E., Diamond, I., and Gordon, A. S. (1991). cAMP-dependent protein kinase regulates inhibition of adenosine transport by ethanol. *Molecular Pharmacology* **40**, 812–817.

Nahoum-Grappe, V. (1995). France. *In International Handbook on Alcohol and Culture*, pp. 75–87. Greenwood Press, Westport CT.

National Institute of Diabetes and Digestive and Kidney Diseases, (2004). *Cirrhosis of the Liver* (NIH publication no. 04-1134), National Institute of Diabetes and Digestive and Kidney Diseases, Bethesda MD.

National Institute on Alcohol Abuse and Alcoholism, (1999). Division of Biometry and Epidemiology, *Apparent Per Capita Alcohol Consumption: National, State, and Regional Trends, 1977–1997* [surveillance report no. 51], National Institute on Drug Abuse, Bethesda MD.

National Institute on Alcohol Abuse and Alcoholism, (2005). *Science Report on the Effects of Moderate Drinking*, National Institute on Alcohol Abuse and Alcoholism, Bethesda MD, in press.

National Institute on Alcohol Abuse and Alcoholism, (1996). *State Trends in Alcohol-Related Mortality, 1979–92* (series title: U.S. Alcohol Epidemiologic Data Reference Manual, vol. 5) [NIH publication no. 96-4174], National Institute on Alcohol Abuse and Alcoholism, Bethesda MD.

Nestler, E. J. (2001). Molecular basis of long-term plasticity underlying addiction. *Nature Reviews Neuroscience* **2**, 119–128.

Nestler, E. J. (2004). Historical review: Molecular and cellular mechanisms of opiate and cocaine addiction. *Trends in Pharmacological Sciences* **25**, 210–218.

Newton, P. M., Orr, C. J., Wallace, M. J., Kim, C., Shin, H. S., and Messing, R. O. (2004). Deletion of N-type calcium channels alters ethanol reward and reduces ethanol consumption in mice. *Journal of Neuroscience* **24**, 9862–9869.

Nie, Z., Madamba, S. G., and Siggins, G. R. (1994). Ethanol inhibits glutamatergic neurotransmission in nucleus accumbens neurons by multiple mechanisms. *Journal of Pharmacology and Experimental Therapeutics* **271**, 1566–1573.

Nie, Z., Schweitzer, P., Roberts, A. J., Madamba, S. G., Moore, S. D., and Siggins, G. R. (2004). Ethanol augments GABAergic transmission in the central amygdala via CRF1 receptors. *Science* **303**, 1512–1514.

Nie, Z., Yuan, X., Madamba, S. G., and Siggins, G. R. (1993). Ethanol decreases glutamatergic synaptic transmission in rat nucleus accumbens in vitro: naloxone reversal. *Journal of Pharmacology and Experimental Therapeutics* **266**, 1705–1712.

Norstrom, T. (2002). *Alcohol in Postwar Europe: Consumption, Drinking Patterns, Consequences and Policy Responses in 15 European Countries*. Almqvist and Wiksell, Stockholm.

Nowak, K. L., McBride, W. J., Lumeng, L., Li, T. K., and Murphy, J. M. (2000b). Involvement of dopamine D2 autoreceptors in the ventral tegmental area on alcohol and saccharin intake of the alcohol-preferring P rat. *Alcoholism: Clinical and Experimental Research* **24**, 476–483.

Nyberg, K., and Allebeck, P. Sweden. (1995). *In International Handbook on Alcohol and Culture*, (D. B. Heath, Ed.), pp. 280–288. Greenwood Press, Westport CT.

Nylander, I., Hyytia, P., Forsander, O., and Terenius, L. (1994). Differences between alcohol-preferring (AA) and alcohol-avoiding (ANA) rats in the prodynorphin and proenkephalin systems. *Alcoholism: Clinical and Experimental Research* **18**, 1272–1279.

O'Farrell, T. J., Kleinke, C. L., and Cutter, H. S. (1998). Sexual adjustment of male alcoholics: changes from before to after receiving alcoholism counseling with and without marital therapy. *Addictive Behaviors* **23**, 419–425.

O'Malley, S. S., and Froehlich, J. C. (2003). Advances in the use of naltrexone: an integration of preclinical and clinical findings. *In Research on Alcoholism Treatment* (series title: *Recent Developments in Alcoholism*, vol. 16), (M. Galanter, Ed.), pp. 217–245. Plenum Press, New York.

O'Malley, S. S., Jaffe, A. J., Chang, G., Schottenfeld, R. S., Meyer, R. E., and Rounsaville, B. (1992). Naltrexone and coping skills therapy for alcohol dependence: A controlled study. *Archives of General Psychiatry* **49**, 881–887.

O'Malley, S. S., Krishnan-Sarin, S., Farren, C., Sinha, R., and Kreek, M. J. (2002). Naltrexone decreases craving and alcohol self-administration in alcohol-dependent subjects and activates the hypothalamo-pituitary-adrenocortical axis. *Psychopharmacology* **160**, 19–29.

Ogilvie, K., Lee, S., and Rivier, C. (1997). Effect of three different modes of alcohol administration on the activity of the rat hypothalamic-pituitary-adrenal axis. *Alcoholism: Clinical and Experimental Research* **21**, 467–476.

Ogilvie, K. M., Lee, S., and Rivier, C. (1998). Divergence in the expression of molecular markers of neuronal activation in the parvocellular paraventricular nucleus of the hypothalamus evoked by alcohol administration via different routes. *Journal of Neuroscience* **18**, 4344–4352.

Olive, M. F., Koenig, H. N., Nannini, M. A., and Hodge, C. W. (2002). Elevated extracellular, C. R. F levels in the bed nucleus of the stria terminalis during ethanol withdrawal and reduction by subsequent ethanol intake. *Pharmacology Biochemistry and Behavior* **72**, 213–220.

Olive, M. F., McGeehan, A. J., Kinder, J. R., McMahon, T., Hodge, C. W., Janak, P.H., and Messing, R. O. (2005). The mGluR5 antagonist 6-methyl-2-(phenylethynyl)pyridine decreases ethanol consumption via a protein kinase C epsilon-dependent mechanism. *Molecular Pharmacology* **67**, 349–355.

Olive, M. F., Mehmert, K. K., Messing, R. O., and Hodge, C. W. (2000). Reduced operant ethanol self-administration and in vivo mesolimbic dopamine responses to ethanol in PKC-epsilon-deficient mice. *European Journal of Neuroscience* **12**, 4131–4140.

Olsen, R. W., Chang, C. S., Li, G., Hanchar, H. J., and Wallner, M. (2004). Fishing for allosteric sites on GABA(A) receptors. *Biochemical Pharmacolology* **68**, 1675–1684.

Ortiz, J., Fitzgerald, L. W., Charlton, M., Lane, S., Trevisan, L., Guitart, X., Shoemaker, W., Duman, R. S., and Nestler, E. J. (1995). Biochemical actions of chronic ethanol exposure in the mesolimbic dopamine system. *Synapse* **21**, 289–298.

Osmanovic, S. S., and Shefner, S. A. (1990). Enhancement of current induced by superfusion of GABA in locus coeruleus neurons by pentobarbital, but not ethanol. *Brain Research* **517**, 324–329.

Overstreet, D. H., Kampov-Polevoy, A. B., Rezvani, A. H., Braun, C., Bartus, R. T., and Crews, F. T. (1999). Suppression of alcohol intake by chronic naloxone treatment in P rats: tolerance development and elevation of opiate receptor binding. *Alcoholism: Clinical and Experimental Research* **23**, 1761–1771.

Overstreet, D. H., Knapp, D. J., Moy, S. S., and Breese, G. R. (2003). A 5-HT1A agonist and a 5-HT2c antagonist reduce social interaction deficit induced by multiple ethanol withdrawals in rats. *Psychopharmacology* **167**, 344–352.

Overstreet, D. H., Knapp, D. J., and Breese, G. R. (2004a). Modulation of multiple ethanol withdrawal-induced anxiety-like behavior by CRF and CRF1 receptors. *Pharmacology Biochemistry and Behavior* **77**, 405–413.

Overstreet, D. H., Knapp, D. J., and Breese, G. R. (2004b). Similar anxiety-like responses in male and female rats exposed to repeated withdrawals from ethanol. *Pharmacology Biochemistry and Behavior* **78**, 459–464.

Palmer, A. A., Low, M. J., Grandy, D. K., and Phillips, T. J. (2003). Effects of a Drd2 deletion mutation on ethanol-induced locomotor stimulation and sensitization suggest a role for epistasis. *Behavior Genetics* **33**, 311–324.

Pandey, S. C., (2004). The gene transcription factor cyclic AMP-responsive element binding protein: role in positive and negative affective states of alcohol addiction. *Pharmacology and Therapeutics* **104**, 47–58.

Pandey, S. C., Lumeng, L., and Li, T. K. (1996). Serotonin2C receptors and serotonin2C receptor-mediated phosphoinositide hydrolysis in the brain of alcohol-preferring and alcohol-nonpreferring rats. *Alcoholism: Clinical and Experimental Research* **20**, 1038–1042.

Pandey, S. C., Zhang, D., Mittal, N., and Nayyar, D. (1999). Potential role of the gene transcription factor cyclic AMP-responsive element binding protein in ethanol withdrawal-related anxiety. *Journal of Pharmacology and Experimental Therapeutics* **288**, 866–878.

Pandey, S. C., Roy, A., and Mittal, N., (2001). Effects of chronic ethanol intake and its withdrawal on the expression and phosphorylation of the creb gene transcription factor in rat cortex. *Journal of Pharmacology and Experimental Therapeutics* **296**, 857–868.

Pandey, S. C., Roy, A., and Zhang, H. (2003). The decreased phosphorylation of cyclic adenosine monophosphate (cAMP) response element binding (CREB) protein in the central amygdala acts as a molecular substrate for anxiety related to ethanol withdrawal in rats. *Alcoholism: Clinical and Experimental Research* **27**, 396–409.

Parker, L. F., and Radow, B. L. Effects of parachlorophenylalanine on ethanol self-selection in the rat. *Pharmacology Biochemistry and Behavior* **4**, 535–540.

Partridge, L. D., and Valenzuela, C. F. (2001). Neurosteroid-induced enhancement of glutamate transmission in rat hippocampal slices. *Neuroscience Letters* **301**, 103–106.

Pellow, S., and File, S. E. (1986). Anxiolytic and anxiogenic drug effects on exploratory activity in an elevated plus-maze: a novel test of anxiety in the rat. *Pharmacology, Biochemistry and Behavior* **24**, 525–529.

Pennington, G. W., and Brien, T. (1964). The use of the breathalyzer in the determination of blood alcohol concentrations. *Journal of the Irish Medical Association* **54**, 107–109.

Perkins, H. W. (2002). Surveying the damage: a review of research on consequences of alcohol misuse in college populations. *Journal of Studies on Alcohol Supplement* **14**, 91–100.

Peugh J, and Belenko, S. (2001). Alcohol, drugs and sexual function: a review. *Journal of Psychoactive Drugs* **33**, 223–232.

Pfeffer, A. O., and Samson, H. H. (1986). Effect of pimozide on home cage ethanol drinking in the rat: dependence on drinking session length. *Drug and Alcohol Dependence* **17**, 47–55.

Phillips, T. J., Crabbe, J. C., Metten, P., and Belknap, J. K. (1994). Localization of genes affecting alcohol drinking in mice. *Alcoholism: Clinical and Experimental Research* **18**, 931–941.

Phillips, T. J., Belknap, J. K., Buck, K. J., and Cunningham, C. L. (1998). Genes on mouse chromosomes 2 and 9 determine variation in ethanol consumption. *Mammalian Genome* **9**, 936–941.

Piano, M. R. (2002). Alcoholic cardiomyopathy: incidence, clinical characteristics, and pathophysiology. *Chest* **121**, 1638–1650.

Pinhas, V. (1980). Sex guilt and sexual control on women alcoholics in early sobriety. *Sexuality and Disability* **3**, 256–272.

Pohorecky, L. A. (1981). The interaction of alcohol and stress: a review. *Neuroscience and Biobehavioral Reviews* **5**, 209–229.

Pollard, G. T., and Howard, J. L. (1979). The Geller–Seifter conflict paradigm with incremental shock. *Psychopharmacology* **62**, 117–121.

Poncelet, M., Maruani, J., Calassi, R., and Soubrie, P. (2003). Overeating, alcohol and sucrose consumption decrease in CB1 receptor deleted mice. *Neuroscience Letters* **343**, 216–218.

Popp, R. L., Lickteig, R. L., and Lovinger, D. M. (1999). Factors that enhance ethanol inhibition of N-methyl-D-aspartate receptors in cerebellar granule cells. *Journal of Pharmacology and Experimental Therapeutics* **289**, 1564–1574.

Prescott, C. A., and Kendler, K. S. (1999). Genetic and environmental contributions to alcohol abuse and dependence in a population-based sample of male twins. *American Journal of Psychiatry* **156**, 34–40.

Preston, W. L. (1969). The validity of the 'breathalyzer'. *Medical Journal of Australia* **1**, 286–289.

Pringsheim, J. (1908). Chemische untersuchungen uber das wesen der alkol toleranz [Chemical investigations on the existence of alcohol tolerance] *Biochemische Zeitschrift* **12**, 143–192.

Proctor, W. R., Soldo, B. L., Allan, A. M., and Dunwiddie, T. V. (1992). Ethanol enhances synaptically evoked $GABA_A$ receptor-mediated responses in cerebral cortical neurons in rat brain slices. *Brain Research* **595**, 220–227.

Quertemont, E., and Grant, K. A. (2002). Role of acetaldehyde in the discriminative stimulus effects of ethanol. *Alcoholism: Clinical and Experimental Research* **26**, 812–817.

Quertemont, E., and Grant, K. A. (2004). Discriminative stimulus effects of ethanol: lack of interaction with taurine. *Behavioral Pharmacology* **15**, 495–501.

Quinlan, J. J., Homanics, G. E., and Firestone, L. L. (1998). Anesthesia sensitivity in mice that lack the beta3 subunit of the gamma-aminobutyric acid type A receptor. *Anesthesiology* **88**, 775–780.

Rabin, R. A. (1993). Ethanol-induced desensitization of adenylate cyclase: role of the adenosine receptor and GTP-binding proteins. *Journal of Pharmacology and Experimental Therapeutics* **264**, 977–983.

Randall, R. D., Lee, S. Y., Meyer, J. H., Wittenberg, G. F., and Gruol, D. L. (1995). Acute alcohol blocks neurosteroid modulation of synaptic transmission and long-term potentiation in the rat hippocampal slice. *Brain Research* **701**, 238–248.

Rasmussen, D. D., Boldt, B. M., Bryant, C. A., Mitton, D. R., Larsen, S. A., and Wilkinson, C. W. (2000). Chronic daily ethanol and withdrawal: 1. Long-term changes in the hypothalamo-pituitary-adrenal axis. *Alcoholism: Clinical and Experimental Research* **24**, 1836–1849.

Rasmussen, D. D., Mitton, D. R., Green, J., and Puchalski, S. (2001). Chronic daily ethanol and withdrawal: 2. Behavioral changes during prolonged abstinence. *Alcoholism: Clinical and Experimental Research* **25**, 999–1005.

Rassnick, S., Pulvirenti, L., and Koob, G. F. (1992). Oral ethanol self-administration in rats is reduced by the administration of dopamine and glutamate receptor antagonists into the nucleus accumbens. *Psychopharmacology* **109**, 92–98.

Rassnick, S., D'Amico, E., Riley, E., and Koob, G. F. (1993a). GABA antagonist and benzodiazepine partial inverse agonist reduce motivated responding for ethanol. *Alcoholism: Clinical and Experimental Research* **17**, 124–130.

Rassnick, S., Heinrichs, S. C., Britton, K. T., and Koob, G. F. (1993b). Microinjection of a corticotropin-releasing factor antagonist into the central nucleus of the amygdala reverses anxiogenic-like effects of ethanol withdrawal. *Brain Research* **605**, 25–32.

Rassnick, S., Pulvirenti, L., and Koob, G. F. (1993c) SDZ 205,152, a novel dopamine receptor agonist, reduces oral ethanol self-administration in rats. *Alcohol* **10**, 127–132.

Rassnick, S., Stinus, L., and Koob, G. F. (1993d). The effects of 6-hydroxydopamine lesions of the nucleus accumbens and the mesolimbic dopamine system on oral self-administration of ethanol in the rat. *Brain Research* **623**, 16–24.

Redmond, E. M., Sitzmann, J. V., and Cahill, P. A. (2000). Potential mechanisms for cardiovascular protective effect of ethanol. *Acta Pharmacologica Sinica* **21**, 385–390.

Reich, T., Edenberg, H. J., Goate, A., Williams, J. T., Rice, J. P., Van Eerdewegh, P., Foroud, T., Hesselbrock, V., Schuckit, M. A., Bucholz, K., Porjesz, B., Li, T. K., Conneally, P. M., Nurnberger, J. I. Jr., Tischfield, J. A., Crowe, R. R., Cloninger, C. R., Wu, W., Shears, S., Carr, K., Crose, C., Willig, C., and Begleiter, H. (1998). Genome-wide search for genes affecting the risk for alcohol dependence. *American Journal of Medical Genetics* **81**, 207–215.

Reid, L. D., and Hunter, G. A. (1984). Morphine and naloxone modulate intake of ethanol. *Alcohol* **1**, 33–37.

Richter, C. P., and Campbell, K. H. (1940). Alcohol taste thresholds and concentrations of solution preferred by rats. *Science* **91**, 507–509.

Ripley, T. L., and Little, H. J. (1995). Ethanol withdrawal hyperexcitability in vitro is selectively decreased by a competitive NMDA receptor antagonist. *Brain Research* **699**, 1–11.

Risinger, F. O., Doan, A. M., and Vickrey, A. C. (1999). Oral operant ethanol self-administration in 5-HT1b knockout mice. *Behavioral Brain Research* **102**, 211–215.

Risinger, F. O., Freeman, P. A., Rubinstein, M., Low, M. J., and Grandy, D. K. (2000). Lack of operant ethanol self-administration in dopamine D2 receptor knockout mice. *Psychopharmacology* **152**, 343–350.

Risinger, F. O., Freeman, P. A., Greengard, P., and Fienberg, A. A. (2001). Motivational effects of ethanol in DARPP-32 knock-out mice. *Journal of Neuroscience* **21**, 340–348.

Riveros-Rosas, H., Julian-Sanchez, A., and Pina, E. (1997). Enzymology of ethanol and acetaldehyde metabolism in mammals. *Archives of Medical Research* **28**, 453–471.

Rivier, C., Bruhn, T., and Vale, W. (1984). Effect of ethanol on the hypothalamic-pituitary-adrenal axis in the rat: role of corticotropin-releasing factor (CRF). *Journal of Pharmacology and Experimental Therapeutics* **229**, 127–131.

Roberto, M., Madamba, S. G., Moore, S. D., Tallent, M. K., and Siggins, G. R. (2003). Ethanol increases GABAergic transmission at both pre- and postsynaptic sites in rat central amygdala neurons. *Proceedings of the National Academy of Sciences USA* **100**, 2053–2058.

Roberto, M., Madamba, S. G., Stouffer, D. G., Parsons, L. H., and Siggins, G. R. (2004a). Increased GABA release in the central amygdala of ethanol-dependent rats. *Journal of Neuroscience* **24**, 10159–10166.

Roberto, M., Schweitzer, P., Madamba, S. G., Stouffer, D. G., Parsons, L. H., and Siggins, G. R. (2004b). Acute and chronic ethanol alter glutamatergic transmission in rat central amygdala: an in vitro and in vivo analysis. *Journal of Neuroscience* **24**, 1594-1603.

Roberts, A. J., Cole, M., and Koob, G. F. (1996). Intra-amygdala muscimol decreases operant ethanol self-administration in dependent rats. *Alcoholism: Clinical and Experimental Research* **20**, 1289–1298.

Roberts, A. J., McArthur, R. A., Hull, E. E., Post, C., and Koob, G. F. (1998). Effects of amperozide, 8-OH-DPAT, and FG 5974 on operant responding for ethanol. *Psychopharmacology* **137**, 25–32.

Roberts, A. J., Heyser, C. J., Cole, M., Griffin, P., and Koob, G. F. (2000a). Excessive ethanol drinking following a history of dependence: Animal model of allostasis. *Neuropsychopharmacology* **22**, 581–594.

Roberts, A. J., McDonald, J. S., Heyser, C. J., Kieffer, B. L., Matthes, H. W. D., Koob, G. F., and Gold, L. H. (2000b).

μ-Opioid receptor knockout mice do not self-administer alcohol. *Journal of Pharmacology and Experimental Therapeutics* **293**, 1002–1008.

Rockman, G. E., Amit, Z., Brown, Z. W., Bourque, C., and Ogren, S. O. (1982). An investigation of the mechanisms of action of 5-hydroxytryptamine in the suppression of ethanol intake. *Neuropharmacology* **21**, 341–347.

Rodriguez, L. A., Plomin, R., Blizard, D. A., Jones, B. C., and McClearn, G. E. (1995). Alcohol acceptance, preference, and sensitivity in mice: II. Quantitative trait loci mapping analysis using BXD recombinant inbred strains. *Alcoholism: Clinical and Experimental Research* **19**, 367–373.

Roy, A., and Pandey, S. C. (2002). The decreased cellular expression of neuropeptide Y protein in rat brain structures during ethanol withdrawal after chronic ethanol exposure. *Alcoholism: Clinical and Experimental Research* **26**, 796–803.

Rubin, H. B., and Henson, D. E., (1976). Effects of alcohol on male sexual responding. *Psychopharmacologia* **47**, 123–134.

Salaspuro, M. P. (2003). Alcohol consumption and cancer of the gastrointestinal tract. *Best Practice and Research: Clinical Gastroenterology* **17**, 679–694.

Salimov, R. M., McBride, W. J., Sinclair, J. D., Lumeng, L., and Li, T. (1996). Performance in the cross-maze and slip funnel tests of four pairs of rat lines selectively bred for divergent alcohol drinking behavior. *Addiction Biology* **1**, 273–280.

Sampson, P. D., Streissguth, A. P., Bookstein, F. L., Little, R. E., Clarren, S. K., Dehaene, P., Hanson, J. W., and Graham, J. M. Jr. (1997). Incidence of fetal alcohol syndrome and prevalence of alcohol-related neurodevelopmental disorder. *Teratology* **56**, 317–326.

Samson, H. H. (1987). Initiation of ethanol-maintained behavior: a comparison of animal models and their implication to human drinking. In *Neurobehavioral Pharmacology* (series title: *Advances in Behavioral Pharmacology*, vol. 6), (T. Thompson, P. B. Dews, and J. E. Barrett, Eds.), pp. 221–248. Lawrence Erlbaum, Hillsdale NJ.

Samson, H. H., and Chappell, A. (2001). Muscimol injected into the medial prefrontal cortex of the rat alters ethanol self-administration. *Physiology and Behavior* **74**, 581–587.

Samson, H. H., and Doyle, T. F. (1985). Oral ethanol self-administration in the rat: effect of naloxone. *Pharmacology Biochemistry and Behavior* **22**, 91–99.

Samson, H. H., Tolliver, G. A., Pfeffer, A. O., Sadeghi, K. G., and Mills, F. G. (1987). Oral ethanol reinforcement in the rat: Effect of the partial inverse benzodiazepine agonist RO15-4513. *Pharmacology Biochemistry and Behavior* **27**, 517–519.

Samson, H. H., Hodge, C. W., Tolliver, G. A., and Haraguchi, M. (1993). Effect of dopamine agonists and antagonists on ethanol-reinforced behavior: The involvement of the nucleus accumbens. *Brain Research Bulletin* **30**, 133–141.

Sanna, E., Serra, M., Cossu, A., Colombo, G., Follesa, P., Cuccheddu, T., Concas, A., and Biggio, G. (1993). Chronic ethanol intoxication induces differential effects on GABAA and NMDA receptor function in the rat brain. *Alcoholism: Clinical and Experimental Research* **17**, 115–123.

Sapru, M. K., Diamond, I., and Gordon, A. S. (1994). Adenosine receptors mediate cellular adaptation to ethanol in NG108-15 cells. *Journal of Pharmacology and Experimental Therapeutics* **271**, 542–548.

Sarkola, T., Makisalo, H., Fukunaga, T., and Eriksson, C. J. (1999). Acute effect of alcohol on estradiol, estrone, progesterone, prolactin, cortisol, and luteinizing hormone in premenopausal women. *Alcoholism: Clinical and Experimental Research* **23**, 976–982.

Sarkola, T., Fukunaga, T., Makisalo, H., and Eriksson, C. J. P. (2000). Acute effect of alcohol on androgens in premenopausal women. *Alcohol and Alcoholism* **35**, 84–90.

Savelieva, K. V., Caudle, W. M., Findlay, G. S., Caron, M. G., and Miller, G. W. (2002). Decreased ethanol preference and consumption in dopamine transporter female knock-out mice. *Alcoholism: Clinical and Experimental Research* **26**, 758–764.

Schinka, J. A., Town, T., Abdullah, L., Crawford, F. C., Ordorica, P. I., Francis, E., Hughes, P., Graves, A. B., Mortimer, J. A., and Mullan, M. (2002). A functional polymorphism within the mu-opioid receptor gene and risk for abuse of alcohol and other substances. *Molecular Psychiatry* **7**, 224–228.

Schulteis, G., Markou, A., Cole, M., and Koob, G. (1995). Decreased brain reward produced by ethanol withdrawal. *Proceedings of the National Academy of Sciences USA* **92**, 5880–5884.

Seitz, H. K., Poschl, G., and Simanowski, U. A. (1998). Alcohol and cancer. *In The Consequences of Alcoholism: Medical, Neuropsychiatric, Economic, Cross-Cultural* (series title: *Recent Developments in Alcoholism*, vol. 14), (M. Galanter, Ed.), pp. 67–95. Plenum Press, New York.

Sellers, E. M., Higgins, G. A., and Sobell, M. (1992a). B. 5-HT and alcohol abuse. *Trends in Pharmacological Sciences* **13**, 69–75.

Sellers, E. M., Higgins, G. A., Tompkins, D. M., and Romach, M. K. (1992b). Serotonin and alcohol drinking. *In Problems of Drug Dependence 1991: Proceedings of the 53rd Annual Scientific Meeting, The College on Problems of Drug Dependence, Inc.* (series title: *NIDA Research Monograph*, vol. 119), (L. S. Harris Ed.), pp. 141–145. National Institute on Drug Abuse, Rockville MD.

Sepinwall, J., Cook, L. (1978). Behavioral pharmacology of antianxiety drugs. *In Biology of Mood and Antianxiety Drugs* (series title: *Handbook of Psychopharmacology*, vol. 13), (L. L. Iversen, S. D. Iversen, and S. H. Snyder., Eds.), pp. 345–393. Plenum Press, New York.

Sereny, G., and Kalant, H. (1965). Comparative clinical evaluation of chlordiazepoxide and promazine in treatment of alcohol-withdrawal syndrome. *British Medical Journal* **5427**, 92–97.

Sessoms-Sikes, J. S., Hamilton, M. E., Liu, L. X., Lovinger, D. M., and Machu, T. K. (2003). A mutation in transmembrane domain II of the 5-hydroxytryptamine(3A) receptor stabilizes channel opening and alters alcohol modulatory actions. *Journal of Pharmacology and Experimental Therapeutics* **306**, 595–604.

Shannon, E. E., Shelton, K. L., Vivian, J. A., Yount, I., Morgan, A. R., Homanics, G. E., and Grant, K. A. (2004). Discriminative stimulus effects of ethanol in mice lacking the gamma-aminobutyric acid type A receptor delta subunit. *Alcoholism: Clinical and Experimental Research* **28**, 906–913.

Shefner, S. A. (1990). Electrophysiological effects of ethanol on brain neurons. *In Biochemistry and Physiology of Substance Abuse*, vol. 2, (R. R. Watson, Ed.), pp. 25–53. CRC Press, Boca Raton.

Shelton, K. L., and Grant, K. A. (2002). Discriminative stimulus effects of ethanol in C57BL/6J and DBA/2J inbred mice. *Alcoholism: Clinical and Experimental Research* **26**, 747–757.

Shen, R. Y., and Chiodo, L. A. (1993). Acute withdrawal after repeated ethanol treatment reduces the number of spontaneously active dopaminergic neurons in the ventral tegmental area. *Brain Research* **622**, 289–293.

Shirley, R. L., Walter, N. A., Reilly, M. T., Fehr, C., and Buck, K. J. (2004). Mpdz is a quantitative trait gene for drug withdrawal seizures. *Nature Neuroscience* **7**, 699–700.

Sieghart, W., and Sperk, G. (2002). Subunit composition, distribution and function of GABA(A) receptor subtypes. *Current Topics in Medicinal Chemistry* **2**, 795–816.

Siems, W., Maul, B., Krause, W., Gerard, C., Hauser, K. F., Hersh, L. B., Fischer, H. S., Zernig, G., and Saria, A. (2000). Neutral endopeptidase and alcohol consumption, experiments in neutral endopeptidase-deficient mice. *European Journal of Pharmacology* **397**, 327–334.

Siggins, G. R., Pittman, Q. J., and French, E. D. (1987). Effects of ethanol on CA1 and CA3 pyramidal cells in the hippocampal slice preparation: an intracellular study. *Brain Research* **414**, 22–34.

Siggins, G. R., Nie, Z., and Madamba, S. B. (1999). A metabotropic hypothesis for ethanol sensitivty of GABAergic and glutamatergic central synapses. *In The 'Drunken' Synapse: Studies of Alcohol-Related Disorders*, (Y. Liu, and W. A. Hunt, Eds.), pp. 135–143. Kluwer Academic/Plenum, New York.

Sillaber, I., Rammes, G., Zimmermann, S., Mahal, B., Zieglgansberger, W., Wurst, W., Holsboer, F., and Spanagel, R. (2002). Enhanced and delayed stress-induced alcohol drinking in mice lacking functional CRH1 receptors. *Science* **296**, 931–933.

Simmonds, M. A. (1991). Modulation of the $GABA_A$ receptor by steroids. *Seminars in the Neurosciences* **3**, 231–239.

Simson, P. E., Criswell, H. E., Johnson, K. B., Hicks, R. E., and Breese, G. R. (1991). Ethanol inhibits NMDA-evoked electrophysiological activity in vivo. *Journal of Pharmacology and Experimental Therapeutics* **257**, 225–231.

Simson, P. E., Criswell, H. E., and Breese, G. R. (1993). Inhibition of NMDA-evoked electrophysiological activity by ethanol in selected brain regions: evidence for ethanol-sensitive and ethanol-insensitive NMDA-evoked responses. *Brain Research* **607**, 9–16.

Sinclair, J. D. (1990). Drugs to decrease alcohol drinking. *Annals of Medicine* **22**, 357–362.

Sinclair, J. D., and Senter, R. J. (1967). Increased preference for ethanol in rats following alcohol deprivation. *Psychonomic Science* **8**, 11–12.

Sinclair, J. D., and Senter, R. J. (1968). Development of an alcohol-deprivation effect in rats, *Quarterly Journal of Studies on Alcohol* **29**, 863–867.

Singletary, K. W., and Gapstur, S. M. (2001). Alcohol and breast cancer: review of epidemiologic and experimental evidence and potential mechanisms. *Journal of the American Medical Association* **286**, 2143–2151.

Slawecki, C. J., Somes, C., and Ehlers, C. L. (1999). Effects of chronic ethanol exposure on neurophysiological responses to corticotropin-releasing factor and neuropeptide Y. *Alcohol and Alcoholism* **34**, 289–299.

Slawecki, C. J., Jimenez-Vasquez, P., Mathe, A. A., and Ehlers, C. L. (2001). Substance, P. and neurokinin levels are decreased in the cortex and hypothalamus of alcohol-preferring (P) rats. *Journal of Studies on Alcohol* **62**, 736–740.

Snell, L. D., Tabakoff, B., and Hoffman, P. L. (1993). Radioligand binding to the N-methyl-D-aspartate receptor/ionophore complex: alterations by ethanol in vitro and by chronic in vivo ethanol ingestion. *Brain Research* **602**, 91–98.

Soini, S. L., Ovaska, T., Honkanen, A., Hyytia, P., and Korpi, E. R. (1998). Brain opioid receptor binding of [3H]CTOP and [3H]naltrindole in alcohol-preferring AA and alcohol-avoiding ANA rats. *Alcohol* **15**, 227–232.

Sokol, R. J., Delaney-Black, V., and Nordstrom, B. (2003). Fetal alcohol spectrum disorder. *Journal of the American Medical Association.* **290**, 2996–2999.

Soldo, B. L., Proctor, W. R., and Dunwiddie, T. V. (1994). Ethanol differentially modulates GABAA receptor-mediated chloride currents in hippocampal, cortical, and septal neurons in rat brain slices. *Synapse* **18**, 94–103.

Soldo, B. L., Proctor, W. R., and Dunwiddie, T. V. (1998). Ethanol selectively enhances the hyperpolarizing component of neocortical neuronal responses to locally applied GABA. *Brain Research* **800**, 187–197.

Sood, B., Delaney-Black, V., Covington, C., Nordstrom-Klee, B., Ager, J., Templin, T., Janisse, J., Martier, S., and Sokol, R. J. (2001). Prenatal alcohol exposure and childhood behavior at age 6 to 7 years: I. Dose-response effect. *Pediatrics* **108**, E34.

Spanagel, R., Hölter, S. M., Allingham, K., Landgraf, R., and Zieglgänsberger, W. (1996). Acamprosate and alcohol: I. Effects on alcohol intake following alcohol deprivation in the rat. *European Journal of Pharmacology* **305**, 39–44.

Spanagel, R., Montkowski, A., Allingham, K., Stohr, T., Shoaib, M., Holsboer, F., and Landgraf, R. (1995). Anxiety: a potential predictor of vulnerability to the initiation of ethanol self-administration in rats. *Psychopharmacology* **122**, 369–373.

Spanagel, R., Pendyala, G., Abarca, C., Zghoul, T., Sanchis-Segura, C., Magnone, M. C., Lascorz, J., Depner, M., Holzberg, D., Soyka, M., Schreiber, S., Matsuda, F., Lathrop, M., Schumann, G., and Albrecht, U. (2005). The clock gene Per2 influences the glutamatergic system and modulates alcohol consumption. *Nature Medicine* **11**, 35–42.

Spanagel, R., Siegmund, S., Cowen, M., Schroff, K. C., Schumann, G., Fiserova, M., Sillaber, I., Wellek, S., Singer, M., and Putzke, J. (2002). The neuronal nitric oxide synthase gene is critically involved in neurobehavioral effects of alcohol. *Journal of Neuroscience* **22**, 8676–8683.

Speisky, M. B., and Kalant, H. (1985). Site of interaction of serotonin and desglycinamide-arginine-vasopressin in maintenance of ethanol tolerance. *Brain Research* **326**, 281–290.

Spillantini, M. G., Crowther, R. A., Jakes, R., Hasegawa, M., and Goedert, M. (1998). Alpha-Synuclein in filamentous inclusions of Lewy bodies from Parkinson's disease and dementia with lewy bodies. *Proceedings of the National Academy of Sciences USA* **95**, 6469–6473.

Stefanini, E., Frau, M., Garau, M. G., Garau, B., Fadda, F., and Gessa, G. L. (1992). Alcohol-preferring rats have fewer dopamine D2 receptors in the limbic system. *Alcohol and Alcoholism* **27**, 127–130.

Steffensen, S. C., Svingos, A. L., Pickel, V. M., and Henriksen, S. J. (1998). Electrophysiological characterization of GABAergic neurons in the ventral tegmental area. *Journal of Neuroscience* **18**, 8003–8815.

Stephens, D. N. (1995). A glutamatergic hypothesis of drug dependence: extrapolations from benzodiazepine receptor ligands. *Behavioral Pharmacology* **6**, 425–446.

Stewart, R. B., Gatto, G. J., Lumeng, L., Li, T. K., and Murphy, J. M. (1993). Comparison of alcohol-preferring (P) and -nonpreferring (NP) rats on tests of anxiety and for the anxiolytic effects of ethanol. *Alcohol* **10**, 1–10.

Strother, W. N., Chernet, E. J., Lumeng, L., Li, T. K., and McBride, W. J. (2001). Regional central nervous system densities of delta-opioid receptors in alcohol-preferring P, alcohol-nonpreferring NP, and unselected Wistar rats. *Alcohol* **25**, 31–38.

Substance Abuse and Mental Health Services Administration. (2000). *National Household Survey on Drug Abuse: Main Findings 1998* (DHHS publication no. [SMA] 00-3381), U.S. Department of Health and Human Services, Rockville MD.

Substance Abuse and Mental Health Services Administration. (2003a). *Emergency Department Trends from the Drug Abuse Warning Network: Final Estimates 1995–2002* (Office of Applied Statistics, DAWN Series D-24, DHHS Publication No. SMA 03-3780), Rockville MD.

Substance Abuse and Mental Health Services Administration. (2003b). *Results from the 2002 National Survey on Drug Use and Health: National Findings* (Office of Applied Studies, NHSDA Series H-22, DHHS Publication No. SMA 03–3836), Rockville MD.

Sundstrom-Poromaa, I., Smith, D. H., Gong, Q. H., Sabado, T. N., Li, X., Light, A., Wiedmann, M., Williams, K., and Smith, S.S. (2002). Hormonally regulated alpha-4-beta-2-delta GABA-A receptors are a target for alcohol. *Nature Neuroscience* **5**, 721–722.

Suzdak, P. D., Glowa, J. R., Crawley, J. N., Schwartz, R. D., Skolnick, P., and Paul, S. M. (1986a). A selective imidazobenzodiazepine antagonist of ethanol in the rat. *Science* **234**, 1243–1247.

Suzdak, P. D., Schwartz, R. D., Skolnick, P., Paul, S. M. (1986b). Ethanol stimulates gamma-aminobutyric acid receptor-mediated chloride transport in rat brain synaptoneurosomes, *Proceedings of the National Academy of Sciences USA* **83**, 4071–4075.

Suzdak, P. D., Schwartz, R. D., Skolnick, P., and Paul, S. M. (1988). Alcohols stimulate gamma-aminobutyric acid receptor-mediated chloride uptake in brain vesicles: correlation with intoxication potency. *Brain Research* **444**, 340–345.

Svensson, L., Engel, J., and Hard, E. (1989). Effects of the 5-HT receptor agonist, 8-OH-DPAT, on ethanol preference in the rat. *Alcohol* **6**, 17–21.

Syvalahti, E. K., Pohjalainen, T., Korpi, E. R., Palvimaki, E. P., Ovaska, T., Kuoppamaki, M., and Hietala, J. (1994). Dopamine D2 receptor gene expression in rat lines selected for differences in voluntary alcohol consumption. *Alcoholism: Clinical and Experimental Research* **18**, 1029–1031.

Szeliga, K. T., and Grant, K. A. (1998). Analysis of the 5-HT2 receptor ligands dimethoxy-4-indophenyl-2-aminopropane and ketanserin in ethanol discriminations. *Alcoholism: Clinical and Experimental Research* **22**, 646–651.

Szumlinski, K. K., Toda, S., Middaugh, L. D., Worley, P. F., and Kalivas, P. W. (2003). Evidence for a relationship between Group 1 mGluR hypofunction and increased cocaine and ethanol sensitivity in Homer2 null mutant mice. *In Glutamate and Disorders of Cognition and Motivation* (series title: *Annals of the New York Academy of Sciences*, vol. 1003), (B. Moghaddam and M.E. Wolf, Eds.), pp. 468–471. New York Academy of Sciences, New York.

Tabakoff, B. (1980). Alcohol tolerance in humans and animals. *In Animal Models in Alcohol Research*, (K. Eriksson, J. D. Sinclair, and K. Kiianmaa, Eds.), pp. 271–292. Academic Press, London.

Tabakoff, B., and Hoffman, P. L. (1992). Alcohol: neurobiology. *In Substance Abuse: A Comprehensive Textbook*, (J. H. Lowinson, P. Ruiz, and R. B. Millman, Eds.), pp. 152–185. 2nd ed., Williams and Wilkins, Baltimore.

Tabakoff, B., Melchior, C. L., and Hoffman, P. L. (1982). Commentary on ethanol tolerance. *Alcoholism: Clinical and Experimental Research* **6**, 252–259.

Thanos, P. K., Rivera, S. N., Weaver, K., Grandy, D. K., Rubinstein, M., Umegaki, H., Wang, G. J., Hitzemann, R., and Volkow, N. D. (2005). Dopamine D_2R DNA transfer in dopamine D_2 receptor-deficient mice: Effects on ethanol drinking. *Life Sciences* **77**, 130–139.

Thatcher-Britton, K., and Koob, G. F. (1986). Alcohol reverses the proconflict effect of corticotropin-releasing factor. *Regulatory Peptides* **16**, 315–320.

Thiele, T. E., Koh, M. T., and Pedrazzini, T. (2002). Voluntary alcohol consumption is controlled via the neuropeptide Y Y1 receptor. *Journal of Neuroscience* **22**, RC208.

Thiele, T. E., Marsh, D. J., St. Marie, L., Bernstein, I. L., and Palmiter, R. D. (1998). Ethanol consumption and resistance are inversely related to neuropeptide Y levels. *Nature* **396**, 366–369.

Thiele, T. E., Naveilhan, P., and Ernfors, P. (2004). Assessment of ethanol consumption and water drinking by NPY Y(2) receptor knockout mice. *Peptides* **25**, 975–983.

Thiele, T. E., Willis, B., Stadler, J., Reynolds, J. G., Bernstein, I. L., and McKnight, G. S. (2000). High ethanol consumption and low sensitivity to ethanol-induced sedation in protein kinase A-mutant mice. *Journal of Neuroscience* **20**, RC75.

Thielen, R. J., McBride, W. J., Chernet, E., Lumeng, L., and Li, T. K. (1997). Regional densities of benzodiazepine sites in the CNS of alcohol–naive P and NP rats. *Pharmacology Biochemistry and Behavior* **57**, 875–882.

Thielen, R. J., McBride, W. J., Lumeng, L., and Li, T. K. (1993). Housing conditions alter GABAA receptor of alcohol-preferring and -nonpreferring rats. *Pharmacology Biochemistry and Behavior* **46**, 723–727.

Thielen, R. J., McBride, W. J., Lumeng, L., and Li, T. K. (1998). GABA(A) receptor function in the cerebral cortex of alcohol-naive P and NP rats. *Pharmacology Biochemistry and Behavior* **59**, 209–214.

Thomson, A. D. (2000). Mechanisms of vitamin deficiency in chronic alcohol misusers and the development of the Wernicke-Korsakoff syndrome. *Alcohol and Alcoholism Supplement* **35** (Suppl. 1), 2–7.

Thorsell, A., Rimondini, R., and Heilig, M. (2002). Blockade of central neuropeptide Y (NPY) Y2 receptors reduces ethanol self-administration in rats. *Neuroscience Letters* **332**, 1–4.

Ticku, M. K., Lowrimore, P., and Lehoullier, P. (1986). Ethanol enhances GABA-induced 36C1-influx in primary spinal cord cultured neurons. *Brain Research Bulletin* **17**, 123–126.

Tomkins, D. M., Higgins, G. A., and Sellers, E. M. (1994). Low doses of the 5-HT1A agonist 8-hydroxy-2-(di-n-propylamino)-tetralin (8-OH DPAT) increase ethanol intake. *Psychopharmacology* **115**, 173-179.

Tomkins, D. M., Joharchi, N., Tampakeras, M., Martin, J. R., Wichmann, J., and Higgins, G. A. (2002). An investigation of the role of 5-HT(2C) receptors in modifying ethanol self-administration behaviour. *Pharmacology Biochemistry and Behavior* **71**, 735-744.

Tomkins, D. M., Le, A. D., and Sellers, E. M. (1995). Effect of the 5-HT3 antagonist ondansetron on voluntary ethanol intake in rats and mice maintained on a limited access procedure. *Psychopharmacology* **117**, 479–485.

Tomkins, D. M., and O'Neill, M. F. (2000). Effect of 5-HT(1B) receptor ligands on self-administration of ethanol in an operant procedure in rats. *Pharmacology Biochemistry and Behavior* **66**, 129–136.

Trevisan, L., Fitzgerald, L. W., Brose, N., Gasic, G. P., Heinemann, S. F., Duman, R. S., and Nestler, E. J. (1994). Chronic ingestion of ethanol up-regulates NMDAR1 receptor subunit immunoreactivity in rat hippocampus. *Journal of Neurochemistry* **62**, 1635–1638.

Trudell, J. R., and Harris, R. A. (2004). Are sobriety and consciousness determined by water in protein cavities? *Alcoholism: Clinical and Experimental Research* **28**, 1–3.

Tsutsumi, M., Lasker, J. M., Shimizu, M., Rosman, A. S., and Lieber, C. S. (1989). The intralobular distribution of ethanol-inducible P450IIE1 in rat and human liver. *Hepatology* **10**, 437–446.

Valdez, G. R., Roberts, A. J., Chan, K., Davis, H., Brennan, M., Zorrilla, E. P., and Koob, G. F. (2002). Increased ethanol self-administration and anxiety-like behavior during acute withdrawal and protracted abstinence: regulation by corticotropin-releasing factor. *Alcoholism: Clinical and Experimental Research* **26**, 1494–1501.

Valdez, G. R., Zorrilla, E. P., Roberts, A. J., and Koob, G. F. (2003). Antagonism of corticotropin-releasing factor attenuates the enhanced responsiveness to stress observed during protracted ethanol abstinence. *Alcohol* **29**, 55–60.

VanDoren, M. J., Matthews, D. B., Janis, G. C., Grobin, A. C., Devaud, L. L., and Morrow, A. L. (2000). Neuroactive steroid 3alpha-hydroxy-5alpha-pregnan-20-one modulates electrophysiological and behavioral actions of ethanol. *Journal of Neuroscience* **20**, 1982–1989.

Victor, M., Adams, R. D., and Collins, G. H. (1971). The Wernicke-Korsakoff syndrome. A clinical and pathological study of 245 patients, 82 with post-mortem examinations. *Contemporary Neurology Series* **7**, 1–206.

Victor, M., Adams, R. D., and Collins, G. H. (1989). *The Wernicke-Korsakoff Syndrome and Related Neurologic Disorders due to Alcoholism and Malnutrition*, 2nd edition, FA Davis, Philadelphia.

Virkkunen, M., and Linnoila, M. (1993). Brain serotonin, type II alcoholism and impulsive violence. *Journal of Studies on Alcoholism Supplement* **11**, 163–169.

Virkkunen, M., Kallio, E., Rawlings, R., Tokola, R., Poland, R. E., Guidotti, A., Nemeroff, C., Bissette, G., Kalogeras, K., Karonen, S. L., and Linnoila, M. (1994). Personality profiles and state aggressiveness in Finnish alcoholic, violent offenders, fire setters, and healthy volunteers. *Archives of General Psychiatry* **51**, 28–33.

Vogel, J. R., Beer, B., and Clody, D. E. (1971). A simple and reliable conflict procedure for testing anti-anxiety agents. *Psychopharmacologia* **21**, 1–7.

Volpicelli, J. R., Alterman, A. I., Hayashida, M., and O'Brien, C. P. (1992). Naltrexone in the treatment of alcohol dependence. *Archives of General Psychiatry* **49**, 876–880.

Wafford, K. A., Burnett, D. M., Leidenheimer, N. J., Burt, D. R., Wang, J. B., Kofuji, P., Dunwiddie, T. V., Harris, R. A., and Sikela, J. M. (1991). Ethanol sensitivity of the GABAA receptor expressed in Xenopus oocytes requires 8 amino acids contained in the gamma 2L subunit. *Neuron* **7**, 27–33.

Wafford, K. A., and Whiting, P. J. (1992). Ethanol potentiation of GABAA receptors requires phosphorylation of the alternatively spliced variant of the gamma 2 subunit. *FEBS Letters* **313**, 113–117.

Wall, T. L., Gallen, C. C., and Ehlers, C. L. (1993). Effects of alcohol on the EEG in Asian men with genetic variations of ALDH2. *Biological Psychiatry* **34**, 91–99.

Waller, M. B., McBride, W. J., Gatto, G. J., Lumeng, L., and Li, T. K. (1984). Intragastric self-infusion of ethanol by ethanol-preferring and -nonpreferring lines of rats. *Science* **225**, 78–80.

Wallner, M., Hanchar, H. J., and Olsen, R. W. (2003). Ethanol enhances alpha 4 beta 3 delta and alpha 6 beta 3 delta gamma-aminobutyric acid type A receptors at low concentrations known to affect humans. *Proceedings of the National Academy of Sciences USA* **100**, 15218–15223.

Walter, H. J., McMahon, T., Dadgar, J., Wang, D., and Messing, R. O. (2000). Ethanol regulates calcium channel subunits by protein kinase C delta-dependent and -independent mechanisms. *Journal of Biological Chemistry* **275**, 25717–25722.

Walters, J. K. (1977). Effects of, P. C. P. A. on the consumption of alcohol, water and other solutions. *Pharmacology Biochemistry and Behavior* **6**, 377–383.

Wan, F. J., Berton, F., Madamba, S. G., Francesconi, W., and Siggins, G. R. (1996). Low ethanol concentrations enhance GABAergic inhibitory postsynaptic potentials in hippocampal pyramidal neurons only after block of GABAB receptors. *Proceedings of the National Academy of Sciences USA* **93**, 5049–5054.

Wand, G. S., and Dobs, A. S. (1991). Alterations in the hypothalamic-pituitary-adrenal axis in actively drinking alcoholics. *Journal of Clinical Endocrinology and Metabolism* **72**, 1290–1295.

Wand, G. S., and Levine, M. A. (1991). Hormonal tolerance to ethanol is associated with decreased expression of the GTP-binding protein, Gs alpha, and adenylyl cyclase activity in ethanol-treated LS mice. *Alcoholism: Clinical and Experimental Research* **15**, 705–710.

Wand, G. S., McCaul, M., Yang, X., Reynolds, J., Gotjen, D., Lee, S., and Ali, A. (2002). The mu-opioid receptor gene polymorphism (A118G) alters HPA axis activation induced by opioid receptor blockade. *Neuropsychopharmacology* **26**, 106–114.

Wang, L., Liu, J., Harvey-White, J., Zimmer, A., and Kunos, G. (2003). Endocannabinoid signaling via cannabinoid receptor 1 is involved in ethanol preference and its age-dependent decline in mice. *Proceedings of the National Academy of Sciences USA* **100**, 1393–1398.

Weathermon, R., and Crabb, D. W. (1999). Alcohol and medication interactions. *Alcohol Research and Health* **23**, 40–54.

Wegelius, K., Honkanen, A., and Korpi, E. R. (1994). Benzodiazepine receptor ligands modulate ethanol drinking in alcohol-preferring rats. *European Journal of Pharmacology* **263**, 141–147.

Weight, F. F. (1992). Cellular and molecular physiology of alcohol actions in the nervous system. *International Review of Neurobiology* **33**, 289–348.

Weiner, J. L., Zhang, L., and Carlen, P. L. (1994). Potentiation of GABAA-mediated synaptic current by ethanol in hippocampal CA1 neurons: possible role of protein kinase C. *Journal of Pharmacology and Experimental Therapeutics* **268**, 1388–1395.

Weiner, J. L., Gu, C., and Dunwiddie, T. V. (1997). Differential ethanol sensitivity of subpopulations of GABAA synapses onto rat hippocampal CA1 pyramidal neurons. *Journal of Neurophysiology* **77**, 1306–1312.

Weiner, J. L., Dunwiddie, T. V., and Valenzuela, C. F. (1999). Ethanol inhibition of synaptically evoked kainate responses in rat hippocampal CA3 pyramidal neurons. *Molecular Pharmacology* **56**, 85–90.

Weinshenker, D., Rust, N. C., Miller, N. S., and Palmiter, R. D. (2000). Ethanol-associated behaviors of mice lacking norepinephrine. *Journal of Neuroscience* **20**, 3157–3164.

Weiss, F., Mitchiner, M., Bloom, F. E., and Koob, G. F. (1990). Free-choice responding for ethanol versus water in alcohol preferring (P) and unselected Wistar rats is differentially modified by naloxone, bromocriptine, and methysergide. *Psychopharmacology* **101**, 178–186.

Weiss, F., Lorang, M. T., Bloom, F. E., and Koob, G. F. (1993). Oral alcohol self-administration stimulates dopamine release in the rat nucleus accumbens: genetic and motivational determinants. *Journal of Pharmacology and Experimental Therapeutics* **267**, 250–258.

Weiss, F., Parsons, L. H., Schulteis, G., Hyytia, P., Lorang, M. T., Bloom, F. E., and Koob, G. F. (1996). Ethanol self-administration restores withdrawal-associated deficiencies in accumbal dopamine and 5-hydroxytryptamine release in dependent rats. *Journal of Neuroscience* **16**, 3474–3485.

Wenger, J. R., Tiffany, T. M., Bombardier, C., Nicholls, K., and Woods, S. C. (1981). Ethanol tolerance in the rat is learned. *Science* **213**, 575–577.

Wetsel, W. C., Khan, W. A., Merchenthaler, I., Rivera, H., Halpern, A. E., Phung, H. M., Negro-Vilar, A., and Hannun, Y. A. (1992). Tissue and cellular distribution of the extended family of protein kinase C isoenzymes. *Journal of Cellular Biology* **117**, 121–133.

White, G., Lovinger, D. M., and Weight, F. F. (1990). Ethanol inhibits NMDA-activated current but does not alter GABA-activated current in an isolated adult mammalian neuron. *Brain Research* **507**, 332–336.

Wilsnack, S. C., Klassen, A. D., and Wilsnack, R. W. (1984). Drinking and reproductive dysfunction among women in a 1981 national survey. *Alcoholism: Clinical and Experimental Research* **8**, 451–458.

Wilson, G. T., and Lawson, D. M. (1976). Expectancies, alcohol, and sexual arousal in male social drinkers. *Journal of Abnormal Psychology* **85**, 587–594.

Wirkner, K., Eberts, C., Poelchen, W., Allgaier, C., and Illes, P. (2000). Mechanism of inhibition by ethanol of NMDA and AMPA receptor channel functions in cultured rat cortical neurons. *Naunyn Schmiedebergs Archives of Pharmacology* **362**, 568–576.

Wolffgramm, J., and Heyne, A. (1995). From controlled drug intake to loss of control: the irreversible development of drug addiction in the rat. *Behavioural Brain Research* **70**, 77–94.

Wong, G., Ovaska, T., and Korpi, E. R. (1996). Brain regional pharmacology of GABA(A) receptors in alcohol-preferring AA and alcohol-avoiding ANA rats. *Addiction Biology* **1**, 263–272.

World Drink Trends: International Beverage Alcohol Consumption and Production Trends. (2002). Produktschap voor Gedistilleerde Dranken, Henley-on-Thames, Oxfordshire, .

Wright, H. I., Gabaler, J. S., and Van Thiel, D. (1991). Effects of alcohol on the male reproductive system. *Alcohol Health and Research World* **15**, 110–114.

Wright, J. M., Peoples, R. W., and Weight, F. F. (1996). Single-channel and whole-cell analysis of ethanol inhibition of NMDA-activated currents in cultured mouse cortical and hippocampal neurons. *Brain Research* **738**, 249–256.

Wu, P. H., Mihic, S. J., Liu, J. F., Le, A. D., and Kalant, H. (1993). Blockade of chronic tolerance to ethanol by the NMDA antagonist, (+)-MK-801. *European Journal of Pharmacology* **231**, 157–164.

Xiao, J. (1995). China. *In International Handbook on Alcohol and Culture*, (D. B. Heath, Ed.), pp. 42–50. Greenwood Press, Westport CT.

Yang, X., Criswell, H. E., Simson, P., Moy, S., and Breese, G. R. (1996). Evidence for a selective effect of ethanol on N-methyl-d-aspartate responses: ethanol affects a subtype of the ifenprodil-sensitive N-methyl-d-aspartate receptors. *Journal of Pharmacology and Experimental Therapeutics* **278**, 114–124.

Yao, L., Arolfo, M. P., Dohrman, D. P., Jiang, Z., Fan, P., Fuchs, S., Janak, P. H., Gordon, A. S., and Diamond, I. (2002). Betagamma Dimers mediate synergy of dopamine D2 and adenosine A2 receptor-stimulated PKA signaling and regulate ethanol consumption. *Cell* **109**, 733–743.

Yi, H., Williams, G. D., Dufour, M. C. (2000). *Trends in Alcohol-Related Fatal Traffic Crashes,* United National Institute on Alcohol Abuse and Alcoholism, *10th Special Report to the U.S. Congress on Alcohol and Health: Highlights from Current Research*, National Institute on Alcohol Abuse and Alcoholism, Bethesda MD.

Yoon, Y. H., Yi, H., Grant, B. F., Stinson, F. S., Dufour, M. C. (2003). *Liver Cirrhosis Mortality in the United States, 1970–2000* (series title: *Surveillance Report*, vol. 63), National Institute on Alcohol Abuse and Alcoholism, Bethesda MD.

Zabik, J. E., Roache, J. D. (1983). 5-hydroxytryptophan-induced conditioned taste aversion to ethanol in the rat. *Pharmacology Biochemistry and Behavior* **18**, 785–790.

Zhang, L., Hosoi, M., Fukuzawa, M., Sun, H., Rawlings, R. R., and Weight, F. F. (2002). Distinct molecular basis for differential sensitivity of the serotonin type 3A receptor to ethanol in the absence and presence of agonist. *Journal of Biological Chemistry* **277**, 46256–46264.

Zhou, F. C., Bledsoe, S., Lumeng, L., and Li, T. K. (1991a). Immunostained serotonergic fibers are decreased in selected brain regions of alcohol-preferring rats. *Alcohol* **8**, 425–431.

Zhou, F. C., Pu, C. F., Lumeng, L., and Li, T. K. (1991b). Fewer number of immunostained serotonergic neurons in raphe of alcohol-preferring rats. *Alcoholism: Clinical and Experimental Research* **15**, 315.

Zhou, F. C., Zhang, J. K., Lumeng, L., and Li, T. K. (1995). Mesolimbic dopamine system in alcohol-preferring rats. *Alcohol* **12**, 403–412.

Zhou, Y., Franck, J., Spangler, R., Maggos, C. E., Ho, A., and Kreek, M. J. (2000). Reduced hypothalamic POMC and anterior pituitary CRF1 receptor mRNA levels after acute, but not chronic, daily 'binge' intragastric alcohol administration. *Alcoholism: Clinical and Experimental Research* **24**, 1575–1582.

Zimmerman, R. S., Warheit, G. J., Ulbrich, P. M., and Auth, J. B. (1990). The relationship between alcohol use and attempts and success at smoking cessation. *Addictive Behaviors* **15**, 197–207.

CHAPTER

6

Nicotine

OUTLINE

DEFINITIONS

Tobacco can be defined as the dried leaves of the cultivated plant *Nicotiana tabacum*, a native of North and South America (*Nicotiana rustica*), and a plant that is a member of the *Solanaceae* (nightshade) family. A wide variety of plants all native to North America were either combined with tobacco or had nicotine-like substances, but the only two widely cultivated were *Nicotiana tabacum* (common tobacco) and *Nicotiana rustica* (Aztec tobacco) (Bentley and Trimen, 1880) (**Fig. 6.1**). The derivation of the word *tobacco* comes from the West Indian (Caribbean) word *tabaco* (Oviedo, 1526), and Spanish *tobaco* (*tobago* or *tobah*) (Fairholt, 1859), which was the pipe or tube in which the Indians smoked the plant. The name was transferred by the Spaniards to the plant itself (Leach, 1972). It is widely used in various products that can be smoked, such as cigars, pipes, and cigarettes, or administered in the nasal or oral cavities, such as with snuff or by chewing tobacco. There are over 4000 chemicals in cigarette smoke, many of which could potentially contribute to the addictive properties of tobacco. However, the consensus is that individuals smoke primarily to experience the acute psychopharmacological properties of nicotine, and that nicotine is a major component in tobacco smoke responsible for addiction (Balfour, 1984; Stolerman, 1991; Stolerman and Jarvis, 1995). Nicotine derives its name from the botanical name *Nicotiana*, which in turn was derived from the name of Jean Nicot de Villemain, the French ambassador to Portugal who introduced tobacco to the French court. Nicot brought tobacco powder via Portugal to Queen Catherine de Medicis after the death of Henri II in 1561 (Pieyre, 1886; Haug, 1961). Catherine de Medicis appreciated the beneficial pleasurable effects of the *poudre américaine* ('American powder'). She developed a taste for it, became an enthusiast, and ensured its popularity first inside, then outside the court. Nicotine was isolated by Posselt and Reimann in 1828 (Investigations of tobacco—nicotia, 1829). It is a highly toxic alkaloid that is derived from

Neurobiology of Addiction, by George F. Koob and Michel Le Moal.
ISBN – 13: 978-0-12-419239-3 ISBN – 10: 0-12-419239-4

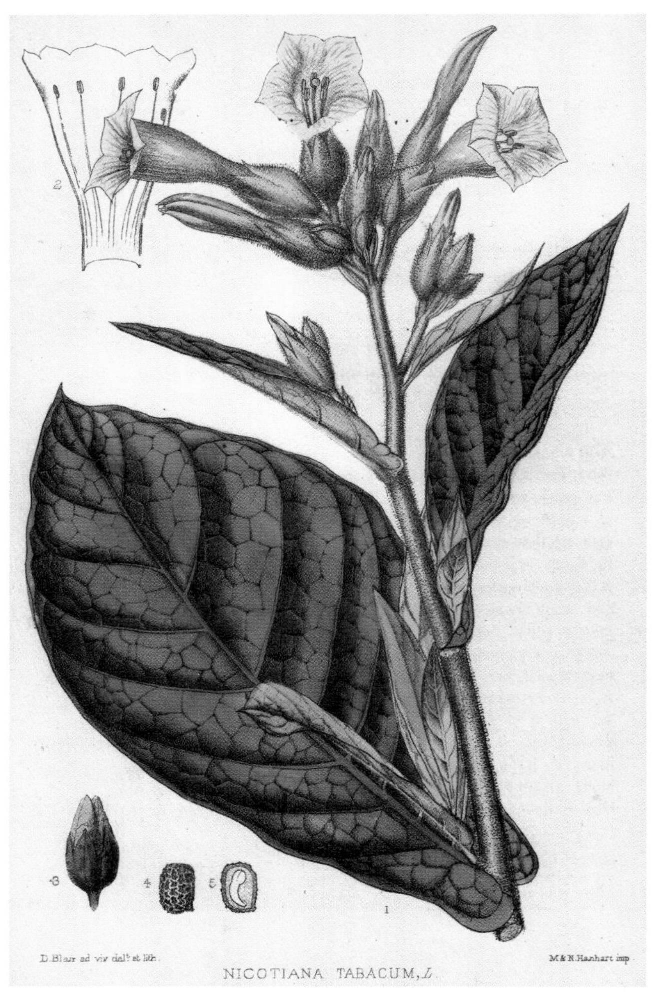

FIGURE 6.1 *Nicotiana tabacum*. Parts of the plant: 1—summit of stem with inflorescence; 2—corolla split open; 3—capsule with persistent calyx; 4—a seed; 5—section of the same (4 and 5 are greatly enlarged). [Reproduced with permission from Bentley and Trimen, 1880.]

tobacco. It is a water soluble, colorless, bitter-tasting in the liquid form, weak base with a pKa of 8.5. Nicotine is not nicotinic acid which is the fat-soluble vitamin B-3, called niacin, used in the treatment of pellagra, a niacin deficiency disorder characterized by cutaneous, gastrointestinal, neurologic, and mental symptoms. Tobacco smoke contains not only nicotine but also carbon monoxide and tar. 'Tar' is defined as what remains after moisture and nicotine are removed and consists largely of aromatic hydrocarbons, many of which are carcinogens (Day, 1967; Hoffman, 1967, 1979; Dalhamn, 1972; Lazar *et al.*, 1974; Battista, 1976; Wynder and Hecht, 1976; Wynder and Hoffman *et al.*, 1976; Hoffman *et al.*, 1979) (**Table 6.1**).

HISTORY OF TOBACCO USE, ABUSE, AND ADDICTION

Tobacco use and cigarette smoking are one of the most popular and persistent forms of drug taking of our modern age (**Fig. 6.2**). The names ascribed to the plant itself varied greatly from culture to culture, including *apooke* in Virginia, *yetl* by the Aztecs, *oyngona* by the Huron, *sayri* by the Peruvians, *kohiha* in the Caribbean, and *cogiaba* or *cohiba* by the Spanish (Fairholt, 1859; Billings, 1875). The use of *Nicotiana tabacum* by indigenous peoples of North and South America can be traced back archeologically and ethnopharmacologically 8000 years (Wilbert, 1987, 1991), and it was used both for medicinal and ceremonial purposes (de Bry and Le Moyne de Morgues, 1591) (**Fig. 6.3**). Most historians contend that European explorers, such as Columbus in 1492 (Billings, 1875; Akehurst, 1968), were the first to record the practice of smoking the dried leaves of the tobacco plant. European travelers to Mexico noted the medical uses of tobacco by the natives:

> 'In this country, *tabaco* cures pain caused by cold; taken in smoke it is beneficial against colds, asthma and coughs; Indians and Negroes use it in powder in their mouths in order to fall asleep and feel no pain' (Stewart, 1967).

Tobacco has been used to prevent fatigue, whiten teeth, treat abscesses, heal wounds, purge nasal passages, relieve thirst, treat syphilis, etc. (Stewart, 1967). English explorers first were made aware of the existence of the plant in Florida in 1565 (Billings, 1875). The plant proliferated to other countries such as India, Portugal, Japan, and Turkey. The first American colonial commercial crop was grown for export in Jamestown, Virginia, in 1612 by John Rolfe, husband of Pocahontas (a benevolent Algonquian Native American who helped save the Jamestown colony by supplying it with food during its hard times). Cultivation extended to Maryland in 1631, and Virginia and Maryland were the main producers through the 1700s. By 1630, over one and a half million pounds of tobacco were being exported from Jamestown every year (Akehurst, 1968). Tobacco now is grown in 120 countries throughout the world and in the United States, including Maryland, Tennessee, and Kentucky.

Tobacco was introduced in Europe in the sixteenth century, and its use has survived significant historical penalties in attempts at prohibition. Smoking tobacco was considered both pleasurable and also a cure for ailments. Tobacco ingestion has fluctuated between smoking, chewing, and snuffing, but one method has often been replaced with another such as when in the early eighteenth century, the British imported *Nicotiana*

TABLE 6.1 Major Toxic Agents in Cigarette Smoke (unaged)

Gas phase

Agent	Concentration per cigarette: Range reported	U.S. cigarettes[1]
Carcinogen		
Dimethylnitrosamine	1–200 ng	13 ng
Ethylmethylnitrosamine	0.1–10 ng	1.8 ng
Eiethylnitrosamine	0–10 ng	1.5 ng
Nitrosopyrrolodine	2–42 ng	11 ng
Other nitrosamines	0–20 ng	nt
Hydrazine	24–43 ng	32 ng
Vinyl chloride	1–16 ng	12 ng
Arsine	nt	nt
Nickel carbonyl	nt	nt
Cocarcinogen		
Formaldehyde	20–90 μg	30 μg
Tumor initiator		
Urethane	10–35 ng	30 ng
Cilia toxic agent		
Formaldehyde	20–90 μg	30 μg
Hydrogen cyanide	30–200 μg	110 μg
Acrolein	25–140 μg	70 μg
Acetaldehyde	18–1400 μg	800 μg
Toxic agent		
Hydrogen cyanide	30–200 μg	110 μg
Nitrogen oxides (NO_x)[2]	10–600 μg	350 μg
Ammonia[3]	10–150 μg	60 μg
Pyridine[3]	9–93 μg	10 μg
Carbon monoxide	2–20 mg	17 mg
Other		
Volatile chlorinated Olefins & nitro-olefins	nt	nt

Particulate phase

Agent	Concentration per cigarette: Range reported	U.S. cigarettes[1]
Carcinogen		
N′-nitrosonornicotine	100–250 ng	250 ng
Polonium-210	0.03–1.3 pCi	nt
Nickel compounds	10–600 ng	nt
Cadmium compounds	9–70 ng	nt
Arsenic	1–25 μg	nt
Cocarcinogen		
Pyrene	50–200 ng	150 ng
Fluoranthene	50–250 ng	170 ng
Benzo(g,h,i)perylene	10–60 ng	30 ng
Naphthalenes	1–10 μg	6 μg
1-methylindoles	0.3–0.9 μg	0.8 μg
9-methylcarbazoles	5–200 ng	100 ng
Catechol	40–460 μg	270 μg
3- & 4-methyl-catechols	30–40 μg	32 μg
Tumor initiator		
Benzo(a)pyrene	8–50 ng	20 ng
5-methylchrysene	0.5–2 ng	0.6 ng
Benzo(j)fluoranthene	5–40 ng	10 ng
Benz(a)anthracene	5–80 ng	40 ng
Dibenz(a,j)acridine	3–10 ng	8 ng
Dibenz(a,h)acridine	nt	nt
Dibenzo(c.g)carbazole	0.7 ng	0.7 ng
Cilia toxic agent		
Phenol	10–200 μg	85 μg
Cresols	10–150 μg	70 μg
Toxic agent		
Nicotine	0.1–2.0 mg	1.5 mg
Minor tobacco alkaloids	10–200 μg	100 μg
Bladder carcinogen		
β-Naphthylamine	0.25 ng	20 ng

[1] 85 mm cigarettes without filter tips bought on the open market 1973–1976.
[2] NO_x > 95% NO; remainder, NO_2.
[3] Not toxic in smoke of blended U.S. cigarettes because pH < 6.5; therefore, ammonia and pyridines are present only in protonated form.
nt: not tested.
[Adapted from Wynder and Hoffman, 1979.]

tabacum from Virginia in the form of snuff for medical use (Fairholt, 1859). By 1726, snuff had nearly eclipsed other forms of tobacco (Fairholt, 1859; Stewart, 1967). In what is probably an apocryphal story, the origin of the cigarette is attributed to serendipity at the siege of Constantinople by the French in the middle of the nineteenth century:

'It is told in Alsace, France, the following story that if apocryphal, has nevertheless the merit to be plausible. At the siege of Constantinople (today's Istanbul) in 1854, an Alsacian Zouave soldier had his pipe pulled out of his teeth by a shell fragment. Not knowing how to smoke his remaining tobacco, he had the idea of rolling it into a tube made of paper. Therefore, it was an accident of war that gave birth to the cigarette, and we all know the comfort it would bring to all soldiers and civilians during the coming wars, whiling away the long hours of anticipation, hunger, depression' (translated from Haug, 1961).

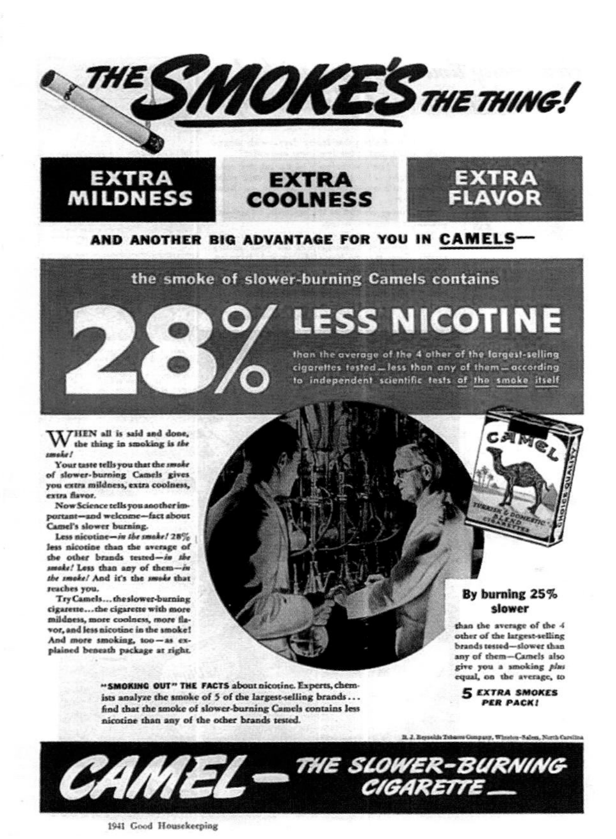

FIGURE 6.2 Camel (R.J. Reynolds Tobacco Company, Winston-Salem, North Carolina) cigarette advertisement, circa 1941. Notice the smoke at the tip of the cigarette—absent now from contemporary advertisements. Notice also the lack of the now ubiquitous Surgeon General's warning, which did not appear on cigarette packages until 1966.

However, reference also is made to another form of smoking, which resembles the cigarette. An early sixteenth-century expedition to Mexico noted that the natives would pack tobacco and liquid ambar (an herb) into a hollow reed which was allowed to smolder on one end, and the smoke inhaled from the other (Akehurst, 1968). According to some, the wrapping evolved from reeds, to corn husks, to papers used for manufacturing cigars in Spain. The Peninsular Wars of the early nineteenth century disseminated these new, smaller cigars from Spain to France. The French renamed it the 'cigarette'. The Crimean War of the mid-nineteenth century introduced it to England, where, in 1856, Robert P. Gloag set up a factory in Walworth for mass production. Aromatic tobaccos were used in these new cigarettes because they were the only types of tobacco suitable for smoking in this form. Flue-cured aromatic tobacco followed by air-cured Burley tobacco in 1864 were introduced for further mildness. These flue-cured and air-cured

FIGURE 6.3 Hand-colored engraving of tobacco smoking as a Floridian Native American health remedy (de Bry and Le Moyne de Morgues, 1591).

tobaccos were substituted for some of the aromatic tobaccos to form the blended cigarette in America in the late 1800s. The first cigarette-making machine was introduced in 1880 (Akehurst, 1968). From then, tobacco and cigarette production accelerated through the end of the century due to the convenience of the smoking vehicle, ease of production, transportation and distribution, mass media advertising, and demand (Akehurst, 1968; Giovino *et al.*, 1995) (**Fig. 6.4**).

Tobacco smoking continues to be a worldwide health problem. The high addictive potential of nicotine is indicated by the vast number of people who habitually smoke and relapse. In 2002, 30 per cent of the U.S. population aged 12 and over (71.5 million) reported past-month use of a tobacco product. 19 per cent of those aged 12 and over (Substance Abuse and Mental Health Services Administration, 2003), and 22.5 per cent of those aged 18 and over (MMWR, 2004), were smokers who smoked every day in the past month (**Fig. 6.5**). There are over 1.3 billion daily smokers worldwide (Guindon and Boisclair, 2003; Thun and da Costa e Silva, 2003). For example, France reports that 36 per cent of its population are current smokers, but this percentage is less than the 40 per cent seen in 1973 (Kopp and Fenoglio, 2000). Daily smoking (current smokers) has declined in the European Union over the past

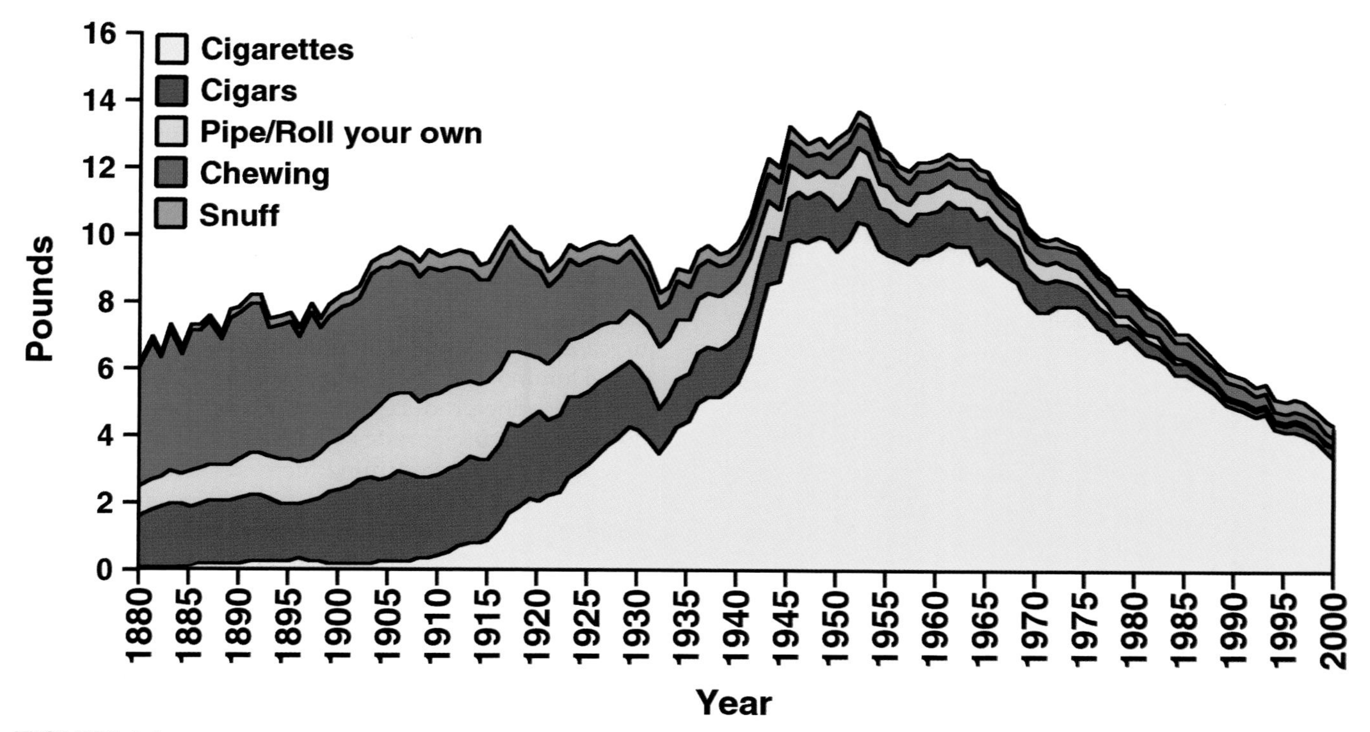

FIGURE 6.4 Trends in per capita consumption of various tobacco products in the United States (in pounds) from 1880 to 2000 among persons aged 18 year or older. The latest data indicate per capita cigarette consumption increased to 3.4 in 2001, increased to 3.5 in 2002, and decreased to 3.3 in 2003. [Data from the U.S. Department of Agriculture *Tobacco Situation and Outlook Report* series; see also Capehart, 2003, for latest data.]

20 years, but smoking prevalence in 1999–2001 still ranges from 19 per cent of the population in Sweden to 40 per cent of the population in Serbia (mean of 30 per cent for 28 countries) (data from the World Health Organization Regional Office for Europe).

The cost to society is significant in terms of health problems that frequently lead to death, medical costs, and human suffering. It led to 4.9 million premature deaths worldwide per year in 2000 (Ezzati and Lopez, 2003) and has been projected to lead to upward of 10 million deaths by 2020 (Mackay and Eriksen, 2002). Tobacco smoking is the leading, *avoidable* cause of disease and premature death in the United States alone, responsible for 440 000 deaths annually and 5.6 million years of potential life lost (Fellows *et al.*, 2002). Smoking is implicated in 67 per cent of death

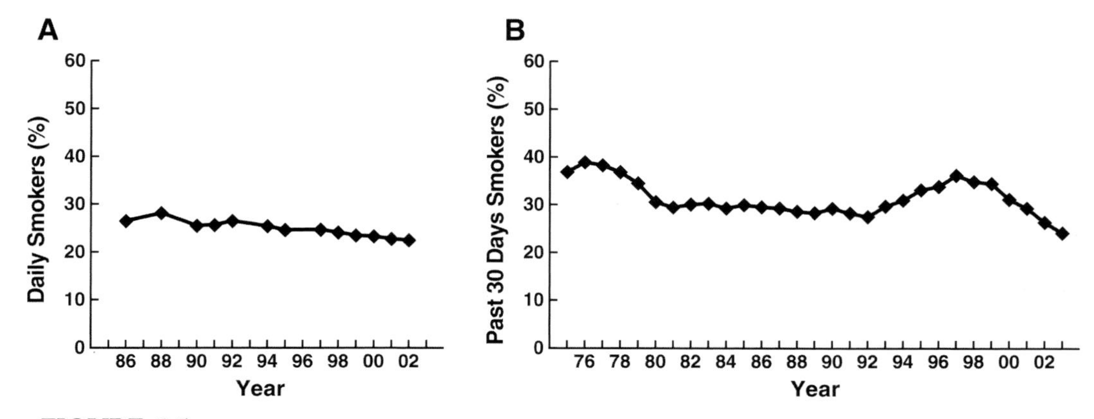

FIGURE 6.5 **(A)** Percentage of United States civilian adults, aged 18 and older, who smoked every day. Data collected from *Morbidity and Mortality Weekly Report* (MMWR, 1987, 1991, 1992, 1993b, 1994, 1996, 1997, 1999–2001, 2002b, 2003b, 2004). **(B)** Percentage of 12th graders who smoked in the past 30 days, from 1976 to 2003. [Data from Substance Abuse and Mental Health Services Administration, 2003.]

from cancer, 78 per cent of deaths from chronic obstructive pulmonary disease, 52 per cent of deaths from respiratory disease (including bronchitis and emphysema), and 16 per cent of deaths from cardiovascular disease (MMWR, 2002a). Of current smokers in 2000, 49 per cent developed bronchitis, 24 per cent emphysema, and 7 per cent cancer, and 13 per cent are likely to experience heart attack and 7 per cent, stroke (MMWR, 2003a). On average, smoking shortens life span by 13 years in males and 15 years in females (MMWR, 2002a). A study sponsored by the World Health Organization and the World Bank estimated that in the United States, smoking-related healthcare expenses accounted for 6 per cent of all annual healthcare costs. Nicotine addiction overall costs the United States $155 billion annually (Centers for Disease Control and Prevention, 2004), and by comparison, $16 billion in France (Kopp and Fenoglio, 2000).

Smokeless tobacco can be defined as either chewing tobacco or using snuff, and is also of significant health concern. Adverse health consequences of smokeless tobacco use include oral (gum and buccal mucosa) cancers in smokeless tobacco users (chewing quid and tobacco) (Critchley and Unal, 2003). Oropharyngeal cancer, dental caries, and cardiovascular disease have been proposed as potential toxic effects, but the data to date are not conclusive (Critchley and Unal, 2003). The number of smokeless tobacco users nearly tripled between 1972 and 1991 (Office of Evaluations and Inspections, 1992). Since 1991, the percentage of past month users has remained fairly constant (range: 3.1–4.0 per cent). In 2002, 3.3 per cent of persons aged 12 and older (7.8 million) used smokeless tobacco within the past month. The majority of users are 18–25 years old and are five-times more likely to be males than females (MMWR, 1993a; Substance Abuse and Mental Health Services Administration, 2003).

MEDICAL USE AND BEHAVIORAL EFFECTS

Evidence indicates that people smoke primarily to experience the psychopharmacological properties of nicotine and that the majority of smokers eventually become dependent upon nicotine when they start as adolescents (Balfour, 1984; Stolerman, 1991). Numerous preclinical studies have demonstrated nicotine's reinforcing properties in many species (Goldberg *et al.*, 1981; Risner and Goldberg, 1983; Fudala *et al.*, 1985; Goldberg and Henningfield, 1988; Corrigall and Coen, 1991; Huston-Lyons and Kornetsky, 1992; Watkins *et al.*, 1999; Markou and Paterson, 2001; Paterson and Markou, 2002).

In humans, nicotine produces positive reinforcing effects, including mild euphoria (Pomerleau and Pomerleau, 1992), increased energy, heightened arousal, reduced stress, reduced anxiety, and reduced appetite (Stolerman and Jarvis, 1995; Benowitz, 1996). Cigarette smokers report that smoking produces arousal, particularly with the first cigarette of the day, and relaxation when under stress (Benowitz, 1988). Nicotine causes an alerting pattern in the electroencephalogram, with decreases in total alpha and theta activity and increases in low-voltage fast activity (Hauser *et al.*, 1958; Domino, 1967; Murphree *et al.*, 1967; Kadoya *et al.*, 1994). Nicotine produces autonomic arousal, and smoking one or two cigarettes increases resting heart rate by about 5–40 beats/min, increases blood pressure 5–20 mm Hg, and increases epinephrine and cortisol (Roth *et al.*, 1945; Herxheimer, 1967; Hill and Wynder, 1974). Most smokers report that smoking is pleasurable (81 per cent), helps them concentrate (63 per cent), calms them down when stressed or upset (90 per cent), and helps them deal with difficult situations (82 per cent) (Etter *et al.*, 2000).

Nicotine paradoxically produces decreases in tension, and an anxiolytic-like effect. The basis for such tension reduction is still unknown but has been linked both to decreases in skeletal muscle tone (Domino, 1973) and the subsequent reduction in muscle tension (Webster, 1964), and possibly its analgesic effect. Called *Nesbitt's paradox*, smokers allowed to smoke during a stressful experience (a session where they received painful shocks to the left forearm and upper arm) showed more arousal (increase in pulse rate) but reported less emotion (more pain endurance, more shocks taken) than smokers not allowed to smoke but simulating smoking (Nesbitt, 1973). These results were interpreted to support the paradox reported by smokers where their physiological arousal is increased but they report themselves calmer and more relaxed (Nesbitt, 1973). A majority of smokers have reported that they smoke to reduce negative affect or to achieve pleasurable relaxation (Ikard *et al.*, 1969; Ikard and Thompkins, 1973).

The effects of nicotine have been assessed under laboratory conditions in human volunteers who were current smokers and had histories of abuse of opioids, psychostimulants, and sedatives, and pressed a lever for intravenous nicotine. The subjects reported that nicotine produced subjective effects similar to those of morphine or cocaine, and the number of injections was inversely related to dose (Henningfield *et al.*, 1983) (**Fig. 6.6**). The positive reinforcing effects of acute nicotine administration through tobacco smoking are considered critically important in the initiation and maintenance of tobacco smoking that ultimately

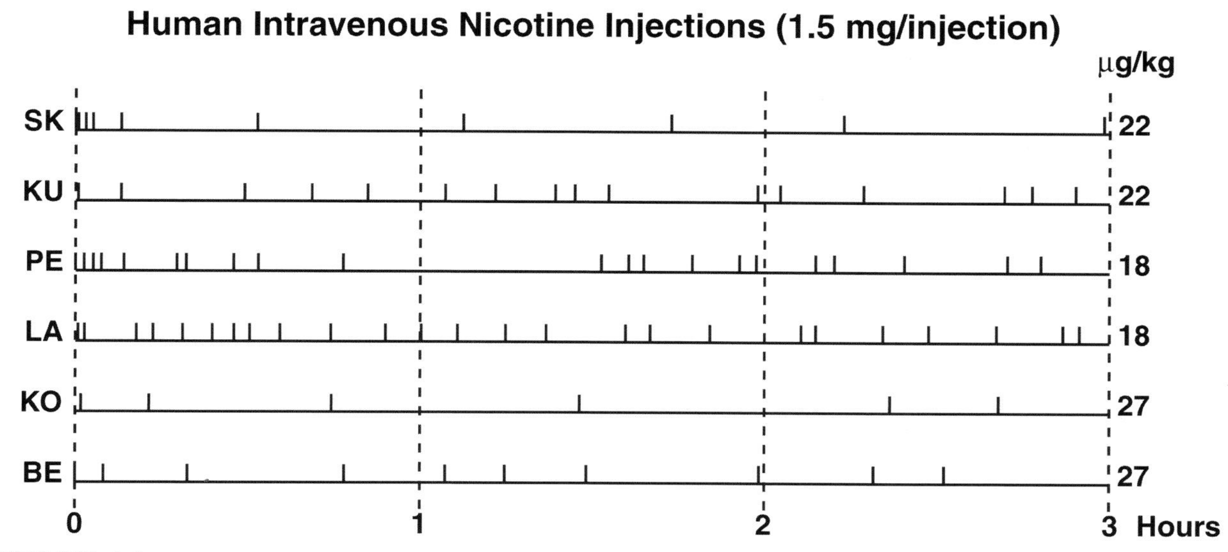

FIGURE 6.6 Pattern of nicotine deliveries obtained during a 3-h session in humans (1.5 mg/injection). Vertical lines indicate a single nicotine self-infusion. The unit dose for humans is indicated on the right side of each record. The number of injections per session were inversely related to the expression of the unit dose. Letters on the left are the initials referring to each of the subjects. [Reproduced with permission from Henningfield *et al.*, 1983.]

leads to dependence (Watkins *et al.*, 2000a; Kenny and Markou 2001). Nevertheless, factors other than nicotine contribute to smoking in human smokers, including sensory and conditioned reinforcing effects of smoking (Crooks and Dwoskin, 1997; Bardo *et al.*, 1999; Jacob *et al.*, 1999; Shahan *et al.*, 1999; Caggiula *et al.*, 2001, 2002). For example, cigarette smoke has monoamine oxidase inhibiting properties which could contribute to its psychotropic effects and addiction potential (Fowler *et al.*, 1996) (see *Imaging* chapter).

Nicotine also decreases appetite, particularly the desire for sweet tasting food and carbohydrates in both rats and humans, (Grunberg *et al.*, 1985; Perkins *et al.*, 1990) and has been linked to the motivation to continue smoking in women and their higher rates of relapse (Kabat and Wynder, 1987; Sorensen and Pechacek, 1987). This appetite suppression is accompanied by decreases in blood insulin (Grunberg *et al.*, 1988; Saah *et al.*, 1994) and changes in serotonin function (Ribeiro *et al.*, 1993). Nicotine also increases metabolism (fat oxidation) (Jensen *et al.*, 1995).

Nicotine is well documented to improve attention and cognition in animals (Hahn *et al.*, 2003; Buccafusco *et al.*, 2005) and learning, reaction time, and problem solving in abstinent smokers (Le Houezec and Benowitz, 1991; Levin *et al.*, 1998) and is particularly effective in enhancing selective attention and enhancing vigilance in performance of repetitive tasks, a classic stimulant effect (Mangan, 1982; Wesnes *et al.*, 1983). There also is evidence that nicotine enhances cognitive performance in non-smokers (Wesnes *et al.*, 1983).

In animals, nicotine lowers brain reward thresholds similar to other drugs of abuse (Huston-Lyons and Kornetsky, 1992) (**Fig. 6.7**). Nicotine sustains intravenous self-administration in animals and in humans, and in rats the dose range is relatively narrow (Watkins *et al.*, 1999) (**Fig. 6.8**). Animals and humans titrate their intake of nicotine (see *Behavioral Mechanism of Action* section). Nicotine in nondependent animals has less efficacy than cocaine as a reinforcer (Risner and Goldberg, 1983) (**Fig. 6.9**). Nicotine also produces locomotor activation, analgesia, appetite suppression, and improvements in learning and memory in rats.

In rats, nicotine produces a stimulant effect that can show tolerance or sensitization depending on the nature of exposure to the drug. In nicotine-naive rats, acute administration of nicotine decreased exploratory locomotor activity, whereas repeated administration of nicotine produced a rapid tolerance to the locomotor-depressant effects, followed by an increase in locomotor activity (Stolerman *et al.*, 1973, 1974; Clarke and Kumar, 1983). Moreover, with repeated administration, sensitization to the locomotor activating effects of nicotine develops (Clarke and Kumar, 1983; Benwell and Balfour, 1992) (**Fig. 6.10**).

Acute administration of nicotine can produce an anxiolytic-like effect in the social interaction test, but high doses can be anxiogenic-like (File *et al.*, 1998). Others have observed an anxiolytic-like effect in potentiated startle (Vale and Green, 1996), elevated plus maze (Brioni *et al.*, 1993), and mirrored

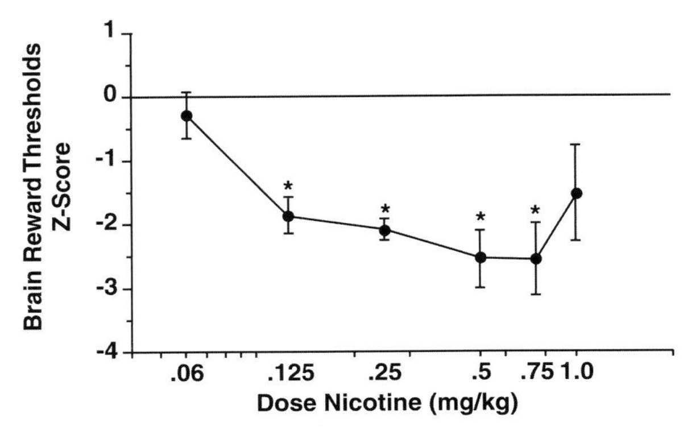

FIGURE 6.7 Mean z-score (standard score) ± SEM changes in reward threshold from pre- to post-drug after administration of various doses of nicotine in male F-344 rats. Saline post- minus predrug threshold is indicated by a z-score of 0. Asterisks (*) indicate doses of nicotine that significantly lowered threshold ($p < 0.025$, two-tailed paired t test). [Reproduced with permission from Huston-Lyons and Kornetsky, 1992.]

chamber (Cao *et al.*, 1993) tests of anxiety-like behavior. Anxiogenic-like responses have been observed after withdrawal from chronic nicotine in rats and mice (Costall *et al.*, 1989; Irvine *et al.*, 1999). However, others have observed an increase in anxiety-like responses as measured by decreased social interaction during daily nicotine self-administration sessions limited to 0.45 mg/kg per day, but no changes in anxiety-like responses during withdrawal from self-administered nicotine (Irvine *et al.*, 2001).

Nicotine has a number of physiological effects mainly attributed to its ability to activate ganglionic receptors in the autonomic nervous system, including the adrenal medulla. However, this activation is transient and is followed by a persistent depression of all autonomic ganglia depending on the dose and history of nicotine intake. In humans (as noted above),

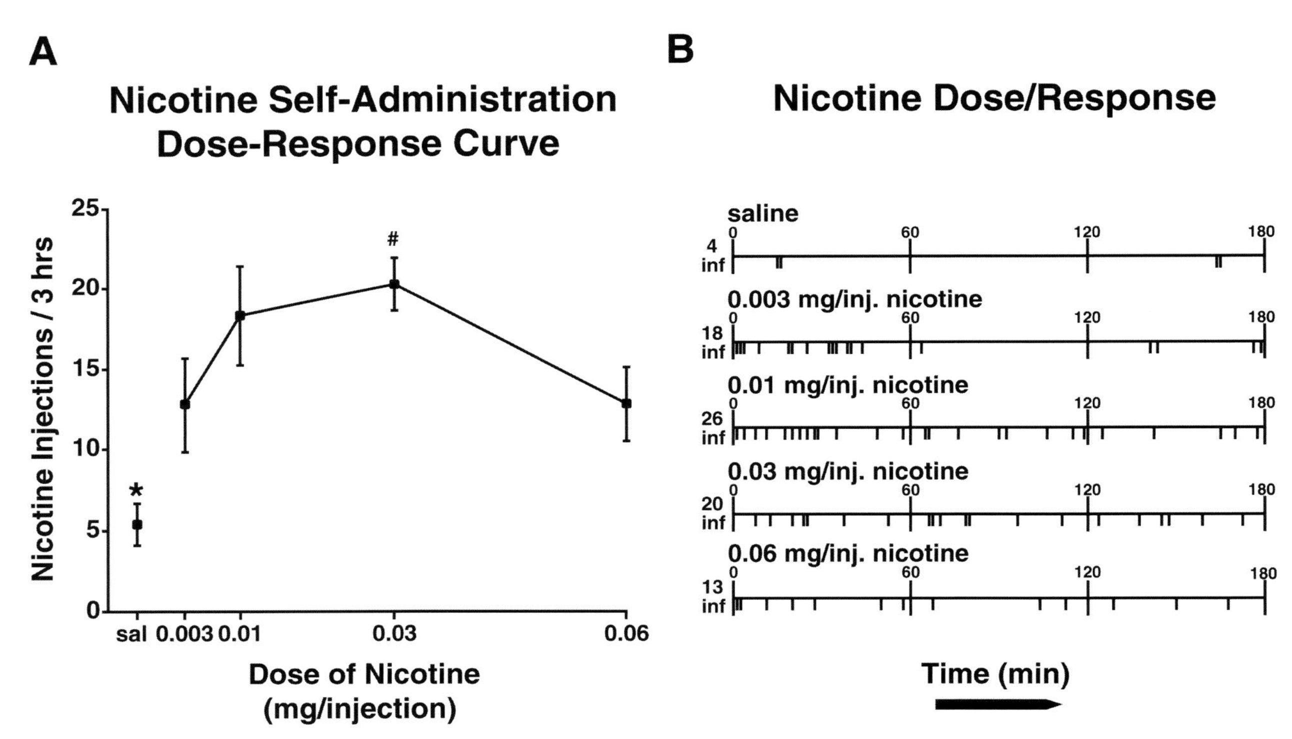

FIGURE 6.8 **(A)** Mean number of nicotine infusions earned during 3-h sessions of intravenous nicotine self-administration in male Wistar rats (n = 5–7). Closed squares represent the mean (± SEM) of the second and third days of 3 days of self-administration at each nicotine dose (0, 0.003, 0.01, 0.03, and 0.06 mg/kg/infusion, i.v.). The square at the saline (sal) dose represents the mean (± SEM) of the second and third day after substitution of saline for nicotine. All rats initially were trained on 0.03 mg/kg/infusion nicotine. The asterisk (*) indicates that responding for saline was significantly different from responding for all nicotine doses ($p < 0.05$) by *post hoc* comparisons after a significant main effect in an analysis of variance. The pound sign (#) indicates that responding for the 0.03 mg/kg dose was significantly higher than responding for the 0.003 and 0.06 mg/kg doses ($p < 0.05$). All nicotine doses refer to free base. **(B)** Event record of responding for nicotine for each dose tested. [Reproduced with permission from Watkins *et al.*, 1999.]

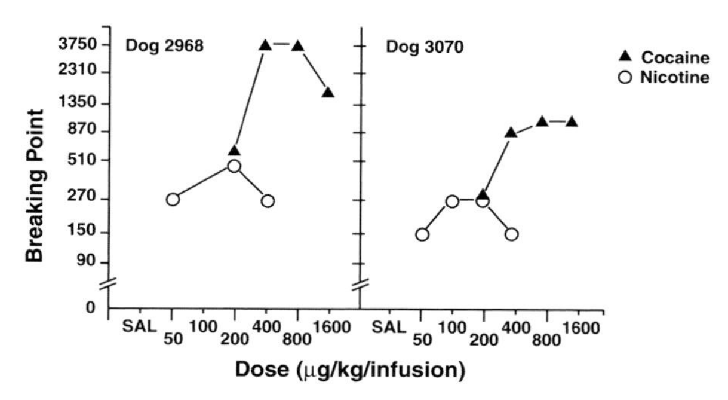

FIGURE 6.9 The highest fixed-ratio completed by two dogs for intravenous infusion of nicotine, cocaine, or saline under a progressive-ratio schedule of reinforcement. Each point represents a single determination at the selected dose and drug. Dog #2968 was tested twice with saline. Note the logarithmic scale used for the ordinate. [Reproduced with permission from Risner and Goldberg, 1983.]

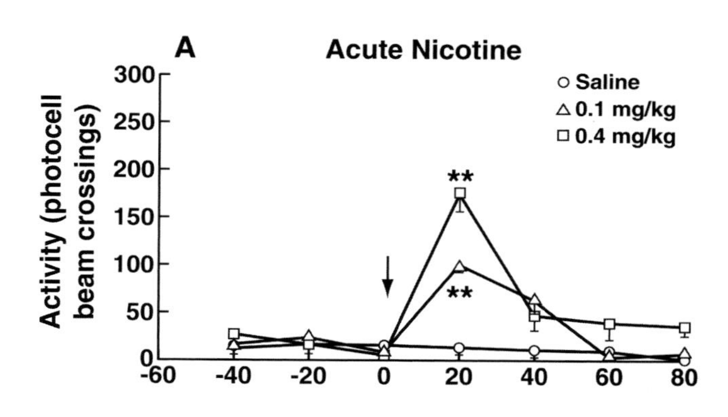

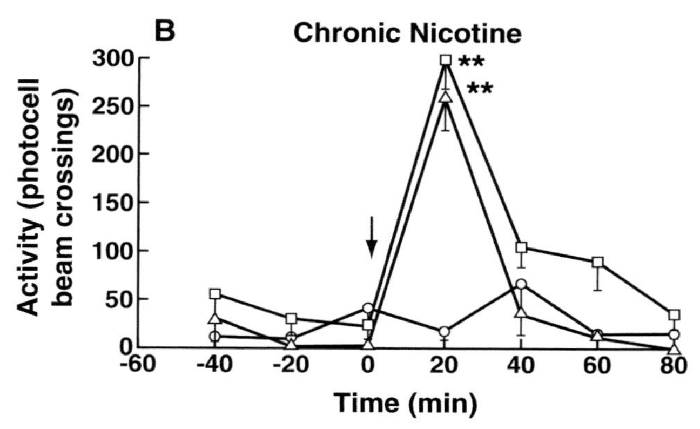

FIGURE 6.10 The effect of acute and subchronic injections of nicotine on spontaneous locomotor activity in rats. Rats were habituated to the testing environment for 80 min prior to injection. **(A)** For acute treatment, subcutaneous injections of saline (n = 6), 0.1 mg/kg nicotine (n = 6), or 0.4 mg/kg nicotine (n = 8) were given at the point indicated by the arrow (time = 0). The results are expressed as mean ± SEM. $^{**}p < 0.01$. **(B)** For chronic treatment, rats were pretreated with daily subcutaneous injections of saline (n = 6), 0.1 mg/kg nicotine (n = 6), or 0.4 mg/kg nicotine (n = 10) for 5 days before the test day. On the test day, the animals were given injections of saline or nicotine (0.1 or 0.4 mg/kg), respectively, at the time indicated by the arrow (time = 0). The results are expressed as mean ± SEM. $^{**}p < 0.01$. [Reproduced with permission from Benwell and Balfour, 1992.]

cigarette smoking increases heart rate and blood pressure (Lee *et al.*, 1983; Henningfield *et al.*, 1985). Nicotine also produces bronchial dilation and stimulation of the salivary glands.

Nicotine causes an initial stimulation of motility in the gastrointestinal tract followed by inhibition (Coulie *et al.*, 2001). Nicotine in a nicotine-naive individual causes vomiting due to activation of the emetic chemoreceptor trigger zone in the area postrema of the medulla and activation of vagal and spinal nerves that form the sensory component of the vomiting reflex (Laffan and Borison, 1957; Beleslin and Krstic, 1987; Srivastava *et al.*, 1991). Thus, nicotine has acute activating effects, reinforcing effects, and anxiolytic effects in man and animals. Nicotine transiently activates the autonomic nervous system. It appears to have less reinforcing effects in nondependent animals than other drugs of abuse, and indeed, initially it is aversive to humans. Factors other than its acute reinforcing effects may contribute to its high addiction potential.

PHARMACOKINETICS

Nicotine when smoked in cigarettes and inhaled, is quickly absorbed in the lungs as the freebase is largely suspended on tiny particles of tar. Nicotine reaches the brain within 8 s, almost as quickly as via an intravenous injection (Benowitz *et al.*, 1988) (**Fig. 6.11**). Tobacco in oral products is more basic and better absorbed via the mouth than cigarette smoke, which is more acidic. Cigarettes contain 1–2 per cent nicotine, or approximately 10–20 mg of nicotine (Caravati *et al.*, 2004). Of the available nicotine in mainstream smoke (25 per cent of total nicotine content of a cigarette), up to 90 per cent of the smoke drawn into the mouth is absorbed by inhalation, and much less is absorbed through the mouth (< 25 per cent) (Armitage *et al.*, 1975). Dependent cigarette smokers tend to titrate their intake over time within the confines of the rapid rise and fall associated with each cigarette, but on an average, approximately 1 mg is delivered per cigarette (Isaac and Rand, 1972) (**Fig. 6.12**), and dependent smokers will maintain a relatively stable blood nicotine level over the course of waking hours (Benowitz and Jacob, 1984) (**Fig. 6.13**). The plasma levels of

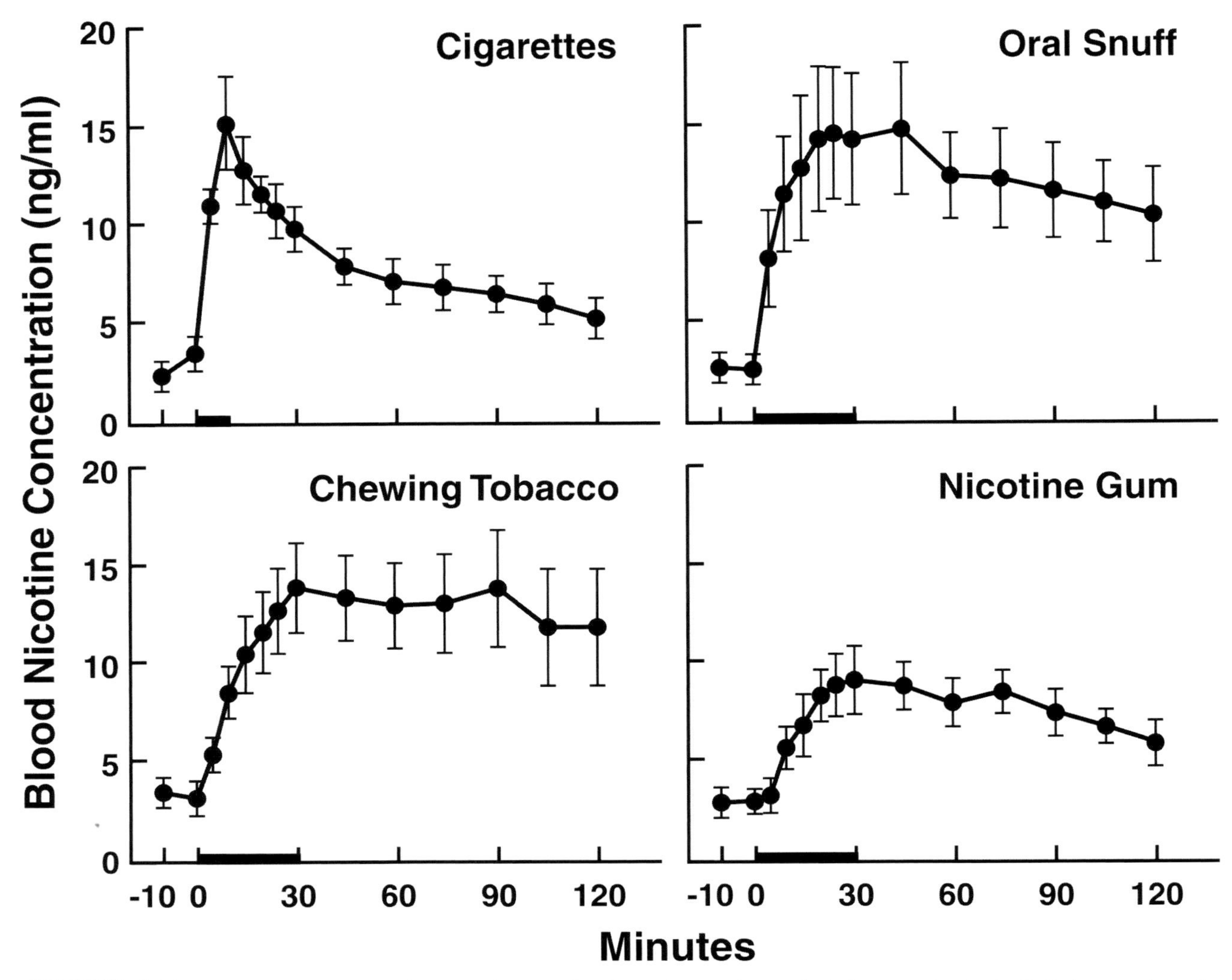

FIGURE 6.11 Blood concentrations of nicotine (mean ± SEM) in subjects who smoked cigarettes for 9 min (1.33 cigarettes), used oral snuff (2.5 g), used chewing tobacco (mean 7.0 g), or chewed nicotine gum (two 2 mg pieces). Shaded bars indicate the period of exposure to tobacco or nicotine gum. [Reproduced with permission from Benowitz *et al.*, 1988.]

nicotine range from 4–6 ng/ml for pipe smoking, to 20–50 ng/ml for cigarette smoking. Tobacco smokers can obtain significant carboxyhemoglobin levels (7–10 per cent) that would be produced by 12 h of tobacco exposure. Formation of carboxyhemoglobin prevents the normal transfer of carbon dioxide and oxygen during blood circulation. These carboxyhemoglobin levels produced by exposure to carbon monoxide are sufficient to meet and exceed the Environmental Protection Agency occupational threshold limit of 50 ppm of carbon monoxide (which produces 5 per cent carboxyhemoglobin). For example, in both males and females with an average cigarette consumption of 36 and 32 cigarettes per day, respectively, who had smoked 21 and 18 cigarettes, respectively, by the time of sampling, venous blood nicotine levels were approximately 32 ng/ml, and the average carboxyhemoglobin levels were 8 per cent (Russell *et al.*, 1980) (**Table 6.2**). Passive absorption of nicotine has been established, and nonsmokers living in the same home with a 40 cigarette per day smoker will obtain urinary cotinine (a nicotine metabolite; see below) levels equivalent to that of smoking approximately 3 cigarettes (Matsukura *et al.*, 1984). The acute fatal dose of oral nicotine of 60 mg has been repeatedly referenced as the commonly accepted lethal dose for an adult, and the earliest reference found by the authors is the statement, 'The fatal dose for a man is about 60 mg' (Sollman, 1926, p. 411). 60 mg has been referenced as the minimal fatal dose (Sollmann, 1926),

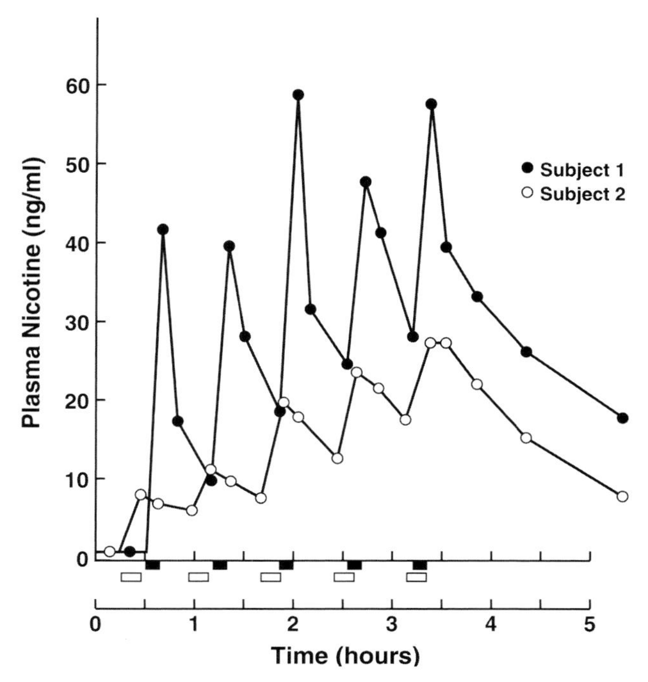

FIGURE 6.12 Effects of five consecutive cigarettes on plasma nicotine concentration in two subjects. 30 min elapsed between the end of one cigarette and the start of the next. The periods of smoking each cigarette are shown as solid bars (Subject 1) and open bars (Subject 2) beneath the abscissa. The curves are brought to coincidence at the end of smoking the fifth cigarette. Blood samples were taken before smoking, at 0.5, 10, and 30 min after the last puff of each cigarette, and at 60 and 120 min after the fifth cigarette. [Reproduced with permission from Isaac and Rand, 1972.]

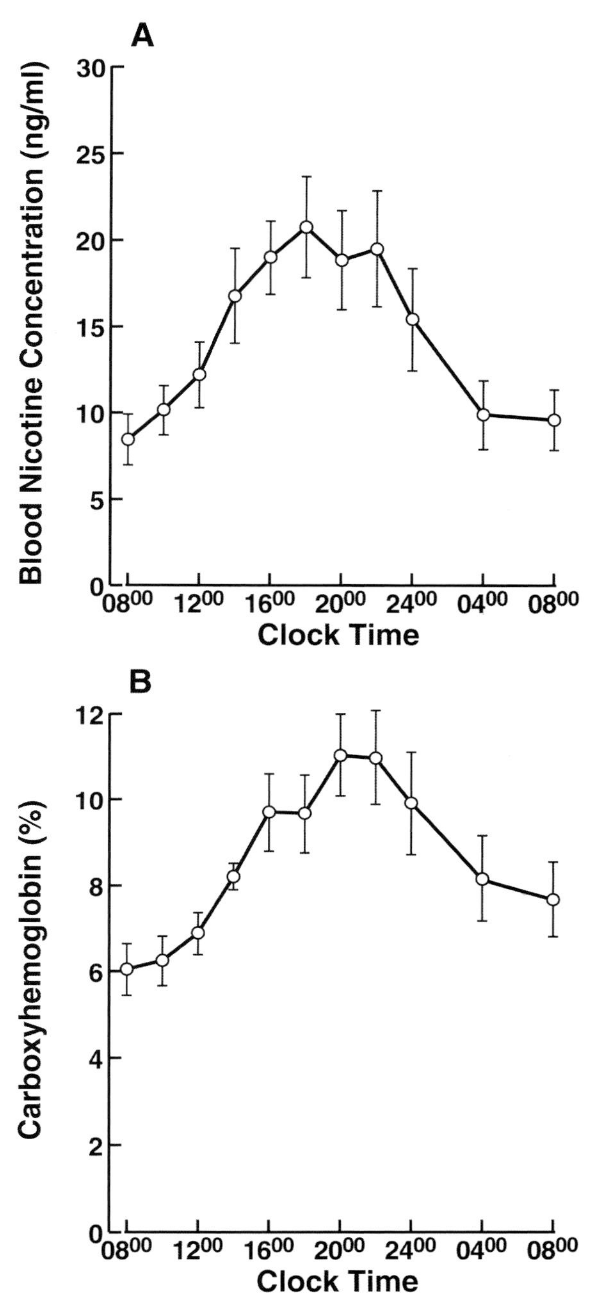

FIGURE 6.13 Venous blood nicotine **(A)** and carboxyhemoglobin **(B)** concentrations throughout the day in 11 habitual smokers smoking approximately 30 cigarettes per day (average yield of 1–2 mg nicotine per cigarette). [Reproduced with permission from Benowitz and Jacob, 1984.]

but individuals have ingested much larger quantities and recovered (Franke and Thomas, 1936; McGuigan, 2004).

Nicotine is largely metabolized in the liver (80–90 per cent), and nicotine and its metabolites, mainly cotinine and nicotine-1′-*N*-oxide, are excreted in the urine. Approximately 4 per cent of nicotine is converted to nicotine *N*-oxide which is largely excreted in urine without further metabolism. 70 per cent of nicotine is converted to cotinine (Benowitz *et al.*, 1983), which is further metabolized (Jacob *et al.*, 1988) (**Fig. 6.14**). The half-life of nicotine is about 2 h after inhalation or parenteral administration, and the half-life of cotinine is 19 h (Russell and Feyerabend, 1978). Cotinine was reported to have some psychoactive effects in that intravenously administered cotinine (30 mg) that produced blood cotinine levels similar to those of smokers (378 mg/ml) significantly decreased tobacco withdrawal-like symptoms in abstinent cigarette smokers (Keenan *et al.*, 1994).

ABUSE AND ADDICTION POTENTIAL

Tobacco smoking typically begins in adolescence, which significantly increases the likelihood to smoke as adults (see *What is Addiction?* chapter).

TABLE 6.2 Cigarette Consumption, Type of Cigarette Smoked (average), and Blood Nicotine and Carboxyhemoglobin Concentrations (averages) in Men and Women

	Men (n = 124)	Women (n = 206)
% smoking filtered cigarettes	61.3%	70.9%
% smoking unfiltered cigarettes	13.7%	1.9%
% smoking low-nicotine cigarettes (<1.0 mg)	25.0%	27.2%
Cigarette consumption per day	36.2 cigarettes	32.6 cigarettes
Cigarette consumption on day of test	20.7 cigarettes	18.2 cigarettes
Tar yield per cigarette	17.3 mg	15.8 mg
Nicotine yield per cigarette	1.3 mg	1.2 mg
Nicotine level in blood	33 ng/ml	32 ng/ml
Carboxyhemoglobin level in blood	7.8%	8.6%

Reproduced with permission from Russell *et al.*, 1980.

Social class, parental, older sibling, and peer smoking habits, type of school, academic achievement, church attendance, and drinking habits have all been related to smoking prevalence (Horn, 1959; McKennell and Thomas, 1967; Salber and Abelin, 1967; Kandel, 1975). Perhaps even more important is leaving school, which doubled the likelihood of being a smoker (Todd, 1969). A hypothetical archetype of the English smoker was described during the 1960s and 1970s as follows:

> 'Taking a number of the societal factors into account, the hypothetical archetype of the English smoker would be a lower working-class male, aged 25–30, educated at a secondary modern school which he left at 15. He would have grown up in a family of smokers where little effort was made to dissuade him. He would prefer chasing girls and regular drinking with smoking friends rather than going to church. Such a person would have a 95 per cent chance of being a smoker as opposed to his contratype with a loading of about 10 per cent' (Russell, 1971).

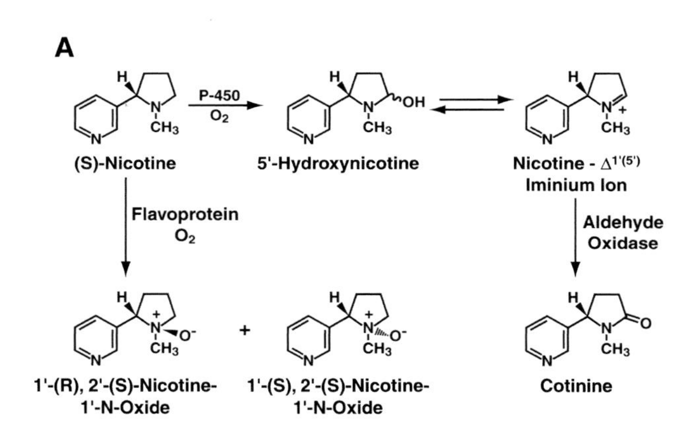

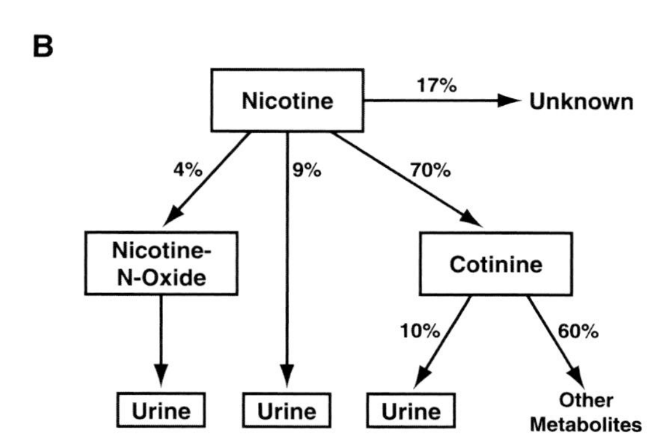

FIGURE 6.14 **(A)** Oxidative metabolism of nicotine. **(B)** Quantitative disposition of nicotine in smokers. [Reproduced with permission from Jacob *et al.*, 1988.]

Adolescents report being able to obtain tobacco easily. Most (95 per cent) are aware of the health risks associated with smoking, but to them, this is of little concern. Three-quarters of adolescents attempt to quit, but only 30 per cent have been able to abstain for more than one month (Chassin *et al.*, 1990; Dappen *et al.*, 1996), and the prevalence of adolescent smoking increases with age: 12 (2 per cent), 13 (5 per cent), 14 (9 per cent), 15 (14 per cent), 16 (22 per cent), and 17 (28 per cent) (Substance Abuse and Mental Health Services Administration, 2003).

As smoking progresses, tolerance develops to the autonomic side effects of smoking. Several trajectories of cigarette use and dependence then follow (**Fig. 6.15**). Smokers can become dependent under a rapid trajectory with exposure to as little as four cigarettes during adolescence, producing a 94 per cent chance of progressing to regular smoking (Russell, 1990). Once regular smoking is established, dependence follows rapidly (McNeill *et al.*, 1986), and smokers

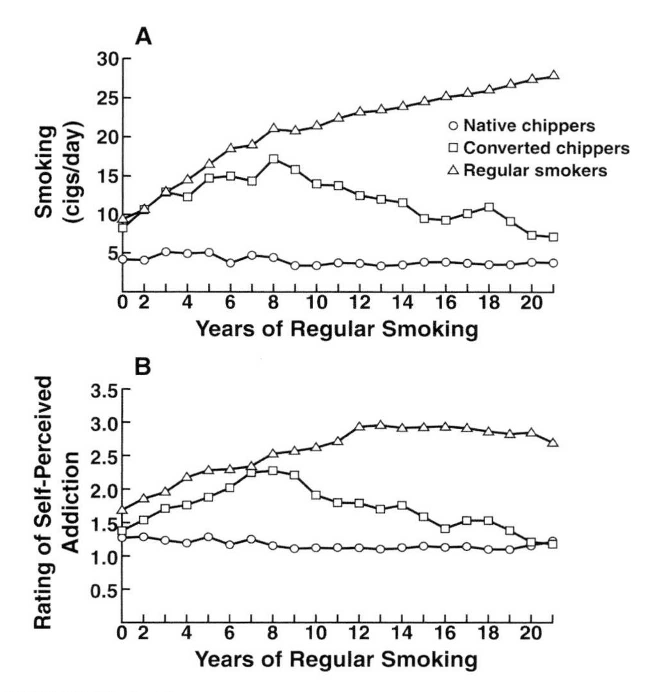

FIGURE 6.15 Different patterns of smoking that evolve over time as measured by **(A)** the number of cigarettes smoked and **(B)** the self-rating of addiction in subjects. Chippers were arbitrarily defined as individuals who smoked 1–5 cigarettes per day and reported no dependence (Shiffman *et al.*, 1994). Chippers reported casual abstinence without any withdrawal symptoms or any evidence of tolerance. Converted chippers were individuals that met the above criteria for chipping but had experienced extended periods of heavy smoking (Shiffman *et al.*, 1994). [Reproduced with permission by Dr. Saul Shiffman].

can become dependent under a rapid trajectory, consistent with an 'exposure model' of addiction (Alexander and Hadaway, 1982).

However, with other drugs of abuse, other trajectories of cigarette use have revealed a category of nondependent smokers called *chippers*. Chippers were defined as those who smoke < 5 cigarettes/day. There were those who remained chippers, and those who became *converted chippers* (those who previously were heavy smokers but currently smoked < 5 cigarettes/day) (Shiffman *et al.*, 1994). In a study of chippers' smoking behavior, smoking history and dependence were compared to matched regular smokers. Chippers smoked < 5 cigarettes/day and did not meet the criteria for Substance Dependence (addiction) on nicotine (Shiffman *et al.*, 1994) (**Table 6.3**). When they first started smoking, regular smokers went through a phase of approximately 2 years when they engaged in light smoking (< 5 cigarettes/day). Once they reached 15 cigarettes/day for at least two years, these regular smokers never returned to chipping (two years at ≤ 5 cigarettes/day). In contrast, in the chipping group chippers smoked on average ≤ 5 cigarettes/day for 16 years. However, some chippers, defined as converted chippers (29 per cent), had started off with a trajectory like regular smokers of accelerating after a brief period of two years, became chippers after a prolonged period of heavy smoking, and then had remained as chippers for six years. The converted chippers showed a lack of

TABLE 6.3 Fagerstrom Tolerance Questionnaire Items

	Chippers		*Regular smokers*		
Variable[a]	M	SD	M	SD	Effect size[b]
Latency to first cigarette of the day (min)	347.4	286.2	18.2	27.8	41.1****
Rate-adjusted latency to first cigarette of the day[c]	124.1	276.3	−16.2	24.4	12.0***
Smoking when awakes during night[d] (monthly frequency)	0.2	0.7	1.8	5.0	32.6**
Most hate to give up first cigarette in the morning (%)	4.6	—	9.4	—	5.0*
Smoking or craving more or less in the morning (1–5 scale)	1.7	1.1	3.4	1.1	38.6****
Difficulty refraining when forbidden (1–5 scale)	1.3	0.6	2.6	1.0	39.7****
Smoking when ill in bed (1–5 scale)	1.1	0.4	2.9	1.1	54.2****

Questionnaire from Fagerstrom, 1978.
[a]Canonical correlation = 0.81, Wilks' $\lambda = 0.35$, $F_{6,130} = 40.4$, $p < 0.0000001$.
[b]Effect size is expressed as percentage of variance accounted.
[c]Adjustment made for differences in smoking frequency; lag to first cigarette compared to average inter-cigarette interval. Positive values indicate smoking later than expected; negative values indicate smoking sooner than expected.
[d]This item is not part of the original Fagerstrom Tolerance Questionnaire, but relates to similar content. Analyzed with nonparametric statistics because of highly skewed distribution.
*$p < 0.01$, **$p < 0.001$, ***$p < 0.00005$, ****$p < 0.0000001$.
[Reproduced with permission from Shiffman *et al.*, 1994.]

dependence and decreased craving profiles similar to nonconverted chippers (Shiffman *et al.*, 1994).

Smoking 3 cigarettes/h is sufficient to maintain a 20-min peak level of nicotine effect more or less continuously. A nicotine abstinence syndrome after chronic nicotine exposure has been characterized in both humans (Shiffman and Jarvik, 1976; Hughes *et al.*, 1991b) and rats (Malin *et al.*, 1992, 1993, 1994; Hildebrand *et al.*, 1997; Epping-Jordan *et al.*, 1998; Watkins *et al.*, 2000b), and has both somatic and motivational components. In humans, acute nicotine withdrawal is characterized by somatic symptoms, such as bradycardia, gastrointestinal discomfort, and increased appetite leading to weight gain, as well as motivational symptoms, including depressed mood, dysphoria, irritability, anxiety, frustration, increased reactivity to environmental stimuli, and difficulty concentrating (Hughes *et al.*, 1991b; American Psychiatric Association, 1994). Anxiety, difficulty concentrating, hunger, irritability, restlessness, weight gain, and decreased heart rate are validated signs of nicotine withdrawal in self-quitters. These symptoms begin 6–12 h after cessation, peak in 1–3 days, and then return to normal within 7–30 days postsmoking cessation, on average lasting 3–4 weeks (Hughes, 1992) (**Fig. 6.16**).

The somatic syndrome associated with nicotine withdrawal has been modeled in rats (Malin *et al.*, 1992, 1993, 1994; Hildebrand *et al.*, 1997; Epping-Jordan *et al.*, 1998; Watkins *et al.*, 2000b) (**Table 6.4**). In rats, the somatic signs of nicotine withdrawal after exposure to 9 mg/kg/day of nicotine hydrogen tartrate salt (3.16 mg/kg nicotine base) via minipump for six days resemble those seen in opiate withdrawal, including the symptoms of abdominal constrictions, facial fasciculation, and ptosis. These somatic signs are both centrally and peripherally mediated (Hildebrand *et al.*, 1997; Malin *et al.*, 1997; Carboni *et al.*, 2000; Watkins *et al.*, 2000a) and are called somatic because they are expressed as responses of the body (e.g., gasps, writhes). This syndrome has been observed after spontaneous nicotine withdrawal, as well as withdrawal precipitated by the nicotinic acetylcholine receptor antagonists mecamylamine and chlorisondamine (Malin *et al.*, 1992, 1993, 1994; Hildebrand *et al.*, 1997; Epping-Jordan *et al.*, 1998; Watkins *et al.*, 2000b) and dihydro-β-erythroidine (Malin *et al.*, 1998; however, see Epping-Jordan *et al.*, 1998).

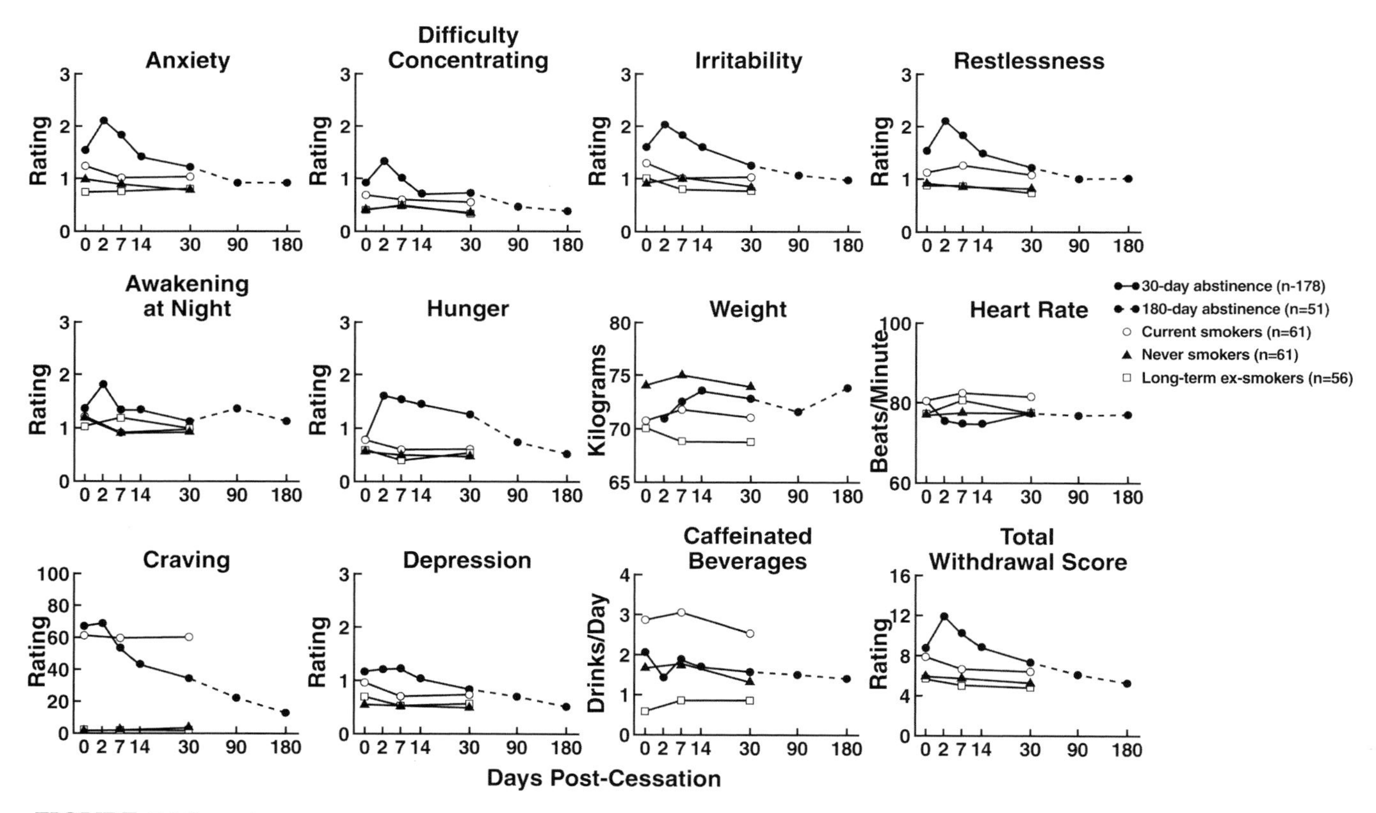

FIGURE 6.16 Self-reported symptoms pre- and postcessation of tobacco use in current smokers, smoking an average of 27 cigarettes per day (nicotine yield: 0.85/cigarette) for an average of 22 years. The 0 value on abscissa indicates precessation. [Reproduced with permission from Hughes, 1992.]

TABLE 6.4 Frequencies of Individual Categories of Abstinence Signs (Mean ± SEM) Observed in Rats over 15 min, 16 h After End of Nicotine Infusion

	Prior infusion rate of nicotine tartrate		
Category	0 mg/kg/day	3 mg/kg/day	9 mg/kg/day
Teeth-chattering/chews	2.6 ± 1.5	7.0 ± 2.5	15.9 ± 4.9*
Gasps/writhes	3.2 ± 1.4	7.2 ± 1.4	17.1 ± 4.6*
Ptosis	0.6 ± 0.4	1.4 ± 0.5	6.5 ± 1.2*
Shakes/tremors	2.0 ± 0.6	7.0 ± 1.8†	4.1 ± 1.7
Yawns, dyspnea, seminal ejaculation	0.1 ± 0.1	0.8 ± 0.3	1.3 ± 0.7††

*$p < 0.01$ vs saline-infused controls (Dunnett's t test).
†$p < 0.05$ vs saline-infused controls (Dunnett's t test).
††$0.05 < p < 0.10$ vs saline-infused controls (Dunnett's t test).
[Reproduced with permission from Malin *et al.*, 1992.]

The motivational measures of nicotine withdrawal have been modeled with intracranial self-stimulation, which has been shown to be a valid and reliable measure of changes in reward associated with acute drug exposure (Kornetsky and Esposito, 1981; Stellar and Rice, 1989; Moolten and Kornetsky, 1990; Huston-Lyons and Kornetsky, 1992; Bauco and Wise, 1994; Baldo *et al.*, 1999). Nicotine produces a pronounced elevation in brain reward thresholds (Epping-Jordan *et al.*, 1998) similar to the elevations in threshold during withdrawal from most drugs of abuse, including cocaine, amphetamine, morphine, ethanol, and Δ^9-tetrahydrocannabinol (Leith and Barrett, 1976; Gardner *et al.*, 1988; Markou and Koob, 1991; Schulteis *et al.*, 1994, 1995; Epping-Jordan *et al.*, 1998; Lin *et al.*, 1999) (**Fig. 6.17**). The elevation in reward thresholds produced by nicotine withdrawal is dependent on the dose of nicotine, with higher doses

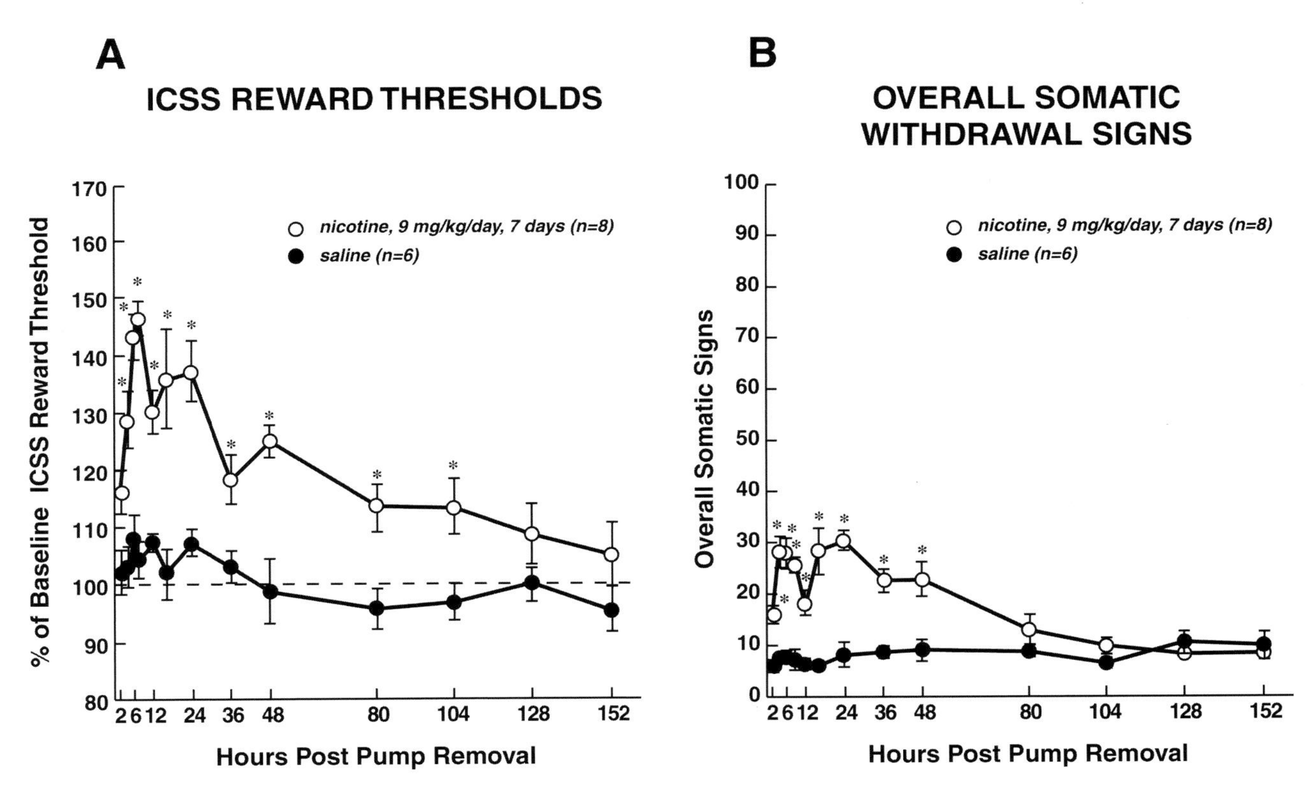

FIGURE 6.17 **(A)** Intracranial self-stimulation (ICSS) reward thresholds, and **(B)** overall somatic withdrawal signs in rats during spontaneous withdrawal after termination of chronic administration of nicotine hydrogen tartrate (9.0 mg/kg/day for 7 days; $n = 8$) or saline ($n = 6$). Asterisks indicate statistically significant differences ($p < 0.05$) between nicotine- and saline-treated groups after removal of the minipumps. [Adapted with permission from Epping-Jordan *et al.*, 1998.]

producing greater elevations in thresholds, and continuous drug administration producing a longer-lasting reward elevation withdrawal than intermittent administration even when the total dose administered was the same (Skjei and Markou, 2003). There was little correlation between somatic signs of withdrawal and reward threshold elevations in this study (Skjei and Markou, 2003), and neuropharmacological dissociations have been observed between the threshold elevations and the somatic signs associated with nicotine withdrawal (Watkins *et al.*, 2000b; Harrison *et al.*, 2001), suggesting that the various aspects of withdrawal are mediated by different substrates.

Enduring symptoms of nicotine withdrawal (protracted abstinence) include continued powerful craving that can last up to 6 months (Shiffman and Jarvik, 1976; Hughes *et al.*, 1984, 1991b; Hughes, 1992). Depression rates did not increase during spontaneous nicotine withdrawal overall. However, subjects with increases in depression were more likely to relapse (West *et al.*, 1989; Covey *et al.*, 1990; Hughes 1992). While the somatic symptoms of withdrawal from chronic intake of drugs of abuse are reported in humans as being unpleasant and annoying, it has been hypothesized that avoidance of the affective components of drug withdrawal, including those associated with nicotine withdrawal, plays a more important role in the maintenance of nicotine dependence than the somatic symptoms of withdrawal (Koob *et al.*, 1993; Markou *et al.*, 1998). Although many smokers attempt to quit smoking and may be successful early on, relapse rates in the long term are high, with only 10–20 per cent of individuals still abstinent after one year (Hunt *et al.*, 1971; Hunt and Bespalec 1974; Hughes *et al.*, 1991a,b, 1992).

Nicotine replacement therapy, including nicotine gum (Schneider *et al.*, 1984) and sublingual nicotine tablets (Molander *et al.*, 2000) reduce the occurrence of withdrawal symptoms in abstinent smokers. The efficacy of nicotine replacement therapy in smoking cessation trials, at least in certain individuals (Fagerstrom 1988; Sachs and Leischow 1991; Fagerstrom *et al.*, 1992), is related to its ability to prevent the onset and reduce the duration of nicotine withdrawal. Thus, treatment of nicotine withdrawal symptoms will facilitate abstinence and increase the percentage of individuals who succeed in quitting smoking permanently (Glassman *et al.*, 1990). 20–30 per cent of smokers that use nicotine replacement therapy remain abstinent after one year compared to 10–20 per cent of those untreated (Hunt *et al.*, 1971; Hunt and Bespalec 1974; Hughes *et al.*, 1991a; 1992).

The strong relationship between withdrawal from smoking and negative affect, including anxiety, frustration, anger, and depressed mood (Pomerleau *et al.*, 1978; Waal-Manning and de Hamel, 1978), has led to the other major therapeutic approach to smoking cessation, that of antidepressant drug treatment. There are estimates indicating that up to 60 per cent of smokers have a history of clinical depression (Hughes *et al.*, 1986; Glassman *et al.*, 1988), and the incidence of major depressive disorder among smokers was twice that of nonsmokers (Glassman *et al.*, 1990). In addition, smokers with a history of clinical depression were significantly less likely to succeed in quitting smoking than smokers without depressive histories (14 per cent versus 28 per cent) (Glassman *et al.*, 1990). It is unknown whether individuals who suffer from depressive symptomatology are more likely to become smokers or whether depressive symptoms are induced or exacerbated by long-term smoking (Markou *et al.*, 1998). Studies with tricyclic antidepressants showed some promise in smoking cessation (Edwards *et al.*, 1989; Prochazka *et al.*, 1998), whereas selective serotonin reuptake inhibitors appeared not to affect smoking behavior in heavy smokers (Sellers *et al.*, 1987). Bupropion (Wellbutrin; Zyban), an 'atypical' antidepressant with actions in facilitating norepinephrine and dopamine but not serotonin neurotransmission (Stahl, 1998), has proven effective in two double-blind, placebo-controlled studies. Twice the number of subjects who received 300 mg of bupropion per day for approximately two months were still abstinent compared to placebo after one year (Hurt *et al.*, 1997; Jorenby *et al.*, 1999).

BEHAVIORAL MECHANISM OF ACTION

The behavioral mechanism of action of nicotine can be described as 'mood titration' and can be defined as 'regulation of an individual's mood by adding known amounts of nicotine over circumscribed periods of time until a given mood state occurs'. Cigarette smokers maintain a given level of nicotine intake over the course of a smoking bout and in dependent smokers over the course of waking hours (see above). It is well documented that when cigarette smokers cut down on the number of cigarettes smoked, they compensate by altering the topography of the behavior (Fredericksen, 1976). Smokers will consume fewer cigarettes more intensely by taking in more smoke per puff, smoking more of each cigarette, and taking more puffs (Ashton and Watson, 1970; Benowitz *et al.*, 1986). Low nicotine-yield cigarettes lead to compensation by increased inhalation or increased numbers of cigarettes (Schachter *et al.*, 1977; Russell *et al.*, 1980; Maron and Fortmann 1987).

Perhaps an even more compelling example of such regulation of nicotine intake is the observation of 'vent blocking' (Kozlowski *et al.*, 1989). To lower smoke and nicotine content of cigarettes, the filter on the end of low-yield smoke cigarettes was ventilated with holes so that each puff was diluted with ambient air. A significant proportion of smokers (40 per cent) were shown to negate the benefit of this ventilation by unconsciously blocking the air holes either by their lips or fingers. This blocking was sufficient to produce tobacco toxins that were similar to individuals smoking regular-yield cigarettes (Kozlowski *et al.*, 1989). Similar titration can be seen when cigarette smokers are allowed to self-administer nicotine intravenously (Henningfield *et al.*, 1983). Subjects were allowed access to 1.5 mg/injection of nicotine. The subjects all self-administered between 18 and 27 μg/kg despite wide variations in the number of injections (see **Fig. 6.6** above). The number of injections per session was inversely related to the μg/kg administered. These data indicate that subjects adjust their intake to compensate for body mass and presumably volume of distribution of the nicotine in the blood (see above).

The *boundary model* of Herman and Kozlowski (1979) provides an explanation for the mood titration behavioral mechanism proposed above. In the boundary model there are three zones of behavioral effects associated with smoking behavior limited by an aversive state of withdrawal when plasma nicotine levels fall below a certain point (*lower boundary*) and the noxious aversive state associated with toxic high doses of nicotine (*upper boundary*) (Herman and Kozlowski, 1979; Kozlowski and Herman, 1984). The zone in between is called the *zone of indifference* to the pharmacological effects which can be very large for a nondependent smoker and very limited for a heavily dependent smoker (Kozlowski and Herman, 1984) (**Fig. 6.18**). A narrow zone of indifference would explain the exquisite regulation of mood hypothesized to be the basis of the behavioral mechanism of mood titration.

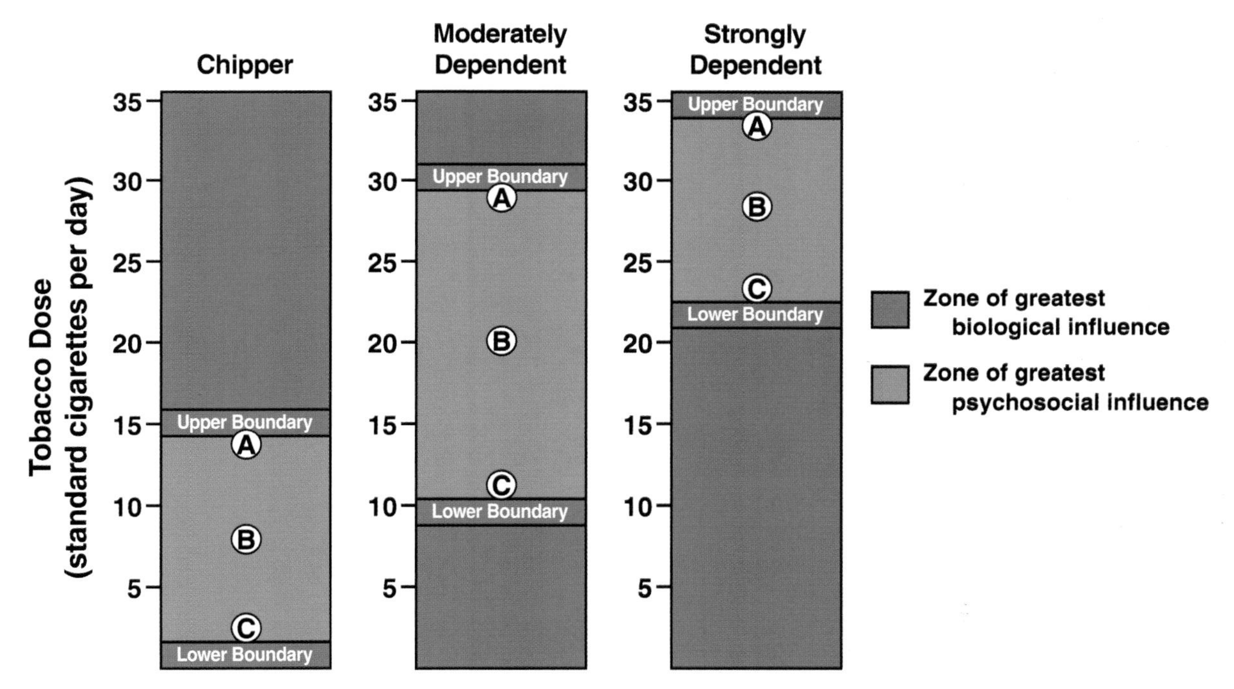

FIGURE 6.18 The 'Boundary Model' of tobacco use. The upper boundary **(A)** marks a zone representing the weighted sum of all the aversive biological consequences of smoking cigarettes. **(B)** represents the relative range of indifference to the pharmacological properties of the drug. It is the plateau of biological reward in which large dose changes have little effect on behavior. This zone demarcates general satisfaction with the drug dose. The lower boundary **(C)** marks a zone representing the weighted sum of all the biologically based pressures to smoke. A, B and C represent different rates of smoking sustained by an individual that vary from high psychosocial pressure to smoke (A) to very low psychosocial pressure (C). [Reproduced with permission from Kozlowski and Herman, 1984.]

NEUROBIOLOGICAL MECHANISM—NEUROCIRCUITRY

Acute Reinforcing and Stimulant-like Actions of Nicotine

Animal studies of the acute reinforcing effects of nicotine using intravenous self-administration have largely focused on activation of nicotinic acetylcholine receptors (nAChR) in the mesocorticolimbic dopaminergic system that project from the ventral tegmental area to the nucleus accumbens and the prefrontal cortex (Corrigall *et al.*, 1992, 1994; Nisell *et al.*, 1995; Pontieri *et al.*, 1996) (**Fig. 6.19**). Other neurochemical systems including cholinergic, glutamatergic, γ-aminobutyric acid (GABA), and opioid peptide systems may modulate nicotine reinforcement processes, but the preponderance of data to date indicate that other neurochemical systems involved in nicotine reinforcement interact with the midbrain dopamine system. Dopamine-independent positive reinforcing effects of nicotine remain to be demonstrated using the self-administration model but have been demonstrated with place preference (see below).

Nicotine produces its central and peripheral actions by binding to the nAChR complex (see below), and the cholinergic input to the mesolimbic dopamine pathway may provide a system through which nicotine may increase dopamine release. Administration of both noncompetitive and competitive nAChR antagonists block nicotine self-administration in the rat,

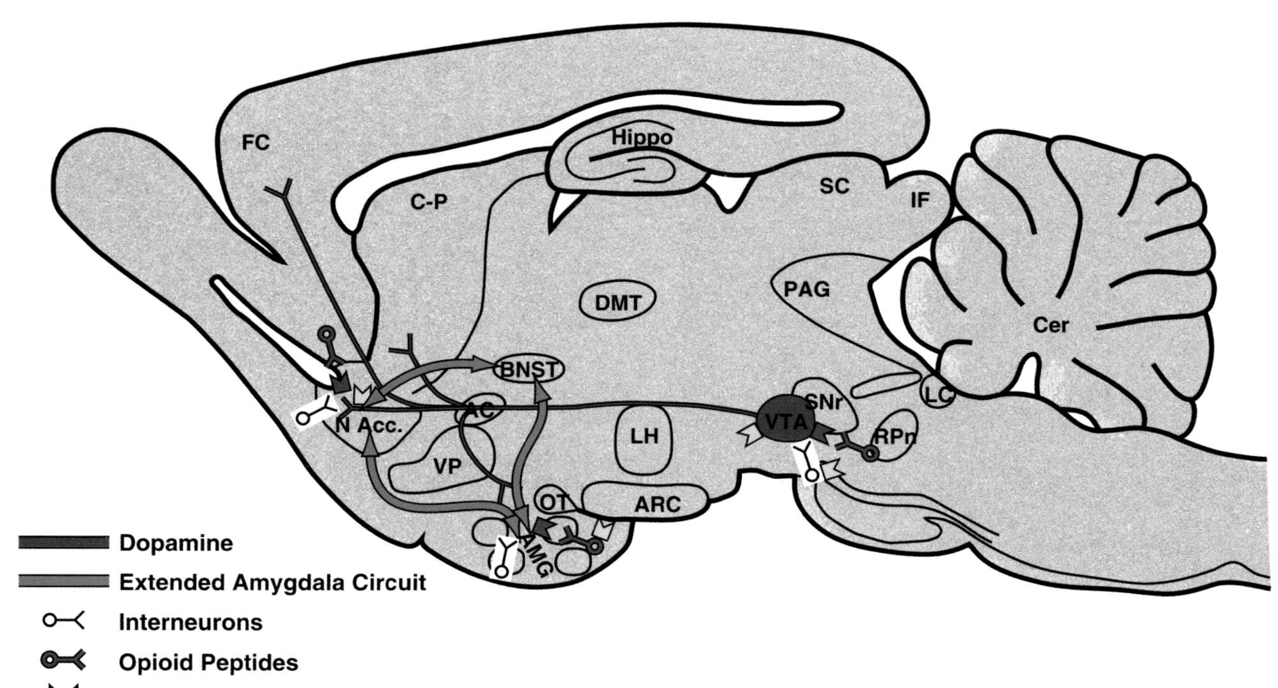

FIGURE 6.19 Sagittal section through a representative rodent brain illustrating the pathways and receptor systems implicated in the acute reinforcing actions of nicotine. Nicotine activates nicotinic acetylcholine receptors in the ventral tegmental area, nucleus accumbens, and amygdala either directly or indirectly via actions on interneurons. Nicotine also may activate opioid peptide release in the nucleus accumbens or amygdala independent of the dopamine system. The blue arrows represent the interactions within the extended amygdala system hypothesized to have a key role in nicotine reinforcement. AC, anterior commissure; AMG, amygdala; ARC, arcuate nucleus; BNST, bed nucleus of the stria terminalis; Cer, cerebellum; C-P, caudate-putamen; DMT, dorsomedial thalamus; FC, frontal cortex; Hippo, hippocampus; IF, inferior colliculus; LC, locus coeruleus; LH, lateral hypothalamus; N Acc., nucleus accumbens; OT, olfactory tract; PAG, periaqueductal gray; RPn, reticular pontine nucleus; SC, superior colliculus; SNr, substantia nigra pars reticulata; VP, ventral pallidum; VTA, ventral tegmental area.

indicating that the activation of nAChRs is involved in the reinforcing actions of nicotine (Corrigall and Coen, 1989; Corrigall *et al.*, 1994; Watkins *et al.*, 1999). Anatomical studies show that the ventral tegmental area, the region containing the cell bodies of the mesolimbic dopamine system, has high levels of nicotinic receptors (Clarke and Pert, 1985).

Evidence for the role of nAChRs in the ventral tegmental area in the positive reinforcing effects of nicotine is the observation that infusions of the competitive nAChR antagonist dihydro-β-erythroidine directly into the ventral tegmental area, but not the nucleus accumbens, produced a significant decrease in nicotine self-administration behavior (Corrigall *et al.*, 1994). Further, 6-hydroxydopamine lesions of the nucleus accumbens, or systemic administration of selective dopamine D_1 or D_2 receptor antagonists attenuated nicotine self-administration (Corrigall and Coen, 1991; Corrigall *et al.*, 1992) (**Fig. 6.20**). Finally, infusion of the nAChR antagonist methyllycaconitine into the ventral tegmental area attenuated nicotine-induced lowering of brain reward thresholds (Panagis *et al.*, 2000). Taken together, these results support the hypothesis that nicotine exerts its primary reinforcing action by activating dopamine neurons along the mesolimbic dopamine pathway.

However, an alternative, opposing hypothesis has been generated using conditioned place preference and direct injections of nicotine into the ventral tegmental area (Laviolette and van der Kooy, 2003). Direct infusions of nicotine into the ventral tegmental area of the rat produced a dose–dependent place aversion and

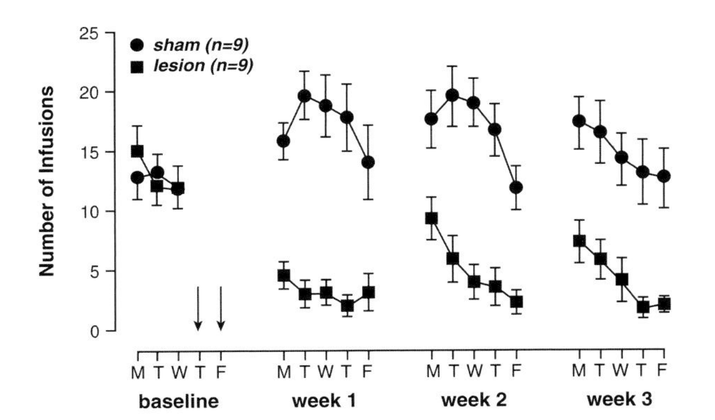

FIGURE 6.20 Effects of bilateral 6-hydroxydopamine lesions or ascorbate vehicle infusions into the dopamine terminal field in the nucleus accumbens on nicotine self-administration in rats. Each rat was lesioned on one of two days, as indicated by the arrows. Notice that although the number of nicotine infusions increases at the beginning of the second and third weeks of testing, this increase is not sustained. Over the 3-week test period, there is a marked difference between control and lesion groups. M, T, W, T, and F = days of the week. [Reproduced with permission from Corrigall *et al.*, 1992.]

then place preference in rats, and the place preference was potentiated by systemic or intra-nucleus accumbens administration of the dopamine receptor antagonist α-flupenthixol (Laviolette and van der Kooy, 2003). The authors interpreted their results to suggest that blockade of mesolimbic dopamine signaling may block the aversive properties of nicotine and thus increase the vulnerability to nicotine's rewarding and addictive properties. They further argue that these results may have bearing on the excessive smoking associated with antipsychotic treatment of schizophrenics. As attractive as the schizophrenia interaction hypothesis is at face value, there is a major difference between these results and intravenous self-administration of nicotine, where dopamine antagonists block the reinforcing effects of nicotine (Corrigall and Coen, 1991). Reconciliation of these positions may require further studies of intravenous nicotine self-administration with a range of doses.

As noted above, a primary site of action for the acute positive reinforcing properties of nicotine appears to be the activation of the mesolimbic dopamine system. Experimental evidence indicates that nicotine induces dopamine release partly by binding directly to nAChRs located within the mesolimbic system, specifically within the ventral tegmental area (Nisell *et al.*, 1994). In the rat brain, nAChRs have been identified on the cell bodies and dendrites of dopamine neurons in the ventral tegmental area, as well as their terminal fields in the nucleus accumbens (Schwartz *et al.*, 1984; Clarke and Pert, 1985; Swanson *et al.*, 1987; Wada *et al.*, 1989). While the presence of nAChRs throughout the mesolimbic dopamine system suggests that any of these sites could mediate the effects of nicotine , it has been hypothesized that nAChRs in the ventral tegmental area play a more important role than those in the nucleus accumbens in mediating the effects of nicotine on dopamine release. Dopamine metabolites were increased by ventral tegmental area infusion of nicotine, but not by nicotine infusion into the nucleus accumbens, an effect likely to reflect the larger area in the terminal field affected by the ventral tegmental area administration (Nisell *et al.*, 1994).

Systemic administration of nicotine has been shown to produce a dose–dependent increase in extracellular dopamine levels in the shell of the nucleus accumbens, a neurochemical effect shared by other drugs that also serve as positive reinforcers (Pontieri *et al.*, 1996, 1998; Nisell *et al.*, 1997). Infusion of the noncompetitive nicotinic antagonist mecamylamine into the ventral tegmental area blocked systemically administered nicotine-induced dopamine release in the nucleus accumbens, while infusion of mecamylamine directly into the nucleus accumbens failed to block dopamine

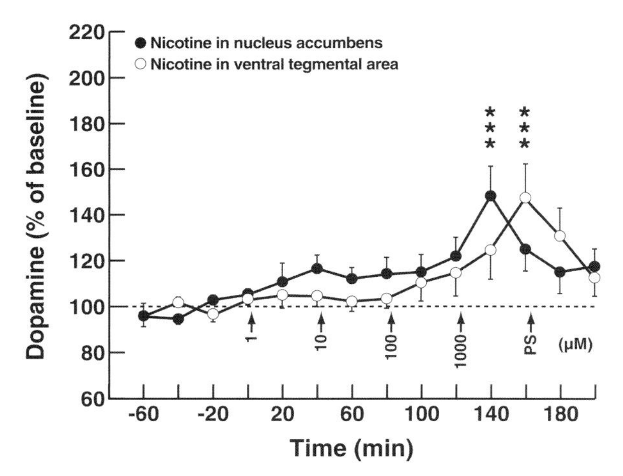

FIGURE 6.21 Temporal changes in extracellular concentrations of dopamine in rats after nicotine infusion (1, 10, 100, 1000 μM, 40 min each concentration) in the nucleus accumbens ($n = 5$) or ventral tegmental area ($n = 7$). Arrows indicate the start of infusion of each drug concentration or perfusion solution (PS). ***$p < 0.001$. [Reproduced with permission from Nisell *et al.*, 1994.]

release (Nisell *et al.*, 1994). Further, nicotine-induced dopamine release from terminals in the nucleus accumbens has been shown to be impulse- and voltage-dependent (Benwell *et al.*, 1993). Nevertheless, direct continuous infusion of nicotine in the ventral tegmental area *and* nucleus accumbens produced an increase in dopamine release in extracellular dopamine in the nucleus accumbens by approximately 50 per cent (Nisell *et al.*, 1994) (**Fig. 6.21**).

The cholinergic innervation from the pedunculopontine tegmental and laterodorsal tegmental nuclei projects to the dopamine neurons in the substantia nigra and ventral tegmental area and may be an important component of the nicotine reward circuit (Tago *et al.*, 1989; Oakman *et al.*, 1995; Azam *et al.*, 2002). Myelinated axons of the medial forebrain bundle (an area supporting high rates of self-stimulation behavior), projecting to the pedunculopontine nucleus, activate cholinergic neurons which are hypothesized to activate dopamine neurons in the ventral tegmental area by stimulating both nicotinic and muscarinic receptors (Yeomans and Baptista, 1997). Stimulation of cholinergic neurons within the pedunculopontine tegmental nucleus by exogenously administered nicotine leads to the release of endogenous acetylcholine which excites dopamine neurons in the substantia nigra and ventral tegmental area, and this activation is blocked by mecamylamine (Clarke *et al.*, 1987). Lesions of the pedunculopontine tegmental nucleus (Corrigall *et al.*, 1994), or direct administration of the nAChR antagonist dihydro-β-erythroidine into this region (Lanca *et al.*, 2000), blocked nicotine self-administration. Similarly, lesions of the pedunculopontine tegmental nucleus blocked the rewarding effects of nicotine as measured by place conditioning when the nicotine was injected directly into the ventral tegmental area (Laviolette *et al.*, 2002) (**Fig. 6.22**).

The mechanism by which nicotine activates the mesolimbic dopamine system also may involve an excitatory role for *N*-methyl-D-aspartate (NMDA) receptors in the ventral tegmental area. Acute administration of nicotine activates nAChRs located presynaptically on glutamatergic terminals, leading to increased evoked glutamate release (McGehee *et al.*, 1995). In turn, through excitatory actions at NMDA receptors on ventral tegmental area dopaminergic neurons, glutamate increases the burst firing of these neurons and subsequent dopamine release in the nucleus accumbens (Chergui *et al.*, 1993; Kalivas *et al.*, 1993; Hu and White, 1996). Most importantly, blockade of NMDA receptors with 2-amino-5-phosphonopentanoic acid injected directly into the ventral tegmental area dose–dependently attenuated nicotine-induced dopamine release in the nucleus accumbens (Schilstrom *et al.*, 1998). Systemic administration of another NMDA antagonist, MK-801 (dizocilpine), also blocked nicotine-induced dopamine release in the nucleus accumbens (Sziraki *et al.*, 1998). These data indicate that activation of excitatory nAChRs on glutamatergic terminals may be a key component of the nicotine mesolimbic dopamine interaction in the ventral tegmental area important for the acute reinforcing properties of nicotine (Kelley, 2002) (**Fig. 6.23**). Blockade of NMDA receptors decreased nicotine intravenous self-administration and nicotine-induced lowering of brain reward thresholds (Kenny and Markou, 2005).

There also are GABAergic inhibitory afferents to dopaminergic ventral tegmental area neurons (Walaas and Fonnum, 1980; Yim and Mogenson, 1980), inhibitory GABAergic interneurons within the ventral tegmental area, and medium spiny GABAergic neurons in the nucleus accumbens that also inhibit mesolimbic dopamine release (Kalivas *et al.*, 1993) and may have a role in nicotine–dopamine interactions. Enhancement of GABAergic neurotransmission through administration of γ-vinyl GABA (also referred to as vigabatrin), an indirect GABA agonist and irreversible inhibitor of GABA transaminase, the primary enzyme involved in GABA metabolism, blocked the reinforcing effects of nicotine as reflected in the conditioned place preference paradigm, and nicotine-induced dopamine increases in the nucleus accumbens (Dewey *et al.*, 1999).

In addition, systemic injections of γ-vinyl GABA that increase GABA levels (Jung *et al.*, 1977; Lippert *et al.*, 1977) decreased nicotine self-administration in rats (Paterson and Markou, 2002). Further, γ-vinyl

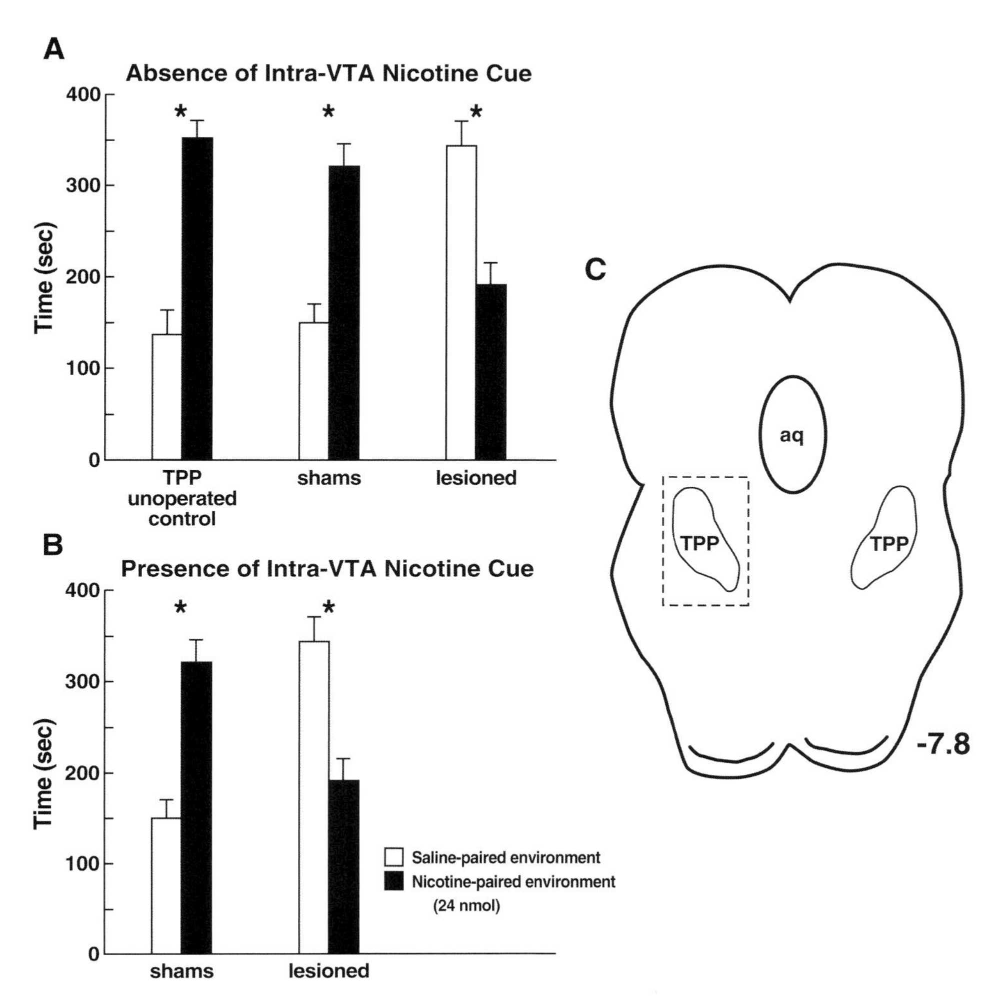

FIGURE 6.22 Motivational effects of intra-VTA nicotine in tegmental pedunculopontine nucleus (TPP; referred to elsewhere in the text as the pedunculopontine tegmental nucleus) in rats that did not have surgery (controls) and sham- and TPP-lesioned rats tested in the presence or absence of an intra-ventral tegmental area (VTA) nicotine cue (24 nmol). Error bars indicate SEM. **(A)** Animals receiving no TPP manipulations (TPP unoperated control; $n = 10$) displayed a robust conditioned place preference for the environment paired with nicotine ($^*p < 0.05$). Similarly, animals with sham lesions of the TPP ($n = 12$) displayed a robust conditioned place preference for the environment paired with intra-VTA nicotine ($^*p < 0.05$). In contrast, animals with excitotoxic bilateral TPP lesions ($n = 12$) demonstrated a reversal in the motivational valence of intra-VTA nicotine and displayed a significant aversion to the environment paired with intra-VTA nicotine ($^*p < 0.05$). **(B)** Both sham- and TPP-lesioned animals were retested in the presence of the intra-VTA nicotine cue. Although sham-lesioned animals continued to display a robust conditioned place preference for the environment paired with intra-VTA nicotine ($^*p < 0.05$), TPP-lesioned animals continued to demonstrate a significant aversion to the environment paired with intra-VTA nicotine ($^*p < 0.05$). **(C)** Schematic of a coronal section showing the approximate anatomical TPP region. aq, aqueduct. [Reproduced with permission from Laviolette *et al.*, 2002.]

GABA administration dose- and time-dependently lowered the nicotine-induced increases in nucleus accumbens dopamine in both naïve and chronically nicotine-treated rats as measured by *in vivo* microdialysis (Dewey *et al.*, 1999). It also abolished nicotine-induced increases in dopamine in the striatum of primates as measured by positron emission tomography (Schiffer *et al.*, 2000). Other evidence suggesting that increased GABAergic transmission through $GABA_B$ receptors is involved in these effects involves the use of selective $GABA_B$ receptor agonists. Systemic or microinjections of baclofen or (3-amino-2[*S*]-hydroxypropyl)-methylphosphinic acid (CGP 44532), two $GABA_B$ receptor agonists, directly into the ventral tegmental area or the pedunculopontine tegmental nucleus, but not into the caudate-putamen, decreased the reinforcing effects of nicotine (Corrigall *et al.*, 2000, 2001; Fattore *et al.*, 2002; Paterson *et al.*, 2004). By contrast, 4,5,6,7-tetrahydroisoxazolo[5,4-c]pyridin-3-ol hydrochloride, a $GABA_A$ agonist, or a weak (+)baclofen enantiomer did not modify nicotine self-administration (Paterson *et al.*, 2004). Taken together, these findings provide support for the hypothesis that GABAergic mechanisms may be involved in modulating nicotine reinforcement (Kelley, 2002) (**Fig. 6.23**). Another possible neuropharmacological interaction of tobacco with the ventral tegmental area may involve activation of dopamine neurons via acetaldehyde, a component of cigarette smoke, and this may be a source of tobacco/alcohol interactions (see *Alcohol* chapter). Acetaldehyde is self-administered intravenously (Myers *et al.*, 1982), intracerebroventricularly (Brown *et al.*, 1980) and directly into the ventral tegmental area (Rodd-Henricks *et al.*, 2002).

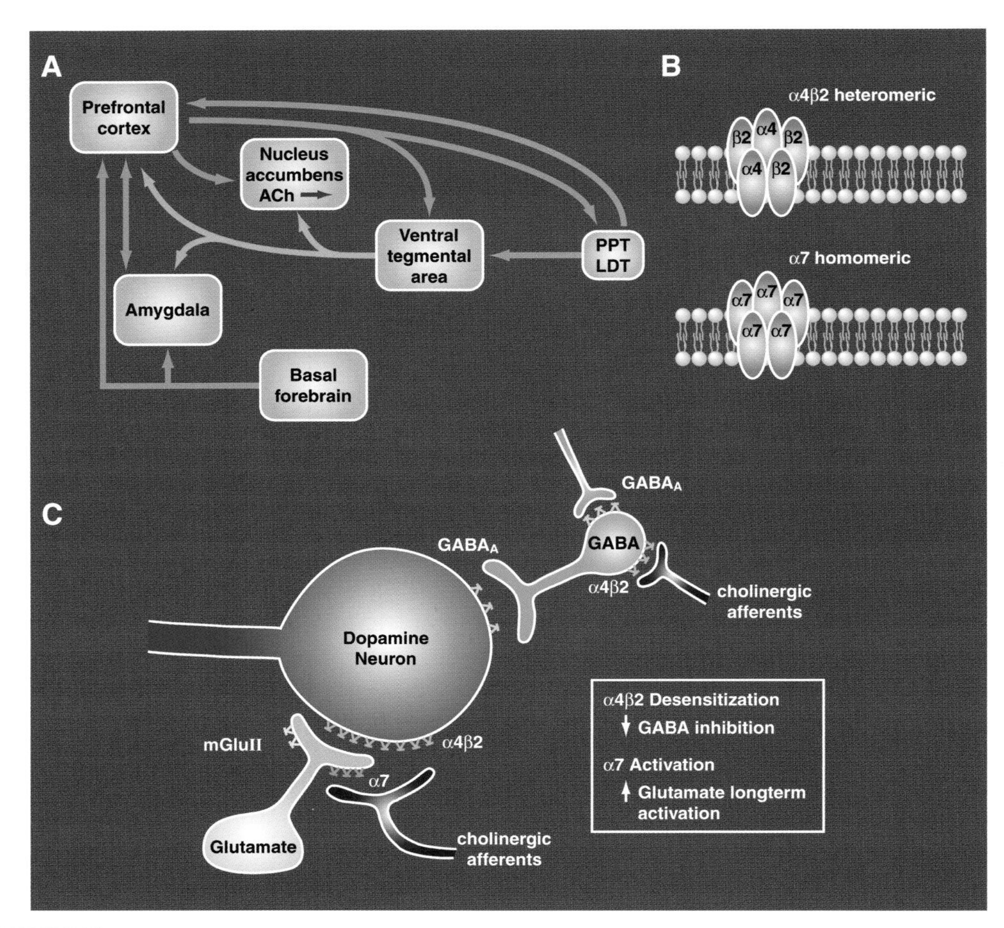

FIGURE 6.23 Interconnecting brain pathways, receptors, and synapses hypothesized to mediate neuroadaptive changes that accompany nicotine addiction. **(A)** Neural systems implicated in addiction and in behavioral sensitization. Dopamine neurons (*pink*), located in the midbrain ventral tegmental area project to limbic forebrain and cortical regions such as the prefrontal cortex, amygdala, and nucleus accumbens. These structures in turn are connected by glutamate pathways (*green*). The cholinergic pathways (*blue*) include the cell bodies in the pedunculopontine nucleus (PPT) and lateral dorsal tegmentum (LDT), located in the pons, that project to the ventral tegmental area and also to cortical regions. A second major cholinergic pathway arises from the basal forebrain and reaches widespread cortical and limbic regions. A third source of forebrain acetylcholine constitutes the intrinsic neurons of the nucleus accumbens, a brain reward region. **(B)** Two different types of neuronal nicotinic receptors commonly found in the cholinergic pathways described above and proposed to be involved in the brain's response to drugs of abuse. [Panels A and B are reproduced with permission from Kelley, 2002.] **(C)** Potential interactions between acetylcholine, glutamate, and dopamine cells in the midbrain. Acetylcholine release in this area could activate both presynaptic nicotinic α_7 receptors, located on glutamatergic terminals (GLU), and postsynaptic $\alpha_4\beta_2$ nicotinic receptors, located on dopamine cell bodies. Activated acetylcholine release by drugs of abuse such as cocaine and amphetamine in the ventral tegmental area (and in other regions) could amplify the excitatory effects of glutamate and dopamine. Complex interactions between the acetylcholine, dopamine, and glutamate systems are likely to set in motion during the addiction process. [Based on Mansvelder and McGehee, 2002.]

Nicotine has long been hypothesized to interact with opioid peptides (Pomerleau and Pomerleau, 1984), and there is some clinical data supporting this hypothesis, but limited evidence to date from intravenous self-administration models. Nicotine stimulates nAChRs within the hypothalamus that in turn activate the proopiomelanocortin peptide group that includes the precursor to β-endorphin (Pomerleau, 1998), but it is likely that activation of pituitary β-endorphin has more of a hormonal role. β-endorphin and adrenocorticotropic hormone are released and regulated simultaneously (Guillemin *et al.*, 1977). Nicotine administered systemically has been shown to increase tissue levels of opioid peptides in the nucleus accumbens (Pierczchala *et al.*, 1987; Houdi *et al.*, 1991). High densities of μ opioid receptors have been identified in the nucleus accumbens, and it has been suggested that nicotine may release endogenous opioid peptides that bind to these receptors

(Tempel and Zukin, 1987; Davenport *et al.*, 1990). Studies in humans have suggested that naltrexone blocked or attenuated the reinforcing effects of nicotine (Karras and Kane, 1980; Gorelick *et al.*, 1988–1989). However, neuropharmacological studies examining the role of opioid receptors in the ventral tegmental area and nucleus accumbens in the acute reinforcing effects of nicotine in rat intravenous self-administration models have been largely negative (Corrigall and Coen, 1991; Corrigall *et al.*, 2000). However, the μ opioid receptor agonist DAMGO suppressed nicotine intake when injected into the pedunculopontine tegmental nucleus (Corrigall *et al.*, 2000, 2002).

More recent studies may provide some explanation for this pattern of results. μ opioid receptor knockout mice showed a blockade of the reinforcing effects of nicotine as measured by place preference, a reduction in the analgesic effects of nicotine, and a reduction in nicotine withdrawal in dependent mice (Berrendero *et al.*, 2002). These results suggest that endogenous opioids may play a role in the reinforcing effects of nicotine, but possibly outside of a direct action on the mesolimbic dopamine system. One hypothesis would be that the reinforcing properties of nicotine are modulated by activation of enkephalin neurons in a reward system parallel to the dopamine system (i.e., dopamine-independent system), as has been hypothesized for other drugs of abuse (Pomerleau and Pomerleau, 1984; Houdi *et al.*, 1991; Koob *et al.*, 1998).

Evidence for the involvement of the serotonergic system in the acute positive reinforcing effects of nicotine is limited. High-affinity nAChRs are located in both the median raphe nucleus and the hippocampus (Benwell *et al.*, 1988; Marks *et al.*, 1992; Alkondon and Albuquerque, 1993; Li *et al.*, 1998), and acute systemic administration of a high nicotine dose increased the release of serotonin in the frontal cortex of rats (Ribeiro *et al.*, 1993). However, subsequent pharmacological studies have provided little support for a role of the serotonin system in acute nicotine reinforcement. Administration of either ICS 205-930 or MDL 72222, two selective serotonin 5-HT_3 receptor antagonists, had no effect on intravenous nicotine self-administration in the rat (Corrigall and Coen, 1994). Furthermore, in a rat model of oral nicotine self-administration, administration of ipsapirone, a 5-HT_{1A} agonist, had no effect on nicotine intake (Mosner *et al.*, 1997). However, the 5-HT_{2C} receptor antagonist RO 60-0178 decreased nicotine self-administration in rats (Grottick *et al.*, 2001).

A cannabinoid CB_1 receptor antagonist also blocked nicotine self-administration (Cohen *et al.*, 2002), and CB_1 knockout mice did not exhibit a conditioned place preference to nicotine (Castane *et al.*, 2002). Rats also developed a reliable conditioned place preference to nicotine, and both the acquisition and expression of such place preference was blocked by a CB_1 antagonist (Le Foll and Goldberg, 2004; Forget *et al.*, 2005). Nicotine-associated cues produced reinstatement of nicotine-seeking behavior in rats, similar to other drugs of abuse. Rats trained to self-administer nicotine intravenously and subjected to extinction in the absence of cues showed reinstatement of responding to a stimulus cue paired with nicotine self-administration, but not to nicotine itself (Le Sage *et al.*, 2004; De Vries *et al.*, 2005). Such cue-induced reinstatement was blocked by a CB_1 antagonist (Cohen *et al.*, 2005a) and a $GABA_B$ agonist (Paterson *et al.*, 2005). Together these results suggest that as with other drugs of abuse there may be a cannabinoid component to nicotine-seeking behavior (De Vries and Schoffelmeer, 2005).

In summary, intravenous nicotine self-administration is blocked by dopamine antagonists and lesions of the midbrain dopamine system. However, place preference produced by nicotine is not blocked by dopamine antagonists but in fact is potentiated by dopamine antagonism. Nicotine increases dopamine release in the nucleus accumbens as measured by *in vivo* microdialysis. Other sites hypothesized to be involved in the acute reinforcing effects of nicotine include the pedunculopontine tegmental nucleus of which lesions block both nicotine self-administration and place preference. Neurochemical systems hypothesized to modulate nicotine reward include glutamate and GABA, with glutamate antagonists blocking nicotine self-administration and dopamine release produced by nicotine, and GABA agonists having a similar effect (Kenny and Markou, 2005). Together these results suggest a combination of glutamate activation and GABA inhibition in the activation of ventral tegmental area dopamine cells by nicotine. Opioid peptides may be involved in the reinforcing effects of nicotine and may be independent of dopamine activation. Endocannabinoids also may have a role in nicotine-seeking behavior.

Nicotine Dependence

As with other drugs of abuse such as opiates and alcohol, the nicotine withdrawal syndrome in non-human animals includes somatic and motivational signs, and the remarkable aspect of nicotine withdrawal is the pronounced and prolonged elevation in brain reward thresholds compared to its efficacy as a reinforcer and the lack of intensity of the somatic withdrawal signs which are relatively mild (see above). This parallels the human condition, where the predominant symptoms of nicotine withdrawal are high irritability, malaise, and high craving. The hypothesized neurochemical basis for the dramatic elevations in brain reward function during nicotine withdrawal includes changes in nicotinic receptor function (see below), decreased dopamine activity, alterations in dopamine-glutamate and dopamine-GABA interactions, changes in serotonergic function, and changes in opioid peptide function. There have been few studies on the effects of chronic nicotine exposure on subsequent nicotine self-administration. Exposure of rats to nicotine during periadolescence increased nicotine self-administration in adult animals (Adriani *et al.*, 2003). Increases in dopamine neuron-specific nicotinic receptor subunits $\alpha5$, $\alpha6$, and $\beta2$ also were observed, suggesting that adolescent exposure may produce enhanced neurobehavioral vulnerability to nicotine (Adriani *et al.*, 2003). Adult rats will self-administer nicotine in continuous access situations (24 h/day) (Valentine *et al.*, 1997; Fu *et al.*, 2001; Brower *et al.*, 2002; Le Sage *et al.*, 2002), but specific neuro-behavioral changes associated with such long-term self-administration have yet to be explored.

Decreases in extracellular levels of dopamine in the nucleus accumbens and amygdala have been observed during precipitated nicotine withdrawal in rats chronically exposed to nicotine (Hildebrand *et al.*, 1998, 1999; Panagis *et al.*, 2000) (**Fig. 6.24**). In spontaneous nicotine withdrawal, tissue levels of dopamine in the nucleus accumbens were reduced approximately 32 per cent compared to saline controls (Fung *et al.*, 1996). A possible explanation for the reduction in dopamine release following chronic nicotine exposure may involve nicotinic receptor desensitization that leads to decreased neuronal firing.

Decreased neuronal firing in the ventral tegmental area has been reported during chronic continuous nicotine infusion (6 mg/kg/day, nicotine base, for 12 days) (Rasmussen and Czachura, 1995). During spontaneous nicotine withdrawal, neuronal firing in the continuously nicotine-treated group returned to baseline (Rasmussen and Czachura, 1995). In animals subjected to chronic nicotine infusion (4 mg/kg/day, nicotine base, for 7 days), a subsequent acute nicotine challenge potentiated the increase in nicotine-induced dopamine release compared to the increase measured after an acute nicotine challenge without previous nicotine exposure (Marshall *et al.*, 1997). The acute challenge, however, was given approximately 20 h after termination of chronic nicotine exposure, possibly sufficient time to effect the recovery of desensitized nAChRs.

Another part of the neuronal circuit that possibly drives nicotine reward during withdrawal may be at the level of pedunculopontine tegmental nucleus cholinergic neurons which terminate on dopamine neurons in the ventral tegmental area (Yeomans *et al.*, 1993; Yeomans and Baptista, 1997; Laviolette *et al.*, 2002) (see **Fig. 6.22**). It may be that after nAChR desensitization and upregulation, in the absence of sufficient agonist to stimulate the receptors, there is reduced cholinergic activation of dopamine neurons. Thus, a reduction in cholinergic input to dopamine neurons along the reward pathway may result in decreased brain reward function. However, these proposed neuroadaptations involving dopamine may only partly contribute to nicotine withdrawal symptomatology.

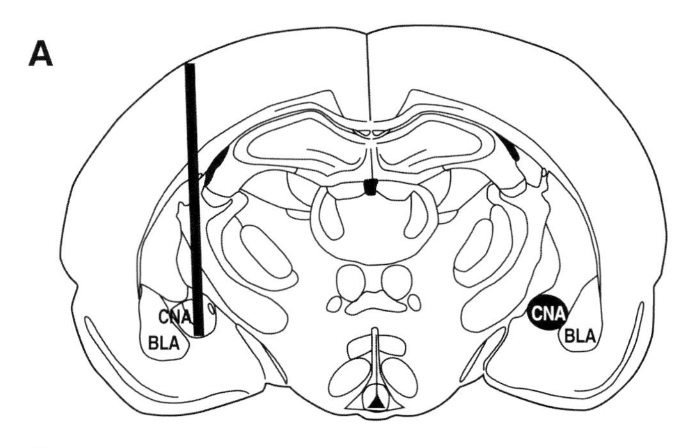

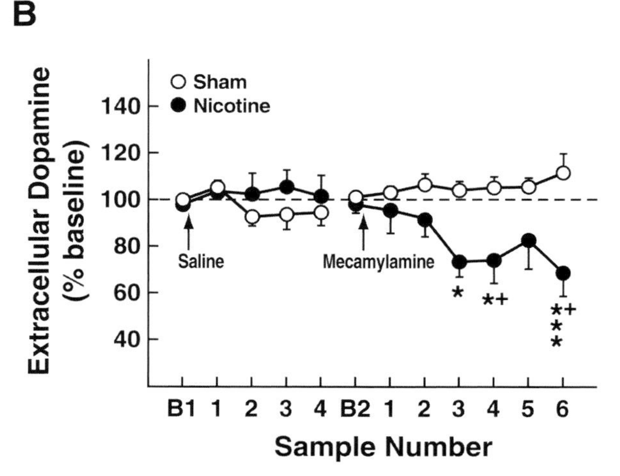

FIGURE 6.24 (A) Representative schematic illustrating the placement of a microdialysis probe in the rat central nucleus of the amygdala (CNA, left side) and the region in which Fos-immunoreactive nuclei were counted (CNA, shaded area, right side). BLA: basolateral amygdala. (B) Temporal changes of extracellular concentrations of dopamine in the central nucleus of the amygdala after systemic injections (as indicated by the arrows) of saline and mecamylamine (1 mg/kg, s.c.) in a group of animals infused for 14 days with nicotine (1, $n = 5$–6) and in control animals implanted with empty minipumps (Sham, $n = 5$–8). B1 indicates baseline 1, which is the sample immediately preceding saline injection (100 per cent in the figure is the average of the two samples preceding saline injection). B2 indicates baseline 2, which is the sample immediately preceding mecamylamine injection (100 per cent is defined as the average of the two samples preceding the mecamylamine injection). $^*p < 0.05$ compared to B2, $^{***}p < 0.05$, [Reproduced with permission from Panagis *et al.*, 2000.]

Alterations in glutamatergic, GABAergic, opioidergic, and serotonergic systems also may contribute to the motivational aspects of nicotine withdrawal. Nicotine withdrawal also is accompanied by increases in the acoustic startle response. The presynaptic Group II metabotropic glutamate receptor agonist LY354740 completely blocked the increased startle response induced by nicotine withdrawal (Helton *et al.*, 1997). Group II receptors are most likely presynaptic, based on the finding that activation of these receptors leads to decreased glutamatergic neurotransmission in limbic areas, such as the hippocampus and the amygdala (Pin and Duvoisin, 1995; Battaglia *et al.*, 1998). The attenuation of a nicotine withdrawal syndrome is presumably mediated by reversing an overexcitation of the glutamatergic system resulting from chronic nicotine administration (**Table 6.5**). However, it is interesting to note that the Group II agonist LY314582 precipitated withdrawal-like elevations of brain reward thresholds in nicotine-dependent, but not control, rats, and the Group II antagonist LY341495 attenuated the elevations of brain reward thresholds in rats undergoing spontaneous nicotine withdrawal (Kenny *et al.*, 2003) (**Fig. 6.23**). These data demonstrate a complex role for Group II metabotropic glutamate receptors in nicotine dependence and provide further evidence for a dissociation of the systems that regulate somatic and motivational components of withdrawal (for review, see Kenny and Markou, 2004).

There also is evidence that glutamate is involved in the sensitization of locomotor activity associated with chronic administration of nicotine. Co-administration with nicotine of NMDA receptor antagonists, such as the noncompetitive antagonist MK-801 (dizocilpine) or the competitive antagonist D-CPP, reduced the development of tolerance to the locomotor depressant effect of nicotine, attenuated the development of tolerance to the aversive stimulus effects of nicotine as measured by conditioned taste aversion, and prevented sensitization to the locomotor activating effects of nicotine (Shoaib *et al.*, 1994, 1997; Shoaib and Stolerman, 1996). Furthermore, pretreatment with MK-801 reduced nicotinic receptor upregulation during chronic exposure, suggesting a role for glutamate in this neuroadaptive response (Shoaib *et al.*, 1997).

Opioid peptide systems also may play a role in nicotine dependence. The somatic signs of nicotine withdrawal have been precipitated by the opiate antagonist naloxone in nicotine-dependent rats (Malin *et al.*, 1993; but see Watkins *et al.*, 2000b), or dansyl-RFamide, an analog of neuropeptide FF, an antiopiate peptide (Malin *et al.*, 1996). Moreover, acute injections of morphine, an opiate agonist, reversed the somatic signs of nicotine withdrawal (Malin *et al.*, 1993). However, naloxone is less effective in precipitating motivational signs of withdrawal in nicotine-dependent rats (Watkins *et al.*, 2000b). Administration of naloxone to nicotine-dependent humans produced dose–dependent increases in self-reported affective and somatic signs of nicotine withdrawal, also suggesting that long-term exposure to nicotine is associated with alterations in endogenous opioid peptide systems (Krishnan-Sarin *et al.*, 1999). Thus, it may be hypothesized that during chronic nicotine exposure there is a release of opioid peptides (Pomerleau and Pomerleau, 1984; Pomerleau and Rosecrans, 1989; Boyadjieva and Sarkar, 1997), which leads to a downregulation of μ opioid receptors or opioid receptor transduction mechanisms. During nicotine abstinence (i.e., in the absence of an agonist), this downregulation of μ opioid receptors or opioid receptor transduction mechanisms may contribute to some, but not all, aspects of nicotine withdrawal.

Evidence also suggests a role of altered serotonin neurotransmission in nicotine withdrawal. Chronic nicotine treatment produced a selective decrease in the concentration of serotonin in the hippocampus (Benwell and Balfour, 1979), and increases in the number of hippocampal 5-HT_{1A} receptors have been measured in chronic smokers. This receptor upregulation may

TABLE 6.5 Changes in Neurotransmitter Function During Various Stages of the Nicotine Self-Administration and Addiction Cycle

Neurotransmitter	Acute administration	Chronic administration	Withdrawal	Protracted abstinence
Nicotinic acetylcholine receptors	↑	↓	↑	↑
Dopamine receptors	↑	↓	↓	nd
Glutamate receptors	↓	↑	↑	nd
Opioid peptides	↑	nd	↓	nd
Serotonin receptors	↑	↓	↓	nd
Corticotropin-releasing factor receptors	↓	↑	↑	nd

nd = not determined.

reflect a reduction in the activity of serotonergic neurons within the median raphe nucleus, which innervates the hippocampus, amygdala, and several other forebrain structures (Benwell *et al.*, 1990). Considering the findings that serotonin deficits have been implicated in depression and anxiety (Coppen, 1967; Young *et al.*, 1985; Delgado *et al.*, 1990, 1991; Markou *et al.*, 1998), it may be hypothesized that during chronic nicotine exposure and nicotine withdrawal, the decreases in serotonin function play a role in the onset of negative affective symptoms, such as depressed mood, impulsivity, and irritability. nAChRs are located in the somatodendritic region in the median raphe nucleus and the terminal fields in the forebrain and may facilitate serotonin release. Chronic nicotine treatment causes nicotinic receptor desensitization, decreased serotonin release, and ultimately increased postsynaptic 5-HT_{1A} receptors to maintain baseline functional activity within the terminal regions. During nicotine abstinence, the decreased serotonin, combined with upregulated postsynaptic 5-HT_{1A} serotonergic receptors, may be hypothesized to contribute to the depressed mood often reported during nicotine withdrawal (Hughes *et al.*, 1991b). In rats withdrawing from nicotine, pretreatment with the 5-HT_{1A} antagonists NAN-190, LY206130, and WAY-100635 significantly reduced the withdrawal-induced increase in the startle response, supporting this hypothesis (Rasmussen *et al.*, 1997). However, administration of p-MPPI, a 5-HT_{1A} receptor antagonist, did not reverse either brain stimulation threshold elevations or the somatic signs of nicotine withdrawal, suggesting that different symptoms of nicotine withdrawal may be mediated by different components of the serotonergic system (Harrison *et al.*, 2001).

Alterations in brain stress systems also may contribute to the motivational symptoms associated with nicotine withdrawal. Specifically, overactivity of the stress hormone corticotropin-releasing factor (CRF) may underlie the symptoms of anxiety, increased stress, and irritability often reported by abstinent smokers. The hypothesis that CRF is activated during nicotine withdrawal is based on the observation that acute withdrawal from nicotine can produce an increase in circulating corticosterone (Benwell and Balfour, 1979) and that CRF is increased in the central nucleus of the amygdala during precipitated nicotine withdrawal (Ghozland *et al.*, 2004). CRF has been shown to be increased during withdrawal from chronic administration of other major drugs of abuse, including cocaine, ethanol, and cannabinoids (Baldwin *et al.*, 1991; Sarnyai *et al.*, 1995; Rodriguez de Fonseca *et al.*, 1997).

While little is known about the role of the extended amygdala in nicotine reinforcement, several studies have investigated the role of the extended amygdala in reinforcement produced by other drugs of abuse (see *Psychostimulants*, *Opioids*, and *Alcohol* chapters). Nicotine may exert similar effects on specific components of the extended amygdala, the dopamine projections to the extended amygdala, or both. nAChR mRNA has been located in neurons throughout the rat amygdala, including the central nucleus of the amygdala (Wada *et al.*, 1989), indicating potential functional nAChRs within these areas. Further, Fos expression and decreased dopamine release have been measured in the central nucleus of the amygdala during precipitated nicotine withdrawal (Panagis *et al.*, 2000), suggesting that alterations occur within the extended amygdala during chronic nicotine exposure. The ability of nicotine to increase dopamine release specifically in the shell and not the core of the nucleus accumbens (Nisell *et al.*, 1994; Pontieri *et al.*, 1996), and the expression of nicotinic receptor gene products in the central nucleus of the amygdala (Wada *et al.*, 1989) leave open the possibility of a potential role of the extended amygdala in the reinforcing effects of nicotine.

In summary, the same neurotransmitter systems hypothesized to mediate the acute reinforcing effects of nicotine are compromised during the development of dependence and this is reflected in decreases in function during precipitated withdrawal (**Table 6.5**). Dopamine and serotonin show decreased activity during withdrawal. In addition, acute withdrawal is accompanied by increases in glutamate activity and recruitment of CRF, and these changes also may contribute to the negative affective state hypothesized to mediate the negative reinforcement associated with nicotine dependence.

NEUROBIOLOGICAL MECHANISM—CELLULAR

Nicotine increases the firing rate of dopaminergic neurons in the ventral tegmental area and substantia nigra, and this activation appears to be via the $\alpha 4\beta 2$ subtype (Clarke *et al.*, 1985; Grenhoff *et al.*, 1986; Pidoplichko *et al.*, 1997; Erhardt *et al.*, 2002; Matsubayashi *et al.*, 2003). The same concentration of nicotine that a smoker would achieve activated and then desensitized nicotinic receptors on midbrain dopamine neurons in a slice preparation, suggesting a mechanism for acute tolerance (Pidoplichko *et al.*, 1997) (**Fig. 6.25**). Nicotine at an intraperitoneal dose of 0.5 mg/kg increased the amount of burst firing in ventral tegmental area neurons, and mecamylamine decreased firing, suggesting a tonic nicotinic inhibition (Grenhoff *et al.*, 1986). Using a whole-cell patch clamp technique, nicotine induced inward currents

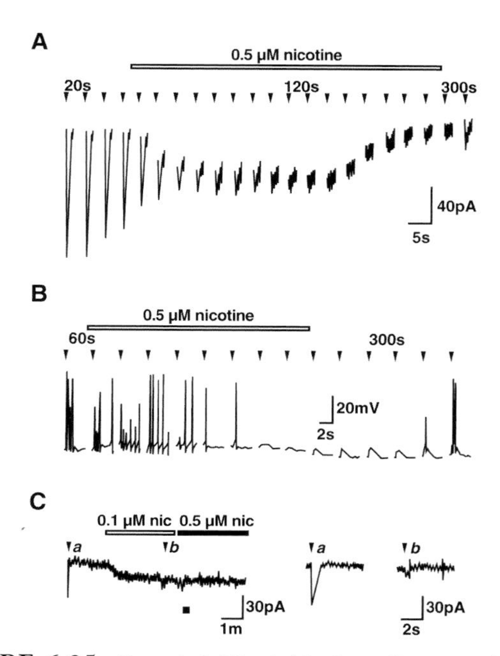

FIGURE 6.25 Extended (19 min) bath applications of 0.5 μM nicotine have multiple effects on dopamine neuron activity in a rat brain slice. **(A)** Under voltage clamp, injections of acetylcholine (1 mM for 40 ms, arrowheads) caused a depolarization of the resting membrane potential (induced inward currents) that were desensitized by bath application of nicotine (solid bar). The injections of acetylcholine were always given every 20 s, but the time interval between displayed traces is initially 20 s, then 120 s, and then 300 s. The current activated by bath-application of nicotine is displayed by a downward displacement of the currents induced by injection of acetylcholine, and at later times in nicotine, the baseline current desensitizes toward its original value. **(B)** Under current clamp, injection of acetylcholine (1 mM for 40 ms) initially depolarized the neuron and caused action potentials, but at later times following nicotine the neuron did not respond to acetylcholine. The injections of acetylcholine were given every 20 s, but the time interval between displayed traces is altered for clarity from 60 to 300 s. **(C)** Left, bath application of 0.1 μM nicotine activated a 17 pA current. After 3 min in 0.1 μM, applying 0.5 μM nicotine activated very little additional current. If there had been no desensitization, the black square marks the average size of the current activated by 0.5 μM nicotine. Acetylcholine pressure injections (1 mM, 30 ms, arrow heads) were applied before (A) and near the end (B) of the 0.1 μM nicotine. Those acetylcholine-induced currents on an expanded time scale (right) illustrate the extent of desensitization. [Reproduced with permission from Pidoplichko *et al.*, 1997.]

reflecting depolarization with an increase in firing (Matsubayashi *et al.*, 2003). However, prolonged exposure to the same concentrations of nicotine that acutely increase firing caused desensitization of the nicotinic receptors and may explain acute tolerance to nicotine's effects (Pidoplichko *et al.*, 1997). At a local circuit level, presynaptic activation of nicotinic acetylcholine receptors on glutamate terminals may mediate some of this activation (Mansvelder and McGehee, 2000; Erhardt *et al.*, 2002). Also, in slice preparations, nicotine enhanced firing of dopamine and nondopamine (presumably GABAergic) neurons, but the non-dopamine neurons desensitized more rapidly, possibly contributing to a more prolonged activation of the dopamine neurons via a disinhibitory effect (Yin and French, 2000; Mansvelder *et al.*, 2002). One hypothesis to explain the overall actions of nicotine in the ventral tegmental area is that nicotine binds to $\alpha 7$ nAChRs on presynaptic glutamate terminals in the ventral tegmental area (Jones and Wonnacott, 2004), and induces a long-term potentiation of excitatory glutamate transmission onto midbrain dopamine neurons. However, the high affinity $\alpha 4\beta 2$ receptors on GABAergic neurons may desensitize quickly, resulting in a concomitant decrease of inhibitory GABA transmission onto dopamine neurons. The combination of effects on glutamate and GABA neurons is hypothesized to produce a powerful and prolonged activation of dopamine neurotransmission in the mesolimbic dopamine system (Mansvelder *et al.*, 2002; Fagen *et al.*, 2003) (**Figs. 6.23** and **6.26**).

Other reward areas activated by nicotine include the facilitation of striatal neuron activity by microiontophoretic application of nicotine that presumably acts by facilitating release of dopamine in the striatum (Yu *et al.*, 2000). This facilitation of dopamine release may be indirect, resulting from a nicotine-evoked increase in glutamate release produced by stimulation of presynaptic glutamate receptors on dopamine-containing terminals (Garcia-Munoz *et al.*, 1996).

Nicotine had mixed effects on dorsal raphe neurons in studies using slice preparations producing increases in firing rates in approximately two thirds of the neurons and decreases in firing in one third of the neurons (Mihailescu *et al.*, 1998, 2002). The stimulatory effects on raphe neurons may be direct. Patch clamp recordings from serotonin-containing neurons in the raphe magnus demonstrated a loss of nicotine-elicited currents in both $\alpha 4$ and $\beta 2$ mutant mice, suggesting a contribution of the $\alpha 4$ and $\beta 2$ subunits to functional receptors in the raphe (Marubio *et al.*, 1999). The inhibitory effects of nicotine were reversed by a 5-HT_{1A} inhibitor, suggesting that these inhibitory effects were mediated through serotonin release (Mihailescu *et al.*, 2002). The excitatory effects of nicotine were potentiated by administration of a $GABA_A$ antagonist, also suggesting an inhibitory effect via GABA interneurons (Mihailescu *et al.*, 2002). In support of this hypothesis, iontophoretic application of nicotine inside the dorsal raphe failed to inhibit firing (Engberg *et al.*, 2000).

Nicotine withdrawal following chronic exposure showed increased sensitivity to a 5-HT_{1A} agonist in an *in vivo* anesthetized single unit preparation, suggesting that the serotonin system also may have a role in the neuroadaptations associated with the development of dependence (Rasmussen and Czachura, 1997).

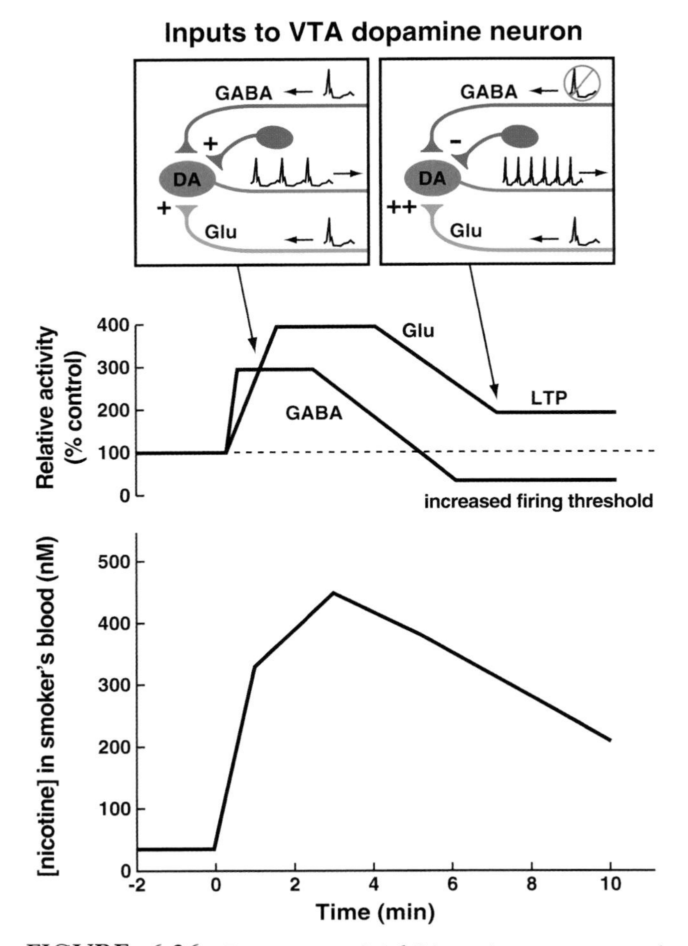

FIGURE 6.26 Excitatory and inhibitory inputs to ventral tegmental area (VTA) dopamine (DA) neurons have a different time course of activity in response to nicotine concentrations in a smoker's arterial blood. The upper diagram depicts the changes in activity of the different ventral tegmental area cell types following nicotine exposure. The middle panel is a schematic of the relative activity of γ-aminobutyric acid (GABA) and glutamate inputs to ventral tegmental area dopamine neurons during a physiologically relevant nicotine concentration profile, as depicted in the lower panel. Time course of nicotine concentrations measured in smoker's arterial blood during cigarette smoking is adapted from Henningfield *et al.* (1993). [Reproduced with permission from Mansvelder *et al.*, 2002.]

More specifically, nicotine withdrawal may increase the inhibitory influence of somatodendritic 5-HT_{1A} autoreceptors located within the raphe nuclei contributing to decreases in serotonin release into forebrain sites (e.g., Benwell and Balfour, 1982) which may contribute to nicotine withdrawal signs. This hypothesis is supported by the observation that a serotonergic antidepressant treatment that combines the selective serotonin reuptake inhibitor fluoxetine and the 5-HT_{1A} receptor antagonist p-MPPI rapidly reversed the elevation in brain-stimulation reward thresholds, but not the somatic signs, observed in rats undergoing nicotine withdrawal (Harrison *et al.*, 2001). Long-term self-administration of nicotine in adults rats profoundly decreased the expression of neural cell adhesion molecules and neurogenesis in the hippocampal dentate gyrus and increased cell death (Abrous *et al.*, 2002).

In summary, nicotine acutely activates the mesolimbic dopamine system by three potential actions: (1) via a direct action on dopamine neurons themselves, (2) via presynaptic activation of glutamate release which shows a long-lasting potentiation, and (3) via activation of GABA interneurons which desensitize more rapidly. The combination of activation of glutamate and deactivation of GABA presumably prolongs the dopaminergic response.

NEUROBIOLOGICAL MECHANISM—MOLECULAR

The initial molecular site of action for the acute reinforcing and dependence-inducing effects of nicotine is the nAChR. nAChRs are cationic ligand-gated ion channels expressed throughout the central nervous system (Lindstrom, 2000). Neuronal nAChRs can be presynaptic or postsynaptic and activate other neurons by increasing the influx of calcium and producing neurotransmitter release. The influx of calcium on postsynaptic neurons also can trigger many cellular processes, including activation of protein kinase and calmodulin-dependent kinase (Lindstrom, 2000; Brunzell *et al.*, 2003). The nAChR is a pentameric structure with at least two ligand-binding sites situated at the interface between subunits (Noda *et al.*, 1983; Numa *et al.*, 1983; Changeaux *et al.*, 1998; Picciotto *et al.*, 1999) (**Fig. 6.28**). Genes encoding nAChR subunits have been identified and cloned in mammals (α1-α10; β1-β9), and several subunits have been found in the central nervous system (α2-α7 and β2-β4) (Wonnacott, 1997; Elgoyhen *et al.*, 2001; Le Novere *et al.*, 2002). These subunits can co-assemble to form functional pentameric receptors (Le Novere and Changeux, 1995; Changeux *et al.*, 1998). α1, β1, γ, δ, and ε subunits are thought to represent the muscle acetylcholine receptor family. α2-6 and β2-4 subunits represent the neuronal receptor family. α7-10 subunits represent the homopentameric acetylcholine receptor family (**Fig. 6.27**). The homomeric acetylcholine receptors, but not other neuronal acetylcholine receptors, bind snake venom toxins like the muscle acetylcholine receptors (Lindstrom, 2000). There appears to be a wide variety of nAChR subunit subtypes with different pharmacological and

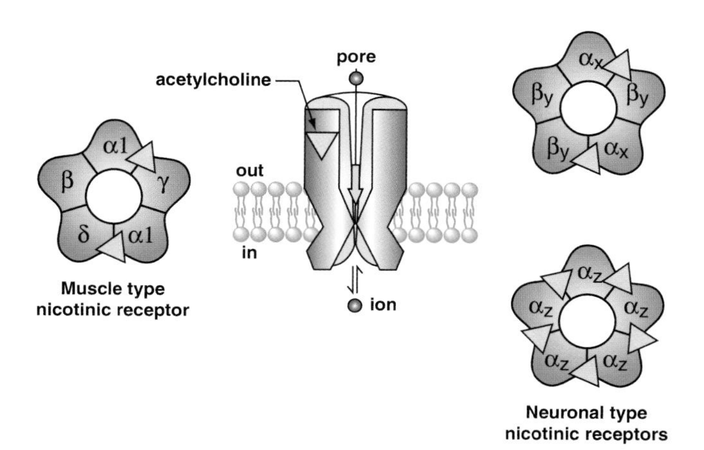

FIGURE 6.27 Schematic diagram of the structural organization of the nicotinic acetylcholine receptor (nAChR). The structure of the muscular receptor (left) of the nAChR is well characterized, and much of what we know about the structure of the neuronal receptor (right) is based on studies of the muscular receptor. The nAChRs cross the membrane, and binding of acetylcholine (or nicotine) causes the receptor to change shape and open a channel that allows ions to flow into and out of the cell. Whereas the nAChR in adult muscle is made up of the α_1, β_1, γ, and δ subunits, there are two different families of nAChRs expressed in neurons. One family is made up of a combination of the α and β subunits (in this subtype, $\alpha_x = \alpha_{2-6}$ and $\beta_y = \beta_{2-4}$), whereas the other family can form active receptors made up of only one type of subunit (in this homomeric subtype, $\alpha_z = \alpha_{7-9}$). The five binding sites on the α_7 subtype are hypothesized based on a symmetry argument. Because all five subunits of the complex are identical, structural elements involved in binding ligand should be identical on each subunit providing five binding sites. [Reproduced with permission from Picciotto *et al.*, 1999.]

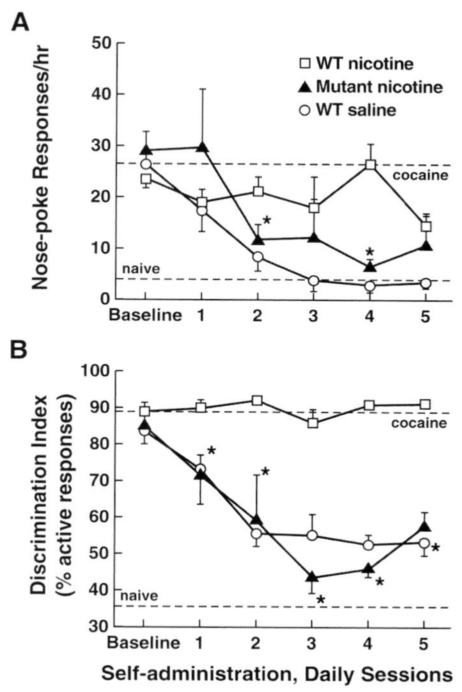

FIGURE 6.28 Intravenous self-administration of cocaine and nicotine in $\beta_2^{-/-}$ mice (Mutant nicotine) and their wildtype (WT) siblings with a history of cocaine self-administration. **(A)** Number of active nosepoke responses per session. Dashed lines represent the average of the last 3 sessions of cocaine self-administration in all groups of mice (cocaine), or of spontaneous nosepoke behavior in naive, sham-operated mice (naive). **(B)** Percentage active responses; that is, the percentage of the number of active responses compared to the sum of active and inactive responses per session. Values of 50 per cent indicate a lack of discrimination between active and inactive detectors. $\beta_2^{-/-}$ mice differed significantly from wildtype mice self-administering nicotine in either active nosepoke ($p < 0.05$) or discrimination index ($p < 0.01$), but did not differ from wildtype mice during saline-induced extinction (ANOVA). The increase in the variability of nosepoke response of $\beta_2^{-/-}$ mice on the first day of nicotine treatment was due to increased responding of some mice, usually interpreted as a transient overresponse to reinforcer devaluation. Asterisks (*) represent post hoc comparisons of the wildtype nicotine group and the $\beta_2^{-/-}$ group (*$p < 0.05$). [Reproduced with permission from Picciotto *et al.*, 1998.]

electrophysiological properties. However, the overlapping patterns of expression of the nAChR subunits and the lack of specific pharmacological agents have limited the identification of functionally specific nicotinic receptor subunit subtypes *in vivo*.

The effects of nicotine on central nAChRs are complex and have been described as 'paradoxical' in that chronic nicotine exposure leads to receptor desensitization and inactivation which is followed by an upregulation in nicotinic receptors during withdrawal (Wonnacott, 1990; Marks *et al.*, 1992; Bhat *et al.*, 1994). Acute administration of nicotine stimulates the nAChR which leads to a brief opening of the ion channel (receptor activation), but then transiently becomes unresponsive to further exposure to agonists (receptor inactivation and desensitization) (Corringer *et al.*, 1998). Consequently, the desensitization leads to an increase in the number of nAChRs (receptor upregulation) (Collins *et al.*, 1990; Wonnacott, 1990; Perry *et al.*, 1999). Although this nicotinic receptor activation, desensitization, and upregulation can be viewed as a neuronal response to maintain the baseline level of synaptic activity within cholinergic and other neurotransmitter systems during chronic nicotine exposure (Dani and Heinemann, 1996; Reitstetter *et al.*, 1999), it is not entirely clear if upregulation reflects an increase in functional receptors (Wonnacott, 1997).

Recombinant DNA technology has provided valuable insights into the role of specific nAChR subunits in nicotine reinforcement and dependence. Knockout mice lacking the $\alpha3$ (Xu *et al.*, 1999a), $\alpha4$ (Marubio *et al.*, 1999; Ross *et al.*, 2000), $\alpha5$ (Orr-Urtreger *et al.*, 1997), $\alpha9$ (Vetter *et al.*, 1999), $\beta2$ (Picciotto *et al.*, 1995), $\beta3$ (Booker *et al.*, 1999), and $\beta4$ (Xu *et al.*, 1999b) subunits have been generated (**Table 6.6**). Phenotypes from these knockout mice have included such

TABLE 6.6 Behavioral and Pharmacological Effects of Nicotinic Acetylcholine Receptor Subunit Knockout Mice

Knockout	Baseline phenotype	Effects of nicotine vs. Wildtype	References
α3	Impaired growth Dilated pupils do not constrict to light Enlarged bladder, dribbling urination, urinary tract infection, bladder stones Lethal postnatal (1 day to 3 months)	Lack of nicotine-induce bladder contractility	Xu *et al.*, 1999a
α4	↑ Locomotor activity ↑ Sniffing ↑ Rearing ↑ Striatal dopamine levels ↓ Exploration in elevated plus maze ↑ Experimental colitis	↑ Recovery from nicotine-induced depression of locomotor activity ↓ Antinociception ↓ Raphe magnus neuronal response ↓ Thalamic neuronal response No nicotine-induced ↑ in dopamine release	Marubio *et al.*, 1999; Ross *et al.*, 2000; Marubio *et al.*, 2003
α5	↓ α3 Receptor expression ↑ Experimental colitis	↑ Seizure activity ↓ Seizure activity —Open field activity Ameliorates colitis	Salas *et al.*, 2003a; Kedmi *et al.*, 2004; Orr-Urtreger *et al.*, 2005
α6	None	Not tested	Champtiaux *et al.*, 2002
α7	↑ Y-maze activity ↑ Open-field activity ↑ Sympathetic response (↑ heart rate) to sodium nitroprusside-induced vasodilation ↓ Appetitive learning	—Somatic signs of withdrawal	Franceschini *et al.*, 2000; Tritto *et al.*, 2004; Bowers *et al.*, 2005; Grabus *et al.*, 2005; Keller *et al.*, 2005
α9	↓ Number of outer hair cells in cochlea ↓ Olivocochlear function Abnormal cochlear terminal morphology	Not tested	Vetter *et al.*, 1999
β2	↑ Passive avoidance latency ↑ Corticosterone levels ↑ Visual response latency ↑ Preference for higher temporal frequencies ↑ Contrast sensitivity ↓ Freezing in tone-conditioned fear (aged mice) ↓ Visual acuity at cortical level ↓ Locomotor activity in familiar environment ↓ Hippocampal pyramidal neurons ↓ Spatial learning Abnormal cholinergic transmission Neural tissue atrophy Neocortical hypotrophy Astrogliosis Microgliosis	↑ Locomotor activity ↑ Body temperature ↑ Passive avoidance latency ↓ Startle ↓ —Nicotine self-administration ↓ Dopamine response in substantia nigra ↓ Dopamine response in ventral tegmental area ↓ Thalamic neuronal response ↓ Raphe magnus neuronal response ↓ Antinociception	Picciotto *et al.*, 1995, 1998; Marubio *et al.*, 1999; Zoli *et al.*, 1999; Caldarone *et al.*, 2000; Rossi *et al.*, 2001; Owens *et al.*, 2003; Tritto *et al.*, 2004; Grubb and Thompson, 2004; King *et al.*, 2004
β3	↑ Open field activity ↓ Prepulse inhibition of startle ↓ Nigrostriatal dopamine release	↑ Striatal dopaminergic transmission	Cui *et al.*, 2003
β4	↑ Heart rate ↑ Exploration in elevated plus maze ↑ Activity in stair case test ↓ α3 receptor expression ↓ Core body temperature Impaired bladder contractility	↓ Seizure activity ↓ Somatic withdrawal signs ↓ Hyperalgesia ↓ Hyperthermic response	Xu *et al.*, 1999b; Salas *et al.*, 2003b, 2004; Kedmi *et al.*, 2004; Sack *et al.*, 2005
α5β4	↓ α3 receptor expression	↓ Seizure activity	Kedmi *et al.*, 2004
α7β2	↑ Rotarod performance ↓ Passive avoidance latency		Marubio and Paylor, 2004
β2β4	Dilated pupils do not contract to light Impaired bladder contractility, enlarged bladder, dribbling urination, urinary tract infection, bladder stones Lethal at 1–3 weeks postnatal	Not tested	Xu *et al.*, 1999b

↓, decreased measure; ↑, increased measure; —, no change.
[Modified and updated from Cordero-Erasquin *et al.*, 2000.]

domains as anxiety, attention, memory, and cognition (Cordero-Erausquin *et al.*, 2000), but the focus here will be on nicotine addiction.

The most widely expressed subtypes of the nAChR in the brain contain $\alpha4$, $\alpha7$, and $\beta2$ subunits (Wada *et al.*, 1989; Flores *et al.*, 1992; Zoli *et al.*, 1998). Various nAChR α and β subunit combinations, including the $\alpha4\beta2$ subtype, are present throughout the mesolimbic pathway, including the ventral tegmental area, prefrontal cortex, amygdala, septal area, and nucleus accumbens (Wada *et al.*, 1989; Marks *et al.*, 1992; Sargent, 1993). The $\alpha4$ subunit appears critical for the slow currents associated with the slow desensitization component of nAChRs (Pidoplichko *et al.*, 1997; Klink *et al.*, 2001), and the release of dopamine produced by systemic nicotine (Marubio *et al.*, 2003).

All high-affinity binding sites for nicotine include the $\beta2$ subunit (Picciotto *et al.*, 1995). Nicotine-induced dopamine release and nicotine activation of dopamine neurons *in vitro* is dependent on the $\beta2$ subunit (Picciotto *et al.*, 1998). Mutant mice lacking the $\beta2$ subunit will not self-administer nicotine (Picciotto *et al.*, 1998) (**Fig. 6.28**), indicating that the $\beta2$ subunit is critically involved in nicotine reinforcement. In contrast, engineering at a single point mutation of the α_4 subunit (Leu9′→Ala9′) in the pore-forming domain produced mice with α_4 receptors that were hypersensitive to nicotine and hypersensitive to the rewarding effects of nicotine (Tapper *et al.*, 2004). Mutant α_4 mice showed a place preference for nicotine at 10 μg/kg, a dose that had no effect in wildtype mice. These mutant mice also showed rapid tolerance to the hypothermic effects of nicotine and rapid sensitization to the locomotor-activating effects of nicotine, again at doses that did not produce these effects in wildtype mice (Tapper *et al.*, 2004).

The high-affinity αCtxMII binding sites that also control dopamine release are lost in $\alpha6$ knockout mice (Champtiaux *et al.*, 2002). An indirect action on dopamine neurons via an activation of $\alpha7$ nicotinic receptors on presynaptic NMDA glutamate neurons which would in turn activate dopamine neurons has been hypothesized based on pharmacological studies (Nomikos *et al.*, 2000), but confirmation will require studies with $\alpha7$ knockout animals (for review, see Drago *et al.*, 2003) (**Table 6.6**). Thus, it appears the main heteromeric nAChR subtypes expressed in dopamine neurons are the $\alpha4\beta2$, $\alpha7$, $\alpha4\alpha5\beta2$, and $\alpha4\alpha6\alpha5\beta2$ subunits (Klink *et al.*, 2001) (**Fig. 6.23** above).

As noted above, during chronic nicotine exposure there is a desensitization of nAChRs in the short-term and an upregulation of nAChR number in the long-term. Since nAChRs may exist in many different functional states within the brain (Changeux *et al.*, 1984; Reitstetter *et al.*, 1999), differential activation could be one driving force for the neuroadaptive changes associated with dependence. The $\alpha2$, $\alpha4$, and $\alpha7$ subunits become inactive and desensitized in the chronic presence of nicotine, while the $\alpha3$ and $\alpha6$ subunits do not show inactivation (Olale *et al.*, 1997), suggesting that some subunits show a greater sensitivity to nicotine than others. Injection of $\alpha3\beta2$ or $\alpha4\beta2$ subunit RNAs in oocytes followed by subsequent nicotine administration indicated that $\alpha4\beta2$ nAChRs desensitize more quickly and recover more slowly than $\alpha3\beta2$ receptors (Hsu *et al.*, 1996). Perinatal exposure to nicotine produced functional deficits in growth, respiration, arousal, and catecholamine biosynthesis. These deficits were similar to pups lacking the β_2-containing nicotinic acetylcholine receptor, suggesting a link between perinatal exposure to nicotine and selective desensitization of the β_2 subunit (Cohen *et al.*, 2005b). Chronic nicotine also increased expression of α_1 and α_2/∂_1 subunits of the L-type high voltage-gated calcium channel (Hayashida *et al.*, 2005). Thus, a differential effect of chronic nicotine exposure on release of various neurotransmitter systems may be explained by the balance of receptor density, desensitization, and functionality.

During nicotine abstinence, such changes in nAChR function may mediate some of the motivational and somatic symptoms associated with nicotine withdrawal. For example, during nicotine abstinence which leads to decreased plasma nicotine levels, the previously desensitized or inactive nAChRs may begin to recover to functional states at different rates depending on the brain region or receptor subtype. During chronic nicotine exposure, upregulation of nAChRs also may occur along nonreward-related cholinergic pathways such as during abstinence. The dysregulation of nAChRs in both reward (desensitized) and nonreward (upregulated) circuits may contribute to motivational or somatic withdrawal symptoms (Dani and Heinemann, 1996). For example, nAChR activation in the ventral tegmental area is followed by receptor desensitization (Pidoplichko *et al.*, 1997). Receptor desensitization and recovery occurred at different rates, suggesting that within the ventral tegmental area, there are multiple types of nAChRs with different activation and desensitization profiles (Pidoplichko *et al.*, 1997). Thus, the development and perpetuation of nicotine addiction may involve self-medication to effectively control the number of functional nAChRs along pathways affected by nicotine (Dani and Heinemann, 1996; Koob *et al.*, 1998).

Smokers report the first cigarette of the day to be the most pleasurable (Russell, 1989), possibly because of nicotine-induced activation of recovered nAChRs in the ventral tegmental area, leading to greater dopamine release than later in the day. Throughout the day, smokers maintain a steady blood nicotine level (Benowitz, 1996) and are exposed to nicotine concentrations which cause nAChR desensitization in the ventral tegmental area (Pidoplichko *et al.*, 1997). If different nAChRs in the ventral tegmental area have different sensitivities to nicotine, as suggested above, it may be that once a steady-state of nicotine is reached, periodic readministration of nicotine engages nAChRs only activated by high nicotine doses (Dani and Heinemann, 1996). Activation of these receptors also would cause dopamine release, thus contributing to the maintenance of cigarette smoking throughout the waking hours. Alternatively, the maintenance of smoking behavior in dependent animals despite the development of nAChR desensitization within the ventral tegmental area may indicate the involvement of parallel reward systems in the positive reinforcing actions of nicotine which extend beyond the mesolimbic dopamine pathway (see above). Few studies to date have explored the neurobiology of

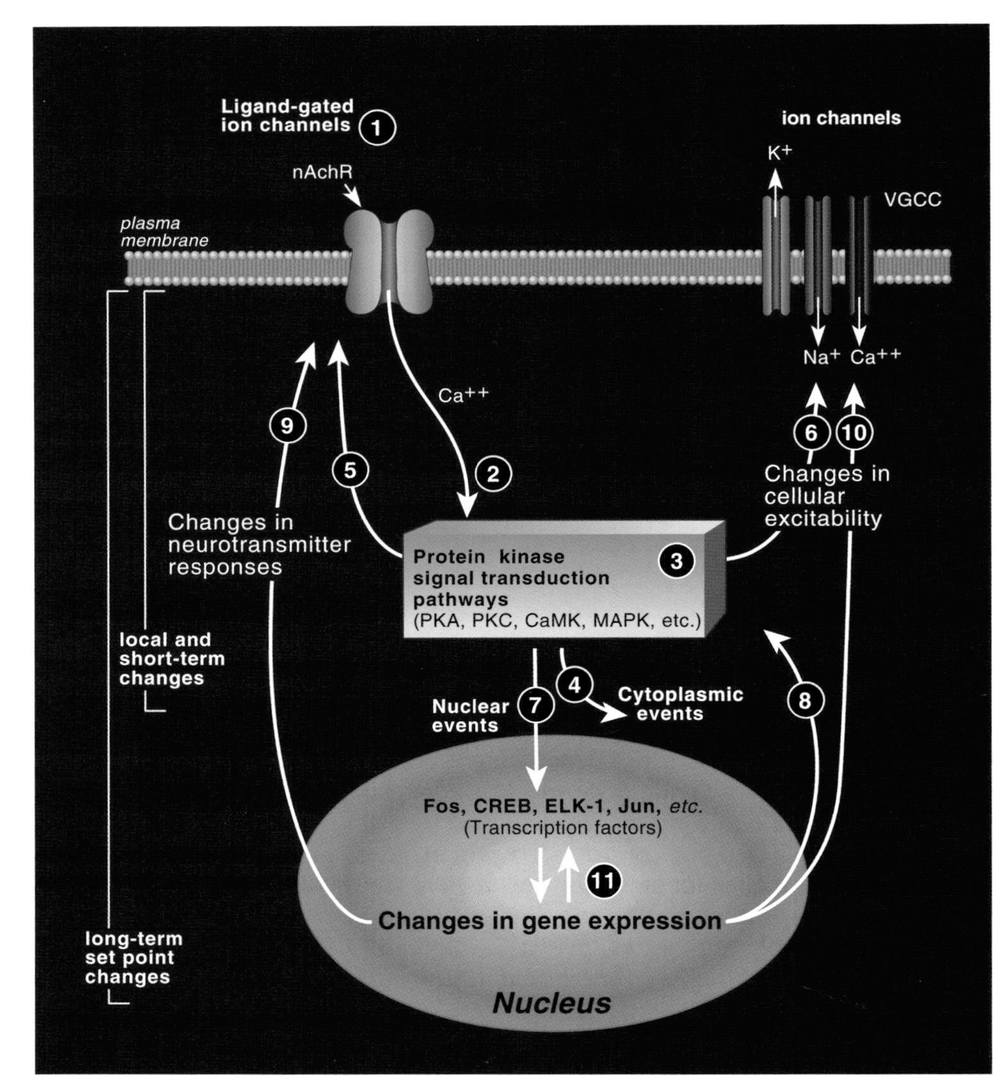

FIGURE 6.29 Molecular mechanisms of nicotine neuroadaptation. Nicotine acts directly on ligand-gated ion channels (①). These receptors modulate the levels of Ca^{2+} (②), which in turn regulate the activity of protein kinase transducers (③). Such protein kinases affect the functions of proteins located in the cytoplasm, plasma membrane, and nucleus (④⑤⑥⑦). Among membrane proteins affected are ligand-gated and voltage-gated ion channels (⑤⑥). Protein kinase transduction pathways also affect the activities of transcription factors (⑦). Some of these factors, like cyclic adenosine monophosphate response element binding protein (CREB), are regulated posttranslationally by phosphorylation; others, like Fos, are regulated transcriptionally; still others, like Jun, are regulated both posttranslationally and/or transcriptionally. While membrane and cytoplasmic changes may be only local (e.g., dendritic domains or synaptic boutons), changes in the activity of transcription factors may result in long-term functional changes. These may include changes in gene expression of proteins involved in signal transduction (⑧) and/or neurotransmission (⑨⑩), resulting in altered neuronal responses. For example, chronic exposure to nicotine has been reported to increase levels of protein kinase A (PKA) (⑧) in the nucleus accumbens. Chronic exposure to nicotine also alters the expression of transcription factors themselves (⑪). CREB expression, for instance, is depressed in the amygdala and prefrontal cortex and increased in the nucleus accumbens and ventral tegmental area. The receptor systems depicted in the figure may not coexist in the same cells. [Modified with permission from Koob *et al.*, 1998.]

the positive reinforcing actions of nicotine in dependent animals.

As noted above, the upregulation of nicotinic receptors with dependence is well documented (Benwell *et al.*, 1988), and may contribute to the neuroadaptation associated with chronic nicotine administration. However, another mechanism of neuroadaptation that may have long-term consequences is alterations in gene transcription via changes in the activity of transcription factors that have been observed with other drugs of abuse (Nestler and Aghajanian, 1997) (**Fig. 6.29**). Nicotine exposure, nicotine self-administration, and withdrawal result in immediate early gene activation in the ventral tegmental area and the projection areas of the mesolimbic dopamine system (Pich *et al.*, 1997; Salminen *et al.*, 1999; Panagis *et al.*, 2000). Nicotine self-administration and nicotine cues also increase Fos-like immunoreactivity in the prefrontal cortex (Pagliusi *et al.*,1996; Pich *et al.*, 1997; Schroeder *et al.*, 2001). Decreased phosphorylated cyclic adenosine monophosphate (cAMP) response element binding protein (CREB) levels in the cingulate gyrus, cerebral cortex, and medial and basolateral amygdala, but not in the central nucleus of the amygdala, were observed 18 h after nicotine withdrawal from 2 mg/kg nicotine injected twice a day for 10 days (Pandey *et al.*, 2001). These decreases in CREB during nicotine withdrawal correlated with increases in anxiety-like responses in the elevated plus maze (Pandey *et al.*, 2001). Others have shown that chronic administration to mice of 200 μg/kg nicotine in 2 per cent saccharin in the drinking water for 30 days, a dose sufficient to upregulate nicotinic receptors (Sparks and Pauly, 1999), resulted in a decrease in phosphorylated CREB in the nucleus accumbens and withdrawal (Brunzell *et al.*, 2003). In contrast, increases were observed in phosphorylated CREB in the prefrontal cortex and ventral tegmental area during chronic nicotine exposure (Brunzell *et al.*, 2003). Phosphorylated extracellular signal-regulated kinase (ERK) decreased in the amygdala during chronic nicotine exposure, but phosphorylated ERK increased in the prefrontal cortex (Brunzell *et al.*, 2003). These results support a role for the ERK/CREB signaling pathway in nicotine dependence through which changes in expression of cAMP-inducible genes may lead to long-term changes in plasticity (Brunzell *et al.*, 2003) (**Fig. 6.30**).

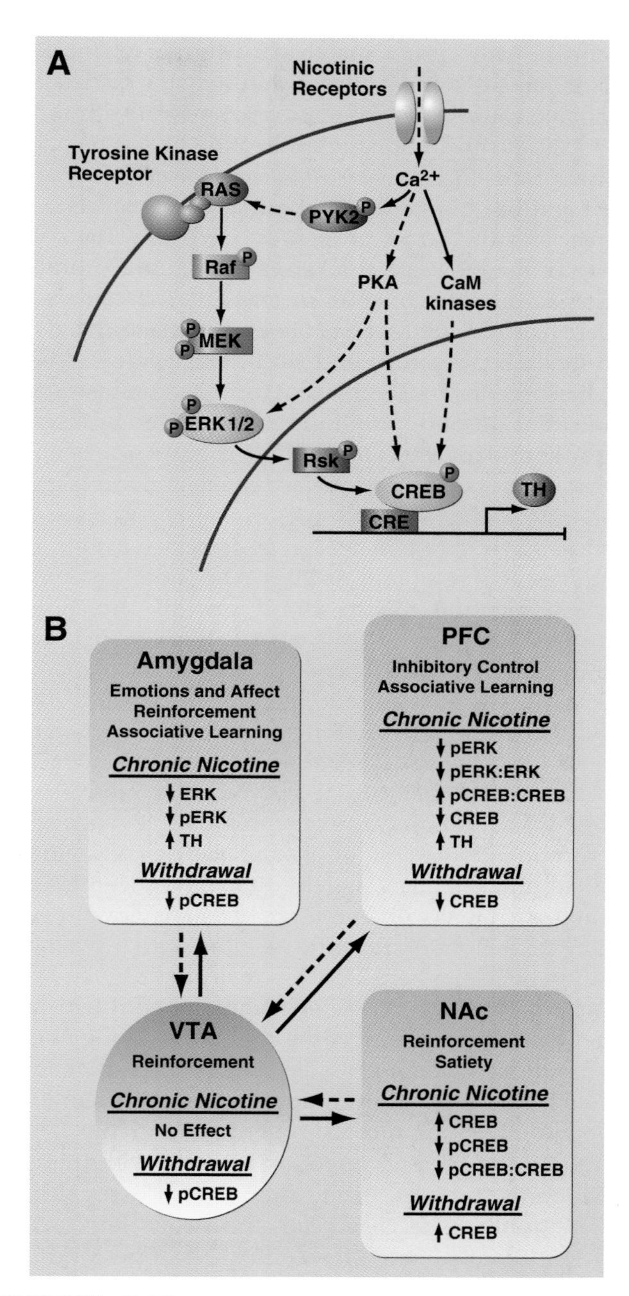

FIGURE 6.30 Signaling pathways related to extracellular signal-regulated kinase (ERK) activation that may be involved in nicotine dependence and withdrawal. **(A)** Nicotine could affect ERK/CREB (cyclic adenosine monophosphate response element binding protein) signaling by direct activation of nicotinic receptors, or indirectly via changes in tyrosine kinase receptor activation. Dashed lines indicate multi-step processes. The PYK2, ERK1/2, CREB, and TH signaling proteins were measured in the current study. **(B)** Summary of changes in response to chronic nicotine exposure and withdrawal in the amygdala, prefrontal cortex (PFC), nucleus accumbens (NAc), and ventral tegmental area (VTA). Each of these brain areas is connected by reciprocal connections to the ventral tegmental area, and thus, changes in signaling within these pathways are likely to alter communication between brain areas involved in drug dependence. [Modified with permission from Brunzell *et al.*, 2003.]

SUMMARY

Although declining in the United States, tobacco addiction remains the most prevalent addiction of all drugs of abuse, with approximately 19 per cent of the U.S. population still daily smokers, and remains the leading avoidable cause of disease and premature death today (440 000 deaths annually). Nicotine is an alkaloid derived from the *Nicotiana tabacum* (common tobacco) plant and has been cultivated for centuries for its ceremonial and medicinal properties. Generally, the dried leaves of the plant are smoked, although oral administration in the form of snuff or chewing tobacco also is widespread. Cigarettes began to be manufactured in the mid-nineteenth century, and since 1940 cigarette smoking, has been the most prevalent form of administration.

Nicotine produces euphoria characterized by arousal and a stimulant effect, including cognitive activation with concomitant autonomic activation and increases in heart rate and blood pressure, but also paradoxically produces a reduction in tension and stress. Nicotine also produces analgesia and decreases appetite. In animal studies, nicotine lowers brain reward thresholds, increases locomotor activity, and is intravenously self-administered. Nicotine can produce anxiolytic-like effects in various animal models of anxiety. It is rapidly absorbed by smoking and by oral administration in pH-modified preparations (chewing tobacco) and has a short half-life of 2 h. Nicotine smoke produces not only high carboxyhemoglobin levels due to the carbon monoxide in smoke but also delivers significant amounts of carcinogens that have been linked to the toxic effects of smoking. The behavioral mechanism of action of nicotine is linked to its mood-titration effects, where smokers maintain a given level of nicotine over the course of a smoking bout which in dependent smokers is over the course of waking hours. The boundary model provides a heuristic explanation for such titration with the upper boundary being the aversive effects of nicotine overdose and the lower boundary the aversive effects of nicotine withdrawal.

Much is known from animal work regarding the neurobiological basis for the acute reinforcing and stimulant effects of nicotine. Nicotine binds to nAChRs and activates the mesolimbic dopamine system, both at the level of the ventral tegmental area and the nucleus accumbens. At the level of the ventral tegmental area, nAChRs are hypothesized to be localized on not only dopaminergic cell bodies but also on GABAergic and glutamatergic afferents. The combination of these actions, on possibly different subunit subtypes of the nAChR, produces a powerful dopamine neuron-facilitating action. There also is evidence of opioid peptide interactions with nicotine reinforcement. μ opioid receptor knockout mice show no nicotine place preference. The molecular site of action for the acute reinforcing effects of nicotine is hypothesized to be the result mainly of interactions with $\alpha4\beta2$ and $\alpha6\beta2$ subunits, with mutant mice lacking the $\beta2$ subunit failing to self-administer nicotine.

Much less is known about the neurobiological bases for the dependence-inducing properties of nicotine. The differential desensitization hypothesis has been proposed that may lead to negative reinforcement to suppress a differential recovery of nicotine receptor effects, and decreases in dopaminergic neurotransmission both at the level of the ventral tegmental area and nucleus accumbens are associated with the malaise of acute withdrawal from nicotine. Loss of glutamate functional activity, serotonin functional activity, and hypothesized opioid peptide activity also are associated with nicotine withdrawal. μ opioid receptor knockout mice do not show the somatic signs of nicotine withdrawal. Preliminary evidence also suggests dysregulation of brain stress neurotransmitters in acute nicotine withdrawal. nAChRs, when activated, trigger changes in intracellular calcium that can activate protein kinase A and with chronic nicotine administration, can alter CREB activity. Decreases in CREB have been observed in the amygdala and prefrontal cortex and increases in CREB in the ventral tegmental area and nucleus accumbens. These changes may alter gene expression and trigger long-term neuroadaptive responses as observed with other drugs of abuse. Future advances in the neurobiology of nicotine addiction will require the same level of neurobiological sophistication applied to better motivational measures of nicotine dependence and relapse with animal models.

REFERENCES

Abrous, D. N., Adriani, W., Montaron, M. F., Aurousseau, C., Rougon, G., Le Moal, M., and Piazza, P. V. (2002). Nicotine self-administration impairs hippocampal plasticity. *Journal of Neuroscience* **22**, 3656–3662.

Adriani, W., Spijker, S., Deroche-Gamonet, V., Laviola, G., Le Moal, M., Smit, A. B., and Piazza, P. V. (2003). Evidence for enhanced neurobehavioral vulnerability to nicotine during periadolescence in rats. *Journal of Neuroscience* **23**, 4712–4716.

Akehurst, B. C. (1968). *Tobacco*, Longmans, London.

Alexander, B. K., and Hadaway, P. F. (1982). Opiate addiction: the case for an adaptive orientation. *Psychological Bulletin* **92**, 367–381.

Alkondon, M., and Albuquerque, E. X. (1993). Diversity of nicotinic acetylcholine receptors in rat hippocampal neurons: I. Pharmacological and functional evidence for distinct structural subtypes. *Journal of Pharmacology and Experimental Therapeutics* **265**, 1455–1473.

American Psychiatric Association (1994). *Diagnostic and Statistical Manual of Mental Disorders*, 4th ed. American Psychiatric Press, Washington DC.

Armitage, A. K., Dollery, C. T., George, C. F., Houseman, T. H., Lewis, P. J., and Turner, D. M. (1975). Absorption and metabolism of nicotine from cigarettes. *British Medical Journal* **4**, 313–316.

Ashton, H., and Watson, D. W. (1970). Puffing frequency and nicotine intake in cigarette smokers. *British Medical Journal* **3**, 679–681.

Azam, L., Winzer-Serhan, U. H., Chen, Y., and Leslie, F. M. (2002). Expression of neuronal nicotinic acetylcholine receptor subunit mRNAs within midbrain dopamine neurons. *Journal of Comparative Neurology* **444**, 260–274.

Baldo, B. A., Koob, G. F., and Markou, A. (1999). Role of adenosine A2 receptors in brain stimulation reward under baseline conditions and during cocaine withdrawal in rats. *Journal of Neuroscience* **19**, 11017–11026.

Baldwin, H. A., Rassnick, S., Rivier, J., Koob, G. F., and Britton, K. T. (1991). CRF antagonist reverses the 'anxiogenic' response to ethanol withdrawal in the rat. *Psychopharmacology* **103**, 227–232.

Balfour, D. J. (1984). Nicotine and the tobacco smoking habit. *In Nicotine and the Tobacco Smoking Habit* (series title: *International Encyclopedia of Pharmacology and Therapeutics*, vol. 114), (D. J. K. Balfour, Ed.), pp. 61–74. Pergamon Press, New York.

Bardo, M. T., Green, T. A., Crooks, P. A., and Dwoskin, L. P. (1999). Nornicotine is self-administered intravenously by rats. *Psychopharmacology* **146**, 290–296.

Battaglia, G., Bruno, V., Ngomba, R. T., Di Grezia, R., Copani, A., and Nicoletti, F. (1998). Selective activation of group-II metabotropic glutamate receptors is protective against excitotoxic neuronal death. *European Journal of Pharmacology* **356**, 271–274.

Battista, S. P. (1976). Cilia toxic components of cigarette smoke. *In Proceedings of the Third World Conference on Smoking and Health, New York, June 2-5, 1975: Volume 1. Modifying the Risk for the Smoker* (E. L. Wynder, D. Hoffmann, and G. B. Gori, Eds.), pp. 517–534. DHEW publ. no. NIH 76-1221, National Cancer Institute, Washington DC.

Bauco, P., and Wise, R. A. (1994). Potentiation of lateral hypothalamic and midline mesencephalic brain stimulation reinforcement by nicotine: Examination of repeated treatment. *Journal of Pharmacology and Experimental Therapeutics* **271**, 294–301.

Beleslin, D. B., and Krstic, S. K. (1987). Further studies on nicotine-induced emesis: nicotinic mediation in area postrema. *Physiology and Behavior* **39**, 681–686.

Benowitz, N. L. (1988). Pharmacologic aspects of cigarette smoking and nicotine addition. *New England Journal of Medicine* **319**, 1318–1330.

Benowitz, N. L. (1996). Pharmacology of nicotine: addiction and therapeutics. *Annual Review of Pharmacology and Toxicology* **36**, 597–613.

Benowitz, N. L., Kuyt, F., Jacob, P. 3rd, Jones, R. T., and Osman, A. L. (1983). Cotinine disposition and effects. *Clinical Pharmacology and Therapeutics* **34**, 604–611.

Benowitz, N. L., Kuyt, F., and Jacob, P. 3rd. (1984). Influence of nicotine on cardiovascular and hormonal effects of cigarette smoking. *Clinical Pharmacology and Therapeutics* **36**, 74–81.

Benowitz, N. L., Jacob, P. 3rd, Kozlowski, L. T., and Yu, L. (1986). Influence of smoking fewer cigarettes on exposure to tar, nicotine, and carbon monoxide. *New England Journal of Medicine* **315**, 1310–1313.

Benowitz, N. L., Porchet, H., Sheiner, L., and Jacob, P., 3rd. (1988). Nicotine absorption and cardiovascular effects with smokeless tobacco use: comparison with cigarettes and nicotine gum. *Clinical Pharmacology and Therapeutics* **44**, 23–28.

Bentley, R., and Trimen, H. (1880). *Medicinal Plants: Being Descriptions with Original Figures of the Principal Plants Employed in Medicine and an Account of the Characters, Properties, and Uses of Their Parts and Products of Medicinal Value*, vol. 3 (no. **191**), J & A Churchill, London.

Benwell, M. E., and Balfour, D. J. (1979). Effects of nicotine administration and its withdrawal on plasma corticosterone and brain 5-hydroxyindoles. *Psychopharmacology* **63**, 7–11.

Benwell, M. E., and Balfour, D. J. (1982). The effects of nicotine administration on 5-HT uptake and biosynthesis in rat brain. *European Journal of Pharmacology* **84**, 71–77.

Benwell, M. E., and Balfour, D. J. (1992). The effects of acute and repeated nicotine treatment on nucleus accumbens dopamine and locomotor activity. *British Journal of Pharmacology* **105**, 849–856.

Benwell, M. E., Balfour, D. J., and Anderson, J. M. (1988). Evidence that tobacco smoking increases the density of (-)-[3H]nicotine binding sites in human brain. *Journal of Neurochemistry* **50**, 1243–1247.

Benwell, M. E., Balfour, D. J., and Anderson, J. M. (1990). Smoking-associated changes in the serotonergic systems of discrete regions of human brain. *Psychopharmacology* **102**, 68–72.

Benwell, M. E., Balfour, D. J., and Lucchi, H. M. (1993). Influence of tetrodotoxin and calcium on changes in extracellular dopamine levels evoked by systemic nicotine. *Psychopharmacology* **112**, 467–474.

Berrendero, F., Kieffer, B. L., and Maldonado, R. (2002). Attenuation of nicotine-induced antinociception, rewarding effects, and dependence in mu-opioid receptor knock-out mice. *Journal of Neuroscience* **22**, 10935–10940.

Bhat, R. V., Marks, M. J., and Collins, A. C. (1994). Effects of chronic nicotine infusion on kinetics of high-affinity nicotine binding. *Journal of Neurochemistry* **62**, 574–581.

Billings, E. R. (1875). *Tobacco: Its History, Varieties, Culture, Manufacture and Commerce, With an Account of its Various Modes of Use, From Its First Discovery Until Now,* American Publishing Company, Hartford CT.

Booker, T. K., Allen, R. S., Marks, M. J., Grady, S. R., Whiteaker, P., Kolman, J., Smith, K. W., and Collins, A. C. (1999). Analysis of the β3 nicotinic acetylcholine receptor subunit in mouse brain using β3 null mutant mice. *In Neuronal Nicotinic Receptors: From Structure to Therapeutics Abstracts*, University of Milan, Italy p. 2.

Bowers, B. J., McClure-Begley, T. D., Keller, J. J., Paylor, R., Collins, A. C., and Wehner, J. M. (2005). Deletion of the alpha 7 nicotinic receptor subunit gene results in increased sensitivity to several behavioral effects produced by alcohol. *Alcoholism: Clinical and Experimental Research* **29**, 295–302.

Boyadjieva, N. I., and Sarkar, D. K. (1997). The secretory response of hypothalamic beta-endorphin neurons to acute and chronic nicotine treatments and following nicotine withdrawal. *Life Sciences* **61**, PL59–PL66.

Brioni, J. D., O'Neill, A. B., Kim, D. J., and Decker, M. W. (1993). Nicotinic receptor agonists exhibit anxiolytic-like effects on the elevated plus-maze test. *European Journal of Pharmacology* **238**, 1–8.

Brower, V. G., Fu, Y., Matta, S. G., and Sharp, B. M. (2002). Rat strain differences in nicotine self-administration using an unlimited access paradigm. *Brain Research* **930**, 12–20.

Brown, Z. W., Amit, Z., and Smith, B. (1980). Intraventricular self-administration of acetaldehyde and voluntary consumption of ethanol in rats. *Behavioral and Neural Biology* **28**, 150–155.

Brunzell, D. H., Russell, D. S., and Picciotto, M. R. (2003). In vivo nicotine treatment regulates mesocorticolimbic CREB and ERK signaling in C57Bl/6J mice. *Journal of Neurochemistry* **84**, 1431–1441.

Buccafusco, J. J., Letchworth, S. R., Bencherif, M., and Lippiello, P. M. (2005). Long-lasting cognitive improvement with nicotinic receptor agonists: mechanisms of pharmacokinetic-pharmacodynamic discordance. *Trends in Pharmacological Sciences* **26**, 352–360.

Caggiula, A. R., Donny, E. C., White, A. R., Chaudhri, N., Booth, S., Gharib, M. A., Hoffman, A., Perkins, K. A., and Sved, A. F. (2001). Cue dependency of nicotine self-administration and smoking. *Pharmacology Biochemistry and Behavior* **70**, 515–530.

Caggiula, A. R., Donny, E. C., White, A. R., Chaudhri, N., Booth, S., Gharib, M. A., Hoffman, A., Perkins, K. A., and Sved, A. F. (2002). Environmental stimuli promote the acquisition of nicotine self-administration in rats. *Psychopharmacology* **163**, 230–237.

Caldarone, B. J., Duman, C. H., and Picciotto, M. R. (2000). Fear conditioning and latent inhibition in mice lacking the high affinity subclass of nicotinic acetylcholine receptors in the brain. *Neuropharmacology* **39**, 2779–2784.

Cao, W., Burkholder, T., Wilkins, L., and Collins, A. C. (1993). A genetic comparison of behavioral actions of ethanol and nicotine in the mirrored chamber. *Pharmacology Biochemistry and Behavior* **45**, 803–809.

Capehart, T. (October 6, 2003). *Tobacco Outlook* (issue no. TBS-255), U.S. Department of Agriculture, Washington DC.

Caravati, E. M., McCowan, C. L., and Marshall, S. W. (2004). Plants. *In Medical Toxicology* (3rd ed.) (R. C. Dart, Ed.), pp. 1671–1713, Lippincott Williams and Wilkins, Philadelphia.

Carboni, E., Bortone, L., Giua, C., and Di Chiara, G. (2000). Dissociation of physical abstinence signs from changes in extracellular dopamine in the nucleus accumbens and in the prefrontal cortex of nicotine dependent rats. *Drug and Alcohol Dependence* **58**, 93–102.

Castane, A., Valjent, E., Ledent, C., Parmentier, M., Maldonado, R., and Valverde, O. (2002). Lack of CB1 cannabinoid receptors modifies nicotine behavioural responses, but not nicotine abstinence. *Neuropharmacology* **43**, 857–867.

Centers for Disease Control and Prevention (2004). *Targeting Tobacco Use: The Nation's Leading Cause of Death*, Centers for Disease Control and Prevention, Atlanta.

Champtiaux, N., Han, Z. Y., Bessis, A., Rossi, F. M., Zoli, M., Marubio, L., McIntosh, J. M., and Changeux, J. P. (2002). Distribution and pharmacology of alpha 6-containing nicotinic acetylcholine receptors analyzed with mutant mice. *Journal of Neuroscience* **22**, 1208–1217.

Changeux, J. P., Devillers-Thiery, A., and Chemouilli, P. (1984). Acetylcholine receptor: an allosteric protein. *Science* **225**, 1335–1345.

Changeux, J. P., Bertrand, D., Corringer, P. J., Dehaene, S., Edelstein, S., Lena, C., Le Novere, N., Marubio, L., Picciotto, M., and Zoli, M. (1998). Brain nicotinic receptors: structure and regulation, role in learning and reinforcement. *Brain Research Reviews* **26**, 198–216.

Chassin, L., Presson, C. C., Sherman, S. J., and Edwards, D. A. (1990). The natural history of cigarette smoking: predicting young-adult smoking outcomes from adolescent smoking patterns. *Health Psychology* **9**, 701–716.

Chergui, K., Charlety, P. J., Akaoka, H., Saunier, C. F., Brunet, J. L., Buda, M., Svensson, T. H., and Chouvet, G. (1993). Tonic activation of NMDA receptors causes spontaneous burst discharge of rat midbrain dopamine neurons in vivo. *European Journal of Neuroscience* **5**, 137–144.

Clarke, P. B., and Pert, A. (1985). Autoradiographic evidence for nicotine receptors on nigrostriatal and mesolimbic dopaminergic neurons. *Brain Research* **348**, 355–358.

Clarke, P. B., Hommer, D. W., Pert, A., and Skirboll, L. R. (1985). Electrophysiological actions of nicotine on substantia nigra single units. *British Journal of Pharmacology* **85**, 827–835.

Clarke, P. B., Hommer, D. W., Pert, A., and Skirboll, L. R. (1987). Innervation of substantia nigra neurons by cholinergic afferents from pedunculopontine nucleus in the rat: neuroanatomical and electrophysiological evidence. *Neuroscience* **23**, 1011–1019.

Clarke, P. B., and Kumar, R. (1983). The effects of nicotine on locomotor activity in non-tolerant and tolerant rats. *British Journal of Pharmacology* **78**, 329–337.

Cohen, C., Kodas, E., and Griebel, G. (2005a). CB(1) receptor antagonists for the treatment of nicotine addiction. *Pharmacology Biochemistry and Behavior* **81**, 387–395.

Cohen, C., Perrault, G., Voltz, C., Steinberg, R., and Soubrie, P. (2002). SR141716, a central cannabinoid (CB(1)) receptor antagonist, blocks the motivational and dopamine-releasing effects of nicotine in rats. *Behavioural Pharmacology* **13**, 451–463.

Cohen, G., Roux, J. C., Grailhe, R., Malcolm, G., Changeux, J. P., and Lagercrantz, H. (2005b). Perinatal exposure to nicotine causes deficits associated with a loss of nicotinic receptor function. *Proceedings of the National Academy of Sciences USA* **102**, 3817–3821.

Collins, A. C., Bhat, R. V., Pauly, J. R., and Marks, M. J. (1990). Modulation of nicotine receptors by chronic exposure to nicotinic agonists and antagonists. *In The Biology of Nicotine Dependence* (series title: *Ciba Foundation Symposium*, vol. 152), (G. Bock, J. Marsh, Eds.), pp. 87–105, John Wiley, New York.

Coppen, A. (1967). The biochemistry of affective disorders. *British Journal of Psychiatry* **113**, 1237–1264.

Cordero-Erausquin, M., Marubio, L. M., Klink, R., and Changeux, J. P. (2000). Nicotinic receptor function: new perspectives from knockout mice. *Trends in Pharmacological Sciences* **21**, 211–217.

Corrigall, W. A., and Coen, K. M. (1989). Nicotine maintains robust self-administration in rats on a limited-access schedule. *Psychopharmacology* **99**, 473–478.

Corrigall, W. A., and Coen, K. M. (1991). Selective dopamine antagonists reduce nicotine self-administration. *Psychopharmacology* **104**, 171–176.

Corrigall, W. A., Franklin, K. B. J., Coen, K. M., and Clarke, P. B. S. (1992). The mesolimbic dopaminergic system is implicated in the reinforcing effects of nicotine. *Psychopharmacology* **107**, 285–289.

Corrigall, W. A., Coen, K. M., and Adamson, K. L. (1994). Self-administered nicotine activates the mesolimbic dopamine system through the ventral tegmental area. *Brain Research* **653**, 278–284.

Corrigall, W. A., Coen, K. M., Adamson, K. L., Chow, B. L., and Zhang, J. (2000). Response of nicotine self-administration in the rat to manipulations of mu-opioid and gamma-aminobutyric acid receptors in the ventral tegmental area. *Psychopharmacology* **149**, 107–114.

Corrigall, W. A., Coen, K. M., Zhang, J., and Adamson, K. L. (2001). GABA mechanisms in the pedunculopontine tegmental nucleus influence particular aspects of nicotine self-administration selectively in the rat. *Psychopharmacology* **158**, 190–197.

Corrigall, W. A., Coen, K. M., Zhang, J., and Adamson, L. (2002). Pharmacological manipulations of the pedunculopontine tegmental nucleus in the rat reduce self-administration of both nicotine and cocaine. *Psychopharmacology* **160**, 198–205.

Corringer, P. J., Bertrand, S., Bohler, S., Edelstein, S. J., Changeux, J. P., and Bertrand, D. (1998). Critical elements determining diversity in agonist binding and desensitization of neuronal nicotinic acetylcholine receptors. *Journal of Neuroscience* **18**, 648–657.

Costall, B., Kelly, M. E., Naylor, R. J., and Onaivi, E. S. (1989). The actions of nicotine and cocaine in a mouse model of anxiety. *Pharmacology Biochemistry and Behavior* **33**, 197–203.

Coulie, B., Camilleri, M., Bharucha, A. E., Sandborn, W. J., and Burton, D. (2001). Colonic motility in chronic ulcerative proctosigmoiditis and the effects of nicotine on colonic motility in patients and healthy subjects. *Alimentary Pharmacology and Therapeutics* **15**, 653–663.

Covey, L. S., Glassman, A. H., and Stetner, F. (1990). Depression and depressive symptoms in smoking cessation. *Comprehensive Psychiatry* **31**, 350–354.

Critchley, J. A., and Unal, B. (2003). Health effects associated with smokeless tobacco: a systematic review. *Thorax* **58**, 435–443.

Crooks, P. A., and Dwoskin, L. P. (1997). Contribution of CNS nicotine metabolites to the neuropharmacological effects of nicotine and tobacco smoking. *Biochemical Pharmacology* **54**, 743–753.

Cui, C., Booker, T. K., Allen, R. S., Grady, S. R., Whiteaker, P., Marks, M. J., Salminen, O., Tritto, T., Butt, C. M., Allen, W. R., Stitzel, J. A., McIntosh, J. M., Boulter, J., Collins, A. C., and Heinemann, S. F. (2003). The beta3 nicotinic receptor subunit: a component of alpha-conotoxin MII-binding nicotinic acetylcholine receptors that modulate dopamine release and related behaviors. *Journal of Neuroscience* **23**, 11045–11053.

Cuthbert, B. N., Holaday, J. W., Meyerhoff, J., and Li, C. H. (1989). Intravenous beta-endorphin: behavioral and physiological effects in conscious monkeys. *Peptides* **10**, 729–734.

Dalhamn, T. (1972). Some factors influencing the respiratory toxicity of cigarette smoke. *Journal of the National Cancer Institute* **48**, 1821–1824.

Dani, J. A., and Heinemann, S., (1996) Molecular and cellular aspects of nicotine abuse. *Neuron* **16**, 905–908.

Dappen, A., Schwartz, R. H., and O'Donnell, R. (1996). A survey of adolescent smoking patterns. *Journal of the American Board of Family Practioners* **9**, 7–13.

Davenport, K. E., Houdi, A. A., and Van Loon, G. R. (1990). Nicotine protects against mu-opioid receptor antagonism by beta-funaltrexamine: evidence for nicotine-induced release of endogenous opioids in brain. *Neuroscience Letters* **113**, 40–46.

Day, T. D. (1967). Carcinogenic action of cigarette smoke condensate on mouse skin. *British Journal of Cancer* **21**, 56–81.

de Bry, T., and Le Moyne de Morgues, J. (1591). *Brevis Narratio Eorvm Qvae in Florida Americae Provicia Gallis Acciderunt [A Brief Narration of Those Things which Befell the French in the Province of Florida in America]*, Francoforti ad Moenvm, Typis I. Wecheli, sumtibus vero T. de Bry, venales reperiutur in fficina S. Feirabedii.

De Vries, T. J., and Schoffelmeer, A. N. M. (2005). Cannabinoid CB1 receptors control conditioned drug seeking. *Trends in Pharmacological Sciences* **26**, 420–426.

De Vries, T. J., de Vries, W., Janssen, M. C., and Schoffelmeer, A. N. (2005). Suppression of conditioned nicotine and sucrose seeking by the cannabinoid-1 receptor antagonist SR141716A. *Behavioural Brain Research* **161**, 164–168.

Delgado, P. L., Charney, D. S., Price, L. H., Aghajanian, G. K., Landis, H., and Heninger, G. R. (1990). Serotonin function and the mechanism of antidepressant action: reversal of antidepressant-induced remission by rapid depletion of plasma tryptophan. *Archives of General Psychiatry* **47**, 411–418.

Delgado, P. L., Price, L. H., Miller, H. L., Salomon, R. M., Licinio, J., Krystal, J. H., Heninger, G. R., and Charney, D. S. (1991). Rapid serotonin depletion as a provocative challenge test for patients with major depression: relevance to antidepressant action and the neurobiology of depression. *Psychopharmacology Bulletin* **27**, 321–330.

Dewey, S. L., Brodie, J. D., Gerasimov, M., Horan, B., Gardner, E. L., and Ashby, C. R. Jr. (1999). A pharmacologic strategy for the treatment of nicotine addiction. *Synapse* **31**, 76–86.

Domino, E. G. (1967). Electroencephalographic and behavioral arousal effects of small doses of nicotine: a neuropsychopharmacological study. *Annals of the New York Academy of Sciences* **142**(1), 216–244.

Domino, E. G. (1973). Neuropsychopharmacology. *In Smoking Behavior: Motives and Incentives*, (W. L. Dunn, Ed.), pp. 5–31, VH Winston, Washington DC.

Drago, J., McColl, C. D., Horne, M. K., Finkelstein, D. I., and Ross, S. A. (2003). Neuronal nicotinic receptors: insights gained from gene knockout and knockin mutant mice. *Cellular and Molecular Life Sciences* **60**, 1267–1280.

Edwards, N. B., Murphy, J. K., Downs, A. D., Ackerman, B. J., and Rosenthal, T. L. (1989). Doxepin as an adjunct to smoking cessation: a double-blind pilot study. *American Journal of Psychiatry* **146**, 373–376.

Elgoyhen, A. B., Vetter, D. E., Katz, E., Rothlin, C. V., Heinemann, S. F., and Boulter, J. (2001). Alpha10: a determinant of nicotinic cholinergic receptor function in mammalian vestibular and cochlear mechanosensory hair cells. *Proceedings of the National Academy of Sciences USA* **98**, 3501–3506.

Engberg, G., Erhardt, S., Sharp, T., and Hajos, M. (2000). Nicotine inhibits firing activity of dorsal raphe 5-HT neurones in vivo. *Naunyn Schmiedebergs Archives of Pharmacology* **362**, 41–45.

Epping-Jordan, M. P., Watkins, S. S., Koob, G. F., and Markou, A. (1998). Dramatic decreases in brain reward function during nicotine withdrawal. *Nature* **393**, 76–79.

Erhardt, S., Schwieler, L., and Engberg, G. (2002). Excitatory and inhibitory responses of dopamine neurons in the ventral tegmental area to nicotine. *Synapse* **43**, 227–237.

Etter, J. F., Humair, J. P., Bergman, M. M., and Perneger, T. V. (2000). Development and validation of theattitudes towards smoking scale (ATS-18). *Addiction* **95**, 613–625.

Ezzati, M., and Lopez, A. D. (2003). Estimates of global mortality attributable to smoking in 2000. *Lancet* **362**, 847–852.

Fagen, Z. M., Mansvelder, H. D., Keath, J. R., and McGehee, D. S. (2003). Short- and long-term modulation of synaptic inputs to brain reward areas by nicotine. *In Glutamate and Disorders of Cognition and Motivation* (series title: *Annals of the New York Academy of Sciences*, vol. 1003), (B. Moghaddam, and M. E. Wolf, Eds.), pp. 185–195, New York Academy of Sciences, New York.

Fagerstrom, K. O. (1978). Measuring degree of physical dependence to tobacco smoking with reference to individualization of treatment. *Addictive Behaviors* **3**, 235–241.

Fagerstrom, K. O. (1988). Efficacy of nicotine chewing gum: a review. *In Nicotine Replacement: A Critical Evaluation* (series title: *Progress in Clinical and Biological Research*, vol. 261), (O. F. Pomerleau, and C. S. Pomerleau, Eds.), pp. 109–128, Liss, New York.

Fagerstrom, K. O., Ramstrom, L. M., and Svensson, T. H. (1992). Health in tobacco control. *Lancet* **339**, 934.

Fairholt, F. W. (1859). *Tobacco: Its History and Associations*, Chapman and Hall, London.

Fattore, L., Cossu, G., Martellotta, M. C., and Fratta, W. (2002). Baclofen antagonizes intravenous self-administration of nicotine in mice and rats. *Alcohol and Alcoholism* **37**, 495–498.

Fellows, J. L., Trosclair, A., and Adams, E. K. (2002). Annual smoking-attributable mortality, years of potential life lost, and economic costs: United States, 1995–1999. *Morbidity and Mortality Weekly Report* **51**, 300–303.

File, S. E., Kenny, P. J., and Ouagazzal, A. M. (1998). Bimodal modulation by nicotine of anxiety in the social interaction test: role of the dorsal hippocampus. *Behavioral Neuroscience* **112**, 1423–1429.

Flores, C. M., Rogers, S. W., Pabreza, L. A., Wolfe, B. B., and Kellar, K. J. (1992). A subtype of nicotinic cholinergic receptor in rat brain is composed of alpha 4 and beta 2 subunits and is up-regulated by chronic nicotine treatment. *Molecular Pharmacology* **41**, 31–37.

Forget, B., Hamon, M., and Thiebot, M. H. (2005). Cannabinoid CB1 receptors are involved in motivational effects of nicotine in rats. *Psychopharmacology* (in press).

Fowler, J. S., Volkow, N. D., Wang, G. J., Pappas, N., Logan, J., MacGregor, R., Alexoff, D., Shea, C., Schlyer, D., Wolf, A. P., Warner, D., Zezulkova, I., and Cilento, R. (1996). Inhibition of monoamine oxidase B in the brains of smokers. *Nature* **379**, 733–736.

Franke, F. E., and Thomas, J. E. (1936). The treatment of acute nicotine poisoning. *Journal of the American Medical Association* **106**, 507–512.

Frederiksen, L. W. (1976). Single-case designs in the modification of smoking. *Addictive Behaviors* **1**, 311–319.

Fu, Y., Matta, S. G., Brower, V. G., and Sharp, B. M. (2001). Norepinephrine secretion in the hypothalamic paraventricular nucleus of rats during unlimited access to self-administered nicotine: An in vivo microdialysis study. *Journal of Neuroscience* **21**, 8979–8989.

Fudala, P. J., Teoh, K. W., and Iwamoto, E. T. (1985). Pharmacologic characterization of nicotine-induced conditioned place preference. *Pharmacology Biochemistry and Behavior* **22**, 237–241.

Fung, Y. K., Schmid, M. J., Anderson, T. M., and Lau, Y. S. (1996). Effects of nicotine withdrawal on central dopaminergic systems. *Pharmacology Biochemistry and Behavior* **53**, 635–640.

Garcia-Munoz, M., Patino, P., Young, S. J., and Groves, P. M. (1996). Effects of nicotine on dopaminergic nigrostriatal axons requires stimulation of presynaptic glutamatergic receptors. *Journal of Pharmacology and Experimental Therapeutics* **277**, 1685–1693.

Gardner, E. L., Paredes, W., Smith, D., Donner, A., Milling, C., Cohen, D., and Morrison, D. (1988). Facilitation of brain stimulation reward by delta-9-tetrahydrocannabinol. *Psychopharmacology* **96**, 142–144.

Ghozland, S., Zorrilla, E., Parsons, L. H., and Koob, G. F. (2004). Mecamylamine increases extracellular CRF levels in the central nucleus of the amygdala or nicotine-dependent rats. *Society for Neuroscience Abstracts* **30**, 708.8.

Giovino, G. A., Henningfield, J. E., Tomar, S. L., Escobedo, L. G., and Slade, J. (1995). Epidemiology of tobacco use and dependence. *Epidemiologic Reviews* **17**, 48–65.

Glassman, A. H., Stetner, F., Walsh, B. T., Raizman, P. S., Fleiss, J. L., Cooper, T. B., and Covey, L. S. (1988). Heavy smokers, smoking cessation, and clonidine. Results of a double-blind, randomized trial. *Journal of the American Medical Association* **259**, 2863–2866.

Glassman, A. H., Helzer, J. E., Covey, L. S., Cottler, L. B., Stetner, F., Tipp, J. E., and Johnson, J. (1990). Smoking, smoking cessation, and major depression. *Journal of the American Medical Association* **264**, 1546–1549.

Goldberg, S. R., and Henningfield, J. E. (1988). Reinforcing effects of nicotine in humans and experimental animals responding under intermittent schedules of i.v. drug injection. *Pharmacology Biochemistry and Behavior* **30**, 227–234.

Goldberg, S. R., Spealman, R. D., and Goldberg, D. M. (1981). Persistent behavior at high rates maintained by intravenous self-administration of nicotine. *Science* **214**, 573–575.

Gorelick, D. A., Rose, J., and Jarvik, M. E. (1988–1989). Effect of naloxone on cigarette smoking. *Journal of Substance Abuse* **1**, 153–159.

Grabus, S. D., Martin, B. R., and Damaj, I. M. (2005). Nicotine physical dependence in the mouse: Involvement of the alpha(7) nicotinic receptor subtype. *European Journal of Pharmacology* **515**, 90–93.

Grenhoff, J., Aston-Jones, G., and Svensson, T. H. (1986). Nicotinic effects on the firing pattern of midbrain dopamine neurons. *Acta Physiologica Scandinavica* **128**, 351–358.

Grottick, A. J., Corrigall, W. A., and Higgins, G. A. (2001). Activation of 5-HT(2C) receptors reduces the locomotor and rewarding effects of nicotine. *Psychopharmacology* **157**, 292–298.

Grubb, M. S., and Thompson, I. D. (2004). Visual response properties in the dorsal geniculate nucleus of mice lacking the beta2 subunit of the nicotinic acetycholine receptor. *Journal of Neuroscience* **24**, 8459–8469.

Grunberg, N. E., Bowen, D. J., Maycock, V. A., and Nespor, S. M. (1985). The importance of sweet taste and caloric content in the effects of nicotine on specific food consumption. *Psychopharmacology* **87**, 198–203.

Grunberg, N. E., Popp, K. A., Bowen, D. J., Nespor, S. M., Winders, S. E., and Eury, S. E. (1988). Effects of chronic nicotine administration on insulin, glucose, epinephrine, and norepinephrine. *Life Sciences* **42**, 161–170.

Guillemin, R., Vargo, T., Rossier, J., Minick, S., Ling, N., Rivier, C., Vale, W., and Bloom, F. (1977). β-endorphin and adrenocorticotropin are selected concomitantly by the pituitary gland. *Science* **197**, 1367–1369.

Guindon, G. E., and Boisclair, D. (2003). *Past, Current and Future Trends in Tobacco Use* (series title: *Economics of Tobacco Control*, vol. 6), World Bank, Washington DC.

Hahn, B., Shoaib, M., and Stolerman, I. P. (2003). Involvement of the prefrontal cortex but not the dorsal hippocampus in the attention-enhancing effects of nicotine in rats. *Psychopharmacology* **168**, 271–279.

Harrison, A. A., Liem, Y. T., and Markou, A. (2001). Fluoxetine combined with a serotonin-1A receptor antagonist reversed reward deficits observed during nicotine and amphetamine withdrawal in rats. *Neuropsychopharmacology* **25**, 55–71.

Haug, H. (1961). *Petite Histoire du Tabac en Alsace*. Strasbourg.

Hauser, H., Schwartz, B., Roth, G., and Bickford, R. (1958). Electroencephalographic changes related to smoking. *Electroencephalography and Clinical Neurophysiology Supplement* **10**, 576.

Hayashida, S., Katsura, M., Torigoe, F., Tsujimura, A., and Ohkuma, S. (2005). Increased expression of L-type high voltage-gated calcium channel alpha1 and alpha2/delta subunits in mouse brain after chronic nicotine administration. *Molecular Brain Research* **135**, 280–284.

Helton, D. R., Tizzano, J. P., Monn, J. A., Schoepp, D. D., and Kallman, M. J. (1997). LY354740: a metabotropic glutamate receptor agonist which ameliorates symptoms of nicotine withdrawal in rats. *Neuropharmacology* **36**, 1511–1516.

Henningfield, J. E., Miyasato, K., and Jasinski, D. R. (1983). Cigarette smokers self-administer intravenous nicotine. *Pharmacology Biochemistry and Behavior* **19**, 887–890.

Henningfield, J. E., Miyasato, K., and Jasinski, D. R. (1985). Abuse liability and pharmacodynamic characteristics of intravenous and inhaled nicotine. *Journal of Pharmacology and Experimental Therapeutics* **234**, 1–12.

Henningfield, J. E., Stapleton, J. M., Benowitz, N. L., Grayson, R. F., and London, E. D. (1993). Higher levels of nicotine in arterial than in venous blood after cigarette smoking. *Drug and Alcohol Dependence* **33**, 23–29.

Herman, C. P., and Kozlowski, L. T. (1979). Indulgence, excess, and restraint: perspectives on consummatory behavior in everyday life. *Journal of Drug Issues* **9**, 185–196.

Herxheimer, A. (1967). Circulatory effects of nicotine aerosol inhalations and cigarette smoking in man. *Lancet* **290(7519)**, 754–755.

Hildebrand, B. E., Nomikos, G. G., Bondjers, C., Nisell, M., and Svensson, T. H. (1997). Behavioral manifestations of the nicotine abstinence syndrome in the rat: peripheral versus central mechanisms. *Psychopharmacology* **129**, 348–356.

Hildebrand, B. E., Nomikos, G. G., Hertel, P., Schilstrom, B., and Svensson, T. H. (1998). Reduced dopamine output in the nucleus accumbens but not in the medial prefrontal cortex in rats displaying a mecamylamine-precipitated nicotine withdrawal syndrome. *Brain Research* **779**, 214–225.

Hildebrand, B. E., Panagis, G., Svensson, T. H., and Nomikos, G. G. (1999). Behavioral and biochemical manifestations of mecamylamine-precipitated nicotine withdrawal in the rat: role of nicotinic receptors in the ventral tegmental area. *Neuropsychopharmacology* **21**, 560–574.

Hill, P., and Wynder, E. L. (1974). Smoking and cardiovascular disease. Effect of nicotine on the serum epinephrine and corticoids. *American Heart Journal* **87**, 491–496.

Hoffman, D., Schmeltz, I., Hecht, S. S., and Wynder, E. L. (1976). Chemical studies on tobacco smoke: XXXIX. On the identification of carcinogens, tumor promoters, and cocarcinogens in tobacco smoke. *In Proceedings of the Third World Conference on Smoking and Health, New York, June 2–5, 1975: Volume 1.* (E. L. Wynder, D. Hoffmann, and G. B. Gori, Eds.), pp. 125–145, *Modifying the Risk for the Smoker* [DHEW publ. no. NIH 76-1221], National Cancer Institute, Washington DC.

Hoffmann, D., Rivenson, A., Hecht, S. S., Hilfrich, J., Kobayashi, N., and Wynder, E. L. (1979). Model studies in tobacco carcinogenesis with the Syrian golden hamster. *Progress in Experimental Tumor Research* **24**, 370–390.

Horn, D., Courts, F. A., Taylor, R. M., and Solomon, E. S. (1959). Cigarette smoking among high school students. *American Journal of Public Health* **49**, 1497–1511.

Houdi, A. A., Pierzchala, K., Marson, L., Palkovits, M., and Van Loon, G. R. (1991). Nicotine-induced alteration in Tyr-Gly-Gly and Met-enkephalin in discrete brain nuclei reflects altered enkephalin neuron activity. *Peptides* **12**, 161–166.

Hsu, Y. N., Amin, J., Weiss, D. S., and Wecker, L. (1996). Sustained nicotine exposure differentially affects alpha 3 beta 2 and alpha 4 beta 2 neuronal nicotinic receptors expressed in Xenopus oocytes. *Journal of Neurochemistry* **66**, 667–675.

Hu, X. T., and White, F. J. (1996). Glutamate receptor regulation of rat nucleus accumbens neurons in vivo. *Synapse* **23**, 208–218.

Hughes, J. R. (1992). Tobacco withdrawal in self-quitters. *Journal of Consulting and Clinical Psychology* **60**, 689–697.

Hughes, J. R., Hatsukami, D. K., Pickens, R. W., Krahn, D., Malin, S., and Luknic, A. (1984). Effect of nicotine on the tobacco withdrawal syndrome. *Psychopharmacology* **83**, 82–87.

Hughes, J. R., Hatsukami, D. K., Mitchell, J. E., and Dahlgren, L. A. (1986). Prevalence of smoking among psychiatric outpatients. *American Journal of Psychiatry* **143**, 993–997.

Hughes, J. R., Gust, S. W., Keenan, R., Fenwick, J. W., Skoog, K., and Higgins, S. T. (1991a). Long-term use of nicotine vs placebo gum. *Archives of Internal Medicine* **151**, 1993–1998.

Hughes, J. R., Gust, S. W., Skoog, K., Keenan, R. M., and Fenwick, J. W. (1991b). Symptoms of tobacco withdrawal. A replication and extension. *Archives of General Psychiatry* **48**, 52–59.

Hunt, W. A., Barnett, L. W., and Branch, L. G. (1971). Relapse rates in addiction programs. *Journal of Clinical Psychology* **27**, 455–456.

Hunt, W. A., and Bespalec, D. A. (1974). An evaluation of current methods of modifying smoking behavior. *Journal of Clinical Psychology* **30**, 431–438.

Hurt, R. D., Sachs, D. P., Glover, E. D., Offord, K. P., Johnston, J. A., Dale, L. C., Khayrallah, M. A., Schroeder, D. R., Glover, P. N., Sullivan, C. R., Croghan, I. T., and Sullivan, P. M. (1997). A comparison of sustained-release bupropion and placebo for smoking cessation. *New England Journal of Medicine* **337**, 1195–1202.

Huston-Lyons, D., and Kornetsky, C. (1992). Effects of nicotine on the threshold for rewarding brain stimulation in rats. *Pharmacology Biochemistry and Behavior* **41**, 755–759.

Ikard, F. F., Green, D. E., and Horn D. (1969). A scale to differentiate between types of smoking as related to the management of affect. *International Journal of the Addictions* **4**, 649–659.

Ikard, F. F., and Tomkins, S. (1973). The experience of affect as a determinant of smoking behavior: a series of validity studies. *Journal of Abnormal Psychology* **81**, 172–181.

Investigations of tobacco—nicotia. (1829). *Quarterly Journal of Science* **2**, 418.

Irvine, E. E., Cheeta, S., and File, S. E. (1999). Time-course of changes in the social interaction test of anxiety following acute and chronic administration of nicotine. *Behavioural Pharmacology* **10**, 691–697.

Irvine, E. E., Bagnalasta, M., Marcon, C., Motta, C., Tessari, M., File, S. E., and Chiamulera, C. (2001). Nicotine self-administration and withdrawal: modulation of anxiety in the social interaction test in rats. *Psychopharmacology* **153**, 315–320.

Isaac, P. F., and Rand, M. J. (1972). Cigarette smoking and plasma levels of nicotine. *Nature* **236**, 308–310.

Jacob, P. 3rd, Benowitz, N. L., and Shulgin, A. T. (1988). Recent studies of nicotine metabolism in humans. *Pharmacology Biochemistry and Behavior* **30**, 249–253.

Jacob, P. 3rd, Yu, L., Shulgin, A. T., and Benowitz, N. L. (1999). Minor tobacco alkaloids as biomarkers for tobacco use: comparison of users of cigarettes, smokeless tobacco, cigars, and pipes. *American Journal of Public Health* **89**, 731–736.

Jensen, E. X., Fusch, C., Jaeger, P., Peheim, E., and Horber, F. F. (1995). Impact of chronic cigarette smoking on body composition and fuel metabolism. *Journal of Clinical Endocrinology and Metabolism* **80**, 2181–2185.

Jones, I. W., and Wonnacott, S. (2004). Precise localization of alpha7 nicotinic acetylcholine receptors on glutamatergic axon terminals in the rat ventral tegmental area. *Journal of Neuroscience* **24**, 11244–11252.

Jorenby, D. E., Leischow, S. J., Nides, M. A., Rennard, S. I., Johnston, J. A., Hughes, A. R., Smith, S. S., Muramoto, M. L., Daughton, D. M., Doan, K., Fiore, M. C., and Baker, T. B. (1999). A controlled trial of sustained-release bupropion, a nicotine patch, or both for smoking cessation. *New England Journal of Medicine* **340**, 685–691.

Jung, M. J., Lippert, B., Metcalf, B. W., Bohlen, P., and Schechter, P. J. (1977). γ-Vinyl GABA (4-amino-hex-5-enoic acid), a new selective irreversible inhibitor of GABA–T: effects on brain GABA metabolism in mice. *Journal of Neurochemistry* **29**, 797–802.

Kabat, G. C., and Wynder, E. L. (1987). Determinants of quitting smoking. *American Journal of Public Health* **77**, 1301–1305.

Kadoya, C., Domino, E. F., and Matsuoka, S. (1994). Relationship of electroencephalographic and cardiovascular changes to plasma nicotine levels in tobacco smokers. *Clinical Pharmacology and Therapeutics* **55**, 370–377.

Kalivas, P. W., Churchill, L., and Klitenick, M. A. (1993). GABA and enkephalin projection from the nucleus accumbens and ventral pallidum to the ventral tegmental area. *Neuroscience* **57**, 1047–1060.

Kandel, D. (1975). Stages in adolescent involvement in drug use. *Science* **190**, 912–914.

Karras, A., and Kane, J. M. (1980). Naloxone reduces cigarette smoking. *Life Sciences* **27**, 1541–1545.

Kedmi, M., Beaudet, A. L., and Orr-Urtreger, A. (2004). Mice lacking neuronal nicotinic acetylcholine receptor beta4-subunit and mice lacking both alpha5- and beta4–subunits are highly resistant to nicotine-induced seizures. *Physiological Genomics* **17**, 221–229.

Keenan, R. M., Hatsukami, D. K., Pentel, P. R., Thompson, T. N., and Grillo, M. A. (1994). Pharmacodynamic effects of cotinine in abstinent cigarette smokers. *Clinical Pharmacology and Therapeutics* **55**, 581–590.

Keller, J. J., Keller, A. B., Bowers, B. J., and Wehner, J. M. (2005). Performance of alpha7 nicotinic receptor null mutants is impaired in appetitive learning measured in a signaled nose poke task. *Behavioural Brain Research* **162**, 143–152.

Kelley, A. E. (2002). Nicotinic receptors: addiction's smoking gun? *Nature Medicine* **8**, 447–449.

Kenny, P. J., and Markou, A. (2001). Neurobiology of the nicotine withdrawal syndrome. *Pharmacology Biochemistry and Behavior* **70**, 531–549.

Kenny, P. J., Gasparini, F., Markou, A., and Group, I. I. (2003). Metabotropic and alpha-amino-3-hydroxy-5-methyl-4-isoxazole propionate (AMPA)/kainate glutamate receptors regulate the deficit in brain reward function associated with nicotine withdrawal in rats. *Journal of Pharmacology and Experimental Therapeutics* **306**, 1068–1076.

Kenny, P. J., and Markou, A., (2004). The ups and downs of addiction: role of metabotropic glutamate receptors. *Trends in Pharmacological Sciences* **25**, 265–272.

Kenny, P. J., and Markou, A. (2005). NMDA receptors gate the hedonic valence of nicotine: role in nicotine self-administration, submitted.

King, S. L., Caldarone, B. J., and Picciotto, M. R. (2004). Beta2-subunit-containing nicotinic acetylcholine receptors are critical for dopamine-dependent locomotor activation following repeated nicotine administration. *Neuropharmacology* **47**(*Suppl.* 1), 132–139.

Klink, R., de Kerchove d'Exaerde, A., Zoli, M., and Changeux, J. P. (2001). Molecular and physiological diversity of nicotinic acetylcholine receptors in the midbrain dopaminergic nuclei. *Journal of Neuroscience* **21**, 1452–1463.

Koob, G. F., Markou, A., Weiss, F., and Schulteis, G. (1993). Opponent process and drug dependence: Neurobiological mechanisms. *Seminars in the Neurosciences* **5**, 351–358.

Koob, G. F., Sanna, P. P., and Bloom, F. E. (1998). Neuroscience of addiction. *Neuron* **21**, 467–476.

Kopp, P., and Fenoglio, P. (2000). *Le cout Social des Drogues Licites (Alcool et Tabac) et Illicites en France*, etude 22, Observatoire Francais des Drogues et des Toxicomanies, Paris.

Kornetsky, C., and Esposito, R. U. (1981). Reward and detection thresholds for brain stimulation: Dissociative effects of cocaine. *Brain Research* **209**, 496–500.

Kozlowski, L. T., and Herman, C. P. (1984). The interaction of psychosocial and biological determinants of tobacco use: more on the boundary model. *Journal of Applied Social Psychology* **14**, 244–256.

Kozlowski, L. T., Wilkinson, D. A., Skinner, W., Kent, C., Franklin, T., and Pope, M. (1989). Comparing tobacco cigarette dependence with other drug dependencies: greater or equal 'difficulty quitting'and 'urges to use,' but less 'pleasure' from cigarettes. *Journal of the American Medical Association* **261**, 898–901.

Krishnan-Sarin, S., Rosen, M. I., and O'Malley, S. S. (1999). Naloxone challenge in smokers. Preliminary evidence of an opioid component in nicotine dependence. *Archives of General Psychiatry* **56**, 663–668.

Laffan, R. J., and Borison, H. L. Emetic action of nicotine and lobeline. *Journal of Pharmacology and Experimental Therapeutics* **121**, 468–476.

Lanca, A. J., Adamson, K. L., Coen, K. M., Chow, B. L., and Corrigall, W. A. (2000). The pedunculopontine tegmental nucleus and the role of cholinergic neurons in nicotine self-administration in the rat: a correlative neuroanatomical and behavioral study. *Neuroscience* **96**, 735–742.

Laviolette, S. R., Alexson, T. O., and van der Kooy, D. (2002). Lesions of the tegmental pedunculopontine nucleus block the rewarding effects and reveal the aversive effects of nicotine in the ventral tegmental area. *Journal of Neuroscience* **22**, 8653–8660.

Laviolette, S. R., and van der Kooy, D. (2003). Blockade of mesolimbic dopamine transmission dramatically increases sensitivity to the rewarding effects of nicotine in the ventral tegmental area. *Molecular Psychiatry* **8**, 50–59.

Lazar, P., Chouroulinkov, I., Izard, C., Moree-Testa, P., and Hemon, D. (1974). Bioassays of carcinogenicity after fractionation of cigarette smoke condensate. *Biomedicine* **20**, 214–222.

Le Foll, B., and Goldberg, S. R. (2004). Rimonabant, a CB1 antagonist, blocks nicotine-conditioned place preferences. *Neuroreport* **15**, 2139–2143.

Le Houezec, J., and Benowitz, N. L. (1991). Basic and clinical psychopharmacology of nicotine. *Clinics in Chest Medicine* **12**, 681–699.

Le Novere, N., and Changeux, J. P. (1995). Molecular evolution of the nicotinic acetylcholine receptor: an example of multigene family in excitable cells. *Journal of Molecular Evolution* **40**, 155–172.

Le Novere, N., Corringer, P. J., and Changeux, J. P. (2002). The diversity of subunit composition in nAChRs: evolutionary origins, physiologic and pharmacologic consequences. *Journal of Neurobiology* **53**, 447–456.

Le Sage, M. G., Burroughs, D., Dufek, M., Keyler, D. E., and Pentel, P. R. (2004). Reinstatement of nicotine self-administration in rats by presentation of nicotine-paired stimuli, but not nicotine priming. *Pharmacology Biochemistry and Behavior* **79**, 507–513.

Le Sage, M. G., Keyler, D. E., Shoeman, D., Raphael, D., Collins, G., and Pentel, P. R. (2002). Continuous nicotine infusion reduces nicotine self-administration in rats with 23-h/day access to nicotine. *Pharmacology Biochemistry and Behavior* **72**, 279–289.

Leach, M. (Ed.) (1972). *Funk and Wagnalls Standard Dictionary of Folklore, Mythology, and Legend*, Funk and Wagnalls, New York.

Lee, L. Y., Morton, R. F., Hord, A. H., and Frazier, D. T. (1983). Reflex control of breathing following inhalation of cigarette smoke in conscious dogs. *Journal of Applied Physiology* **54**, 562–570.

Leith, N. J., and Barrett, R. J. (1976). Amphetamine and the reward system: Evidence for tolerance and postdrug depression. *Psychopharmacologia* **46**, 19–25.

Lemaire, I., Tseng, R., and Lemaire, S. (1978). Systemic administration of beta-endorphin: potent hypotensive effect involving a serotonergic pathway. *Proceedings of the National Academy of Sciences USA* **75**, 6240–6242.

Levin, E. D., Conners, C. K., Silva, D., Hinton, S. C., Meck, W. H., March, J., and Rose, J. E. (1998). Transdermal nicotine effects on attention. *Psychopharmacology* **140**, 135–141.

Li, X., Rainnie, D. G., McCarley, R. W., and Greene, R. W. (1998). Presynaptic nicotinic receptors facilitate monoaminergic transmission. *Journal of Neuroscience* **18**, 1904–1912.

Lin, D., Koob, G. F., and Markou, A. (1999). Differential effects of withdrawal from chronic amphetamine or fluoxetine administration on brain stimulation reward in the rat: interactions between the two drugs. *Psychopharmacology* **145**, 283–294.

Lindstrom, J. M. (2000). The structure of neuronal nicotinic receptors. *In Neuronal Nicotinic Receptors* (series title: *Handbook of Experimental Pharmacology*, vol. 144), (F. Clementi, D. Fornasari, and C. Gotti, Eds.), pp. 101–162. Springer, New York.

Lippert, B., Metcalf, B. W., Jung, M. J., and Casara, P. (1977). 4-amin-hex-5-enoic acid, a selective catalytic inhibitor of 4-aminobutyric-acid aminotransferase in mammalian brain. *European Journal of Biochemistry* **74**, 441–445.

Mackay, J., and Eriksen, M. (2002). *The Tobacco Atlas*. World Health Organization, Geneva.

Malin, D. H., Lake, J. R., Newlin-Maultsby, P., Roberts, L. K., Lanier, J. G., Carter, V. A., Cunningham, J. S., and Wilson, O. B. (1992). Rodent model of nicotine abstinence syndrome. *Pharmacology Biochemistry and Behavior* **43**, 779–784.

Malin, D. H., Lake, J. R., Carter, V. A., Cunningham, J. S., and Wilson, O. B. (1993). Naloxone precipitates nicotine abstinence syndrome in the rat. *Psychopharmacology* **112**, 339–342.

Malin, D. H., Lake, J. R., Carter, V. A., Cunningham, J. S., Hebert, K. M., Conrad, D. L., and Wilson, O. B. (1994). The nicotine antagonist mecamylamine precipitates nicotine abstinence syndrome in the rat. *Psychopharmacology* **115**, 180–184.

Malin, D. H., Lake, J. R., Short, P. E., Blossman, J. B., Lawless, B. A., Schopen, C. K., Sailer, E. E., Burgess, K., and Wilson, O. B. (1996). Nicotine abstinence syndrome precipitated by an analog of neuropeptide FF. *Pharmacology Biochemistry and Behavior* **54**, 581–585.

Malin, D. H., Lake, J. R., Schopen, C. K., Kirk, J. W., Sailer, E. E., Lawless, B. A., Upchurch, T. P., Shenoi, M., and Rajan, N. (1997). Nicotine abstinence syndrome precipitated by central but not peripheral hexamethonium. *Pharmacology Biochemistry and Behavior* **58**, 695–699.

Malin, D. H., Lake, J. R., Upchurch, T. P., Shenoi, M., Rajan, N., and Schweinle, W. E. (1998). Nicotine abstinence syndrome precipitated by the competitive nicotinic antagonist dihydro-β-erythroidine. *Pharmacology Biochemistry and Behavior* **60**, 609–613.

Mangan, G. L. (1982). The effects of cigarette smoking on vigilance performance. *Journal of General Psychology* **106**, 77–83.

Mansvelder, H. D., and McGehee, D. S. (2000). Long-term potentiation of excitatory inputs to brain reward areas by nicotine. *Neuron* **27**, 349–357.

Mansvelder, H. D., and McGehee, D. S. (2002). Cellular and synaptic mechanisms of nicotine addiction. *Journal of Neurobiology* **53**, 606–617.

Mansvelder, H. D., Keath, J. R., and McGehee, D. S. (2002). Synaptic mechanisms underlie nicotine-induced excitability of brain reward areas. *Neuron* **33**, 905–919.

Markou, A., and Koob, G. F. (1991). Post-cocaine anhedonia: An animal model of cocaine withdrawal. *Neuropsychopharmacology* **4**, 17–26.

Markou, A., Kosten, T. R., and Koob, G. F. (1998). Neurobiological similarities in depression and drug dependence: A self-medication hypothesis. *Neuropsychopharmacology* **18**, 135–174.

Markou, A., and Paterson, N. E. (2001). The nicotinic antagonist methyllycaconitine has differential effects on nicotine self-administration and nicotine withdrawal in the rat. *Nicotine and Tobacco Research* **3**, 361–373.

Marks, M. J., Pauly, J. R., Gross, S. D., Deneris, E. S., Hermans-Borgmeyer, I., Heinemann, S. F., and Collins, A. C. (1992). Nicotine binding and nicotinic receptor subunit RNA after chronic nicotine treatment. *Journal of Neuroscience* **12**, 2765–2784.

Maron, D. J., and Fortmann, S. P. (1987). Nicotine yield and measures of cigarette smoke exposure in a large population: are lower-yield cigarettes safer? *American Journal of Public Health* **77**, 546–549.

Marshall, D. L., Redfern, P. H., and Wonnacott, S. (1997). Presynaptic nicotinic modulation of dopamine release in the three ascending pathways studied by in vivo microdialysis: comparison of naive and chronic nicotine-treated rats. *Journal of Neurochemistry* **68**, 1511–1519.

Marubio, L. M., del Mar Arroyo-Jimenez, M., Cordero-Erausquin, M., Lena, C., Le Novere, N., de Kerchove d'Exaerde, A., Huchet, M., Damaj, M. I., and Changeux, J. P. (1999). Reduced antinociception in mice lacking neuronal nicotinic receptor subunits. *Nature* **398**, 805–810.

Marubio, L. M., Gardier, A. M., Durier, S., David, D., Klink, R., Arroyo-Jimenez, M. M., McIntosh, J. M., Rossi, F., Champtiaux, N., Zoli, M., and Changeux, J. P. (2003). Effects of nicotine in the dopaminergic system of mice lacking the alpha4 subunit of neuronal nicotinic acetylcholine receptors. *European Journal of Neuroscience* **17**, 1329–1337.

Marubio, L. M., and Paylor R. (2004). Impaired passive avoidance learning in mice lacking central neuronal nicotinic acetylcholine receptors. *Neuroscience* **129**, 575–582.

Matsubayashi, H., Amano, T., Seki, T., Sasa, M., and Sakai, N. (2003). Electrophysiological characterization of nicotine-induced excitation of dopaminergic neurons in the rat substantia nigra. *Journal of Pharmacological Sciences* **93**, 143–148.

Matsukura, S., Taminato, T., Kitano, N., Seino, Y., Hamada, H., Uchihashi, M., Nakajima, H., and Hirata, Y. (1984). Effects of environmental tobacco smoke on urinary cotinine excretion in nonsmokers: Evidence for passive smoking. *New England Journal of Medicine* **311**, 828–832.

McGehee, D. S., Heath, M. J., Gelber, S., Devay, P., and Role, L. W. (1995). Nicotine enhancement of fast excitatory synaptic transmission in CNS by presynaptic receptors. *Science* **269**, 1692–1696.

McGuigan, M. A. (2004). Nicotine. *In Medical Toxicology*, 3rd ed. (R. C. Dart, Ed.), pp. 601–604. Lippincott Williams and Wilkins, Philadelphia.

McKennell, A. C., and Thomas, R. K. (1967). *Adults' and Adolescents' Smoking Habits and Attitudes*, Cox and Sharland, London.

McNeill, A. D., West, R. J., Jarvis, M., Jackson, P., and Bryant, A. (1986). Cigarette withdrawal symptoms in adolescent smokers. *Psychopharmacology* **90**, 533–536.

Mihailescu, S., Palomero-Rivero, M., Meade-Huerta, P., Maza-Flores, A., and Drucker-Colin, R. (1998). Effects of nicotine and mecamylamine on rat dorsal raphe neurons. *European Journal of Pharmacology* **360**, 31–36.

Mihailescu, S., Guzman-Marin, R., Dominguez Mdel, C., and Drucker-Colin, R. (2002). Mechanisms of nicotine actions on dorsal raphe serotoninergic neurons. *European Journal of Pharmacology* **452**, 77–82.

MMWR. (1987). Cigarette smoking in the United States, 1986. *Morbidity and Mortality Weekly Report* **36**, 581–585.

MMWR. (1991). Cigarette smoking among adults: United States, 1988. *Morbidity and Mortality Weekly Report* **40**, 757–759, 765 [erratum: **41**(20):367].

MMWR. (1992). Cigarette smoking among adults: United States, 1990. *Morbidity and Mortality Weekly Report* **41**, 354–355, 361–362.

MMWR. (1993a). Use of smokeless tobacco among adults: United States, 1991. *Morbidity and Mortality Weekly Report* **42**, 263–266 [erratum: **42**(19):382].

MMWR. (1993b). Cigarette smoking among adults: United States, 1991. *Morbidity and Mortality Weekly Report* **42**, 230–233 [erratum: **42**(13):255].

MMWR. (1994). Cigarette smoking among adults—United States, 1992, and changes in the definition of current cigarette smoking. *Morbidity and Mortality Weekly Report* **43**, 342–346 [erratum: **43**(43), 801–803].

MMWR. (1996). Cigarette smoking among adults: United States, 1994. *Morbidity and Mortality Weekly Report* **45**, 588–590.

MMWR. Cigarette smoking among adults: United States, 1995. *Morbidity and Mortality Weekly Report* **46**, 1217–1220.

MMWR. (1997). Cigarette smoking among adults: United States, 1997. *Morbidity and Mortality Weekly Report* 1999, **48**, 993–996.

MMWR. (2000). Cigarette smoking among adults: United States, 1998. *Morbidity and Mortality Weekly Report* **49**, 881–884.

MMWR. (2001). Cigarette smoking among adults: United States, 1999. *Morbidity and Mortality Weekly Report* **50**, 869–873 [erratum: **50**(47):1066].

MMWR. (2002a). Annual smoking-attributable mortality, years of potential life lost, and economic costs: United States, 1995–1999. *Morbidity and Mortality Weekly Report* **51**, 300–303.

MMWR. (2002b). Cigarette smoking among adults: United States, 2000. *Morbidity and Mortality Weekly Report* **51**, 642–645.

MMWR. (2003a). Cigarette smoking-attributable morbidity: United States, 2000. *Morbidity and Mortality Weekly Report* **52**, 842–844.

MMWR. (2003b). Cigarette smoking among adults: United States, 2001. *Morbidity and Mortality Weekly Report* **52**, 953–956 [erratum: **52**(42), 1025].

MMWR. (2004). Cigarette smoking among adults: United States, 2002. *Morbidity and Mortality Weekly Report* **53**, 427–431.

Molander, L., Lunell, E., and Fagerstrom, K. O. (2000). Reduction of tobacco withdrawal symptoms with a sublingual nicotine tablet: a placebo controlled study. *Nicotine and Tobacco Research* **2**, 187–191.

Moolten, M., Kornetsky, C. (1990). Oral self-administration of ethanol and not experimenter-administered ethanol facilitates rewarding electrical brain stimulation. *Alcohol* **7**, 221–225.

Mosner, A., Kuhlman, G., Roehm, C., and Vogel, W. H. (1997). Serotonergic receptors modify the voluntary intake of alcohol and morphine but not of cocaine and nicotine by rats. *Pharmacology* **54**, 186–192.

Murphree, H. B., Pfeiffer, C. C., and Price, L. M. (1967). Electroencephalographic changes in man following smoking. *In The Effects of Nicotine and Smoking on the Central Nervous System* (series title: *Annals of the New York Academy of Sciences*, vol. 142[1]), (H. B. Murphree, Ed.), pp. 245–260. New York Academy of Sciences, New York.

Myers, W. D., Ng, K. T., and Singer, G. (1982). Intravenous self-administration of acetaldehyde in the rat as a function of schedule, food deprivation and photoperiod. *Pharmacology Biochemistry and Behavior* **17**, 807–811.

Nesbitt, P. D. (1973). Smoking, physiological arousal, and emotional response. *Journal of Personality and Social Psychology* **25**, 137–144.

Nisell, M., Nomikos, G. G., and Svensson, T. H. (1994). Systemic nicotine–induced dopamine release in the rat nucleus accumbens is regulated by nicotinic receptors in the ventral tegmental area. *Synapse* **16**, 36–44.

Nisell, M., Nomikos, G. G., and Svensson, T. H. (1995). Nicotine dependence, midbrain dopamine systems and psychiatric disorders. *Pharmacology and Toxicology* **76**, 157–162.

Nisell, M., Marcus, M., Nomikos, G. G., and Svensson, T. H. (1997). Differential effects of acute and chronic nicotine on dopamine output in the core and shell of the rat nucleus accumbens. *Journal of Neural Transmission* **104**, 1–10.

Nestler, E. J., and Aghajanian, G. K. (1997). Molecular and cellular basis of addiction. *Science* **278**, 58–63.

Noda, M., Takahashi, H., Tanabe, T., Toyosato, M., Kikyotani, S., Hirose, T., Asai, M., Takashima, H., Inayama, S., Miyata, T., and Numa, S. (1983). Primary structures of beta- and delta-subunit precursors of Torpedo californica acetylcholine receptor deduced from cDNA sequences. *Nature* **301**, 251–255.

Nomikos, G. G., Schilstrom, B., Hildebrand, B. E., Panagis, G., Grenhoff, J., and Svensson, T. H. (2000). Role of alpha7 nicotinic receptors in nicotine dependence and implications for psychiatric illness. *Behavioural Brain Research* **113**, 97–103.

Numa, S., Noda, M., Takahashi, H., Tanabe, T., Toyosato, M., Furutani, Y., and Kikyotani, S. (1983). Molecular structure of the nicotinic acetylcholine receptor. *Cold Spring Harbor Symposia on Quantitative Biology* **48**(Pt. 1), 57–69.

Oakman, S. A., Faris, P. L., Kerr, P. E., Cozzari, C., and Hartman, B. K. (1995). Distribution of pontomesencephalic cholinergic neurons projecting to substantia nigra differs significantly from those projecting to ventral tegmental area. *Journal of Neuroscience* **15**, 5859–5869.

Office of Evaluations and Inspections (1992). *Spit Tobacco and Youth* [DHHS publication no. (OEI-06)92-00500)], Office of the Inspector General, Washington DC.

Olale, F., Gerzanich, V., Kuryatov, A., Wang, F., and Lindstrom, J. (1997). Chronic nicotine exposure differentially affects the function of human alpha-3, alpha-4, and alpha-7 neuronal nicotinic receptor subtypes. *Journal of Pharmacology and Experimental Therapeutics* **283**, 675–683.

Orr-Urtreger, A., Goldner, F. M., Saeki, M., Lorenzo, I., Goldberg, L., De Biasi, M., Dani, J. A., Patrick, J. W., and Beaudet, A. L. (1997). Mice deficient in the alpha7 neuronal nicotinic acetylcholine receptor lack alpha-bungarotoxin binding sites and hippocampal fast nicotinic currents. *Journal of Neuroscience* **17**, 9165–9171.

Orr-Urtreger, A., Kedmi, M., Rosner, S., Karmeli, F., and Rachmilewitz, D. (2005). Increased severity of experimental colitis in alpha5 nicotinic acetylcholine receptor subunit-deficient mice. *Neuroreport* **16**, 1123–1127.

Oviedo, G. F. (1526). *Historia General y Natural de las Indias*, R. de Petras, Toledo.

Owens, J. C., Balogh, S. A., McClure-Begley, T. D., Butt, C. M., Labarca, C., Lester, H. A., Picciotto, M. R., Wehner, J. M., and Collins, A. C. (2003). Alpha4beta2* nicotinic acetylcholine receptors modulate the effects of ethanol and nicotine on the acoustic startle response. *Alcoholism: Clinical and Experimental Research* **27**, 1867–1875.

Pagliusi, S. R., Tessari, M., DeVevey, S., Chiamulera, C., and Pich, E. M. (1996). The reinforcing properties of nicotine are associated with a specific patterning of c-fos expression in the rat brain. *European Journal of Neuroscience* **8**, 2247–2256.

Panagis, G., Hildebrand, B. E., Svensson, T. H., and Nomikos, G. G. (2000). Selective c-fos induction and decreased dopamine release in the central nucleus of amygdala in rats displaying a mecamylamine-precipitated nicotine withdrawal syndrome. *Synapse* **35**, 15–25.

Pandey, S. C., Roy, A., Xu, T., and Mittal, N. (2001). Effects of protracted nicotine exposure and withdrawal on the expression and phosphorylation of the CREB gene transcription factor in rat brain. *Journal of Neurochemistry* **77**, 943–952.

Paterson, N. E., and Markou, A. (2002). Increased GABA neurotransmission via administration of gamma-vinyl GABA decreased nicotine self-administration in the rat. *Synapse* **44**, 252–253.

Paterson, N. E., Froestl, W., and Markou, A. (2004). The GABA-B receptor agonists baclofen and CGP44532 decreased nicotine self-administration in the rat. *Psychopharmacology* **172**, 179–186.

Paterson, N. E., Froestl, W., and Markou, A. (2005). Repeated administration of the GABAB receptor agonist CGP44532 decreased nicotine self-administration, and acute administration decreased cue-induced reinstatement of nicotine-seeking in rats. *Neuropsychopharmacology* **30**, 119–128.

Perkins, K. A., Epstein, L. H., Stiller, R. L., Fernstrom, M. H., Sexton, J. E., and Jacob, R. G. (1990). Perception and hedonics of sweet and fat taste in smokers and nonsmokers following nicotine intake. *Pharmacology Biochemistry and Behavior* **35**, 671–676.

Perry, D. C., Davila-Garcia, M. I., Stockmeier, C. A., and Kellar, K. J. (1999). Increased nicotinic receptors in brains from smokers: membrane binding and autoradiography studies. *Journal of Pharmacology and Experimental Therapeutics* **289**, 1545–1552.

Picciotto, M. R., Zoli, M., Lena, C., Bessis, A., Lallemand, Y., Le Novere, N., Vincent, P., Pich, E. M., Brulet, P., and Changeux, J. P. (1995). Abnormal avoidance learning in mice lacking functional high-affinity nicotine receptor in the brain. *Nature* **374**, 65–67.

Picciotto, M. R., Zoli, M., Rimondini, R., Lena, C., Marubio, L. M., Pich, E. M., Fuxe, K., and Changeux, J. P. (1998). Acetylcholine receptors containing the beta2 subunit are involved in the reinforcing properties of nicotine. *Nature* **391**, 173–177.

Picciotto, M. R., Zoli, M., and Changeux, J. P. (1999). Use of knock-out mice to determine the molecular basis for the actions of nicotine. *Nicotine and Tobacco Research* **1** (*Suppl.* 2), s121–s125.

Pich, E. M., Pagliusi, S. R., Tessari, M., Talabot-Ayer, D., Hooft van Huijsduijnen, R., and Chiamulera, C. (1997). Common neural substrates for the addictive properties of nicotine and cocaine. *Science* **275**, 83–86.

Pidoplichko, V. I., DeBiasi, M., Williams, J. T., and Dani, J. A. (1997). Nicotine activates and desensitizes midbrain dopamine neurons. *Nature* **390**, 401–404.

Pierzchala, K., Houdi, A. A., and Van Loon, G. R. (1987). Nicotine-induced alterations in brain regional concentrations of native and cryptic Met- and Leu-enkephalin. *Peptides* **8**, 1035–1043.

Pieyre, A. (1886). *Memoire sur Jean Nicot*, A. Catelan, Nimes, France.

Pin, J. P., and Duvoisin, R. (1995). The metabotropic glutamate receptors: structure and functions. *Neuropharmacology* **34**, 1–26.

Pomerleau, O. F. (1998). Endogenous opioids and smoking: a review of progress and problems. *Psychoneuroendocrinology* **23**, 115–130.

Pomerleau, O., Adkins, D., and Pertschuk, M. (1978). Predictors of outcome and recidivism in smoking cessation treatment. *Addictive Behaviors* **3**, 65–70.

Pomerleau, O. F., and Pomerleau, C. S. (1984). Neuroregulators and the reinforcement of smoking: towards a biobehavioral explanation. *Neuroscience and Biobehavioral Reviews* **8**, 503–513.

Pomerleau, O. F., and Rosecrans, J. (1989). Neuroregulatory effects of nicotine. *Psychoneuroendocrinology* **14**, 407–423.

Pomerleau, C. S., and Pomerleau, O. F. (1992). Euphoriant effects of nicotine in smokers. *Psychopharmacology* **108**, 460–465.

Pontieri, F. E., Tanda, G., Orzi, F., and Di Chiara, G. (1996). Effects of nicotine on the nucleus accumbens and similarity to those of addictive drugs. *Nature* **382**, 255–257.

Pontieri, F. E., Passarelli, F., Calo, L., and Caronti, B. (1998). Functional correlates of nicotine administration: similarity with drugs of abuse. *Journal of Molecular Medicine* **76**, 193–201.

Prochazka, A. V., Weaver, M. J., Keller, R. T., Fryer, G. E., Licari, P. A., and Lofaso, D. (1998). A randomized trial of nortriptyline for smoking cessation. *Archives of Internal Medicine* **158**, 2035–2039.

Rasmussen, K., and Czachura, J. F. (1995). Nicotine withdrawal leads to increased firing rates of midbrain dopamine neurons. *Neuroreport* **7**, 329–332.

Rasmussen, K., and Czachura, J. F. (1997). Nicotine withdrawal leads to increased sensitivity of serotonergic neurons to the 5-HT1A agonist 8-OH-DPAT. *Psychopharmacology* **133**, 343–346.

Rasmussen, K., Kallman, M. J., and Helton, D. R. (1997). Serotonin-1A antagonists attenuate the effects of nicotine withdrawal on the auditory startle response. *Synapse* **27**, 145–152.

Reitstetter, R., Lukas, R. J., and Gruener, R. (1999). Dependence of nicotinic acetylcholine receptor recovery from desensitization on the duration of agonist exposure. *Journal of Pharmacology and Experimental Therapeutics* **289**, 656–660.

Ribeiro, E. B., Bettiker, R. L., Bogdanov, M., and Wurtman, R. J. (1993). Effects of systemic nicotine on serotonin release in rat brain. *Brain Research* **621**, 311–318.

Risner, M. E., and Goldberg, S. R. (1983). A comparison of nicotine and cocaine self-administration in the dog: fixed-ratio and progressive-ratio schedules of intravenous drug infusion. *Journal of Pharmacology and Experimental Therapeutics* **224**, 319–326.

Rodd-Henricks, Z. A., Melendez, R. I., Zaffaroni, A., Goldstein, A., McBride, W. J., and Li, T. K. (2002). The reinforcing effects of acetaldehyde in the posterior ventral tegmental area of alcohol-preferring rats. *Pharmacology Biochemistry and Behavior* **72**, 55–64.

Rodriguez de Fonseca, F., Carrera, M. R. A., Navarro, M., Koob, G. F., and Weiss, F. (1997). Activation of corticotropin-releasing factor in the limbic system during cannabinoid withdrawal. *Science* **276**, 2050–2054.

Ross, S. A., Wong, J. Y., Clifford, J. J., Kinsella, A., Massalas, J. S., Horne, M. K., Scheffer, I. E., Kola, I., Waddington, J. L., Berkovic, S. F., and Drago, J. (2000). Phenotypic characterization of an alpha 4 neuronal nicotinic acetylcholine receptor subunit knock-out mouse. *Journal of Neuroscience* **20**, 6431–6441.

Rossi, F. M., Pizzorusso, T., Porciatti, V., Marubio, L. M., and Maffei, L. (2001). Changeux, JP. Requirement of the nicotinic acetylcholine receptor beta 2 subunit for the anatomical and functional development of the visual system. *Proceedings of the National Academy of Sciences USA* **98**, 6453–6458.

Roth, G. M., McDonald, J. B., and Sheard, C. (1945). The effect of smoking cigarets [sic] and the intravenous administration of nicotine on the heart and peripheral blood vessels. *Medical Clinics of North America* **29**, 949–957.

Russell, M. A. (1971). Cigarette smoking: natural history of a dependence disorder. *British Journal of Medical Psychology* **44**, 1–16.

Russell, M. A. (1989). Subjective and behavioural effects of nicotine in humans: some sources of individual variation. *In: Nicotinic Receptors in the CNS: Their Role in Synaptic Transmission* (series title: *Progress in Brain Research*, vol. 79), (A. Nordberg, Ed.), pp. 289–302. Elsevier, Amsterdam.

Russell, M. A., and Feyerabend, C. (1978). Cigarette smoking: a dependence on high-nicotine boli. *Drug Metabolism Reviews* **8**, 29–57.

Russell, M. A., Jarvis, M., Iyer, R., and Feyerabend, C. (1980). Relation of nicotine yield of cigarettes to blood nicotine concentrations in smokers. *British Medical Journal* **280**, 972–976.

Russell, M. A. H., (1990). The nicotine addiction trap: a 40-year sentence for four cigarettes. *British Journal of Addiction* **85**, 293–300.

Saah, M. I., Raygada, M., and Grunberg, N. E. (1994). Effects of nicotine on body weight and plasma insulin in female and male rats. *Life Sciences* **55**, 925–931.

Sachs, D. P., and Leischow, S. J. (1991). Pharmacologic approaches to smoking cessation. *Clinics in Chest Medicine* **12**, 769–791 [erratum: 1992, **13**:ix–xi].

Sack, R., Gochberg-Sarver, A., Rozovsky, U., Kedmi, M., Rosner, S., and Orr-Urtreger, A. (2005). Lower core body temperature and attenuated nicotine-induced hypothermic response in mice lacking the beta4 neuronal nicotinic acetylcholine receptor subunit. *Brain Research Bulletin* **66**, 30–36.

Salas, R., Orr-Urtreger, A., Broide, R. S., Beaudet, A., Paylor, R., and De Biasi, M. (2003a). The nicotinic acetylcholine receptor subunit alpha 5 mediates short-term effects of nicotine in vivo. *Molecular Pharmacology* **63**, 1059–1066.

Salas, R., Pieri, F., Fung, B., Dani, J. A., and De Biasi, M. (2003b). Altered anxiety-related responses in mutant mice lacking the beta4 subunit of the nicotinic receptor. *Journal of Neuroscience* **23**, 6255–6263.

Salas, R., Pieri, F., and De Biasi, M. (2004). Decreased signs of nicotine withdrawal in mice null for the beta4 nicotinic acetylcholine receptor subunit. *Journal of Neuroscience* **24**, 10035–10039.

Salber, E. J., and Abelin, T. (1967). Smoking behavior of Newton school children: 5-year follow-up. *Pediatrics* **40**, 363–372.

Salminen, O., Seppa, T., Gaddnas, H., and Ahtee, L., (1999). The effects of acute nicotine on the metabolism of dopamine and the expression of Fos protein in striatal and limbic brain areas of rats during chronic nicotine infusion and its withdrawal. *Journal of Neuroscience* **19**, 8145–8151.

Sargent, P. B. (1993). The diversity of neuronal nicotinic acetylcholine receptors. *Annual Review of Neuroscience* **16**, 403–443.

Sarnyai, Z., Biro, E., Gardi, J., Vecsernyes, M., Julesz, J., and Telegdy, G. (1995). Brain corticotropin-releasing factor mediates 'anxiety-like' behavior induced by cocaine withdrawal in rats. *Brain Research* **675**, 89–97.

Schachter, S., Silverstein, B., Kozlowski, L. T., Perlick, D., Herman, C. P., and Liebling, B. (1977). Studies of the interaction of psychological and pharmacological determinants of smoking. *Journal of Experimental Psychology: General* **106**, 3–12.

Schiffer, W. K., Gerasimov, M. R., Bermel, R. A., Brodie, J. D., and Dewey, S. L. (2000). Stereoselective inhibition of dopaminergic activity by gamma vinyl-GABA following a nicotine or cocaine challenge: a PET/microdialysis study. *Life Sciences* **66**, PL169–PL173.

Schilstrom, B., Nomikos, G. G., Nisell, M., Hertel, P., and Svensson, T. H. (1998). N-methyl-D-aspartate receptor antagonism in the ventral tegmental area diminishes the systemic nicotine–induced dopamine release in the nucleus accumbens. *Neuroscience* **82**, 781–789.

Schneider, N. G., Jarvik, M. E., and Forsythe, A. B. (1984). Nicotine vs. placebo gum in the alleviation of withdrawal during smoking cessation. *Addictive Behaviors* **9**, 149–156.

Schroeder, B. E., Binzak, J. M., and Kelley, A. E. (2001). A common profile of prefrontal cortical activation following exposure to nicotine- or chocolate-associated contextual cues. *Neuroscience* **105**, 535–545.

Schulteis, G., Markou, A., Gold, L. H., Stinus, L., and Koob, G. F. (1994). Relative sensitivity to naloxone of multiple indices of opiate withdrawal: A quantitative dose-response analysis. *Journal of Pharmacology and Experimental Therapeutics* **271**, 1391–1398.

Schulteis, G., Markou, A., Cole, M., and Koob, G. (1995). Decreased brain reward produced by ethanol withdrawal. *Proceedings of the National Academy of Sciences USA* **92**, 5880–5884.

Schwartz, R. D., Lehmann, J., and Kellar, K. J. (1984). Presynaptic nicotinic cholinergic receptors labeled by [3H]acetylcholine on catecholamine and serotonin axons in brain. *Journal of Neurochemistry* **42**, 1495–1498.

Sellers, E. M., Naranjo, C. A., and Kadlec, K. (1987). Do serotonin uptake inhibitors decrease smoking? Observations in a group of heavy drinkers. *Journal of Clinical Psychopharmacology* **7**, 417–420.

Shahan, T. A., Bickel, W. K., Madden, G. J., and Badger, G. J. (1999). Comparing the reinforcing efficacy of nicotine containing and de-nicotinized cigarettes: a behavioral economic analysis. *Psychopharmacology* **147**, 210–216.

Shiffman, S. M., and Jarvik, M. E. (1976). Smoking withdrawal symptoms in two weeks of abstinence. *Psychopharmacology* **50**, 35–39.

Shiffman, S., Paty, J. A., Kassel, J. D., Gnys, M., and Zettler-Segal, M. (1994). Smoking behavior and smoking history of tobacco chippers. *Experimental and Clinical Psychopharmacology* **2**, 126–142.

Shoaib, M., Benwell, M. E., Akbar, M. T., Stolerman, I. P., and Balfour, D. J. (1994). Behavioural and neurochemical adaptations to nicotine in rats: influence of NMDA antagonists. *British Journal of Pharmacology* **111**, 1073–1080.

Shoaib, M., and Stolerman, I. P. (1996). The NMDA antagonist dizocilpine (MK-801) attenuates tolerance to nicotine in rats. *Journal of Psychopharmacology* **10**, 214–218.

Shoaib, M., Schindler, C. W., Goldberg, S. R., and Pauly, J. R. (1997). Behavioural and biochemical adaptations to nicotine in rats: influence of MK801, an NMDA receptor antagonist. *Psychopharmacology* **134**, 121–130.

Skjei, K. L., and Markou, A. (2003). Effects of repeated withdrawal episodes, nicotine dose, and duration of nicotine exposure on the severity and duration of nicotine withdrawal in rats. *Psychopharmacology* **168**, 280–292.

Sollmann, T. H. (1926). *A Manual of Pharmacology and Its Applications to Therapeutics and Toxicology*, W.B. Saunders, Philadelphia.

Sorensen, G., and Pechacek, T. F. (1987). Attitudes toward smoking cessation among men and women. *Journal of Behavioral Medicine* **10**, 129–137.

Sparks, J. A., and Pauly, J. R. (1999). Effects of continuous oral nicotine administration on brain nicotinic receptors and responsiveness to nicotine in C57Bl/6 mice. *Psychopharmacology* **141**, 145–153.

Srivastava, E. D., Russell, M. A., Feyerabend, C., Masterson, J. G., and Rhodes J. (1991). Sensitivity and tolerance to nicotine in smokers and nonsmokers. *Psychopharmacology* **105**, 63–68.

Stahl, S. M. (1998). Basic psychopharmacology of antidepressants, part 1: Antidepressants have seven distinct mechanisms of action. *Journal of Clinical Psychiatry* **59** (Suppl. 4), 5–14.

Stellar, J. R., and Rice, M. B. (1989). Pharmacological basis of intracranial self-stimulation reward. *In The Neuropharmacological Basis of Reward* (series title: *Topics in Experimental Psychopharmacology*, vol. 1), (J. M. Liebman, and S. J. Cooper, Eds.), pp. 14–65. Clarendon Press, Oxford.

Stewart, G. G. (1967). A history of the medicinal use of tobacco, 1492–1860. *Medical History* **11**, 228–268.

Stolerman, I. P. (1991). Behavioural pharmacology of nicotine: multiple mechanisms. *British Journal of Addiction* **86**, 533–536.

Stolerman, I. P., Fink, R., and Jarvik, M. E. (1973). Acute and chronic tolerance to nicotine measured by activity in rats. *Psychopharmacologia* **30**, 329–342.

Stolerman, I. P., Bunker, P., and Jarvik, M. E. (1974). Nicotine tolerance in rats; role of dose and dose interval. *Psychopharmacologia* **34**, 317–324.

Stolerman, I. P., and Jarvis, M. J. (1995). The scientific case that nicotine is addictive. *Psychopharmacology* **117**, 2–10.

Substance Abuse and Mental Health Services Administration. (2003). *Results from the 2002 National Survey on Drug Use and Health: National Findings* (Office of Applied Studies, NHSDA Series H–22, DHHS Publication No. SMA 03–3836), Rockville MD.

Swanson, L. W., Simmons, D. M., Whiting, P. J., and Lindstrom, J. (1987). Immunohistochemical localization of neuronal nicotinic receptors in the rodent central nervous system. *Journal of Neuroscience* **7**, 3334–3342.

Sziraki, I., Sershen, H., Benuck, M., Hashim, A., and Lajtha, A. (1998). Receptor systems participating in nicotine-specific effects. *Neurochemistry International* **33**, 445–457.

Tago, H., McGeer, P. L., McGeer, E. G., Akiyama, H., and Hersh, L. B. (1989). Distribution of choline acetyltransferase immunopositive structures in the rat brainstem. *Brain Research* **495**, 271–297.

Tapper, A. R., McKinney, S. L., Nashmi, R., Schwarz, J., Deshpande, P., Labarca, C., Whiteaker, P., Marks, M. J., Collins, A. C., and Lester, H. A. (2004). Nicotine activation of alpha4* receptors: sufficient for reward, tolerance, and sensitization. *Science* **306**, 1029–1032.

Tempel, A., and Zukin, R. S. (1987). Neuroanatomical patterns of the mu, delta, and kappa opioid receptors of rat brain as determined by quantitative in vitro autoradiography. *Proceedings of the National Academy of Sciences USA* **84**, 4308–4312.

Thun, M. J., and da Costa e Silva, V. L. (2003). Introduction and overview of global tobacco surveillance. *In Tobacco Control Country Profiles*, 2nd ed., (O. Shafey, S. Dolwick, and G. E. Guindon, Eds.), pp. 7–12. American Cancer Society, Atlanta.

Todd, G. F. (1969). *Statistics of Smoking in the United Kingdom*, Tobacco Research Council, London.

Tritto, T., McCallum, S. E., Waddle, S. A., Hutton, S. R., Paylor, R., Collins, A. C., and Marks, M. J. (2004). Null mutant analysis of responses to nicotine: deletion of beta2 nicotinic acetylcholine receptor subunit but not alpha7 subunit reduces sensitivity to nicotine-induced locomotor depression and hypothermia. *Nicotine and Tobacco Research* **6**, 145–158.

Vale, A. L., and Green, S. (1996). Effects of chlordiazepoxide, nicotine and d-amphetamine in the rat potentiated startle model of anxiety. *Behavioural Pharmacology* **7**, 138–143.

Valentine, J. D., Hokanson, J. S., Matta, S. G., and Sharp, B. M. (1997). Self-administration in rats allowed unlimited access to nicotine. *Psychopharmacology* **133**, 300–304.

Vetter, D. E., Liberman, M. C., Mann, J., Barhanin, J., Boulter, J., Brown, M. C., Saffiote-Kolman, J., Heinemann, S. F., and Elgoyhen, A. B. (1999). Role of alpha9 nicotinic ACh receptor subunits in the development and function of cochlear efferent innervation. *Neuron* **23**, 93–103.

Waal-Manning, H. J., and de Hamel, F. A. (1978). Smoking habit and psychometric scores: a community study. *New Zealand Medical Journal* **88**, 188–191.

Wada, E., Wada, K., Boulter, J., Deneris, E., Heinemann, S., Patrick, J., and Swanson, L. W. (1989). Distribution of alpha 2, alpha 3, alpha 4, and beta 2 neuronal nicotinic receptor subunit mRNAs in the central nervous system: a hybridization histochemical study in the rat. *Journal of Comparative Neurology* **284**, 314–335.

Walaas, I., and Fonnum, F. (1980). Biochemical evidence for gamma-aminobutyrate containing fibres from the nucleus accumbens to the substantia nigra and ventral tegmental area in the rat. *Neuroscience* **5**, 63–72.

Watkins, S. S., Epping-Jordan, M. P., Koob, G. F., and Markou, A. (1999). Blockade of nicotine self-administration with nicotinic antagonists in rats. *Pharmacology Biochemistry and Behavior* **62**, 743–751.

Watkins, S. S., Koob, G. F., and Markou, A. (2000a). Neural mechanisms underlying nicotine addiction: Acute positive reinforcement and withdrawal. *Nicotine and Tobacco Research* **2**, 19–37.

Watkins, S. S., Stinus, L., Koob, G. F., and Markou, A. (2000b). Reward and somatic changes during precipitated nicotine withdrawal in rats: Centrally and peripherally mediated effects. *Journal of Pharmacology and Experimental Therapeutics* **292**, 1053–1064.

Webster, D. D. (1964). The dynamic quantitation of spasticity with automated integrals of passive motion resistance. *Clinical Pharmacology and Therapeutics* **5**, 900–908.

Wesnes, K., Warburton, D. M., and Matz, B. (1983). Effects of nicotine on stimulus sensitivity and response bias in a visual vigilance task. *Neuropsychobiology* **9**, 41–44.

West, R., Hajek, P., and Belcher, M. (1989). Time course of cigarette withdrawal symptoms while using nicotine gum. *Psychopharmacology* **99**, 143–145.

Wilbert, J. (1987). *Tobacco Shamanism in South America*, Yale University Press, New Haven, CT.

Wilbert, J. (1991). Does pharmacology corroborate the nicotine therapy and practices of South American shamanism? *Journal of Ethnopharmacology* **32**, 179–186.

Wonnacott, S. (1990). The paradox of nicotinic acetylcholine receptor upregulation by nicotine. *Trends in Pharmacological Sciences* **11**, 216–219.

Wonnacott, S. (1997). Presynaptic nicotinic ACh receptors. *Trends in Neurosciences* **20**, 92–98.

Wynder, E. L., and Hoffmann, D. (1967). *Tobacco and Tobacco Smoke: Studies in Experimental Carcinogenesis*, Academic Press, New York.

Wynder, E. L., and Hecht, S. (Eds.) (1976). *Lung Cancer* (series title: *UICC Technical Report Series*, vol. 25) International Union Against Cancer, Geneva.

Wynder, E. L., and Hoffmann, D. (1979). Tobacco and health: a societal challenge. *New England Journal of Medicine* **300**, 894–903.

Xu, W., Gelber, S., Orr-Urtreger, A., Armstrong, D., Lewis, R. A., Ou, C. N., Patrick, J., Role, L., De Biasi, M., and Beaudet, A. L. (1999a). Megacystis, mydriasis, and ion channel defect in mice lacking the alpha3 neuronal nicotinic acetylcholine receptor. *Proceedings of the National Academy of Sciences USA* **96**, 5746–5751.

Xu, W., Orr-Urtreger, A., Nigro, F., Gelber, S., Sutcliffe, C. B., Armstrong, D., Patrick, J. W., Role, L. W., Beaudet, A. L., and De Biasi, M. (1999b). Multiorgan autonomic dysfunction in mice lacking the beta2 and the beta4 subunits of neuronal nicotinic acetylcholine receptors. *Journal of Neuroscience* **19**, 9298–9305.

Yeomans, J. S., Mathur, A., and Tampakeras, M. (1993). Rewarding brain stimulation: role of tegmental cholinergic neurons that activate dopamine neurons. *Behavioral Neuroscience* **107**, 1077–1087.

Yeomans, J., and Baptista, M., (1997). Both nicotinic and muscarinic receptors in ventral tegmental area contribute to brain-stimulation reward. *Pharmacology Biochemistry and Behavior* **57**, 915–921.

Yim, C. Y., and Mogenson, G. J. (1980). Electrophysiological studies of neurons in the ventral tegmental area of Tsai. *Brain Research* **181**, 301–313.

Yin, R., and French, E. D. (2000). A comparison of the effects of nicotine on dopamine and non-dopamine neurons in the rat ventral tegmental area: an in vitro electrophysiological study. *Brain Research Bulletin* **51**, 507–514.

Young, S. N., Smith, S. E., Pihl, R. O., and Ervin, F. R. (1985). Tryptophan depletion causes a rapid lowering of mood in normal males. *Psychopharmacology* **87**, 173–177.

Yu, H., Matsubayashi, H., Amano, T., Cai, J., and Sasa, M. (2000). Activation by nicotine of striatal neurons receiving excitatory input from the substantia nigra via dopamine release. *Brain Research* **872**, 223–226.

Zoli, M., Lena, C., Picciotto, M. R., and Changeux, J. P. (1998). Identification of four classes of brain nicotinic receptors using beta2 mutant mice. *Journal of Neuroscience* **18**, 4461–4472.

Zoli, M., Picciotto, M. R., Ferrari, R., Cocchi, D., and Changeux, J. P. (1999). Increased neurodegeneration during ageing in mice lacking high-affinity nicotine receptors. *EMBO Journal* **18**, 1235–1244.

CHAPTER

7

Cannabinoids

OUTLINE

DEFINITIONS

Cannabis is a highly adaptive annual plant that grows throughout the temperate and tropical zones of the world. Cannabis in the form of hemp likely originated in Central Asia or near the Alai and Tian Shan mountain ranges extending from Kyrgyzstan through Tajikistan to the China–Mongolia border and was a familiar agricultural crop from the beginning of civilization. *Hemp* is defined by the Merriam-Webster dictionary as a tall, widely cultivated Asian herb (*Cannabis sativa*) of the mulberry family with tough phloem fiber used for cords and ropes. Archeologists have found evidence of hemp plant use as far back as the late Neolithic era, dating back at least 6000 years (the New Stone Age) (Cheng, 1966; Chang, 1968; Li, 1974b; Clarke and Watson, 2002). *Ma* is the Chinese word for *hemp*, and *ta ma* is the Chinese word for psychoactive hemp (Read, 1936; Li, 1974a; Abel, 1980) (**Fig. 7.1**). Reference to psychoactive hemp (*ta ma*) as a medicinal herb dates to 2838 BC with Emperor Shen Nung's compilation of the *Pen Ts'ao* (or *The Herbal*), a kind of herbal standard (Li, 1974a,b; Wong and Wu, 1985). Emperor Nung experimented with various herbs and listed *ta ma* as medicinal (Wong and Wu, 1985; Read, 1936): '...ma-fen (fruits of hemp)...if taken in excess will produce hallucinations (literally ''seeing devils''). If taken over a long term, it makes one communicate with spirits and lightens one's body' (Li, 1974a,b). Hua T'o, a Chinese physician from the 2nd century AD, used an oral preparation of cannabis called *ma-fei-san* (hemp-boiling compound, combined with wine) to anesthetize patients undergoing abdominal surgery (Li, 1974b). The Swedish botanist Carl Linnaeus named the plant *Cannabis sativa* in 1753 (Zhou and Bartholomew, 2003). Though accounts of the etymology of the word *cannabis* differ, it may have derived from the Greek and Latin *kannabis*, the Assyrian *kunnapu* (a way to produce smoke) (Maykut, 1985) or perhaps from the Sanskrit *cana* (hemp or cane) (Booth, 2003). The word *kan* referred to hemp or cane in many ancient languages and *bis* can be linked to the word *aromatic*; thus, cannabis is the 'fragrant cane' (Booth, 2003).

Neurobiology of Addiction, by George F. Koob and Michel Le Moal.
ISBN – 13: 978-0-12-419239-3 ISBN – 10: 0-12-419239-4

FIGURE 7.1 Evolvement of the Chinese character for hemp, or *ma*, from the archaic *chuan* script (dating from 1766–1122 BC during the Shang Dynasty), to the contemporary cursive *hsing* script (emerging in the 3rd century AD) (Li, 1974a). The ideogram is composed of the *madare* radical (top and left) which represents a tilted roof and is used in the characters for words such as *house, shop, to live*, etc. Under the 'roof' are two small characters for *tree* which by themselves mean 'small forest'. Literally, the character for *hemp* expresses the idea of a 'small forest in or at one's house', or a 'domestic forest'. The part beneath and to the right of the straight lines represents hemp fibers dangling from a rack. The horizontal and vertical lines represent the home in which they are drying (Abel, 1980). The pictograph at the bottom combines *big* (*ta*) with *ma* to form the Chinese ideogram (*ta ma*) for psychoactive marijuana, *Cannabis sativa* (Read, 1936).

Marijuana is the dry shredded green or brown mixture of flowers, stems, seeds, and leaves of the hemp plant *Cannabis sativa*. It has been suggested that the word *marijuana* derived from the Spanish *Mariguana*, one of the islands forming the Bahamas (Partridge, 1958), though others have suggested that it derived from the Spanish prenomes *Maria* (Mary) and *Juana* (Jane). This claim, however, is unsubstantiated (Hendrickson, 1987). The dried mixture can be smoked like a cigarette (termed a *joint*) or in a pipe (*bong*) or in a cigar where the tobacco has been removed from the inside (*blunt*). Other preparations include hash or hashish, the dried sticky resin of the flowers of the female plant, or hash oil, a sticky black liquid.

The marijuana plant has numerous chemical constituents, but its major active constituent responsible for its main pharmacological effect is (-)Δ^9-6a, 10a-trans-tetrahydrocannabinol (THC) (Gaoni and Mechoulam, 1964), or simply referred to here as Δ^9-THC. The term *cannabinoid* originally referred to Δ^9-THC and related phytocannabinoids of the marijuana plant *Cannabis sativa* with a typical c-21 structure and any products transformed from these structures (Razdan, 1986). A broader definition based on pharmacology and chemistry 'encompasses kindred structures, or any other compound that affects cannabinoid receptors' (Pate, 2002) or 'all ligands of the cannabinoid receptor and related compounds including endogenous ligands of the receptors and a large number of synthetic cannabinoid analogues' (Grotenhermen, 2003).

> 'The dried Hemp plant which has flowered, and *from which the resin has not been removed*, is called *Gunjah*. It sells from twelve annas to one rupee seer, in the Calcutta bazars, and yields to alcohol twenty per 100 of resinous extract, composed of the resin (*churrus*) and green colouring matter (*Chloro-phille*). Distilled with a large quantity of water, traces of essential oil pass over, and the distilled liquor has the powerful narcotic odour of the plant. The *gunjah* is sold for smoking chiefly. The bundles of *gunjah* are about two feet long, and three inches in diameter, and contain 24 to 36 plants. The colour is dusky green—the odour agreeably narcotic—the whole plant resinous and adhesive to the touch. The larger leaves and capsules without the stalks are called ''*Bangh Subjee or Sidhee*''. They are used for making an intoxicating drink, for smoking, and in the conserve or confection termed *Majoon*. *Bang* is cheaper than *gunjah*, and though less powerful, is sold at such a low price, that for one pice enough can be purchased to intoxicate an experienced person' (O'Shaughnessy, 1838).

In the plant, Δ^9-THC content is highest in the oil from the flowering tops and lowest in the seeds, declining in concentration in the following order: flowering tops, bracts, leaves, stems, roots, and seeds of the plant (Doorenbos *et al.*, 1971) (**Table 7.1**). Cannabis is able to survive in very hot, arid climates due to the resin film which protects it from losing moisture due to evaporation. This sticky coating is called hashish:

> 'The means by which cannabis accomplishes this amazing feat is by producing a thick, sticky resin that coats its leaves and flowers. This protective canopy prevents life-sustaining moisture from disappearing

TABLE 7.1 Levels of Cannabinoids in Various Plant Parts

	Fetterman *et al.*, 1971a	ElSohly and Holley 1983	Fetterman *et al.*, 1971b	Fairbairn *et al.*, 1971
Δ⁹-THC content				
Bracts*	0.37%	—	3.7%	—
Buds†	—	3.6–4.6%	—	—
Flowers	—	—	1.6%	1.56%
Leaves	0.32%	—	1.0–1.4%	0.83–1.56%
Stems	0.02%	—	0.89%	0.07–0.11%
Roots	0.002%	—	—	—
Seeds	0.00057%	—	0.01%	—
Sinsemilla	—	3.6–4.6%	—	—
Hashish	—	2.4–3.4%	2.1%	—
Hash oil	—	11.5–21.6%	10%	—
Cannabidiol content				
Bracts	5.55%	—	0.15%	—
Leaves	1.6%	—	0.079–0.085%	—
Stems	0.19%	—	0.055%	—
Roots	0.015%	—	—	—
Seeds	0.00887%	—	trace	—
Hashish	—	—	9.8%	3.5%
Hash oil	—	—	0.88%	—
Cannabinol content				
Bracts	0.038%	—	0.18%	—
Leaves	0.088%	—	0.047–0.051%	—
Stems	trace	—	0.076%	—
Roots	0.002%	—	nt	—
Seeds	trace	—	0.01%	—
Hashish	—	—	3.5%	—
Hash oil	—	—	3.5%	—

Note that variability exists due to methods of analysis, species of plant, and plant origin (see Fetterman *et al.*, 1971a).
*A bract is a leaflike plant part, usually small, located just below a flower, a flower stalk, or an inflorescence.
†Buds include flowering tops, smaller leaves and seeds.

into the dry air. But this thick stocky resin is not an ordinary goo. It is the stuff that dreams are made of, the stuff that holds time suspended in limbo, the stuff that makes men forgetful, makes them both sad and deliriously happy, makes them ravenously hungry or completely disinterested in food. It is a god to some and a devil to others. It is all of these things and more. This resin, this shield against the sun, this sticky goo... hashish' (Abel, 1980).

The term *hashish* has an even more interesting derivation and has been related to a particular sect of one of the main branches of Islam (Shiite branch) known as the Nizari Ismaili. Led by a famous Islamic dissident named Hasan ibn-Sabah (1050–1124), the movement extended into the period of the Crusades and was marked by terrorist-like secret assassinations of prominent leaders within both Islam and Christianity. Known as the Hashshahin or Heyssessini, there are two accounts that the followers of this sect may have used cannabis, one by Marco Polo (though he only referred to the drug as an unidentified potion; Yule, 1875) and another by twelfth-century friar Arnold von Lubeck (who did refer to it as *hemp per se*) (Lewin, 1964). Though the state of intoxication with cannabis does not lend itself to acts of violence (quite the contrary; see below) the legend begot the name and was embellished by the writings of Marco Polo (Yule, 1875). Eventually an *assassin* came to be known as a perfidious murderer, and hashish came to be considered a drug that turned normal individuals into assassins. The myth continues to some extent even in modern times (Booth, 2003). It has been implied that the word *hashish* itself may have derived from the Arabic *asas*, or 'foundation' (which applied to Islam's religious leaders) and *hassas*, meaning either 'to kill' or the followers of Hasan ibn-Sabah (Abel, 1980).

The French naturalist Jean Baptiste Lamarck argued that the hemp plant grown in Europe was sufficiently different from that grown in India to be a different species. The plant grown for fiber use was termed *Cannabis sativa* and that grown for psychoactive properties was called *Cannabis indica* (Lamarck, 1783; Booth, 2003). A Russian botanist, Janischevski, recognized a third wild species in Asia called *Cannabis ruderalis* (Janichevsky, 1924). The relationships between an individual plant and its environment are complex and determine the representative phenotypes of the species; therefore, the genetic plasticity of *Cannabis sativa* enables wide phenotypic variability in adapting to diverse conditions (Merlin, 1972).

The taxonomy of cannabis continues to be in flux, with discussion often referring to either two or three species. The two-species formulation includes *Cannabis sativa* (not very psychoactive and used mainly for fiber) and *Cannabis indica* (psychoactive) (Zhukovskii and Hudson, 1962; Clarke and Watson, 2002). Another formulation supports three species (*Cannabis sativa, Cannabis indica*, and *Cannabis ruderalis*) (Schultes *et al.*, 1974). A recent formulation argues for two species of *Cannabis sativa* to include all wild, hemp, and drug cannabis races and *Cannabis indica* to include cannabis races used for hashish production (also termed *Cannabis afghanica*) (Clarke and Watson, 2002) (**Fig. 7.2**). For growing hemp for fiber or seed (sativa or indica) both male and female plants are left undisturbed until harvest. However, marijuana cultivators in the 1970s in North America and Europe began to grow *sinsemilla* (Spanish for 'without seed'). Sinsemilla preparations can effectively be implemented by eliminating staminate (male) plants from the fields and keeping only the unfertilized pistillate (female) plants to mature for later harvest. The female plants continue to produce flowers, are high in resin glands, and thus have an increased Δ^9-THC content. More recently, indoor plant growing has become popular for growing cannabis for medical purposes. Here, plants are reproduced vegetatively by rooting cuttings of only female plants, producing uniform crops of seedless females (Clarke and Watson, 2002). Marijuana prepared from the dried flowering tops and leaves has Δ^9-THC concentrations ranging from 0.5–5.0 per cent. Hashish can range up to 15–20 per cent. Hash oil is made by soaking cannabis leaves and flowering tops in a solvent like isopropanol. The plant material is then removed, and the isopropanol containing the cannabinoids is heated to allow the isopropanol to evaporate leaving the pure hash oil. It has the highest Δ^9-THC concentration of any marijuana preparation, reaching up to 50–60 per cent. Selective breeding has resulted in special varieties of marijuana such as sinsemilla and 'Netherweed' (Dutch hemp) that may have concentions of as high as 20 per cent (World Health Organization, 1997).

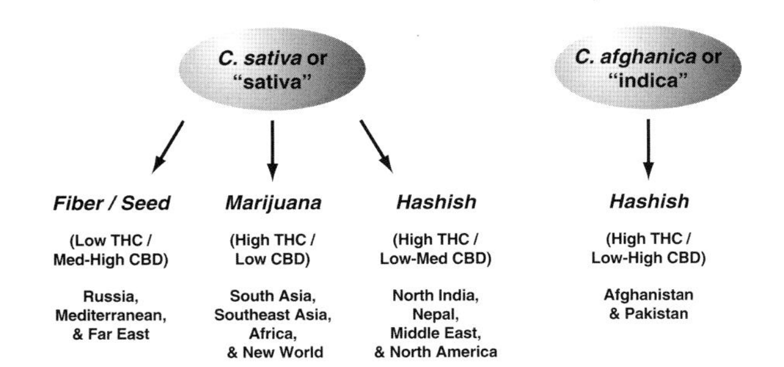

FIGURE 7.2 The four major *Cannabis* gene pools. Most modern medical cannabis varieties are a blend of traditional *sativa* marijuana varieties with *indica* hashish varieties. The traditional *Cannabis* gene pools originate either from *Cannabis sativa*, which comprises the vast majority of naturally occurring hemp and drug land races, or from *Cannabis afghanica* from Afghanistan and Pakistan, which is commonly called 'indica' and has become a component in many modern drug *Cannabis* cultivars. Also, all taxonomists recognize the species *Cannabis sativa*. Schultes *et al.* (1974) divided *Cannabis* into three species: *Cannabis sativa, Cannabis indica*, and *Cannabis ruderalis*. Clarke and Watson (2002) consider *Cannabis sativa* to describe all wild hemp and drug *Cannabis* races with the possible exception of the races used for hashish production in Afghanistan and Pakistan which they term *Cannabis afghanica* and others refer to as *indica*. THC, Δ^9-tetrahydrocannabinol; CBD, cannabidiol. [Adapted from Clarke and Watson, 2002.]

It should be noted that there are two numbering systems for cannabinoids. In the formal numbering system for pyran compounds (the dibenzopyran nomenclature), the main active ingredient is numbered Δ^9-THC. However, not all cannabinoids are pyran compounds, and hence a second nomenclature using a biogenetic basis exists (monoterpenoid nomenclature). In this older numbering system, Δ^9-THC is actually referred to as Δ^1-THC (which readers will find to be the case when referencing older literature) (Ellenhorn and Barceloux, 1988; Grotenhermen, 2003).

Cannabis contains a total of 66 phytocannabinoids with several different subclasses of compounds within this grouping (ElSohly, 2002) (**Table 7.2**). However, over 483 total natural components have been isolated (for reviews, see Turner *et al.*, 1980; Ross and ElSohly, 1995) (**Table 7.3**). Of the 66 unique cannabinoids in the *Cannabis* plant (ElSohly, 2002), three are known to be psychoactive: Δ^9-THC (Gaoni and Mechoulam, 1964; Grunfeld and Edery, 1969), Δ^8-THC (67 per cent as a potent isomer of Δ^9-THC) (Grunfeld and Edery, 1969), and propyl homologue of Δ^9-tetrahydrocannabidivarin (25 per cent as potent as Δ^9-THC) (Gill *et al.*, 1970; Mechoulam, 1970). However, the latter two probably do not contribute to its psychological or physiological effects, representing considerably less than 10 per cent of samples (Mechoulam, 1970; Mason and McBay, 1985; Turner *et al.*, 1973). Cannabis also

TABLE 7.2 Average Cannabinoid Concentrations in 35312 Cannabis Preparations Confiscated in the United States, 1980–1997

	Concentration (%)			
	Δ^9-THC	Cannabidiol	Cannabichromene	Cannabinol
Marijuana	3.1	0.3	0.2	0.3
Sinsemilla	8.0	0.6	0.2	0.2
Hashish	5.2	4.2	0.4	1.7
Hashish oil	15.0	2.7	1.1	4.1

Data from ElSohly *et al.*, 2000.

frequently contains cannabinoid acids in various amounts relative to Δ^9-THC and are devoid of psychotropic effects (Dewey, 1986): Δ^9-tetrahydrocannabinolic acid A (6a,7,8,10a-tetrahydro-1-hydroxy-6,6,9-trimethyl-3-pentyl-6H-benzo[*c*]chromene-2-carb oxylic acid) (Fetterman *et al.*, 1971a,b; Turner *et al.*, 1973) and Δ^9-tetrahydrocannabinolic acid B (6a,7,8,10a-tetrahydro-1-hydroxy-6,6,9-trimethyl-3-pentyl-6H-benzo[*c*]chromene-4-carboxylic acid) (Mechoulam *et al.*, 1969; Turner *et al.*, 1974) which readily decarboxylate to yield Δ^9-THC. Cannabichromene has slight Δ^9-THC-like effects (Isbell *et al.*, 1967a,b; Davis and Hatoum, 1983). Cannabidiol and cannabinol appear to be largely psychotropically inactive (Isbell *et al.*, 1967a,b; Hollister, 1973; Karniol *et al.*, 1974, 1975; Grotenhermen, 2003), although weak effects of cannabinol have been reported in man (Perez-Reyes *et al.*, 1973), and anticonvulsant effects have been reported for cannabidiol in mice (Karler *et al.*, 1973). Cannabidiol does not bind to cannabinoid CB_1 or CB_2 receptors, but it has antiarthritic properties in a murine model of arthritis (Malfait *et al.*, 2000) and may have anxiolytic-like effects (Zuardi *et al.*, 2002), antipsychotic-like effects in animal models, and anti-Δ^9-THC-like effects (Dalton *et al.*, 1976; Zuardi *et al.*, 1982).

The fiber type of cannabis plant has very little Δ^9-THC (Avico *et al.*, 1985; Ross *et al.*, 2000; Clarke and Watson, 2002). Dronabinol (Marinol) is the (-)*trans*-isomer of Δ^9-THC and is dissolved in sesame oil in capsules of 2.5, 5.0, and 10 mg (Medical Economics Company, 2004).

TABLE 7.3 Cannabis Consituents

Chemical class	Known constituents in cannabis
Cannabinoids	66
Nitrogenous compounds	27
Amino acids	18
Proteins, glycoproteins, and enzymes	11
Sugars and related compounds	34
Hydrocarbons	50
Simple alcohols	7
Simple aldehydes	12
Simple ketones	13
Simple acids	21
Fatty acids	22
Simple esters and lactones	13
Steroids	11
Terpenes	120
Noncannabinoid phenols	25
Flavonoids	21
Vitamins	1
Pigments	2
Elements	9
Total	***483***

Reproduced with permission from ElSohly, 2002.

HISTORY OF CANNABINOID USE, ABUSE, AND ADDICTION

Cannabis is the most commonly used illicit drug in the United States (Anthony *et al.*, 1994; Substance Abuse and Mental Health Services Administration, 2003) with 15 million current users (approximately 6 per cent of the U.S. population), and in 2002, over 14 million Americans aged 12 and older used marijuana at least once in the past month (Substance Abuse and Mental Health Services Administration, 2003), with approximately 3 million daily or almost daily users. Approximately 34 per cent of Americans 12 years or older (76.3 million) have tried marijuana at some point in their lives (Substance Abuse and Mental Health Services Administration, 2000) and most first-time use occurs in adolescence (Gruber and Pope, 2002). The average age of first cannabis use in the United States was 14 in 1997 (Substance Abuse and Mental Health Services Administration, 1996), and in a report of adolescents referred for conduct and substance abuse problems, cannabis, tobacco, and alcohol showed high overall prevalence and rapid progression from first use to regular use (Crowley *et al.*, 1998) (**Fig. 7.3**).

In the past ten years in the United States, there has been a steady increase in the prevalence of cannabis use particularly among young people and an increase in the number of patients entering treatment for cannabis-related problems. Long-term trends showed increases in marijuana use in the 1960s and 1970s, declines in the 1980s, and increases again in the 1990s. In the mid-1960s only 5 per cent of young adults aged 18–25 had ever used marijuana, but this increased to 54 per cent in 1982 (Substance Abuse and Mental Health Services Administration, 2003) (**Fig. 7.4**). Although the rate for 'ever used' in young adults declined somewhat, it never dropped below 43 per cent and had reached 54 per cent in 2002.

Approximately 10 per cent of individuals who ever use cannabis become daily users, which corresponds to estimates of rates of Substance Dependence on cannabis and the prevalence of a withdrawal syndrome (Anthony *et al.*, 1994; Wiesbeck *et al.*, 1996; Johnston *et al.*, 1999). For some time a common belief was that marijuana abuse and dependence rarely occurred as a primary problem, and many believed that cannabis did not produce a true dependence syndrome. Starting in the 1980s, however, there has been an increasing number of individuals who cannot stop smoking cannabis on their own and have sought treatment primarily for cannabis dependence (Zweben and O'Connell, 1992). Two-thirds of current illicit users of marijuana used only marijuana (Substance Abuse and Mental Health Services Administration, 2003). However, in a study of adolescent marijuana-dependent subjects, multiple dependencies were the prevalent condition with each youth, on an average being dependent on 3.2 substances (Crowley *et al.*, 1998). When subjects present for treatment, various rationales for discontinuing marijuana use are given, including fear of the physical consequences of smoking cannabis, difficulty expressing emotions, difficulty experiencing feelings of intimacy and closeness with a partner, and dissatisfaction with a failure in achieving life goals (Roffman and Barnhart, 1987; Zweben and O'Connell, 1992). Chronic users are described as having a mild boredom, lack of zest, or a low level of depression that resolves during marijuana abstinence (Zweben and O'Connell, 1992). Researchers found that subjects readily respond to advertisements for treatment of cannabis dependence, and the majority were not abusing other substances (Budney *et al.*, 1998; Copeland *et al.*, 2001). The demand for treatment of cannabis-related problems at substance abuse programs doubled in the United States between 1992 and 1998 (Substance Abuse and Mental Health Services Administration, 2000). In 1999, there were 220 000 cannabis-related admissions to publicly funded substance abuse treatment programs in the United States, representing 14 per cent of all admissions. Two thirds of these admissions were subjects 12–25 years of age. In 2002, the number seeking treatment jumped to 974 000 for persons aged 12 and older

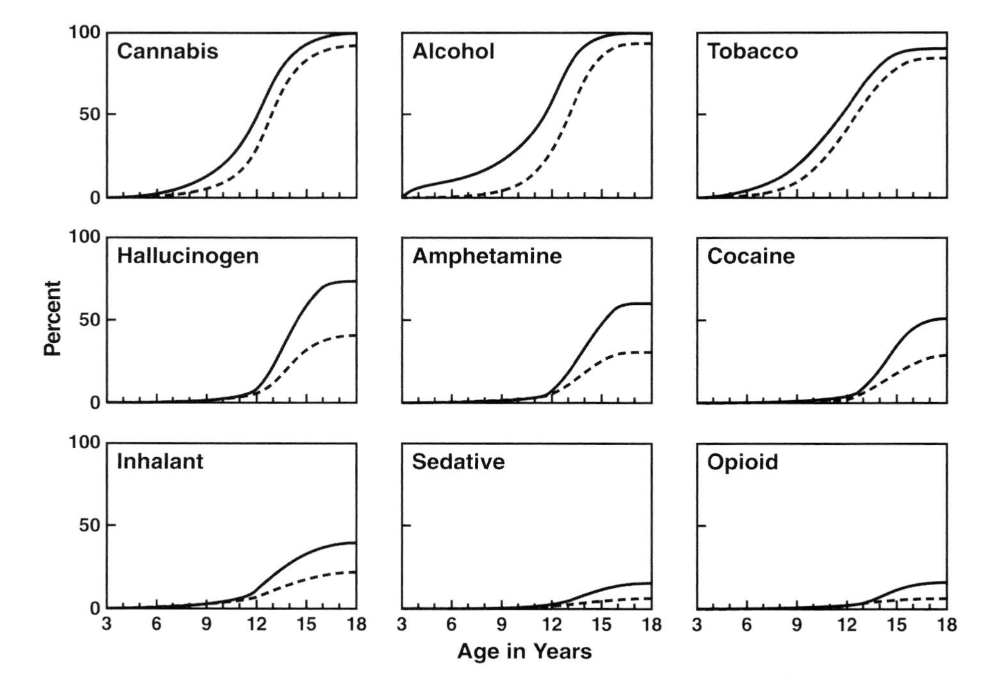

FIGURE 7.3 Cumulative percentage of 218 subjects aged 13–19 who had used the drugs shown either once (solid line) or at least monthly (dashed line) by various ages. Subjects for this study were patients admitted to a substance abuse treatment program between 1991 and 1994. All patients had: (1) significant antisocial problems, diagnosed substance problems, and diagnosed conduct disorder, (2) been judged by clinical staff not to be currently psychotic, mentally retarded, homicidal, suicidal, or a current arson risk, (3) no physical illness which would prevent participation in active, group-oriented treatment, (4) written, informed consent from a parent or guardian, (5) assent from the youth. [Data from Crowley *et al.*, 1998.]

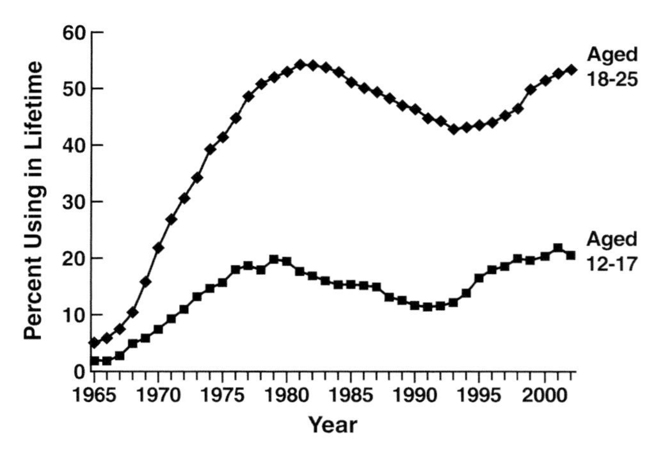

FIGURE 7.4 Percentage of persons aged 12–17 and 18–25 who had ever used marijuana through the year 2000. [from Substance Abuse and Mental Health Services Administration, 2003.]

(Substance Abuse and Mental Health Services Administration, 2003).

Substance Dependence on cannabis has a characteristic clinical course that can be identified, is predictable, and has elements common to Substance Dependence on other drugs of abuse (Miller and Gold, 1989). Typically, a user will begin with social use, often enjoying the pleasurable effects of cannabis in a social situation while learning to use the drug to enhance such effects (learning to inhale) (see *Case Studies* in appendix). Subsequently, marijuana use increases to a point where the drug is no longer used socially, and users begin smoking whenever possible (for example, on the way to school or even between classes). Tolerance develops, and more and more cannabis is smoked presumably due to tolerance. Performance in social settings and occupational functioning (for adolescents, school work) declines, and denial and resistance to considering cannabis a problem develop. The most important criteria for diagnosing Substance Dependence on cannabis has been argued to be that cannabis is affecting one's ability to function (Miller and Gold, 1989). Often abnormal cannabis use is attributed to a result of a given problem instead of the cause of it. Difficulty with sleep, depression, and stress fall into these categories. While both social users and cannabis-dependent subjects may use marijuana for the same reasons, 'the reason addicts use marijuana *abnormally* is because marijuana apparently provides an unusual reinforcement to those with a vulnerability to marijuana not possessed by others' (Miller and Gold, 1989).

Two clinical forms of dependence have been described (**Table 7.4**). One of the forms involves an individual who self-administers cannabis several times per day with an interval of approximately 2–4 h except during sleep and who escalates their intake. Such subjects present with significant impairment in social and occupational functioning and often seek treatment. The other form of dependence presents at mandatory screening and involves individuals who usually self-administer cannabis every 24–36 h. While presenting with impairment in social and occupational functioning, these individuals have less withdrawal and less perceived dependence (Tennant *et al.*, 1986) (see *Case Studies* in appendix).

Ultimately, most of the *Diagnostic and Statistical Manual of Mental Disorders-IV* criteria for Substance Dependence are met by those individuals diagnosed with Substance Dependence on cannabis (Miller and Gold, 1989; American Psychiatric Association, 1994; Crowley *et al.*, 1998). Preoccupation with marijuana is represented by a persistent presence of marijuana in the individual's pattern of living and choices of activities (DSM-IV criteria #3 and #5), and compulsivity is shown by the continued use in the presence of, or in spite of, marijuana-related consequences (DSM-IV criteria #6 and #7). Relapse or propensity to relapse (DSM-IV criterion #4) is reflected in a return to marijuana use after a period of abstinence and may be a confirmation of a suspected diagnosis (Miller and

TABLE 7.4 Two Clinical Forms of Marijuana Dependence

	Frequency of self-administration	Likely dependence metabolites	Usual treatment referral	Self-perceived dependence	Severity of withdrawal symptoms	Relapse rate
Type One	Multiple times each day	Δ^9-Tetrahydrocannabinol 11-Hydroxy-Δ^9-tetrahydrocannabinol	Voluntary; self-referred	Significant	Moderate	High
Type Two	Every 24–48 h	11-nor-Δ9-Tetrahydrocannabinol-carboxylic acid	Involuntary; detected by mandatory screening	Minor-moderate	Mild	High

Reproduced with permission from Tennant *et al.*, 1986.

Gold, 1989). In a study of adolescents referred for substance and conduct problems, the most frequently observed criteria were much time getting, using, or recovering from use (DSM-IV criteria #5 and #7; 80 per cent), continued use despite problems in social and occupational functioning (DSM-IV criterion #6; 70–80 per cent), or tolerance and withdrawal (DSM-IV criteria #1 and #2; approximately 70 per cent) (Crowley *et al.*, 1998) (**Fig. 7.5**).

Treatment of cannabis dependence has relied almost exclusively on behavioral therapies. Controlled clinical trials of outpatient treatment have yielded responsiveness and success rates similar to those for Substance Dependence on other drugs of abuse (McRae *et al.*, 2003). Cognitive behavioral therapy (Stephens *et al.*, 1994), motivational enhancement therapy (Budney *et al.*, 2000), relapse prevention support groups (Stephens *et al.*, 1994), social support groups, and motivational interviewing (Copeland *et al.*, 2001) all have shown significant reductions in marijuana use compared to baseline, but no particular therapy has stood out as significantly more effective than any other (McRae *et al.*, 2003). In a large scale, multi-site trial, preliminary results suggested that at 4 months, approximately 23 per cent of an extended cognitive behavioral therapy combined with a motivational enhancement therapy treatment group remained abstinent compared to approximately 9 per cent of a brief treatment group and approximately 4 per cent for a delayed treatment group, suggesting some semblance of a dose response (Stephens *et al.*, 2002; McRae *et al.*, 2003). No pharmacotherapies have been systematically evaluated for Substance Dependence on cannabis but there is some suggestion that serotonin antidepressants (Cornelius *et al.*, 1999) and anxiolytics (McRae *et al.*, 2004) may have promise (McRae *et al.*, 2003).

MEDICAL USE AND BEHAVIORAL EFFECTS

Medical uses of marijuana were ubiquitous for centuries in Asia, Southeast Asia, and India, and included use as an intoxicant and tonic. Hemp was used to expel flatulence, to excite appetite, and to induce eloquence (O'Shaughnessy, 1839). Both stimulant and sedative effects were described as dose-related, with higher doses producing more sedative-like effects (O'Shaughnessy, 1839). Noted in these early explorations of the medical use of cannabis was its ability to cause profound 'narcoticism' which was effective in treating a case of infantile convulsions, but also resulted in a 'singular form of delirium which the incautious use of the hemp preparations often occasions, especially among young men first commencing the practice'. Reflecting a state of intoxication, the user is described as having 'a strange balancing gait, perpetual giggling, expressions of cunning and merriment, increased libido and a voracious appetite' (O'Shaughnessy, 1839). Possible indications for cannabis preparations are numerous and can be found in a number of reviews (Mechoulam, 1986; Mathre, 1997; Joy and Benson, 1999; Iversen, 2000; Grotenhermen, 2002) and at least one author has outlined a hierarchy of therapeutic effects (**Table 7.5**). Cannabinoids have two accepted medical uses in the United States. Dronabinol (Marinol) is approved for use in refractory nausea and vomiting associated with cancer chemotherapy and for appetite loss in anorexia of HIV/AIDS patients (Department of Justice, 1999). Potential medical effects that have been relatively less well confirmed include amelioration of spasticity due to spinal cord injury and multiple sclerosis. Cannabinoids are effective analgesics and have shown some effectiveness in chronic pain conditions.

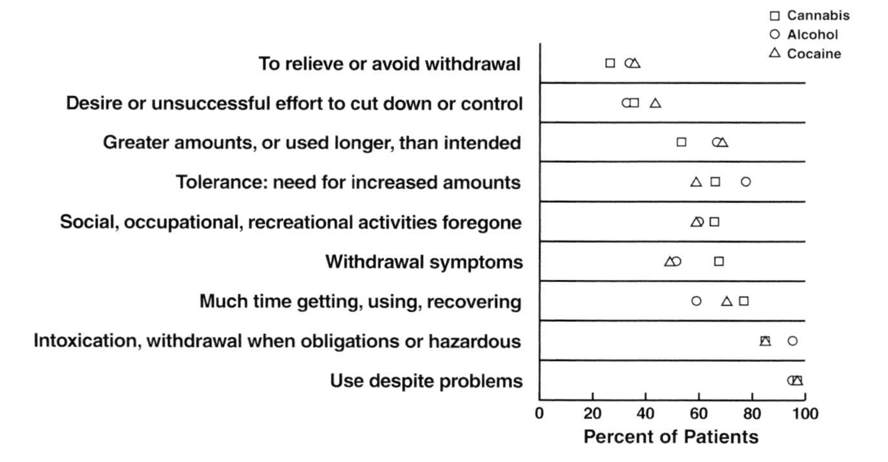

FIGURE 7.5 Percentage of patients with various DSM-III-R dependence symptoms, among those dependent on cannabis (n = 180), alcohol (n = 186), or cocaine (n = 51). [Reproduced with permission from Crowley *et al.*, 1998.]

TABLE 7.5 Medical Uses of Cannabinoids

Well-confirmed clinical effects	
Refractory nausea/vomiting	Sallan *et al.*, 1980; Lane *et al.*, 1991; Abrahamov *et al.*, 1995; Dansak, 1997; Soderpalm *et al.*, 2001
Anorexia appetite loss	Beal *et al.*, 1997
HIV/AIDS/cancer cachexia	Plasse *et al.*, 1991; Beal *et al.*, 1995, 1997; Jatoi *et al.*, 2002; Nauck and Klaschik, 2004
Less well-confirmed clinical effects	
Spasticity due to spinal cord injury	Petro, 1980; Malec *et al.*, 1982; Maurer *et al.*, 1990; Brenneisen *et al.*, 1996
Multiple sclerosis	Petro, 1980; Petro and Ellenberger, 1981; Ungerleider *et al.*, 1987; Meinck *et al.*, 1989; Martyn *et al.*, 1995; Brenneisen *et al.*, 1996; Killestein *et al.*, 2003; Pryce *et al.*, 2003; Zajicek *et al.*, 2003; Pryce and Baker, 2005 (but see Killestein *et al.*, 2002; van Oosten *et al.*, 2004; Fox *et al.*, 2004)
Neurogenic pain/ neuropathy/allodynia	Noyes *et al.*, 1975a,b; Petro, 1980; Maurer *et al.*, 1990; Elsner *et al.*, 2001; Ibrahim *et al.*, 2003; Karst *et al.*, 2003; Kehl *et al.*, 2003; Lynch and Clark, 2003; Wade *et al.*, 2003; Ware *et al.*, 2003; AIDS Patient Care STDS, 2004; AIDS Reader, 2004; De Vry *et al.*, 2004; Notcutt *et al.*, 2004; (but see Buggy *et al.*, 2003; Naef *et al.*, 2003; Attal *et al.*, 2004)
Movement disorders (Tourette's syndrome; dystonia; dyskinesia)	Clifford, 1983; Sandyk and Awerbuch, 1988; Hemming and Yellowlees, 1993; Muller-Vahl *et al.*, 1999, 2002a,b, 2003b; Sieradzan *et al.*, 2001; Fox *et al.*, 2002; Gauter *et al.*, 2004 (but see Muller-Vahl *et al.*, 2003a)
Bronchodilation effects	Tashkin *et al.*, 1974; Williams *et al.*, 1976; Hartley *et al.*, 1978
Glaucoma	Hepler and Frank, 1971; Crawford and Merritt, 1979; Merritt *et al.*, 1980, 1981; Porcella *et al.*, 2001
Largely unexplored, suggested potential clinical effects	
Epilepsy	Cunha *et al.*, 1980; Carlini and Cunha, 1981; Karler and Turkanis, 1981; Wallace *et al.*, 2003; Lorenz, 2004
Hiccups	Gilson and Busalacchi, 1998
Bipolar disorder	Grinspoon and Bakalar, 1998
Alzheimer's disease	Volicer *et al.*, 1997
Alcohol dependence	Mikuriya, 1970
Basic research	
Amyloid formation	Milton, 2002; Iuvone *et al.*, 2004
Opiate withdrawal	Yamaguchi *et al.*, 2001
Ischemia	Bar-Joseph *et al.*, 1994; Vered *et al.*, 1994; Belayev *et al.*, 1995a,b; Leker *et al.*, 1999; Nagayama *et al.*, 1999; Louw *et al.*, 2000; Sinor *et al.*, 2000; Hampson, 2002; Di Filippo *et al.*, 2004
Hypertension	Ralevic and Kendall, 2001; Wagner *et al.*, 2001
Neoplasms	Jacobsson *et al.*, 2001; Sanchez *et al.*, 2001; Portella *et al.*, 2003; Massi *et al.*, 2004; Velasco *et al.*, 2004
Diarrhea	Izzo *et al.*, 2000, 2003; Tyler *et al.*, 2000
	Calignano *et al.*, 2000
Sleep apnea	Carley *et al.*, 2002
Colonic inflammation	Massa *et al.*, 2004
Irritable bowel syndrome	Camilleri *et al.*, 2004
Cough	Patel *et al.*, 2003
Huntington's disease	Lastres-Becker *et al.*, 2003 (but see Consroe *et al.*, 1991)
Optic nerve damage	Zalish and Lavie, 2003

Cannabinoids also have been shown to be effective in treating asthma and in lowering intraocular pressure in the treatment of glaucoma. Cannabinoids have been hypothesized for use in the treatment of movement disorders, including dystonias and tardive dyskinesia. Other proposed actions of cannabinoids that are largely unconfirmed range from treatment of allergies and inflammation to epilepsy and psychiatric disorders (Grotenhermen, 2003).

Cannabis is reinforcing in humans as demonstrated by numerous anecdotal and laboratory studies. The subjective effects reported include euphoria and mood swings characterized by initial feelings of 'happiness' or sudden talkativeness, or a dreaming or lolling state, or general activation and hyporeactivity. The user can report feeling fuzzy, dizzy, sleepy and in a dream-like state (Miller and Gold, 1989). A user can feel more friendly toward others and finds more pleasure in the company of others. In such a group setting, smoking cannabis can produce talkativeness among subjects, with much contagiousness of laughing and joking with a particular high-pitched giggly laughter. In studies of self-reported effects of cannabis in regular human cannabis users, both in naturalistic studies

TABLE 7.6 Ten Most Frequently Reported Effects of Marijuana Intoxication (based on open-ended question)

Berke (1974) (n = 522)	%	Goode (1970) (n = 191)	%	Atha (1998) (n = 2794)	%
Enhanced relaxation	25.7	More relaxed	46.1	Relaxation	25.9
Happiness	16.1	Senses more perceptive	36.1	Insight/personal growth	8.7
Appreciation of music	15.5	Think deeper	31.4	Pain relief	6.1
Visual illusions/hallucinations	13.4	Laugh much more	28.8	Antidepressant/happy	4.9
Enhanced insight into others	11.9	Time slowed down	23.0	Respiratory benefit	2.4
Hunger/appetite enhanced	11.9	Become more withdrawn	22.0	Creativity	2.3
Heightened sense perception	11.7	Feels nice, pleasant	20.9	Socializing	2.0
Elation	11.5	Mind wanders	20.9	Sensory perception	1.6
Colors brighter	11.1	Feel dizzy, giddy	20.4	Improved sleep	1.6

Reproduced with permission from Green *et al.*, 2003.

and laboratory-based studies, the most frequently reported effect was relaxation. Enhanced mood (happiness or laughing more), sensory alterations, enhanced appetite and greater insight/thinking also ranked high (**Table 7.6**). There are large individual differences in response to cannabis, and the effects are heavily influenced by expectancies (Aarons *et al.*, 2001). Indeed, decreased talkativeness and sociability are often reported, and these effects may be dose-related and more likely to occur with higher doses (Tart, 1971; Green *et al.*, 2003). A characteristic effect of marijuana intoxication compared to alcohol intoxication is 'less noisy and boisterous at parties than when drunk or tipsy on alcohol' (Tart, 1971). Sociability can go either way, with self-reports of 'I become more sociable' (more likely at lower levels of intoxication) and 'I become less sociable' (at higher levels of intoxication) (Tart, 1971) (**Fig. 7.6**). In human laboratory studies, where subjects who smoked at least 4 times per week (average 6.6 times per week) participated in a 16-day residential study, self-administration of marijuana cigarettes varied as a function of the Δ^9-THC content. Marijuana cigarettes with Δ^9-THC concentrations of 2.2 and 3.9 per cent were self-administered more than placebo cigarettes (Haney *et al.*, 1997). Subjects reported significant increases in ratings of *high*, *stimulated*, and *good drug effect*, but also measures of *forgetful* and *can't concentrate* were increased, and performance on a digit-symbol substitution task, divided attention task, rapid information task, and math task were decreased. Some of these effects were greater in the subjects smoking the 3.9 per cent versus 2.2 per cent cigarettes (Haney *et al.*, 1997).

Subjects may have a form of disinhibition in that there is an inclination to increase motor activity, and behavior is impulsive. However, paradoxically, even the simplest tasks appear to require enormous effort such that users generally seek situations where no

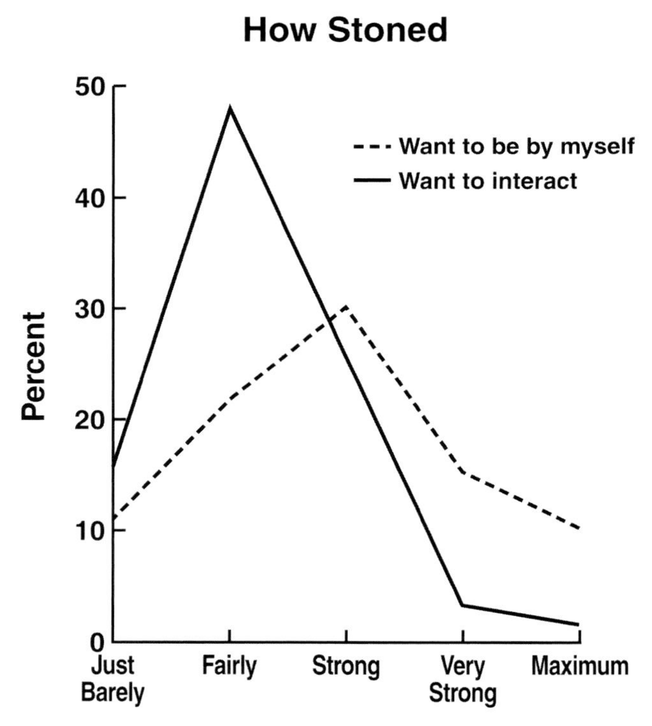

FIGURE 7.6 Effect of level of marijuana intoxication (how "stoned") on sociability. 'I become more sociable' is a common effect, but its converse, 'I become less sociable; I want to be by myself' is just as common. The latter effect occurs at higher levels of intoxication than the former. At a party, for instance, the people sitting by themselves often may be more intoxicated than the ones conversing. The intoxication level data will not always add up to 100 per cent because of variable numbers of respondents skipping various questions and/or due to rounding errors. In interpreting the graph, note that the percentage of users plotted at each level is the percentage indicating that level as their minimal level of intoxication for experiencing that particular effect. Thus, a drop in the curve with increasing minimal level of intoxication does not mean that fewer users experience that effect at higher levels, but that fewer give a higher level as their minimal level for experiencing that effect. [Reproduced with permission from Tart, 1971.]

physical effort is required. Thus, an individual may be disinhibited, but incoordination and clumsiness prevent the subject from attempting many activities. Following is an early clinical description of acute intoxication:

> 'Walking becomes effortless. The paresthesias and changes in bodily sensations help to give an astounding feeling of lightness to the limbs and body. Elation continues: he laughs uncontrollably and explosively for brief periods of time without at times the slightest provocation: if there is a reason it quickly fades, the point of the joke is lost immediately. Speech is rapid, flighty, the subject has the impression that his conversation is witty, brilliant; ideas flow quickly' (Bromberg, 1934).

There are also psychedelic-like effects associated with marijuana intoxication. Subjects report an increased sensitivity to sound and a keener appreciation of rhythm and timing. Perception of time often is slowed down with an exaggeration of the sense of time. Perceptions of space may be broadened, and proximal objects may appear distant. There are some visual hallucination-like effects which are, as with psychedelics such as LSD, mostly illusionary transformations of the outer world. There can be flashes of light or amorphous forms of vivid color which evolve into geometric figures, shapes, or faces. The depth of color is striking with an apparent increase in auditory faculties (**Table 7.7**). There also may be a subjective feeling of unreality that borders on depersonalization with various sensations of lightness or heaviness in the body and a sensation of floating in air or walking on waves. One adolescent subject reported:

> 'Man, when I'm up on the weed, I'm really livin'. I float up and up and up until I'm miles above the earth. Then, Baby, I begin to come apart. My fingers leave my hands, my hands leave my wrists, my arms and legs leave my body and I just flooooooat all over the universe' (Bloomquist, 1967).

TABLE 7.7 Behavioral Effects of Marijuana Intoxication

- Euphoria
- Perceptual changes
  - Feelings of 'unreality' Lightness or heaviness of the body
  - Time slowed down
  - Distorted spatial vision
  - More vivid colors
  - More intense sounds
  - Synesthesia (at high doses)
- Sedation
- Motor impairment
- Decreased mental performance
- Increased appetite
- Analgesia

PHARMACOKINETICS

Cannabis is used primarily by the smoking route of administration, although oral marijuana is active at producing both subjective effects and increases in heart rate (Chait and Zacny, 1992). After inhalation, Δ^9-THC is detectable in plasma within seconds, with peak concentrations being measured 3–10 min after smoking and with bioavailability ranging from 10 to 35 per cent depending on the experience of the smoker (Chait and Zacny, 1992; Grotenhermen, 2003) (**Fig. 7.7**). Oral administration results in low bioavailability (6–7 per cent) presumably due to numerous factors but including an extensive first-pass liver metabolism (Grotenhermen, 2003). In addition, the effects of oral Δ^9-THC are significantly delayed, up to 60–120 min. These routes of administration correlate well with their relative effectiveness in producing the subjective 'high' associated with marijuana intoxication (Ohlsson *et al.*, 1980; Hollister *et al.*, 1981) (**Fig. 7.8**). Other potential effective routes of administration include rectal (bioavailability of 13.5 per cent) (ElSohly *et al.*, 1991), sublingual (Guy and Flint, 2000), transdermal (Touito *et al.*, 1988), and opthalmic (bioavailability of 6–40 per cent) (Chiang *et al.*, 1983). Inhalation of Δ^9-THC produces dose-dependent peak blood levels; significantly less peak blood levels are reached after oral administration (**Table 7.8**).

Δ^9-THC has a peculiar distribution because of its high lipophilicity compared to other drugs of abuse (Grotenhermen, 2003) and as a result, rapidly enters highly vascularized tissues (Ho *et al.*, 1970). However, after intravenous administration, only 1 per cent is estimated to enter the brain at the time of peak behavioral effects (Gill and Jones, 1972). Significant accumulation of cannabinoids occurs later in less vascularized tissues and body fat which is the major long-term storage site resulting in concentration ratios of 10:1 (fat:other tissues) though the exact composition of the cannabinoids is unknown (Kreuz and Axelrod, 1973; Harvey *et al.*, 1982; Grotenhermen, 2003).

THC is broken down in the liver by enzymes of the cytochrome P450 system (Matsunaga *et al.*, 1995). In humans, hydroxylation leads to 11-OH-THC and oxidation to THC-COOH and ultimately glucuronidate (Frytak *et al.*, 1984) (**Fig. 7.9**). The elimination half-life for Δ^9-THC is estimated to be in the range of 20–60 h (Lemberger *et al.*, 1971; Hunt and Jones, 1980; Ohlsson

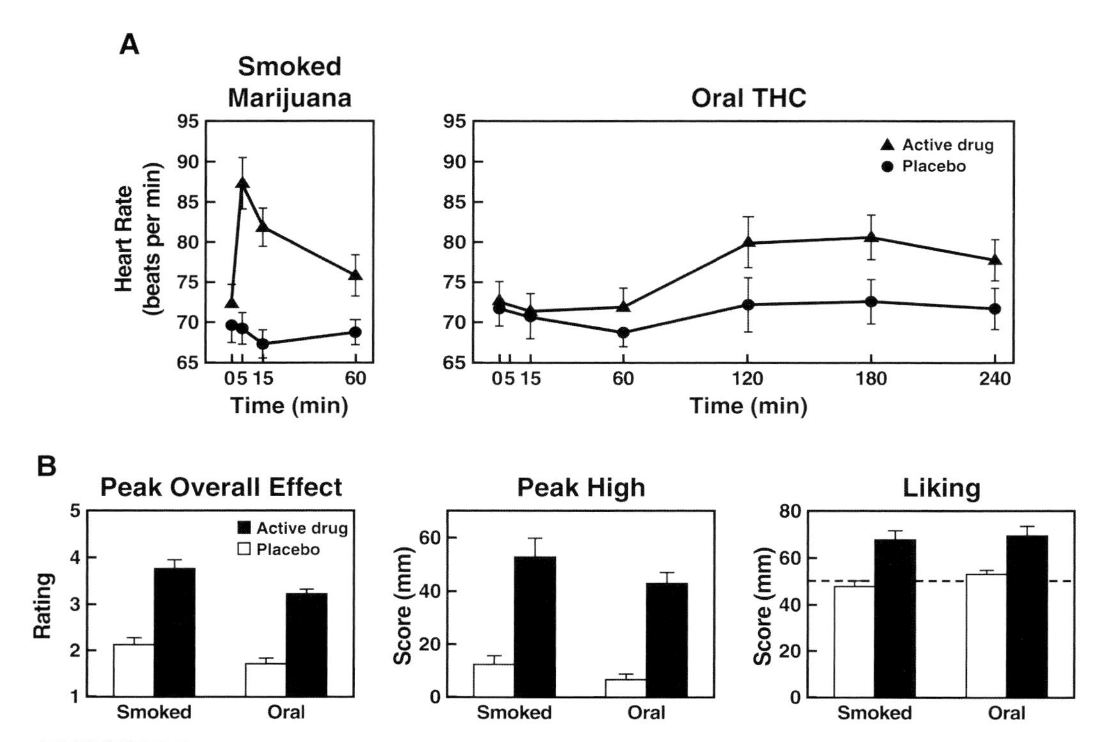

FIGURE 7.7 Physiological and subjective effects of smoked versus oral administration of Δ^9-tetrahydrocannabinol. Smoked marijuana cigarettes contained 2.3–3.6 per cent Δ^9-THC. Oral administration was via dronabinol (Marinol) capsules (2.5, 5, and 10 mg). **(A)** Group mean heart rate ± SEM before and after active drug or placebo. **(B)** Group mean scores ± SEM from the end-of-session questionnaire after active drug or placebo. [Modified with permission from Chait and Zacny, 1992.]

et al., 1982; Wall *et al.*, 1983). The half-life in chronic users is shorter (28 h) than nonusers (57 h) (Lemberger *et al.*, 1971). The elimination half-life for the metabolites of Δ^9-THC is longer than that of the parent compound, ranging up to 5–6 days. Δ^9-THC is excreted as acid metabolites (Grotenhermen, 2003), and these metabolites can be detected in urine for 27 days on average in chronic users, with positive results for cannabinoids in urine at 20 ng/ml or above using the Syva Enzyme Multiple Immunoassay Technique for as long as 46 days (Ellis *et al.*, 1985).

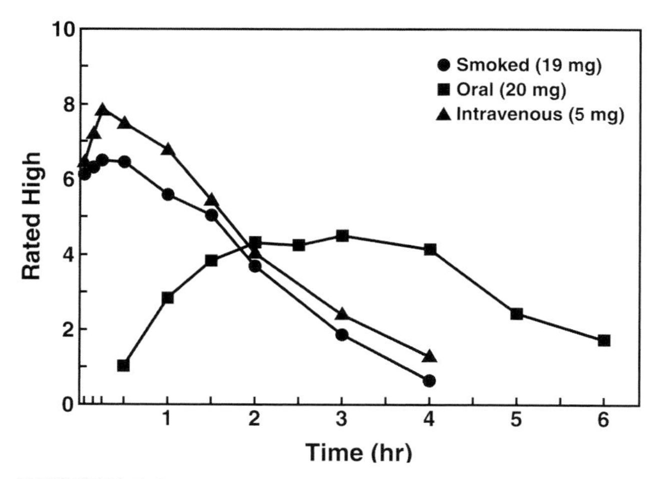

FIGURE 7.8 Time course of subjective high after smoked (19 mg), oral (20 mg), and intravenous (5 mg) administration of Δ^9-THC. [Adapted from Hollister *et al.*, 1981.]

CANNABINOID TOLERANCE

Tolerance develops readily to most of the effects of cannabinoids in humans and is largely attributed to pharmacodynamic (brain neuroadaptation), not pharmacokinetic (metabolic or distribution) changes (Hunt and Jones, 1980; Maykut, 1985). Numerous studies have reported tolerance to cannabis intoxication in humans as measured by either changes in the increases in heart rate or the subjective high (Williams *et al.*, 1946; Dornbush *et al.*, 1972; Fink *et al.*, 1976; Jones and Benowitz, 1976; Jones *et al.*, 1981; Chait, 1989). In an inpatient study of 30 subjects smoking marijuana, the mean heart rate and subjective 'high' decreased over a 94-day smoking period (Nowlan and Cohen, 1977) (**Fig. 7.10**). Here, subjects smoked 2.2 marijuana cigarettes/day (light smokers) to 8.5 marijuana

TABLE 7.8 Peak Plasma Levels of Δ^9-THC after Smoked vs Oral Administration

Route	Dose Δ^9-THC	Mean peak plasma level (range)	Reference
Smoked	16 mg	84.3 (50.0–129.0) ng/ml	Huestis *et al.*, 1992
	34 mg	162.2 (76.0–267.0) ng/ml	Huestis *et al.*, 1992
Oral	2.5 mg	2.01 (0.6–12.5) ng/ml	Timpone *et al.*, 1997
	15 mg	3.9 (2.7–6.3) ng/ml	Frytak *et al.*, 1984
	20 mg	6.0 (4.4–11.0) ng/ml	Ohlsson *et al.*, 1980
	30 mg	15.5 ng/ml*	Haney *et al.*, 2003

*peak at 1 h.

cigarettes/day (heavy smokers) with greater decreases in subjective 'high' in the heavy smokers (Nowlan and Cohen, 1977). Tolerance to the subjective effects but not effects on food intake have been observed in residential studies (Haney *et al.*, 1997, 1999a; Hart *et al.*, 2002). Tolerance also has been observed to the effects of Δ^9-THC on intraocular pressure (Flach, 2002), cardiovascular measures other than heart rate (Sidney, 2002), sedative effects (Kirk and de Wit, 1999), autonomic and sleep changes (Jones *et al.*, 1981).

CANNABINOID WITHDRAWAL

Cannabis withdrawal in humans has long been described anecdotally but not generally accepted by the medical community in that it has not been included in the *Diagnostic and Statistical Manual of Mental Disorders* (American Psychiatric Association, 2000). However, an accumulation of data from both inpatient and outpatient studies has led to a proposal for criteria for cannabis withdrawal in humans (**Table 7.9**). The most common symptoms associated with cannabis withdrawal were decreased appetite/weight loss, irritability, nervousness, anxiety, anger, aggression, restlessness, and sleep disturbances (Budney *et al.*, 2003). A substantial percentage of heavy marijuana users (16 per cent) showed these symptoms (Weisbeck *et al.*, 1996; Haney *et al.*, 1999a,b; Kouri and Pope, 2000; Budney *et al.*, 2001). Inpatient studies showed increased ratings of anxiety, irritability, decreased food intake, sleep disturbances, and depressed mood (Cohen *et al.*, 1976; Jones and Benowitz,

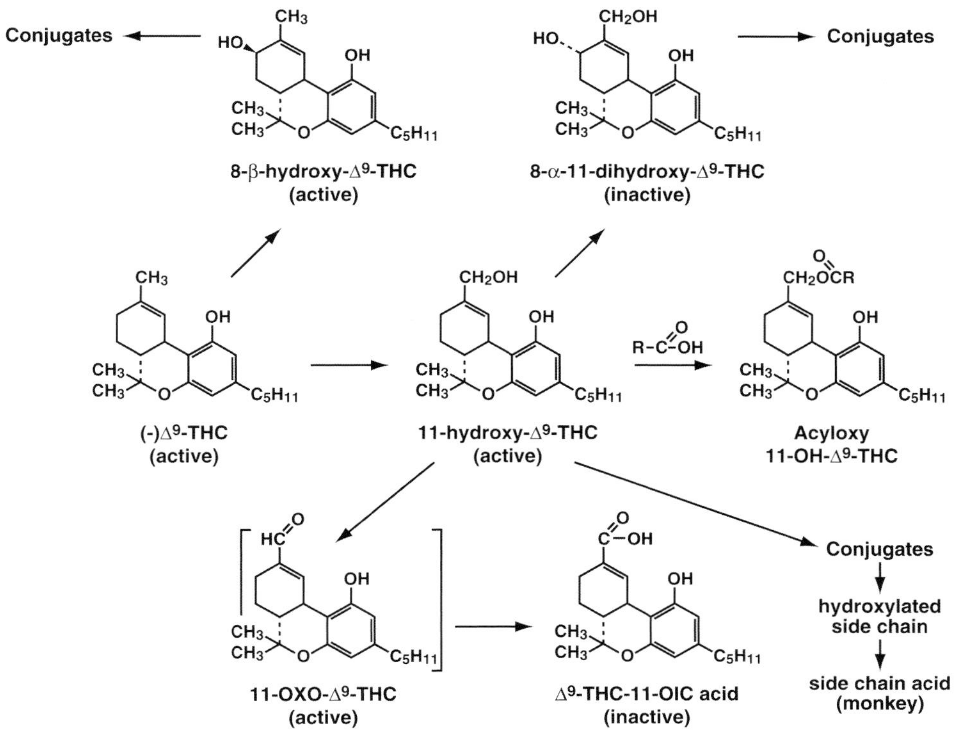

FIGURE 7.9 Metabolic pathways of Δ^9-THC. [Reproduced with permission from Frytak *et al.*, 1984.]

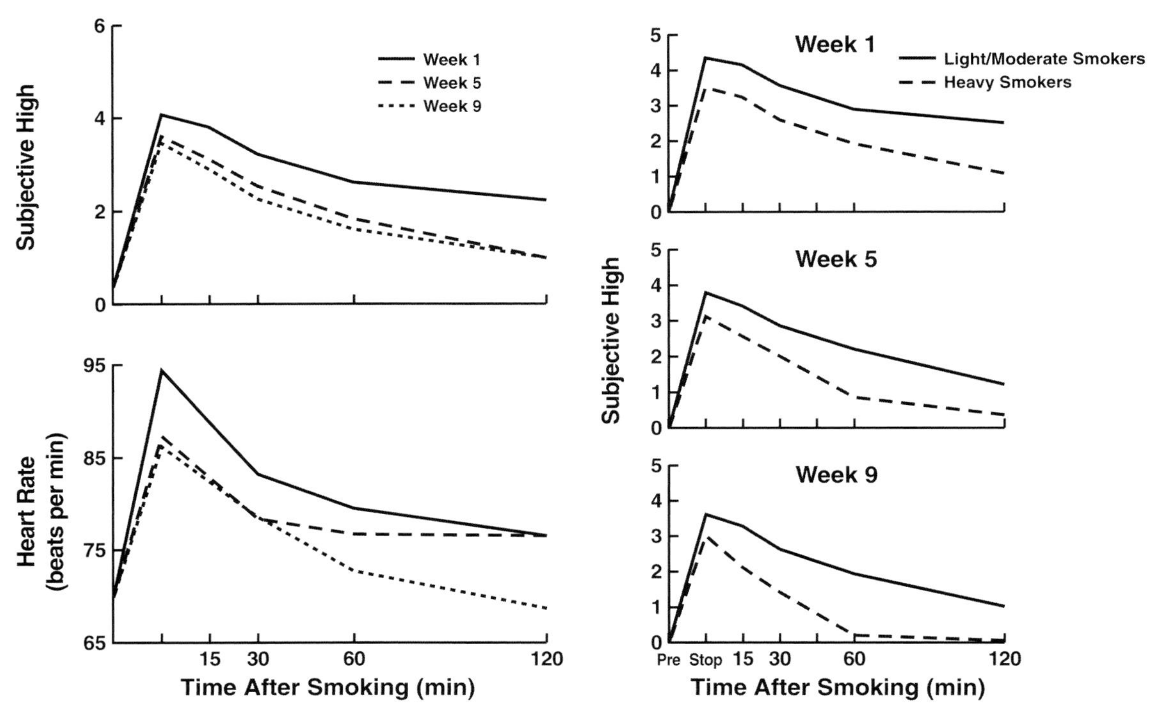

FIGURE 7.10 Tolerance to the subjective and physiological measures of marijuana intoxication. Left graphs: Mean heart rate and subjective high after smoking one marijuana cigarette after 1, 5, or 9 weeks of smoking at least one 900 mg marijuana cigarette per day. Right graphs: Comparison of light-to-moderate smokers to heavy smokers in ratings of subjective high. [Reproduced with permission from Nowlan and Cohen, 1977.]

1976; Georgotas and Zeidenberg, 1979; Mendelson *et al.*, 1984; Haney *et al.*, 1999a,b). Outpatient studies in adolescents and adults seeking treatment of cannabis dependence have shown that the majority of them reported histories of marijuana withdrawal (Stephens *et al.*, 1993; Budney *et al.*, 1998, 1999; Copeland *et al.*, 2001). Onset of withdrawal typically occurs within 1–3 days. Peak effects are experienced between days 2 and 6, and most symptoms last 4–14 days (Budney *et al.*, 2003) (**Fig. 7.11**). The long onset of Δ^9-THC withdrawal appears to be directly related to the long half-life and slow decline of blood Δ^9-THC levels (see *Pharmacokinetics* above).

Some clinicians have argued for the existence of a protracted abstinence state where the body is reconstituting to a normal or predrug state that can be quite prolonged, and may last up to 15–18 months (Zweben and O'Connell, 1992). Mild flu-like symptoms may occur a week or more later, and other subjective effects of protracted abstinence include cognitive deficits and sleep disturbances (Tennant, 1986). During this recovery, subjects report behaviors as subtle as being able to better sustain concentration when doing visualizations or meditations within a treatment session, engaging in more difficult reading material, and being less accident prone (Zweben and O'Connell, 1992).

In preclinical work, cannabinoid withdrawal syndromes have been described in both rats and mice. Rats show a variety of somatic withdrawal signs, including wet dog shakes, scratching, facial rubbing, ptosis, mastication, hunched posture, and ataxia (Kaymakcalan *et al.*, 1977; Aceto *et al.*, 1996). In mice exposed to chronic administration of Δ^9-THC (twice daily for 5 days of 20 mg/kg i.p.), administration of the CB_1 receptor antagonist SR14716A precipitated a robust withdrawal syndrome (Hutcheson *et al.*, 1998) (**Fig. 7.12**). Signs in order of prevalence included wet dog shakes, facial rubbing, ptosis, hunched posture, front paw tremor, piloerection, and ataxia. The signs of wet dog shakes, ptosis, and hunched posture are similar to the signs observed in rats (Aceto *et al.*, 1996). These signs also are common with opioid withdrawal. Spontaneous somatic withdrawal from Δ^9-THC has not been observed, although a mild somatic withdrawal syndrome has been seen during abstinence from the synthetic cannabinoid WIN 55,212-2 (Aceto *et al.*, 2001). Some motivational signs of withdrawal such as elevations in reward thresholds as measured by brain

TABLE 7.9 Proposed Withdrawal Symptom List for DSM Consideration

Symptom	Controlled studies reporting symptom	All studies reporting symptom
Common		
Decreased appetite/weight loss	6/6	11/11
Irritability	6/6	10/10
Nervousness/anxiety	5/6	6/6
Anger/aggression	3/4	8/8
Restlessness	3/4	9/10
Sleep difficulty/strange dreams	3/5	9/11
Less common/equivocal		
Depressed mood	3/5	5/8
Stomach ache/physical pain	3/5	4/7
Chills	1/4	4/7
Shakiness	1/4	3/6
Sweating	1/4	2/5

Reproduced with permission from Budney *et al.*, 2003.

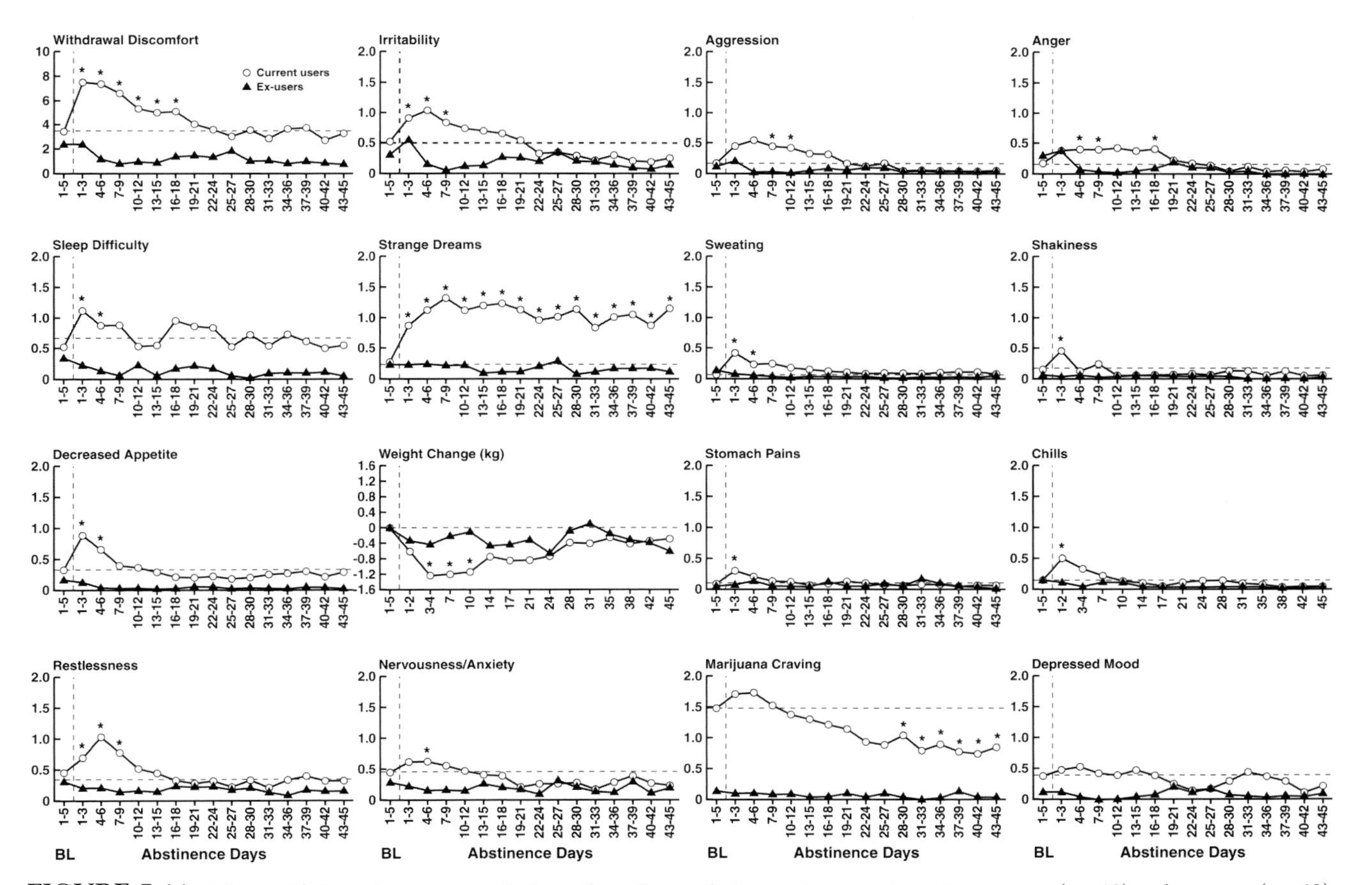

FIGURE 7.11 Mean withdrawal scores over 45 days of marijuana abstinence in current marijuana users ($n = 18$) and ex-users ($n = 12$). Ranges overall: 0–3. Range for Withdrawal Discomfort: 0–36. The value of the baseline (BL) data point reflects the mean of Days 1, 3, and 5. The dotted horizontal line represents the baseline mean score. Asterisks indicate significant difference between specific 3-day abstinence periods and baseline mean ($*p < 0.05$), or between body weight at specific abstinence days and baseline ($*p < 0.05$). [Reproduced with permission from Budney *et al.*, 2003.]

stimulation reward have been observed (Gardner and Vorel, 1998) (**Fig. 7.13**). Acute administration of Δ^9-THC itself can produce place aversions at certain doses in rats and mice (Parker and Gillies, 1995; McGregor *et al.*, 1996; Hutcheson *et al.*, 1998), but to date there are no reported studies of place aversion during spontaneous or precipitated cannabinoid withdrawal. However, administration of the CB_1 antagonist SR141716A in rats receiving long-term cannabinoid agonist treatment (HU-210) resulted in anxiety-like responses in the defensive withdrawal test (Rodriguez de Fonseca *et al.*, 1997).

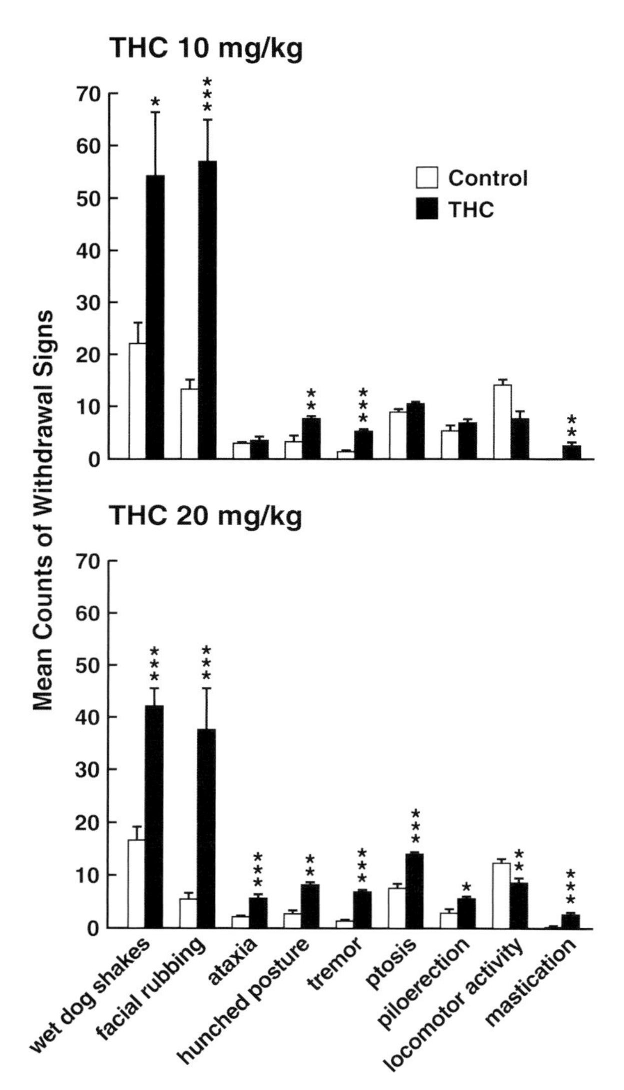

FIGURE 7.12 Effects of the CB_1 cannabinoid receptor antagonist SR141716A (10 mg/kg) after five days of twice-daily administration of either 10 mg/kg or 20 mg/kg Δ^9-THC in mice. Data are expressed as mean ± SEM incidents of counted and checked physical withdrawal signs observed during the 45 min immediately after SR141716A administration. $^*p < 0.05$, $^{**}p < 0.01$, $^{***}p < 0.001$ compared to vehicle. [Reproduced with permission from Hutcheson *et al.*, 1998.]

PATHOLOGY AND PSYCHOPATHOLOGY

Acute administration of cannabis produces, in most users, a pleasant experience associated with intoxication and the subjective high (see above). However, some individuals with existing psychopathology, and particularly naive individuals who have ingested cannabis unknowingly, can experience anxiety and panic reactions (Naditch, 1974; Noyes *et al.*, 1975a; Bloodworth, 1985). Sometimes such reactions can include dysphoria, paranoia, depersonalization, and psychosis (Binitie, 1975; Hall and Solowij, 1998; American Psychiatric Association, 2000; Johns, 2001).

There are a number of physiological effects in the cardiovascular system produced by cannabis, including an increase in heart rate that, in vulnerable individuals, may contribute to pathology. Δ^9-THC increases heart rate, slightly increases supine blood pressure, produces postural hypotension, and increases cardiac output. Tolerance develops to the increased heart rate and blood pressure (Benowitz and Jones, 1975; Nowlan and Cohen, 1977). While the cardiovascular effects of marijuana are not associated with serious health risks in young healthy users (although occasional cardiac events have been reported), marijuana smoking by people with cardiovascular disease poses health risks (Jones, 2002). Smoking of cannabis by patients with angina decreased performance during exercise and precipitated chest pains sooner compared to healthy subjects (Aronow and Cassidy, 1974).

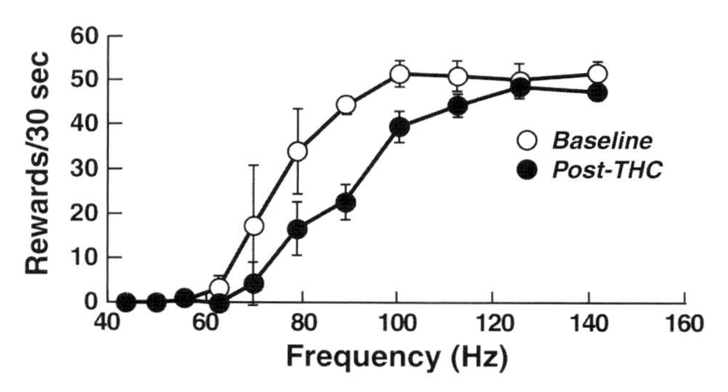

FIGURE 7.13 Diminished brain-stimulation reward during withdrawal from an acute 1.0 mg/kg dose of Δ^9-THC in rats. Diminished brain reward is experimentally equivalent to a right shift in the mean rate-frequency electrical brain stimulation reward functions of the medial forebrain bundle. Withdrawal from Δ^9-THC significantly shifted the reward function to the right. Although animals were tested at 16 different pulse frequencies from 25 to 141 Hz, only the central portion of each mean rate-frequency brain stimulation reward function is depicted. Rewards per 30 s were asymptotic for all animals below 50 Hz and above 120 Hz. [Reproduced with permission from Gardner and Vorel, 1998.]

The adverse effects on the respiratory tract of smoking cannabis are hypothesized to be similar to those of smoking tobacco cigarettes, but firm conclusions are limited by lack of animal models and concomitant tobacco use and a relatively short period of prevalence of marijuana smoking in Western society (30–35 years) (Van Hoozen and Cross, 1997). Common untoward respiratory effects reported by abusers are cough, dyspnea, sore throat, nasal congestion, and bronchitis (Van Hoozen and Cross, 1997). The long-term effects on the respiratory tract are likely to be similar to those of tobacco because of the similarities in smoke composition between cannabis and tobacco (**Table 7.10**). While the number of cannabis cigarettes smoked may be less than the number of tobacco cigarettes smoked in a chronic smoker, several characteristics of marijuana smoking are likely to increase the burden of tar and carbon monoxide. Marijuana cigarettes are not usually filtered, and smokers tend to smoke to the very end of the butt length which increases tar, carbon monoxide, and Δ^9-THC delivery to the lungs (Tashkin *et al.*, 1991). Also, marijuana smokers tend to inhale larger puff volumes, draw more deeply, and hold the smoke longer in their lungs (Wu *et al.*, 1988). In subjects with both tobacco and marijuana experience, smoking marijuana was associated with a five times greater increase in carboxyhemoglobin (Wu *et al.*, 1988). Marijuana smoking is associated with decrements in pulmonary function, but it is unknown to date if abnormalities in pulmonary fuction will progress to chronic obstructive pulmonary disease with continued use (Van Hoozen and Cross, 1997).

Cannabis use at intoxicating doses impairs psychomotor performance in any situation which requires perceptual, cognitive, and psychomotor functioning, including driving an automobile or flying an airplane. Mental and motor performance, including response speed, physical work capacity, fine hand–eye coordination, complex tracking, divided attention tasks, visual information processing, altered time sense, and impaired short-term memory, particularly in complex tasks, are dose-related, beginning with impairments after 4–16 puffs on a 3.55 per cent Δ^9-THC cigarette yielding 63–188 ng/ml (Heishman *et al.*, 1997) or after smoking a 1 g marijuana cigarette (2.0 or 3.5 per cent Δ^9-THC) (Kelly *et al.*, 1993) dose ranges (**Fig. 7.14**). Automobile driving or airplane flying ability using a computer simulator have shown dose–related deficits even after doses as low as 5–10 mg and up to 24 h after smoking (Janowsky *et al.*, 1976a,b; Janowsky *et al.*, 1976a,b; Leirer *et al.*, 1991; Brookoff *et al.*, 1994) (**Table 7.11**). From the perspective of behavioral pathology, marijuana, even at low (100 μg/kg smoked delivered) to moderate (200 μg/kg smoked delivered) doses, negatively affected driving performance in real traffic situations with reduced capacity to avoid collision if confronted with the sudden need for evasive action (Robb and O'Hanlon, 1999). The combination of moderate doses of alcohol with moderate doses of marijuana resulted in more dramatic decrements in performance (Robbe and O'Hanlon, 1999; Lamers and Ramaekers, 2000). The actual extent to which cannabis ingestion contributes to road accidents is controversial (Ashton, 1999), but in many countries cannabis is the most common drug reported to be detected in individuals involved in reckless driving or traffic accidents with or without alcohol. Significant percentages of impaired drivers or drivers involved in fatal accidents in the United Kingdom, Canada, Europe, New Zealand, and Australia have been reported (Ashton, 1999), with approximately 47 per cent of 1842 impaired drivers

TABLE 7.10 Comparison of Marijuana and Tobacco Smoke Constituents

Constituent	Marijuana	Tobacco
Whole smoke		
Burning rate	11.6 mm/min/g	5.7 mm/min/g
pH (3rd to 10th puffs)	6.56–6.58	6.14–6.02
Moisture	10.3%	11.1%
Particulate phase		
Total particulate (per puff)	1.6 mg	2.4 mg
Phenol	76.8 μg	39.0 μg
o-Cresol	17.9 μg	24.0 μg
m-Cresol + p-Cresol	54.4 μg	65.0 μg
2,4-Dimethylphenol + 2,5-Dimethylphenol	6.8 μg	14.4 μg
Naphthalene	3.0 mg	1.2 mg
Benz(a)anthracene	75 μg	43 μg
Benz(a)pyrene	31 μg	22 μg
Nicotine	—	2.85 mg
Δ^9-Tetrahydrocannabinol	820 μg	—
Cannabinol	400 μg	—
Cannabidiol	190 μg	—
Gas phase		
Carbon monoxide (per cigarette)	2600 ppm	4100 ppm
Ammonia	228 μg	198 μg
Hydrogen cyanide	532 μg	498 μg
Isoprene	83 μg	310 μg
Acetaldehyde	1.20 mg	0.98 mg
Acetone	443 μg	578 μg
Acrolein	92 μg	85 μg
Acetonitrile	132 μg	123 μg
Benzene	76 μg	67 μg
Toluene	112 μg	108 μg
Dimethylnitrosamine	75 μg	84 μg
Methylnitrosamine	27 μg	30 μg

Reproduced with permission from Huber *et al.*, 1991.

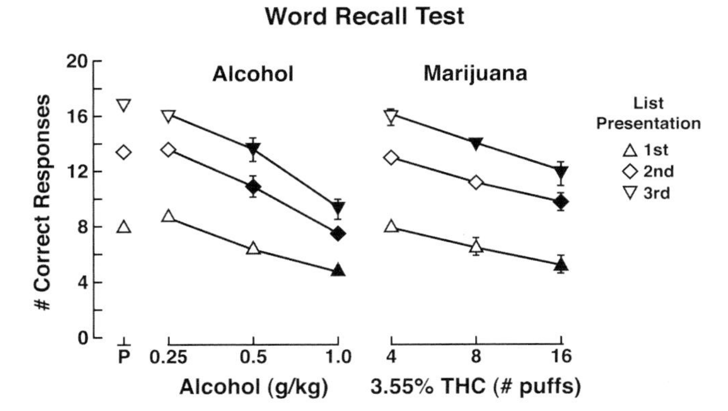

FIGURE 7.14 Number of correct responses on the word recall test as a function of list presentation order and placebo, alcohol (0.25, 0.5, 1.0 g/kg) and marijuana (3.55 per cent Δ^9-THC, equivalent to 34 mg; Huestis *et al.*, 1992). Each data point represents the mean ± SEM of five measurements made at 0, 30, 60, 90, and 120 min post-dosing for five subjects. If the SEM is not visible, it was less than the radius of the symbol. Filled symbols indicate significant difference from placebo ($p < 0.05$; Tukey's HSD test). [Reproduced with permission from Heishman *et al.*, 1997.]

(driving while intoxicated) being suspected of cannabis use by drug recognition experts, with most of the opinions confirmed by chemical tests (Preusser *et al.*, 1992). About 33 per cent tested positive for marijuana in a group not impaired by alcohol (Brookoff *et al.*, 1994). These results suggested that cannabis with or without other sedative hypnotics including alcohol and benzodiazepines can contribute to automobile accidents.

Daily or chronic use of cannabis at intoxicating doses can lead to chronic impairments in social and occupational functioning, including ineffectiveness in school, sports, work, and learning to initiate and sustain healthy relationships (Gruber and Pope, 2002). The state of chronic intoxication increases risk-taking

TABLE 7.11 Effects of Cannabis Which Impair Driving and Piloting Skills

Effects
Slowed complex reaction time
Poor detection of peripheral light stimuli
Poor oculomotor tracking
Space and time distortion
Impaired coordination
Brake and accelerator errors
Poor speed control
Poor judgment
Increased risks in overtaking
Impaired attention, especially for divided attention tasks
Impaired short-term memory
Additive effects with alcohol and other drugs

Reproduced with permission from Ashton, 1999.

behaviors as a result of disinhibition. These can include participating in unprotected sex, exposing a fetus to cannabis, driving while intoxicated, or riding with an intoxicated driver (Brookoff *et al.*, 1994; Graves and Leigh, 1995; National Highway Traffic Safety Administration, 2000).

Perhaps even more striking is the lack of motivation, direction, ambition, or even ability to hold a coherent conversation. This group of effects was termed 'amotivational syndrome' (McGlothlin and West, 1968; Kupfer *et al.*, 1973; Cohen, 1980a,b; Hendin and Haas, 1985; Reilly *et al.*, 1998; Gruber and Pope, 2002). Characterized by diminished goal-directed behavior, apathy, and an inability to master new problems, the syndrome often has been used to explain poor school performance in adolescents, personality deterioration, and a general decrease in function. Most discussions of this topic recognize this syndrome as a state of chronic intoxication rather than a neurotoxic effect of marijuana ingestion. The 'cannabis syndrome' or 'amotivational syndrome' may reflect an interaction between the pharmacological effects of the drug and psychological and social factors associated with adolescence, including self-medication of existing psychopathology (Millman and Sbriglio, 1986) (see *Case Studies* in appendix). The cannabis syndrome is more likely to occur in high-dose compulsive users than controlled low-dose users, and often remits with cessation of use.

Acute cannabis intoxication can lead to a well-documented acute transient psychotic episode in some individuals (Mayor's Committee on Marijuana, 1944; Isbell *et al.*, 1967a,b; Jones, 1971; Leweke *et al.*, 1999; D'Souza *et al.*, 2004; Hall and Degenhardt, 2004). Such psychotic episodes are characterized by delusions, loosening of associations, and marked illusions. Cannabis also can produce a short-term exacerbation or recurrence of pre-existing psychotic symptoms (Mathers and Ghodse, 1992; D'Souza *et al.*, 2004). Although controversial, more and more evidence is accumulating that cannabis can contribute to the causes of a functional psychotic illness or schizophrenia (Arseneault *et al.*, 2004; Hall and Degenhardt, 2004). While causality is difficult to define, much less prove, several findings point to a causal relationship (Thornicroft, 1990; Hall and Degenhardt, 2004). Cross-sectional national surveys have found that rates of cannabis use are approximately 2 times higher among subjects diagnosed with schizophrenia than among the general population (Grech *et al.*, 1998; van Os *et al.*, 2002). Daily cannabis use has been shown to double the risk of reporting psychotic symptoms (Tien and Anthony, 1990). In a series of prospective studies in Sweden, The Netherlands, and New Zealand where cannabis use in adolescence was examined relative to later adult psychotic symptoms, evidence was presented to support the hypothesis that 'cannabis use is an independent risk factor for the emergence of psychosis in psychosis-free persons' (van Os *et al.*, 2002). Cannabis use in adolescence most probably preceded schizophrenia (Andreasson *et al.*, 1987; Arseneault *et al.*, 2002, 2004; van Os *et al.*, 2002; Zammit *et al.*, 2002; Fergusson *et al.*, 2003) (**Table 7.12**). Such an association between cannabis use in adolescence and adult psychosis appears to persist even after controlling for numerous social, gender, age, and ethnic group factors (Arseneault *et al.*, 2004). Overall, cannabis use conveys a two-fold risk of later schizophrenia or schizophreniform disorder, and calculations indicate that without cannabis use there would be 8 per cent less schizophrenia in the general population (Arseneault *et al.*, 2004).

BEHAVIORAL MECHANISM OF ACTION

Cannabis and its active ingredient Δ^9-THC have behavioral effects that cross between two drug classes: sedative-hypnotics and psychedelics. Sedative hypnotic drugs, as discussed in the *Alcohol* chapter, 'disinhibited' behavior (i.e., individuals show a release of inhibitions in situations where normally they might be socially constrained). Alcohol is widely used as a 'social lubricant' at social events to promote conversation and social interaction. The disinhibition associated with alcohol is often mistaken for a psychostimulant effect, and alcohol is often labeled as a stimulant at lower doses as a result. In contrast, the disinhibition produced by cannabis is a more cognitive disinhibition with an overlay of a pronounced decreased motivation to exert energy that significantly limits any actual disinhibited behavior that would resemble the stimulant-like effect of alcohol. Indeed, the increased psychedelic-like perceptual effects, decreased motivation, and impaired cognitive functioning associated with increased doses of Δ^9-THC, which lead to a unique behavioral mechanism of action that begs the question of the functional role of endogenous cannabinoids.

One hypothesis is that endogenous cannabinoids as neuromodulators have a functional role in the brain to temper excessive arousal and excessive cognitive function but at the same time, or even as a result, increase responses to novelty and as such, increase perceptual function and facilitate hedonic processes. Hypotheses regarding the role of cannabis range from marijuana inducing variability of information processing by higher brain structures that leads to retardation of habituation of classical reinforcers and induces novel experiences (Feeney, 1976) to cannabinoids amplifying hedonic aspects of feeding (Harrold and Williams, 2003). Thus, the behavioral mechanism of action of cannabinoids could be proposed as a perceptual disinhibition of both external and internal cues/states without a motivational disinhibition at least as would be manifest by psychomotor activation. This *perceptual disinhibition* can be pleasant in the appropriate external context of stimuli of positive valence that can range from the external sensory modalities (visual and auditory and touch) to the taste modality (sweet or particularly palatable food), or even unpleasant in situations with stimuli of negative emotional valence.

NEUROBIOLOGICAL MECHANISM—NEUROCIRCUITRY

Evidence for acute reinforcing effects of Δ^9-THC comes from studies of drug discrimination, brain stimulation reward, place preference, and intravenous self-administration. In drug discrimination, rats and monkeys will discriminate Δ^9-THC, and a correlation can be established between the drugs that generalize to Δ^9-THC and drugs that bind to the cannabinoid receptors (Balster and Prescott, 1992; Gold *et al.*, 1992). Cannabinoid agonists substitute for Δ^9-THC in drug discrimination studies (Balster and Prescott, 1992; Wiley *et al.*, 1993, 1995a,b; Barrett *et al.*, 1995), and these effects can be blocked by cannabinoid receptor antagonists (Wiley *et al.*, 1995b). Two classes of drugs have been identified to substitute for Δ^9-THC: bicyclic compounds (Gold *et al.*, 1992) and aminoalkylindoles (Compton *et al.*, 1992; Gold *et al.*, 1992) (**Fig. 7.15**). No cross substitution has been observed with a wide variety of different neuropharmacological agents, including opioids, anticonvulsants, antipsychotics, serotonergic drugs, psychostimulants, and psychedelics (Balster and Prescott, 1992; Wiley and Martin, 1999). Partial substitution has been observed with diazepam, phencyclidine, and methylenedioxymethamphetamine (Barrett *et al.*, 1995). Also, only very high doses of anandamide (a CB_1 receptor agonist) show substitution (Wiley *et al.*, 1998), whereas other more stable analogs of anandamide show substitution but only with low doses of Δ^9-THC (Jarbe *et al.*, 1998, 2000).

Most drugs of abuse have acute reinforcing effects as measured by brain stimulation reward, place preference, and self-administration. Reward thresholds as measured by intracranial self-stimulation are decreased by Δ^9-THC administration in rats upon acute administration (Gardner and Lowinson, 1991; Lepore *et al.*,

TABLE 7.12 Epidemiological Studies on Cannabis Use and Schizophrenia

Study	Design (n)	Year of enrollment	Gender	Number of participants	Age of cannabis users	Diagnostic criteria	Definition of cannabis use	Outcome (n)	Adjusted risk (odds ratio $p < 0.05$)	Reference
Swedish conscript cohort	Conscript cohort (50 000)	1969–1970	Male	45 570	18	ICD-8	Used cannabis > 50 times	Inpatient admission for schizophrenia (*n* = 246; 0.5%)	2.3 (1.0–5.3)	Andreasson *et al.*, 1987
	Conscript cohort (50 000)	1969–1970	Male	50 053	18	ICD-8/9	Used cannabis > 50 times	Hospital admission for schizophrenia (*n* = 362; 0.7%)	3.1 (1.7–5.5)	Zammit *et al.*, 2002
Netherland Mental Health Survey and Incidence Study	Population-based (7076)	1996	Male/ Female	4104	18–64	BPRS	Cannabis use at baseline (age 16–17)	Any level of psychotic symptoms (*n* = 33; 0.9%)	2.76 (1.2–6.5)	van Os *et al.*, 2002
								Pathology level of psychotic symptoms (*n* = 10; 0.3%)	24.17 (5.44–107.5)	
								Need for care (*n* = 7; 0.2%)	12.01 (2.4–64.3)	
Christchurch Study	Birth cohort (1265)	1977	Male/ Female	1011	21	SCL-90	DSM-IV cannabis dependent at age 21	Psychotic symptoms (NA)	1.8 (1.2–2.6)	Fergusson *et al.*, 2003
Dunedin Study	Birth cohort (1037)	1972–1973	Male/ Female	759	15–18	DSM-IV	Users by age 15 and continued at age 18	Schizophreniform disorder symptoms	$\beta = 6.56$ (4.8–8.34)	Arseneault *et al.*, 2002
								Schizophreniform disorder diagnosis (*n* = 25; 3.3%)	3.12 (0.7–13.3)	
									Overall Risk for Psychosis: 2.34 (1.69–2.96)	

ICD-8, *International Statistical Classification of Diseases and Related Health Problems*, 8th and 9th revisions.
BPRS, Brief Psychiatric Rating Scale.
SCL-90, 90-item Symptom Checklist.
DSM-IV, *Diagnostic and Statistical Manual of Mental Disorders*, 4th edition.
NA, not available.
Note: The adjusted odds ratios included in the calculation of the *Overall Risk for Psychosis* are underlined.
[Adapted from Arseneault *et al.*, 2004.]

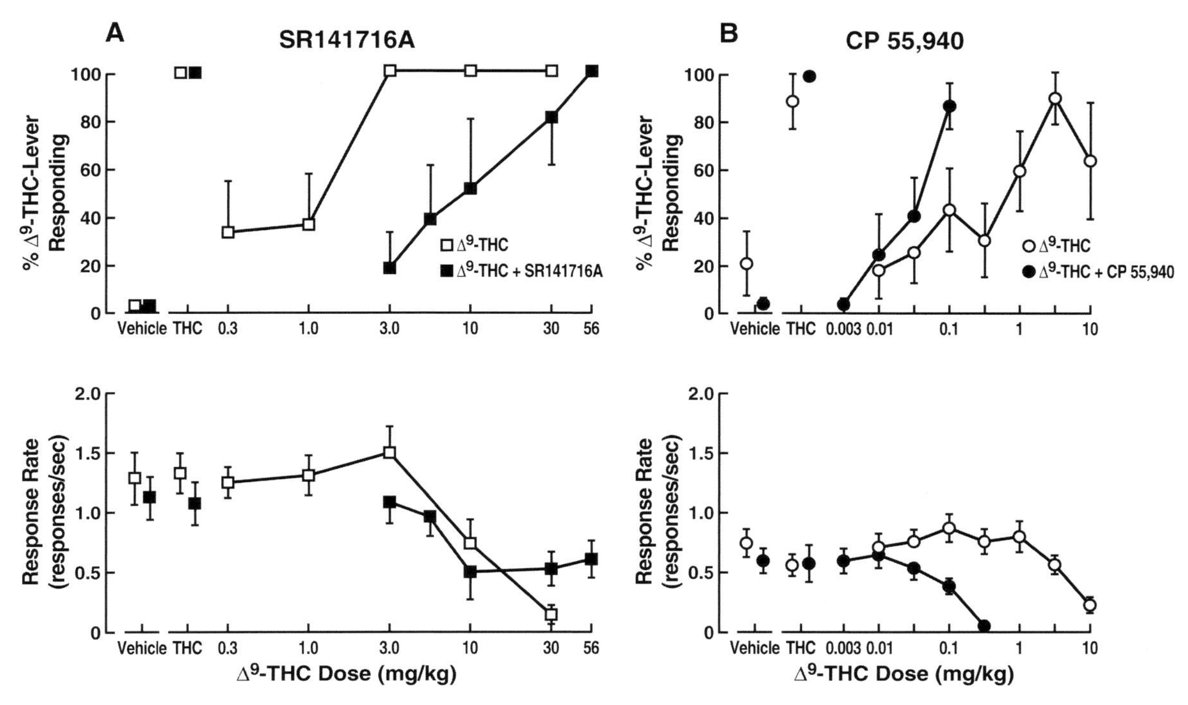

FIGURE 7.15 Percentage of Δ^9-THC lever responding and response rate in rats trained to discriminate Δ^9-THC (3.0 mg/kg, i.p.) from vehicle as a function of dose of Δ^9-THC and **(A)** Δ^9-THC + SR141716A (a CB_1 receptor antagonist; 1 mg/kg; $n = 4$–6) and **(B)** CP 55,940 (a CB_1 receptor agonist; $n = 6$–8). Corresponding vehicle and Δ^9-THC control tests were conducted before each dose-effect curve determination. Note that in Panel B only 4 rats were included for the highest dose of Δ^9-THC in the mean of percentage Δ^9-THC lever responding. [Panel A reproduced with permission from Wiley *et al.*, 1995b. Panel B reproduced with permission from Gold *et al.*, 1992.]

1996; Gardner *et al.*, 1988). This decrease in thresholds is similar to that observed with all other major drugs of abuse and has been measured with both rate-frequency and reward-threshold paradigms (Gardner *et al.*, 1988). This reward enhancement is blocked by administration of the opioid antagonist naloxone (Gardner, 2002) (**Fig. 7.16**).

Δ^9-THC produces both place aversions and place preferences in rats and mice depending on the dose and experience of the animals. Place aversions have been observed with acute administration of moderate to high doses (Parker and Gillies, 1995; McGregor *et al.*, 1996; Sanudo-Pena *et al.*, 1997; Chaperon *et al.*, 1998). Place preferences have been observed in rats with low doses (Lepore *et al.*, 1995) (**Fig. 7.17**) and mice by pre-exposure to Δ^9-THC (Valjent and Maldonado, 2000). The potent synthetic cannabinoid CP 55,940 produced a significant place preference that was reversed by administration of the opioid antagonist naloxone (Braida *et al.*, 2001a).

Early studies were marked by failures to observe reliable cannabinoid self-administration in laboratory animals, but studies with synthetic cannabinoid agonists and low doses of Δ^9-THC have shown robust intravenous and intracerebroventricular self-administration (Tanda and Goldberg, 2003). Intravenous self-administration of a synthetic Δ^9-THC agonist, WIN 55,212-2, has been observed in mice (Martellotta *et al.*, 1998; Ledent *et al.*, 1999) and rats (Fattore *et al.*, 2001), and intracerebroventricular self-administration of the cannabinoid agonist CP 55,940 in rats has been reported (Braida *et al.*, 2001b).

Intravenous self-administration of Δ^9-THC has been observed in squirrel monkeys with previous drug experience (Tanda *et al.*, 2000) and naive animals (Justinova *et al.*, 2003) (**Fig. 7.18**). Low doses of Δ^9-THC were employed, and the Δ^9-THC was dissolved in a vehicle containing sufficient emulsifier Tween-80 (0.4–1.0 per cent) and ethanol (0.4–1.0 per cent) to produce a clear solution that could be injected rapidly (0.2 ml/200 ms). Squirrel monkeys with unlimited access to food and water were allowed a wide range of doses of Δ^9-THC (1.0–8.0 μg/kg/injection) on a fixed-ratio 10, timeout 60 s schedule (10 lever presses were required to produce each 200 ms Δ^9-THC infusion with a 60 s timeout period after each injection) (Tanda *et al.*, 2000;

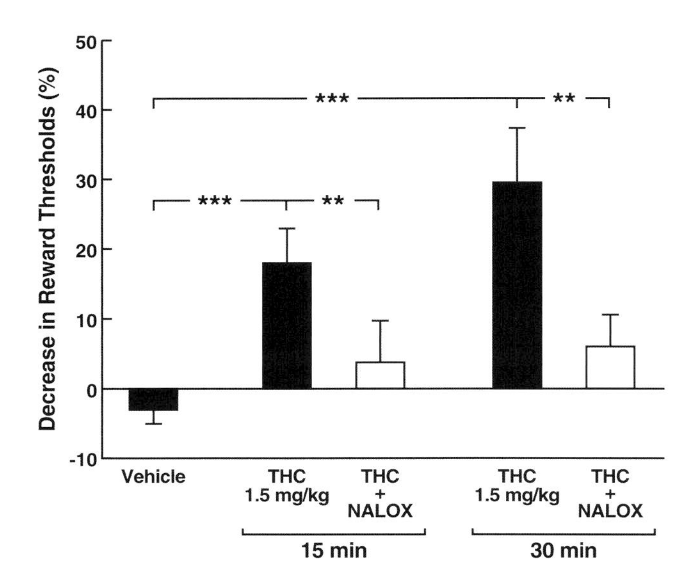

FIGURE 7.16 Cannabinoid-induced enhancement of brain reward in rats (as indicated by a decrease in intracranial self-stimulation reward thresholds), and attenuation of cannabinoid-induced brain-reward enhancement by the opioid antagonist naloxone. THC, Δ^9-tetrahydrocannabinol; NALOX, naloxone. Asterisks indicate statistical significance levels for the comparisons indicated (** $p < 0.01$; *** $p < 0.005$; **** $p < 0.001$). Data from Gardner *et al.*, 1988, 1989. [Reproduced with permission from Gardner, 2002.]

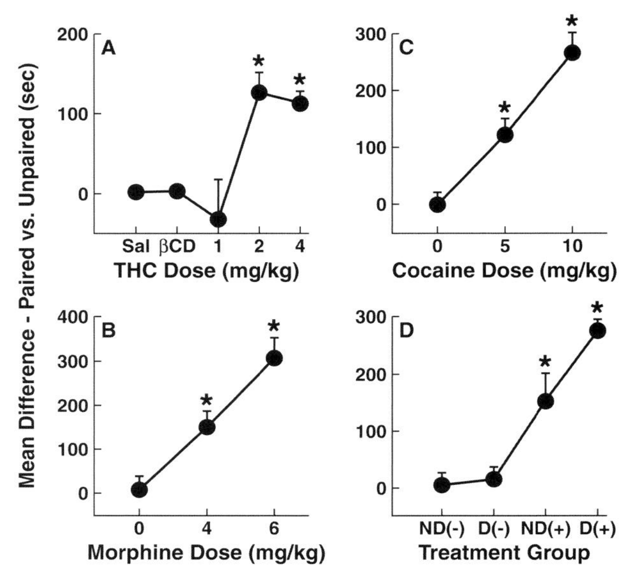

FIGURE 7.17 Means (± SEM) for conditioned place preference scores following Δ^9-THC, morphine, cocaine, or food place conditioning in rats. Panel A shows the abrupt step-up dose-response relationship observed following 1.0, 2.0, or 4.0 mg/kg Δ^9-THC administration. SAL denotes saline administration. β-CD denotes β-cyclodextrin administration (the vehicle for Δ^9-THC). Panels B and C show the dose-response relationships observed following 4.0 and 6.0 mg/kg pairings with morphine, or with 5.0 and 10.0 mg/kg cocaine. Panel D shows the effects of food deprivation on the shift in preference scores following food reward conditioning. A significant dose-effect relationship was observed with increases in deprivation level. Asterisks (*) indicate statistically significant differences between means ($p < 0.05$). [Reproduced with permission from Lepore *et al.*, 1995.]

Tanda and Goldberg, 2003). The self-administration of Δ^9-THC and synthetic cannabinoid was blocked by the CB_1 antagonist SR141716A (Martelotta *et al.*, 1998; Tanda *et al.*, 2000; Braida *et al.*, 2001b).

The acute reinforcing effects of Δ^9-THC have been hypothesized to involve activation of the mesocorticolimbic dopamine system (Ng Cheong Ton *et al.*, 1988; Malone and Taylor, 1999; Chen *et al.*, 1990a,b, 1991). Δ^9-THC-induced facilitation of dopamine has been observed with both *in vivo* microdialysis and *in vivo* voltammetry and is calcium-dependent and naloxone-reversible (Ng Cheong Ton *et al.*, 1988; Chen *et al.*, 1990b; Gardner *et al.*, 1990a; Tanda *et al.*, 1997). Δ^9-THC selectively increases the release of dopamine in the shell of the nucleus accumbens (Tanda *et al.*, 1997), an effect also observed with all major drugs of abuse (**Fig. 7.19**). Similar increases in extracellular dopamine in the nucleus accumbens have been observed with the synthetic cannabinoid agonist WIN 55,212-2 (Tanda *et al.*, 1997).

Another potential neuropharmacological mechanism for the acute reinforcing effects of Δ^9-THC is the release of endogenous opioid peptides. Intravenous self-administration of Δ^9-THC and intracerebroventricular self-administration of CP 55,940 are blocked by administration of the opioid antagonist naltrexone, suggesting a role for opioid peptides in the reinforcing effects of Δ^9-THC (Braida *et al.*, 2001b; Justinova *et al.*, 2004) (**Fig. 7.20**). These results are consistent with cannabinoids increasing the synthesis and release of endogenous opioid peptides (Manzanares *et al.*, 1998; Valverde *et al.*, 2001). Δ^9-THC has been shown to increase β-endorphin levels in the ventral tegmental area (Solinas *et al.*, 2003). Another potential mechanism is that both cannabinoid and opioid receptors are colocalized on medium spiny γ-aminobutyric acid (GABA) efferent neurons in the nucleus accumbens. Stimulation of either cannabinoid or opioid receptors inhibit GABAergic neurotransmission (Hoffman and Lupica, 2001). Thus, as with other drugs of abuse, the sites of action of Δ^9-THC may involve both actions at the ventral tegmental area and the nucleus accumbens, and possibly the extended amygdala (**Fig. 7.21**).

Δ^9-THC produces a reliable spontaneous withdrawal syndrome in humans and a robust precipitated withdrawal syndrome in animals (see above). The most characteristic somatic signs in rodents are a combination of opioid and sedative hypnotic signs such as wet

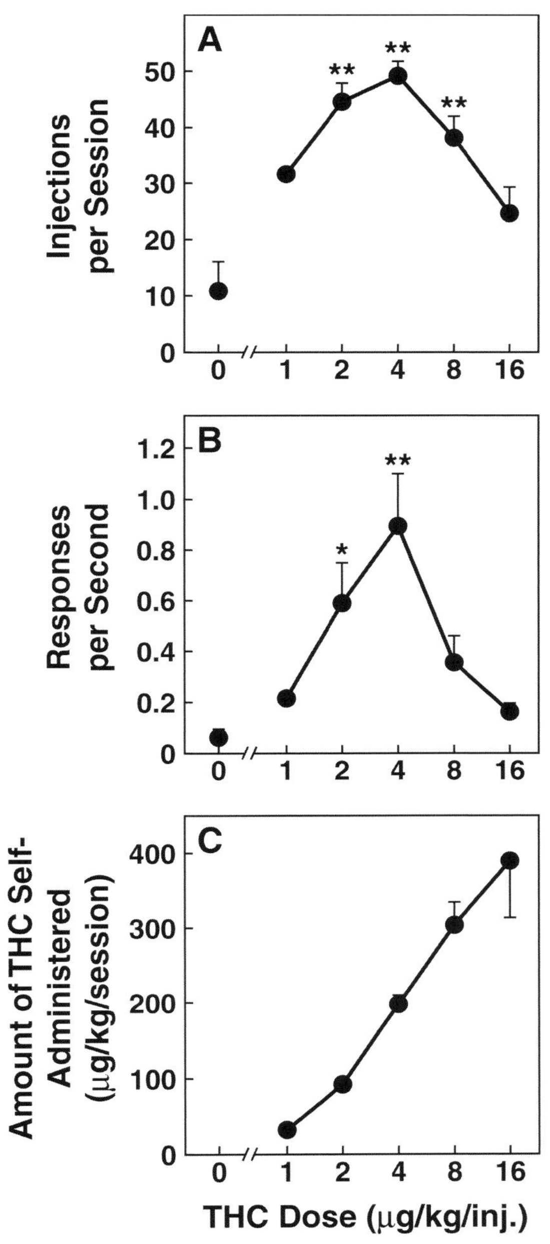

FIGURE 7.18 Δ^9-THC dose–response curves in squirrel monkeys with no history of exposure to other drugs ($n = 3$). **(A)** Numbers of injections per session. **(B)** Overall rates of responding in the presence of a green light signaling Δ^9-THC availability. **(C)** Total Δ^9-THC intake per session. Each panel is presented as a function of injection dose of Δ^9-THC. Each symbol represents the mean (±SEM) of the last three sessions under each Δ^9-THC injection dose condition and under a vehicle condition from three monkeys, with the exception of the values for the 1 μg/kg per injection dose, which represent means from two monkeys. Asterisks (*) indicate statistically significant differences from vehicle conditions (*$p < 0.05$, **$p < 0.01$, compared to the vehicle conditions after significant one-way ANOVA for repeated measures main effect, Dunnett's test). [Reproduced with permission from Justinova *et al.*, 2003.]

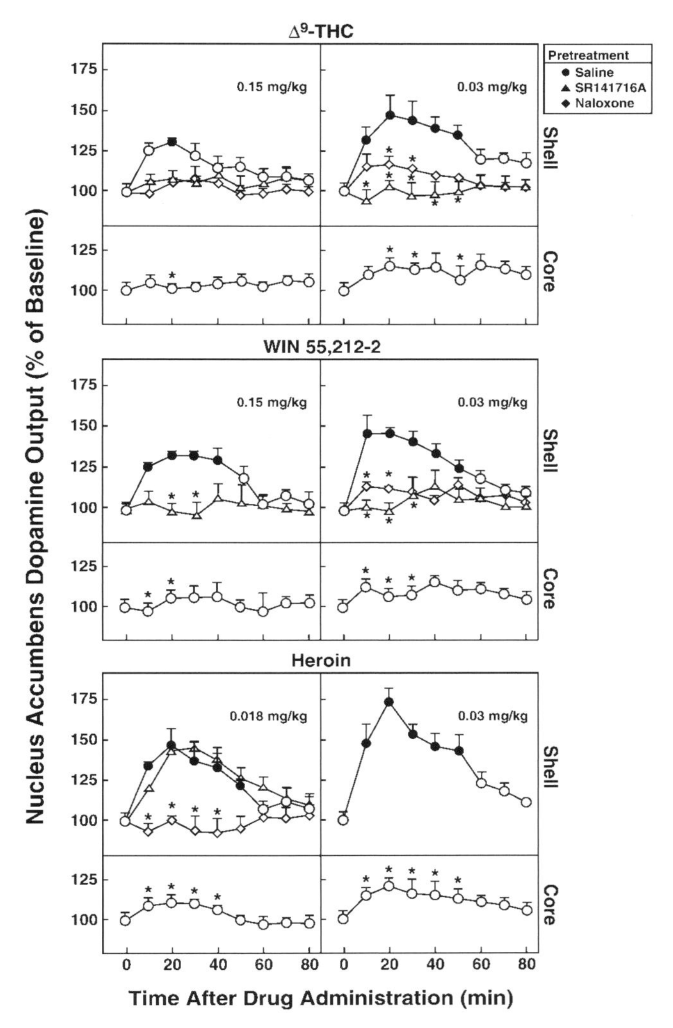

FIGURE 7.19 Effect of intravenous Δ^9-THC, WIN 55212-2, and heroin on dialysate dopamine in the shell (upper panels) and core (lower panels) of the nucleus accumbens. Rats were pretreated with saline (circles), SR141716A (triangles; 1 mg/kg, s.c.), or naloxone (diamonds; 0.1 mg/kg, i.p.). Results are means ± SEM of the amount of dopamine in 10 min dialysate samples, expressed as a percentage of baseline values. Solid symbols: $p < 0.05$ compared to baseline values. *$p < 0.05$ compared to the corresponding value obtained in the shell of saline-pretreated controls. [Reproduced with permission from Tanda *et al.*, 1997.]

dog shakes, head shakes, front paw and body tremor, ptosis, piloerection, and grooming behavior. However, precipitated cannabinoid withdrawal also produces some evidence of motivational withdrawal. Precipitated Δ^9-THC withdrawal disrupts operant behavior in rats (Beardsley and Martin, 2000) and produces anxiety-like effects in animal models of anxiety (Rodriguez de Fonseca *et al.*, 1997). Perhaps more compelling, spontaneous withdrawal from an acute injection of Δ^9-THC can produce an elevation in brain reward thresholds (Gardner and Vorel, 1998) (**Fig. 7.13**).

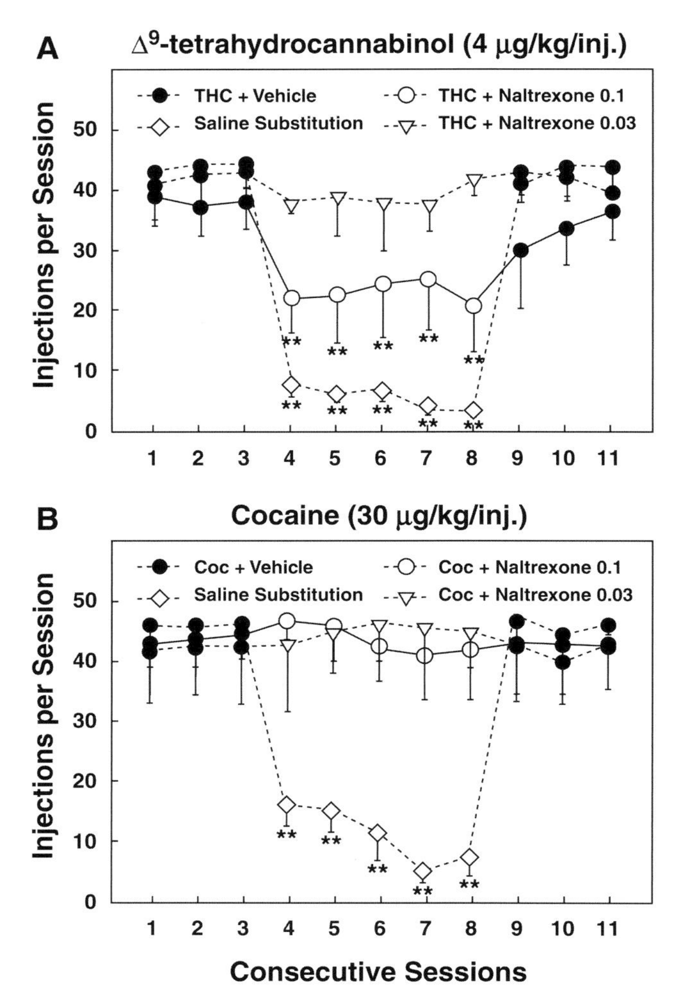

FIGURE 7.20 Effects of 0.03 and 0.1 mg/kg naltrexone on responding maintained by **(A)** Δ9-THC and **(B)** cocaine (Coc) in monkeys over consecutive sessions and extinction of self-administration behavior by substitution of saline injections for injections of (A) Δ9-THC or (B) cocaine. Numbers are shown of injections per session during Δ9-THC (4 μg/kg/inj) and cocaine (30 μg/kg/inj) self-administration sessions after pretreatment with vehicle (sessions 1–3 and 9–11) or naltrexone (sessions 4–8). Numbers also are shown of injections per session during self-administration sessions when saline was substituted for Δ9-THC or cocaine (sessions 4-8). Symbols represent the means (± SEM) of injections per session from four (Δ9-THC) and three (cocaine) monkeys. $^{**}p < 0.01$, *post hoc* comparisons with the last Δ9-THC session before naltrexone pretreatment or saline substitution (session 3) after significant one-way ANOVA for repeated measures main effect, Dunnett's test. [Reproduced with permission from Justinova *et al.*, 2004.]

A neuropharmacological mechanism that may be involved in the motivational withdrawal syndrome is decreased activity in the mesolimbic dopamine system and activation of brain and pituitary adrenal corticotropin-releasing factor (CRF) systems in the extended amygdala. Precipitated withdrawal from chronic cannabinoids decreases the firing of ventral tegmental area dopamine neurons (Diana *et al.*, 1998a,b) and decreases in extracellular dopamine levels in the nucleus accumbens (Tanda *et al.*, 1999) (**Fig. 7.22**). The anxiogenic-like effects of precipitated Δ9-THC withdrawal are blocked by a CRF antagonist, and precipitated withdrawal increases extracellular levels of CRF in the central nucleus of the amygdala (Rodriguez de Fonseca *et al.*, 1997) (**Fig. 7.23**).

Significant interactions between cannabinoid and opioid reinforcement and dependence provided a basis for investigating the role of endogenous cannabinoid systems in drug dependence in general. Administration of the cannabinoid CB_1 receptor antagonist SR141716A blocked morphine self-administration in mice and heroin self-administration in rats (Navarro *et al.*, 1998). Acute administration of SR141716A also blocked the expression of morphine-induced place preference (Navarro *et al.*, 1998). SR141716A precipitated both somatic and motivational opioid withdrawal in morphine-dependent rats (Navarro *et al.*, 1998). However, this interaction is bidirectional because the opioid antagonist naloxone induced partial cannabinoid withdrawal in rats treated chronically with the synthetic CB_1 agonist HU-210 (Kaymakcalan *et al.*, 1977; Navarro *et al.*, 1998) (**Fig. 7.24**). Precipitated Δ9-THC withdrawal is decreased in preproenkephalin knockout mice dependent on Δ9-THC (Valverde *et al.*, 2000b) but not altered in prodynorphin knockout mice (Zimmer *et al.*, 2001). This appears to be an interaction with μ opioid receptors in that μ knockout mice chronically treated with Δ9-THC also showed a blunted Δ9-THC-precipitated withdrawal (Lichtman *et al.*, 2001). Together, these results suggest that endogenous opioid peptides are important for the withdrawal syndrome associated with Δ9-THC and that there is some endogenous cannabinoid tone that contributes to opioid dependence.

The discovery of endogenous cannabinoid substances in the brain has a rich history that began with the synthesis and radiolabeling of potent bicyclic cannabinoids such as CP 55,940 (Melvin and Johnson, 1987) that allowed identification and characterization of a cannabinoid receptor in rat brain membranes (Devane *et al.*, 1988). Data revealed a single binding site that was saturable and reversible, and the pharmacological potency of cannabinoids correlated well with their affinity for the cannabinoid binding site (Compton *et al.*, 1993) (**Fig. 7.25**). The cannabinoid receptor distribution as determined with autoradiography in brain sections using radiolabeled [^{3}H]CP 55,940 showed the most dense binding in the basal ganglia and the cerebellum with intermediate levels in the hippocampus, amygdala, and cortex. Low levels of

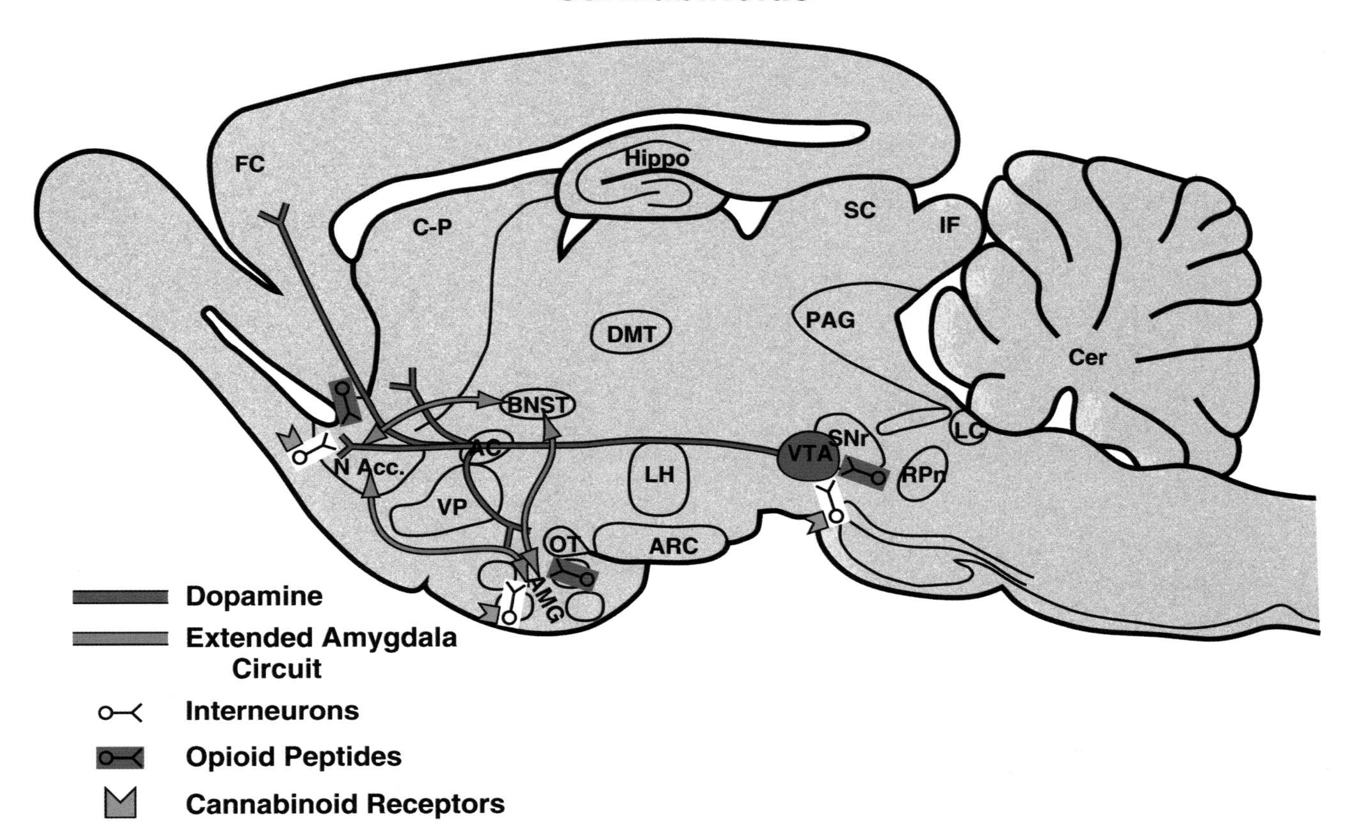

FIGURE 7.21 Sagittal section through a representative rodent brain illustrating the pathways and receptor systems implicated in the acute reinforcing actions of cannabinoids. Cannabinoids activate cannabinoid CB_1 receptors in the ventral tegmental area, nucleus accumbens, and amygdala via direct actions on interneurons. Cannabinoids facilitate the release of dopamine in the nucleus accumbens via an action either in the ventral tegmental area or the nucleus accumbens, but also are hypothesized to activate elements independent of the dopamine system. Endogenous cannabinoids may interact with postsynaptic elements in the nucleus accumbens involving dopamine and/or opioid peptide systems. The blue arrows represent the interactions within the extended amygdala system hypothesized to have a key role in cannabinoid reinforcement. AC, anterior commissure; AMG, amygdala; ARC, arcuate nucleus; BNST, bed nucleus of the stria terminalis; Cer, cerebellum; C-P, caudate-putamen; DMT, dorsomedial thalamus; FC, frontal cortex; Hippo, hippocampus; IF, inferior colliculus; LC, locus coeruleus; LH, lateral hypothalamus; N Acc., nucleus accumbens; OT, olfactory tract; PAG, periaqueductal gray; RPn, reticular pontine nucleus; SC, superior colliculus; SNr, substantia nigra pars reticulata; VP, ventral pallidum; VTA, ventral tegmental area.

cannabinoid receptors were found in brainstem areas corresponding to the low levels of lethality of cannabinoids (Herkenham *et al.*, 1990; Freund *et al.*, 2003) (**Fig. 7.26**).

The identification of a brain binding site and the early work showing that cannabinoids inhibited adenylate cyclase led to the hypothesis that the cannabinoid receptor was G-protein-linked (Howlett and Fleming, 1984). The serendipitous cloning of the cannabinoid receptor isolated from a rat brain library of clones that had homology with other G-protein receptors confirmed this hypothesis (Matsuda *et al.*, 1990). The heretofore unidentified clone did not bind to traditional agonists of G-proteins, but the mRNA distribution of the receptor clone paralleled that of the distribution of the cannabinoid receptor. Chinese hamster ovary cells transfected with the cannabinoid clone were cannabinoid-responsive, and Δ^9-THC and synthetic analogs dose-dependently inhibited forskolin-stimulated accumulation of cyclic adenosine

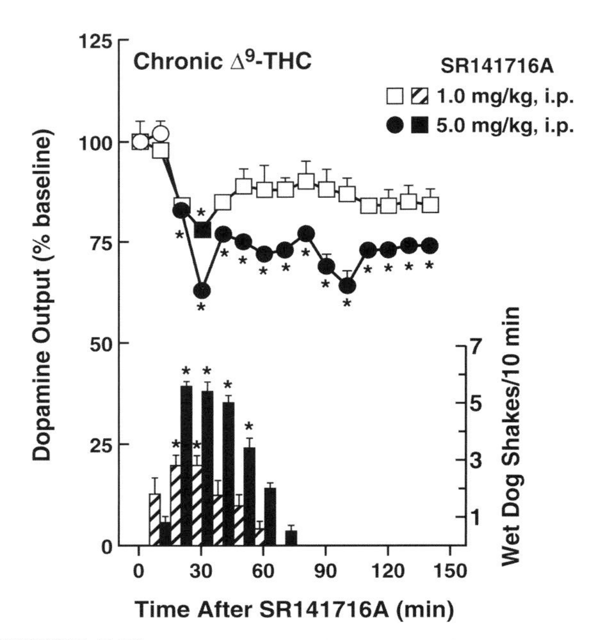

FIGURE 7.22 Time course of the effects of the administration of the CB_1 cannabinoid receptor antagonist SR141716A on dopamine output in dialysates from the nucleus accumbens shell and on wet dog shakes of rats chronically treated with Δ^9-THC. Dopamine in dialysates is expressed as percentage of baseline values (mean ± SEM). Bars represent the number of wet dog shakes (mean ± S.E.M.) observed every 10 min after the administration of SR141716A. Filled symbols: $p < 0.05$ compared to baseline. $^*p < 0.05$ compared to the corresponding time point of chronic saline-injected rats challenged with SR141716A. [Reproduced with permission from Tanda *et al.*, 1999.]

monophosphate (cAMP) in this preparation (Matsuda *et al.*, 1990). The rank order potency of cannabinoids to inhibit adenylate cyclase in transfected cells correlated with their rank order potency for psychoactive effects (Matsuda *et al.*, 1990) (**Fig. 7.27**). This cannabinoid receptor was deemed the CB_1 receptor and belongs to a G-protein-coupled receptor subfamily that includes the adrenocorticotropin receptor (Mountjoy *et al.*, 1992).

The discovery of a cannabinoid receptor immediately raised the hypothesis of potential endogenous ligands for the receptor, and a search of lipid extracts from porcine brain yielded the isolation and discovery of an arachidonic acid derivative, arachidonoylethanolamide (later termed anandamide) that bound to the CB_1 receptor (Devane *et al.*, 1992) (**Fig 7.28**). Anandamide was shown to bind competitively to the CB_1 receptor (Smith *et al.*, 1994), inhibit adenylate cyclase, inhibit voltage-sensitive calcium channels, and produce behavioral and pharmacological effects similar to Δ^9-THC such as antinociception, hypomotility, catalepsy, and hypothermia (Crawley *et al.*, 1993; Fride and Mechoulam, 1993; Adams *et al.*, 1998). However, relative to cannabinoids, anandamide produces only weak and transient behavioral effects *in vivo* probably due to its rapid catabolism (Smith *et al.*, 1994; Willoughby *et al.*, 1997), and in mice its effects were not blocked by pretreatment with the selective CB_1 antagonist SR141716A (Adams *et al.*, 1998).

Subsequently, a second endocannabinoid was isolated: 2-arachidonylglycerol (2-AG) (Sugiura *et al.*, 1995). 2-AG is an anandamide analog with a glycerol backbone (Sugiura *et al.*, 1999), possesses high affinity for the CB_1 receptor in synaptosomal membranes, and is found in the brain in amounts 1000 times higher than anandamide (Sugiura *et al.*, 1995) (**Fig. 7.29**). The endocannabinoids are lipid-like compounds that are synthesized and released from neurons, bind to cannabinoid receptors, activate transduction mechanisms, and have a reuptake system (Yamamoto and Takada, 2000) (**Fig. 7.30**). The endocannabinoids do not fulfill all of the requirements of 'classical' neurotransmitters. They are not synthesized in the cytosol of neurons or stored in synaptic vesicles to be secreted by exocytosis following excitation of nerve terminals by action potentials. Rather, endocannabinoids are synthesized when required in response to depolarization by receptor-stimulated cleavage of membrane lipid precursors and released from cells immediately after their production (Piomelli *et al.*, 2000), fulfilling more a role of a neuromodulator (Piomelli *et al.*, 2000) (**Fig. 7.31**). Once released, endocannabinoids act on cannabinoid receptors or are taken back into cells via an energy-independent transport system (Beltramo *et al.*, 1997). Once inside cells, both anandamide and 2-AG can be hydrolyzed (**Fig. 7.30**). Such a nonsynaptic release mechanism and rapid breakdown of both anandamide and 2-AG suggest that these compounds may locally regulate the effects of primary neurotransmitters (Piomelli *et al.*, 2000; Elphick and Egertova, 2001) (**Fig. 7.32**).

Advances in the understanding of the neuropharmacology of endocannabinoids have been facilitated by the identification of drugs that can block the formation, reuptake, and inactivation of both anandamide and 2-AG (**Fig. 7.32**). Anandamide reuptake is blocked by the anandamide transport inhibitor *N*-(4-hydroxyphenyl)-arachidonamide (AM 404) and as such the effects of exogenous anandamide are potentiated (Beltramo *et al.*, 1997). AM 404 has been shown to elevate levels of circulating anandamide and to decrease locomotor activity (Gonzalez *et al.*, 1999; Piomelli *et al.*, 2000). There also are inhibitors of anandamide amidohydrolase which can increase the levels of anandamide by blocking its breakdown (**Fig. 7.32**).

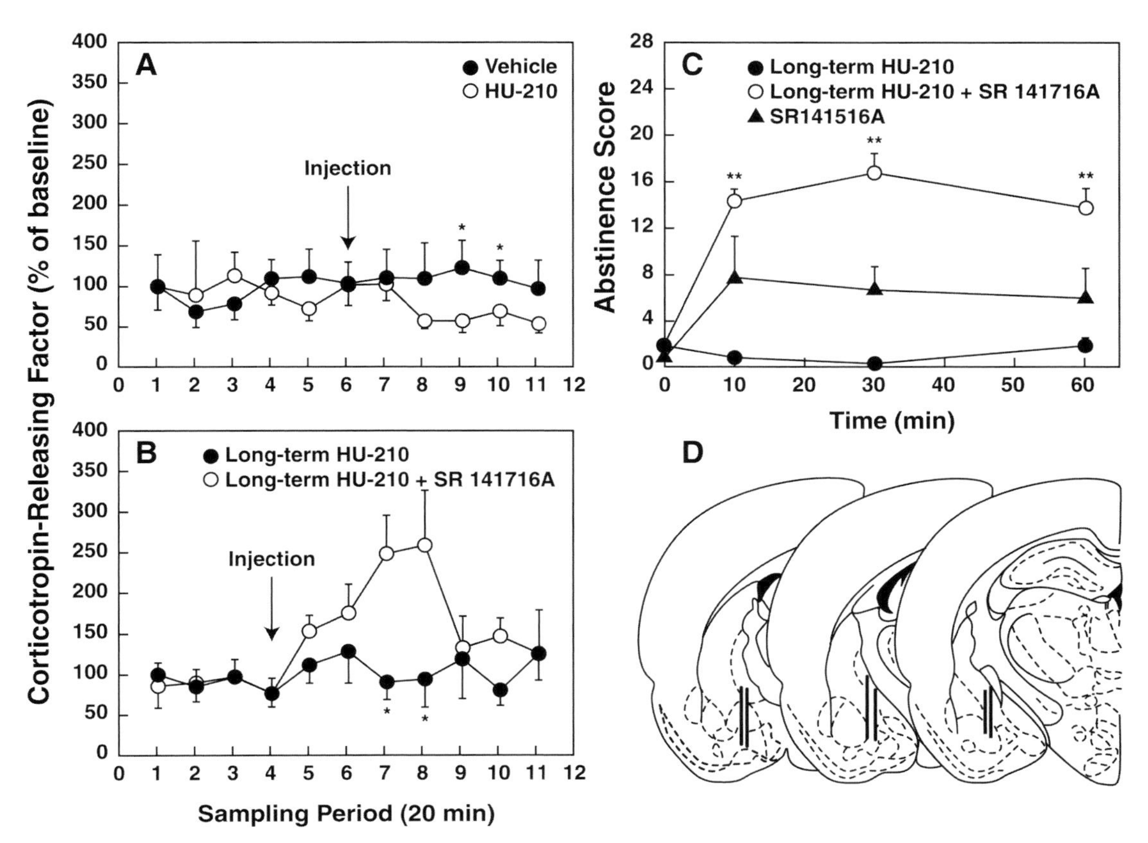

FIGURE 7.23 Effects of acute and chronic administration of, and precipitated withdrawal from, a cannabinoid receptor agonist on corticotropin-releasing factor (CRF) levels in the amygdala. **(A)** Effects of a single injection of the cannabinoid receptor agonist HU-210 (100 mg/kg) in rats on CRF release from the central nucleus of the amygdala. Statistical analysis (one-way ANOVA for repeated measures) revealed that HU-210 lowered CRF release (*$p < 0.05$, $n = 7$). Vehicle injections did not alter CRF efflux ($p = 0.93$, $n = 6$). Administration of the CB_1 cannabinoid receptor antagoist SR141716A did not modify CRF release. **(B)** Effects of SR141716A (3 mg/kg) on CRF release from the central nucleus of the amygdala in animals pretreated for 14 days with HU-210 (100 mg/kg). Cannabinoid withdrawal induced by SR141716A was associated with increased CRF release (*$p < 0.005$, $n = 8$). Vehicle injections did not alter CRF efflux ($F_{11,66} = 0.69$, not significant, $n = 7$). Data in (A) and (B) were standardized by transforming dialysate CRF concentrations into percentages of baseline values based on averages of the first four fractions. **(C)** Mean ± SEM of summed cannabinoid withdrawal scores 0, 10, 30, and 60 min after injection of SR141716A in rats treated for 14 days with HU-210 or its vehicle. SR141716A induced a mild behavioral syndrome in drug-naive rats receiving long-term pretreatment with vehicle (SR141716A) and a clear withdrawal syndrome in animals pretreated with HU-210 (long-term HU-210 + SR141716A; **$p < 0.0001$). Rats pretreated with cannabinoid (long-term HU-210) that received vehicle on the test day did not exhibit withdrawal signs. Drug-naive control animals that received vehicle injections were indistinguishable from the long-term HU-210 treatment group, and cannabinoid-naive rats did not exhibit observable changes in behavior after a single injection of HU-210. **(D)** Anatomical location of the active region of microdialysis probes (outer diameter, 0.5 mm) in animals subjected to SR141716A-induced cannabinoid withdrawal. [Reproduced with permission from Rodriguez de Fonseca *et al.*, 1997.]

Another enzyme implicated in regulating anandamide function is fatty acid amide hydrolase (FAAH), a membrane-associated hydrolase that is enriched in the brain and known to catalyze hydrolysis of another fatty acid signaling molecule, oleamide (Cravatt *et al.*, 1995). FAAH hydrolyzes anandamide *in vitro* (Cravatt *et al.*, 1996), and mice lacking FAAH have a significant inability to degrade anandamide. These mice also show an enhanced response to exogenous administration of anandamide, including hypoactivity, analgesia, catalepsy, and hypothermia (Cravatt *et al.*, 2001). Monoglyceride lipase hydrolyzes 2-AG (Dinh *et al.*, 2002; Saario *et al.*, 2004). In the brain, FAAH is localized in the somato-dendritic areas of neurons that are postsynaptic to CB_1-expressing axons, suggesting that FAAH may participate in cannabinoid signaling mechanisms by inactivating locally released endocannabinoids (Egertova *et al.*, 1998) (**Fig. 7.32**).

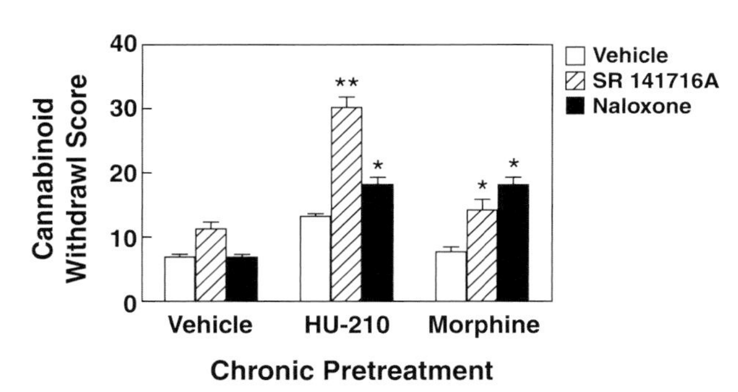

FIGURE 7.24 Naloxone administration (1 mg/kg, i.p.) induced a partial cannabinoid withdrawal syndrome in male rats chronically exposed to either the cannabinoid receptor agonist HU-210 (100 μg/kg for 14 days) or to morphine (2 pellets of 75 mg morphine base implanted s.c. for 72 h). The CB_1 cannabinoid receptor antagonist SR141716A (3 mg/kg) induced a partial cannabinoid withdrawal syndrome in morphine-dependent animals. Asterisks (*) indicate statistically significant differences from saline-treated animals ($^{*}p < 0.05$; $^{**}p < 0.01$; Newman-Keuls test; n = 9–15 per group). [Reproduced with permission from Navarro *et al.*, 2001.]

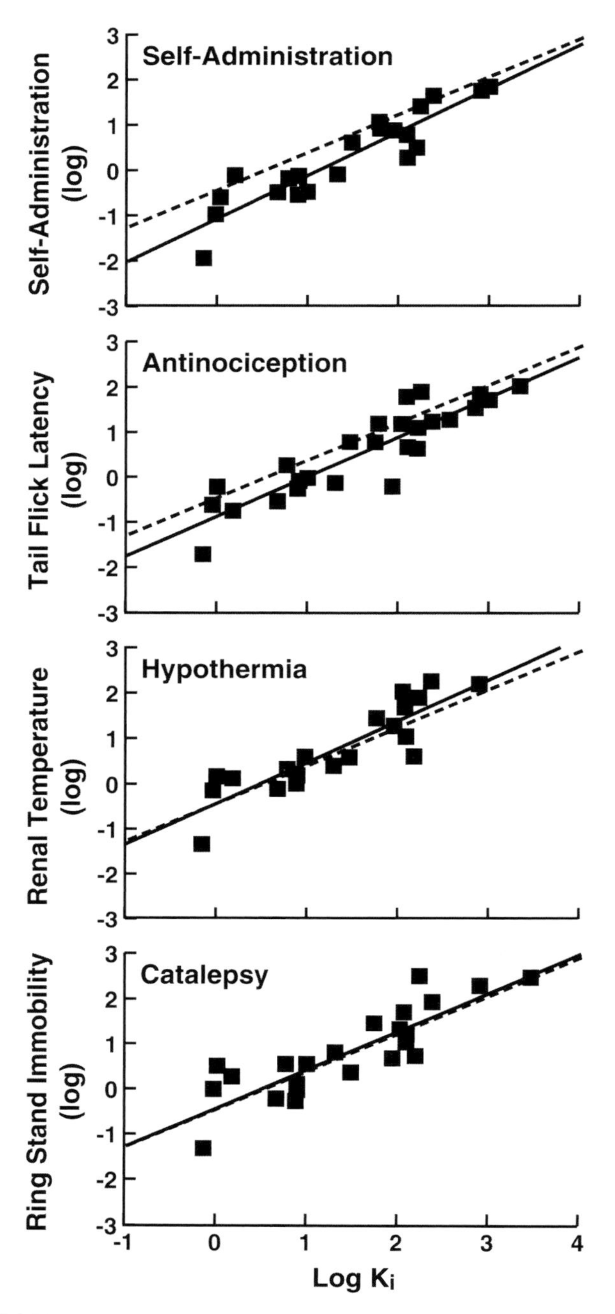

FIGURE 7.25 Linear correlations of log ED50 values (μmol/kg) for cannabinoids. The potency of these analogs administered intravenously in the mouse were assessed on four pharmacological measures, including self-administration, antinociception, hypothermia, and catalepsy. The dashed line represents the linear regression that would be obtained if the mean ED50 value of the four measures was plotted instead of that shown for each individual behavior. [Reproduced with permission from Compton *et al.*, 1993.]

NEUROBIOLOGICAL MECHANISM—CELLULAR

Much work has focused on the interactions of cannabinoids with the circuitry of the basal ganglia because of the high density of cannabinoid receptors in the basal ganglia and the complex effects of cannabinoids on movement (low dose activation and high dose inhibition of movement). The CB_1 receptor is expressed by the axons of striatal GABAergic medium spiny projection neurons (Matsuda *et al.*, 1993) (**Fig. 7.26**), and this results in high concentrations of CB_1 receptors in the globus pallidus and substantia nigra pars reticulata (Herkenham *et al.*, 1991). Cannabinoids inhibit GABA-mediated inhibitory postsynaptic potentials in the cell bodies of striatal medium-spiny neurons, possibly by decreasing presynaptic GABA release from the medium-spiny axon collaterals themselves (Szabo *et al.*, 1998). This ultimately results in a disinhibitory effect, increasing the firing rate of striatal medium-spiny neurons. Similar presynaptic effects have been hypothesized both for GABA and glutamate neurotransmission in the substantia nigra pars reticulata (Miller and Walker, 1998; Szabo *et al.*, 2000) and hippocampus (Freund *et al.*, 2003) (**Fig. 7.33**).

The well documented analgesic effects of cannabinoids may involve supraspinal, spinal, and even peripheral effects on sensory neurons (Elphick and Egertova, 2001). One likely target for the supraspinal analgesic effects is the periaqueductal gray (Martin *et al.*, 1995). In the periaqueductal gray, electrical

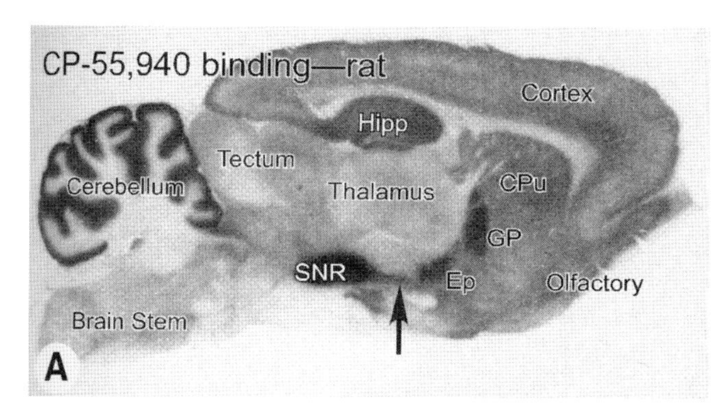

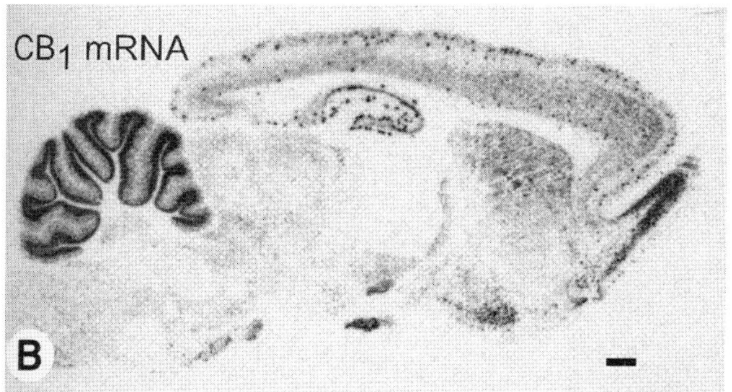

FIGURE 7.26 Autoradiographic film images showing cannabinoid receptor localization **(A)** in rat marked by the tritiated ligand CP 55,940 in an *in vitro* binding assay described by Herkenham *et al.* (1990). Sagittal slide-mounted section of rat brain hybridized with a CB_1-specific oligonucleotide probe **(B)** shows locations of neurons that express the mRNA at this level. High levels of receptor protein are visible in the basal ganglia structures globus pallidus (GP), entopeduncular nucleus (Ep), and substantia nigra pars reticulata (SNR). High binding also is seen in the cerebellum. Moderate binding is found in the hippocampus (Hipp), cortex, and caudate putamen (CPu). Low binding is seen in the brain stem and thalamus. Note that the GP, Ep, and SNR do not contain CB_1 mRNA-expressing cells (B). This is because the receptors in these areas are on axons (large arrows in Panel A) and terminals, and the mRNA-expressing cells of origin reside in the caudate and putamen. [Reproduced with permission from Freund *et al.*, 2003.]

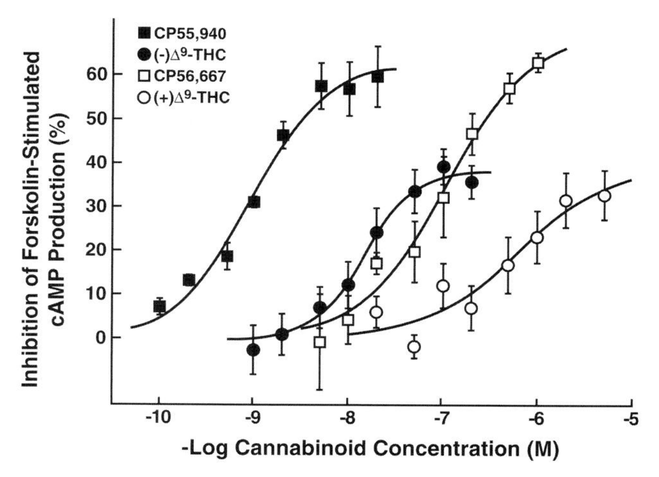

FIGURE 7.27 Cannabinoid-induced inhibition of forskolin-stimulated cyclic adenosine monophosphate (cAMP) production in SKR6-transfected CHO-K1 cells isolated from a rat cerebral cortex cDNA library. Stereoselective inhibition by Δ^9-THC and CP 55,940. Data represent the average percentage inhibition ± SEM of cAMP accumulation for three to five experiments, each performed in triplicate. Curves were generated using the GraphPad InPlot nonlinear regression analysis program. Cannabinoids did not significantly inhibit cAMP accumulation in nontransfected cells (data not shown). EC50 values (mean nM ± SEM) for the inhibition of stimulated cAMP accumulation were: (−)Δ^9-THC, 13.5 ± 2.7; (+)Δ^9-THC, 773 ± 187; CP 55,940, 0.87 ± 0.20; CP 56,667, 96.3 ± 7.1; 11-OHΔ^9-THC, 8.9 ± 1.8; nabilone, 16.6 ± 4.9; Δ^8-THC, 274 ± 8.4. Cannabinoid-induced inhibition of cAMP accumulation also was observed in transfected cells in which cAMP production was stimulated by the peptide hormone calcitonin gene-related peptide instead of forskolin. [Reproduced with permission from Matsuda *et al.*, 1990.]

stimulation of the dorsal and lateral parts produced cannabinoid-mediated analgesia accompanied by a marked increase in the release of anandamide in the periaqueductal gray, suggesting that endogenous anandamide may contribute to this analgesia (Walker *et al.*, 1999) (**Fig. 7.34**). As in the basal ganglia, cannabinoids in the periaqueductal gray have been shown to inhibit electrically induced inhibitory and excitatory postsynaptic potentials. The inhibition of inhibitory postsynaptic potentials may be via inhibition of presynaptic GABA release, and the inhibition of excitatory postsynaptic potentials may be via inhibition of glutamate release (Vaughan *et al.*, 2000). Similar actions may occur in the rostral ventromedial medulla that are relevant to analgesia (Vaughan *et al.*, 1999).

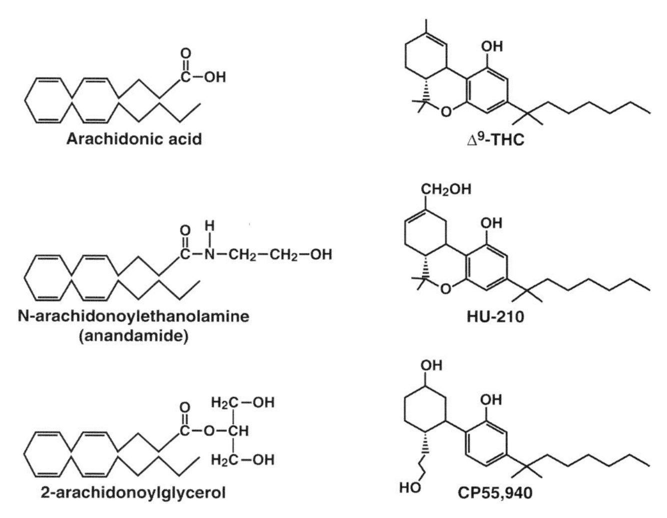

FIGURE 7.28 Chemical structures of 2-arachidonoylglycerol, structural analogs, and synthetic cannabinoids.

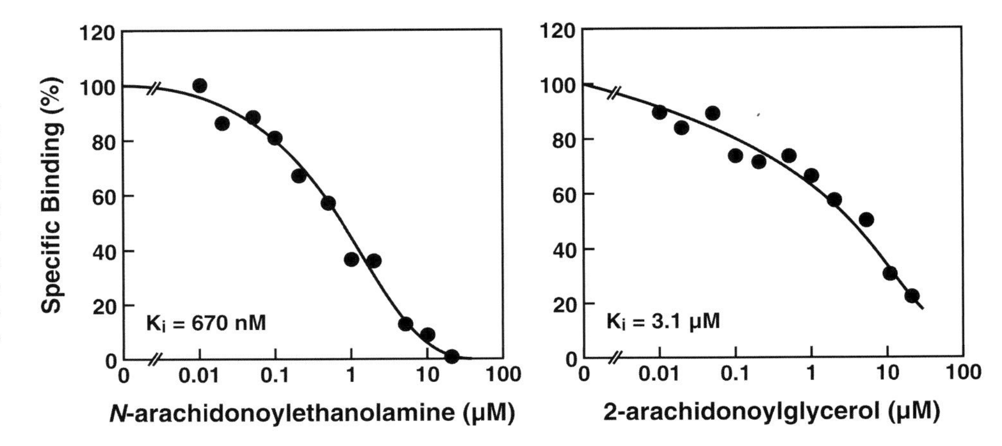

FIGURE 7.29 Effects of *N*-arachidonoylethanolamine (anandamide) and 2-arachidonoylglycerol on the specific binding of [^{3}H]CP 55, 940 to rat brain synaptosomal membranes at 0°C. The data are the means of four (*N*-arachidonoylethanolamine) or five (2-arachidonoylglycerol) separate experiments, each performed in quadruplicate. [Reproduced with permission from Sugiura *et al.*, 1995.]

As with the association between the basal ganglia and movement, and the periaqueductal gray and pain, the effects of cannabinoids on reward-related structures are hypothesized to involve disinhibitory effects. Using *in vivo* unit recording approaches, Δ^9-THC and synthetic cannabinoids have been shown to increase neuronal firing of dopamine neurons in ventral tegmental area (Melis *et al.*, 1996; French., 1997; French *et al.*, 1997; Gifford *et al.*, 1997; Diana *et al.*, 1998a; Gessa *et al.*, 1998; Cheer *et al.*, 2003). This activation is blocked by CB_1 receptor antagonists, but not by the opioid antagonist naloxone (French, 1997). This appears to be a presynaptic action as was observed in the basal ganglia and periaqueductal gray. In whole-cell patch clamp recordings in slices of the ventral tegmental area, the cannabinoid agonists WIN 55,212-2 and CP 55,940 reduced the amplitude of $GABA_A$ receptor-mediated inhibitory postsynaptic potentials (Szabo *et al.*, 2002). Such an inhibition of a GABAergic inhibitory input of dopaminergic neurons could be a mechanism for increasing the firing rate of ventral tegmental area neurons *in vivo*.

During withdrawal from chronic administration of cannabinoids, there is a decrease in firing of ventral tegmental dopamine neurons. Extracellular recordings from antidromically identified mesolimbic dopamine neurons showed decreased activity to both precipitated and spontaneous cannabinoid withdrawal (Diana *et al.*, 1998b).

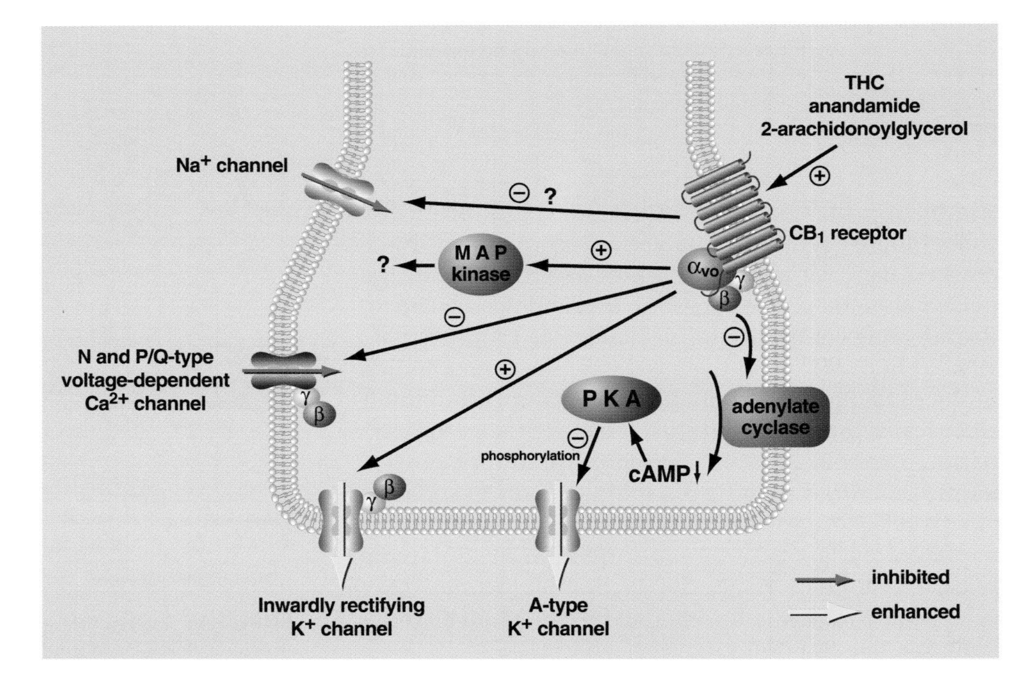

FIGURE 7.30 Schematic illustration of signal transduction mechanisms stimulated by CB_1 receptors in a presynaptic nerve terminal. This schematic is based on Ameri *et al.* (1999) and modified by Yamamoto and Takada (2000). [Reproduced with permission from Yamamoto and Takada, 2000.]

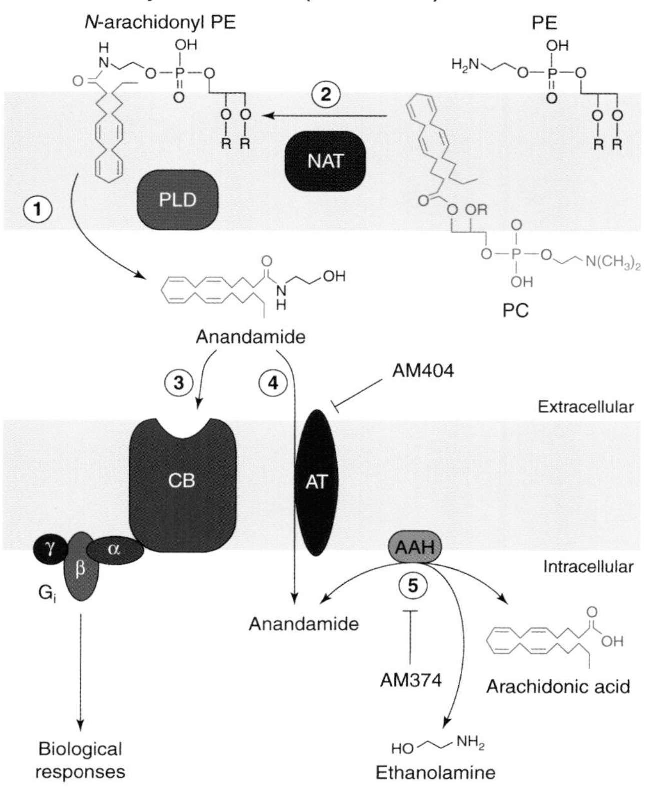

FIGURE 7.31. *N-arachidonoylethanolamine (anandamide)*: Formation and inactivation of anandamide. Anandamide can be generated by hydrolysis of *N*-arachidonyl phosphatidylethanolamine (*N*-arachidonyl PE), which is catalyzed by phospholipase D(PLD) ①. The synthesis of *N*-arachidonyl PE, depleted during anandamide formation, might be mediated by *N*-acyl transferase activity (NAT) ②, which detaches an arachidonate moiety (red) from the *sn*-1 position of phospholipids such as phosphatidylcholine (PC) and transfers it to the primary amino group of PE. The membrane localizations of PLD and NAT are speculative. Newly formed anandamide can be released into the extracellular space, where it can activate G-protein-coupled cannabinoid (CB) receptors located on neighboring cells ③ or on the same cells that produce anandamide (not shown). Anandamide release in the external milieu has been demonstrated both *in vitro* and *in vivo*. Anandamide can be removed from its sites of action by carrier-mediated transport (anandamide transport, AT) ④, which can be inhibited by AM404. Transport into cells can be followed by hydrolysis catalyzed by a membrane-bound anandamide amidohydrolase (AAH, also called fatty acid amide hydrolase) ⑤, which can be inhibited by AM374. Arachidonic acid produced during the AAH reaction can be rapidly reincorporated into phospholipid and is unlikely to undergo further metabolism. *In vitro*, AAH also can act in reverse, catalyzing the formation of anandamide from arachidonic acid and ethanolamine. The physiological significance of this reaction in anandamide formation is unclear. Abbreviation *R* indicates a fatty acid group. *2-arachidonylglycerol*: Formation and inactivation of 2-arachidonylglycerol (2-AG). Hydrolysis of phosphatidylinositol(4,5)-bisphosphate [PtdIns(4,5)P_2] by phospholipase C (PLC) produces the second messengers 1,2-diacyl-glycerol (DAG) and inositol (1,4,5)-trisphosphate [Ins(1,4,5)P_3] ①. DAG serves as a substrate for DAGlipase (DAGL), which catalyzes the production of 2-AG ②. This pathway also gives rise to free arachidonic acid. 2-AG can be released into the external milieu, measured *in vitro*, allowing it to interact with cannabinoid receptors, and its effects can be terminated by uptake into cells (not shown). However, extracellular release of 2-AG has not yet been reported *in vivo*. Intracellular 2-AG can be hydrolyzed to arachidonic acid and glycerol by an uncharacterized esterase such as monoacylglycerol lipase (MAGL) ③. [Reproduced with permission from Piomelli *et al.*, 2000.]

In the extended amygdala, cannabinoids also potentiate GABA-mediated postsynaptic inhibitory potentials. Using whole-cell patch clamp recordings in slices of the lateral and basal nuclei of the amygdala, the cannabinoid agonists WIN 55,212-2 and CP 55,940 reduced the amplitude of $GABA_A$ receptor-mediated inhibitory postsynaptic potentials (Katona *et al.*, 2001). Single-cell recordings in anesthetized rats have revealed that cannabinoids inhibit firing of interneurons in the basolateral amygdala as well as projection neurons from the shell of the nucleus accumbens (Pistis *et al.*, 2004). Using extracellular recordings with stimulation of basolateral or prefrontal cortex afferents, cannabinoids also were shown to inhibit excitatory inputs to the shell of the nucleus accumbens similar to other drugs of abuse (Pistis *et al.*, 2002). Similarly, in nucleus accumbens slice preparations, cannabinoids strongly inhibited stimulus-evoked glutamate-mediated neurotransmission suggesting a presynaptic localization (Robbe *et al.*, 2001). This inhibition has been hypothesized to be a negative feedback loop mediated via a direct presynaptic action on vesicular release mechanisms (Robbe *et al.*, 2003) (**Fig. 7.35**). Others have argued for cell-autonomous regulation in a subtype of cortical interneurons (Glickfeld and Scanziani, 2005). Endocannabinoids also may be involved in long-term synaptic plasticity such as long-term depression in the nucleus accumbens and striatum (Gerdeman *et al.*, 2002; Robbe *et al.*, 2002, 2003).

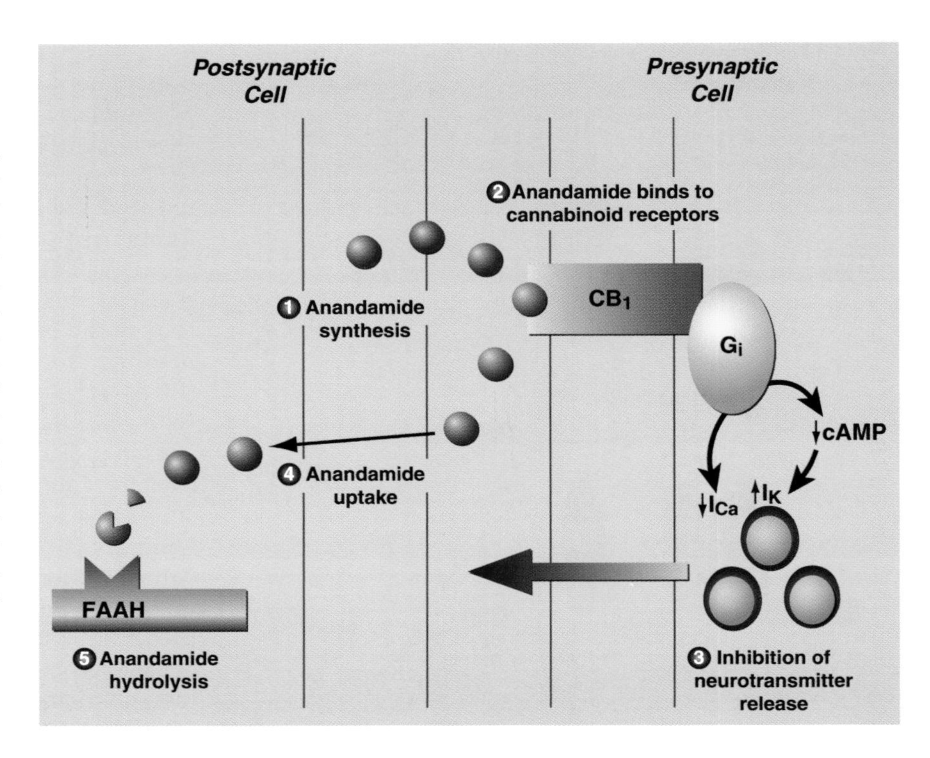

FIGURE 7.32 A model of synaptic cannabinoid signaling in which anandamide functions as a retrograde signaling molecule that modulates (inhibits) the release of classical anterograde transmitters by presynaptic terminals. Anandamide is synthesized and released by the postsynaptic cell ① and then diffuses into the synaptic cleft where it binds to and activates presynaptic CB_1-type cannabinoid receptors ②. Activated CB_1 receptors cause inhibition of neurotransmitter release from presynaptic terminals via G-protein-mediated mechanisms involving activation of K^+ channels or inhibition of Ca^{2+} channels ③. Anandamide dissociates from CB_1 receptors and then, following uptake into the postsynaptic cell ④, is hydrolyzed intracellularly by fatty acid amide hydrolase (FAAH) ⑤. [Reproduced with permission from Elphick and Egertova, 2001.]

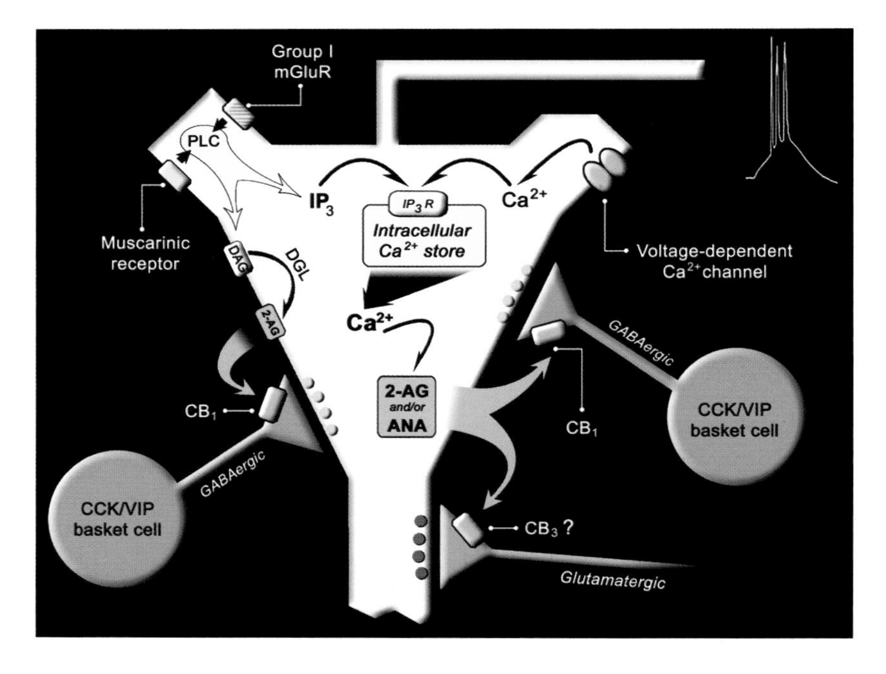

FIGURE 7.33 Schematic diagram of endocannabinoid-mediated retrograde synaptic signaling. The possible physiological mechanisms that may trigger endocannabinoid synthesis and release from hippocampal pyramidal neurons are outlined (similar mechanisms are likely to operate in most brain areas where endocannabinoid signaling takes place). The large Ca^{2+} transient required for endocannabinoid synthesis likely involves Ca^{2+} mobilization from intracellular stores upon activation of the inositol 1,4,5-trisphosphate (IP_3) system (metabotropic receptors) and voltage-dependent Ca^{2+} channels (burst firing). Another root independent of intracellular Ca^{2+} transients is illustrated on the left of the schematized pyramidal cell body. Activation of phospholipase C (PLC) via group I metabotropic glutamate (mGluR) or muscarinic cholinergic receptors will produce, in addition to IP3, 1,2-diacylglycerol (DAG), which likely remains in the plasma membrane. This then could be converted to 2-arachidonoylglycerol (2-AG) by the enzyme 1,2-diacylglycerol lipase (DGL) still within the membrane, which may ensure a rapid diffusion into the extracellular space. The released endocannabinoids act on CB_1 receptors located on axon terminals of γ-aminobutyric acid (GABA) interneurons that contain CCK, or on a new cannabinoid receptor subtype (CB_3?) expressed by glutamatergic axons. Activation of CB_1 reduces GABA release via G_i-mediated blocking of N-type Ca^{2+} channels, whereas the new receptor likely reduces glutamate release via a similar mechanism. ANA, anandamide; CCK/VIP, cholecystokinin/vasoactive intestinal peptide. [Reproduced with permission from Freund *et al.*, 2003.]

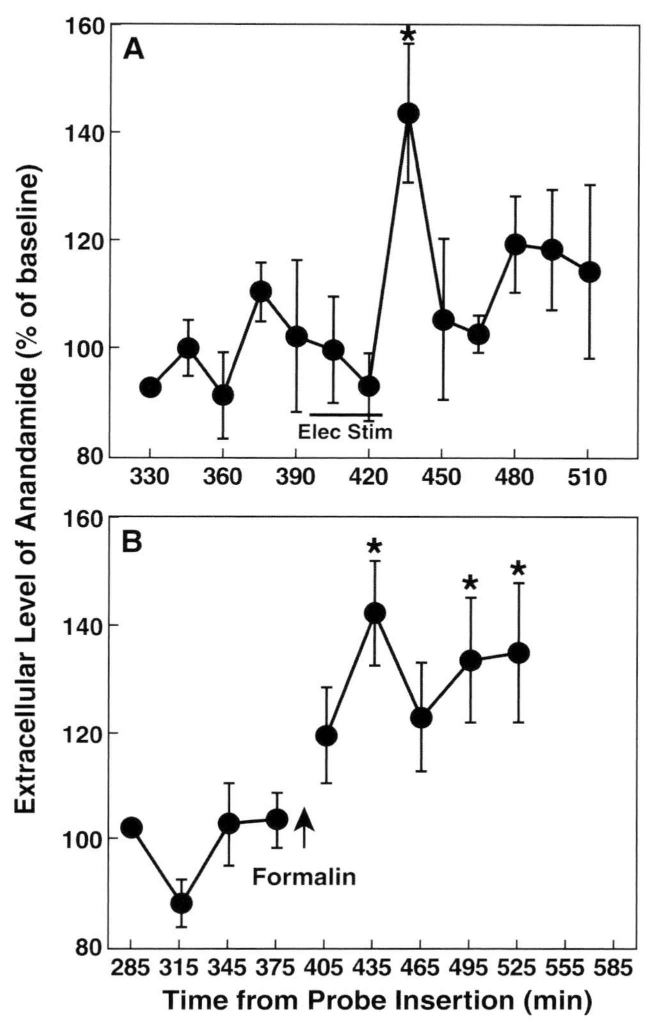

FIGURE 7.34 The release of anandamide in the periaqueductal gray of the rat stimulated by electrical depolarization or pain. **(A)** Increased extracellular levels of anandamide after electrical stimulation of the periaqueductal gray in urethane-anesthetized rats. After the establishment of stable baseline values, electrical stimulation (bipolar 0.1 ms/1 mA, 60 Hz, 5 s trains with 5 s rest intervals) was delivered for 30 min. Microdialysis samples were collected in 15 min intervals and analyzed by high performance liquid chromatography with detection by atmospheric pressure chemical ionization—liquid chromatography/mass spectrometry, with selected ion monitoring mode at molecular weight 348.3 ($n = 5$, $p < 0.05$, in repeated measures ANOVA). The asterisks (*) mark points that are significantly different from the baseline average by *post hoc* test ($p < 0.05$). The delay in measurement presumably reflects the time needed to produce anandamide in the extracellular space with sufficient overflow to achieve recovery by microdialysis. **(B)** Increased extracellular levels of anandamide in the periaqueductal gray after induction of prolonged pain in urethane-anesthetized rats. After the establishment of a stable baseline, a 4 per cent 150 μl formalin solution was injected subcutaneously in both hindpaws of the rat. The samples shown span 30 min intervals ($n = 6$; $p < 0.001$, repeated measures ANOVA). [Reproduced with permission from Walker *et al.*, 1999.]

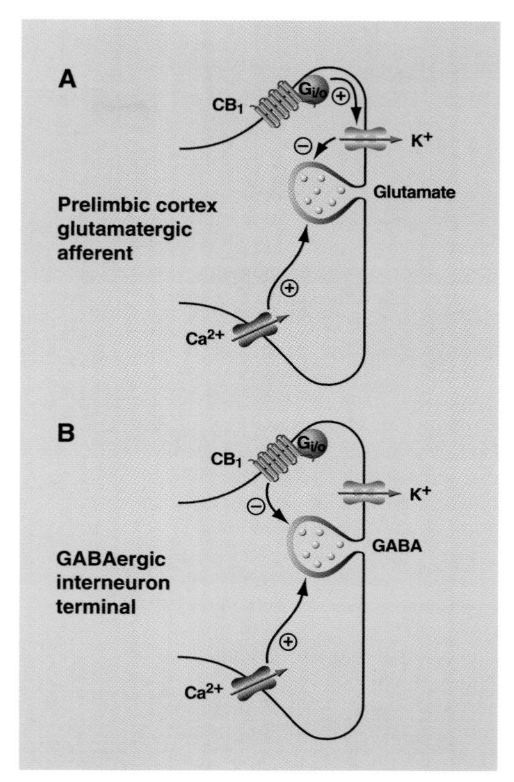

FIGURE 7.35 Schematic view of the presynaptic actions of CB_1 receptors in the nucleus accumbens. **(A)** At glutamatergic terminals, CB_1 receptors reduce synaptic release via activation of presynaptic K^+ channels. **(B)** At γ-aminobutyric acid (GABA) terminals, CB_1 receptors reduce synaptic release downstream of presynaptic Ca^{2+} channels. [Reproduced with permission from Robbe *et al.*, 2003.]

NEUROBIOLOGICAL MECHANISM—MOLECULAR

As noted above, early evidence for a receptor-mediated action of cannabinoids was based on the synthesis and radiolabeling of potent bicyclic cannabinoids and the establishment of saturable binding in brain membranes. The initial site of Δ^9-THC binding was hypothesized to be the CB_1 receptor which is widely distributed throughout the brain, but is particularly concentrated in the extrapyramidal motor system of the rat (Herkenham *et al.*, 1990). The CB_1 receptor was first cloned and localized to brain structures, and subsequently the CB_2 receptor was cloned from macrophages and spleen and localized peripherally (Munro *et al.*, 1993). The CB_1 receptor appears to mediate the psychoactive effects of cannabinoids. CB_2 receptors appear to be involved in the immunosuppressive effects of cannabinoids (Buckley *et al.*, 2000).

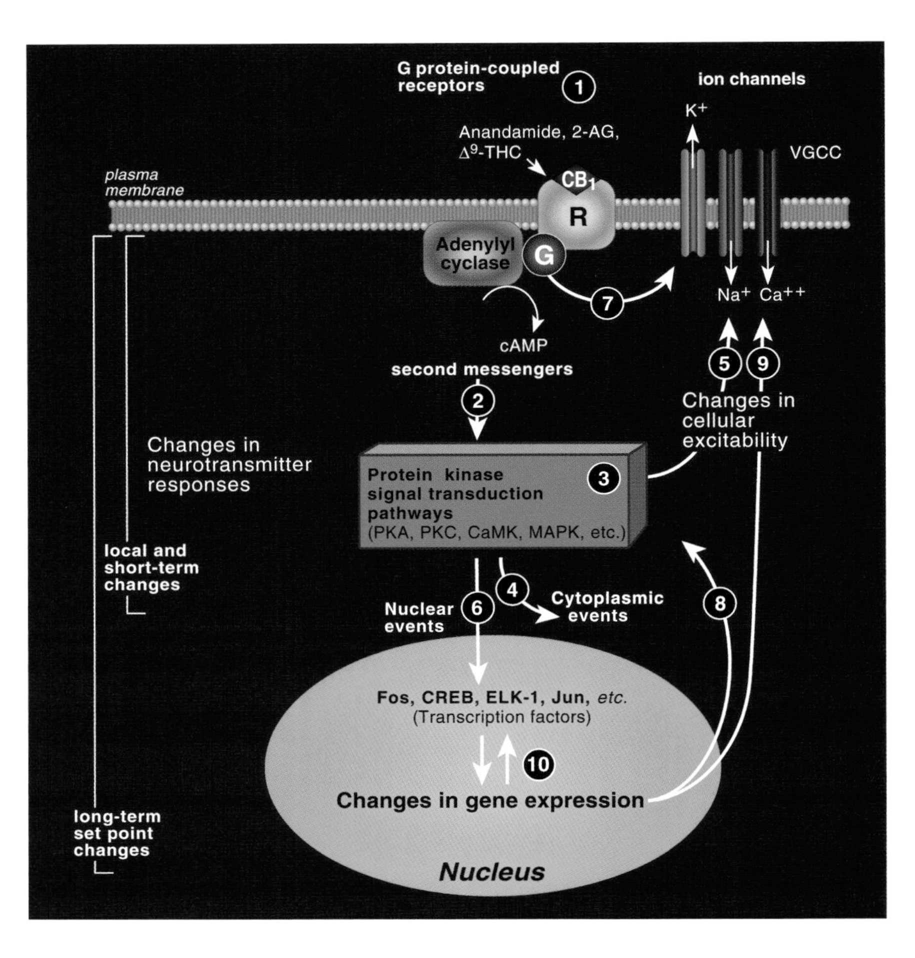

FIGURE 7.36 Molecular mechanisms of neuroadaptation. Δ^9-THC, by acting on neurotransmitter systems, affects the phenotypic and functional properties of neurons through the general mechanisms outlined in the diagram. Cannabinoids act on a G-protein-coupled receptor (R), the cannabinoid CB_1 receptor (①). The CB_1 receptor also is activated by endogenous cannabinoids such as anandamide. This receptor modulates (inhibits) the levels of second messengers like cAMP and Ca^{2+} (②), which in turn regulate the activity of protein kinase transducers (③). Such protein kinases affect the functions of proteins located in the cytoplasm, plasma membrane, and nucleus (④) (⑤⑥). Among membrane proteins affected are ligand-gated and voltage-gated ion channels (⑤). Protein kinase transduction pathways affect the activities of transcription factors (⑥). G_i and G_o proteins can regulate potassium and calcium channels directly through their $\beta\gamma$ subunits(⑦). Some of these factors, like CREB, are regulated post-translationally by phosphorylation. Others, like Fos, are regulated transcriptionally. Still other, like Jun, are regulated both posttranslationally and/or transcriptionally. While membrane and cytoplasmic changes may be only local (e.g., dentritic domains or synaptic boutons), changes in the activity of transcription factors may result in long-term functional changes. These may include changes in gene expression of proteins involved in signal transduction (⑧) and/or neurotransmission (⑨), resulting in altered neuronal responses. Chronic exposure to Δ^9-THC also alters the expression of transcription factors themselves (⑩). [Modified with permission from Koob *et al.*, 1998.]

Both the CB_1 and CB_2 receptors are coupled to the G-proteins G_i and G_o, and this interaction produces an inhibition of adenylate cyclase (Howlett and Fleming, 1984; Howlett *et al.*, 1986; Koob *et al.*, 1998) (**Fig. 7.36**). This leads to a subsequent reduction in cAMP and an increase in mitogen-activated protein kinase (Ameri, 1999) (**Fig. 7.30**). Cannabinoids also act to enhance A-type potassium channels (enhanced outward potassium current) (Deadwyler *et al.*, 1993), inhibit voltage-activated N-type calcium channels (Caulfield and Brown, 1992; Mackie and Hill, 1992), and inhibit presynaptic P/Q calcium channels (Sullivan, 1999; Yamamoto and Takada, 2000). Thus, activation of CB_1 receptors can lead to inhibition of presynaptic release via several molecular pathways, including inhibition of adenylyl cyclase, and thus reduction in cAMP and a decrease in protein kinase A-mediated phosphorylation of A-type potassium channels (enhancement of A-Type currents), or a direct inhibition of P/Q and N-type calcium channels by G-protein inhibition.

Tetanic stimulation mimicking naturally occurring frequencies of prefrontal cortex afferents to the nucleus accumbens induced a presynaptic long-term depression dependent on endogenous cannabinoids and CB_1 receptors (Robbe *et al.*, 2003). Similar effects were observed in striatum via cortical stimulation, and it was suggested that the endocannabinoids are synthesized postsynaptically but act presynaptically in the striatum as a retrograde messenger (Gerdeman *et al.*, 2002).

CB_1 receptor knockout mice studies have confirmed a critical role for the CB_1 receptor in the behavioral effects of cannabinoids. Knockout mice showed no analgesia or locomotor activity when challenged with Δ^9-THC, no intravenous self-administration of WIN 55,212-2, and no precipitated cannabinoid withdrawal when administered Δ^9-THC chronically (Ledent *et al.*, 1999) (**Fig. 7.37**). However, CB_1 knockout mice retain their response to anandamide (Di Marzo *et al.*, 2000b) and methanandamide (Baskfield *et al.*, 2004), suggesting that anandamide-like endocannabinoids may have actions in the brain outside the CB_1 receptor (Di Marzo *et al.*, 2000a). Other *in vitro* and *in vivo* studies have shown anandamide and some synthetic cannabinoids retain activity in CB_1 knockout mice, again suggesting that anandamide may act at non CB_1 and nonCB_2 sites (Di Marzo *et al.*, 2000b; Wiley and Martin, 2002). One hypothesis is that the effects of anandamide may be mediated by the vanilloid receptor. The vanilloid receptor is a ligand-gated receptor activated by anandamide at concentrations 10–20 times higher than those that activate the CB_1 receptor (Zygmunt *et al.*, 1999; Smart *et al.*, 2000).

In support of the pharmacological studies with cannabinoid antagonists, studies with knockout mice have shown significant interactions between the reinforcing and dependence-producing effects of opioids and the cannabinoid system in CB_1 knockout mice. Morphine-induced conditioned place preference (Martin *et al.*, 2000; but see Rice *et al.*, 2002), intravenous morphine self-administration (Ledent *et al.*, 1999; Cossu *et al.*, 2001), and morphine-induced dopamine release in the nucleus accumbens (Mascia *et al.*, 1999) are eliminated in CB_1 knockout mice. Naloxone-precipitated withdrawal also is decreased in CB_1 knockout mice (Lichtman *et al.*, 2001). Emphasizing the cross talk between cannabinoid and opioid systems, Δ^9-THC place preference is blocked in μ opioid receptor knockout mice, and Δ^9-THC place aversions are blocked in κ opioid receptor knockout mice (Ghozland *et al.*, 2002). Also, the cannabinoid withdrawal syndrome is decreased in μ knockout mice (Lichtman *et al.*, 2001), in double μ and δ knockout mice (Castane *et al.*, 2003), and in preproenkephalin knockout mice (Valverde *et al.*, 2000b).

Cannabinoid knockout mice show blockade of analgesia associated with cannabinoids (Ledent *et al.*, 1999; Zimmer *et al.*, 2001), and there are reports of decreased locomotor activity, hypoalgesia and catalepsy in these mice (Zimmer *et al.*, 1999). Opioid antinociception is not modified in CB_1 knockout mice (Valverde *et al.*, 2000a). In addition, the antinociception produced by Δ^9-THC is not modified in μ, δ or κ opioid receptor knockout mice (Ghozland *et al.*, 2002). These observations, combined with the extensive reports that cannabinoids combined with morphine can produce a greater than additive effect (Cichewicz, 2004), suggest that the interaction may occur at a neurocircuitry level rather than as a result of direct receptor interaction.

Knockout studies have provided evidence for cannabinoid systems in the reinforcing actions of other drugs of abuse and in feeding. Voluntary ethanol consumption and dopamine release in the nucleus accumbens are reduced in cannabinoid CB_1 receptor knockout mice (Hungund *et al.*, 2003; Wang *et al.*, 2003). Nicotine place preference is blocked (Castane *et al.*, 2002), and anxiogenic-like responses in animal models of anxiety are observed in CB_1 knockout mice (Haller *et al.*, 2002; Martin *et al.*, 2002). CB_1 knockout mice are slightly hypophagic and lean (Ravinet Trillou *et al.*, 2004). These mice also have reduced plasma insulin and reduced plasma leptin levels and resistance to high fat diet-induced obesity, suggesting a role for cannabinoid systems in high fat diet-induced obesity.

SUMMARY

Marijuana is the dry shredded green or brown mixture of flowers, stems, seeds, and leaves of the hemp plant *Cannabis sativa*, and cannabinoids were defined originally as the phytocannabinoids of the hemp plant *Cannabis sativa*. As with the opioids, later definitions were more inclusive, defining cannabinoids as 'all ligands of the cannabinoid receptor and related compounds including endogenous ligands of the receptors and a large number of synthetic cannabinoid analogues' (Grotenhermen, 2003). Preparations of marijuana have a rich history in terms of medical use and myth, with the fiber of the plant having a major use in rope and the more psychoactive varieties having medicinal and intoxicating effects. There are over 60 different cannabinoids in marijuana, but the primary active ingredient is Δ^9-THC. Cannabinoids, at present, have two accepted medical uses in the United States and many potential therapeutic uses. Dronabinol (Marinol; the oral capsule preparation of Δ^9-THC) is approved for use in refractory nausea and vomiting associated with cancer chemotherapy and for appetite loss in anorexia of HIV/AIDS patients. Marijuana also is the most commonly used illicit drug in the United States, with estimates of approximately 15 million current users. Of those people who ever used marijuana, approximately 10 per cent ultimately will meet the criteria for Substance Dependence on marijuana (American Psychiatric Association, 1994). Marijuana produces intoxicating effects when ingested, usually by inhalation, that include euphoria and mood swings characterized by initial feelings of 'happiness' or sudden talkativeness, a dreaming or lolling state, or general activation and hyperactivity.

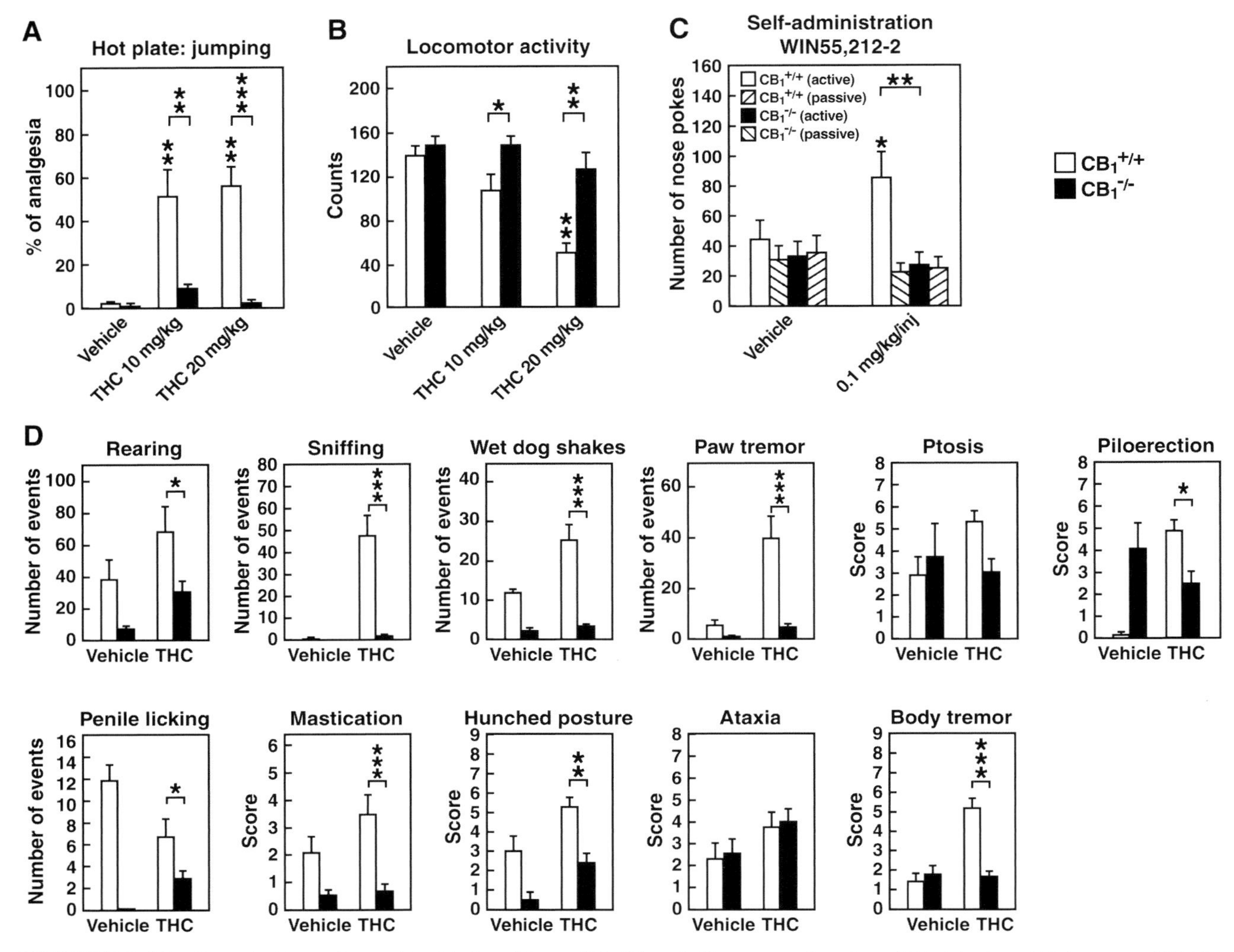

FIGURE 7.37 Central effects of cannabinoids in wildtype ($CB_1^{+/+}$) and CB_1 receptor knockout ($CB_1^{-/-}$) mice. **(A)** Latency + SEM for escape jumping in the hot-plate test (n = 10). **(B)** Spontaneous activity in locomotor activity boxes (number of photocell counts within 10 min + SEM; n = 10). For (A) and (B), an intraperitoneal injection of Δ^9- THC (or vehicle alone) was made 20 min before measurements. **(C)** Self-administration of WIN 55,212-2. Injection of agonist or vehicle to active and passive mice was coupled to the nose-poke behavior (+ SEM) of the active mouse (n = 8 for WIN 55,212-2; n = 4 for vehicle). **(D)** Signs (+ SEM) reflecting Δ^9-THC withdrawal were monitored (n = 5–15). Statistical significance was measured between genotypes and against vehicle for drug-treated group [t test for (A) and (B) and (D), and Neuman-Keuls test for (C)]. [Reproduced with permission from Ledent *et al.*, 1999.]

Higher doses are associated with psychedelic-like effects such as increased sensitivity to sound and a keener appreciation of rhythm and timing. Perception of time often is slowed with an exaggeration of the sense of time. Perceptions of space may be broadened, and near objects may appear distant. There are some visual 'hallucination-like' effects which are, as with psychedelics like lysergic acid diethylamide, mostly illusionary transformations of the outer world.

Following ingestions, Δ^9-THC has a peculiar distribution in the body, sequestering initially in vascularized tissue and later in fat rich tissues. It has a long terminal half-life of approximately 20-60 h, but metabolites can be measured in urine for weeks after ingestion. Tolerance develops to the intoxicating and physiological effects of cannabis, and a withdrawal syndrome has been defined in humans and animals. In humans, the most common symptoms associated with cannabis withdrawal are decreased appetite/ weight loss, irritability, nervousness and anxiety, anger and aggression, restlessness, and sleep disturbances. Pathological effects of intoxication or chronic high dose use include the dangers of using an automobile or flying an airplane while intoxicated, health risks for cardiac problems for those with pre-existing cardiac disease, anxiety and panic reactions, exacerbation of schizophrenic-like symptoms in individuals who are vulnerable to develop schizophrenia, and some evidence of precipitation of schizophrenic-like syndromes. In addition, marijuana smoke is hypothesized to have the same potential toxicity as cigarette smoke vis a vis lung function. The behavioral mechanism of action of cannabinoids is hypothesized to be a perceptual disinhibition of both external and internal cues/states without a motivational disinhibition, at least as would be manifest by psychomotor activation. This perceptual disinhibition can be pleasant or unpleasant depending on internal or external context.

In animals, Δ^9-THC and other cannabinoids have reinforcing properties in models of brain stimulation reward, place preference, and self-administration. Tolerance is observed in animals, and both somatic and motivational withdrawal syndromes have been observed. Significant evidence exists for a role of the mesolimbic dopamine system and opioid peptide systems in the ventral tegmental area and basal forebrain in the acute reinforcing effects of cannabinoids and the motivational effects of cannabinoid withdrawal. In addition, acute withdrawal from cannabinoids is associated with an activation of CRF in the extended amygdala as with other drugs of abuse. The discovery of high potency ligands and a cannabinoid receptor in brain led to the identification of endocannabinoids, such as anandamide and 2-AG, that act as neuromodulators in the central nervous system regulating brain excitability through local actions on presynaptic GABA and glutamate neurons. As with alcohol and opioids, such disinhibition of specific GABAergic and glutamatergic systems may provide the activation of reward neurotransmitters associated with the acute reinforcing effects of cannabinoids. Molecular studies have identified two cannabinoid receptors: the CB_1 receptor in the central nervous system and the CB_2 receptor in the periphery localized to macrophages and spleen. Molecular genetic studies have shown that knockout of the CB_1 receptor blocks the acute reinforcing and dependence-inducing effects of cannabinoids but also blunts some of the effects of opioids, further supporting cross talk between opioid and cannabinoid systems. Cannabinoids acting through the CB_1 receptor inhibit adenylate cyclase which in turn inhibits protein kinase A leading to an activation of A-type potassium channels but at the same time inhibits P/Q-type voltage-dependent calcium channels directly. Such combined inhibition has been hypothesized to lead to a presynaptic long-term depression in the basal ganglia that may be of significance for neuroadaptive plasticity at the cellular level in this system.

The explosion of research in the field of cannabinoid neurobiology will most certainly provide not only insights into the neurobiology of cannabinoid addiction but also into the role of the endogenous cannabinoids in normal adaptive function.

REFERENCES

Aarons, G. A., Brown, S. A., Stice, E., and Coe, M. T. (2001). Psychometric evaluation of the marijuana and stimulant effect expectancy questionnaires for adolescents. *Addictive Behaviors* **26**, 219–236.

Abel, E. L. (1980). *Marihuana: The First Twelve Thousand Years.* Plenum Press, New York.

Abrahamov, A., Abrahamov, A., and Mechoulam, R. (1995). An efficient new cannabinoid antiemetic in pediatric oncology. *Life Sciences* **56**, 2097–2102.

Aceto, M. D., Scates, S. M., Lowe, J. A., and Martin, B. R. (1996). Dependence on delta 9-tetrahydrocannabinol: studies on precipitated and abrupt withdrawal. *Journal of Pharmacology and Experimental Therapeutics* **278**, 1290–1295.

Aceto, M. D., Scates, S. M., and Martin, B. B. (2001). Spontaneous and precipitated withdrawal with a synthetic cannabinoid, WIN 55212-2. *European Journal of Pharmacology* **416**, 75–81.

Adams, I. B., Compton, D. R., and Martin, B. R. (1998). Assessment of anandamide interaction with the cannabinoid brain receptor: SR 141716A antagonism studies in mice and autoradiographic analysis of receptor binding in rat brain. *Journal of Pharmacology and Experimental Therapeutics* **284**, 1209–1217.

AIDS Patient Care STDS. (2004). Marijuana effective for polyneuropathy. *AIDS Patient Care STDS* **18**, 251.

AIDS Reader. (2004). Marijuana eases HIV-related nerve pain. *AIDS Reader* **14**, 164–165.

Ameri, A. (1999). The effects of cannabinoids on the brain. *Progress in Neurobiology* **58**, 315–348.

American Psychiatric Association. (1994). *Diagnostic and Statistical Manual of Mental Disorders*, 4th ed. American Psychiatric Press, Washington DC.

American Psychiatric Association. (2000). *Diagnostic and Statistical Manual of Mental Disorders*, 4th ed. text revision, American Psychiatric Press, Washington DC.

Andreasson, S., Allebeck, P., Engstrom, A., and Rydberg, U. (1987). Cannabis and schizophrenia. A longitudinal study of Swedish conscripts. *Lancet* **2(8574)**, 1483–1486.

Anthony, J. C., Warner, L. A., and Kessler, R. C. (1994). Comparative epidemiology of dependence on tobacco, alcohol, controlled substances, and inhalants: Basic findings from the National Comorbidity Survey. *Experimental and Clinical Psychopharmacology* **2**, 244–268.

Aronow, W. S., and Cassidy, J. (1974). Effect of marihuana and placebo-marihuana smoking on angina pectoris. *New England Journal of Medicine* **291**, 65–67.

Arseneault, L., Cannon, M., Poulton, R., Murray, R., Caspi, A., and Moffitt, T. E. (2002). Cannabis use in adolescence and risk for adult psychosis: longitudinal prospective study. *British Medical Journal* **325**, 1212–1213.

Arseneault, L., Cannon, M., Witton, J., and Murray, R. M. (2004). Causal association between cannabis and psychosis: examination of the evidence. *British Journal of Psychiatry* **184**, 110–117.

Ashton, C. H. (1999). Adverse effects of cannabis and cannabinoids. *British Journal of Anaesthesia* **83**, 637–649.

Attal, N., Brasseur, L., Guirimand, D., Clermond-Gnamien, S., Atlami, S., and Bouhassira, D. (2004). Are oral cannabinoids safe and effective in refractory neuropathic pain? *European Journal of Pain* **8**, 173–177.

Avico, U., Pacifici, R., and Zuccaro, P. (1985). Variations of tetrahydrocannabinol content in cannabis plants to distinguish the fibre-type from drug-type plants. *Bulletin on Narcotics* **37**, 61–65.

Balster, R. L., and Prescott, W. R. (1992). Delta 9-tetrahydrocannabinol discrimination in rats as a model for cannabis intoxication. *Neuroscience and Biobehavioral Reviews* **16**, 55–62.

Bar-Joseph, A., Berkovitch, Y., Adamchik, J., and Biegon, A. (1994). Neuroprotective activity of HU-211, a novel NMDA antagonist, in global ischemia in gerbils. *Molecular and Chemical Neuropathology* **23**, 125–135.

Barrett, R. L., Wiley, J. L., Balster, R. L., and Martin, B. R. (1995). Pharmacological specificity of delta 9-tetrahydrocannabinol discrimination in rats. *Psychopharmacology* **118**, 419–424.

Baskfield, C. Y., Martin, B. R., and Wiley, J. L. (2004). Differential effects of delta9-tetrahydrocannabinol and methanandamide in CB1 knockout and wild-type mice. *Journal of Pharmacology and Experimental Therapeutics* **309**, 86–91.

Beal, J. E., Olson, R., Laubenstein, L., Morales, J. O., Bellman, P., Yangco, B., Lefkowitz, L., Plasse, T. F., and Shepard, K. V. (1995). Dronabinol as a treatment for anorexia associated with weight loss in patients with AIDS. *Journal of Pain and Symptom Management* **10**, 89–97.

Beal, J. E., Olson, R., Lefkowitz, L., Laubenstein, L., Bellman, P., Yangco, B., Morales, J. O., Murphy, R., Powderly, W., Plasse, T. F., Mosdell, K. W., and Shepard, K. V. (1997). Long-term efficacy and safety of dronabinol for acquired immunodeficiency syndrome-associated anorexia. *Journal of Pain and Symptom Management* **14**, 7–14.

Beardsley, P. M., and Martin, B. (2000). Effects of the cannabinoid CB(1) receptor antagonist, SR141716A, after delta(9)-tetrahydrocannabinol withdrawal. *European Journal of Pharmacology* **387**, 47–53.

Belayev, L., Bar-Joseph, A., Adamchik, J., and Biegon, A. (1995a). HU-211, a nonpsychotropic cannabinoid, improves neurological signs and reduces brain damage after severe forebrain ischemia in rats. *Molecular and Chemical Neuropathology* **25**, 19–33.

Belayev, L., Busto, R., Zhao, W., and Ginsberg, M. D. (1995b). HU-211, a novel noncompetitive, N-methyl-D-aspartate antagonist, improves neurological deficit and reduces infarct volume after reversible focal cerebral ischemia in the rat. *Stroke* **26**, 2313–2319.

Beltramo, M., Stella, N., Calignano, A., Lin, S. Y., Makriyannis, A., and Piomelli, D. (1997). Functional role of high-affinity anandamide transport, as revealed by selective inhibition. *Science* **277**, 1094–1097.

Benowitz, N. L., and Jones, R. T. (1975). Cardiovascular effects of prolonged delta-9-tetrahydrocannabinol ingestion. *Clinical Pharmacology and Therapeutics* **18**, 287–297.

Binitie, A. (1975). Psychosis following ingestion of hemp in children. *Psychopharmacologia* **44**, 301–302.

Bloodworth, R. C. (1985). Medical problems associated with marijuana abuse. *Psychiatric Medicine* **3**, 173–184.

Bloomquist, E. R. (1967). Marijuana: social benefit or social detriment? *California Medicine* **106**, 346–353.

Booth, M. (2003). *Cannabis: A History*. Doubleday, London.

Braida, D., Pozzi, M., Cavallini, R., and Sala, M. (2001a). Conditioned place preference induced by the cannabinoid agonist CP 55 ,940: interaction with the opioid system. *Neuroscience* **104**, 923–926.

Braida, D., Pozzi, M., Parolaro, D., and Sala, M. (2001b). Intracerebral self-administration of the cannabinoid receptor agonist CP 55940 in the rat: interaction with the opioid system. *European Journal of Pharmacology* **413**, 227–234.

Brenneisen, R., Egli, A., Elsohly, M. A., Henn, V., and Spiess, Y. (1996). The effect of orally and rectally administered delta 9-tetrahydrocannabinol on spasticity: a pilot study with 2 patients. *International Journal of Clinical Pharmacology and Therapeutics* **34**, 446–452.

Bromberg, W. (1934). Marihuana intoxication: a clinical study of Cannabis sativa intoxication. *American Journal of Psychiatry* **91**, 303–330.

Brookoff, D., Cook, C. S., Williams, C., and Mann, C. S. (1994). Testing reckless drivers for cocaine and marijuana. *New England Journal of Medicine* **331**, 518–522.

Buckley, N. E., McCoy, K. L., Mezey, E., Bonner, T., Zimmer, A., Felder, C. C., Glass, M., and Zimmer, A. (2000). Immunomodulation by cannabinoids is absent in mice deficient for the cannabinoid CB(2) receptor. *European Journal of Pharmacology* **396**, 141–149.

Budney, A. J., Radonovich, K. J., Higgins, S. T., and Wong, C. J. (1998). Adults seeking treatment for marijuana dependence: a comparison with cocaine-dependent treatment seekers. *Experimental and Clinical Psychopharmacology* **6**, 419–426.

Budney, A. J., Novy, P. L., and Hughes, J. R. (1999). Marijuana withdrawal among adults seeking treatment for marijuana dependence. *Addiction* **94**, 1311–1322.

Budney, A. J., Higgins, S. T., Radonovich, K. J., and Novy, P. L. (2000). Adding voucher-based incentives to coping skills and motivational enhancement improves outcomes during treatment for marijuana dependence. *Journal of Consulting and Clinical Psychology* **68**, 1051–1061.

Budney, A. J., Hughes, J. R., Moore, B. A., and Novy, P. L. (2001). Marijuana abstinence effects in marijuana smokers maintained in their home environment. *Archives of General Psychiatry* **58**, 917–924.

Budney, A. J., Moore, B. A., Vandrey, R. G., and Hughes, J. R. (2003). The time course and significance of cannabis withdrawal. *Journal of Abnormal Psychology* **112**, 393–402.

Buggy, D. J., Toogood, L., Maric, S., Sharpe, P., Lambert, D. G., and Rowbotham, D. J. (2003). Lack of analgesic efficacy of oral delta-9-tetrahydrocannabinol in postoperative pain. *Pain* **106**, 169–172.

Calignano, A., Katona, I., Desarnaud, F., Giuffrida, A., La Rana, G., Mackie, K., Freund, T. F., and Piomelli, D. (2000). Bidirectional control of airway responsiveness by endogenous cannabinoids. *Nature* **408**, 96–101.

Camilleri, M. (2004). Treating irritable bowel syndrome: overview, perspective and future therapies. *British Journal of Pharmacology* **141**, 1237–1248.

Carley, D. W., Paviovic, S., Janelidze, M., and Radulovacki, M. (2002). Functional role for cannabinoids in respiratory stability during sleep. *Sleep* **25**, 391–398.

Carlini, E. A., and Cunha, J. M. (1981). Hypnotic and antiepileptic effects of cannabidiol. *Journal of Clinical Pharmacology* **21**(8–9 Suppl), 417s–427s.

Castane, A., Valjent, E., Ledent, C., Parmentier, M., Maldonado, R., and Valverde, O. (2002). Lack of CB1 cannabinoid receptors modifies nicotine behavioural responses, but not nicotine abstinence. *Neuropharmacology* **43**, 857–867.

Castane, A., Robledo, P., Matifas, A., Kieffer, B. L., and Maldonado, R. (2003). Cannabinoid withdrawal syndrome is reduced in double mu and delta opioid receptor knockout mice. *European Journal of Neuroscience* **17**, 155–159 [erratum: **17**, 427].

Caulfield, M. P., and Brown, D. A. (1992). Cannabinoid receptor agonists inhibit Ca current in NG108–15 neuroblastoma cells via a pertussis toxin-sensitive mechanism. *British Journal of Pharmacology* **106**, 231–232.

Chait, L. D. (1989). Delta-9-tetrahydrocannabinol content and human marijuana self-administration. *Psychopharmacology* **98**, 51–55.

Chait, L. D., and Zacny, J. P. (1992). Reinforcing and subjective effects of oral delta 9-THC and smoked marijuana in humans. *Psychopharmacology* **107**, 255–262.

Chang, K. (1968). *The Archaeology of Ancient China*. pp. 111–112. Yale University Press, New Haven CT.

Chaperon, F., Soubrie, P., Puech, A. J., and Thiebot, M. H. (1998). Involvement of central cannabinoid (CB1) receptors in the establishment of place conditioning in rats. *Psychopharmacology* **135**, 324–332.

Cheer, J. F., Kendall, D. A., Mason, R., and Marsden, C. A. (2003). Differential cannabinoid-induced electrophysiological effects in rat ventral tegmentum. *Neuropharmacology* **44**, 633–641.

Chen, J., Paredes, W., Lowinson, J. H., and Gardner, E. L. (1990a). Delta 9-tetrahydrocannabinol enhances presynaptic dopamine efflux in medial prefrontal cortex. *European Journal of Pharmacology* **190**, 259–262.

Chen, J. P., Paredes, W., Li, J., Smith, D., Lowinson, J., and Gardner, E. L. (1990b). Delta 9-tetrahydrocannabinol produces naloxone-blockable enhancement of presynaptic basal dopamine efflux in nucleus accumbens of conscious, freely-moving rats as measured by intracerebral microdialysis. *Psychopharmacology* **102**, 156–162.

Chen, J. P., Paredes, W., Lowinson, J. H., and Gardner, E. L. (1991). Strain-specific facilitation of dopamine efflux by delta 9-tetrahydrocannabinol in the nucleus accumbens of rat: An in vivo microdialysis study. *Neuroscience Letters* **129**, 136–140.

Cheng, T. (1966). *New Light on Prehistoric China* (series title: *Archaeology in China*, vol. 1 Suppl. 1), W. Heffer and Sons, Cambridge.

Chiang, C. W., Barnett, G., and Brine, D. (1983). Systemic absorption of delta 9-tetrahydrocannabinol after ophthalmic administration to the rabbit. *Journal of Pharmaceutical Sciences* **72**, 136–138.

Cichewicz, D. L. (2004). Synergistic interactions between cannabinoid and opioid analgesics. *Life Sciences* **74**, 1317–1324.

Clarke, R. C., and Watson, D. P. (2002). Botany of natural *Cannabis* medicines. *In Cannabis and Cannabinoids: Pharmacology, Toxicology and Therapeutic Potential* (F. Grotenhermen, and E. Russo, Eds.), pp. 3–13. Haworth Integrative Healing Press, New York.

Clifford, D. B. (1983). Tetrahydrocannabinol for tremor in multiple sclerosis. *Annals of Neurology* **13**, 669–671.

Cohen, S. (1980a). Cannabis: impact on motivation, Part 1. *Drug Abuse and Alcoholism Newsletter*, vol. 9(10).

Cohen, S. (1980b). Cannabis: impact on motivation, Part 2. *Drug Abuse and Alcoholism Newsletter*, vol. 10(1).

Cohen, S., Lessin, P. J., Hahn, P. M., and Tyrrell, E. D. (1976). A 94–day cannabis study. *In Pharmacology of Marihuana*, vol. 2, (M. C. Braude, and S. Szara, Eds.), pp. 621–626. Raven Press, New York.

Compton, D. R., Johnson, M. R., Melvin, L. S., and Martin, B. R. (1992). Pharmacological profile of a series of bicyclic cannabinoid analogs: classification as cannabimimetic agents. *Journal of Pharmacology and Experimental Therapeutics* **260**, 201–209.

Compton, D. R., Rice, K. C., De Costa, B. R., Razdan, R. K., Melvin, L. S., Johnson, M. R., and Martin, B. R. (1993). Cannabinoid structure-activity relationships: correlation of receptor binding and in vivo activities. *Journal of Pharmacology and Experimental Therapeutics* **265**, 218–226.

Consroe, P., Laguna, J., Allender, J., Snider, S., Stern, L., Sandyk, R., Kennedy, K., and Schram, K. (1991). Controlled clinical trial of cannabidiol in Huntington's disease. *Pharmacology Biochemistry and Behavior* **40**, 701–708.

Copeland, J., Swift, W., Roffman, R., and Stephens, R. (2001). A randomized controlled trial of brief cognitive-behavioral interventions for cannabis use disorder. *Journal of Substance Abuse Treatment* **21**, 55–64.

Cornelius, J. R., Salloum, I. M., Haskett, R. F., Ehler, J. G., Jarrett, P. J., Thase, M. E., and Perel, J. M. (1999). Fluoxetine versus placebo for the marijuana use of depressed alcoholics. *Addictive Behaviors* **24**, 111–114.

Cossu, G., Ledent, C., Fattore, L., Imperato, A., Bohme, G. A., Parmentier, M., and Fratta, W. (2001). Cannabinoid, CB1 receptor knockout mice fail to self-administer morphine but not other drugs of abuse. *Behavioural Brain Research* **118**, 61–65.

Cravatt, B. F., Prospero-Garcia, O., Siuzdak, G., Gilula, N. B., Henriksen, S. J., Boger, D. L., and Lerner, R. A. (1995). Chemical characterization of a family of brain lipids that induce sleep. *Science* **268**, 1506–1509.

Cravatt, B. F., Giang, D. K., Mayfield, S. P., Boger, D. L., Lerner, R. A., and Gilula, N. B. (1996). Molecular characterization of an enzyme that degrades neuromodulatory fatty-acid amides. *Nature* **384**, 83–87.

Cravatt, B. F., Demarest, K., Patricelli, M. P., Bracey, M. H., Giang, D. K., Martin, B. R., and Lichtman, A. H. (2001). Supersensitivity to anandamide and enhanced endogenous cannabinoid signaling in mice lacking fatty acid amide hydrolase. *Proceedings of the National Academy of Sciences USA* **98**, 9371–9376.

Crawford, W. J., and Merritt, J. C. (1979). Effects of tetrahydrocannabinol on arterial and intraocular hypertension. *International Journal of Clinical Pharmacology and Biopharmacy* **17**, 191–196.

Crawley, J. N., Corwin, R. L., Robinson, J. K., Felder, C. C., Devane, W. A., and Axelrod, J. (1993). Anandamide, an endogenous ligand of the cannabinoid receptor, induces hypomotility and hypothermia in vivo in rodents. *Pharmacology Biochemistry and Behavior* **46**, 967–972.

Crowley, T. J., Macdonald, M. J., Whitmore, E. A., and Mikulich, S. K. (1998). Cannabis dependence, withdrawal, and reinforcing effects among adolescents with conduct symptoms and substance use disorders. *Drug and Alcohol Dependence* **50**, 27–37.

Cunha, J. M., Carlini, E. A., Pereira, A. E., Ramos, O. L., Pimentel, C., Gagliardi, R., Sanvito, W. L., Lander, N., and Mechoulam, R. (1980). Chronic administration of cannabidiol to healthy volunteers and epileptic patients. *Pharmacology* **21**, 175–185.

D'Souza, D. C., Cho, H. S., Perry, E. B., and Krystal, J. H. (2004). Cannabinoid 'model psychosis, dopamine-cannabinoid interactions and implications for schizophrenia. *In Marijuana and Madness: Psychiatry and Neurobiology* (D. J. Castle, and R. Murray, Eds.), pp. 142–165. Cambridge University Press, Cambridge.

Dalton, W. S., Martz, R., Lemberger, L., Rodda, B. E., and Forney, R. B. (1976). Influence of cannabidiol on delta-9-tetrahydrocannabinol effects. *Clinical Pharmacology and Therapeutics* **19**, 300–309.

Dansak, D. A. (1997). As an antiemetic and appetite stimulant in cancer patients. *In Cannabis in Medical Practice: A Legal, Historical and Pharmacological Overview of the Therapeutic Use of Marijuana* (M. L. Mathre, Ed.), McFarland, Jefferson NC.

Davis, W. M., and Hatoum, N. S. (1983). Neurobehavioral actions of cannabichromene and interactions with delta 9-tetrahydrocannabinol. *General Pharmacology* **14**, 247–252.

De Vry, J., Denzer, D., Reissmueller, E., Eijckenboom, M., Heil, M., Meier, H., and Mauler, F. (2004). 3-[2-Cyano-3-(trifluoromethyl) phenoxy]phenyl-4,4,4-trifluoro-1-butanesulfonate (BAY 59-3074): a novel cannabinoid CB1/CB2 receptor partial agonist with antihyperalgesic and antiallodynic effects. *Journal of Pharmacology and Experimental Therapeutics* **310**, 620–632.

Deadwyler, S. A., Hampson, R. E., Bennett, B. A., Edwards, T. A., Mu, J., Pacheco, M. A., Ward, S. J., and Childers, S. R. (1993). Cannabinoids modulate potassium current in cultured hippocampal neurons. *Receptors and Channels* **1**, 121–134.

Department of Justice. (1999). Drug Enforcement Administration, Schedules of controlled substances: rescheduling of the Food and Drug Administration approved product containing synthetic dronabinol [(-)Δ^9-(trans)-tetrahydrocannabinol] in sesame oil and encapsulated in soft gelatin capsules from schedule II to schedule III. Final rule. *Federal Register* **64**, 35928–35930.

Devane, W. A., Dysarz, F. A. 3rd, Johnson, M. R., Melvin, L. S., and Howlett, A. C. (1988). Determination and characterization of a cannabinoid receptor in rat brain. *Molecular Pharmacology* **34**, 605–613.

Devane, W. A., Hanus, L., Breuer, A., Pertwee, R. G., Stevenson, L. A., Griffin, G., Gibson, D., Mandelbaum, A., Etinger, A., and Mechoulam, R. (1992). Isolation and structure of a brain constituent that binds to the cannabinoid receptor. *Science* **258**, 1946–1949.

Dewey, W. L. (1986). Cannabinoid pharmacology. *Pharmacological Reviews* **38**, 151–178.

Di Filippo, C., Rossi, F., Rossi, S., and D'Amico, M. (2004). Cannabinoid CB2 receptor activation reduces mouse myocardial ischemia-reperfusion injury: involvement of cytokine/chemokines and PMN. *Journal of Leukocyte Biology* **75**, 453–459.

Di Marzo, V., Breivogel, C., Bisogno, T., Melck, D., Patrick, G., Tao, Q., Szallasi, A., Razdan, R.K., and Martin, B. R. (2000a). Neurobehavioral activity in mice of N-vanillyl-arachidonyl-amide. *European Journal of Pharmacology* **406**, 363–374.

Di Marzo, V., Breivogel, C. S., Tao, Q., Bridgen, D. T., Razdan, R. K., Zimmer, A. M., Zimmer, A., and Martin, B. R. (2000b). Levels, metabolism, and pharmacological activity of anandamide in CB(1) cannabinoid receptor knockout mice: evidence for non-CB(1), non-CB(2) receptor-mediated actions of anandamide in mouse brain. *Journal of Neurochemistry* **75**, 2434–2444.

Diana, M., Melis, M., and Gessa, G. L. (1998a). Increase in meso-prefrontal dopaminergic activity after stimulation of CB1 receptors by cannabinoids. *European Journal of Neuroscience* **10**, 2825–2830.

Diana, M., Melis, M., Muntoni, A. L., and Gessa, G. L. (1998b). Mesolimbic dopaminergic decline after cannabinoid withdrawal. *Proceedings of the National Academy of Sciences USA* **95**, 10269–10273.

Dinh, T. P., Freund, T. F., and Piomelli, D. (2002). A role for monoglyceride lipase in 2-arachidonoylglycerol inactivation. *Chemistry and Physics of Lipids* **121**, 149–158.

Doorenbos, N. J., Fetterman, P. S., Quimby, M. W., and Turner, C. E. (1971). Cultivation, extraction, and analysis of *Cannabis sativa L. In Marijuana: Chemistry, Pharmacology, and Patterns of Social Use* (series title: *Annals of the New York Academy of Sciences*, vol. 191), (A. J. Singer, Ed.), pp. 3–14. New York Academy of Sciences, New York.

Dornbush, R. L., Clare, G., Zaks, A., Crown, P., Volavka, J., and Fink, M. (1972). 21–day administration of marijuana in male volunteers. *In Current Research in Marijuana* (M. F. Lewis, Ed.), pp. 115–128. Academic Press, New York.

Egertova, M., Giang, D. K., Cravatt, B. F., and Elphick, M. R. (1998). A new perspective on cannabinoid signalling: complementary localization of fatty acid amide hydrolase and the, CB1 receptor in rat brain. *Proceedings of the Royal Society of London (Series B: Biological Sciences* **265**, 2081–2085.

Ellenhorn, M. J., and Barceloux, D. G. (1988). *Medical Toxicology: Diagnosis and Treatment of Human Poisoning* Elsevier, New York.

Ellis, G. M. Jr., Mann, M. A., Judson, B. A., Schramm, N. T., and Tashchian, A. (1985). Excretion patterns of cannabinoid metabolites after last use in a group of chronic users. *Clinical Pharmacology and Therapeutics* **38**, 572–578.

Elphick, M. R., and Egertova, M. (2001). The neurobiology and evolution of cannabinoid signaling. *Philosophical Transactions: Series B. Biological Sciences* **356**, 381–408.

Elsner, F., Radbruch, L., and Sabatowski, R. (2001) Tetrahydrocannabinol for treatment of chronic pain. *Schmerz*, **15**, 200–204. [German]

ElSohly, M. A., and Holley, J. H. (1983). *Potency Monitoring Project* (series title: *University of Mississippi Quarterly Report*, vol. 5), University of Mississippi, Jackson MS.

ElSohly, M. A., Stanford, D. F., Harland, E. C., Hikal, A. H., Walker, L. A., Little, T. L. Jr., Rider, J. N., and Jones, A. B. (1991). Rectal bioavailability of delta-9-tetrahydrocannabinol from the hemisuccinate ester in monkeys. *Journal of Pharmaceutical Sciences* **80**, 942–945.

ElSohly, M. A., Ross, S. A., Mehmedic, Z., Arafat, R., Yi, B., and Banahan, B. F. 3rd ed. (2000). Potency trends of delta9-THC and other cannabinoids in confiscated marijuana from 1980–1997. *Journal of Forensic Sciences* **45**, 24–30.

ElSohly, M. A. (2002). Chemical constituents of *Cannabis*. *In Cannabis and Cannabinoids: Pharmacology, Toxicology and Therapeutic Potential* (F. Grotenhermen, and E. Russo, Eds.), pp. 27–36. Haworth Integrative Healing Press, New York.

Fairbairn, J. W., Liebmann, J. A., and Simic, S. (1971). The tetrahydrocannabinol content of cannabis leaf. *Journal of Pharmacy and Pharmacology* **23**, 558–559.

Fattore, L., Cossu, G., Martellotta, C. M., and Fratta, W. (2001). Intravenous self-administration of the cannabinoid CB1 receptor agonist WIN 55,212–2 in rats. *Psychopharmacology* **156**, 410–416.

Feeney, D. M. (1976). The marijuana window: a theory of cannabis use. *Behavioral Biology* **18**, 455–471.

Fergusson, D. M., and Horwood, L. J., and Swain-Campbell, N. R. (2003). Cannabis dependence and psychotic symptoms in young people. *Psychological Medicine* **33**, 15–21.

Fetterman, P. S., Doorenbos, N. J., Keith, E. S., and Quimby, M. W. (1971a). A simple gas liquid chromatography procedure for determination of cannabinoidic acids in Cannabis sativa L. *Experientia* **27**, 988–990.

Fetterman, P. S., Keith, E. S., Waller, C. W., Guerrero, O., Doorenbos, N. J., and Quimby, M. W. (1971b). Mississippi-grown Cannabis sativa L: preliminary observation on chemical definition of phenotype and variations in tetrahydrocannabinol content versus age, sex, and plant part. *Journal of Pharmaceutical Science* **60**, 1246–1249.

Fink, M., Volavka, J., Panayiotopoulos, C. P., and Stefanis, C. (1976). Quantitative, EEG studies of marihuana, Δ^9-tetrahydrocannabinol and hashish in man. *In Pharmacology of Marihuana*, vol. 1, (M. C. Braude, and S. Szara, Eds.), pp. 383–391. Raven Press, New York.

Flach, A. J. (2002). Delta-9-tetrahydrocannabinol (THC) in the treatment of end-stage open-angle glaucoma. *Transactions of the American Ophthalmological Society* **100**, 215–222.

Fox, P., Bain, P. G., Glickman, S., Carroll, C., and Zajicek, J. (2004). The effect of cannabis on tremor in patients with multiple sclerosis. *Neurology* **62**, 1105–1109.

Fox, S.H., Kellett, M., Moore, A. P., Crossman, A. R., and Brotchie, J. M. (2002). Randomised, double-blind, placebo-controlled trial to assess the potential of cannabinoid receptor stimulation in the treatment of dystonia. *Movement Disorders* **17**, 145–149.

French, E. D. (1997). Delta9-tetrahydrocannabinol excites rat VTA dopamine neurons through activation of cannabinoid CB1 but not opioid receptors. *Neuroscience Letters* **226**, 159–162.

French, E. D., Dillon, K., and Wu, X. (1997). Cannabinoids excite dopamine neurons in the ventral tegmentum and substantia nigra. *Neuroreport* **8**, 649–652.

Freund, T. F., Katona, I., and Piomelli, D. (2003). Role of endogenous cannabinoids in synaptic signaling. *Physiological Reviews* **83**, 1017–1066.

Fride, E., and Mechoulam, R. (1993). Pharmacological activity of the cannabinoid receptor agonist, anandamide, a brain constituent. *European Journal of Pharmacology* **231**, 313–314.

Frytak, S., Moertel, C. G., and Rubin, J. (1984). Metabolic studies of delta-9-tetrahydrocannabinol in cancer patients. *Cancer Treatment Reports* **68**, 1427–1431.

Gaoni, Y., and Mechoulam, R. (1964). Isolation, structure, and partial synthesis of an active constituent of hashish. *Journal of the American Chemical Society* **86**, 1646–1647.

Gardner, E. L. (2002). Addictive potential of cannabinoids: the underlying neurobiology. *Chemistry and Physics of Lipids* **121**, 267–290.

Gardner, E. L., Paredes, W., Smith, D., Donner, A., Milling, C., Cohen, D., and Morrison, D. (1988). Facilitation of brain stimulation reward by delta-9-tetrahydrocannabinol. *Psychopharmacology* **96**, 142–144.

Gardner, E. L., Paredes, W., Smith, D., and Zukin, R. S. (1989). Facilitation of brain stimulation reward by Δ^9-tetrahydrocannabinol is mediated by an endogenous opioid mechanism. *In Progress in Opioid Research* (series title: *Advances in the Biosciences*, vol. 75), (J. Cros, J. C. Meunier, and M. Hamon, Eds.), pp. 671–674. Pergamon Press, Oxford.

Gardner, E. L., Chen, J., Paredes, W., Smith, D., Li, J., and Lowinson, J. (1990a). Enhancement of presynaptic dopamine efflux in brain by delta-9-tetrahydrocannabinol is mediated by an endogenous opioid mechanism. *In New Leads in Opioid Research* (J. M. van Ree, A. H. Mulder, V. M. Wiegant, and T. B. van Wimersma Greidanus, Eds.), pp. 243–245. Elsevier, New York.

Gardner, E. L., and Lowinson, J. H. (1991). Marijuana's interaction with brain reward systems: update 1991. *Pharmacology Biochemistry and Behavior* **40**, 571–580.

Gardner, E. L., and Vorel, S. R. (1998). Cannabinoid transmission and reward-related events. *Neurobiology of Disease* **5**, 502–533.

Gauter, B., Rukwied, R., and Konrad, C. (2004). Cannabinoid agonists in the treatment of blepharospasm-a case report study. *Neuroendocrinology Letters* **25**, 45–48.

Georgotas, A., and Zeidenberg, P. (1979). Observations on the effects of four weeks of heavy marihuana smoking on group interaction and individual behavior. *Comprehensive Psychiatry* **20**, 427–432.

Gerdeman, G. L., Ronesi, J., and Lovinger, D. M. (2002). Postsynaptic endocannabinoid release is critical to long-term depression in the striatum. *Nature Neuroscience* **5**, 446–451.

Gessa, G. L., Melis, M., Muntoni, A. L., and Diana, M. (1998). Cannabinoids activate mesolimbic dopamine neurons by an action on cannabinoid CB1 receptors. *European Journal of Pharmacology* **341**, 39–44.

Ghozland, S., Matthes, H. W., Simonin, F., Filliol, D., Kieffer, B. L., and Maldonado, R. (2002). Motivational effects of cannabinoids are mediated by mu-opioid and kappa-opioid receptors. *Journal of Neuroscience* **22**, 1146–1154.

Gifford, A. N., Gardner, E. L., and Ashby, C. R. Jr. (1997). The effect of intravenous administration of delta-9-tetrahydrocannabinol on the activity of A10 dopamine neurons recorded in vivo in anesthetized rats. *Neuropsychobiology* **36**, 96–99.

Gill, E. W., Paton, W. D., and Pertwee, R. G. (1970). Preliminary experiments on the chemistry and pharmacology of cannabis. *Nature* **228**, 134–136.

Gill, E. W., and Jones, G. (1972). Brain levels of delta1-tetrahydrocannabinol and its metabolites in mice: correlation with behaviour, and the effect of the metabolic inhibitors SKF 525A and piperonyl butoxide. *Biochemical Pharmacology* **21**, 2237–2248.

Gilson, I., and Busalacchi, M. (1998). Marijuana for intractable hiccups. *Lancet* **351**, 267.

Glickfeld, L. L., and Scanziani, M. (2005). Self-administering cannabinoids. *Trends in Neurosciences* **28**, 341–343.

Gold, L. H., Balster, R. L., Barrett, R. L., Britt, D. T., and Martin, B. R. (1992). A comparison of the discriminative stimulus properties of delta-9-tetrahydrocannabinol and CP 55 940 in rats and rhesus monkeys. *Journal of Pharmacology and Experimental Therapeutics* **262**, 479–486.

Gonzalez, S., Romero, J., de Miguel, R., Lastres-Becker, I., Villanua, M. A., Makriyannis, A., Ramos, J. A., and Fernandez-Ruiz, J. J. (1999). Extrapyramidal and neuroendocrine effects of AM404, an inhibitor of the carrier-mediated transport of anandamide. *Life Sciences* **65**, 327–336.

Graves, K. L., and Leigh, B. C. (1995). The relationship of substance use to sexual activity among young adults in the United States. *Family Planning Perspectives* **27**, 18–22, 33.

Grech, A., Takei, N., and Murray, R. M. (1998). Comparison of cannabis use in psychotic patients and controls in London and Malta. *Schizophrenia Research* **29**, 22.

Green, B., Kavanagh, D., and Young, R. (2003). Being stoned: a review of self-reported cannabis effects. *Drug and Alcohol Review* **22**, 453–460.

Grinspoon, L., and Bakalar, J. B. (1998). The use of cannabis as a mood stabilizer in bipolar disorder: anecdotal evidence and the need for clinical research. *Journal of Psychoactive Drugs* **30**, 171–177.

Grotenhermen, F. (2002). Review of therapeutic effects. *In Cannabis and Cannabinoids: Pharmacology, Toxicology, and Therapeutic Potential* (F. Grotenhermen, and E. Russo, Eds.), pp. 123–142. Haworth Integrative Healing Press, New York.

Grotenhermen, F. (2003). Pharmacokinetics and pharmacodynamics of cannabinoids. *Clinical Pharmacokinetics* **42**, 327–360.

Gruber, A. J., and Pope, H. G. Jr. (2002). Marijuana use among adolescents. *Pediatric Clinics of North America* **49**, 389–413.

Grunfeld, Y., and Edery, H. (1969). Psychopharmacological activity of the active constituents of hashish and some related cannabinoids. *Psychopharmacologia* **14**, 200–210.

Guy, G. W., and Flint, M. E. (2000). A phase one study of sublingual Cannabis based medicinal extracts. *In 2000 Symposium on the Cannabinoids*, Burlington VT, International Cannabinoid Research Society, abstract# 115.

Hall, W., and Degenhardt, L. (2004). Is there a specific 'cannabis psychosis'? *In Marijuana and Madness: Psychiatry and Neurobiology* (D. J. Castle, and R. Murray, Eds.), pp. 89–100. Cambridge University Press, Cambridge.

Hall, W., and Solowij, N. (1998). Adverse effects of cannabis. *Lancet* **352**, 1611–1616.

Haller, J., Bakos, N., Szirmay, M., Ledent, C., and Freund, T. F. (2002). The effects of genetic and pharmacological blockade of the CB1 cannabinoid receptor on anxiety. *European Journal of Neuroscience* **16**, 1395–1398.

Hampson, A. (2002). Cannabinoids as neuroprotectants against ischemia. *In Cannabis and Cannabinoids: Pharmacology, Toxicology and Therapeutic Potential* (F. Grotenhermen, and E. Russo, Eds.), pp. 101–109. Haworth Integrative Healing Press, New York.

Haney, M., Comer, S. D., Ward, A. S., Foltin, R. W., and Fischman, M. W. (1997). Factors influencing marijuana self-administration by humans. *Behavioural Pharmacology* **8**, 101–112.

Haney, M., Ward, A. S., Comer, S. D., Foltin, R. W., and Fischman, M. W. (1999a). Abstinence symptoms following oral THC administration to humans. *Psychopharmacology* **141**, 385–394.

Haney, M., Ward, A. S., Comer, S. D., Foltin, R. W., and Fischman, M. W. (1999b). Abstinence symptoms following smoked marijuana in humans. *Psychopharmacology* **141**, 395–404.

Haney, M., Bisaga, A., and Foltin, R. W. (2003). Interaction between naltrexone and oral THC in heavy marijuana smokers. *Psychopharmacology* **166**, 77–85.

Harrold, J. A., and Williams, G. (2003). The cannabinoid system: a role in both the homeostatic and hedonic control of eating? *British Journal of Nutrition* **90**, 729–734.

Hart, C. L., Ward, A. S., Haney, M., Comer, S. D., Foltin, R. W., and Fischman, M. W. (2002). Comparison of smoked marijuana and oral Delta(9)-tetrahydrocannabinol in humans. *Psychopharmacology* **164**, 407–415.

Hartley, J. P., Nogrady, S. G., and Seaton A. (1978). Bronchodilator effect of delta1-tetrahydrocannabinol. *British Journal of Clinical Pharmacology* **5**, 523–525.

Harvey, D. J., Leuschner, J. T., and Paton, W. D. (1982). Gas chromatographic and mass spectrometric studies on the metabolism and pharmacokinetics of delta 1-tetrahydrocannabinol in the rabbit. *Journal of Chromatography* **239**, 243–250.

Heishman, S. J., Arasteh, K., and Stitzer, M. L. (1997). Comparative effects of alcohol and marijuana on mood, memory, and performance. *Pharmacology Biochemistry and Behavior* **58**, 93–101.

Hemming, M., and Yellowlees, P. M. (1993). Effective treatment of Tourette's syndrome with marijuana. *Journal of Psychopharmacology* **7**, 389–391.

Hendin, H., and Haas, A. P. (1985). The adaptive significance of chronic marijuana use for adolescents and adults. *Advances in Alcohol and Substance Abuse* **4**, 99–115.

Hendrickson, R. (1987). *The Facts on File Encyclopedia of Word and Phrase Origins*. Facts on File Publications, New York.

Hepler, R. S., and Frank, I. R. (1971). Marihuana smoking and intraocular pressure. *Journal of the American Medical Association* **217**, 1392.

Herkenham, M., Lynn, A. B., Little, M. D., Johnson, M. R., Melvin, L. S., de Costa, B. R., and Rice, K. C. (1990). Cannabinoid receptor localization in brain. *Proceedings of the National Academy of Sciences USA* **87**, 1932–1936.

Herkenham, M., Lynn, A. B., Johnson, M. R., Melvin, L. S., de Costa, B. R., and Rice, K. C. (1991). Characterization and localization of cannabinoid receptors in rat brain: a quantitative in vitro autoradiographic study. *Journal of Neuroscience* **11**, 563–583.

Ho, B. T., Fritchie, G. E., Kralik, P. M., Englert, L. F., McIsaac, W. M., and Idanpaan-Heikkila, J. (1970). Distribution of tritiated-1 delta 9-tetrahydrocannabinol in rat tissues after inhalation. *Journal of Pharmacy and Pharmacology* **22**, 538–539.

Hoffman, A. F., and Lupica, C. R. (2001). Direct actions of cannabinoids on synaptic transmission in the nucleus accumbens: a comparison with opioids. *Journal of Neurophysiology* **85**, 72–83.

Hollister, L. E. (1973). Cannabidiol and cannabinol in man. *Experientia* **29**, 825–826.

Hollister, L. E., Gillespie, H. K., Ohlsson, A., Lindgren, J. E., Wahlen, A., and Agurell, S. (1981). Do plasma concentrations of delta 9-tetrahydrocannabinol reflect the degree of intoxication? *Journal of Clinical Pharmacology* **21**(8–9 Suppl), 171s–177s.

Howlett, A. C., and Fleming, R. M. (1984). Cannabinoid inhibition of adenylate cyclase. Pharmacology of the response in neuroblastoma cell membranes. *Molecular Pharmacology* **26**, 532–538.

Howlett, A. C., Qualy, J. M., and Khachatrian, L. L. (1986). Involvement of Gi in the inhibition of adenylate cyclase by cannabimimetic drugs. *Molecular Pharmacology* **29**, 307–313.

Huber, G. L., and First, M. W. (1991). Grubner, O., Marijuana and tobacco smoke gas-phase cytotoxins. *Pharmacology Biochemistry and Behavior* **40**, 629–636.

Huestis, M. A., Henningfield, J. E., and Cone, E. J. (1992). Blood cannabinoids. I. Absorption of THC and formation of 11-OH-THC and THCCOOH during and after smoking marijuana. *Journal of Analytical Toxicology* **16**, 276–282.

Hungund, B. L., Szakall, I., Adam, A., Basavarajappa, B. S., and Vadasz, C. (2003). Cannabinoid CB1 receptor knockout mice exhibit markedly reduced voluntary alcohol consumption and lack alcohol-induced dopamine release in the nucleus accumbens. *Journal of Neurochemistry* **84**, 698–704.

Hunt, C. A., and Jones, R. T. (1980). Tolerance and disposition of tetrahydrocannabinol in man. *Journal of Pharmacology and Experimental Therapeutics* **215**, 35–44.

Hutcheson, D. M., Tzavara, E. T., Smadja, C., Valjent, E., Roques, B. P., Hanoune, J., and Maldonado, R. (1998). Behavioural and biochemical evidence for signs of abstinence in mice chronically treated with delta-9-tetrahydrocannabinol. *British Journal of Pharmacology* **125**, 1567–1577.

Ibrahim, M. M., Deng, H., Zvonok, A., Cockayne, D. A., Kwan, J., Mata, H. P., Vanderah, T. W., Lai, J., Porreca, F., Makriyannis, A., and Malan, T. P. Jr. (2003). Activation of CB2 cannabinoid receptors by AM1241 inhibits experimental neuropathic pain: pain inhibition by receptors not present in the CNS. *Proceedings of the National Academy of Sciences USA* **100**, 10529–10533.

Isbell, H., Gorodetzsky, C. W., Jasinski, D., Claussen, U., von Spulak, F., and Korte, F. (1967a). Effects of (-)delta-9-trans-tetrahydrocannabinol in man. *Psychopharmacologia* **11**, 184–188.

Isbell, H., Jasinski, D. J., Gorodetzky, C. W., Korte, F., Claussen, U., Haage, M., Sieper, H., and von Spulak, F. (1967b). Studies on tetrahydrocannabinol: I. Method of assay in human subjects and results with crude extracts, purified tetrahydrocannabinols and synthetic compounds. *Bulletin on Problems of Drug Dependence* **29**, 4832–4846.

Iuvone, T., Esposito, G., Esposito, R., Santamaria, R., Di Rosa, M., and Izzo, A. A. (2004). Neuroprotective effect of cannabidiol, a nonpsychoactive component from Cannabis sativa, on beta-amyloid-induced toxicity in PC12 cells. *Journal of Neurochemistry* **89**, 134–141.

Iversen, L. L. (2000). *The Science of Marijuana*. Oxford University Press, New York.

Izzo, A. A., Capasso, F., Costagliola, A., Bisogno, T., Marsicano, G., Ligresti, A., Matias, I., Capasso, R., Pinto, L., Borrelli, F., Cecio, A., Lutz, B., Mascolo, N., and Di Marzo, V. (2003). An endogenous cannabinoid tone attenuates cholera toxin-induced fluid accumulation in mice. *Gastroenterology* **125**, 765–774.

Izzo, A. A., Pinto, L., Borrelli, F., Capasso, R., Mascolo, N., and Capasso, F. (2000). Central and peripheral cannabinoid modulation of gastrointestinal transit in physiological states or during the diarrhoea induced by croton oil. *British Journal of Pharmacology* **129**, 1627–1632.

Jacobsson, S. O., Wallin, T., and Fowler, C. J. (2001). Inhibition of rat C6 glioma cell proliferation by endogenous and synthetic cannabinoids. Relative involvement of cannabinoid and vanilloid receptors. *Journal of Pharmacology and Experimental Therapeutics* **299**, 951–959.

Janischevsky, D. E. (1924). Cannabis specimens in the drossy places of South-East Russia. *Ucen. Zap. Saratovsk, Gosud. Cernysevskgo Univ [Scientific Proceedings of the State Chernyshevsky University of Saratov]* **2**, 3–17. [Russian]

Janowsky, D. S., Meacham, M. P., Blaine, J. D., Schoor, M. and Bozzetti, L. P. (1976a). Marijuana effects on simulated flying ability. *American Journal of Psychiatry* **133**, 384–388.

Janowsky, D. S., Meacham, M. P., Blaine, J. D., Schoor, M., and Bozzetti, L. P. (1976b). Simulated flying performance after marihuana intoxication. *Aviation, Space, and Environmental Medicine* **47**, 124–128.

Jarbe, T. U., Lamb, R. J., Makriyannis, A., Lin, S., and Goutopoulos, A. (1998). Delta9-THC training dose as a determinant for (R)-methanandamide generalization in rats. *Psychopharmacology* **140**, 519–522.

Jarbe, T. U., Lamb, R. J., Lin, S., and Makriyannis, A. (2000). Delta9-THC training dose as a determinant for (R)-methanandamide generalization in rats: a systematic replication. *Behavioural Pharmacology* **11**, 81–86.

Jatoi, A., Windschitl, H. E., Loprinzi, C. L., Sloan, J. A., Dakhil, S. R., Mailliard, J. A., Pundaleeka, S., Kardinal, C. G., Fitch, T. R., Krook, J. E., Novotny, P. J., and Christensen, B. (2002). Dronabinol versus megestrol acetate versus combination therapy for cancer-associated anorexia: a North Central Cancer Treatment Group study. *Journal of Clinical Oncology* **20**, 567–573.

Johns, A. (2001). Psychiatric effects of cannabis. *British Journal of Psychiatry* **178**, 116–122.

Johnston, L. D., O'Malley, P. M., and Bachman, J. G. (1999). *National Survey Results on Drug Use from the Monitoring the Future Study, 1975–1998* (NIH publication no. 99-4660-61). National Institute on Drug Abuse, Bethesda, MD.

Jones, R. T. (1971). Tetrahydrocannabinol and the marijuana-induced social 'high,' or the effect of the mind on marijuana. *In Marijuana: Chemistry, Pharmacology, and Patterns of Social Use* (series title: *Annals of the New York Academy of Sciences*, vol. 191), (A. J. Singer, Ed.), pp. 155–165. New York Academy of Sciences, New York.

Jones, R. T. (2002). Cardiovascular system effects of marijuana. *Journal of Clinical Pharmacology* **42**(11 Suppl.), 58s–63s.

Jones, R. T., and Benowitz, N. (1976). The 3–day trip: clinical studies of cannabis tolerance and dependence. *In Pharmacology of Marihuana*, vol. 2, (M. C. Braude, and S. Szara, Eds.), pp. 627–642. Raven Press, New York.

Jones, R. T., Benowitz, N. L., and Herning, R. I. (1981). Clinical relevance of cannabis tolerance and dependence. *Journal of Clinical Pharmacology* **21**(8–9 Suppl), 143s–152s.

Joy, S. J. Jr., Benson, J. A. Jr. (Eds.). (1999). *Marijuana and Medicine: Assessing the Science Base*, National Academy Press, Washington DC.

Justinova, Z., Tanda, G., Redhi, G. H., and Goldberg, S. R. (2003). Self-administration of delta9-tetrahydrocannabinol (THC) by drug naive squirrel monkeys. *Psychopharmacology* **169**, 135–140.

Justinova, Z., Tanda, G., Munzar, P., and Goldberg, S. R. (2004). The opioid antagonist naltrexone reduces the reinforcing effects of delta(9)-tetrahydrocannabinol (THC) in squirrel monkeys. *Psychopharmacology* **173**, 186–194.

Karler, R., Cely, W., and Turkanis, S. A. (1973). The anticonvulsant activity of cannabidiol and cannabinol. *Life Sciences* **13**, 1527–1531.

Karler, R., and Turkanis, S. A. (1981). The cannabinoids as potential antiepileptics. *Journal of Clinical Pharmacology* **21**(8–9 Suppl), 437s–448s.

Karniol, I. G., Takahashi, R. N., and Musty, R. E. (1974). Effects of delta9-tetrahydrocannabinol and cannabinol on operant performance in rats. *Archives Internationales de Pharmacodynamie et de Therapie* **212**, 230–237.

Karniol, I. G., Shirakawa, I., Takahashi, R. N., Knobel, E., and Musty, R. E. (1975). Effects of delta9-tetrahydrocannabinol and cannabinol in man. *Pharmacology* **13**, 502–512.

Karst, M., Salim, K., Burstein, S., Conrad, I., Hoy, L., and Schneider, U. (2003). Analgesic effect of the synthetic cannabinoid CT-3 on chronic neuropathic pain: a randomized controlled trial. *Journal of the American Medical Association* **290**, 1757–1762.

Katona, I., Rancz, E. A., Acsady, L., Ledent, C., Mackie, K., Hajos, N., and Freund, T. F. (2001). Distribution of CB1 cannabinoid receptors in the amygdala and their role in the control of GABAergic transmission. *Journal of Neuroscience* **21**, 9506–9518.

Kaymakcalan, S., Ayhan, I. H., and Tulunay, F. C. (1977). Naloxone-induced or postwithdrawal abstinence signs in delta9-tetrahydrocannabinol-tolerant rats. *Psychopharmacology* **55**, 243–249.

Kehl, L. J., Hamamoto, D. T., Wacnik, P. W., Croft, D. L., Norsted, B. D., Wilcox, G. L., and Simone, D. A. (2003). A cannabinoid agonist differentially attenuates deep tissue hyperalgesia in animal models of cancer and inflammatory muscle pain. *Pain* **103**, 175–186.

Kelly, T. H., Foltin, R. W., Emurian, C. S., and Fischman, M. W. (1993). Performance-based testing for drugs of abuse: dose and time profiles of marijuana, amphetamine, alcohol, and diazepam. *Journal of Analytical Toxicology* **17**, 264–272.

Killestein, J., Hoogervorst, E. L., Reif, M., Kalkers, N. F., Van Loenen, A. C., Staats, P. G., Gorter, R. W., Uitdehaag, B. M., and Polman, C. H. (2002). Safety, tolerability, and efficacy of orally administered cannabinoids in MS. *Neurology* **58**, 1404–1407.

Killestein, J., Hoogervorst, E. L., Reif, M., Blauw, B., Smits, M., Uitdehaag, B. M., Nagelkerken, L., and Polman, C. H. (2003). Immunomodulatory effects of orally administered cannabinoids in multiple sclerosis. *Journal of Neuroimmunology* **137**, 140–143.

Kirk, J. M., and de Wit, H. (1999). Responses to oral delta9-tetrahydrocannabinol in frequent and infrequent marijuana users. *Pharmacology Biochemistry and Behavior* **63**, 137–142.

Koob, G. F., Sanna, P. P., and Bloom, F. E. (1998). Neuroscience of addiction. *Neuron* **21**, 467–476.

Kouri, E. M., and Pope, H. G. Jr. (2000). Abstinence symptoms during withdrawal from chronic marijuana use. *Experimental and Clinical Psychopharmacology* **8**, 483–492.

Kreuz, D. S., and Axelrod, J. (1973). Delta-9-tetrahydrocannabinol: localization in body fat. *Science* **179**, 391–393.

Kupfer, D. J., Detre, T., Koral, J., and Fajans, P. (1973). A comment on the "amotivational syndrome" in marijuana smokers. *American Journal of Psychiatry* **130**, 1319–1322.

Lamarck, J. P. (1783). *Encyclopedie Methodique: Botanique*, Panckoucke, Paris.

Lamers, C. T. J., and Ramaekers, J. G. (2000). *Visual Search and Urban City Driving Under the Influence of Marijuana and Alcohol* [DOT HS 809 020], National Highway Traffic Safety Administration, Washington DC.

Lane, M., Vogel, C. L., Ferguson, J., Krasnow, S., Saiers, J. L., Hamm, J., Salva, K., Wiernik, P. H., Holroyde, C. P., Hammill, S., Shepard, K., and Plasse, T. (1991). Dronabinol and prochlorperazine in combination for treatment of cancer chemotherapy-induced nausea and vomiting. *Journal of Pain and Symptom Management* **6**, 352–359.

Lastres-Becker, I., Bizat, N., Boyer, F., Hantraye, P., Brouillet, E., and Fernandez-Ruiz, J. (2003). Effects of cannabinoids in the rat model of Huntington's disease generated by an intrastriatal injection of malonate. *Neuroreport* **14**, 813–816.

Ledent, C., Valverde, O., Cossu, G., Petitet, F., Aubert, J. F., Beslot, F., Bohme, G. A., Imperato, A., Pedrazzini, T., Roques, B. P., Vassart, G., Fratta, W., and Parmentier, M. (1999). Unresponsiveness to cannabinoids and reduced addictive effects of opiates in CB1 receptor knockout mice. *Science* **283**, 401–404.

Leirer, V. O., Yesavage, J. A., and Morrow, D. G. (1991). Marijuana carry-over effects on aircraft pilot performance. *Aviation, Space, and Environmental Medicine* **62**, 221–227.

Leker, R. R., Shohami, E., Abramsky, O., and Ovadia, H. (1999). Dexanabinol; a novel neuroprotective drug in experimental focal cerebral ischemia. *Journal of Neurological Science* **162**, 114–119.

Lemberger, L., Tamarkin, N. R., Axelrod, J., and Kopin, I. J. (1971). Delta-9-tetrahydrocannabinol: metabolism and disposition in long-term marihuana smokers. *Science* **173**, 72–74.

Lepore, M., Vorel, S. R., Lowinson, J., and Gardner, E. L. (1995). Conditioned place preference induced by delta 9-tetrahydrocannabinol: Comparison with cocaine, morphine, and food reward. *Life Sciences* **56**, 2073–2080.

Lepore, M., Liu, X., Savage, V., Matalon, D., and Gardner, E. L. (1996). Genetic differences in delta 9-tetrahydrocannabinol-induced facilitation of brain stimulation reward as measured by a rate-frequency curve-shift electrical brain stimulation paradigm in three different rat strains. *Life Sciences* **58**, PL365–PL372.

Leweke, F. M., Schneider, U., Thies, M., Munte, T. F., and Emrich, H. M. (1999). Effects of synthetic delta9-tetrahydrocannabinol on binocular depth inversion of natural and artificial objects in man. *Psychopharmacology* **142**, 230–235.

Lewin, L. (1964). *Phantastica: Narcotic and Stimulating Drugs, Their Use and Abuse*, Dutton, New York.

Li, H. L. (1974a). The origin and use of cannabis in Eastern Asia: linguistic-cultural implications. *Economic Botany* **28**, 293–301.

Li, H. L. (1974b). An archaeological and historical account of cannabis in China. *Economic Botany* **28**, 437–448.

Lichtman, A. H., Sheikh, S. M., Loh, H. H., and Martin, B. R. (2001). Opioid and cannabinoid modulation of precipitated withdrawal in delta(9)-tetrahydrocannabinol and morphine-dependent mice. *Journal of Pharmacology and Experimental Therapeutics* **298**, 1007–1014.

Lorenz, R. (2004). On the application of cannabis in paediatrics and epileptology. *Neuroendocrinology Letters* **25**, 40–44.

Louw, D. F., Yang, F. W., and Sutherland, G. R. (2000). The effect of delta-9-tetrahydrocannabinol on forebrain ischemia in rat. *Brain Research* **857**, 183–187.

Lynch, M. E., and Clark, A. J. (2003). Cannabis reduces opioid dose in the treatment of chronic non-cancer pain. *Journal of Pain and Symptom Management* **25**, 496–498.

Mackie, K., and Hille, B. (1992). Cannabinoids inhibit N-type calcium channels in neuroblastoma-glioma cells. *Proceedings of the National Academy of Sciences USA* **89**, 3825–3829.

Malec, J., Harvey, R. F., and Cayner, J. J. (1982). Cannabis effect on spasticity in spinal cord injury. *Archives of Physical Medicine and Rehabilitation* **63**, 116–118.

Malfait, A. M., Gallily, R., Sumariwalla, P. F., Malik, A. S., Andreakos, E., Mechoulam, R., and Feldmann, M. (2000). The nonpsychoactive cannabis constituent cannabidiol is an oral anti-arthritic therapeutic in murine collagen-induced arthritis. *Proceedings of the National Academy of Sciences USA* **97**, 9561–9566.

Malone, D. T., and Taylor, D. A. (1999). Modulation by fluoxetine of striatal dopamine release following Delta9-tetrahydrocannabinol: a microdialysis study in conscious rats. *British Journal of Pharmacology* **128**, 21–26.

Manzanares, J., Corchero, J., Romero, J., Fernandez-Ruiz, J. J., Ramos, J. A., and Fuentes, J. A. (1998). Chronic administration of cannabinoids regulates proenkephalin mRNA levels in selected regions of the rat brain. *Molecular Brain Research* **55**, 126–132.

Martellotta, M. C., Cossu, G., Fattore, L., Gessa, G. L., and Fratta, W. (1998). Self-administration of the cannabinoid receptor agonist WIN 55 212–2 in drug-naive mice. *Neuroscience* **85**, 327–330.

Martin, B. R., Beletskaya, I., Patrick, G., Jefferson, R., Winckler, R., Deutsch, D. G., Di Marzo, V., Dasse, O., Mahadevan, A., and Razdan, R. K. (2000). Cannabinoid properties of methylfluorophosphonate analogs. *Journal of Pharmacology and Experimental Therapeutics* **294**, 1209–1218.

Martin, W. J., Patrick, S. L., Coffin, P. O., Tsou, K., and Walker, J. M. (1995). An examination of the central sites of action of cannabinoid-induced antinociception in the rat. *Life Sciences* **56**, 2103–2109.

Martin, M., Ledent, C., Parmentier, M., Maldonado, R., and Valverde, O. (2002). Involvement of CB1 cannabinoid receptors in emotional behaviour. *Psychopharmacology* **159**, 379–387.

Martyn, C. N., Illis, L. S., and Thom, J. (1995). Nabilone in the treatment of multiple sclerosis. *Lancet* **345**, 579.

Mascia, M. S., Obinu, M. C., Ledent, C., Parmentier, M., Bohme, G. A., Imperato, A., and Fratta, W. (1999). Lack of morphine-induced dopamine release in the nucleus accumbens of cannabinoid CB(1) receptor knockout mice. *European Journal of Pharmacology* **383**, R1–R2.

Mason, A. P., and McBay, A. J. (1985). Cannabis: pharmacology and interpretation of effects. *Journal of Forensic Sciences* **30**, 615–631.

Massa, F., Marsicano, G., Hermann, H., Cannich, A., Monory, K.., Cravatt, B. F., Ferri, G. L., Sibaev, A., Storr, M., and Lutz, B. (2004). The endogenous cannabinoid system protects against colonic inflammation. *Journal of Clinical Investigation* **113**, 1202–1209.

Massi, P., Vaccani, A., Ceruti, S., Colombo, A., Abbracchio, M. P., and Parolaro, D. (2004). Antitumor effects of cannabidiol, a nonpsychoactive cannabinoid, on human glioma cell lines. *Journal of Pharmacology and Experimental Therapeutics* **308**, 838–845.

Mathers, D. C., and Ghodse, A. H. (1992). Cannabis and psychotic illness. *British Journal of Psychiatry* **161**, 648–653.

Mathre, M. L. (Ed.) (1997). *Cannabis in Medical Practice: A Legal, Historical and Pharmacological Overview of the Therapeutic Use of Marijuana*, McFarland, Jefferson NC.

Matsuda, L. A., Lolait, S. J., Brownstein, M. J., Young, A. C., and Bonner, T. I. (1990). Structure of a cannabinoid receptor and functional expression of the cloned cDNA. *Nature* **346**, 561–564.

Matsuda, L. A., Bonner, T. I., and Lolait, S. J. (1993). Localization of cannabinoid receptor mRNA in rat brain. *Journal of Comparative Neurology* **327**, 535–550.

Matsunaga, T., Iwawaki, Y., Watanabe, K., Yamamoto, I., Kageyama, T., and Yoshimura, H. (1995). Metabolism of delta 9-tetrahydrocannabinol by cytochrome P450 isozymes purified from hepatic microsomes of monkeys. *Life Sciences* **56**, 2089–2095.

Maurer, M., Henn, V., Dittrich, A., and Hofmann, A. (1990). Delta-9-tetrahydrocannabinol shows antispastic and analgesic effects in a single case double-blind trial. *European Archives of Psychiatry and Clinical Neuroscience* **240**, 1–4.

Maykut, M. O. (1985). Health consequences of acute and chronic marihuana use. *Progress in Neuropsychopharmacology and Biological Psychiatry* **9**, 209–238.

Mayor's Committee on Marijuana (1944). *The Marihuana Problem in the City of New York: Sociological, Medical, Psychological and Pharmacological Studies*, Jacques Cattell Press, Lancaster, PA.

McGlothlin, W. H., and West, L. J. (1968). The marihuana problem: an overview. *American Journal of Psychiatry* **125**, 126–134.

McGregor, I. S., Issakidis, C. N., and Prior, G. (1996). Aversive effects of the synthetic cannabinoid CP 55 940 in rats. *Pharmacology Biochemistry and Behavior* **53**, 657–664.

McRae, A. L., Budney, A. J., and Brady, K. T. (2003). Treatment of marijuana dependence: a review of the literature. *Journal of Substance Abuse Treatment* **24**, 369–376.

McRae, A. L., Sonne, S. C., Brady, K. T., Durkalski, V., and Palesch, Y. (2004). A randomized, placebo-controlled trial of buspirone for the treatment of anxiety in opioid-dependent individuals. *American Journal on Addictions* **13**, 53–63.

Mechoulam, R. (1970). Marihuana chemistry. *Science* **168**, 1159–1166.

Mechoulam, R. (Ed.) (1986). *Cannabinoids as Therapeutic Agents*, CRC Press, Boca Raton.

Mechoulam, R., Ben-Zvi, Z., Yagnitinsky, B., and Shani, A. (1969). A new tetrahydrocannabinolic acid. *Tetrahedron Letters* **28**, 2339–2341.

Medical Economics Company. (2004). *Physicians' Desk Reference*, 58th ed. Medical Economics Company, Oradell NJ.

Meinck, H. M., Schonle, P. W., and Conrad, B. (1989). Effect of cannabinoids on spasticity and ataxia in multiple sclerosis. *Journal of Neurology* **236**, 120–122.

Melis, M., Muntoni, A. L., Diana, M., and Gessa, G. L. (1996). The cannabinoid receptor agonist WIN 55 212-2 potently stimulates dopaminergic neuronal activity in the mesolimbic system. *Behavioural Pharmacology* **7**(*Suppl.* 1), 68.

Melvin, L. S., and Johnson, M. R. (1987). Structure-activity relationships of tricyclic and nonclassical bicyclic cannabinoids. *In Structure-Activity Relationships of the Cannabinoids* (series title: *NIDA Research Monograph*, vol. 79), (R. S. Rapaka, and A. Makriyannis, Eds.), pp. 31–47. National Institute on Drug Abuse, Rockville, MD.

Mendelson, J. H., Mello, N. K., Lex, B. W., and Bavli, S. (1984). Marijuana withdrawal syndrome in a woman. *American Journal of Psychiatry* **141**, 1289–1290.

Merlin, M. D. (1972). *Man and Marijuana: Some Aspects of Their Ancient Relationship*, Fairleigh Dickinson University Press, Rutherford, WI.

Merritt, J. C., Crawford, W. J., Alexander, P. C., Anduze, A. L., and Gelbart, S. S. (1980). Effect of marihuana on intraocular and blood pressure in glaucoma. *Ophthalmology* **87**, 222–228.

Merritt, J. C., Olsen, J. L., Armstrong, J. R., and McKinnon, S. M. (1981). Topical delta 9-tetrahydrocannabinol in hypertensive glaucomas. *Journal of Pharmacy and Pharmacology* **33**, 40–41.

Mikuriya, T. H. (1970). Cannabis substitution: an adjunctive therapeutic tool in the treatment of alcoholism. *Medical Times* **98**, 187–191.

Miller, A. S., and Walker, J. M. (1998). Local effects of cannabinoids on spontaneous activity and evoked inhibition in the globus pallidus. *European Journal of Pharmacology* **352**, 199–205.

Miller, N. S., and Gold, M. S. (1989). The diagnosis of marijuana (cannabis) dependence. *Journal of Substance Abuse Treatment* **6**, 183–192.

Millman, R. B., and Sbriglio, R. (1986). Patterns of use and psychopathology in chronic marijuana users. *Psychiatric Clinics of North America* **9**, 533–545.

Milton, N. G. (2002). Anandamide and noladin ether prevent neurotoxicity of the human amyloid-beta peptide. *Neuroscience Letters* **332**, 127–130.

Mountjoy, K. G., Robbins, L. S., Mortrud, M. T., and Cone, R. D. (1992). The cloning of a family of genes that encode the melanocortin receptors. *Science* **257**, 1248–1251.

Muller-Vahl, K. R., Schneider, U., Kolbe, H., and Emrich, H. M. (1999). Treatment of Tourette's syndrome with delta-9-tetrahydrocannabinol. *American Journal of Psychiatry* **156**, 495.

Muller-Vahl, K. R., Kolbe, H., Schneider, U., and Emrich, H. M. (2002a). Movement disorders. *In Cannabis and Cannabinoids: Pharmacology, Toxicology, and Therapeutic Potential* (F. Grotenhermen, and E. Russo, Eds.), pp. 205–214. Haworth Integrative Healing Press, New York.

Muller-Vahl, K. R., Schneider, U., Koblenz, A., Jobges, M., Kolbe, H., Daldrup, T., and Emrich, H. M. (2002b). Treatment of Tourette's syndrome with Delta 9-tetrahydrocannabinol (THC): a randomized crossover trial. *Pharmacopsychiatry* **35**, 57–61.

Muller-Vahl, K. R., Prevedel, H., Theloe, K., Kolbe, H., Emrich, H. M., and Schneider, U. (2003a). Treatment of Tourette syndrome with delta-9-tetrahydrocannabinol (delta 9-THC): no influence on neuropsychological performance. *Neuropsychopharmacology* **28**, 384–388.

Muller-Vahl, K. R., Schneider, U., Prevedel, H., Theloe, K., Kolbe, H., Daldrup, T., and Emrich, H. M. (2003b). Delta 9-tetrahydrocannabinol (THC) is effective in the treatment of tics in Tourette syndrome: a 6-week randomized trial. *Journal of Clinical Psychiatry* **64**, 459–465.

Munro, S., Thomas, K. L., and Abu-Shaar, M. (1993). Molecular characterization of a peripheral receptor for cannabinoids. *Nature* **365**, 61–65.

Naditch, M. P. (1974). Acute adverse reactions to psychoactive drugs, drug usage, and psychopathology. *Journal of Abnormal Psychology* **83**, 394–403.

Naef, M., Curatolo, M., Petersen-Felix, S., Arendt-Nielsen, L., Zbinden, A., and Brenneisen, R. (2003). The analgesic effect of oral delta-9-tetrahydrocannabinol (THC), morphine, and a THC-morphine combination in healthy subjects under experimental pain conditions. *Pain* **105**, 79–88.

Nagayama, T., Sinor, A. D., Simon, R. P., Chen, J., Graham, S. H., Jin, K., and Greenberg, D. A. (1999). Cannabinoids and neuroprotection in global and focal cerebral ischemia and in neuronal cultures. *Journal of Neuroscience* **19**, 2987–2995.

National Highway Traffic Safety Administration. (2000). Marijuana and alcohol combined severely impede driving performance. *Annals of Emergency Medicine* **35**, 398–399.

Nauck, F., and Klaschik, E. (2004). Cannabinoids in the treatment of the cachexia-anorexia syndrome in palliative care patient. *Schmerz* **18**, 197–202. [German]

Navarro, M., Chowen, J., Carrera, M. R. A., Del Arco, I., Villanua, M. A., Martin, Y., Roberts, A. J., Koob, G. F., and Rodriguez de Fonseca F. (1998). CB-1 cannabinoid receptor antagonist-induced opiate withdrawal in morphine-dependent rats. *Neuroreport* **9**, 3397–3402.

Navarro, M., Carrera, M. R. A., Fratta, W., Valverde, O., Cossu, G., Fattore, L., Chowen, J. A., Gomez, R., Del Arco, I., Villanua, M. A., Maldonado, R., Koob, G. F., and Rodriguez de Fonseca, F. (2001). Functional interaction between opioid and cannabinoid receptors in drug self-administration. *Journal of Neuroscience* **21**, 5344–5350.

Ng Cheong Ton, J. M., Gerhardt, G. A., Friedemann, M., Etgen, A. M., Rose, G. M., Sharpless, N. S., and Gardner, E. L. (1988). The effects of delta 9-tetrahydrocannabinol on potassium-evoked release of dopamine in the rat caudate nucleus: an in vivo electrochemical and in vivo microdialysis study. *Brain Research* **451**, 59–68.

Nowlan, R., and Cohen, S. (1977). Tolerance to marijuana: heart rate and subjective "high". *Clinical Pharmacology and Therapeutics* **22**, 550–556.

Notcutt, W., Price, M., Miller, R., Newport, S., Phillips, C., Simmons, S., and Sansom, C. (2004). Initial experiences with medicinal extracts of cannabis for chronic pain: results from 34 'N of 1' studies. *Anaesthesia* **59**, 440–452.

Noyes, R. Jr., Brunk, S. F., Avery, D. A., and Canter, A. C. (1975a). The analgesic properties of delta-9-tetrahydrocannabinol and codeine. *Clinical Pharmacology and Therapeutics* **18**, 84–89.

Noyes R. Jr., Brunk, S. F., Baram, D. A., and Canter, A. (1975b). Analgesic effect of delta-9-tetrahydrocannabinol. *Journal of Clinical Pharmacology* **15**, 139–143.

O'Shaughnessy, W. B. (1838–1840). On the preparations of the Indian hemp, or gunjah, *Transactions of the Medical and Physical Society of Bengal*, pp. 421–461.

Ohlsson, A., Lindgren, J. E., Wahlen, A., Agurell, S., Hollister, L. E., and Gillespie, H. K. (1980). Plasma delta-9 tetrahydrocannabinol concentrations and clinical effects after oral and intravenous administration and smoking. *Clinical Pharmacology and Therapeutics* **28**, 409–416.

Ohlsson, A., Lindgren, J. E., Wahlen, A., Agurell, S., Hollister, L. E., and Gillespie, H. K. (1982). Single dose kinetics of deuterium labelled delta 1-tetrahydrocannabinol in heavy and light cannabis users. *Biomedical Mass Spectrometry* **9**, 6–10.

Parker, L. A., and Gillies, T. (1995). THC-induced place and taste aversions in Lewis and Sprague-Dawley rats. *Behavioral Neuroscience* **109**, 71–78.

Partridge, E. (1958). *Origins: A Short Etymological Dictionary of Modern English*, Macmillan, New York.

Pate, D. W. (2002). Taxonomy of cannabinoids. *In Cannabis and Cannabinoids: Pharmacology, Toxicology, and Therapeutic Potential* (F. Grotenhermen, and E. Russo, Eds.), pp. 15–26. Haworth Integrative Healing Press, New York.

Patel, H. J., Birrell, M. A., Crispino, N., Hele, D. J., Venkatesan, P., Barnes, P. J., Yacoub, M. H., and Belvisi, M. G. (2003). Inhibition of guinea-pig and human sensory nerve activity and the cough reflex in guinea-pigs by cannabinoid (CB2) receptor activation. *British Journal of Pharmacology* **140**, 261–268.

Perez-Reyes, M., Timmons, M. C., Davis, K. H., and Wall, E. M. (1973). A comparison of the pharmacological activity in man of intravenously administered delta9-tetrahydrocannabinol, cannabinol, and cannabidiol. *Experientia* **29**, 1368–1369.

Petro, D. J. (1980). Marihuana as a therapeutic agent for muscle spasm or spasticity. *Psychosomatics* **21,** 81–85.

Petro, D. J., and Ellenberger, C. Jr. (1981). Treatment of human spasticity with delta 9-tetrahydrocannabinol. *Journal of Clinical Pharmacology* **21**(8–9 *Suppl.*), 413s–416s.

Piomelli, D., Giuffrida, A., Calignano, A., and Rodriguez de Fonseca, F. (2000). The endocannabinoid system as a target for therapeutic drugs. *Trends in Pharmacological Sciences* **21**, 218–224.

Pistis, M., Muntoni, A. L., Pillolla, G., and Gessa, G. L. (2002). Cannabinoids inhibit excitatory inputs to neurons in the shell of the nucleus accumbens: an in vivo electrophysiological study. *European Journal of Neuroscience* **15**, 1795–1802.

Pistis, M., Perra, S., Pillolla, G., Melis, M., Gessa, G. L., and Muntoni, A. L. (2004). Cannabinoids modulate neuronal firing in the rat basolateral amygdala: evidence for CB1- and non-CB1-mediated actions. *Neuropharmacology* **46**, 115–125.

Plasse, T. F., Gorter, R. W., Krasnow, S. H., Lane, M., Shepard, K. V., and Wadleigh, R. G. (1991). Recent clinical experience with dronabinol. *Pharmacology Biochemistry and Behavior* **40**, 695–700.

Porcella, A., Maxia, C., Gessa, G. L., and Pani, L. (2001). The synthetic cannabinoid WIN55212-2 decreases the intraocular pressure in human glaucoma resistant to conventional therapies. *European Journal of Neuroscience* **13**, 409–412.

Portella, G., Laezza, C., Laccetti, P., De Petrocellis, L., Di Marzo, V., and Bifulco, M. (2003). Inhibitory effects of cannabinoid CB1 receptor stimulation on tumor growth and metastatic spreading: actions on signals involved in angiogenesis and metastasis. *FASEB Journal* **17**, 1771–1773.

Preusser, D. F., Ulmer, R. G., and Preusser, C. W. (1992). *Evaluation of the Impact of the Drug Evaluation and Classification Program on Enforcement and Adjudication*, U.S. National Highway Traffic Safety Administration, Washington DC.

Pryce, G., Ahmed, Z., Hankey, D. J., Jackson, S. J., Croxford, J. L., Pocock, J. M., Ledent, C., Petzold, A., Thompson, A. J., Giovannoni, G., Cuzner, M. L., and Baker, D. (2003). Cannabinoids inhibit neurodegeneration in models of multiple sclerosis. *Brain* **126**, 2191–2202.

Pryce, G., and Baker, D. (2005). Emerging properties of cannabinoid medicines in management of multiple sclerosis. *Trends in Neurosciences* **28**, 272–276.

Ralevic, V., and Kendall, D. A. (2001). Cannabinoid inhibition of capsaicin-sensitive sensory neurotransmission in the rat mesenteric arterial bed. *European Journal of Pharmacology* **418**, 117–125.

Ravinet Trillou, C., Delgorge, C., Menet, C., Arnone, M., and Soubrie, P. (2004). CB1 cannabinoid receptor knockout in mice leads to leanness, resistance to diet-induced obesity and enhanced leptin sensitivity. *International Journal of Obesity and Related Metabolic Disorders* **28**, 640–648.

Razdan, R. K. (1986). Structure-activity relationships in cannabinoids. *Pharmacological Reviews* **38**, 75–149.

Read B. E. (1936). Chinese Medicinal Plants from the Pen Ts'ao Kang Mu A.D. 1596, *Peking National History Bulletin*, Peiping.

Reilly, D., Didcott, P., Swift, W., and Hall, W. (1998). Long-term cannabis use: characteristics of users in an Australian rural area. *Addiction* **93**, 837–846.

Rice, O. V., Gordon, N., and Gifford, A. N. (2002). Conditioned place preference to morphine in cannabinoid CB1 receptor knockout mice. *Brain Research* **945**, 135–138.

Robbe, H. W. J., and O'Hanlon, J. F. (1999). *Marijuana, Alcohol and Actual Driving Performance* (DOT HS 808 939), National Highway Traffic Safety Administration, Washington DC.

Robbe, D., Alonso, G., Duchamp, F., Bockaert, J., and Manzoni, O. J. (2001). Localization and mechanisms of action of cannabinoid receptors at the glutamatergic synapses of the mouse nucleus accumbens. *Journal of Neuroscience* **21**, 109–116.

Robbe, D., Kopf, M., Remaury, A., Bockaert, J., and Manzoni, O. J. (2002). Endogenous cannabinoids mediate long-term synaptic depression in the nucleus accumbens. *Proceedings of the National Academy of Sciences USA* **99**, 8384–8388.

Robbe, D., Alonso, G., and Manzoni, O. J. (2003). Exogenous and endogenous cannabinoids control synaptic transmission in mice nucleus accumbens. *In Glutamate and Disorders of Cognition and Motivation* (series title: *Annals of the New York Academy of Sciences*, vol. 1003), (B. Moghaddam, and M. E. Wolf, Eds.), pp. 212–225. New York Academy of Sciences, New York.

Rodriguez de Fonseca, F., Carrera, M. R. A., Navarro, M., Koob, G. F., and Weiss, F. (1997). Activation of corticotropin-releasing factor in the limbic system during cannabinoid withdrawal. *Science* **276**, 2050–2054.

Roffman, R. A., and Barnhart, R. (1987). Assessing need for marijuana dependence treatment through an anonymous telephone interview. *International Journal of the Addictions* **22**, 639–651.

Ross, S., and ElSohly, M. A. (1995). Constituents of Cannabis sativa L: XXVIII. A review of the natural constituents: 1980–1994. *Zagazig Journal of Pharmaceutical Sciences* **4**, 1–10.

Ross, S. A., Mehmedic, Z., Murphy, T. P., and Elsohly, M. A. (2000). GC-MS analysis of the total delta9-THC content of both drug- and fiber-type cannabis seeds. *Journal of Analytical Toxicology* **24**, 715–717.

Sallan, S. E., Cronin, C., Zelen, M., and Zinberg, N. E. (1980). Antiemetics in patients receiving chemotherapy for cancer: a randomized comparison of delta-9-tetrahydrocannabinol and prochlorperazine. *New England Journal of Medicine* **302**, 135–138.

Sanchez, C., de Ceballos, M. L., del Pulgar, T. G., Rueda, D., Corbacho, C., Velasco, G., Galve-Roperh, I., Huffman, J. W., Ramon y Cajal, S., and Guzman, M. (2001). Inhibition of glioma growth in vivo by selective activation of the CB(2) cannabinoid receptor. *Cancer Research* **61**, 5784–5789.

Sandyk, R., and Awerbuch, G. (1988). Marijuana and Tourette's syndrome. *Journal of Clinical Psychopharmacology* **8**, 444–445.

Sanudo-Pena, M. C., Tsou, K., Delay, E. R., Hohman, A. G., Force, M., and Walker, J. M. (1997). Endogenous cannabinoids as an aversive or counter-rewarding system in the rat. *Neuroscience Letters* **223**, 125–128.

Saario, S. M., Savinainen, J. R., Laitinen, J. T., Jarvinen, T., and Niemi, R. (2004). Monoglyceride lipase-like enzymatic activity is responsible for hydrolysis of 2-arachidonoylglycerol in rat cerebellar membranes. *Biochemical Pharmacology* **67**, 1381–1387.

Schultes, R. E., Klein, W. M., Plowman, T., and Lockwood, T. E. (1974). Cannabis: an example of taxonomic neglect. *Botanical Museum Leaflets, Harvard University* **23**, 337–367.

Sidney, S. (2002). Cardiovascular consequences of marijuana use. *Journal of Clinical Pharmacology* **42**(11 *Suppl.*), 64s–70s.

Sieradzan, K. A., Fox, S. H., Hill, M., Dick, J. P., Crossman, A. R., and Brotchie, J. M. (2001). Cannabinoids reduce levodopa-induced dyskinesia in Parkinson's disease: a pilot study. *Neurology* **57**, 2108–2111.

Sinor, A. D., Irvin, S. M., and Greenberg, D. A. (2000). Endocannabinoids protect cerebral cortical neurons from in vitro ischemia in rats. *Neuroscience Letters* **278**, 157–160.

Smart, D., Gunthorpe, M. J., Jerman, J. C., Nasir, S., Gray, J., Muir, A. I., Chambers, J. K., Randall, A. D., and Davis, J. B. (2000). The endogenous lipid anandamide is a full agonist at the human vanilloid receptor (hVR1). *British Journal of Pharmacology* **129**, 227–230.

Smith, P. B., Compton, D. R., Welch, S. P., Razdan, R. K., Mechoulam, R., and Martin, B. R. (1994). The pharmacological activity of anandamide, a putative endogenous cannabinoid, in mice. *Journal of Pharmacology and Experimental Therapeutics* **270**, 219–227.

Substance Abuse and Mental Health Services Administration (1996). *National Household Survey on Drug Abuse: Population Estimates 1995* (DHHS publication no. (SMA) 96–3095), U.S. Department of Health and Human Services, Rockville, MD.

Soderpalm, A. H., Schuster, A., and de Wit, H. (2001). Antiemetic efficacy of smoked marijuana: subjective and behavioral effects on nausea induced by syrup of ipecac. *Pharmacology Biochemistry and Behavior* **69**, 343–350.

Solinas, M., Panlilio, L. V., Antoniou, K., Pappas, L. A., and Goldberg, S. R. (2003). The cannabinoid CB1 antagonist N-piperidinyl-5-(4-chlorophenyl)-1-(2,4-dichlorophenyl)-4-methylpyrazole-3-carboxamide (SR-141716A) differentially alters the reinforcing effects of heroin under continuous reinforcement, fixed ratio, and progressive ratio schedules of drug self-administration in rats. *Journal of Pharmacology and Experimental Therapeutics* **306**, 93–102.

Stephens, R. S., Roffman, R. A., and Simpson, E. E. (1993). Adult marijuana users seeking treatment. *Journal of Consulting and Clinical Psychology* **61**, 1100–1104.

Stephens, R. S., Roffman, R. A., and Simpson, E. E. (1994). Treating adult marijuana dependence: a test of the relapse prevention model. *Journal of Consulting and Clinical Psychology* **62**, 92–99.

Stephens, R. S., Babor, T. F., Kadden, R., and Miller, M. (2002). The Marijuana Treatment Project: rationale, design and participant characteristics. *Addiction* **97**(*Suppl*. 1), 109–124.

Substance Abuse and Mental Health Services Administration (2000). *National Household Survey on Drug Abuse: Main Findings 1998* (DHHS publication no. (SMA) 00–3381), U.S. Department of Health and Human Services, Rockville, MD.

Substance Abuse and Mental Health Services Administration (2003). *Results from the 2002 National Survey on Drug Use and Health: National Findings* (Office of Applied Studies, NHSDA Series H–22, DHHS Publication No. SMA 03–3836), Rockville, MD.

Sugiura, T., Kondo, S., Sukagawa, A., Nakane, S., Shinoda, A., Itoh, K., Yamashita, A., and Waku, K. (1995). 2-Arachidonoylglycerol: a possible endogenous cannabinoid receptor ligand in brain. *Biochemical and Biophysical Research Communications* **215**, 89–97.

Sugiura, T., Kodaka, T., Nakane, S., Miyashita, T., Kondo, S., Suhara, Y., Takayama, H., Waku, K., Seki, C., Baba, N., and Ishima, Y. (1999). Evidence that the cannabinoid CB1 receptor is a 2-arachidonoylglycerol receptor: Structure-activity relationship of 2-arachidonoylglycerol, ether-linked analogues, and related compounds. *Journal of Biological Chemistry* **274**, 2794–2801.

Sullivan, J. M. (1999). Mechanisms of cannabinoid-receptor-mediated inhibition of synaptic transmission in cultured hippocampal pyramidal neurons. *Journal of Neurophysiology* **82**, 1286–1294.

Szabo, B., Dorner, L., Pfreundtner, C., Norenberg, W., and Starke, K. (1998). Inhibition of GABAergic inhibitory postsynaptic currents by cannabinoids in rat corpus striatum. *Neuroscience* **85**, 395–403.

Szabo, B., Wallmichrath, I., Mathonia, P., and Pfreundtner, C. (2000). Cannabinoids inhibit excitatory neurotransmission in the substantia nigra pars reticulata. *Neuroscience* **97**, 89–97.

Szabo, B., Siemes, S., and Wallmichrath, I. (2002). Inhibition of GABAergic neurotransmission in the ventral tegmental area by cannabinoids. *European Journal of Neuroscience* **15**, 2057–2061.

Tanda, G., Pontieri, F. E., and Di Chiara, G. (1997). Cannabinoid and heroin activation of mesolimbic dopamine transmission by a common mu1 opioid receptor mechanism. *Science* **276**, 2048–2050.

Tanda, G., Loddo, P., and Di Chiara, G. (1999). Dependence of mesolimbic dopamine transmission on delta9-tetrahydrocannabinol. *European Journal of Pharmacology* **376**, 23–26.

Tanda, G., Munzar, P., and Goldberg, S. R. (2000). Self-administration behavior is maintained by the psychoactive ingredient of marijuana in squirrel monkeys. *Nature Neuroscience* **3**, 1073–1074.

Tanda, G., and Goldberg, S. R. (2003). Cannabinoids: reward, dependence, and underlying neurochemical mechanisms: a review of recent preclinical data. *Psychopharmacology* **169**, 115–134.

Tart, C. T. (1971). *On Being Stoned: A Psychological Study of Marijuana Intoxication*, Science and Behavior Books, Palo Alto, CA.

Tashkin, D. P., Shapiro, B. J., and Frank, I. M. (1974). Acute effects of smoked marijuana and oral delta9-tetrahydrocannabinol on specific airway conductance in asthmatic subjects. *American Review of Respiratory Disease* **109**, 420–428.

Tashkin, D. P., Gliederer, F., Rose, J., Change, P., Hui, K. K., Yu, J. L., and Wu, T. C. (1991). Effects of varying marijuana smoking profile on deposition of tar and absorption of CO and delta-9-THC. *Pharmacology Biochemistry and Behavior* **40**, 651–656.

Tennant, F. S. (1986). The clinical syndrome of marijuana dependence. *Psychiatric Annals* **16**, 225–234.

Thornicroft, G. (1990). Cannabis and psychosis: Is there epidemiological evidence for an association? *British Journal of Psychiatry* **157**, 25–33 (erratum: **157**, 460).

Tien, A. Y., and Anthony, J. C. (1990). Epidemiological analysis of alcohol and drug use as risk factors for psychotic experiences. *Journal of Nervous and Mental Disease* **178**, 473–480.

Timpone, J. G., Wright, D. J., Li, N., Egorin, M. J., Enama, M. E., Mayers, J., and Galetto, G. (1997). The safety and pharmacokinetics of single-agent and combination therapy with megestrol acetate and dronabinol for the treatment of HIV wasting syndrome. *AIDS Research and Humam Retroviruses* **13**, 305–315.

Touitou, E., Fabin, B., Dany, S., and Almog, S. (1988). Transdermal delivery of tetrahydrocannabinol. *International Journal of Pharmaceutics* **43**, 9–16.

Turner, C. E., Hadley, K. W., Fetterman, P. S., Doorenbos, N. J., Quimby, M. W., and Waller, C. (1973). Constituents of Cannabis sativa L: IV. Stability of cannabinoids in stored plant material. *Journal of Pharmaceutical Sciences* **62**, 1601–1605.

Turner, C. E., Hadley, K. W., Henry J., and Mole, M. L. (1974). Constituents of Cannabis sativa L. VII: use of silyl derivatives in routine analysis. *Journal of Pharmaceutical Sciences* **63**, 1872–1876.

Turner, C. E., Elsohly, M. A., and Boeren, E. G. (1980). Constituents of Cannabis sativa L. XVII. A review of the natural constituents. *Journal of Natural Products* **43**, 169–234.

Tyler, K., Hillard, C. J., and Greenwood-Van Meerveld, B. (2000). Inhibition of small intestinal secretion by cannabinoids is CB1 receptor-mediated in rats. *European Journal of Pharmacology* **409**, 207–211.

Ungerleider, J. T., Andyrsiak, T., Fairbanks, L., Ellison, G. W., and Myers, L. W. (1987). Delta-9-THC in the treatment of spasticity associated with multiple sclerosis. *Advances in Alcohol and Substance Abuse* **7**, 39–50.

Valjent, E., and Maldonado, R. (2000). A behavioural model to reveal place preference to delta 9-tetrahydrocannabinol in mice. *Psychopharmacology* **147**, 436–438.

Valverde, O., Ledent, C., Beslot, F., Parmentier, M., and Roques, B. P. (2000a). Reduction of stress-induced analgesia but not of exogenous opioid effects in mice lacking CB1 receptors. *European Journal of Neuroscience* **12**, 533–539.

Valverde, O., Maldonado, R., Valjent, E., Zimmer, A. M., and Zimmer, A. (2000b). Cannabinoid withdrawal syndrome is reduced in preproenkephalin knock-out mice. *Journal of Neuroscience* **20**, 9284–9289.

Valverde, O., Noble, F., Beslot, F., Dauge, V., Fournie-Zaluski, M. C., and Roques, B. P. (2001). Delta9-tetrahydrocannabinol releases and facilitates the effects of endogenous enkephalins: reduction in morphine withdrawal syndrome without change in rewarding effect. *European Journal of Neuroscience* **13**, 1816-1824.

Van Hoozen, B. E., and Cross, C. E. (1997). Marijuana. Respiratory tract effects. *Clinical Reviews in Allergy and Immunology* **15**, 243–269.

van Oosten, B. W., Killestein, J., Mathus-Vliegen, E. M., and Polman, C. H. (2004). Multiple sclerosis following treatment with a cannabinoid receptor-1 antagonist. *Multiple Sclerosis* **10**, 330–331.

van Os, J., Bak, M., Hanssen, M., Bijl, R. V., de Graaf, R., and Verdoux, H. (2002). Cannabis use and psychosis: a longitudinal population-based study. *American Journal of Epidemiology* **156**, 319–327.

Vaughan, C. W., McGregor, I. S., and Christie, M. J. (1999). Cannabinoid receptor activation inhibits GABAergic neurotransmission in rostral ventromedial medulla neurons in vitro. *British Journal of Pharmacology* **127**, 935–940.

Vaughan, C. W., Connor, M., Bagley, E. E., and Christie, M. J. (2000). Actions of cannabinoids on membrane properties and synaptic transmission in rat periaqueductal gray neurons in vitro. *Molecular Pharmacology* **57**, 288–295.

Velasco, G., Galve-Roperh, I., Sanchez, C., Blazquez, C., and Guzman, M. (2004). Hypothesis: cannabinoid therapy for the treatment of gliomas? *Neuropharmacology* **47**, 315–323.

Vered, M., Bar-Joseph, A., Belayev, L., Berkovich, Y., and Biegon, A. (1994). Anti-ischemia activity of HU-211, a nonpsychotropic synthetic cannabinoid. *Acta Neurochirurgica Supplement* **60**, 335–337.

Volicer, L., Stelly, M., Morris, J., McLaughlin, J., and Volicer, B. J. (1997). Effects of dronabinol on anorexia and disturbed behavior in patients with Alzheimer's disease. *International Journal of Geriatric Psychiatry* **12**, 913–919.

Wade, D. T., Robson, P., House, H., Makela, P., and Aram, J. (2003). A preliminary controlled study to determine whether whole-plant cannabis extracts can improve intractable neurogenic symptoms. *Clinical Rehabilitation* **17**, 21–29.

Wagner, J. A., Jarai, Z., Batkai, S., and Kunos, G. (2001). Hemodynamic effects of cannabinoids: coronary and cerebral vasodilation mediated by cannabinoid CB(1) receptors. *European Journal of Pharmacology* **423**, 203–210.

Walker, J. M., Huang, S. M., Strangman, N. M., Tsou, K., and Sanudo-Pena, M. C. (1999). Pain modulation by release of the endogenous cannabinoid anandamide. *Proceedings of the National Academy of Sciences USA* **96**, 12198–12203.

Wall, M. E., Sadler, B. M., Brine, D., Taylor, H., and Perez-Reyes, M. (1983). Metabolism, disposition, and kinetics of delta-9-tetrahydrocannabinol in men and women. *Clinical Pharmacology and Therapeutics* **34**, 352–363.

Wallace, M. J., Blair, R. E., Falenski, K. W., Martin, B. R., and DeLorenzo, R. J. (2003). The endogenous cannabinoid system regulates seizure frequency and duration in a model of temporal lobe epilepsy. *Journal of Pharmacology and Experimental Therapeutics* **307**, 129–137.

Wang, L., Liu, J., Harvey-White, J., Zimmer, A., and Kunos, G. (2003). Endocannabinoid signaling via cannabinoid receptor 1 is involved in ethanol preference and its age-dependent decline in mice. *Proceedings of the National Academy of Sciences USA* **100**, 1393–1398.

Ware, M. A., Doyle, C. R., Woods, R., Lynch, M. E., and Clark, A. J. (2003). Cannabis use for chronic noncancer pain: results of a prospective survey. *Pain* **102**, 211–216.

Wiesbeck, G. A., Schuckit, M. A., Kalmijn, J. A., Tipp, J. E., Bucholz, K. K., and Smith, T. L. (1996). An evaluation of the history of a marijuana withdrawal syndrome in a large population. *Addiction* **91**, 1469–1478.

Wiley, J. L., Barrett, R. L., Britt, D. T., Balster, R. L., and Martin, B. R. (1993). Discriminative stimulus effects of delta 9-tetrahydrocannabinol and delta-9-11-tetrahydrocannabinol in rats and rhesus monkeys. *Neuropharmacology* **32**, 359–365.

Wiley, J. L., Huffman, J. W., Balster, R. L., and Martin, B. R. (1995a). Pharmacological specificity of the discriminative stimulus effects of delta-9-tetrahydrocannabinol in rhesus monkeys. *Drug and Alcohol Dependence* **40**, 81–86.

Wiley, J. L., Lowe, J. A., Balster, R. L., and Martin, B. R. (1995b). Antagonism of the discriminative stimulus effects of delta-9-tetrahydrocannabinol in rats and rhesus monkeys. *Journal of Pharmacology and Experimental Therapeutics* **275**, 1–6.

Wiley, J. L., Ryan, W. J., Razdan, R. K., and Martin, B. R. (1998). Evaluation of cannabimimetic effects of structural analogs of anandamide in rats. *European Journal of Pharmacology* **355**, 113–118.

Wiley, J. L., and Martin, B. R. (1999). Effects of SR141716A on diazepam substitution for delta9-tetrahydrocannabinol in rat drug discrimination. *Pharmacology Biochemistry and Behavior* **64**, 519–522.

Wiley, J. L., and Martin, B. R. (2002). Cannabinoid pharmacology: implications for additional cannabinoid receptor subtypes. *Chemistry and Physics of Lipids* **121**, 57–63.

Williams, E., Himmelsbach, C., Wikler, A., Rubley, D., and Lloyd, B. (1946). Studies in marihuana and pyrahexyl compound. *Public Health Reports* **61**, 1059–1083.

Williams, S. J., Hartley, J. P., and Graham, J. D. (1976). Bronchodilator effect of delta1-tetrahydrocannabinol administered by aerosol of asthmatic patients. *Thorax* **31**, 720–723.

Willoughby, K. A., Moore, S. F., Martin, B. R., and Ellis, E. F. (1997). The biodisposition and metabolism of anandamide in mice. *Journal of Pharmacology and Experimental Therapeutics* **282**, 243–247.

Wong, K. C., and Wu, L. T. (1985). *History of Chinese Medicine: Being a Chronicle of Medical Happenings in China from Ancient Times to the Present Period*, 2nd ed. Southern Materials Center, Taipei.

World Health Organization. (1997). *Cannabis: A Health Perspective and Research Agenda*, World Health Organization, Geneva.

Wu, T. C., Tashkin, D. P., Djahed, B., and Rose, J. E. (1988). Pulmonary hazards of smoking marijuana as compared with tobacco. *New England Journal of Medicine* **318**, 347–351.

Yamaguchi, T., Hagiwara, Y., Tanaka, H., Sugiura, T., Waku, K., Shoyama, Y., Watanabe, S., and Yamamoto, T. (2001). Endogenous cannabinoid, 2-arachidonoylglycerol, attenuates naloxone-precipitated withdrawal signs in morphine-dependent mice. *Brain Research* **909**, 121–126.

Yamamoto, T., and Takada, K. (2000). Role of cannabinoid receptor in the brain as it relates to drug reward. *Japanese Journal of Pharmacology* **84**, 229–236.

Yule, H., (Ed.) (1875). *The Book of Ser Marco Polo, the Venetian, Concerning the Kingdoms and Marvels of the East*, 2nd ed. John Murray, London.

Zajicek, J., Fox, P., Sanders, H., Wright, D., Vickery, J., Nunn, A., Thompson, A. UK MS Research Group. (2003). Cannabinoids for treatment of spasticity and other symptoms related to multiple sclerosis (CAMS study): multicentre randomised placebo-controlled trial. *Lancet* **362**, 1517–1526.

Zalish, M., and Lavie, V. (2003). Dexanabinol (HU-211) has a beneficial effect on axonal sprouting and survival after rat optic nerve crush injury. *Vision Research* **43**, 237–242.

Zammit, S., Allebeck, P., Andreasson, S., Lundberg, I., and Lewis, G. (2002). Self reported cannabis use as a risk factor for schizophrenia in Swedish conscripts of 1969: historical cohort study. *British Medical Journal* **325**, 1199.

Zhou, Z., and Bartholomew, B. (2003). Cannabaceae. *Flora of China* **5**, 74–75.

Zhukovskii, P. M., and Hudson, P. S. (1962). *Cultivated Plants and Their Wild Relatives*, Commonwealth Agricultural Bureaux, Bucks, England.

Zimmer, A., Zimmer, A. M., Hohmann, A. G., Herkenham, M., and Bonner, T. I. (1999). Increased mortality, hypoactivity, and hypoalgesia in cannabinoid CB1 receptor knockout mice. *Proceedings of the National Academy of Sciences USA* **96**, 5780–5785.

Zimmer, A., Valjent, E., Konig, M., Zimmer, A. M., Robledo, P., Hahn, H., Valverde, O., and Maldonado, R. (2001). Absence of delta-9-tetrahydrocannabinol dysphoric effects in dynorphin-deficient mice. *Journal of Neuroscience* **21**, 9499–9505.

Zuardi, A. W., Shirakawa, I., Finkelfarb, E., and Karniol, I. G. (1982). Action of cannabidiol on the anxiety and other effects produced by delta 9-THC in normal subjects. *Psychopharmacology* **76**, 245–250.

Zuardi, A. W., Guimaraes, F. S., Guimaraes, V. M. C., and Del Bel, E. A. (2002). Cannabidiol: possible therapeutic application. *In Cannabis and Cannabinoids: Pharmacology, Toxicology, and Therapeutic Potential* (F. Grotenhermen, and E. Russo, Eds.), pp. 359–369. Haworth Integrative Healing Press, New York.

Zweben, J. E., and O'Connell, K. (1992). Strategies for breaking marijuana dependence. *Journal of Psychoactive Drugs* **24**, 165–171.

Zygmunt, P. M., Petersson, J., Andersson, D. A., Chuang, H., Sorgard, M., Di Marzo, V., Julius, D., and Hogestatt, E. D. (1999). Vanilloid receptors on sensory nerves mediate the vasodilator action of anandamide. *Nature* **400**, 452–457.

C H A P T E R

8

Imaging

OUTLINE

INTRODUCTION

Recent advances in technology and methodology in brain imaging render it possible in awake, conscious subjects to obtain detailed images of the brain's structures and to produce images of functions corresponding to mental processes in normal conditions. Anatomical and functional imaging provides a dynamic way to facilitate major advances in the ever-debated relationship between structure and function. Recently, enormous progress has been made in brain imaging that ultimately may provide sensitive and quantitative approaches to bridge the gap between neurology and psychiatry (Laureys *et al.*, 2002) to advance our understanding of the pathophysiology of drug abuse and addiction.

Brain imaging methods are based largely on indices of hemodynamics, which are reflected in the distribution of radiotracers or on nuclear magnetic resonance (a quantitative chemical analysis of biological structures based on nuclear magnetic relaxation times). For radiotracer studies, radioactively labeled drugs are injected into the body, and their distributions are measured for a certain period of time, and this procedure allows the assessment of blood flow, metabolism in a particular region, or neurochemical reactions. Two radiotracer methods are generally used: position emission tomography (PET) and single photon emission computed tomography (SPECT) (Hitzemann *et al.*, 2002). Nuclear magnetic resonance imaging involves three methods. Magnetic resonance imaging (MRI) measures radio signals that differ based on proton composition of tissue and provides detailed images of brain regions. Functional MRI (fMRI) measures the change in magnetic fields associated with the ratio of oxygenated to deoxygenated hemoglobin and thus can detect functionally induced changes in blood oxygenation and allows the visualization of neural activities with high spatial and temporal resolution.

Neurobiology of Addiction, by George F. Koob and Michel Le Moal.
ISBN – 13: 978-0-12-419239-3 ISBN – 10: 0-12-419239-4

Magnetic resonance spectroscopy (MRS) allows the measurement of chemical products without the radiation problems of radiotracer use—but at the cost of less sensitivity. As noted by Magistretti *et al.* (1999), the basic principle of brain imaging was formulated by Sherrington more than a century ago when he suggested that neuronal activity and energy metabolism were tightly coupled. PET and fMRI detect brain activity indirectly by signals reflecting brain energy consumption. Energy is delivered to the brain by the oxydation of glucose from the blood, and PET monitors changes in blood flow, glucose utilization, or oxygen consumption; fMRI signals reflect the degree of blood oxygenation and flow. The methods now are implemented in more and more psychiatric institutions but largely for research purposes. The progress in this field is so rapid that any chapter on the application of imaging methods in any given field of clinical neuroscience will remain preliminary (Koretsky, 2004).

Imaging data optimally should be compared and analyzed consistently by the scientific community. New technologies provide quantification and statistical voxel-based treatments and statistical parametric mapping by computer-assisted analyses. It is important to note that hundreds of megabytes of data are assimilated per session to generate relatively few pieces of functional data. For studies in psychopathology and addiction research, the data are acquired digitally and are amenable to computer-assisted analysis, a domain of intensive development, with ever-larger storage capacities, increasing speed, and more complex statistical analyses. Neuroimaging papers provide attractive color-encoded images but also require examination of the main data in tables and graphs where the mean and variance of the measures are reported (Seibyl *et al.*, 1999). Research in humans is not without challenges. Differences within a sample group and between sample groups sometimes make it difficult for data to be directly compared. For example, inherent individual differences in brain structure, age, sex differences, polydrug abuse, duration of dependence, the state of the subject when undergoing the measures (whether still drug dependent, in treatment, or in withdrawal) are all parameters contributing to group differences between studies, independent of the hypothesis under test.

Other neurophysiological methods that are used in correlation with imaging techniques in the field of addiction include electroencephalography (EEG), event-related potentials, and magneto-encephalography (MEG). There is no universal noninvasive method. A comparison of these different methods in terms of temporal resolution and spatial resolution is represented in **Fig. 8.1** (Laureys *et al.*, 2002).

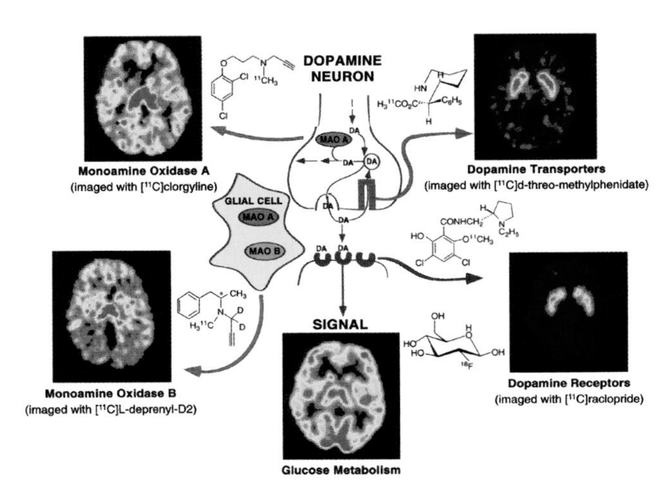

FIGURE 8.1 Diagram of the dopamine (DA) synapse along with images generated by labeling different molecules that participate in the transmission of dopamine signals. The images for the different molecules (dopamine transporters, dopamine D_2 receptors, glucose metabolism, monoamine oxidase-A and-B[MAO A, MAO, B]) were obtained for the same brain level. Notice that although some of the images are visually similar, the information they convey reflects different molecular targets separated from each other by the synaptic cleft, perhaps no more than 50 nm apart, as is the case for the dopamine transporters and the D_2 receptors. The various ligands are [^{11}C] methylphenidate (dopamine transporter), [^{11}C]raclopride (D_2 dopamine receptor), 2-[^{18}F]fluoro-2-deoxy-D-glucose (FDG) (metabolism), [^{11}C]deprenyl (monoamine oxidase [MAO]-B) and [^{11}C]clorgyline (MAO-A). [Brain sections reproduced with permission from Volkow et al., 1999. Chemical structures reproduced with permission from Hitzemann et al., 2002.]

BASIC TECHNICAL PRINCIPLES OF NEUROIMAGING

Biological imaging methods lie at the interface of physics, mathematics, computer sciences, and engineering (Seibyl *et al.*, 1999). Positron emission tomography (PET) derives from the well-known tissue autoradiography techniques developed in the mid-twentieth century, the introduction of X-ray computed tomography (CT) in the 1970s, and the introduction of radioisotopes emitting positrons (Laureys *et al.*, 2002). PET is, by its very nature, a molecular imaging technology. A positron combines immediately with an electron, and they annihilate each other immediately and emit high-energy gamma rays. PET technology has rapidly developed to capture markers of brain activity, such as oxygen use, blood flow, drug uptake, and glucose metabolism. Most PET imaging studies of functional brain activity assess synaptic activity by measuring regional changes in regional cerebral blood flow (rCBF) or the regional metabolic rates for glucose. The radioactive tracers are produced by a cyclotron and administered into the blood stream. The disintegration of particles from the short half-life positron-emitting radionucleotides is recorded by many sensors distributed around the subject's head. The scanner detects the rays produced as positron annihilation occurs. Images are reconstructed to reflect both location and concentration of the isotope for a given place of the head, and the information depends on the isotope. The radioisotopes of oxygen (^{14}O and ^{15}O), nitrogen (^{13}N), and carbon (^{11}C) are all positron emitters and can be administered to a subject and detected externally. There is no positron emitter for hydrogen, so ^{18}fluorine is used as a hydrogen substitute. Deoxyglucose tracers also are used, such as 2-[^{18}F]fluoro-2-deoxy-D-glucose (FDG) (Fowler *et al.*, 2004). Isotopes integrate organic molecules for natural moieties from water to L-DOPA or receptor ligands. PET locates activity changes in spatial resolution of 6 mm for all the different parts of the brain (**Fig. 8.1**). This method has been used to explore the neural substrates of higher functions, and while progressively being replaced by fMRI, PET remains necessary for molecular, receptor, neurotransmitter, and gene imaging in physiology and pathology (Phelps, 2000). Recent developments in PET with increased detector crystals and three-dimensional image acquisition have enhanced both sensitivity and resolution significantly, with commercially available PET now providing a resolution of 2.0–2.5 mm (Chatziioannou *et al.*, 1999).

SPECT also uses radiotracers to detect individual photons (low-energy gamma rays) (Laureys *et al.*, 2002). The scanner consists of two or three cameras rotating around the head, and multiple angular projections are captured and reconstructed. SPECT tracers are more limited but longer lasting. They include ^{99m}technetium with a 6 h half-life and ^{123}iodine with a 13 h half-life. Generally, the agents are soluble in lipids, making brain uptake rapid, and they are longer lasting. Because of this, a delay can be introduced between injection and scanning. Spatial and temporal performances are inferior to PET (**Fig. 8.1**), but SPECT is less expensive and does not require a cyclotron on-site.

A continuing difficulty for radiotracer imaging of molecular targets is the relatively small window of the appropriate combination of lipophilicity, molecular weight, and affinity that allows the molecule to readily cross the blood brain barrier. There is no major difficulty in developing the proper radiotracers, in that almost all candidate ligands contain carbon and hydrogen. Positron-emitting nuclides ^{11}C and ^{18}F can be incorporated as an isotopic variant (^{11}C for ^{12}C) or an atomic substitute. As a result, the available probes, synthesized as analogues of agents active at synaptic sites of neurotransmission, have permitted the measurement of pre-, post-, and intrasynaptic targets. However, little progress has been made in measuring intracellular signal transduction or gene expression (Fumita and Innis, 2002).

MRI corresponds to a vast variety of techniques that do not use ionizing radiation (Laureys *et al.*, 2002). The method is based on the fact that spinning atoms (hydrogen nuclei, or protons) are perturbed in a precise manner under radio wave pulses applied to tissue. Protons under the magnetic field act as little compass needles. There is an emission of radio signals depending on the state and the number of the atoms. The signal is bright or dark depending on the type of tissue and the pulse sequence. The strength of the signal—emitted when nuclei return to equilibrium—depends on proton density in tissues and the relaxation times, longitudinal or transverse, that make different biological parameters bright or dark.

In the early 1990s, it was demonstrated that fMRI was able to detect small magnetic fields, such as instantaneous changes in blood oxygenation in the human brain. During brain activation, there is an increase in blood flow, and more glucose is consumed, but there is no alteration in the amount of oxygen. However, during sudden efforts, the brain works through anaerobic metabolism, and additional blood flow leads to more oxygen in the draining venous system. The blood oxygen level-dependent (BOLD) signal, the ratio of oxygenated-to-deoxygenated hemoglobin, changes in relation to oxygenated blood increase and to a reduction of the levels of paramagnetic deoxygenated hemoglobin. The regional changes are translated into magnetic resonance signals, used as markers of functional activation (i.e., fMRI), and mapped directly onto high resolution ultra-fast scans. In practice, because there is no natural baseline following the BOLD signal in traditional fMRI, rest periods are assumed to represent the baseline, a zero-activity level to be compared with activation tasks or drug intake. However, rest is not without brain activity and is frequently associated with significant cognitive activity, in particular in the frequently explored limbic and cortical regions (Stark and Squire, 2001). fMRI has a resolution of 1–2 mm, and with new developments (e.g., echo-planar imaging) it monitors changes in blood flow-induced signals in real-time, faster than $H_2^{15}O$-PET which has a time resolution of approximately 45 s. Note, however, that time resolution is on the order of milliseconds with EEG and MEG (**Fig. 8.1**).

Increased interest in molecular imaging is increasing the number of processes that can be imaged in the brain (Koretsky *et al.*, 2004). One of the latest advancements of MRI technology has been diffusion tensor imaging (DTI) that measures the diffusibility of water molecules and renders visible the preferential orientation of the movements of these water molecules into tissue. The orientation is equally distributed in all directions, such as in gray matter, and is either isotropic (equally behaving) or anisotropic (not equally behaving in all directions) (Makris *et al.*, 2002). The DTI method provides a new approach to tractography for exploring various white matter fiber pathways that may be involved in cognitive function (Makris *et al.*, 2002).

The problem raised by significant individual differences in brain structure has prompted imaging scientists to promote a separate MRI-based neuroanatomy (Evans, 2002). The International Consortium for Brain Mapping (ICBM) is attempting to create a probabilistic human brain atlas (Mazziotta *et al.*, 1995). Fully automated techniques that produce accurate neuroanatomic segmentation in MRI data sets are necessary for answering questions regarding cross-sectional variability and detection of abnormalities in groups (Evans, 2002). Once the MRI image has been segmented, each voxel (three-dimensional pixels) in an image space carries an anatomical label and a measure of the confidence in that label. This information detects subtle neuroanatomical or neuropathologic changes, comparing one subject to a group or making intragroup comparisons or longitudinal studies in a single subject or in a group (Evans, 2002). Another problem that limits the general use of MRI is that many subjects find remaining motionless in a narrow tunnel uncomfortable—some subjects even become claustrophobic.

In summary, imaging techniques differ in their sensitivity for measuring brain changes. A major advantage of the PET and SPECT radiotracer techniques is their extraordinarily high sensitivity (10^{-9}–10^{-12} M) compared to MRI (10^{-4} M) or MRS (10^{-3}–10^{-5} M). MRI detection of gandolinium occurs at concentrations of approximately 10^{-4} M, and MRS measures γ-aminobutyric acid (GABA) and glutamine at about 10^{-3} M. This power cannot compete with [^{11}C]NNC756 PET which uses a conventional bismuth germanate-base scintillator which can measure extrastriatal dopamine D_1 receptors at a concentration of about 10^{-9} M, a concentration approaching physiological ranges considering that many molecules relevant to psychiatric disorders are present at concentrations of less than 10^{-8} M (Fumita and Innis, 2002).

Application of neuroimaging methods to substance use disorders has developed rapidly over the last 10 years (for review, see Fowler *et al.*, 1999, 2004; Hitzemann *et al.*, 2002). In practice, from a pharmacokinetic approach, PET techniques are used to measure the absolute uptake, regional distribution, and kinetics of drugs of abuse. These techniques can also monitor activation patterns in response to drug interventions by measuring rCBF or glucose metabolism. However, the temporal resolution does not match that of fMRI, and each activation measure requires a separate radiotracer injection for rCBF and clearance between interventions (**Fig. 8.1**). fMRI provides a better temporal potential and allows one to track the duration of regionally generated signals in parallel with the course and duration of the functional–behavioral effects of a drug (Breiter *et al.*, 1997).

From a pharmacodynamic approach, serial studies can be performed in a given subject to evaluate the effect of an acute drug challenge (Fowler *et al.*, 1999). By means of rCBF, enzyme activity, and transporter occupancy, neurotransmitter function can be assessed with PET and SPECT. The brain regions most sensitive are determined, and the relations between metabolism, rCBF, and behavior can be established. Most drugs of abuse acutely decrease regional brain glucose metabolism, yielding a value that reflects the average of the changes that occurs over the 35–45 min of the uptake period of FDG. Data from rCBF provide an indirect index of neuronal activity in physiological or pathophysiological conditions, but can be variable depending on the direct vasoactive properties of a given drug. Imaging specialists are aware that drug-induced cerebrovascular changes and/or pathologies may confound interpretation of regional activation. The temporal resolution of rCBF is better than that of metabolism measures. For example, SPECT studies of rCBF have been able to demonstrate subclinical neurovascular sequelae associated with chronic cocaine use (Volkow *et al.*, 1988; Holman *et al.*, 1993). Moreover, [$^{15}O_2H_2$] PET has been demonstrated to be useful for exploring drug toxicity and rCBF abnormalities due to vasoconstrictor or platelet aggregation in cocaine abuse.

The pharmacodynamics of drug action can be evaluated with PET and SPECT (Hitzemann et al., 2002) (**Fig. 8.2**). Determining transporter and receptor occupancy by drugs is the first step in assessing drug-induced changes in synaptic function. The ability to measure dopamine transporter occupancy allows an assessment of the relationship between the amount of occupancy and the level of a drug-induced behavioral effect, such as intoxication or 'high', and conversely, the efficacy of dopamine transporter-blocking agents in preventing the 'high'. A similar conceptual framework guides the parallel studies of receptor binding (Fowler *et al.*, 1999). Binding of dopamine receptor ligands is hypothesized to be sensitive to the endogenous concentrations of dopamine (Volkow *et al.*, 1997b).

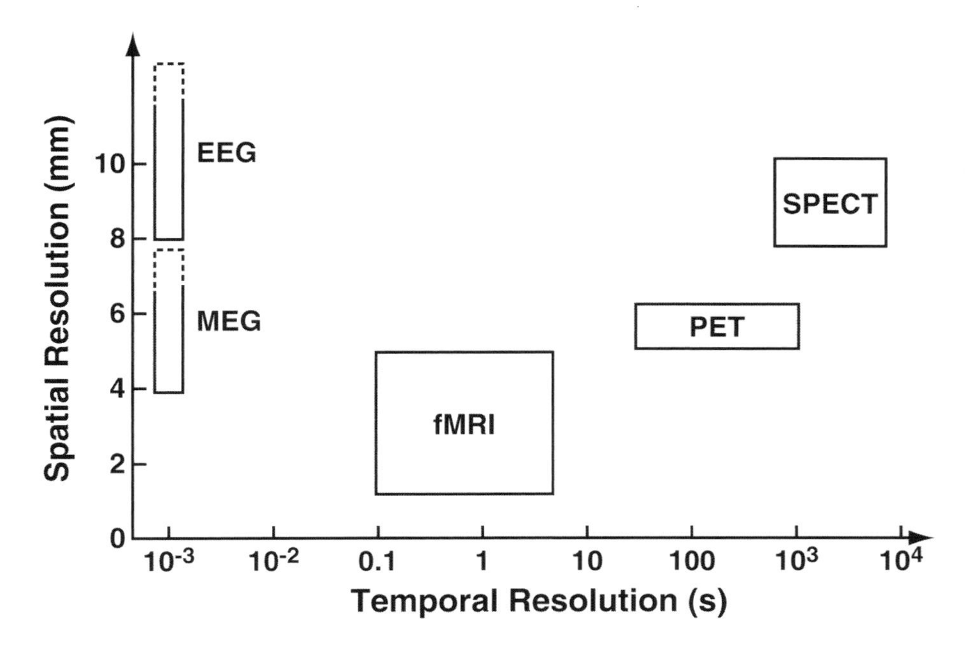

FIGURE 8.2 Approximation of the resolution in time and space of the most commonly employed functional neuroimaging techniques based on measurements of hemodynamic (functional magnetic resonance imaging, positron emission tomography, single photon emission computed tomography) and electrical (electroencephalography and magnetoencephalography) activity of the brain. [Reproduced with permission from Laureys *et al.*, 2002.]

The ligands typically used are [^{11}C]raclopride for PET studies and [^{123}I]-S-(−)-N-[(1-ethyl-2-pyrrolidinyl) methyl]-2-hydroxy-3-iodo-6-methoxybenzamide ([^{123}I]IBZM) for SPECT. In the synapse, dopamine competes with the radioligand so that increases in dopamine are reflected in drug-induced decreases in radiotracer binding relative to baseline. Hypothetically, synaptic neurotransmitter function can be followed over the course of intoxication and dependence.

BRAIN IMAGING OF DRUG ADDICTION

Psychostimulants

Research with imaging techniques on the effects of cocaine, the most studied psychostimulant, began in the early 1990s. Cocaine hydrochloride injected intravenously into polydrug abusers decreased regional cerebral metabolic rates as measured by FDG PET, including decreases in all neocortical areas, striatum, and midbrain (London *et al.*, 1990b). In contrast, using fMRI, acute infusion of cocaine selectively increased or decreased brain metabolic activity, depending on the brain region, in cocaine-dependent subjects abstinent for 18 h. The brain was imaged for 5 min before and 13 min after infusion of either cocaine (0.6 mg/kg) or saline, while the subjects rated scales of *rush*, *high*, *low*, or *craving* (Breiter *et al.*, 1997) (**Fig. 8.3**). The *high* only appeared in the drug group, followed by dysphoria 11 min postinfusion, while peak *craving* occurred 12 min postinfusion. Prior to both infusions (saline and cocaine), within 5 min a positive signal change was observed in the ventral region of the nucleus accumbens and subcallosal cortex. After the cocaine infusion, the signal increased in some areas, such as the nucleus accumbens, subcallosal cortex, caudate putamen, thalamus, insula, hippocampus, cingulate, lateral prefrontal cortex, temporal cortex, ventral tegmentum, and pons. The signal decreased in some subjects in the amygdala, temporal pole, and medial prefrontal cortex. Saline infusion produced a limited set of significant activations, such as frontal and temporal-occipital cortices matching some of the activation seen in the cocaine group. The activation correlated with the degree of emotional ratings (Breiter *et al.*, 1997) (**Fig. 8.3**).

In contrast, detoxified cocaine abusers showed major decreases in brain metabolism in the orbitofrontal and cingulate areas. FDG metabolic studies of brain glucose metabolism showed decreased metabolism in several regions of the frontal lobes, mostly in the orbitofrontal cortex and cingulate gyri in detoxified cocaine abusers months after cocaine use (Volkow *et al.*, 1993b) (**Fig. 8.4**). Consistent with these results, in a cognitive situation requiring inhibition, a GO/NO-GO task with fMRI, cocaine users had significant hypoactivity in the cingulate, presupplementary motor, and insula brain regions (Kaufman *et al.*, 2003). This attenuated response in the presence of comparable activation levels in other task-related cortical areas, such as the prefrontal cortex, parietal cortex, and putamen suggest specific cortical functional deficits in cocaine abusers (Kaufman *et al.*, 2003).

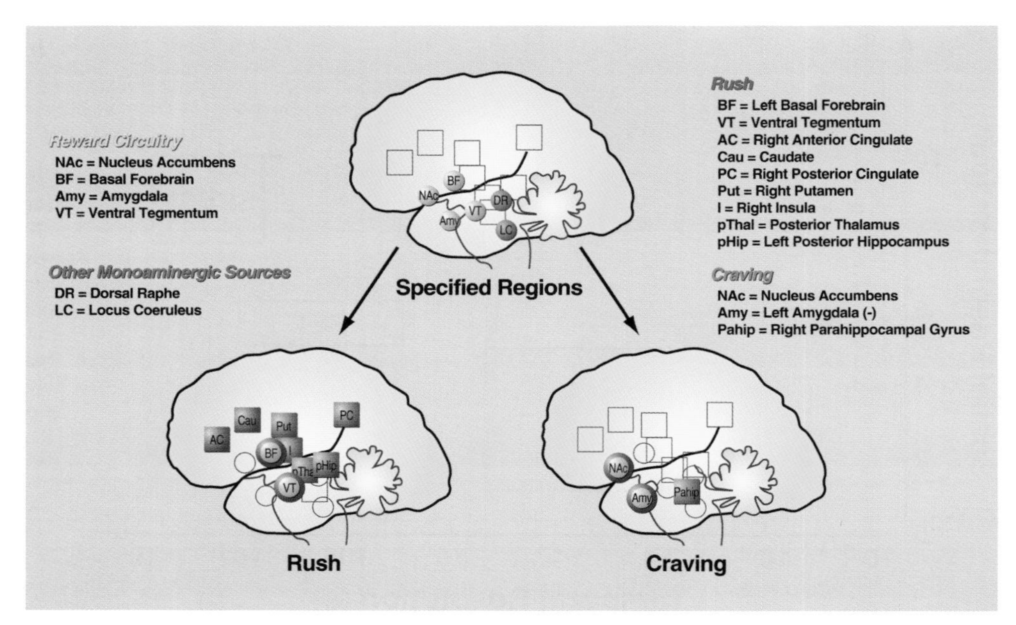

FIGURE 8.3 Summary schematic of limbic and paralimbic brain regions that correlate with euphoria (*brown*) versus those regions that correlate with craving (*green*). Above these summary schematics is a schematic of the brain regions (*yellow*) predicted to be active after an infusion of cocaine. Two other brainstem monoaminergic regions, potentially encompassed in a pontine activation seen in baseline versus postinfusion comparison, are illustrated in *blue*. This pontine activation correlated with behavioral ratings for 'rush.' [Reproduced with permission from Breiter *et al.*, 1997.]

Some of the metabolic changes may be lateralized. Cocaine abusers, tested weeks and months after their last drug intake, had significantly lower metabolic activity in 16 of 21 left frontal regions and 8 of 21 right frontal regions, and these effects persisted after 3–4 months of detoxification (Volkow *et al.*, 1992b) (**Fig. 8.5**).

Addicts often attribute relapse to intense desire or craving that may arise in an environment associated with drug use (Wikler, 1948; Childress *et al.*, 1992; Drummond *et al.*, 1995). The role of drug-related cues has been demonstrated in laboratory settings. A study with FDG PET showed increased glucose metabolism in cortical and limbic regions implicated in several

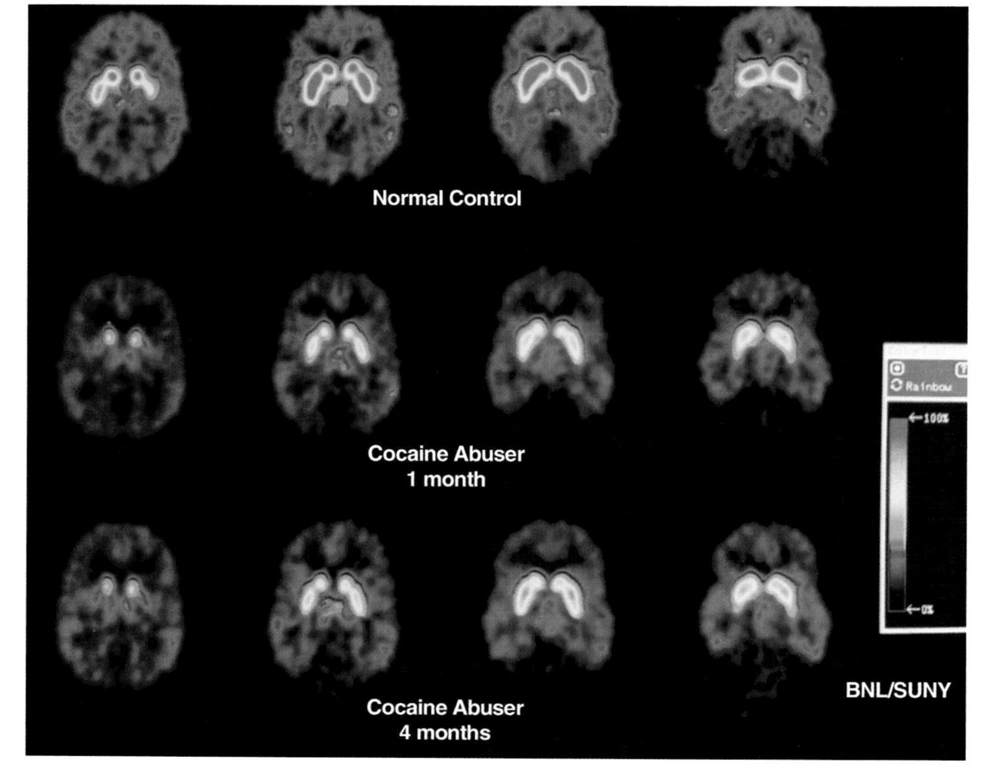

FIGURE 8.4 [^{18}F]*N*-methylspiroperidol images in a normal control subject and in a cocaine abuser tested 1 month and 4 months after the last cocaine use. The images correspond to the four sequential planes where the basal ganglia are located. The color scale has been normalized to the injected dose. Notice the lower uptake of the tracer in the cocaine abuser compared to the normal control. Notice also the persistence of the decreased uptake even after 4 months of cocaine discontinuation. BNL, Brookhaven National Laboratory; SUNY, State University of New York. [Reproduced with permission from Volkow *et al.*, 1993b.]

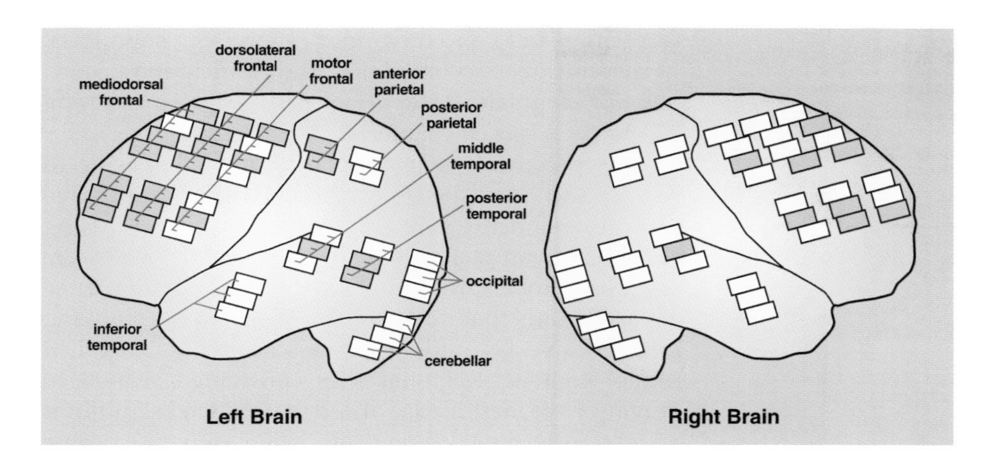

FIGURE 8.5 Comparison between regional metabolic values of normal controls and cocaine abusers tested 1–6 weeks after their last use of cocaine. Cocaine abusers, tested weeks and months after their last drug intake, had significantly lower metabolic activity in 16 of 21 left frontal regions and 8 of 21 right frontal regions, and these effects persisted after 3–4 months of detoxification. Blue areas represent regions which were significantly lower ($p < 0.05$) than those of normal controls (Student two-tailed t-test). [Reproduced with permission from Volkow *et al.*, 1992b.]

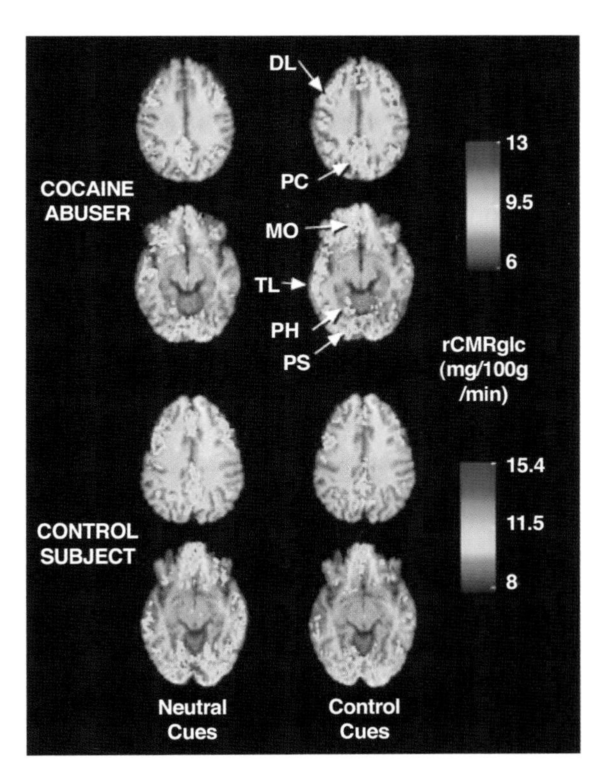

FIGURE 8.6 Representative images of relative regional cerebral glucose metabolism (rCMRglc) in selected subjects from the cocaine group (Upper) and the control group (Lower) during the neutral (Left) and cocaine-related (Right) stimulus sessions. Two levels of the brain in each subject are displayed. In each case, a pseudocolored metabolic positron emission tomography image was superimposed on a structural magnetic resonance image. Arrows indicate regions which exhibited significant increases in rCMRglc in the cocaine group (COC): DL, dorsolateral prefrontal cortex; PC, precuneus; PS, peristriate cortex; MO, medial orbitofrontal cortex; TL, temporal lobe; PH, parahippocampal gyrus. Other areas that exhibited metabolic increases in the subject shown, however, did not manifest increases in other subjects of the COC group. In contrast to the metabolic increases in the COC group, control subjects exhibited a tendency for a decrease in rCMRglc. [Reproduced with permission from Grant *et al.*, 1996.]

forms of memory in long-term cocaine abusers exposed to cocaine-related cues and drug paraphernalia (Grant *et al.*, 1996) (**Fig. 8.6**). Importantly, correlations were found between metabolic increases in the dorsolateral prefrontal cortex, medial temporal lobe (amygdala), and cerebellum with self-reports of craving. These regions correspond to a distributed network integrating emotional and cognitive aspects that link environmental cues with drug craving (see *Neurobiological Theories of Addiction* chapter). In another study linking craving to exposure to cocaine-related cues in cocaine abusers using FDG PET, cocaine cues resulted in left hemisphere activation of the lateral amygdala, lateral orbitofrontal cortex, and rhinal cortex, and right hemispheric activation of the dorsolateral prefrontal cortex and cerebellum (Bonson *et al.*, 2002). The intensity of the activation in these regions (with the exception of the cerebellum) was correlated with craving. Deactivation occurred in the left ventral pole and left medial prefrontal cortex (Bonson *et al.*, 2002).

Using [$H_2{}^{15}O$] PET as a tracer, rCBF was measured in detoxified cocaine users and compared to naive controls who were exposed to cocaine-related videos or neutral scenes (Childress *et al.*, 1999) (**Fig. 8.7**). Results indicated that during cocaine videos, users experienced craving and showed a pattern of increases in rCBF in the amygdala and anterior cingulate cortex and decreases in rCBF in the basal ganglia relative to their responses to the neutral stimuli. However, the two groups did not differ in their responses in the dorsolateral prefrontal cortex, cerebellum, thalamus, or visual cortex. Similar data have been obtained from comparable protocols (Kilts *et al.*, 2001). Craving-related activation corresponded to a network of limbic, paralimbic, and striatal brain regions, and included other regions involved in stimulus–reward associations

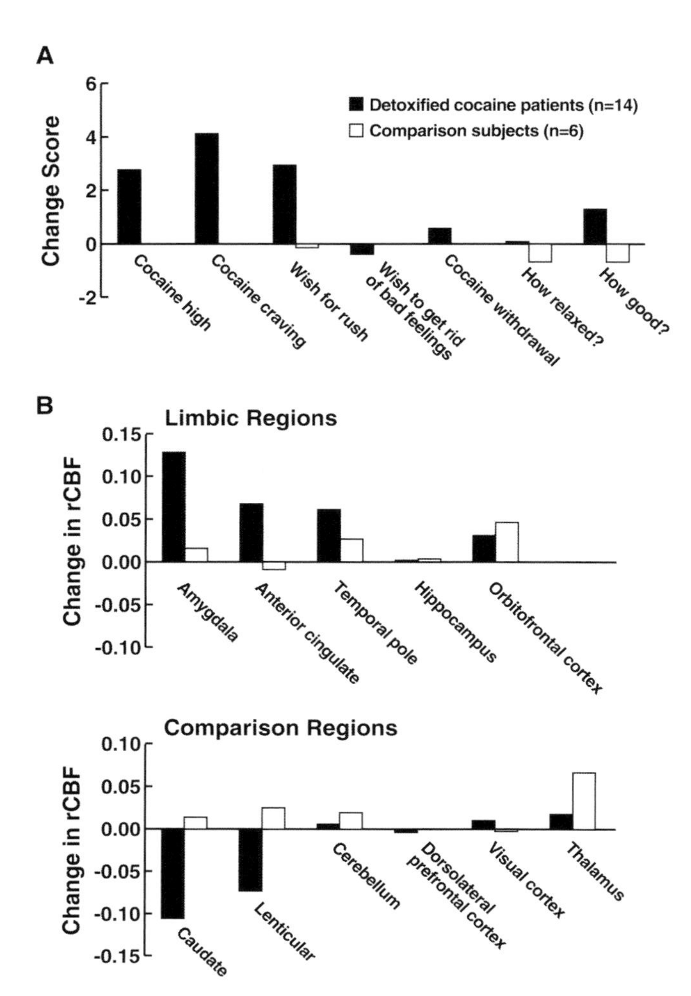

FIGURE 8.7 **(A)** Changes in subjective responses to a cocaine-related video among detoxified cocaine patients and cocaine-naive comparison subjects. Scores represent the change from resting baseline for items self-rated on a 10-point (0 to 9) scale. Changes in *Cocaine high, Cocaine craving,* and *Wish for rush* differed both from the patients' own baseline responses and from the responses of the comparison subjects. **(B and C)** Changes in relative regional cerebral blood flow (rCBF) in limbic and comparison brain regions of detoxified cocaine patients and cocaine-naive comparison subjects in response to a cocaine-related video. Values represent the change in regional cerebral blood flow (rCBF) between a non-drug-related (nature) video and a cocaine-related video. rCBF in the cocaine patients showed a pattern of differential limbic increases and basal ganglia decreases in response to the cocaine video. This pattern did not occur in comparison subjects without a cocaine history. For these analyses, the hippocampus included the adjacent entorhinal cortex. The orbitofrontal cortex included the rectal gyrus. The visual cortices included both primary and association cortices. **(B)** There were significant changes in rCBF in response to the cocaine video for the amygdala and the anterior cingulate both within the patient group and between the patients and the comparison subjects. Within the cocaine group, there was also a significant change in rCBF for the temporal pole. **(C)** Within the cocaine group, there were significant reductions in rCBF in response to the cocaine video for the caudate and the lenticular nuclei. There also was a significant difference in caudate rCBF between the patients and the comparison subjects. [Reproduced with permission from Childress et al., 1999.]

(e.g., amygdala), incentive motivation (e.g., subcallosal gyrus/nucleus accumbens), and anticipation (e.g., anterior cingulate cortex) (**Fig. 8.8**) (see *Neurobiological Theories of Addiction* chapter).

This pattern of activation may not be specific to drugs and may be involved in other appetitive responses and for natural rewards. In a study using fMRI (Garavan *et al.*, 2000), current cocaine users were compared to controls in a protocol of viewing three separate films that portrayed: (1) individuals smoking crack cocaine, (2) outdoor nature scenes, and (3) explicit sexual content. Regional sites considered related to craving were identified as those that showed significant activation in the cocaine users when viewing cocaine videos. These regions showed significantly greater activation when contrasted with comparison subjects viewing the cocaine film (population specificity) and cocaine users viewing the neutral scenes (content specificity). The regions identified were largely lateralized to the left side of the brain, and included the frontal lobe, parietal lobe, insula, and anterior/posterior cingulate. Of the 13 regions identified as putative craving sites, only the anterior cingulate, right inferior parietal lobule, caudate, and lateral dorsal nucleus showed significantly greater activation during the cocaine film than during the sex film in the cocaine users. These data suggested that cocaine cues activate similar substrates as naturally evocative stimuli in the cocaine users (see also Breiter *et al.*, 2001). However, cocaine users showed a smaller neuroanatomical response to the sex film in frontal, cingulate, and parietal cortices and cerebellum than the control subjects, suggesting some impairment in responses to natural rewards.

Because of decades of animal data elucidating a role for dopamine in the psychostimulant actions of cocaine, imaging studies of the neuropharmacology of cocaine dependence have focused primarily on dopamine transmission. The relationship between dopamine transporter blockade and occupancy by cocaine and the subjective effects of the drug has been explored in regular cocaine abusers with PET using [^{11}C]cocaine as the transporter ligand (Volkow *et al.*, 1996b, 1997a). Cocaine (0.3–0.6 mg/kg, i.v., doses commonly abused) blocks between 60 and 77 per cent of the transporters, and the magnitude of the self-reported 'high' was correlated with the degree of dopamine transporter occupancy in the striatum (Volkow *et al.*, 1997a) (**Fig. 8.9**). It was shown that at least 47 per cent of the transporters need to be blocked for subjects to perceive cocaine effects (Volkow *et al.*, 1997a). The time-course for the 'high' parallels that of cocaine concentration within the striatum at large. However, in an earlier study, pre-blocking the

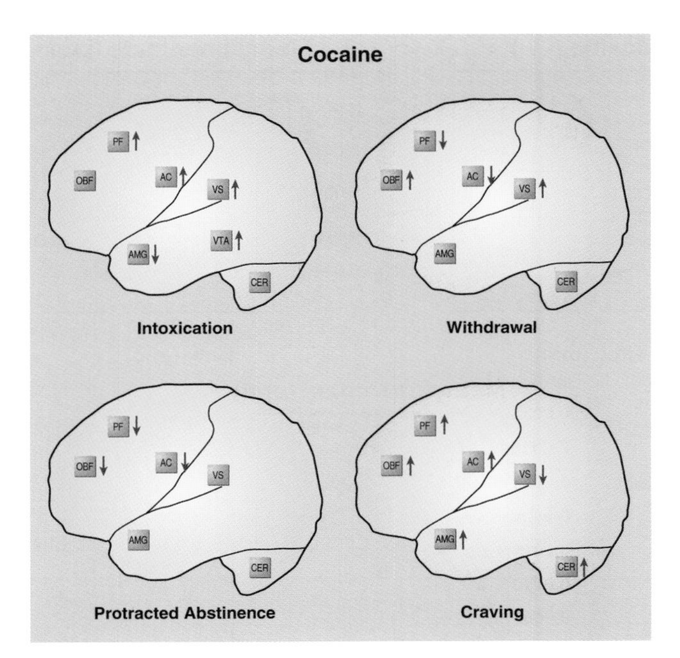

FIGURE 8.8 Changes in brain metabolic activity associated with different stages of the cocaine addiction cycle in humans: intoxication (Breiter *et al.*, 1997), withdrawal (Volkow *et al.*, 1992b, 1993b), protracted abstinence/craving (Volkow *et al.*, 1992b, 1993b; Grant *et al.*, 1996; Breiter *et al.*, 1997; Childress *et al.*, 1999).

dopamine transporter with administration of methylphenidate that occupied up to 80 per cent of dopamine transporter did not block the 'high' from a subsequent dose of methylphenidate given 60 min later (Volkow *et al.*, 1996b). Methylphenidate binds to the dopamine transporter and blocks dopamine uptake but has little abuse potential. Almost total dopamine transporter occupancy would be required for a medication to block cocaine's reinforcing effects (Volkow *et al.*, 1996b, 1997a).

The question of why methylphenidate is not an abused drug and cocaine is an abused drug may be linked to their brain pharmacokinetics. Both drugs have similar affinity for the dopamine transporter. PET studies showed that both drugs administered intravenously had a rapid and high uptake in the brain, a similar regional distribution, and a similar time course of 'high', but the half-life clearance of methylphenidate from the striatum was far slower than that of cocaine (> 90 min for methylphenidate versus 20 min for cocaine). Also, methylphenidate given orally has a much slower uptake and clearance than methylphenidate given intravenously (Hitzemann *et al.*, 2002) (**Fig. 8.10**).

Experimental studies have shown that a history of cocaine abuse decreases dopamine receptors of the D_2 family as measured with PET combined with [^{18}F]methylspiroperidol (Volkow *et al.*, 1993b) or [^{11}C]raclopride (Volkow *et al.*, 1993a). Cocaine abusers detoxified for one week to several months had significantly lower values of D_2 binding in striatum compared to normal subjects (Volkow *et al.*, 1990a, 1993b). Detoxified cocaine abusers also had reduced dopamine release induced by intravenous methylphenidate in the striatum, including the nucleus accumbens, as measured by [^{11}C]raclopride with PET, and showed a reduced 'high' with methylphenidate challenge compared to controls

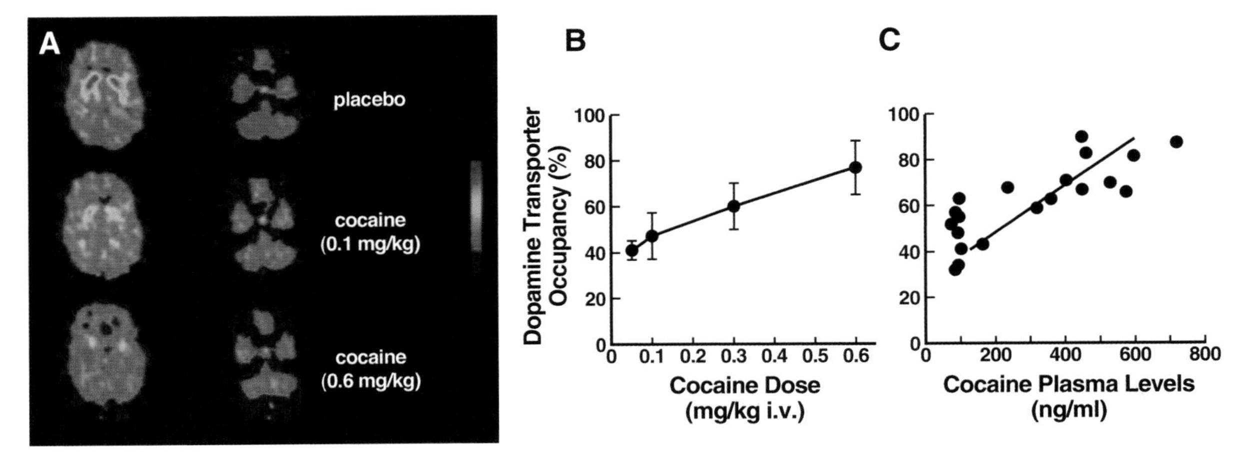

FIGURE 8.9 **(A)** Distribution volume images at the level of the striatum and the cerebellum obtained with [^{11}C]cocaine for a subject tested at baseline (placebo) and with 0.1 and 0.6 mg/kg (i.v.) cocaine doses. **(B)** Correlation between dopamine transporter occupancy by cocaine and doses of cocaine (mg/kg). Values correspond to means and standard deviations. **(C)** Correlation between dopamine transporter occupancy and plasma concentration of cocaine (ng/ml). [Reproduced with permission from Volkow *et al.*, 1997a.]

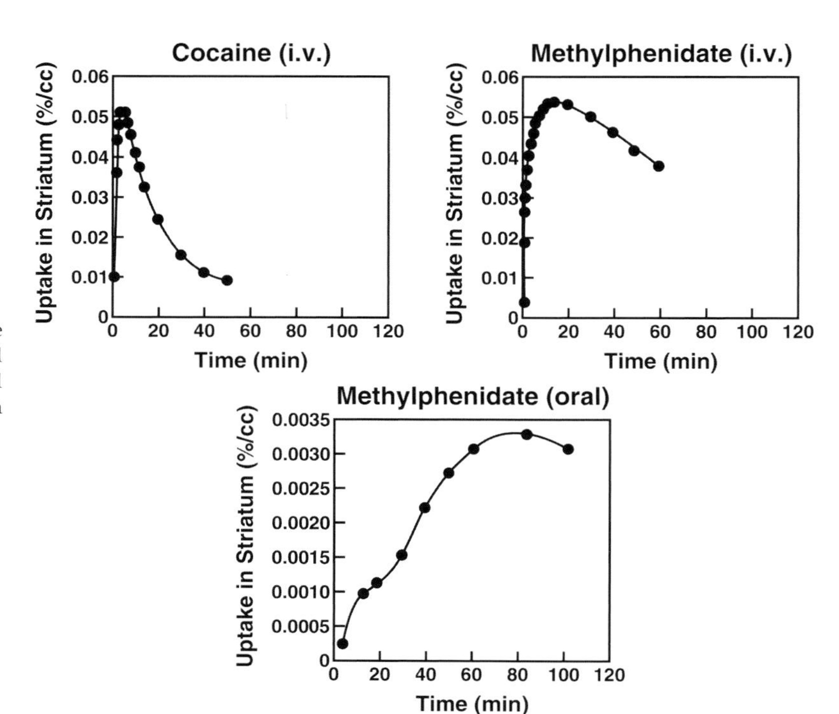

FIGURE 8.10 Time courses for the striatal uptake of intravenously administered [11C]cocaine and [11C]methylphenidate and orally administered [11C]methylphenidate. [Reproduced with permission from Hitzemann *et al.*, 2002.]

(Volkow *et al.*, 1993b, 1997b) (**Fig. 8.4**). As a result, decreased dopaminergic function was postulated to underlie cocaine dependence (Volkow *et al.*, 1997b). These findings were a major challenge to the hypothesis that addiction to cocaine involved an enhanced striatal dopamine response to cocaine and/or an enhanced induction of euphoria (Volkow *et al.*, 1997b). However, cocaine addicts showed an enhanced dopamine response to intravenous methylphenidate in the thalamus that was associated with craving that may reflect activation of the thalamo-orbitofrontal system (Volkow *et al.*, 1997b).

Activity at D_2 receptors also may predict potential reinforcing responses of psychostimulants. Low levels of dopamine receptors have been hypothesized to be a factor that predisposes subjects to use drugs as a means of compensating for the decrease in activation of reward circuits activated by these receptors (Blum *et al.*, 1996). Some support for this hypothesis can be found in imaging studies with methylphenidate. Intravenous [11C]raclopride and PET have been used to measure D_2 receptor levels in the striatum (the marker binds to D_2 and D_3 receptors, but there are few D_3 receptors in the striatum) in non-drug users administered methylphenidate (0.5 mg/kg, i.v.) (Volkow *et al.*, 1999c). Methylphenidate binds to the dopamine transporter and blocks dopamine uptake but has little abuse potential. Subjects who reportedly 'liked' the effects of methylphenidate had significantly lower D_2 receptor levels than subjects who disliked its effects, and the higher the receptor level, the more intense were the drug's unpleasant effects (Volkow *et al.*, 1999c) (**Fig. 8.11**).

To test whether increasing dopamine activity could reverse the metabolic decrements that were linked previously to decreased dopamine D_2 receptors in cocaine abusers, subjects underwent two FDG PET scans, one after two sequential placebo injections, and one after two intravenous methylphenidate injections. D_2 receptor activity also was measured, as above, to evaluate the relation of dopamine activity to methylphenidate-induced metabolic changes. Data showed that methylphenidate produced significant increases in metabolism in the anterior cingulate cortex. The increases in the right orbitofrontal cortex and right striatum were associated with craving, and those in the prefrontal cortex were associated with mood changes. The data also showed that although methylphenidate increased metabolism in the anterior cingulate, it only increased metabolism in the orbitofrontal cortex and striatum for the subjects in whom it enhanced craving and mood, respectively (Volkow *et al.*, 1999b) (**Fig. 8.12**). Thus, dopamine enhancement *per se* was not sufficient to increase orbitofrontal or prefrontal brain activity. The authors hypothesized that the observed activation of the orbitofrontal cortex and striatum may be related to compulsive drug administration in addicted subjects because these regions have been proposed to mediate the salience of reinforcing stimuli (Volkow *et al.*, 1999b).

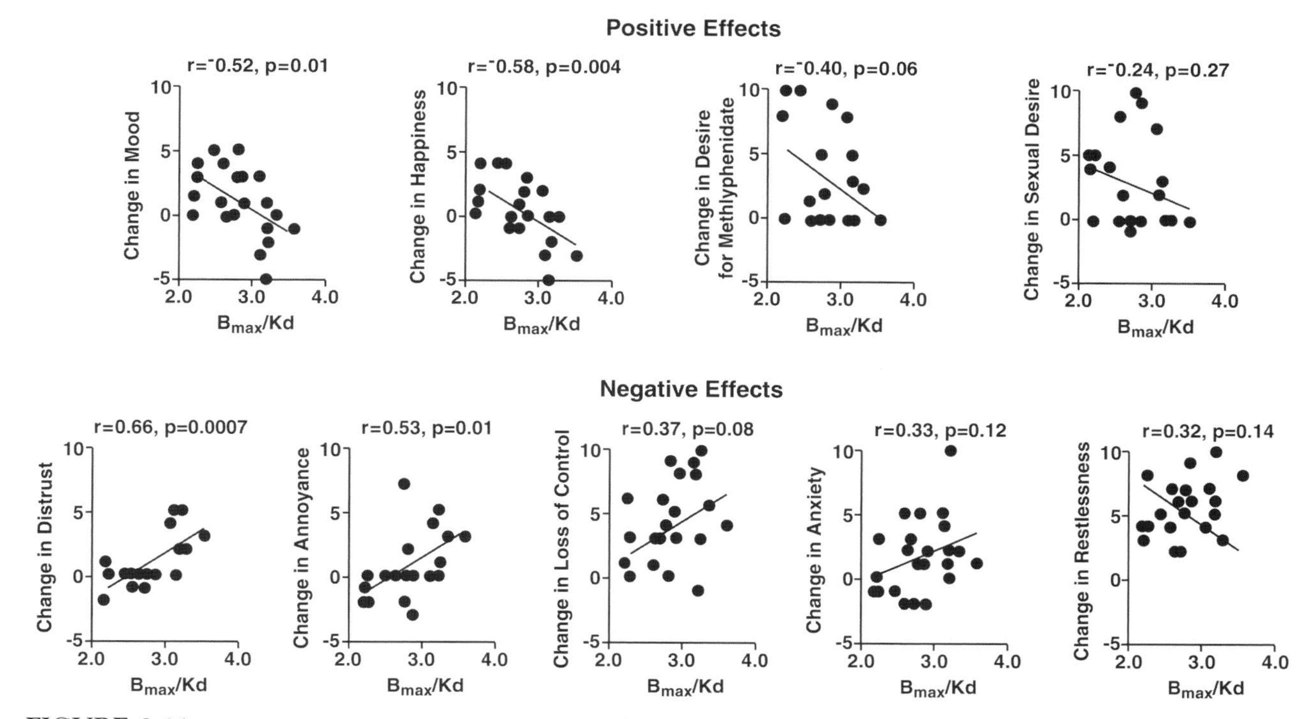

FIGURE 8.11 Regression slopes and correlations between the behavioral effects of methylphenidate (minus baseline ratings) and the measures of dopamine D_2 receptor levels (B_{max}/Kd) in 23 healthy male subjects. [Reproduced with permission from Volkow *et al.*, 1999c.]

Brain γ-aminobutyric acid (GABA) levels in cocaine-dependent subjects with and without alcohol use disorder also differed from healthy subjects using two-dimensional MRS (Ke *et al.*, 2004). The cocaine-dependent subjects without a history of alcohol use had lower GABA levels in the left prefrontal lobe than the controls, and those with an alcohol history had lower levels than the other two groups, suggesting cocaine dependence alone decreased GABA levels. In cocaine addicts subjected to studies of μ opioid system function with [^{11}C]carfentanil and PET 1–4 days postcocaine, there was increased μ opioid receptor binding in the striatum, thalamus, anterior cingulate, frontal cortex, and temporal cortex, suggesting possible decreases in endogenous opioid production (Zubieta *et al.*, 1996). Similar changes in μ opioid receptor binding have been observed in animal studies following 'binge' administration of cocaine (Unterwald *et al.*, 1992).

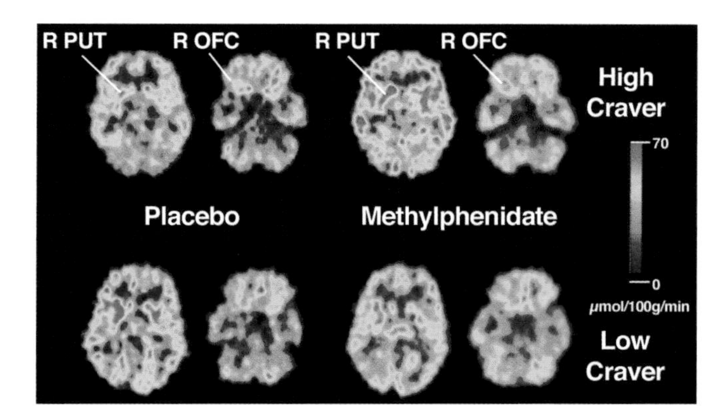

FIGURE 8.12 Brain metabolic images at the level of the striatum and the orbitofrontal cortex after placebo and after methylphenidate for a cocaine abuser who experienced high levels of craving and for a cocaine abuser who experienced low levels of craving after methylphenidate. Notice that the cocaine abuser who experienced craving had an increase in metabolism in the right orbitofrontal cortex (R OFC) and in the right striatum, including the right caudate and right putamen (R PUT). [Reproduced with permission from Volkow *et al.*, 1999b.]

The functional changes described above as resulting from cocaine abuse also may be accompanied by structural changes (Strickland *et al.*, 1998). Focal structural analysis by using brain imaging can reveal long-term selective modifications. MRI and voxel-based morphometry in cocaine-dependent subjects 3–5 days after the last cocaine use showed a decrease in gray matter concentration in the ventromedial orbitofrontal, anterior cingulate, anteroventral insular, and superior temporal cortices (regions involved in cognitive and inhibitory processes and emotional valence), compared to controls. The average percentage decrease in gray matter concentration ranged from 5 to 11 per cent. White matter concentration did not differ between groups (Franklin *et al.*, 2002). In general, with some methodological differences, the same results were found in cocaine abusers abstinent for 20 days (Matochik *et al.*, 2003). In another study among cocaine-dependent individuals, a negative correlation was observed between the volume of the lateral ventricles as measured by MRI and respective ratings of the 'high' experienced just after an intravenous cocaine infusion, and a negative correlation was observed between frontal cortex cerebrospinal fluid volume and euphoria 30 min after cocaine infusion, indicating that larger lateral ventricular volumes are associated with a decrease in response to intravenous cocaine (Bartzokis *et al.*, 2000).

Amphetamines in general have been less investigated than cocaine. Methamphetamine abuse has raised much concern about potential neurotoxicity to the human brain because of preclinical data and neurological effects in humans (Davidson *et al.*, 2001; Cadet *et al.*, 2003). Brain imaging of methamphetamine abusers during 4–7 days of abstinence showed higher ratings of depression and anxiety than controls, accompanied by significant decreases in regional glucose metabolism with FDG PET in anterior cingulate and insula cortices, but higher activity in orbitofrontal cortex, amygdala, and ventral striatum (London *et al.*, 2004). A PET study where brain glucose metabolism was measured in 15 detoxified methamphetamine abusers 2 weeks to 5 months from last use revealed lower activity in the thalamus and striatum and higher functional activity in the parietal cortex (Volkow *et al.*, 2001b). Methamphetamine-dependent subjects also have been compared by using fMRI with age and education-matched controls (Paulus *et al.*, 2002). A two-choice prediction task and a two-choice response task were employed during the imaging session. The dependent subjects exhibited cognitive deficits during decision-making correlated with orbitofrontal and dorsolateral prefrontal cortex dysfunction. Methamphetamine addicts abstinent for 8 months and evaluated with perfusion MRI and neuropsychological testing showed slower responses on a working memory task and decreases in relative rCBF measures in striatum, insula cortex, parietal cortex, occipital cortex, and temporal cortex, suggesting persistent physiological changes during protracted abstinence (Chang *et al.*, 2002).

Methamphetamine abusers had lower levels of dopamine D_2 receptors in striatum as measured by $[^{11}C]$raclopride with PET, and this decrease in D_2 receptors was associated with decreases in metabolic activity of the orbitofrontal cortex (Volkow *et al.*, 2001a). In preclinical studies, methamphetamine induces long-lasting depletions in dopamine and serotonin, and neuropathological deficits have been shown. It is difficult to translate data from laboratory to humans due to the large doses frequently administered in animals (McCann and Ricaurte, 2004). Nonetheless, a postmortem study of 12 methamphetamine abusers (Wilson *et al.*, 1996) and an imaging study months after withdrawal (McCann *et al.*, 1998) reported reductions in dopamine transporters in the brain, especially in the caudate, suggesting that the drug affected dopamine terminals. PET scans with $[^{11}C]$WIN35 428 of abstinent methamphetamine users showed that levels of the dopamine transporter were persistently reduced in former addicts for months after withdrawal, especially in the caudate (McCann *et al.*, 1998). PET scans of detoxified methamphetamine abusers after administration of $[^{11}C]$-D-threo-methylphenidate, a dopamine transporter ligand, revealed a mean dopamine transporter reduction of 27 per cent in the caudate and 21 per cent in the putamen compared to control subjects (Volkow *et al.*, 2001d). The reduction was evident even in abusers who had been detoxified for at least 11 months (Volkow *et al.*, 2001d). The dopamine transporter reduction was correlated with motor slowing and memory impairment, and the decrement in dopamine transporter site availability presumably reflected presynaptic pathology (Volkow *et al.*, 2001d).

In chronic methamphetamine abusers tested at least 4–7 days after the last methamphetamine injection, severe structural deficits were demonstrated using MRI and surface-based computational image analyses to map regional abnormalities (Thompson *et al.*, 2004). Cortical maps revealed severe gray matter deficits in the cingulate, limbic, and paralimbic cortices in abusers (11.3 per cent below control). Moreover, they had, on average, 7.8 per cent smaller hippocampal volume and significant white matter hypertrophy (7 per cent). Hippocampal deficits correlated with memory performance on a word recall test. It was concluded that the drug damage to the cingulate and limbic cortices provoked reactive gliosis, secondary to neuronal damage, which in turn caused white matter hypertrophy. The loss of dopamine transporters in methamphetamine abusers does recover with protracted abstinence (Volkow *et al.*, 2001c), but changes such as local perfusion abnormalities, as measured by ^{99m}technetium-hexamethylpropylenamine oxime (^{99m}Tc-HMPAO) and SPECT, may not recover (Iyo *et al.*, 1997). Long-term neurotoxicity also was observed in methamphetamine users abstinent for 4 months as measured by *N*-acetylaspartate levels using $[^{1}H]$MRS (Ernst *et al.*, 2000). *N*-acetylaspartate was reduced significantly in striatum and frontal cortex compared to controls, suggesting long-term neuronal damage in abstinent methamphetamine users. Subjects with a history of methamphetamine addiction and tested at least 4–7 days after the last methamphetamine dose also showed structural abnormalities as measured by high-resolution MRI in cingulate, limbic, and paralimbic cortices, and hippocampus, and the hippocampal deficits correlated with memory performance impairments (Thompson *et al.*, 2004).

A large percentage of subjects in medical care and treatment for drug abuse are polydrug abusers. In male polydrug abusers (cocaine, alcohol, marijuana, heroin) abstinent from drugs (except nicotine) for at least 15 days before MRI scanning, total volumes for right and left prefrontal lobes were significantly smaller than in normal volunteers, and the deficits were localized in gray but not in white matter (Liu *et al.*, 1998) (**Fig. 8.13**).

From a neurobiological perspective, these long-term decrements in brain metabolism and rCBF suggest a slow process of recovery in addiction. A functional downregulation of the dopamine system parallels a reduction in the euphoric response and an increase in drug craving. Significant decreases in orbitofrontal cortex, cingulate cortex, and frontal cortical function also parallel the development of dependence. Craving for psychostimulant was associated with a reactivation of some of the same structures, such as prefrontal cortex, orbitofrontal cortex, and anterior cingulate, but also with the recruitment of the amygdala. Structural deficits also are observed with long-term methamphetamine abusers and may persist for months. Altogether, the data for cocaine and methamphetamine suggest that significant neuroadaptation takes place in a striato-thalamo-orbitofrontal network (**Table 8.1**).

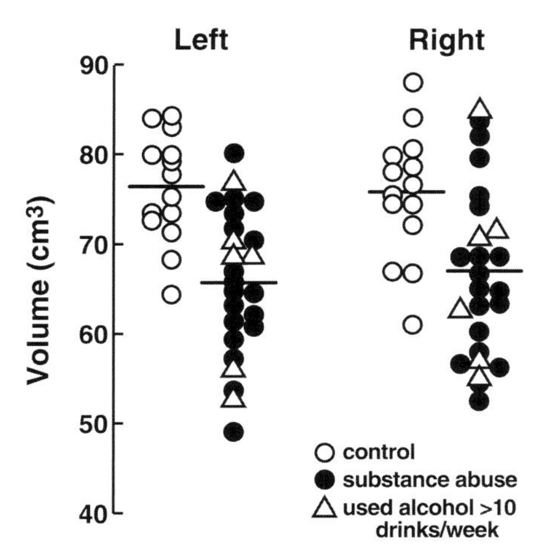

FIGURE 8.13 Absolute volumes of prefrontal lobe in controls and substance abusers. Drugs included alcohol, marijuana, heroin, and cocaine. Each dot represents data for a single research subject. Horizontal lines indicate means for each group. Substance abusers with an average use of alcohol of more than 10 drinks per week are plotted as triangles. [Reproduced with permission from Liu *et al.*, 1998.]

In summary, cocaine and methamphetamine have pronounced effects on brain metabolic activity at all three stages of the addiction cycle that reflect changes in several critical areas: nucleus accumbens/ventral striatum, amygdala, anterior cingulate, prefrontal cortex, and orbitofrontal cortex (**Table 8.1**). Cocaine may acutely activate brain metabolism in parts of the brain reward system (ventral tegmental area, nucleus accumbens, anterior cingulate, prefrontal cortex) but decrease metabolism in cortical regions depending on the imaging measure used. In contrast, during cocaine and methamphetamine withdrawal, there are major decreases in brain metabolism in orbitofrontal cortex, anterior cingulate cortex, and prefrontal cortex. In laboratory situations associated with craving, cocaine produces brain metabolic activation in amygdala, anterior cingulate, prefrontal cortex, and orbitofrontal cortex.

Cocaine and methamphetamine also significantly alter dopaminergic activity in these same regions as measured by imaging studies. The intoxication with cocaine requires a large number of dopamine transporters to be occupied, but the magnitude of the self-reported 'high' correlated with the degree of occupancy of the dopamine transporter. Chronic administration of cocaine and methamphetamine is associated with a decrease in dopamine D_2 receptors in the striatum during withdrawal, and low levels of D_2 receptors have been associated with increased rewarding effects of methylphenidate in nondrug users. Similar results have been observed in cynomologous macaque monkeys where, after 3 months of social housing, subordinate monkeys showed greater self-administration of cocaine, and lower D_2 dopamine receptor binding, than dominant monkeys (Morgan *et al.*, 2002). Decreases in dopamine transporter have been observed in chronic methamphetamine abusers during withdrawal. Finally, methylphenidate administration to cocaine abusers could reverse the decreases in metabolic activity in anterior cingulate and also reverse the decreases in metabolic activity in orbitofrontal cortex, prefrontal cortex, and striatum in subjects in whom it enhanced craving and mood. These results have been interpreted as supporting the hypothesis that decreased dopamine function has a major role in psychostimulant addiction and that this decrease is linked to the disruption of function in the anterior cingulate, prefrontal cortex, and orbitofrontal cortex.

TABLE 8.1 Summary of Imaging Data

	Neurobiological characteristics	
Stage of addiction process	Behavioral and cognitive	Neurobiological mechanisms
Drug Intoxication	↑ Drug reinforcement and salience attribution ↑ Conditioned response ↑ Memories	↑ Reward circuitry (mesolimbic regions) ↑ Amygdala ↑ Hippocampus ↑ Thalamus
Bingeing and Relapse	↑ Loss of control ↑ Response disinhibition	↑ Reward circuitry (ventral tegmental area, nucleus accumbens) ↑ Prefrontal cortex
Acute Withdrawal	↑ Dysthymia ↓ Reward	↓ Reward circuitry ↑ Anterior cingulate cortex ↓ Prefrontal cortex ↓ Striatum ↓ Dopamine D_2 receptors
Protracted Abstinence	↑ Negative affective state ↑ Vulnerability	↓ Prefrontal cortex ↓ Orbitofrontal cortex ↓ Cingulate gyrus ↓ Dopamine D_2 receptors
Craving	↑ Drug expectation of drug effects ↑ Urge to abuse ↑ Negative affective state ↑ Self-medication	↑ Cingulate gyrus ↑ Prefrontal cortex ↑ Orbitofrontal cortex

Alcohol

As soon as functional neuroimaging became available, given the medical importance and the cognitive deficits associated with the neuropathology of alcohol abuse and dependence, neuroimaging of alcohol abuse was investigated. Early studies (Berglund and Ingvar, 1976; Berglund and Risberg, 1977) showed decreased blood flow mainly in fronto-temporal cortical regions. These observations were consistent with the hypothesis that the limbic-basal ganglia and thalamo-cortical networks were involved in the loss of control in alcohol intake associated with alcoholism (Modell *et al.*, 1990). Thus, chronic alcoholism was hypothesized to have effects on blood flow and cerebral metabolism independent of the neuropathological defects due to Korsakoff's syndrome (Joyce *et al.*, 1994) or in the absence of marked anatomical changes (Wang *et al.*, 1993).

Acute alcohol intake decreased brain glucose metabolism as measured by FDG PET in cortex and cerebellum, and this inhibition was more pronounced in alcoholics alcohol-free for 10–15 days prior to imaging (Volkow *et al.*, 1990b). In nonalcoholics, alcohol at doses that produced a 'high' decreased metabolic activity in occipital cortex but increased metabolic activity in the left basal ganglia (Wang *et al.*, 2000). When basic functional and perceptual capacities are considered, brain imaging techniques also can reveal significant brain changes even after two doses of alcohol. The Motor-Free Visual Perception Test—Revised was used with event-related fMRI to study brain activation patterns in healthy controls when sober or after two doses of alcohol sufficient to produce intoxication (Calhoun *et al.*, 2004). This test provides measures of overall visual processing ability and incorporates different cognitive elements, including visual discrimination, spatial relationship, and mental rotation. Alcohol resulted in a dose-dependent decrease in fMRI activation amplitude over much of the visual perception network and a decrease in the maximum contrast-to-noise ratio in the lingua gyrus. Significant dose–response activation decreases were observed in anterior and posterior cingulate, precuneus, and middle frontal areas (Calhoun *et al.*, 2004). In a study with ^{99m}Tc-HMPAO SPECT and morphometric analysis by CT scan, alcoholics showed a reduction in rCBF in the frontal lobe which was related to recent alcohol intake, reflected by functional impairments, and was independent of frontal lobe atrophy (Nicolas *et al.*, 1993).

These changes normalized after two months of abstinence.

Studies of alcoholics during acute and chronic withdrawal have shown decreases in brain metabolism that resolve over time, with more recovery the longer the detoxification. Brain metabolic activity measures using FDG PET showed decreased regional metabolic levels in frontal, parietal, and temporal cortex when tested 6–32 days after alcohol discontinuation (Volkow *et al.*, 1992a). In general, the deficit in brain metabolism (again, in frontal, orbitofrontal, and parietal cortices) began to disappear after the second week of abstinence and returned to normal after two months (Nicolas *et al.*, 1993; Volkow *et al.*, 1994) (**Fig. 8.14**). Decrements persisted in the basal ganglia and temporal cortex. A survey of PET and SPECT studies that investigated and compared short-term and long-term (from 10 months to 10.5 years) abstinent alcohol abusers showed that deficits in brain metabolism, blood flow, and cognitive performance improved with the passage of time for most subjects (Dupont *et al.*, 1996; Johnson-Greene *et al.*, 1997; Harris *et al.*, 1999; Gansler *et al.*, 2000).

Neurobiological correlates of cue-induced craving also have been demonstrated in alcoholic patients. Adolescents aged 14–17 with alcohol use disorders were compared to matched controls who were infrequent drinkers (Tapert *et al.*, 2003). Subjects were shown pictures of alcoholic and nonalcoholic beverage advertisements during BOLD fMRI, while self-report craving measures were obtained before and after cue exposure. Subjects with alcohol use disorders displayed significantly more BOLD responses than controls to alcohol pictures relative to neutral beverage pictures in the left hemisphere, including frontal and limbic regions (i.e., ventral anterior cingulate, prefrontal cortex, orbital gyrus, and subcallosal cortex) and other areas, such as the inferior frontal gyrus, paracentral lobule, parahippocampus, amygdala, fusiform gyrus, temporal lobe, hypothalamus, posterior cingulate, precuneus, cuneus, and angular gyrus. Interestingly, controls showed more BOLD response to alcohol pictures in right frontal regions than alcohol use disorder subjects (Tapert *et al.*, 2003). In a study conducted with fMRI on alcohol-dependent young women, a greater BOLD response to drug-related cues was observed in subcallosal, anterior cingulate, left prefrontal, and bilateral insular regions (Tapert *et al.*, 2004) (**Fig. 8.15**).

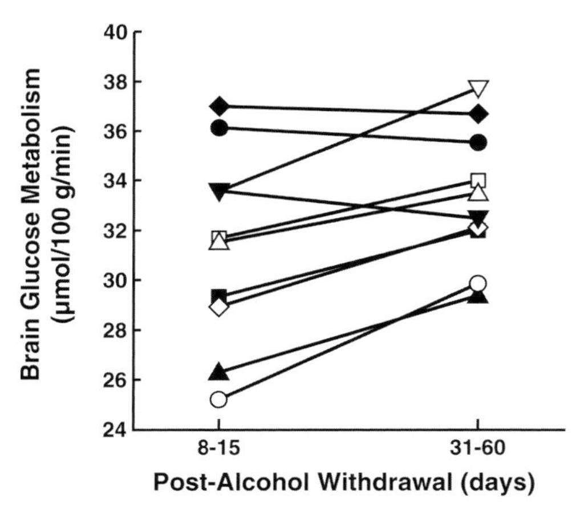

FIGURE 8.14 Individual values for global brain metabolism in alcoholic subjects at 8–15 days and 31–60 days after alcohol withdrawal (n = 10). Symbols represent individual subjects. [Reproduced with permission from Volkow *et al.*, 1994.]

Neurological correlates with cue-induced craving can be influenced by therapeutic interventions. Alcohol-dependent patients who had undergone detoxification were presented with alcohol odor, and fMRI was used to map cerebral responses. Patients underwent standardized behavioral therapy with classic psychopharmacological intervention, and all subjects were again submitted to fMRI to evaluate the effect of therapy on the functional correlates of craving (Schneider *et al.*, 2001). Data showed that, before treatment, activation was primarily located in the subcortical-limbic region of the right amygdala/hippocampal area and in the cerebellum. After treatment, activation was first found in the superior temporal sulcus, while subcortical or cerebellar activation was no longer present. The presence of emotional aspects of craving suggested amygdala activation (Schneider *et al.*, 2001) (see *Psychostimulants* section above and *Neurobiological Theories of Addiction* chapter).

Imaging techniques also are suited for tracking potential vulnerability in subjects at risk for drug abuse. Multiple studies have demonstrated that the amplitude of the P300 component of the event-related potential is smaller in high-risk children and adolescent offspring of alcoholics than in control subjects (for review, see Polich *et al.*, 1994). As an example, MRI was used to measure cerebral, amygdala, and hippocampal volumes in high-risk adolescents (Hill *et al.*, 2001). High-risk adolescents and young adults showed reduced right amygdala volume compared to control subjects. In another study, when fMRI images were collected during performance of a visual oddball task, two areas had significantly lower activation in the offspring of male alcoholics with high risk than in controls—the bilateral inferior parietal lobule and the bilateral inferior frontal gyrus—suggesting a dysfunctional frontoparietal circuit deficiency in the rehearsal component of the working memory system (Rangaswamy *et al.*, 2004).

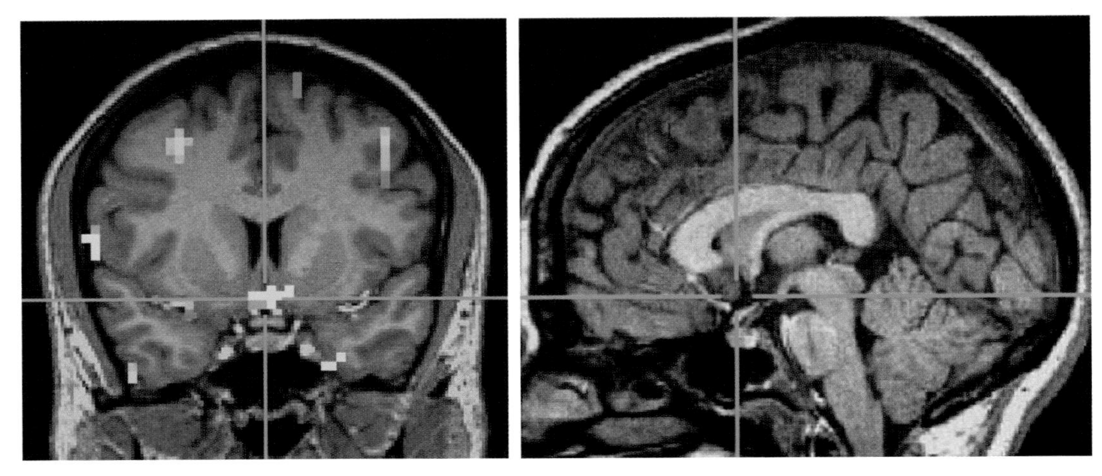

FIGURE 8.15 Coronal view of blood oxygen level-dependent (BOLD) functional magnetic resonance imaging results showing between-group t-test of BOLD response differences during exposure to alcohol words relative to neutral words. *Yellow* and *orange* indicate where alcohol-dependent subjects had more alcohol-word response than normal control participants relative to neutral words (shown here: subcallosal cortex, anterior cingulate, left inferior frontal gyrus, and right uncus). *Blue* shows where normal controls had more alcohol-word response than alcohol-dependent subjects relative to neutral words (shown here: right superior frontal, right and left middle frontal, left inferior frontal, and left middle temporal). Group $p < 0.025$, voxel clusters > 243 μl, $n = 17$. Thumbnail image on the right shows location of coronal slice. [Reproduced with permission from Tapert *et al.*, 2004.]

Besides functional changes, alcohol impacts the brain by direct toxic actions of its metabolic by-products and by secondary nutritional deficiencies. Brain weight studies showed that uncomplicated alcoholics (drinking more than 80 g of alcohol per day for more than 15 years) had a significantly reduced brain weight (Harper and Kril, 1993), and this reduction in brain weight was largely accounted for by a reduction in white matter volume (Harper *et al.*, 1985). Brain lesions and atrophy were present in a majority of chronic alcoholics, but atrophy was only partially reversible during abstinence (Mann *et al.*, 1995; Kril and Harper, 1997). Neuropathologists have shown alcohol-related neuronal loss in specific regions of the cerebral cortex (superior frontal association cortex, hypothalamus, and cerebellum) (Harper, 1998; Mann *et al.*, 2001) with a 20 per cent reduction in numbers of neurons in the superior frontal cortex (Kril *et al.*, 1989). Differences in frontal cortical neurons, thalamic neurons, cerebellum, and white matter volume were significant (**Table 8.2**).

TABLE 8.2 **Summary of Quantitative Neuropathological Data (% of control) in Different Groups of Alcoholic Patients**

Region	Uncomplicated alcoholic (%)	Chronic Wernicke (%)	Korsakoff psychosis (%)
Brain shrinkage (↑ pericerebral space)	36	77	77
↓ Frontal cortical neurons	77	80	84
↓ Cortical neuronal dendrites	81	NA	NA
↓ White matter volume	98	79	83
↓ Mamillary body neuronal number	98	53	32
↓ Thalamic neurons (mediodorsal)	100	52	36
↓ Thalamic neurons (anterior principal)	100	86	47
↓ Basal forebrain neurons	100	76	79
↓ Median raphe neurons	95	30	NA
↓ Dorsal raphe neurons	98	36	NA
↓ Cerebellar vermis Purkinje cells	95	57	NA

Uncomplicated Alcoholic: no liver disease or Wernicke-Korsakoff syndrome.
Chronic Wernicke: nonamnestic Wernicke-Korsakoff syndrome.
Korsakoff Psychosis: amnestic Wernicke-Korsakoff.
100% indicates no variation from control data.
NA = not available.
[Reproduced with permission from Mann *et al.*, 2001.]

Brain imaging revealed significantly smaller brain volumes in alcoholic subjects recruited from an inpatient program and sober for three weeks (Hommer *et al.*, 2001). Both men and women alcoholics had smaller volumes of gray and white matter as well as greater volumes of sulcal and ventricle cerebrospinal fluid than controls. The differences between alcoholic women and nonalcoholic women were greater than the differences between alcoholic men and nonalcoholic men (Hommer *et al.*, 2001). Cortical volume loss has been observed in both white matter and gray matter using MRI (Jernigan *et al.*, 1991; Pfefferbaum *et al.*, 1992).

Numerous neuroimaging and postmortem studies have linked the significant evidence for alcohol-induced abnormalities in frontal lobes and cerebellum to alcohol-related impairment in cognitive function (Harper and Kril, 1989; Pfefferbaum *et al.*, 1997; Kubota *et al.*, 2001). *In vivo* imaging with MRI has shown abnormalities throughout the cortex in an older alcoholic group (Pfefferbaum *et al.*, 1997). Abnormalities in cerebellar function in alcoholics are well documented (Sullivan, 2000; Sullivan *et al.*, 2002), and the link of the cerebellum to some classic frontal lobe functions has suggested the hypothesis of an overall frontocerebellar circuit deficit in alcoholism (Sullivan *et al.*, 2003) (see *Neurobiological Theories of Addiction* chapter).

Another classic syndrome observed in chronic alcoholics is central pontine myelinolysis. In a study with approximately 600 alcoholic patients examined by MRI, 11 were found to have pontine lesions with noticeable shapes, and for three of them the lesions were in the middle cerebellar peduncles, with secondary Wallerian degeneration of the pontocerebellar tract (Uchino *et al.*, 2003). Some deficits may be more subtle. Regions that support the articulatory control system of working memory may require a compensatory increase to maintain the same level of performance in alcoholics compared to controls (Desmond *et al.*, 2003). Alcoholics performed at comparable levels as matched controls in a Sternberg verbal working memory task, but nevertheless showed greater activation with fMRI in the left frontal and right superior cerebellar regions compared to controls.

Alcohol may affect the brains of men and women differently. Alcoholism has been traditionally thought to be a 'male disease,' due to a classic higher prevalence among men. Studies of sex differences are particularly difficult, largely because of methodological problems of controlling for consumption and differences in the metabolism and absorption of alcohol. Female blood alcohol levels tend to be higher after ingesting the same quantity as men (see *Alcohol* chapter), and this may contribute to the increased damage to liver and heart in females. In one study, female alcoholics showed a greater loss of gray matter (11 per cent) than male alcoholics (6 per cent) (Hommer *et al.*, 2001), suggesting that the female brain may be more sensitive to alcohol. However, other studies using MRI have shown greater effects in male alcoholics, sober for several months, compared to female alcoholic outpatients. Both groups were matched on age and sobriety (2–15 months). Women, regardless of alcoholism diagnosis, had less cortical gray and white matter and smaller third ventricles than men. There was a striking increase in the volume of cortical sulci and a decrease in brain and third ventricle volume in men but not in women (Pfefferbaum *et al.*, 2001). A blunted response also has been observed in the changes in metabolic activity produced by alcohol ingestion sufficient to produce intoxication in women, compared to men, using FDG PET (Wang *et al.*, 2003). Males had consistently larger decreases in brain metabolism due to alcohol intoxication than females, but females had greater self-reports of intoxication, both of which were not due to differences in blood alcohol levels (Wang *et al.*, 2003). These results suggest the possibility of gender differences in response to the acute responses and neuropathological effects of alcohol, but caution must be considered because of serious confounds in how much alcohol is ingested and the pattern of ingesting alcohol.

It is well documented that dangerous effects on the brain can occur with alcohol exposure during brain development to form a major component of fetal alcohol spectrum disorders (O'Leary, 2004) (see *Alcohol* chapter). The brains of children and adults with histories of heavy prenatal alcohol exposure have been examined by using quantitative structural MRI. Besides microcephaly, structural abnormalities exist in the cerebellum, corpus callosum, and basal ganglia. More specifically, accompanying the reduced size and displacement in three-dimensional space of the corpus callosum, increased gray matter densities were noted in the perisylvian regions in both hemispheres, and altered gray matter asymmetry in portions of the temporal lobes. A narrowing in the temporal region and reduced brain growth in portions of the frontal regions also were observed long after the prenatal toxic insult (Riley *et al.*, 2004), and these changes are consistent with the neurocognitive symptoms described in these patients.

The acute reinforcing effects of alcohol are hypothesized to be due to interactions with the $GABA_A$ receptor, but also with complex effects on many other ligandgated ion channels, including *N*-methyl-D-aspartate (NMDA), nicotine acetylcholine (nACh), serotonin (5-HT), dopamine, and opioid peptides (Lewis, 1996) (see *Alcohol* chapter). Benzodiazepine drugs have a binding site on the benzodiazepine $GABA_A$ receptor, and alcohol has been hypothesized to have specific alcohol-receptive elements within the same complex (Lewis, 1996) (see *Alcohol* chapter), suggesting that some of the alcohol effects are perhaps mediated by actions on the benzodiazepine $GABA_A$ ionophore receptor complex within the GABA system. An acute challenge with lorazepam, a benzodiazepine drug that decreases brain glucose metabolism, as does alcohol, produced a blunted response in alcoholics in cingulate orbitofrontal cortex that continued after acute withdrawal (Volkow *et al.*, 1997c). Benzodiazepine receptor distribution volume or availability has been evaluated using [^{123}I]iomazenil SPECT and is decreased significantly in detoxified alcoholics in frontal, parietal, temporal, and anterior cingulate cortices, and in cerebellum (Abi-Dargham *et al.*, 1998; Lingford-Hughes *et al.*, 1998, 2000).

Mesocorticolimbic dopamine systems also have been hypothesized to contribute to the mechanisms mediating alcohol abuse. The effect of an oral dose of alcohol (1 ml/kg of 95 per cent USP alcohol) has been tested in healthy volunteers by using [^{11}C]raclopride PET scans co-registered with MRI (Boileau *et al.*, 2003). A significant reduction in [^{11}C]raclopride binding potential, bilaterally in the ventral striatum-nucleus accumbens, was observed in the alcohol condition compared to the orange juice condition, indicative of increased extracellular dopamine during intoxication. Moreover, the magnitude of the change in binding correlated with the alcohol-induced increase in heart rate, a marker of the psychostimulant effect of the drug and also correlated with a personality dimension of impulsiveness (Boileau *et al.*, 2003) (**Fig. 8.16**).

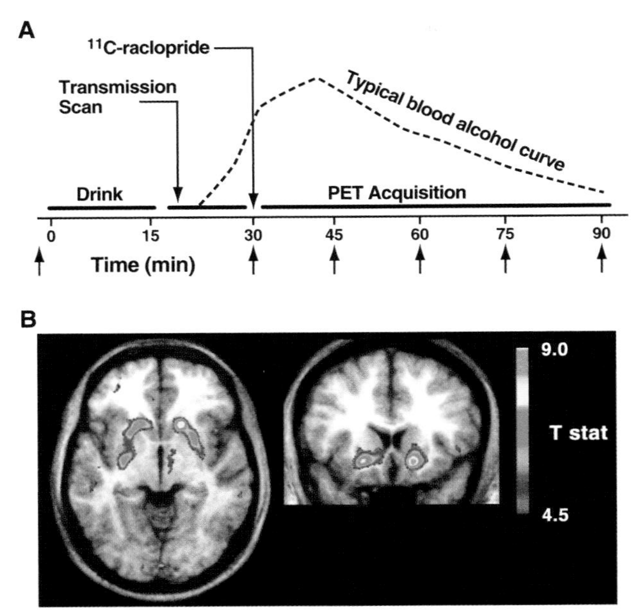

FIGURE 8.16 **(A)** A study designed to assess whether an oral dose of alcohol in humans would lead to dopamine release in the ventral striatum. The vertical arrows indicate the time points of blood sampling, subjective mood assessments, and physiological measurement. **(B)** Statistical *t*-map of the change in [^{11}C]raclopride binding potential induced by an acute oral dose of alcohol (1 ml/kg) in healthy volunteers (n = 6). Color clusters superimposed on the average magnetic resonance imaging from all subjects depict a significant change in binding potential in the ventral striatum. PET, positron emission tomography. [Reproduced with permission from Boileau *et al.*, 2003.]

By using PET and SPECT with D_2 or D_2/D_3 receptor ligands, data showed that these receptors, as seen with other drugs of abuse, seem to be lower in alcoholism in temporal cortex and striatum (Volkow *et al.*, 1996c; Kuikka *et al.*, 2000). The dopamine transporter also is either not changed, or is lower, in alcoholics as shown by [^{123}I]βCIT SPECT and returns to normal long after withdrawal (Laine *et al.*, 1999).

The serotonin transporter also is reduced as a function of the duration of the illness (Heinz *et al.*, 1998). The serotonin transporter has been investigated using PET in various brain regions of abstinent or recovering alcoholics compared to social drinkers. In subjects with a history of alcoholism, total distribution volume of the radioligand [^{11}C]McN5652, which labels the serotonin transporter, was lower in the 11 regions studied and, in particular, in the midbrain, thalamus, amygdala, pons, cingulate gyrus, frontal cortex, and cerebellum (Szabo *et al.*, 2004). Hommer *et al.*, (1997) examined the effects of intravenously administered m-chlorophenylpiperazine, a mixed serotonin agonist/antagonist that can induce craving for alcohol, particularly in male alcoholics in conjunction with PET. In healthy volunteers, m-chlorophenylpiperazine produced increased glucose utilization in 15 regions, particularly in basal ganglia circuitry involving the orbitofrontal cortex and prefrontal cortex, but the responses in orbitofrontal cortex, prefrontal cortex, and basal ganglia were totally blunted in alcoholics. Results suggested that a serotonergic challenge can activate orbitofrontal-striatal circuits, but this activation is blunted in alcoholics (Hommer *et al.*, 1997). In a study using [^{123}I]βCIT SPECT in alcoholics, serotonin dysfunction was associated with anxiety and depression which increased with the risk of relapse. A significant reduction of 30 per cent in the availability of brainstem serotonin transporter was found in alcoholics which was correlated with anxiety and depression during withdrawal (Heinz *et al.*, 1998).

Animal and human studies also have suggested that the endogenous opioid system may be linked to alcohol dependence and craving. Regional brain μ opioid receptor binding potential was measured by using [^{11}C]carfentanil PET in male alcohol-dependent subjects undergoing withdrawal and compared to matched controls (Bencherif *et al.*, 2004). Data showed a lower μ opioid receptor binding potential compared to controls, and μ receptor binding in the right dorsal lateral prefrontal, right anterior frontal, and right parietal cortices, was associated with higher reported craving. Lower binding also was correlated with negative affective states. The involvement of the endogenous opioid peptide system in alcohol dependence has received support from studies using naltrexone, an opioid antagonist used in pharmacotherapy. The effect of a naltrexone challenge was tested in chronic alcoholic patients during detoxification by measuring rCBF with SPECT. In the baseline condition, patients showed lower SPECT blood flow values than controls in the left orbitofrontal cortex and frontal cortex. After naltrexone administration, a significant rCBF decrease was found in the left basal ganglia and left medial temporal region, but no other areas showed changes. This result suggested that increased activity in the basal ganglia and medial temporal region may be driven by craving and may be opioid-dependent, and supported the authors' hypothesis of a role of emotional memory and an obsessive–compulsive component in craving (Catafau *et al.*, 1999).

In summary, acute alcohol intoxication decreases brain metabolic activity in the orbitofrontal cortex, anterior cingulate, and prefrontal cortex, but increases brain metabolic activity in the ventral striatum. During acute withdrawal and protracted abstinence, there is a decrease in prefrontal cortex and a decrease in cerebellar functional activity. Chronic alcoholism causes decrements in brain metabolism and rCBF in association with cognitive deficits, results that are consistent with the effects of other drugs of abuse. Benzodiazepine $GABA_A$ receptors also are decreased, as are D_2 dopamine receptors, dopamine transporters, and serotonin transporters in chronic alcoholics. During craving for alcohol in alcoholics, there is activation of the orbitofrontal, anterior cingulate, and prefrontal cortices, and the striatal system (**Figs. 8.8** and **8.17**).

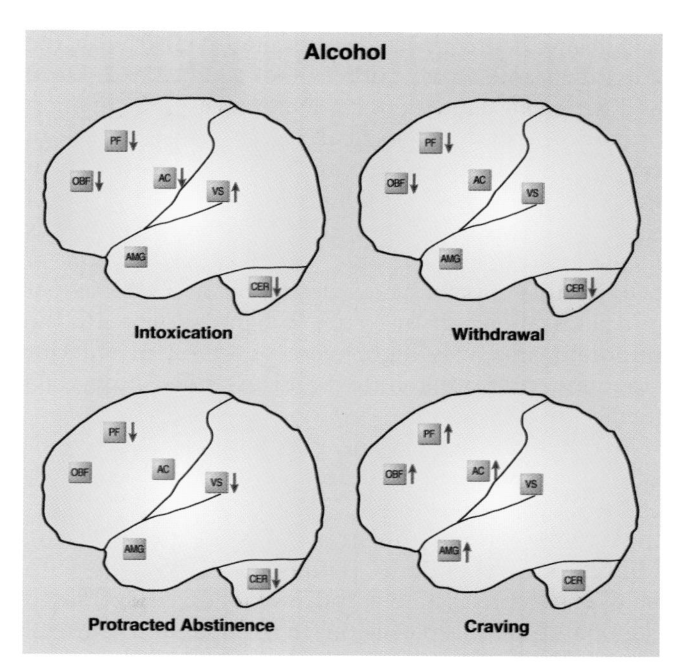

FIGURE 8.17 Changes in brain metabolic activity associated with different stages of the alcohol addiction cycle in humans: intoxication (Volkow *et al.*, 1990b; Wang *et al.*, 2000, 2003; Calhoun *et al.*, 2004), withdrawal (Berglund and Risberg, 1977), protracted abstinence (Berglund and Ingvar, 1976; Volkow *et al.*, 1992a, 1994), craving (Schneider *et al.*, 2001; Tapert *et al.*, 2004).

Opioids

Opioid dependence has not been explored with brain imaging to the extent that one would expect. An early study showed that morphine (30 mg, i.m.) administered to 12 polydrug abusers decreased cerebral glucose metabolism in whole brain and six cortical areas as measured by FDG PET (London *et al.*, 1990a). Cerebral blood flow with [$H_2{}^{15}O$] PET in opioid addicts participating in a methadone program explored how brain regions involved in the reward circuitry reacted to prototypical rewards (Martin-Soelch *et al.*, 2001). During a visual–spatial recognition task with delayed response, three conditions were presented: nonsense feedback, nonmonetary reinforcement, and monetary reward. In control subjects, activation increased in regions associated with visual and motor performance and in the terminal areas of corticolimbic circuits in response to both monetary and nonmonetary reward. In opioid addicts, reinforcement elicited rCBF increases in fewer regions than in control subjects, and typical regions associated with reward, such as the striatum and the orbitofrontal cortex were activated only by monetary reward. The lack of activation in dopaminergic and performance-related regions in addicts was accompanied by poorer performance. It appears that the addicts under methadone treatment process feedback through a different neural pathway than healthy subjects.

When opioid addicts received heroin or placebo or are presented videos showing heroin injection scenes or neutral films, PET scans showed that the effects of heroin and the drug video were centered on the periaqueductal grey and ventral tegmentum, regions which are interconnected and connect also with the amygdala and limbic cortices (Sell *et al.*, 1999). Drug and drug cue-related effects share the same activation of subcortical and cortical regions implicated in reward-related behavior, specifically activation of the periaqueductal grey, ventral tegmental area, dorsal raphe, and insula. Significant modulatory effects, as measured by a psychophysiological interaction of activity that discriminates positively between salient and nonsalient visual cues when activity in the midbrain area was high, were observed in anterior cingulate, prefrontal cortex, and extended amygdala (Sell *et al.*, 1999).

In withdrawal situations, rCBF is decreased as indicated by SPECT mainly in frontal, temporal, and parietal cortices at one week, but less at three weeks, after the last opioid administration (Rose *et al.*, 1996). Methadone-dependent patients challenged with naloxone to precipitate withdrawal showed decreased regional cerebral perfusion in frontal and parietal cortices and increased regional cerebral perfusion in the thalamus compared to controls (Krystal *et al.*, 1995). However, after adjusting for placebo administration, opioid withdrawal was associated with only relatively small increases in right temporal cortex activity and a trend toward decreased frontal cortex activity (Krystal *et al.*, 1995).

When audiotaped autobiographical scripts of an episode of craving were presented to opioid-dependent subjects abstinent for months, compared to presentation of a mental stimulus, rCBF increased in left medial prefrontal and left anterior cingulate cortices in parallel to craving for drug (Daglish *et al.*, 2001). Similar to that observed in animal studies, the few experiments designed to explore whether opioid receptor binding is affected by opioid dependence have not provided reproducible effects (see *Opioids* chapter).

In summary, acute administration of opioids and drug-related cues activate the periaqueductal gray, ventral tegmental area, and extended amygdala. Natural rewards have less of an impact on brain metabolism in these areas in opioid-dependent subjects. Craving during protracted abstinence is associated with medial prefrontal and anterior cingulate cortex activation.

Nicotine

The use of, and dependence on, nicotine is largely via cigarettes. Despite the world-wide public health problem related to tobacco use, neuroimaging studies are not abundant. The acute effects of nicotine administration and smoking were investigated with [$H_2{}^{15}O$] PET in a group of healthy volunteer cigarette smokers (Rose *et al.*, 2003). The subjects were exposed to three conditions in a single session: smoking a nicotine-containing cigarette, smoking a denicotinized cigarette, or receiving intravenous nicotine injections in conjunction with smoking a denicotinized cigarette. In another experiment, other subjects received the nicotinic antagonist mecamylamine (10 mg) or placebo orally after smoking nicotine-containing cigarettes or denicotinized cigarettes. Nicotine increased rCBF in the left frontal region and decreased it in the left amygdala. rCBF in the right hemisphere reticular system was related to nicotine dose in an inverted U-shaped pattern and to self-reports of craving. The effects of mecamylamine was opposite to those of nicotine (Rose *et al.*, 2003).

The effect of nicotine itself on brain imaging in active cigarette smokers also has been explored. Nicotine was administered intravenously at different doses (0.0, 0.75, 1.50, and 2.25 mg/70 kg body weight) and fMRI was used for brain imaging (Stein *et al.*, 1998). The behavioral parameters and subjective drug effects increased dose-dependently. The drug induced a dose-dependent increase in neuronal activity in a set of brain regions that included the frontal lobes, amygdala, cingulate, and nucleus accumbens. In another study, smokers who were abstinent overnight before the study received an intranasal spray dose of 1–2 mg nicotine or placebo 3 min before FDG PET (Domino *et al.*, 2000). The left inferior frontal gyrus, left posterior cingulate, and right thalamus were shown to be activated. The thalamus is a region of high cholinergic receptor density, but other regions with a known high density of receptors were not activated. These brain metabolism data have some correspondence with rCBF studies using [$H_2{}^{15}O$] PET. Decreases in rCBF were observed in the anterior temporal cortex and in the right amygdala, and increases in rCBF in the right thalamus in 12 h abstinent smokers receiving a similar intranasal dose of nicotine spray (Zubieta *et al.*, 2001).

Nicotine (12 μg/kg body weight) was administered subcutaneously to healthy nonsmoking males, and neural correlates of cognitive tasks, hypothesized to activate the prefrontal-parietal areas, were examined using fMRI (Kumari *et al.*, 2003) (**Fig. 8.18**). Nicotine improved attention, arousal, and behavioral performance and produced an increased response in the anterior cingulate, superior frontal cortex, and superior parietal cortex during the task. At rest, nicotine produced increased responses in the parahippocampal gyrus, cerebellum, and medial occipital lobe. Intravenous nicotine (1.5 mg) reduced regional cerebral

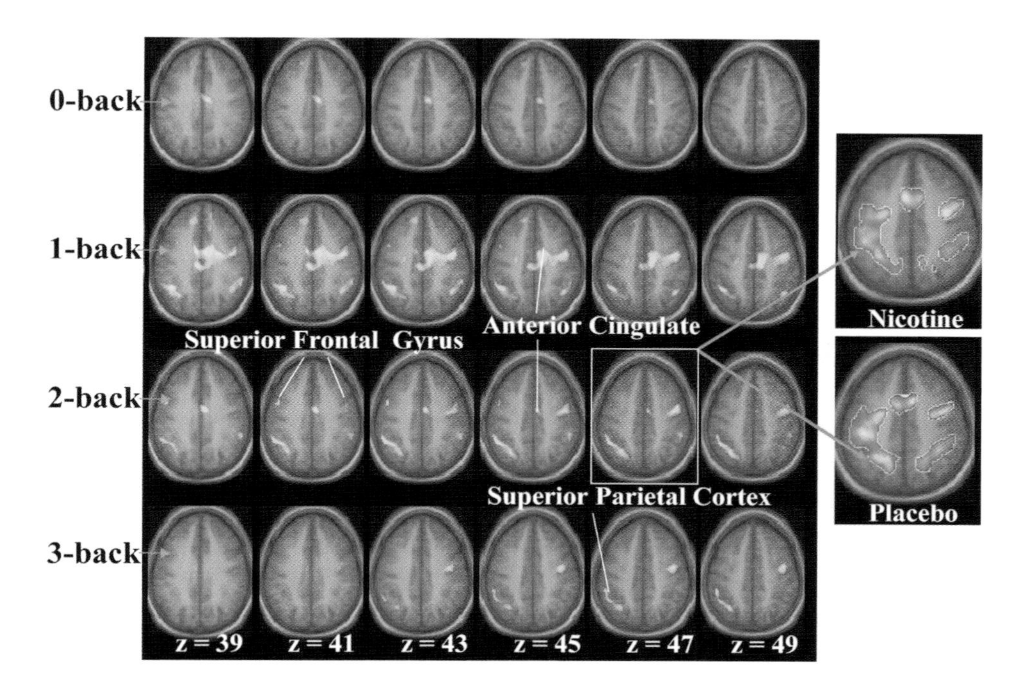

FIGURE 8.18 Nicotine-related modulations at each working memory load. The significant differences between nicotine and placebo activations (paired *t*-test) for 0-back, 1-back, 2-back, and 3-back versus rest contrasts are shown superimposed on the average structural image. Six transverse slices are shown from each condition with their associated Talairach *z* coordinates. The images have been thresholded at $p < 0.05$ uncorrected, although most regions are significant at $p < 0.05$ corrected within a 5 mm sphere located within the regions of interest. The left hemisphere is shown on the left of each slice. Increased activation is demonstrated in the anterior cingulate (0-back minus rest, 1-back minus rest, and 2-back minus rest), superior frontal cortex (bilateral for 1-back minus rest and 2-back minus rest; right side only for 3-back minus rest), and superior parietal cortex (bilateral for 1-back minus rest and 2-back minus rest; left side only for 3-back minus rest third row). The inset panel shows the generic maps (one sample *t*-test) under nicotine and placebo for the 2-back minus rest comparison from which the difference map is constructed. [Reproduced with permission from Kumari *et al.*, 2003.]

metabolic rates for glucose using FDG PET in most of the 30 brain regions examined. Nine regions had bilateral effects that reached significance, including the frontal cortex, anterior cingulate, insula, and striatum (Stapleton *et al.*, 2003).

Using [$H_2{}^{15}O$] PET, smokers showed a differential response to monetary and nonmonetary rewards compared to nonsmokers as was seen in opioid addicts (Martin-Solch *et al.*, 2001). Nonsmokers showed activation in response to nonmonetary rewards and monetary rewards in frontal cortex, orbitofrontal cortex, and cingulate gyrus, whereas smokers showed such activation only to monetary but not nonmonetary rewards (Martin-Solch *et al.*, 2001). It is now clear that cigarette smoke inhalation and nicotine are not completely equivalent (Fowler *et al.*, 1996a,b, 1998, 2000). Cigarette smokers show a marked inhibition of both monoamine oxidase (MAO)-A and -B (Fowler *et al.*, 1996b) (**Fig. 8.19**). MAO breaks down neurotransmitter amines, such as dopamine, serotonin, and norepinephrine, as well as amines from exogenous sources. MAO can be imaged *in vivo* using [^{11}C]clorgyline and [^{11}C]L-deprenyl-D_2 and PET, respectively. The smoking-associated decrease in enzyme activity for both forms was about 40–50 per cent relative to nonsmokers and former smokers. However, nicotine does not inhibit MAO at physiologically relevant levels (Fowler *et al.*, 1998); further, a single cigarette does not decrease MAO activity (Fowler *et al.*, 1999). From these observations, it was hypothesized that smoke from cigarettes and the associated inhibition of MAO may be associated with enhanced activity of dopamine, potentially involved in incentive motivation. Additionally, there may be decreased production of hydrogen peroxide, which is lethal for dopamine cells. Thus, MAO inhibition may account for the lower incidence of Parkinson's disease in smokers and for the higher rate of smoking in individuals with depression and schizophrenia.

Nicotine increased dopamine release in the striatum using [^{11}C]raclopride PET in dependent smokers who smoked compared to dependent smokers who did not smoke. Also, a subject who had smoked while on a substantial dose of haloperidol (40 mg/day) had less upregulation of dopamine D_2 receptors as measured by [^{11}C]raclopride PET than a nonsmoker on a much

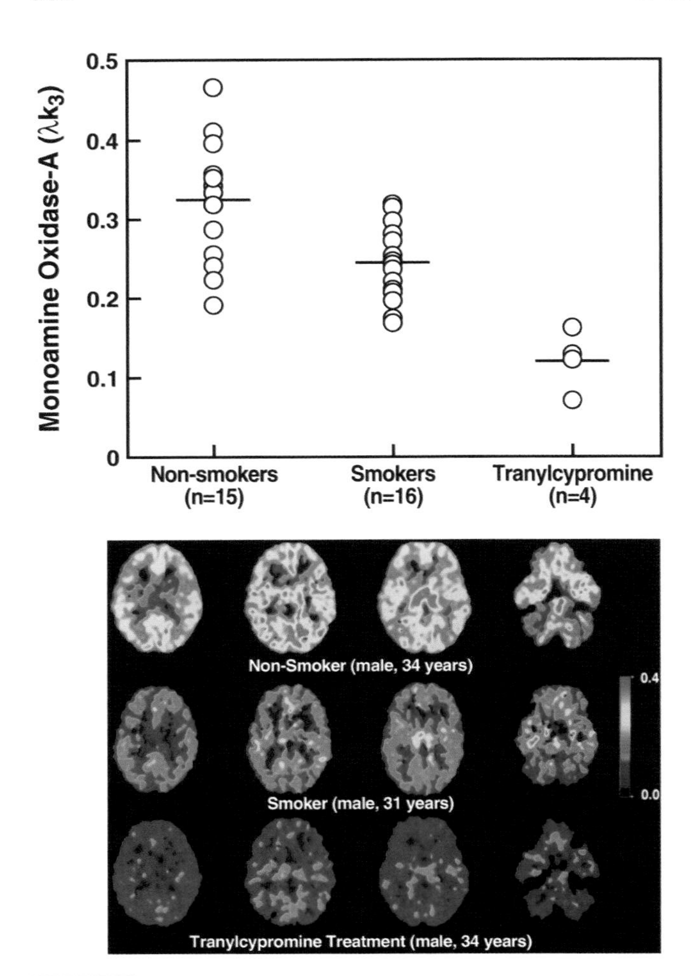

FIGURE 8.19 **(A)** Comparison of monoamine oxidase A levels in the thalamus (as expressed by the model term λk_3) for nonsmokers (n = 15), smokers (n = 16), and nonsmokers who were treated with the monoamine oxidase inhibitor tranylcypromine (n = 4). **(B)** Pixel-by-pixel images of the model term which is a function of monoamine oxidase A activity for a nonsmoker (top row), a smoker (middle row), and the same nonsmoker after treatment with tranylcypromine (bottom row). The same four planes of the brain are shown for each subject and correspond to brain sections at 5.8 cm (level of the occipital cortex and the lateral ventricles), 5.1 cm (level of the occipital cortex and the lateral ventricles), 4.5 cm (level of the thalamus), and 1 cm (level of the lower temporal poles and the cerebellum) above the canthomeatal line (proceeding from left to right). The color scale represents values of λk_3 (scales from 0.4 [red] to 0 [black]). The values of λk_3 for the thalamus are 0.319, 0.212, and 0.128 $cc_{brain}/ml_{plasma}/min$ for the nonsmoker, the smoker and the nonsmoker treated with tranylcypromine, respectively. The corresponding K_1 values are 0.434, 0.421, 0.404 ml/cc/min. [Reproduced with permission from Fowler *et al.*, 1996b.]

lower dose of haloperidol (10 mg/day) suggesting that chronic nicotine counteracted the effects of haloperidol in producing D_2 hypersensitivity (Silvestri *et al.*, 2004). However, another interpretation is that nicotine dependence is producing decreases in D_2 receptors as do other drugs of abuse.

In summary, nicotine administration in smokers and nonsmokers is associated with activation of the frontal cortex, anterior cingulate, amygdala, and ventral striatum. Tobacco smoking also significantly decreases brain MAO activity, suggesting a contribution of smoking to brain function independent of nicotine itself. Smoking also suppresses reward system activation to natural rewards as is observed with opioid addiction.

Marijuana

Acute exposure to marijuana has consistently resulted in increased brain activity (Loeber and Yurgelun-Todd, 1999). Among the first studies using FDG PET, intravenous Δ^9-tetrahydrocannabinol (THC) administration (2 mg) to marijuana abusers increased brain metabolic activity in the orbitofrontal cortex, prefrontal cortex, and basal ganglia (Volkow *et al.*, 1991, 1996a). Baseline cerebellar metabolism was lower in marijuana smokers. Intravenous administration of Δ^9-THC also has been shown to increase rCBF in frontal, temporal, parietal, anterior cingulate, and insula cortices, and striatum and thalamus in a number of studies (Mathew and Wilson, 1993; Volkow *et al.*, 1996a; Mathew *et al.*, 1997).

The effects of smoking marijuana on rCBF as measured by $[H_2{}^{15}O]$ PET and auditory attention task performance also have been assessed in a double-blind, placebo-controlled study in recreational users. Measures were made before and after smoking of marijuana or placebo cigarettes. The behavioral performance on the attention task was not significantly altered, but heart rate and blood pressure increased dramatically. rCBF increases were noted in orbitofrontal cortex, medial frontal cortex, insula cortex, temporal poles, anterior cingulate, and cerebellum, and increases in the paralimbic regions were hypothesized to be related to the drug mood-related effects. Reduced rCBF was observed in temporal lobe auditory regions, visual cortex, and regions controlling attentional processes, such as the thalamus and parietal and frontal lobes. However, no change was noted in the nucleus accumbens, basal ganglia, or hippocampus where high quantities of cannabinoid receptors exist (O'Leary *et al.*, 2000, 2002) (**Fig. 8.20**).

The effects of marijuana on brain perfusion and internal timing (i.e., a self-paced counting task) have been assessed using $[H_2{}^{15}O]$ PET in chronic daily users (for years) and recreational users before and after smoking (O'Leary *et al.*, 2003). Marijuana smoking increased rCBF in ventral forebrain and cerebellar cortex in both groups, but resulted in significantly less frontal lobe activation in chronic users. Counting rate for the self-paced counting task increased after

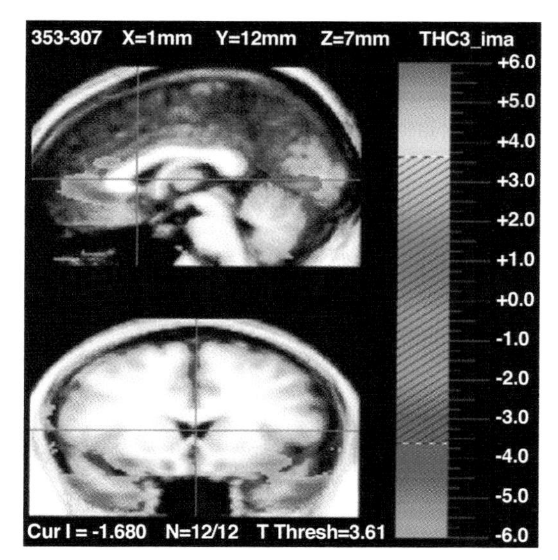

FIGURE 8.20 Positron emission tomography *t*-map image for the presmoking minus postsmoking marijuana analysis (in color) overlaid on an average magnetic resonance image for 12 subjects. Subjects were performing the same dichotic task prior to and following smoking. The two views represent sagittal and coronal views of the unthresholded *t*-map, with *t* values ranging from −6.0 (purple) to +6.0 (red) as illustrated on the pallet on the right overlaid on the average magnetic resonance image, but with the *t* values thresholded at a value of 3.61 (uncorrected $p < .0005$). Radiological convention is followed, and the crosshairs are at the same location in both views. The sagittal view illustrates the increase in anterior cingulate regional cerebral blood flow (rCBF) and the decrease in rCBF in occipital lobe after smoking marijuana, and additionally illustrates a region in the inferior cerebellum that has higher rCBF following smoking. The coronal view illustrates bilateral regions of the temporal poles, insula, and orbitofrontal cortex that have significantly higher rCBF after smoking marijuana. Note that the regions of increased rCBF in the coronal view are lateral and ventral to the basal ganglia. [Reproduced with permission from O'Leary *et al.*, 2002.]

smoking in both groups, as did self-paced tapping. Both measures correlated with rCBF in the cerebellum (O'Leary *et al.*, 2003). Abstinence from chronic marijuana use showed lower rCBF measures than nonsmokers at baseline (Tunving *et al.*, 1986; Volkow *et al.*, 1996a). Decreased prefrontal cortex and increased anterior cingulate cortex activation in chronic marijuana users was observed using fMRI after a 30-day washout period (Loeber and Yurgelun-Todd, 1999). Chronic marijuana users also showed decreased cerebellar metabolism (Volkow *et al.*, 1991, 1996a), and these changes, combined with the decreases in prefrontal cortex activity suggested a fronto-ponto-cerebellar dysfunction with chronic marijuana use, not unlike that observed with chronic alcohol use (Loeber and Yurgelun-Todd, 1999) (see *Alcohol* and *Neurobiological Theories of Addiction* chapters). Abstinent heavy marijuana users abstinent for 25 days showed hypoactivity in the anterior cingulate and prefrontal cortices, and hyperactivity in the hippocampus using [$H_2{}^{15}O$] PET (Eldreth *et al.*, 2004).

The brain effects of chronic marijuana use, namely on tissue volume or morphology, also have been explored. It has been hypothesized that the age of first consumption of marijuana may play a differential role on brain morphological and functional effects. In a study of 57 subjects using MRI and PET, it was found that age of first use was important for whole brain, gray matter, white matter, ventricle volume, and global rCBF measures. When subjects started using marijuana before age 17 compared to those who started later, they had smaller whole brain and gray matter. Males had significantly higher rCBF, and both sexes were physically smaller in height and weight, suggesting effects of marijuana on gonadal and pituitary hormones (Wilson *et al.*, 2000). In another study using PET, no evidence of cerebral atrophy or global or regional changes in tissue volumes were found in adult users who nevertheless had lower ventricular volumes (Block *et al.*, 2000). The investigators also showed lower rCBF in a large region of posterior cerebellum in frequent users who were abstinent for 26 h. Heavy marijuana users, after 20 days of abstinence, also showed lower gray matter density in the parahippocampus gyrus, precentral gyrus, and thalamus, and higher white matter density around the parahippocampal gyrus using voxel-based morphometry with MRI (Matochik *et al.*, 2005). The authors hypothesized that these changes may be related to neurocognitive deficits observed with heavy marijuana use (Bolla *et al.*, 2002).

In summary, intoxication with marijuana, as with other drugs of abuse, produces increases in brain metabolism in orbitofrontal cortex, medial prefrontal cortex, anterior cingulate, basal ganglia, and cerebellum. There is some evidence for tolerance to the effects in the frontal cortex in chronic users. Chronic users show decreased brain metabolism in frontal cortex and cerebellum, and heavy long-term use of marijuana is associated with structural changes in cortex and hippocampus that may be related to neurocognitive deficits associated with heavy marijuana use (Bolla *et al.*, 2002).

INTEGRATION OF IMAGING STUDIES IN HUMANS WITH THE NEUROCIRCUITRY OF ADDICTION

Drug Neuroimaging: A Discipline in Progress

Drug addiction is a human disease, and brain imaging provides a unique approach to studying its pathophysiology by noninvasive *in vivo* methods at a given stage of the disease process. PET, SPECT, and BOLD with fMRI allow the researcher to image brain activity usually related to oxygen or glucose use with relatively good spatial and temporal resolution. fMRI can resolve structures as small as 1–2 mm and view brain activity in time-blocks as small as 2–3 s. This time resolution may appear rather crude relative to the millisecond speed of neuronal activity. However, while imaging techniques reveal the activity of clusters of brain cells through a sustained time domain in association with drug intake, a given behavior, or a task, one cannot image the directional flow of information through the brain. Nevertheless, using these techniques, researchers can correlate regional activities in a subject who self-administers a drug, or is in a withdrawal state, or performs a behavioral act, or is re-experiencing images of drug use (George and Bohning, 2002).

The use of imaging in humans might also help to elucidate the human and social conditions that lead to vulnerability toward developing dependence that cannot be properly modeled in animals. Similarly, the various comorbid psychiatric disorders that may contribute to deficits observed in addiction cannot be studied easily in animals. Obviously, from studies in human drug users, it is not possible to discern whether brain activity alterations observed, in the long term, after cessation of cocaine intake (Volkow *et al.*, 1993b) are due to chronic abuse or to a pre-existing functional change (Porrino and Lyons, 2000). Prospective and long-term longitudinal studies could address these issues but remain challenges for future imaging research.

Many neurotransmitter systems are the target of the various drugs of abuse. The dopamine system has been a focal point for the preponderance of neurobiological studies on drug addiction (see previous chapters). Dopamine neurons also have been the target of many of the imaging studies devoted to neurotransmitters, transporters, and receptors, presumably because of both a limited conceptual framework (see *Neurobiological Theories of Addiction* and *Drug Addiction: Transition from Neuroadaptation to Pathophysiology* chapters) and the limited availability of research tools. Acute administration of cocaine and alcohol has been shown to enhance dopamine release in the ventral striatum as measured by imaging studies. However, rapid tolerance develops to this enhanced release, and acute withdrawal and protracted withdrawal are associated with decreases in dopamine D_2 receptors for both drugs. Even more intriguing are data showing that low levels of D_2 receptors predict the rewarding effects of psychostimulants.

In animal studies, based largely on work with psychostimulant drugs, the mesolimbic dopamine circuits (i.e., mesoaccumbens and mesoamygdala) have been associated with the acute reinforcing effects of drugs and with the emotional and motivational changes associated with drug withdrawal. The mesocortical dopamine circuits, again, largely based on work with psychostimulant drugs (e.g., mesoprefrontal, orbitofrontal, and mesoanterior cingulate cortices), have been associated with incentive salience, drug expectation, compulsive intake, loss of control, conscious experiences of drug intoxication, and craving. These projections are hypothesized to act in parallel to facilitate the processing of information and also to interact with each other through feedback mechanisms (Le Moal and Simon, 1991; Le Moal, 1995; Jentsch and Taylor, 1999) (see *Neurobiological Theories of Addiction* and *Drug Addiction: Transition from Neuroadaptation to Pathophysiology* chapters).

Integration of Prefrontal Cortex and Orbitofrontal Cortex Pathophysiology in the Course of Addiction

Changes in the function of the orbitofrontal and medial prefrontal cortices in the course of addiction appear to be a common observation in imaging studies. The prefrontal cortex is central for brain signals that convey knowledge derived from prior experience. Many motivated behaviors are learned and depend on cognitive systems that implement the 'rules of the game' needed to achieve a given goal in a given situation, and thus the frontal cortex helps synthesize a range of information that lays the foundation for complex forms of behavior in humans (Miller, 2000). Prediction is an important aspect of prefrontal physiology, in particular expectation/uncertainty about reward outcome and magnitude (Fiorillo, 2004). The prefrontal top–down control process may be dysregulated via striato-thalamo-orbitofrontal circuits and, as such, contribute to drive-disruptions and compulsions (Miller and Cohen, 2001).

The orbitofrontal cortex receives information from the nucleus accumbens via the ventral pallidum and subsequently the mediodorsal nucleus of the thalamus, and in turn the cortex sends dense projections to the nucleus accumbens. Moreover, the orbitofrontal cortex receives direct inputs from the amygdala, and indirect inputs via the thalamus, cingulate gyrus, and hippocampus (Haber *et al.*, 1965; Ray and Price, 1993; Carmichael and Price, 1995; Volkow and Fowler, 2000). This circuit represents a set of interconnected regions driving drug-seeking and compulsive behaviors (Breiter and Rosen, 1999; Porrino and Lyons, 2000; Goldstein and Volkow, 2002) (see *Neurobiological Theories of Addiction* chapter). In functional terms, the circuit involving the orbitofrontal cortex has been hypothesized to process information about rewarding properties of stimuli and pharmacological reinforcers, attention, and learning stimulus/reinforcement associations, salience attribution, inhibition of common or irrelevant responses, and conditioned responses associated with drugs of abuse (Volkow and Fowler, 2000). The circuit promotes the value and prediction of natural reward in general and of risk-taking under natural conditions.

Acute administration of psychostimulants produced increases in brain metabolism in orbitofrontal cortex, prefrontal cortex, and cingulate cortex (Breiter *et al.*, 1997; Volkow *et al.*, 1999b). In contrast, detoxified cocaine abusers and alcoholics showed major decreases in brain metabolism in the orbitofrontal, prefrontal, and cingulate cortices (Volkow *et al.*, 1993b). During exposure to stimuli that elicit craving in humans there is a reactivation of the prefrontal cortex, orbitofrontal cortex, and anterior cingulate, and a recruitment of the amygdala (Grant *et al.*, 1996; Childress *et al.*, 1999). Similar effects have been observed with alcohol, nicotine, and marijuana.

When one attempts to assimilate these results with rodent studies, problems of definitions and limitations of the cortical areas arise. Early attempts in rats to define a region homologous to the prefrontal region in primates were based on the connection with the mediodorsal nucleus of the thalamus (Leonard, 1969). Later, it became evident that given the complexity of the numerous thalamic projections, it was not possible to base conclusions simply on thalamic connectivities, and functional considerations became necessary (Kolb, 1984, 1990). However, the prefrontal–cortical correspondences between rats and primates are still not clear. From a cytoarchitectural point of view, the rat prefrontal cortex includes dorsal and ventral cingulate, prelimbic, infralimbic, medial orbitofrontal, lateral orbitofrontal, and dorsal and ventral agranular insular cortices (Zilles and Wree, 1995), with distinct connectivities and physiological and behavioral functions (Van Eden, 1986; Groenewegen,1988; Berendse *et al.*, 1992). This anatomical diversity is not accounted for in most rodent studies.

Studies in rodents have implicated the prefrontal cortex in drug addiction, particularly in psychostimulant addiction. Thus, disruption of the medial prefrontal cortex causes a possible loss of incentive-motivational properties of cocaine-associated cues, a loss of controlling influences on behavior (inhibition) after lesioning, and an increase of perseverative drug-seeking behavior (Porrino and Lyons, 2000) (see *Psychostimulants* and *Neurobiological Theories of Addiction* chapters). The dorsal prelimbic region projects to the nucleus accumbens core, whereas the ventral prelimbic and infralimbic cortices project to the shell, and that may explain some of the functional differences of these two parts of the frontal cortex in abuse (Barrot *et al.*, 2000). Whether or not these cortico-subcortial subsystems are organized in parallel remains to be determined.

Human studies currently do not allow longitudinal prospective studies. Conversely, animal studies can address specific questions, such as long-term consequences of abuse. Metabolic mapping studies using 2-[^{14}C]deoxyglucose was applied to rats during self-administration sessions, and rates of energy metabolism were compared to those of rats receiving yoked-saline infusions (Macey *et al.*, 2004). Previous work with acute cocaine administration in rats had shown extensive elevations in cerebral metabolism throughout the mesolimbic–ventral striatal brain systems (Porrino *et al.*, 1988; Porrino, 1993; Thomas *et al.*, 1996; Zocchi *et al.*, 2001). However, the pattern changed with chronic exposure to cocaine where reductions in cerebral metabolism have been observed in these same regions (Hammer and Cooke, 1994). Rats allowed to self-administer cocaine for different periods (5 and 30 days) showed decreases in brain metabolic activity that became more widespread with the longer exposures (Macey *et al.*, 2004). Following 5 days of cocaine self-administration, decreases were observed in the nucleus accumbens, but after 30 days, decreases extended throughout the striatum and included the central nucleus of the amygdala, anterior cingulate, and prefrontal cortex. The changes in the brain in 30-day self-administration animals were accompanied by a loss of locomotor sensitization (Macey *et al.*, 2004). Because these effects were not present in the initial stages, but developed over time, the authors concluded that they are unlikely to be necessary for cocaine reinforcement, but likely to be necessary to the processes related to compulsive drug-seeking behavior and craving, or other motivational states associated with escalating drug use. These conclusions have been confirmed for subcortical regions with functional studies in rats (Ahmed and Koob, 1998, 1999, 2004; Ahmed *et al.*, 2000, 2003; Deroche-Gamonet *et al.*, 2004; Vanderschuren and Everitt, 2004) (see *Neurobiological Theories of Addiction* chapter).

Primate imaging studies have been conducted with the same purpose, to compare acute and chronic effects of cocaine (Porrino and Lyons, 2000) (**Fig. 8.21**). Rates of glucose utilization (local cerebral glucose metabolism) following an acute administration of cocaine in rhesus monkeys with long-term cocaine

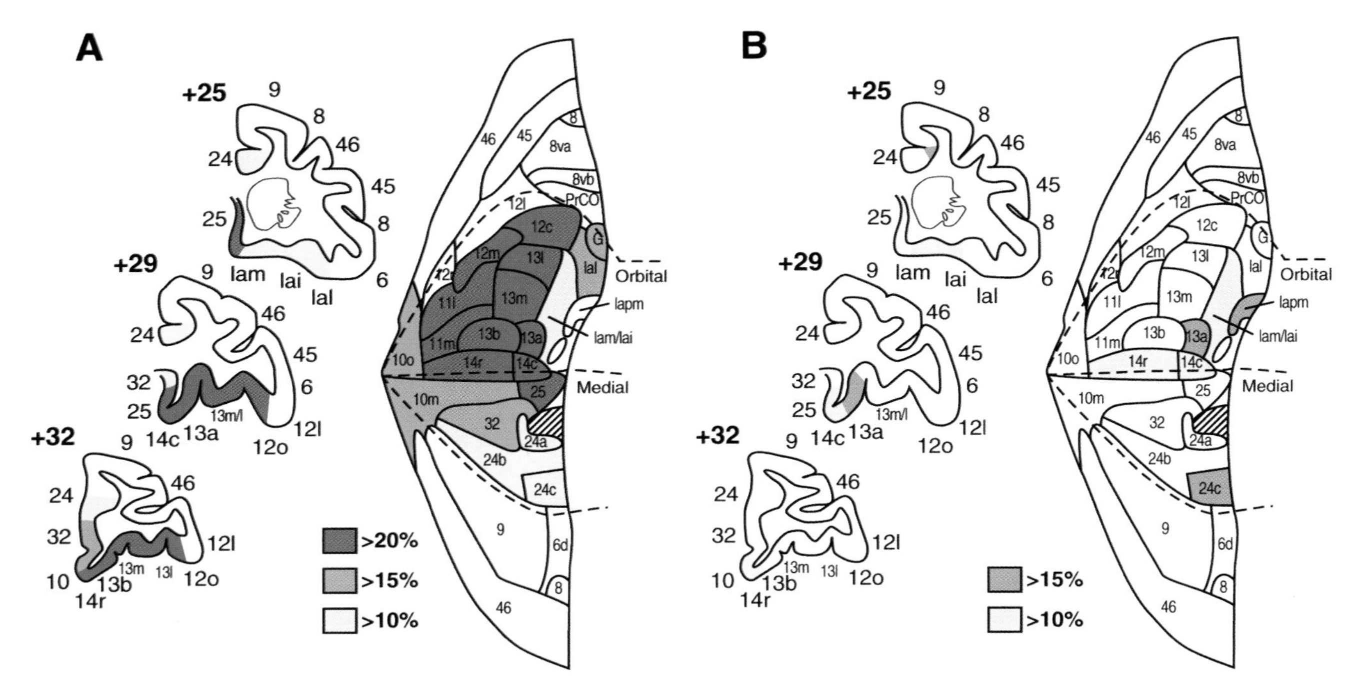

FIGURE 8.21 Distribution of changes in rates of local cerebral glucose utilization in the prefrontal cortex as assessed by the 2-[^{14}C]deoxyglucose method, accompanying the **(A)** acute intravenous administration of cocaine (1 mg/kg) in drug-naive cynomologous monkeys, or **(B)** chronic cocaine (1 mg/kg) in rhesus monkeys with 18–22 months of cocaine self-administration experience. In acute administration (A), the most intense changes were concentrated in areas 13 and 14. The magnitude of the effect diminished gradually more medially in areas 25, 32, and 24a (anterior cingulate). Moderate changes also were found rostrally in areas 10 and 11 and caudally in the anterior insular cortex. In self-administration (B), the most intense changes were concentrated in the caudomedial part of area 13, the anterior insular cortex (within the caudal regions of the orbitofrontal cortex), and area 14. Shown on the left of each panel are three coronal sections adapted from the atlas of Szabo and Cowan (1984). On the right is a representative unfolded two-dimensional map of the prefrontal cortex adapted from Carmichael and Price (1994). The color-coding indicates the differing degrees of decreases in glucose utilization. Cytoarchitectonic areas are defined according to Carmichael and Price (1996). [Reproduced with permission from Porrino and Lyons, 2000.]

self-administration histories were compared to rates of glucose metabolism following acute self-administration of cocaine in naive control monkeys (Lyons *et al.*, 1996; Porrino and Lyons, 2000). The experiment in the animals with a history of self-administration lasted for 18–22 months. The subjects (rhesus monkeys) were allowed to self-administer cocaine intravenously in 4 h sessions daily. Average daily intake was around 1.35 mg/kg, and lifetime intake ranged between 431 and 588 mg/kg. On the last day, monkeys made a single lever press for one infusion of 1.0 mg/kg cocaine, and the 2-[^{14}C]deoxyglucose method started immediately thereafter. Rates of glucose utilization in the monkeys with a long history of cocaine self-administration were compared to rates in cocaine-naive controls used in other experiments. In naive monkeys, an injection of cocaine produced a widespread decrease in metabolic activity in the orbitofrontal cortex, medial prefrontal cortex, and anterior cingulate. With chronic exposure, there were striking differences compared to acute administration. Chronic exposure led to a far more restricted pattern of deactivation in the orbitofrontal cortex and medial prefrontal cortex (**Fig. 8.21**). Functional changes were more concentrated within the caudal aspects of the network as defined above: anterior insular cortex (Iam, Iai, Iapm), the caudal portion of area 13 (13a), along the length of the gyrus rectus (14c, 14r), and in the anterior cingulate (24c). Functional changes also were more concentrated within the striatum, with the most intense changes occurring in the shell of the nucleus accumbens, but also spread to include more of the dorsal striatum (Porrino *et al.*, 2004).

In a more detailed examination, based on the cytoarchitectonic boundaries described by Carmichael and Price (1994, 1996), the authors proposed that the orbitofrontal cortex and prefrontal cortex can be subdivided into two distinct networks: (1) an orbitofrontal network, including areas in the posterior, lateral, and central parts of the orbitofrontal cortex, and (2) a medial prefrontal network including areas along the medial wall and the gyrus rectus (Lyons *et al.*, 1996; Porrino and Lyons, 2000). In naive monkeys, an injection of cocaine produced functional changes in the medial and orbitofrontal cortex that were greatest in the regions corresponding to areas 11, 12, and 14, and to a lesser extent within the surrounding areas (i.e., area 10 rostrally, caudally in anterior insula, areas Iam, Iai, Iapm, Iapl, and area G), medially in area 14 and 25, 32, 24a, 24b, and 24c of the anterior cingulate. In the monkeys with an extensive history of cocaine self-administration, changes in the medial and orbitofrontal cortex were restricted to areas 13a, 14, and 24c (**Fig. 8.21**). Therefore, the effects of an acute injection were localized to anterior limbic cortical areas. Chronic administration of cocaine, in particular, produced functional consequences in the caudal medial part of the prefrontal cortex, the caudal regions of the orbitofrontal cortex, and more dorsal caudal parts of the striatum (Porrino and Lyons, 2000; Porrino *et al.*, 2004).

The changes in metabolic activity in primates with a history of no cocaine or chronic cocaine (Lyons *et al.*, 1996; Porrino and Lyons, 2000) raised two interesting considerations. The differential activation in acute versus chronic drug use revealed a modular organization of selective cortico-subcortical networks, both at functional-pathophysiological and anatomical levels. The portions of the orbitofrontal cortex and medial prefrontal cortex affected by chronic cocaine predominantly innervate the shell, whereas the changes after acute administration were found in prefrontal regions which innervate both the accumbal core and shell. The restriction of functional changes to the projection fields of the shell in cocaine-experienced monkeys parallels the changes within the infralimbic and ventral prelimbic cortices of rats after prolonged drug experience. These portions of rodent medial prefrontal cortex also send selective projections to the shell (Graham and Porrino, 1995; Porrino and Lyons, 2000). Although data from metabolic studies do not directly show a relationship between prefrontal metabolic activity and cognitive processes, they nevertheless suggest that the drug may modify the higher order processing of converging sensory and visceral information integrated in these areas (Porrino and Lyons, 2000). In rodents, primates, and humans, functionality might follow patterns of cortico-striatal connectivity. Future refinement of imaging techniques in humans may allow further exploration of the heterogeneous nature of the prefrontal cortex.

Nondrug and Related Rewards and Compulsions Activate Prefrontal-related Circuit

The interpretation of brain imaging data meets the crucial question of the causal relationship between drug, intoxication, and dependence and what is observed in the limbic and striato-thalamo-cortical networks. However, a basic observation from real life in humans is that some subjects are vulnerable to addiction-like syndromes independent of drugs (e.g., gambling, compulsive eating, compulsive sex, compulsive shopping, and so on). Thus, an inherent or acquired propensity to have a prefrontal dysfunction (loss of control), or to have a change in reward function (higher sensitivity to rewards, or some sort of 'self-medication' for the brain in an allostatic state) may transcend the reinforcer involved. In brief, for genetic and/or environmental reasons leading to predisposition and vulnerability, the same brain circuits implicated in drugs of abuse also may be in a pathophysiological state outside the range of homeostasis (i.e., in a preclinical state and prone to engage in disease). These considerations open another question—whether the general features revealed by brain imaging are found in other pathologies or behavioral reactivities behind the obvious and important fact that each agent or object, or each genetic–environmental characteristic, adds a specific signature on the neural networks described above.

Pleasant and unpleasant emotions versus neutral emotions increased rCBF in the medial prefrontal cortex, thalamus, hypothalamus, and midbrain (Lane *et al.*, 1997). In male and female subjects viewing erotic film excerpts versus emotional mental films, bilateral BOLD signal increased in medial prefrontal, orbitofrontal, insular, anterior cingulate, and occipito-temporal cortices as well as in the amygdala and ventral striatum. Only male subjects had increased signals in thalamus and hypothalamus (Redoute *et al.*, 2000; Karama *et al.*, 2002). Pleasantness or aversiveness have been tested while eating chocolate compared to an aversive saline solution. These different pleasant or unpleasant stimuli, representing different levels of affective stimuli, activate the prefrontal and orbitofrontal cortices, cingulate gyrus, and amygdala (Lane *et al.*, 1997; Zald *et al.*, 1998; Francis *et al.*, 1999; Small *et al.*, 2001) (**Fig. 8.22**).

Rewarding stimuli with cultural loading also modulate reward circuitry. Photographs of sports cars rated more attractive than small cars which activated orbitofrontal cortex, anterior cingulate, occipital regions, and ventral striatum (Erk *et al.*, 2002). Decision-making in risk-taking tasks, in the case of financial reinforcers or of expectancy and experience of monetary gains and losses, induced strong emotional reactions associated with cognitive processes and outcomes of decisions (Mellers *et al.*, 1997, 1999). Hemodynamic responses increased monotonically with monetary value in the orbitofrontal region, extended amygdala (nucleus accumbens), and hypothalamus (Breiter *et al.*, 2001). The same pattern was found with various imaging techniques and in similar experimental situations (i.e., activation of orbitofrontal cortex and dorsolateral prefrontal cortex, anterior cingulate, amygdala, and nucleus accumbens) (Thut *et al.*, 1997; Ernst *et al.*, 2002; Akitsuki *et al.*, 2003; Arana *et al.*, 2003; Elliot *et al.*, 2003). Choosing between small, likely rewards and large, unlikely rewards also activates the orbital prefrontal cortex as measured by [$H_2{}^{15}O$] PET (Rogers *et al.*, 1999). Pleasurable music, relaxation responses, and meditation as well as subjective experience of pain or social exclusion and self-reported distress tivate prefrontal and anterior cingulate cortices, and for each condition other regions are activated specific to the situation (Lazar *et al.*, 2000; Blood and Zatorre, 2001; Coghill *et al.*, 2003; Eisenberger *et al.*, 2003; **Fig. 8.23**).

Experimental neuropsychological studies also have been performed in normal subjects or subjects with brain regions using gambling tasks with financial rewards and penalties, risk-taking, and incentives for decision-making cognition. These studies have helped refine the role of subregions within the frontal lobe in relation to limbic cortices. Subjects with pathological gambling or gambling addiction were presented with videotaped scenarios of gambling, with happy or sad content, and fMRI was used to assess brain function during viewing (Potenza *et al.*, 2003b). The gambling scenes triggered gambling urges and decreased activity in frontal and orbitofrontal cortices, caudate-basal ganglia regions, and thalamus. Distinct patterns of regional brain activity were observed in specific

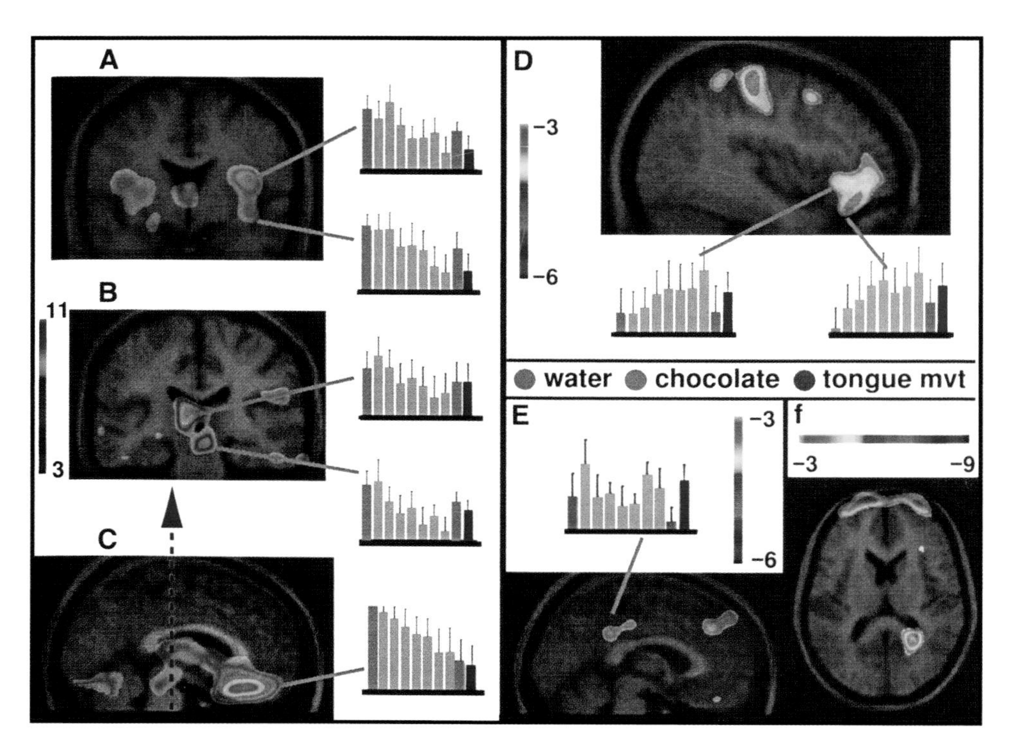

FIGURE 8.22 Cortical regions demonstrating significant regional cerebral blood flow (rCBF) correlations with affective rating for the question of how much they would like or not like to have another piece of chocolate. Regression analyses were used to correlate rCBF from averaged positron emission tomography (PET) data (Choc 1 minus Choc 7) with affective ratings taken immediately after these scans. Correlations are shown as *t* statistic images superimposed on corresponding averaged magnetic resonance imaging scans. The *t* statistic ranges for each set of images are coded by color bars, one in each box. Bar graphs represent normalized CBF in an 8 mm radius surrounding the peak. The y-axis corresponds to normalized activity, and the bars along the x-axis represent scans. The three colors represent scan type. Purple bars represent water PET scans. Turquoise bars represent chocolate PET scans. Blue bars represent tongue movement scans. Each bar graph corresponds to activations indicated by a turquoise line. **(A)** Coronal section taken at $y = 1$ showing the decrease in rCBF in the primary gustatory area (bilaterally in the anterior insula/frontal operculum and in the right ventral insula). **(B)** Coronal section taken at $y = -26$ showing decreases in rCBF in the left thalamus and medial midbrain (possibly corresponding to the ventral tegmental area). **(C)** Sagittal section taken at $x = -1$ showing decreases in rCBF in the subcallosal region, thalamus and midbrain. **(D)** Sagittal section taken at $x = 42$ showing the increase in rCBF in the right caudolateral orbitofrontal cortex. Activation is also evident in the motor and premotor areas. **(E)** Sagittal section taken at $x = 8$ showing an increase in rCBF in the posterior cingulate gyrus (peak at 8, −30, 45) in subtraction analysis *Choc 1 – water-post*. This was the only region where CBF was consistently greater in affective scans regardless of valence, compared to the neutral chocolate scan (Choc 4) and the two water baseline scans (water-pre and water-post). **(F)** Horizontal section at $z = 12$ showing an increase in rCBF in the retrosplenial cortex (area 30) that correlated with affective rating, but not the affective rating when scan order was covaried out of the regression analysis. [Reproduced with permission from Small *et al.*, 2001.]

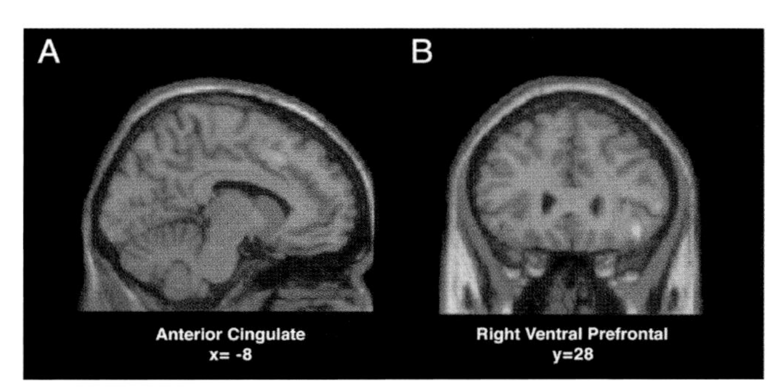

FIGURE 8.23 **(A)** Increased activity in anterior cingulate cortex during social exclusion relative to inclusion. **(B)** Increased activity in right ventral prefrontal cortex during social exclusion relative to inclusion. [Reproduced with permission from Eisenberger *et al.*, 2003.]

temporal epochs of videotape viewing. Differences were localized to the ventral anterior cingulate during the final period of gambling videotape viewing, corresponding to the presentation of the most provocative gambling stimuli. These results support the hypothesis that decreased activity in the ventromedial prefrontal cortex is associated with impulse dysregulation (Best *et al.*, 2002). Pathological gamblers subjected to the Stroop test demonstrated decreased activity in the left ventromedial prefrontal cortex, again reflecting changes in a brain region implicated in poor impulse control (Potenza *et al.*, 2003a). Cocaine abusers abstinent for 25 days showed alterations in normalized rCBF measured with [$H_2{}^{15}O$] PET in the orbitofrontal cortex during the Iowa Gambling Task, which measures the ability to weigh short-term rewards against long-term losses. Cocaine abusers showed greater activation during the task in the right orbitofrontal cortex and less activation in the right dorsolateral prefrontal cortex and left medial prefrontal cortex compared to the control group. There was a positive correlation with the performance and the activation in the right orbitofrontal cortex in both groups. The amount of cocaine self-administered prior to the 25 days of enforced abstinence was negatively correlated with activation in the left orbitofrontal cortex. These data point to persistent functional abnormalities in the prefrontal network involved in decision-making and anticipation of reward (Bolla *et al.*, 2003). In another study with abstinent cocaine abusers subjected to the Iowa Gambling Task, no deficits were observed, but there was decreased resting dorsolateral prefrontal cortex rCBF as measured by [^{99m}Tc-HMPAO] SPECT. This decreased rCBF correlated with Iowa Gambling Task performance in both groups, suggesting that cocaine-related changes in dorsolateral prefrontal cortex were associated with impaired decision-making (Adinoff *et al.*, 2003).

Obsessive-compulsive disorder (OCD) provides an interesting parallel to drug addiction that reflects the neurocircuitry for compulsive behavior. Human brain imaging has contributed comprehensively to its neuropathological basis (for review, see Baxter, 1999), and these processes may help to understand some aspects of addiction symptomatology. Symptoms of anxiety and depression are common in OCD. During obsession or compulsion manifestations, abnormal functional activities are observed in orbitofrontal cortex, striatum, cingulate gyrus, and thalamus, and all are disinhibited (Baxter, 1999) (**Fig. 8.24**). The cortico-basal

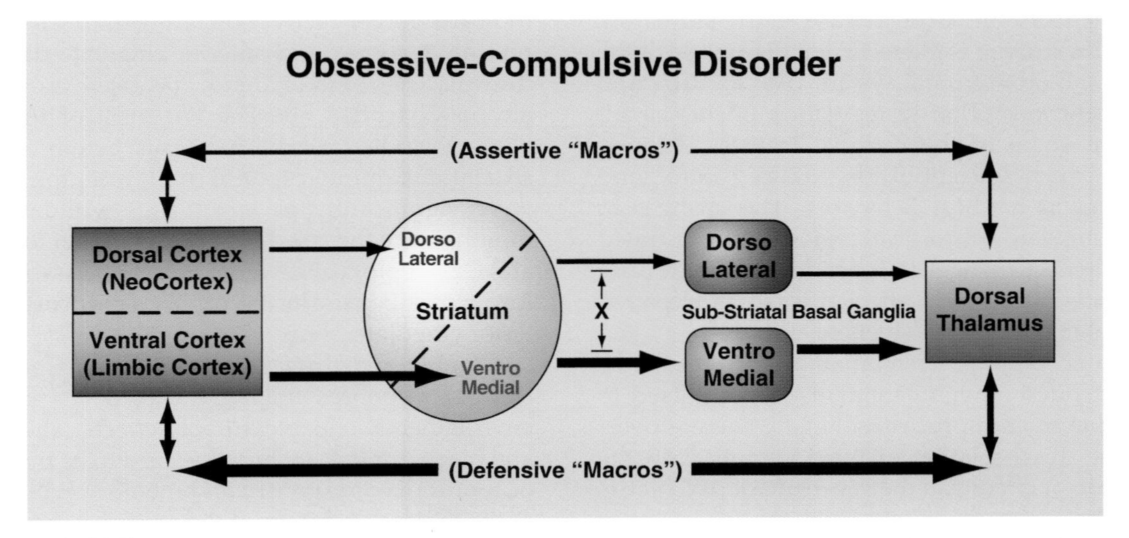

FIGURE 8.24 Schema depicting the relative effects of direct versus indirect system tone in obsessive–compulsive disorder. In obsessive–compulsive disorder, there is relatively increased tone in the direct basal ganglia pathways that predominate in outflow from the ventral medial striatum compared to control tone. A 'macro' is defined as behavioral routines, or complex sets of interrelated behaviors that must be precisely choreographed for semiautomatic implementation in specific situations to be effective. [Reproduced with permission from Baxter, 1999.]

ganglion-pallidal-thalamic system is organized as partially segregated subchannels according to topographically distributed afferents from cortex to dorsal and ventral striatum. Given the dorsolateral-to-ventromedial distributions of various structures and neurotransmitters and receptors, associations with neocortex and limbic regions have different degrees of effects on direct versus indirect basal ganglia systems efferent to the striatum. Anatomical and clinical observations, to which imaging has contributed, indicate that the dorsal association neocortex may be more related to moment-to-moment multiple-contingent behaviors that require moderation of impulses and cognitive control (Miller, 2000), whereas the ventromedial limbic systems are more involved in impulsive behaviors (Baxter, 1999). Each system seems to mutually inhibit each other to influence thalamic tone (Baxter, 1999). When major depression is superimposed on OCD, there is a decrease in dorsal tone, but limbic ventral tone increases, leading to increased drive between thalamus and ventral prelimbic cortex and finally a worsening of OCD and depression. Serotonin transmission has long been hypothesized to be involved in OCD and depression physiopathology. A recent review on the neuronal circuitry underlying mood and anxiety disorders again puts emphasis on changes in serotonin neurotransmission in the prefrontal cortex, amygdala, hippocampus, and nucleus accumbens (Davidson, 2002).

In summary, brain imaging studies show common features for all drugs of abuse, despite discrepancies in some details, where a pattern of regional change appears in most of the studies as representative of the long-term effects of drug abuse and addiction. However, also striking is the recurrent disruption of the striato-thalamo-orbitofrontal system with all drug and nondrug addictions. The dysregulation of this circuit may explain some of the cardinal features defining dependence, such as compulsion, loss of control, and perseveration in harmful behaviors. The changes in prefrontal-cingulate-subcortical regions demonstrated by imaging reflect changes in functions for which drug dependence is particularly affected, such as impaired decision-making and impulsiveness. The dysregulated circuits also encompass dysregulation of ventral striatal and extended amygdala circuits hypothesized to be involved in reward processes. Finally, there is overlap with the dysregulation of these circuits with the dysregulation of circuits in the spectrum of psychiatric disorders that span from 'atypical' impulse control disorders (e.g., gambling) to compulsive disorders (e.g., OCD).

SUMMARY

Brain imaging has been used to discriminate between the stages of addiction. While there may be some discrepancies with this overall synthesis that may be attributed to potential methodological issues, such as vasoactive responses and different temporal courses of drug effects, the following emphasizes the more salient and common responses (**Figs. 8.8** and **8.17**). The prefrontal cortex, orbitofrontal cortex, cingulate gyrus, and ventral striatum are involved in the intoxication process, and they are activated by acute administration of drugs of abuse in humans according to previous drug experiences. The rCBF is higher in the prefrontal cortex during nicotine, alcohol, and marijuana intoxication, and the right prefrontal activation with alcohol is associated with euphoria. Mapping studies with fMRI to measure BOLD responses showed activation of the prefrontal and cingulate regions during cocaine or nicotine intoxication and correlated with drug reinforcement properties. The short-term intoxication after a single drug dosing is hypothesized to be accompanied by increased dopamine activity in various regions, in particular the nucleus accumbens and frontal cortex for cocaine and alcohol.

Drug withdrawal is a clinical state that comprises irritability, negative affect, and dysphoria, and these symptoms are hypothesized to contribute to relapse (see *Neurobiological Theories of Addiction* chapter). Many transmitter systems are involved in this functional disruption (Koob and Le Moal, 2001). rCBF and glucose metabolism measurements after the last cocaine intake indicate lower activation in orbitofrontal cortex, prefrontal cortex, and anterior cingulate during withdrawal. The same evidence has been provided from alcohol-detoxified abusers. For most of the drugs of abuse, including psychostimulants, heroin, and alcohol, striatal dopamine response or D_2 receptor availability were significantly lower during withdrawal, even months since the last drug use. The lower levels of striatal D_2 receptors also were found to be associated with lower metabolism in the orbitofrontal cortex and anterior cingulate gyrus.

Craving and drug expectation are associated with learned responses linking the drug, its conditioned environments, and pleasurable effects. The amygdala and hippocampus are hypothesized to be the substrates of conditioned responses and memory, with activation of the thalamo-orbitofrontal circuit and anterior cingulate gyrus hypothesized to underlie the craving experience in humans. Higher levels of activation are shown in the orbitofrontal cortex, dorsolateral prefrontal cortex, and cingulate gyrus in situations linked to craving. Activation of the amygdala also is recruited during craving as measured by rCBF and BOLD. Increases in amygdala activity have been reported when drug-related videotapes are presented to addicts, but no increase is observed in naive subjects. Many cortical regions activated are hypothesized to be involved in anticipation processes.

Imaging studies have not yet investigated one of the most important questions related to human addiction. Why, under apparently similar circumstances, do some people succumb to drug abuse and addiction, but others do not? The abnormalities found in the orbitofrontal-prefrontal cortex 'craving circuit' may precede the first drug consumption or may result from the drug effects themselves. Similarly, the neurobiological mechanisms and imbalances between many neurotransmitters in the reward and stress systems, including mesocortical and mesolimbic dopamine systems, opioid peptides, and stress hormones may underlie potential vulnerability to drug use or result from the drug effects. Finally, the compulsive drug-seeking behavior may activate the ventral striatum–ventral pallidum–thalamus–orbitofrontal cortex compulsive circuit, and this circuit also may be a substrate for vulnerability (for details on these three circuits, see *Neurobiological Theories of Addiction* chapter).

REFERENCES

Abi-Dargham, A., Krystal, J. H., Anjilvel, S., Scanley, B. E., Zoghbi, S., Baldwin, R. M., Rajeevan, N., Ellis, S., Petrakis, I. L., Seibyl, J. P., Charney, D. S., Laruelle, M., and Innis, R. B. (1998). Alterations of benzodiazepine receptors in type II alcoholic subjects measured with SPECT and [123I]iomazenil. *American Journal of Psychiatry* **155**, 1550–1555.

Adinoff, B., Devous, M. D. Sr., Cooper, D. B., Best, S. E., Chandler, P., Harris, T., Cervin, C. A., and Cullum, C. M. (2003). Resting regional cerebral blood flow and gambling task performance in cocaine-dependent subjects and healthy comparison subjects. *American Journal of Psychiatry* **160**, 1892–1894.

Ahmed, S. H., and Koob, G. F. (1998). Transition from moderate to excessive drug intake: Change in hedonic set point. *Science* **282**, 298–300.

Ahmed, S. H., and Koob, G. F. (1999). Long-lasting increase in the set point for cocaine self-administration after escalation in rats. *Psychopharmacology* **146**, 303–312.

Ahmed, S. H., Walker, J. R., and Koob, G. F. (2000). Persistent increase in the motivation to take heroin in rats with a history of drug escalation. *Neuropsychopharmacology* **22**, 413–421.

Ahmed, S. H., Lin, D., Koob, G. F., and Parsons, L. H. (2003). Escalation of cocaine self-administration does not depend on altered cocaine-induced nucleus accumbens dopamine levels. *Journal of Neurochemistry* **86**, 102–113.

Ahmed, S. H., and Koob, G. F. (2004). Changes in response to a dopamine antagonist in rats with escalating cocaine intake. *Psychopharmacology* **172**, 450–454.

Akitsuki, Y., Sugiura, M., Watanabe, J., Yamashita, K., Sassa, Y., Awata, S., Matsuoka, H., Maeda, Y., Matsue, Y., Fukuda, H., and Kawashima, R. (2003). Context-dependent cortical activation in response to financial reward and penalty: an event-related fMRI study. *Neuroimage* **19**, 1674–1685.

Arana, F. S., Parkinson, J. A., Hinton, E., Holland, A. J., Owen, A. M., and Roberts, A. C. (2003). Dissociable contributions of the human amygdala and orbitofrontal cortex to incentive motivation and goal selection. *Journal of Neuroscience* **23**, 9632–9638.

Barrot, M., Marinelli, M., Abrous, D. N., Rouge-Pont, F., Le Moal, M., and Piazza, P. V. (2000). The dopaminergic hyper-responsiveness of the shell of the nucleus accumbens is hormone-dependent. *European Journal of Neuroscience* **12**, 973–979.

Bartzokis, G., Beckson, M., Lu, P. H., Edwards, N., Rapoport, R., Wiseman, E., and Bridge, P. (2000). Increased CSF volumes are associated with diminished subjective responses to cocaine infusion. *Neuropsychopharmacology* **23**, 468–473.

Baxter, L. R. (1999). Functional imaging of brain systems mediating obsessive-compulsive disorder. *In Neurobiology of Mental Illness* (D. S. Charney, E. J. Nestler, and B. S. Bunney, Eds.), pp. 534–547. Oxford University Press, New York.

Bencherif, B., Wand, G. S., McCaul, M. E., Kim, Y. K., Ilgin, N., Dannals, R. F., and Frost, J. J. (2004). Mu-opioid receptor binding measured by [11C]carfentanil positron emission tomography is related to craving and mood in alcohol dependence. *Biological Psychiatry* **55**, 255–262.

Berendse, H. W., Galis-de Graaf, Y., and Groenewegen, H. J. (1992). Topographical organization and relationship with ventral striatal compartments of prefrontal corticostriatal projections in the rat. *Journal of Comparative Neurology* **316**, 314–347.

Berglund, M., and Ingvar, D. H. (1976). Cerebral blood flow and its regional distribution in alcoholism and in Korsakoff's psychosis. *Journal of Studies on Alcohol* **37**, 586–597.

Berglund, M., and Risberg, J. (1977). Regional cerebral blood flow during alcohol withdrawal related to consumption and clinical symptomatology. *Acta Neurologica Scandinavica Supplementum* **64**, 480–481.

Best, M., Williams, J. M., and Coccaro, E. F. (2002). Evidence for a dysfunctional prefrontal circuit in patients with an impulsive aggressive disorder. *Proceedings of the National Academy of Sciences USA* **99**, 8448–8453.

Block, R. I., O'Leary, D. S., Ehrhardt, J. C., Augustinack, J. C., Ghoneim, M. M., Arndt, S., and Hall, J. A. (2000). Effects of frequent marijuana use on brain tissue volume and composition. *Neuroreport* **11**, 491–496.

Blood, A. J., and Zatorre, R. J. (2001). Intensely pleasurable responses to music correlate with activity in brain regions implicated in reward and emotion. *Proceedings of the National Academy of Sciences USA* **98**, 11818–11823.

Blum, K., Cull, J. G., Braverman, E. R., and Comings, D. E. (1996). Reward deficiency syndrome. *American Scientist* **84**, 132–145.

Boileau, I., Assaad, J. M., Pihl, R. O., Benkelfat, C., Leyton, M., Diksic, M., Tremblay, R. E., and Dagher, A. (2003). Alcohol promotes dopamine release in the human nucleus accumbens. *Synapse* **49**, 226–231.

Bolla, K. I., Brown, K., Eldreth, D., Tate, K., and Cadet, J. L. (2002). Dose-related neurocognitive effects of marijuana use. *Neurology* **59**, 1337–1343.

Bolla, K. I., Eldreth, D. A., London, E. D., Kiehl, K. A., Mouratidis, M., Contoreggi, C., Matochik, J. A., Kurian, V., Cadet, J. L., Kimes, A. S.,

Funderburk, F. R., and Ernst, M. (2003). Orbitofrontal cortex dysfunction in abstinent cocaine abusers performing a decision-making task. *Neuroimage* **19**, 1085–1094.

Bonson, K. R., Grant, S. J., Contoreggi, C. S., Links, J. M., Metcalfe, J., Weyl, H. L., Kurian, V., Ernst, M., and London, E. D. (2002). Neural systems and cue-induced cocaine craving. *Neuropsychopharmacology* **26**, 376–386.

Breiter, H. C., Gollub, R. L., Weisskoff, R. M., Kennedy, D. N., Makris, N., Berke, J. D., Goodman, J. M., Kantor, H. L., Gastfriend, D. R., Riorden, J. P., Mathew, R. T., Rosen, B. R., and Hyman, S. E. (1997). Acute effects of cocaine on human brain activity and emotion. *Neuron* **19**, 591–611.

Breiter, H. C., and Rosen, B. R. (1999). Functional magnetic resonance imaging of brain reward circuitry in the human. *In Advancing from the Ventral Striatum to the Extended Amygdala: Implications for Neuropsychiatry and Drug Abuse* (series title: *Annals of the New York Academy of Sciences*, vol. 877) (J. F. McGinty, Ed.), pp. 523–547. New York Academy of Sciences, New York.

Breiter, H. C., Aharon, I., Kahneman, D., Dale, A., and Shizgal, P. (2001). Functional imaging of neural responses to expectancy and experience of monetary gains and losses. *Neuron* **30**, 619–639.

Cadet, J. L., Jayanthi, S., and Deng, X. (2003). Speed kills: cellular and molecular bases of methamphetamine-induced nerve terminal degeneration and neuronal apoptosis. *FASEB Journal* **17**, 1775–1788.

Calhoun, V. D., Altschul, D., McGinty, V., Shih, R., Scott, D., Sears, E., and Pearlson, G. D. (2004). Alcohol intoxication effects on visual perception: an fMRI study. *Human Brain Mapping* **21**, 15–26 [erratum: **21**, 298–299].

Carmichael, S. T., and Price, J. L. (1994). Architectonic subdivision of the orbital and medial prefrontal cortex in the macaque monkey. *Journal of Comparative Neurology* **346**, 366–402.

Carmichael, S. T., and Price, J. L. (1995). Limbic connections of the orbital and medial prefrontal cortex in macaque monkeys. *Journal of Comparative Neurology* **363**, 615–641.

Carmichael, S. T., and Price, J. L. (1996). Connectional networks within the orbital and medial prefrontal cortex of macaque monkeys. *Journal of Comparative Neurology* **371**, 179–207.

Catafau, A. M., Etcheberrigaray, A., Perez de los Cobos, J., Estorch, M., Guardia, J., Flotats, A., Berna, L., Mari, C., Casas, M., and Carrio, I. (1999). Regional cerebral blood flow changes in chronic alcoholic patients induced by naltrexone challenge during detoxification. *Journal of Nuclear Medicine* **40**, 19–24.

Chang, L., Ernst, T., Speck, O., Patel, H., DeSilva, M., Leonido-Yee, M., and Miller, E. N. (2002). Perfusion MRI and computerized cognitive test abnormalities in abstinent methamphetamine users. *Psychiatry Research* **114**, 65–79.

Chatziioannou, A. F., Cherry, S. R., Shao, Y., Silverman, R. W., Meadors, K., Farquhar, T. H., Pedarsani, M., and Phelps, M. E. (1999). Performance evaluation of microPET: a high-resolution lutetium oxyorthosilicate PET scanner for animal imaging. *Journal of Nuclear Medicine* **40**, 1164–1175.

Childress, A. R., Ehrman, R., Rohsenow, D. J., Robbins, S. J., and O'Brien, C. P. (1992). Classically conditioned factors in drug dependence. *In Substance Abuse: A Comprehensive Textbook*, 2nd ed. (J. H. Lowinson, P. Ruiz, and R. B. Millman, Eds.), pp. 56–69. Williams and Wilkins, Baltimore.

Childress, A. R., Mozley, P. D., McElgin, W., Fitzgerald, J., Reivich, M., and O'Brien, C. P. (1999). Limbic activation during cue-induced cocaine craving. *American Journal of Psychiatry* **156**, 11–18.

Coghill, R. C., McHaffie, J. G., and Yen, Y. F. (2003). Neural correlates of interindividual differences in the subjective experience of pain. *Proceedings of the National Academy of Sciences USA* **100**, 8538–8542.

Daglish, M. R., Weinstein, A., Malizia, A. L., Wilson, S., Melichar, J. K., Britten, S., Brewer, C., Lingford-Hughes, A., Myles, J. S., Grasby, P., and Nutt, D. J. (2001). Changes in regional cerebral blood flow elicited by craving memories in abstinent opiate-dependent subjects. *American Journal of Psychiatry* **158**, 1680–1686.

Davidson, C., Gow, A. J., Lee, T. H., and Ellinwood, E. H. (2001). Methamphetamine neurotoxicity: necrotic and apoptotic mechanisms and relevance to human abuse and treatment. *Brain Research Reviews* **36**, 1–22.

Davidson, R. J. (2002). Activation paradigms in affective and cognitive neuroscience: probing the neuronal circuitry underlying mood and anxiety disorders. *In Neuropsychopharmacology: The Fifth Generation of Progress* (K. L. Davis, D. Charney, J. T. Coyle, and C. Nemeroff, Eds.), pp. 373–381. Lippincott Williams and Wilkins, New York.

Deroche-Gamonet, V., Belin, D., and Piazza, P. V. (2004). Evidence for addiction-like behavior in the rat. *Science* **305**, 1014–1017.

Desmond, J. E., Chen, S. H., DeRosa, E., Pryor, M. R., Pfefferbaum, A., and Sullivan, E. V. (2003). Increased frontocerebellar activation in alcoholics during verbal working memory: an fMRI study. *Neuroimage* **19**, 1510–1520.

Domino, E. F., Minoshima, S., Guthrie, S. K., Ohl, L., Ni, L., Koeppe, R. A., Cross, D. J., and Zubieta, J. (2000). Effects of nicotine on regional cerebral glucose metabolism in awake resting tobacco smokers. *Neuroscience* **101**, 277–282.

Drummond, D. C., Tiffany, S. T., Glautier, S., and Remington, B. (Eds.) (1995). *Addictive Behaviour: Cue Exposure Theory and Practice*. Wiley, New York.

Dupont, R. M., Rourke, S. B., Grant, I., Lehr, P. P., Reed, R. J., Challakere, K., Lamoureux, G., and Halpern, S. (1996). Single photon emission computed tomography with iodoamphetamine-123 and neuropsychological studies in long-term abstinent alcoholics. *Psychiatry Research* **67**, 99–111.

Eisenberger, N. I., Lieberman, M. D., and Williams, K. D. (2003). Does rejection hurt? An FMRI study of social exclusion. *Science* **302**, 290–292.

Eldreth, D. A., Matochik, J. A., Cadet, J. L., and Bolla, K. I. (2004). Abnormal brain activity in prefrontal brain regions in abstinent marijuana users. *Neuroimage* **23**, 914–920.

Elliott, R., Newman, J. L., Longe, O. A., and Deakin, J. F. (2003). Differential response patterns in the striatum and orbitofrontal cortex to financial reward in humans: a parametric functional magnetic resonance imaging study. *Journal of Neuroscience* **23**, 303–307.

Erk, S., Spitzer, M., Wunderlich, A. P., Galley, L., and Walter, H. (2002). Cultural objects modulate reward circuitry. *Neuroreport* **13**, 2499–2503.

Ernst, M., Bolla, K., Mouratidis, M., Contoreggi, C., Matochik, J. A., Kurian, V., Cadet, J. L., Kimes, A. S., and London, E. D. (2002). Decision-making in a risk-taking task: a PET study. *Neuropsychopharmacology* **26**, 682–691.

Ernst, T., Chang, L., Leonido-Yee, M., and Speck, O. (2000). Evidence for long-term neurotoxicity associated with methamphetamine abuse: a 1H MRS study. *Neurology* **54**, 1344–1349.

Evans, A. C. (2002). Automated 3D analysis of large brain MRI databases. *In Neuropsychopharmacology: The Fifth Generation of Progress* (K. L. Davis, D. Charney, J. T. Coyle, and C. Nemeroff, Eds.), pp. 301–313. Lippincott Williams and Wilkins, New York.

Fiorillo, C. D. (2004). The uncertain nature of dopamine. *Molecular Psychiatry* **9**, 122–123.

Fowler, J. S., Volkow, N. D., Wang, G. J., and Ding, Y. S. (2004). 2-deoxy-2-[18F]fluoro-D-glucose and alternative radiotracers for positron emission tomography imaging using the human brain as a model. *Seminars in Nuclear Medicine* **34**, 112–121.

Fowler, J. S., Volkow, N. D., Wang, G. J., Pappas, N., Logan, J., MacGregor, R., Alexoff, D., Shea, C., Schlyer, D., Wolf, A. P., Warner, D., Zezulkova, I., and Cilento, R. (1996a). Inhibition of monoamine oxidase B in the brains of smokers. *Nature* **379**, 733–736.

Fowler, J. S., Volkow, N. D., Wang, G. J., Pappas, N., Logan, J., Shea, C., Alexoff, D., MacGregor, R. R., Schlyer, D. J., Zezulkova, I., and Wolf, A. P. (1996b). Brain monoamine oxidase A inhibition in cigarette smokers. *Proceedings of the National Academy of Sciences USA* **93**, 14065–14069.

Fowler, J. S., Volkow, N. D., Wang, G. J., Pappas, N., Logan, J., MacGregor, R., Alexoff, D., Wolf, A. P., Warner, D., Cilento, R., and Zezulkova, I. (1998). Neuropharmacological actions of cigarette smoke: brain monoamine oxidase B (MAOB) inhibition. *Journal of Addictive Diseases* **17**, 23–34.

Fowler, J. S., Wang, G. J., Volkow, N. D., Franceschi, D., Logan, J., Pappas, N., Shea, C., MacGregor, R. R., and Garza, V. (1999). Smoking a single cigarette does not produce a measurable reduction in brain MAOB in non-smokers. *Nicotine and Tobacco Research* **1**, 325–329.

Fowler, J. S., Volkow, N. D., Malison, R., and Gatley, S. J. (1999). Neuroimaging studies of substance abuse disorders. *In Neurobiology of Mental Illness* (D. S. Charney, E. J. Nestler, and B. S. Bunney, Eds.), pp. 616–626. Oxford University Press, New York.

Fowler, J. S., Wang, G. J., Volkow, N. D., Franceschi, D., Logan, J., Pappas, N., Shea, C., MacGregor, R. R., and Garza, V. (2000). Maintenance of brain monoamine oxidase B inhibition in smokers after overnight cigarette abstinence. *American Journal of Psychiatry* **157**, 1864–1866.

Francis, S., Rolls, E. T., Bowtell, R., McGlone, F., O'Doherty, J., Browning, A., Clare, S., and Smith, E. (1999). The representation of pleasant touch in the brain and its relationship with taste and olfactory areas. *Neuroreport* **10**, 453–459.

Franklin, T. R., Acton, P. D., Maldjian, J. A., Gray, J. D., Croft, J. R., Dackis, C. A., O'Brien, C. P., and Childress, A. R. (2002). Decreased gray matter concentration in the insular, orbitofrontal, cingulate, and temporal cortices of cocaine patients. *Biological Psychiatry* **51**, 134–142.

Fumita, M., and Innis, R. B. (2002). In vivo molecular imaging: ligand development and research applications. *In Neuropsychopharmacology: The Fifth Generation of Progress* (K. L. Davis, D. Charney, J. T. Coyle, and C. Nemeroff, Eds.), pp. 411–425. Lippincott Williams and Wilkins, New York.

Gansler, D. A., Harris, G. J., Oscar-Berman, M., Streeter, C., Lewis, R. F., Ahmed, I., and Achong, D. (2000). Hypoperfusion of inferior frontal brain regions in abstinent alcoholics: a pilot SPECT study. *Journal of Studies on Alcohol* **61**, 32–37.

Garavan, H., Pankiewicz, J., Bloom, A., Cho, J. K., Sperry, L., Ross, T. J., Salmeron, B. J., Risinger, R., Kelley, D., and Stein, E. A. (2000). Cue-induced cocaine craving: neuroanatomical specificity for drug users and drug stimuli. *American Journal of Psychiatry* **157**, 1789–1798.

George, M. S., and Bohning, D. E. (2002). Measuring brain connectivity with functional imaging and transcranial magnetic stimulation. *In Neuropsychopharmacology: The Fifth Generation of Progress* (K. L. Davis, D. Charney, J. T. Coyle, and C. Nemeroff, Eds.), pp. 393–410. Lippincott Williams and Wilkins, New York.

Goldstein, R. Z., and Volkow, N. D. (2002). Drug addiction and its underlying neurobiological basis: neuroimaging evidence for the involvement of the frontal cortex. *American Journal of Psychiatry* **159**, 1642–1652.

Graham, J. H., and Porrino, L. J. (1995). Neuroanatomical substrates of cocaine self-administration. *In The Neurobiology of Cocaine: Cellular and Molecular Mechanisms* (R. P. Hammer, Jr., Ed.), pp. 3–14. CRC Press, Boca Raton, FL.

Grant, S., London, E. D., Newlin, D. B., Villemagne, V. L., Liu, X., Contoreggi, C., Phillips, R. L., Kimes, A. S., and Margolin, A. (1996). Activation of memory circuits during cue-elicited cocaine craving. *Proceedings of the National Academy of Sciences USA* **93**, 12040–12045.

Groenewegen, H. J. (1988). Organization of the afferent connections of the mediodorsal thalamic nucleus in the rat, related to the mediodorsal-prefrontal topography. *Neuroscience* **24**, 379–431.

Haber, B., Kohl, H., and Pscheidt, G. R. (1965). Supernatant-particulate distribution of exogenous serotonin in rat brain homogenates. *Biochemical Pharmacology* **14**, 1–6.

Hammer, R. P. Jr., and Cooke, E. S. (1994). Gradual tolerance of metabolic activity is produced in mesolimbic regions by chronic cocaine treatment, while subsequent cocaine challenge activates extrapyramidal regions of rat brain. *Journal of Neuroscience* **14**, 4289–4298.

Harper, C. (1998). The neuropathology of alcohol-specific brain damage, or does alcohol damage the brain? *Journal of Neuropathology and Experimental Neurology* **57**, 101–110.

Harper, C. G., Kril, J. J., and Holloway, R. L. (1985). Brain shrinkage in chronic alcoholics: a pathological study. *British Medical Journal* **290**, 501–504.

Harper, C., and Kril, J. (1989). Patterns of neuronal loss in the cerebral cortex in chronic alcoholic patients. *Journal of Neurological Science* **92**, 81–89.

Harper, C. G., and Kril, J. J. (1993). Neuropathological changes in alcoholics. *In Alcohol-Induced Brain Damage* (series title: *NIAAA Research Monograph*, vol. 22) (W. A. Hunt, and S. J. Nixon, Eds.), pp. 39–69. National Institute on Alcohol Abuse and Alcoholism, Rockville, MD.

Harris, G. J., Oscar-Berman, M., Gansler, A., Streeter, C., Lewis, R. F., Ahmed, I., and Achong, D. (1999). Hypoperfusion of the cerebellum and aging effects on cerebral cortex blood flow in abstinent alcoholics: a SPECT study. *Alcoholism: Clinical and Experimental Research* **23**, 1219–1227.

Heinz, A., Ragan, P., Jones, D. W., Hommer, D., Williams, W., Knable, M. B., Gorey, J. G., Doty, L., Geyer, C., Lee, K. S., Coppola, R., Weinberger, D. R., and Linnoila, M. (1998). Reduced central serotonin transporters in alcoholism. *American Journal of Psychiatry* **155**, 1544–1549.

Hill, S. Y., De Bellis, M. D., Keshavan, M. S., Lowers, L., Shen, S., Hall, J., and Pitts, T. (2001). Right amygdala volume in adolescent and young adult offspring from families at high risk for developing alcoholism. *Biological Psychiatry* **49**, 894–905.

Hitzemann, R., Volkow, N., Fowler, J., and Wang, G. J. (2002). Neuroimaging and substance abuse. *In Biological Psychiatry*, vol. 1, (H. D'Haenen, J. A. den Boer, and P. Willner, Eds.), pp. 523–535. John Wiley, Chichester.

Holman, B. L., Mendelson, J., Garada, B., Teoh, S. K., Hallgring, E., Johnson, K. A., and Mello, N. K. (1993). Regional cerebral blood flow improves with treatment in chronic cocaine polydrug users. *Journal of Nuclear Medicine* **34**, 723–727.

Hommer, D., Andreasen, P., Rio, D., Williams, W., Ruttimann, U., Momenan, R., Zametkin, A., Rawlings, R., and Linnoila, M. (1997). Effects of m-chlorophenylpiperazine on regional brain glucose utilization: a positron emission tomographic comparison of alcoholic and control subjects. *Journal of Neuroscience* **17**, 2796–2806.

Hommer, D., Momenan, R., Kaiser, E., and Rawlings, R. (2001). Evidence for a gender-related effect of alcoholism on brain volumes. *American Journal of Psychiatry* **158**, 198–204.

Iyo, M., Namba, H., Yanagisawa, M., Hirai, S., Yui, N., and Fukui, S. (1997). Abnormal cerebral perfusion in chronic methamphetamine abusers: a study using 99MTc-HMPAO and SPECT. *Progress in Neuropsychopharmacology and Biological Psychiatry* **21**, 789–796.

Jentsch, J. D., and Taylor, J. R. (1999). Impulsivity resulting from frontostriatal dysfunction in drug abuse: Implications for the control of behavior by reward-related stimuli. *Psychopharmacology* **146**, 373–390.

Jernigan, T. L., Butters, N., DiTraglia, G., Schafer, K., Smith, T., Irwin, M., Grant, I., Schuckit, M., and Cermak, L. S. (1991). Reduced cerebral grey matter observed in alcoholics using magnetic resonance imaging. *Alcoholism: Clinical and Experimental Research* **15**, 418–427.

Johnson-Greene, D., Adams, K. M., Gilman, S., Koeppe, R. A., Junck, L., Kluin, K. J., Martorello, S., and Heumann, M. (1997). Effects of abstinence and relapse upon neuropsychological function and cerebral glucose metabolism in severe chronic alcoholism. *Journal of Clinical and Experimental Neuropsychology* **19**, 378–385.

Joyce, E. M., Rio, D. E., Ruttimann, U. E., Rohrbaugh, J. W., Martin, P. R., Rawlings, R. R., and Eckardt, M. J. (1994). Decreased cingulate and precuneate glucose utilization in alcoholic Korsakoff's syndrome. *Psychiatry Research* **54**, 225–239.

Karama, S., Lecours, A. R., Leroux, J. M., Bourgouin, P., Beaudoin, G., Joubert, S., and Beauregard, M. (2002). Areas of brain activation in males and females during viewing of erotic film excerpts. *Human Brain Mapping* **16**, 1–13.

Kaufman, J. N., Ross, T. J., Stein, E. A., and Garavan, H. (2003). Cingulate hypoactivity in cocaine users during a GO-NOGO task as revealed by event-related functional magnetic resonance imaging. *Journal of Neuroscience* **23**, 7839–7843.

Ke, Y., Streeter, C. C., Nassar, L. E., Sarid-Segal, O., Hennen, J., Yurgelun-Todd, D. A., Awad, L. A., Rendall, M. J., Gruber, S. A., Nason, A., Mudrick, M. J., Blank, S. R., Meyer, A. A., Knapp, C., Ciraulo, D. A., and Renshaw, P. F. (2004). Frontal lobe GABA levels in cocaine dependence: a two-dimensional, J-resolved magnetic resonance spectroscopy study. *Psychiatry Research* **130**, 283–293.

Kilts, C. D., Schweitzer, J. B., Quinn, C. K., Gross, R. E., Faber, T. L., Muhammad, F., Ely, T. D., Hoffman, J. M., and Drexler, K. P. (2001). Neural activity related to drug craving in cocaine addiction. *Archives of General Psychiatry* **58**, 334–341.

Kolb, B. (1984). Functions of the frontal cortex of the rat: a comparative review. *Brain Research* **320**, 65–98.

Kolb, B. (1990). Animal models for human PFC-related disorders. *In The Prefrontal Cortex: Its Structure, Function, and Pathology* (series title: *Progress in Brain Research*, vol. 85) (H. B. M. Uylings, C. G. van Eden, J. P. C. de Bruin, M. A. Corner, and M. G. P. Feenstra, Eds.), pp. 501–519. Elsevier, New York.

Koob, G. F., and Le Moal, M. (2001). Drug addiction, dysregulation of reward, and allostasis. *Neuropsychopharmacology* **24**, 97–129.

Koretsky, A. P. (2004). New developments in magnetic resonance imaging of the brain. *NeuroRx* **1**, 155–164.

Kril, J. J., and Harper, C. G. (1989). Neuronal counts from four cortical regions of alcoholic brains. *Acta Neuropathologica* **79**, 200–204.

Kril, J. J., Halliday, G. M., Svoboda, M. D., and Cartwright, H. (1997). The cerebral cortex is damaged in chronic alcoholics. *Neuroscience* **79**, 983–998.

Krystal, J. H., Woods, S. W., Kosten, T. R., Rosen, M. I., Seibyl, J. P., van Dyck, C. C., Price, L. H., Zubal, I. G., Hoffer, P. B., and Charney, D. S. (1995). Opiate dependence and withdrawal: preliminary assessment using single photon emission computerized tomography (SPECT). *American Journal of Drug and Alcohol Abuse* **21**, 47–63.

Kubota, M., Nakazaki, S., Hirai, S., Saeki, N., Yamaura, A., and Kusaka, T. (2001). Alcohol consumption and frontal lobe shrinkage: study of 1432 non-alcoholic subjects. *Journal of Neurology, Neurosurgery and Psychiatry* **71**, 104–106.

Kuikka, J. T., Repo, E., Bergstrom, K. A., Tupala, E., and Tiihonen, J. (2000). Specific binding and laterality of human extrastriatal dopamine D2/D3 receptors in late onset type 1 alcoholic patients. *Neuroscience Letters* **292**, 57–59.

Kumari, V., Gray, J. A., Ffytche, D. H., Mitterschiffthaler, M. T., Das, M., Zachariah, E., Vythelingum, G. N., Williams, S. C., Simmons, A., and Sharma, T. (2003). Cognitive effects of nicotine in humans: an fMRI study. *Neuroimage* **19**, 1002–1013.

Laine, T. P., Ahonen, A., Rasanen, P., and Tiihonen, J. (1999). Dopamine transporter availability and depressive symptoms during alcohol withdrawal. *Psychiatry Research* **90**, 153–157.

Lane, R. D., Reiman, E. M., Bradley, M. M., Lang, P. J., Ahern, G. L., Davidson, R. J., and Schwartz, G. E. (1997). Neuroanatomical correlates of pleasant and unpleasant emotion. *Neuropsychologia* **35**, 1437–1444.

Laureys, S., Peigneux, P., and Goldman, S. (2002). Brain imaging. *In Biological Psychiatry*, vol. 1, (H. D'Haenen, J. A. den Boer, and P. Willner, Eds.), pp. 155–166. John Wiley, Chichester.

Lazar, S. W., Bush, G., Gollub, R. L., Fricchione, G. L., Khalsa, G., and Benson, H. (2000). Functional brain mapping of the relaxation response and meditation. *Neuroreport* **11**, 1581–1585.

Le Moal, M. (1995). Mesocorticolimbic dopaminergic neurons: Functional and regulatory roles. *In Psychopharmacology: The Fourth Generation of Progress* (F. E. Bloom, and D. J. Kupfer, Eds.), pp. 283–294. Raven Press, New York.

Le Moal, M., and Simon, H. (1991). Mesocorticolimbic dopaminergic network: functional and regulatory roles. *Physiological Reviews* **71**, 155–234.

Leonard, C. M. (1969). The prefrontal cortex of the rat. I. Cortical projection of the mediodorsal nucleus: II. Efferent connections. *Brain Research* **12**, 321–343.

Lewis, M. J. (1996). Alcohol reinforcement and neuropharmacological therapeutics. *Alcohol and Alcoholism* **31**(*Suppl.* 1), 17–25.

Lingford-Hughes, A. R., Acton, P. D., Gacinovic, S., Suckling, J., Busatto, G. F., Boddington, S. J., Bullmore, E., Woodruff, P. W., Costa, D. C., Pilowsky, L. S., Ell, P. J., Marshall, E. J., and Kerwin, R. W. (1998). Reduced levels of GABA-benzodiazepine receptor in alcohol dependency in the absence of grey matter atrophy. *British Journal of Psychiatry* **173**, 116–122.

Lingford-Hughes, A. R., Acton, P. D., Gacinovic, S., Boddington, S. J., Costa, D. C., Pilowsky, L. S., Ell, P. J., Marshall, E. J., and Kerwin, R. W. (2000). Levels of gamma-aminobutyric acid-benzodiazepine receptors in abstinent, alcohol-dependent women: preliminary findings from an 123I-iomazenil single photon emission tomography study. *Alcoholism: Clinical and Experimental Research* **24**, 1449–1455.

Liu, X., Matochik, J. A., Cadet, J. L., and London, E. D. (1998). Smaller volume of prefrontal lobe in polysubstance abusers: a magnetic resonance imaging study. *Neuropsychopharmacology* **18**, 243–252.

Loeber, R. T., and Yurgelun-Todd, D. A. (1999). Human neuroimaging of acute and chronic marijuana use: implications for frontocerebellar dysfunction. *Human Psychopharmacology: Clinical and Experimental* **14**, 291–304.

London, E. D., Broussolle, E. P., Links, J. M., Wong, D. F., Cascella, N. G., Dannals, R. F., Sano, M., Herning, R., Snyder, F. R., and Rippetoe, L. R. (1990a). Morphine-induced metabolic changes in human brain. Studies with positron emission tomography and [fluorine 18]fluorodeoxyglucose. *Archives of General Psychiatry* **47**, 73–81.

London, E. D., Cascella, N. G., Wong, D. F., Phillips, R. L., Dannals, R. F., Links, J. M., Herning, R., Grayson, R., Jaffe, J. H., and Wagner, H. N. Jr. (1990b). Cocaine-induced reduction of glucose utilization in human brain: A study using positron emission tomography and [fluorine 18]-fluorodeoxyglucose. *Archives of General Psychiatry* **47**, 567–574.

London, E. D., Simon, S. L., Berman, S. M., Mandelkern, M. A., Lichtman, A. M., Bramen, J., Shinn, A. K., Miotto, K., Learn, J., Dong, Y., Matochik, J. A., Kurian, V., Newton, T., Woods, R., Rawson, R., and Ling, W. (2004). Mood disturbances and regional cerebral metabolic abnormalities in recently abstinent methamphetamine abusers. *Archives of General Psychiatry* **61**, 73–84.

Lyons, D., Friedman, D. P., Nader, M. A., and Porrino, L. J. (1996). Cocaine alters cerebral metabolism within the ventral striatum and limbic cortex of monkeys. *Journal of Neuroscience* **16**, 1230–1238.

Macey, D. J., Rice, W. N., Freedland, C. S., Whitlow, C. T., and Porrino, L. J. (2004). Patterns of functional activity associated with cocaine self-administration in the rat change over time. *Psychopharmacology* **172**, 384–392.

Magistretti, P. J., Pellerin, L., Rothman, D. L., and Shulmanm, R. G. (1999). Energy on demand. *Science* **283**, 496–497.

Makris, N., Papadimitriou, G. M., Worth, A. J., Jenkins, B. G., Garrido, L., Sorensen, A. G., Wedeen, V. J., Tuch, D. S., Wu, O., Cudkowicz, M. E., Caviness, V. S. Jr., Rosen, B. R., and Kennedy, D. N. (2002). Diffusion tensor imaging. *In Neuropsychopharmacology: The Fifth Generation of Progress* (K. L. Davis, D. Charney, J. T. Coyle, and C. Nemeroff, Eds.), pp. 357–371. Lippincott Williams and Wilkins, New York.

Mann, K., Mundle, G., Strayle, M., and Wakat, P. (1995). Neuroimaging in alcoholism: CT and MRI results and clinical correlates. *Journal of Neural Transmission: General Section* **99**, 145–155.

Mann, K., Agartz, I., Harper, C., Shoaf, S., Rawlings, R. R., Momenan, R., Hommer, D. W., Pfefferbaum, A., Sullivan, E. V., Anton, R. F., Drobes, D. J., George, M. S., Bares, R., Machulla, H. J., Mundle, G., Reimold, M., and Heinz, A. (2001). Neuroimaging in alcoholism: ethanol and brain damage. *Alcoholism: Clinical and Experimental Research* **25**(5 *Suppl.* ISBRA), 104s–109s.

Martin-Soelch, C., Chevalley, A. F., Kunig, G., Missimer, J., Magyar, S., Mino, A., Schultz, W., and Leenders, K. L. (2001). Changes in reward-induced brain activation in opiate addicts. *European Journal of Neuroscience* **14**, 1360–1368.

Martin-Solch, C., Magyar, S., Kunig, G., Missimer, J., Schultz, W., and Leenders, K. L. (2001). Changes in brain activation associated with reward processing in smokers and nonsmokers. A positron emission tomography study. *Experimental Brain Research* **139**, 278–286.

Mathew, R. J., and Wilson, W. H. (1993). Acute changes in cerebral blood flow after smoking marijuana. *Life Sciences* **52**, 757–767.

Mathew, R. J., Wilson, W. H., Coleman, R. E., Turkington, T. G., and DeGrado, T. R. (1997). Marijuana intoxication and brain activation in marijuana smokers. *Life Sciences* **60**, 2075–2089.

Matochik, J. A., London, E. D., Eldreth, D. A., Cadet, J. L., and Bolla, K. I. (2003). Frontal cortical tissue composition in abstinent cocaine abusers: a magnetic resonance imaging study. *Neuroimage* **19**, 1095–1102.

Matochik, J. A., Eldreth, D. A., Cadet, J. L., and Bolla, K. I. (2005). Altered brain tissue composition in heavy marijuana users. *Drug and Alcohol Dependence* **77**, 23–30.

Mazziotta, J. C., Toga, A. W., Evans, A., Fox, P., and Lancaster, J. (1995). A probabilistic atlas of the human brain: theory and rationale for its development. The International Consortium for Brain Mapping (ICBM). *Neuroimage* **2**, 89–101.

McCann, U. D., Wong, D. F., Yokoi, F., Villemagne, V., Dannals, R. F., and Ricaurte, G. A. (1998). Reduced striatal dopamine transporter density in abstinent methamphetamine and methcathinone users: evidence from positron emission tomography studies with [11C]WIN-35,428. *Journal of Neuroscience* **18**, 8417–8422.

McCann, U. D., and Ricaurte, G. A. (2004). Amphetamine neurotoxicity: accomplishments and remaining challenges. *Neuroscience and Biobehavioral Reviews* **27**, 821–826.

Mellers, B. A., Schwartz, A., Ho, K., and Ritov, I. (1997). Decision affect theory: emotional reactions to the outcomes of risky options. *Psychological Science* **8**, 423–429.

Mellers, B., Schwartz, A., and Ritov, I. (1999). Emotion-based choice. *Journal of Experimental Psychology: General* **128**, 332–345.

Miller, E. K. (2000). The prefrontal cortex and cognitive control. *Nature Reviews Neuroscience* **1**, 59–65.

Miller, E. K., and Cohen, J. D. (2001). An integrative theory of prefrontal cortex function. *Annual Review of Neuroscience* **24**, 167–202.

Modell, J. G., Mountz, J. M., and Beresford, T. P. (1990). Basal ganglia / limbic striatal and thalamocortical involvement in craving and loss of control in alcoholism. *Journal of Neuropsychiatry and Clinical Neurosciences* **2**, 123–144.

Nicolas, J. M., Catafau, A. M., Estruch, R., Lomena, F. J., Salamero, M., Herranz, R., Monforte, R., Cardenal, C., and Urbano-Marquez, A. (1993). Regional cerebral blood flow-SPECT in chronic alcoholism: relation to neuropsychological testing. *Journal of Nuclear Medicine* **34**, 1452–1459.

O'Leary, C. M. (2004). Fetal alcohol syndrome: diagnosis, epidemiology, and developmental outcomes. *Journal of Paediatrics and Child Health* **40**, 2–7.

O'Leary, D. S., Block, R. I., Flaum, M., Schultz, S. K., Boles Ponto, L. L., Watkins, G. L., Hurtig, R. R., Andreasen, N. C., and Hichwa, R. D. (2000). Acute marijuana effects on rCBF and cognition: a PET study. *Neuroreport* **11**, 3835–3841.

O'Leary, D. S., Block, R. I., Koeppel, J. A., Flaum, M., Schultz, S. K., Andreasen, N. C., Ponto, L. B., Watkins, G. L., Hurtig, R. R., and Hichwa, R. D. (2002). Effects of smoking marijuana on brain perfusion and cognition. *Neuropsychopharmacology* **26**, 802–816.

O'Leary, D. S., Block, R. I., Turner, B. M., Koeppel, J., Magnotta, V. A., Ponto, L. B., Watkins, G. L., Hichwa, R. D., and Andreasen, N. C. (2003). Marijuana alters the human cerebellar clock. *Neuroreport* **14**, 1145–1151.

Paulus, M. P., Hozack, N. E., Zauscher, B. E., Frank, L., Brown, G. G., Braff, D. L., and Schuckit, M. A. (2002). Behavioral and functional neuroimaging evidence for prefrontal dysfunction in methamphetamine-dependent subjects. *Neuropsychopharmacology* **26**, 53–63.

Pfefferbaum, A., Lim, K. O., Zipursky, R. B., Mathalon, D. H., Rosenbloom, M. J., Lane, B., Ha, C. N., and Sullivan, E. V. (1992). Brain gray and white matter volume loss accelerates with aging in chronic alcoholics: a quantitative MRI study. *Alcoholism: Clinical and Experimental Research* **16**, 1078–1089.

Pfefferbaum, A., Sullivan, E. V., Mathalon, D. H., and Lim, K. O. (1997). Frontal lobe volume loss observed with magnetic resonance imaging in older chronic alcoholics. *Alcoholism: Clinical and Experimental Research* **21**, 521–529.

Pfefferbaum, A., Rosenbloom, M., Deshmukh, A., and Sullivan, E. (2001). Sex differences in the effects of alcohol on brain structure. *American Journal of Psychiatry* **158**, 188–197.

Phelps, M. E. (2000). Inaugural article: positron emission tomography provides molecular imaging of biological processes. *Proceedings of the National Academy of Sciences USA* **97**, 9226–9233.

Polich, J., Pollock, V. E., and Bloom, F. E. (1994). Meta-analysis of P300 amplitude from males at risk for alcoholism. *Psychological Bulletin* **115**, 55–73.

Porrino, L. J. (1993). Functional consequences of acute cocaine treatment depend on route of administration. *Psychopharmacology* **112**, 343–351.

Porrino, L. J., Domer, F. R., Crane, A. M., and Sokoloff, L. (1988). Selective alterations in cerebral metabolism within the mesocorticolimbic dopaminergic system produced by acute cocaine administration in rats. *Neuropsychopharmacology* **1**, 109–118.

Porrino, L. J., and Lyons, D. (2000). Orbital and medial prefrontal cortex and psychostimulant abuse: studies in animal models. *Cerebral Cortex* **10**, 326–333.

Porrino, L. J., Daunais, J. B., Smith, H. R., and Nader, M. A. (2004). The expanding effects of cocaine: studies in a nonhuman primate model of cocaine self-administration. *Neuroscience and Biobehavioral Reviews* **27**, 813–820.

Potenza, M. N., Leung, H. C., Blumberg, H. P., Peterson, B. S., Fulbright, R. K., Lacadie, C. M., Skudlarski, P., and Gore, J. C. (2003a). An fMRI Stroop task study of ventromedial prefrontal cortical function in pathological gamblers. *American Journal of Psychiatry* **160**, 1990–1994.

Potenza, M. N., Steinberg, M. A., Skudlarski, P., Fulbright, R. K., Lacadie, C. M., Wilber, M. K., Rounsaville, B. J., Gore, J. C., and Wexler, B. E. (2003b). Gambling urges in pathological gambling: a functional magnetic resonance imaging study. *Archives of General Psychiatry* **60**, 828–836.

Rangaswamy, M., Porjesz, B., Ardekani, B. A., Choi, S. J., Tanabe, J. L., Lim, K. O., and Begleiter, H. (2004). A functional MRI study of visual oddball: evidence for frontoparietal dysfunction in subjects at risk for alcoholism. *Neuroimage* **21**, 329–339.

Ray, J. P., and Price, J. L. (1993). The organization of projections from the mediodorsal nucleus of the thalamus to orbital and medial prefrontal cortex in macaque monkeys. *Journal of Comparative Neurology* **337**, 1–31.

Redoute, J., Stoleru, S., Gregoire, M. C., Costes, N., Cinotti, L., Lavenne, F., Le Bars, D., Forest, M. G., and Pujol, J. F. (2000). Brain processing of visual sexual stimuli in human males. *Human Brain Mapping* **11**, 162–177.

Riley, E. P., McGee, C. L., and Sowell, E. R. (2004). Teratogenic effects of alcohol: a decade of brain imaging. *American Journal of Medical Genetics: C. Seminars in Medical Genetics* **127**, 35–41.

Rogers, R. D., Owen, A. M., Middleton, H. C., Williams, E. J., Pickard, J. D., Sahakian, B. J., and Robbins, T. W. (1999). Choosing between small, likely rewards and large, unlikely rewards activates inferior and orbital prefrontal cortex. *Journal of Neuroscience* **19**, 9029–9038.

Rose, J. E., Behm, F. M., Westman, E. C., Mathew, R. J., London, E. D., Hawk, T. C., Turkington, T. G., and Coleman, R. E. (2003). PET studies of the influences of nicotine on neural systems in cigarette smokers. *American Journal of Psychiatry* **160**, 323–333.

Rose, J. S., Branchey, M., Buydens-Branchey, L., Stapleton, J. M., Chasten, K., Werrell, A., and Maayan, M. L. (1996). Cerebral perfusion in early and late opiate withdrawal: a technetium-99m-HMPAO SPECT study. *Psychiatry Research* **67**, 39–47.

Schneider, F., Habel, U., Wagner, M., Franke, P., Salloum, J. B., Shah, N. J., Toni, I., Sulzbach, C., Honig, K., Maier, W., Gaebel, W., and Zilles, K. (2001). Subcortical correlates of craving in recently abstinent alcoholic patients. *American Journal of Psychiatry* **158**, 1075–1083.

Seibyl, J. P., Scanley, E., Krystal, J. H., and Innis, R. B. (1999). Neuroimaging methodologies. *In Neurobiology of Mental Illness* (D. S. Charney, E. J. Nestler, and B. S. Bunney, Eds.), pp. 170–189. Oxford University Press, New York.

Sell, L. A., Morris, J., Bearn, J., Frackowiak, R. S., Friston, K. J., and Dolan, R. J. (1999). Activation of reward circuitry in human opiate addicts. *European Journal of Neuroscience* **11**, 1042–1048.

Silvestri, S., Negrete, J. C., Seeman, M. V., Shammi, C. M., and Seeman, P. (2004). Does nicotine affect D2 receptor upregulation? A case-control study. *Acta Psychiatrica Scandinavica* **109**, 313–317.

Small, D. M., Zatorre, R. J., Dagher, A., Evans, A. C., and Jones-Gotman, M. (2001). Changes in brain activity related to eating chocolate: from pleasure to aversion. *Brain* **124**, 1720–1733.

Stapleton, J. M., Gilson, S. F., Wong, D. F., Villemagne, V. L., Dannals, R. F., Grayson, R. F., Henningfield, J. E., and London, E. D. (2003). Intravenous nicotine reduces cerebral glucose metabolism: a preliminary study. *Neuropsychopharmacology* **28**, 765–772.

Stark, C. E., and Squire, L. R. (2001). When zero is not zero: the problem of ambiguous baseline conditions in fMRI. *Proceedings of the National Academy of Sciences USA* **98**, 12760–12766.

Stein, E. A., Pankiewicz, J., Harsch, H. H., Cho, J. K., Fuller, S. A., Hoffmann, R. G., Hawkins, M., Rao, S. M., Bandettini, P. A., and Bloom, A. S. (1998). Nicotine-induced limbic cortical activation in the human brain: a functional MRI study. *American Journal of Psychiatry* **155**, 1009–1015.

Strickland, T. L., Miller, B. L., Kowell, A., and Stein, R. (1998). Neurobiology of cocaine-induced organic brain impairment: contributions from functional neuroimaging. *Neuropsychology Review* **8**, 1–9.

Sullivan, E. V. (2000). Human brain vulnerability to alcoholism: evidence from neuroimaging studies. *In Review of NIAAA's Neuroscience and Behavioral Research Portfolio* (series title: *NIAAA Research Monograph*, vol. 34), (A. Noronha, M. Eckardt, and K. Warren, Eds.), pp. 473–508. National Institute on Alcohol Abuse and Alcoholism, Bethesda, MD.

Sullivan, E. V., Desmond, J. E., Lim, K. O., and Pfefferbaum A. (2002). Speed and efficiency but not accuracy or timing deficits of limb movements in alcoholic men and women. *Alcoholism: Clinical and Experimental Research* **26**, 705–713.

Sullivan, E. V., Harding, A. J., Pentney, R., Dlugos, C., Martin, P. R., Parks, M. H., Desmond, J. E., Chen, S. H., Pryor, M. R., De Rosa, E., and Pfefferbaum, A. (2003). Disruption of fronto-cerebellar circuitry and function in alcoholism. *Alcoholism: Clinical and Experimental Research* **27**, 301–309.

Szabo, J., and Cowan, W. M. (1984). A stereotaxic atlas of the brain of the cynomolgus monkey (*Macaca fascicularis*). *Journal of Comparative Neurology* **222**, 265–300.

Szabo, Z., Owonikoko, T., Peyrot, M., Varga, J., Mathews, W. B., Ravert, H. T., Dannals, R. F., and Wand, G. (2004). Positron emission tomography imaging of the serotonin transporter in subjects with a history of alcoholism. *Biological Psychiatry* **55**, 766–771.

Tapert, S. F., Cheung, E. H., Brown, G. G., Frank, L. R., Paulus, M. P., Schweinsburg, A. D., Meloy, M. J., and Brown, S. A. (2003). Neural response to alcohol stimuli in adolescents with alcohol use disorder. *Archives of General Psychiatry* **60**, 727–735.

Tapert, S. F., Brown, G. G., Baratta, M. V., and Brown, S. A. (2004). fMRI BOLD response to alcohol stimuli in alcohol dependent young women. *Addictive Behaviors* **29**, 33–50.

Thomas, W. L. Jr., Cooke, E. S., and Hammer, R. P. Jr. (1996). Cocaine-induced sensitization of metabolic activity in extrapyramidal circuits involves prior dopamine D1-like receptor stimulation. *Journal of Pharmacology and Experimental Therapeutics* **278**, 347–353.

Thompson, P. M., Hayashi, K. M., Simon, S. L., Geaga, J. A., Hong, M. S., Sui, Y., Lee, J. Y., Toga, A. W., Ling, W., and London, E. D. (2004). Structural abnormalities in the brains of human subjects who use methamphetamine. *Journal of Neuroscience* **24**, 6028–6036.

Thut, G., Schultz, W., Roelcke, U., Nienhusmeier, M., Missimer, J., Maguire, R. P., and Leenders, K. L. (1997). Activation of the human brain by monetary reward. *Neuroreport* **8**, 1225–1228.

Tunving, K., Thulin, S. O., Risberg, J., and Warkentin, S. (1986). Regional cerebral blood flow in long-term heavy cannabis use. *Psychiatry Research* **17**, 15–21.

Uchino, A., Yuzuriha, T., Murakami, M., Endoh, K., Hiejima, S., Koga, H., and Kudo, S. (2003). Magnetic resonance imaging of sequelae of central pontine myelinolysis in chronic alcohol abusers. *Neuroradiology* **45**, 877–880.

Unterwald, E. M., Horne-King, J., and Kreek, M. J. (1992). Chronic cocaine alters brain mu opioid receptors. *Brain Research* **584**, 314–318.

Van Eden, C. G. (1986). Development of connections between the mediodorsal nucleus of the thalamus and the prefrontal cortex in the rat. *Journal of Comparative Neurology* **244**, 349–359.

Vanderschuren, L. J., and Everitt, B. J. (2004). Drug seeking becomes compulsive after prolonged cocaine self-administration. *Science* **305**, 1017–1019.

Volkow, N. D., Mullani, N., Gould, K. L., Adler, S., and Krajewski, K. (1988). Cerebral blood flow in chronic cocaine users: a study with positron emission tomography. *British Journal of Psychiatry* **152**, 641–648.

Volkow, N. D., Fowler, J. S., Wolf, A. P., Schlyer, D., Shiue, C. Y., Alpert, R., Dewey, S. L., Logan, J., Bendriem, B., and Christman, D. (1990a). Effects of chronic cocaine abuse on postsynaptic dopamine receptors. *American Journal of Psychiatry* **147**, 719–724.

Volkow, N. D., Hitzemann, R., Wolf, A. P., Logan, J., Fowler, J. S., Christman, D., Dewey, S. L., Schlyer, D., Burr, G., Vitkun, S., and Hirschowitz, J. (1990b). Acute effects of ethanol on regional brain glucose metabolism and transport. *Psychiatry Research* **35**, 39–48.

Volkow, N. D., Gillespie, H., Mullani, N., Tancredi, L., Grant, C., Ivanovic, M., and Hollister, L. (1991). Cerebellar metabolic activation by delta-9-tetrahydro-cannabinol in human brain: a study with positron emission tomography and 18F-2-fluoro-2-deoxyglucose. *Psychiatry Research* **40**, 69–78.

Volkow, N. D., Hitzemann, R., Wang, G. J., Fowler, J. S., Burr, G., Pascani, K., Dewey, S. L., and Wolf, A. P. (1992a). Decreased brain metabolism in neurologically intact healthy alcoholics. *American Journal of Psychiatry* **149**, 1016–1022.

Volkow, N. D., Hitzemann, R., Wang, G. J., Fowler, J. S., Wolf, A. P., Dewey, S. L., and Handlesman, L. (1992b). Long-term frontal brain metabolic changes in cocaine abusers. *Synapse* **11**, 184–190 [erratum: **12**, 86].

Volkow, N. D., Fowler, J. S., Wang, G. J., Dewey, S. L., Schlyer, D., MacGregor, R., Logan, J., Alexoff, D., Shea, C., and Hitzemann, R. (1993a). Reproducibility of repeated measures of carbon-11-raclopride binding in the human brain. *Journal of Nuclear Medicine* **34**, 609–613 [erratum: **34**, 838].

Volkow, N. D., Fowler, J. S., Wang, G. J., Hitzemann, R., Logan, J., Schlyer, D. J., Dewey, S. L., and Wolf, A. P. (1993b). Decreased dopamine D2 receptor availability is associated with reduced frontal metabolism in cocaine abusers. *Synapse* **14**, 169–177.

Volkow, N. D., Wang, G. J., Hitzemann, R., Fowler, J. S., Overall, J. E., Burr, G., and Wolf, A. P. (1994). Recovery of brain glucose metabolism in detoxified alcoholics. *American Journal of Psychiatry* **151**, 178–183.

Volkow, N. D., Gillespie, H., Mullani, N., Tancredi, L., Grant, C., Valentine, A., and Hollister, L. (1996a). Brain glucose metabolism in chronic marijuana users at baseline and during marijuana intoxication. *Psychiatry Research* **67**, 29–38.

Volkow, N. D., Wang, G. J., Fowler, J. S., Gatley, S. J., Ding, Y. S., Logan, J., Dewey, S. L., Hitzemann, R., and Lieberman, J. (1996b). Relationship between psychostimulant-induced 'high' and dopamine transporter occupancy. *Proceedings of the National Academy of Sciences USA* **93**, 10388–10392.

Volkow, N. D., Wang, G. J., Fowler, J. S., Logan, J., Hitzemann, R., Ding, Y. S., Pappas, N., Shea, C., and Piscani, K. (1996c). Decreases in dopamine receptors but not in dopamine transporters in alcoholics. *Alcoholism: Clinical and Experimental Research* **20**, 1594–1598.

Volkow, N. D., Wang, G. J., Fischman, M. W., Foltin, R. W., Fowler, J. S., Abumrad, N. N., Vitkun, S., Logan, J., Gatley, S. J., Pappas, N., Hitzemann, R., and Shea, C. E. (1997a). Relationship between subjective effects of cocaine and dopamine transporter occupancy. *Nature* **386**, 827–830.

Volkow, N. D., Wang, G. J., Fowler, J. S., Logan, J., Gatley, S. J., Hitzemann, R., Chen, A. D., Dewey, S. L., and Pappas, N. (1997b). Decreased striatal dopaminergic responsiveness in detoxified cocaine-dependent subjects. *Nature* **386**, 830–833.

Volkow, N. D., Wang, G. J., Overall, J. E., Hitzemann, R., Fowler, J. S., Pappas, N., Frecska, E., and Piscani, K. (1997c). Regional brain metabolic response to lorazepam in alcoholics during early and late alcohol detoxification. *Alcoholism: Clinical and Experimental Research* **21**, 1278–1284.

Volkow, N. D., Fowler, J. S., Ding, Y. S., Wang, G. J., and Gatley, S. J. (1999a). Imaging the neurochemistry of nicotine actions: studies with positron emission tomography. *Nicotine and Tobacco Research* **1(Suppl. 2)**, s127–s132.

Volkow, N. D., Wang, G. J., Fowler, J. S., Hitzemann, R., Angrist, B., Gatley, S. J., Logan, J., Ding, Y. S., and Pappas, N. (1999b). Association of methylphenidate-induced craving with changes in right striato-orbitofrontal metabolism in cocaine abusers: implications in addiction. *American Journal of Psychiatry* **156**, 19–26.

Volkow, N. D., Wang, G. J., Fowler, J. S., Logan, J., Gatley, S. J., Gifford, A., Hitzemann, R., Ding, Y. S., and Pappas, N. (1999c). Prediction of reinforcing responses to psychostimulants in humans by brain dopamine D2 receptor levels. *American Journal of Psychiatry* **156**, 1440–1443.

Volkow, N. D., and Fowler, J. S. (2000). Addiction, a disease of compulsion and drive: Involvement of the orbitofrontal cortex. *Cerebral Cortex* **10**, 318–325.

Volkow, N. D., Chang, L., Wang, G. J., Fowler, J. S., Ding, Y. S., Sedler, M., Logan, J., Franceschi, D., Gatley, J., Hitzemann, R., Gifford, A., Wong, C., and Pappas, N. (2001a). Low level of brain dopamine D2 receptors in methamphetamine abusers: association with metabolism in the orbitofrontal cortex. *American Journal of Psychiatry* **158**, 2015–2021.

Volkow, N. D., Chang, L., Wang, G. J., Fowler, J. S., Franceschi, D., Sedler, M. J., Gatley, S. J., Hitzemann, R., Ding, Y. S., Wong, C., and Logan, J. (2001b). Higher cortical and lower subcortical metabolism in detoxified methamphetamine abusers. *American Journal of Psychiatry* **158**, 383–389.

Volkow, N. D., Chang, L., Wang, G. J., Fowler, J. S., Franceschi, D., Sedler, M., Gatley, S. J., Miller, E., Hitzemann, R., Ding, Y. S., and Logan, J. (2001c). Loss of dopamine transporters in methamphetamine abusers recovers with protracted abstinence. *Journal of Neuroscience* **21**, 9414–9418.

Volkow, N. D., Chang, L., Wang, G. J., Fowler, J. S., Leonido-Yee, M., Franceschi, D., Sedler, M. J., Gatley, S. J., Hitzemann, R., Ding, Y. S., Logan, J., Wong, C., and Miller, E. N. (2001d). Association of dopamine transporter reduction with psychomotor impairment in methamphetamine abusers. *American Journal of Psychiatry* **158**, 377–382.

Wang, G. J., Volkow, N. D., Roque, C. T., Cestaro, V. L., Hitzemann, R. J., Cantos, E. L., Levy, A. V., and Dhawan, A. P. (1993). Functional importance of ventricular enlargement and cortical atrophy in healthy subjects and alcoholics as assessed with PET, MR imaging, and neuropsychologic testing. *Radiology* **186**, 59–65.

Wang, G. J., Volkow, N. D., Franceschi, D., Fowler, J. S., Thanos, P. K., Scherbaum, N., Pappas, N., Wong, C. T., Hitzemann, R. J., and Felder, C. A. (2000). Regional brain metabolism during alcohol intoxication. *Alcoholism: Clinical and Experimental Research* **24**, 822–829.

Wang, G. J., Volkow, N. D., Fowler, J. S., Franceschi, D., Wong, C. T., Pappas, N. R., Netusil, N., Zhu, W., Felder, C., and Ma, Y. (2003). Alcohol intoxication induces greater reductions in brain metabolism in male than in female subjects. *Alcoholism: Clinical and Experimental Research* **27**, 909–917.

Wikler, A. (1948). Recent progress in research on the neurophysiologic basis of morphine addiction. *American Journal of Psychiatry* **105**, 329–338.

Wilson, J. M., Kalasinsky, K. S., Levey, A. I., Bergeron, C., Reiber, G., Anthony, R. M., Schmunk, G. A., Shannak, K., Haycock, J. W., and Kish, S. J. (1996). Striatal dopamine nerve terminal markers in human, chronic methamphetamine users. *Nature Medicine* **2**, 699–703.

Wilson, W., Mathew, R., Turkington, T., Hawk, T., Coleman, R. E., and Provenzale, J. (2000). Brain morphological changes and early marijuana use: a magnetic resonance and positron emission tomography study. *Journal of Addictive Diseases* **19**, 1–22.

Zald, D. H., Lee, J. T., Fluegel, K. W., and Pardo, J. V. (1998). Aversive gustatory stimulation activates limbic circuits in humans. *Brain* **121**, 1143–1154.

Zilles, K., and Wree, A. (1995). Cortex: area-1 and laminar structure. *In The Rat Nervous System* (G. Paxinos, Ed.), pp. 649–680. Academic Press, San Diego.

Zocchi, A., Conti, G., and Orzi, F. (2001). Differential effects of cocaine on local cerebral glucose utilization in the mouse and in the rat. *Neuroscience Letters* **306**, 177–180.

Zubieta, J. K., Gorelick, D. A., Stauffer, R., Ravert, H. T., Dannals, R. F., and Frost, J. J. (1996). Increased mu opioid receptor binding detected by PET in cocaine-dependent men is associated with cocaine craving. *Nature Medicine* **2**, 1225–1229.

Zubieta, J., Lombardi, U., Minoshima, S., Guthrie, S., Ni, L., Ohl, L. E., Koeppe, R. A., and Domino, E. F. (2001). Regional cerebral blood flow effects of nicotine in overnight abstinent smokers. *Biological Psychiatry* **49**, 906–913.

CHAPTER

9

Neurobiological Theories of Addiction

OUTLINE

Neurobiology of Addiction, by George F. Koob and Michel Le Moal.
ISBN – 13: 978-0-12-419239-3 ISBN – 10: 0-12-419239-4

INTRODUCTION

The purpose of the present chapter is to outline and summarize the different neurobiological theories of addiction at three levels of analysis—neurocircuitry, cellular, and molecular—independent of any particular class of drugs. There is a historical overlay provided that forms an undercurrent within the reviews herein, and an attempt has been made to represent the latest version of each author perspective. Representative diagrams and summaries from all the original sources are provided for the reader's use. Finally, an attempt is made to provide a generic circuit of addiction with the main frameworks of reward, behavioral output, and craving represented. The generic circuit is not all-inclusive and undoubtedly has left out specific brain regions that may be of importance in addiction. However, it is hoped that it will provide a heuristic framework for future work.

NEUROCIRCUITRY HYPOTHESES OF ADDICTION—DOPAMINE AND REWARD

Mesolimbic Dopamine Reward Hypothesis of Addiction

Wise (1980)

One of the original theories of the actions of drugs of abuse on the brain reward system conceptualized an action on a critical dopaminergic synapse where all reward sites were argued to be afferent to a critical dopaminergic synapse. Two components of the brain reward system were outlined: the high-frequency-sensitive, fast-conducting myelinated fibers of the medial forebrain bundle, and the ventral tegmental dopamine system hypothesized to synapse directly on the dopamine link. At that time, Wise did not specify the mesolimbic dopamine system. Rather, most references to specific dopamine pathways in the discussion were to the ventral tegmentum or the tegmental-striatal projections (Wise, 1980). Later discussions came to focus on the mesolimbic component *per se* (see below).

Data generated to support this theory at the time included evidence that amphetamine and cocaine act directly in the dopamine synapse in the terminal areas of the mesolimbic dopamine system, and evidence that opiates act at the dopamine cell bodies or dopamine synapse. Alcohol, barbiturates, and benzodiazepines were speculated to act via a naloxone-reversible inhibition of noradrenergic function, which disinhibited rather than directly excited the dopamine reward link (Wise, 1980) (**Fig. 9.1**). Dopamine antagonists were shown to block amphetamine and cocaine reward (Yokel and Wise, 1975, 1978; de Wit and Wise, 1977) and food reward (Wise *et al.*, 1978a,b). All major drugs of abuse were shown to facilitate brain stimulation reward and by extrapolation, were linked to their activation of the dopamine system. Wise wrote, 'It is attractive to consider the possibility that opiate reward, like brain stimulation, food, and stimulant reward, ultimately activates a common dopaminergic substrate' (Wise, 1980).

This theory had a profound effect on the neurobiology of drug abuse, and more than any other theory, has guided research in this area. Dopamine receptor blockade was argued to interfere with all rewards (tested up until the publication of Wise, 1980) and the model developed had as a central element, the dopamine neuron and its efferents. In addition, at least one afferent link was added to illustrate how the target neurons for brain stimulation reward and psychomotor stimulant reward were linked, thus yielding the now famous 'two-neuron' theory of reward. Wise did argue that the specific anatomy of the medial forebrain bundle connection was not known (and to some extent

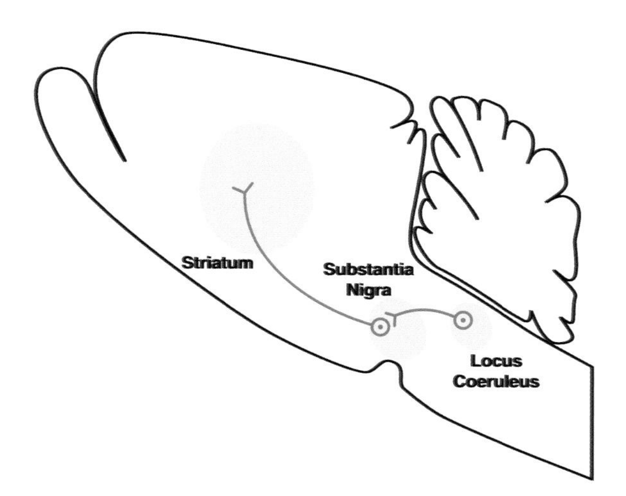

FIGURE 9.1 Early neurocircuitry diagram of drug reward for opiates (Wise, 1980). *'Suggested sites of potential interaction of opiates with brain reward circuitry. Opiate receptor fields are shaded in the region of the striatal dopamine terminal field, the tegmental dopamine cell region, and the region of the locus coeruleus, which is thought to inhibit reward circuitry, perhaps by an inhibitory synapse on the dopamine cells themselves. Opiates inhibit locus coeruleus firing; their actions in the tegmentum and striatum are not yet understood, and may be either pre- or postsynaptic in either region. Thus, opiates may act on, or either afferent or efferent to, the dopamine cells implicated in reward function.'* [Reproduced with permission from Wise, 1980.]

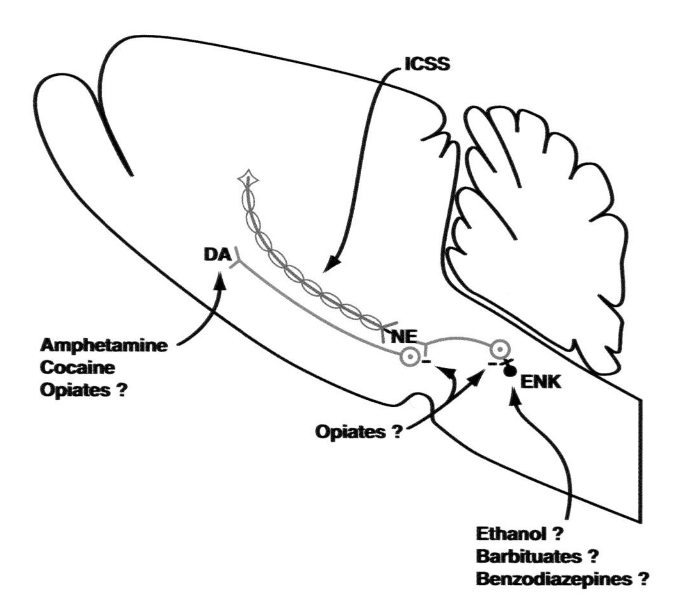

FIGURE 9.2 This diagram illustrates a summary model of current candidates for brain reward circuitry related to drugs of abuse and brain stimulation reward (Wise, 1980). *'The dopamine neuron is thought to be at least one synapse efferent to the directly activated fiber system in brain stimulation reward, which is shown as myelinated. Amphetamine and cocaine are known to act at the dopamine link, presumably in the synapse though perhaps at tegmental autoreceptors. Opiates might act at any level of the diagrammed model. Alcohol, barbiturates, and benzodiazepines are speculated to link through inputs to an opiate receptor to a noradrenergic inhibitory control over the dopamine cells; current evidence for this particular site of anxiolytic action is suggestive at best, but some disinhibitory links with the reward system must be taken as a serious possibility in current models of reward circuitry.'* ENK, enkephalin; DA, dopamine; NE, norepinephrine; ICSS, intracranial self-stimulation. [Reproduced with permission from Wise, 1980.]

is *still* unknown) and that drugs of abuse may enter the system to activate the dopamine system at different sites, a hypothesis that has persisted to this day (Wise, 1980) (**Fig. 9.2**). Thus, Wise wrote, 'While the rewarding and reward-facilitating effects of opiates, benzodiazepines, alcohol, and barbiturates may be mediated at the level of the dopamine neuron, it seems more likely that these agents interact with the dopamine link in reward circuitry through its afferents, either by exciting dopaminergic activity directly or by causing disinhibition' (Wise, 1980).

In summary, the focal point of drug reward—and by extrapolation reward per se*—had become inextricably linked to activation of the midbrain dopamine neurons and remains so today.*

Psychomotor Stimulant Theory of Addiction

Wise and Bozarth (1987)

The thesis of this conceptual framework was based on the premise that addiction was synonymous with operant reinforcement and specified that independent psychomotor stimulant properties were predictors of whether a drug will be reinforcing in an operant situation. In short, 'the crux of the theory is that the reinforcing effects of drugs, and thus their addiction liability, can be predicted from their ability to induce psychomotor activation' (Wise and Bozarth, 1987). Building on the Glickman and Schiff (1967) formulation that both approach behaviors and positive reinforcement can be elicited by activation of the medial forebrain bundle, Wise and Bozarth (1987) argued that all drugs that are positive reinforcers should elicit forward locomotion.

Much evidence was marshaled to support this hypothesis from the domain of behavioral pharmacology. Wise and Bozarth argued that all drugs that have addiction potential, such as amphetamine, cocaine, nicotine, caffeine, opiates, barbiturates, alcohol, benzodiazepines, cannabis, and phencyclidine, have psychomotor stimulant actions, and these psychostimulant actions are due to activation of central dopaminergic systems (Wise and Bozarth, 1987). Wise and Bozarth further argued that even the drugs of abuse with central nervous system depression as a dominant effect have stimulant properties, which are mediated by the same brain mechanism that mediates the psychostimulant, and by extrapolation, the addictive properties of drugs of abuse with psychomotor stimulation as the dominant effect. The increased locomotor activity of all these drugs was hypothesized to be due to activation of the mesolimbic dopamine system.

The final conceptual argument of the psychomotor theory of addiction was that a common biological mechanism played homologous roles in psychomotor stimulation and positive reinforcement. Forward locomotion was proposed to be the unconditioned response to all positive reinforcers and that the medial forebrain bundle mediated such a response. The midbrain dopamine systems were thought to be critical for the reinforcing effects of brain stimulation reward (Fouriezos *et al.*, 1978; Esposito *et al.*, 1979), and the mesolimbic dopamine system also was thought to be critical not only for the psychomotor stimulation produced by all drugs of abuse, but also for their reinforcing actions. Most of the data at the time were generated with psychomotor stimulants and opioids (Yokel and Wise, 1975, 1976; Bozarth and Wise, 1981).

Subsequent studies have largely discredited the psychomotor theory of addiction, but the theory had a major effect in that, more than any other theory, the focus was placed on the mesolimbic dopamine system as being critical for the positive reinforcing effects of drugs of abuse. Evidence against the theory today includes observations that in the rat, very little psychomotor activation, if any, is observed with administration of alcohol and cannabis at doses that are regularly self-administered. In addition, locomotor activation and self-administration produced by opiates, alcohol, and phencyclidine in rats and mice can be observed in the absence of the mesolimbic dopamine system (Pettit *et al.*, 1984; Amalric and Koob, 1985; Vaccarino *et al.*, 1986; Cunningham *et al.*, 1992; Rassnick *et al.*, 1993b; Carlezon and Wise, 1996), suggesting that there are reinforcing effects of these drugs independent of the mesolimbic dopamine system. Despite powerful arguments to suggest that dopamine is only one *part* of reward circuitry, there are persistent arguments for a central role for the mesolimbic dopamine system in the positive reinforcement of drugs, and other arguments that positive reinforcement is directly equated with addiction.

In summary, the psychomotor theory of addiction, in its literal form, is largely discredited. However, it had a major impact of further propagating the theory that the mesolimbic dopamine system was the critical substrate for the acute positive reinforcing effects of drugs of abuse.

Mesolimbic Dopamine Reward Hypothesis of Addiction: Update

Wise (2002)

A more recent version of addiction neurocircuitry with a focus on the mesolimbic dopamine system has moved the focus from forward locomotion to 'neuroadaptations associated with the learning of the drug seeking habit' (Wise, 2002). The argument put forth by Wise (2002) is that the memories of early drug experiences are 'stamped in' by the same reinforcement process that stamps in ordinary habits (nondrug habits) via weaker incentives. The theoretical framework has moved to a neuroadaptational perspective where the brain neuroadaptations are again designated to be within the domain of the mesolimbic dopamine system but are argued to be the 'neuroadaptations of habit formation'.

The mesolimbic dopamine system via its cortical inputs (glutamatergic afferents) and nucleus accumbens output (γ-aminobutyric acid [GABA]ergic efferents) is hypothesized to comprise a major portion of the endogenous circuitry through which the pleasures of the flesh come to shape the habits of an animal (Wise, 2002) (**Fig. 9.3**). This neurocircuit strikingly resembles the neurocircuitry outlined much earlier (Koob, 1992b) (see below), but again, with a critical dopaminergic link in series with reward. Wise argues that the mesolimbic dopamine system is activated trans-synaptically by the normal pleasures of life or is activated directly by drugs of abuse or electrical brain stimulation reward. Relatively little is new in this conceptualization except that the 'forward locomotion' of Wise and Bozarth (1987) is replaced by the augmentation of the 'consolidation—by stamping in—the still active memory traces of the exteroceptive (reward-associated) and interoceptive (response feedback) stimuli that led to the behavior that preceded activation of the system'.

The segue of the function of the mesolimbic dopamine system, from forward locomotion and pure reward to consolidation of memory, appears to depend on the critical observations made in primates by Schultz and co-workers that the mesolimbic dopamine system is more activated by the distant sensory message that predicts reward rather than the receipt of reward (Ljungberg *et al.*, 1992; Schultz *et al.*, 1993, 1997; Mirenowicz and Schultz, 1994; Schultz, 2002). Monkeys trained to respond to fruit juice delivered into their mouth showed activation of midbrain dopamine neurons initially to the juice itself, but with repeated testing the response of the neurons became less associated with the juice and more associated with the stimuli that predicted the presentation of the stimulus

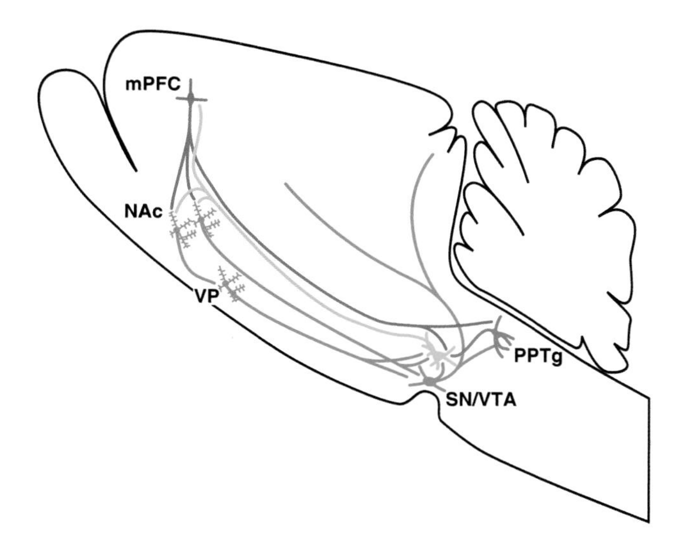

FIGURE 9.3 Selected elements and neurocircuitry of brain reward related to drugs of abuse and brain stimulation reward (Wise, 2002). *'The mesolimbic dopamine system is in gold. Amphetamine and cocaine are rewarding because they act at the dopamine transporter to elevate nucleus accumbens (NAc) dopamine levels; nicotine is rewarding because of actions on nicotinic cholinergic receptors, expressed at both the cell bodies and the terminals of the mesolimbic system, that result in elevated dopamine release in NAc. Dopamine in NAc inhibits the output neurons of NAc. The normal cholinergic input to these receptors in the VTA is from the pedunculo-pontine tegmental nucleus (PPTg) and the latero-dorsal pontine tegmental nucleus; these nuclei send branching projections to several basal forebrain targets (not shown). Rewarding electrical stimulation of the lateral hypothalamus is thought to be rewarding because it activates fibers to PPTg. The excitatory amino acid (glutamate) projections of medial prefrontal cortex (mPFC) are in blue. Projections from this and other cortical areas that receive mesolimbic dopamine input (amygdala, hippocampus) also project to NAc; amygdala also projects to the substantia nigra and ventral tegmental area (SN/VTA). Phencyclidine is rewarding because it blocks NMDA-type glutamate receptors in NAc and mPFC. Blockade of NMDA receptors in NAc reduces the excitatory input to the GABAergic output neurons. Electrical stimulation of mPFC is rewarding because it causes glutamate release in VTA and dopamine release in NAc. Two subsets of GABAergic projection neurons exit NAc; one projects to the ventral pallidum (VP) and the other to the SN/VTA. GABAergic neurons in VP also project to SN/VTA. Most of the GABAergic projection to SN synapses again on GABAergic neurons; these, in turn, project to the pedunculo-pontine tegmental nucleus, the deep layers of the superior colliculus, and the dorsomedial thalamus. Heroin and morphine have two rewarding actions: inhibition of GABAergic cells that normally hold the mesolimbic dopamine system under inhibitory control (thus morphine disinhibits the dopamine system) and inhibition of output neurons in NAc. Alcohol and cannabis act by unknown mechanisms to increase the firing of the mesolimbic dopamine system and are apparently rewarding for that reason. The habit-forming effects of barbiturates and benzodiazepines appear to be triggered at one or more of the GABAergic links in the circuitry, not necessarily through feedback links to the dopamine system. Caffeine appears to be rewarding through some independent circuitry.'* [Reproduced with permission from Wise, 2002.]

and could be reinstated by presentation of a novel incentive such as a slice of apple (Ljungberg *et al.*, 1992). The authors concluded that these data provided 'evidence for the involvement of dopamine neurons in arousing, motivational, and behavioral activating processes that determine behavioral reactivity *without* encoding specific information about the behavioral reaction' (Ljungberg *et al.*, 1992). Wise concluded that these data suggest that it is the receipt of reward predictors (promise of reward) that produces the most arousal, and thus these reward predictors are conditioned rewards, not primary rewards and thus are rewards only because of previous learning. The activation of the midbrain dopamine system serves to establish the response habits that are followed by its activation caused by either the normal pleasures of life or directly by intravenous drugs, or electrical brain stimulation (i.e., a form of consolidation, stamping in of the still-active memory traces of the stimuli that led to the behavior that preceded activation of the system). This neo-dopamine reward theory strikingly resembles the argument put forth years earlier of a role for midbrain dopamine systems in incentive motivation and the activation associated with the presentation of incentives (Koob, 1992a; Salamone, 1992, 1994) with the addition of a new role for dopamine to 'stamp in' the memory of the association between previously neutral stimuli and the incentive (see Ranje and Ungerstedt, 1977, for an earlier version of the dopamine learning hypothesis).

In summary, from the perspective of drug abuse, Wise has moved his dopamine theory of primary drug reward to a more important role for midbrain dopamine to establish response habits that are followed by its activation caused by drugs of abuse. As such, Wise has moved the role of dopamine more from a primary reward function to the preoccupation/anticipation or 'craving' stage.

NEUROCIRCUITRY THEORIES OF ADDICTION—EXECUTIVE FUNCTION

Motive Circuits: Prefrontal Cortex/Ventral Striatal Hypotheses of Addiction

Jentsch and Taylor (1999)

The thesis of this conceptual framework is that regions of the frontal cortex are involved in inhibitory response control that are directly affected by long-term exposure to drugs of abuse. The resulting frontal cortical cognitive dysfunction produces an inability to inhibit unconditioned or conditioned responses elicited by drugs (Jentsch and Taylor, 1999). Drug-seeking behavior is hypothesized to be due to two related phenomena: increased incentive motivational qualities of the drug and drug-associated stimuli due to limbic/amygdalar dysfunction, and impaired inhibitory control due to frontal cortical dysfunction.

The authors postulated that drugs and drug-associated stimuli rely on dopaminergic function in the nucleus accumbens, particularly the shell of the nucleus accumbens, to modulate behavioral output. They further postulated that the amygdala and prefrontal cortex contribute to learning about the associations between drugs and external and internal cues and also thus may contribute to impulsivity (Jentsch and Taylor, 1999) (**Fig. 9.4**). The amygdala is linked to the control of behavior by reward-related stimuli in that the amygdala has a role in mediating incentive learning and in mediating the incentive value of conditioned stimuli. Excitotoxic lesions of the central nucleus of the amygdala block amphetamine-induced potentiation of conditioned reward (Robledo *et al.*, 1996), blunt the suppression in behavior produced by a fear stimulus (Killcross *et al.*, 1997), and also prevent the development of autonomic responses to primary or secondary conditioned stimuli (Kapp *et al.*, 1979; Gentile *et al.*, 1986; Iwata *et al.*, 1986; LeDoux *et al.*, 1988). In contrast, excitotoxic lesions of the basolateral amygdala produce impairments in the ability of stimuli to affect instrumental responding (Killcross *et al.*, 1997). Excitotoxic lesions of the basolateral amygdala but not the central nucleus of the amygdala block cocaine-seeking behavior under a second-order schedule of reinforcement (Whitelaw *et al.*, 1996) and also block cue-induced reinstatement of drug responding in rats (Meil and See, 1997) (**Table 9.1**). However, excitotoxic lesions of the basolateral amygdala do *not* block acquisition of heroin-seeking behavior under a second-order schedule (Alderson *et al.*, 2000).

Many of these effects of drugs of abuse in the amygdala were hypothesized by the authors to be facilitated

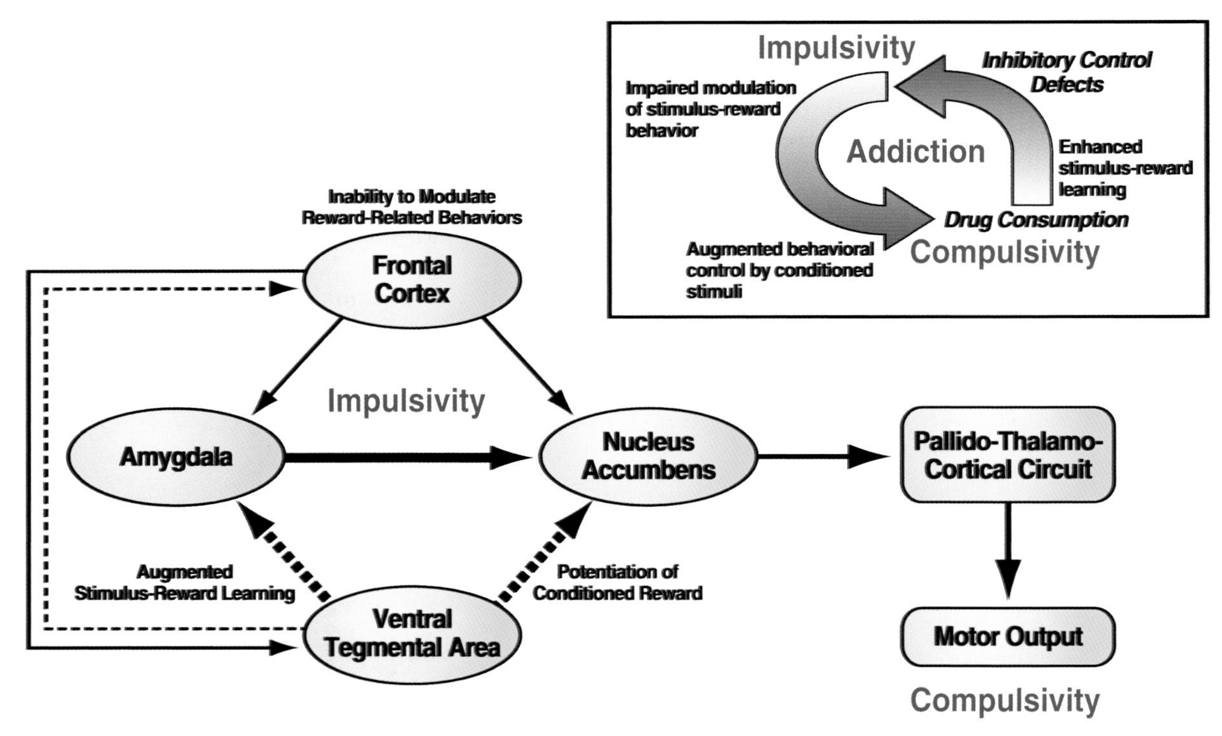

FIGURE 9.4 Model for the functional synergism between augmented conditioned reward and deficits in the ability to modulate reward-related behavior at the cognitive level in drug addiction (Jentsch and Taylor, 1999). *'Increased stimulus-reward learning and conditioned reward, produced by subcortical dopaminergic hyperactivity (bold dashed line arrows), may result in an augmented impulse to seek drugs. Drug-induced impulsivity and cue-elicited drug-seeking may be mutually reinforcing, as repeated drug consumption may progressively augment impulsivity, resulting in greater susceptibility to subsequent relapse. Moreover, the inhibitory control functions of the frontostriatal system that act to modulate reward-related behaviors likewise may be impaired with chronic drug use due to altered frontal cortical dopaminergic function (thin dashed-line arrow) and impaired frontal cortical modulation of subcortical systems by descending corticostriatal projections (thin single-line arrows). [The] theoretical model (see insert to figure) predicts that compulsive drug-seeking behavior may be seen as a functional synergism between augmented conditioned reward and deficits in the ability to modulate reward-related behaviors at a cognitive level.'* [Reproduced with permission from Jentsch and Taylor, 1999.]

TABLE 9.1 **Behavioral Effects Relevant to Drug of Abuse Action of Lesions of Basolateral and Central Nuclei of the Amygdala**

Functional Effect	Reference
Central nucleus of the amygdala	
↓ Amphetamine-induced potentiation of conditioned reward	Robledo *et al.*, 1996; Killcross *et al.*, 1997
↓ Naloxone-precipitated withdrawal-induced place aversion	Watanabe *et al.*, 2002
↓ Morphine antinociception	Manning and Mayer, 1995; Manning, 1998
↓ Anxiety-like effects of benzodiazepines	Yadin *et al.*, 1991
Basolateral nucleus of the amygdala	
↓ Cocaine-seeking behavior under a second-order schedule of reinforcement	Whitelaw *et al.*, 1996
↓ Cue-induced reinstatement of drug responding in rats	Meil and See, 1997
↓ Modulation of intracranial self-stimulation thresholds by drug-associated cues	Hayes and Gardner, 2004
↓ Cue-induced and cocaine-induced reinstatement in a discriminative stimulus task	Yun and Fields, 2003
↓ Acquisition and extinction of cocaine conditioned place preference	Fuchs *et al.*, 2002
↓ Conditioned withdrawal from opiates	Schulteis *et al.*, 2000
↓ Self-administration of ethanol	Moller *et al.*, 1997

by the activation of dopaminergic substrates within the amygdala. For example, intracerebral administration of amphetamine into the amygdala facilitates the acquisition of stimulus-reward associations (Hitchcott *et al.*, 1997), and psychostimulants in general facilitate memory consolidation through actions in the amygdala (Cestari *et al.*, 1996). Thus, Jentsch and Taylor hypothesized that during drug self-administration, the synaptic release of dopamine increases in the amygdala and helps lay down the associations between the rewarding qualities of the drug and exteroceptive stimuli (Jentsch and Taylor, 1999). In the context of drug addiction, an increase in dopamine release within the nucleus accumbens produced by repeated administration of a drug of abuse results in increased responding for a conditioned reinforcer, and the acquisition of stimulus-reward associations is facilitated by drug-induced neuroadaptations within the amygdala (Jentsch and Taylor, 1999) (**Fig. 9.4**).

However, Jentsch and Taylor (1999) assign an equally important role, or a possibly primary role, to the frontal cortex in the impulsivity of drug dependence. They argued that internal motivational states to seek food, water, and sex and other primary reinforcers are regulated by an active inhibitory control mechanism in the frontal striatal systems. This inhibitory control mechanism would transiently suppress rapid conditioned responses and reflexes so that slower cognitive processes can guide behavior. Lesions of the frontal cortex can lead to marked cognitive impairments including disinhibition (Milner, 1982) and a preferential response for immediate small rewards over delayed rewards where calculation of future outcome is not possible (Damasio, 1996). Lesions of the dorsolateral prefrontal, lateral orbitofrontal, or ventromedial frontal cortex resulted in increased perseveration and deficits in inhibition (Iversen and Mishkin, 1970; Ridley *et al.*, 1993). In addition, there is strong evidence for a role for dopamine mediating some of these same cognitive functions in the frontal cortex (Blanc *et al.*, 1980; Simon *et al.*, 1980; Herman *et al.*, 1982; Sawaguchi and Goldman-Rakic, 1991). Thus, an impairment in inhibitory control combined with the progressive enhancement of the conditioned reinforcing effects of drug-associated stimuli would represent a state where reward-related stimuli could dominate responding.

Chronic high dose administration of amphetamine, cocaine, and even cannabinoids can lead to reductions in basal frontal cortex dopamine transmission (Ricaurte *et al.*, 1980; Robinson and Becker, 1986; Karoum *et al.*, 1990; Jentsch *et al.*, 1998). Similar effects have been observed with phencyclidine and have shown a correlation between performance impairments in cognitive function and dopaminergic hypofunction in the dorsolateral, prefrontal, and ventromedial frontal cortex (Jentsch *et al.*, 1999). Also, consistent with these results, one study showed tolerance to the footshock-enhanced release of dopamine into the medial prefrontal cortex one week after chronic administration of cocaine (Sorg and Kalivas, 1993). However, others have observed increased dopamine utilization in the frontal cortex following chronic administration of amphetamine (Robinson *et al.*, 1985) and an increase in footshock-enhanced release of dopamine in the medial prefrontal cortex one week after chronic administration of D-amphetamine (Hamamura and Fibiger, 1993). The dysregulation of dopaminergic neurotransmission within the prefrontal cortex was hypothesized to underlie the loss of inhibition of reward-seeking behavior and the increased susceptibility to drug-induced relapse (Jentsch and Taylor, 1999). Finally, Jentsch and Taylor (1999) hypothesized that the inhibitory modulation of reward-seeking behavior may depend critically upon the corticostriatal projections from the medial frontal cortex to the caudate nucleus and nucleus accumbens core and shell. Dysfunction of the frontal cortex or hypofunction of dopamine activity in the frontal cortex can activate subcortical dopamine systems (Louilot *et al.*, 1989; Piazza *et al.*, 1991a; for review, see Le Moal and Simon, 1991). Decreased dopamine in the prefrontal cortex induced locomotor activation due to an activation of dopamine in the nucleus accumbens (Tassin *et al.*, 1978). This functional interaction between the prefrontal cortex and nucleus accumbens is highlighted by the observation that animals more vulnerable to acquiring intravenous drug self-administration showed reduced dopaminergic activity in the prefrontal cortex (Piazza *et al.*, 1991a). Such hypoactivity also may contribute to locomotor sensitization (Banks and Gratton, 1995), and by extrapolation, presumably compulsive drug self-administration.

In summary, in the Jentsch and Taylor model, compulsive drug-seeking behavior is a functional synergism between deficits in the frontal striatal system driven by a cortical dopamine hypofunction and augmented conditioned reward, presumably driven by subcortical (nucleus accumbens and amygdala) dopamine activation. The primary focus of this hypothesis is on changes in function in the prefrontal cortex which may underlie the loss of inhibition of reward-seeking behavior.

Frontal Cortex Dysfunction, Cognitive Performance, and Executive Function: Disruption of Frontocerebellar Circuitry and Function in Alcoholism

Pfefferbaum, Sullivan, Mathalon, and Lim (1997)
Sullivan, Harding, Pentney, Dlugos, Martin, Parks, Desmond, Chen, Pryor, De Rosa, and Pfefferbaum (2003)

Classical neuropsychological behaviors that are typical of frontal lobe dysfunction characterize alcoholics and include impaired judgment, blunted affect, poor insight, social withdrawal, reduced motivation, and attentional deficits (Parsons *et al.*, 1987; for reviews, see Oscar-Berman and Hutner, 1993; Sullivan *et al.*, 2000). Neuroimaging and neuropathological studies have provided significant evidence for alcohol-induced abnormalities in frontal lobes and cerebellum that are of particular relevance to these alcohol-related impairments in cognitive function (Harper and Kril, 1989; Pfefferbaum *et al.*, 1997; Kubota *et al.*, 2001). More formal testing has raised the hypothesis that the frontal lobe component of these deficits can be impairments in executive function and short-term or working memory (Oscar-Berman and Hutner 1993; Parsons, 1993; Sullivan *et al.*, 1993). Postmortem studies (Harper and Kril, 1989) and *in vivo* magnetic resonance imaging (MRI) have shown frontal cortex abnormalities in alcoholism (Pfefferbaum *et al.*, 1997) (**Fig. 9.5**). Deficits in balance and gait in alcoholics have been linked to abnormalities in cerebellar function (Sullivan *et al.*, 2000b, 2002), and the cerebellum may be involved in some classic frontal lobe functions, suggesting the hypothesis of an overall frontocerebellar circuitry deficit in alcoholism (Sullivan *et al.*, 2003).

The link of cognitive impairment from these abnormalities to frontal cortex were predicted by neuropsychological studies revealing alcohol-dependent changes in frontal lobe function (Oscar-Berman and Hutner, 1993; Sullivan *et al.*, 2000a,b). These performance deficits correlated with resting frontal lobe metabolism (Noel *et al.*, 2001). Perhaps more compelling were the data showing that tasks that produce robust frontal lobe activation may be especially sensitive to alcohol-related changes in brain function. Different patterns of frontal lobe activation were observed in alcoholics compared to controls while performing visual attentional and working memory tasks (Pfefferbaum *et al.*, 1997). Similarly, in a function MRI (fMRI) study, alcoholic and non-alcoholic subjects showed no difference in performance of a working memory task, but fMRI activation was greater in the alcoholics for the high load condition suggesting that alcoholics may require more extensive activation of frontal lobe function to maintain normal performance in such cognitive tasks (Fama *et al.*, 2004).

Brain Circuitry for Addiction from Brain Imaging Studies

Volkow, Fowler, and Wang (2003)

Brain imaging studies of drug addiction have largely explored human addicts using imaging technologies, such as positron emission tomography (PET) and fMRI. PET imaging is based on the use of radiotracers labeled with isotopes, which can measure in the brain at very low concentrations, and as such, PET can measure labeled compounds that selectively bind to receptors, transporters, or enzymes (see *Imaging* chapter). The fMRI is based on the changes in magnetic properties in brain tissue. During brain activation, there is an excess of arterial blood delivered to a given region and a change in the ratio of deoxyhemoglobin to oxyhemoglobin which have different magnetic properties (see *Imaging* chapter).

Based on the results of brain imaging studies, Volkow and associates have proposed four circuits that are

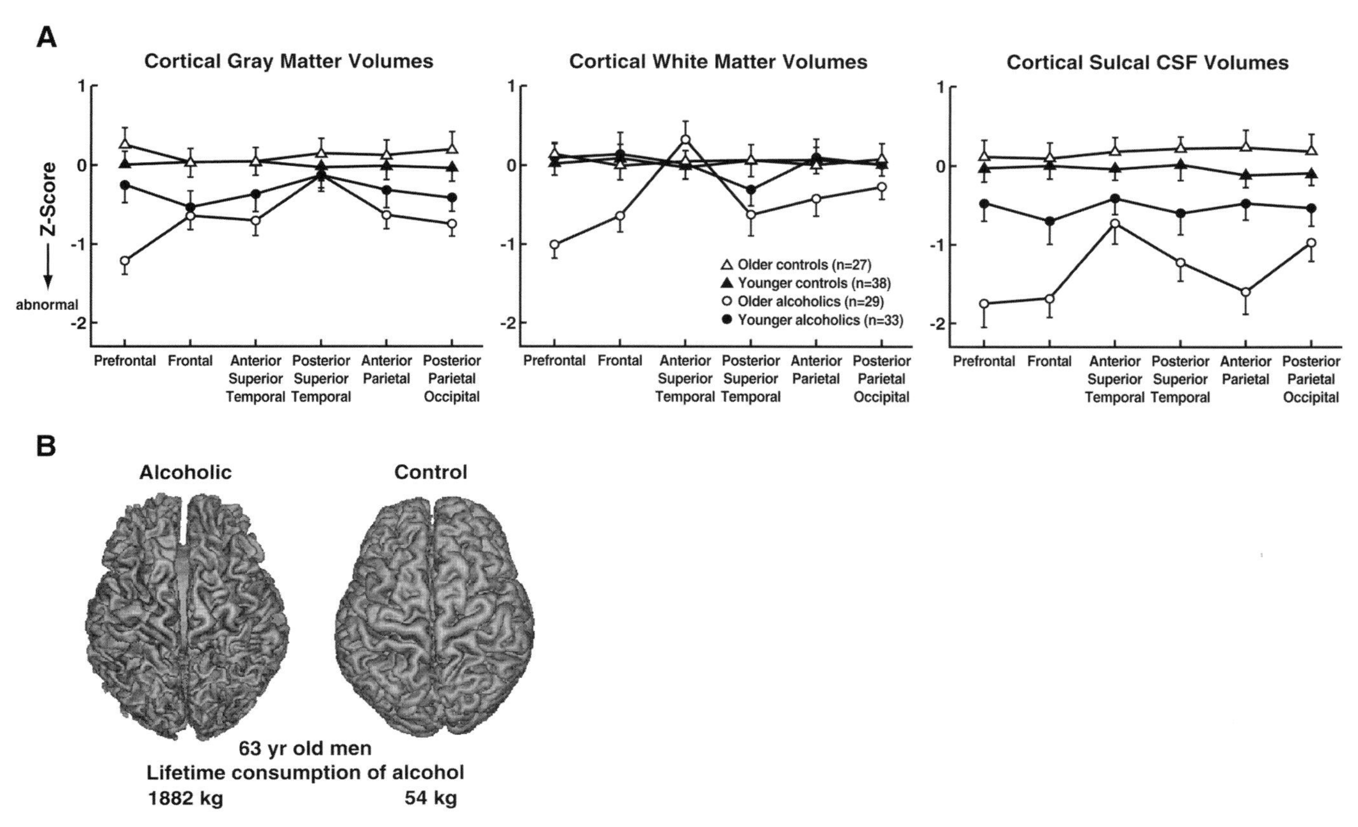

FIGURE 9.5 **(A)** Volume profiles of (left) cortical gray matter, (middle) cortical white matter, and (right) cortical sulcal cerebrospinal fluid of two control and two alcoholic groups. For all measures, lower scores reflect greater abnormality. The two alcoholic groups had distinctly different profiles on three magnetic resonance imaging measures. Only the older alcoholic group had a cortical gray matter deficit in the prefronal region. The white matter deficit of the older alcoholic group was greatest in the prefrontal, frontal, and posterior superior temporal regions. Both alcoholic groups had widespread regional sulcal enlargement. The enlargement of the older group was especially pronounced in the prefrontal, frontal, and parietal regions. **(B)** Three-dimensional renderings of the brains of two men at age 63. To produce these images, the scalp, skull, and cerebrospinal fluid were digitally peeled away to reveal the grooves (i.e., sulci) and ridges (i.e., gyri) marking each brain's external surface. These images of living brains are comparable to postmortem photographs of brains for pathological study. The control brain is from a healthy social drinker who has an estimated lifetime consumption of 54 kg of pure alcohol. The alcoholic brain is from a male alcoholic who reported a history of heavy drinking over the past 32 years, with an estimated lifetime consumption of 1882 kg of pure alcohol. The shriveled appearance (i.e., wider sulci and narrower gyri) of the alcoholic brain sharply contrasts with the control brain's relatively plump appearance (i.e., well-filled gyri and narrower sulci) and reflects the tissue shrinkage associated with heavy drinking. Tissue shrinkage has widened the interhemispheric fissure in the alcoholic brain, exposing the bundle of fibers connecting the two hemispheres (i.e., corpus callosum). In contrast, the corpus callosum can barely be seen in the control brain. [Taken with permission from Pfefferbaum et al., 1995.]

disrupted in drug addiction: (1) reward, localized to the nucleus accumbens and ventral pallidum, (2) motivation/drive, localized to the orbitofrontal cortex and subcallosal cortex, (3) memory and learning, localized to the amygdala and hippocampus, and (4) control, localized to the prefrontal cortex and anterior cingulate gyrus (Volkow *et al.*, 2003) (**Fig. 9.6**). In the reward circuit, imaging studies in subjects who are both drug abusers and subjects who are not drug abusers have shown a normal acute response of the dopamine system to drugs of abuse. Both groups showed increases in the extracellular concentrations of dopamine in the striatum, and the subjects who had the greatest increases in dopamine were the ones who experienced the subjective effects associated with 'euphoria' most intensely (Laruelle *et al.*, 1995; Volkow *et al.*, 1999b; Drevets *et al.*, 2001). The route of administration providing the fastest drug uptake produced the greatest subjective effects (Volkow *et al.*, 1999b, 2001). However, subjects who became drug abusers or became drug addicted showed long-lasting decreases in the number of dopamine D_2 receptors in the striatum compared to controls, and cocaine abusers have reduced dopamine release in response to a pharmacological challenge (Volkow *et al.*, 1997, 2002). The authors concluded that the decrease in D_2 receptors and decreases in dopamine system activity would result in decreased sensitivity of reward circuits to

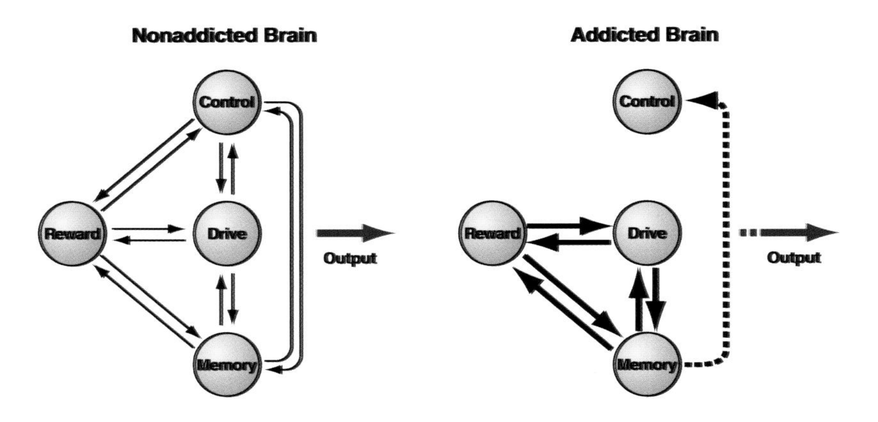

FIGURE 9.6 Model proposing a network of four circuits involved with addiction, reward, motivation/drive, memory and control (Volkow *et al.*, 2003). *'These circuits work together and change with experience. Each is linked to an important concept: saliency (reward), internal state (motivation/drive), learned associations (memory), and conflict resolution (control). During addiction, the enhanced value of the drug in the reward, motivation, and memory circuits overcomes the inhibitory control exerted by the prefrontal cortex, thereby favoring a positive-feedback loop initiated by the consumption of the drug and perpetuated by the enhanced activation of the motivation/drive and memory circuits.'* [Reproduced with permission from Volkow *et al.*, 2003.]

stimulation by natural reinforcers (Volkow *et al.*, 2003), which would put subjects at greater risk for seeking drug stimulation to temporarily activate these reward circuits (Volkow *et al.*, 2002). The hypothesis of decreased reward processing in addiction is supported by imaging studies in opioid-, cocaine-, and tobacco-dependent subjects (Volkow *et al.*, 1997, 2000; Martin-Solch *et al.*, 2001).

Imaging studies also have shown disruption of the motivation/drive circuit in addiction (Volkow *et al.*, 2003). The orbitofrontal cortex is hyperactive in active cocaine abusers (Volkow *et al.*, 1991), during intoxication (Volkow *et al.*, 1999a), during presentation of cocaine-associated cues (Grant *et al.*, 1996; Volkow *et al.*, 1999a; Wang *et al.*, 1999), and during presentation of cigarette-associated cues (Brody *et al.*, 2002), but hypoactive during withdrawal in addicted subjects (Volkow *et al.*, 1992b; Adinoff *et al.*, 2001). Since increased orbitofrontal activation is associated with obsessive–compulsive disorder, activation of this structure may contribute to the compulsive nature of drug intake (Volkow *et al.*, 2003). Amygdala/prefrontal (anterior cingulate) systems may be involved in conditioning or how neutral stimuli paired with drug-taking acquire reinforcing properties and motivational salience (see animal studies below). Structures, such as the amygdala and anterior cingulate have been shown to be activated during intoxication and during craving induced by drug exposure or videos (Grant *et al.*, 1996; Childress *et al.*, 1999; Kilts *et al.*, 2001).

In summary, Volkow and colleagues hypothesized a control circuit in drug addiction based on one of the most robust findings from imaging studies of drug addicts, showing abnormalities in the prefrontal cortex and anterior cingulate gyrus. Disruption of the prefrontal cortex would impair the inhibitory control and decision making that leads drug addicts to choose immediate rewards over delayed reward and could contribute to the loss of control over intake (Goldstein and Volkow, 2002).

Impaired Response Inhibition and Salience Attribution Syndrome of Addiction

Goldstein and Volkow (2002)
Volkow, Fowler, and Wang (2003)

In an elaboration of the role of the frontal cortex and orbitofrontal cortex in the motivational effects of drug addiction, Goldstein and Volkow (2002) have conceptualized drug addiction as a syndrome of impaired response inhibition and salience attribution (I-RISA). The authors argued that four clusters of behaviors are interconnected in a positive feedback loop (drug reinforcement, craving, bingeing, and withdrawal), and activity of the prefrontal circuits and the subcortical reward pathway are differentially represented at each stage (Goldstein and Volkow, 2002) (**Fig. 9.7**). During the drug intoxication stage, strong positive and negative reinforcement effects are strengthened through repeated self-administration of the drug, giving more attribution to incentive salience of the drug at the expense of less powerful reinforcers. During the drug intoxication stage, the prefrontal and orbitofrontal cortex are activated in human drug addicts challenged with drugs (Volkow *et al.*, 1988, 1996; Ingvar *et al.*, 1998; Nakamura *et al.*, 2000). During the relapse and bingeing stage, high levels of brain activation also have been observed in the prefrontal cortex and orbitofrontal cortex. Higher levels of brain activation as measured by glucose metabolism or cerebral blood flow have been seen in frontolimbic areas in drug abusers exposed to stimuli associated with drugs (Grant *et al.*, 1996; Maas *et al.*, 1998; Childress *et al.*, 1999; Wang *et al.*, 1999; Garavan *et al.*, 2000; Wexler *et al.*, 2001).

However, during the relapse and bingeing stage, the response inhibition system is hypothesized to be impaired because of impaired salience attribution leading to response disinhibition or impulsive responding, again to immediately salient drug-related rewards

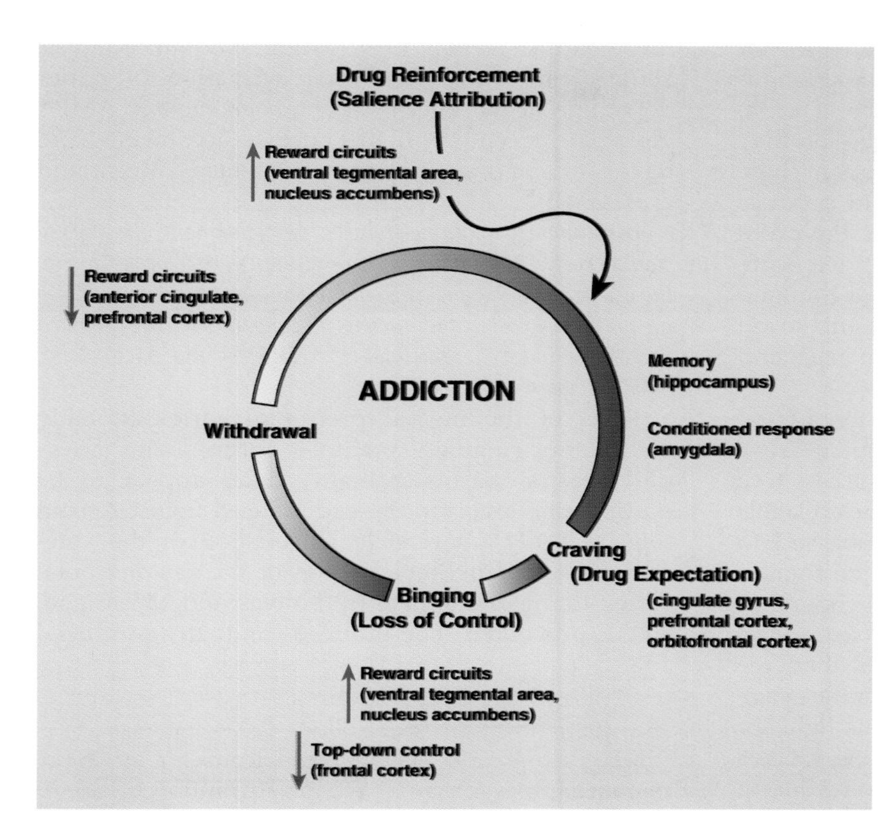

FIGURE 9.7 Integrative model of brain and behavior describing the Impaired Response Inhibition and Salience Attribution (I-RISA) syndrome of drug addiction (Goldstein and Volkow, 2002). *'The mesolimbic dopamine circuit, which includes the nucleus accumbens, amygdala, and hippocampus, has been traditionally associated with the acute reinforcing effects of a drug and with the memory and conditioned responses that have been linked to craving. It is also likely to be involved in the emotional and motivational changes seen in drug abusers during withdrawal. The mesocortical dopamine circuit, which includes the prefrontal cortex, orbitofrontal cortex, and anterior cingulate, is likely to be involved in the conscious experience of drug intoxication, drug incentive salience, drug expectation/craving, and compulsive drug administration. Note the circular nature of this interaction: the attribution of salience to a given stimulus, which is a function of the orbitofrontal cortex, depends on the relative value of a reinforcer compared to simultaneously available reinforcers, which requires knowledge of the strength of the stimulus as a reinforcer, a function of the hippocampus and amygdala. Consumption of the drug in turn will further activate cortical circuits (orbitofrontal cortex and anterior cingulate) in proportion to the dopamine stimulation by favoring the target response and decreasing nontarget-related background activity. The activation of these interacting circuits may be indispensable for maintaining the compulsive drug administration observed during bingeing and to the vicious circle of addiction'.* [Reproduced with permission from Goldstein and Volkow, 2002.]

(Goldstein and Volkow, 2002). During drug withdrawal, particularly protracted abstinence, brain metabolism is lower in the orbitofrontal cortex, frontal cortex, and anterior cingulate gyrus in drug abusers/addicts than in normal controls (Volkow *et al.*, 1992a,b, 1994). Dysthymia is a key symptom of abstinence and is hypothesized to be produced by a lower sensitivity of the brain reward systems induced by adaptational changes in response to repeated activation by drugs of abuse.

The Goldstein and Volkow (2002) and Volkow *et al.*, (2003) models conceptualize addiction as a state that is initiated by taking drugs that are highly effective rewards compared to other stimuli, and this state triggers a series of adaptations in the *reward, motivation/drive, memory*, and *control* circuits of the brain (Volkow *et al.*, 2003) (**Fig. 9.6**). As a result, there is an enhanced and permanent saliency value for the drug, superimposed on a reward deficit for nondrug rewards, and a loss of inhibitory control, favoring the emergence of compulsive drug administration. The authors further argued that their model leads to suggestions for treatment strategies, such as decreasing the rewarding value of drugs while increasing the value of nondrug reinforcers, weakening learned drug associations, and strengthening cognitive function specifically in the frontal control circuit (Volkow *et al.*, 2003).

In summary, Volkow and colleagues conceptualized addiction as a disruption of incentive salience attribution for normal rewards with a redirection of salience to drug rewards and accompanied by deficits in response inhibition and a lower sensitivity of reward function during withdrawal.

NEUROCIRCUITRY THEORIES OF ADDICTION—RELAPSE

Brain Circuitry of Reinstatement of Drug-seeking

Kalivas and McFarland (2003)
Shaham, Shalev, Lu, DeWit, and Stewart (2003)

The thesis of this conceptual framework, based almost exclusively on animal models, is that reinstatement of drug-seeking behavior by the drug itself (drug priming), drug cues, or stressors converge on the medial prefrontal cortex (anterior cingulate cortex) and output through the core of the nucleus accumbens (Kalivas and

McFarland, 2003). A neurochemical subhypothesis is that the effects of such priming stimuli are conveyed by a glutamatergic projection from the anterior cingulate to the nucleus accumbens core, and that as a result, changes in glutamate system neurotransmission may be a key component of vulnerability to relapse. Three distinct types of stimuli are postulated to produce a related interoceptive state that increases the probability of executing drug-seeking behavior: exposure to a pharmacological stimulus that induces a common component of the drug experience, exposure to an environmental stimulus associated with the drug, and exposure to stressors.

The animal models used to explore the neurobiological bases of reinstatement typically utilize a paradigm where rats are allowed limited-access to a drug (usually cocaine) until they have achieved stable responding, and then the animals are subjected to prolonged extinction procedures. Typically, the animals are trained to self-administer intravenous cocaine in daily 2-h sessions (McFarland *et al.*, 2004) (see *Animal Models of Addiction* chapter). The animals then are subjected to extinction where responding on the active lever results in illumination of the stimulus light for 20 s and infusion of saline instead of cocaine. Once a criteria of <10 per cent of the average responding during the maintenance of self-administration is achieved, the animals are tested for reinstatement following passive administration of drug (cocaine-primed reinstatement), cues (discrete and/or contextual cues associated with drug administration), or stressors.

The dorsal prefrontal cortex, core of the nucleus accumbens, and ventral pallidum, but not the basolateral amygdala, appear to play a critical role in drug-induced reinstatement (such as that primed by cocaine) (Grimm and See, 2000; McFarland and Kalivas, 2001) (**Figs. 9.8** and **9.9**). Cocaine-primed reinstatement was not altered by blockade of dopamine receptors in the core of the nucleus accumbens (McFarland and Kalivas, 2001), but was blocked by inactivation of D_1 receptors in the shell of the nucleus accumbens (Anderson *et al.*, 2003) and blockade of dopamine D_1/D_2 receptors in the dorsal prefrontal cortex prevented cocaine-primed reinstatement (McFarland and Kalivas, 2001). Cocaine-primed reinstatement also was blocked by inhibiting ionotropic α-amino-3-hydroxy-5-methyl-4-isoxazole propionic acid (AMPA)/kainate glutamate receptors but not *N*-methyl-D-aspartate (NMDA) glutamate receptors in the core of the nucleus accumbens. In fact, NMDA antagonists themselves can reinstate cocaine-seeking behavior (Cornish and Kalivas, 2000; Park *et al.*, 2002). Both dopamine agonists and glutamate agonists administered directly into the nucleus accumbens provoked reinstatement in rats (Cornish *et al.*, 1999; McFarland and Kalivas, 2001). Thus, it appears that activation of either dopamine in the nucleus accumbens shell or AMPA glutamate receptors in the nucleus accumbens core have a role in drug-primed reinstatement (McFarland *et al.*, 2003).

In summary, cocaine-induced reinstatement involves a frontal cortex (cingulate cortex) glutamatergic projection to the nucleus accumbens core, and dopamine projections both to the prefrontal cortex and nucleus accumbens shell are involved in cocaine-induced reinstatement (Kalivas and McFarland, 2003; Shaham et al.*, 2003).*

Parts of the medial prefrontal cortex (notably the anterior cingulate/prelimbic cortices) and amygdala (rostral basolateral amygdala) appear to be particularly involved in cue-induced reinstatement (**Figs. 9.9–9.11**). Reversible inactivation of the anterior cingulate/prelimbic region of the rat prefrontal cortex (termed cingulate by Kalivas and McFarland, 2003) prevented cue-induced and drug-induced

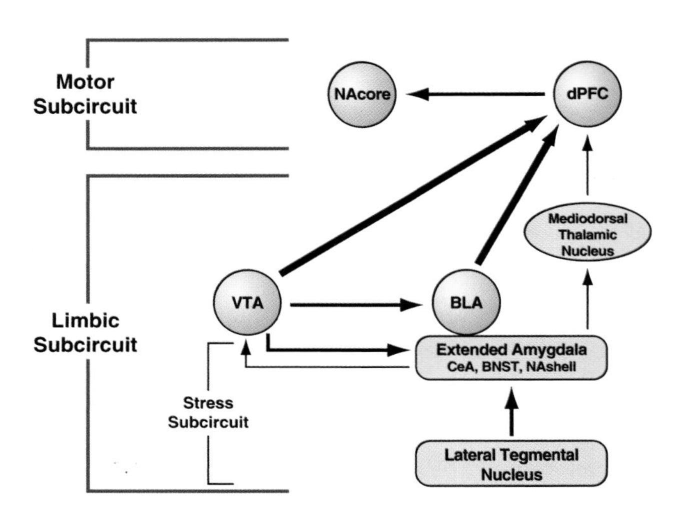

FIGURE 9.8 A schematic illustration of the cocaine reinstatement circuit according to Kalivas and McFarland (2003). *'Schematic illustration of the hypothetical circuitry involved in primed reinstatement. It is proposed that the motor subcircuit containing the projection from the anterior cingulate/prelimbic region of the dorsal prefrontal cortex (dPFC) to the core of the nucleus accumbens (NAcore) corresponds to a final common pathway for all priming stimuli. The remainder of the circuit is termed the limbic subcircuit and contains the connected nuclei shown to be distinct to a class of priming stimulus (e.g., cue, stress, and drug). This includes the VTA and basolateral amygdala (BLA) for drug and cue priming, and adrenergic projections from the lateral tegmental nucleus (LTN) to the extended amygdala for stress-primed reinstatement. The lighter lines correspond to anatomical connections that might theoretically link stress priming to the dPFC, but for which no evidence currently exists. BNST bed nucleus of the stria terminalis, CeA central amygdala nucleus, NAshell shell of the nucleus accumbens.'* [Reproduced with permission from Kalivas and McFarland, 2003.]

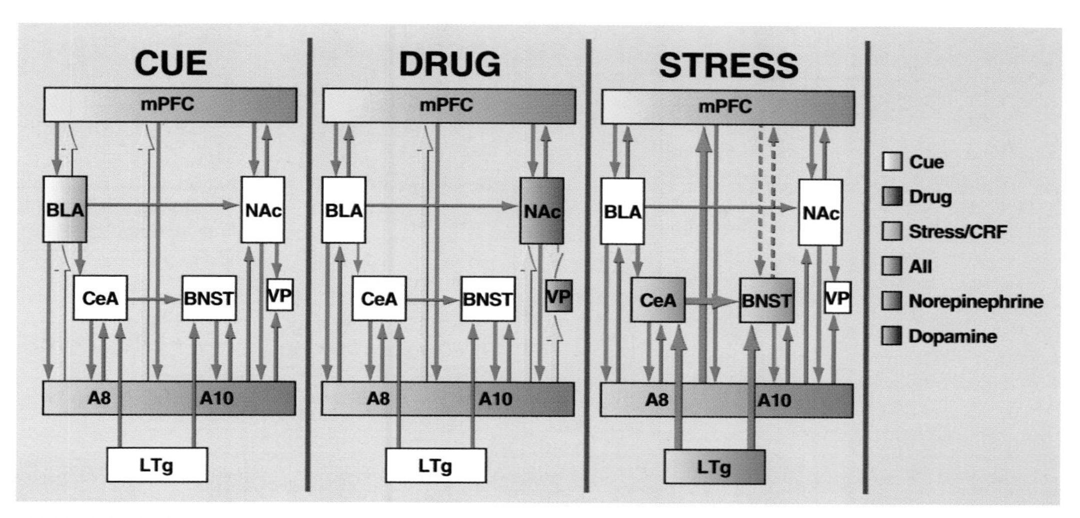

FIGURE 9.9 Hypothetical brain circuits critical for the induction of reinstatement of drug-seeking behavior by cues, drugs, and footshock stressor (Shaham *et al.*, 2003). *'Most of the data described [in this figure] is from studies with cocaine-trained rats. In each case, fat arrows indicate pathways that may be involved in reinstatement; thin arrows indicate some of the existing direct anatomical connections and dashed arrows indicate indirect connections. Cue-induced reinstatement: the effect of cocaine cues on reinstatement is blocked by systemic and intra-BLA injections of D_1-like receptor antagonists and by reversible inactivation of the BLA and the dorsal mPFC. Drug-induced reinstatement: the effect of cocaine priming on reinstatement is blocked by systemic, intra-mPFC and possibly by intra-NAc shell injections of DA receptor antagonists. This effect of cocaine priming is also blocked by reversible inactivation of the dorsal mPFC, NAc core, VP and the VTA. In contrast, the effect of cocaine on reinstatement is mimicked by systemic injections of D_2-like receptor agonists, by activation of VTA DA neurons with morphine or excitatory amino acid agonists, and by infusions of cocaine, DA, and amphetamine into the NAc and mPFC. Footshock stress-induced reinstatement: the effect of footshock on reinstatement is blocked by systemic and intra-BNST, but not intra-amygdala, injections of a CRF receptor antagonist. This effect of footshock stress is also blocked by systemic and ventricular, but not intra-LC, injections of α_2-adrenoceptor agonists (which reduce NA cell firing and release), by intra-BNST and intra-amygdala infusions of α_2-adrenoceptor antagonists, and by 6-OHDA lesions of the VNAB projections arising from the LTg. In addition, the effect of footshock on reinstatement is attenuated by disrupting the CRF projections from the CeA to the BNST by infusions of a CRF antagonist into the BNST and TTX into the contralateral CeA. 6-OHDA 6-hydroxydopamine, A8 and A10 dopamine (DA) cell groups, BLA basolateral amygdala, BNST bed nucleus of the stria terminalis, CeA central amygdala, CRF corticotropin-releasing factor, LC locus coeruleus, LTg noradrenaline (NA) cell groups of the lateral tegmental nuclei, mPFC medial prefrontal cortex, NAc nucleus accumbens, TTX tetrodotoxin, VNAB ventral noradrenergic bundle, VTA ventral tegmental area, VP ventral pallidum.'* [Reproduced with permission from Shaham *et al.*, 2003.]

reinstatement (McFarland and Kalivas, 2001). Inactivation of the rostral parts of the basolateral amygdala also blocked cue-induced reinstatement (Kantak *et al.*, 2002). In addition, inactivation of the ventral tegmental area or nucleus accumbens core reversed cocaine-seeking under a second-order schedule of reinforcement (Di Ciano and Everitt, 2004a), and inactivation of the lateral orbitofrontal cortex also blocked cue-induced reinstatement (Di Ciano and Everitt, 2004a; Fuchs *et al.*, 2004). Blockade of dopamine receptors in the basolateral amygdala blocked cue-induced reinstatement (See *et al.*, 2001) and cocaine-seeking behavior in a second-order schedule (Di Ciano and Everitt, 2004b). Discriminative stimuli that predict availability of cocaine and reinstate cocaine-seeking behavior increased dopamine and glutamate release in the amygdala and nucleus accumbens (Weiss *et al.*, 2000). These same discriminative stimuli increased Fos activation in the basolateral amygdala and medial prefrontal cortex, and Fos activation and reinstatement were reversed by dopamine D_1 receptor antagonism (Weiss *et al.*, 2000; Ciccocioppo *et al.*, 2001). Blockade of ionotropic AMPA/kainate, but not NMDA glutamate receptors in the core of the nucleus accumbens prevented drug-seeking behavior in response to a cocaine-associated cue in a second-order schedule (Di Ciano and Everitt, 2001), and the core of the nucleus accumbens is preferentially innervated by the rostral basolateral amygdala.

In summary, cue-induced reinstatement appears to be dependent on activation of the basolateral amygdala and medial prefrontal cortex (cingulate cortex) via a glutamatergic innervation of the nucleus accumbens core (Kalivas and McFarland, 2003; Shaham et al., *2003). Dopamine*

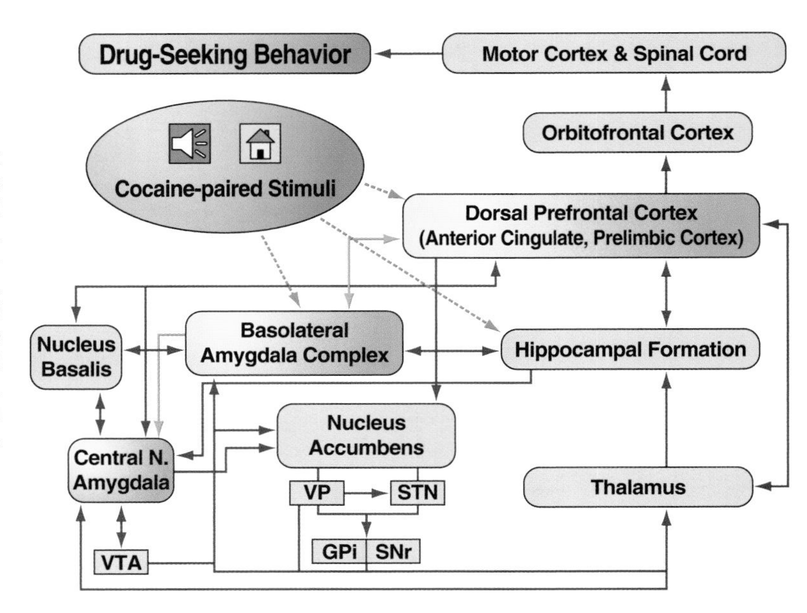

FIGURE 9.10 The conditioned-cued reinstatement of drug-seeking behavior neurocircuit as proposed by See *et al.*, (2003). *'The motivational properties of drug-associated stimuli are processed by a network of brain structures which, based on the results of our lesion and pharmacological inactivation experiments, include the basolateral amygdala complex, central nucleus of the amygdala, anterior cingulate cortex, and prelimbic cortex. Note: A number of nuclei and connections are omitted for the sake of clarity.'* GPi, globus pallidus internal segment; SNr, substantia nigra pars reticulata; STN, subthalamic nucleus; VP, ventral pallidum; VTA ventral tegmental area. [Reproduced with permission from See *et al.*, 2003.]

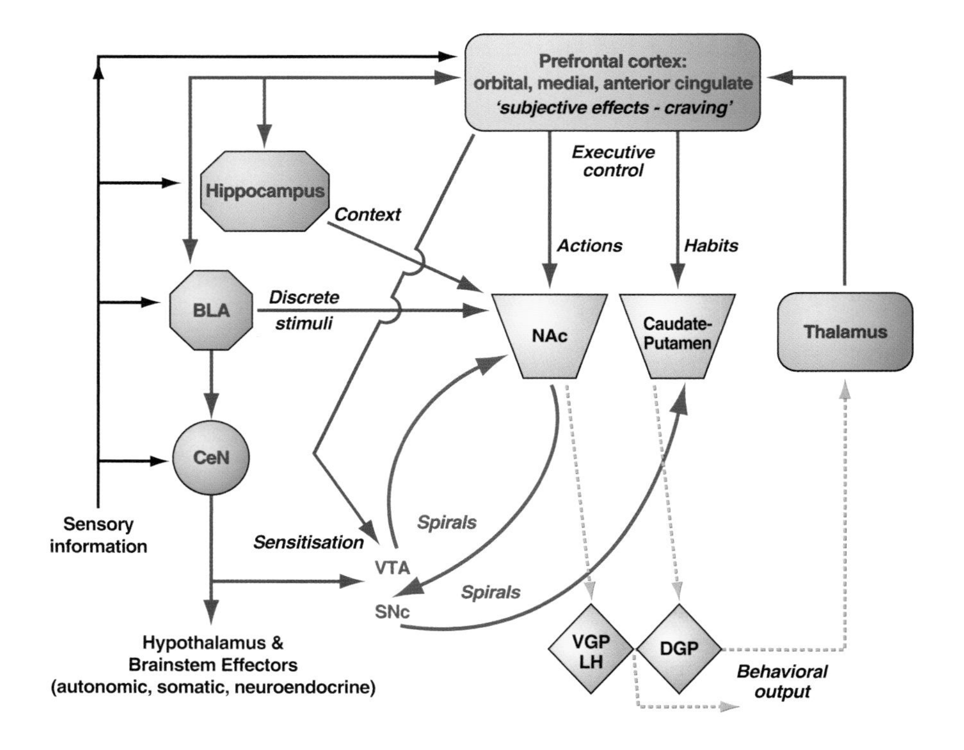

FIGURE 9.11 A neural systems perspective of psychomotor stimulant addiction as proposed by Everitt and Wolf (2002). *'A highly schematic representation of limbic cortical-ventral striatopallidal circuitry that tentatively localizes particular functions: (1) sensitization—ventral tegmental area and also the nucleus accumbens, via glutamate–dopamine interactions; (2) processing of discrete and contextual drug-associated conditioned stimuli—basolateral amygdala and hippocampal formation, respectively; (3) goal-directed actions (''action-outcome'' associations)—nucleus accumbens; (4) ''habits'' (stimulus-response learning)—dorsal striatum. Both 3 and 4 involve interactions between cortical afferents and striatal processes modulated by dopamine. (5) ''Executive control''—prefrontal cortical areas; (6) subjective processes, such as craving, activate areas such as orbital and anterior cingulate cortex, as well as temporal lobe structures including the amygdala, in functional imaging studies; (7) ''behavioral output'' is intended to subsume ventral and dorsal striatopallidal outflow via both brainstem structures and re-entrant thalamocortical loop circuitry; (8) ''spirals'' refers to the serial, spiraling interactions between the striatum and midbrain dopamine neurons that are organized in a ventral-to-dorsal progression (Haber* et al., *2000); (9) green arrows indicate glutamatergic pathways; yellow arrows indicate GABAergic pathways; red arrows indicate dopaminergic pathways. The transmitter used by central amygdala neurons is less certain but is probably glutamate and also possible neuropeptides.'* BLA, basolateral amygdala; CeN, central nucleus of the amygdala; VTA, ventral tegmental area; SNc, substantia nigra pars compacta. [Reproduced with permission from Everitt and Wolf, 2002.]

innervation of the basolateral amygdala and prefrontal cortex also may have a role in cue-induced reinstatement.

Inactivation of the central nucleus of the amygdala and the bed nucleus of the stria terminalis (BNST) prevented stress-induced reinstatement (Shaham *et al.*, 2000a), and antagonists of corticotropin-releasing factor (CRF) injected into central nucleus of the amygdala and lateral BNST also block stress-induced reinstatement, but injections into the basolateral amygdala do not block stress-induced reinstatement (see **Fig. 9.9**). Microinjection of noradrenergic antagonists into both the central nucleus of the amygdala and the lateral BNST also block stress-induced reinstatement (Leri *et al.*, 2002). Transient inactivation of the prefrontal cortex, extended amygdala, nucleus accumbens, and ventral tegmental area also can prevent stress-induced reinstatement (McFarland and Kalivas, 2001; Capriles *et al.*, 2003; Kalivas and McFarland, 2003; Shaham *et al.*, 2003) (**Figs. 9.8** and **9.9**).

In summary, there is hypothesized to be an important role for the extended amygdala (central nucleus of the amygdala and lateral BNST) in stress-induced reinstatement and a possible link between the extended amygdala and the prefrontal cortex-nucleus accumbens projection hypothesized to be critical for drug- and cue-induced reinstatement (Shaham et al., *2003; Stewart, 2000).*

Kalivas and McFarland divide the neurocircuitry involved in drug-, cue-, and stress- induced reinstatement of cocaine-seeking behavior into three subcircuits (Kalivas and McFarland, 2003) (**Fig. 9.8**). The frontal cortex (anterior cingulate) projection to the core of the nucleus accumbens is referred to as the motor subcircuit. The circuitry distinct for each mode of priming is described as the limbic subcircuit. The stress subcircuit is embedded within the limbic subcircuit. The neurocircuitry of drug-induced reinstatement is described as a series of connections from the ventral tegmental area to the prefrontal cortex (anterior cingulate) and then onto the core of the nucleus accumbens. The neurocircuitry of cue-induced reinstatement is described as a series of connections from the ventral tegmental area to the basolateral amygdala, and then from the basolateral amygdala to the prefrontal cortex (anterior cingulate), and then to the core of the nucleus accumbens. The neurocircuitry of stress-induced reinstatement is described as a series of adrenergic and CRF inputs to the extended amygdala which in turn activate the prefrontal cortex (anterior cingulate), which then projects to the core of the nucleus accumbens. One hypothesis postulated by the authors is that afferents from the extended amygdala are modulating GABAergic projections from the ventral tegmental area to the prefrontal (cingulate) cortex (McFarland and Kalivas, 2003). Another possibility may be a projection from the extended amygdala to the mediodorsal thalamus and then to the prefrontal (cingulate) cortex.

The key component of this conceptualization is the hypothesized role of the prefrontal (cingulate) cortex-to-nucleus accumbens glutamatergic projection as a critical subcircuit in driving drug-, cue-, and stress-induced reinstatement and in initiating motivated behavior in general (McFarland and Kalivas, 2003). While there is a common focus on the frontal cortex with the Jentsch and Taylor (1999), McFarland and Kalivas (2003), Shaham *et al.*, (2003), and Volkow *et al.*, (2003) conceptual frameworks, one area of significant discrepancy is that the reinstatement neurocircuitry hypothesis posits an activation of the prefrontal cortex, whereas the imaging (Volkow *et al.*, 2003) and cognitive-behavioral hypotheses posit a hypoactivity in the prefrontal cortex (hypofrontality). Kalivas and McFarland suggest an explanation via an increased signal-to-noise ratio, but another issue that needs exploration is that all of the reviewed animal models of reinstatement to date have involved nondependent animal models, and most of the imaging studies in humans have involved cocaine-dependent subjects. Data in reinstatement models where there is a history of dependence may shed new light on the actual valence of changes within these circuits so critical for guiding motivated behavior.

Drug Addiction and Relapse: Amygdala and Corticostriatopallidal Circuits

Everitt and Wolf (2002)
Fuchs, Ledford, and McLaughlin (2003)

Relapse to drugs of abuse has long been associated with a nebulous concept termed 'craving' which involves not only memories for the positive reinforcing effects of drugs but also drug-opposite associated effects such as the motivational components of withdrawal (Tiffany and Carter, 1998). Nevertheless, evidence is clear that drug-related stimuli can trigger increased motivation toward drug-seeking and drug-taking (Carter and Tiffany, 1999). See and colleagues argued that through a process of associative learning, previously neutral stimuli come to acquire incentive-motivational properties after repeated pairing with the drug (See *et al.*, 2003). Everitt and Wolf further proposed that at a systems level, progressive strengthening or the 'consolidation' of behavior paralleling the progression to addiction may be a form of habit learning (Robbins and Everitt, 1999; Everitt and Wolf, 2002). In such habit learning, it is further hypothesized that voluntary control over drug use is lost, and the propensity to relapse is high and readily precipitated by exposure to drug-associated stimuli. While studying cocaine-seeking behavior, both frameworks have identified corticostriatal pallidal systems as key components, with a focus on the basolateral amygdala for cue-induced relapse and cue-controlled cocaine-seeking and a role for the prefrontal cortex in loss of inhibitory control mechanisms hypothesized to reflect the development of locomotor sensitization in animals and impulsivity in humans (Everitt and Wolf, 2002; See *et al.*, 2003) (**Figs. 9.10** and **9.11**).

The evidence for a role of the basolateral amygdala in cue-induced reinstatement of cocaine-seeking behavior is overwhelming. Inactivation of the basolateral amygdala with excitotoxic lesions, tetrodotoxin, or lidocaine all blocked the reinstatement of cocaine-seeking by cues paired with cocaine self-administration in animal models of cue-induced reinstatement (Meil and See, 1997; Grimm and See, 2000; Kantak *et al.*, 2002) (**Table 9.1**). Typically, various stimuli that were paired with drug administration in rats trained to self-administer cocaine are then presented in the absence of the drug after extinction, and the amount of responding for these stimuli are used as a measure of cue-induced reinstatement (see *Animal Models of Addiction* chapter). These lesions do not block subsequent cocaine self-administration, but the lesions block the ability of cocaine-paired stimuli to reinstate extinguished lever responding. Intra-basolateral amygdala administration of tetrodotoxin also blocked the acquisition of cue-induced cocaine-seeking behavior as well as the expression of reinstatement (Kruzich and See, 2001). Lesions of the central nucleus of the amygdala only blocked the expression of reinstatement. The dorsomedial prefrontal cortex (also labeled the anterior cingulate cortex or prelimbic cortex) (See *et al.*, 2003) has been hypothesized to act in concert with the basolateral amygdala during the process of conditioned cued reinstatement (See *et al.*, 2003). Tetrodotoxin lesions of the dorsomedial prefrontal cortex also significantly attenuated conditioned cued reinstatement produced by cocaine-paired stimuli (McLaughlin and See, 2003). The neuropharmacological basis of cue-induced reinstatement of cocaine-seeking behavior appears to depend on dopamine D_1 receptor elements in the basolateral amygdala but not D_2 receptors or glutamate AMPA or NMDA receptors (See *et al.*, 2001). However, systemic administration of both D_1 and D_2 antagonists blocked cue-induced reinstatement (Weiss *et al.*, 2001b). Presentation of a stimulus that predicts cocaine availability increased extracellular dopamine levels in the basolateral amygdala and also increased basolateral amygdala c-fos activity (Weiss *et al.*, 2000; Ciccocioppo *et al.*, 2001). Finally, imaging studies in humans have shown that cocaine-related cues activate the amygdala (Childress *et al.*, 1999; Bonson *et al.*, 2002).

Consistent with the results of cue-induced reinstatement, excitotoxic lesions of the basolateral amygdala and medial prefrontal cortex also blocked the acquisition of cocaine-seeking behavior under a second-order schedule of reinforcement (Whitelaw *et al.*, 1996; Weissenborn *et al.*, 1997). In addition, inactivation of the nucleus accumbens core (Ito *et al.*, 2004) and administration of an AMPA, but not an NMDA receptor antagonist in the core of the nucleus accumbens (Di Ciano and Everitt, 2001) blocked cocaine-seeking behavior as measured in a second-order schedule. Consistent with the results observed with the second-order schedule, intra-nucleus accumbens core administration of an AMPA receptor antagonist blocked conditioned place preference to psychostimulants (Layer *et al.*, 1993; Kaddis *et al.*, 1995). Further reinforcing the role of corticostriatal glutamate projections are neurochemical studies showing that discrete cocaine-related stimuli increase nucleus accumbens glutamate levels but decrease basal nucleus accumbens glutamate levels (Hotsenpiller *et al.*, 2001). The potentiation of conditioned reinforcement associated with psychostimulants also depends critically on the basolateral and central nuclei of the amygdala and their interactions with the shell of the nucleus accumbens (Burns *et al.*, 1993; for review, see Everitt *et al.*, 2000).

It should be noted that another glutamate projection from the ventral subiculum has been implicated in cocaine reinstatement (Vorel *et al.*, 2001). Stimulation of the ventral subiculum induced a long-lasting dopamine release in the nucleus accumbens and reinstated cocaine-seeking behavior (Vorel *et al.*, 2001). Lesions of the ventral subiculum also blocked the facilitation of responding for conditioned reinforcers produced by intra-nucleus accumbens injection of D-amphetamine (Burns *et al.*, 1993). Thus, glutamatergic inputs into the nucleus accumbens via the ventral subiculum may amplify information that also is provided by the basolateral amygdala (Everitt *et al.*, 2000).

However, discriminated approach to appetitive Pavlovian stimuli in an autoshaping procedure appears to depend more on the central nucleus of the amygdala and the anterior cingulate core of the nucleus accumbens system, but not the basolateral nucleus of the amygdala (Gallagher *et al.*, 1990; Parkinson *et al.*, 2000a,b). In such a procedure, presentation of a visual stimulus is followed by food, and the animal develops a conditioned response of approaching the food-predicting conditioned stimulus before retrieving the

TABLE 9.2 Effects of Lesions of Specific Forebrain Sites on Stimulus-Reward Learning

	Stimulus-Reward Conditioned Response (NAc amphetamine)	Stimulus-Reward Conditioned Response	Pavlovian Approach	Pavlovian Instrumental Transfer	Reference
Basolateral amygdala	↓	↓	—	—	Cador *et al.*, 1989; Burns *et al.*, 1993; Parkinson *et al.*, 2000a; Hall *et al.*, 2001
Central amygdala	↓	—	↓	↓	Gallagher *et al.*, 1990; Robledo *et al.*, 1996; Parkinson *et al.*, 2000a; Hall *et al.*, 2001
Nucleus accumbens shell	↓	—	—	—	Parkinson *et al.*, 1999, 2000b; Hall *et al.*, 2001
Nucleus accumbens core	—	—	↓	↓	Parkinson *et al.*, 1999, 2000b; Hall *et al.*, 2001
Medial prefrontal cortex	—	—	↓	nd	Burns *et al.*, 1993; Parkinson *et al.*, 2000b; Chudasama and Robbins, 2003
Subiculum	↓	—	nd	nd	Burns *et al.*, 1993

nd, not determined.

primary reward. Similarly, the potentiation of instrumental behavior by noncontingent presentations of Pavlovian stimuli also depends on the central nucleus of the amygdala and the core of the nucleus accumbens but not the basolateral amygdala (Hall *et al.*, 2001) (**Table 9.2**).

Such potentiation of instrumental responding by noncontingent presentations of Pavlovian stimuli is augmented by increased dopamine in the nucleus accumbens shell and has been linked to the sensitized conditioned salience associated with locomotor sensitization (Robinson and Berridge, 1993), and locomotor sensitization also appears to depend on activation of the prefrontal cortex and basolateral amygdala. Lesions of the prefrontal cortex, basolateral amygdala, and ventral tegmental area blocked both the induction and expression of locomotor sensitization to repeated administration of psychostimulant drugs (Wolf *et al.*, 1995; Pierce *et al.*, 1998; Li *et al.*, 1999), and the mechanism is hypothesized to be mediated through a glutamate projection to the ventral tegmental area (Wolf, 1998).

Complex interactions were proposed to attempt to explain how locomotor sensitization leads to a loss of inhibitory control mechanisms and the development of impulsivity (Robbins and Everitt, 1999). For example, it was proposed that in the sensitized state, loss of inhibitory tone in the prefrontal cortex may lead to prefrontal disinhibition of the basolateral amygdala which in turn could enhance the basolateral amygdala–nucleus accumbens role in drug-conditioned reward (Everitt and Wolf, 2002) and impulsivity (Jentsch and Taylor, 1999; Robbins and Everitt, 1999). In contrast, compulsivity was hypothesized to arise from consolidation of habitual timulus–response drug-seeking through engagement of corticostriatal loops operating through both the dorsal and ventral striatum (nucleus accumbens core) (Robbins and Everitt, 1999; Everitt *et al.*, 2001). These changes that lead from impulsivity to compulsivity may be linked, as these authors and others argued that the transition from voluntary drug-seeking to a compulsive habit also may depend on the disruption of executive control provided by prefrontal cortex descending influences on striatal mechanisms (Jentsch and Taylor 1999; Volkow *et al.*, 2003). What is somewhat difficult to reconcile is how frontostriatal dysfunction—which in animal studies simultaneously blocks cue-induced reinstatement, cocaine-induced reinstatement, and locomotor sensitization—'acts synergistically with sensitization of stimulus–response mechanisms to produce compulsive drug-seeking behavior' (Everitt and Wolf, 2002). This conundrum becomes particularly salient when one notes that cocaine locomotor sensitization is accompanied by a reduction in the magnitude of AMPA receptor-mediated quantal synaptic events and long-term depression in the nucleus accumbens shell and ventral tegmental area (Thomas *et al.*, 2001; Thomas and Malenka, 2003) (see cellular section below). Interestingly, the authors ultimately end up suggesting that the decreased excitability of the nucleus accumbens associated with repeated cocaine administration produced by loss of prefrontal cortex executive control and loss of glutamate tone in the nucleus accumbens could be specifically related to 'withdrawal phenomena,' such as elevated reward thresholds (Markou and Koob, 1991) and anhedonia or dysphoria (Koob and Le Moal, 2001) that also may contribute to persistent cocaine-seeking and relapse.

In summary, Everitt and Wolf (2002) and See et al.*, (2003) present a framework where cue-induced reinstatement depends on a circuit involving the basolateral amygdala, the prefrontal cortex, and the core of the nucleus accumbens (ventral striatal/ventral pallidal loops). Other stimulus–reward associations as measured by approach to appetitive Pavlovian stimuli and Pavlovian instrumental transfer also involve activation of the nucleus accumbens core. The psychostimulant-induced potentiation of conditioned reinforcement involves the basolateral and central nuclei of the amygdala, but via the shell of the nucleus accumbens. Impulsivity is linked to sensitization of dopamine systems via a loss of inhibitory control from the prefrontal cortex, and compulsivity is hypothesized to develop from consolidation of stimulus response drug-seeking through engagement of corticostriatal loops operating through both the ventral and dorsal striatum.*

NEUROCIRCUITRY THEORIES OF ADDICTION—REWARD AND STRESS

Drugs of Abuse: Anatomy, Pharmacology, and Function of Reward Pathways

Koob (1992)

In one of the earliest attempts to explore the interaction of drugs of abuse with reward pathways, a midbrain-forebrain-extrapyramidal circuit with a focus on the nucleus accumbens was proposed (Koob, 1992b) (**Fig. 9.12**). Three neurochemical systems were hypothesized to be involved in the initial reinforcing (or rewarding) actions of drugs of abuse: dopamine, opioid, and GABA. For indirect sympathomimetics such as cocaine and amphetamines, the mesolimbic dopamine system was proposed as critical and supported by numerous neuropharmacological studies. For example, neurotoxin-specific lesions of the mesolimbic dopamine system blocked the self-administration of cocaine and amphetamine (Lyness *et al.*, 1979; Roberts *et al.*, 1980) but not the self-administration of heroin (Pettit *et al.*, 1984) or alcohol (Rassnick *et al.*, 1993b). For opioids, the opioid receptors were hypothesized to be a critical first step in the reinforcing actions of opioid drugs with a predominant role for the μ opioid receptor and for sites both pre- and postsynaptic to the mesolimbic dopamine system in the nucleus accumbens and ventral tegmental area. For alcohol, the $GABA_A$ receptor was hypothesized to be an initial site of action in the reinforcing actions of alcohol with a predominant role for the $GABA_A$ receptor in the nucleus accumbens and amygdala (Hyytia and Koob, 1995).

Based on this synthesis, Koob (1992b) proposed an early neurobiological circuit for drug reward. The starting point for the reward circuit was the medial forebrain bundle which is composed of myelinated fibers connecting the olfactory tubercle and nucleus accumbens with the hypothalamus and ventral tegmental area (Nauta and Haymaker, 1969), and of ascending monoamine pathways such as the mesocorticolimbic dopamine system. Drug reward was hypothesized to depend on dopamine release in the nucleus accumbens for cocaine and amphetamine, opioid peptide receptor activation in the ventral tegmental area, and nucleus accumbens for opiates, and $GABA_A$ receptors in the amygdala for alcohol. Even at this time, the independence of both opioid and alcohol reward from a critical role for dopamine was noted and was used as a basis to argue for multiple independent neurochemical elements in drug reward. It was further argued that the nucleus accumbens was situated strategically to receive important limbic information from the amygdala, frontal cortex, and hippocampus that could be converted to motivational action via its connections with the extrapyramidal motor system. Finally, it was noted that the nucleus accumbens was not a homogeneous structure and that the 'shell' part (medial and ventral) may be part of an extended amygdala system (see below), while the core resembled more the corpus striatum (Alheid and Heimer, 1988; Groenewegen *et al.*, 1990).

In summary, an early neurobiological circuit for drug reward included a midbrain-forebrain-extrapyramidal circuit with a focus on the nucleus accumbens and multiple parallel neurochemical sites of action for different drugs of abuse.

Neural Substrates of Alcohol Self-administration: Neurobiology of High Alcohol Drinking Behavior in Rodents

McBride and Li (1998)

Based on an extensive series of studies with animals selectively bred for high alcohol consumption, a number of key brain regions and neurochemical systems were identified that contribute to circuitry important for alcohol reward (McBride and Li, 1998) (**Fig. 9.13**). A key focus of this circuitry framework is

A

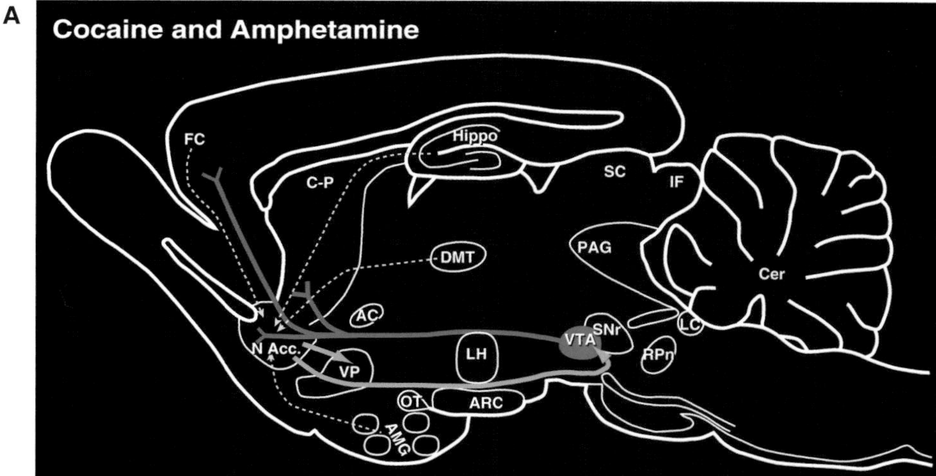

B

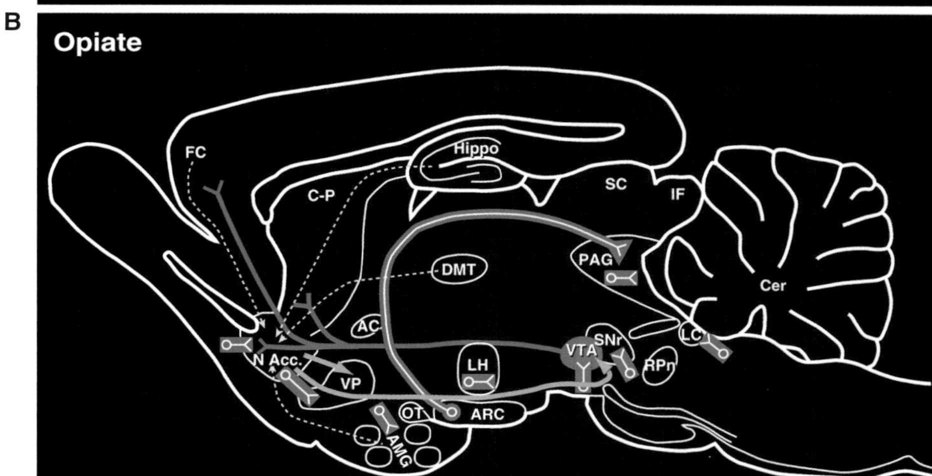

C

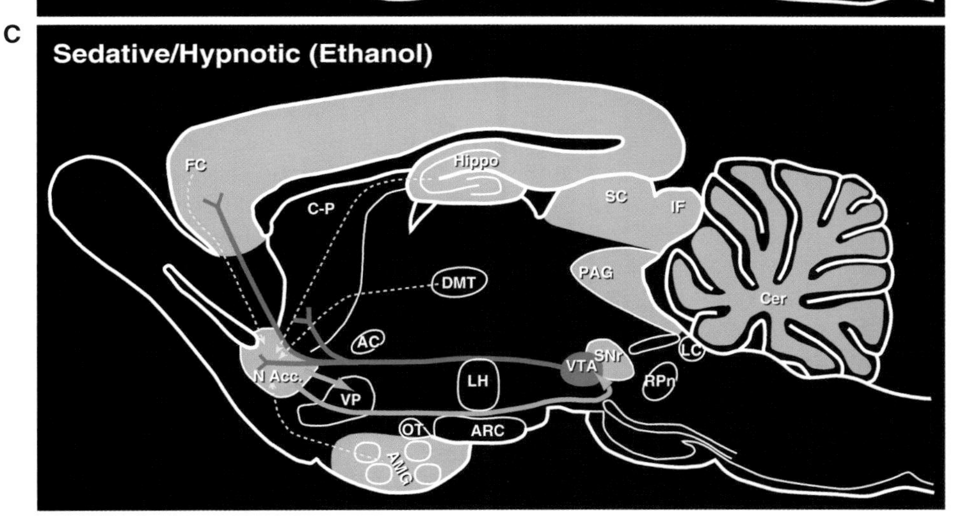

FIGURE 9.12 Sagittal rat brain sections illustrating proposed cocaine and amphetamine (A), opioid peptide (B), and sedative/hypnotic (C) drug reward neurocircuits. **(A)** *The cocaine and amphetamine reward circuit includes a limbic-extrapyramidal motor interface. Yellow indicates limbic afferents to the nucleus accumbens and orange represents efferents from the nucleus accumbens thought to be involved in psychomotor stimulant reward. Red indicates projections of the mesocorticolimbic dopamine system thought to be a critical substrate for psychomotor stimulant reward. This system originates in the A10 cell group of the ventral tegmental area and projects to the nucleus accumbens, olfactory tubercle, and ventral striatal domains of the caudate putamen.* **(B)** *These opioid peptide systems (green) include local enkephalin circuits (short segments) and the hypothalamic midbrain β-endorphin circuit (long segment). The system is superimposed on the neural reward circuit shown in* (A). **(C)** *Approximate distribution of $GABA_A$ receptor complexes (blue) determined by the relative distribution of both [^{3}H]flumazenil binding and expression of the α, β, and γ subunits of the $GABA_A$ receptor (Sequier* et al.*, 1988; Shivers* et al.*, 1989; Olsen* et al.*, 1990). This distribution is superimposed on the neural reward circuit shown in* (A). *AC, anterior commissure; AMG, amygdala; ARC, arcuate nucleus; Cer, cerebellum; C-P, caudate putamen; DMT, dorsomedial thalamus; FC, frontal cortex; Hippo, hippocampus; IF, inferior colliculus; LC, locus coeruleus; LH, lateral hypothalamus; N Acc, nucleus accumbens; OT, olfactory tubercle; PAG, periaqueductal gray; RPn, reticular pontine nucleus; SC, superior colliculus; SNr, substantia nigra pars reticulata; VP, ventral pallidum; VTA, ventral tegmental area.* [Reproduced with permission from Koob, 1992b.]

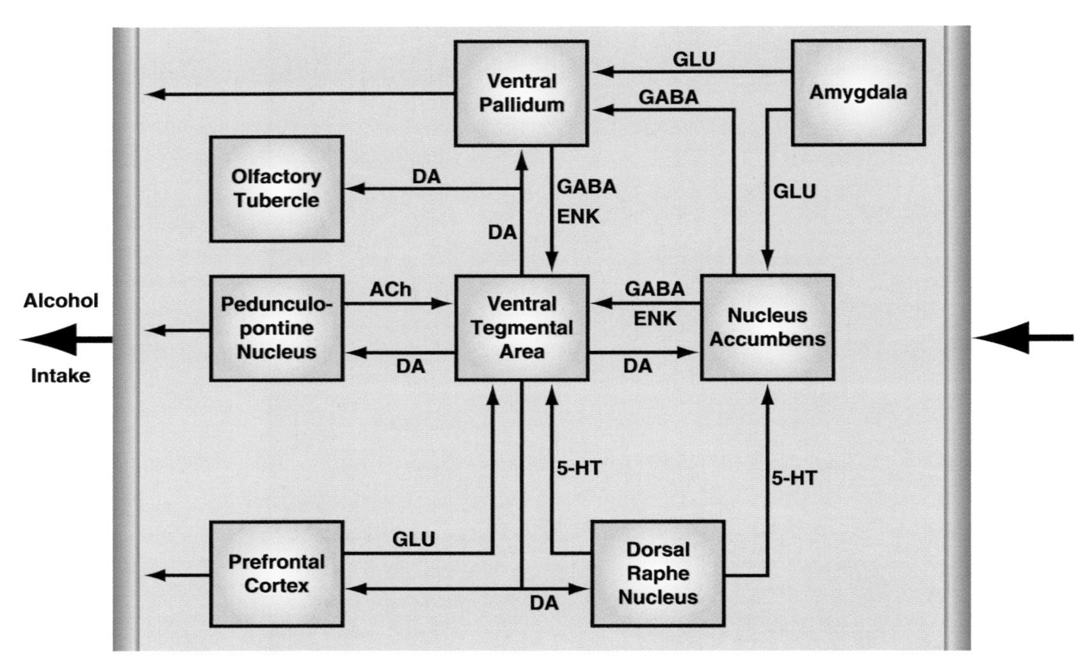

FIGURE 9.13 A hypothetical simplified scheme of neuronal pathways and CNS regions mediating alcohol drinking (McBride and Li, 1998): *'The ventral tegmental area and its dopamine (DA) projections to the nucleus accumbens, ventral pallidum, prefrontal cortex, and olfactory tubercle play key roles in regulating alcohol intake. The dorsal raphe nucleus serotonin (5-HT) system serves to regulate the activity of the ventral tegmental area—nucleus accumbens dopamine pathway. Furthermore, the ventral tegmental area is regulated by γ-aminobutyric acid (GABA) and enkephalinergic (ENK) projections from the ventral pallidum and nucleus accumbens, a cholinergic (ACh) input from the pedunculopontine nucleus, and a glutamatergic (GLU) pathway from the prefrontal cortex. The ventral pallidum receives a major GABA input from the nucleus accumbens. Both of these limbic regions are regulated by glutamate projections from the amygdala. In addition to these interactions, inputs from other CNS regions influence the activity of these limbic structures in regulating alcohol intake.'* [Reproduced with permission from McBride and Li, 1998.]

the ventral tegmental area and dopamine projections to the nucleus accumbens and olfactory tubercle, prefrontal cortex, and ventral pallidum in mediating alcohol drinking. Ethanol was hypothesized to activate ventral tegmental area neurons and causes release of dopamine in the nucleus accumbens (see *Alcohol* chapter). Further evidence supporting a role for the ventral tegmental area as one site mediating the reinforcing actions of alcohol came from microinjections and intracranial self-administration studies. Alcohol-preferring P rats, but not nonpreferring NP rats, self-administered 50–200 mg per cent ethanol directly into the ventral tegmental area (Gatto *et al.*, 1994; Rodd-Henricks *et al.*, 2000), and this self-administration was dependent on dopamine activation (Rodd *et al.*, 2004, 2005). Acetaldehyde also was self-administered into the ventral tegmental area (Rodd-Henricks *et al.*, 2002), and salsolinol was self-administered into the nucleus accumbens (Rodd *et al.*, 2003). Together, these results suggested that in alcohol-preferring rats, alcohol and its metabolites may interact with the mesolimbic dopamine system to produce some of its reinforcing effects.

Blocking $GABA_A$ receptors in the ventral tegmental area also blocked ethanol intake of P rats at doses that did not alter saccharin consumption (Nowak *et al.*, 1998) and was interpreted as GABA antagonists mimicking the effects of alcohol by activating dopamine firing, and thus substituting for the effects of alcohol (McBride and Li, 1998). Similar effects in decreasing alcohol responding were observed with microinjection of a benzodiazepine inverse agonist into the ventral tegmental area (June *et al.*, 1998a,b), and similar interactions have been hypothesized for the decrease in alcohol intakes in P rats resulting from activating muscarinic acetylcholine receptors in the ventral tegmental area (Katner *et al.*, 1997).

Other key sites for alcohol reinforcement included the nucleus accumbens, pedunculopontine tegmental nucleus, and midbrain raphe (McBride and Li, 1998) (**Fig. 9.13**). Microinjection into the nucleus accumbens of both GABA agonists and antagonists blocked alcohol self-administration (Hodge *et al.*, 1995), and local injection of a benzodiazepine inverse agonist decreased alcohol responding in P rats (June *et al.*, 1998a,b). Suppression of serotonergic function in the raphe nuclei increased alcohol intake in Wistar rats (Tomkins *et al.*, 1994) (see *Alcohol* chapter for a discussion of the role of serotonin in alcoholism). The authors argued for a role of brain sites within the extended amygdala (central nucleus of the amygdala, BNST, and the transition zone in the shell of the nucleus accumbens) in mediating the reinforcing effects of alcohol and implicated several neurotransmitter systems, including GABA, serotonin, CRF, and neuropeptide Y (see *Alcohol* chapter). Finally, extensive neurochemical studies on alcohol-prefering P rats, sP rats, and HAD rats (see *Animal Models of Addiction* chapter) showed inherent differences in the function of the mesolimbic dopamine system, endogenous opioid peptide systems, GABAergic systems, and serotonin systems consistent with the neuropharmacological studies of alcohol reward (see *Alcohol* chapter). Thus, there was a convergence of data from neuropharmacological studies and genetic animal models for a role of the circuit elements outlined in **Fig. 9.13** in the reinforcing effects of alcohol (McBride and Li, 1998).

In summary, significant evidence was marshaled to support the hypothesis that a neurocircuit with focal points in the ventral tegmental area, nucleus accumbens, and extended amygdala mediates the acute reinforcing effects of alcohol. Neuropharmacological data from alcohol-preferring rats, combined with inherent neurochemical changes in alcohol-preferring rats, have identified four key neurochemical systems—dopamine, serotonin, GABA, and opioid peptide—in the acute reinforcing actions of alcohol. Changes in these systems at specific points in the neurocircuit may convey vulnerability to the excessive alcohol consumption of alcoholism.

Stress, Dysregulation of Drug Reward Pathways, and Drug Addiction

Piazza and Le Moal (1996)
Shaham, Shalev, Lu, DeWit, and Stewart (2003)
Aston-Jones and Harris (2004)
Kreek and Koob (1998)

Drugs of abuse acutely activate the hypothalamic-pituitary-adrenal response to stress, and as dependence develops ultimately engage brain stress systems. These basic observations have led to the hypothesis that the brain and brain pituitary stress systems have a role in the initial vulnerability to drugs of abuse (Piazza and Le Moal, 1996; Kreek and Koob, 1998), the development of dependence to drugs of abuse (Kreek and Koob, 1998), and the vulnerability to stress-induced relapse (Shaham *et al.*, 2003). Stressors facilitated the acquisition of cocaine and amphetamine self-administration, and food restriction increased self-administration of most drugs of abuse (Carroll and Meisch, 1984; Piazza and Le Moal, 1998) (**Table 9.3**). Rats bred for increased basal exploration of a novel environment and a high initial corticosterone response were much more likely to self-administer psychostimulant drugs (Piazza and Le Moal, 1996). There was a positive correlation between the locomotor response to novelty, behavioral reactivity to stress, and the amount of amphetamine that was self-administered by individual rats (Piazza *et al.*, 1989, 1991). Rats receiving repeated injections of corticosterone acquired cocaine self-administration at a lower dose of cocaine than did rats that were administered vehicle (Mantsch *et al.*, 1998). Administration of corticosterone caused rats that would not self-administer cocaine at low doses to self-administer cocaine (Piazza *et al.*, 1991b). Glucocorticoids have been shown to facilitate the locomotor activation associated with drugs of abuse (Marinelli *et al.*, 1994, 1997b), and acute blockade of corticosterone secretion decreased the psychomotor response to cocaine (Marinelli *et al.*, 1997a). The enhanced propensity to self-administer drugs of abuse produced by stressors was linked to increased activation of the mesolimbic dopamine system mediated by glucocorticoid release (Piazza and Le Moal, 1998) (**Fig. 9.14**). Glucocorticoids facilitated dopamine-dependent behaviors by modulating dopamine transmission in the ventral striatum. Suppression of glucocorticoids by adrenalectomy reduced extracellular concentrations of dopamine in the ventral striatum, the shell in particular (Barrot *et al.*, 2000), both in basal conditions and after psychostimulant administration. These effects were reversed by corticosterone replacement and are selectively mediated by glucocorticoid receptors (Barrot *et al.*, 2000).

The initial responses of the hypothalamic-pituitary-adrenal system to drugs change dramatically when the animals were experimenter-administered or self-administered binge-like amounts of psychostimulant drugs (Zhou *et al.*, 1996, 2003; Mantsch *et al.*, 2003). There were large increases in adrenocorticotropic hormone and corticosterone in rats during an acute binge (Zhou *et al.*, 1996, 2003). However, the increase showed tolerance in the chronic binge stage and then a reactivation during acute withdrawal and dysregulation that persists during protracted abstinence (Mantsch *et al.*, 2003; Zhou *et al.*, 2003).

TABLE 9.3 Stressors that increase drug self-administration

Stressor	Approaches to the study of drug self-administration				Reference
	Acquisition	Dose–response	Progressive-ratio	Reinstatement	
Food restriction	↑ Psychostimulants ↑ Opiates ↑ Alcohol	↑ Psychostimulants ↑ Opiates ↑ Alcohol			Carroll and Meisch, 1984 Shaham and Stewart, 1994
Tail pinch	↑ Amphetamine				Piazza *et al.*, 1990
Footshock	↑ Cocaine		↑ Heroin	↑ Heroin ↑ Cocaine	Goeders and Guerin, 1994 Shaham and Stewart, 1994 Shaham and Stewart, 1995 Erb *et al.*, 1996
Restraint	↑ Morphine				Shaham, 1993
Social aggression	↑ Cocaine	↑ Cocaine			Haney *et al.*, 1995 Miczek and Mutschler, 1996
Social competition	↑ Amphetamine				Maccari *et al.*, 1991
Social isolation	↑ Opiates ↑ Alcohol	↑ Cocaine* ↑ Heroin			Piazza *et al.*, 1991c Hadaway *et al.*, 1979 Alexander *et al.*, 1981 Marks-Kaufman and Lewis, 1984 Schenck *et al.*, 1987, 1990 Bozarth *et al.*, 1989 Wolffgramm and Heyne, 1991
Witnessing stress	↑ Cocaine				Ramsey and Van Ree, 1993
Prenatal stress	↑ Amphetamine				Deminiere *et al.*, 1992

↑, Depending on the type of self-administration used: facilitation of acquisition; upward shift of the dose–response curve; higher break point; induction of responding on the device previously associated with the infusion of the drug.
*, Note that for social isolation, slightly higher (Boyle *et al.*, 1991), equal (Bozarth *et al.*, 1989), and lower (Phillips *et al.*, 1994) sensitivities to the reinforcing effects of cocaine also have been reported.
[Reproduced with permission from Piazza and Le Moal, 1998.]

At the same time, escalation in drug intake either with extended access or dependence-induction produced an activation of the brain stress system and norepinephrine stress systems outside of the hypothalamus in the extended amygdala (Koob, 1999, 2003b; Aston-Jones and Harris, 2004). Acute withdrawal from cocaine (Richter and Weiss, 1999), alcohol (Merlo-Pich *et al.*, 1995; Olive *et al.*, 2002), nicotine (Ghozland *et al.*, 2004), cannabinoids (Rodriguez de Fonseca *et al.*, 1997), and opioids (Weiss *et al.*, 2001a) produced increases in CRF release as measured by *in vivo* microdialysis in the central nucleus of the amygdala and in some cases, in the BNST. CRF antagonists blocked the increased anxiety-like or aversive responses associated with alcohol (Baldwin *et al.*, 1991; Rassnick *et al.*, 1993a), precipitated opiate withdrawal (Heinrichs *et al.*, 1995; Stinus *et al.*, 2005), and cocaine withdrawal (Sarnyai *et al.*, 1995; Basso *et al.*, 1999). More recent studies have shown that these increases in CRF may have motivational significance in that a competitive CRF antagonist blocked the excessive drinking in dependent rats but not baseline nondependence drinking (Valdez *et al.*, 2002). A competitive CRF_1 antagonist blocked the development of place aversion to precipitated opiate withdrawal (Stinus *et al.*, 2005).

Norepinephrine projections to the BNST, which are known to activate CRF systems, also have been shown to be critically involved in the motivational aspects of opiate withdrawal (Aston-Jones *et al.*, 1999; Delfs *et al.*, 2000; Aston-Jones and Harris, 2004). Microinjection of β-adrenergic receptor antagonists into the BNST blocked opiate withdrawal-induced conditioned place aversion (Delfs *et al.*, 2000). The origin of the norepinephrine projections for this effect appears to be from the ventral noradrenergic bundle from the noradrenergic cell groups of the caudal medulla, in that lesions of this ventral system but not the dorsal noradrenergic bundle from the locus coeruleus also blocked opiate withdrawal-induced place aversions (Delfs *et al.*, 2000). These results suggest that the brain stress systems involving CRF and norepinephrine may be activated during the development of dependence and contribute to the motivation for excessive drug-seeking associated with dependence (Koob and Le Moal, 2004) (**Fig. 9.15**). Other antistress neural systems such as neuropeptide Y may be dysregulated as a 'brake' on the stress system, which would further exacerbate a dysregulation of the brain stress systems (Heilig *et al.*, 1994).

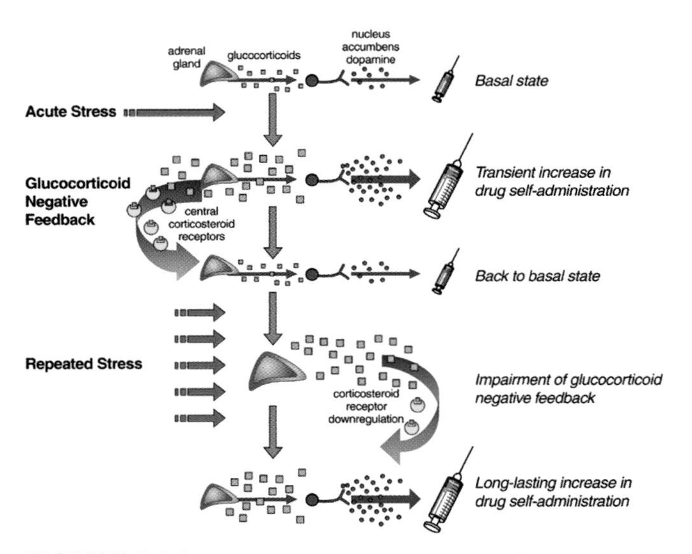

FIGURE 9.14 Possible pathophysiological mechanisms for the increase in drug self-administration induced by acute and repeated stress as hypothesized by Piazza and Le Moal (1998). *'The interactions of two biological systems are schematically represented by the secretion of glucocorticoids from the adrenal gland in yellow and the release of dopamine from the meso-accumbens dopaminergic projection in red. In basal conditions, glucocorticoid secretion and dopamine release are low, as is sensitivity to drugs of abuse. An acute stress causes an increase in glucocorticoid secretion, which, by enhancing the release of dopamine, results in an increase in the sensitivity to the reinforcing effects of drugs of abuse, and thus in an increase in self-administration. Negative feedback rapidly returns the glucocorticoid system to basal levels. The repeated exposure to stress and consequent repeated increase in the concentrations of glucocorticoids is hypothesized to progressively impair glucocorticoid negative feedback via decreasing the number of central corticosteroid receptors in the hippocampus. This results in a long-lasting increase in the secretion of glucocorticoids and increased in the release of dopamine in the nucleus accumbens and may explain why, after repeated stress, an increase in the sensitivity to drugs is found also long (weeks) after the end of the stressor exposure.'* [Modified with permission from Piazza and Le Moal, 1998.]

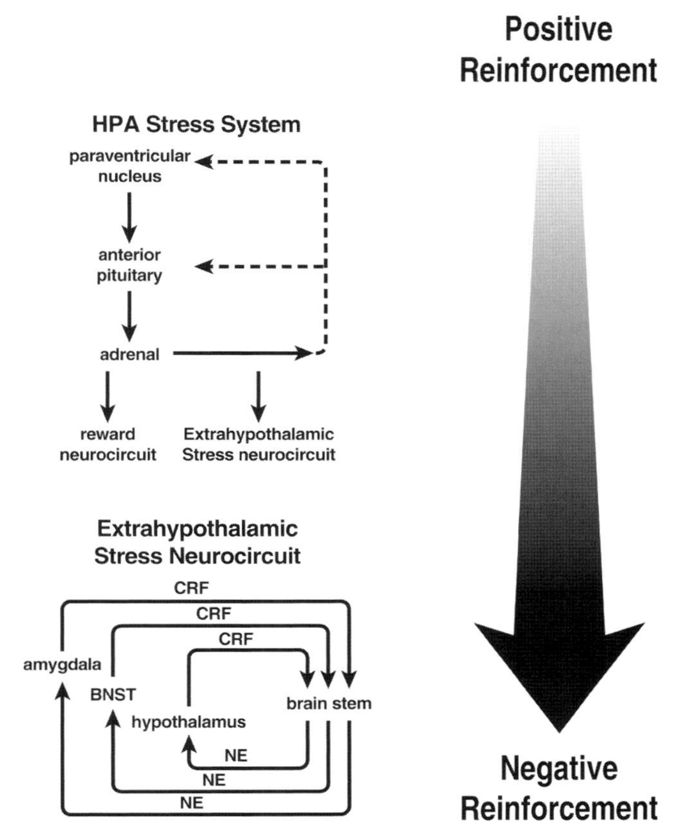

FIGURE 9.15 Hypothalamic-pituitary-adrenal and brain stress circuits hypothesized to be recruited at different stages of the addiction cycle as addiction moves from positive reinforcement to negative reinforcement (Shaham *et al.*, 2003; Aston-Jones and Harris, 2004; Koob and Kreek, 2005). The top circuit refers to the HPA axis which (a) feeds back to regulate itself, (b) activates the brain reward neurocircuit, and (c) facilitates the extrahypothalamic stress neurocircuit. The bottom circuit refers to the brain stress circuits in feed-forward loops. Here, corticotropin-releasing factor in the extended amygdala drives norepinephrine systems in the pons-medulla of the brain stem, which in turn drives CRF in the extended amygdala (see Koob, 1999). [Adapted from Koob and Le Moal, 2004.]

Some of these dysregulations of the CRF and norepinephrine systems have been shown to persist during protracted abstinence and can be reinstated by the reinstatement of an acute stressor (Aston-Jones and Harris, 2004; Koob and Kreek, 2005). Rats made dependent on alcohol and tested 2–6 weeks post-acute withdrawal showed an enhanced anxiety-like response in the elevated plus maze that was blocked by intracerebroventricular administration of a competitive CRF antagonist (Valdez *et al.*, 2003). Parallel results have been obtained in studies with chronic morphine and place preference where chronic morphine treatment produced an enhanced place preference for morphine weeks after acute withdrawal, and this enhanced place preference was accompanied by increased anxiety-like responses and by increases in Fos staining in the medial prefrontal cortex (cingulate cortex), basolateral amygdala, and ventral part of the BNST (Harris and Aston-Jones, 2003). Aston-Jones and Harris suggested that the conditioned release of norepinephrine in the BNST in response to stressors may elevate anxiety which then augments the reward value of drugs through negative reinforcement mechanisms (Aston-Jones and Harris, 2004; Koob and Le Moal, 2004) (**Fig. 9.15**).

The same brain stress systems have been strongly implicated in stress-induced reinstatement of drug-seeking behavior. Clear evidence has been generated showing that rats previously trained to self-administer cocaine and heroin and then extinguished will reinstate their responding if exposed immediately prior to the testing session with a mild footshock (Shaham and Stewart, 1995; Erb *et al.*, 1996; Le *et al.*, 1998; Ahmed *et al.*, 2000; Martin-Fardon *et al.*, 2000) (**Table 9.3**). This stress-induced reinstatement was not blocked by removal of glucocorticoids (Shaham *et al.*, 1997), but was blocked with pretreatment of CRF antagonists administered directly into the brain (Shaham *et al.*, 1997; Erb *et al.*, 1998; Lu *et al.*, 2000, 2001). This effect appeared to be mediated by CRF_1 receptors (Shaham *et al.*, 1998; Lu *et al.*, 2000, 2001). The brain site responsible for the actions of CRF antagonists on cocaine reinstatement appeared to be the ventral BNST since infusions of CRF into this area reinstate responding, and local administration of CRF antagonists into the ventral BNST blocked footshock-induced reinstatement, though administration into the central nucleus of the amygdala was without effect (Erb and Stewart, 1999). However, reversible activation of both the central nucleus of the amygdala and the BNST with tetrodotoxin blocked the footshock-induced reinstatement of heroin responding (Shaham *et al.*, 2000a). An asymmetric lesion procedure to functionally disconnect the CRF-containing pathway from the central nucleus of the amygdala to the BNST significantly reduced footshock-induced reinstatement, suggesting that an important origin of the CRF terminals in the BNST for cocaine-induced reinstatement was the central nucleus of the amygdala (Erb *et al.*, 2001).

Noradrenergic functional antagonists also blocked footshock-induced reinstatement (Erb *et al.*, 2000; Shaham *et al.*, 2000b; Highfield *et al.*, 2001). The brain sites for these effects appeared to be the ventral noradrenergic bundle projections to the BNST. Neurotoxin-specific lesions of the ventral noradrenergic bundle attenuated footshock-induced reinstatement of heroin responding (Shaham *et al.*, 2000b), and local injection of a β-adrenergic receptor antagonist into the BNST also blocked footshock-induced reinstatement in cocaine-trained rats (Leri *et al.*, 2002). The exact interaction between the noradrenergic ventral bundle projections and the intrinsic CRF systems of the central nucleus of the amygdala and BNST are not known but may involve activation by the noradrenergic pathway of CRF to the BNST pathway (Shalev *et al.*, 2002), which in turn may activate noradrenergic systems in the ventral medulla (Koob, 1999). Such a feed-forward system has been hypothesized for CRF/noradrenergic interactions for anxiety and stress-like responses, and also may become activated during the development of dependence.

In summary, the hypothalamic-pituitary-adrenal hormonal stress system and the brain stress systems are engaged by drugs of abuse and may contribute to not only the initial vulnerability to take drugs but also to the development of dependence and vulnerability to relapse. The hypothalamic-pituitary-adrenal hormonal stress system appears to have an important role in the initiation of drug-seeking and in the maintenance of drug-taking behavior. However, the brain extrahypothalamic stress systems appear to have a more important role in the motivational effects of both acute withdrawal and protracted abstinence and stress-induced reinstatement (Piazza and Le Moal, 1996; Kreek and Koob, 1998; Shaham et al., *2003; Aston-Jones and Harris, 2004).*

Neurocircuits in the Extended Amygdala as a Common Substrate for Dysregulation of Reward and Stress Function in Addiction

Koob and Le Moal (2001)

While changes in reward function are components of other conceptual frameworks in the neurocircuitry of addiction, for Koob and Le Moal (2001) the primary 'deficit' is a neuroadaptational shift in how rewards are processed. A loss of positive reinforcement and a recruitment of negative reinforcement are hypothesized to take place within a specific basal forebrain '*uber*-circuit' termed the extended amygdala. The extended amygdala has been identified by neuroanatomical studies (Alheid and Heimer, 1988; Koob *et al.*, 1998) as a separate entity within the basal forebrain, and has been hypothesized to be a common neural circuit for the reinforcing actions of drugs (Alheid and Heimer, 1988). The term *extended amygdala* originally was described by Johnston (1923) and represents a macrostructure that is composed of several basal forebrain structures: the lateral and medial BNST, the central and medial nuclei of the amygdala, the area termed the sublenticular substantia innominata, and a transition zone that forms the medial posterior part of the nucleus accumbens (e.g., shell) (Heimer and Alheid, 1991; Alheid *et al.*, 1995). These structures have similarities in morphology, immunohistochemistry, and connectivity (Alheid and Heimer 1988; Heimer and Alheid, 1991) (**Fig. 9.16**).

However, further examination of this anatomical system reveals two major divisions: the *central division* and the *medial division* (see **Fig. 9.15**). These two divisions have important differences in structure and afferent and efferent connections (Alheid *et al.*, 1995) that may be of heuristic value. The *central division* of the extended amygdala includes the central nucleus of the amygdala, the central sublenticular extended amygdala, the lateral BNST, and a transition area in the medial and caudal portions of the nucleus accumbens. These structures are within the central division and are largely defined by their network of intrinsic connections and extensive connections to the lateral hypothalamus (Koob, 2003a). The *medial division* of the extended amygdala includes the medial BNST, medial nucleus of the amygdala, and the medial sublenticular extended amygdala. These structures have been largely defined as the *medial division* by their network of intrinsic associative connections and extensive

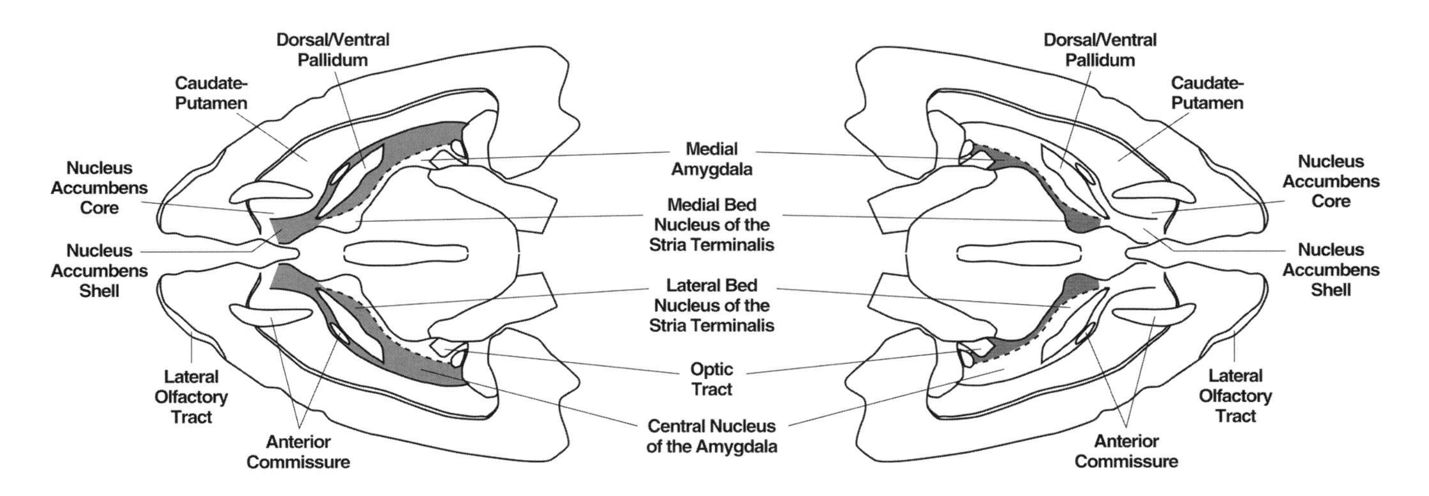

Central Division

Neuropharmacological Component
- Dopamine terminals
- Norepinephrine terminals
- CRF terminals and cell bodies
- NPY terminals
- Galanin cell bodies

Afferent/Efferent
- Receives cortical information from prefrontal cortex and insular cortex
- Sends information to lateral hypothalamus, ventral tegmental area, and pedunculopontine tegmental nucleus

Functional Attributes
- Regulates pituitary adrenal axis
- Major role in mediating reinforcing actions of drugs of abuse and ethanol

Medial Division

Neuropharmacological Component
- Vasopressin terminals
- Sex hormone receptors

Afferent/Efferent
- Receives olfactory information from infralimbic cortex, entorhinal cortex and subiculum
- Sends information to ventral striatum, ventromedial hypothalamus and mesencephalic central gray

Functional Attributes
- Sexually dimorphic – larger in males than females
- Major role in mediating male and female sexual behavior
- Regulates sympathetic and physiological responses

FIGURE 9.16 Neuroanatomical connections of the extended amygdala relevant for the positive reinforcing effects of drugs and the negative reinforcing effects of addiction. Neuroanatomical studies reveal two major divisions: the *central division* and the *medial division*. These two divisions have important differences in structure and afferent and efferent connections. The *central division* of the extended amygdala includes the central nucleus of the amygdala, the central sublenticular extended amygdala, the lateral BNST, and a transition area in the medial and caudal portions of the nucleus accumbens. These structures are within the central division and are largely defined by their network of intrinsic connections and extensive connections to the lateral hypothalamus. The *medial division* of the extended amygdala includes the medial BNST, medial nucleus of the amygdala, and the medial sublenticular extended amygdala. These structures have been largely defined as the *medial division* by their network of intrinsic associative connections and extensive relations to the medial hypothalamus (Alheid *et al.*, 1995). Evidence suggests that the *central division* may be involved in receiving cortical information and regulating the hypothalamic-pituitary-adrenal axis (Gray *et al.*, 1993), whereas the medial division may be more involved in sympathetic and physiological responses and receiving olfactory information (Lesur *et al.*, 1989; Pompei *et al.*,1991; Nijsen *et al.*, 2001). Most neuropharmacological effects resulting from motivational manipulations associated with the reinforcing effects of drugs of abuse have been in the central nucleus of the amygdala and the lateral nucleus of the BNST. [Brain sections modified with permission from Heimer and Alheid, 1991.]

relations to the medial hypothalamus (Alheid *et al.*, 1995). The lateral BNST which forms a key element of the *central division* of the extended amygdala has high amounts of dopamine and norepinephrine terminals, CRF terminals, CRF cell bodies, neuropeptide Y (NPY), terminals, and galanin cell bodies and receives afferents from the prefrontal cortex, insular cortex, and amygdalopiriform area. The medial BNST, in contrast, contains high amounts of vasopressin, is sexually dimorphic, and receives afferents from structures such as infralimbic cortex, entorhinal cortex, and subiculum (Allen *et al.*, 1984; Phelix and Paull, 1990; Gray and Magnuson, 1992; McDonald *et al.*, 1999; Dong *et al.*, 2001; Kozicz, 2001). Evidence suggests that the *central division* may be involved in receiving cortical information and regulating the hypothalamic-pituitary-adrenal axis (Gray *et al.*, 1993), whereas the *medial division* may be more involved in sympathetic and physiological responses, and in receiving olfactory information (Lesur *et al.*, 1989; Pompei *et al.*, 1991; Nijsen *et al.*, 2001). Most motivational manipulations resulting in modification of the reinforcing effects of drugs of abuse have been in the central nucleus of the amygdala and the lateral nucleus of the BNST.

Specific sites within the extended amygdala and selective neurochemical and neuropharmacological actions have been identified for both the acute positive reinforcing effects of drugs of abuse and in the negative reinforcement associated with drug abstinence. Microinjections of dopamine D_1 antagonists directly into the shell of the nucleus accumbens, the central nucleus of the amygdala (Caine *et al.*, 1995), and the BNST (Epping-Jordan *et al.*, 1998) were particularly effective in blocking cocaine self-administration. *In vivo* microdialysis studies showed a selective activation of dopaminergic transmission in the shell of the nucleus accumbens in response to acute administration of virtually all major drugs of abuse (Pontieri *et al.*, 1995, 1996; Tanda *et al.*, 1997). In addition, the acute reinforcing effects of alcohol are blocked by administration of GABAergic and opioidergic competitive antagonists into the central nucleus of the amygdala (Hyytia and Koob, 1995; Heyser *et al.*, 1999), while lesions of the cell bodies within this structure markedly suppressed alcohol self-administration (Moller *et al.*, 1997) (**Table 9.1**). Thus, activation of dopamine and opioid peptides and potential contributions from other neurotransmitters, such as GABA and serotonin, and neuromodulators, such as endocannabinoids, are importantly involved in the acute reinforcing effects of drugs of abuse (**Table 9.4**, binge/intoxication stage).

A role for the involvement of the extended amygdala in the aversive stimulus effects of drug withdrawal includes changes in opioidergic, GABAergic, and CRF neurotransmission during acute withdrawal. There was enhanced sensitivity of alcohol-dependent rats to GABA agonists during acute withdrawal (Roberts *et al.*, 1996), and the CRF systems in the central nucleus of the amygdala were activated during acute cocaine, alcohol, opioid, Δ^9-tetrahydrocannabinol, and nicotine withdrawal as measured by *in vivo* microdialysis and neuropharmacological probes (Heinrichs *et al.*, 1995; Merlo-Pich *et al.*, 1995; Ghozland *et al.*, 2004). Similar effects have been observed with alcohol and opioids in the lateral BNST (Olive *et al.*, 2002; Aston-Jones and Harris, 2004) (**Table 9.4**, withdrawal/negative affect stage). However, decreases in the function of dopamine, serotonin, and opioid peptides, as well as recruitment of brain stress systems such as CRF, are hypothesized to contribute to a shift in reward set point that characterizes the transition to the addictive state.

Thus, acute withdrawal from drugs of abuse produces opponent process-like changes in reward neurotransmitters in specific elements of reward circuitry associated with the extended amygdala as well as recruitment of brain stress systems that motivationally oppose the hedonic effects of drugs of abuse. Such changes in these brain systems associated with the development of motivational aspects of withdrawal are hypothesized to be a major source of neuroadaptive changes that drive and maintain addiction. All of these changes are hypothesized to be focused on a dysregulation of function within the neurocircuitry of the basal forebrain macrostructure of the extended amygdala.

In summary, the extended amygdala circuit for drug addiction puts a major focus on the role of the shell of the nucleus accumbens, BNST and central nucleus of the amygdala for the changes in hedonic tone associated with the transition from positive reinforcement to negative reinforcement as addiction develops (Koob and Le Moal, 2001).

TABLE 9.4 Neurochemical Systems in the Extended Amygdala Involved in the Motivational Effects of Different Stages of the Addiction Cycle

Stage of Addiction Cycle	Neurochemical System	Functional Effect
Binge/Intoxication	↑ Dopamine	Euphoria
	↑ Opioid peptides	Euphoria
	↑ GABA	Antianxiety
	↑ NPY	Antistress
Withdrawal/Negative Affect	↓ Dopamine	'Dysphoria'
	↓ Serotonin	'Dysphoria'
	↓ GABA	Anxiety
	↓ NPY	Antistress
	↑ Dynorphin	'Dysphoria'
	↑ CRF	Stress
	↑ Norepinephrine	Stress
Preoccupation/Anticipation		
Cue-induced craving	↑ Dopamine	'Craving'
	↑ Opioid peptides	'Craving'
	↓ Glutamate	'Craving'
Residual negative affective state	↑ Dynorphin	'Dysphoria'
	↑ CRF	Stress
	↑ Norepinephrine	Stress
	↓ GABA	Anxiety, panic attacks

Differential Role of the Nucleus Accumbens Core and Shell in Addiction

Ito, Robbins, and Everitt (2004)
Di Chiara (2002)

The nucleus accumbens has long been considered a key element for the locomotor activation and reinforcing

TABLE 9.5 **Differential Neurochemical and Lesion Effects in the Shell and Core of the Nucleus Accumbens Related to Drug Addiction**

	Nucleus Accumbens	
	Shell	Core
Acute drug-induced activation of dopamine		
Morphine	↑↑	↑
Heroin	↑↑↑	↑
Cocaine	↑↑↑	↑
Amphetamine	↑↑↑	↑↑
Δ^9-Tetrahydrocannabinol	↑↑	↑
Nicotine	↑↑	—
Nondrug-induced activation of dopamine		
Acute Fonzies	↑	↑
Chronic Fonzies	—	↑↑
Cell body-specific excitotoxic lesion effects		
Locomotor activity	↓	↑
Amphetamine locomotor activity	↓	↑
Amphetamine potentiation of conditioned response	↓	—
Discriminated approach to Pavlovian conditioned response	—	↓
Second-order schedule for drug reward	—	↓

— = not measured.

actions of psychostimulants and other drugs of abuse (Di Chiara and North, 1992; Koob, 1992b), and the neuroadaptational changes associated with development of addiction and the reinstatement of drug-seeking behavior after extinction (see previous chapters and previous discussions in this chapter). However, the development of the extended amygdala concept where the shell or medial part of the nucleus accumbens forms a transition zone within the extended amygdala (Alheid and Heimer, 1988) and the strong resemblance of the core of the nucleus accumbens to striatum and its intersection with ventral striatal-ventral pallidal loops (Groenewegen *et al.*, 1993; Haber *et al.*, 2000) has led to an intense interest in a differential role of the shell and core of the nucleus accumbens in addiction (Di Chiara, 2002; Ito *et al.*, 2004).

Di Chiara and colleagues argued that the incentive-sensitization and allostatic-counteradaptive theories do not account for two basic properties of motivational disturbances associated with drug addiction: the narrowing of behavior toward drug incentives versus nondrug incentives and the irreversibility of such a focus on drug reinforcement. To address this issue the shell of the nucleus accumbens was hypothesized to mediate the strengthening of stimulus-drug associations. This hypothesis was based on the observation that most, if not all, drugs of abuse (Pontieri *et al.*, 1995, 1996; Tanda *et al.*, 1997; Tanda and Di Chiara, 1998; Cadoni and Di Chiara, 1999) share with nondrug rewards (e.g., Fonzies and chocolate) (Bassareo and Di Chiara, 1997, 1999a,b; Barrot *et al.*, 1999, 2000; Bassareo *et al.*, 2002) the ability to stimulate dopamine transmission in the shell of the nucleus accumbens (**Table 9.5**). Much less activation was observed in the core of the nucleus accumbens with both drug and nondrug rewards. Nondrug rewards also increased dopamine release in the prefrontal cortex. The activation of dopamine release showed habituation with repeated administration of nondrug (food) rewards in the nucleus accumbens *shell* (Bassareo and Di Chiara, 1997, 1999a,b), but not with repeated administration of nondrug (food) rewards in the nucleus accumbens core or prefrontal cortex. Indeed, the activation appeared to show sensitization in these two structures (Bassareo and Di Chiara, 1997, 1999a,b; Bassareo *et al.*, 2002). Based on these results, Di Chiara hypothesized that the properties of dopamine responsiveness in the nucleus accumbens shell suggests a role for dopamine in associative stimulus-reward learning. Release of dopamine in the nucleus accumbens shell by unpredicted primary appetitive stimuli may serve to associate the discriminative properties of the rewarding stimulus with its biological outcome (e.g., the taste of food). In contrast, Di Chiara argued that the nucleus accumbens core and prefrontal cortex have a role in the expression of motivation to seek a reward and in converting motivation to action (Di Chiara, 1999).

Thus, Di Chiara argued that drugs of abuse share the effects on a nondrug reward in activating the shell of the nucleus accumbens, but this activation in the *shell* of the nucleus accumbens does not undergo habituation as with nondrug (food) rewards, imparting much more stimulus-reward learning and contributing to the compulsive components of drug-seeking behavior (Di Chiara, 1999). Repeated drug exposure also induced a sensitization of drug-induced stimulation of dopamine neurotransmission in the core of the nucleus accumbens that was hypothesized to mediate instrumental performance involved in drug-seeking (Cadoni and Di Chiara, 1999; Di Chiara, 2002).

Based on these observations, Di Chiara argued for a differential role for the dopamine system in addiction depending on the stage of the addiction cycle (Di Chiara, 1999, 2002). In the stage of controlled use, the reinforcing properties of the drug facilitate further exposure to the drug via Pavlovian incentive learning related to a release of dopamine in the shell. With repeated drug exposure, the subject enters the stage of drug abuse where repeated association of drug reward and drug-related stimuli occurs via the presence of a nonhabituating stimulation of dopamine neurotransmission in the shell of the nucleus accumbens. Sensitization of dopamine neurotransmission in the core of the nucleus accumbens is hypothesized to begin in this stage. Subsequently, the tolerance and dependence stage develops, which is characterized by a negative emotional state, and the actions of the drug are amplified by a potentiation of the dopamine-releasing properties of the drug (presumably superimposed on a decreased baseline) (see *Psychostimulants* chapter). In the postaddiction stage, abstinence and sensitization disappear, but Pavlovian associations remain as powerful incentives for reinstatement of drug self-administration (relapse).

Everitt, Robbins and colleagues initiated a parallel series of studies on the differential role of the shell and core of the nucleus accumbens on drug-seeking in addiction involving selective lesions of the core and shell, and sophisticated behavioral procedures for separating motivational components of drug reward, notably drug-seeking and drug-taking. In a study of cocaine self-administration (Ito *et al.*, 2004), animals were trained on a second-order schedule of reinforcement where response requirements for cocaine and a conditioned reinforcer were progressively increased. Under this schedule, rats were required to make *y* responses to obtain a single presentation of a 2 s light conditioned stimulus (conditioned reinforcer), while completion of *x* of these responses resulted in the delivery of cocaine, illumination of the light conditioned stimulus for 20 s, retraction of both levers, and turning off of the house light during a 20 s timeout period. Ultimately, the *y* was increased from 1 to 10, and the *x* was increased from 5 to 10, yielding a schedule of FR10 (FR10:S) in daily 2-h sessions. Excitotoxic lesions of the shell and core of the nucleus accumbens had no effect on the acquisition of responding for cocaine on a continuous reinforcement schedule, but lesions of the *core* of the nucleus accumbens profoundly impaired the acquisition of drug-seeking behavior that was maintained by drug-associated conditioned reinforcers (Ito *et al.*, 2004) (**Table 9.5**).

In contrast, selective excitotoxic lesions of the shell of the nucleus accumbens failed to alter the acquisition of drug-seeking behavior that was maintained by drug-associated conditioned reinforcers. Similar effects were observed with selective nucleus accumbens core lesions on Pavlovian approach behavior (Parkinson *et al.*, 1999) (**Table 9.5**), Pavlovian-to-instrumental transfer (Hall *et al.*, 2001), and conditioned reinforcement (Parkinson *et al.*, 1999). Taken together with data implicating the basolateral amygdala in similar effects on cocaine self-administration maintained on a second-order schedule (Whitelaw *et al.*, 1996), the nucleus accumbens *core* was hypothesized to be a component of reward-related information derived from conditioned reinforcers (Ito *et al.*, 2004) or essential for the influence of associative stimulus-reward information on goal-directed action.

In contrast, selective excitotoxic lesions of the shell of the nucleus accumbens failed to alter the acquisition of drug-seeking behavior that was maintained by drug-associated conditioned reinforcers and failed to alter Pavlovian approach behavior (Parkinson *et al.*, 1999), Pavlovian-to-instrumental transfer (Hall *et al.*, 2001), and Pavlovian conditioned reinforcement (Parkinson *et al.*, 1999) (**Table 9.5**). Shell lesions, however, produced hypoactivity, attenuated the psychostimulant effects of amphetamine (Parkinson *et al.*, 1999), and decreased nucleus accumbens-administered amphetamine potentiation of conditioned reinforcement (Parkinson *et al.*, 1999). Similar effects on amphetamine potentiation of conditioned reinforcement have been observed with lesions of the central nucleus of the amygdala (Robledo *et al.*, 1996) and the ventral subiculum (Burns *et al.*, 1993). These results suggested that the shell of the nucleus accumbens may potentiate ongoing instrumental responding in the presence of motivationally significant stimuli (Parkinson *et al.*, 1999) or the psychostimulant actions of stimulant drugs.

Reconciling the microdialysis position (Di Chiara, 2002) and the excitotoxic lesion position (Ito *et al.*, 2004) may require reassessment of how one defines a global term such as *incentive motivation*. For Di Chiara, incentive motivation is inferred by attribution of motivational effects to changes in dopamine function in the shell of the nucleus accumbens that change with nondrug reward but show resistance to change with drug rewards. In contrast, Ito, Robbins, and Everitt (2004) have operationally defined the impact of conditioned reinforcement on cocaine-seeking behavior as being disrupted by lesions of the nucleus accumbens core, but not the shell, using a second-order schedule. In contrast, using their models, the nucleus accumbens shell is more involved in the response-invigorating effects of psychomotor stimulant drugs. Overall, the results from the lesion work and the microdialysis work could be reconciled with a reversal of roles for the core and shell of the nucleus accumbens as proposed by Di Chiara (2002). Indeed, the observation that the dopamine response in the core of the nucleus accumbens sensitizes with repeated exposure to drugs of abuse could provide a neurochemical basis for the stimulus-reward association facilitation previously attributed to the shell of the nucleus accumbens.

In summary, based on microdialysis data, Di Chiara proposed that the shell of the nucleus accumbens is involved in the incentive motivational properties of drugs of abuse via the enhancement of stimulus-reward associations, while the core of the nucleus accumbens was more involved in the performance components of compulsive drug-taking. Based on lesion data and a sophisticated behavioral paradigm for exploring drug-seeking versus drug-taking, Ito, Robbins, and Everitt proposed that the shell of the nucleus accumbens was more likely to be involved in the activational (psychostimulant) component of drug effects and that the core of the nucleus accumbens was critical for imparting the conditioned reinforcing properties to previously neutral stimuli. Reconciliation of these two positions may require some reinterpretation of the neurochemical changes in the shell and core of the nucleus accumbens given the weight of evidence from the literature on differential roles of the inputs to the shell and core subregions of the nucleus accumbens on drug-seeking and drug-taking behavior (Di Chiara, 2002; Ito et al.*, 2004).*

CELLULAR HYPOTHESES OF ADDICTION

Synaptic Plasticity in the Mesolimbic Dopamine System and Drugs of Abuse

Hyman and Malenka (2001)
Thomas and Malenka (2003)

The basic underlying thesis of the mesolimbic dopamine synaptic plasticity conceptual framework is that the compulsive characteristics of drug abuse and its persistence even after abstinence are based on a pathological disruption of molecular and physiological mechanisms involved in memory. Several premises guide this conceptual framework. First, the mesolimbic dopamine system originating in the ventral tegmental area and terminating in the nucleus accumbens is viewed as the major, though not exclusive, substrate for reward and reinforcement of both drugs of abuse and natural rewards. Second, repeated exposure to drugs is argued to increase their rewarding and locomotor-activating properties (Robinson and Berridge, 1993). This 'sensitization' is hypothesized to be mediated by changes in the mesolimbic dopamine system specifically with induction of sensitization dependent on functional changes in the ventral tegmental area (Carlezon and Nestler, 2002), whereas the long-term maintenance of sensitization involves adaptations in the nucleus accumbens (Kalivas, 1995; Everitt and Wolf, 2002). Third, synaptic plasticity has been demonstrated in the mesolimbic dopamine system in that long-term potentiation (LTP) and long-term depression (LTD) can be elicited at excitatory synapses on medium spiny neurons in the nucleus accumbens (Kombian and Malenka, 1994), and excitatory synapses on dopamine neurons in the ventral tegmental area also undergo both LTP and LTD (Bonci and Malenka, 1999).

The similarities between the LTP and LTD observed in the nucleus accumbens and those observed in the CA1 region of the hippocampus have led to the hypothesis that associative synapse-specific plasticity can take place in the mesolimbic dopamine system and link this 'reward' system to a role in the development of addiction (Hyman and Malenka, 2001; Robinson and Kolb, 2004) (**Fig. 9.17**). High frequency tetanic stimulation of presynaptic fibers induced LTP but low-frequency stimulation during modest depolarization of the postsynaptic cell-induced LTD. Both LTP and LTD in the nucleus accumbens require activation of NMDA glutamate receptors, and the excitatory input appears to derive from prelimbic cortical afferents (Pennartz *et al.*, 1993; Kombian and Malenka, 1994; Thomas *et al.*, 2000). LTP in the ventral tegmental area also is NMDA receptor-dependent (Bonci and Malenka, 1999). LTD is generated in the ventral tegmental area by activation of voltage-dependent calcium channels and does not require activation of NMDA receptors (Bonci and Malenka, 1999; Jones *et al.*, 2000). However, LTD is blocked by dopamine or amphetamine acting through dopamine D_2 receptors (Jones *et al.*, 2000), providing an additional means by which psychostimulants can potentiate excitatory synaptic transmission.

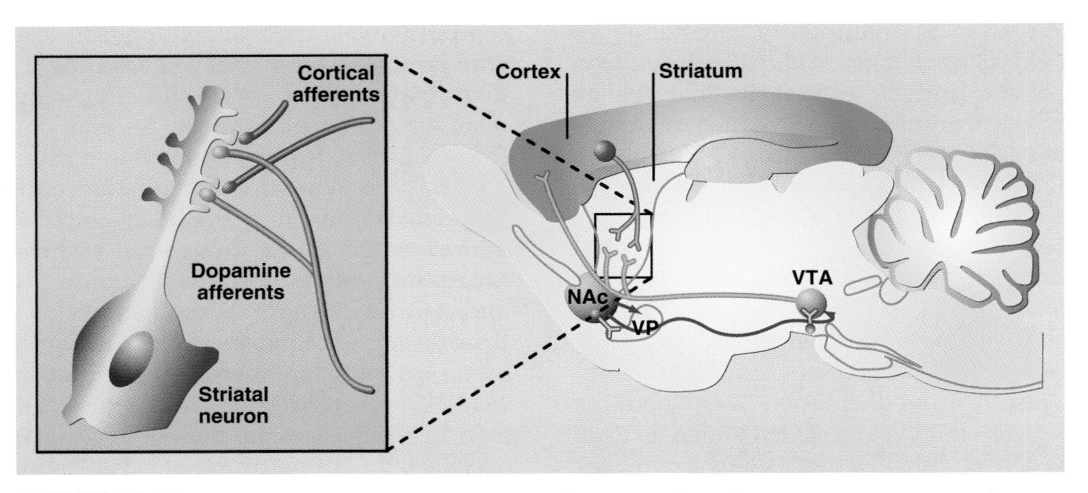

FIGURE 9.17 Dopamine–glutamate interactions in the striatum hypothesized to be involved in the reorganization of neurocircuitry of drug dependence (Hyman and Malenka, 2001). *'Approximately 95 per cent of neurons in the dorsal striatum and nucleus accumbens (NAc) are medium-sized spiny projection neurons, which use γ-aminobutyric acid (GABA) as their main neurotransmitter. These neurons receive glutamatergic projections from the cerebral cortex, which form well-defined synapses on the heads of dendritic spines. Dopaminergic axons from the midbrain pass by the necks of spines, where they release neurotransmitters; however, dopamine receptors are widely distributed on the cell membrane, including the soma. Changes in synaptic strength may result from a change in neurotransmitter release, neurotransmitter receptors, or receptor-mediated signaling. Alternatively, changes in the intrinsic excitability of neurons might follow changes in the properties or numbers of voltage-dependent ion channels. A third possibility is that morphological changes, such as the generation of new synaptic connections or pruning away of existing ones, might be initiated by various forms of synaptic plasticity.'* VP, ventral pallidum; VTA, ventral tegmental area; NAc, nucleus accumbens. [Reproduced with permission from Hyman and Malenka, 2001.]

The demonstration of synaptic plasticity at excitatory synapses in the source and terminal areas of the mesolimbic dopamine system were argued to be evidence for a role of synaptic plasticity in addiction. In support of this hypothesis, a single *in vivo* administration of cocaine caused a marked potentiation in *synaptic strength* in the ventral tegmental area in midbrain slices using whole-cell recording techniques that was due to an upregulation of AMPA glutamate receptors. Similar to locomotor sensitization to cocaine, the potentiation lasted up to 5 days postcocaine (Ungless *et al*., 2001). This potentiation did not occur on ventral tegmental area GABAergic neurons or on hippocampal CA1 neurons. The resemblance to sensitization was further argued by the observation that, like sensitization, the cocaine-induced synaptic potentiation was blocked by co-administration of cocaine with an NMDA antagonist. Also, a similar synaptic potentiation was observed with many different drugs of abuse, including amphetamine, morphine, alcohol, and nicotine (Saal *et al*., 2003). An acute stressor exposure also caused a robust increase in excitatory synaptic strength in the ventral tegmental area (Saal *et al*., 2003).

In contrast, pre-exposure to cocaine *in vivo* produced decreases in synaptic strength in neurons from prefrontal cortical afferents in slices of the nucleus accumbens shell, but not the core. In fact, a microdialysis study showed that a single injection of cocaine (15 mg/kg) resulted in an initial increase (24 h) and then a long-term decrease in glutamate release in the dorsal striatum for up to 14 days post-cocaine (McKee and Meshul, 2005). The amplitude of excitatory postsynaptic potentials and LTD were decreased at these excitatory synapses from prelimbic cortical afferents (Thomas *et al*., 2001). The authors argued that the development of an LTD-like process in the nucleus accumbens contributes to the reorganization of the neural circuitry that underlies behavioral sensitization to cocaine and thus, may be important for the development of addiction. These results suggested to the authors that increasing synaptic drive onto the mesolimbic dopamine cells enhances the motivational significance of the drugs themselves, as well as stimuli closely associated with drug-seeking. They further argued that the results showing LTP and LTD and changes in synaptic strength with drug pre-exposure show that persistent drug-induced behavioral changes, such as sensitization, probably occur because of the ability of drugs to elicit long-lasting changes in synaptic weights in crucial brain circuits, notably the mesolimbic dopamine circuit (Hyman and Malenka, 2001; Thomas and Malenka, 2003).

In summary, the development of drug addiction is conceptualized as being reflected in plasticity both in the ventral tegmental area and nucleus accumbens as demonstrated by drug-induced changes in LTP and LTD, increases in synaptic strength of the mesolimbic dopamine system, and decreases in synaptic strength of prefrontal nucleus accumbens synapses that parallel the development of locomotor sensitization (Hyman and Malenka, 2001; Thomas and Malenka, 2003).

Narrowing of Neuronal Activity by Changes in Signal and Background Firing of Nucleus Accumbens Neurons during Cocaine Self-administration

Peoples and Cavanaugh (2003)
Peoples, Lynch, Lesnock, and Gangadhar (2004)

Extracellular recordings from the nucleus accumbens during cocaine self-administration in awake, freely moving animals have revealed a population of nucleus accumbens neurons that exhibit phasic excitatory responses that are time-locked to drug-related events, specifically drug-related responding for the drug and cues that predict drug delivery (Carelli *et al.*, 1993; Chang *et al.*, 1994; Peoples and West, 1996; Uzwiak, 1997; Janak *et al.*, 1999; Peoples and Cavanaugh, 2003; Peoples *et al.*, 2004). These phasic changes have been linked to drug reward-related events. During the course of a self-administration session, the phasic nucleus accumbens neurons typically decrease their firing rate as do the majority of nucleus accumbens neurons (Peoples *et al.*, 1998, 1999). However, the phasic firing associated with drug-seeking appears to be less sensitive to the inhibitory effects of cocaine on firing. During intravenous cocaine self-administration, neurons that showed an excitatory phasic response to cocaine did not diminish their firing rate during the self-administration session but did show large decreases in background firing compared to predrug baselines (Peoples and Cavanaugh, 2003). Prior to the recording session, rats were trained to self-administer cocaine daily in 6 h or 80 injections of cocaine for 12–17 days. Recording sessions began with a 20 min, nondrug baseline recording session and then a 6-h drug session. Lever-press firing patterns were defined as a significant increase or decrease in average firing rate within ± 3 s of the lever press, relative to firing during the –12 to –9 s prepress period. The background period (control) was defined as –12 to –4 s prepress for all neurons, whether there was a lever-press or not. Changes in signal and background firing were evaluated relative to two periods: the 30 s period that preceded the onset of the self-administration session, and the 30 s period before the first press. The results showed that in neurons that exhibited an excitatory response (phasic increase in firing rate around the time of the lever press relative to background termed 'lever press neurons') there was an increase in response rate to cocaine infusion (relative to the 30 s presession), but a time-related decline in firing in the background periods (relative to the 30 s presession) (Peoples and Cavanaugh, 2003) (**Fig. 9.18**).

One interpretation of these results by the authors was focused on dopamine release in the nucleus accumbens by cocaine, where mechanisms were amplifed that normally contribute to reward-related learning. This abnormal strengthening in turn was hypothesized to contribute to compulsive drug-seeking. However, there was no evidence of a change of absolute magnitude of excitatory reward-related nucleus accumbens firing; rather, there was a differential inhibition of signal and background firing such that there was a net enhancement of drug reward-related signals relative to background firing. Dopamine input to the nucleus accumbens has been proposed to modulate feed-forward inhibition and as a result, contribute to selective activation of neural ensembles relevant to a particular behavioral situation (Pennartz *et al.*, 1994). The differential inhibition of drug reward-related firing and background firing thus may reflect differential modulation of ensemble activity in the form of filtering, where only certain neurons and ensembles are activated. This filtering in turn could contribute to learning by narrowing the ensembles of neurons in the nucleus accumbens to those that mediate the strengthening of particular associations.

An alternative interpretation was that during drug self-administration sessions, dopamine-mediated drug effects may weaken synaptic connections and neural responses that are involved in transmission of signals unrelated to reward, and a highly selective sparing of neural responses related to drug reward. This in turn could lead to a progressive decline in the throughput of nondrug reward-related signals in circuits involving the nucleus accumbens such as the cortico-striato-pallido-thalamo-cortical loops. Such a narrowing of information flow also could potentially contribute to the progressive narrowing in the behavioral repertoire, decreased reward function and dysphoria, and cognitive deficits associated with the development of drug addiction (Rogers *et al.*, 1999; Grant *et al.*, 2000; Koob and Le Moal, 2001; Volkow *et al.*, 2003).

In summary, Peoples and colleagues argued that a narrowing of the behavioral repertoire to drug-related rewards rather than natural rewards during the development of addiction is dependent on phasic response neurons in the

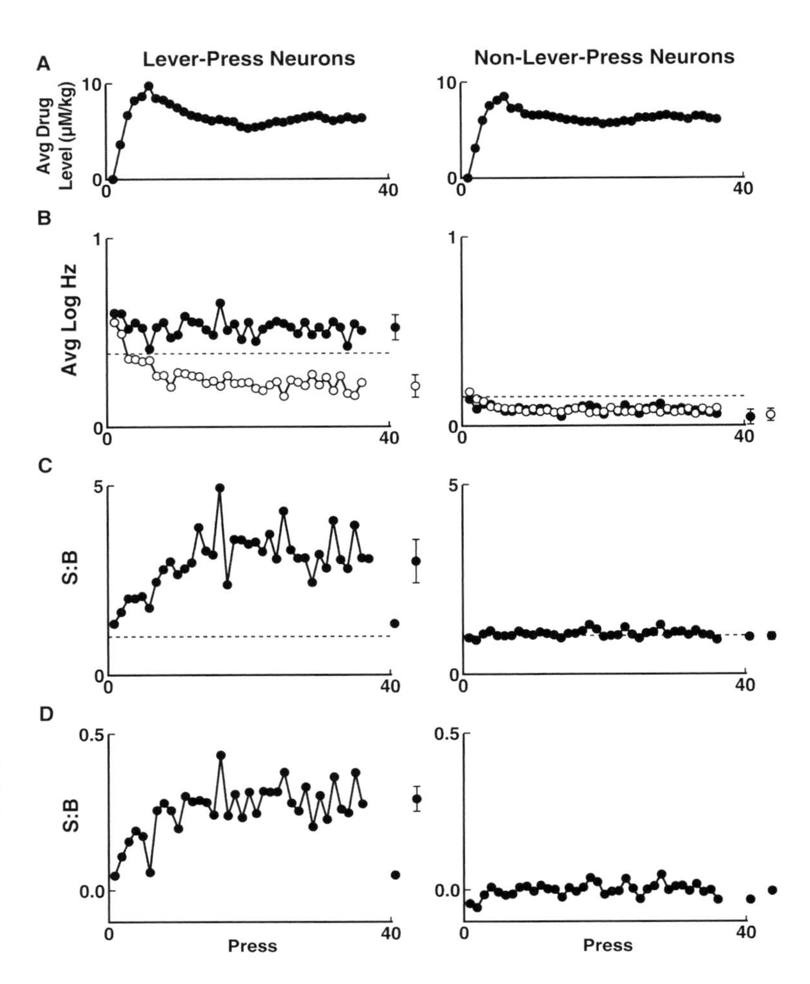

FIGURE 9.18 Changes in signal and background for tonically inhibited neurons in the nucleus accumbens during extracellular recordings of intravenous cocaine self-administration sessions. Rats were trained to intravenously self-administer cocaine in 6-h (or 80 infusion) sessions. A recording session started with a 20 min nondrug baseline recording period, followed by the typical daily self-administration session and ended by a 40 min nondrug recovery period. '*Left and right: tonically inhibited lever-press neurons (n = 14) and nonlever-press neurons (n = 31), respectively.* (A) *average drug level at the time of each press is plotted as a function of press number.* (B) *average log Hz during the signal (closed circles) and background (open circles) periods is plotted as a function of lever-press number. Dashed line, average firing rate during the 30 s pre-*S^D. *To the right of the plot, standard error bars are shown for the median average log Hz during presses 11–36 for the signal (closed circles, left) and background (open circles, right) periods.* (C) *average signal-to-background ratio (S:B) is plotted as a function of lever-press number. Dotted line, a signal to background ratio of 1. To the right of the plot, standard error bars are shown for the first press (light gray bars, left) and the press at which the median ratio during presses 11–36 was observed (black bars, right).* (D) *average difference between signal and background (S-B) plotted as a function of lever-press number. To the right of the plot, standard error bars are shown for the first press (light gray bars, left) and the press at which the median S-B difference during presses 11–36 was observed (black bars, right).*' [Reproduced with permission from Peoples and Cavanaugh, 2003.]

nucleus accumbens responding differentially to repeated cocaine compared to nonphasic neurons, with a stable response to cocaine in phasic neurons superimposed on a time-related decline in background firing (Peoples and Cavanaugh, 2003; Peoples et al., *2004).*

MOLECULAR HYPOTHESES OF ADDICTION

Molecular Basis of Long-term Plasticity Underlying Addiction

Nestler (2001, 2004)

Acknowledging that all drugs of abuse share some common neurocircuitry actions, namely inhibition of medium spiny neurons in the nucleus accumbens either through dopamine or other G_i-coupled receptors, Nestler has focused on identifying the molecular and cellular mechanisms involved (Nestler, 2001). The conceptual framework is that repeated perturbation of intracellular signal transduction pathways leads to changes in nuclear function and altered rates of transcription of particular target genes. Altered expression of such genes would lead to altered activity of the neurons where such changes occur, and ultimately, to changes in neural circuits in which those neurons operate (Nestler, 2001) (**Fig. 9.19**).

A transcription factor particularly implicated in the plasticity associated with the addiction cycle is cyclic adenosine monophosphate (cAMP) response element binding protein (CREB). CREB regulates the transcription of genes that contain a cAMP response element site within the regulatory regions and can be found ubiquitously in genes expressed in the central nervous system such as those encoding neuropeptides, synthetic enzymes for neurotransmitters, signaling proteins, and other transcription factors. CREB can be phosphorylated by not only protein kinase A but also by protein kinases regulated by growth factors putting it at a point of convergence for several intracellular messenger pathways that can regulate the expression of genes (Nestler, 2001) (**Fig. 9.19**).

Much work in the addiction field has shown that activation of CREB in the nucleus accumbens is a consequence of chronic exposure to opiates and

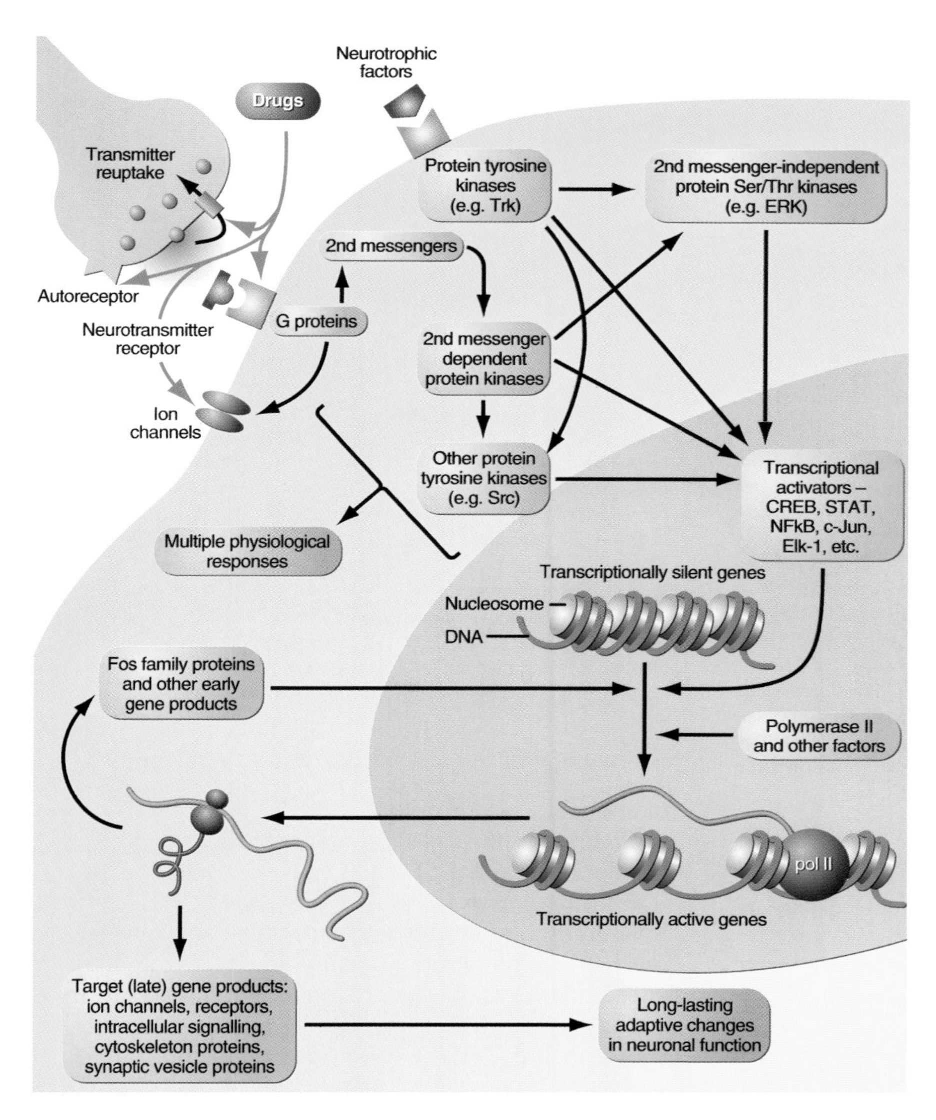

FIGURE 9.19 Regulation of gene expression by drugs of abuse (Nestler, 2001). *'The rate of expression of a particular gene is controlled by its location within nucleosomes and by the activity of the transcriptional machinery (Carey and Smale, 2000). A nucleosome is a tightly wound span of DNA that is bound to histones and other nuclear proteins. Transcription requires the unwinding of a nucleosome, which makes the gene accessible to a basal transcription complex. This complex consists of RNA polymerase (pol II, which transcribes the new RNA strand) and numerous regulatory proteins (some of which unwind nucleosomes through histone acetylation). Transcription factors bind to specific sites (response elements; also called promoter or enhancer elements) that are present within the regulatory regions of certain genes, and thereby increase or decrease the rate at which they are transcribed. Transcription factors act by enhancing or inhibiting the activity of the basal transcription complex, in some cases by altering nucleosomal structure through changes in the histone acetylation of the complex. Regulation of transcription factors is the best-understood mechanism by which changes in gene expression occur in the adult brain (Nestler* et al., *1993; Nestler and Aghajanian, 1997; Berke and Hyman, 2000). Most transcription factors are regulated by phosphorylation. Accordingly, by causing repeated perturbation of synaptic transmission and hence of protein kinases or protein phosphatases, repeated exposure to a drug of abuse would lead eventually to changes in the phosphorylation state of particular transcription factors such as CREB that are expressed under basal conditions. This would lead to the altered expression of their target genes. Among such target genes are those for additional transcriptional factors (such as c-Fos), which—through alterations in their levels— would alter the expression of additional target genes. Drugs of abuse could conceivably produce stable changes in gene expression through regulation of many other types of nuclear proteins, but such actions have not yet been shown.'* [Reproduced with permission from Nestler, 2001.]

cocaine (Terwilliger *et al.*, 1991; Turgeon *et al.*, 1997; Shaw-Lutchman *et al.*, 2002, 2003). Such an activation was first recognized because CREB is activated as a consequence of upregulation of the cAMP pathway, an upregulation long associated with chronic administration of opiates (Collier and Francis, 1975; Sharma *et al.*, 1975; Nestler, 2004) (**Fig. 9.20**). Opioids acutely inhibit adenylyl cyclase via G_i-coupled receptors, and an upregulation of the cAMP pathway is the result of a compensatory homeostatic response of cells in various sites of the central nervous system to persistent opioid inhibition, including the nucleus accumbens (Terwilliger *et al.*, 1991). Mice with mutations in the CREB gene show decreased development of opioid dependence as measured by physical signs of opioid withdrawal (Maldonado *et al.*, 1996). Upregulation of the cAMP pathway has been observed in various brain regions and has been hypothesized to have a role not only in the physical signs of opioid withdrawal, but also in the motivational aspects of opioid withdrawal (Nestler, 2001) (**Table 9.6**).

Similar upregulation of the cAMP pathway in the nucleus accumbens has been observed with other drugs of abuse (Ortiz *et al.*, 1995). *In vitro*, in a NG108-15 neuroblastoma x glioma hybrid cell line, acute and chronic alcohol also increased cAMP levels via activation

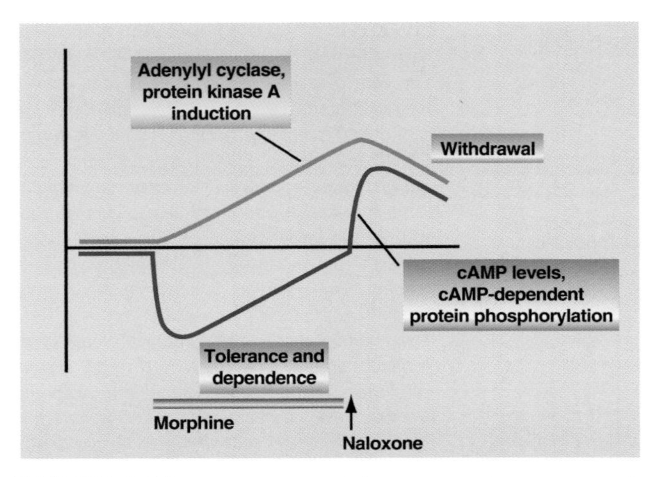

FIGURE 9.20 Upregulation of the cAMP pathway as a mechanism of opiate tolerance and dependence (Nestler, 2004). *'Opiates acutely inhibit the functional activity of the cAMP pathway (indicated by cellular levels of cAMP and cAMP-dependent protein phosphorylation). With continued opiate exposure, functional activity of the cAMP pathway gradually recovers, and increases far above control levels following removal of the opiate (e.g., by administration of the opioid receptor antagonist naloxone). These changes in the functional state of the cAMP pathway are mediated via the induction of adenylyl cyclase and protein kinase A (PKA) in response to chronic administration of opiates. Induction of these enzymes accounts for the gradual recovery in the functional activity of the cAMP pathway that occurs during chronic opiate exposure (tolerance and dependence) and activation of the cAMP pathway observed on removal of opiate (withdrawal).'* [Reproduced with permission from Nestler, 2004.]

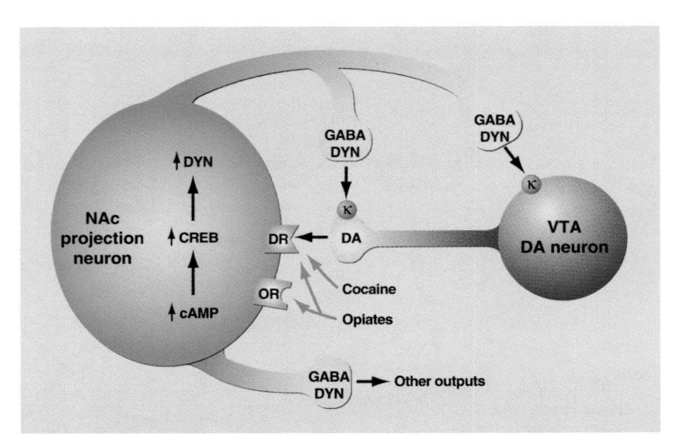

FIGURE 9.21 Regulation of CREB by drugs of abuse (Nestler, 2001). *'The figure shows a dopamine (DA) neuron of the ventral tegmental area (VTA) innervating a class of γ-aminobutyric acid (GABA) projection neuron from the nucleus accumbens (NAc) that expresses dynorphin (DYN). Dynorphin constitutes a negative feedback mechanism in this circuit: dynorphin, released from terminals of the NAc neurons, acts on κ-opioid receptors located on nerve terminals and cell bodies of the DA neurons to inhibit their functioning. Chronic exposure to cocaine or opiates upregulates the activity of this negative feedback loop through upregulation of the cAMP pathway, activation of CREB and induction of dynorphin.'* cAMP, cyclic adenosine monophosphate; CREB, cyclic adenosine monophosphate response element binding protein; DR, dopamine receptor; OR, opioid receptor. [Reproduced with permission from Nestler, 2001.]

TABLE 9.6 Upregulation of the cAMP Pathway and Opiate Addiction

Site of upregulation	Functional consequence
Locus coeruleus*	Physical dependence and withdrawal
Ventral tegmental area†	Dysphoria during early withdrawal periods
Periaqueductal grey	Dysphoria during early withdrawal periods
	Physical dependence and withdrawal
Nucleus accumbens	Dysphoria during early withdrawal periods
Amygdala	Conditioned aspects of addiction?
Dorsal horn of spinal cord	Tolerance to opiate-induced analgesia
Myenteric plexus of gut	Tolerance to opiate-induced reductions in intestinal motility
	Increases in motility during withdrawal

*The cyclic adenosine monophosphate (cAMP) pathway is upregulated within the principal noradrenergic neurons located in this region.
†Indirect evidence indicates that the cAMP pathway may be upregulated within γ-aminobutyric acid (GABA) neurons that innervate the dopamine and serotonin cells located in the ventral tegmental area and periaqueductal grey, respectively. During withdrawal, the upregulated cAMP pathway would become fully functional and could contribute to a state of dysphoria by increasing the activity of the GABA neurons, which would then inhibit the dopamine and serotonin neurons (Rasmussen *et al.*, 1990; Delfs *et al.*, 2000).
[Reproduced with permission from Nestler, 2001.]

of adenosine A_2 receptors, leading to phosphorylation of CREB (Sapru *et al.*, 1994; Constantinescu *et al.*, 1999). Nestler has argued, 'There is now compelling evidence that upregulation of the cAMP pathway and CREB in this brain region (nucleus accumbens) represents a mechanism of "motivational tolerance and dependence": these molecular adaptations decrease an individual's sensitivity to the rewarding effects of subsequent drug exposures (tolerance) and impair the reward pathway (dependence) so that after removal of the drug the individual is left in an amotivational, depressed-like state' (Carlezon *et al.*, 1998; Self *et al.*, 1998; Pliakas *et al.*, 2001; Walters and Blendy, 2001; Barrot *et al.*, 2002; Newton *et al.*, 2002). The effects of upregulation of the cAMP pathway and CREB also contributed to increased expression of dynorphin induced by exposure to drugs of abuse (Carlezon *et al.*, 1998). Dynorphin is a κ agonist and causes dysphoric-like responses. It is hypothesized that CREB increases the gain on the dynorphin-mediated negative feedback circuit in the nucleus accumbens and contributes to the aversive states associated with acute withdrawal (Nestler, 2001) (**Fig. 9.21**). However, regulation of the cAMP pathway, CREB, and dynorphin are relatively short-lived and may not explain long-term changes associated with the propensity to relapse (Nestler, 2004) (**Fig. 9.20**).

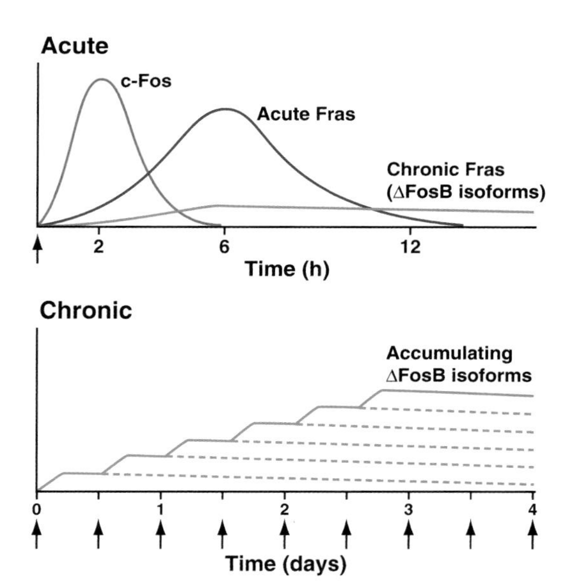

FIGURE 9.22 Regulation of FosB by drugs of abuse (Nestler, 2001). *'The top graph shows the several waves of induction of Fos family proteins in the NAc after a single exposure (black arrow) to a drug of abuse. These proteins include c-Fos and several Fras (fos-related antigens; for example, FosB, Fra-1, Fra-2). All of these proteins of the Fos family are unstable. By contrast, isoforms of FosB are highly stable and therefore persist in the brain long after drug exposure. Because of this stability, FosB accumulates with repeated drug exposures, as shown in the bottom graph.'* [Reproduced with permission from Nestler, 2001.]

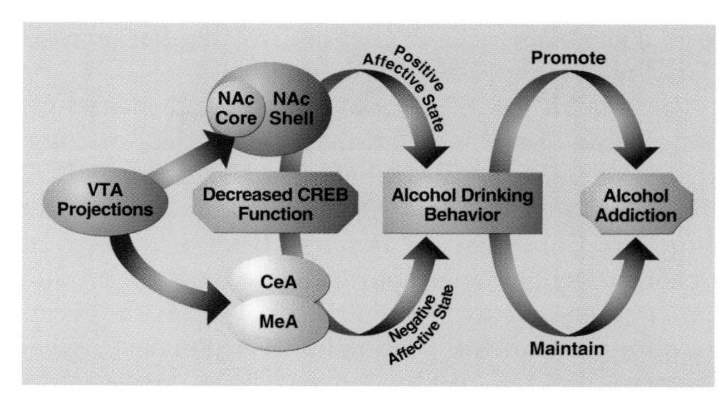

FIGURE 9.23 Proposed molecular model in the extended amygdala for the positive and negative affective states of alcohol addiction (Pandey, 2004). *'The decreased cyclic AMP-responsive element binding protein (CREB) function in the shell of nucleus accumbens (NAc), which receives dopaminergic projections from ventral tegmental area (VTA), might be associated with euphoric, rewarding (positive affective state) properties of alcohol use. The decreased CREB function in the central and medial nucleus of amygdala (CeA and MeA, respectively) might be involved in a negative affective state related to alcohol withdrawal, or as a result of pre-existing conditions.'* [Reproduced with permission from Pandey, 2004.]

The molecular changes associated with long-term changes in brain function as a result of chronic exposure to drugs of abuse also have been linked to changes in other transcription factors, such as members of the Fos family. Such factors that can change gene expression and produce long-term changes in protein expression, and as a result presumably neuronal function. While acute administration of drugs of abuse can cause a rapid (on the order of hours) activation of members of the Fos family, such as c-fos, FosB, Fra-1, and Fra-2 in the nucleus accumbens (Hope *et al.*, 1994; Moratalla *et al.*, 1996), other transcription factors, isoforms of ΔFosB, accumulate over longer periods of time (days) with repeated drug administration (Nestler, 2001) (**Fig. 9.22**). Data suggest that increased ΔFosB may mediate enhanced locomotor and rewarding responses to drugs of abuse (Kelz *et al.*, 1999). Inducible expression of ΔFosB increased cocaine-seeking behavior (Colby *et al.*, 2003). Knockouts of ΔFosB have not supported a role for ΔFosB in increased sensitivity to drugs of abuse (Hiroi *et al.*, 1997), but such knockouts lack both products of the gene: ΔFosB and full-length FosB. Interestingly, transgenic mice that express a dominant-negative mutant form of c-jun which selectively antagonizes the transcriptional effects of ΔFosB showed reduced responses to drugs of abuse (Peakman *et al.*, 2003; McClung *et al.*, 2004). Nestler argued that ΔFosB may be a sustained molecular switch that helps to initiate and maintain a state of addiction (Nestler *et al.*, 2001). How changes in ΔFosB which can last for days can translate into structural changes linked to chronic administration of drugs such as a reduction in the density of dendritic spines of medium spiny neurons in the nucleus accumbens (Robinson and Kolb, 1997, 1999), and whether such molecular and structural changes actually underlie the vulnerability to relapse, remain challenges for future work.

In summary, chronic administration of drugs of abuse leads to a short-term upregulation of the cAMP pathway and CREB that may represent a mechanism of tolerance and dependence, but long-term changes in transcription factors such as ΔFosB may maintain the state of addiction and be the basis for vulnerability to relapse (Nestler, 2001, 2004).

The Gene Transcription Factor CREB: Role in Positive and Negative Affective States of Alcohol Addiction

Pandey (2004)

The conceptual framework explored here is that signaling of the gene transcription factor CREB in the extended amygdala has a role in the positive and negative affective states associated with alcohol addiction (Pandey, 2004) (**Fig. 9.23**). Both increased positive affective states and negative affective states are hypothesized to promote and maintain alcohol addiction, and form a key element in the hedonic dysregulation associated with compulsive drinking.

The cAMP-protein kinase A second-messenger system has long been implicated in alcohol tolerance and dependence (Tabakoff and Hoffman, 1992; Diamond and Gordon, 1997), but in an opposite direction from that of opiates. Brief exposures to alcohol potentiated receptor-activated cAMP production, and chronic exposure to alcohol decreased receptor-stimulated cAMP production (Gordon *et al.*, 1992). Decreased cAMP production has been considered a cellular model for alcohol dependence since stimulated cAMP levels are abnormally low in NG108-15 neural cell cultures after alcohol withdrawal but return to normal levels when alcohol is reintroduced (Gordon *et al.*, 1986). CREB phosphorylation also has been shown to be lower in the cortex during alcohol withdrawal (Pandey *et al.*, 1999, 2001). Decreased CREB phosphorylation and decreases in CaMKIV protein levels also have been observed in the central nucleus of the amygdala during alcohol withdrawal and have been linked to the increased anxiety-like responses associated with acute alcohol withdrawal (Pandey *et al.*, 2003). No changes were observed during alcohol exposure *per se*. Infusion of a protein kinase A activator, S_p-cAMPS (S_p-adenosine 3′,5′-cyclic monophosphorothioate triethylammonium salt), into the central nucleus of the amygdala normalized CREB phosphorylation and prevented the development of anxiety-like responses during alcohol withdrawal (Pandey *et al.*, 2003). Others have shown that CREB phosphorylation is decreased in the nucleus accumbens during voluntary alcohol exposure and remained decreased up to 72 h into withdrawal (Li *et al.*, 2003). Consistent with these results, decreased CREB phosphorylation after administration of a protein kinase A inhibitor into the central nucleus of the amygdala produced anxiety-like responses (Zhang and Pandey, 2003). Also consistent with these results is the observation that CREB mutant mice, which are deficient in transcriptionally active α and β CREB isoforms, have increased anxiety-like responses (Graves *et al.*, 2002; Pandey *et al.*, 2004; Valverde *et al.*, 2004).

Preclinical and clinical studies suggest a relationship between high anxiety-like states and greater alcohol intake (see *Alcohol* chapter) and as such, decreased CREB phosphorylation in the central nucleus of the amygdala may be responsible for increased anxiety-like responses in alcohol-dependent animals and thus increased drinking in dependent animals possibly through actions on the NPY systems which is a CREB target gene (Chance *et al.*, 2000). Normalization of CREB phosphorylation by infusing the protein kinase A activator S_p-cAMPS into the central nucleus of the amygdala that prevented the anxiety-like effects in rats during alcohol withdrawal (Pandey *et al.*, 2003) and also reversed the decreased expression of NPY during alcohol withdrawal (Zhang and Pandey, 2003). Since CREB regulates the expression of the gene encoding NPY, and decreased NPY is associated with anxiety and excessive drinking, it was further hypothesized that the decreased function of CREB in the central nucleus of the amygdala may regulate both anxiety and alcohol intake via decreased expression of NPY (Pandey, 2003). Similar results have been observed in the shell of the nucleus accumbens during acute nicotine withdrawal where there were reductions in total CREB and phosphorylated CREB (Pluzarev and Pandey, 2004).

What is difficult to resolve is how decreased CREB phosphorylation in the central nucleus of the amygdala and nucleus accumbens yields decreased NPY, to yield increased anxiety, to yield increased drinking for alcohol, while increased CREB phosphorylation in the nucleus accumbens with cocaine and opioid leads to increased dynorphin gain which leads to dysphoria and increased cocaine and morphine intake (Nestler, 2001, 2004). One possibility is that there are simply differences between different drugs of abuse, and while CREB changes occur during neuroadaptation to all drugs of abuse, the valence of these changes differ between drugs. Another possibility is that there are differential molecular changes in different parts of the extended amygdala circuitry. Yet another possibility is that the anxiety-like responses associated with enhanced alcohol drinking during acute withdrawal reflects decreased CREB function in the central nucleus of the amygdala (Pandey *et al.*, 2003), but the increased CREB function in the nucleus accumbens occurs during active drug-taking to decrease sensitivity to the rewarding effects of the drug and to promote tolerance (Carlezon *et al.*, 1998; Nestler, 2004). Clearly, the relationship of these molecular changes to the motivational effects of drug addiction require further research.

In summary, Pandey argued that CREB under-expression in the central nucleus of the amygdala promotes anxiety-like responses, possibly through decreased function of NPY, that in turn promotes excessive drinking, but decreased CREB function in the nucleus accumbens promotes an increase in the intake of drugs to achieve an increase in reward (Pandey, 2004).

Glutamate Systems in Cocaine Addiction

Kalivas, McFarland, Bowers, Szumlinski, Xi and Baker (2003)
Kalivas (2004)

As noted earlier, Kalivas and colleagues and other groups have shown that reversible inhibition of

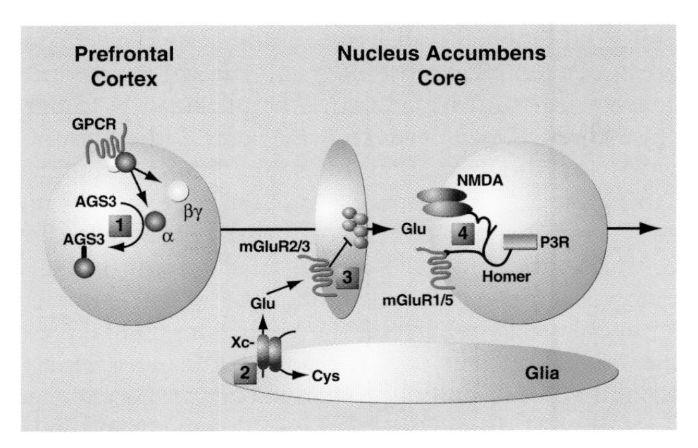

FIGURE 9.24 Illustration of the neuroadaptations produced by repeated cocaine in the glutamatergic projection from the prefrontal cortex to the nucleus accumbens core (Kalivas *et al.*, 2003). *'The numbers refer to specific adaptations described in the text. (1) Increased activator of G-protein signaling 3 (AGS3) results in reduced signaling through G_i [protein] coupled receptors (GPCR) because AGS3 competes with β for G_i binding. (2) The rate of cystine-glutamate exchange (Xc-) is reduced in the nucleus accumbens, which decreases the basal extracellular level of glutamate. (3) Reduced extracellular glutamate combined with desensitization of mGluR2/3 decreases inhibitory tone on synaptic glutamate release. (4) Reduced levels of Homer1bc compromise signaling in the postsynaptic density between group I mGluR (mGluR1/5), NMDA receptors and IP3 receptors.'* AGS3, activator of G-protein signaling 3; Cys, cystine; Glu, glutamate; mGluR, metabotropic glutamate receptor; NMDA, *N*-methyl-D-aspartate. [Reproduced with permission from Kalivas *et al.*, 2003.]

the prefrontal cortex prevents the reinstatement of drug-seeking induced by cocaine, stress, or cocaine-associated cues (McFarland and Kalivas, 2001). The frontal cortex activation involved in reinstatement has been linked to a glutamatergic pathway to the core of the nucleus accumbens (McFarland *et al.*, 2003). With this neurocircuitry as a background, Kalivas and co-workers have identified four neuroadaptations of potential importance in mediating the dysregulation of glutamatergic neurotransmission associated with cocaine-primed reinstatement of drug-seeking (Kalivas *et al.*, 2003) (see (1) of **Fig. 9.24**).

One neuroadaptation involves protein regulation of a G-protein-linked receptor to increase glutamate release in prefrontal cortex neurons. Activator of G-protein signaling 3 (AGS3) is a protein that regulates G-protein signaling by reducing the access of $G_{i\alpha}$ to receptor and effector proteins. Withdrawal from repeated cocaine administration produces an increase in AGS3 in the prefrontal cortex and nucleus accumbens core and is accompanied by reduced signaling of presynaptic $G_{i\alpha}$-coupled receptors, notably metabotropic glutamate 2/3 receptors (mGluR2/3) and an increase in cocaine-primed glutamate release in the nucleus accumbens core (Xi *et al.*, 2002; Kalivas *et al.*, 2003) (see (1) of **Fig. 9.24**). Support for a role of enhanced glutamate release in this pathway associated with cocaine-induced reinstatement is the observation that infusion of antisense oligonucleotides to AGS3 into the prefrontal cortex prevented cocaine-induced reinstatement in rats subjected to extinction, and increasing AGS3 facilitated the ability of cocaine to elevate glutamate levels in the nucleus accumbens (Bowers *et al.*, 2004). Thus, cocaine-induced elevation in AGS3 in the prefrontal cortex is associated with enhanced release of prefrontal-accumbens glutamate and reinstatement of cocaine self-administration.

A second proposed neuroadaptation in glutamate neurotransmission associated with repeated administration of cocaine is a decrease in the cystine–glutamate exchanger function (Kalivas *et al.*, 2003) (see (2) of **Fig. 9.24**). The cystine–glutamate exchanger is a heterodimer that is hypothesized to have a primary role in regulating extracellular levels of glutamate in the nucleus accumbens. Repeated cocaine administration decreases the rate of cystine–glutamate exchange and thus reduces extracellular basal levels of glutamate, and administration of the procysteine drug *N*-acetylcysteine normalizes extracellular levels of glutamate and blocks reinstatement by a cocaine priming (Baker *et al.*, 2003).

A third proposed neuroadaptation is that the reduced basal extracellular glutamate levels reduce the tone at mGluR2/3 autoreceptors and thus facilitate phasic synaptic release of glutamate during cocaine-primed reinstatement (Kalivas *et al.*, 2003) (see (3) of **Fig. 9.24**).

A fourth proposed neuroadaptation in glutamate neurotransmission is reduced expression of post-synaptic glutamate receptors caused by a reduction in Homer proteins, notably Homer 1bc and Homer 2 (Kalivas *et al.*, 2003) (see (4) of **Fig. 9.24**). Homer 2 knockout mice are more sensitive to the motor stimulant and rewarding effects of cocaine (Ghasemzadeh *et al.*, 2003; Szumlinski *et al.*, 2004). Thus, reduced Homer proteins resulted in reduced metabotropic glutamate 1/5 receptor signaling which is hypothesized to cause enhanced responsiveness to cocaine. It should be noted that deletion of the mGluR5 gene or mGluR5 antagonists blocked cocaine reinforcement, suggesting that the above effects are indeed mediated by the mGluR1 receptor (Chiamulera *et al.*, 2001; McGeehan and Olive, 2003; Kalivas, 2004).

Thus, the Kalivas model is focused on molecular changes within a specific neurotransmitter system (glutamate) within a specific neurocircuit (prefrontal cortex to nucleus accumbens). Prolonged decreases in tonic glutamate in the prefrontal cortex-nucleus accumbens core pathway, combined with desensitization of mGluR2/3 receptors, results in a loss of feedback control of dynamic synaptic glutamate release.

Lower basal levels of glutamate, combined with increased release of synaptic glutamate in the nucleus accumbens core in response to activation of prefrontal cortical afferents results in a hypothesized amplified signal, and at the behavioral level, results in increased drive for drug-seeking. A variety of postsynaptic glutamate receptor functional changes also have been described, including a decrease in both NMDA and mGluR1 receptor function. However, the exact nature of this relationship remains to be resolved. How the molecular changes associated with prefrontal glutamate dysfunction in preclinical reinstatement models are related with clinical observations of prefrontal cortical hypofunction in established drug addicts, and how such changes relate to vulnerability to drug addiction and relapse remain to be established.

In summary, the Kalivas model emphasizes molecular changes within the glutamate projection from the prefrontal cortex to the nucleus accumbens. Lower basal levels of glutamate caused by a decreased cystine–glutamate exchange, combined with augmented cocaine-induced glutamate release due to reduced tone of mGluR2/3 autoreceptors and an increase of the AGS3 protein, are hypothesized to promote cocaine reinstatement (*Kalivas* et al., *2003; Kalivas, 2004*).

SYNTHESIS: COMMON ELEMENTS OF MOST NEUROBIOLOGICAL MODELS OF ADDICTION

The neural elements comprising each of the neurocircuitry models described above have several common elements that include a reward/stress circuit, a behavioral output (compulsivity) circuit, and a drug- and cue-induced reinstatement (craving) circuit that can be integrated into an overall heuristic model (**Fig. 9.25**). Most models include a key role for some component of reward usually involving the ventral tegmental area and the nucleus accumbens and sometimes the central nucleus of the amygdala. Some have argued for a key role for the basal forebrain suprastructure of the extended amygdala in this reward function (**Fig. 9.25**). The ventral striatal-pallidal-thalamic loops have long been hypothesized to play a key role in translating motivation to action. The prefrontal cortex is hypothesized to have a key role in drug-induced reinstatement and the basolateral amygdala in cue-induced reinstatement. Stress neurocircuitry involving brain stem-basal forebrain loops is implicated not only in the negative affect associated with acute and protracted abstinence from drugs of abuse but also stress-induced relapse.

The ensemble of subcircuits outlined in **Fig. 9.25** not only incorporates most elements linked to various components of drug addiction emphasized by different theoretical perspectives, but also provides a framework for the evolution of neurocircuitry changes during the development of addiction. For example, the acute reinforcing effects of drugs of abuse that comprise the binge/intoxication stage most likely involve actions with an emphasis on the extended amygdala reward system, the ventral tegmental area, the arcuate nucleus of the hypothalamus, and ventral striatal-ventral pallidal-thalamic-cortical loops (see light gray of drug-associated reinforcement and light blue of behavioral output of **Fig. 9.25**). In contrast, the symptoms of acute withdrawal important for addiction, such as negative affect and increased anxiety associated with the negative affect stage, most likely involve decreases in function of the extended amygdala reward system but a recruitment of brain stress neurocircuitry (see dark gray of **Fig. 9.25**). The preoccupation/anticipation or craving stage involves key afferent projections to the extended amygdala and nucleus accumbens, specifically the prefrontal cortex for drug-induced reinstatement and the basolateral amygdala for cue-induced reinstatement (see yellow of drug- and cue-induced reinstatement in **Fig. 9.25**). Unreconciled at the present time is how the activation of prefrontal cortex executive system important for cue- and drug-priming-induced reinstatement, but also known to be severely compromised in established drug addicts, also can be responsible for elements of drug-induced craving (see *Drug Addiction: Transition from Neuroadaptation to Pathophysiology* chapter).

Note also that the development of addiction is hypothesized to involve not only neurocircuitry changes with emphasis on different elements but also to involve changes in how rewards are processed. Initial activation of the extended amygdala and ventral striatal-ventral pallidal-thalamic loops may represent positive reinforcement, but during the development of addiction, the positive reinforcement gives way to negative reinforcement with recruitment of the stress neurocircuits (**Fig. 9.25**). Understanding how both of these hypothesized changes evolve in the extended amygdala to effect increased motivation for drug use under the construct of negative reinforcement represents a challenge for future research.

Clearly, much emphasis at the cellular and molecular levels has focused on neuroplasticity in the origin and terminal areas of the mesocorticolimbic dopamine system, as this system has long been associated with the acute reinforcing effects of drugs of abuse. However, the acute reinforcing effects of drugs of abuse other than indirect sympathomimetic psychostimulants have

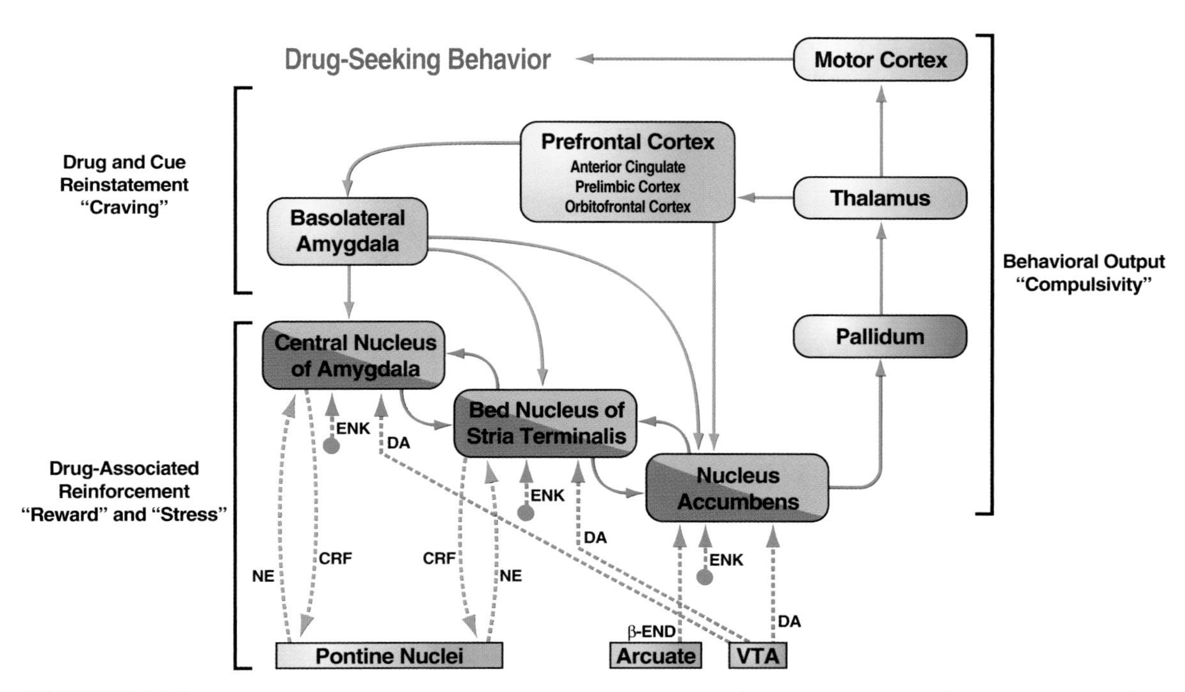

FIGURE 9.25 Key common neurocircuitry elements in drug-seeking behavior of addiction. Three major circuits that underlie addiction can be distilled from the literature. A drug-reinforcement circuit ('reward' and 'stress') is comprised of the extended amygdala including the central nucleus of the amygdala, the bed nucleus of the stria terminalis, and the transition zone in the shell of the nucleus accumbens. Multiple modulator neurotransmitters are hypothesized, including dopamine and opioid peptides for reward and corticotropin-releasing factor and norepinephrine for stress. The extended amygdala is hypothesized to mediate integration of rewarding stimuli or stimuli with positive incentive salience and aversive stimuli or stimuli with negative aversive salience. During acute intoxication, valence is weighted on processing rewarding stimuli, and during the development of dependence aversive stimuli come to dominate function. A drug- and cue-induced reinstatement ('craving') neurocircuit is comprised of the prefrontal (anterior cingulate, prelimbic, orbitofrontal) cortex and basolateral amygdala with a primary role hypothesized for the basolateral amygdala in cue-induced craving and a primary role for the medial prefrontal cortex in drug-induced craving, based on animal studies. Human imaging studies have shown an important role for the orbitofrontal cortex in craving (see text). A drug-seeking circuit ('compulsive') circuit is comprised of the nucleus accumbens, ventral pallidum, thalamus, and orbitofrontal cortex. The nucleus accumbens has long been hypothesized to have a role in translating motivation to action and forms an interface between the reward functions of the extended amygdala and the motor functions of the ventral striatal-ventral pallidal-thalamic-cortical loops. The striatal-pallidal-thalamic loops reciprocally move from prefrontal cortex to orbitofrontal cortex to motor cortex, ultimately leading to drug-seeking behavior. Note that for the sake of simplicity, other structures are not included, such as the hippocampus (which presumably mediates context-specific learning, including that associated with drug actions). Also note that dopamine and norepinephrine both have widespread innervation of cortical regions and may modulate function relevant to drug addiction in those structures. DA dopamine, ENK enkephalin, CRF corticotropin-releasing factor, NE norepinephrine, β-END β-endorphin.

been shown to be independent of the mesocorticolimbic dopamine system. Rodents continue to show rewarding effects of heroin, alcohol, and nicotine despite inactivation of the mesocorticolimbic dopamine system (Pettit *et al.*, 1984; Rassnick *et al.*, 1993b; Laviolette and Van der Kooy, 2003). While arguments can be made for residual dopamine mediating such effects or a role for other dopamine systems, a reasonable hypothesis is that elements independent of the dopamine systems can activate reward circuits outlined in **Fig. 9.25**. At a minimum, many hypotheses regarding the neural substrates for the acute reinforcing effects of drugs of abuse designate targets other than a direct action on dopamine cells as the initial sites of action for the acute reinforcing effects of drugs of abuse. These include interneurons that interact with either the cell bodies of the mesolimbic dopamine system or the medium spiny neurons of the nucleus accumbens that are postsynaptic to the mesolimbic dopamine system.

Nevertheless, despite accumulating evidence suggesting that the role of the midbrain dopamine system is more to facilitate responding to salient incentives than to mediate primary reward, the simplistic view continues to be *drug reward = activation of the mesocorticolimbic dopamine system*. As a result, the focus of most cellular studies to date has been on the neuroplasticity of the mesolimbic dopamine system and its projection areas. Such studies have raised a number of potential cellular targets that could mediate long-term changes associated with the development of the channeling of behavior toward drug-seeking that is the hallmark of the compulsive drug-taking and loss of control over drug-taking in addiction. These changes range from evidence of short-term changes in synaptic strength within the dopamine system or its targets, to development of LTP and LTD in the ventral tegmental area and nucleus accumbens (See Kombian and Malenka, 1994; Bonci and Malenka, 1999; Hyman and Malenka, 2001; Thomas *et al.*, 2001; Saal *et al.*, 2003; Thomas and Malenka, 2003), to reductions in nondrug-related reward firing in the nucleus accumbens (Peoples and West, 1996; Uzwiak *et al.*, 1997; Peoples *et al.*, 1998, 1999, 2004; Peoples and Cavanaugh, 2003). A challenge for future research will be to move out from under the constraints of the mesocorticolimbic dopamine system to other neurocircuits and targets (see also Caine, 1998; Nestler, 2004).

At a molecular level, chronic exposure or repeated exposure to drugs of abuse produce short-term and long-term changes in second-messenger systems, transcription factors (Nestler *et al.*, 1993, 2001; Nestler and Aghajanian, 1997; Nestler, 2001), the cystine–glutamate exchanger, and proteins related to glutamate neurotransmission involved in G-protein signaling and Homer protein. However, the mesocorticolimbic system remains a focal point, and changes in other reward-related structures have largely been unexplored, but see Pandey *et al.*, (2004) for studies on CREB in the amygdala. Perhaps more importantly, causal relationships of these molecular and cellular changes have not yet been extensively linked to motivational measures of dependence (see *Animal Models of Addiction* and *Drug Addiction: Transition from Neuroadaptation to Pathophysiology* chapters).

Thus, the circuitry outlined in **Fig. 9.25** represents the neural substrates clearly linked to the acute reinforcing effects of drugs of abuse, the neural systems implicated in the neuroadaptational changes associated with the transition to addiction, and the residual neurobiological changes associated with relapse. The extended amygdala is well situated to process emotional stimuli and weigh positive and negative hedonic valence. The ventral striatal-ventral pallidal-thalamic-cortical loops are well situated to translate motivation to action and 'stamp-in' stimulus-reward associations. The prefrontal cortex (anterior cingulate) has a critical role in executive function and drug- and cue-primed reinstatement, and the basolateral amygdala appears to be critical for cue-induced reinstatement (and conditioned withdrawal) and thus, may be a critical substrate for processing the powerful associations of previously neutral stimuli with positive and negative reinforcing properties of drug addiction. Note that this circuitry provides a role for the mesocorticolimbic dopamine system and the opioid peptide systems in modulating the basic circuitry, but does not place the addiction process in series with either system. In the case of the mesolimbic dopamine system, much evidence has accumulated suggesting a role in facilitating the incentive motivational function or the activational aspects of motivation (Salamone and Correa, 2002). Much less is known about the modulatory effects of opioid peptides, with the exception perhaps of alcohol reinforcement, where opioid antagonists clearly limit the reinforcing properties of alcohol in both rodents and humans (Reid and Hunter, 1984; O'Malley, 1996; Sharpe and Samson, 2001; O'Malley *et al.*, 2002).

Note also that the model presented in **Fig. 9.25** is a minimalistic model, with what is hoped to be of heuristic value. Other structures can easily be added that obviously contribute to some aspects of addiction, such as the hippocampus and hypothalamus. However, the structures included are the structures most represented in the reviews outlined above. Emphasis obviously varies from investigator to investigator, with some emphasizing the critical role for the dorsal prefrontal cortex (anterior cingulate), the extended amygdala, or the ventral striatal-ventral pallidal-thalamic-cortical loops. What will be critical for the future of the neuroscience of addiction is to identify how molecular changes, and in turn cellular changes, weight specific neurocircuits to convey vulnerability to addiction. This challenge remains largely unexplored.

REFERENCES

Adinoff, B., Devous, M. D. Sr., Best, S. M., George, M. S., Alexander, D., and Payne, K. (2001). Limbic responsiveness to procaine in cocaine-addicted subjects. *American Journal of Psychiatry* **158**, 390–398.

Ahmed, S. H., Walker, J. R., and Koob, G. F. (2000). Persistent increase in the motivation to take heroin in rats with a history of drug escalation. *Neuropsychopharmacology* **22**, 413–421.

Alderson, H. L., Robbins, T. W., and Everitt, B. J. (2000). The effects of excitotoxic lesions of the basolateral amygdala on the acquisition of heroin-seeking behaviour in rats. *Psychopharmacology* **153**, 111–119.

Alexander, B. K., Beyerstein, B. L., Hadaway, P. F., and Coambs, R. B. (1981). Effect of early and later colony housing on oral ingestion of morphine in rats. *Pharmacology Biochemistry and Behavior* **15**, 571–576.

Alheid, G. F., and Heimer, L. (1988). New perspectives in basal forebrain organization of special relevance for neuropsychiatric disorders: The striatopallidal, amygdaloid, and corticopetal components of substantia innominata. *Neuroscience* **27**, 1–39.

Alheid, G. F., De Olmos, J. S., and Beltramino, C. A. (1995). Amygdala and extended amygdala. *In The Rat Nervous System* (G. Paxinos, Ed.), pp. 495–578. Academic Press, San Diego.

Allen, Y. S., Roberts, G. W., Bloom, S. R., Crow, T. J., and Polak, J. M. (1984). Neuropeptide Y in the stria terminalis: evidence for an amygdalofugal projection. *Brain Research* **321**, 357–362.

Amalric, M., and Koob, G. F. (1985). Low doses of methylnaloxonium in the nucleus accumbens antagonize hyperactivity induced by heroin in the rat. *Pharmacology Biochemistry and Behavior* **23**, 411–415.

Anderson, S. M., Bari, A. A., and Pierce, R. C. (2003). Administration of the D1-like dopamine receptor antagonist SCH-23390 into the medial nucleus accumbens shell attenuates cocaine priming-induced reinstatement of drug-seeking behavior in rats. *Psychopharmacology* **168**, 132–138.

Aston-Jones, G., and Harris, G. C. (2004). Brain substrates for increased drug seeking during protracted withdrawal, Neuropharmacology **47**(Suppl 1.), 167–179.

Aston-Jones, G., Delfs, J. M., Druhan, J., and Zhu, Y. (1999). The bed nucleus of the stria terminalis. A target site for noradrenergic actions in opiate withdrawal. *In Advancing from the Ventral Striatum to the Extended Amygdala: Implications for Neuropsychiatry and Drug Abuse* (series title: *Annals of the New York Academy of Sciences*, vol. 877), (J. F. McGinty, Ed.), pp. 486–498. New York Academy of Sciences, New York.

Baker, D. A., McFarland, K., Lake, R. W., Shen, H., Tang, X. C., Toda, S., and Kalivas, P. W. (2003). Neuroadaptations in cystine-glutamate exchange underlie cocaine relapse. *Nature Neuroscience* **6**, 743–749.

Baldwin, H. A., Rassnick, S., Rivier, J., Koob, G. F., and Britton, K. T. (1991). CRF antagonist reverses the 'anxiogenic' response to ethanol withdrawal in the rat. *Psychopharmacology* **103**, 227–232.

Banks, K. E., and Gratton, A. (1995). Possible involvement of medial prefrontal cortex in amphetamine-induced sensitization of mesolimbic dopamine function. *European Journal of Pharmacology* **282**, 157–167.

Barrot, M., Marinelli, M., Abrous, D. N., Rougé-Pont, F., Le Moal, M., and Piazza, P. V. (1999). Functional heterogeneity in dopamine release and in the expression of Fos-like proteins within the rat striatal complex. *European Journal of Neuroscience* **11**, 1155–1166.

Barrot, M., Marinelli, M., Abrous, D. N., Rougé-Pont, F., Le Moal, M., and Piazza, P. V. (2000). The dopaminergic hyper-responsiveness of the shell of the nucleus accumbens is hormone-dependent. *European Journal of Neuroscience* **12**, 973–979.

Barrot, M., Olivier, J. D., Perrotti, L. I., DiLeone, R. J., Berton, O., Eisch, A. J., Impey, S., Storm, D. R., Neve, R. L., Yin, J. C., Zachariou, V., and Nestler, E. J. (2002). CREB activity in the nucleus accumbens shell controls gating of behavioral responses to emotional stimuli. *Proceedings of the National Academy of Sciences USA* **99**, 11435–11440.

Bassareo, V., and Di Chiara, G. (1997). Differential influence of associative and nonassociative learning mechanisms on the responsiveness of prefrontal and accumbal dopamine transmission to food stimuli in rats fed ad libitum. *Journal of Neuroscience* **17**, 851–861.

Bassareo, V., and Di Chiara, G. (1999a). Modulation of feeding-induced activation of mesolimbic dopamine transmission by appetitive stimuli and its relation to motivational state. *European Journal of Neuroscience* **11**, 4389–4397.

Bassareo, V., and Di Chiara, G. (1999b). Differential responsiveness of dopamine transmission to food-stimuli in nucleus accumbens shell/core compartments. *Neuroscience* **89**, 637–641.

Bassareo, V., De Luca, M. A., and Di Chiara, G. (2002). Differential Expression of Motivational Stimulus Properties by Dopamine in nucleus accumbens shell versus core and prefrontal cortex. *Journal of Neuroscience* **22**, 4709–4719.

Basso, A. M., Spina, M., Rivier, J., Vale, W., and Koob, G. F. (1999). Corticotropin-releasing factor antagonist attenuates the 'anxiogenic-like' effect in the defensive burying paradigm but not in the elevated plus-maze following chronic cocaine in rats. *Psychopharmacology* **145**, 21–30.

Berke, J. D., and Hyman, S. E. (2000). Addiction, dopamine, and the molecular mechanisms of memory. *Neuron* **25**, 515–532.

Blanc, G., Herve, D., Simon, H., Lisoprawski, A., Glowinski, J., and Tassin, J. P. (1980). Response to stress of mesocortico-frontal dopaminergic neurones in rats after long-term isolation. *Nature* **284**, 265–267.

Bonci, A., and Malenka, R. C. (1999). Properties and plasticity of excitatory synapses on dopaminergic and GABAergic cells in the ventral tegmental area. *Journal of Neuroscience* **19**, 3723–3730.

Bonson, K. R., Grant, S. J., Contoreggi, C. S., Links, J. M., Metcalfe, J., Weyl, H. L., Kurian, V., Ernst, M., and London, E. D. (2002). Neural systems and cue-induced cocaine craving. *Neuropsychopharmacology* **26**, 376–386.

Bowers, M. S., McFarland, K., Lake, R. W., Peterson, Y. K., Lapish, C. C., Gregory, M. L., Lanier, S. M., and Kalivas, P. W. (2004). Activator of G protein signaling 3: a gatekeeper of cocaine sensitization and drug seeking. *Neuron* **42**, 269–281.

Boyle, A. E., Gill, K., Smith, B. R., and Amit, Z. (1991). Differential effects of an early housing manipulation on cocaine-induced activity and self-administration in laboratory rats. *Pharmacology Biochemistry and Behavior* **39**, 269–274.

Bozarth, M. A., and Wise, R. A. (1981). Heroin reward is dependent on a dopaminergic substrate. *Life Sciences* **29**, 1881–1886.

Bozarth, M. A., Murray, A., and Wise, R. A. (1989). Influence of housing conditions on the acquisition of intravenous heroin and cocaine self-administration in rats. *Pharmacology Biochemistry and Behavior* **33**, 903–907.

Brody, A. L., Mandelkern, M. A., London, E. D., Childress, A. R., Lee, G. S., Bota, R. G., Ho, M. L., Saxena, S., Baxter, L. R. Jr., Madsen, D., and Jarvik, M. E. (2002). Brain metabolic changes during cigarette craving. *Archives of General Psychiatry* **59**, 1162–1172.

Burns, L. H., Robbins, T. W., and Everitt, B. J. (1993). Differential effects of excitotoxic lesions of the basolateral amygdala, ventral subiculum and medial prefrontal cortex on responding with conditioned reinforcement and locomotor activity potentiated by intra-accumbens infusions of D-amphetamine. *Behavioural Brain Research* **55**, 167–183.

Cadoni, C., and Di Chiara, G. (1999). Reciprocal changes in dopamine responsiveness in the nucleus accumbens shell and core and in the dorsal caudate-putamen in rats sensitized to morphine. *Neuroscience* **90**, 447–455.

Cador, M., Robbins, T. W., and Everitt, B. J. (1989). Involvement of the amygdala in stimulus-reward associations: Interaction with the ventral striatum. *Neuroscience* **30**, 77–86.

Caine, S. B. (1998). Cocaine abuse: hard knocks for the dopamine hypothesis? *Nature Neuroscience* **1**, 90–92.

Caine, S. B., Heinrichs, S. C., Coffin, V. L., and Koob, G. F. (1995). Effects of the dopamine D1 antagonist SCH 23390 microinjected into the accumbens, amygdala or striatum on cocaine self-administration in the rat. *Brain Research* **692**, 47–56.

Capriles, N., Rodaros, D., Sorge, R. E., and Stewart, J. (2003). A role for the prefrontal cortex in stress- and cocaine-induced reinstatement of cocaine seeking in rats. *Psychopharmacology* **168**, 66–74.

Carey, M., and Smale, S. T. (2000). *Transcriptional Regulation in Eukaryotes: Concepts, Strategies, and Techniques* Cold Spring Harbor Laboratory Press, Cold Spring Harbor, NY.

Carlezon, W. A. Jr., and Nestler, E. J. (2002). Elevated levels of GluR1 in the midbrain: a trigger for sensitization to drugs of abuse? *Trends in Neurosciences* **25**, 610–615.

Carlezon, W. A. Jr., and Wise, R. A. (1996). Rewarding actions of phencyclidine and related drugs in nucleus accumbens shell and frontal cortex. *Journal of Neuroscience* **16**, 3112–3122.

Carlezon, W. A. Jr., Thome, J., Olson, V. G., Lane-Ladd, S. B., Brodkin, E. S., Hiroi, N., Duman, R. S., Neve, R. L., and Nestler, E. J. (1998). Regulation of cocaine reward by CREB. *Science* **282**, 2272–2275.

Carelli, R. M., King, V. C., Hampson, R. E., and Deadwyler, S. A. (1993). Firing patterns of nucleus accumbens neurons during cocaine self-administration in rats. *Brain Research* **626**, 14–22.

Carroll, M. E., and Meisch, R. A. (1984). Increased drug-reinforced behavior due to food deprivation. *Advances in Behavioral Pharmacology* **4**, 47–88.

Carter, B. L., and Tiffany, S. T. (1999). Meta-analysis of cue-reactivity in addiction research. *Addiction* **94**, 327–340.

Cestari, V., Mele, A., Oliverio, A., and Castellano, C. (1996). Amygdala lesions block the effect of cocaine on memory in mice. *Brain Research* **713**, 286–289.

Chance, W. T., Sheriff, S., Peng, F., and Balasubramaniam, A. (2000). Antagonism of NPY-induced feeding by pretreatment with cyclic AMP response element binding protein antisense oligonucleotide. *Neuropeptides* **34**, 167–172.

Chang, J. Y., Sawyer, S. F., Lee, R. -S, and Woodward, D. J. (1994). Electrophysiological and pharmacological evidence for the role of the nucleus accumbens in cocaine self-administration in freely moving rats. *Journal of Neuroscience* **14**, 1224–1244.

Chiamulera, C., Epping-Jordan, M. P., Zocchi, A., Marcon, C., Cottiny, C., Tacconi, S., Corsi, M., Orzi, F., and Conquet, F. (2001). Reinforcing and locomotor stimulant effects of cocaine are absent in mGluR5 null mutant mice. *Nature Neuroscience* **4**, 873–874.

Childress, A. R., Mozley, P. D., McElgin, W., Fitzgerald, J., Reivich, M., and O'Brien, C. P. (1999). Limbic activation during cue-induced cocaine craving. *American Journal of Psychiatry* **156**, 11–18.

Chudasama, Y., and Robbins, T. W. (2003). Dissociable contributions of the orbitofrontal and infralimbic cortex to pavlovian autoshaping and discrimination reversal learning: further evidence for the functional heterogeneity of the rodent frontal cortex. *Journal of Neuroscience* **23**, 8771–8780.

Ciccocioppo, R., Sanna, P. P., and Weiss, F. (2001). Cocaine-predictive stimulus induces drug-seeking behavior and neural activation in limbic brain regions after multiple months of abstinence: reversal by D(1) antagonists. *Proceedings of the National Academy of Sciences USA* **98**, 1976–1981.

Colby, C. R., Whisler, K., Steffen, C., Nestler, E. J., and Self, D. W. (2003). Striatal cell type-specific overexpression of DeltaFosB enhances incentive for cocaine. *Journal of Neuroscience* **23**, 2488–2493.

Collier, H. O., and Francis, D. L. (1975). Morphine abstinence is associated with increased brain cyclic AMP. *Nature* **255**, 159–162.

Constantinescu, A., Diamond, I., and Gordon, A. S. (1999). Ethanol-induced translocation of cAMP-dependent protein kinase to the nucleus: mechanism and functional consequences. *Journal of Biological Chemistry* **274**, 26985–26991.

Cornish, J. L., and Kalivas, P. W. (2000). Glutamate transmission in the nucleus accumbens mediates relapse in cocaine addiction. *Journal of Neuroscience* **20**, RC89.

Cornish, J. L., Duffy, P., and Kalivas, P. W. (1999). A role for nucleus accumbens glutamate transmission in the relapse to cocaine-seeking behavior. *Neuroscience* **93**, 1359–1367.

Cunningham, C. L., Malott, D. H., Dickinson, S. D., and Risinger, F. O. (1992). Haloperidol does not alter expression of ethanol-induced conditioned place preference. *Behavioural Brain Research* **50**, 1–5.

Damasio, A. R. (1996). The somatic marker hypothesis and the possible functions of the prefrontal cortex. *Philosophical Transactions: Series B. Biological Sciences* **351**, 1413–1420.

de Wit, H., and Wise, R. A. (1977). Blockade of cocaine reinforcement in rats with the dopamine receptor blocker pimozide, but not with the noradrenergic blockers phentolamine and phenoxybenzamine. *Canadian Journal of Psychology* **31**, 195–203.

Delfs, J. M., Zhu, Y., Druhan, J. P., and Aston-Jones, G. (2000). Noradrenaline in the ventral forebrain is critical for opiate withdrawal-induced aversion. *Nature* **403**, 430–434.

Deminière, J. M., Piazza, P. V., Guegan, G., Abrous, N., Maccari, S., Le Moal, M., and Simon, H. (1992). Increased locomotor response to novelty and propensity to intravenous amphetamine self-administration in adult offspring of stressed mothers. *Brain Research* **586**, 135–139.

Di Chiara, G. (1999). Drug addiction as dopamine-dependent associative learning disorder. *European Journal of Pharmacology* **375**, 13–30.

Di Chiara, G. (2002). Nucleus accumbens shell and core dopamine: differential role in behavior and addiction. *Behavioural Brain Research* **137**, 75–114.

Di Chiara, G., and North, R. A. (1992). Neurobiology of opiate abuse. *Trends in Pharmacological Sciences* **13**, 185–193.

Di Ciano, P., and Everitt, B. J. (2001). Dissociable effects of antagonism of NMDA and AMPA/KA receptors in the nucleus accumbens core and shell on cocaine-seeking behavior. *Neuropsychopharmacology* **25**, 341–360.

Di Ciano, P., and Everitt, B. J. (2004a). Contribution of the ventral tegmental area to cocaine-seeking maintained by a drug-paired conditioned stimulus in rats. *European Journal of Neuroscience* **19**, 1661–1667.

Di Ciano, P., and Everitt, B. J. (2004b). Direct interactions between the basolateral amygdala and nucleus accumbens core underlie cocaine-seeking behavior by rats. *Journal of Neuroscience* **24**, 7167–7173.

Diamond, I., and Gordon, A. S. (1997). Cellular and molecular neuroscience of alcoholism. *Physiological Reviews* **77**, 1–20.

Dong, H. W., Petrovich, G. D., and Swanson, L. W. (2001). Topography of projections from amygdala to bed nuclei of the stria terminalis. *Brain Research Reviews* **38**, 192–246.

Drevets, W. C., Gautier, C., Price, J. C., Kupfer, D. J., Kinahan, P. E., Grace, A. A., Price, J. L., and Mathis, C. A. (2001). Amphetamine-induced dopamine release in human ventral striatum correlates with euphoria. *Biological Psychiatry* **49**, 81–96.

Epping-Jordan, M. P., Markou, A., and Koob, G. F. (1998). The dopamine, D-1 receptor antagonist SCH 23390 injected into the dorsolateral bed nucleus of the stria terminalis decreased cocaine reinforcement in the rat. *Brain Research* **784**, 105–115.

Erb, S., and Stewart, J. (1999). A role for the bed nucleus of the stria terminalis, but not the amygdala, in the effects of corticotropin-releasing factor on stress-induced reinstatement of cocaine seeking. *Journal of Neuroscience* **19**, RC35.

Erb, S., Shaham, Y., and Stewart, J. (1996). Stress reinstates cocaine-seeking behavior after prolonged extinction and a drug-free period. *Psychopharmacology* **128**, 408–412.

Erb, S., Shaham, Y., and Stewart, J. (1998). The role of corticotropin-releasing factor and corticosterone in stress- and cocaine-induced relapse to cocaine seeking in rats. *Journal of Neuroscience* **18**, 5529–5536.

Erb, S., Hitchcott, P. K., Rajabi, H., Mueller, D., Shaham, Y., and Stewart, J. (2000). Alpha-2 adrenergic receptor agonists block stress-induced reinstatement of cocaine seeking. *Neuropsychopharmacology* **23**, 138–150.

Erb, S., Salmaso, N., Rodaros, D., and Stewart, J. (2001). A role for the CRF-containing pathway from central nucleus of the amygdala to bed nucleus of the stria terminalis in the stress-induced reinstatement of cocaine seeking in rats. *Psychopharmacology* **158**, 360–365.

Esposito, R. U., Faulkner, W., and Kornetsky, C. (1979). Specific modulation of brain stimulation reward by haloperidol. *Pharmacology Biochemistry and Behavior* **10**, 937–940.

Everitt, B. J., and Wolf, M. E. (2002). Psychomotor stimulant addiction: a neural systems perspective. *Journal of Neuroscience* **22**, 3312–3320 [erratum: **22**(16):1a].

Everitt, B. J., Cardinal, R. N., Hall, J., Parkinson, J. A., and Robbins, T. W. (2000). Differential involvement of amygdala subsystems in appetitive conditioning and drug addiction. *In The Amygdala: A Functional Analysis* (J. P. Aggleton, Ed.), pp. 353–390. Oxford University Press, New York.

Everitt, B. J., Dickinson, A., and Robbins, T. W. (2001). The neuropsychological basis of addictive behaviour. *Brain Research Reviews* **36**, 129–138.

Fama, R., Pfefferbaum, A., and Sullivan, E. V. (2004). Perceptual learning in detoxified alcoholic men: contributions from explicit memory, executive function, and age. *Alcoholism: Clinical and Experimental Research* **28**, 1657–1665.

Fouriezos, G., Hansson, P., and Wise, R. A. (1978). Neuroleptic-induced attenuation of brain stimulation reward in rats. *Journal of Comparative and Physiological Psychology* **92**, 661–671.

Fuchs, R. A., Weber, S. M., Rice, H. J., and Neisewander, J. L. (2002). Effects of excitotoxic lesions of the basolateral amygdala on cocaine-seeking behavior and cocaine conditioned place preference in rats. *Brain Research* **929**, 15–25.

Fuchs, R. A., Evans, K. A., Parker, M. P., and See, R. E. (2004). Differential involvement of orbitofrontal cortex subregions in conditioned cue-induced and cocaine-primed reinstatement of cocaine seeking in rats. *Journal of Neuroscience* **24**, 6600–6610.

Gallagher, M., Graham, P. W., and Holland, P. C. (1990). The amygdala central nucleus and appetitive Pavlovian conditioning: lesions impair one class of conditioned behavior. *Journal of Neuroscience* **10**, 1906–1911.

Garavan, H., Pankiewicz, J., Bloom, A., Cho, J. K., Sperry, L., Ross, T. J., Salmeron, B. J., Risinger, R., Kelley, D., and Stein, E. A. (2000). Cue-induced cocaine craving: neuroanatomical specificity for drug users and drug stimuli. *American Journal of Psychiatry* **157**, 1789–1798.

Gatto, G. J., McBride, W. J., Murphy, J. M., Lumeng, L., and Li, T. K. (1994). Ethanol self-infusion into the ventral tegmental area by alcohol-preferring rats. *Alcohol* **11**, 557–564.

Gentile, C. G., Jarrell, T. W., Teich, A., McCabe, P. M., and Schneiderman, N. (1986). The role of amygdaloid central nucleus in the retention of differential Pavlovian conditioning of bradycardia in rabbits. *Behavioural Brain Research* **20**, 263–273.

Ghasemzadeh, M. B., Permenter, L. K., Lake, R., Worley, P. F., and Kalivas, P. W. (2003). Homer1 proteins and AMPA receptors modulate cocaine-induced behavioural plasticity. *European Journal of Neuroscience* **18**, 1645–1651.

Ghozland, S., Zorrilla, E., Parsons, L. H., and Koob, G. F. (2004). Mecamylamine increases extracellular CRF levels in the central nucleus of the amygdala or nicotine-dependent rats, Society for Neuroscience Abstracts, abstract# 708.8.

Glickman, S. E., and Schiff, B. B. (1967). A biological theory of reinforcement. *Psychological Review* **74**, 81–109.

Goeders, N. E., and Guerin, G. F. (1994). Non-contingent electric footshock facilitates the acquisition of intravenous cocaine self-administration in rats. *Psychopharmacology* **114**, 63–70.

Goldstein, R. Z., and Volkow, N. D. (2002). Drug addiction and its underlying neurobiological basis: neuroimaging evidence for the involvement of the frontal cortex. *American Journal of Psychiatry* **159**, 1642–1652.

Gordon, A. S., Collier, K., and Diamond, I. (1986). Ethanol regulation of adenosine receptor-stimulated cAMP levels in a clonal neural cell line: an in vitro model of cellular tolerance to ethanol. *Proceedings of the National Academy of Sciences USA* **83**, 2105–2108.

Gordon, A. S., Mochly-Rosen, D., and Diamond, I. (1992). Alcoholism: a possible G protein disorder. *In G proteins, signal transduction and disease* (G. Milligan, and M. J. O. Wakelam, Eds.), pp. 191–216. Academic Press, London.

Grant, S., London, E. D., Newlin, D. B., Villemagne, V. L., Liu, X., Contoreggi, C., Phillips, R. L., Kimes, A. S., and Margolin, A. (1996). Activation of memory circuits during cue-elicited cocaine craving. *Proceedings of the National Academy of Sciences USA* **93**, 12040–12045.

Grant, S., Contoreggi, C., and London, E. D. (2000). Drug abusers show impaired performance in a laboratory test of decision making. *Neuropsychologia* **38**, 1180–1187.

Graves, L., Dalvi, A., Lucki, I., Blendy, J. A., and Abel, T. (2002). Behavioral analysis of CREB alphadelta mutation on a B6/129 F1 hybrid background. *Hippocampus* **12**, 18–26.

Gray, T. S., and Magnuson, D. J. (1992). Peptide immunoreactive neurons in the amygdala and the bed nucleus of the stria terminalis project to the midbrain central gray in the rat. *Peptides* **13**, 451–460.

Gray, T. S., Piechowski, R. A., Yracheta, J. M., Rittenhouse, P. A., Bethea, C. L., and Van de Kar, L. D. (1993). Ibotenic acid lesions in the bed nucleus of the stria terminalis attenuate conditioned stress-induced increases in prolactin, ACTH and corticosterone. *Neuroendocrinology* **57**, 517–524.

Grimm, J. W., and See, R. E. (2000). Dissociation of primary and secondary reward-relevant limbic nuclei in an animal model of relapse. *Neuropsychopharmacology* **22**, 473–479.

Groenewegen, H. J., Berendse, H. W., Wolters, J. G., and Lohman, A. H. (1990). The anatomical relationship of the prefrontal cortex with the striatopallidal system, the thalamus and the amygdala: evidence for a parallel organization. *In The Prefrontal Cortex: Its Structure, Function, and Pathology* (series title: *Progress in Brain Research*, vol. 85), (H. B. M. Uylings, C. G. van Eden, J. P. C. de Bruin, M. A. Corner, and M. G. P. Feenstra, Eds.), pp. 95–116. Elsevier, New York.

Groenewegen, H. J., Berendse, H. W., and Haber, S. N. (1993). Organization of the output of the ventral striatopallidal system in the rat: ventral pallidal efferents. *Neuroscience* **57**, 113–142.

Haber, S. N., Fudge, J. L., and McFarland, N. R. (2000). Striatonigrostriatal pathways in primates form an ascending spiral from the shell to the dorsolateral striatum. *Journal of Neuroscience* **20**, 2369–2382.

Hadaway, P. F., Alexander, B. K., Coambs, R. B., and Beyerstein, B. (1979). The effect of housing and gender on preference for morphine-sucrose solutions in rats. *Psychopharmacology* **66**, 87–91.

Hall, J., Parkinson, J. A., Connor, T. M., Dickinson, A., and Everitt, B. J. (2001). Involvement of the central nucleus of the amygdala and nucleus accumbens core in mediating Pavlovian influences on instrumental behaviour. *European Journal of Neuroscience* **13**, 1984–1992.

Hamamura, T., and Fibiger, H. C. (1993). Enhanced stress-induced dopamine release in the prefrontal cortex of amphetamine-sensitized rats. *European Journal of Pharmacology* **237**, 65–71.

Haney, M., Maccari, S., Le Moal, M., Simon, H., and Piazza, P. V. (1995). Social stress increases the acquisition of cocaine self-administration in male and female rats. *Brain Research* **698**, 46–52.

Harper, C., and Kril, J. (1989). Patterns of neuronal loss in the cerebral cortex in chronic alcoholic patients. *Journal of Neurological Science* **92**, 81–89.

Harris, G. C. (2003). Aston-Jones, G. (2003). Enhanced morphine preference following prolonged abstinence: association with increased Fos expression in the extended amygdala. *Neuropsychopharmacology* **28**, 292–299.

Hayes, R. J., and Gardner, E. L. (2004). The basolateral complex of the amygdala mediates the modulation of intracranial self-stimulation threshold by drug-associated cues. *European Journal of Neuroscience* **20**, 273–280.

Heilig, M., Koob, G. F., Ekman, R., and Britton, K. T. (1994). Corticotropin-releasing factor and neuropeptide Y: role in emotional integration. *Trends in Neurosciences* **17**, 80–85.

Heimer, L., and Alheid, G. (1991). Piecing together the puzzle of basal forebrain anatomy. *In The Basal Forebrain: Anatomy to Function* (series title: *Advances in Experimental Medicine and Biology*, vol. 295), (T. C. Napier, P. W. Kalivas, and I. Hanin, Eds.), pp. 1–42. Plenum Press, New York.

Heinrichs, S. C., Menzaghi, F., Schulteis, G., Koob, G. F., and Stinus, L. (1995). Suppression of corticotropin-releasing factor in the amygdala attenuates aversive consequences of morphine withdrawal. *Behavioural Pharmacology* **6**, 74–80.

Herman, J. P., Guillonneau, D., Dantzer, R., Scatton, B., Semerdjian-Rouquier, L., and Le Moal, M. (1982). Differential effects of inescapable footshocks and of stimuli previously paired with inescapable footshocks on dopamine turnover in cortical and limbic areas of the rat. *Life Sciences* **30**, 2207–2214.

Heyser, C. J., Roberts, A. J., Schulteis, G., and Koob, G. F. (1999). Central administration of an opiate antagonist decreases oral ethanol self-administration in rats. *Alcoholism: Clinical and Experimental Research* **23**, 1468–1476.

Highfield, D., Yap, J., Grimm, J. W., Shalev, U., and Shaham, Y. (2001). Repeated lofexidine treatment attenuates stress-induced, but not drug cues-induced reinstatement of a heroin-cocaine mixture (speedball) seeking in rats. *Neuropsychopharmacology* **25**, 320–331.

Hiroi, N., Brown, J. R., Haile, C. N., Ye, H., Greenberg, M. E., and Nestler, E. J. (1997). FosB mutant mice: Loss of chronic cocaine induction of Fos-related proteins and heightened sensitivity to cocaine's pyschomotor and rewarding effects. *Proceedings of the National Academy of Sciences USA* **94**, 10397–10402.

Hitchcott, P. K., Harmer, C. J., and Phillips, G. D. (1997). Enhanced acquisition of discriminative approach following intra-amygdala d-amphetamine. *Psychopharmacology* **132**, 237–246.

Hodge, C. W., Chappelle, A. M., and Samson, H. H. (1995). GABAergic transmission in the nucleus accumbens is involved in the termination of ethanol self-administration in rats. *Alcoholism: Clinical and Experimental Research* **19**, 1486–1493.

Hope, B. T., Nye, H. E., Kelz, M. B., Self, D. W., Iadarola, M. J., Nakabeppu, Y., Duman, R. S., and Nestler, E. J. (1994). Induction of a long-lasting AP-1 complex composed of altered Fos-like proteins in brain by chronic cocaine and other chronic treatments. *Neuron* **13**, 1235–1244.

Hotsenpiller, G., Giorgetti, M., and Wolf, M. E. (2001). Alterations in behaviour and glutamate transmission following presentation of stimuli previously associated with cocaine exposure. *European Journal of Neuroscience* **14**, 1843–1855.

Hyman, S. E., and Malenka, R. C. (2001). Addiction and the brain: the neurobiology of compulsion and its persistence. *Nature Reviews Neuroscience* **2**, 695–703.

Hyytia, P., and Koob, G. F. (1995). GABA-A receptor antagonism in the extended amygdala decreases ethanol self-administration in rats. *European Journal of Pharmacology* **283**, 151–159.

Ingvar, M., Ghatan, P. H., Wirsen-Meurling, A., Risberg, J., Von Heijne, G., Stone-Elander, S., and Ingvar, D. H. (1998). Alcohol activates the cerebral reward system in man. *Journal of Studies on Alcohol* **59**, 258–269.

Ito, R., Robbins, T. W., and Everitt, B. J. (2004). Differential control over cocaine-seeking behavior by nucleus accumbens core and shell. *Nature Neuroscience* **7**, 389–397.

Iversen, S. D., and Mishkin, M. (1970). Perseverative interference in monkeys following selective lesions of the inferior prefrontal convexity. *Experimental Brain Research* **11**, 376–386.

Iwata, J., LeDoux, J. E., Meeley, M. P., Arneric, S., and Reis, D. J. (1986). Intrinsic neurons in the amygdaloid field projected to by the medial geniculate body mediate emotional responses conditioned to acoustic stimuli. *Brain Research* **383**, 195–214.

Janak, P. H., Chang, J. Y., and Woodward, D. J. (1999). Neuronal spike activity in the nucleus accumbens of behaving rats during ethanol self-administration. *Brain Research* **817**, 172–184.

Jentsch, J. D., and Taylor, J. R. (1999). Impulsivity resulting from frontostriatal dysfunction in drug abuse: Implications for the control of behavior by reward-related stimuli. *Psychopharmacology* **146**, 373–390.

Jentsch, J. D., Verrico, C. D., Le, D., and Roth, R. H. (1998). Repeated exposure to delta 9-tetrahydrocannabinol reduces prefrontal cortical dopamine metabolism in the rat. *Neuroscience Letters* **246**, 169–172.

Jentsch, J. D., Taylor, J. R., Elsworth, J. D., Redmond, D. E. Jr., and Roth, R. H. (1999). Altered frontal cortical dopaminergic transmission in monkeys after subchronic phencyclidine exposure: involvement in frontostriatal cognitive deficits. *Neuroscience* **90**, 823–832.

Johnston, J. B. (1923). Further contributions to the study of the evolution of the forebrain. *Journal of Comparative Neurology* **35**, 337–481.

Jones, S., Kornblum, J. L., and Kauer, J. A. (2000). Amphetamine blocks long-term synaptic depression in the ventral tegmental area. *Journal of Neuroscience* **20**, 5575–5580.

June, H. L., Eggers, M. W., Warren-Reese, C., DeLong, J., Ricks-Cord, A., Durr, L. F., and Cason, C. R. (1998a). The effects of the novel benzodiazepine receptor inverse agonist Ru 34000 on ethanol-maintained behaviors. *European Journal of Pharmacology* **350**, 151–158.

June, H. L., Torres, L., Cason, C. R., Hwang, B. H., Braun, M. R., and Murphy, J. M. (1998b). The novel benzodiazepine inverse agonist RO19-4603 antagonizes ethanol motivated behaviors: neuropharmacological studies. *Brain Research* **784**, 256–275.

Kaddis, F. G., Uretsky, N. J., and Wallace, L. J. (1995). DNQX in the nucleus accumbens inhibits cocaine-induced conditioned place preference. *Brain Research* **697**, 76–82.

Kalivas, P. W. (1995). Interactions between dopamine and excitatory amino acids in behavioral sensitization to psychostimulants. *Drug and Alcohol Dependence* **37**, 95–100.

Kalivas, P. W. (2004). Glutamate systems in cocaine addiction. *Current Opinion in Pharmacology* **4**, 23–29.

Kalivas, P. W., and McFarland, K. (2003). Brain circuitry and the reinstatement of cocaine-seeking behavior, *Psychopharmacology* **168**, 44–56.

Kalivas, P. W., McFarland, K., Bowers, S., Szumlinski, K., Xi, Z. X., and Baker, D. (2003). Glutamate transmission and addiction to cocaine. *In Glutamate and Disorders of Cognition and Motivation* (series title: *Annals of the New York Academy of Sciences*, vol. 1003), (B. Moghaddam, and M. E. Wolf, Eds.), pp. 169–175. New York Academy of Sciences, New York.

Kantak, K. M., Black, Y., Valencia, E., Green-Jordan, K., and Eichenbaum, H. B. (2002). Dissociable effects of lidocaine inactivation of the rostral and caudal basolateral amygdala on the maintenance and reinstatement of cocaine-seeking behavior in rats. *Journal of Neuroscience* **22**, 1126–1136.

Kapp, B. S., Frysinger, R. C., Gallagher, M., and Haselton, J. R. (1979). Amygdala central nucleus lesions: effect on heart rate conditioning in the rabbit. *Physiology and Behavior* **23**, 1109–1117.

Karoum, F., Suddath, R. L., and Wyatt, R. J. (1990). Chronic cocaine and rat brain catecholamines: long-term reduction in hypothalamic and frontal cortex dopamine metabolism. *European Journal of Pharmacology* **186**, 1–8.

Katner, S. N., McBride, W. J., Lumeng, L., Li, T. K., and Murphy, J. M. (1997). Alcohol intake of P rats is regulated by muscarinic receptors in the pedunculopontine nucleus and VTA. *Pharmacology Biochemistry and Behavior* **58**, 497–504.

Kelz, M. B., Chen, J., Carlezon, W. A. Jr., Whisler, K., Gilden, L., Beckmann, A. M., Steffen, C., Zhang, Y. J., Marotti, L., Self, D. W., Tkatch, T., Baranauskas, G., Surmeier, D. J., Neve, R. L., Duman, R. S., Picciotto, M. R., and Nestler, E. J. (1999). Expression of the transcription factor deltaFosB in the brain controls sensitivity to cocaine. *Nature* **401**, 272–276.

Killcross, S., Robbins, T. W., and Everitt, B. J. (1997). Different types of fear-conditioned behaviour mediated by separate nuclei within amygdala. *Nature* **388**, 377–380.

Kilts, C. D., Schweitzer, J. B., Quinn, C. K., Gross, R. E., Faber, T. L., Muhammad, F., Ely, T. D., Hoffman, J. M., and Drexler, K. P. (2001). Neural activity related to drug craving in cocaine addiction. *Archives of General Psychiatry* **58**, 334–341.

Kombian, S. B., and Malenka, R. C. (1994). Simultaneous LTP of non-NMDA- and LTD of NMDA-receptor-mediated responses in the nucleus accumbens. *Nature* **368**, 242–246.

Koob, G. F. (1992a). Dopamine, addiction and reward. *Seminars in the Neurosciences* **4**, 139–148.

Koob, G. F. (1992b). Drugs of abuse: anatomy, pharmacology, and function of reward pathways. *Trends in Pharmacological Sciences* **13**, 177–184.

Koob, G. F. (1999). Corticotropin-releasing factor, norepinephrine and stress. *Biological Psychiatry* **46**, 1167–1180.

Koob, G. F. (2003a). Alcoholism: allostasis and beyond. *Alcoholism: Clinical and Experimental Research* **27**, 232–243.

Koob, G. F. (2003b). Neuroadaptive mechanisms of addiction: studies on the extended amygdala. *European Neuropsychopharmacology* **13**, 442–452.

Koob, G. F., and Kreek, M. J. (2005). Stress, dysregulation of drug reward pathways, and drug dependence: an updated perspective commemorating the 30th anniversary of the National Institute on Drug Abuse, submitted.

Koob, G. F., and Le Moal, M. (2001). Drug addiction, dysregulation of reward, and allostasis. *Neuropsychopharmacology* **24**, 97–129.

Koob, G. F., and Le Moal, M. (2004). Drug addiction and allostasis. *In Allostasis, Homeostasis, and the Costs of Physiological Adaptation* (J. Schulkin, Ed.), pp. 150–163. Cambridge University Press, New York.

Koob, G. F., Sanna, P. P., and Bloom, F. E. (1998). Neuroscience of addiction. *Neuron* **21**, 467–476.

Kozicz, T. (2001). Axon terminals containing tyrosine hydroxylase- and dopamine-beta-hydroxylase immunoreactivity form synapses with galanin immunoreactive neurons in the lateral division of the bed nucleus of the stria terminalis in the rat. *Brain Research* **914**, 23–33.

Kreek, M. J., and Koob, G. F. (1998). Drug dependence: Stress and dysregulation of brain reward pathways. *Drug and Alcohol Dependence* **51**, 23–47.

Kruzich, P. J., and See, R. E. (2001). Differential contributions of the basolateral and central amygdala in the acquisition and expression of conditioned relapse to cocaine-seeking behavior. *Journal of Neuroscience* **21**: RC155.

Kubota, M., Nakazaki, S., Hirai, S., Saeki, N., Yamaura, A., and Kusaka, T. (2001). Alcohol consumption and frontal lobe shrinkage: study of 1432 non-alcoholic subjects. *Journal of Neurology, Neurosurgery and Psychiatry* **71**, 104–106.

Laruelle, M., Abi-Dargham, A., van Dyck, C. H., Rosenblatt, W., Zea-Ponce, Y., Zoghbi, S. S., Baldwin, R. M., Charney, D. S., Hoffer, P. B., and Kung, H. F. (1995). SPECT imaging of striatal dopamine release after amphetamine challenge. *Journal of Nuclear Medicine* **36**, 1182–1190.

Laviolette, S. R., and van der Kooy, D. (2003). Blockade of mesolimbic dopamine transmission dramatically increases sensitivity to the rewarding effects of nicotine in the ventral tegmental area. *Molecular Psychiatry* **8**, 50–59.

Layer, R. T., Uretsky, N. J., and Wallace, L. J. (1993). Effects of the AMPA/kainate receptor antagonist DNQX in the nucleus accumbens on drug-induced conditioned place preference. *Brain Research* **617**, 267–273.

Le, A. D., Quan, B., Juzytch, W., Fletcher, P. J., Joharchi, N., and Shaham, Y. (1998). Reinstatement of alcohol-seeking by priming injections of alcohol and exposure to stress in rats. *Psychopharmacology* **135**, 169–174.

Le Moal, M., and Simon, H. (1991). Mesocorticolimbic dopaminergic network: functional and regulatory roles. *Physiological Reviews* **71**, 155–234.

LeDoux, J. E., Iwata, J., Cicchetti, P., and Reis, D. J. (1988). Different projections of the central amygdaloid nucleus mediate autonomic and behavioral correlates of conditioned fear. *Journal of Neuroscience* **8**, 2517–2529.

Leri, F., Flores, J., Rodaros, D., and Stewart, J. (2002). Blockade of stress-induced but not cocaine-induced reinstatement by infusion of noradrenergic antagonists into the bed nucleus of the stria terminalis or the central nucleus of the amygdala. *Journal of Neuroscience* **22**, 5713–5718.

Lesur, A., Gaspar, P., Alvarez, C., and Berger, B. (1989). Chemoanatomic compartments in the human bed nucleus of the stria terminalis. *Neuroscience* **32**, 181–194.

Li, Y., Hu, X. T., Berney, T. G., Vartanian, A. J., Stine, C. D., Wolf, M. E., and White, F. J. (1999). Both glutamate receptor antagonists and prefrontal cortex lesions prevent induction of cocaine sensitization and associated neuroadaptations. *Synapse* **34**, 169–180.

Li, J., Li, Y. H., and Yuan, X. R. (2003). Changes of phosphorylation of cAMP response element binding protein in rat nucleus accumbens after chronic ethanol intake: naloxone reversal. *Acta Pharmacologica Sinica* **24**, 930–936.

Ljungberg, T., Apicella, P., and Schultz, W. (1992). Responses of monkey dopamine neurons during learning of behavioral reactions. *Journal of Neurophysiology* **67**, 145–163.

Louilot, A., Le Moal, M., and Simon, H. (1989). Opposite influences of dopaminergic pathways to the prefrontal cortex or the septum on the dopaminergic transmission in the nucleus accumbens: an in vivo voltammetric study. *Neuroscience* **29**, 45–56.

Lu, L., Ceng, X., and Huang, M. (2000). Corticotropin-releasing factor receptor type I mediates stress-induced relapse to opiate dependence in rats. *Neuroreport* **11**, 2373–2378.

Lu, L., Liu, D., and Ceng, X. (2001). Corticotropin-releasing factor receptor type 1 mediates stress-induced relapse to cocaine-conditioned place preference in rats. *European Journal of Pharmacology* **415**, 203–208.

Lyness, W. H., Friedle, N. M., and Moore, K. E. (1979). Destruction of dopaminergic nerve terminals in nucleus accumbens: Effect on d-amphetamine self-administration. *Pharmacology Biochemistry and Behavior* **11**, 553–556.

Maas, L. C., Lukas, S. E., Kaufman, M. J., Weiss, R. D., Daniels, S. L., Rogers, V. W., Kukes, T. J., and Renshaw, P. F. (1998). Functional magnetic resonance imaging of human brain activation during cue-induced cocaine craving. *American Journal of Psychiatry* **155**, 124–126.

Maccari, S., Piazza, P. V., Deminière, J. M., Lemaire, V., Mormède, P., Simon, H., Angelucci, L., and Le Moal, M. (1991). Life events-induced decrease of corticosteroid type I receptors is associated with reduced corticosterone feedback and enhanced vulnerability to amphetamine self-administration. *Brain Research* **547**, 7–12.

Maldonado, R., Blendy, J. A., Tzavara, E., Gass, P., Roques, B. P., Hanoune, J., and Schutz, G. (1996). Reduction of morphine abstinence in mice with a mutation in the gene encoding CREB. *Science* **273**, 657–659.

Manning, B. H. (1998). A lateralized deficit in morphine antinociception after unilateral inactivation of the central amygdala. *Journal of Neuroscience* **18**, 9453–9470.

Manning, B. H., and Mayer, D. J. (1995). The central nucleus of the amygdala contributes to the production of morphine antinociception in the formalin test. *Pain* **63**, 141–152.

Mantsch, J. R., Saphier, D., and Goeders, N. E. (1998). Corticosterone facilitates the acquisition of cocaine self-administration in rats: opposite effects of the type II glucocorticoid receptor agonist dexamethasone. *Journal of Pharmacology and Experimental Therapeutics* **287**, 72–80.

Mantsch, J. R., Yuferov, V., Mathieu-Kia, A. M., Ho, A., and Kreek, M. J. (2003). Neuroendocrine alterations in a high-dose, extended-access rat self-administration model of escalating cocaine use. *Psychoneuroendocrinology* **28**, 836–862.

Marinelli, M., Piazza, P. V., Deroche, V., Maccari, S., Le Moal, M., and Simon, H. (1994). Corticosterone circadian secretion differentially facilitates dopamine-mediated psychomotor effect of cocaine and morphine. *Journal of Neuroscience* **14**, 2724–2731.

Marinelli, M., Rougé-Pont, F., De Jesus-Oliveira, C., Le Moal, M., and Piazza, P. V. (1997a). Acute blockade of corticosterone secretion decreases the psychomotor stimulant effects of cocaine. *Neuropsychopharmacology* **16**, 156–161.

Marinelli, M., Rougé-Pont, F., Deroche, V., Barrot, M., De Jesus-Oliveira, C., Le Moal, M., and Piazza, P. V. (1997b). Glucocorticoids and behavioral effects of psychostimulants. I: locomotor response to cocaine depends on basal levels of glucocorticoids. *Journal of Pharmacology and Experimental Therapeutics* **281**, 1392–1400.

Markou, A., and Koob, G. F. (1991). Post-cocaine anhedonia: An animal model of cocaine withdrawal. *Neuropsychopharmacology* **4**, 17–26.

Marks-Kaufman, R., and Lewis, M. J. (1984). Early housing experience modifies morphine self-administration and physical dependence in adult rats. *Addictive Behaviors* **9**, 235–243.

Martin-Fardon, R., Ciccocioppo, R., Massi, M., and Weiss, F. (2000). Nociceptin prevents stress-induced ethanol- but not cocaine-seeking behavior in rats. *Neuroreport* **11**, 1939–1943.

Martin-Solch, C., Magyar, S., Kunig, G., Missimer, J., Schultz, W., and Leenders, K. L. (2001). Changes in brain activation associated with reward processing in smokers and nonsmokers. A positron emission tomography study. *Experimental Brain Research* **139**, 278–286.

McBride, W. J., and Li, T. K. (1998). Animal models of alcoholism: Neurobiology of high alcohol-drinking behavior in rodents. *Critical Reviews in Neurobiology* **12**, 339–369.

McClung, C. A., Ulery, P. G., Perrotti, L. I., Zachariou, V., Berton, O., and Nestler, E. J. (2004). DeltaFosB: a molecular switch for long-term adaptation in the brain. *Molecular Brain Research* **132**, 146–154.

McDonald, A. J., Shammah-Lagnado, S. J., Shi, C., and Davis, M. (1999). Cortical afferents to the extended amygdala. *In Advancing from the Ventral Striatum to the Extended Amygdala: Implications for Neuropsychiatry and Drug Abuse* (series title: *Annals of the New York Academy of Sciences*, vol. 877), (J. F. McGinty, Ed.), pp. 309–338. New York Academy of Sciences, New York.

McFarland, K., and Kalivas, P. W. (2001). The circuitry mediating cocaine-induced reinstatement of drug-seeking behavior. *Journal of Neuroscience* **21**, 8655–8663.

McFarland, K., Lapish, C. C., Kalivas, P. W. (2003) Prefrontal glutamate release into the core of the nucleus accumbens mediates cocaine-induced reinstatement of drug-seeking behavior, *Journal of Neuroscience* **23**, 3531–3537.

McFarland, K., Davidge, S. B., Lapish, C. C., and Kalivas, P. W. (2004). Limbic and motor circuitry underlying footshock-induced reinstatement of cocaine-seeking behavior. *Journal of Neuroscience* **24**, 1551–1560.

McGeehan, A. J., and Olive, M. F. (2003). The mGluR5 antagonist MPEP reduces the conditioned rewarding effects of cocaine but not other drugs of abuse. *Synapse* **47**, 240–242.

McKee, B. L., and Meshul, C. K. (2005). Time-dependent changes in extracellular glutamate in the rat dorsolateral striatum following a single cocaine injection. *Neuroscience* **133**, 605–613.

McLaughlin, J., and See, R. E. (2003). Selective inactivation of the dorsomedial prefrontal cortex and the basolateral amygdala attenuates conditioned-cued reinstatement of extinguished cocaine-seeking behavior in rats. *Psychopharmacology* **168**, 57–65.

Meil, W. M., and See, R. E. (1997). Lesions of the basolateral amygdala abolish the ability of drug associated cues to reinstate responding during withdrawal from self-administered cocaine. *Behavioural Brain Research* **87**, 139–148.

Merlo-Pich, E., Lorang, M., Yeganeh, M., Rodriguez de Fonseca, F., Raber, J., Koob, G. F., and Weiss, F. (1995). Increase of extracellular corticotropin-releasing factor-like immunoreactivity levels in the amygdala of awake rats during restraint stress and ethanol withdrawal as measured by microdialysis. *Journal of Neuroscience* **15**, 5439–5447.

Miczek, K. A., and Mutschler, N. H. (1996). Activational effects of social stress on IV cocaine self-administration in rats. *Psychopharmacology* **128**, 256–264.

Milner, B. (1982). Some cognitive effects of frontal-lobe lesions in man. *Philosophical Transactions of the Royal Society of London B Biological Sciences* **298**, 211–226.

Mirenowicz, J., and Schultz, W. (1994). Importance of unpredictability for reward responses in primate dopamine neurons. *Journal of Neurophysiology* **72**, 1024–1027.

Moller, C., Wiklund, L., Sommer, W., Thorsell, A., and Heilig, M. (1997). Decreased experimental anxiety and voluntary ethanol consumption in rats following central but not basolateral amygdala lesions. *Brain Research* **760**, 94–101.

Moratalla, R., Elibol, B., Vallejo, M., and Graybiel, A. M. (1996). Network-level changes in expression of inducible Fos-Jun proteins in the striatum during chronic cocaine treatment and withdrawal. *Neuron* **17**, 147–156.

Nakamura, H., Tanaka, A., Nomoto, Y., Ueno, Y., and Nakayama, Y. (2000). Activation of fronto-limbic system in the human brain by cigarette smoking: evaluated by a CBF measurement. *Keio Journal of Medicine* **49**(*Suppl.* 1), A122–A124.

Nauta, J. H., and Haymaker, W. (1969). Hypothalamic nuclei and fiber connections. *In The Hypothalamus* (W. Haymaker, E. Anderson, and W. J. H. Nauta, Eds.), pp. 136–209. Charles C. Thomas, Springfield IL.

Nestler, E. J. (2001). Molecular basis of long-term plasticity underlying addiction. *Nature Reviews Neuroscience* **2**, 119–128.

Nestler, E. J. (2004). Historical review: Molecular and cellular mechanisms of opiate and cocaine addiction. *Trends in Pharmacological Sciences* **25**, 210–218.

Nestler, E. J., and Aghajanian, G. K. (1997). Molecular and cellular basis of addiction. *Science* **278**, 58–63.

Nestler, E. J., Hope, B. T., and Widnell, K. L. (1993). Drug addiction: A model for the molecular basis of neural plasticity. *Neuron* **11**, 995–1006.

Nestler, E. J., Barrot, M., and Self, D. W. (2001). DeltaFosB: a sustained molecular switch for addiction. *Proceedings of the National Academy of Sciences USA* **98**, 11042–11046.

Newton, S. S., Thome, J., Wallace, T. L., Shirayama, Y., Schlesinger, L., Sakai, N., Chen, J., Neve, R., Nestler, E. J., and Duman, R. S. (2002). Inhibition of cAMP response element-binding protein or dynorphin in the nucleus accumbens produces an antidepressant-like effect. *Journal of Neuroscience* **22**, 10883–10890.

Nijsen, M. J., Croiset, G., Diamant, M., De Wied, D., and Wiegant, V. M. (2001). CRH signalling in the bed nucleus of the stria terminalis is involved in stress-induced cardiac vagal activation in conscious rats. *Neuropsychopharmacology* **24**, 1–10.

Noel, X., Paternot, J., Van der Linden, M., Sferrazza, R., Verhas, M., Hanak, C., Kornreich, C., Martin, P., De Mol, J., Pelc, I., and Verbanck, P. (2001). Correlation between inhibition, working memory and delimited frontal area blood flow measure by 99mTc-Bicisate SPECT in alcohol-dependent patients. *Alcohol and Alcoholism* **36**, 556–563.

Nowak, K. L., McBride, W. J., Lumeng, L., Li, T. K., and Murphy, J. M. (1998). Blocking GABA(A) receptors in the anterior ventral tegmental area attenuates ethanol intake of the alcohol-preferring P rat. *Psychopharmacology* **139**, 108–116.

O'Malley, S. S. (1996). Opioid antagonists in the treatment of alcohol dependence: clinical efficacy and prevention of relapse. *Alcohol and Alcoholism* **31**(*Suppl.* 1), 77–81.

O'Malley, S. S., Krishnan-Sarin, S., Farren, C., Sinha, R., and Kreek, M. J. (2002). Naltrexone decreases craving and alcohol self-administration in alcohol-dependent subjects and activates the hypothalamo-pituitary-adrenocortical axis. *Psychopharmacology* **160**, 19–29.

Olive, M. F., Koenig, H. N., Nannini, M. A., and Hodge, C. W. (2002). Elevated extracellular CRF levels in the bed nucleus of the stria terminalis during ethanol withdrawal and reduction by subsequent ethanol intake. *Pharmacology Biochemistry and Behavior* **72**, 213–220.

Olsen, R. W., McCabe, R. T., and Wamsley, J. K. (1990). GABAA receptor subtypes: autoradiographic comparison of GABA, benzodiazepine, and convulsant binding sites in the rat central nervous system. *Journal of Chemical Neuroanatomy* **3**, 59–76.

Ortiz, J., Fitzgerald, L. W., Charlton, M., Lane, S., Trevisan, L., Guitart, X., Shoemaker, W., Duman, R. S., and Nestler, E. J. (1995). Biochemical actions of chronic ethanol exposure in the mesolimbic dopamine system. *Synapse* **21**, 289–298.

Oscar-Berman, M., and Hutner, N. (1993). Frontal lobe changes after chronic alcohol ingestion. *In Alcohol-Induced Brain Damage* (series title: *NIAAA Research Monograph*, vol. 22), (W. A. Hunt, and S. J. Nixon, Eds.), pp. 121–156. National Institute on Alcohol Abuse and Alcoholism, Bethesda MD.

Pandey, S. C. (2003). Anxiety and alcohol abuse disorders: a common role for CREB and its target, the neuropeptide Y gene. *Trends in Pharmacological Sciences* **24**, 456–460.

Pandey, S. C. (2004). The gene transcription factor cyclic AMP-responsive element binding protein: role in positive and negative affective states of alcohol addiction. *Pharmacology and Therapeutics* **104**, 47–58.

Pandey, S. C., Zhang, D., Mittal, N., and Nayyar, D. (1999). Potential role of the gene transcription factor cyclic AMP-responsive element binding protein in ethanol withdrawal-related anxiety. *Journal of Pharmacology and Experimental Therapeutics* **288**, 866–878.

Pandey, S. C., Roy, A., and Mittal, N. (2001). Effects of chronic ethanol intake and its withdrawal on the expression and phosphorylation of the creb gene transcription factor in rat cortex. *Journal of Pharmacology and Experimental Therapeutics* **296**, 857–868.

Pandey, S. C., Roy, A., and Zhang, H. (2003). The decreased phosphorylation of cyclic adenosine monophosphate (cAMP) response element binding (CREB) protein in the central amygdala acts as a molecular substrate for anxiety related to ethanol withdrawal in rats. *Alcoholism: Clinical and Experimental Research* **27**, 396–409.

Pandey, S. C., Roy, A., Zhang, H., and Xu, T. (2004). Partial deletion of the cAMP response element-binding protein gene promotes alcohol-drinking behaviors. *Journal of Neuroscience* **24**, 5022–5030.

Park, W. K., Bari, A. A., Jey, A. R., Anderson, S. M., Spealman, R. D., Rowlett, J. K., and Pierce, R. C. (2002). Cocaine administered into the medial prefrontal cortex reinstates cocaine-seeking behavior by increasing AMPA receptor-mediated glutamate transmission in the nucleus accumbens. *Journal of Neuroscience* **22**, 2916–2925.

Parkinson, J. A., Olmstead, M. C., Burns, L. H., Robbins, T. W., and Everitt, B. J. (1999). Dissociation in effects of lesions of the nucleus accumbens core and shell on appetitive pavlovian approach behavior and the potentiation of conditioned reinforcement and locomotor activity by D-amphetamine. *Journal of Neuroscience* **19**, 2401–2411.

Parkinson, J. A., Robbins, T. W., and Everitt, B. J. (2000a). Dissociable roles of the central and basolateral amygdala in appetitive emotional learning. *European Journal of Neuroscience* **12**, 405–413.

Parkinson, J. A., Willoughby, P. J., Robbins, T. W., and Everitt, B. J. (2000b). Disconnection of the anterior cingulate cortex and nucleus accumbens core impairs Pavlovian approach behavior: further evidence for limbic cortical-ventral striatopallidal systems. *Behavioral Neuroscience* **114**, 42–63.

Parsons, O. (1993). Impaired neuropsychological cognitive functioning in sober alcoholcs. *In Alcohol-Induced Brain Damage* (series title: *NIAAA Research Monograph*, vol. 22), (W. A. Hunt, and S. J. Nixon, Eds.), pp. 173–194. National Institute on Alcohol Abuse and Alcoholism, Bethesda, MD.

Parsons, O. A., Butters, N., and Nathan, P. E. (Eds.). (1987). *Neuropsychology of Alcoholism: Implications for Diagnosis and Treatment*, Guilford Press, New York.

Peakman, M. C., Colby, C., Perrotti, L. I., Tekumalla, P., Carle, T., Ulery, P., Chao, J., Duman, C., Steffen, C., Monteggia, L., Allen, M. R., Stock, J. L., Duman, R. S., McNeish, J. D., Barrot, M., Self, D. W., Nestler, E. J., and Schaeffer, E. (2003). Inducible, brain region-specific expression of a dominant negative mutant of c-Jun in transgenic mice decreases sensitivity to cocaine. *Brain Research* **970**, 73–86.

Pennartz, C. M., Ameerun, R. F., Groenewegen, H. J., and Lopes da Silva, F. H. (1993). Synaptic plasticity in an in vitro slice preparation of the rat nucleus accumbens. *European Journal of Neuroscience* **5**, 107–117.

Pennartz, C. M., Groenewegen, H. J., and Lopes da Silva, F. H. (1994). The nucleus accumbens as a complex of functionally distinct neuronal ensembles: an integration of behavioural, electrophysiological and anatomical data. *Progress in Neurobiology* **42**, 719–761.

Peoples, L. L., and Cavanaugh D. (2003). Differential changes in signal and background firing of accumbal neurons during cocaine self-administration. *Journal of Neurophysiology* **90**, 93–1010.

Peoples, L. L., and West, M. O. (1996). Phasic firing of single neurons in the rat nucleus accumbens correlated with the timing of intravenous cocaine self-administration. *Journal of Neuroscience* **16**, 3459–3473.

Peoples, L. L., Uzwiak, A. J., Guyette, F. X., and West, M. O. (1998). Tonic inhibition of single nucleus accumbens neurons in the rat: a predominant but not exclusive firing pattern induced by cocaine self-administration sessions. *Neuroscience* **86**, 13–22.

Peoples, L. L., Uzwiak, A. J., Gee, F., Fabbricatore, A. T., Muccino, K. J., Mohta, B. D., and West, M. O. (1999). Phasic accumbal firing may contribute to the regulation of drug taking during intravenous cocaine self-administration sessions. *In Advancing from the Ventral Striatum to the Extended Amygdala: Implications for Neuropsychiatry and Drug Abuse* (series title: *Annals of the New York Academy of Sciences*, vol. 877), (J. F. McGinty, Ed.), pp. 781–787. New York Academy of Sciences, New York.

Peoples, L. L., Lynch, K. G., Lesnock, J., and Gangadhar, N. (2004). Accumbal neural responses during the initiation and maintenance of intravenous cocaine self-administration. *Journal of Neurophysiology* **91**, 314–323.

Pettit, H. O., Ettenberg, A., Bloom, F. E., and Koob, G. F. (1984). Destruction of dopamine in the nucleus accumbens selectively attenuates cocaine but not heroin self-administration in rats. *Psychopharmacology* **84**, 167–173.

Pfefferbaum, A., Sullivan, E. V., Mathalon, D. H., and Lim, K. O. (1997). Frontal lobe volume loss observed with magnetic resonance imaging in older chronic alcoholics. *Alcoholism: Clinical and Experimental Research* **21**, 521–529.

Phelix, C. F., and Paull, W. K. (1990). Demonstration of distinct corticotropin releasing factor-containing neuron populations in the bed nucleus of the stria terminalis: a light and electron microscopic immunocytochemical study in the rat. *Histochemistry* **94**, 345–364.

Phillips, G. D., Howes, S. R., Whitelaw, R. B., Wilkinson, L. S., Robbins, T. W., and Everitt, B. J. (1994). Isolation rearing enhances the locomotor response to cocaine and a novel environment, but impairs the intravenous self-administration of cocaine. *Psychopharmacology* **115**, 407–418.

Piazza, P. V., and Le Moal, M. (1996). Pathophysiological basis of vulnerability to drug abuse: Role of an interaction between stress, glucocorticoids, and dopaminergic neurons. *Annual Review of Pharmacology and Toxicology* **36**, 359–378.

Piazza, P. V., and Le Moal, M. (1998). The role of stress in drug self-administration. *Trends in Pharmacological Sciences* **19**, 67–74.

Piazza, P. V., Deminière, J. M., Le Moal, M., and Simon, H. (1989). Factors that predict individual vulnerability to amphetamine self-administration. *Science* **245**, 1511–1513.

Piazza, P. V., Deminière, J. M., Le Moal, M., and Simon, H. (1990). Stress- and pharmacologically-induced behavioral sensitization increases vulnerability to acquisition of amphetamine self-administration. *Brain Research* **514**, 22–26.

Piazza, P. V., Deminière, J. M., Maccari, S., Le Moal, M., Mormède, P., and Simon, H. (1991a). Individual vulnerability to drug self-administration: action of corticosterone on dopaminergic systems as a possible pathophysiological mechanism. *In The Mesolimbic Dopamine System: From Motivation to Action*, (P. Willner, and J. Scheel-Kruger, Eds.), pp. 473–495. John Wiley and Sons, Chichester.

Piazza, P. V., Maccari, S., Deminière, J. M., Le Moal, M., Mormede, P., and Simon, H. (1991b). Corticosterone levels determine individual vulnerability to amphetamine self-administration. *Proceedings of the National Academy of Sciences USA* **88**, 2088–2092.

Piazza, P. V., Rougé-Pont, F., Deminière, J. M., Kharoubi, M., Le Moal, M., and Simon, H. (1991c). Dopaminergic activity is reduced in the prefrontal cortex and increased in the nucleus accumbens of rats predisposed to develop amphetamine self-administration. *Brain Research* **567**, 169–174.

Pierce, R. C., Reeder, D. C., Hicks, J., Morgan, Z. R., and Kalivas, P. W. (1998). Ibotenic acid lesions of the dorsal prefrontal cortex disrupt the expression of behavioral sensitization to cocaine. *Neuroscience* **82**, 1103–1114.

Pliakas, A. M., Carlson, R. R., Neve, R. L., Konradi, C., Nestler, E. J., and Carlezon, W. A. Jr. (2001). Altered responsiveness to cocaine and increased immobility in the forced swim test associated with elevated cAMP response element-binding protein expression in nucleus accumbens. *Journal of Neuroscience* **21**, 7397–7403.

Pluzarev, O., and Pandey, S. C. (2004). Modulation of CREB expression and phosphorylation in the rat nucleus accumbens during nicotine exposure and withdrawal. *Journal of Neuroscience Research* **77**, 884–891.

Pompei, P., Tayebaty, S. J., De Caro, G., Schulkin, J., and Massi, M. (1991). Bed nucleus of the stria terminalis: site for the antinatriorexic action of tachykinins in the rat. *Pharmacology Biochemistry and Behavior* **40**, 977–981.

Pontieri, F. E., Tanda, G., and Di Chiara, G. (1995). Intravenous cocaine, morphine, and amphetamine preferentially increase extracellular dopamine in the "shell" as compared with the "core" of the rat nucleus accumbens. *Proceedings of the National Academy of Sciences USA* **92**, 12304–12308.

Pontieri, F. E., Tanda, G., Orzi, F., and Di Chiara, G. (1996). Effects of nicotine on the nucleus accumbens and similarity to those of addictive drugs. *Nature* **382**, 255–257.

Ramsey, N. F., and Van Ree, J. M. (1993). Emotional but not physical stress enhances intravenous cocaine self-administration in drug-naive rats. *Brain Research* **608**, 216–222.

Ranje, C., and Ungerstedt, U. (1977). Lack of acquisition in dopamine denervated animals tested in an underwater Y-maze. *Brain Research* **134**, 95–111.

Rasmussen, K., Beitner-Johnson, D. B., Krystal, J. H., Aghajanian, G. K., and Nestler, E. J. (1990). Opiate withdrawal and the rat locus coeruleus: behavioral, electrophysiological, and biochemical correlates. *Journal of Neuroscience* **10**, 2308–2317.

Rassnick, S., Heinrichs, S. C., Britton, K. T., and Koob, G. F. (1993a). Microinjection of a corticotropin-releasing factor antagonist into the central nucleus of the amygdala reverses anxiogenic-like effects of ethanol withdrawal. *Brain Research* **605**, 25–32.

Rassnick, S., Stinus, L., and Koob, G. F. (1993b). The effects of 6-hydroxydopamine lesions of the nucleus accumbens and the mesolimbic dopamine system on oral self-administration of ethanol in the rat. *Brain Research* **623**, 16–24.

Reid, L. D., and Hunter, G. A. (1984). Morphine and naloxone modulate intake of ethanol. *Alcohol* **1**, 33–37.

Ricaurte, G. A., Schuster, C. R., and Seiden, L. S. (1980). Long-term effects of repeated methylamphetamine administration on dopamine and serotonin neurons in the rat brain: a regional study. *Brain Research* **193**, 153–163.

Richter, R. M., and Weiss, F. (1999). In vivo CRF release in rat amygdala is increased during cocaine withdrawal in self-administering rats. *Synapse* **32**, 254–261.

Ridley, R. M., Clark, B. A., Durnford, L. J., and Baker, H. F. (1993). Stimulus-bound perseveration after frontal ablations in marmosets. *Neuroscience* **52**, 595–604.

Robbins, T. W., and Everitt, B. J. (1999). Drug addiction: bad habits add up. *Nature* **398**, 567–570.

Roberts, D. C. S., Koob, G. F., Klonoff, P., and Fibiger, H. C. (1980). Extinction and recovery of cocaine self-administration following 6-hydroxydopamine lesions of the nucleus accumbens. *Pharmacology Biochemistry and Behavior* **12**, 781–787.

Roberts, A. J., Cole, M., and Koob, G. F. (1996). Intra-amygdala muscimol decreases operant ethanol self-administration in dependent rats. *Alcoholism: Clinical and Experimental Research* **20**, 1289–1298.

Robinson, T. E., and Becker, J. B. (1986). Enduring changes in brain and behavior produced by chronic amphetamine administration: a review and evaluation of animal models of amphetamine psychosis. *Brain Research* **396**, 157–198.

Robinson, T. E., and Berridge, K. C. (1993). The neural basis of drug craving: An incentive-sensitization theory of addiction. *Brain Research Reviews* **18**, 247–291.

Robinson, T. E., and Kolb, B. (1997). Persistent structural modifications in nucleus accumbens and prefrontal cortex neurons produced by previous experience with amphetamine. *Journal of Neuroscience* **17**, 8491–8497.

Robinson, T. E., and Kolb, B. (1999). Morphine alters the structure of neurons in the nucleus accumbens and neocortex of rats. *Synapse* **33**, 160–162.

Robinson, T. E., and Kolb, B. (2004). Structural plasticity associated with exposure to drugs of abuse. *Neuropharmacology* **47(Suppl. 1)**, 33–46.

Robinson, T. E., Becker, J. B., Moore, C. J., Castaneda, E., and Mittleman G. (1985). Enduring enhancement in frontal cortex dopamine utilization in an animal model of amphetamine psychosis. *Brain Research* **343**, 374–377.

Robledo, P., Robbins, T. W., and Everitt, B. J. (1996). Effects of excitotoxic lesions of the central amygdaloid nucleus on the potentiation of reward-related stimuli by intra-accumbens amphetamine. *Behavioral Neuroscience* **110**, 981–990.

Rodd, Z. A., Bell, R. L., Zhang, Y., Goldstein, A., Zaffaroni, A., McBride, W. J., and Li, T. K. (2003). Salsolinol produces reinforcing effects in the nucleus accumbens shell of alcohol-preferring (P) rats. *Alcoholism: Clinical and Experimental Research* **27**, 440–449.

Rodd, Z. A., Melendez, R. I., Bell, R. L., Kuc, K. A., Zhang, Y., Murphy, J. M., and McBride, W. J. (2004). Intracranial self-administration of ethanol within the ventral tegmental area of male Wistar rats: evidence for involvement of dopamine neurons. *Journal of Neuroscience* **24**, 1050–1057.

Rodd, Z. A., Bell, R. L., Zhang, Y., Murphy, J. M., Goldstein, A., Zaffaroni, A., Li, T. K., and McBride, W. J. (2005). Regional heterogeneity for the intracranial self-administration of ethanol and acetaldehyde within the ventral tegmental area of alcohol-preferring (P) rats: involvement of dopamine and serotonin. *Neuropsychopharmacology* **30**, 330–338.

Rodd-Henricks, Z. A., McKinzie, D. L., Crile, R. S., Murphy, J. M., and McBride, W. J. (2000). Regional heterogeneity for the intracranial self-administration of ethanol within the ventral tegmental area of female Wistar rats. *Psychopharmacology* **149**, 217–224.

Rodd-Henricks, Z. A., Melendez, R. I., Zaffaroni, A., Goldstein, A., McBride, W. J., and Li, T. K. (2002). The reinforcing effects of acetaldehyde in the posterior ventral tegmental area of alcohol-preferring rats. *Pharmacology Biochemistry and Behavior* **72**, 55–64.

Rodriguez de Fonseca, F., Carrera, M. R. A., Navarro, M., Koob, G. F., and Weiss, F. (1997). Activation of corticotropin-releasing factor in the limbic system during cannabinoid withdrawal. *Science* **276**, 2050–2054.

Rogers, R. D., Everitt, B. J., Baldacchino, A., Blackshaw, A. J., Swainson, R., Wynne, K., Baker, N. B., Hunter, J., Carthy, T., Booker, E., London, M., Deakin, J. F., Sahakian, B. J., and Robbins, T. W. (1999). Dissociable deficits in the decision-making cognition of chronic amphetamine abusers, opiate abusers, patients with focal damage to prefrontal cortex, and tryptophan-depleted normal volunteers: evidence for monoaminergic mechanisms. *Neuropsychopharmacology* **20**, 322–339.

Rosenbloom, M. J., Pfefferbaum, A., and Sullivan, E. V. (1995). Structural brain alterations associated with alcoholism. *Alcohol Health and Research World* **19**, 266–272.

Saal, D., Dong, Y., Bonci, A., and Malenka, R. C. (2003). Drugs of abuse and stress trigger a common synaptic adaptation in dopamine neurons. *Neuron* **37**, 577–582 [erratum: **38**, 359].

Salamone, J. D. (1992). Complex motor and sensorimotor functions of striatal and accumbens dopamine: involvement in instrumental behavior processes. *Psychopharmacology* **107**, 160–174.

Salamone, J. D. (1994). The involvement of nucleus accumbens dopamine in appetitive and aversive motivation. *Behavioural Brain Research* **61**, 117–133.

Salamone, J. D., and Correa, M. (2002). Motivational views of reinforcement: implications for understanding the behavioral functions of nucleus accumbens dopamine. *Behavioural Brain Research* **137**, 3–25.

Sapru, M. K., Diamond, I., and Gordon, A. S. (1994). Adenosine receptors mediate cellular adaptation to ethanol in NG108–15 cells. *Journal of Pharmacology and Experimental Therapeutics* **271**, 542–548.

Sarnyai, Z., Biro, E., Gardi, J., Vecsernyes, M., Julesz, J., and Telegdy, G. (1995). Brain corticotropin-releasing factor mediates "anxiety-like" behavior induced by cocaine withdrawal in rats. *Brain Research* **675**, 89–97.

Sawaguchi, T., and Goldman-Rakic, P. S. (1991). D1 dopamine receptors in prefrontal cortex: involvement in working memory. *Science* **251**, 947–950.

Schenk, S., Lacelle, G., Gorman, K., and Amit, Z. (1987). Cocaine self-administration in rats influenced by environmental conditions: implications for the etiology of drug abuse. *Neuroscience Letters* **81**, 227–231.

Schenk, S., Gorman, K., and Amit, Z. (1990). Age-dependent effects of isolation housing on the self-administration of ethanol in laboratory rats. *Alcohol* **7**, 321–326.

Schulteis, G., Ahmed, S. H., Morse, A. C., Koob, G. F., and Everitt, B. J. (2000). Conditioning and opiate withdrawal: The amygdala links neutral stimuli with the agony of overcoming drug addiction. *Nature* **405**, 1013–1014.

Schultz, W. (2002). Getting formal with dopamine and reward. *Neuron* **36**, 241–263.

Schultz, W., Apicella, P., and Ljungberg, T. (1993). Responses of monkey dopamine neurons to reward and conditioned stimuli during successive steps of learning a delayed response task. *Journal of Neuroscience* **13**, 900–913.

Schultz, W., Dayan, P., and Montague, P. R. (1997). A neural substrate of prediction and reward. *Science* **275**, 1593–1599.

See, R. E., Kruzich, P. J., and Grimm, J. W. (2001). Dopamine, but not glutamate, receptor blockade in the basolateral amygdala attenuates conditioned reward in a rat model of relapse to cocaine-seeking behavior. *Psychopharmacology* **154**, 301–310.

See, R. E., Fuchs, R. A., Ledford, C. C., and McLaughlin, J. (2003). Drug addiction, relapse, and the amygdala (series title: *Annals of the New York Academy of Sciences*, vol. 985), (P. Shinnick-Gallagher, A. Pitkanen, A. Shekhar, and L. Cahill, Eds.), pp. 294–307. New York Academy of Sciences, New York.

Self, D. W., Genova, L. M., Hope, B. T., Barnhart, W. J., Spencer, J. J., and Nestler, E. J. (1998). Involvement of cAMP-dependent protein kinase in the nucleus accumbens in cocaine self-administration and relapse of cocaine-seeking behavior. *Journal of Neuroscience* **18**, 1848–1859.

Sequier, J. M., Richards, J. G., Malherbe, P., Price, G. W., Mathews, S., and Mohler, H. (1988). Mapping of brain areas containing RNA homologous to cDNAs encoding the alpha and beta subunits of the rat GABAA gamma-aminobutyrate receptor. *Proceedings of the National Academy of Sciences USA* **85**, 7815–7819.

Shaham, Y. (1993). Immobilization stress-induced oral opioid self-administration and withdrawal in rats: role of conditioning factors and the effect of stress on 'relapse' to opioid drugs. *Psychopharmacology* **111**, 477–485.

Shaham, Y., and Stewart, J. (1994). Exposure to mild stress enhances the reinforcing efficacy of intravenous heroin self-administration in rats. *Psychopharmacology* **114**, 523–527.

Shaham, Y., and Stewart, J. (1995). Stress reinstates heroin-seeking in drug-free animals: an effect mimicking heroin, not withdrawal. *Psychopharmacology* **119**, 334–341.

Shaham, Y., Funk, D., Erb, S., Brown, T. J., Walker, C. D., and Stewart, J. (1997). Corticotropin-releasing factor, but not corticosterone, is involved in stress-induced relapse to heroin-seeking in rats. *Journal of Neuroscience* **17**, 2605–2614.

Shaham, Y., Erb, S., Leung, S., Buczek, Y., and Stewart, J. (1998). CP-154,526, a selective, non-peptide antagonist of the corticotropin-releasing factor1 receptor attenuates stress-induced relapse to drug seeking in cocaine- and heroin-trained rats. *Psychopharmacology* **137**, 184–190.

Shaham, Y., Erb, S., and Stewart, J. (2000a). Stress-induced relapse to heroin and cocaine seeking in rats: a review. *Brain Research Reviews* **33**, 13–33.

Shaham, Y., Highfield, D., Delfs, J., Leung, S., and Stewart, J. (2000b). Clonidine blocks stress-induced reinstatement of heroin seeking in rats: an effect independent of locus coeruleus noradrenergic neurons. *European Journal of Neuroscience* **12**, 292–302.

Shaham, Y., Shalev, U., Lu, L., De Wit, H., and Stewart, J. (2003). The reinstatement model of drug relapse: history, methodology and major findings. *Psychopharmacology* **168**, 3–20.

Shalev, U., Grimm, J. W., and Shaham, Y. (2002). Neurobiology of relapse to heroin and cocaine seeking: a review. *Pharmacological Reviews* **54**, 1–42.

Sharma, S. K., Klee, W. A., and Nirenberg, M. (1975). Dual regulation of adenylate cyclase accounts for narcotic dependence and tolerance. *Proceedings of the National Academy of Sciences USA* **72**, 3092–3096.

Sharpe, A. L., and Samson, H. H. (2001). Effect of naloxone on appetitive and consummatory phases of ethanol self-administration. *Alcoholism: Clinical and Experimental Research* **25**, 1006–1011.

Shaw-Lutchman, T. Z., Barrot, M., Wallace, T., Gilden, L., Zachariou, V., Impey, S., Duman, R. S., Storm, D., and Nestler, E. J. (2002). Regional and cellular mapping of cAMP response element-mediated transcription during naltrexone-precipitated morphine withdrawal. *Journal of Neuroscience* **22**, 3663–3672.

Shaw-Lutchman, T. Z., Impey, S., Storm, D., and Nestler, E. J. (2003). Regulation of CRE-mediated transcription in mouse brain by amphetamine. *Synapse* **48**, 10–17.

Shivers, B. D., Killisch, I., Sprengel, R., Sontheimer, H., Kohler, M., Schofield, P. R., and Seeburg, P. H. (1989). Two novel GABAA receptor subunits exist in distinct neuronal subpopulations. *Neuron* **3**, 327–337.

Simon, H., Scatton, B., and Le Moal, M. (1980). Dopaminergic A10 neurones are involved in cognitive functions. *Nature* **286**, 150–151.

Sorg, B. A., and Kalivas, P. W. (1993). Effects of cocaine and footshock stress on extracellular dopamine levels in the medial prefrontal cortex. *Neuroscience* **53**, 695–703.

Stewart, J. (2000). Pathways to relapse: the neurobiology of drug- and stress-induced relapse to drug-taking. *Journal of Psychiatry and Neuroscience* **25**, 125–136.

Stinus, L., Cador, M., Zorrilla, E. P., and Koob, G. F. (2005). Buprenorphine and a CRF1 antagonist block the acquisition of opiate withdrawal-induced conditioned place aversion in rats. *Neuropsychopharmacology* **30**, 90–98.

Sullivan, E. V. (2000). Human brain vulnerability to alcoholism: evidence from neuroimaging studies. *In Review of NIAAA's Neuroscience and Behavioral Research Portfolio* (series title: *NIAAA Research Monograph*, vol. 34), (A. Noronha, M. Eckardt, and K. Warren, Eds.), pp. 473–508. National Institute on Alcohol Abuse and Alcoholism, Bethesda, MD.

Sullivan, E. V., Mathalon, D. H., Zipursky, R. B., Kersteen-Tucker, Z., Knight, R. T., and Pfefferbaum, A. (1993). Factors of the Wisconsin Card Sorting Test as measures of frontal-lobe function in schizophrenia and in chronic alcoholism. *Psychiatry Research* **46**, 175–199.

Sullivan, E. V., Rosenbloom, M. J., Lim, K. O., and Pfefferbaum, A. (2000a). Longitudinal changes in cognition, gait, and balance in abstinent and relapsed alcoholic men: relationships to changes in brain structure. *Neuropsychology* **14**, 178–188.

Sullivan, E. V., Rosenbloom, M. J., and Pfefferbaum, A. (2000b). Pattern of motor and cognitive deficits in detoxified alcoholic men. *Alcoholism: Clinical and Experimental Research* **24**, 611–621.

Sullivan, E. V., Desmond, J. E., Lim, K. O., and Pfefferbaum, A. (2002). Speed and efficiency but not accuracy or timing deficits of limb movements in alcoholic men and women. *Alcoholism: Clinical and Experimental Research* **26**, 705–713.

Sullivan, E. V., Harding, A. J., Pentney, R., Dlugos, C., Martin, P. R., Parks, M. H., Desmond, J. E., Chen, S. H., Pryor, M. R., De Rosa, E., and Pfefferbaum, A. (2003). Disruption of frontocerebellar circuitry and function in alcoholism. *Alcoholism: Clinical and Experimental Research* **27**, 301–309.

Szumlinski, K. K., Dehoff, M. H., Kang, S. H., Frys, K. A., Lominac, K. D., Klugmann, M., Rohrer, J., Griffin, W. 3rd, Toda, S., Champtiaux, N. P., Berry, T., Tu, J. C., Shealy, S. E., During, M. J., Middaugh, L. D., Worley, P. F., and Kalivas, P. W. (2004). Homer proteins regulate sensitivity to cocaine. *Neuron* **43**, 401–413.

Tabakoff, B., and Hoffman, P. L. (1992). Alcohol: neurobiology. *In Substance Abuse: A Comprehensive Textbook*, 2nd ed., (J. H. Lowinson, and P. Ruiz, Eds.), pp. 152–185. Williams and Wilkins, Baltimore.

Tanda, G., and Di Chiara, G. (1998). A dopamine-mu1 opioid link in the rat ventral tegmentum shared by palatable food (Fonzies) and non-psychostimulant drugs of abuse. *European Journal of Neuroscience* **10**, 1179–1187.

Tanda, G., Pontieri, F. E., Frau, R., and Di Chiara, G. (1997). Contribution of blockade of the noradrenaline carrier to the increase of extracellular dopamine in the rat prefrontal cortex by amphetamine and cocaine. *European Journal of Neuroscience* **9**, 2077–2085.

Tassin, J. P., Stinus, L., Simon, H., Blanc, G., Thierry, A. M., Le Moal, M., Cardo, B., and Glowinski, J. (1978). Relationship between the locomotor hyperactivity induced by A10 lesions and the destruction of the fronto-cortical dopaminergic innervation in the rat. *Brain Research* **141**, 267–281.

Terwilliger, R. Z., Beitner-Johnson, D., Sevarino, K. A., Crain, S. M., and Nestler, E. J. (1991). A general role for adaptations in G-proteins and the cyclic AMP system in mediating the chronic actions of morphine and cocaine on neuronal function. *Brain Research* **548**, 100–110.

Thomas, M. J., and Malenka, R. C. (2003). Synaptic plasticity in the mesolimbic dopamine system. *Transactions of the Royal Society of London B Biological Sciences* **358**, 815–819.

Thomas, M. J., Malenka, R. C., and Bonci, A. (2000). Modulation of long-term depression by dopamine in the mesolimbic system. *Journal of Neuroscience* **20**, 5581–5586.

Thomas, M. J., Beurrier, C., Bonci, A., and Malenka, R. C. (2001). Long-term depression in the nucleus accumbens: a neural correlate of behavioral sensitization to cocaine. *Nature Neuroscience* **4**, 1217–1223.

Tiffany, S. T., and Carter, B. L. (1998). Is craving the source of compulsive drug use? *Journal of Psychopharmacology* **12**, 23–30.

Tomkins, D. M., Sellers, E. M., and Fletcher, P. J. (1994). Median and dorsal raphe injections of the 5-HT1A agonist, 8-OH-DPAT, and the GABAA agonist, muscimol, increase voluntary ethanol intake in Wistar rats. *Neuropharmacology* **33**, 349–358.

Turgeon, S. M., Pollack, A. E., and Fink, J. S. (1997). Enhanced CREB phosphorylation and changes in c-Fos and FRA expression in striatum accompany amphetamine sensitization. *Brain Research* **749**, 120–126.

Ungless, M. A., Whistler, J. L., Malenka, R. C., and Bonci, A. (2001). Single cocaine exposure in vivo induces long-term potentiation in dopamine neurons. *Nature* **411**, 583–587.

Uzwiak, A. J., Guyette, F. X., West, M. O., and Peoples, L. L. (1997). Neurons in accumbens subterritories of the rat: phasic firing time-locked within seconds of intravenous cocaine self-infusion. *Brain Research* **767**, 363–369.

Vaccarino, F. J., Amalric, M., Swerdlow, N. R., and Koob, G. F. (1986). Blockade of amphetamine but not opiate-induced locomotion following antagonism of dopamine function in the rat. *Pharmacology Biochemistry and Behavior* **24**, 61–65.

Valdez, G. R., Roberts, A. J., Chan, K., Davis, H., Brennan, M., Zorrilla, E. P., and Koob, G. F. (2002). Increased ethanol self-administration and anxiety-like behavior during acute withdrawal and protracted abstinence: regulation by corticotropin-releasing factor. *Alcoholism: Clinical and Experimental Research* **26**, 1494–1501.

Valdez, G. R., Zorrilla, E. P., Roberts, A. J., and Koob, G. F. (2003). Antagonism of corticotropin-releasing factor attenuates the enhanced responsiveness to stress observed during protracted ethanol abstinence. *Alcohol* **29**, 55–60.

Valverde, O., Mantamadiotis, T., Torrecilla, M., Ugedo, L., Pineda, J., Bleckmann, S., Gass, P., Kretz, O., Mitchell, J. M., Schutz, G., and Maldonado, R. (2004). Modulation of anxiety-like behavior and morphine dependence in CREB-deficient mice. *Neuropsychopharmacology* **29**, 1122–1133.

Volkow, N. D., Mullani, N., Gould, K. L., Adler, S., and Krajewski, K. (1988). Cerebral blood flow in chronic cocaine users: a study with positron emission tomography. *British Journal of Psychiatry* **152**, 641–648.

Volkow, N. D., Fowler, J. S., Wolf, A. P., Hitzemann, R., Dewey, S., Bendriem, B., Alpert, R., and Hoff, A. (1991). Changes in brain glucose metabolism in cocaine dependence and withdrawal. *American Journal of Psychiatry* **148**, 621–626.

Volkow, N. D., Hitzemann, R., Wang, G. J., Fowler, J. S., Burr, G., Pascani, K., Dewey, S. L., and Wolf, A. P. (1992a). Decreased brain metabolism in neurologically intact healthy alcoholics. *American Journal of Psychiatry* **149**, 1016–1022.

Volkow, N. D., Hitzemann, R., Wang, G. J., Fowler, J. S., Wolf, A. P., Dewey, S. L., and Handlesman, L. (1992b). Long-term frontal brain metabolic changes in cocaine abusers. *Synapse* **11**, 184–190 [erratum: **12**, 86].

Volkow, N. D., Wang, G. J., Hitzemann, R., Fowler, J. S., Overall, J. E., Burr, G., and Wolf, A. P. (1994). Recovery of brain glucose metabolism in detoxified alcoholics. *American Journal of Psychiatry* **151**, 178–183.

Volkow, N. D., Gillespie, H., Mullani, N., Tancredi, L., and Grant, C., Valentine, A., Hollister, L. (1996). Brain glucose metabolism in chronic marijuana users at baseline and during marijuana intoxication. *Psychiatry Research* **67**, 29–38.

Volkow, N. D., Wang, G. J., Fowler, J. S., Logan, J., Gatley, S. J., Hitzemann, R., Chen, A. D., Dewey, S. L., and Pappas, N. (1997). Decreased striatal dopaminergic responsiveness in detoxified cocaine-dependent subjects. *Nature* **386**, 830–833.

Volkow, N. D., Wang, G. J., Fowler, J. S., Hitzemann, R., Angrist, B., Gatley, S. J., Logan, J., Ding, Y. S., and Pappas, N. (1999a). Association of methylphenidate-induced craving with changes in right striato-orbitofrontal metabolism in cocaine abusers: Implications in addiction. *American Journal of Psychiatry* **156**, 19–26.

Volkow, N. D., Wang, G. J., Fowler, J. S., Logan, J., Gatley, S. J., Wong, C., Hitzemann, R., and Pappas, N. R. (1999b). Reinforcing effects of psychostimulants in humans are associated with increases in brain dopamine and occupancy of D(2) receptors. *Journal of Pharmacology and Experimental Therapeutics* **291**, 409–415.

Volkow, N. D., Wang, G. J., Fowler, J. S., Franceschi, D., Thanos, P. K., Wong, C., Gatley, S. J., Ding, Y. S., Molina, P., Schlyer, D., Alexoff, D., Hitzemann, R., and Pappas, N. (2000). Cocaine abusers show a blunted response to alcohol intoxication in limbic brain regions. *Life Sciences* **66**, PL161–PL167.

Volkow, N. D., Wang, G., Fowler, J. S., Logan, J., Gerasimov, M., Maynard, L., Ding, Y., Gatley, S. J., Gifford, A., and Franceschi, D. (2001). Therapeutic doses of oral methylphenidate significantly increase extracellular dopamine in the human brain. *Journal of Neuroscience* **21**, RC121.

Volkow, N. D., Fowler, J. S., and Wang, G. J. (2002). Role of dopamine in drug reinforcement and addiction in humans: results from imaging studies. *Behavioural Pharmacology* **13**, 355–366.

Volkow, N. D., Fowler, J. S., and Wang, G. J. (2003). The addicted human brain: insights from imaging studies. *Journal of Clinical Investigation* **111**, 1444–1451.

Vorel, S. R., Liu, X., Hayes, R. J., Spector, J. A., and Gardner, E. L. (2001). Relapse to cocaine-seeking after hippocampal theta burst stimulation. *Science* **292**, 1175–1178.

Walters, C. L., and Blendy, J. A. (2001). Different requirements for cAMP response element binding protein in positive and negative reinforcing properties of drugs of abuse. *Journal of Neuroscience* **21**, 9438–9444.

Wang, G. J., Volkow, N. D., Fowler, J. S., Cervany, P., Hitzemann, R. J., Pappas, N. R., Wong, C. T., and Felder, C. (1999). Regional brain metabolic activation during craving elicited by recall of previous drug experiences. *Life Sciences* **64**, 775–784.

Watanabe, T., Yamamoto, R., Maeda, A., Nakagawa, T., Minami, M., and Satoh, M. (2002). Effects of excitotoxic lesions of the central or basolateral nucleus of the amygdala on naloxone-precipitated withdrawal-induced conditioned place aversion in morphine-dependent rats. *Brain Research* **958**, 423–428.

Weiss, F., Maldonado-Vlaar, C. S., Parsons, L. H., Kerr, T. M., Smith, D. L., and Ben-Shahar, O. (2000). Control of cocaine-seeking behavior by drug-associated stimuli in rats: Effects on recovery of extinquished operant-responding and extracellular dopamine levels in amygdala and nucleus accumbens. *Proceedings of the National Academy of Sciences USA* **97**, 4321–4326.

Weiss, F., Ciccocioppo, R., Parsons, L. H., Katner, S., Liu, X., Zorrilla, E. P., Valdez, G. R., Ben-Shahar, O., Angeletti, S., and Richter, R. R. (2001a). Compulsive drug-seeking behavior and relapse: neuroadaptation, stress, and conditioning factors. *In The Biological Basis of Cocaine Addiction* (series title: *Annals of the New York Academy of Sciences*, vol. 937), (Quinones-Jenab, Ed.), pp. 1–26. New York Academy of Sciences, New York.

Weiss, F., Martin-Fardon, R., Ciccocioppo, R., Kerr, T. M., Smith, D. L., and Ben-Shahar, O. (2001b). Enduring resistance to extinction of cocaine-seeking behavior induced by drug-related cues. *Neuropsychopharmacology* **25**, 361–372.

Weissenborn, R., Robbins, T. W., and Everitt, B. J. (1997). Effects of medial prefrontal or anterior cingulate cortex lesions on responding for cocaine under fixed-ratio and second-order schedules of reinforcement in rats. *Psychopharmacology* **134**, 242–257.

Wexler, B. E., Gottschalk, C. H., Fulbright, R. K., Prohovnik, I., Lacadie, C. M., Rounsaville, B. J., and Gore, J. C. (2001s). Functional magnetic resonance imaging of cocaine craving. *American Journal of Psychiatry* **158**, 86–95.

Whitelaw, R. B., Markou, A., Robbins, T. W., and Everitt, B. J. (1996). Excitotoxic lesions of the basolateral amygdala impair the acquisition of cocaine-seeking behaviour under a second-order schedule of reinforcement. *Psychopharmacology* **127**, 213–224.

Wise, R. A. (1980). Action of drugs of abuse on brain reward systems. *Pharmacology Biochemistry and Behavior* **13**(*Suppl.* 1), 213–223.

Wise, R. A. (2002). Brain reward circuitry: insights from unsensed incentives. *Neuron* **36**, 229–240.

Wise, R. A., and Bozarth, M. A. (1987). A psychomotor stimulant theory of addiction. *Psychological Review* **94**, 469–492.

Wise, R. A., Spindler, J., deWit, H., and Gerberg, G. J. (1978a). Neuroleptic-induced 'anhedonia' in rats: pimozide blocks reward quality of food. *Science* **201**, 262–264.

Wise, R. A., Spindler, J., and Legault, L. (1978b). Major attenuation of food reward with performance-sparing doses of pimozide in the rat. *Canadian Journal of Psychology* **32**, 77–85.

Wolf, M. E. (1998). The role of excitatory amino acids in behavioral sensitization to psychomotor stimulants. *Progress in Neurobiology* **54**, 679–720.

Wolf, M. E., Dahlin, S. L., Hu, X. T., Xue, C. J., and White, K. (1995). Effects of lesions of prefrontal cortex, amygdala, or fornix on behavioral sensitization to amphetamine: comparison with N-methyl-D-aspartate antagonists. *Neuroscience* **69**, 417–439.

Wolffgramm, J., and Heyne, A. (1991). Social behavior, dominance, and social deprivation of rats determine drug choice. *Pharmacology Biochemistry and Behavior* **38**, 389–399.

Yokel, R. A., and Wise, R. A. (1975). Increased lever pressing for amphetamine after pimozide in rats: Implications for a dopamine theory of reward. *Science* **187**, 547–549.

Yokel, R. A., and Wise, R. A. (1976). Attenuation of intravenous amphetamine reinforcement by central dopamine blockade in rats. *Psychopharmacology* **48**, 311–318.

Yokel, R. A., and Wise, R. A. (1978). Amphetamine-type reinforcement by dopaminergic agonists in the rat. *Psychopharmacology* **58**, 289–296.

Yun, I. A., and Fields, H. L. (2003). Basolateral amygdala lesions impair both cue- and cocaine-induced reinstatement in animals trained on a discriminative stimulus task. *Neuroscience* **121**, 747–757.

Xi, Z. X., Ramamoorthy, S., Baker, D. A., Shen, H., Samuvel, D. J., and Kalivas, P. W. (2002). Modulation of group II metabotropic glutamate receptor signaling by chronic cocaine. *Journal of Pharmacology and Experimental Therapeutics* **303**, 608–615.

Yadin, E., Thomas, E., Strickland, C. E., and Grishkat, H. L. (1991). Anxiolytic effects of benzodiazepines in amygdala-lesioned rats. *Psychopharmacology* **103**, 473–479.

Zhang, H., and Pandey, S. C. (2003). Effects of PKA modulation on the expression of neuropeptide Y in rat amygdaloid structures during ethanol withdrawal. *Peptides* **24**, 1397–1402.

Zhou, Y., Spangler, R., LaForge, K. S., Maggos, C. E., Ho, A., and Kreek, M. J. (1996). Corticotropin-releasing factor and type 1 corticotropin-releasing factor receptor messenger RNAs in rat brain and pituitary during "binge"-pattern cocaine administration and chronic withdrawal. *Journal of Pharmacology and Experimental Therapeutics* **279**, 351–358.

Zhou, Y., Spangler, R., Ho, A., and Kreek, M. J. (2003). Increased CRH mRNA levels in the rat amygdala during short-term withdrawal from chronic 'binge' cocaine. *Molecular Brain Research* **114**, 73–79.

C H A P T E R

10

Drug Addiction: Transition from Neuroadaptation to Pathophysiology

OUTLINE

COMMON NEUROBIOLOGICAL ELEMENTS IN ADDICTION

Definitions and Animal Models

Drug addiction, also known as Substance Dependence ('big D'), is a chronically relapsing disorder characterized by: (1) compulsion to seek and take the drug, (2) loss of control in limiting intake, and (3) emergence of a negative emotional state (e.g., dysphoria, anxiety, irritability) when access to the drug is prevented (defined here as dependence with a 'little d'). Clinically, the occasional but limited use of an abusable drug is distinct from escalated drug use and the emergence of chronic Substance Dependence. An important goal of current neurobiological research is to understand the neuropharmacological and neuroadaptive mechanisms within specific neurocircuits that mediate the transition from occasional, controlled drug use and the loss of behavioral control over drug-seeking and drug-taking that defines chronic addiction.

Much of the recent progress in understanding the mechanisms of addiction has been derived from the study of animal models of addiction on specific drugs, such as stimulants, opioids, alcohol, nicotine, and Δ^9-tetrahydrocannabinol. While no animal model of addiction fully emulates the human condition, animal models do permit investigation of specific elements of

Neurobiology of Addiction, by George F. Koob and Michel Le Moal.
ISBN – 13: 978-0-12-419239-3 ISBN – 10: 0-12-419239-4

the process of drug addiction. Such elements can be defined by models of different systems, models of psychological constructs such as positive and negative reinforcement, and models of different stages of the addiction cycle. While much focus in animal studies has been on the synaptic sites and transductive mechanisms in the central nervous system on which drugs of abuse act initially to produce their positive reinforcing effects, new animal models of components of the negative reinforcing effects of dependence have been developed and are beginning to be used to explore how the nervous system adapts to drug use. In addition, there are models of craving involving drug-, cue- and stress-induced reinstatement of drug-seeking behavior. Finally, animal models have been designed to address the transition to addiction ranging from compulsive drug-taking, as measured by escalation in drug intake (Ahmed and Koob, 1998), or drug-seeking despite negative or aversive consequences (Deroche-Gamonet *et al.*, 2004; Vanderschuren and Everitt, 2004). These models also have begun to address specific elements of the diagnostic criteria used by the American Psychiatric Association and the World Health Organization to define addiction or Substance Dependence (World Health Organization, 1992; American Psychiatric Association, 1994). The neurobiological mechanisms of addiction that are involved in various stages of the addiction cycle have a specific focus on certain brain circuits, and the neurochemical changes associated with those circuits, during the transition from drug-taking to drug addiction and how those changes persist in the vulnerability to relapse.

Neurocircuitry of the Acute Rewarding Effects of Drugs of Abuse

A key element of drug addiction is how the brain reward system changes with the development of addiction, and one must understand the neurobiological bases for acute drug reward to understand how these systems change with the development of addiction. A principal focus of research on the neurobiology of the positive reinforcing effects of drugs of abuse has been the origins and terminal areas of the mesocorticolimbic dopamine system. There is compelling evidence for the importance of this system in psychostimulant reward. This specific circuit has been broadened to include the many neural inputs and outputs that interact with the ventral tegmental area and the basal forebrain, and as such has been termed by some as the mesolimbic reward system, and so broadened involves interactions with other drugs of abuse. More recently, specific components of the basal forebrain that have been identified with drug reward have focused on the 'extended amygdala'. The extended amygdala includes the central nucleus of the amygdala, the bed nucleus of the stria terminalis, and a transition zone in the medial (shell) part of the nucleus accumbens. As the neural circuits for the reinforcing effects of drugs of abuse have evolved, the role of neurotransmitters/neuromodulators also has evolved. Four of these systems have been identified as having an important role in the acute reinforcing effects of drugs of abuse: the mesolimbic dopamine system, the opioid peptide system, the γ-aminobutyric acid (GABA) system, and the endocannabinoid system (**Table 10.1**). While some have argued for a drug reward system in series with the mesolimbic dopamine system, others have argued that as one moves away from the psychostimulant drugs, such as cocaine and amphetamines, the importance of dopamine diminishes, and other neurochemical systems may have drug reward functions independent of dopamine in the extended amygdala.

Neurocircuitry of the Withdrawal/Negative Affect Stage of the Addiction Cycle

The neural substrates and neuropharmacological mechanisms for the negative motivational effects of drug withdrawal may involve disruption of the same neurochemical systems and neurocircuits implicated in the positive reinforcing effects of drugs of abuse (**Table 10.2**). Repeated administration of psychostimulants produces an initial facilitation of dopamine and glutamate neurotransmission (Ungless *et al.*, 2001; Vorel *et al.*, 2002). However, chronic administration leads to decreases in dopaminergic and glutamatergic neurotransmission in the nucleus accumbens during acute withdrawal, opposite responses of opioid receptor transduction mechanisms in the nucleus accumbens during opioid withdrawal, changes in GABAergic neurotransmission during alcohol withdrawal, and differential regional changes in nicotinic acetylcholine receptor function during nicotine withdrawal. Decreases in reward neurotransmitter function have been hypothesized to contribute significantly to the negative motivational state associated with acute drug abstinence and may trigger long-term biochemical changes that contribute to the clinical syndrome of protracted abstinence and vulnerability to relapse.

Different neurochemical systems involved in stress modulation also may be engaged within the neurocircuitry of the brain stress systems in an attempt to overcome the chronic presence of the perturbing drug and to restore normal function despite the presence of the drug. Both the hypothalamic–pituitary–adrenal (HPA) axis and the brain stress system mediated by

TABLE 10.1 Neurobiological substrates for the acute reinforcing effects of drugs of abuse

Drug of Abuse	Neurotransmitter	Site
Cocaine and amphetamines	Dopamine γ-Aminobutyric acid	Nucleus accumbens Amygdala
Opioids	Opioid peptides Dopamine Endocannabinoids	Nucleus accumbens Ventral tegmental area
Nicotine	Dopamine γ-Aminobutyric acid Opioid peptides	Nucleus accumbens Ventral tegmental area Amygdala
Δ^9-Tetrahydrocannabinol	Endocannabinoids Opioid peptides Dopamine	Nucleus accumbens Ventral tegmental area
Alcohol	Dopamine Opioid peptides γ-Aminobutyric acid Glutamate Endocannabinoids	Nucleus accumbens Ventral tegmental area Amygdala

corticotropin-releasing factor (CRF) are dysregulated by chronic administration of drugs of abuse, with a common response of dysregulated adrenocorticotropic hormone (ACTH) and corticosterone and extended amygdala CRF during acute and protracted withdrawal from all major drugs of abuse. Acute withdrawal from drugs of abuse also may increase the release of norepinephrine in the bed nucleus of the stria terminalis and decrease levels of neuropeptide Y (NPY) in the amygdala.

These results suggest not only a change in the function of neurotransmitters associated with the acute reinforcing effects of drugs of abuse during the development of dependence, such as dopamine, opioid peptides, serotonin and GABA, but also recruitment of the CRF and norepinephrine brain stress system and dysregulation of the NPY brain antistress system. Additionally, activation of the brain stress systems may not only contribute to the negative motivational state associated with acute abstinence but may also contribute to the vulnerability to stressors observed during protracted abstinence in humans.

TABLE 10.2 Neurotransmitters Implicated in the Motivational Effects of Withdrawal From Drugs of Abuse

Neurotransmitter	Functional Effect
↓ Dopamine	'Dysphoria'
↓ Serotonin	'Dysphoria'
↓ γ-Aminobutyric acid	Anxiety, panic attacks
↓ Neuropeptide Y	Antistress
↑ Dynorphin	'Dysphoria'
↑ Corticotropin-releasing factor	Stress
↑ Norepinephrine	Stress

The neuroanatomical entity termed the extended amygdala thus may represent not only a common anatomical substrate for acute drug reward but also a common neuroanatomical substrate for the negative effects on reward function produced by stress that help drive compulsive drug administration. As stated above, the extended amygdala is composed of the bed nucleus of the stria terminalis, the central nucleus of the amygdala, and a transition zone in the medial subregion of the nucleus accumbens (shell of the nucleus accumbens). Each of these regions has certain cytoarchitectural and circuitry similarities. The extended amygdala receives numerous afferents from limbic structures, such as the basolateral amygdala and hippocampus, and sends efferents to the medial part of the ventral pallidum and to the lateral hypothalamus, thus further defining the specific brain areas that interface classical limbic (emotional) structures with the extrapyramidal motor system.

Neurocircuitry of the Preoccupation/Anticipation (Craving) Stage of the Addiction Cycle

Animal models of 'craving' involve the use of drug-, cue-, and stress-induced reinstatement in animals acquiring drug self-administration and then are subjected to extinction for responding to the drug. Most evidence from animal studies suggests that drug-induced reinstatement is localized to the medial prefrontal cortex–nucleus accumbens–ventral pallidum circuit mediated in part by the neurotransmitter glutamate in the prefrontal cortex–nucleus accumbens circuit. In contrast, neuropharmacological and neurobiological studies using animal models for cue-induced reinstatement involve the basolateral amygdala as a critical substrate, with a possible feed-forward mechanism through the prefrontal cortex involved in drug-induced reinstatement. Stress-induced reinstatement of drug-related responding in animal models appears to depend on the activation of both CRF and norepinephrine in elements of the extended amygdala (central nucleus of the amygdala and bed nucleus of the stria terminalis).

The hippocampus may also be a part of the circuitry of cognitive pathophysiology associated with addiction. The hippocampus has been hypothesized to have a role in contextual cues associated with 'craving' (Selden *et al.*, 1991; Everitt and Wolf, 2002), and stimulation of the hippocampus can induce reinstatement of drug-seeking behavior (Vorel *et al.*, 2001). Even more intriguing is the discovery that there is continuous neurogenesis in adult rat hippocampus (for review, see Abrous *et al.*, 2005). Such plasticity can be influenced by drugs of abuse. Rats trained to self-administer nicotine had a dose-dependent decrease in neurogenesis, a decrease in the production of cell adhesion molecules, and an increase in cellular apoptosis in the dentate gyrus of the hippocampus (Abrous *et al.*, 2002). How these changes impact the cognitive inflexibility associated with drug addiction and whether they generalize to all drugs of abuse are challenges for future research.

In summary, three neurobiological circuits have been identified from animal studies that have heuristic value for the study of the neurobiological changes associated with the development and persistence of drug dependence. The acute reinforcing effects of drugs of abuse comprising the binge/intoxication stage most likely involve actions with an emphasis on the extended amygdala reward system and inputs from the ventral tegmental area (dopamine), the arcuate nucleus of the hypothalamus (β-endorphin), local opioid peptide circuits (enkephalins), local inputs from GABA neurons, and endocannabinoid modulation. In contrast, the effects associated with the withdrawal/negative affect stage, such as increased anxiety, most likely involve not only decreases in function of the extended amygdala reward system, but also recruitment of brain stress neurocircuitry within that system (CRF, norepinephrine, and glucocorticoids). The effects of the preoccupation/anticipation (craving) stage involve key afferent projections to the extended amygdala and nucleus accumbens, specifically the prefrontal cortex for drug-induced reinstatement and the basolateral amygdala for cue-induced reinstatement. Compulsive drug-seeking behavior is hypothesized to be driven by activated ventral striatal–ventral pallidal–thalamic–cortical loops.

Brain Imaging Circuits Involved in Human Addiction

Brain imaging studies using positron emission tomography with ligands for measuring oxygen utilization or glucose metabolism or using magnetic resonance imaging techniques have provided dramatic insights into the neurocircuitry changes in the human brain associated with the development, maintenance, and even vulnerability to addiction. These imaging results, overall, have a striking resemblance to the neurocircuitry identified by animal studies. During acute intoxication with drugs, such as cocaine, methamphetamine, nicotine, alcohol, and Δ^9-tetrahydrocannabinol, there are changes in the orbitofrontal cortex, prefrontal cortex, anterior cingulate, extended amygdala, and ventral striatum. These changes are reflected in activation as measured by functional magnetic resonance imaging, and decreases in oxygen utilization as measured by positron emission tomography (London *et al.*, 1990a,b; Breiter *et al.*, 1997). The changes during acute intoxication are often accompanied by an increase in the availability of dopamine, particularly in the striatal regions.

During acute and chronic withdrawal there is a reversal of these changes with decreases in metabolic activity in the orbitofrontal cortex, prefrontal cortex, and anterior cingulate, and decreases in basal dopamine activity in the ventral striatum and prefrontal cortex. Cue-induced reinstatement, in limited studies, appears to involve a reactivation of these same circuits, much like acute intoxication, but with the additional recruitment of activation in the amygdala. In summary, two strongly represented markers for active substance dependence in humans across drugs of different neuropharmacological actions are decreases in prefrontal cortex metabolic activity and decreases in brain dopamine D_2 receptors that are hypothesized to reflect decreases in brain dopamine function.

Cellular Targets within Brain Circuits Associated with Addiction

Acute administration of virtually all drugs of abuse activates firing of neurons in key elements of the reward circuits outlined earlier. Acute administration of psychostimulants, opioids, alcohol, and nicotine all activate neurons in the nucleus accumbens and the central nucleus of the amygdala. With psychostimulants, *in vivo* recordings of freely moving animals self-administering drugs have shown activation or inhibition of firing in the nucleus accumbens not only in response to delivery of the drug reinforcer itself, but also in anticipation of drug delivery. For many drugs of abuse, changes in both GABA and glutamate neuronal activity mediate the cellular actions of acute administration of drugs of abuse in the nucleus accumbens and central nucleus of the amygdala via disinhibition. Another common element of the acute actions of drugs of abuse is the activation of dopamine neurons in the ventral tegmental area. All three sites—nucleus accumbens, central nucleus of the amygdala, and ventral tegmental area—have been implicated in the neuroplasticity associated with chronic administration of drugs of abuse, with opposite changes in cellular function relative to the acute effects of drugs of abuse during acute withdrawal (decreases in firing in ventral tegmental area dopamine neurons; opposite changes in GABA and glutamate activity in the nucleus accumbens), but also some evidence of residual plasticity associated with a history of drug exposure.

Molecular Targets within the Brain Circuits Associated with Addiction

All drugs of abuse share some common neurocircuitry actions, namely inhibition of medium spiny neurons in the nucleus accumbens either through ligand-gated ion channel receptors, metabotropic G-protein-coupled receptors, or other effector targets. The search at the molecular level has led to examining how repeated perturbation of intracellular signal transduction pathways leads to changes in nuclear function and altered rates of transcription of particular target genes. Altered expression of such genes would lead to altered activity of the neurons where such changes occur, and ultimately to changes in neural circuits in which those neurons operate.

Two transcription factors in particular have been implicated in the plasticity associated with addiction: cyclic adenosine monophosphate response element binding protein (CREB) and ΔFosB. CREB regulates the transcription of genes that contain a CRE site (cAMP response element) within their regulatory regions. CREB is ubiquitous in genes expressed in the central nervous system such as those encoding neuropeptides, synthetic enzymes for neurotransmitters, signaling proteins, and other transcription factors. CREB can be phosphorylated not only by protein kinase A but also by protein kinases regulated by growth factors placing it at a point of convergence for several intracellular messenger pathways that can regulate the expression of genes.

Much work in the addiction field has shown that activation of CREB in the nucleus accumbens, one component of the reward circuit, is a consequence of chronic exposure to opioids and cocaine, and deactivation of CREB in the central nucleus of the amygdala, another part of the reward circuit, is a consequence of chronic exposure to alcohol and nicotine. The activation of CREB in the nucleus accumbens is linked to an increase in the κ opioid receptor binding dynorphin, which has been hypothesized to produce dysphoric-like responses. Nestler (2004) has argued that there is compelling evidence that upregulation of the cAMP pathway and CREB in this brain region (nucleus accumbens) represents a mechanism of 'motivational tolerance and dependence'. These molecular adaptations decrease an individual's sensitivity to the rewarding effects of subsequent drug exposures (tolerance) and impair the reward pathway (dependence), so that after removal of the drug, the individual is left in an amotivational, dysphoric, or depressed-like state. In contrast, decreased CREB phosphorylation has been observed in the central nucleus of the amygdala during alcohol and nicotine withdrawal and has been linked to decreased NPY function and consequently the increased anxiety-like responses associated with withdrawal from chronic alcohol. These changes are not necessarily mutually exclusive and point to transduction mechanisms that could produce neurochemical changes in the neurocircuits outlined above as important for breaks with reward homeostasis in addiction.

Chronic exposure to drugs of abuse also has been linked to more prolonged changes in other transcription factors that can alter gene expression and produce long-term changes in protein expression and as a result neuronal function. While acute administration of drugs of abuse can cause a rapid (on the order of hours) activation of members of the Fos family, such as c-fos, FosB, Fra-1, and Fra-2 in the nucleus accumbens, other transcription factors, such as isoforms of ΔFosB, accumulate over longer periods of time (days) with repeated drug administration. Animals with activated ΔFosB have exaggerated sensitivity to the rewarding effects of drugs of abuse. ΔFosB has been characterized as a sustained molecular 'switch' that helps to initiate and maintain a state of addiction. How changes in ΔFosB that last for days can translate into vulnerability to relapse remains a challenge for future work.

Molecular genetic animal models have provided a convergence of data to support the neuropharmacological substrates identified in neurocircuitry studies. Rats bred for high alcohol preference show high voluntary consumption of alcohol, increased anxiety-like responses, and numerous neuropharmacological phenotypes, such as decreased dopaminergic activity and decreased NPY activity. In a preferring (P) and nonpreferring (nP) cross, a quantitative trait locus was identified on Chromosome 4, a region to which the gene for NPY has been mapped. In the inbred P and nP quantitative trait loci analyses, loci on Chromosomes 3, 4, and 8 have been identified which correspond to loci near the genes for the dopamine D_2 and serotonin $5HT_{1B}$ receptors. Quantitative trait loci analyses have also led to the identification of the *Mpdz* gene important for alcohol withdrawal.

Advances in molecular biology provide the ability to systematically inactivate the genes that control the expression of proteins that make up receptors or neurotransmitter/neuromodulators in the central nervous system using gene knockout and/or transgenic approaches. While such an approach does not guarantee that these genes are the ones that are vulnerable in the human population, they do provide viable candidates for exploring the genetic basis of endophenotypes associated with addiction.

Notable positive results with gene knockout studies in mice have focused on knockout of the μ opioid receptor, which eliminates opioid, nicotine, and cannabinoid reward and alcohol drinking in mice. Opioid reinforcement as measured by place preference or self-administration is absent in μ knockout mice, and there is no development of somatic signs of dependence to morphine in these mice. To date, all morphine effects tested, including analgesia, hyperlocomotion, respiratory depression, and inhibition of gastrointestinal transit, are abolished in μ knockout mice.

Selective deletion of the genes for expression of different dopamine receptor subtypes and the dopamine transporter has revealed significant effects to challenges with psychomotor stimulants. Dopamine D_1 receptor knockout mice show no response to D_1 agonists or antagonists and show a blunted response to the locomotor-activating effects of cocaine and amphetamine. D_1 knockout mice are also impaired in their acquisition of intravenous cocaine self-administration compared to wildtype mice. D_2 knockout mice have severe motor deficits and blunted responses to psychostimulants and opioids, but the effects on psychostimulant reward are less consistent.

OVERALL CONCLUSIONS—NEUROBIOLOGY OF ADDICTION

Much progress in neurobiology has provided a heuristic neurocircuitry framework with which to identify the neurobiological and neuroadaptive mechanisms involved in the development of drug addiction. The brain reward system implicated in the development of addiction is comprised of key elements of a basal forebrain macrostructure termed the extended amygdala and its connections. Neuropharmacological studies in animal models of addiction have provided evidence for the dysregulation of specific neurochemical mechanisms in specific brain reward neurochemical systems in the extended amygdala (dopamine, opioid peptides, GABA, and endocannabinoids). There is also recruitment of brain stress systems (CRF and norepinephrine) and dysregulation of brain antistress systems (NPY) in the extended amygdala associated with the positive reinforcing effects of drugs of abuse that contribute to the negative motivational state associated with drug abstinence. The changes in the reward and stress systems during the development of dependence are hypothesized to remain outside of a homeostatic state, and as such help convey the vulnerability for relapse in addiction. Additional neurobiological and neurochemical systems have been implicated in animal models of relapse with the prefrontal cortex and basolateral amygdala (and glutamate systems therein) being implicated in drug- and cue-induced relapse, respectively. The brain stress systems in the extended amygdala are directly implicated in stress-induced relapse. Electrophysiological studies have identified several common elements of acute drug reward, including activation of mesolimbic dopamine neurons and drug reward sensitive neurons in the nucleus accumbens and amygdala. Both GABAergic and glutamatergic systems are directly or indirectly altered by acute administration of drugs of abuse in the ventral tegmental area or nucleus accumbens and provide a key substrate for the cellular neuroplasticity associated with chronic drug administration. Genetic studies in animals, to date, suggest roles for the genes encoding the neurochemical elements involved in brain reward (dopamine, opioid peptide), withdrawal (*Mpdz*), and stress (NPY) systems in the vulnerability to addiction. Molecular studies have identified transduction (cAMP-protein kinase system) and transcription factors that may mediate dependence-induced reward and stress dysregulation (CREB) and chronic vulnerability changes (ΔFosB) in neurocircuitry associated with the development and maintenance of addiction. Human imaging studies reveal similar neurocircuits involved in acute intoxication, chronic drug dependence, and vulnerability to relapse. While no exact imaging results necessarily predict addiction, two salient changes that cut across different drugs in established and unrecovered substance-dependent individuals are decreases in prefrontal cortex function and decreases in brain dopamine D_2 receptors. No biochemical markers are sufficiently specific to predict a given stage of the addiction cycle, but changes in certain intermediate early genes with chronic drug exposure in animal models show the possibility of long-term changes in specific brain regions that may be common to all drugs of abuse.

HOMEOSTASIS VERSUS ALLOSTASIS IN ADDICTION

An overall conceptual theme championed by the authors of this book is that drug addiction represents a break with homeostatic brain regulatory mechanisms that regulate the emotional state of the animal. However, the view that drug addiction represents a simple break with homeostasis is not sufficient to explain a number of key elements of addiction. Drug addiction, as with other chronic physiological disorders such as high blood pressure worsens over time, is subject to significant environmental influences, and leaves a residual neuroadaptive trace that allows rapid 'readdiction' even months and years after detoxification and abstinence. These characteristics of drug addiction have led us to reconsider drug addiction as more than simply a homeostatic dysregulation of hedonic function and executive function, but rather as a dynamic break with homeostasis of these systems that has been termed allostasis.

Allostasis is a physiological concept that differs from homeostasis and is argued to provide a more parsimonious explanation of the neuroadaptive changes that occur in the brain reward and stress systems to drive the pathological condition of addiction (Koob and Le Moal, 2001). Allostasis as a physiological concept was developed originally by neurobiologist Peter Sterling and epidemiologist James Eyer to explain the basis for changes in patterns of human morbidity and mortality associated with the Baby Boom generation of individuals born after World War II. This generation reached an age where major causes of death were renal, cerebral, and cardiovascular diseases, and the single largest contributor to these diseases was hypertension. These researchers could find no obvious disease explanation for these findings but noted that hypertension was most prevalent where social disruption was the greatest (Sterling and Eyer, 1988). They conceptualized that the brain could be the only possible link between such social psychological and physiological phenomena and as a result argued for a different form of brain–body regulation than *homeostasis*, that of *allostasis*.

Claude Bernard is credited with being the first to suggest that the internal milieu of the body is critically important for establishing and maintaining stable states within the body. This concept was termed *homeostasis* by Walter Cannon (1929). *Homeostasis* can be defined as 'preserving constancy in the internal environment'. Bernard argued as early as 1859–1860 that there are two environments that impact on an organism: a *general milieu* which is the outside world of inanimate and animate objects, and the *internal milieu* where the elements of a living body find an optimal climate for operation. Originally described largely for the circulatory system, Bernard later argued for involvement of the lymphatic systems as well (Bernard, 1865). A central thesis of Bernard's view of homeostasis was that not only was an organism required to respond to outside stimuli to regulate energy by bodily adjustments, but that the body also maintained the internal milieu remarkably constant in the face of such challenges, making the organism, at some level, independent of exterior challenges. Fundamentally, Homeostatic regulation is local at the level of a given organ, and this differs significantly from an allostatic perspective. Finally, he argued that all the vital mechanisms of the body have but one goal, that of maintaining the conditions of the life of the internal milieu constant (Bernard, 1865). Fundamentally for Bernard and Cannon, the regulations operated at a local cellular-organ level, and the 'fluids' had an essential role.

Later, Cannon added the intervention of the adrenal–medulla–sympathetic system and related hormones, transported by the 'fluids' (Cannon, 1929). However, Bernard and Cannon considered the concept of physiological adaptation within a rather short timeframe, even if they were aware that stability also had to be considered in the long-term and that an organism would need to mobilize resources for the long-term.

Bernard had a peculiar view of the relationship between the normal and pathological states and considered them to be two parts of the same process (Bernard, 1865). Cannon argued that homeostasis corresponded to a 'condition which may vary but which is relatively constant' (Cannon, 1932). Both argued that the defense of the body was through physiological mechanisms under the control of negative feedback. This model of homeostasis reflecting the restoration of internal parameters to a given set point has guided much of twentieth-century physiology and medicine. Later conceptualizations emphasized that a rigid set point did not fit with medical and physiological observations and did not fit with consequences of long-term or heavy demands on the organism (stability versus flexibility). It was necessary to have biological flexibility in the context of environmental changes and demands and to anticipate demands—the concepts of 'predictive homeostasis' (Moore-Ede, 1986) or 'rheostasis' (Mrosovsky, 1990). Behavior also was invoked to regulate the internal milieu, presumably via the brain, to organize whole-body physiology and adjust for conflicting motivations. The key problem was to account appropriately for the transition from normal to pathological (Canguilhem, 1978), and the new concept of allostasis has proven to be useful in this regard.

For example, in homeostasis the bodily system returns physiological parameters to a specific 'set point'. Allostasis is defined as 'stability through change'. Allostasis is far more complex than homeostasis and in contrast has several unique characteristics that differ from homeostasis (Sterling and Eyer, 1988) (**Table 10.3**). Allostasis involves a feed-forward mechanism rather than the negative feedback mechanisms of homeostasis. A feed-forward mechanism has many advantages because in homeostasis when increased need produces a signal, negative feedback can correct the need, but the time required may be long and resources may not be available. However, in allostasis there is a continuous re-evaluation of need and continuous readjustment of all parameters toward new set points. Thus, there is a fine matching of resources to needs.

However, if demand continues unabated, the set point remains adjusted out of the homeostatic range. Another difference with allostasis versus homeostasis is that allostasis can anticipate altered need, and the system can make adjustments in advance. In homeostasis, increased need triggers a signal for negative feedback mechanisms. In this type of situation the break with homeostasis could get very large before

TABLE 10.3 Homeostasis versus Allostasis

Homeostasis	Allostasis
Normal set point	Changing set point
Physiologic equilibrium	Compensated equilibrium
No anticipation of demand	Anticipation of demand
No adjustment based on history	Adjustment based on history
Adjustment carries no price	Adjustment and accommodation carry a price
No pathology	Leads to pathology

a correction is made, particularly if resources are not available. Homeostatic systems do not change according to experience, whereas allostatic systems can be hypothesized to use past experience to anticipate demand (Sterling and Eyer, 1988). Finally, allostatic regulation is hypothesized to be driven by the brain. The brain evaluates the pressure of the environment, anticipates these demands, and is in an elevated arousal state. Allostatic regulation implicates many organs and involves the whole brain and body rather than a simple local feedback (Sterling and Eyer, 1988).

Yet, it is precisely this ability to mobilize resources quickly and use feed-forward mechanisms that leads to an allostatic state and an ultimate cost of an allostatic mechanism that is known as *allostatic load* (McEwen, 1998). An *allostatic state* can be defined as a state of chronic deviation of the regulatory system from its normal (homeostatic) operating level. *Allostatic load* can be defined as the 'long-term cost of allostasis that accumulates over time and reflects the accumulation of damage that can lead to pathological states'. Allostatic load is the consequence of repeated deviations from homeostasis that takes on the form of changed set points that require more and more energy to defend. Such allostatic adjustments have been hypothesized to move to the level of allostatic load by four situations (McEwen, 2000). First, the challenge may be frequent, usually environmental. Second, the allostatic change may not habituate to repeated challenge. Third, there may be an inability to shut off the allostatic processes. Fourth, there may be inadequate allostatic responses that trigger compensatory responses in other systems (McEwen, 2000).

The failure of allostatic change to habituate or not to shut off is inherent in a feed-forward system that is in place for rapid, anticipated challenge to homeostasis. However, the very physiological mechanism that allows rapid response to environmental challenge becomes the engine of pathology if adequate time or resources are not available to shut off the response. Thus, for example, chronically elevated blood pressure is 'appropriate' in an allostasis model to meet environmental demand of chronic arousal but is 'certainly not healthy' (Sterling and Eyer, 1988). Another example of such a feed-forward system is illustrated in the interaction between CRF and norepinephrine in the brainstem and basal forebrain that could lead to pathological anxiety (Koob, 1999). Allostatic mechanisms also have been hypothesized to be involved in maintaining a functioning brain reward system that has relevance for the pathology of addiction (Koob and Le Moal, 2001). There are two components hypothesized to adjust to challenges to the brain produced by drugs of abuse: overactivation of brain reward transmitters and circuits and recruitment of the brain antireward or brain stress systems. Repeated challenges, as in the case of drugs of abuse, lead to attempts to maintain stability by the brain via molecular, cellular, and neurocircuitry changes, but at a cost. The residual deviation from normal brain reward threshold regulation is termed an *allostatic state*. This state represents a combination of chronic elevation of reward set point fueled by dysregulation of reward circuits, loss of executive control, and facilitation of stimulus response associations, all of which lead to the compulsivity of drug-seeking and drug-taking (see next section).

DRUG-SEEKING TO ADDICTION—AN ALLOSTATIC VIEW

Allostasis, as a concept, also may apply to any number of pathological states that are challenged by external and internal events. Aging and major affective disorders have been argued to reflect lifetime stress, and the brain's response to stress has all of the features of an allostatic mechanism (Carroll, 2002). There is increased activity of the HPA axis, impaired glucocorticoid feedback and regulation, circadian abnormalities, and decreased neurogenesis in the dentate gyrus with other brain system projection losses (Duman *et al.*, 1997). Such effects of chronic stress are seen with aging. In addition, it is argued that depression is an established outcome of stress, and an ongoing depressive episode itself can constitute chronic stress. One can argue, therefore, that depression also fits an allostasis model (Carroll, 2002). The neuroendocrinology of human depression closely resembles that of chronic stress in the laboratory, including increased HPA axis activity, reduced glucocorticoid feedback, and dysregulated diurnal rhythms or cortisol (Checkley, 1996). Closing the loop, there is a strong relationship between the number of depressive symptoms exhibited by subjects, and premature mortality and depression are associated with many cardiovascular risk factors (Carroll, 2002; Gold *et al.*, 2005). Similar connections have been made between allostatic changes in brain stress systems and post-traumatic stress disorder and anxiety disorders (Schulkin *et al.*, 1994; Lindy and Wilson, 2001). One can reasonably see how both developmental and genetic domains can modify allostatic load that may determine vulnerability to pathology (McEwen, 2000). The challenge for future research will be to explore how these brain stress systems integrate with brain motivational systems and to broaden allostasis as a concept from its current main focus on the activation of the HPA axis.

Vulnerability to Drug-seeking—Individual Differences

Individual differences, either via genetic or environmental factors, at critical periods may cause a predisposition to initially self-administer drugs of abuse, and a large body of evidence exists for individual vulnerability to drug reward and by extrapolation to addiction (for review, see Piazza and Le Moal, 1996; Piazza *et al.*, 1998). The nature of the acute impact of the effects of drugs of abuse on neuronal circuits of reinforcement also can contribute to the subsequent neuroadaptations that form an allostatic state in the brain reward systems. Individual differences in the propensity to develop drug intake have been demonstrated readily in the rat (Deminière *et al.*, 1989; Piazza *et al.*, 1989) as well as in the propensity to manifest many other adaptive biological responses (Piazza *et al.*, 1989, 1993; Hooks *et al.*, 1994a; Spanagel *et al.*, 1995).

A classic operational design to identify individual differences is to differentiate animals on the basis of their reactivity to a stressful event (for instance, their locomotor reactivity to novelty) and to divide them into high reactive (HR) and low reactive (LR) groups. Systematic studies from different models, including responses to novelty and responses to stressors, have led to the demonstration of increased drug intake across the full dose–effect function in HR rats (Rougé-Pont *et al.*, 1993, 1995; Deroche *et al.*, 1997; Piazza *et al.*, 2000). HR animals have an increased reactivity of the stress axis, particularly the HPA axis. Levels of corticosterone 2 h after exposure to a stressor are positively correlated with the amount of drug consumed when it is presented for the first time (i.e., acquisition) to an HR subject. Moreover, genetic inactivation of the glucocorticoid receptor gene in the brain reduces the excessive drug response not only in the HR animals but also in animals after long-term exposure to cocaine (Deroche-Gamonet *et al.*, 2003). Finally, the levels of the stress hormone before drug administration were correlated with the extent of self-administration (Piazza *et al.*, 1991a; Goeders and Guerin, 1994). A vulnerable phenotype for drug-taking, inherent and/or acquired through life experience, implied new drug reward set points and has led to the exploration of the interaction of the HPA axis with the mesolimbic dopamine system.

The propensity of HR rats to develop drug intake, compared to LRs, has been correlated with other drug-dependent responses within the mesolimbic region (Piazza *et al.*, 1989, 1991a; Hooks *et al.*, 1991, 1992a,b,c; Exner and Clark, 1993). HR rats, independent of drug administration, show an increase of dopamine utilization in the nucleus accumbens and a decrease in dopamine utilization in the prefrontal cortex (Piazza *et al.*, 1991b). HR rats also show a lower number of dopamine D_1 and D_2 receptors in the nucleus accumbens (Hooks *et al.*, 1994b). Note that the changes in D_2 receptors are similar to those reported in human drug addicts after the development of addiction (Volkow *et al.*, 1993).

Progressive changes in the HPA axis are observed during the transition from acute to chronic administration of drugs of abuse. Acute administration of most drugs of abuse in animals activates the HPA axis (see earlier chapters), but with cocaine and alcohol these acute changes are blunted with repeated administration (Rivier *et al.*, 1984; Zhou *et al.*, 1996, 2003). During withdrawal, there are, again, increases in the activity of the HPA axis for most drugs of abuse (see earlier chapters).

There are clear interactions between stress, glucocorticoids, and mesocorticolimbic dopaminergic neurons, and between dopaminergic neurons and vulnerability to drugs of abuse. Glucocorticoid receptors are localized in brain monoaminergic neurons, particularly in the ventral tegmental area (Harfstrand *et al.*, 1986), although direct cellular interactions between stress hormones and dopamine neurons have been difficult to document. However, glucocorticoid receptors have pivotal regulatory roles in many regions of the brain (de Kloet, 1991; Joels and de Kloet, 1992, 1994), and glucocorticoids can interact with dopamine reward circuitry in the basal forebrain that may be independent of direct glucocorticoid–dopamine interactions. More specifically, glucocorticoids modulate the transmission of the neuropeptides dynorphin, enkephalin, tachykinin, and neurotensin, especially in the basal ganglia and the nucleus accumbens (Chao and McEwen, 1990; Ahima *et al.*, 1992; for review, see Angulo and McEwen, 1994; Schoffelmeer *et al.*, 1996).

Increased corticosterone secretion or a higher sensitivity to the central effects of the hormone, either genetically present in certain individuals or induced by stress, increases the vulnerability to develop intake of drugs of abuse, and may have a role in dependence and relapse via an enhancement of the activity of mesencephalic dopaminergic neurons. Dopaminergic hyperresponsiveness in forebrain structures involved in regulation of motivation, such as the shell of the nucleus accumbens in the extended amygdala, is glucocorticoid-dependent (Barrot *et al.*, 2000), and this effect is state-dependent in that it varies with nutritional status and arousal status (Piazza *et al.*, 1996). In addition, high circulating levels of glucocorticoids can feedback to shut off the HPA axis but can 'sensitize' the CRF systems in the central nucleus of the amygdala known to be involved in behavioral responses to stressors (Lee *et al.*, 1994; Schulkin *et al.*, 1994; Shepard *et al.*, 2000). These central CRF systems are well documented to contribute to behavioral responses to stressors (see Heinrichs and Koob, 2004). CRF, when injected intracerebroventricularly, is aversive and produces place aversions (Cador *et al.*, 1992) and taste aversions (Heinrichs *et al.*, 1991) and raises brain stimulation reward thresholds (Macey *et al.*, 2000). Thus, first there is an activation of brain reward systems possibly through an interaction with the HPA axis, but ultimately this activation can lead to activation of brain stress systems. Engagement of the brain stress systems, in contrast to the HPA axis, may contribute to the negative mood state that dissipates with time, but with repeated administration of drug grows larger with time (or fails to return to normal homeostatic baseline), setting up a negative reinforcement mechanism (see Motivational view of dependence section below).

Thus, the HPA axis and glucocorticoids, which are linked to high responsivity to novelty, also may be involved in adaptations in many parts of the neuraxis, including the basal ganglia/extended amygdala systems, the stress axis hypophyseal systems, and cortical systems that contribute to the shift from homeostasis to pathophysiology associated with drug abuse. These results and the interaction with stress which drives the HPA axis suggest that the HPA axis can play an important role in facilitating both reward and brain stress neurochemical systems implicated in the development of addiction. Thus, alteration in the HPA axis may contribute to a vulnerable state that facilitates the transition from drug use to abuse by initially increasing the probability of drug-taking (initial allostatic state). Subsequently, the HPA axis during chronic drug administration and withdrawal may 'sensitize' the between-system opponent processes that promote addiction such as recruitment of the brain CRF systems (secondary allostatic state).

Vulnerability to Drug-seeking—Environmental Factors

Environmental factors in adults, both intrinsic and extrinsic, also are important determinants of self-administration behavior, particularly during acquisition of the behavior and during reinstatement of drug-taking following extinction (Le Sage *et al.*, 1999). Food deprivation increased drug-maintained behavior, and this generalized to different species, routes of administration, and reinforcement schedules (de la Garza and Johanson, 1987). Food restriction increased cocaine self-administration during unlimited access to cocaine during acquisition and reinstatement (Carroll *et al.*, 1979; Carroll, 1985). Nondrug reinforcers, such as those concurrently available during acquisition and maintenance of cocaine self-administration, prevented acquisition and decreased maintenance of self-administration behavior (Carroll *et al.*, 1989; Carroll and Lac, 1993; Carroll and Rodefer, 1993). Nondrug reinforcers in a clinical setting also reduced cocaine intake (Higgins *et al.*, 1994). How these intrinsic and extrinsic factors interact with the allostasis model of addiction proposed here remains a challenge for future work.

Environmental events during critical periods of development can produce enduring neuroendocrinological and neurodevelopmental changes that could influence drug reward responsivity and propensity to addiction (Moyer *et al.*, 1978; Fride and Weinstock, 1989; Henry *et al.*, 1994). Prenatal stress has been found to have long-term effects on the activity of the dopamine system and on dopamine-related behaviors (Moyer *et al.*, 1978; Fride and Weinstock, 1989). Moreover, there is evidence that prenatal stress increases and prolongs corticosterone secretion in response to stress (Henry *et al.*, 1994). Self-administration of stimulants has been studied in the offspring of mothers submitted to a restraint stress procedure during the last week of pregnancy (Maccari *et al.*, 1991; Deminière *et al.*, 1992). These animals also were tested for locomotor reactivity to novelty and to psychostimulants, since these behaviors are characteristically enhanced in animals spontaneously predisposed to self-administration of drugs of abuse (Piazza *et al.*, 1989). Prenatallly stressed animals had an increased and more rapid locomotor reactivity to amphetamine, particularly over the first hour of testing. Furthermore, prenatal stress influenced the propensity to develop amphetamine self-administration. While control and stressed animals did not differ during the first day of testing, animals in the prenatal stress group showed a higher intake of amphetamine on subsequent days.

Although the development of an organism presumably carries a strong genetic component, the organism's early environment has long-lasting influence. Both components shape psychobiological temperaments and are at the origin of individual differences. Moreover, both components can contribute equally to vulnerabilities for neurodegenerative processes and ultimately deleterious life events. Prenatal and postnatal events modify the activity of the HPA axis (Caldji *et al.*, 1998; Ladd *et al.*, 2000), and maternal glucocorticoids have a major role in the development of endocrine function in the offspring. In fact, high levels of maternal glucocorticoids during prenatal stress have marked long-term repercussions on the efficiency of the offspring's HPA negative feedback mechanisms. Thus, a modification of corticosterone secretion via changes in HPA axis activity could be a biological substrate of the long-term behavioral effects of prenatal and postnatal events that could contribute to individual differences in vulnerability to primary allostasis in the brain reward system (see above).

Motivational View of Dependence

Motivational changes associated with acute withdrawal reflect opponent-process-like changes in the reward circuitry activated by the acute reinforcing effects of drugs of abuse (see previous chapters). Brain systems associated with the development of motivational aspects of withdrawal have been hypothesized to be a major source of potential allostatic changes that drive and maintain addiction. Acute withdrawal from the chronic use of drugs of abuse has long been associated with physical signs, and the manifestation of these physical signs varies with each drug of abuse. However, the manifestation of physical signs of withdrawal are only one of a constellation of physical and motivational symptoms associated with addiction (see *What is Addiction?* chapter), and the physical symptoms of withdrawal may be less relevant to the motivation to take drugs than earlier conceptualizations proposed (Himmelsbach, 1942, 1943). For example, patients relapse long after physical signs of withdrawal have dissipated; and depending on how one defines physical symptoms, some drugs of abuse such as cocaine do not produce obvious physical symptoms. Clearly, physical withdrawal symptoms vary with the drug of abuse and may have motivational properties and discriminative stimulus properties (signal the onset of dysphoria) with drugs such as alcohol and opioids, but much less of a role with psychostimulant drugs. Finally, it should be noted that the dualism of physical or somatic signs of withdrawal versus psychological signs may be false if one considers withdrawal as a physiological response in the brain.

However, a common element of acute withdrawal observed with all drugs of abuse is a dysphoric state, including various negative emotions, such as dysphoria, depression, irritability, and anxiety, and has been termed 'motivational withdrawal' or, more operationally, the aversive stimulus effects associated with drug withdrawal. These emotional states may have major motivational significance in contributing to the maintenance of drug addiction. For example, cocaine withdrawal in humans in the outpatient setting is characterized by severe depressive symptoms combined with irritability, anxiety, and anhedonia lasting several hours to several days (i.e., the 'crash') and may be one of the motivating factors in the maintenance of the cocaine-dependence cycle (Gawin and Kleber, 1986). Inpatient studies of cocaine withdrawal in cocaine-dependent subjects have shown similar changes in mood and anxiety states, but they generally are much less severe (Weddington *et al.*, 1991). Opioid withdrawal is characterized by severe dysphoria, and alcohol withdrawal produces pronounced dysphoria and anxiety. Recent studies using animal models of reward and anxiety have provided measures of behavioral changes that can be linked to emotional states associated with withdrawal from all major drugs of abuse, including opioids, psychostimulants, alcohol, and nicotine. Using the technique of intracranial self-stimulation to measure reward thresholds throughout the course of drug dependence, reward thresholds are increased during withdrawal from chronic administration of all major drugs of abuse (reflecting a decrease in reward), and some of these changes can persist for up to a week postdrug (see previous chapters).

The important contribution of motivational withdrawal to addiction was part of the original conceptualizations of Wikler (1973) but was overlooked when physical withdrawal was removed as a critical component of the nosology of addiction (transition from DSM-III to DSM-IIIR criteria; see American Psychiatric Association, 1980, 1987). An extension of the original Wikler conceptualization is the hypothesis that acute withdrawal shifts the hedonic state of the animal to a negative position that raises the salience of re-exposure to the drug or to drug cues. For example, previous experience with heroin withdrawal in rats provides a motivational state that enhances the incentive value of the drug (Hutcheson *et al.*, 2001). Rats trained to self-administer heroin were subjected to induction of dependence (8–30 mg/kg, i.p., of morphine twice daily for 3 days) and allowed to self-administer heroin during spontaneous opioid withdrawal. Rats that had previously taken heroin when in opioid withdrawal subsequently showed greater drug-seeking behavior when again in withdrawal compared to a group of rats that had no experience with withdrawal (Hutcheson *et al.*, 2001). The authors argued that 'withdrawal functions as a motivational state to enhance the incentive value of the drug' (Hutcheson *et al.*, 2001), and a distinction could be made between incentive-salience theories (Robinson and Berridge, 1993) and opponent-process theories (Solomon and Corbit, 1974; Koob *et al.*, 1989). Similar results have been observed repeatedly with alcohol self-administration during withdrawal, where alcohol self-administration increases with repeated withdrawal experience (Roberts *et al.*, 1996, 2000).

The allostasis hypothesis, as an extension of opponent-process theory, in fact predicts a strong role of a negative affective state in not only negative reinforcement, but also in enhancing the relative change in motivational value of administration of drugs of abuse superimposed upon that negative state (Ahmed and Koob, 1998, 2005; Koob and Le Moal, 2001; Koob, 2004). Indeed, increased intake of cocaine, opioids, and alcohol associated with prolonged access (escalation models) is associated with higher break points on a progressive-ratio schedule of reinforcement, suggesting an increased motivation to obtain drug at a time point when brain reward thresholds are elevated (decreased reward or elevated reward set point) (see previous chapters for details and references).

Thus, the hypothesis is that the *b process* (see *Neurobiological Theories of Addiction* chapter) of the opponent-process represents both a neuroadaptive response to the drug and becomes an allostatic state that contributes to the pathophysiology of addiction when there is not sufficient time between drug administrations for the system to fully recuperate (**Fig. 10.1**) (see **Figs. 1.11–1.13** in *What is Addiction?* chapter). The hysteresis between drug level and recovery of the *b process* (blood levels of the drug continue to be elevated but there is no euphoria, only the dysphoria of the *b process*) not only explains acute hedonic apparent tolerance, but also how the allostatic state continues to build much like the development of hypertension. However, what makes this allostatic state most powerful in driving compulsive behavior is not only the loss of activity in reward systems and the gain of activity in stress systems, but also the loss of prefrontal inhibitory executive control which removes any brake on the impulse circuit driving continued drug-seeking despite negative consequences (see Neurocircuitry of Compulsive Drug-Seeking and Drug-Taking section below).

The neural substrates and neuropharmacological mechanisms for the negative emotional-motivational effects of drug withdrawal have been proposed to involve two major components: disruption of the same neural systems implicated in the positive reinforcing effects of drugs of abuse and recruitment of dysregulated brain stress systems. As such, these effects may reflect changes in the activity of the same extended amygdala system implicated in the positive reinforcing effects of drugs and can last up to 72 h depending on the drug and dose administered (see above).

The activation of brain and pituitary stress systems may be another common response to repeated administration of drugs of abuse that may be involved in the negative emotional state associated with acute

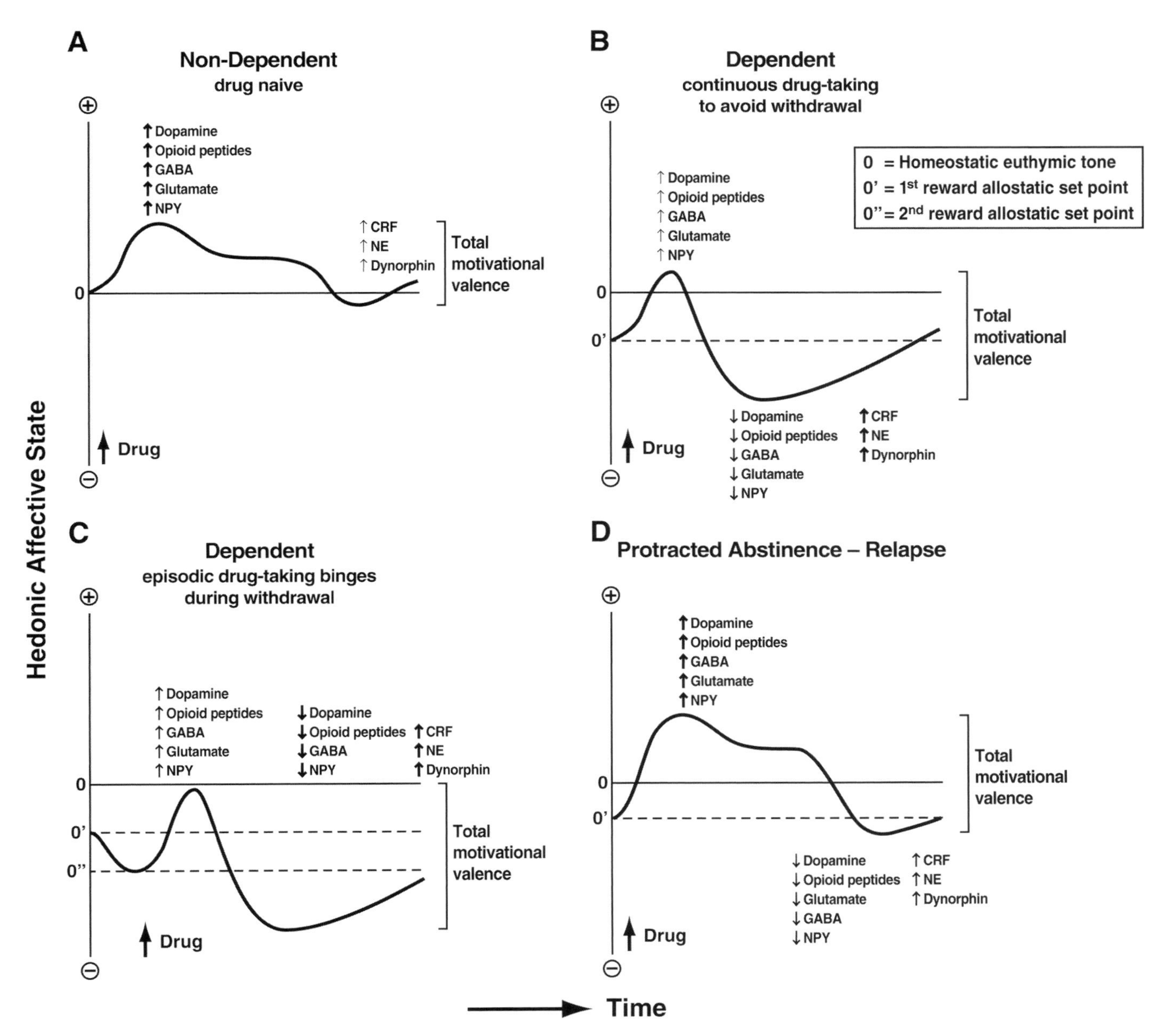

FIGURE 10.1 Koob and Le Moal's latest conceptualization of the hedonic responses associated with drug intake at various stages of drug addiction correlated with changes in neurotransmitter systems within the extended amygdala circuitry hypothesized to mediate drug reward (arbitrarily defined as dopamine, opioid peptides, γ-aminobutyric acid [GABA], glutamate, and neuropeptide Y [NPY]) and anti-reward (arbitrarily defined as corticotropin-releasing factor [CRF], norepinephrine [NE], and dynorphin). **(A)** The hedonic response to an acute drug administration in a drug-naive individual with activity in neurotransmitter systems involved in reward predominating with a minor anti-reward opponent process-like response. **(B)** The hedonic response to an acute drug administration in a drug-Dependent (big 'D'; see *What is Addiction*? chapter) individual while taking drug regularly. Initial activity in neurotransmitter systems involved in reward is followed by a decrease in function of neurotransmitter systems involved in reward and a major anti-reward opponent process-like response. *0'* (zero prime) refers to the change in hedonic set point produced by chronic dysregulation of reward neurotransmitters and chronic recruitment of anti-reward neurotransmitters. This change in hedonic set point is the allostatic state of reward dysregulation conceptualized in Koob and Le Moal (2001). **(C)** The hedonic response to an acute drug administration in a drug-Dependent individual during withdrawal. A major anti-reward opponent process-like response at the beginning of the time course is followed by modest activity in neurotransmitter systems involved in reward triggered by a drug administration during withdrawal. *0'* (zero prime) refers to the change in hedonic set point associated with the development of Dependence while still taking drug. *0"* (zero double-prime) refers to the hedonic set point during peak withdrawal after cessation of drug taking. **(D)** The hedonic response to an acute drug administration in a formerly drug-Dependent individual during protracted abstinence. Note that previously a drug-Dependent individual was hypothesized to remain at a residual *0'* state termed *protracted abstinence*. A robust activity in neurotransmitter systems involved in reward triggered by a drug administration is followed by an exaggerated anti-reward opponent process-like response afterward (i.e., dysregulation of reward neurotransmitters and recruitment of anti-reward neurotransmitters) that drives the subject back to below *0'*. *Total motivational valence* refers to the combined motivation for compulsive drug use driven by both positive reinforcement (the most positive state above the *0* euthymic set point) and negative reinforcement (movement from the most negative state to *0* set point). The magnitude of a response is designated by the thickness of the arrows. The large upward arrow *Drug* at the bottom of each panel refers to drug administration. The total time scale is estimated to be approximately 8 h.

withdrawal (see above). HPA function is activated during drug dependence and during acute withdrawal from drugs of abuse in humans. Dysregulation of the HPA axis can persist even past acute withdrawal (Kreek *et al.*, 1984; Kreek, 1987). However, the CRF function in extrahypothalamic regions outside of the HPA axis also is activated during acute withdrawal from cocaine, alcohol, opioids, Δ^9-tetrahydrocannabinol, and nicotine, and thus may mediate some of the behavioral responses to stress associated with acute abstinence (Koob *et al.*, 1994; Heinrichs *et al.*, 1995; Rodriguez de Fonseca *et al.*, 1997; Richter and Weiss, 1999; Ghozland *et al.*, 2004). Rats treated repeatedly with cocaine, opioids, nicotine, and alcohol show significant anxiety-like responses following cessation of chronic drug administration which are reversed with intracerebroventricular administration of a CRF antagonist (Rassnick *et al.*, 1993; Sarnyai *et al.*, 1995; Schulteis *et al.*, 1998b; Basso *et al.*, 1999; Knapp *et al.*, 2004). Microinjections into the central nucleus of the amygdala of lower doses of the same CRF antagonist also reversed the anxiogenic-like effects of alcohol withdrawal (Rassnick *et al.*, 1993), and similar doses of this CRF antagonist injected into the amygdala reversed the aversive effects of opioid withdrawal (Heinrichs *et al.*, 1995). Studies using *in vivo* microdialysis have shown that rats withdrawn from chronic alcohol, withdrawn from chronic cocaine, and precipitously withdrawn from chronic cannabinoids show increases in the release of CRF from the central nucleus of the amygdala (Cummings *et al.*, 1983; Merlo-Pich *et al.*, 1995; Rodriguez de Fonseca *et al.*, 1997; Olive *et al.*, 2002; Ghozland *et al.*, 2004). Indeed, one could speculate that the profound activation of both the HPA axis and central CRF systems during the development of drug dependence, particularly during a binge of drug-taking, represents the ultimate activation of the HPA axis and contributes to a subsequent sensitization of central CRF that cannot return to homeostatic levels. Thus, a cascade of events could be hypothesized to develop in the following manner: (1) reward system activation in a binge, with HPA axis activation; (2) downregulation of dopamine/opioid peptide systems at the end of a binge; (3) continued dysregulation of reward systems during acute withdrawal; (4) HPA activation and central CRF activation during acute withdrawal that may persist into protracted abstinence (see **Fig. 10.2**).

NPY, a 36 amino acid member of the pancreatic polypeptide family, also has been implicated in the neuroadaptations associated with the development of drug addiction. NPY is abundantly present in brain areas implicated in alcohol and drug dependence, such as the ventral striatum and the amygdala (de Quidt and Emson, 1986). Acute effects of NPY are remarkably similar to those of alcohol in producing a suppression of anxiety-like responses, sedation (Heilig *et al.*, 1994), and anticonvulsant actions (Vezzani *et al.*, 1999). A quantitative trait locus contributing to the phenotype of alcohol-preferring P rats has been found within a chromosomal region containing the NPY gene (Carr *et al.*, 1998). Furthermore, in several brain areas, the central expression of NPY differs between alcohol-preferring and -nonpreferring rats. Among these differences are suppressed levels of NPY in the central amygdala, also seen in the high alcohol drinking (HAD) line of rats, suggesting that NPY within this structure might play a role in the regulation of alcohol intake (Hwang *et al.*, 1999). A decrease in NPY-like immunoreactivity has been observed in the amygdala of rats with a history of alcohol dependence (Ehlers *et al.*, 1998). Furthermore, the electrophysiological response to intracerebroventricular NPY differs between P and NP rats (Ehlers *et al.*, 1999) and after chronic alcohol exposure (Slawecki *et al.*, 1999). The anxiogenic responses of alcohol withdrawal after chronic alcohol exposure have been extensively linked to a downregulation of phospho-CREB and decreased NPY activity in the central nucleus of the amygdala (Pandey *et al.*, 2003) (see *Alcohol* chapter for details). Further evidence of a causal relation between NPY expression and alcohol intake has been suggested by the inverse relationship between NPY-expression and alcohol intake in NPY transgenic and mutant mice, respectively (Thiele *et al.*, 1998). NPY_1 receptor knockout mice also show an anxiogenic-like profile in animal models of anxiety (Palmiter *et al.*, 1998), but NPY-overexpressing mice do not show an anxiolytic-like profile (Thiele *et al.*, 1998). One hypothesis is that dysregulation of both the CRF and NPY systems contributes significantly to the motivational basis of continued alcohol-seeking behavior during alcohol dependence and where the CRF system is hypothesized to become sensitized with repeated withdrawal, the NPY system is hypothesized to show tolerance, both contributing to an allostatic state (Valdez and Koob, 2004).

NPY also may be involved in dependence on drugs other than alcohol. NPY expression in the ventral striatum is suppressed following prolonged treatment with cocaine, possibly related to the anhedonic state present during cocaine withdrawal (Wahlestedt *et al.*, 1991). Conversely, withdrawal from opioids is antagonized by central NPY (Woldbye *et al.*, 1998). Overall, the role of NPY in dependence may be best viewed perhaps as an inhibitor of neuronal excitability (Palmiter *et al.*, 1998), and it is important to note that NPY and CRF have been hypothesized to have reciprocal actions in mediating behavioral responses to stressors (Heilig *et al.*, 1994). Thus, one may speculate that during acute withdrawal, and perhaps

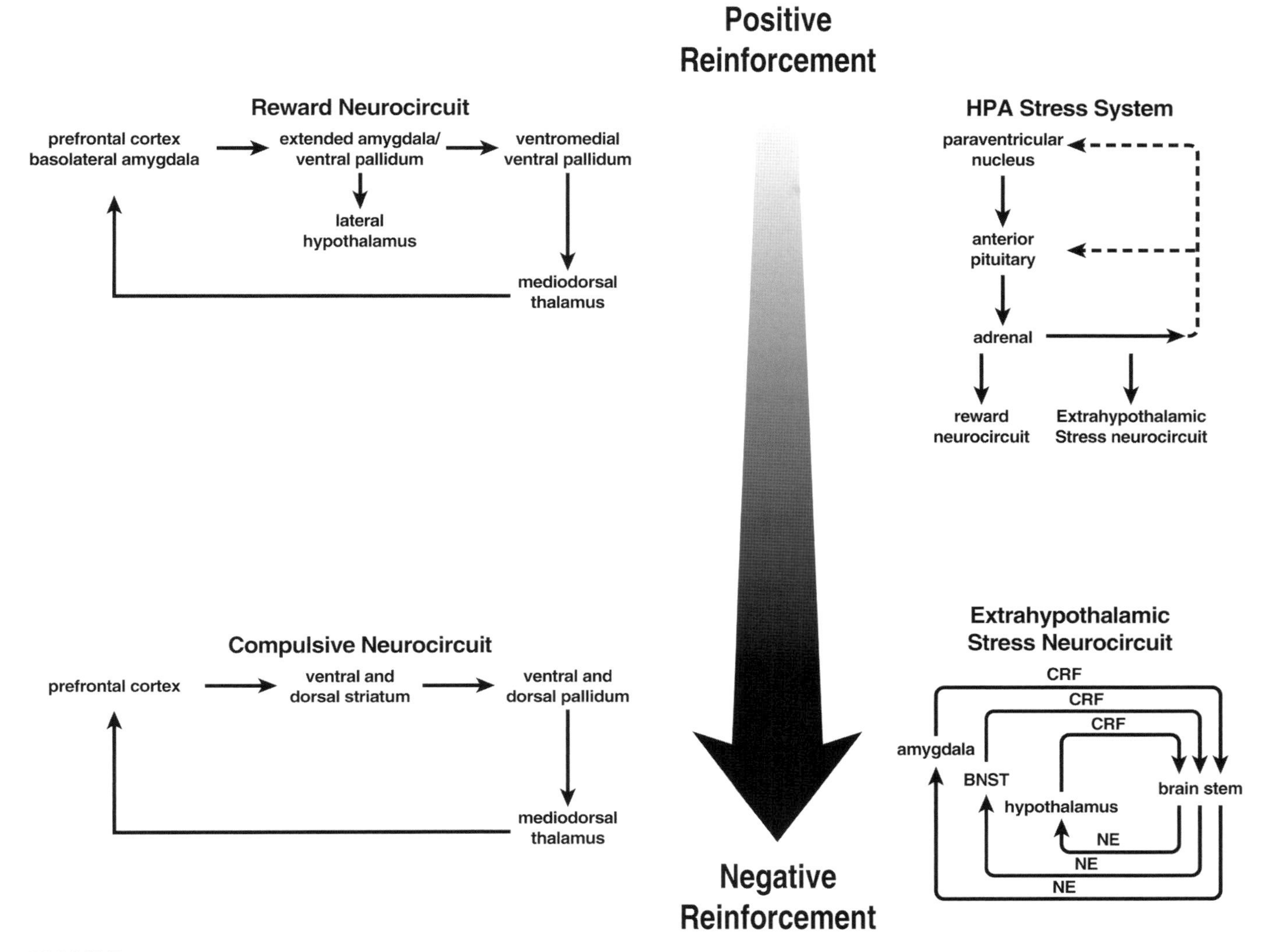

FIGURE 10.2 Brain circuits hypothesized to be recruited at different stages of the addiction cycle as addiction moves from positive reinforcement to negative reinforcement. The top left diagram illustrates an increase in the inactivity of a brain reward system circuit with a focus on the extended amygdala and an increase in the drug- and cue-induced reinstatement circuit with a focus on the prefrontal cortex and basolateral amygdala which both drive positive reinforcement and 'impulsivity.' The bottom left diagram illustrates a decrease in the brain reward circuit and an increase in the behavioral output or 'compulsivity' circuit, both involved in driving negative reinforcement and "compulsivity." The top right diagram refers to the hypothalamic-pituitary-adrenal axis which (1) feeds back to regulate itself, (2) activates the brain reward neurocircuit, and (3) facilitates the extrahypothalamic stress neurocircuit. The bottom right diagram refers to the brain stress circuits in feed-forward loops. β-END, β-endorphin; CRF, corticotropin-releasing factor; DA, dopamine; ENK, enkephalin; HPA, hypothalamic-pituitary-adrenal axis; NE, norepinephrine. [Adapted from Koob and Le Moal, 2004.]

extended to protracted abstinence, decreases in NPY activity may accompany increases in CRF activity further potentiating the neurochemical brain stress system contribution to allostasis in the reward system. As such, NPY falls into the category of a reward transmitter whose function is compromised during the development of dependence.

Recruitment of nonstress, antireward systems involving other neurotransmitter systems has been hypothesized to contribute to the motivational changes associated with chronic administration of drugs of abuse and may represent another contribution to the allostatic state. These include the neuropeptides dynorphin, neuropeptide FF (NPFF), and more recently orphanin FQ, and these antireward neurotransmitters may be natural counteradaptive mechanisms that are activated to limit impulsive reward-seeking behavior. Dynorphin peptides appear to decrease dopamine activity via a presynaptic action on κ opioid receptors in the nucleus accumbens, and κ agonists produce aversive effects in rodents and humans (Cole *et al.*, 1995; Carlezon *et al.*, 1998; Newton *et al.*, 2002; Nestler, 2004). Moreover, dynorphin transmission is modulated by glucocorticoids, and this modulation could represent an indirect action upon dopamine neurons at the terminal projection level (for review, see Angulo and McEwen, 1994).

Antiopioid activities have been hypothesized for NPFF (previously termed F8Fa) based on the effects of intracerebroventricular injection of NPFF-related peptides. Administration of NPFF attenuated morphine- and stress-induced analgesia (Kavaliers, 1990), and precipitated morphine withdrawal (Malin *et al.*, 1990). More compelling, NPFF antagonists can increase both morphine- and stress-induced analgesia, reverse morphine tolerance (Lake *et al.*, 1992), and attenuate the naloxone-precipitated withdrawal syndrome in morphine-dependent rats. An NPFF antagonist also blocked some aspects of nicotine withdrawal (Malin *et al.*, 1996).

Antiopioid-like effects have been reported with administration of the orphan receptor binding peptide orphanin FQ (nociceptin) (Mogil *et al.*, 1996), although more recent evidence suggests that orphanin FQ produces more anxiolytic-like effects (Jenck *et al.*, 1997, 2000; Ciccocioppo *et al.*, 2002). Orphanin FQ blocked stress-induced anorexia (Ciccocioppo *et al.*, 2002), blocked alcohol intake in alcohol-preferring rats (Ciccocioppo *et al.*, 1999), and blocked cue-induced (Ciccocioppo *et al.*, 2004) and stress-induced (Martin-Fardon *et al.*, 2000) reinstatement of alcohol-seeking behavior.

Thus, chronic drug self-administration sets up two additional major components of allostasis within the brain reward systems during a binge/withdrawal cycle. There is hypofunctioning of neurotransmitter systems involved in positive reinforcement and a recruitment of neurotransmitter systems involved in negative emotional states that provide the motivation for negative reinforcement. Hypothetically, counter-regulatory processes within the reward domain could acutely limit drug intake. However, if the initial counter-regulatory response cannot exactly balance the activational forces involved in an allostatic response because the anti-reward processes develop more slowly and dissipate more slowly, such hysteresis could lead to further drug intake and the further exaggeration of allostatic-like changes in the counter-regulatory processes where the antireward effects develop slowly (Koob and Le Moal, 2001) (see **Figs 10.1** and **1.11–1.13** from *What is Addiction?* chapter). One could envision genetically and environmentally based vulnerability at both ends of the counter-adaptive process. Thus, these antireward systems are hypothesized to be abnormally activated during the development of dependence, and thus contribute to the allostatic state of reward dysfunction in addiction.

THE ALLOSTATIC VIEW VERSUS INCENTIVE SENSITIZATION VIEW

Repeated exposure to many drugs of abuse results in a progressive and enduring enhancement in the motor stimulant effects elicited by a subsequent challenge, and this phenomenon, termed 'behavioral sensitization,' has been thought to underlie some aspects of drug addiction (for review, see Vanderschuren and Kalivas, 2000) (see *What is Addiction?* chapter). This phenomenon has been observed and well characterized for various drugs (Babbini *et al.*, 1975; Eichler and Antelman, 1979; Bartoletti *et al.*, 1983; Segal and Schuckit, 1983; Kolta *et al.*, 1985), and sensitization gets larger with the passage of time (Antelman *et al.*, 1980, 1983, 2000). Moreover, stress and stimulant effects on sensitization are interchangeable, with one sensitizing the animal to the other and vice versa (Antelman *et al.*, 1980). Finally, passive stimulant administration induced neuroendocrine effects and sensitization in the same manner as stress (Vanderschuren *et al.*, 1999).

The incentive-sensitization theory of addiction has the central premise that 'addictive drugs enduringly alter [nucleus accumbens]-related brain systems that mediate a basic incentive-motivational function, the attribution of incentive salience' (Robinson and Berridge, 2003). Accordingly, with repeated (usually intermittent) administration of drugs of abuse, these neural circuits become hypersensitive to specific drug effects and to drug-associated stimuli. This hypersensitivity of the nucleus accumbens circuitry is further hypothesized to cause pathological *wanting* to take drugs. *Wanting* is defined as activation of incentive-salience processes (Robinson and Berridge, 2003). Neurobehavioral sensitization, and by extrapolation *wanting*, is operationally defined as either locomotor sensitization or increased psychomotor activation with repeated intermittent administration of drugs (Robinson and Berridge, 1993) or increased incentive salience for a cue, paired with sucrose reward, produced by a drug of abuse after repeated exposure to the drug of abuse (Wyvell and Berridge, 2001). Robinson and Berridge further argue that this *wanting* can be dissociated from hedonic *liking*. *Liking* represents the hedonic aspects of drug use and is operationally defined in rodents by species-specific facial movements reflecting positive reinforcing substances such as sucrose (Berridge, 2000). Neural sensitization by drugs increases only *wanting*. Thus, Robinson and Berridge argue that the incentive-sensitization process is the fundamental process in the transition to addiction and relapse. It is specifically sensitization of incentive salience attribution to representation of drug cues and drug-taking that causes the pursuit of drugs and persisting vulnerability to relapse in an addict. Individuals are guided to incentive stimuli by the influence of Pavlovian stimulus–stimulus associations in motivational systems, which are psychologically separable from the symbolic cognitive systems that mediate conscious desire, declarative expectancies of reward, and act–outcome representations.

Numerous investigations have shown that sensitization is a complex process arising from many cellular changes in many brain regions, depending on the specific drug's molecular targets, and on possible involvements of learned associations. In brief, Vanderschuren and Kalivas (2000) have summarized the main neuropharmacological aspects as follows: (1) Common substrates for the induction of sensitization with drugs of abuse are activation of glutamate transmission, at *N*-methyl-D-aspartate receptors especially, and an action within the ventral tegmental region; (2) A role for dopamine in the induction of sensitization is only seen in amphetamine sensitization, due largely to its robust capacity to release mesoaccumbens dopamine; (3) Other regions, such as the nucleus accumbens and prefrontal cortex, are involved in the induction of cocaine and morphine sensitization, but not amphetamine sensitization; (4) Enhanced dopamine transmission and an action in the nucleus accumbens is associated with the expression of sensitization by all drugs; (5) Glutamatergic cortical and allocortical systems appear to be more important for the expression of cocaine sensitization than amphetamine or morphine sensitization (Vanderschuren and Kalivas, 2000) (**Table 10.4**). The pathophysiological changes within these networks involved in the induction and expression of behavioral sensitization after repeated exposure to drugs, in particular in the prefrontal cortex, are hypothesized to cause modification of behavioral strategies described in drug addiction, including impulsive behaviors and deficits in the ability to inhibit and shift behaviors that are characteristic of human dependence (Logue *et al.*, 1992; Dias *et al.*, 1996; Madden *et al.*, 1997; Weissenborn *et al.*, 1997; Jentsch and Taylor, 1999; Petry and Casarella, 1999; Richards *et al.*, 1999; Roberts *et al.*, 2000).

However, there are a number of predictions that are made by incentive sensitization theory that are not validated in the addiction process. In animal studies, the literature linking increased self-administration to the sensitization process is not overwhelming. There is some data supporting the hypothesis that pre-exposure to psychostimulant drugs or stress can enhance acquisition of self-administration, but linking this observation to drug addiction makes a number of untenable assumptions, one of which is that simple drug-seeking or drug-taking behavior is addiction. In contrast, when animals with a history of extended access escalate their intake of cocaine there is a loss of sensitization (Ben-Shahar *et al.*, 2004).

In human studies, one prominent prediction of an incentive sensitization view would be that with repeated use, addicts would take less drug (i.e., if the subject is sensitized, then more nucleus accumbens dopamine activation should occur for a given dose,

TABLE 10.4 Neurotransmitters and Brain Sites Involved in the Induction and Expression of Sensitization to Psychostimulants and Opioids

	Amphetamine	Cocaine	Opioids
Induction of sensitization			
Dopamine neurotransmission	Yes	No	No
Glutamate neurotransmission	Yes	Yes	Yes
Ventral tegmental area	Yes	Yes	Yes
Nucleus accumbens	No	Yes	Yes/No
Prefrontal cortex	Yes/No	Yes	Yes/No
Long-term expression of sensitization			
Dopamine neurotransmission	Yes	Yes	Yes
Glutamate neurotransmission	Yes/No	Yes	Not known
Ventral tegmental area	No	No	Yes
Nucleus accumbens	Yes	Yes	Yes
Prefrontal cortex	No	Yes	Not known

Yes or No refers to either a predominant body of evidence, or lack thereof, indicating involvement of these neurotransmitters and brain sites. Yes/No indicates roughly equivalent sizes of supporting data.
[Reproduced with permission from Vanderschuren and Kalivas, 2000.]

and a smaller dose therefore would be required). The evidence from human studies of addiction is overwhelmingly one of tolerance to the pleasurable or hedonic effects of drugs (see previous chapters). Indeed, with psychostimulants such tolerance is readily observable within one self-administration session (see *Psychostimulants*, *Opioids*, and *Alcohol* chapters). Perhaps equally problematic for the neural sensitization hypothesis are data from imaging studies in human cocaine addicts showing tolerance to the 'high' associated with intravenous methylphenidate challenge in cocaine addicts actively using cocaine, and decreased release of dopamine in the striatum (Volkow *et al.*, 1997). In fact, it was metabolic activity in the rectal gyrus (medial orbitofrontal cortex) that increased with methylphenidate challenge and correlated with cocaine craving (Volkow *et al.*, 1999, 2004).

Similarly in animal models of escalation in drug intake there is evidence of *upward* shifts in the dose–effect function for cocaine self-administration in the animals showing escalation in drug intake compared to controls (Ahmed and Koob, 1998). This escalation is correlated with a dramatic elevation in basal reward thresholds (Ahmed *et al.*, 2002). From a theoretical perspective, an allostatic decrease in reward function provides the most parsimonious explanation of escalated levels of cocaine self-administration and associated vertical shifts upward in the dose–injection function observed with extended access to the drug (Ahmed and Koob, 1998, 1999, 2004, 2005; Manstch *et al.*, 2004). Reward sensitization would predict a *leftward* shift of the dose–injection function. In an incentive sensitization model, there is no logical connection between variations in the incentive effect of cocaine and variations in cocaine intake. The incentive sensitization model fails to predict how the level of drug intake changes in response to increased drug availability (Robinson and Berridge, 2003; Vezina, 2004). Preexposure to psychostimulants is argued to enhance the psychomotor response, drug intake, and midbrain dopaminergic function (Vezina *et al.*, 2002; Vezina, 2004). In contrast, in the allostasis model, a simulation of the decrease in baseline reward function seen in animals with escalating cocaine intake produces both between-system tolerance and increased motivation for the rewarding effects of cocaine (Ahmed and Koob, 2005). More specifically, in a computer simulation of cocaine self-administration that integrates pharmacokinetics, pharmacodynamics, and motivational factors, the simulation of an allostatic decrease in reward system responsivity exacerbates the initial error that drives cocaine self-administration, thereby increasing both the intake of, and motivation for, the drug (Ahmed and Koob, 2005).

Thus, increased motivation to take drugs is an inherent part of both theories, but the derivation of the sources of reinforcement differ significantly. In the Robinson and Berridge model, increased dopaminergic activity in the nucleus accumbens produces neural sensitization which is hypothesized to increase incentive salience, and though the authors eschew the hedonic interpretation, others extrapolate this increased incentive salience to increased drug-seeking and drug-taking (Vezina, 2004). In the allostasis model, decreased reward function through within-system and between-system tolerance (e.g., recruitment of brain antireward systems) drives the enhanced drug-seeking via a negative reinforcement mechanism. One common element to both theories is the enhanced motivation to seek drugs, but one major difference between both theories is the emphasis on the midbrain dopamine systems in the Berridge and Robinson theory and the relatively minor role of the midbrain dopamine systems in the allostasis theory.

Robinson and Berridge's most exploited criticism of opponent-process (allostasis) theory is that in animal studies, acute precipitated withdrawal from drugs of abuse has not been shown to have motivational significance either in reinstatement or conditioning models (Robinson and Berridge, 1993). However, most of the studies cited use psychomotor stimulants or opioids and have not noted that to make an association between a negative state and a stimulus requires pairing of the negative state with previously neutral stimuli. When such pairings are explicitly made, animals exhibit place aversions to opioid withdrawal and conditioned withdrawal responses (Schulteis *et al.*, 1998a; Stinus *et al.*, 2000, 2005). Such stimuli have been shown recently to have motivational significance in opioid-dependent rats (Walker *et al.*, 2003). However, even more telling are recent studies in an animal model of alcoholism and opioid dependence. Alcohol-dependent animals self-administer significantly more alcohol during acute withdrawal if the animals have a history of alcohol self-administration and have repeated experience with access to alcohol under conditions where they experience withdrawal (Roberts *et al.*, 1996, 2000; O'Dell *et al.*, 2004). Opioid-dependent animals during opioid withdrawal, with a history of opioid withdrawal, showed greater drug-seeking compared to rats with no history. Perhaps even more compelling is the observation that anti-stress doses of a competitive CRF antagonist injected intracerebroventricularly blocked the increased self-administration of alcohol in dependent animals but had no effect in nondependent animals, and this effect appeared to be mediated in the extended amygdala (Valdez *et al.*, 2003; Reiter-Funk *et al.*, 2005).

These results conform well to a negative reinforcement, opponent-process model and have very little consistency with a dopaminergic incentive-sensitization model.

Finally, it is not at all clear that incentive sensitization has any validity for dependence models. Animals with prolonged access to cocaine escalate their intake of cocaine but lose their locomotor sensitization, suggesting in fact that locomotor sensitization may be valid only for initial exposure to the drug. This is what was argued in Fig. 4B (see **Fig. 1.4** here) of Koob and Le Moal (1997) where sensitization was hypothesized to impact only on the initial use of the drug but had little or nothing to do with the development of dependence (Koob and Le Moal, 1997).

A NONDOPAMINE-CENTRIC VIEW OF ADDICTION

Dopamine neurons are activated and dopamine utilization is increased in the synaptic cleft as a result of active intake or passive administration of drugs of abuse (see previous chapters). This effect is hypothesized to be one of the primary and unconditioned actions of drugs of abuse and also of stress, activation, exploration, novelty, natural rewards, reward expectation, and cognitive, learning, and intentionality processes. The activation of dopamine neurons by such factors has allowed attribution of general functional roles or more holistic constructs necessary for survival. 'Reward neuron' or 'reward system' are labels currently used (e.g., 'reward probability and uncertainty' neurons) (Wise, 1980, 2002; Wise and Bozarth, 1987; Fiorillo *et al.*, 2003; Fiorillo, 2004).

Ventral mesencephalic dopamine cells are a part of the upper region of the reticular formation and as such form part of an activation system engaged by sensations and signals of the environment (Le Moal and Olds, 1979). These neurons project to more than 20 regions, including limbic, striatal, epithalamic, and cortical, and some have a branched organization (Thierry *et al.*, 1983). Although the detailed organization is still not fully understood, most, if not all, of these modulated regions project back to the dopamine subnuclei, and exert feedback control (Louilot *et al.*, 1985, 1986, 1989; Louilot and Le Moal, 1994). Numerous studies have shown that the function of a given projection of the midbrain dopamine systems, as determined by lesion or stimulation studies, is reflected more by the known functions of the terminal area receiving the projection than by any intrinsic property of the dopamine neurons themselves (Simon *et al.*, 1980; Le Moal and Simon, 1991).

Drug reward and reinforcement have been intrinsically linked to dopamine activation and have evolved into the concept of a 'dopamine code,' in which dopamine neurons code for a given function (i.e., drug reward). However, it is significant that the most robust role attributed to dopamine in the brain (where lesions of the nigrostriatal system induce Parkinson's disease symptoms) (Carlsson, 1987) has not engaged the labeling of the dopamine neurons as 'motor neurons.' What is clear is that dopamine in the basal ganglia is necessary for complex psychomotor activities to be engaged and that it is a part of complex integrated circuitries through which a given function is elaborated (Alexander *et al.*, 1986; Le Moal, 1995).

Our position is that reward is a complex function represented by a set of interrelated regions and circuitries, one component of which is the mesocorticolimbic dopamine system. There is confusion between the structure–function mechanism (termed 'reward,' if it exists) and elements contributing to the mechanisms (with *mechanism* being defined by the Oxford English Dictionary as 'structure or arrangement of parts that work together as the parts of a machine do'). From this perspective, dopamine can be conceptualized as 'oil in the machine.' Dopamine allows the appropriate functioning of complex circuits that it innervates, but itself does not have a functional attribute.

An additional oversimplification regarding the function of dopamine results from a conception of a neuron insulated from its cellular environment and from the inter-neuronal regulations in which it participates. As an example, a pronounced norepinephrine–dopamine interaction has been demonstrated in the frontal cortex and nucleus accumbens through a long series of studies developed by Tassin, Glowinski and colleagues (Hervé *et al.*, 1982; Tassin *et al.*, 1982). A selective blockade of α_1 adrenergic receptors in the prefrontal cortex prevents the stimulant-induced activation and release of dopamine in the nucleus accumbens (Blanc *et al.*, 1994; Darracq *et al.*, 1998). In mice lacking α_{1b} adrenergic receptors, psychostimulant-induced locomotor activation is suppressed (Drouin *et al.*, 2002) and dopamine release is absent (Auclair *et al.*, 2002). Conversely, pharmacological stimulation of this receptor increases dopamine-mediated locomotor responses (Villegier *et al.*, 2003). The blockade of cortical α_1 adrenergic receptors induces the expression of acute and sensitized locomotor responses to morphine (Drouin *et al.*, 2001). Moreover, serotonin 5-HT_{2A} and α_{1b} adrenergic receptors control a common neural pathway responsible for the release of dopamine in the nucleus accumbens by psychostimulants (Auclair *et al.*, 2004). Thus, dopamine release is dramatically modified by another widely distributed monoamine system at the same terminal regions independent of activity at the mesocorticolimbic dopamine cell bodies in the ventral tegmental area.

In addition, as amply discussed in previous chapters, drug reward or drug reinforcement are largely independent of a critical, essential role for the mesocorticolimbic dopamine system. Rats continue to self-administer heroin and alcohol in the absence of the mesocorticolimbic dopamine system, and place preference studies show robust place preferences to morphine and nicotine in the presence of massive dopamine receptor blockade (see earlier chapters). Thus, dopamine is *not* essential for acute drug reward. A role for a multiple-component system in the initiation of drug self-administration is further exemplified by the results with dopamine transporter knockout mice where the primary clearance mechanism for released dopamine was lacking (Giros *et al.*, 1996). The mutants also had increased extracellular dopamine, decreased dopamine stores, and decreased dopamine receptors. The mice were hyperactive and did not respond to passive cocaine administration, but they did self-administer cocaine. A serotonin connection was hypothesized to explain these results (Giros *et al.*, 1996; Rocha *et al.*, 1997; Caine, 1998).

Accumulating evidence also suggests that dopamine is not required for nondrug reward. In a study in which dopamine release in the nucleus accumbens core and shell was measured with precise voltammetric techniques during self-stimulation, it was shown that if dopamine activation is a necessary condition for brain stimulation reward, evoked dopamine release is actually not observed during brain stimulation reward and is even diminished (Garris *et al.*, 1999). Also, mice completely lacking tyrosine hydroxylase, such that they cannot make dopamine, demonstrated the ability to learn to consume sweet solutions and showed a preference for sucrose and saccharin. Dopamine was not required for animals to find sweet tastes of sucrose or saccharin rewarding (Cannon and Palmiter, 2003; Cannon and Bseikri, 2004).

Finally, the emphasis on dopamine is a major problem for the incentive-sensitization theory. While psychostimulant drugs produce sensitization and promote incentive salience in operant situations, drugs in the sedative-hypnotic class such as alcohol do not produce locomotor activation or sensitization in rats, nor have there been any systematic studies of facilitation by opioids or alcohol of incentive sensitization. The authors know of no studies of opioids or alcohol on operant incentive salience using sucrose reward. Similarly, while locomotor sensitization has been observed with opioid drugs, and there is cross-sensitization to psychomotor stimulants, tolerance is a more prevalent response for opioids in doses and regimens that are self-administered by addicts. The lack of generalizability of the incentive-salience component of sensitization to other drugs of abuse correlates well with the less than essential role for midbrain dopamine systems in the acute reinforcing effects of drugs of abuse. As noted above, while cocaine and amphetamine self-administration are virtually abolished with neurotoxin-specific destruction of the mesolimbic dopamine system, alcohol and heroin self-administration are largely unaffected (see previous chapters).

In summary, the midbrain dopamine systems have a critical role in facilitating neuronal circuits in the regions that they innervate, and these regions determine the functional attributes of dopamine. Dopamine activity is critical for only *one* class of drugs of abuse, psychostimulant indirect sympathomimetics (e.g., cocaine and amphetamines). Dopamine may have a role in the actions of other drugs of abuse, but the overemphasis on dopaminergic function in the study of the neurobiology of drug addiction has limited progress in the field and has limited the translation between animal models and the human condition (see also Nestler, 2004).

PAIN AND ADDICTION

The neurobiology of addiction may have some parallels to the field of pain research. Two major types of pain have been conceptualized: pain sensation and affective pain (Melzack and Casey, 1968; Price, 2002). Affective pain has a strong emotional or affective component, and much research now suggests that affective pain is mediated by many of the same neurobiological structures involved in addiction. Affective pain has been hypothesized to involve both the moment-to-moment unpleasantness of pain and 'secondary pain affect' which is the emotional feeling directed toward long-term implications of having pain (i.e., 'suffering') (Price, 2002). Affective pain is an important component of the chronic or persistent pain syndrome. Persistent pain is associated significantly with anxiety and depressive disorders, as is addiction (Huyser and Parker, 1999; Wilson *et al.*, 2001; McWilliams *et al.*, 2003). Persistent pain arises initially from pathological conditions, such as injury and inflammation, and sensitization subsequently occurs that is defined by enhanced responsiveness to incoming signals by primary afferent nerve fibers and neurons in the peripheral and central nervous systems (i.e., peripheral and central sensitization). Addiction also can be considered a type of chronic pain syndrome characterized by emotional pain, dysphoria, stress, anxiety, and interpersonal difficulties. Drugs can be argued to be sources of self-medication for such emotional pain (Khantzian *et al.*, 1974; Khantzian, 1985, 1990, 1995, 1997; Khantzian and Wilson, 1993) (see *What is Addiction?* chapter).

A principal ascending spinal pathway critical for sensory processing of pain is the lateral spinothalamic pathway which projects from nociceptive neurons to the dorsal horn of the spinal cord, to the ventroposterior lateral thalamus, and to the primary and secondary somatosensory cortex. Other ascending pathways include the spino-parabrachio-amygdaloid and spino-parabrachial-hypothalamic pathways (Bernard and Besson, 1990). Both of these pathways have been implicated in mediating the affective dimension of pain (Price, 2002).

There has been substantial evidence to suggest that the amygdala has a key role in pain modulation and emotional responses to pain (Neugebauer *et al.*, 2004). The amygdala is anatomically situated to receive highly processed affective and cognitive information (Aggleton, 2000; Le Doux, 2000), and pain-related information is well documented to be conveyed to the lateral, basolateral, and central nuclei of the amygdala via extensions of the spinohypothalamic and spinothalamic pain pathways (Millan, 1999; Le Doux, 2000). The central nucleus of the amygdala receives nociceptive-specific information via the spino-parabrachial-amygdaloid pain pathway (Gauriau and Bernard, 2002) and direct projections from the spinal cord (Wang *et al.*, 1999). Other projections to and from the central nucleus of the amygdala have been reviewed in the *Neurobiological Theories of Addiction* chapter, but include extensive afferent projections from the cortex and thalamus and extensive efferent projections to other components of the extended amygdala, thalamus, hypothalamus, and brainstem. A wide variety of pain stimuli elicit both activational and deactivational changes in metabolic activity in the amygdala as demonstrated in human neuroimaging studies (Neugebauer *et al.*, 2004).

Consistent with these observations, it appears that the amygdala is critically involved in pain enhancement (hyperalgesia) and pain reduction (analgesia) (Neugebauer *et al.*, 2004). The interaction of the amygdala is dependent on the context, the affective state, and the pain state. For example, the amygdala is hypothesized to be involved both in conditioned analgesia (Helmstetter, 1992) and the enhancement of pain responses with persistent pain (Manning, 1998; Rhudy and Meagher, 2003). A hypothetical model of the interaction between emotions and pain mediated by the amygdala has been proposed where negative emotions such as stress and fear that produce analgesia activate descending inhibitory pathways, and negative affective states such as anxiety and depression enhance facilitatory pathways (Neugebauer *et al.*, 2004). Positive emotions that decrease the unpleasantness of pain were hypothesized to inhibit amygdala-induced pain facilitation (Neugebauer *et al.*, 2004).

The interaction of pain and addiction may provide insights into the very nature of the addiction process. Tolerance and sensitization processes—two apparently opposite phenomena—can be modified concomitantly by one given biological process. The fact that opioids produce not only analgesia but also long-lasting hyperalgesia suggests that tolerance to the analgesic effect of an opioid could be the result in part of an actual sensitization of pronociceptive systems (Célèrier *et al.*, 2001; Laulin *et al.*, 2002; Rivat *et al.*, 2002). First, both magnitude and duration of heroin-induced delayed hyperalgesia increase with intermittent opiate (e.g., heroin) administration, leading to an apparent decrease in the analgesic effectiveness of a given heroin dose. Second, a small dose of heroin, ineffective at triggering delayed hyperalgesia in drug-naive rats, induced an enhancement in pain sensitivity for several days after a series of heroin administrations. These data are in agreement with a sensitization hypothesis (i.e., sensitization is the secondary effect). Third, the effectiveness of naloxone to precipitate hyperalgesia in rats that had recovered their predrug nociceptive value after single or repeated heroin administrations indicates that heroin-deprived rats were in a new biological state associated with a high-level balance between opioid-dependent analgesia systems and pronociceptive systems. Fourth, the *N*-methyl-D-aspartate receptor antagonist MK-801 prevented both long-lasting enhancement in pain sensitivity and naloxone-precipitated hyperalgesia. Thus, tolerance, sensitization, and one of the opioid withdrawal symptoms (hyperalgesia) are hypothesized to result from a single neuroadaptive process (i.e., a new allostatic equilibrium) (Célèrier *et al.*, 2001).

There are numerous additional parallels between the hypothesized modulation of chronic or persistent pain by a key element of the extended amygdala, the central nucleus of amygdala (Neugebauer *et al.*, 2004), and the modulation of addiction. As outlined above and in the *Neurobiological Theories of Addiction* chapter, the central nucleus of the amygdala has been implicated in the positive reinforcing effects of drugs of abuse and the aversive stimulus state hypothesized to drive the negative reinforcement of drug withdrawal. The central nucleus of the amygdala has been implicated in Pavlovian stimulus–reward associations

(Everitt *et al.*, 2000) (see *Neurobiological Theories of Addiction* chapter) and has been hypothesized to contribute to the formation of habit-based learning important for addiction (Robbins and Everitt, 1999; Everitt and Wolf, 2002). Even the hypothesized role of glucocorticoids in sensitizing the brain stress system described earlier in contributing to the allostatic state associated with dependence has parallels with pain modulation. Implants of corticosterone into the central nucleus of the amygdala increased indices of anxiety-like responses and produced hypersensitivity to noxious visceral stimuli (Greenwood-Van Meerveld *et al.*, 2001). Enhanced glutamatergic function in the central nucleus of the amygdala is associated with the sensitization of persistent pain (Neugebauer *et al.*, 2003). These parallels make a compelling argument for further exploration of the interaction of pain and addiction and more specifically the interaction of the affective pain, of persistent pain, and persistent emotional pain hypothesized to drive addiction.

NEUROCIRCUITRY OF COMPULSIVE DRUG-SEEKING AND DRUG-TAKING

Drug addiction is a chronic relapsing disorder characterized by compulsive drug-taking behavior with impairment in social and occupational functioning. From a psychiatric perspective, drug-seeking, drug abuse, and addiction have aspects of both impulse control disorders and compulsive disorders (see *What is Addiction?* chapter). In the development of drug addiction, individuals are hypothesized to move from an impulsive disorder to a compulsive disorder where there is a shift from positive reinforcement driving the motivated behavior to negative reinforcement driving the motivated behavior. Drug addiction has been conceptualized as a disorder that progresses from impulsivity to compulsivity in a collapsed cycle of addiction comprised of three stages: (1) preoccupation/anticipation; (2) binge/intoxication; and (3) withdrawal/negative affect (Koob and Le Moal, 1997). Thus, addiction is conceptualized as a cycle of spiraling dysregulation of brain reward systems that progressively increases, resulting in the compulsive use of drugs.

Three basic circuits have been identified from animal studies in the *Neurobiological Theories of Addiction* chapter as key common elements of the neurobiology of drug-seeking behavior: drug-associated reinforcement ('reward and stress'), drug- and cue-induced reinstatements ('craving'), and behavioral output ('compulsivity'). The challenge remains to determine how these three circuits change with the development of dependence, how they account for the impulsivity associated with drug use and abuse, and how they account for the compulsivity of addiction (**Figs. 10.2** and **10.3**). An additional challenge is to integrate these three circuits with the limited knowledge to date from human imaging studies. Such research may inform the human imaging studies, and, reciprocally, human imaging studies may reach a level of sophistication to allow them to inform the animal studies.

As defined in the *What is Addiction?* chapter, impulsivity can be associated with increased tension before performing an act, relief or pleasure upon performing the act, and little, if any, withdrawal or negative affect after the act. As such, impulsivity follows a positive reinforcement model. In contrast, compulsivity involves anxiety and distress before performing an act, some relief of the anxiety and stress upon completing the act, but significant distress and negative affect after the act that sets up even more anxiety and distress and recruits substantial guilt—and the cycle continues. Compulsivity best fits a negative reinforcement model.

Clearly, drug-seeking behavior in nondependent rats has many of the characteristics of the impulsivity model. Rats will self-administer a drug with limited access and with limited evidence of brain reward changes after self-administration. This drug self-administration has multiple neurotransmitter contributions at multiple sites, but a strong argument can be made for important roles for the elements of the extended amygdala (shell of the nucleus accumbens, central nucleus of the amygdala, and bed nucleus of the stria terminalis) and the ventral tegmental area. Some would argue that activation of the mesocorticolimbic dopamine system is critical for acute drug reward (Wise, 2002), but we would argue for a more limited role with the exception of psychostimulant drugs. Other candidate neurochemical systems would include GABA, glutamate, opioid peptides, endocannabinoids, and even glucocorticoids (Koob, 1992; Koob and Le Moal, 2001). Limited evidence in human imaging studies suggests that acute administration of drugs of abuse also changes brain metabolic activity in the ventral striatum (Volkow *et al.*, 2003).

Cues can be paired with drug-seeking, and such cues can obtain motivational value in nondependent animals (e.g., cue-induced reinstatement) and the drug itself (i.e., drug-induced reinstatement), or stressors (stress-induced reinstatement) can have motivational value. The circuitry mediating these effects in the rat has focused on key elements in the basolateral amygdala (i.e., cue-induced reinstatement), prefrontal cortex (i.e., drug-induced reinstatement), and extended amygdala (i.e., stress-induced reinstatement) (Everitt and Wolf, 2002; Kalivas and McFarland, 2003; See *et al.*, 2003; Shaham *et al.*, 2003). From human imaging studies, and from animal studies, it is clear that an initial vulnerability to drug-taking can be conveyed by a hypoactive frontal cortex and consequently low levels of executive function or high levels of impulsivity. Such hypofrontality, combined with individual vulnerability, can contribute significantly to the initial allostatic state (see Drug-Seeking to Addiction—An Allostatic View section).

However, outlining how the neurobiological circuits change with the transition to compulsive behavior presents several challenges for future research. For example, to a large extent how the 'craving' circuits change in the dependent animal is still largely unknown, and this question remains a challenge for future research. Until recently, few animal models existed for excessive drug intake, and the focus of neurobiological research was either on nondependent animals or withdrawal syndromes.

One window into the changes that occur in the brain circuits outlined above can be derived from exploring the neurobiological basis of acute withdrawal. Several common neurobiological elements have been identified that account for the motivational components of acute withdrawal from drugs of abuse. Increases in brain reward thresholds have been observed with all major drugs of abuse during acute withdrawal and presumably reflect an altered functioning of the neurochemical systems within the medial forebrain bundle. Some of the neurochemical changes associated with virtually all drugs of abuse include decreases in dopamine activity in the nucleus accumbens (Koob and Le Moal, 2001), decreases in glutamate activity in the prefrontal cortex and nucleus accumbens (Kalivas, 2004), increases in dynorphin activity in the nucleus accumbens (Nestler, 2004), increases in CRF in the extended amygdala (Koob and Kreek, 2005), and increases in norepinephrine in the extended amygdala during acute withdrawal (Aston-Jones and Harris, 2004). At the electrophysiological level, there is evidence of a decrease in firing of ventral tegmental area neurons and a quiescent nucleus accumbens during acute withdrawal (Everitt and Wolf, 2002; Thomas and Malenka, 2003). At the molecular level, there are changes in gene transcription factors, such as CREB in the nucleus accumbens (Nestler, 2004) and amygdala (Pandey, 2004). Human imaging studies have identified two common components of acute withdrawal from chronic administration of drugs in human addicts: a decrease in prefrontal cortex/anterior cingulate metabolic activity and a decrease in striatal and cortical dopamine activity as measured by decreases in D_2 receptor activity (Volkow *et al.*, 2003). All these brain activity and neurochemical/biochemical changes point to mechanisms for the dysregulation of emotion characteristic of addicts as the hallmark of the allostatic state involved in addiction.

Thus, to a large extent the changes that occur in the brain during acute drug-taking in the nondependent state are mirrored by the opposite changes during acute withdrawal from the dependent state. This would be of no surprise, even to early theoreticians in the field (Himmelsbach, 1943), but what is new since the time of Himmelsbach is that these neuroadaptive changes occur in brain circuits that have important relevance for the regulation of emotion and a tendency toward organized activity in the environment (i.e., motivation) (Hebb, 1949). One exception to this observation is that the hypofrontality and activation of the HPA and brain stress axes, both components of initial vulnerability to impulsivity, continue to 'sensitize' with the development of dependence such that the action part of compulsive drug-seeking is disinhibited.

The critical question then becomes how the activation of reward pathways (positive reinforcement) transfer information of previously neutral stimuli (reward–stimulus associations; conditioned reinforcement), loss of executive function (hypofrontality), and a compromised reward system (hyperactivity of the anti-reward circuits; negative reinforcement) to converge in the drug addict to produce compulsive behavior. Our argument would be that all four components combine to produce a very powerful drive for drug-seeking behavior that not only channels behavior away from natural rewards but also persists to explain the powerful urges to relapse so characteristic of drug addiction. This view has elements close to the incentive-motivational view argued by Hutcheson *et al.* (2001) where the reinforcing effects of heroin gain motivational value with a history of withdrawal, and close to the impaired response inhibition and salience attribution view argued by Goldstein and Volkow (2002). They argued that craving, bingeing, withdrawal, and drug reinforcement are interconnected in a positive feedback loop and depend on the functioning of the prefrontal cortex. What differs in our current perspective is the *emphasis on a powerful drive from the 'dark side', a negative motivational state (dysphoria, anxiety, distress, guilt, etc.)* that combines with a powerful positive reinforcer (or incentive) that has gained strength through environmental associations and dysregulated executive function due to hypofrontality to produce the compulsion associated with addiction.

The narrowing of the behavioral repertoire toward drug-seeking (beginning of compulsive behavior) may begin early in the process, significantly before the manifestation of a major physical withdrawal syndrome or even motivational signs of withdrawal (Deroche-Gamonet *et al.*, 2004; Vanderschuren and Everitt, 2004), but it should be noted that throughout this book there has been ample evidence that the motivational component of withdrawal begins early in the process of drug-taking. In humans and rats, this occurs within one bout of smoked or intravenous cocaine (see *What is Addiction?* chapter). Similarly, acute tolerance and withdrawal has been observed for opioids and alcohol (see earlier chapters). Human imaging studies also support the hypothesis that the drug cues activate the anterior cingulate, prefrontal cortex, orbitofrontal cortex, and the amygdala in detoxified human addicts. Metabolic mapping studies in primates show that brain changes may involve a narrowing to specific subcomponents of the prefrontal–orbitofrontal areas with a spread to the ventral striatum (Porrino and Lyons, 2000) (see *Imaging* chapter). Similarly, limited electrophysiological studies show that with chronic administration of cocaine, the phasic response neurons in the nucleus accumbens maintain their response to cocaine despite a time-related decline in background firing (Peoples and Cavanaugh, 2003). One could argue that the induction of ΔFosB in ventral striatal regions might also be a basic biochemical reflection of compulsivity (Nestler, 2004).

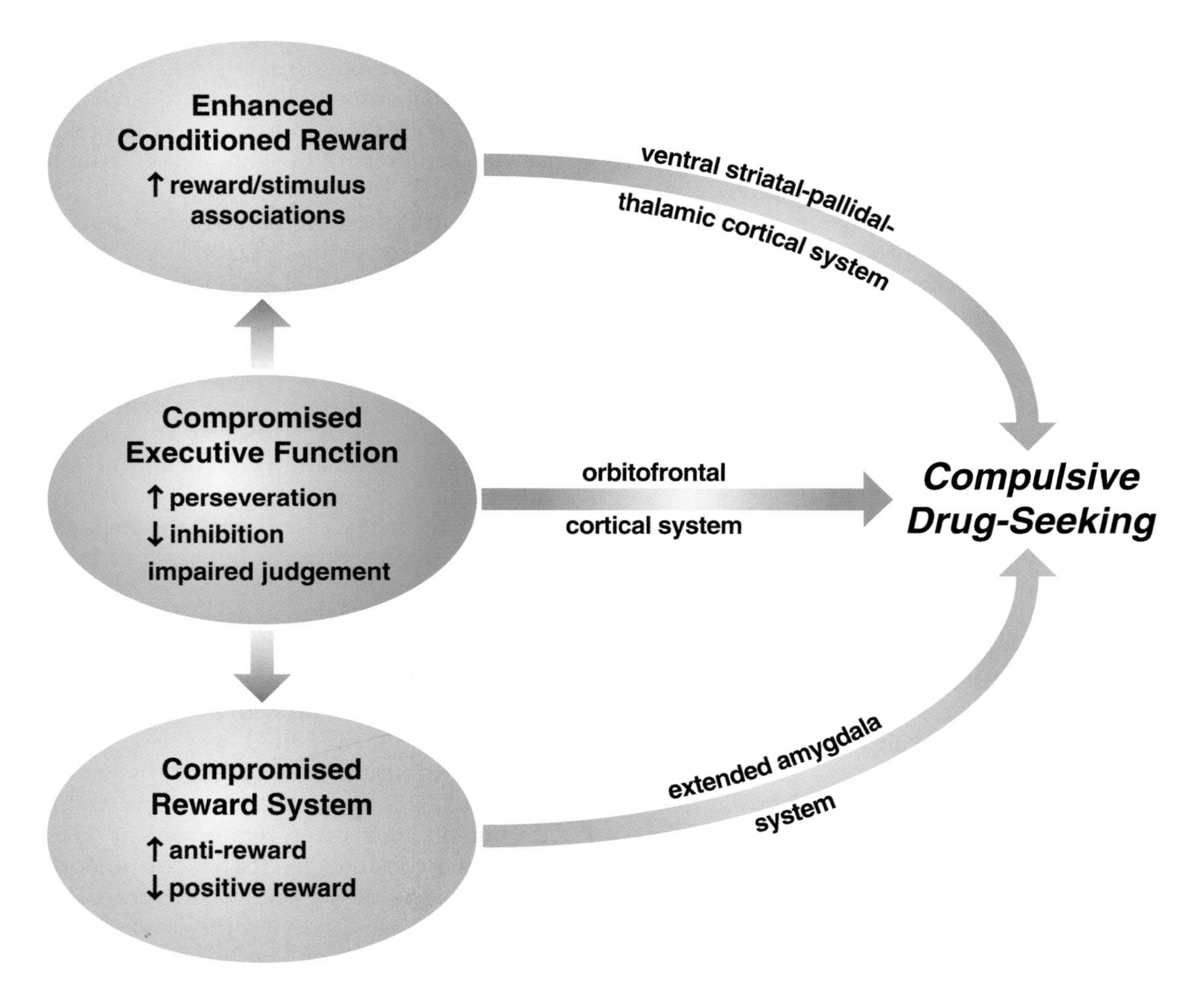

FIGURE 10.3 Three components contributing to compulsive drug-seeking in addiction that are engendered by neuroadaptive changes in three specific brain circuits. A key element of compulsive drug-seeking is hypothesized to be due to compromised executive function mediated by basal hypoactivity and situational hyperactivity in the orbitofrontal system based on both animal and human studies. Such changes in the orbitofrontal system are hypothesized to impact on, or combine with, neuroplasticity in both enhanced conditioned reward (stimulus-reward associations) mediated via the ventral striatal-pallidal-thalamic cortical system and to also impact on, or combine with, a compromised primary reward system mediated via extended amygdala circuitry. Combined, these neuroadaptations provide a powerful impetus for compulsive drug-seeking.

What is still unknown is how basal hypofunction of the reward circuit, which leads to motivational withdrawal and basal hypofunction in the prefrontal cortex circuits, which consequently leads to loss of executive function (Pfefferbaum *et al.*, 1997; Jentsch and Taylor, 1999; Sullivan *et al.*, 2003), interfaces with a 'hot-wired' craving circuit. One possibility is that these are independent neuroadaptive changes that have additive effects in an incentive-motivational model (pushing-drive state combined with pulling-incentive state). Another possibility is a direct link such that the overactivity in the reward pathways during drug-taking has engaged Pavlovian and instrumental learning processes through associative mechanisms (Robbins and Everitt, 1999; Everitt and Wolf, 2002) that persist even when the basic reward system is compromised and, in fact, helps continue the allostatic process. Such an argument has parallels with obsessive–compulsive disorder and affective disorder where engagement of negative affect not only worsens the symptoms but also produces a cognitive inflexibility that perpetuates the negative state. Reversing the molecular/biochemical factors that feed both the hypofunctioning reward systems, the hypofrontality that leads to impulsiveness, and the 'hot-wired' craving circuits may be the key to treatment and prevention of drug addiction.

IMPLICATIONS OF THE ALLOSTATIC VIEW FOR MOTIVATION AND PSYCHOPATHOLOGY

The allostatic view discussed in this chapter has implications not only for drug addiction, but also for motivational processes in general. *Motivation* can be defined as stimulation that arouses activity of a particular kind (Hebb, 1949), and presumably some kind of reward or reinforcement system is the key to any motivational conceptual framework. Such a natural reward system, however, may have a biological limitation, where excessive engagement of a brain reward system may leave an organism vulnerable to environmental challenge, be it a sudden change in weather or the appearance of a predator. Under such a formulation, brain reward is limited, and the studies with drugs of abuse suggest that this limitation involves not only dysregulation of brain transmitters important for directing appetitively motivated behavior but also recruitment of brain stress systems that limit appetitively motivated behavior.

A biological perspective of the brain's reward system, even in the context of nondrug states, suggests that it may be a limited resource (Koob and Le Moal, 1997). An hedonic system with limited energy was conceptualized by Carl Jung where the psyche was regarded as a relatively closed system (Jung, 1948). Jung described the psyche as a closed system near that of 'entropy'. In entropy, the system is closed and no energy from the outside can be transferred into it. The term 'libido' was used to describe a limited general life instinct or psychic energy. One can expend the psychic energy or hedonic resource rapidly in a binge of compulsive behavior but at the risk of entering into the spiraling dysregulation of the addiction cycle, thereby triggering allostatic mechanisms and ultimately allostatic load. A more regulated 'hedonic Calvinistic' approach, where the brain's reward system is allowed the time and resources to return to a near homeostatic set point, would prevent the development of an allostatic state and the subsequent spiraling distress associated with an addiction-like cycle (**Fig. 10.3**) (Koob and Le Moal, 1997).

REFERENCES

Abrous, D. N., Adriani, W., Montaron, M. F., Aurousseau, C., Rougon, G., Le Moal, M., and Piazza, P. V. (2002). Nicotine self-administration impairs hippocampal plasticity. *Journal of Neuroscience* **22**, 3656–3662.

Abrous, D. N., Koehl, M., and Le Moal, M. (2005). Adult neurogenesis from precursors to network and physiology. *Physiological Reviews*, in press.

Aggleton, J. P. (Ed.) (2000). *The Amygdala: A Functional Analysis*, Oxford University Press, New York.

Ahima, R. S., Garcia, M. M., and Harlan, R. E. (1992). Glucocorticoid regulation of preproenkephalin gene expression in the rat forebrain. *Molecular Brain Research* **16**, 119–127.

Ahmed, S. H., and Koob, G. F. (1998). Transition from moderate to excessive drug intake: Change in hedonic set point. *Science* **282**, 298–300.

Ahmed, S. H., and Koob, G. F. (1999). Long-lasting increase in the set point for cocaine self-administration after escalation in rats. *Psychopharmacology*, **146**, 303–312.

Ahmed, S. H., and Koob, G. F. (2004). Changes in response to a dopamine antagonist in rats with escalating cocaine intake. *Psychopharmacology* **172**, 450–454.

Ahmed, S. H., and Koob, G. F. (2005). The transition to drug addiction: a negative reinforcement model based on an allostatic decrease in reward function. *Psychopharmacology* **180**, 473–490.

Ahmed, S. H., Kenny, P. J., Koob, G. F., and Markou, A. (2002). Neurobiological evidence for hedonic allostasis associated with escalating cocaine use. *Nature Neuroscience* **5**, 625–626.

Alexander, G. E., DeLong, M. R., and Strick, P. L. (1986). Parallel organization of functionally segregated circuits linking basal ganglia and cortex. *Annual Review of Neuroscience* **9**, 357–381.

American Psychiatric Association (1980). *Diagnostic and Statistical Manual of Mental Disorders*, 3rd ed. American Psychiatric Press, Washington, DC.

American Psychiatric Association (1987). *Diagnostic and Statistical Manual of Mental Disorders*, 3rd ed. revised, American Psychiatric Press, Washington, DC.

American Psychiatric Association (1994). *Diagnostic and Statistical Manual of Mental Disorders*, 4th ed. American Psychiatric Press, Washington, DC.

Angulo, J. A., and McEwen, B. S. (1994). Molecular aspects of neuropeptide regulation and function in the corpus striatum and nucleus accumbens. *Brain Research Reviews* **19**, 1–28.

Antelman, S. M., Eichler, A. J., Black, C. A., and Kocan, D. (1980). Interchangeability of stress and amphetamine in sensitization. *Science* **207**, 329–331.

Antelman, S. M., DeGiovanni, L. A., Kocan, D., Perel, J. M., and Chiodo, L. A. (1983). Amitriptyline sensitization of a serotonin-mediated behavior depends on the passage of time and not repeated treatment. *Life Sciences* **33**, 1727–1730.

Antelman, S. M., Levine, J., and Gershon, S. (2000). Time-dependent sensitization: the odyssey of a scientific heresy from the laboratory to the door of the clinic. *Molecular Psychiatry* **5**, 350–356.

Aston-Jones, G., and Harris, G. C. (2004). Brain substrates for increased drug seeking during protracted withdrawal. *Neuropharmacology* **47**(Suppl 1.), 167–179.

Auclair, A., Cotecchia, S., Glowinski, J., and Tassin, J. P. (2002). D-amphetamine fails to increase extracellular dopamine levels in mice lacking alpha 1b-adrenergic receptors: relationship between functional and nonfunctional dopamine release. *Journal of Neuroscience* **22**, 9150–9154.

Auclair, A., Drouin, C., Cotecchia, S., and Glowinski, J. (2004). 5-HT2A and alpha1b-adrenergic receptors entirely mediate dopamine release, locomotor response and behavioural sensitization to opiates and psychostimulants. *European Journal of Neuroscience* **20**, 3073–3084.

Babbini, M., Gaiardi, M., and Bartoletti, M. (1975). Persistence of chronic morphine effects upon activity in rats 8 months after ceasing the treatment. *Neuropharmacology* **14**, 611–614.

Barrot, M., Marinelli, M., Abrous, D. N., Rougé-Pont, F., Le Moal, M., and Piazza, P. V. (2000). The dopaminergic hyper-responsiveness of the shell of the nucleus accumbens is hormone-dependent. *European Journal of Neuroscience* **12**, 973–979.

Bartoletti, M., Gaiardi, M., Gubellini, G., Bacchi, A., and Babbini, M. (1983). Long-term sensitization to the excitatory effects of morphine: a motility study in post-dependent rats. *Neuropharmcology* **22**, 1193–1196.

Basso, A. M., Spina, M., Rivier, J., Vale, W., and Koob, G. F. (1999). Corticotropin-releasing factor antagonist attenuates the 'anxiogenic-like' effect in the defensive burying paradigm but not in the elevated plus-maze following chronic cocaine in rats. *Psychopharmacology* **145**, 21–30.

Ben-Shahar, O., Ahmed, S. H., Koob, G. F., and Ettenberg, A. (2004). The transition from controlled to compulsive drug use is associated with a loss of sensitization. *Brain Research* **995**, 46–54.

Bernard, C. (1865). *Introduction a l'Etude de la Médecine Expérimentale [Introduction to the Study of Experimental Medicine]*. JB Baillierre, New York.

Bernard, J. F., and Besson, J. M. (1990). The spino(trigemino) pontoamygdaloid pathway: electrophysiological evidence for an involvement in pain processes. *Journal of Neurophysiology* **63**, 473–490.

Berridge, K. C. (2000). Measuring hedonic impact in animals and infants: microstructure of affective taste reactivity patterns. *Neuroscience and Biobehavioral Reviews* **24**, 173–198.

Blanc, G., Trovero, F., Vezina, P., Hervé, D., Godeheu, A. M., Glowinski, J., and Tassin, J. P. (1994). Blockade of prefronto-cortical alpha 1-adrenergic receptors prevents locomotor hyperactivity induced by subcortical D-amphetamine injection. *European Journal of Neuroscience* **6**, 293–298.

Breiter, H. C., Gollub, R. L., Weisskoff, R. M., Kennedy, D. N., Makris, N., Berke, J. D., Goodman, J. M., Kantor, H. L., Gastfriend, D. R., Riorden, J. P., Mathew, R. T., Rosen, B. R., and Hyman, S. E. (1997). Acute effects of cocaine on human brain activity and emotion. *Neuron* **19**, 591–611.

Cador, M., Ahmed, S. H., Koob, G. F., Le Moal, M., and Stinus, L. (1992). Corticotropin-releasing factor induces a place aversion independent of its neuroendocrine role. *Brain Research* **597**, 304–309.

Caine, S. B. (1998). Cocaine abuse: hard knocks for the dopamine hypothesis? *Nature Neuroscience* **1**, 90–92.

Caldji, C., Tannenbaum, B., Sharma, S., Francis, D., Plotsky, P. M., and Meaney, M. J. (1998). Maternal care during infancy regulates the development of neural systems mediating the expression of fearfulness in the rat. *Proceedings of the National Academy of Sciences USA* **95**, 5335–5340.

Canguilhem, G. (1978). *The Normal and the Pathological*. Zone Books, New York.

Cannon, W. B. (1929). Organization for physiological homeostasis. *Physiological Reviews* **9**, 399–431.

Cannon, W. B. (1932). *The Wisdom of the Body*. WW Norton and Company, New York.

Cannon, C. M., and Bseikri, M. R. (2004). Is dopamine required for natural reward? *Physiology and Behavior* **81**, 741–748.

Cannon, C. M., and Palmiter, R. D. (2003). Reward without dopamine. *Journal of Neuroscience* **23**, 10827–10831.

Carlezon, W. A. Jr., Thome, J., Olson, V. G., Lane-Ladd, S. B., Brodkin, E. S., Hiroi, N., Duman, R. S., Neve, R. L., and Nestler, E. J. (1998). Regulation of cocaine reward by CREB. *Science* **282**, 2272–2275.

Carlsson, A. (1987). Monoamines of the central nervous system: a historical perspective. *In Psychopharmacology: The Third Generation of Progress* (H. Y. Meltzer, Ed.), pp. 39–48. Raven Press, New York.

Carr, L. G., Foroud, T., Bice, P., Gobbett, T., Ivashina, J., Edenberg, H., Lumeng, L., and Li, T. K. (1998). A quantitative trait locus for alcohol consumption in selectively bred rat lines. *Alcoholism: Clinical and Experimental Research* **22**, 884–887.

Carroll, M. E. (1985). The role of food deprivation in the maintenance and reinstatement of cocaine-seeking behavior in rats. *Drug and Alcohol Dependence* **16**, 95–109.

Carroll, B. J. (2002). Ageing, stress and the brain. *In Endocrine Facets of Ageing* (series title: *Novartis Foundation Symposium*, vol. 242), (D. J. Chadwick, and J. A. Goode, Eds.), pp. 26–45. Wiley, New York.

Carroll, M. E., and Lac, S. T. (1993). Autoshaping i.v. cocaine self-administration in rats: Effects of nondrug alternative reinforcers on acquisition. *Psychopharmacology* **110**, 5–12.

Carroll, M. E., and Rodefer, J. S. (1993). Income alters choice between drug and an alternative nondrug reinforcer in monkeys. *Experimental and Clinical Psychopharmacology* **1**, 110–120.

Carroll, M. E., France, C. P., and Meisch, R. A. (1979). Food deprivation increases oral and intravenous drug intake in rats. *Science* **205**, 319–321.

Carroll, M. E., Lac, S. T., and Nygaard, S. L. (1989). A concurrently available nondrug reinforcer prevents the acquisition or decreases the maintenance of cocaine-reinforced behavior. *Psychopharmacology* **97**, 23–29.

Célèrier, E., Laulin, J. -P., Corcuff, J. -B., Le Moal, M., and Simonnet, G. (2001). Progressive enhancement of delayed hyperalgesia induced by repeated heroin administration: a sensitization process. *Journal of Neuroscience* **21**, 4074–4080.

Chao, H. M., and McEwen, B. S. (1990). Glucocorticoid regulation of preproenkephalin messenger ribonucleic acid in the rat striatum. *Endocrinology* **126**, 3124–3130.

Checkley, S. (1996). The neuroendocrinology of depression and chronic stress. *British Medical Bulletin* **52**, 597–617.

Ciccocioppo, R., Panocka, I., Polidori, C., Regoli, D., and Massi, M. (1999). Effect of nociceptin on alcohol intake in alcohol-preferring rats. *Psychopharmacology* **141**, 220–224.

Ciccocioppo, R., Biondini, M., Antonelli, L., Wichmann, J., Jenck, F., and Massi, M. (2002). Reversal of stress- and CRF-induced anorexia in rats by the synthetic nociceptin/orphanin FQ receptor agonist, Ro 64–6198. *Psychopharmacology* **161**, 113–119.

Ciccocioppo, R., Economidou, D., Fedeli, A., Angeletti, S., Weiss, F., Heilig, M., and Massi, M. (2004). Attenuation of ethanol self-administration and of conditioned reinstatement of alcohol-seeking behaviour by the antiopioid peptide nociceptin/orphanin FQ in alcohol-preferring rats. *Psychopharmacology* **172**, 170–178.

Cole, R. L., Konradi, C., Douglass, J., and Hyman, S. E. (1995). Neuronal adaptation to amphetamine and dopamine: molecular mechanisms of prodynorphin gene regulation in rat striatum. *Neuron* **14**, 813–823.

Cummings, S., Elde, R., Ells, J., and Lindall, A. (1983). Corticotropin-releasing factor immunoreactivity is widely distributed within the central nervous system of the rat: An immunohistochemical study. *Journal of Neuroscience* **3**, 1355–1368.

Darracq, L., Blanc, G., Glowinski, J., and Tassin, J. P. (1998). Importance of the noradrenaline-dopamine coupling in the locomotor activating effects of D-amphetamine. *Journal of Neuroscience* **18**, 2729–2739.

de Kloet, E. R. (1991). Brain corticosteroid receptor balance and homeostatic control. *Frontiers in Neuroendocrinology* **12**, 95–164.

de la Garza, R., and Johanson, C. E. (1987). The effects of food deprivation on the self-administration of psychoactive drugs. *Drug and Alcohol Dependence* **19**, 17–27.

de Quidt, M. E., and Emson, P. C. (1986). Distribution of neuropeptide Y-like immunoreactivity in the rat central nervous system: II Immunohistochemical analysis. *Neuroscience* **18**, 545–618.

Deminière, J. M., Piazza, P. V., Le Moal, M., and Simon, H. (1989). Experimental approach to individual vulnerability to psychostimulant addiction. *Neuroscience and Biobehavioral Reviews* **13**, 141–147.

Deminière, J. M., Piazza, P. V., Guegan, G., Abrous, N., Maccari, S., Le Moal, M., and Simon, H. (1992). Increased locomotor response to novelty and propensity to intravenous amphetamine self-administration in adult offspring of stressed mothers. *Brain Research* **586**, 135–139.

Deroche, V., Caine, S. B., Heyser, C. J., Polis, I., Koob, G. F., and Gold, L. H. (1997). Differences in the liability to self-administer intravenous cocaine between C57BL/6 × SJL and BALB/cByJ mice. *Pharmacology Biochemistry and Behavior* **57**, 429–440.

Deroche-Gamonet, V., Sillaber, I., Aouizerate, B., Izawa, R., Jaber, M., Ghozland, S., Kellendonk, C., Le Moal, M., Spanagel, R., Schutz, G., Tronche, F., and Piazza, P. V. (2003). The glucocorticoid receptor as a potential target to reduce cocaine abuse. *Journal of Neuroscience* **23**, 4785–4790.

Deroche-Gamonet, V., Belin, D., and Piazza, P. V. (2004). Evidence for addiction-like behavior in the rat. *Science* **305**, 1014–1017.

Dias, R., Robbins, T. W., and Roberts, A. C. (1996). Dissociation in prefrontal cortex of affective and attentional shifts. *Nature* **380**, 69–72.

Drouin, C., Blanc, G., Trovero, F., Glowinski, J., and Tassin, J. P. (2001). Cortical alpha 1-adrenergic regulation of acute and sensitized morphine locomotor effects. *Neuroreport* **12**, 3483–3486.

Drouin, C., Darracq, L., Trovero, F., Blanc, G., Glowinski, J., Cotecchia, S., and Tassin, J. P. (2002). Alpha1b-adrenergic receptors control locomotor and rewarding effects of psychostimulants and opiates. *Journal of Neuroscience* **22**, 2873–2884.

Duman, R. S., Heninger, G. R., and Nestler, E. J. (1997). A molecular and cellular theory of depression. *Archives of General Psychiatry* **54**, 597–606.

Ehlers, C. L., Li, T. K., Lumeng, L., Hwang, B. H., Somes, C., Jimenez, P., and Mathe, A. A. (1998). Neuropeptide Y levels in ethanol-naive alcohol-preferring and nonpreferring rats and in Wistar rats after ethanol exposure. *Alcoholism: Clinical and Experimental Research* **22**, 1778–1782.

Ehlers, C. L., Somes, C., Lumeng, L., and Li, T. K. (1999). Electrophysiological response to neuropeptide Y (NPY): In alcohol-naive preferring and non-preferring rats. *Pharmacology Biochemistry and Behavior* **63**, 291–299.

Eichler, A. J., and Antelman, S. M. (1979). Sensitization to amphetamine and stress may involve nucleus accumbens and medial frontal cortex. *Brain Research* **176**, 412–416.

Everitt, B. J., and Wolf, M. E. (2002). Psychomotor stimulant addiction: a neural systems perspective. *Journal of Neuroscience* **22**, 3312–3320 [erratum: **22**(16):1a].

Everitt, B. J., Cardinal, R. N., Hall, J., Parkinson, J. A., and Robbins, T. W. (2000). Differential involvement of amygdala subsystems in appetitive conditioning and drug addiction. *In The Amygdala: A Functional Analysis* (J. P. Aggleton, Ed.), pp. 353–390. Oxford University Press, New York.

Exner, E., and Clark, D. (1993). Behaviour in the novel environment predicts responsiveness to D-amphetamine in the rat: A multivariate approach. *Behavioural Pharmacology* **4**, 47–56.

Fiorillo, C. D. (2004). The uncertain nature of dopamine. *Molecular Psychiatry* **9**, 122–123.

Fiorillo, C. D., Tobler, P. N., and Schultz, W. (2003). Discrete coding of reward probability and uncertainty by dopamine neurons. *Science* **299**, 1898–1902.

Fride, E., and Weinstock, M. (1989). Alterations in behavioral and striatal dopamine asymmetries induced by prenatal stress. *Pharmacology Biochemistry and Behavior* **32**, 425–430.

Garris, P. A., Kilpatrick, M., Bunin, M. A., Michael, D., Walker, Q. D., and Wightman, R. M. (1999). Dissociation of dopamine release in the nucleus accumbens from intracranial self-stimulation. *Nature* **398**, 67–69.

Gauriau, C., and Bernard, J. F. (2002). Pain pathways and parabrachial circuits in the rat. *Experimental Physiology* **87**, 251–258.

Gawin, F. H., and Kleber, H. D. (1986). Abstinence symptomatology and psychiatric diagnosis in cocaine abusers: Clinical observations. *Archives of General Psychiatry* **43**, 107–113.

Ghozland, S., Zorrilla, E., Parsons, L. H., and Koob, G. F. (2004). Mecamylamine increases extracellular CRF levels in the central nucleus of the amygdala or nicotine-dependent rats. *Society for Neuroscience Abstracts* **30**, 708.8.

Giros, B., Jaber, M., Jones, S. R., Wightman, R. M., and Caron, M. G. (1996). Hyperlocomotion and indifference to cocaine and amphetamine in mice lacking the dopamine transporter. *Nature* **379**, 606–612.

Goeders, N. E., and Guerin, G. F. (1994). Non-contingent electric footshock facilitates the acquisition of intravenous cocaine self-administration in rats. *Psychopharmacology* **114**, 63–70.

Gold, H. K., Ronsaville, D. S., Esler, M., Alesci, S., Masood, A., Licinio, J., Geracioti, T. D. Jr., Perini, G., De Bellis, M. D., Holmes, C., Vgontzas, A. N., Charney, D. S., Chrousos, G. P., McCann, S. M., and Klin, M. A. (2005). Cardiac implications of increased arterial entry and reversible 24-h central and peripheral norepinephrine levels in melancholia. *Proceedings of the National Academy of Sciences USA* **102**, 8303–8308.

Goldstein, R. Z., and Volkow, N. D. (2002). Drug addiction and its underlying neurobiological basis: neuroimaging evidence for the involvement of the frontal cortex. *American Journal of Psychiatry* **159**, 1642–1652.

Greenwood-Van Meerveld, B., Gibson, M., Gunter, W., Shepard, J., Foreman, R., and Myers, D. (2001). Stereotaxic delivery of corticosterone to the amygdala modulates colonic sensitivity in rats. *Brain Research* **893**, 135–142.

Harfstrand, A., Fuxe, K., Cintra, A., Agnati, L. F., Zini, I., Wikstrom, A. C., Okret, S., Yu, Z. Y., Goldstein, M., Steinbusch, H., Verhofstad, A., and Gustafsson, J. -A. (1986). Glucocorticoid receptor immunoreactivity in monoaminergic neurons of rat brain. *Proceedings of the National Academy of Sciences USA* **83**, 9779–9783.

Hebb, D. O. (1949). *Organization of Behavior: A Neuropsychological Theory*. Wiley, New York.

Heilig, M., Koob, G. F., Ekman, R., and Britton, K. T. (1994). Corticotropin-releasing factor and neuropeptide Y: role in emotional integration. *Trends in Neurosciences* **17**, 80–85.

Heinrichs, S. C., and Koob, G. F. (2004). Corticotropin-releasing factor in brain: a role in activation, arousal, and affect regulation. *Journal of Pharmacology and Experimental Therapeutics* **311**, 427–440.

Heinrichs, S. C., Britton, K. T., and Koob, G. F. (1991). Both conditioned taste preference and aversion induced by corticotropin-releasing factor. *Pharmacology Biochemistry and Behavior* **40**, 717–721.

Heinrichs, S. C., Menzaghi, F., Schulteis, G., Koob, G. F., and Stinus, L. (1995). Suppression of corticotropin-releasing factor in the amygdala attenuates aversive consequences of morphine withdrawal. *Behavioural Pharmacology* **6**, 74–80.

Helmstetter, F. J. (1992). The amygdala is essential for the expression of conditional hypoalgesia. *Behavioral Neuroscience* **106**, 518–528.

Henry, C., Kabbaj, M., Simon, H., Le Moal, M., and Maccari, S. (1994). Prenatal stress increases the hypothalamo-pituitary-adrenal axis response in young and adult rats. *Journal of Neuroendocrinology* **6**, 341–345.

Hervé, D., Blanc, G., Glowinski, J., and Tassin, J. P. (1982). Reduction of dopamine utilization in the prefrontal cortex but not in the nucleus accumbens after selective destruction of noradrenergic fibers innervating the ventral tegmental area in the rat. *Brain Research* **237**, 510–516.

Higgins, S. T., Bickel, W. K., and Hughes, J. R. (1994). Influence of an alternative reinforcer on human cocaine self-administration. *Life Sciences* **55**, 179–187.

Himmelsbach, C. K. (1942). Clinical studies of drug addiction: Physical dependence, withdrawal and recovery. *Archives of Internal Medicine* **69**, 766–772.

Himmelsbach, C. K. (1943). Can the euphoric, analgetic, and physical dependence effects of drugs be separated? IV. With reference to physical dependence. *Federation Proceedings* **2**, 201–203.

Hooks, M. S., Jones, G. H., Smith, A. D., Neill, D. B., and Justice, J. B. Jr. (1991). Response to novelty predicts the locomotor and nucleus accumbens dopamine response to cocaine. *Synapse* **9**, 121–128.

Hooks, M. S., Colvin, A. C., Juncos, J. L., and Justice, J. B. Jr. (1992a). Individual differences in basal and cocaine-stimulated extracellular dopamine in the nucleus accumbens using quantitative microdialysis. *Brain Research* **587**, 306–312.

Hooks, M. S., Jones, G. H., Liem, B. J., and Justice, J. B. Jr. (1992b). Sensitization and individual differences to IP amphetamine, cocaine, or caffeine following repeated intracranial amphetamine infusions. *Pharmacology Biochemistry and Behavior* **43**, 815–823.

Hooks, M. S., Jones, G. H., Neill, D. B., and Justice, J. B. Jr. (1992c). Individual differences in amphetamine sensitization: Dose-dependent effects. *Pharmacology Biochemistry and Behavior* **41**, 203–210.

Hooks, M. S., Jones, G. H., Juncos, J. L., Neill, D. B., and Justice, J. B. (1994a). Individual differences in schedule-induced and conditioned behaviors. *Behavioural Brain Research* **60**, 199–209.

Hooks, M. S., Juncos, J. L., Justice, J. B. Jr., Meiergerd, S. M., Povlock, S. L., Schenk, J. O., and Kalivas, P. W. (1994b). Individual locomotor response to novelty predicts selective alterations in D1 and D2 receptors and mRNAs. *Journal of Neuroscience* **14**, 6144–6152.

Hutcheson, D. M., Everitt, B. J., Robbins, T. W., and Dickinson, A. (2001). The role of withdrawal in heroin addiction: enhances reward or promotes avoidance? *Nature Neuroscience* **4**, 943–947.

Huyser, B. A., and Parker, J. C. (1999). Negative affect and pain in arthritis. *Rheumatic Diseases Clinics of North America* **25**, 105–121.

Hwang, B. H., Zhang, J. K., Ehlers, C. L., Lumeng, L., and Li, T. K. (1999). Innate differences of neuropeptide Y (NPY) in hypothalamic nuclei and central nucleus of the amygdala between selectively bred rats with high and low alcohol preference. *Alcoholism: Clinical and Experimental Research* **23**, 1023–1030.

Jenck, F., Moreau, J. L., Martin, J. R., Kilpatrick, G. J., Reinscheid, R. K., Monsma, F. J. Jr., Nothacker, H. P., and Civelli, O. (1997). Orphanin FQ acts as an anxiolytic to attenuate behavioral responses to stress. *Proceedings of the National Academy of Sciences USA* **94**, 14854–14858.

Jenck, F., Wichmann, J., Dautzenberg, F. M., Moreau, J. L., Ouagazzal, A. M., Martin, J. R., Lundstrom, K., Cesura, A. M., Poli, S. M., Roever, S., Kolczewski, S., Adam, G., and Kilpatrick, G. (2000). A synthetic agonist at the orphanin FQ/nociceptin receptor ORL1: anxiolytic profile in the rat. *Proceedings of the National Academy of Sciences USA* **97**, 4938–4943.

Jentsch, J. D., and Taylor, J. R. (1999). Impulsivity resulting from frontostriatal dysfunction in drug abuse: Implications for the control of behavior by reward-related stimuli. *Psychopharmacology* **146**, 373–390.

Joels, M., and de Kloet, E. R. (1992). Control of neuronal excitability by corticosteroid hormones. *Trends in Neurosciences* **15**, 25–30.

Joels, M., and de Kloet, E. R. (1994). Mineralocorticoid and glucocorticoid receptors in the brain: Implications for ion permeability and transmitter systems. *Progress in Neurobiology* **43**, 1–36.

Jung, C. G. (1948). *Die Beziehungen der Psychotherapie zur Seelsorge [The Structure and Dynamics of the Psyche]*. Rascher, Zurich.

Kalivas, P. W. (2004). Glutamate systems in cocaine addiction. *Current Opinion in Pharmacology* **4**, 23–29.

Kalivas, P. W., and McFarland, K. (2003). Brain circuitry and the reinstatement of cocaine-seeking behavior. *Psychopharmacology* **168**, 44–56.

Kavaliers, M. (1990). Inhibitory influences of mammalian FMRFamide (Phe-Met-Arg-Phe-amide)-related peptides on nociception and morphine- and stress-induced analgesia in mice. *Neuroscience Letters* **115**, 307–312.

Khantzian, E. J. (1985). The self-medication hypothesis of affective disorders: focus on heroin and cocaine dependence. *American Journal of Psychiatry* **142**, 1259–1264.

Khantzian, E. J. (1990). Self-regulation and self-medication factors in alcoholism and the addictions: similarities and differences. *In Combined Alcohol and Other Drug Dependence* (series title: *Recent Developments in Alcoholism*, vol. 8), (M. Galanter, Ed.), pp. 255–271. Plenum Press, New York.

Khantzian, E. J. (1995). The 1994 distinguished lecturer in substance abuse. *Journal of Substance Abuse Treatment* **12**, 157–165.

Khantzian, E. J. (1997). The self-medication hypothesis of substance use disorders: a reconsideration and recent applications. *Harvard Review of Psychiatry* **4**, 231–244.

Khantzian, E. J., and Wilson, A. (1993). Substance abuse, repetition, and the nature of addictive suffering. *In Hierarchical Concepts in Psychoanalysis: Theory, Research, and Clinical Practice* (A. Wilson, and J. E. Gedo, Eds.), pp. 263–283. Guilford Press, New York.

Khantzian, E. J., Mack, J. E., and Schatzberg, A. F. (1974). Heroin use as an attempt to cope: clinical observations. *American Journal of Psychiatry* **131**, 160–164.

Knapp, D. J., Overstreet, D. H., Moy, S. S., and Breese, G. R. (2004). SB242084, flumazenil, and CRA1000 block ethanol withdrawal-induced anxiety in rats. *Alcohol* **32**, 101–111.

Kolta, M. G., Shreve, P., De Souza, V., and Uretsky, N. J. (1985). Time course of the development of the enhanced behavioral and biochemical responses to amphetamine after pretreatment with amphetamine. *Neuropharmacology* **24**, 823–829.

Koob, G. F. (1992). Drugs of abuse: anatomy, pharmacology, and function of reward pathways. *Trends in Pharmacological Sciences* **13**, 177–184.

Koob, G. F. (1999). Corticotropin-releasing factor, norepinephrine and stress. *Biological Psychiatry* **46**, 1167–1180.

Koob, G. F. (2004). Allostatic view of motivation: implications for psychopathology. *In Motivational Factors in the Etiology of Drug Abuse* (series title: *Nebraska Symposium on Motivation*, vol. 50), (R. A. Bevins, and M. T. Bardo, Eds.), pp. 1–18. University of Nebraska Press, Lincoln, NE.

Koob, G. F., and Kreek, M. J. (2005). Stress, dysregulation of drug reward pathways, and drug dependence: an updated perspective commemorating the 30th anniversary of the National Institute on Drug Abuse, submitted.

Koob, G. F., and Le Moal, M. (1997). Drug abuse: Hedonic homeostatic dysregulation. *Science* **278**, 52–58.

Koob, G. F., and Le Moal, M. (2001). Drug addiction, dysregulation of reward, and allostasis. *Neuropsychopharmacology* **24**, 97–129.

Koob, G. F., and Le Moal, M. (2004). Drug addiction and allostasis. *In Allostasis, Homeostasis, and the Costs of Physiological Adaptation* (J. Schulkin, Ed.), pp. 150–163. Cambridge University Press, New York.

Koob, G. F., Stinus, L., Le Moal, M., and Bloom, F. E. (1989). Opponent process theory of motivation: neurobiological evidence from studies of opiate dependence. *Neuroscience and Biobehavioral Reviews* **13**, 135–140.

Koob, G. F., Heinrichs, S. C., Menzaghi, F., Pich, E. M., and Britton, K. T. (1994). Corticotropin releasing factor, stress and behavior. *Seminars in the Neurosciences* **6**, 221–229.

Kreek, M. J. (1987). Multiple drug abuse patterns and medical consequences. *In Psychopharmacology: The Third Generation of Progress* (H. Y. Meltzer, Ed.), pp. 1597–1604. Raven Press, New York.

Kreek, M. J., Ragunath, J., Plevy, S., Hamer, D., Schneider, B., and Hartman, N. (1984). ACTH, cortisol and beta-endorphin response to metyrapone testing during chronic methadone maintenance treatment in humans. *Neuropeptides* **5**, 277–278.

Ladd, C. O., Huot, R. L., Thrivikraman, K. V., Nemeroff, C. B., Meaney, M. J., and Plotsky, P. M. (2000). Long-term behavioral and neuroendocrine adaptations to adverse early experience. *In The Biological Basis for Mind Body Interactions* (series title: *Progress in Brain Research*, vol. 122), (E. A. Mayer, and C. B. Saper, Eds.), pp. 81–103. Elsevier, New York.

Lake, J. R., Hebert, K. M., Payza, K., Deshotel, K. D., Hausam, D. D., Witherspoon, W. E., Arcangeli, K. A., and Malin, D. H. (1992). Analog of neuropeptide FF attenuates morphine tolerance. *Neuroscience Letters* **146**, 203–206.

Laulin, J. P., Maurette, P., Corcuff, J. B., Rivat, C., Chauvin, M., and Simonnet, G. (2002). The role of ketamine in preventing fentanyl-induced hyperalgesia and subsequent acute morphine tolerance. *Anesthesia and Analgesia* **94**, 1263–1269.

Le Doux, J. E. (2000). Emotion circuits in the brain. *Annual Review of Neuroscience* **23**, 155–184.

Le Moal, M. (1995). Mesocorticolimbic dopaminergic neurons: Functional and regulatory roles. *In Psychopharmacology: The Fourth Generation of Progress* (F. E. Bloom, and D. J. Kupfer, Eds.), pp. 283–294. Raven Press, New York.

Le Moal, M., and Olds, M. E, (1979). Peripheral auditory input to the midbrain limbic area and related structures. *Brain Research* **167**, 1–17.

Le Moal, M., and Simon, H. (1991). Mesocorticolimbic dopaminergic network: functional and regulatory roles. *Physiological Reviews* **71**, 155–234.

Le Sage, M. G., Stafford, D., and Glowa, J. R. (1999). Preclinical research on cocaine self-administration: Environmental determinants and their interaction with pharmacological treatment. *Neuroscience and Biobehavioral Reviews* **23**, 717–741.

Lee, Y., Schulkin, J., and Davis, M. (1994). Effect of corticosterone on the enhancement of the acoustic startle reflex by corticotropin releasing factor (CRF). *Brain Research* **666**, 93–98.

Lindy, J. D., and Wilson, J. P. (2001). An allostatic approach to the psychodynamic understanding of PTSD. *In Treating Psychological Trauma and PTSD* (J.P. Wilson, M. J. Friedman, and J. D. Lindy, Eds.), pp. 125–138. Guilford Press, New York.

Logue, A. W., Tobin, H., Chelonis, J. J., Wang, R. Y., Geary, N., and Schachter, S. (1992). Cocaine decreases self-control in rats: a preliminary report. *Psychopharmacology* **109**, 245–247.

London, E. D., Broussolle, E. P., Links, J. M., Wong, D. F., Cascella, N. G., Dannals, R. F., Sano, M., Herning, R., Snyder, F. R., and Rippetoe, L. R. (1990a). Morphine-induced metabolic changes in human brain. Studies with positron emission tomography and [fluorine 18]fluorodeoxyglucose. *Archives of General Psychiatry* **47**, 73–81.

London, E. D., Cascella, N. G., Wong, D. F., Phillips, R. L., Dannals, R. F., Links, J. M., Herning, R., Grayson, R., Jaffe, J. H., and Wagner, H. N. Jr. (1990b). Cocaine-induced reduction of glucose utilization in human brain: A study using positron emission tomography and [fluorine 18]-fluorodeoxyglucose. *Archives of General Psychiatry* **47**, 567–574.

Louilot, A., and Le Moal, M. (1994). Lateralized interdependence between limbicotemporal and ventrostriatal dopaminergic transmission. *Neuroscience* **59**, 495–500.

Louilot, A., Simon, H., Taghzouti, K., and Le Moal, M. (1985). Modulation of dopaminergic activity in the nucleus accumbens following facilitation or blockade of the dopaminergic transmission in the amygdala: a study by in vivo differential pulse voltammetry. *Brain Research* **346**, 141–145.

Louilot, A., Le Moal, M., and Simon, H. (1986). Differential reactivity of dopaminergic neurons in the nucleus accumbens in response to different behavioral situations. An in vivo voltammetric study in free moving rats. *Brain Research* **397**, 395–400.

Louilot, A., Le Moal, M., and Simon, H. (1989). Opposite influences of dopaminergic pathways to the prefrontal cortex or the septum on the dopaminergic transmission in the nucleus accumbens: an in vivo voltammetric study. *Neuroscience* **29**, 45–56.

Maccari, S., Piazza, P. V., Deminière, J. M., Lemaire, V., Mormède, P., Simon, H., Angelucci, L., and Le Moal, M. (1991). Life events-induced decrease of corticosteroid type I receptors is associated with reduced corticosterone feedback and enhanced vulnerability to amphetamine self-administration. *Brain Research* **547**, 7–12.

Macey, D. J., Koob, G. F., and Markou, A. (2000). CRF and urocortin decreased brain stimulation reward in the rat: reversal by a CRF receptor antagonist. *Brain Research* **866**, 82–91.

Madden, G. J., Petry, N. M., Badger, G. J., and Bickel, W. K. (1997). Impulsive and self-control choices in opioid-dependent patients and non-drug-using control participants: drug and monetary rewards. *Experimental and Clinical Psychopharmacology* **5**, 256–262.

Malin, D. H., Lake, J. R., Hammond, M. V., Fowler, D. E., Rogillio, R. B., Brown, S. L., Sims, J. L., and Leecraft, B. M. (1990). Yang HYT, FMRF-NH2-like mammalian octapeptide: Possible role in opiate dependence and abstinence. *Peptides* **11**, 969–972.

Malin, D. H., Lake, J. R., Short, P. E., Blossman, J. B., Lawless, B. A., Schopen, C. K., Sailer, E. E., Burgess, K., and Wilson, O. B. (1996). Nicotine abstinence syndrome precipitated by an analog of neuropeptide FF. *Pharmacology Biochemistry and Behavior* **54**, 581–585.

Manning, B. H. (1998). A lateralized deficit in morphine antinociception after unilateral inactivation of the central amygdala. *Journal of Neuroscience* **18**, 9453–9470.

Mantsch, J. R., Yuferov, V., Mathieu-Kia, A. M., Ho, A., and Kreek, M. J. (2004). Effects of extended access to high versus low cocaine doses on self-administration, cocaine-induced reinstatement and brain mRNA levels in rats. *Psychopharmacology* **175**, 26–36.

Martin-Fardon, R., Ciccocioppo, R., Massi, M., and Weiss, F. (2000). Nociceptin prevents stress-induced ethanol- but not cocaine-seeking behavior in rats. *Neuroreport* **11**, 1939–1943.

McEwen, B. S. (1998). Stress, adaptation, and disease: Allostasis and allostatic load. *In Neuroimmunomodulation: Molecular Aspects, Integrative Systems, and Clinical Advances* (series title: *Annals of the New York Academy of Sciences*, vol. 840), (S. M. McCann, J. M. Lipton, E. M. Sternberg, G. P. Chrousos, P. W. Gold, and C. C. Smith, Eds.), pp. 33–44. New York Academy of Sciences, New York.

McEwen, B. S. (2000). Allostasis and allostatic load: Implications for neuropsychopharmacology. *Neuropsychopharmacology* **22**, 108–124.

McWilliams, L. A., Cox, B. J., and Enns, M. W. (2003). Mood and anxiety disorders associated with chronic pain: an examination in a nationally representative sample. *Pain* **106**, 127–133.

Melzack, R., and Casey, K. L. (1968). Sensory, motivational, and central control determinants of pain. *In The Skin Senses* (D. R. Kenshalo, Ed.), pp. 423–439. Thomas, Springfield IL.

Merlo-Pich, E., Lorang, M., Yeganeh, M., Rodriguez de Fonseca, F., Raber, J., Koob, G. F., and Weiss, F. (1995). Increase of extracellular corticotropin-releasing factor-like immunoreactivity levels in the amygdala of awake rats during restraint stress and ethanol withdrawal as measured by microdialysis. *Journal of Neuroscience* **15**, 5439–5447.

Millan, M. J. (1999). The induction of pain: an integrative review. *Progress in Neurobiology* **57**, 1–164.

Mogil, J. S., Grisel, J. E., Zhangs, G., Belknap, J. K., and Grandy, D. K. (1996). Functional antagonism of mu-, delta- and kappa-opioid antinociception by orphanin FQ. *Neuroscience Letters* **214**, 131–134.

Moore-Ede, M. C. (1986). Physiology of the circadian timing system: predictive versus reactive homeostasis. *American Journal of Physiology* **250**, R737–R752.

Moyer, J. A., Herrenkohl, L. R., and Jacobowitz, D. M. (1978). Stress during pregnancy: Effect on catecholamines in discrete brain regions of offspring as adults. *Brain Research* **144**, 173–178.

Mrosovsky, N. (1990). *Rheostasis: The Physiology of Change*. Oxford University Press, New York.

Nestler, E. J. (2004). Historical review: Molecular and cellular mechanisms of opiate and cocaine addiction. *Trends in Pharmacological Sciences* **25**, 210–218.

Neugebauer, V., Li, W., Bird, G. C., Bhave, G., and Gereau, R. W. 4th ed. (2003). Synaptic plasticity in the amygdala in a model of arthritic pain: differential roles of metabotropic glutamate receptors 1 and 5. *Journal of Neuroscience* **23**, 52–63.

Neugebauer, V., Li, W., Bird, G. C., and Han, J. S. (2004). The amygdala and persistent pain. *Neuroscientist* **10**, 221–234.

Newton, S. S., Thome, J., Wallace, T. L., Shirayama, Y., Schlesinger, L., Sakai, N., Chen, J., Neve, R., Nestler, E. J., and Duman, R. S. (2002). Inhibition of cAMP response element-binding protein or dynorphin in the nucleus accumbens produces an antidepressant-like effect. *Journal of Neuroscience* **22**, 10883–10890.

O'Dell, L. E., Roberts, A. J., Smith, R. T., and Koob, G. F. (2004). Enhanced alcohol self-administration after intermittent versus continuous alcohol vapor exposure. *Alcoholism: Clinical and Experimental Research* **28**, 1676–1682.

Olive, M. F., Koenig, H. N., Nannini, M. A., and Hodge, C. W. (2002). Elevated extracellular CRF levels in the bed nucleus of the stria terminalis during ethanol withdrawal and reduction by subsequent ethanol intake. *Pharmacology Biochemistry and Behavior* **72**, 213–220.

Palmiter, R. D., Erickson, J. C., Hollopeter, G., Baraban, S. C., and Schwartz, M. W. (1998). Life without neuropeptide Y. *Recent Progress in Hormone Research* **53**, 163–199.

Pandey, S. C. (2004). The gene transcription factor cyclic AMP-responsive element binding protein: role in positive and negative affective states of alcohol addiction. *Pharmacology and Therapeutics* **104**, 47–58.

Pandey, S. C., Roy, A., and Zhang, H. (2003). The decreased phosphorylation of cyclic adenosine monophosphate (cAMP) response element binding (CREB) protein in the central amygdala acts as a molecular substrate for anxiety related to ethanol withdrawal in rats. *Alcoholism: Clinical and Experimental Research* **27**, 396–409.

Peoples, L. L., and Cavanaugh, D. (2003). Differential changes in signal and background firing of accumbal neurons during cocaine self-administration. *Journal of Neurophysiology* **90**, 993–1010.

Petry, N. M., and Casarella, T. (1999). Excessive discounting of delayed rewards in substance abusers with gambling problems. *Drug and Alcohol Dependence* **56**, 25–32.

Pfefferbaum, A., Sullivan, E. V., Mathalon, D. H., and Lim, K. O. (1997). Frontal lobe volume loss observed with magnetic resonance imaging in older chronic alcoholics. *Alcoholism: Clinical and Experimental Research* **21**, 521–529.

Piazza, P. V., and Le Moal, M. (1996). Pathophysiological basis of vulnerability to drug abuse: Role of an interaction between stress, glucocorticoids, and dopaminergic neurons. *Annual Review of Pharmacology and Toxicology* **36**, 359–378.

Piazza, P. V., Deminière, J. M., Le Moal, M., and Simon, H. (1989). Factors that predict individual vulnerability to amphetamine self-administration. *Science* **245**, 1511–1513.

Piazza, P. V., Maccari, S., Deminière, J. M., Le Moal, M., Mormède, P., and Simon, H. (1991a). Corticosterone levels determine individual vulnerability to amphetamine self-administration. *Proceedings of the National Academy of Sciences USA* **88**, 2088–2092.

Piazza, P. V., Rougé-Pont, F., Deminière, J. M., Kharoubi, M., Le Moal, M., and Simon, H. (1991b). Dopaminergic activity is reduced in the prefrontal cortex and increased in the nucleus accumbens of rats predisposed to develop amphetamine self-administration. *Brain Research* **567**, 169–174.

Piazza, P. V., Deroche, V., Deminière, J. M., Maccari, S., Le Moal, M., and Simon, H. (1993). Corticosterone in the range of stress-induced levels possesses reinforcing properties: Implications for sensation-seeking behaviors. *Proceedings of the National Academy of Sciences USA* **90**, 11738–11742.

Piazza, P. V., Rougé-Pont, F., Deroche, V., Maccari, S., Simon, H., and Le Moal, M. (1996). Glucocorticoids have state-dependent stimulant effects on the mesencephalic dopaminergic transmission. *Proceedings of the National Academy of Sciences USA* **93**, 8716–8720.

Piazza, P. V., Deroche, V., Rougé-Pont, F., and Le Moal, M. (1998). Behavioral and biological factors associated with individual vulnerability to psychostimulant abuse. *In Laboratory Behavioral Studies of Vulnerability to Drug Abuse* (series title: *NIDA Research Monograph*, vol. 169), (C. L. Wetherington, and J. L. Falk, Eds.), pp. 105–133. National Institute on Drug Abuse, Rockville, MD.

Piazza, P. V., Deroche-Gamonent, V., Rougè-Pont, F., and Le Moal, M. (2000). Vertical shifts in self-administration dose-response functions predict a drug-vulnerable phenotype predisposed to addiction. *Journal of Neuroscience* **20**, 4226–4232.

Porrino, L. J., and Lyons, D. (2000). Orbital and medial prefrontal cortex and psychostimulant abuse: studies in animal models. *Cerebral Cortex* **10**, 326–333.

Price, D. D. (2002). Central neural mechanisms that interrelate sensory and affective dimensions of pain. *Molecular Interventions* **2**, 392–403.

Rassnick, S., Heinrichs, S. C., Britton, K. T., and Koob, G. F. (1993). Microinjection of a corticotropin-releasing factor antagonist into the central nucleus of the amygdala reverses anxiogenic-like effects of ethanol withdrawal. *Brain Research* **605**, 25–32.

Reiter-Funk, C., O'Dell, L. E., and Koob, G. F. (2005). Escalation of ethanol self-administration during acute ethanol withdrawal: regulation by corticotropin releasing factor in the extended amygdala. *Alcoholism: Clinical and Experimental Research* **29(5 Suppl)**, 11.

Rhudy, J. L., and Meagher, M. W. (2003). Negative affect: effects on an evaluative measure of human pain. *Pain* **104**, 617–626.

Richards, J. B., Sabol, K. E., and de Wit, H. (1999). Effects of methamphetamine on the adjusting amount procedure, a model of impulsive behavior in rats. *Psychopharmacology* **146**, 432–439.

Richter, R. M., and Weiss, F. (1999). In vivo CRF release in rat amygdala is increased during cocaine withdrawal in self-administering rats. *Synapse* **32**, 254–261.

Rivat, C., Laulin, J. -P., Corcuff, J. B., Célèrier, E., Pain, L., and Simonnet, G. (2002). Fentanyl enhancement of carrageenan-induced long-lasting hyperalgesia in rats: Prevention by the N-methyl-D-aspartate receptor antagonist ketamine. *Anesthesiology* **96**, 381–391.

Rivier, C., Bruhn, T., and Vale, W. (1984). Effect of ethanol on the hypothalamic-pituitary-adrenal axis in the rat: role of corticotropin-releasing factor (CRF). *Journal of Pharmacology and Experimental Therapeutics* **229**, 127–131.

Robbins, T. W., and Everitt, B. J. (1999). Drug addiction: bad habits add up. *Nature* **398**, 567–570.

Roberts, A. J., Cole, M., and Koob, G. F. (1996). Intra-amygdala muscimol decreases operant ethanol self-administration in dependent rats. *Alcoholism: Clinical and Experimental Research* **20**, 1289–1298.

Roberts, A. J., Heyser, C. J., Cole, M., Griffin, P., and Koob, G. F. (2000). Excessive ethanol drinking following a history of dependence: Animal model of allostasis. *Neuropsychopharmacology* **22**, 581–594.

Robinson, T. E., and Berridge, K. C. (1993). The neural basis of drug craving: An incentive-sensitization theory of addiction. *Brain Research Reviews* **18**, 247–291.

Robinson, T. E., and Berridge, K. C. (2003). Addiction. *Annual Review of Psychology* **54**, 25–53.

Rocha, B. A., Ator, R., Emmett-Oglesby, M. W., and Hen, R. (1997). Intravenous cocaine self-administration in mice lacking 5-HT1B receptors. *Pharmacology Biochemistry and Behavior* **57**, 407–412.

Rodriguez de Fonseca, F., Carrera, M. R. A., Navarro, M., Koob, G. F., and Weiss, F. (1997). Activation of corticotropin-releasing factor in the limbic system during cannabinoid withdrawal. *Science* **276**, 2050–2054.

Rougé-Pont, F., Piazza, P. V., Kharouby, M., Le Moal, M., and Simon, H. (1993). Higher and longer stress-induced increase in dopamine concentrations in the nucleus accumbens of animals predisposed to amphetamine self-administration: A microdialysis study. *Brain Research* **602**, 169–174.

Rougé-Pont, F., Marinelli, M., Le Moal, M., Simon, H., and Piazza, P. V. (1995). Stress-induced sensitization and glucocorticoids: II. Sensitization of the increase in extracellular dopamine induced by cocaine depends on stress-induced corticosterone secretion. *Journal of Neuroscience* **15**, 7189–7195.

Sarnyai, Z., Biro, E., Gardi, J., Vecsernyes, M., Julesz, J., and Telegdy, G. (1995). Brain corticotropin-releasing factor mediates 'anxiety-like' behavior induced by cocaine withdrawal in rats. *Brain Research* **675**, 89–97.

Schoffelmeer, A. N., Voorn, P., Jonker, A. J., Wardeh, G., Nestby, P., Vanderschuren, L. J., De Vries, T. J., Mulder, A. H., and Tjon, G. H. (1996). Morphine-induced increase in D-1 receptor regulated signal transduction in rat striatal neurons and its facilitation by glucocorticoid receptor activation: Possible role in behavioral sensitization. *Neurochemistry Research* **21**, 1417–1423.

Schulkin, J., McEwen, B. S., and Gold, P. W. (1994). Allostasis, amygdala, and anticipatory angst. *Neuroscience and Biobehavioral Reviews* **18**, 385–396.

Schulteis, G., Stinus, L., Risbrough, V. B., and Koob, G. F. (1998a). Clonidine blocks acquisition but not expression of conditioned opiate withdrawal in rats. *Neuropsychopharmacology* **19**, 406–416.

Schulteis, G., Yackey, M., Risbrough, V., and Koob, G. F. (1998b). Anxiogenic-like effects of spontaneous and naloxone-precipitated opiate withdrawal in the elevated plus-maze. *Pharmacology Biochemistry and Behavior* **60**, 727–731.

See, R. E., Fuchs, R. A., Ledford, C. C., and McLaughlin, J. (2003). Drug addiction, relapse, and the amygdala. *In The Amygdala in Brain Function: Basic and Clinical Approaches* (series title: *Annals of the New York Academy of Sciences*, vol. 985), (P. Shinnick-Gallagher, A. Pitkanen, A. Shekhar, and L. Cahill, Eds.), pp. 294–307. New York Academy of Sciences, New York.

Segal, D. S., and Schuckit, M. A. (1983). Animal models of stimulant-induced psychosis. *In Stimulants: Neurochemical, Behavioral, and Clinical Perspectives* (I. Creese, Ed.), pp. 131–167. Raven Press, New York.

Selden, N. R., Everitt, B. J., Jarrard, L. E., and Robbins, T. W. (1991). Complementary roles for the amygdala and hippocampus in aversive conditioning to explicit and contextual cues. *Neuroscience* **42**, 335–350.

Shaham, Y., Shalev, U., Lu, L., de Wit, H., and Stewart, J. (2003). The reinstatement model of drug relapse: history, methodology and major findings. *Psychopharmacology* **168**, 3–20.

Shepard, J. D., Barron, K. W., and Myers, D. A. (2000). Corticosterone delivery to the amygdala increases corticotropin-releasing factor mRNA in the central amygdaloid nucleus and anxiety-like behavior. *Brain Research* **861**, 288–295.

Simon, H., Scatton, B., and Le Moal, M. (1980). Dopaminergic A10 neurones are involved in cognitive functions. *Nature* **286**, 150–151.

Slawecki, C. J., Somes, C., and Ehlers, C. L. (1999). Effects of chronic ethanol exposure on neurophysiological responses to corticotropin-releasing factor and neuropeptide Y. *Alcohol and Alcoholism* **34**, 289–299.

Solomon, R. L., and Corbit, J. D. (1974). An opponent-process theory of motivation: I. Temporal dynamics of affect. *Psychological Reviews* **81**, 119–145.

Spanagel, R., Montkowski, A., Allingham, K., Stohr, T., Shoaib, M., Holsboer, F., and Landgraf, R. (1995). Anxiety: a potential predictor of vulnerability to the initiation of ethanol self-administration in rats. *Psychopharmacology* **122**, 369–373.

Sterling, P., and Eyer, J. (1988). Allostasis: A new paradigm to explain arousal pathology. *In Handbook of Life Stress, Cognition and Health* (S. Fisher, J. Reason, Eds.), pp. 629–649. John Wiley, Chichester.

Stinus, L., Caille, S., and Koob, G. F. (2000). Opiate withdrawal-induced place aversion lasts for up to 16 weeks. *Psychopharmacology* **149**, 115–120.

Stinus, L., Cador, M., Zorrilla, E. P., and Koob, G. F. (2005). Buprenorphine and a CRF1 antagonist block the acquisition of opiate withdrawal-induced conditioned place aversion in rats. *Neuropsychopharmacology* **30**, 90–98.

Sullivan, E. V., Harding, A. J., Pentney, R., Dlugos, C., Martin, P. R., Parks, M. H., Desmond, J. E., Chen, S. H., Pryor, M. R., De Rosa, E., and Pfefferbaum, A. (2003). Disruption of frontocerebellar circuitry and function in alcoholism. *Alcoholism: Clinical and Experimental Research* **27**, 301–309.

Tassin, J. P., Simon, H., Hervé, D., Blanc, G., Le Moal, M., Glowinski, J., and Bockaert, J. (1982). Non-dopaminergic fibres may regulate dopamine-sensitive adenylate cyclase in the prefrontal cortex and nucleus accumbens. *Nature* **295**, 696–698.

Thiele, T. E., Marsh, D. J., St. Marie, L., Bernstein, I. L., and Palmiter, R. D. (1998). Ethanol consumption and resistance are inversely related to neuropeptide Y levels. *Nature* **396**, 366–369.

Thierry, A. M., Chevalier, G., Ferron, A., and Glowinski, J. (1983). Diencephalic and mesencephalic efferents of the medial prefrontal cortex in the rat: electrophysiological evidence for the existence of branched axons. *Experimental Brain Research* **50**, 275–282.

Thomas, M. J., and Malenka, R. C. (2003). Synaptic plasticity in the mesolimbic dopamine system. *Transactions of the Royal Society of London B Biological Sciences* **358**, 815–819.

Ungless, M. A., Whistler, J. L., Malenka, R. C., and Bonci, A. (2001). Single cocaine exposure in vivo induces long-term potentiation in dopamine neurons. *Nature* **411**, 583–587.

Valdez, G. R., and Koob, G. F. (2004). Allostasis and dysregulation of corticotropin-releasing factor and neuropeptide Y systems: implications for the development of alcoholism. *Pharmacology Biochemistry and Behavior* **79**, 671–689.

Valdez, G. R., Zorrilla, E. P., Roberts, A. J., and Koob, G. F. (2003). Antagonism of corticotropin-releasing factor attenuates the enhanced responsiveness to stress observed during protracted ethanol abstinence. *Alcohol* **29**, 55–60.

Vanderschuren, L. J., and Everitt, B. J. (2004). Drug seeking becomes compulsive after prolonged cocaine self-administration. *Science* **305**, 1017–1019.

Vanderschuren, L. J., and Kalivas, P. W. (2000). Alterations in dopaminergic and glutamatergic transmission in the induction and expression of behavioral sensitization: a critical review of preclinical studies. *Psychopharmacology* **151**, 99–120.

Vanderschuren, L. J., Schmidt, E. D., De Vries, T. J., Van Moorsel, C. A., Tilders, F. J., and Schoffelmeer, A. N. (1999). A single exposure to amphetamine is sufficient to induce long-term behavioral, neuroendocrine, and neurochemical sensitization in rats. *Journal of Neuroscience* **19**, 9579–9586.

Vezina, P. (2004). Sensitization of midbrain dopamine neuron reactivity and the self-administration of psychomotor stimulant drugs. *Neuroscience and Biobehavioral Reviews* **27**, 827–839.

Vezina, P., Lorrain, D. S., Arnold, G. M., Austin, J. D., and Suto, N. (2002). Sensitization of midbrain dopamine neuron reactivity promotes the pursuit of amphetamine. *Journal of Neuroscience* **22**, 4654–4662.

Vezzani, A., Sperk, G., and Colmers, W. F. (1999). Neuropeptide Y: Emerging evidence for a functional role in seizure modulation. *Trends in Neurosciences* **22**, 25–30.

Villegier, A. S., Drouin, C., Bizot, J. C., Marien, M., Glowinski, J., Colpaert, F., and Tassin, J. P. (2003). Stimulation of postsynaptic alpha1b- and alpha2-adrenergic receptors amplifies dopamine-mediated locomotor activity in both rats and mice. *Synapse* **50**, 277–284.

Volkow, N. D., Fowler, J. S., Wang, G. J., Hitzemann, R., Logan, J., Schlyer, D. J., Dewey, S. L., and Wolf, A. P. (1993). Decreased dopamine D2 receptor availability is associated with reduced frontal metabolism in cocaine abusers. *Synapse* **14**, 169–177.

Volkow, N. D., Wang, G. J., Fowler, J. S., Logan, J., Gatley, S. J., Hitzemann, R., Chen, A. D., Dewey, S. L., and Pappas, N. (1997). Decreased striatal dopaminergic responsiveness in detoxified cocaine-dependent subjects. *Nature* **386**, 830–833.

Volkow, N. D., Wang, G. J., Fowler, J. S., Hitzemann, R., Angrist, B., Gatley, S. J., Logan, J., Ding, Y. S., and Pappas., N. (1999). Association of methylphenidate-induced craving with changes in right striato-orbitofrontal metabolism in cocaine abusers: Implications in addiction. *American Journal of Psychiatry* **156**, 19–26.

Volkow, N. D., Fowler, J. S., and Wang, G. J. (2003). The addicted human brain: insights from imaging studies. *Journal of Clinical Investigation* **111**, 1444–1451.

Volkow, N. D., Wang, G. J., Ma, J., Fowler, J. S., Wong, C., Logan, J., and Ding, Y. S. (2004). Activation of the rectal gyrus by intravenous methylphenidate in cocaine addicted subjects but not in controls. *Neuropsychopharmacology* **29**(Suppl. 1), s2.

Vorel, S. R., Liu, X., Hayes, R. J., Spector, J. A., and Gardner, E. L. (2001). Relapse to cocaine-seeking after hippocampal theta burst stimulation. *Science* **292**, 1175–1178.

Vorel, S. R., Ashby, C. R. Jr., Paul, M., Liu, X., Hayes, R., Hagan, J. J., Middlemiss, D. N., Stemp, G., and Gardner, E. L. (2002). Dopamine D3 receptor antagonism inhibits cocaine-seeking and cocaine-enhanced brain reward in rats. *Journal of Neuroscience* **22**, 9595–9603.

Wahlestedt, C., Karoum, F., Jaskiw, G., Wyatt, R. J., Larhammar, D., Ekman, R., and Reis, D. J. (1991). Cocaine-induced reduction of brain neuropeptide Y synthesis dependent on medial prefrontal cortex. *Proceedings of the National Academy of Sciences USA* **88**, 2078–2082.

Walker, J. R., Chen, S. A., Moffitt, H., Inturrisi, C. E., and Koob, G. F. (2003). Chronic opioid exposure produces increased heroin self-administration in rats. *Pharmacology Biochemistry and Behavior* **75**, 349–354.

Wang, C. C., Willis, W. D., and Westlund, K. N. (1999). Ascending projections from the area around the spinal cord central canal: a *Phaseolus vulgaris* leucoagglutinin study in rats. *Journal of Comparative Neurology* **415**, 341–367.

Weddington, W. W. Jr., Brown, B. S., Haertzen, C. A., Hess, J. M., Mahaffey, J. R., Kolar, A. F., and Jaffe, J. H. (1991). Comparison of amantadine and desipramine combined with psychotherapy for treatment of cocaine dependence. *American Journal of Drug and Alcohol Abuse* **17**, 137–152.

Weissenborn, R., Robbins, T. W., and Everitt, B. J. (1997). Effects of medial prefrontal or anterior cingulate cortex lesions on responding for cocaine under fixed-ratio and second-order schedules of reinforcement in rats. *Psychopharmacology* **134**, 242–257.

Wikler, A. (1973). Dynamics of drug dependence: Implications of a conditioning theory for research and treatment. *Archives of General Psychiatry* **28**, 611–616.

Wilson, K. G., Mikail, S. F., D'Eon, J. L., and Minns, J. E. (2001). Alternative diagnostic criteria for major depressive disorder in patients with chronic pain. *Pain* **91**, 227–234.

Wise, R. A. (1980). Action of drugs of abuse on brain reward systems. *Pharmacology Biochemistry and Behavior* **13**(Suppl. 1), 213–223.

Wise, R. A. (2002). Brain reward circuitry: insights from unsensed incentives. *Neuron* **36**, 229–240.

Wise, R. A., and Bozarth, M. A. (1987). A psychomotor stimulant theory of addiction. *Psychological Review* **94**, 469–492.

Woldbye, D. P., Klemp K., and Madsen, T. M. (1998). Neuropeptide Y attenuates naloxone-precipitated morphine withdrawal via Y5-like receptors. *Journal of Pharmacology and Experimental Therapeutics* **284**, 633–636.

World Health Organization (1992). *International Statistical Classification of Diseases and Related Health Problems*, 10th revision, World Health Organization, Geneva.

Wyvell, C. L., and Berridge, K. C. (2001). Incentive sensitization by previous amphetamine exposure: increased cue-triggered 'wanting' for sucrose reward. *Journal of Neuroscience* **21**, 7831–7840.

Zhou, Y., Spangler, R., LaForge, K. S., Maggos, C. E., Ho, A., and Kreek, M. J. (1996). Corticotropin-releasing factor and type 1 corticotropin-releasing factor receptor messenger, RNAs in rat brain and pituitary during 'binge'-pattern cocaine administration and chronic withdrawal. *Journal of Pharmacology and Experimental Therapeutics* **279**, 351–358.

Zhou, Y., Spangler, R., Ho, A., and Kreek, M. J. (2003). Increased, C. R.H mRNA levels in the rat amygdala during short-term withdrawal from chronic 'binge' cocaine. *Molecular Brain Research* **114**, 73–79.

APPENDIX

1

Psychostimulants

From: Wesson, D.R., and Smith, D.E. (1977). Cocaine: its use for central nervous system stimulation including recreational and medical uses. ***In Cocaine: 1977*****, (series title:** ***NIDA Research Monograph*****, vol. 13), (R.C. Petersen, and R.C. Stillman, Eds.), pp. 137–152. National Institute on Drug Abuse, Rockville MD.**

A 24-year-old white female worked as a secretary and had no history of significant prior depression. She periodically used cocaine by the intranasal route, and, at one point, was given an unusually large quantity of cocaine. (The material was professionally analyzed by a street drug analysis laboratory in the San Francisco Bay Area, which found it to contain 93 percent cocaine.) After 'snorting' five to eight lines of the cocaine per night for several days, she began waking up feeling depressed. To overcome this depression, and go to work, she snorted one to two lines of cocaine in the morning. By the end of the second week, she was progressively developing severe anxiety, depression and increasing irritability which was interfering with her interpersonal relationships. Her concern over this drug-induced depression and anxiety faded in approximately two days and she re-established her usual level of positive affect and mood. Since this occurrence, she periodically uses cocaine in social-recreational settings; however, she is careful to keep her dosage at a low enough level to avoid recurrence of this drug-induced depression.

From: Wesson, D.R., and Smith, D.E. (1977). Cocaine: its use for central nervous system stimulation including recreational and medical uses. ***In Cocaine: 1977*****, (series title:** ***NIDA Research Monograph*****, vol. 13), (R.C. Petersen, and R.C. Stillman, Eds.), pp. 137–152. National Institute on Drug Abuse, Rockville MD.**

A 19-year-old white male had been experimenting with a variety of drugs, including snorting cocaine for approximately two years. He had used cocaine exclusively in a recreational setting and indicated that he found nothing but pleasure in the drug experience and never had any problem, nor had he escalated his dose. One day a group of friends were injecting cocaine and they persuaded him to try this route of administration. As he had had no difficulty with cocaine previously, and as a result of curiosity and peer group pressure, he decided that he would experiment with injection. Following the intense stimulation and rush he became acutely anxious and frightened. Upon arrival at the Medical Section of the Haight-Ashbury Free Medical Clinic, he was found to have a very rapid pulse rate as well as a hyperventilation administered anxiety [sic]. He was treated with 10 mg of i.v. Valium(R) administered slowly with reassurance. To cease the carpopedal spasms he had developed as a consequence of his hyperventilation syndrome, he was told to breathe into a paper bag which increased his carbon dioxide levels. The acute anxiety and its subsequent sequelae faded in approximately three hours. Follow-up indicated no recurrences or further experimentation with intravenous injection of cocaine by this individual. This use of intravenous sedative hypnotic medication is controversial and some critics of this approach use oral medication only while others stress reassurance alone without medication. We would recommend intravenous Valium(R) only after nonpharmacological intervention has failed.

From: Wesson, D.R., and Smith, D.E. (1977). Cocaine: its use for central nervous system stimulation including recreational and medical uses. ***In Cocaine: 1977*****, (series title:** ***NIDA Research Monograph*****, vol. 13), (R.C. Petersen, and R.C. Stillman, Eds.), pp. 137–152. National Institute on Drug Abuse, Rockville MD.**

A 31-year-old white male law student in his fourth year of law school had a long history of experimental drug use including alcohol (his first drug), marihuana and LSD; but at no time had he abused a psychoactive drug. Approximately two years ago he was introduced to cocaine in a social setting by a group of friends and fellow law students. He became a regular recreational user of cocaine and in a social setting during an evening would chop up and snort between 10 and 20 lines of cocaine in the usual fashion. (Often, as with this case, cocaine is used in a recreational setting along with alcohol and marihuana.) With this law student, the pattern of recreational cocaine use continued for some time, but moved to a more daily pattern when he found that the inhalation of cocaine stimulated his performance and ability to study at night, something he found desirable because he had begun to prepare for the bar examinations. One evening, a female friend with whom he was periodically having sexual relations produced a needle and syringe and indicated that the injection of cocaine produced a pleasurable, orgasmic-like 'rush'. The law student injected the cocaine simultaneously with his female sexual acquaintance and found the orgasmic 'rush' quite desirable. Over a several-month period he escalated his intravenous cocaine use on a daily basis, injecting from approximately 10 p.m. until 7 a.m., on a 15 minute to 1 hour repeated schedule, using approximately 2 g of cocaine per night. Despite the fact that the law student was independently wealthy as a result of a family inheritance, he found that he was rapidly consuming his inheritance as his cocaine habit was costing him $50–150 per day. As a consequence he began dealing cocaine to his friends in order to help support his own habit. While the injection of cocaine involved both male and female figures, he would almost invariably inject with a woman in a sexual context, although he reported that as he became more deeply involved with cocaine, his libido dropped dramatically; for both he and his female sexual partners, the orgasmic effects of the cocaine injection became a substitute for actual sexual intercourse. One evening he injected a female friend in his usual fashion (he would first inject the woman and then himself). She suddenly had a series of seizures, became comatose, required mouth-to-mouth resuscitation and was subsequently transported to an emergency room. During this particular cocaine run, he also experienced the first evidence of a cocaine psychosis, with auditory and visual hallucinations and extreme paranoia. The negative effects both to himself and to his girlfriend were quite shocking because he had believed cocaine to be as free of adverse consequences as marihuana. Because of these two episodes, he decided to quit cocaine use and seek treatment. During the 'withdrawal period' he experienced difficulty sleeping and a severe drug-induced depression associated with anxiety that lasted for approximately one week. Most depressive symptoms gradually abated; however, the anxiety continued along with an urge to use cocaine late in the evening at the time for his previous cocaine runs. To help with the anxiety, depression and sleep disorder, 10 mg of Valium(R) p.o, was administered each night. As there was no evidence of a prolonged underlying depression which preceded the cocaine abuse or that lasted following the 'fade out' period of the drug-induced depression, no tricyclic antidepressants were administered. He made a decision to self-medicate the lethargy and reactive depression with the intranasal use of cocaine which he resumed on a daily basis. He expressed great surprise at the toxic effects of cocaine, but was also quite ambivalent about whether he would completely discontinue cocaine.

From: Wesson, D.R., and Smith, D.E. (1977). Cocaine: its use for central nervous system stimulation including recreational and medical uses. *In Cocaine: 1977*, (series title: *NIDA Research Monograph*, vol. 13), (R.C. Petersen, and R.C. Stillman, Eds.), pp. 137–152. National Institute on Drug Abuse, Rockville MD.
A 35-year-old white female architect intermittently snorted cocaine in a social-recreational setting, but was introduced to a mixture of cocaine and heroin by a rock musician. She found the heroin prolonged and altered the effects of cocaine which she snorted, and produced an encouraging drug effect which was somewhat like an 'acid trip'. She snorted this mixture of cocaine and heroin each evening for approximately two weeks, but found that it was disrupting her work, producing a drug-induced depression and raising concerns in her mind about becoming addicted to heroin since she found the combination so pleasurable. As a result, she discontinued her use and had a moderate drug-induced depression which she thought might be a mild abstinence syndrome. Neither required treatment or medical intervention. Even when examining what appear to be specific drug patterns, it cannot be assumed that the desired effects are the same for all individuals.

From: Wesson, D.R., and Smith, D.E. (1977). Cocaine: its use for central nervous system stimulation including recreational and medical uses. *In Cocaine: 1977*, (series title: *NIDA Research Monograph*, vol. 13), (R.C. Petersen, and R.C. Stillman, Eds.), pp. 137–152. National Institute on Drug Abuse, Rockville MD.
Interviews were done with 15 high-dose intravenous cocaine and heroin abusers in a notorious drug dealing house in the Mission District of San Francisco. Of the 15 individuals, 4 indicated that intravenous injection of cocaine was the drug of choice because of the 'orgasmic rush' they received when they injected, and the fact that it stimulated them to stay active. They mixed cocaine with heroin in order to calm some of its side-effects and to enhance the pleasurable

effects of the cocaine. The other 11 individuals indicated that the nodding, stoned effect of heroin was much more attractive to them and they did not like the 'speedy' effect of the cocaine. They used cocaine combined with heroin only occasionally to have some diversity in their drug experience. Despite the fact that all 15 had experienced adverse drug effects with both cocaine and heroin, none had any interest in stopping their intravenous drug abuse.

From: Khantzian, E.J. (2003). Understanding addictive vulnerability: an evolving psychodynamic perspective. ***Neuro-Psychoanalysis*** **5, 5–21.**
The following case illustrates many of the themes that this report highlights. It is apparent that the patient who is presented suffered with intense and extreme emotions. He also suffered with self-esteem problems and difficult interpersonal relations. His impulsive and risky behaviors also revealed poor self-care in that he invariably failed to anticipate the consequences of the dangers of his actions and activities. The case also reveals the short-term adaptive effects of addictive drugs and how thay can relieve emotional suffering as well as counter certain resticting personality characteristics.

Arnold: When Arnold first came to see me he was 29 years old, and he had just started treatment in a methadone maintenance program. I followed him in psychotherapy over the subsequent ten years, during which time he ultimately achieved abstinence from his dependency on all narcotics. Following him over this period provided the opportunity to appreciate the nature of his suffering and his personality organization. The long-term therapy also allowed him to be observed both during times when he was actively using, as well as an extended time when he abstained from all drugs, thus making it possible to compare how the drugs affected him, as well as how he behaved and reacted when he was free of drugs. The opiates had become his drug-of-choice over the past five years before he came to see me, but, starting in early adolescence, he heavily used and abused sedatives and stimulants (amphetamines). His long-standing shaky self-esteem and tenuous capacity to relate to others lent a self-effacing and reticent quality to his interpersonal dealings, including the way he related to me in his psychotherapy. These qualities gave an appealing and likeable aspect to his personality, combining elements of charm and vulnerability. Nevertheless, by history, and based on his day-to-day encounters and activities reported in therapy, it was clear he had another side. Arnold could be ruthless and sadistic and disavow any need for help and care from others. In contrast to his generally gentle and solicitous veneer, he revealed a penchant to be aggressive if not violent as he disclosed his keen interest in active and risky involvements in athletics, martial arts, and speedy high-performance motorcycles and automobiles. As a result of his violent and risky behaviors he frequently suffered injuries of various sorts, many of which I witnessed when he came for his psychotherapy. He was often quick to play down or dismiss with bravado the nature or seriousness of his injuries.

In his adolescence he used both depressants and amphetamines to overcome his shyness and restricted emotional life. The depressants would disinhibit him, and he could more easily relate to his peers. He said that the amphetamines made him feel powerful and helped him to overcome feelings of vulnerability and weakness in social situations and contact sports. As he continued to use amphetamines, he realized that they also helped to counter his low self-esteem and inertia, aspects of which indicated a long-standing depression. The progressive reliance on the stimulants empowered him and frequently resulted in brutal, punishing fights. As time went on, however, he realized that amphetamines caused him enormous dysphoria and fear, and especially how much they heightened his sadism, whether this involved beating up a person or an emerging cruelty to his pet cats. As he approached his mid-twenties and the uncontrollable violence and rage was interfering with his friendships, work, and life in general he discovered and subsequently became dependent on heroin. In contrast to the amphetamines and sedatives, he was immediately impressed with the containing and calming effect of narcotics; he was aware of a marked diminution in his rage and aggressivity, and he felt more organized, in control, and able to work.

In the course of his therapy, he and I better appreciated how much the extremes and flip-flopping of his emotions derived from his growing-up years. During those years he was both shamed and devalued, being subjected to verbal and physical abuse by his mother, a person who apparently had her own problems with aggressivity and impulse control.

H.C. William Stewart Halsted, 1852–1922 (1922) ***Science*****, 56, 461–464.**
Professor Halsted, certainly one of the most cultivated, and regarded by many as the most eminent surgeon of his time, in view of the character of his contributions, died at noon on Thursday, the seventh of September, in the Johns Hopkins Hospital, of which he had been surgeon-in-chief since soon after its opening. At that time, in 1889, neither he nor his clinical colleagues, Osler and Kelly, had as yet turned forty.

A man of unique personality, shy, something of a recluse, fastidious in his tastes and in his friendships, an aristocrat in his breeding, scholarly in his habits, the victim for many years of indifferent health, he nevertheless was one of the few American surgeons who may be considered to have established a school of surgery, comparable, in a sense, to the school of Billroth in Vienna. He had few of the qualities supposed to accompany what the world regards as a successful surgeon. Over-modest about his work, indifferent to matters of priority, caring little for the gregarious gatherings of medical men, unassuming, having little interest in private practice, he spent his medical life avoiding patients—even students, when this was possible—and, when health permitted, working in clinic and laboratory at the solution of a succession of problems which aroused his interest. He had that rare from of imagination which sees problems, and the technical ability combined with persistence which enabled him to attack them with promise of a successful issue. Many of his contributions, not only to his craft but to the science of medicine in general, were fundamental in character and of enduring importance.

During his last few years in New York he undertook an anatomico-surgical investigation on the anaesthetizing effect of the then little-known and newly introduced drug, cocaine. In this research, which had been begun in 1885, he was the first to utilize for surgical purposes the principle of nerve blocking, and was accustomed to demonstrate to dentists how painless extractions or even more extensive operations on the jaws might thus be carried out. He was the first, also, at this time, to demonstrate spinal anaesthesia by introducing the drug into the lumbar meninges. In the course of these studies he used himself as a subject, injecting his own peripheral nerves in order to map out the areas of anaesthesia, and, unaware of the danger he was running, contracted an habituation to the drug, from which, with the help of a devoted professional friend, he effectually broke himself.

It was natural enough that cocaine was subsequently abhorred by him, and after Schleich's solution came to be generally employed as a local anaesthetic he usually preferred to infiltrate with salt solution alone, which has certain anaesthetizing properties, rather than use even the diluted drug. Fifteen years later when the writer of this note, as Dr. Halsted's resident surgeon, stumbled anew upon the principle of nerve blocking for operations on hernia and published a paper on the subject, he was utterly unaware that his chief had ever made studies with cocaine of any sort, so reticent was he about this particular matter and so little did questions of priority interest him. It has remained for the dentists to call attention to his original work on regional anaesthesia, and a few months before his death they made due public acknowlegment of what Dr. Halsted himself had never laid claim to, and the knowledge of which he had even withheld, at least until recent years, from his house officers.

From: Crowe, S. J. (1957). ***Halsted of Johns Hopkins: The Man and this Men,*** **Charles C. Thomas, Springfield, IL. p. 29.**

Doctor Halsted and the others who injected their own nerves in the preliminary experiments with cocaine unfortunately but understandably fell victims to the 'shadow side' of the drug. That Halsted had the perseverance and character to control and eventually overcome this addiction and to return to a splendid productive life is one of his crowning glories. It required about a year in a hospital in Providence, Rhode Island, for Halsted to control his craving for the drug.

From: Ageyev, M. (1983) [translated by Heim, M.H. (1998)]. ***Novel with Cocaine.*** **Northwestern University Press, Evanston IL.**

During the long nights and long days I spent under the influence of cocaine in Yag's room I came to see that what counts in life is not the events that surround one but the reflection of those events in one's consciousness. Events may change, but insofar as the changes are not reflected in one's consciousness their result is nil. Thus, for example, a man basking in the aura of his riches will continue to feel himself a millionaire so long as he is unaware that the bank where he keeps his capital has gone under; a man basking in the aura of his offspring will continue to feel himself a father until he learns that his child has been run over. Man lives not by the events surrounding him, therefore, but by the reflection of those events in his consciousness.

All of the man's life—his work, his deeds, his will, his physical and mental prowess—is completely and utterly devoted to, and fixed on, bringing about one or another event in the external world, though not so much to experience the event in itself as to experience the reflection of the event on his consciousness. And if, to take it all a step further, everything a man does he does to bring about only those events which, when reflected in his consciousness, will make him feel happiness and joy, then what he spontaneously reveals thereby is nothing less than the basic mechanism behind his life and the life of every man, evil and cruel or good and kind.

One man does everything in his power to overthrow the tsar, another to overthrow the revolutionary junta; one man wishes to strike it rich, another gives his fortune to the poor. Yet what do these contrasts show but the diversity of human activity, which serves at best (and not in every case) as a kind of individual personality index. The *reason* behind human activity, as diverse as that activity may be, is always one: man's need to bring about events in the external world which, when reflected in his consciousness, will make him feel happiness.

So it was in my insignificant life as well. The road to the external event was well marked: I wished to become a rich and famous lawyer. It would seem I had only to take the road and follow it to the end, especially since I had much to recommend me (or so I tried to convince myself). But oddly enough, the more time I spent making my way towards the cherished goal, the more often I would stretch out on the couch in my dark room and imagine I was what I intended to become, my penchant for sloth and reverie persuading me that there was no point in laying out so great an expenditure of time and energy to bring the external events to fruition when my happiness would be all the stronger if the events leading up to it came about rapidly and unexpectedly.

But such was the force of habit that even in my dreams of happiness I thought chiefly of the event rather than the feeling of happiness, certain that the event (should it but occur) would lead to the happiness I desired. I was incapable of divorcing the two.

The problem was that before I first came in contact with cocaine I assumed that happiness was an entity, while in fact all human happiness consists of a clever fusion of two elements: (1) the physical feeling of happiness, and (2) the external event providing the psychic impetus for that feeling. Not until I first tried cocaine did I see the light; not until then did I see that the external event I had dreamed of bringing about—the result I had been slaving day and night for and yet might never manage to achieve—the external event was essential only insofar as I needed its reflection to make me feel happy. What if, as I was convinced, a tiny speck of cocaine could provide my organism with instantaneous happiness on a scale I had never dreamed of before? Then the need for any event whatever disappeared and, with it, the need for expending great amounts of work, time, and energy to bring it about.

Therein lay the power of cocaine—in its ability to produce a feeling of physical happiness psychically independent of all external events, even when the reflection of the events in my consciousness would otherwise have produced feelings of grief, depression, and despair. And it was that property of the drug that exerted so terribly strong an attraction on me that I neither could nor would oppose or resist it. The only way I could have done so was if the feeling of happiness had come less from bringing about the external event than from the work, the effort, the energy invested in bringing it about. But that was a kind of happiness I had never known.

Of course, everything I have said thus far about cocaine must be understood only as the opinion of someone who has only just begun to take the drug and not as a general statement. The neophyte does indeed believe that the main property of cocaine is its ability to make him feel happy, much as the mouse, before it is caught, believes that the main property of mousetraps is to provide him with lard.

The most awful aftereffect of cocaine, and one that followed the hours of euphoria without fail, was the agonizing reaction which doctors call depression and which took hold of me the instant I finished the last grain of powder. It would go on for what seemed an eternity–though by the clock it was only three or four hours–and consisted of the deepest, darkest misery imaginable. True, my mind knew that it would be over in a few hours, but my body could not believe it.

It is a well-known fact that the more a person is ruled by his emotions the less capable he is of lucid observation. The feelings I experienced under the spell of cocaine were so potent that my power of self observation dwindled to a state found only in certain of mental illnesses; my 'feeling I' grew to such proportions that my 'self-observing I' all but ceased operation. There being nothing left to bridle my feelings, they poured out with total abandon—in my face, in my movements, in everything I did. But the moment the cocaine was gone and the misery took over, I began to see myself for what I was: indeed, the misery consisted largely in seeing myself as I had been while under the influence of the drug.

Slumped over as if nauseous, the nails of one hand digging into the palm of the other, I recalled every sinister, shameful detail. Standing frozen by the door of Yag's silent room at night, trembling with the idiotic but insuperable fear that someone was creeping along the passageway about to burst in on me and peer into my frightful eyes. And stealing slowly, ever so slowly, up to the dark, blindless window, certain that the moment I turned my back someone would glower at me through it, yet perfectly aware it was on the second floor. And turning off the lamp and its almost audible glare, which seemed to invite intruders. And lying on the couch, straining my neck to keep my head from touching the pillow and waking the entire house with the racket. And staring into the vibrating red darkness, my eyes aching with terror at the imminent prospect of being gouged. And striking match after match, my hand so numb with cold and horror that the effort appeared doomed until, after a long hiss, one would indeed take fire, and my body recoiled as the match dropped to the couch. And pulling myself up every ten minutes for a new fix, feeling for the packet, scraping the cocaine with the dull end of a steel pen-point, though even when quivering directly beneath my nostril (lifted there in the dark by a hand growing scrawnier with every night) the pen-point failed to give me my sniff: still wet from the last time round, it had moistened the cocaine, which then hardened, and all that came through was an acrid smell of rust. And answering countless calls of the bladder, forcing myself each time to overcome the panic-stricken immobility of my body and use the chamber pot there in the room, gritting my chattering teeth as I listened to the monstrous sound I made for all the building to hear, and then, sticky from a particularly pungent, fetid sweat, climbing back on the frozen mountain of a couch and falling into a state of stupefaction until the next urgent call roused me. And watching day break and objects take on shapes again, a process that did not in the least relax the muscles, which, longing for the protective covering of night and shunning the light that exposes face and eyes, contracted even more. And licking off the rusty pen-point by the morning light, delighting in the dry rush of a fresh fix from a new packet—the slight dizzy spell, the nausea-cum-bliss—then grieving at the first sounds of people in other flats awakening. And, finally, the knock at the door, returning at long, rhythmic intervals, a cough, which, though it racked the body, was necessary to dislodge the tongue, and then my voice quaking with happiness (despite the anguish) as I muttered through my teeth, 'Who's there? What do you want? Who's there?' and suddenly another knock, insistent, implacable, but from a new direction: it was the sound of wood being chopped in the courtyard.

Each time I came to the end of a session I would have visions, fanciful reconstructions of what I had just been through, how I had looked and behaved, and with the visions grew the certainty that soon, very soon—if not tomorrow then next week, if not next week then next month or year—I would end up in an insane asylum. And yet I kept increasing the dose, taking as many as three and a half grams and prolonging the effect for periods of over twenty-four hours. On the one hand, I had an insatiable desire for the drug; on the other, I merely desired to postpone its even more ominous aftereffects.

Whether because I had increased the dosage or because any poison gives the organism a rude shock—and perhaps for both reasons—the shell my cocaine bliss presented to the world eventually began to crumble. I was possessed by the strangest of manias within an hour of the first sniff. Sometimes I would run out of matches and start searching for another box, moving the furniture away from the walls, emptying out all the drawers, carrying on with great pleasure for hours at a stretch, yet knowing all the while there was not a match in the room; at other times I would be obsessed with some dire apprehension, yet have no idea what or whom I so dreaded, and crouch in abject fear, crouch—again, for hours—by the door, torn between the unbearable need for a new fix, which meant going back to the couch, and the terrible risk involved in leaving the door unguarded for even a moment; at yet other times—and these had begun to grow in frequency lately—I would be set upon by all my manias at once, and then my nerves would be strained to the breaking point.

One night, while everyone was asleep and I had my ear to the door keeping vigil, I heard a sudden resonant noise—the sort of bang one sometimes hears in the night—followed immediately by a long wail. It was a moment or two before I realized that it was I who had wailed and my hand that was clamped over my mouth.

Here end the notes of Vadim Maslennikov or, rather, here they break off. Maslennikov was brought, delirious, to our hospital during the terrible frosts of January 1919. Once he had regained consciousness and we could make a preliminary examination, we learned from the patient that he was a cocaine addict, that he had tried many times to break the habit, and that by dint of great effort he had succeeded in doing so for a month or two at a time. In the end, however, he always returned to it. According to his own confession, addiction had grown more painful of late, tending to irritate the psychic apparatus rather then exhilarate it. To be more explicit, if during the early stages of addiction cocaine promoted precision and acuity of consciousness, it now elicited incoherence, which, when coupled with a concomitant sense of anxiety, tended to produce hallucinations.

When the head physician asked him why he kept returning to the drug when he knew in advance that, no matter what the dosage, it would subject him to psychic torture, Maslennikov compared his state to Gogol's. Like Gogol, he said in a trembling voice, like Gogol, who, while working on the second part of *dead souls,* knew that the creative forces of his earlier years had dried up, that every attempt to revive them moved him further from them, yet returned day after day to the tortures of creation, (for without the euphoria, the combustion of creation life had no meaning for him)—like Gogol, he Maslennikov, continued to succumb to his obsession even though it promised him nothing but despair.

Maslennikov exhibited all the symptoms of addiction: extensive damage to the intestinal tract, debility, chronic insomnia, apathy, cachexia, jaundice, and a number of nervous disorders, apparently of psychic origin, the precise diagnosis of which required further observation.

Clearly it made no sense to keep such a patient in a military hospital such as ours. He was informed of his imminent discharge by our chief physician, an extraordinarily kind man, who, obviously upset at not being able to help him, added that what he, Maslennikov, needed was not so much a hospital as a good psychiatric sanatorium. But gaining admission to such a sanatorium would not be easy. In our new socialist era admission depended less on illnesses presented than on services rendered to the Revolution or, failing same, the likelihood of services to be rendered in the future.

Maslennikov listened morosely, a swollen eyelid giving his face a sinister cast. When asked solicitously by the head physician whether he had any relatives or friends who might be of use, have connections, he answered he had not. After a long pause he added that his mother was deceased and that his old nanny, who had made heroic sacrifices to help him through these difficult times, was herself badly in need of help. The only friends he could name were several former classmates: a certain Stein, who had recently left the country, and two others, whose whereabouts he no longer knew, namely, Yegorov and Burkewitz.

The moment he mentioned the name Burkewitz everyone present exchanged glances.

'Comrade Burkewitz?' the head physician asked. 'Why, he's our direct superior. One word from him and you're saved.'

Maslennikov asked us all kinds of questions, apparently worried it was all a misunderstanding or a case of someone with the same name. He was extremely agitated and, I think, pleased when we convinced him that our Comrade Burkewitz was indeed the one he knew. The head physician informed him that the department run by Comrade Burkewitz was located in the same street as the hospital but that, since there was little chance of his finding anyone in at so late an hour, he would have to wait until the next morning to see him. He then offered Maslennikov a bed for the night, but Maslennikov declined and left the hospital.

Next morning, shortly after eleven, three errand boys from Comrade Burkewitz's department carried him in. It was too late to save him. All we could do was establish that death was due to cardiac arrest from acute cocaine poisoning and that it was unmistakably premeditated: the drug had been dissolved in water and swallowed.

The following were found in the inner breast pocket of his jacket: (1) a small calico pouch with ten silver five-kopeck pieces sewn into it, and (2) a manuscript with two words scribbled in large, jittery letters on the front page: *Burkewitz refuses*.

APPENDIX

2

Opioids

From: De Quincey, T. (1986). *Confessions of an English Opium-Eater.* Penguin Books, New York [first appeared in *London Magazine,* October 1821 – December 1822].

It is so long since I first took opium, that if it had been a trifling incident in my life, I might have forgotten its date: but cardinal events are not to be forgotten; and from circumstances connected with it, I remember that it must be referred to the autumn of 1804. During that season I was in London, having come thither for the first time since my entrance at college. And my introduction to opium arose in the following way. From an early age I had been accustomed to wash my head in cold water at least once a day: being suddenly seized with toothache, I attributed it to some relaxation caused by an accidental intermission of that practice; jumped out of bed: plunged my head into a basin of cold water; and with hair thus wetted went to sleep. The next morning, as I need hardly say, I awoke with excruciating rheumatic pains of the head and face, from which I had hardly any respite for about twenty days. On the twenty-first day, I think it was, and on a Sunday, that I went out into the streets; rather to run away, if possible from my torments, than with any distinct purpose. By accident I met a college acquaintance who recommended opium. Opium! dread agent of unimaginable pleasure and pain! I had heard of it as I had of manna or of ambrosia, but no further: how unmeaning a sound was it at that time! what solemn chords does it now strike upon my heart! what heart-quaking vibrations of sad and happy remembrances! Reverting for a moment to these, I feel a mystic importance attached to the minutest circumstances connected with the place and the time, and the man (if man he was) that first laid open to me the paradise of opium-eaters. It was a Sunday afternoon, wet and cheerless: and a duller spectacle this earth of ours has not to show than a rainy Sunday in London.

Arrived at my lodgings, it may be supposed that I lost not a moment in taking the quantity prescribed. I was necessarily ignorant of the whole art and mystery of opium-taking: and, what I took, I took under every disadvantage. But I took it: —and in an hour, oh! heavens! what a revulsion! what an upheaving, from its lowest depths, of the inner spirit! what an apocalypse of the world within me! That my pains had vanished, was now a trifle in my eyes: —this negative effect was swallowed up in the immensity of those positive effects which had opened before me—in the abyss of divine enjoyment thus suddenly revealed. Here was a panacea—a φαgμαχον νηπcενθεζ [a drug to banish grief] for all human woes: here was the secret of happiness, about which philosophers had disputed for so many ages, at once discovered: happiness might now be bought for a penny, and carried in the waistcoat pocket: portable ecstasies might be had corked up in a pint bottle: and peace of mind could be sent down in gallons by the mail coach. But, if I talk in this way, the reader will think I am laughing: and I can assure him, that nobody will laugh long who deals much with opium: its pleasures even are of a grave and solemn complexion; and in his happiest state, the opium-eater cannot present himself in the character of *l'Allegro* even then, he speaks and thinks as becomes *Il Penseroso*. Never-theless, I have a very reprehensible way of jesting at times in the midst of my own misery: and, unless when I am checked by some more powerful feelings, I am afraid I shall be guilty of this indecent practice even in these annals of suffering or enjoyment. The reader must allow a little to my infirm nature in this respect: and with a few indulgences of that sort, I shall endeavour to be as grave, if not drowsy, as fits a theme like opium, so antimercurial as it really is, and so drowsy as it is falsely reputed.

The elevation of spirits produced by opium is necessarily followed by a proportionate depression, and that the natural and even immediate consequence of opium is torpor and stagnation, animal and mental. The first of these errors I shall content myself with simply denying; assuring my reader, that for ten years, during which I took opium at intervals, the day succeeding to that on which I allowed myself this luxury was always a day of unusually good spirits.

Thus I have shown that opium does not, of necessity, produce inactivity or torpor; but that, on the contrary, it often led me into markets and theatres. Yet, in candour, I will admit that markets and theatres are not the appropriate haunts of the opium-eater, when in the divinest state incident to his enjoyment. In that state, crowds become an oppression to him; music even, too sensual and gross. He naturally seeks solitude and silence, as indispensable conditions of those trances, or profoundest reveries, which are the crown and consummation of what opium can do for human nature.

One day a Malay knocked at my door. What business a Malay could have to transact amongst English mountains, I cannot conjecture: but possibly he was on his road to a sea-port about forty miles distant.

He lay down upon the floor for about an hour, and then pursued his journey. On his departure, I presented him with a piece of opium. To him, as an Orientalist, I concluded that opium must be familiar: and the expression of his face convinced me that it was. Nevertheless, I was struck with some little consternation when I saw him suddenly raise his hand to his mouth, and (in the school-boy phrase) bolt the whole, divided into three pieces, at one mouthful. The quantity was enough to kill three dragoons and their horses: and I felt some alarm for the poor creature: but what could be done? I had given him the opium in compassion for his solitary life, on recollecting that if he had travelled on foot from London, it must be nearly three weeks since he could have exchanged a thought with any human being. I could not think of violating the laws of hospitality, by having him seized and drenched with an emetic, and thus frightening him into a notion that we were going to sacrifice him to some English idol. No: there was clearly no help for it: —he took his leave: and for some days I felt anxious—but as I never heard of any Malay being found dead, I became convinced that he was used to opium: and that I must have done him the service I designed, by giving him one night of respite from the pains of wandering.

However, as some people, in spite of all laws to the contrary, will persist in asking what became of the opium-eater, and in what state he now is, I answer for him thus: The reader is aware that opium had long ceased to found its empire on spells of pleasure; it was solely by the tortures connected with the attempt to abjure it, that it kept its hold.

I saw that I must die if I continued the opium: I determined, therefore, if that should be required, to die in throwing it off. How much I was at that time taking I cannot say; for the opium which I used had been purchased for me by a friend who afterwards refused to let me pay him; so that I could not ascertain even what quantity I had used within the year. I apprehend, however, that I took it very irregularly; and that I varied from about fifty or sixty grains, to 150 a-day. My first task was to reduce it to forty, to thirty, and, as fast as I could, to twelve grains.

I triumphed: but think not, reader, that therefore my sufferings were ended; nor think of me as of one sitting in a *dejected* state. Think of me as of one, even when four months had passed still agitated, writhing, throbbing, palpitating, shattered.

APPENDIX

3

Alcohol

From: Yamamoto, H., Tanegashima, A., Hosoe, H., and Fukunaga, T. (2000). Fatal acute alcohol intoxication in an ALDH2 heterozygote: a case report, *Forensic Science International* 112, 201–207.

A 25-year-old man was found dead in his bedroom by his brother who came home from his work at about 18:30 on November 30th. There were empty snap-out sheets of flunitrazepam corresponding to 52 tablets in the trash at his bedside. The deceased had telephoned to some party-phone lines, and made an appointment with a lady. Arriving at the place to meet her at around 2:20 in the same day, he was severely hit and kicked by ten boys and girls, all junior high school students. They also took about 50,000 yen from him. He reported the incident to the police. He apparently went home and went to bed at around 9:10. He had attended a mental hospital every second week for the treatment of depression and been given flunitrazepam tablets.

Autopsy was performed about 12 h after death.

The multiple bruises and/or abrasions on the whole body, the slight subdural hemorrhage, and the multiple fractures of the both ribs observed in the deceased were considered to have been caused by an act of violence. None of these injuries, however, could be the responsible cause for his death.

The ethanol concentrations in the peripheral blood, heart blood, and urine, however, were relatively high: 2.00, 1.97, and 2.04 mg/ml, respectively. The blood alcohol concentration found in this case (2.00 mg/ml) [200 mg%] could be attained when about 700 ml of Japanese sake (ethanol concentration: 16% (v/v)) is consumed. It can, therefore, be presumed that he was severely intoxicated when he died. Moreover, because the alcohol concentration of blood was equal to that of urine, it is considered that the deceased died within relatively short time (1–2 h) after drinking.

Fatal blood alcohol concentrations have generally been reported to be in the range 2.25–6.39 mg/ml. Because the blood alcohol concentration in the present case did not reach what is usually considered to be the fatal level, we did not initially conclude that the cause of death was acute alcohol intoxication.

Further consideration and the information about the [alcohol dehydrogenase 2 (ALDH2)] genotype of the deceased, however, did lead to that conclusion. There has been considerable research concerning the individual differences in the sensitivity to alcohol. It has become clear that those who possess *ALDH2*2* gene, homozygously or heterozygously, are more sensitive than people who are homozygous for the active ALDH2 gene (*ALDH2*1/1*). Since the ALDH2 genotype of the deceased in this case was *ALDH2*1/2* (heterozygote), he would have been particularly sensitive to ethanol and should not have drunk the large quantity of it. As there are no established postmortem methods of AcH [acetaldehyde] measurement that account for artifactural formation and the loss after death, allowing estimates of the level while alive to be made, it is certain given the genotype and the ethanol concentration, that the deceased must have had very high AcH levels.

Since there was marked congestion in each organ, darkish red fluid blood without clotting was present in the atria, and subserosal petechial hemorrhages were observed in each organ (the so-called triad for an acute death), it was considered that the deceased died suddenly. No severe pathological changes accounting for the death were observed except for congestion in each of the organs. It is suspected that suppression of the central nervous system by alcohol and hypotension due to dilatation of peripheral blood vessels by a high concentration of AcH led to a functional standstill of the brainstem.

In conclusion, the cause of death was determined to be acute ethanol intoxication including AcH poisoning. There have previously been no reports identifying the genotype of the alcohol metabolizing enzymes as a contributing factor concerning a decision about the cause of death. We showed here that acute alcohol drinking could be fatal in a ALDH2 deficient individual even at a rather moderate blood alcohol concentration.

From: *San Diego Union Tribune*, Thursday, June 19, 1997.

Girl downs quart of liquor on dare, dies

Orland Park, Ill. —A high school cheerleader who downed a quart of 107-proof liquor on a dare passed out and died after she was dropped off at a friend's house to sleep it off.

Authorities said 16-year-old Elizabeth Wakulich might have been saved if someone had taken her to a hospital sooner.

Wakulich was out with friends early Monday when she answered a challenge to drink a bottle of schnapps with an alcohol content of more than 53 percent.

Wakulich was pronounced dead with a blood-alcohol level .38 percent, nearly four times the legal limit for driving in Illinois. The medical examiner estimated her level was closer to .60 shortly after the binge. A level of .40 to .50 can kill.

Associated Press

From: Knapp, C. (1996). *Drinking: A Love Story*. Dial Press, New York.

I Drank.

I drank Fumé Blanc at the Ritz-Carlton Hotel, and I drank double shots of Johnnie Walker Black on the rocks at a dingy Chinese restaurant across the street from my office, and I drank at home. For a long time I drank expensive red wine, and I learned to appreciate the subtle differences between a silky Merlot and a tart Cabernet Sauvignon and a soft, earthy Beaucastel from the south of France, but I never really cared about those nuances because, honestly, they were beside the point. Toward the end I kept two bottles of Cognac in my house: the bottle for show, which I kept on the counter, and the real bottle, which I kept in the back of a cupboard beside an old toaster. The level of liquid in the show bottle was fairly consistent, decreasing by an inch or so, perhaps less, each week. The liquid in the real bottle disappeared quickly, sometimes within days. I was living alone at the time, when I did this, but I did it anyway and it didn't occur to me not to: it was always important to maintain appearances.

I drank when I was happy and I drank when I was anxious and I drank when I was bored and I drank when I was depressed, which was often. I started to raid my parents' liquor cabinet the year my father was dying. He'd be in the back of their house in Cambridge, lying in the hospital bed in their bedroom, and I'd steal into the front hall bathroom and pull out a bottle of Old Grand-dad that I'd hidden behind the toilet. It tasted vile—the bottle must have been fifteen years old—but my father was dying, dying very slowly and gradually from a brain tumor, so I drank it anyway and it helped.

A love story. Yes: this is a love story.

It's about passion, sensual pleasure, deep pulls, lust, fears, yearning hungers. It's about needs so strong they're crippling. It's about saying goodbye to something you can't fathom living without.

I loved the way drink made me feel, and I loved its special power of deflection, its ability to shift my focus away from my own awareness of self and onto something else, something less painful than my own feelings. I loved the sounds of drink: the slide of a cork as it eased out of a wine bottle, the distinct glug-glug of booze pouring into a glass, the clatter of ice cubes in a tumbler. I loved the rituals, the camaraderie of drinking with others, the warming, melting feeling of ease and courage it gave me.

Our introduction was not dramatic; it wasn't love at first sight, I don't even remember my first taste of alcohol. The relationship developed gradually, over many years, time punctuated by separations and reunions. Anyone who's ever shifted from general affection and enthusiasm for a lover to outright obsession knows what I mean: the relationship is just there, occupying a small corner of your heart, and then you wake up one morning and some indefinable tide has turned forever and you can't go back. You *need* it; it's a central part of who you are.

Still, I look in the mirror sometimes and think, What happened? I have the CV of a model citizen or a gifted child, not a common drunk. Hometown: Cambridge, Massachusetts, backyard of Harvard University. Education: Brown University, class of '81, magna cum laude. Parents: esteemed psychoanalyst (dad) and artist (mom), both devoted and insightful and keenly intelligent.

In other words, nice person, from a good, upper-middle-class family. I look and I think, What *happened*?

Of course, there is no simple answer. Trying to describe the process of becoming an alcoholic is like trying to describe air. It's too big and mysterious and pervasive to be defined. Alcohol is everywhere in your life, omnipresent, and you're both aware and unaware of it almost all the time; all you know is you'd die without it, and there is no simple reason why this happens, no single moment, no physiological event that pushes a heavy drinker across a concrete line into alcoholism. It's a slow, gradual, insidious, elusive *becoming*.

When you love somebody, or something, it's amazing how willing you are to overlook the flaws. Around that same time, in my thirties, I started to notice that tiny blood vessels had burst all along my nose and cheeks. I started to dry-heave in the mornings, driving to work in my car. A tremor in my hands developed, then grew worse, then persisted for longer periods, all day sometimes.

I did my best to ignore all this. I struggled to ignore it, the way a woman hears coldness in a lover's voice and struggles, mightily and knowingly, to misread it.

The phrase is *high-functioning alcoholic*. Smooth and ordered on the outside; roiling and chaotic and desperately secretive underneath, but not noticeably so, never noticeably so. I remember sitting down in my cubicle that morning, my leg propped up on a chair, and thinking: *I wonder if she knows. I wonder if anyone can tell by looking at me that something is wrong*. I used to wonder that *a lot*, that last year or two of drinking—*Something is different about me*, I'd think, sitting in an editorial meeting and looking around at everyone else, at their clear eyes and well-rested expressions. *Can anybody see it?* The wondering itself made me anxious, chipping away at the edges of denial.

Perception versus reality. Outside versus inside. I never missed a day of work because of drinking, never called in sick, never called it quits and went home early because of a hangover. But inside I was falling apart. The discrepancy was huge.

Beneath my own witty, professional façade were oceans of fear, whole rivers of self-doubt. I once heard alcoholism described in an AA meeting, with eminent simplicity, as 'fear of life', and that seemed to sum up the condition quite nicely. I, for example, had spent half my professional life as a reporter who lived in secret terror of the most basic aspects of the job, of picking up the phone and calling up strangers to ask questions. Inside, I harbored a long list of qualities that made my own skin crawl: a basic fragility; a feeling of hypersensitivity to other peoples' reactions, as though some piece of my soul might crumble if you looked at me the wrong way; a sense of being essentially inferior and unprotected and scared. Feelings of fraudulence are familiar to scores of people in and out of the working world—the highly effective, well-defended exterior cloaking the small, insecure person inside—but they're epidemic among alcoholics. You hide behind the professional persona all day; then you leave the office and hide behind the drink.

Sometimes, in small flashes, I'd be aware of this. One night after work, on my way to a bar to meet a friend for drinks, a sentence popped into my head. I thought: *This is the real me, this person driving in the car*. I was anxious. My teeth were clenched, partly from spending a long day hunched over the computer and partly from the physical sensation of wanting a drink badly, and I was aware of an undercurrent of fear deep in my gut, a barely definable sensation that the ground beneath my feet wasn't solid or real. I think I understood in that instant that I'd created two versions of myself: the working version, who sat at the desk and pounded away at the keyboard, and the restaurant version, who sat at a table and pounded away at white wine. In between, for five or ten minutes at a stretch, the real version would emerge: the fearful version, tense and dishonest and uncertain. I rarely allowed her to emerge for long. Work—all that productive, effective, focused work—kept her distracted and submerged during the day. And drink—anesthetizing and constant—kept her too numb to feel at night.

My reflection in the mirror looked awful: my skin was pale and my face was drawn and I had large dark circles under my eyes. I had on a scoop-necked sweater and I could see little burst blood vessels all over my chest, red marks that looked like the beginnings of a rash. My twin sister Becca, a doctor, would see those periodically and tell me she thought they were alcohol related, and I always thought, *Nonsense: they're from too much sun, back when I was younger*. In any event, I looked like hell, and somewhere inside I understood that if I kept this up, kept drinking and working and flailing around like this, I'd die, slowly, but literally kill myself.

Around the time that Wicky died, I started taking those little quizzes about drug and alcohol abuse that you sometimes find in women's magazines or pamphlets at the doctor's office, and I started answering a lot of the questions positively. Do you find yourself having a drink or two before you go to a party where you know alcohol will be served, just to 'get yourself in the mood?' Yes. Do you find yourself gulping drinks? Um ... check. Do you drink more when you're under stress? Sure. But some of the questions seemed a little obvious, even kind of stupid. Do you drink alone? Well, of course I drink alone; I live alone. What kind of a question is that!

The knowledge that some people can have enough while you never can is the single most compelling piece of evidence for a drinker to suggest that alcoholism is, in fact, a disease, that it has powerful physiological roots, that the alcoholic's body simply responds differently to liquor than a nonalcoholic's. Once I started to drink, I simply did not know how or when to stop: the feeling of need kicked in, so pervasively that stopping didn't feel like an option. My friend Bill explains it this way to his mother, who has a hard time wrapping her mind around the disease concept of alcoholism and who holds fast to the belief that he could have controlled his drinking if only he'd exerted enough will. He says, 'Mom, next time you have diarrhea, try controlling that.' Crude, perhaps, but he gets the point across.

The need is more than merely physical: It's psychic and visceral and multilayered. There's a dark fear to the feeling of wanting that wine, that vodka, that bourbon: a hungry, abiding fear of being without, being exposed, without your armor. In meetings you often hear people say that, by definition, an addict is someone who seeks physical solutions to emotional or spiritual problems. I suppose that's an intellectual way of describing that brand of fear, and the instinctive response that accompanies it: there's a sense of deep need, and the response is a grabbiness, a compulsion to latch on to something outside yourself in order to assuage some deep discomfort.

And there it was again, the connection: Expressional + Drink = Openness. At heart alcoholism feels like the accumulation of dozens of such connections, dozens of tiny fears and hungers and rages, dozens of experiences and memories that collect in the bottom of your soul, coalescing over many many many drinks into a single liquid solution.

You drink long and hard enough and your life gets messy. Your relationships (with nondrinkers, with yourself) become strained. Your work suffers. You run into financial trouble, or legal trouble, or trouble with the police. Rack up enough pain and the old math—Discomfort + Drink = No Discomfort—ceases to suffice; feeling 'comfortable' isn't good enough anymore. You're after something deeper than a respite from shyness, or a break from private fears and anger. So after a while you alter the equation, make it stronger and more complete. Pain + Drink = Self-Obliteration.

From: Goodwin, D.W. (1981). Alcoholism: The Facts. Oxford University Press, New York.

I am David. I am an alcoholic. I have always been an alcoholic. I will always be an alcoholic. I cannot touch alcohol. It will destroy me. It is like an allergy—not a real allergy—but *like* an allergy.

I had my first drink at sixteen. I got drunk. For several years I drank every week or so with the boys. I didn't always get drunk, but I know now that alcohol affected me differently than other people. I looked forward to the times I knew I could drink. I drank for the glow, the feeling of confidence it gave me. But maybe that's why my friends drank too. They didn't become alcoholics. Alcohol seemed to satisfy some specific need I had, which I can't describe. True, it made me feel good, helped me forget my troubles, but that wasn't it. What was it? I don't know, but I know I liked it, and after a time, I more than liked it, I needed it. Of course, I didn't realize it. It was maybe ten or fifteen years before I realized it, *let myself* realize it.

My need was easy to hide from myself and others (maybe I'm kidding myself about the others). I only associated with people who drank. I married a woman who drank. There were always reasons to drink. I was low, tense, tired, mad, happy. I probably drank as often because I was happy as for any other reason. And occasions for drinking—when drinking was appropriate, expected—were endless. Football games, fishing, trips, parties, holidays, birthdays, Christmas, or merely Saturday night. Drinking became interwoven with everything pleasurable—food, sex, social life. When I stopped drinking, these things, for a time, lost all interest for me, they were so tied to drinking. I don't think I will ever enjoy them as much as I did when drinking. But if I had kept drinking, I wouldn't be here to enjoy them. I would be dead.

So, drinking came to dominate my life. By the time I was 25 I was drinking every day, usually before dinner, but sometimes after dinner (if there was a 'reason'), and more on weekends, starting in the afternoon. By 30, I drank all weekend, starting with a beer or Bloody Mary in the morning, and drinking off and on, throughout the day, beer or wine or vodka, indiscriminately. The goal always, was to maintain a glow, not enough, I hoped, that people would notice, but a glow. When five o'clock came, I thought, well, now it's cocktail hour and I would have my two of three scotches or martinis before dinner as I did on nonweekend nights. After dinner I might nap, but just as often felt a kind of wakeful calm and power and happiness that I've never experienced any other time. These were the dangerous moments. I called friends, boring them with drunken talk; arranged parties, decided impulsively to drive to a bar. In one year, at the age of 33, I had three accidents, all on Saturday night, and was charged with drunken driving once (I kept my licence, but barely). My friends became fewer, reduced to other heavy drinkers and barflies. I fought with my wife, blaming her for *her drinking*, and once or twice hit her (or so she said—like many things I did while drinking, there was no memory afterward).

And by now I was drinking at noontime, with the lunch hour stretching longer and longer. I began taking off whole afternoons, going home potted. I missed mornings at work because of drinking the night before, particularly Monday mornings. And I began drinking weekday mornings to get going. Vodka and orange juice. I thought Vodka wouldn't smell (it did). It usually lasted until an early martini luncheon, and I then suffered through until cocktail hour, which came earlier and earlier.

By now I was hooked and knew it, but desperately did not want others to know it. I had been sneaking drinks for years—slipping out to the kitchen during parties and such—but now I began hiding alcohol, in my desk, bedroom—car glove compartment, so it would never be far away, ever. I grew panicky even thinking I might not have alcohol when I needed it, which was just about always.

For years, I drank and had very little hangover, but now the hangovers were gruesome, I felt physically bad—headachy, nauseous, weak—but the mental part was the hardest. I loathed myself. I was waking early and thinking what a mess I was, how I had hurt so many others and myself. The words 'guilty' and 'depression' sound superficial in trying to describe how I felt. The loathing was almost physical–a dead weight that could be lifted in only one way, and that was by having a drink, so I drank, morning after morning. After two or three, my hands were steady, I could hold some breakfast down, and the guilt was gone, or almost.

Despite everything, others knew. There was the odour, the rheumy eyes, and flushed face. There was missing work and not working well when there. Fights with wife, increasingly physical. She kept threatening to leave and finally did. My boss gave me a leave of absence after an embarrassed remark about my 'personal problems'. At some point I was without wife, home, or job. I had nothing to do but drink. The drinking was now steady, days on end. I lost appetite and missed meals (besides, money was short), I awoke at night, sweating and shaking, and had a drink, I awoke in the morning vomiting and had a drink. It couldn't last. My ex-wife found me in my apartment shaking and seeing things, and got me in the hospital. I dried out, left, and went back to drinking. I was hospitalized again, and this time stayed dry for six months. I was nervous and couldn't sleep, but got some of my confidence back and found a part-time job. Then my ex-boss offered my job back and I celebrated by having a drink. The next night I had two drinks. In a month I was drinking as much as ever and again unemployed. That was three years ago. I've had two big binges since then but don't drink other times. I think about alcohol and miss it. Life is grey and monotonous. The joy and gaiety are gone. But drinking will kill me. I know this and have stopped—for now.

From: Frey, J. (2003). *A Million Little Pieces*. Doubleday, New York pp. 4–11.

We head north to the Cabin. Along the way I learn that my parents, who live in Tokyo, have been in the States for the last two weeks on business. At four a.m. they received a call from a friend of mine who was with me at a hospital and had tracked them down in a hotel in Michigan. He told them that I had fallen face first down a fire escape and that he thought they should find me some help. He didn't know what I was on, but he knew there was a lot of it and he knew it was bad. They had driven to Chicago during the night.

We drive on and after a few hard silent minutes, we arrive. We get out of the car and we go into the house and I take a shower because I need it. When I get out there are some fresh clothes sitting on my bed. I put them on and I go to my parents' room. They are up drinking coffee and talking but when I come in they stop.

Hi.

Mom starts crying again and she looks away. Dad looks at me.

Feeling better?

No.

You should get some sleep.

I'm gonna.

Good.

I look at my Mom. She can't look back. I breathe.

I just.

I look away.

I just, you know.

I look away. I can't look at them.

I just wanted to say thanks. For picking me up.

Dad smiles. He takes my Mother by the hand and they stand and they come over to me and they give me a hug. I don't like it when they touch me so I pull away.

Good night.

Good night, James. We love you.

I turn and I leave their room and I close their door and I go to the kitchen. I look through the cabinets and I find an unopened half-gallon bottle of whiskey. The first sip brings my stomach back up, but after that it's all right. I go to my room and I drink and I smoke some cigarettes and I think about her. I drink and I smoke and I think about her and at a certain point blackness comes and my memory fails me.

Back in the car with a headache and bad breath. We're heading north and west to Minnesota. My father made some calls and got me into a clinic and I don't have any other options, so I agree to spend some time there and for now I'm fine with it. It's getting colder.

I want to run or die or get fucked up. I want to be blind and dumb and have no heart. I want to crawl in a hole and never come out. I want to wipe my existence straight off the map. Straight off the fucking map. I take a deep breath. We enter a small Waiting Room. They're gonna check you in now. We stand and we move toward a small room where a man sits behind a desk with a computer. He meets us at the door.

You ready to get started?

I don't smile.

Sure.

He gets up and I get up and we walk down a hall. He talks and I don't. The doors are always open here, so if you want to leave, you can. Substance use is not allowed and if you're caught using or possessing, you will be sent home. You are not allowed to say anything more than hello to any women aside from doctors, nurses or staff members. If you violate this rule, you will be sent home. There are other rules, but those are the only ones you need to know right now.

We walk through a door into the Medical Wing. There are small rooms and doctors and nurses and a pharmacy. The cabinets have large steel locks. He shows me to a room. It has a bed and a desk and a chair and a closet and a window. Everything is white.

He stands at the door and I sit on the bed.

A nurse will be here in a few minutes to talk with you.

Fine.

You feel okay?

No, I feel like shit.

It'll get better.

Yeah.

Trust me.

Yeah.

The man leaves and he shuts the door and I'm alone. My feet bounce, I touch my face, I run my tongue along my gums. I'm cold and getting colder. I hear someone scream.

The door opens and a nurse walks into the room. She wears white, all white, and she is carrying a clipboard. She sits in the chair by the desk.

Hi, James.

Hi.

I need to ask you some questions.

All right.

I also need to check your blood pressure and your pulse.

All right.

What type of substances do you normally use?

Alcohol.

Every day?

Yes.

What time do you start drinking?

When I wake up.

She marks it down.

How much per day?

As much as I can.

How much is that?

Enough to make myself look like I do.

She looks at me. She marks it down.

Do you use anything else?

Cocaine.

How often?

Every day.

She marks it down.

How much?

As much as I can.

She marks it down.

In what form?

Lately crack, but over the years, in every form that it exists.

She marks it down.

Anything else?

Pills, acid, mushrooms, meth, PCP and glue.

Marks it down.

How often?

When I have it.

How often?

A few times a week.

Marks it down.

She moves forward and draws out a stethoscope.

How are you feeling?

Terrible.

In what way?

In every way.

She reaches for my shirt.

Do you mind?

No.

She lifts my shirt and she puts the stethoscope to my chest. She listens.

Breathe deeply.

She listens.

Good. Do it again.

She lowers my shirt and she pulls away and she marks it down.

Thank you.

I smile.

Are you cold?

Yes.

She has a blood pressure gauge.

Do you feel nauseous?

Yes.

She straps it on my arm and it hurts.

When was the last time you used?

She pumps it up.

A little while ago.

What and how much?

I drank a bottle of vodka.

How does that compare to your normal daily dosage?

It doesn't.

She watches the gauge and the dials move and she marks it down and she removes the gauge.

I'm gonna leave for a little while, but I'll be back.

I stare at the wall.

We need to monitor you carefully and we will probably need to give you some detoxification drugs.

I see a shadow and I think it moves but I'm not sure.

You're fine right now, but I think you'll start to feel some things.

I see another one. I hate it.

If you need me, just call.

I hate it.

She stands up and she smiles and she puts the chair back and she leaves. I take off my shoes and I lie under the blankets and I close my eyes and I fall asleep.

I wake and I start to shiver and I curl up and I clench my fists. Sweat runs down my chest, my arms, the backs of my legs. It stings my face.

I sit up and I hear someone moan. I see a bug in the corner, but I know it's not there. The walls close in and expand they close in and expand and I can hear them. I cover my ears but it's not enough.

I stand. I look around me. I don't know anything. Where I am, why, what happened, how to escape. My name, my life.

I curl up on the floor and I am crushed by images and sounds. Things I have never seen or heard or ever knew existed. They come from the ceiling, the door, the window, the desk, the chair, the bed, the closet. They're coming from the fucking closet. Dark shadows and bright lights and flashes of blue and yellow and red as deep as the red of my blood. They move toward me and they scream at me and I don't know what they are but I know they're helping the bugs. They're screaming at me.

I start shaking. Shaking shaking shaking. My entire body is shaking and my heart is racing and I can see it pounding through my chest and I'm sweating and it stings. The bugs crawl onto my skin and they start biting me and I try to kill them. I claw at my skin, tear at my hair, start biting myself. I don't have any teeth and I'm biting myself and there are shadows and bright lights and flashes and screams and bugs bugs bugs. I am lost. I am completely fucking lost.

I scream.

I piss on myself.

I shit my pants.

The nurse returns and she calls for help and men in white come in and they put me on the bed and they hold me there. I try to kill the bugs but I can't move so they live. In me. On me. I feel the stethoscope and the gauge and they stick a needle in my arm and they hold me down.

I am blinded by blackness.

I am gone.

APPENDIX

4

Nicotine

From: Bain, J. Jr. (1903). *Tobacco Leaves.* H.M. Caldwell Company, Boston.

Edwin Booth was a fierce smoker. His favourite was a pipe, not a cigar. He smoked in his dressing-room, between acts, in his own room, constantly, and I am not sure that he did not smoke in bed. He loved tobacco as another man might love food and drink. His system was full of nicotine, for he overdid it, and he would be alive to-day if he had been a moderate smoker, as would General Grant...

...The fiercest smoker whom I have ever known was the late Francis Saltus, the marvellous linguist, musician, composer, writer, and traveller. He would smoke (surely) fifty cigarettes a day. You talk about fellows smoking in bed and between courses at a dinner? Well, Frank Saltus would smoke between *mouthfuls*. I have seen him smoke fifty cigarettes in a day, while turning off two or three hundred dialogues ('squibs', he called them) for the papers and magazines. He was a wonder, look at him how you will, and some day the world will know it.

Elihu Root thinks that a cigar after breakfast is the smoke of the day, and there are many smokers who will agree with him. He is reported as saying: 'My breakfast is a very simple meal, and consists of a cup of coffee or chocolate and a roll. When I have finished it, I light my cigar. I find that it assists me in my work. It does not aid me in the creation of ideas so much, nor in reading or actual writing; but when I want to prepare my plans for the day, when I want to arrange and put in shape the work I have before me, I find that smoking is a valuable assistant. I never smoke a large cigar in the morning, and usually do not prolong the smoke beyond the time it takes me to arrange my day's programme. Altogether I should say that I smoke five cigars a day. I have smoked steadily for the past thirty years, and during the first ten years I smoked a pipe. It has been my experience that smoking relieved me at any time when I felt overworked. Consequently, if I find at any time of day that my brain is getting tired, and that my ideas are getting muddled, I stop and light a cigar. I don't think that smoking has a sedative effect upon me, but it composes my thoughts and soothes me to some extent.'

APPENDIX

5

Cannabinoids

From: Tennant F. S. Jr. (1986). The clinical syndrome of marijuana dependence. *Psychiatric Annals.* 16, 225–234.

Case 1: Marijuana Dependence: Voluntary Admission to Treatment

MV was a 25-year-old male who presented with the complaint that he could not 'stop marijuana by myself.' He was a 12-year user having begun marijuana smoking at 13 years of age. He had used marijuana daily for about five years and was using two to three joints per day at the time of admission to outpatient treatment. The patient was married and held a regular job as a warehouse superintendent. He claimed he was having considerable conflicts with his wife and employer. In addition, he had noticed in the two months just prior to admission that he occasionally heard voices that were not real, did not always have total 'control over his mind', and had some thoughts of suicide. He denied use of any other drug or excessive alcohol intake. His treatment admission breath alcohol was negative, and his urine contained marijuana metabolite, but no other abusable drug. The patient was administered desipramine, 25 mg, three times per day and was given weekly psychotherapy for approximately six months. During the first ten days of treatment, he reported insomnia, abdominal cramps, diaphoresis, tachycardia, and anxiety. These symptoms subsided, and he submitted a urine void of marijuana approximately 30 days after admission. Most of the thought disturbances noted above disappeared after about two to six weeks of treatment. He denied any marijuana use during the six months after entering treatment, and he submitted monthly urine tests that showed no marijuana.

Case 2: Volunteer with Unsuccessful Treatment

JS was an 18-year-old male who voluntarily presented because he 'wanted to stop'. Consumption of drugs consisted of about 3.5 grams of marijuana and hashish per day for one year prior to seeking treatment. He self-administered every one to four hours while awake and complained of chronic cough, anorexia, depression, and weight loss. When he had tried abruptly to cease marijuana by himself, he had hallucinations, depression, and anergy. Urinalysis testing revealed the presence of marijuana, but no other drugs. The patient entered a counseling program, but received no medications. Only one return appointment was kept, and he was lost to followup.

Case 3: Mandatory Work-Site Detection and Referral

HS was a 37-year-old male salesperson. He was reported to the management of his company to be a marijuana user who also sold it to other employees while on company premises. A mandatory urine test revealed the presence of marijuana metabolite, and in order to retain employment he was required to undergo withdrawal and enter a periodic urine-testing program. Upon interview, he stated that he had used marijuana every evening for approximately 22 years. He believed this habit had not been injurious to himself until approximately three months prior to treatment when he began to notice some defects in his short-term memory. Physical examination was normal. Plasma analysis showed there to be 148 ng/ml of 11-OH THC and 80 ng/ml of THC-C. He was administered desipramine, 25 mg, three times per day and tyrosine. During the first three weeks following cessation of marijuana, he reported mild insomnia, depression, anergy, and craving. Urine analysis showed no marijuana metabolite after about 30 days. After six weeks of abstinence, he reported improvement of short-term memory and improved job performance.

Case 4: Delayed Withdrawal Symptoms

A 27-year-old male was admitted to a day-treatment program for marijuana dependence. He had been identified at work for being 'under the influence' on more than one occasion and was, therefore, referred for treatment. Drug consumption consisted of intermittent cocaine use and daily use of about one marijuana joint. He perceived that he had been 'addicted' to marijuana for about 15 years, and that he had skipped marijuana use on very few days during this time. A physical examination was normal except for mild nasal-septum inflammation and a swollen uvula. Urine analysis showed the presence of marijuana metabolite and marijuana plasma analysis by high performance liquid chromatography (HPLC) showed no 11-OH-THC and THC-C to be 8 ng/ml. A 24-hour urine specimen showed secretion of 2-methoxy-4-hydroxy-phenyglycol (MHPG) to be 143.0 MCG/24 hours (normal is 1164 to 2216). Since his continued employment was dependent upon attending the day-care program until his urine was void of all drugs, compliance with treatment and testing procedures was good. Withdrawal medication consisted of desipramine, 25 mg, administered three times per day, and the amino acid, tyrosine. On the third treatment day, his urine still contained metabolite, and his plasma contained 3 ng/ml of THC-C. On the eighth day of attendance, he complained of a flu-like illness consisting of nausea, vomiting, diaphoresis, chills, myalgia, anorexia, and insomnia. The patient did not relate these symptoms temporally to his marijuana use, since he had ceased use eight days previous. Plasma analysis showed no detectable presence of 11-OH-THC or THC-C, but marijuana metabolite was still present in urine at this time. The apparent withdrawal symptoms resolved within 48 hours. Marijuana metabolite remained in his urine until the 34th day of treatment.

Index